International Conference on Modern Technologies in Industrial Engineering (ModTech 2015)

AA003309

IOP Conference Series: Materials Science and Engineering
Volume 95

Mamaia, Romania
17 – 20 June 2015

Editors:

Emil Oanta
Radu Comaneci
Constantin Carausu
Marek Placzek
Viorel Cohal
Pavel Topala
Dumitru Nedelcu

ISBN: 978-1-5108-1466-0
ISSN: 1757-8981

Printed from e-media with permission by:

Curran Associates, Inc.
57 Morehouse Lane
Red Hook, NY 12571

proceedings.com

Some format issues inherent in the e-media version may also appear in this print version.

Copyright© (2015) by the Institute of Physics
All rights reserved. The material featured in this book is subject to IOP copyright protection, unless otherwise indicated.

Printed by Curran Associates, Inc. (2016)

For permission requests, please contact the Institute of Physics at the address below.

Institute of Physics
Dirac House, Temple Back
Bristol BS1 6BE UK

Phone: 44 1 17 929 7481
Fax: 44 1 17 920 0979

techtracking@iop.org

Additional copies of this publication are available from:

Curran Associates, Inc.
57 Morehouse Lane
Red Hook, NY 12571 USA
Phone: 845-758-0400
Fax: 845-758-2634
Email: curran@proceedings.com
Web: www.proceedings.com

International Conference on Modern Technologies in Industrial Engineering (ModTech 2015)

IOP Conference Series: Materials Science and Engineering
Volume 95

Mamaia, Romania
17 – 20 June 2015

IOP
Institute of Physics

Table of contents

Volume 95

Modern Technologies in Industrial Engineering (ModTech2015)
17–20 June 2015, Mamaia, Romania

Accepted papers received: 7 October 2015
Published online: 3 November 2015

Preface

Papers

1 - Advanced Materials and Modern Technologies for Manufacturing Processes

3 - New Materials and Modern Technologies in Marine Engineering

IOP Conf. Series: Materials Science and Engineering **95** (2015) 011001 doi:10.1088/1757-899X/95/1/011001

Preface

The dominant feature of the current stage of society development is the update, refinement and innovation of the technological processes and products whose ultimate goal is to satisfy the market requirements. New and modern technologies should be considered in terms of their applicability in industry while the materials can lead to an increase in the quality of the end products.

Replacing the existing technologies with innovative and eco-efficient technologies can contribute to an added value increase in the production of new materials.

Materials are one of the most dynamic and prospective fields, with applications in all other fields. The development of new advanced materials and technologies shall contribute to the procurement of a wide range of reliable products, with competitive prices and worldwide performance, high sensitivity and functionality, user-friendly and reduced energy consumption, for different industrial applications.

Research in the field of advanced/intelligent materials supposes a fundamental, experimental, laboratory and technological research and its approach has to be linked to the application. This involves, even for the niche fields, complex projects which result in scientific issues in top journals, patents and functional models.

The third edition of ModTech International Conference was held in Mamaia, Romania, between June 17-20, with the Professional Association in Modern Manufacturing Technologies, ModTech, as main organizer, and the Constanta Maritime University, Constanta-Romania, Silesian University of Technology, Gliwice-Poland, the Technical University of Chisinau-Republic of Moldova and the Donetsk National Technical University, Donetsk-Ukraine as co-organizers. The ModTech2015 International Conference brought together representatives of technology and materials manufacturers, various universities, professional associations and research institutes that exchanged the latest knowledge on the conference topics. This edition was attended by 140 participants from 17 countries. The authors and co-authors were from various countries worldwide, namely: Sweden, China, Switzerland, Romania, Serbia, Germany, Netherlands, Belgium, France, South Korea, Taiwan, Poland, USA, Slovenia, Turkey, Republic of Moldova, Russia, Finland, Japan, Ukraine, Portugal, Uzbekistan, Iraq, Italy, India, Slovenia.

The Keynote Speakers were as follows: Prof. Esteban Broitman - Linkoping University, Sweden; Prof. Ziyi Ge - NIMTE, Chinese Academy of Sciences, Ningbo, China; Prof. Thomas Graule - EMPA, Switzerland; prof. Razvan Tamas - Constanta Maritime University, Romania; Prof. Rainer Gadow - University of Stuttgart, Germany; Prof. Marcel Van de Voorde - DELFT University of Technology, Netherlands; Prof. Chris Lacor - Vrije University, Brussel, Belgium; Prof. Fiqiri Hodaj - National Polytechnique Institute of Grenoble, France; Prof. Hong Seok Park - University of Ulsan, South Korea; Prof. Der-Jang Liaw - National Taiwan University of Science and Technology, Taiwan; Prof. Petrica Vizureanu - Gheorghe Asachi Technical University of Iasi, Romania.

The main publications of ModTech2015 International Conference are as follows: IOP Conference Series: Materials Science and Engineering, United Kingdom, Indian Journal of Engineering & Materials Sciences (IJEMS) and International Journal of Modern Manufacturing Technologies (IJMMT).

Editors,

Emil Oanta, Radu Comaneci, Constantin Carausu, Marek Placzek, Viorel Cohal, Pavel Topala, Dumitru Nedelcu

Content from this work may be used under the terms of the Creative Commons Attribution 3.0 licence. Any further distribution of this work must maintain attribution to the author(s) and the title of the work, journal citation and DOI.
Published under licence by IOP Publishing Ltd

Conference Committees

Honorary Presidents

Professor Violeta-Vali Ciucur, Ph.D., Rector, Constanta Maritime University, Romania.
Professor Octavian Pruteanu, Ph.D., DHC, Vicepresident ModTech Professional Association Iasi, Romania.
Professor Jerzy Świder, Ph.D., Director of the Institute of Engineering Processes Automation and Integrated Manufacturing Systems, Silesian University of Technology, Gliwice, Poland.
Professor Olexander Minayev, Ph.D., Rector, Donetsk National Technical University, Ukraine.
Academician Dr. Hab. Prof. Univ. Ion Bostan Rector, Technical University of Moldova, Republic of Moldova.

Presidents

Professor Mat. Eliodor Constantinescu, Ph.D., Vice-Rector, Constanta Maritime University, Romania.
Professor Dumitru Nedelcu, Ph.D., President of ModTech Professional Association Iasi, Romania.
Professor Dr. Hab. inż. Arkadiusz Mężyk, Dean of Mechanical Engineering Faculty, Silesian University of Technology, Gliwice, Poland.
Professor Ilia Navka, Ph.D., Vice-Rector, Donetsk National Technical University, Ukraine.
Dr. Hab. Prof. Univ. Valerian Dorogan, Vice-Rector, Technical University of Moldova, Republic of Moldova.

Presidents of Organizing Committee

Associate Professor Ec. Costel Stanca, Ph.D., Dean of Navigation and Naval Faculty, Constanta Maritime University, Romania.
Lecturer Ciprian Ciofu, Ph.D., Member of Executive Council of ModTech Professional Association Iasi, Romania.
Lecturer Marek Płaczek, Ph.D., Faculty of Mechanical Engineering, Silesian University of Technology, Gliwice, Poland.
Professor Alexander Mikhaylov, Ph.D., Donetsk National Technical University, Ukraine.
Dr. Associate Professor Alexei Toca, Dean of Faculty of Engineering and Management in Machine Manufacturing, Republic of Moldova.

Scientific Committee

Dumitru AMARANDEI, "Stefan cel Mare" University of Suceava, Romania
Andrzej BAIER, Silesian University of Technology, Gliwice, Poland
Mihaela BANU, University of Michigan, Ann Arbor, USA
Peter BIKFALVI, University of Miskolc, Hungary
Radhey Shyam BENIWAL, CSIR-NISCAIR, New Delhi, India
Leandru Gheorghe BUJOREANU, "Gheorghe Asachi" Technical University of Iasi, Romania
Konstantinos -Dionysios BOUZAKIS, "Aristotelis" University of Salonic, Greece

Ion BOSTAN, Technical University of Moldova, Chisinau, Republic of Moldova

IOP Conf. Series: Materials Science and Engineering **95** (2015) 011001 doi:10.1088/1757-899X/95/1/011001

Nicolae BÂLC, Technical University of Cluj-Napoca, Romania
Florin BLAGA, University of Oradea, Romania
Gheorghe BRABIE, "Vasile Alexandri" University of Bacau, Romania
Paul Dan BRÂNDASU, "Lucian Blaga"University of Sibiu, Romania
Andrzej BUCHACZ, Silesian University of Technology, Gliwice, Poland
Doru BARDAC, "Politehnica" University of Bucharest, Romania
Ioan CARCEA, "Gheorghe Asachi" Technical University of Iasi, Romania
Constantin CARAUSU, "Gheorghe Asachi" Technical University of Iasi, Romania
Gabriela CARJA, "Gheorghe Asachi" Technical University of Iasi, Romania
Tudor CHERECHES, Military Academy of Bucharest, Romania
Ciprian CIOFU, "Gheorghe Asachi" Technical University of Iasi, Romania
Violeta-Vali CIUCUR, Constanta Maritime University, Romania
Francisco CHINESTA, Ecole Centrale Nantes, France
Laurentiu CHIOTOROIU, Jade University of Applied Sciences, Elsfleth, Germany
Viorel COHAL "Gheorghe Asachi" Technical University of Iasi, Romania
Gheorghe CRETU "Gheorghe Asachi" Technical University of Iasi, Romania
Ioan DOROFTEI, "Gheorghe Asachi" Technical University of Iasi, Romania
Laurentiu Dan GHENGHEA, "Gheorghe Asachi" Technical University of Iasi, Romania
António GONÇALVES-COELHO, University Nova de Lisboa, Portugal
Joao Paulo DAVIM, University of Aveiro, Portugal
George DRAGHICI, "Politehnica" University of Timisoara, Romania
Cristian DOICIN, "Politehnica" University of Bucharest, Romania
Manuel Alberto M. FERREIRA, ISCTE-IUL, Lisboa, Portugal
Catalin FETECAU, "Dunarea de Jos" University of Galati, Romania
Hidehiro KAMIYA, TAT, BASE, Tokyo, Japan
Natalia KAMANINA, Vavilov State Optical Institute, St. Petersburg, Russia
Ziyi GE, Ningbo Institute of Material Technology & Engineering, Chinese Academy of Sciences, Ningbo, China
Mihai GOANTA, "Dunarea de Jos" University of Galati, Romania
Nil GULER, Istanbul Technical University, Turkey
Traian GRAMESCU, "Gheorghe Asachi" Technical University of Iasi, Romania
Huran LIU, Zhejiang University of Science and Technology, China
Alexandru EPUREANU, "Dunarea de Jos" University of Galati, Romania
Cornel IOANA Politechnical Institute, Grenoble, France
Romeo IONESCU, "Stefan cel Mare" University of Suceava, Romania
Gabriel KOST, Silesian University of Technology, Gliwice, Poland
David HUI, University of New Orleans, USA
Vasile KRIVTOV, University of Michigan, USA
Chris LACOR, Vrije University, Brussels, Belgium
Peter LORENZ, University of Applied Sciences Saarbrücken, Germany
Nouras LUPULESCU, "Transilvania" University of Brasov, Romania
Olexander MINAYEV, Donetsk National Technical University, Ukraine
Sarwoko MANGKOEDIHARDJO, Tenth of November Institute of Technology, Surabaya, Indonesia
Alexandr MALIARENCO, National Technical University of Belarus, Belarus
Sergiu MAZURU, Technical University of Moldova, Chisinau, Republic of Moldova
Arkadiusz MEZYK, Silesian University of Technology, Gliwice, Poland
Boyan MEDNIKAROV, Naval Academy, Varna, Bulgaria

Alexandru MIHAILOV, UTN Donetsk, Ukraine
Olivera MILOSEVIC, Institute of Technical Sciences, Belgrade, Serbia
Sevasti MITSI, "Aristotelis" University of Salonic, Greece
Corneliu MUNTEANU, "Gheorghe Asachi" Technical University of Iasi, Romania
António MOURÃO, University Nova de Lisboa, Portugal
Ilia NAVCA, National Technical University of Donetsk, Ukraine
Makio NAITO, JWRI, Osaka University, Japan
Vasile NASTASESCU, Military Academy of Bucharest, Romania
Dumitru NEDELCU, "Gheorghe Asachi" Technical University of Iasi, Romania
Nicolae OANCEA, "Dunarea de Jos" Unversity of Galati, Romania
Emil OANTA, Constanta Maritime University, Romania
Constantin OPREAN, "Lucian Blaga" University of Sibiu, Romania
Cezar OPRISAN, "Gheorghe Asachi" Technical University of Iasi, Romania
Viorel PAUNOIU, "Dunarea de Jos" University of Galati, Romania
Marcel Sabin POPA, Technical University of Cluj-Napoca, Romania
Remon Pop ILIEV, UOIT University, Oshawa, Canada
Mihaela PICU, "Dunarea de Jos" University of Galati, Romania
Marek PŁACZEK, Silesian University of Technology, Gliwice, Poland
Octavian PRUTEANU, "Gheorghe Asachi" Technical University of Iasi, Romania
Serban RAICU, "Politehnica" University of Bucharest, Romania
Akos REDEY, University of Pannonia, Veszprem, Hungary
Robert O. RITCHIE, University of California, Berkeley, USA
Loredana SANTO, "Tor Vergata" University of Rome, Italy
Francisco SANTOS, University of Valladolid, Spain
Mamoru SENNA, Keio University, Japan
Bozena SKOŁUD, Silesian University of Technology, Gliwice, Poland
Costel STANCA, Constanta Maritime University, Romania
Nadezda STEVULOVA, Technical University of Kosice, Slovakia
Andrey N. STRELETSKII, N.N. Semenov Institute, Moscow, Russia
Carol SCHNAKOVSZKY, “Vasile Alecsandri” University of Bacau, Romania
Anatolii SUSLOV, Technical University of Briansk, Russia
Janos SZAVA, "Transilvania" University of Brasov, Romania
Jerzy SWIDER, Silesian University of Technology, Gliwice, Poland
Razvan TAMAS, Constanta Maritime University, Romania
Alexei TOCA, Technical University of Moldova, Chisinau, Republic of Moldova
Pavel TOPALA, "Alecu Russo" University of Balti, Republic of Moldova
Gabriel VASILE, CNRS, Grenoble, France
Petrica VIZUREANU, Technical University "Gheorghe Asachi" of Iasi, Romania
Aurelian VLASE, "Politehnica" University of Bucharest, Romania
Andrzej WRÓBEL, Silesian University of Technology, Gliwice, Poland
Remus ZAGAN, Constanta Maritime University, Romania

Peer review statement

All papers published in this volume of *IOP Conference Series: Materials Science and Engineering* have been peer reviewed through processes administered by the proceedings Editors. Reviews were conducted by expert referees to the professional and scientific standards expected of a proceedings journal published by IOP Publishing.

Content from this work may be used under the terms of the Creative Commons Attribution 3.0 licence. Any further distribution of this work must maintain attribution to the author(s) and the title of the work, journal citation and DOI.
Published under licence by IOP Publishing Ltd

IOP Conf. Series: Materials Science and Engineering **95** (2015) 012001 doi:10.1088/1757-899X/95/1/012001

Numerical investigation of the thermal behavior of heated natural composite materials

S M Qasim[1], F Abbas[2] Mohammed and R Hashim[3]
[1]AL- Mustansiriya University College of Engineering Iraq – Baghdad
[2]Midland Refinery Company, Iraq - Baghdad

E-mail: qasim602006@yahoo.com

Abstract. In the present work numerical investigation was carried out for laminar natural convection heat transfer from natural composite material (NCM). Three types of natural materials such as seed dates, egg shells, and feathers are mixed separately with polyester resin. Natural materials are added with different volume fraction (10%, 20%, and 30%) are heated with different heat flux (1078W/m^2, 928W/m^2, 750W/m^2, 608W/m^2, and 457W/m^2) at (vertical, inclined, and horizontal) position. Continuity and Navier-Stocks equations are solved numerically in three dimensions using ANSYS FLUENT package 12.1 software commercial program. Numerical results showed the temperature distribution was affected for all types at volume fraction 30% and heat flux is 1078 W/m^2, for different position. So, shows that the plumes and temperature behavior are affected by the air and the distance from heat source. Numerical results showed acceptable agreement with the experimental previous results.

1. Introduction

Natural convection heat transfer depends on the geometry of the surface as well as its orientation. It also depends on the variation of temperature on the surface and the thermo physical properties of the fluid involved. The heat transfer from generators, transformers and electronic circuits in the electrical and electronic applications is natural convection heat transfer [1,2]. Hassan and Mohammed [3], studied experimentally of laminar natural convection heat transfer from rectangular plates at horizontal, vertical and inclined heated at constant temperature. The plate fabricated from brass dimensions (0.504 m X 0.2 m) heated by hot water upward to constant temperature. They obtained that the mean of heat transfer in vertical position is higher than horizontal in both cases. Pera and Gebhart [4], studied numerically and experimentally the boundary layer flow for laminar natural convection heat transfer at semi-infinite horizontal plates and inclined at constant temperature and heat flux, with range of Prandtl number ($0.1<Pr<10$). Miyamoto et al.[5], studied numerically the heat transfer from rectangular surfaces at vertical and horizontal position. It is heated with constant heat flux and temperature. They obtained the relation between the mean Nusslet number and Grashof number at Prandtl number ($Pr = 0.72$). Rafah [6], studied experimentally the natural convection heat transfer from isothermal heated upward – facing square horizontal plates and has square hole at range of Grashof number is ($0.5 \times 10^6 < Gr < 6.28 \times 10^6$) and for range of temperatures (T_w=40, 61, 82.5, and 112.8℃). Fujino and Honda[7], described the features of PWFA (Plastic Waste / Fly Ash) composite, which is made mostly from plastic waste and fly ash, is one of the materials developed for the purpose of recycling. A small amount of glass fiber as reinforcement and a fire retardant was

Content from this work may be used under the terms of the Creative Commons Attribution 3.0 licence. Any further distribution of this work must maintain attribution to the author(s) and the title of the work, journal citation and DOI.
Published under licence by IOP Publishing Ltd

added in the ingredient of the composite. The thermal conductivity data for the specimen (100×100) mm^2 is determined and it decreased with increasing the specimen temperature. Faieza [8], studied the effect of volume fraction on the thermal conductivity of some composite materials consisting of epoxy/carbon fibers and epoxy/Kevlar fibers experimentally using Lee's disc method. The results showed that the composites when reinforced with carbon fiber have a higher thermal conductivity than Kevlar composites. Also, for both composites the thermal conductivity increases as the volume fraction increase. In the present work numerical study of laminar natural convection heat transfer from NCM with different heat flux at different position.

2. Physical Model

The heat flux can be determined by the following law [9];

$$q = \frac{Q}{A_s} \tag{1}$$

where q is heat flux(W/m^2)

Heat transfer coefficient can be computed as follows:

$$I.P = Q = Q_{cond.} + Q_{conv.} + Q_{rad} \tag{2}$$

$$Q_{cond.} = \frac{(T_w - T_x)}{R_{ins.}} \tag{3}$$

$$R_{ins.} = \frac{L_{ins.1}}{K_{ins.1} A_{ins.1}} + \frac{L_{ins.2}}{K_{ins.2} A_{ins.2}} \tag{4}$$

$$Q_{rad.} = F A_s \varepsilon\sigma \left(T_w^4 - T_{surr}^4\right) \tag{5}$$

Where F=1,Ɛ=0.9

In the present work Q_{cond} and $Q_{rad.}$ are neglected.

$$Q_{conv.} = \bar{h} A_s \left(T_w - T_\infty\right) \tag{6}$$

Where A_s is area of element (m^2), h is average heat transfer coefficient (W/m^2K),and Q_{conv} is heat transfer by convection(W).

$$\bar{h} = \frac{Q_{conv.}}{A_s (T_w - T_\infty)} \tag{7}$$

The properties of air can be evaluated at reference temperature

$$T_e = T_w - 0.25\,(T_w - T_\infty) \tag{8}$$

Where T_w and T_∞ are wall temperature and ambient temperature respectively (C).

The coefficient of expansion (β) is determined as:

$$T_f = 0.5\,(T_w + T_\infty) \tag{9}$$

$$\beta = \frac{1}{T_f} \tag{10}$$

where β; is expansion volumetric coefficient (1/K)

3. Mathematical Model and Numerical Solution

The geometry of physical model is consists of a solid plate represents natural composite materials plate and have dimensions of (30cm x20cm x0.6 cm). The natural composite materials plates are setting in horizontal, vertical, and inclined with different angles. They are heated with different constant heat flux. A SOLID WORK PREMIUM 2012 is used to draw the geometry of plate as shown in figure 1. ANSYS FLUENT 12.1 package has been conducted for performing numerical simulation across the NCM plate using three dimensional model. The assumptions for the present work are; steady state, Newtonian fluid, incompressible laminar flow, three dimensional, free convection, external flow. The conservation equations for continuity, momentum, and energy can be written as follows [9]:

Continuity Equation:

$$\frac{\partial u}{\partial x} + \frac{\partial v}{\partial y} + \frac{\partial w}{\partial z} = 0 \tag{11}$$

Momentum Equations (x, y, z components):

$$\rho\left(u\frac{\partial u}{\partial x} + v\frac{\partial u}{\partial y} + w\frac{\partial u}{\partial z}\right) = \mu\left(\frac{\partial_u^2}{\partial x^2} + \frac{\partial_u^2}{\partial y^2} + \frac{\partial_u^2}{\partial z^2}\right) - \frac{\partial p}{\partial x} + \rho g_x \tag{12}$$

$$\rho\left(u\frac{\partial v}{\partial x} + v\frac{\partial v}{\partial y} + w\frac{\partial v}{\partial z}\right) = \mu\left(\frac{\partial_v^2}{\partial x^2} + \frac{\partial_v^2}{\partial y^2} + \frac{\partial_v^2}{\partial z^2}\right) - \frac{\partial p}{\partial y} + \rho g_y \tag{13}$$

$$\rho\left(u\frac{\partial w}{\partial x} + v\frac{\partial w}{\partial y} + w\frac{\partial w}{\partial z}\right) = \mu\left(\frac{\partial_w^2}{\partial x^2} + \frac{\partial_w^2}{\partial y^2} + \frac{\partial_w^2}{\partial z^2}\right) - \frac{\partial p}{\partial z} + \rho g_z \tag{14}$$

Energy Equation:

$$u\frac{\partial T}{\partial x} + v\frac{\partial T}{\partial y} + w\frac{\partial T}{\partial z} = \alpha\left(\frac{\partial^2 T}{\partial x^2} + \frac{\partial^2 T}{\partial y^2} + \frac{\partial^2 T}{\partial z^2}\right) + \beta T\left(\frac{dp}{dt} + u\frac{dp}{dx} + v\frac{dp}{dy} + w\frac{dp}{dz}\right) + \mu\phi + q''' \tag{15}$$

Unstructured mesh is used to discretized the computational domain by using the finite volume scheme. The model was meshed by GAMBIT software and figure 2 shows the mesh. In the present study, an average of three million cells is used. The maximum number of iterations performed to get the solver terminate are 500 iteration as shown in figure 3. Boundary conditions are specified for each zone of computational domain as shown in figure 4.

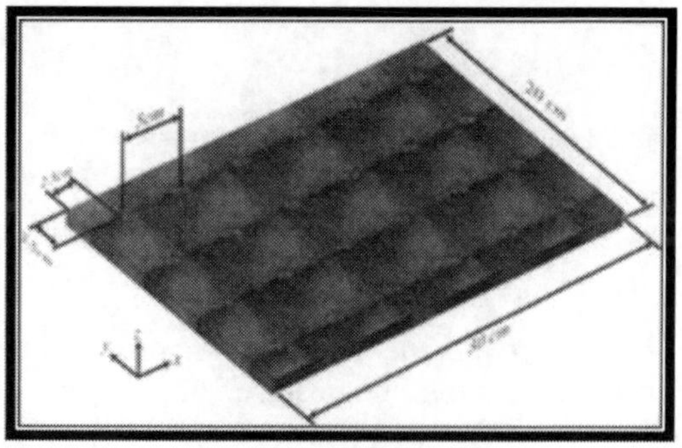

Figure 1. Plate geometry of NCM.

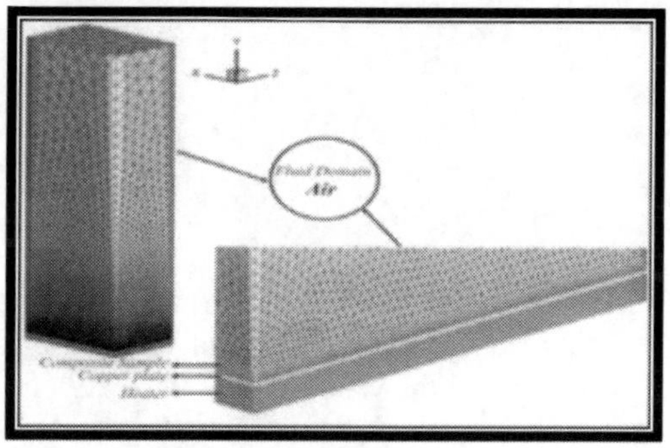

Figure 2. Tetrahedral meshing of the samples and the fluid Domain (Air).

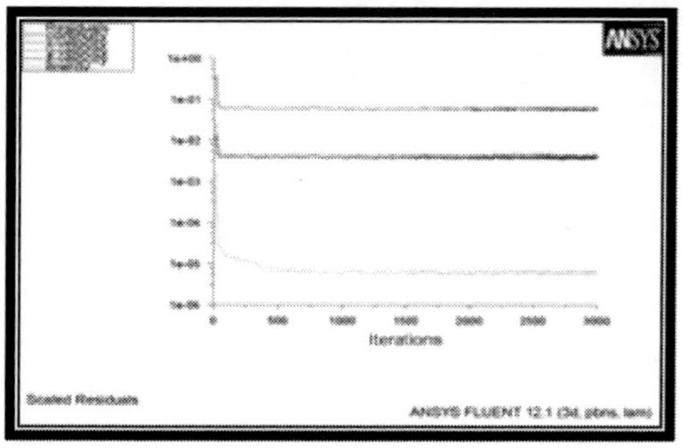

Figure 3. Convergence history for continuity, momentum, and energy equations.

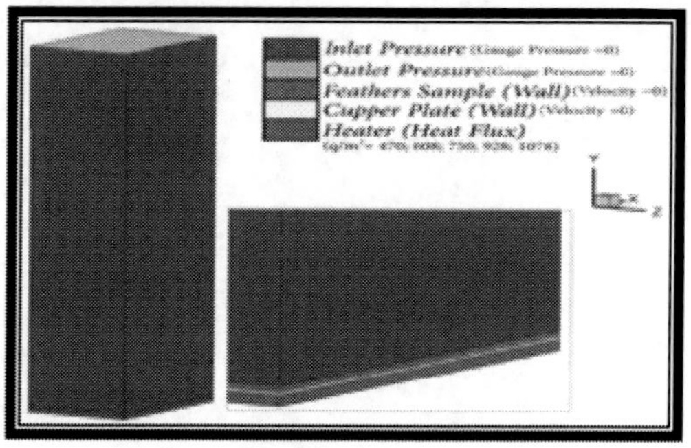

Figure 4. Boundary conditions of the computational domain.

4. Results and Discussion

In order to validate present numerical simulation, a comparison has been carried out with experimental results achieved by Mohammed [10]. Good agreement of behavior between these results has been found which indicates the acceptable validation of present simulation as shown in figure 5 and figure 6. Figures show that the temperature is concerned at mid of NCM plate for different volume of fraction and different type of NCM at horizontal position. But the temperature concerned at the leading edge of plate for vertical position because of air effect.

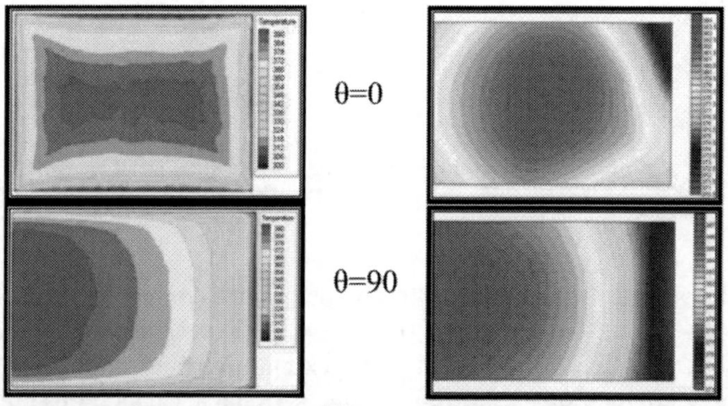

Numerical results (present work) **Experimental results [10]**

Figure 5. Validation of numerical results with reference results for egg shells composite material plate , q=1078W/m^2, V_f=30%, temperatures at (K).

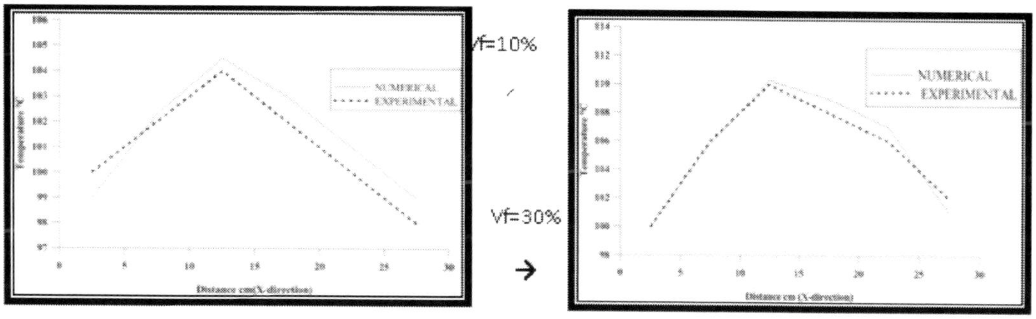

Figure 6. Comparison between numerical and reference experimental results [10] for seed dates composite materials, $q=1078 W/m^2$, $\theta=0^{\circ}$, (X=12.5 cm, Y=12.5 cm).

Figure 7 and figure 8 show the distribution of temperature for two types of NCM (seed dates and feathers) at different positions. Same behavior show in previous figures but different heating temperature. This is because of different properties for NCM (thermal conductivity, specific heat, and thermal diffusivity), which causes difference in the behavior of temperatures. Figures 9, 10 and 11 show the temperature contours at middle Y- plane of three types of NCM with 30% volume of fraction. NCM plates are heated with constant heat flux (1078 W/m^2) at horizontal position. Figures show different shape of plumes above the plate. This is because of the distribution of temperature related with different properties. Figure 12 shows that the heat transfer from the plate is increases with increasing the angle of position and the heat flux depending on the properties on NCM.

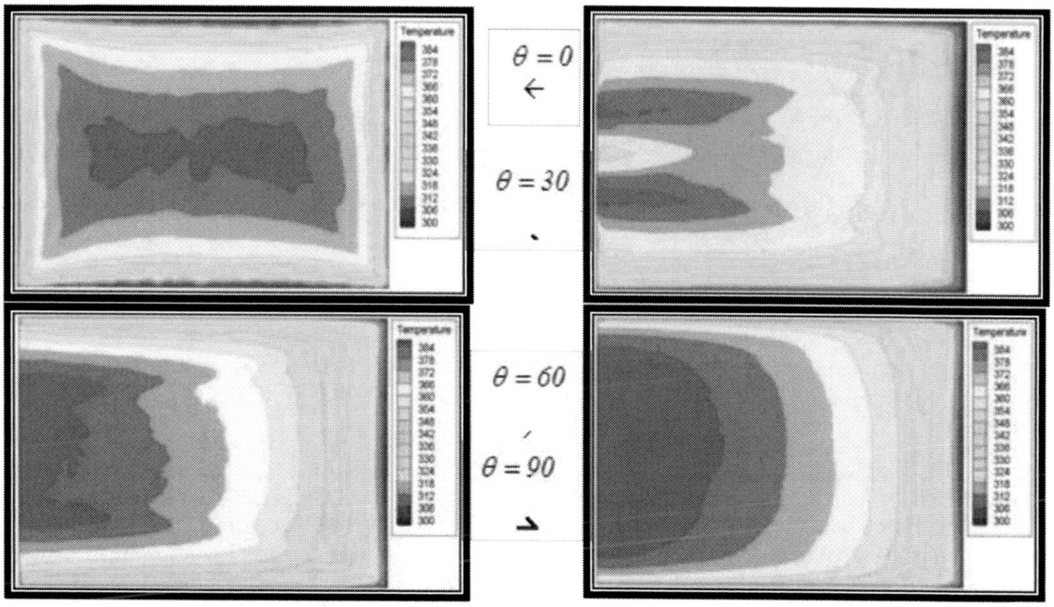

Figure 7. Temperature contours at the plate for seed dates composite material, $q=1078 W/m^2$, $V_f=30\%$.

IOP Conf. Series: Materials Science and Engineering **95** (2015) 012001 doi:10.1088/1757-899X/95/1/012001

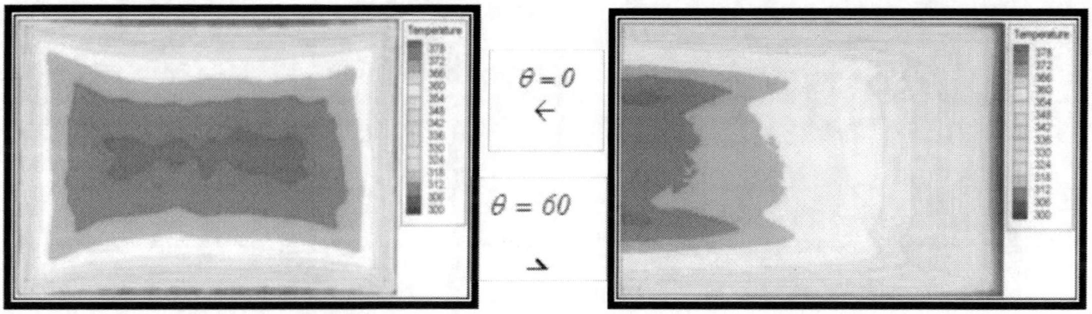

Figure 8. Temperature contours at the plate for feathers composite material, q=1078W/m^2, V_f=30%.

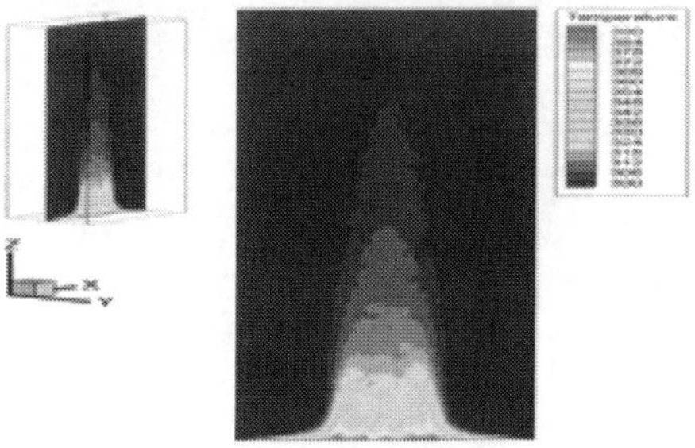

Figure 9. Temp. contour at XZ-plane for egg shells composite material q=1078W/m^2, V_f =30%, θ=0°.

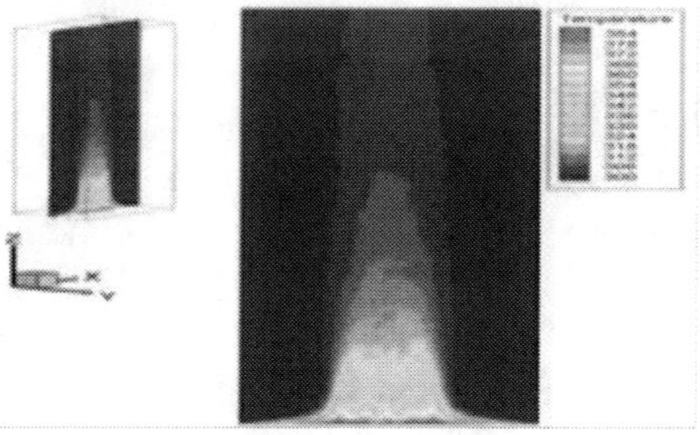

Figure 10. Temp. contour at XZ-plane for seed dates composite material q=1078W/m^2, V_f=30%, θ=0°.

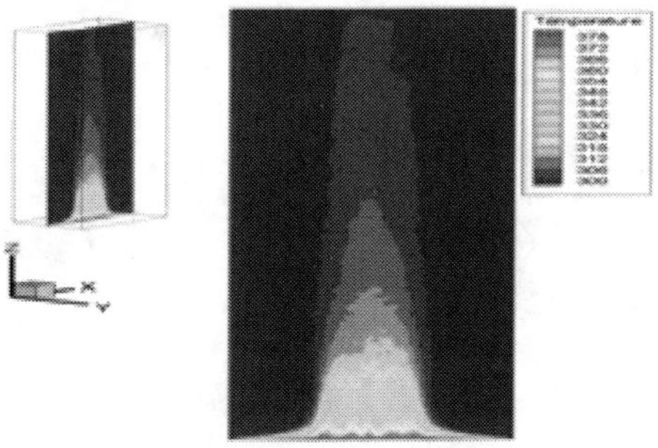

Figure 11. Temp. contour at XZ-plane for feathers composite material q=1078W/m^2, V_f =30%, θ=0°.

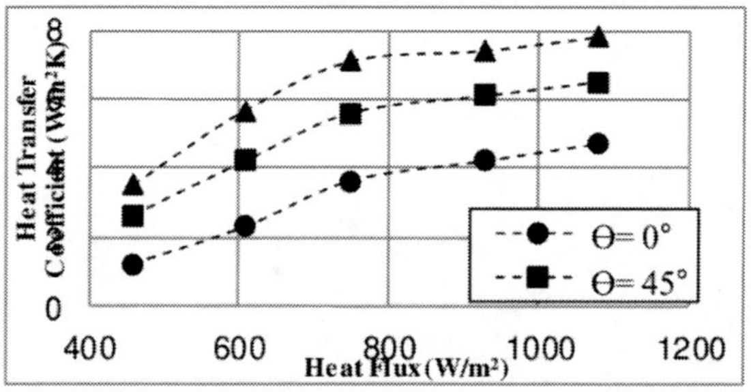

Figure 12. Effect of heat flux on heat transfer coefficient for seed dates composite material V_f=30%.

5. Conclusions

The main concluded points of this study may be summarized as follows:

- The temperature is concentrated at the central of NCM plate in horizontal position. But it is start to deviate toward the edge of plate at any angle of inclination especially vertical position.

-Increasing the volume of fraction of (seed dates, egg shells, and feathers) in the composite materials lead to different properties for NCM such as (thermal conductivity, specific heat, and thermal diffusivity).

-The temperature distribution is different due to the properties of NCM.

- The heat transfer coefficient increases due to the increases in heat flux for different type of NCM and position of plate.

References

[1] Yasmeen H 2005 Numerical Study for a Three Dimensional laminar natural convection heat transfer from an isothermal heated horizontal and inclined square plate and with circle hole *M Sc thesis* (University of Technology)
[2] Yunus A Ç 2002 Heat Transfer *McGraw-Hill* Sixth edition
[3] Hassan K and Mohammed A 1970 Natural Convection from Isothermal Flat Surfaces *Int J Heat and Mass Transfer* **13** pp 1873-1886
[4] Pera L and Gebhart B 1973 Natural Convection Boundary Layer Flow Over Horizontal and Slightly Inclined Surfaces *Int J Heat and Mass Transfer* **16**
[5] Miyamoto M , Katoh Y , Kurim J , Kurihara S and Yamashita K 1985 Free Convection Heat Transfer from Vertical and Horizontal Short Plates *Int J Heat and Mass Transfer* **28(9)** pp 1733-1745
[6] Rafah A 2002 Instruction System to Study Free Convection Heat Transfer from Isothermal Horizontal Square Flat Surfaces *M Sc thesis* U O T
[7] Fujino J and Honda T 2002 Experimental Investigation on the Thermal Conductivity Measurement of the Fly Ash Dispersed Plastic Composite Using Guarded Hot Plate Apparatus (Fukuoka University: Nanakuma Jonan-ku Japan)
[8] Faieza M 2006 Thermal and Mechanical Investigation of Fiber Reinforced Epoxy *Ph D thesis* (University of Technology)
[9] Holman J P 2010 Heat Transfer *McGraw-Hill* tenth edition
[10] Mohammed R 2013 Experimental and Numerical Investigation for laminar- natural convection heat transfer from Natural Composite Materials *M Sc thesis* (Al-Mustansiriya University College of Engineering)

The effect of reconditioning techniques by welding on the quality of deposits on crankshafts, case study SMAW

O R Chivu[1], **D T Cicic**[1], **C Rontescu**[1], **I M Vasile**[1] **and C Petriceanu**[1]
[1]University POLITEHNICA of Bucharest, Faculty of Engineering and Management of Technological Systems, Materials Technology and Welding Department, Romania

E-mail: virlan_oana@yahoo.co.uk

Abstract. Currently, we are searching for a range of solutions for repairing the crankshafts that had snapped during operation. The paper deals with the extension of the two methods for reconditioning by welding in the energy industry in the field of repairing the crankshafts in the automotive industry. The results obtained through the application of methods for reconditioning Weld Toe Tempering Technique and filling layers of sidings, which was used as a welding procedure SMAW. Qualitative and quantitative comparisons between the results of the two methods based on the criterion of rigidity are carried out.

1. Introduction

The defective metal products in conjunction with the price of new products led to the search for and development of various technological processes which help to restore the original dimensions and form of damaged product [1].

Another principle that underlies the choice of repairing by welding and replacement of defective components is that we can get the repair parts with the original superior properties (which have not withstood the conditions of use).

The success of repairing operation by welding depends mostly of the conditions under which it will carry out the analysis of the information related to the pieces to be repaired. One of the reasons for performing this analysis is to establish the causes of cracking for repairing by welding situation, or wear of parts in case of filling by welding.

Once the decision to repair by welding was taken is necessary to determine the causes which have led to the faults from the following possible causes: accident, improper use, abuse and operating conditions over the maximum limit allowed.

In the event that damage has occurred due to accident or operating conditions values being exceeded, the renovation will aim only restoring the shape and characteristics of the component.

The work aims to offer solutions for repairing by welding in a situation in which the non-compliance occurs, which may endanger the safety of operation, on the counterweight on the crankshaft from automotive industry [2].

2. Techniques of reconditioning

Worldwide, there are several methods of repair by welding for various products obtained from various ferrous materials or non-ferrous. In the present paper, two such techniques are used, Weld Toe Tempering Technique and filling on layers sides the counterweight components on a crankshaft from automotive industry.

Content from this work may be used under the terms of the Creative Commons Attribution 3.0 licence. Any further distribution of this work must maintain attribution to the author(s) and the title of the work, journal citation and DOI.
Published under licence by IOP Publishing Ltd

IOP Conf. Series: Materials Science and Engineering **95** (2015) 012002 doi:10.1088/1757-899X/95/1/012002

The technique of Weld Toe Tempering Technique, figure 1, involves filling a layer or an additional channel at the surface of the bead to ensure the recovery effect of the previous interlayer ZIT filed.

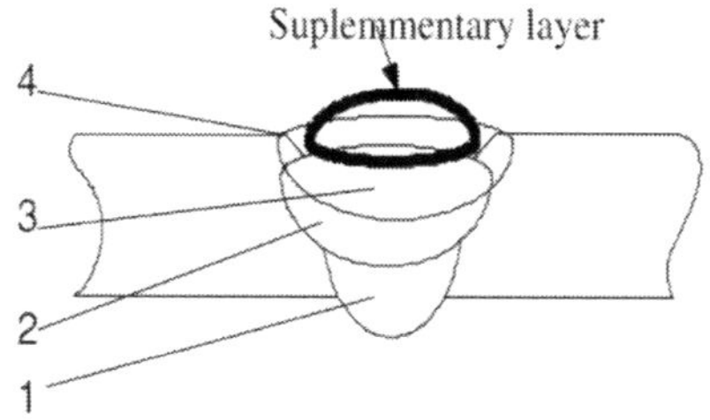

Figure 1. The application of the technique "of an extra layer" in the repairing by welding: 1 - root layer; 2 - fill layer; 3 - layer sealing; 4 - layer of heat treatment.

To illustrate the effect of extra layer, in figure 2 is represented the section by two layers of a welded joint. From this figure it can be seen that the second layer provides the heat necessary to carry out a "thermal self-treatment" the first row microstructure, [3,4,5].

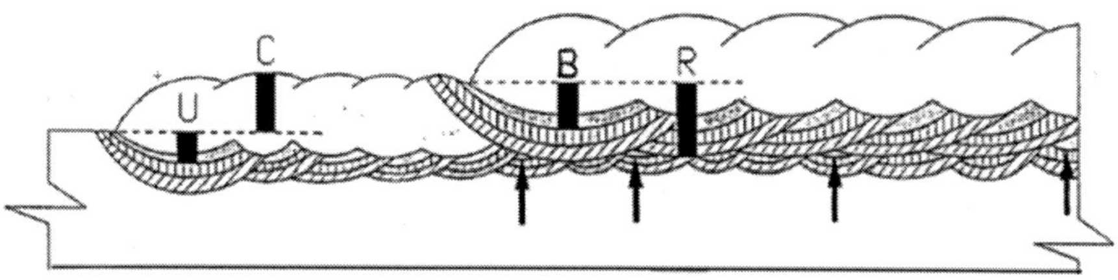

Figure 2. The Section through 2 layers of a welded joint [5]. B - Boundary Layer fusion II; U - Boundary Layer fusion I; C - Median value of aggradation layer 1; R - the maximum value of the finishing layer depth 2; ▭ The stitching material; ▨ The overheating layer of the ZIT; ▥ The layer with the fine film grain of the ZIT; ▧ Partial transformation area of the ZIT.

The technique of filling layers on the side (left-right), figure 3, involves obtaining similar effect on the ZIT, as in the previous case, but with a higher consumption of filler material. In this way, basically, we get the same effect on both sides [6].

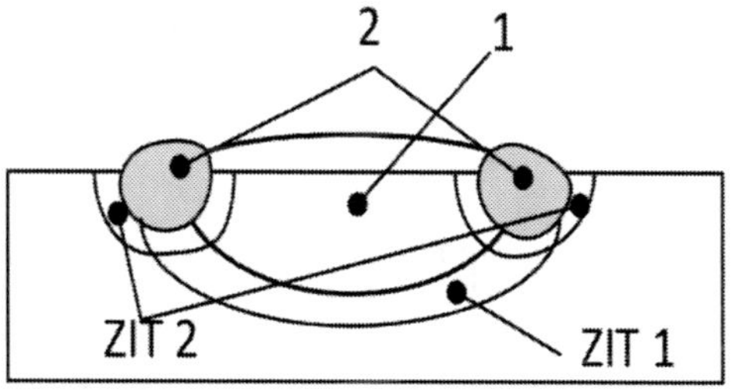

Figure 3. The principle of the technique of lateral filling of additional layers. 1 - filled layer in order to repair; 2 additional layers arranged along the left-right of the reparation; ZIT 1 – the heat influenced area due to layer 1; ZIT 2 – the heat - affected area due to layer 2.

3. Experimental data

The basic material from which counterweight was made is cast iron, type EN-GJS-600-3 according to DIN EN 1564: 2012 [7], whose chemical composition and mechanical properties are given in tables 1 and 2.

Table 1. Standard chemical composition of EN-GJS-600-3 material.

Material designation	C [%]	Si [%]	Mn [%]	P [%]	S [%]
EN-GJS-600-3	2.5-3.6	1.8-2.8	0.3- 0.7	≤0.08	≤0.02

Table 2. Material properties measured on test pieces according to DIN EN 1563:2012.

Material designation	Tensile strength Rm [N/mm^2]	0,2% Proof stress Rp0,2 [N/mm^2]	Elongation A [%]	Microstructure
EN-GJS-600-3	600	370	3	Pearlite/ ferrite

The purpose of the application of the techniques of reconditioning exposed above, on the surface of the counterweight a channel on all the piece has been performed with the dimensions of 16 mm width and depth of 5 mm, figure 4, which simulates the removal of a range with nonconformities which will be subject of repairing by welding. Realization of the channel took place through mechanical processing so as not to influence the thermal structure of the material.

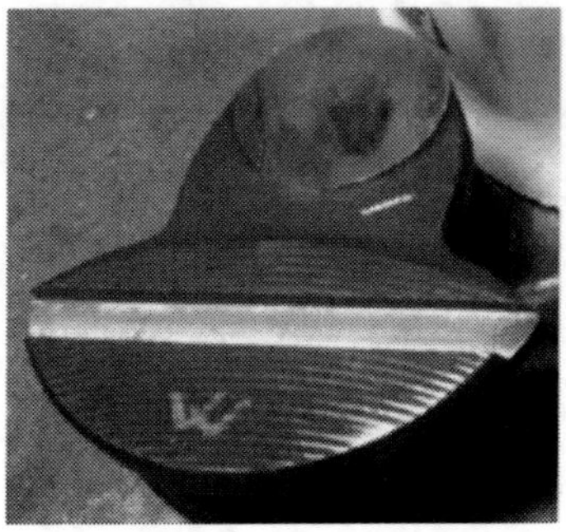

Figure 4. The channel achieved.

3 samples were completed as follows:
- the first sample, coded P1, repaired by welding by normal filling of layers, figure 5.a.;
- the second sample, coded P2, which has filed a supplementary layer by technique Weld Toe Tempering Technique, figure 5.b.;
- the third sample, coded P3, obtained by filling on the side of extra cords, figure 5.c.

a. b. c.

Figure 5. Tests results.

When filling the completed channels crossing, we used the following parameters of reconditioning by welding [8,9]:

- normal filling of the layers: Is = 140A. Ua = 20 V;
- the application of Weld Toe Tempering Technique: coatings filling Is = 140A. Ua = 20 V, extra layer filling Is = 100A and Ua = 18V;
- for filling on the side of extra cords: the filling of layers Is = 140A. UA = 20 V, lateral filing two additional crossings Is = 100A and Ua = 18V.

4. Results

After the end of the experiments, the samples were subjected to examination of optical-visual and with penetrant liquids, not finding any non-conformity.

The reconditioned counterweights by welding have been debited in the central area to obtain samples that were subjected to macroscopic examination and measurement of rigidity. After processing, the samples are shown in figure 6, numbered in the order of the filling layers, in the channel provided for in the counterweight, as follows: 1 - the first layer, 2 - the second layer and so on.

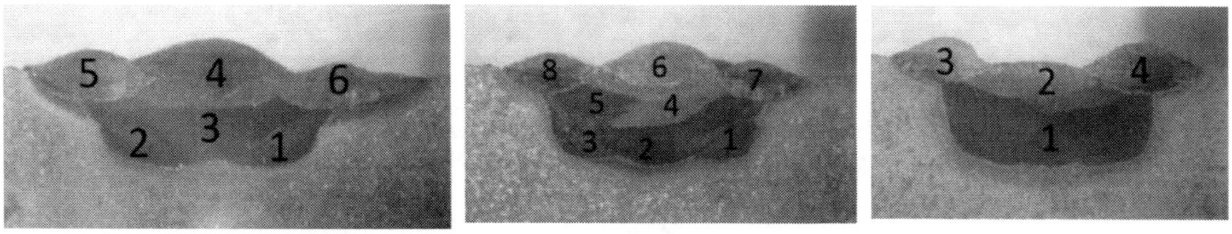

a) the usual technical filling b) Weld Toe Tempering Technique c) sided layers filling

Figure 6. Samples for microscopic examination and measurement of resulted hardness after appropriate processing.

After the macroscopic examination we took the hardness values HV 0.5, in the areas shown in figure 7. The values obtained are shown in table 3 and the graphs showing changes in hardness are contained in figure 8-15, for all areas surveyed.

Table 3. Hardness values - all samples.

Area	Localization	Item No.	Sample			Area	Localization	Item No	Sample		
			P1	P2	P3				P1	P2	P3
S1	ZIT 1	1	689	496	723	The bottom area	Bottom cord	36	331	337	327
		2	653	524	726			37	348	368	314
		3	677	508	737			38	329	355	328
		4	647	494	760			39	347	350	315
		5	611	562	728			40	341	376	316
	ZIT 2	6	535	414	466		ZIT	41	508	330	441
		7	527	374	402			42	428	333	402
		8	437	376	382			43	503	392	426
		9	420	365	375			44	492	354	412
		10	409	382	392			45	415	325	391
S2	ZIT 1	11	632	471	752	Superior Cords	Cord 1	21	261	618	431
		12	665	561	762			22	273	610	427
		13	687	527	740			23	240	584	370
		14	644	591	710			24	250	598	466
		15	615	524	751			25	275	616	434
	ZIT 2	16	520	337	387		Cord 2	26	251	505	253
		17	470	340	353			27	233	491	259
		18	399	312	351			28	238	508	248
		19	347	327	334			29	245	474	255
		20	374	356	367			30	238	500	250

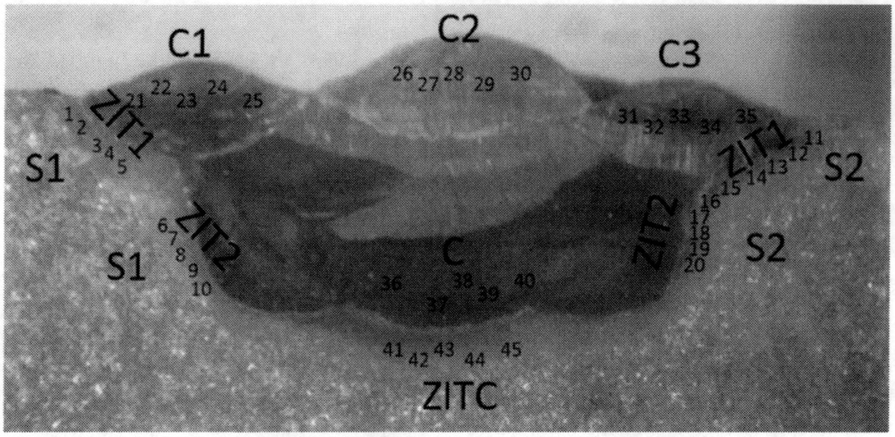

Figure 7. Hardness measurement are:

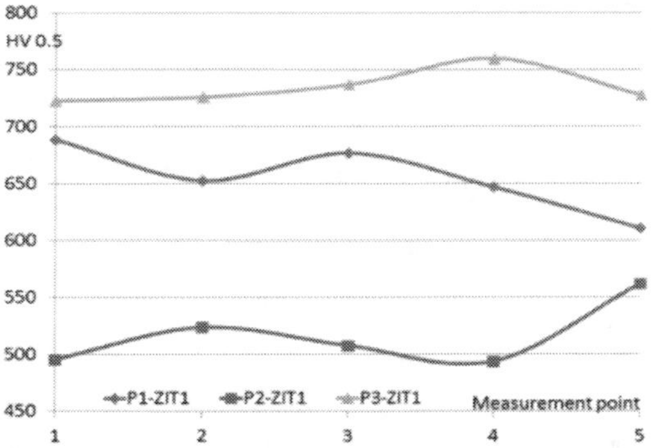

Figure 8. Variation of hardness-S1-ZIT1 Area.

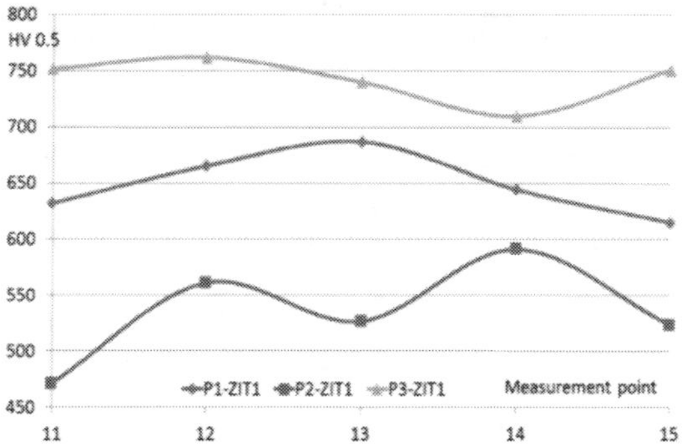

Figure 9. Variation of hardness-S2-ZIT1 Area.

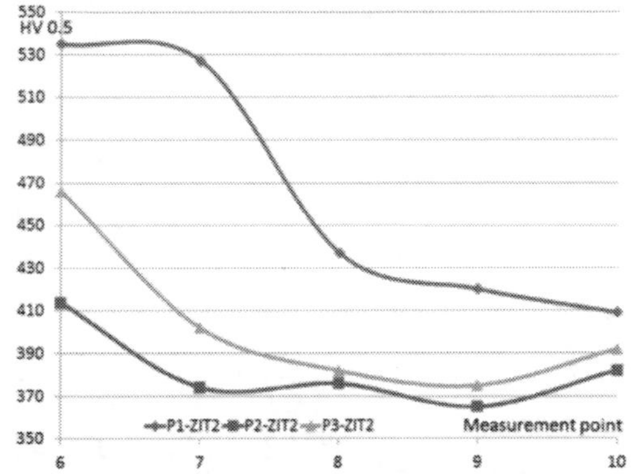

Figure 10. Variation of hardness - S1-ZIT2 Area.

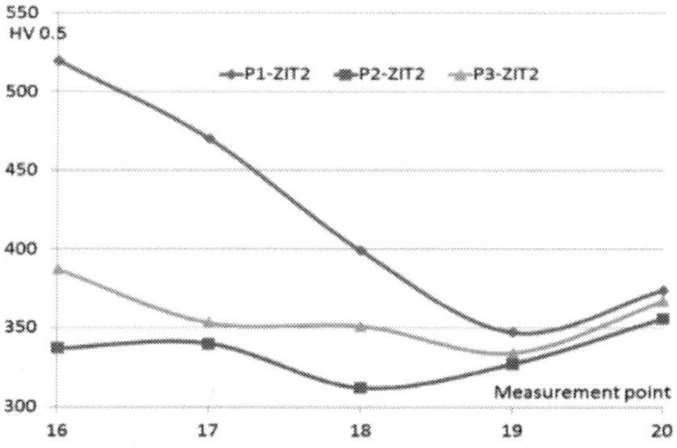

Figure11. Variation of hardness - S2-ZIT2 Area.

Analysing the hardness obtained in the area ZIT 1, top area, figure 8 and figure 9, of all the samples, it can be seen that the higher values were obtained in the sample P3, a fact explained by the filling of side layers.

In of case ZIT-2, at the bottom area, figure 10 and figure 11, the maximum hardness values were obtained for the sample results through normal filling by welding, P1, a fact explained by the swinging movement of the layers filled.

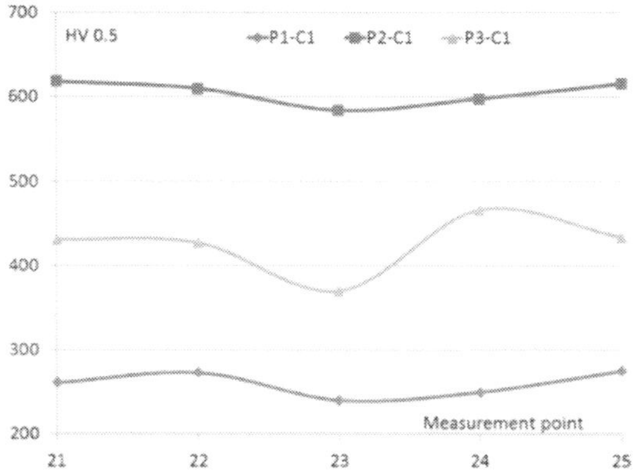

Figure 12. Variation of hardness - C1 Area.

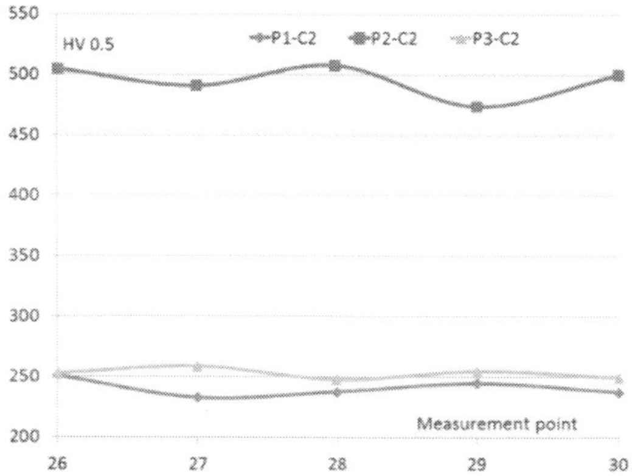

Figure13. Variation of hardness - Area C2.

From the analysis of the hardness values obtained in the upper layers, coded as C1 and C2, figure 12 and figure 13, it can be seen that the same phenomenon occurs, downwards, the resulted hardness values are: sample P2, sample P3 and sample P1.

5. Conclusions

From the research conducted and the results obtained we can draw the following conclusions:

- SMAW procedure can be applied to good effect in counterweights repair of the crankshaft in the automotive industry;
- the results, based on the criterion of rigidity, permits the extension methods for reconditioning by welding in the energy industry and other industrial products;
- the hardness largely depends on the order of transitions filling;
- we can note the extra layer filling importance, or the extra layer, on the hardness of the reconditioned area.

Acknowledge

The work has been funded by the Sectoral Operational Programme Human Resources Development 2007-2013 of the Ministry of European Funds through the Financial Agreement POSDRU/159/1.5/S/134398.

All measurements were carried out in an accredited laboratory LAMET, University Polytechnic of Bucharest, Faculty of engineering and management of technological systems, Materials and welding Technology Department.

References

[1] Cicic D T and Iacobescu G 2014 *Informatizarea şi Optimizarea Proceselor de Sudare* (Bucharest: Politehnica Press Publishing House) pp 150-153
[2] Information on http://www.ustudy.in, accessed: 15.01.2015
[3] Allen P J 1997 Cold weld repair, development and application *Proc. of Welding and Repair Technology* pp 122-127
[4] Cicic D T, Solomon G, Iacobescu G and Rontescu C 2011 Method of establishing the optimum technique for reconditioning by welding, taking into account the consumption of filler metals and electrical power *U.P.B. Sci. Bull.* D 73 **1** pp 85-98
[5] Edgley J S and Pitrun M 1993 The Temper Bead Weld Repair of the Hazelwood Steam Drums, *Pressure Vessels and Pipework Conf.* (Sydney)
[6] Information on http://www.twi-global.com, accessed: 15.01.2015
[7] Vaško M and Vaško A 2014 Correlation between charge composition and fatigue properties of nodular cast irons, *Applied Mechanics and Materials* 474, pp. 291-296
[8] Consumables for manual and automatic welding *ESAB Welding Handbook*, eighth edition
[9] Information on http://www.esabna.com , accessed: 15.01.2015

IOP Conf. Series: Materials Science and Engineering **95** (2015) 012003 doi:10.1088/1757-899X/95/1/012003

Qualitative and quantitative comparisons on reconditioning by welding of crankshafts from auto industry

O R Chivu (Vîrlan)[1], C Rontescu[1], D T Cicic[1], I M Vasile[1] and C Petriceanu[1]
[1]Politehnica University of Bucharest, Faculty of Engineering and Management of Technological Systems, Materials Technology and Welding Department, Romania

E-mail: virlan_oana@yahoo.co.uk

Abstract. One of the goals of modern society is represented by reducing the cost for the maintenance of cars. One of the pieces that break down most often on the engine block is the crankshaft, in various areas of it. Due to the fact that the price of a crankshaft is very high, specialists seek solutions for repairing and not replacing them. In this study, it is presented a comparison in terms of hardness obtained at recovering the counterweight by welding by using two methods of welding, WIG and SMAW through various techniques: normal, WTO, lateral depositing.

1. Introduction

The present work aims to offer solutions for repairing by welding in a situation in which the non-compliance occurs, which may endanger the safety of operation, on the counterweight on the crankshaft from automotive industry, shown in figure 1.

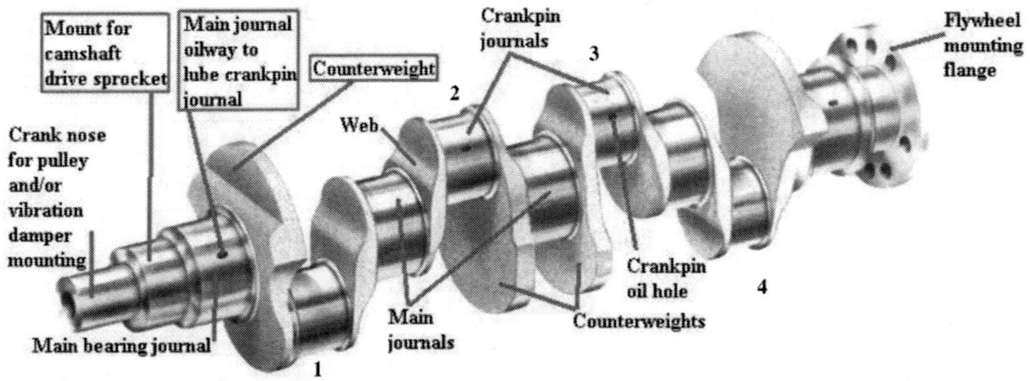

Figure 1. The components of a crankshaft [1]

For rebuilding the geometric dimensions, the favorite repair processes of crankshaft are welding and metallization.

The metallization process presents as main disadvantages, the high cost and the thorough preparation of the piece.

The main problem when applying the technology of repairing by welding is represented by the

Content from this work may be used under the terms of the Creative Commons Attribution 3.0 licence. Any further distribution of this work must maintain attribution to the author(s) and the title of the work, journal citation and DOI.
Published under licence by IOP Publishing Ltd

overheating of the support material which leads to irreversible mechanical and structural transformations.

In order to perform the experiments we opted for 2 repairing techniques, Weld Toe Tempering Technique and depositing lateral layers on counterweight components on a crankshaft from automotive industry, where we used 2 welding processes with electric arc SMAW and WIG.

Weld Toe Tempering Technique involves depositing a layer or an additional passage (supplementary) at the surface of the cord to ensure the recovery effect of the interlayer HAZ deposited above [2,3]. The technique seeks mainly to improve resilience of heat affected zone (HAZ/ZIT) by reducing its enlargement. The additional layer deposited must not form a liquid metal bath with the base material [4].

The depositing technique of lateral layers (left-right), aims to achieve similar effects on the ZIT as and in the previous case, but with a higher consumption of added material by depositing and subsequent removal of the layers [5].

Both techniques seek to improve the structure and hardness of the heat affected zone that appears when using welding processes [6].

2. Experimental data

The basic material from which the counterweight was made is a cast iron, type EN-GJS-600-3 according to DIN EN 1564: 2012.

Regarding the purpose of applying the techniques of reconditioning exposed above the surface of the counterweight, a channel has been made with all the dimensions of 16 mm width and depth of 5 mm, figure 2, which simulates the removal of an area with nonconformities, which will be subject to repairing by welding. Realizing the channel took place through mechanical processing with continuous cooling in order not to influence the thermal structure of the material.

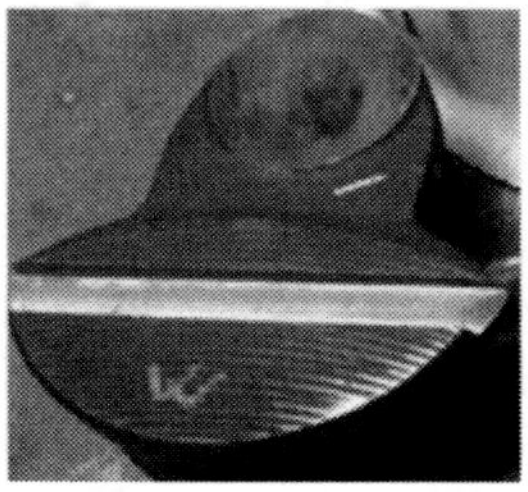

Figure 2. The achieved channel.

The parameters used in the experiments are indicated in table 1.

Table 1. The parameters used in the experiments.

<table>
<tr><th rowspan="2">No. crt.</th><th rowspan="2" colspan="2">Parameter</th><th colspan="2">Process</th></tr>
<tr><th>SMAW</th><th>WIG</th></tr>
<tr><td>1.</td><td colspan="2">Filler material</td><td>E10-UM-60-CZ</td><td>WSG-3GZ-5-T</td></tr>
<tr><td>2.</td><td colspan="2">Filler material diameter [mm]</td><td>3.25</td><td>2.4</td></tr>
<tr><td rowspan="3">3.</td><td rowspan="3">The intensity of welding [A]</td><td>Layers</td><td colspan="2">140</td></tr>
<tr><td>Additional Layer Depositing</td><td colspan="2">100</td></tr>
<tr><td>Lateral Layers Depositing</td><td colspan="2">100</td></tr>
<tr><td rowspan="3">4.</td><td rowspan="3">ARC voltage [V]</td><td>Depositing Layers</td><td>22</td><td>14</td></tr>
<tr><td>Additional Layer Depositing</td><td>24</td><td>16</td></tr>
<tr><td>Lateral Layers Depositing</td><td>22</td><td>14</td></tr>
</table>

3. Results and discussions

The encoding of samples was: P1- the resulting normal deposit sample using the SMAW procedure; P2 – the resulting sample in Weld Toe Tempering Technique using the SMAW procedure; P3 – the sample resulted by depositing lateral layers using the SMAW procedure; P1'- the resulted sample by normal depositing using the WIG procedure; P2'- the resulted sample through the Weld Toe Tempering Technique using the WIG procedure; P3'- the resulted sample by depositing lateral layers using the WIG procedure;

After completing the experiments, the test results, figure 3, were subjected to examination of optical-visual and penetrant liquids, not finding any non-conformity.

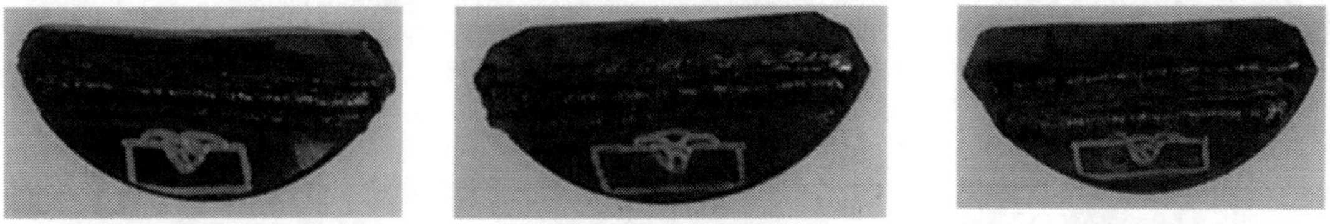

a) The usual technical depositing – SMAW sample P1 | b) Weld Toe Tempering Technique – SMAW sample P2 | c) Depositing lateral layers - SMAW sample P3

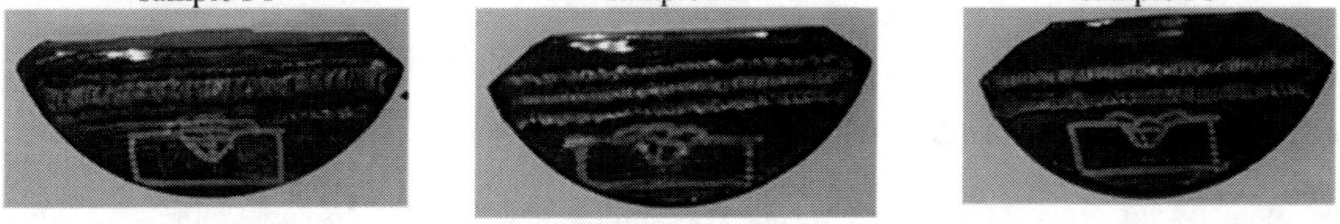

d) Weld Toe Tempering Technique – WIG sample P2' | e) Weld Toe Tempering Technique – WIG sample P2' | f) Depositing lateral layers WIG sample P3'

Figure 3. Image of resulting samples.

a) The usual technical-depositing SMAW - sample P1 | b) Weld Toe Tempering Technique - SMAW- sample P2 | c) Depositing lateral layers – SMAW - sample P3

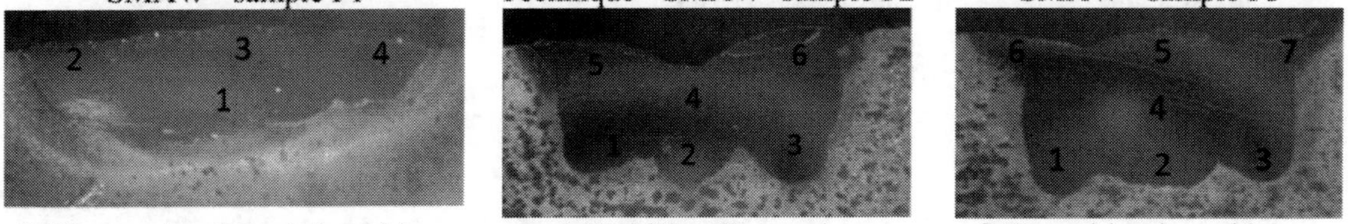

d) The usual technical-depositing - WIG - sample P1' | e) Weld Toe Tempering Technique - WIG - sample P2' | f) Depositing lateral layers WIG - sample P3'

Figure 4. Samples for microscopic examination and measurement of hardness resulted after appropriate processing.

The reconditioned welding counterweights have been debited in the central area to obtain samples that were subjected to macroscopic examination and measurement of hardness. The samples after processing are shown in figure 4, in which it is indicated with numbers, the order of the layers in the channel provided for in the counterweight, as follows: 1 - the first layer, 2 - the second layer and so on.

After the macroscopic examination we took the hardness values HV0.5, in the areas shown in the figure 5. The values obtained are shown in table 2 and the charts, some of the areas analyzed, showing the variation of hardness can be found in figure 6-8.

Table 2. Hardness Values HV0.5 – all samples.

Area	Localization	No. Points.	Samples - SMAW			Samples - WIG		
			P1	P2	P3	P1'	P2'	P3'
S1	ZIT 1	1	689	496	723	434	700	584
		2	653	524	726	439	620	581
		3	677	508	737	460	618	590
		4	647	494	760	498	600	588
		5	611	562	728	495	655	515
	ZIT 2	6	535	414	466	311	333	324
		7	527	374	402	317	338	333
		8	437	376	382	315	356	326
		9	420	365	375	316	341	331
		10	409	382	392	321	343	331
S2	ZIT 1	11	632	471	752	337	762	521
		12	665	561	762	314	776	578
		13	687	527	740	340	718	505
		14	644	591	710	337	728	589
		15	615	524	751	373	770	526
	ZIT 2	16	520	337	387	335	376	345
		17	470	340	353	347	393	360
		18	399	312	351	342	372	355
		19	347	327	334	321	382	335
		20	374	356	367	320	351	337
Superior Cords	Cord 1	21	261	618	431	291	380	370
		22	273	610	427	260	386	359
		23	240	584	370	250	367	358
		24	250	598	466	262	372	357
		25	275	616	434	248	364	336
	Cord 2	26	251	505	253	253	299	269
		27	233	491	259	207	301	281
		28	238	508	248	235	290	273
		29	245	474	255	206	297	283
		30	238	500	250	215	308	271
	Cord 3	31	259	457	324	240	461	353
		32	280	501	386	232	493	376
		33	268	492	358	271	439	355
		34	278	556	385	261	519	346
		35	261	546	366	240	501	362
Bottom Part	Bottom Cord	36	331	337	327	369	406	398
		37	348	368	314	351	420	410
		38	329	355	328	364	399	380
		39	347	350	315	358	400	383
		40	341	376	316	357	391	367
	ZIT	41	508	330	441	427	352	414
		42	428	333	402	420	358	372
		43	503	392	426	431	363	373
		44	492	354	412	447	329	366
		45	415	325	391	423	340	374

After the macroscopic examination we took the hardness values HV0.5, in the areas shown in the figure 5. The values obtained are shown in table 2 and the charts, some of the areas analyzed, showing the variation of hardness can be found in figures 6-8.

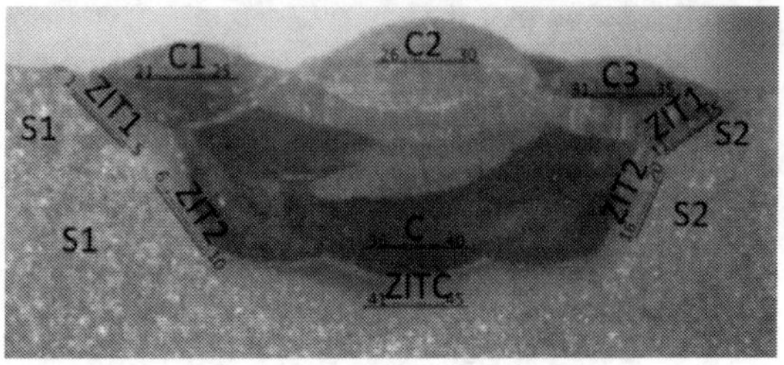

Figure 5. Hardness measurement areas.

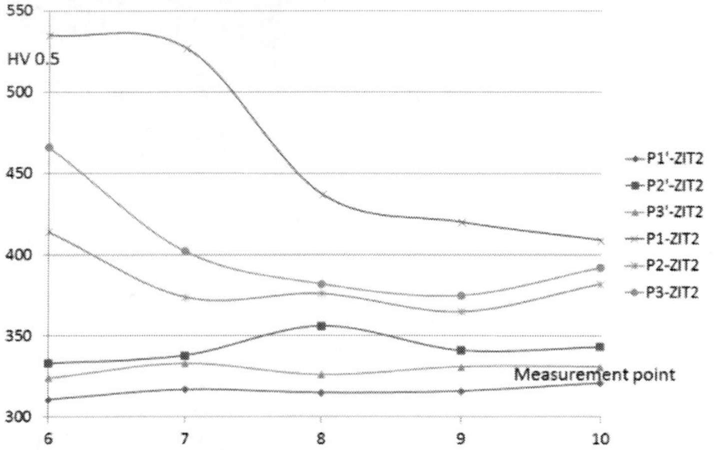

Figure 6. Variation of hardness – S1 - ZIT2 Area.

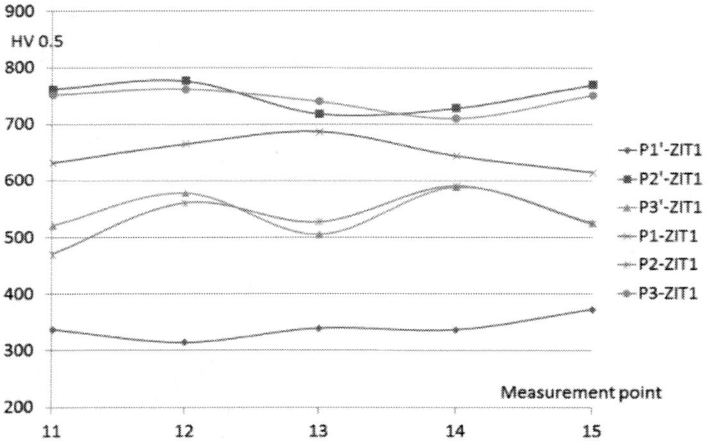

Figure 7. Variation of hardness – S2 - ZIT1 Area.

In case of ZIT2, S1 area, in the lower part of the figure 6, it can be seen that the maximum hardness values were obtained for the samples resulted through normal depositing by welding, P1 and the minimum values are obtained when using the WIG welding process and the normal technique of depositing. The low values obtained are explicable by lower heat into the material.

From analyzing the hardness obtained in area ZIT 1, area S2, top, figure 7, it can be seen that the highest values were obtained in the sample P2', and the minimum values as one would expect in the case of the P1'.

From analysing the hardness values obtained in the median area, encoded with C2, figure 8 shows the maximum values obtained in the case of the sample P2 in the implementation of which we used the SMAW process of welding and Weld Toe Tempering Technique and the minimum values in the case of the P1' realized by the procedure of normal depositing WIG.

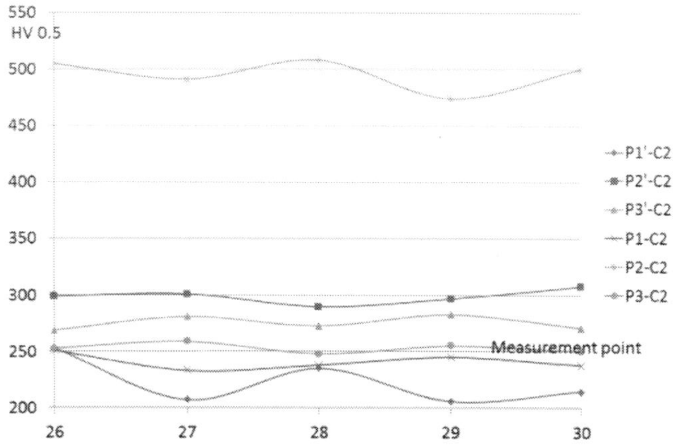

Figure 8. Variation of hardness - Area C2.

4. Conclusions

On the basis of the information in the paper, the following conclusions may be drawn:

- Both welding processes can be used for reconditioning by welding of the elements associated with crankshafts in the automotive industry;
- The layers layout mode to restore the constructive shape, has a primary effect on hardness;
- Using the WIG process presents the major advantage of the introduction of a smaller heat quantities in the reconditioned part, so a more beneficial influence on hardness;
- The variations in hardness, for all 6 samples, are not linear;
- It is necessary to conduct research on the structure for a better correlation between the influence of techniques, procedures and delimitation of their applications.

Acknowledge

The work has been funded by the Sectoral Operational Programme Human Resources Development 2007-2013 of the Ministry of European Funds through the Financial Agreement POSDRU/159/1.5/S/134398. All measurements were carried out in an accredited laboratory LAMET, University Polytechnic of Bucharest, Faculty of Engineering and Management of Technological Systems, Materials and Welding Technology Department.

References

[1] Information on http://www.ustudy.in, accessed: 15.01.2015
[2] Allen P J 1997 Cold weld repair, development and application *Proc. Welding and Repair Tech.* pp 122-127
[3] Cicic D T, Solomon G, Iacobescu G and Rontescu C 2011 Method of establishing the optimum technique for reconditioning by welding, taking into account the consumption of filler metals and electrical power U.P.B. Sci. Bull. Series D Vol.73 **1** pp 85-98
[4] Information on http://www.gowelding.com/met/temper.htm, accessed: 15.01.2015
[5] Information on http://www.twi-global.com, accessed: 15.01.2015
[6] D.T. Cicic D T and Iacobescu G 2014 *Informatizarea şi optimizarea proceselor de sudare* (Bucharest: Politehnica Press Publishing House) pp 154-156

IOP Conf. Series: Materials Science and Engineering **95** (2015) 012004 doi:10.1088/1757-899X/95/1/012004

The surface quality of AWJ cut parts as a function of abrasive material reusing rate

C Schnakovszky[1], **E Herghelegiu**[1], **M C Radu**[1] **and N C Tampu**[1]
[1]Vasile Alecsandri University of Bacau, Calea Marasesti 157, 600115, Bacau, Romania

E-mail: scarol@ub.ro

Abstract. Abrasive water jet cutting (AWJ) has been extensively used during the last years to process a large variety of materials since it offers important advantages as a good quality of the processed surface, without heat affected zones, low environmental impact (no emission of dust or other compounds that endanger the health of the user), small induced mechanical stresses etc. The main disadvantage is the high cost of processing (cost of equipment and consumables). In view of this, the effects of reusing the abrasive material on the quality of processed surface are investigated in this paper. Two steel materials were used: OL 37 (S 235) with large applicability in machine building industry and 2P armor steel used in the arms industry. The reusing rate of the garnet abrasive material was: 0%, 20%, 40%, 60%, 80% and 100%. The quality of processed surface was quantified by the following parameters: width at the jet inlet (Li), width at the jet outlet (Lo), inclination angle (α), deviation from perpendicularity (u) and roughness (Ra).

1. Introduction

The need to process different hard and super hard materials that cannot be processed by conventional processing methods led to the appearance of unconventional technologies. These technologies have been developed greatly in recent decades, as they are used at large scale in different industries. A great advantage of the unconventional technology is the lack of contact between the tool and the surface to be processed; the processing is done using agents such as electricity, water, air, chemicals etc [1].

The newest unconventional technology is water jet cutting (the first reliable ultra-high pressure pump was developed in the early 1970s). However, while water jets could easily cut soft materials, they were not effective in cutting harder materials such as metals. This capability was achieved by adding an abrasive to the water jets. In 1979, Dr. Mohamed Hashish, from the Flow International Cooperation Company, introduced for the first time abrasive water jet cutting by adding an abrasive material into a jet of pure water [2]. Therefore, two subcategories of water jet cutting can be considered: (1) water jet cutting without abrasive material, which is used to process non-metallic materials (wood, plastic, glass, leather, etc.) as well as different products in the food industry since it prevents oxidation and (2) water jet cutting with abrasive material, which is used to cut a wide range of materials including hardened tool steel, titanium, all types of alloys, stone, ceramics etc., based on a proper combination of a durable abrasive water jet nozzle and a reliable high pressure pump [3, 4].

Water jet technology has many advantages such as: is an ecological technology; does not emit toxic fumes or dust that are harmful to the environment or machine operator; can process various metallic and non-metallic materials; can process different geometries of parts; there is no heat release during

Content from this work may be used under the terms of the Creative Commons Attribution 3.0 licence. Any further distribution of this work must maintain attribution to the author(s) and the title of the work, journal citation and DOI.
Published under licence by IOP Publishing Ltd

processing; there is a very low loss of material, having a small cutting width; does not leave burrs or rough surfaces [5, 6].

However, the technology has disadvantages, too: in packet processing cannot be performed; high cost of equipment and replaceable components (water nozzle, focusing tube, cutting head, filter, etc.); large thicknesses cannot be processed accurately, as the shape of cut becomes conic and appears grooves at the bottom of surface; materials that degrade on contact with water cannot be processed [7].

In recent years, many researches have been conducted for process optimization in order to obtain a better quality at a low price. Following this trend, the current paper investigates the effects of reusing abrasive material on the quality of processed surface. In view of this two types of steel were used: OL 37 (S 235) steel, with large applicability in machine building industry and 2P armor steel with low-temperature-tempered martensite structure (Russia), used in the arms industry. Different reusing rates of abrasive material were set for the experiments: 0%, which means only virgin abrasive, 20%, 40%, 60%, 80% and 100% reused abrasive material. The parameters used to quantify the quality of processed surface, according to the ISO/WD/TC44 N 1770 standard were: width of the processed surface at the jet inlet (Li), width of the processed surface at the jet outlet (Lo), inclination angle (α), deviation from perpendicularity (u) and roughness (Ra).

2. Experimental methodology

The experimental tests were performed on a Hydro-Jet Eco 0615 water jet cutting machine using plates made by the above mentioned two materials, with the thickness of 6.5 mm. The mechanical characteristics of materials are presented in table 1 and table 2, respectively. The process parameters were those presented in table 3.

Table 1. Mechanical characteristics of the OL 37 carbon steel (S 235).

Steel grade	Yield stress, Rp0,2 [MPa]	Tensile stress, Rm [MPa]	Elongation, A [%]
S 235 (OL 37)	240	360 - 440	25

Table 2. Mechanical characteristics of armor steel with low-temperature-tempered martensite structure (Russia).

Steel Grade	Alloy System	Carbon, % by weight	Tensile stress (average), Rm [MPa]	Hardness, [HB]	Similar standard
2P	Si-Mn-Mo	≤ 0.29	1550	444 – 514	(MIL 46100)

Table 3. Process parameters.

Parameter	Unit	Value
Abrasive mixture quantity, Q	[g/min]	300
Pressure, P	[MPa]	150
Distance between cutting head and surface, h	[mm]	3
Feed rate, Vf	[mm/min]	100
Orifice diameter, Do	[mm]	0.35
Focusing tube diameter, Df	[mm]	1.02
Length of focusing tube, Lf	[mm]	76.2

The abrasive material as a function of the reusing rate is presented in figure 1. It can be seen that the dimension of abrasive grains is getting smaller as the reusing rate is higher.

The measurements of width of the processed surface at the jet inlet (Li), width of the processed surface at the jet outlet (Lo), deviation from perpendicularity (u), inclination angle (α) and surface macrostructure were performed on a Leica MZ75 microscope, with PC interface. The surface roughness (Ra) was measured with a Mitutoyo SJ 201 digital roughness device.

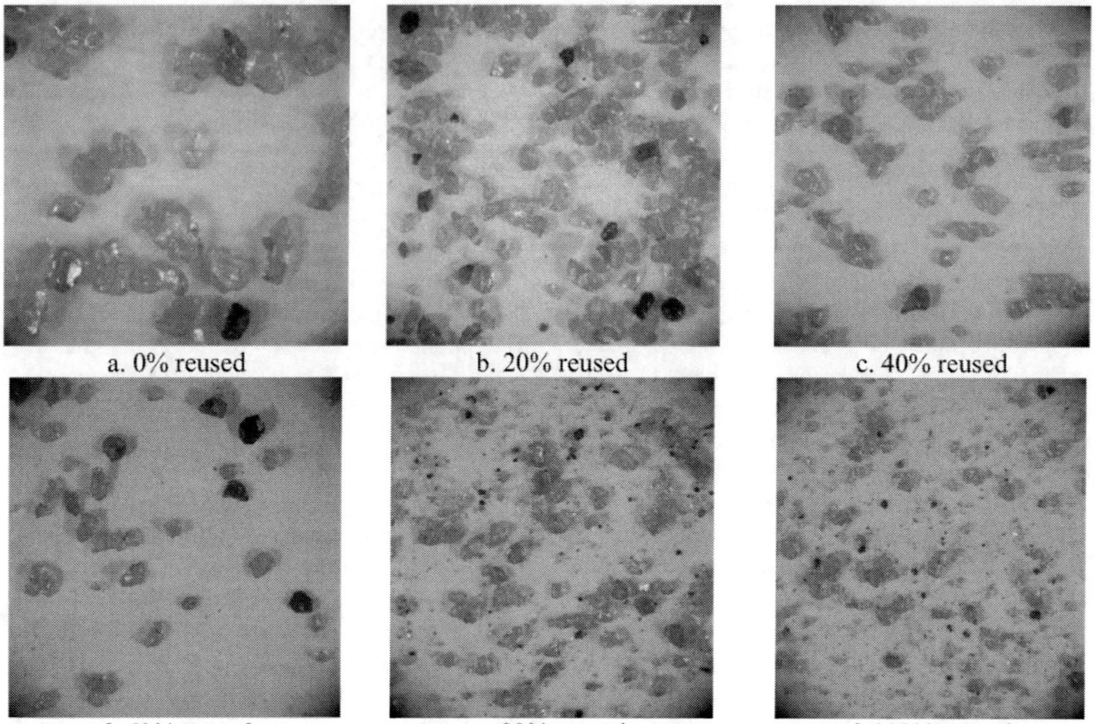

a. 0% reused b. 20% reused c. 40% reused
d. 60% reused e. 80% reused f. 100% reused

Figure 1. The abrasive material used for the experimental tests.

3. Results and discussion

The influence of the reusing rate of abrasive material on the surface quality of plates made by OL 37 (S235) carbon steel and 2P armor steel, cut by abrasive water jet technology, is presented in the graphics below.

The macrostructures of OL 37 (S235) material processed at different percentages of reused abrasive is presented in figure 2. Note that the increasing of reused abrasive percentage leads to a macrostructures without striations. In case of 2P armor steel cut at different rates of reused abrasive material, some small striations resulted on the cut surface when virgin grains of abrasive were used. As the percentage of reused abrasive increased, the striations were getting bigger, the worst surface quality being obtained at a reusing rate of 60%. However, the best quality of processed surface resulted for a reusing rate of abrasive material of 80% (figure 3).

a. 0% reused b. 20% reused c. 40% reused d. 60% reused e. 80% reused f. 100% reused

Figure 2. Macrostructures of OL 37 (S235) steel processed by AWJ at different rates of reused abrasive.

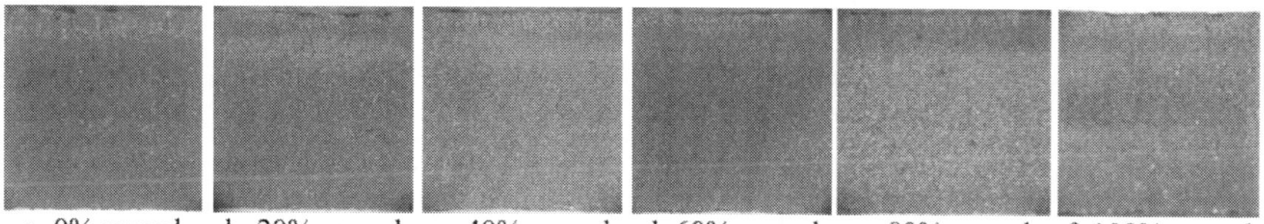

Figure 3. Macrostructures of 2P armor steel processed by AWJ at different rates of reused abrasive.

The results concerning the influence of abrasive material reusing rate on the width of processed surface at the jet inlet (Li) are graphically shown in figure 4. The graph analysis highlights that a higher percentage of reused abrasive material leads to a smaller value of the Li parameter, in case of both materials, due to the small size of the abrasive grains.

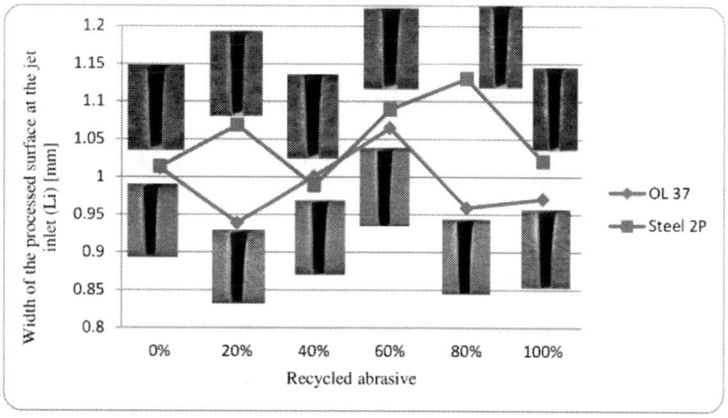

Figure 4. Width of the processed surface at the jet inlet (Li) as a function of the reused abrasive.

The influence of abrasive material reusing rate on the width of processed surface at the jet outlet (Lo) is graphically presented in figure 5. It can be seen that the percentage of reused abrasive has an almost insignificant influence on the Lo parameter till the value of 60%. After this value, the width of processed surface at the jet outlet decreases as the percentage of reused abrasive increases.

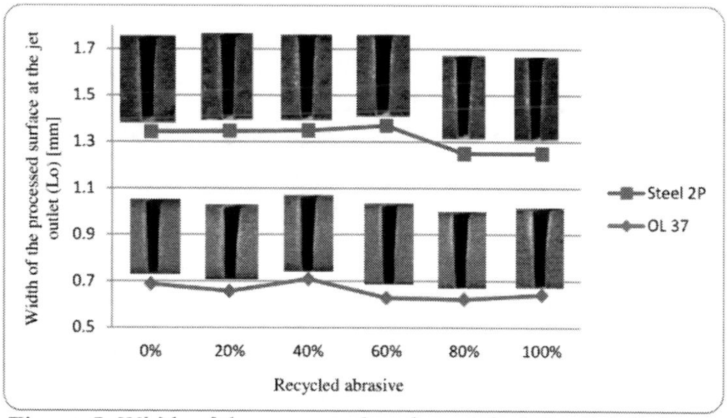

Figure 5. Width of the processed surface at the jet outlet (Lo) as a function of the reused abrasive.

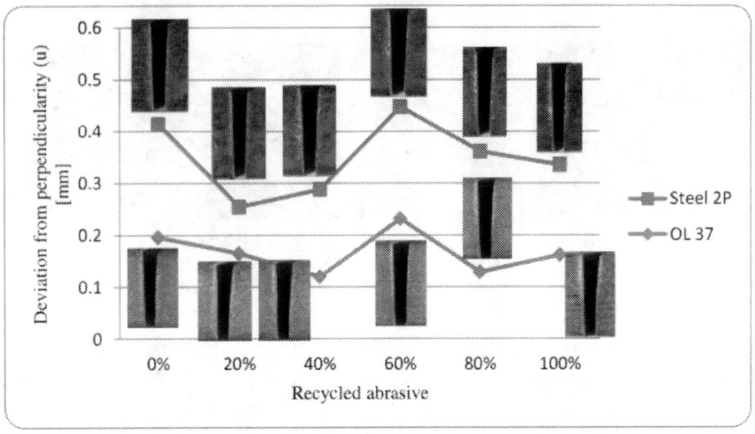

Figure 6. Deviations from perpendicularity (u) as a function of the reused abrasive.

Figure 7 summarize the results concerning the influence of reused abrasive percentage on the inclination angle of processed surface. It can be noticed that the angle of inclination is constant at 3°, for both materials, up to a reusing rate of abrasive of 80%. Over 80%, the inclination angle increases for both materials, more pronounced in case of OL 37 (S235) carbon steel.

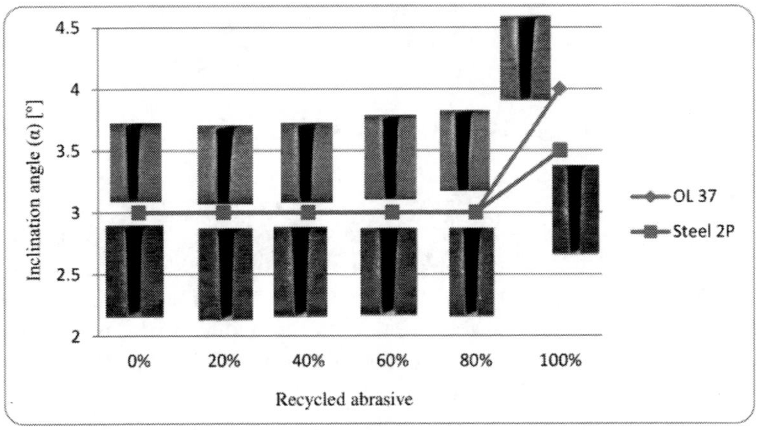

Figure 7. Inclination angle of cut (α) as a function of the reused abrasive.

The influence of abrasive material reusing rate on the surface roughness is presented in figure 8. It can be seen a decreasing trend of roughness as the abrasive is reused in larger percentages, excepting the 40% reusing rate when an increase of roughness is observed for both materials, more pronounced in case of 2P armor steel.

4. Conclusions

The aim of the current paper was to investigate the effects of reusing abrasive material on the quality of surfaces cut by AWJ. As result of the study it was found that reusing of abrasive material leads to a good quality of the processed surface, the processing productivity being slightly affected. The best quality of processed surface, quantified in terms of parameters specified by the ISO/WD/TC44 N 1770 standard, was obtained for reusing rates of abrasive material between 60% and 80%. This means a decreased consumption of virgin abrasive material, which in turn results in a lower cost of processing per unit length (economic benefits) as well as lower power consumption, lower transport costs and fewer related problems (environmental benefits).

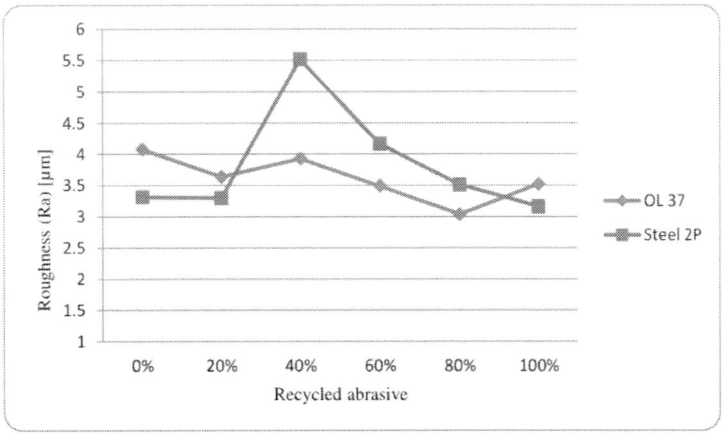

Figure 8. Roughness of the processed surface (Ra) as a function of the reused abrasive.

As concern the two analysed materials, it was observed that in case of 2P armor steel all parameters that quantify the surface quality had higher values compared to those obtained for OL 37 (S235) carbon steel, excepting the inclination angle that, after a constant value of 3oC for both materials, was bigger for carbon steel.

Acknowledgement
This paper was performed through the Partnerships program, in PN II priority areas, developed with MEN-UEFISCDI (Romanian National Authority for Scientific Research) support, project no. 294/2014.

References

[1] Herghelegiu E, Schnakovszky C, Radovanovic M and Raveica C I 2012 Comparative study on the dimensional accuracy and surface quality of plates cut through unconventional methods *Proc. Int. Conf. Modern Technologies, Quality and Innovation-ModTech 2012* (Sinaia) pp 433-436

[2] Radovanovic M, Dasic P and Stefanek M 2005 Some possibilities for determining abrasive water jet cutting parameters *Proc. 13th Int. Sci. Conf. CO-MAT-TECH* pp 969-976

[3] Radovanovic M 2004 Precision cutting by abrasive waterjet Sci. *Conf. with Int. Participation Manufacturing and Management in 21st Century* (Ohrid, Republic of Macedonia) pp 94 – 99

[4] Parikh P J and Lam S S 2009 Parameter estimation for abrasive water jet machining process using neural networks *Int. J. Adv. Manuf. Technol.* **40** pp 497–502

[5] Herghelegiu E 2010 Experimental determination of the cutting speed in case of the water jet processing of different material *Proc. Int. Conf. Modern Technologies, Quality and Innovation-ModTech 2012* (Sinaia) pp 327-330

[6] Schnakovszky C, Herghelegiu E and Tampu C 2014 The metal sheets processed by AWJ Analysis of the surface quality *App. Mech. Mat.* **837** pp 201-205

[7] Schnakovszky C, Herghelegiu E, Radu C and Cristea I 2014 The influence of the feed rate on the quality of surfaces processed by AWJ at high pressures *App. Mech. Mat.* **837** pp 196-200

IOP Conf. Series: Materials Science and Engineering **95** (2015) 012005 doi:10.1088/1757-899X/95/1/012005

Experimental determinations of the eigenmodes for composite bars made with carbon and Kevlar-carbon fibers

C M Miriţoiu[1], **M M Stănescu**[2], **C O Burada**[3], **D Bolcu**[3] **and V Roşca**[3]
[1]Faculty of Mechanics, University of Craiova, 165 Calea Bucuresti, 200620, Craiova, Romania
[2]Department of Applied Mathematics, University of Craiova, 13 AI Cuza, 200396, Craiova, Romania
[3]Faculty of Mechanics, University of Craiova, 165 Calea Bucuresti, 200620, Craiova, Romania

E-mail: miritoiucosmin@yahoo.com

Abstract. For modal identification, the single-point excitation method has been widely used in modal tests and it consists in applying a force in a given point and recording the vibratory structure response in all interest points, including the excitation point. There will be presented the experimental recordings for the studied bars (with Kevlar-carbon or carbon fibers), the frequency response function in Cartesian and polar coordinates. By using the frequency response functions we determine the eigenparameters for each bar. We present the final panel of the eigenmodes (with the damping factors, eigenfrequencies and critical damping) for each considered bar. Using the eigenfrequency of the first determined eigenmode, the bars stiffness has been determined. The presented bars can be used in practical engineering for: car or bus body parts, planes body parts, bullet-proof vests, reinforcements for sandwich beams, and so on.

1. Introduction
Many papers present various studies regarding Kevlar or carbon-fibers studies. In this sense, in [1] there is presented a study regarding the interfacial bonding under bending and uniaxial compression for sandwich structures made by carbon-fiber and aluminum honeycomb with and without Kevlar-fiber interfacial toughening. In [2] there is presented a research on impacted sandwich composites with Kevlar/hybrid and carbon face sheets subjected to different temperatures. Testing was performed to determine bending and core shear stresses, maximum energy absorption, and absorbing energy, moment parameter, performance parameter and compression strength after impact. Other studies regarding these fibers can also be found, but not limited, in [3, 8].

The modal identification is characterized by theoretical and experimental procedures used to determine the system eigenmodes parameters. The procedures used for tests can be divided in two important big groups: single-point excitation and multiple-point excitation. In this paper, starting from the theoretical background of modal identification, it is established an experimental method used to determine the eigenmodes for some composite bars with Kevlar-carbon and carbon fibers with epoxy resin. The epoxy resin is RESOLTECH 1050 type and its hardener is RESOLTECH 1058 type (having a mixed density of 1.11 g/cm^3, mixed viscosity of 633 MPa·s at 23^{0}C, 5% elongation to break and 6% flexion to break). There was used the single-point excitation method to determine the bars

Content from this work may be used under the terms of the Creative Commons Attribution 3.0 licence. Any further distribution of this work must maintain attribution to the author(s) and the title of the work, journal citation and DOI.
Published under licence by IOP Publishing Ltd

eigenmodes. This method has been widely used in modal testing and it consists in applying a force in a given point and recording the vibratory response in all interest points, including the excitation point, if needed. Although the single-point excitation requires a minimum of equipment, it needs a laborious analysis to perform extensive result processing in order to interpret the dynamic behavior of the structure under test.

2. The modal parameters identification procedure

The theoretical procedure used in this paper has been previously presented by the authors in papers like [10] and [11]. So, according to [9], [10] and [11], if a mechanical system is described by n concentrated mass points jointed with elastic elements (with the next characteristics: s_k stiffness and μ_k damping) and can have n degrees of freedom, and if it is loaded with an $\{\Theta(t)\}$ external excitation, the motion equations can be given by equation (1).

$$[G]\cdot\left\{\ddot{v}(t)\right\}+[M]\cdot\left\{\dot{v}(t)\right\}+[S]\cdot\{v(t)\}=\{\Theta(t)\} \quad (1)$$

In the equation (1) we have marked with: [G], [M] and [S] the matrices of mass, damping and stiffness; {v(t)} with its first and second derivatives are the displacement, velocity and acceleration vectors; {Θ(t)} is the generalized forces vector.

The equation (2) is used to determine the system response at external excitation. Its response is processed as a sum of n modal contributions due to each separate degree of freedom (as described in equation (2)).

$$\sum_{k=1}^{N}\frac{\{\vartheta^k\}\{\vartheta^k\}^T\cdot\{\Theta(\omega)\}}{-a_k(\mu_k-i(\omega-\nu_k))}+\sum_{k=1}^{N}\frac{\{\overline{\vartheta}^k\}\{\overline{\vartheta}^k\}^T\cdot\{\Theta(\omega)\}}{-\overline{a}_k(\mu_k-i(\omega+\nu_k))}=\{v(\omega)\} \quad (2)$$

In the equation (2) we have marked with: {v(ω)} is the displacement Fourier transform; $\{\vartheta^k\}$ and $\{\overline{\vartheta}^k\}$ is the eigenvector and its complex conjugate k order; μ_k is the damping factor k order; ν_k is the damped natural frequency k order; a_k and $\overline{a}_k$ are the norm constants of eigenvector; ω is the external excitation frequency.

In the experimental modal identification, the modal vectors can be replaced by two constants. Their value can be determined with equation (3).

$$2\cdot U_{ij}^k=\frac{\xi_i^k\xi_j^k}{a_k}+\frac{\overline{\xi_i^k\xi_j^k}}{\overline{a}_k},2\cdot iV_{ij}^k=\frac{\xi_i^k\xi_j^k}{a_k}-\frac{\overline{\xi_i^k\xi_j^k}}{\overline{a}_k}, \quad (3)$$

According to the modal identification literature, we may insert the system admittance parameter as the ratio between the displacement response and the force excitation. This parameter may be calculated by using the equation (4).

$$\sum_{k=1}^{n} - \frac{U_{ij}^{k} + iV_{ij}^{k}}{\mu_k + i(-\omega + \nu_k)} - \sum_{k=1}^{n} \frac{U_{ij}^{k} + iV_{ij}^{k}}{\mu_k + i(-\omega - \nu_k)} = \Psi_{ij}(\omega) \quad , \tag{4}$$

The concept of discrete system with concentrated mass in n material points was used in the approximations adopted during the mathematical model [9, 10, 11]. In order to obtain an accurate approximation of the real system by the discrete one, n must converge to infinity. Because of experimental and processing technique and of the necessary time for data processing, this is impossible. The frequencies domain is limited to a reasonable width in practical applications, which is obtained by the major resonances of the analysed equipment and the frequency domain of the application goal. In these conditions, the sum from equation (4) is reduced to several components marked in the following with n, too. The contribution of superior and inferior modes are included in two corrections factors known as residual flexibility S'ij (for superior modes) and inferior modal admittance $-(M^{'}_{ij} \cdot \omega^2)^{-1}$ (for inferior modes) (according to [9, 10, 11]). Using these notations, the equation 4 will have the form like in equation (5).

$$\sum_{k=1}^{n} - \frac{U_{ij}^{k} + iV_{ij}^{k}}{\mu_k + i(-\omega + \nu_k)} - \sum_{k=1}^{n} \frac{U_{ij}^{k} + iV_{ij}^{k}}{\mu_k + i(-\omega - \nu_k)} + S^{'}_{ij} - \left(M^{'}_{ij} \cdot \omega^2\right)^{-1} = \Psi_{ij}(\omega) \quad . \tag{5}$$

The modal identification of a system with n degrees of freedom assumes determination of $4n$ modal parameters: μ_k, ν_k, $U^{'}_{ij}$, $V^{'}_{ij}$. These are intrinsic characteristics of the system, independent of the external conditions.

3. The modal identification for the chosen composite bars

In figure 1 there is presented the experimental setup and the analyzed composite bars (a built-in end and the other one is free with an accelerometer placed at the middle of the free length). The bars are marked with 1, 2 and 3 according to figure 1. The samples have the thickness of 0,002 m, the width 0.05 m and the density: 1090 kg/m (sample 1), 1170 kg/m (sample 2) and 1260 kg/m (sample 3).

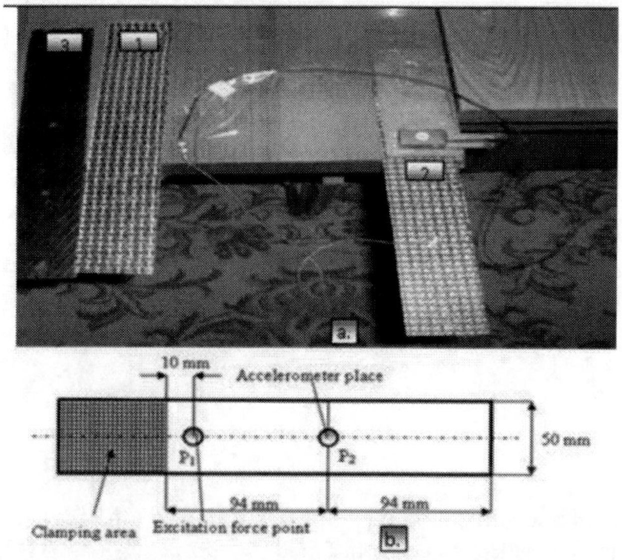

Figure 1. Experimental setup: a) the studied samples; b) a scheme with the experimental setup.

In order to identify the eigenmodes, the next steps were followed:

- we have marked with 1 and 2 the samples with Kevlar-carbon fibers (the difference between them is the direction of fiber application, as shown in figure 1) and with 3 the sample with carbon fiber;
- the experimental recordings are determined (the impact force made with an impact hammer with the sensitivity 1,020 pC/ms^{-2} and the vibratory response with an accelerometer with 0,04 pC/ms^{-2} sensitivity);
- the frequency response functions (abbreviated as FRF in the next paragraphs) are determined in both Cartesian and Nyquist coordinates;
- the resonances approximate position and the initial modal parameters μ_k and ν_k calculus (where k is the number of the modal parameter);
- the first stage identification of modal parameters μ_k, ν_k, U_{ij}, V_{ij}, S'_{ij} and $-(M'_{ij})^{-1}$ on limited frequency domains (the identification is made by using linear procedures, determining those modal parameters that, inserted in relation (5), generate theoretical characteristics which approximate with minimal error the experimentally determined frequency response function);
- the final stage of modal parameters μ_k, ν_k, U_{ij}, V_{ij}, S'_{ij} and $-(M'_{ij})^{-1}$ over the entire domain (the identification is made by using nonlinear procedures of recursive approximation, determining those modal parameters that, inserted in equation (5), generate theoretical characteristics which approximate with minimal error the experimentally determined frequency response function).

The tests are performed in order to determine the bars eigenmodes, eigenfrequencies, the damping factors and the critical damping for each eigenmode.

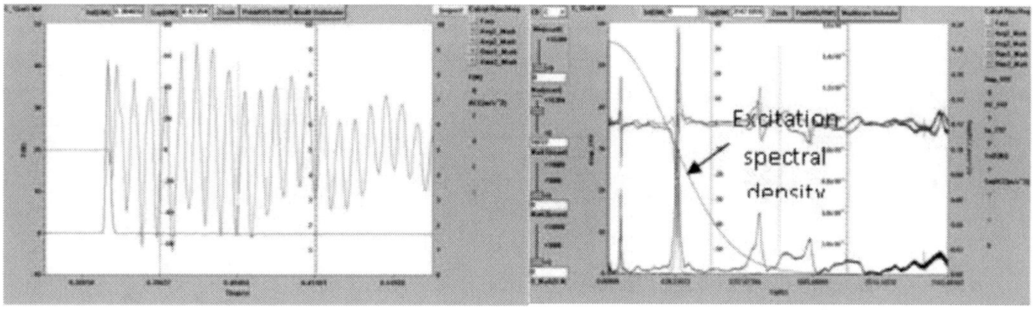

a. b.

Figure 2. Sample 1: a) Experimental recordings; b) FRF in Cartesian coordinates.

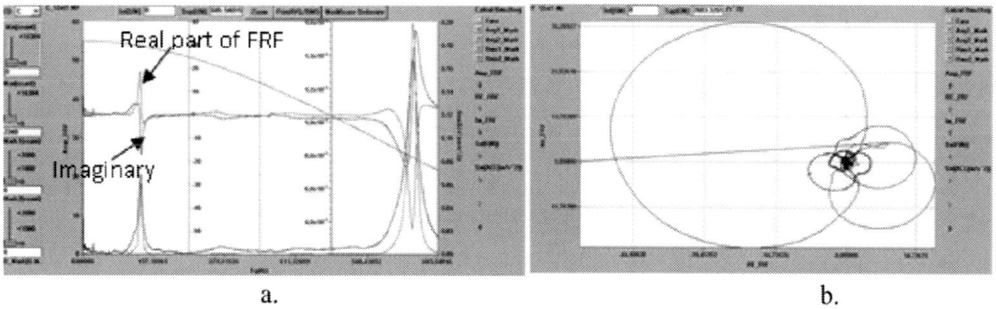

a. b.

Figure 3. Sample 1: a) Real and imaginary parts of FRF; b) FRF in polar coordinates.

Table 1. The modal parameters obtained for the considered samples.

Sample	Eigenmode				
1	μ (miu) [-(Ns/m)/kg]	ν (niu) [1/s]	U	V	ζ (zita) [(Ns/m)/kg]

	-2.223	111.589	40.873	-26.600	0.019
	-5.922	636.327	-312.888	71.232	0.009
	-11.893	1383.824	173.138	49.069	0.008
	-1.969	94.378	42.941	-0.230	0.020
2	-4.103	539.016	-9.932	179.545	0.007
	-8.641	1254.753	61.749	-32.395	0.006
	-1.698	105.322	49.348	-22.632	0.016
3	-5.461	601.732	-159.533	123.668	0.009
	-14.943	2138.874	569.156	326.391	0.006

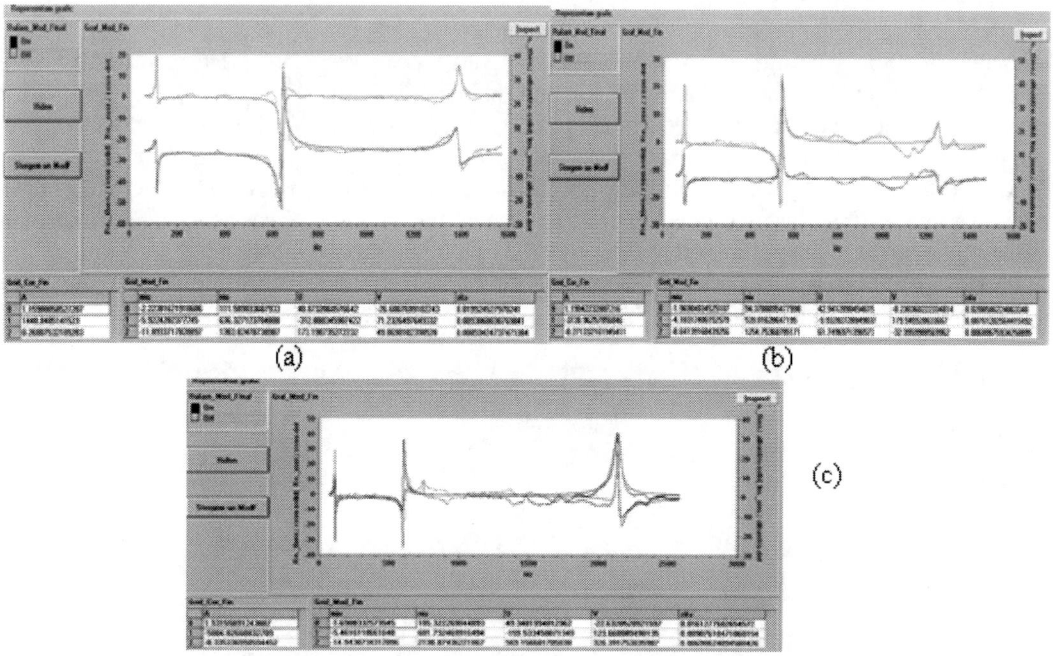

Figure 4. Final panel of the modal parameters: a) Sample 1; b) Sample 2; c) Sample 3.

Using the relation (6) we have determined the loss factor for each eigenmode.

$$\eta = 2\cdot\zeta \tag{6}$$

We have obtained the next values:

- sample 1: 0.038 (first eigenmode); 0.018 (second eigenmode); 0.016 (third eigenmode);
- sample 2: 0.040 (first eigenmode); 0.014 (second eigenmode); 0.012 (third eigenmode);
- sample 3: 0.032 (first eigenmode); 0.018 (second eigenmode); 0.012 (third eigenmode).

Using relation (7), from [13], we have determined the bars flexural rigidity.

$$\nu_\xi \approx 0{,}1591549\cdot\frac{(EI)^{0.5}\cdot 4.694}{(\rho A)^{0.5}\cdot l^2} \tag{7}$$

4. Conclusions

In this paper, the modal parameters were determined for three types of composite bars marked in this way: *sample 1 and 2* – bars with Kevlar-carbon fibers with different orientation and *sample 3* – a bar with carbon fiber. All the fibers were combined with epoxy resin. The single-point excitation method was used. For all the bars, from the experimental setup, three eigenmodes were determined. The above presented experiment shows many distinct eigenmodes in a large frequency domain: from 94.378 s^{-1} up to 2138.874 s^{-1}. The modal identification method has the advantage that it is non-destructive and can be used in the case of complex structures built from modern (like composite materials, shape memory alloys, and so on) or classical (like steel, aluminum or cast iron structures) materials. But this method also presents some disadvantages because it can be used for some mechanical parameters calculus (for example, the ones that characterize the breakage of a certain structure). We have determined the loss factor using the critical damping value (ζ (zita)) and the bars stiffness. In practical engineering, these types of composites structures can be used for: car or bus body parts, planes body parts, bullet-proof vests, reinforcements for sandwich beams, and so on.

Acknowledgement
This work was supported by the strategic grant POSDRU/159/1.5/S/133255, Project ID 133255 (2014), co-financed by the European Social Fund within the Sectorial Operational Program Human Resources Development 2007-2013.

References

[1] Shi S, Sun Z, et al. 2014 Carbon-fiber and aluminum-honeycomb sandwich composites with and without Kevlar-fiber interfacial toughening *Composites: Part A* **67** pp 102-110
[2] Salehi-Khojin A and Mahinfalah M 2007 Temperature effects on Kevlar/hybrid and carbon fiber composite sandwiches under impact loading *Composites structures* **78** pp 197-206
[3] Sargianis J and Suhr J 2012 Core material effect on wave number and vibrational damping characteristics in carbon fiber sandwich composites *Composites Science and Technology* **72** pp 1493-1499
[4] Wong Y C, Ruan D, et al. 2014 The influence of magnetic field on ballistic performance of aramid fibre and ultrahigh molecular weight polyethylene *Materials and Design* **64** pp360-365
[5] Gustin J, Joneson A, et al. 2005 Low velocity impact of combination Kevlar/carbon fiber sandwich composites *Composite Structures* **69** pp 396-406
[6] Caserta G D, L. Ianucci, et. Al 2011 Shock absorption performance of a motorbike helmet with honeycomb reinforced liner *Composite Structures* **93** pp 2748-2759
[7] Mamivand M and Liaghat G H 2010 A model for ballistic impact on multi-layer fabric targets *International Journal of Impact Engineering* **37** pp 806-812
[8] Ong C W and Boey C W 2011 Advanced layered personnel armor *International Journal of Impact Engineering* **38** pp 369-383
[9] Edwins D 1987 Modal Testing. Theory and Practice *Bruel&Kjaer*
[10] Manea I, Bolcu D and Miriţoiu C 2007 Software for mechanical systems modal identification *Annals of the University of Craiova. Mechanical Series* **1** pp 191-199
[11] Miriţoiu C, Bolcu D, et. al 2014 Experimental Determinations of the Eigenmodes for Sandwich Bars With Different Core Reinforced with Metal Fabric *Applied Mechanics and Materials* **658** pp 255-260.
[12] Miriţoiu C, Ilincioiu D, et al. 2011 A Comparison Between the Modal Parameters Obtained by two Different Accelerometers *The 5TH International Conference on Manufacturing Science and Education* **1** pp 39-43
[13] Burada C O and Miriţoiu C M , et. al 2014 Experimental determinations of the damping factor and stiffness for new sandwich platbands with different reinforcement and core *Romanian Journal of Materials* **44** pp 405-413

Thermal analysis and microstructural characterization of Mg-Al-Zn system alloys

M Król, T Tański and W Sitek[1]

[1]Institute of Engineering Materials and Biomaterials, Faculty of Mechanical Engineering, Silesian University of Technology, ul. Konarskiego 18a, 44-100 Gliwice, Poland

E-mail: mariusz.krol@polsl.pl

Abstract. The influence of Zn amount and solidification rate on the characteristic temperature of the evaluation of magnesium dendrites during solidification at different cooling rates (0.6-2.5°C) were examined by thermal derivative analysis (TDA). The dendrite coherency point (DCP) is presented with a novel approach based on second derivative cooling curve. Solidification behavior was examined via one thermocouple thermal analysis method. Microstructural assessments were described by optical light microscopy, scanning electron microscopy and energy dispersive X-ray spectroscopy. These studies showed that utilization of d^2T/dt^2 vs. the time curve methodology provides for analysis of the dendrite coherency point

1. Introduction

The magnesium alloys are an attractive alternative to aluminum alloys for lightweight applications. Mg alloys are widely applied to the production of engines and gearbox housings. However, the magnesium alloys compared to aluminum alloys are inferior casting properties. Castability depends significantly on the dendrite coherency point that represents the temperature, time and the percentage of solid state at which an interlocking solid network forms during solidification. An increase in the solid state at coherence may improve the casting properties of the alloy and reduce casting defects. The thermal analysis technique is a technique used for long time, both in ferrous and nonferrous industries to assess the quality of the melt prior to casting. However, obtaining the appropriate microstructure of a standard cup does not guarantee that the microstructure is correct in real parts which may solidify in a very various cooling rates [1, 2].

The solidification characteristics can be investigated through various thermal analysis techniques. There are standardized techniques such as differential thermal analysis (DTA), thermo gravimetric analysis (TGA), differential scanning calorimetry (DSC) and computer aided cooling curve analysis (CA-CCA) otherwise known as the thermal derivative analysis (TDA). Due to its ease of use and low-cost, TDA is much more appropriate for industrial applications compared to other techniques and TDA method has been successfully applied in investigations of solidification sequences of magnesium, aluminum and zinc alloys in recent years [2-5].

The aim of the present work is to promote thermal analysis of one centre cooling curve for the determination of the dendrite coherency characteristics such as temperature, time, instantaneous solidification rate and fraction solid. The first derivative of the cooling curve is plotted vs. the temperature and time and the thermal characteristics of all metallurgical reactions, including the DCP

Content from this work may be used under the terms of the Creative Commons Attribution 3.0 licence. Any further distribution of this work must maintain attribution to the author(s) and the title of the work, journal citation and DOI.
Published under licence by IOP Publishing Ltd

are determined. The thermal analysis results of the present work will also provide data support for thermodynamic description of Mg-Al-Zn system.

2. Experimental procedure

2.1. Fabrication and materials preparation

The study was carried out on samples of MCMgAl6Zn1 and MCMgAl6Zn3 magnesium alloys. Chemical compositions of the tested materials are given in table 1. Experiments were carried out applying cylindrical samples with a diameter 18mm and a length 20mm. Samples were in the axis the hole to place a super sensitive K type thermocouple positioned at the centre of the test sample to gather the thermal information about the process and monitoring the processing temperatures.

Table 1. Chemical composition of examined magnesium alloys.

The mass concentration of main elements, %							
Alloy	Al	Zn	Mn	Si	Fe	Mg	Rest
MCMgAl6Zn1	5.92	0.49	0.15	0.037	0.007	93.33	0.0613
MCMgAl6Zn3	5.7	2.5	0.22	-	0.0025	91.5	0.0775

2.2. Thermal analysis

The *Thermal Derivative Analysis* during the cycle of melting and solidification was performed using the Universal Metallurgical Simulator and Analyser (UMSA) [6]. Each thermal test was repeated at least three times. Cooling curves have been used to assess the behaviour of solidification process and to determine any specific temperatures. The temperature vs. time and first derivative vs. temperature were calculated and graphed. The test samples were heated to 700±2°C and isothermally maintained at that temperature for 90 seconds to stabilize the conditions of melt. Then the test samples were solidified to ambient temperature with three different cooling rates. Argon was used as a protective gas in order to enhance the cooling rate. The solid fraction was estimated by the calculating the area between the first derivative of the cooling curve and the so-called baseline (BL) [1-3].

2.3. Microstructure examinations

The observations of the examined cast magnesium alloys have been made on the light microscope LEICA MEF4A as well as on the electron scanning microscope Opton DSM-940. The X-ray qualitative and quantitative microanalysis and the analysis of a surface distribution of cast elements have been made on the Opton DSM-940 scanning microscope with the Oxford EDS LINK ISIS dispersive radiation spectrometer at the accelerating voltage of 15 kV

3. Investigation results

3.1. Microstructure analysis

The representative microstructures presented in figure 1 and figure 2 confirms the results acquired by thermal derivative analysis presented in figure 3 and 4 represents a representative SEM structure of samples cut from the middle part of MCMgAl6Zn1 and MCMgAl6Zn3 alloy. It may be noted as a microstructure in which the divorced eutectic phase spread along primary magnesium dendrite and grain boundaries. The microstructures illustrated in figure 1 presents polygonal Mg_2Si particles built-in the edge of γ phase what is confirmed by EDS quantitative analysis. This state observed in entire microstructure of the MCMgAl6Zn1 cast sample. This means that Mg_2Si phase precipitated before eutectic transformation temperature. The SEM observation of MCMgAl6Zn1 magnesium alloy discloses the occurrence of white Al_8Mn_5 intermetallic in needle shapes mostly located at grain boundaries or in interdendritic areas. Detection this type of the intermetallic particle with this type of shape next to the γ phase can be indicated its simultaneous precipitation with Mg_2Si at lower temperature. The XRD pattern is given in figure 1 confirm the coexistence of Al_8Mn_5 and Mg_2Si particles with the main phases of α and γ in the microstructure of MCMgAl6Zn1.

For MCMgAl6Zn3 alloy, SEM analysis presents a microstructure in which the divorced eutectic phase spread along primary magnesium dendrite and grain boundaries. The SEM observation of MCMgAl6Zn3 magnesium alloy reveals the presence of $MgZn_2$ intermetallic precipitation mainly situated at grain boundaries regions (figure 2). Detection this type of intermetallic particles next to the γ phase can be indicated its simultaneous precipitation with $Mg_{17}Al_{12}$ at lower temperature. The XRD pattern given in figure 2 confirms the coexistence of Al_8Mn_5 and $MgZn_2$ particles with the main phases of α and γ in the microstructure MCMgAl6Zn3 cast magnesium alloy. For both analysed cast magnesium alloys can be observed that the increase in cooling rate decreases the size and shape of precipitated particles.

a)

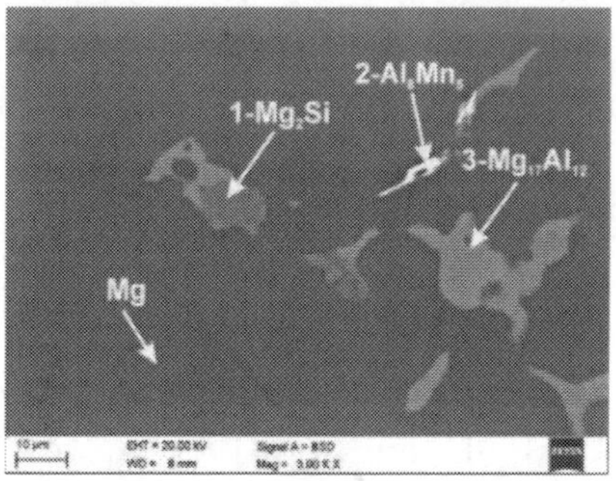

1)

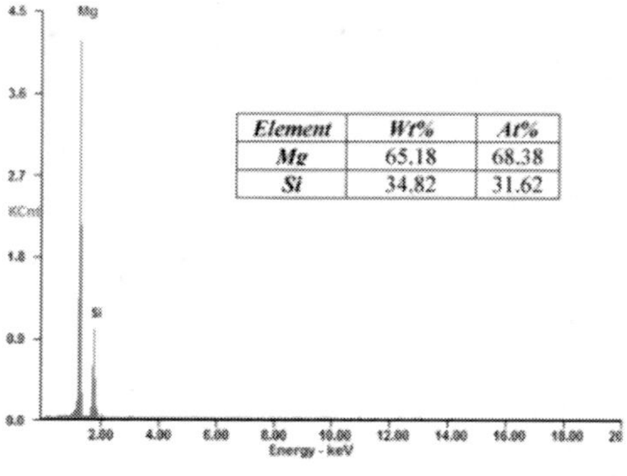

Element	Wt%	At%
Mg	65.18	68.38
Si	34.82	31.62

2)

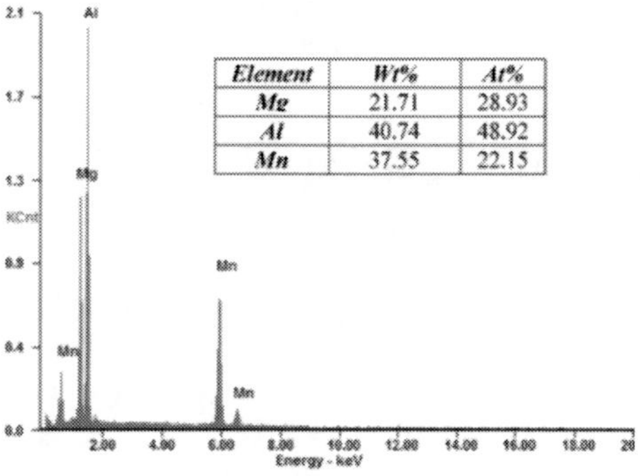

Element	Wt%	At%
Mg	21.71	28.93
Al	40.74	48.92
Mn	37.55	22.15

3)

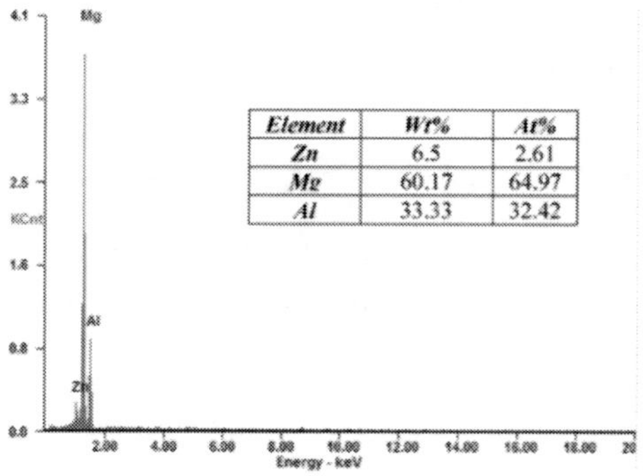

Element	Wt%	At%
Zn	6.5	2.61
Mg	60.17	64.97
Al	33.33	32.42

Figure 1. Representative SEM micrographs of an MCMgAl6Zn1 alloy solidify at 0.6°C/s cooling rate with EDS quantitative analysis results at marked points

a)

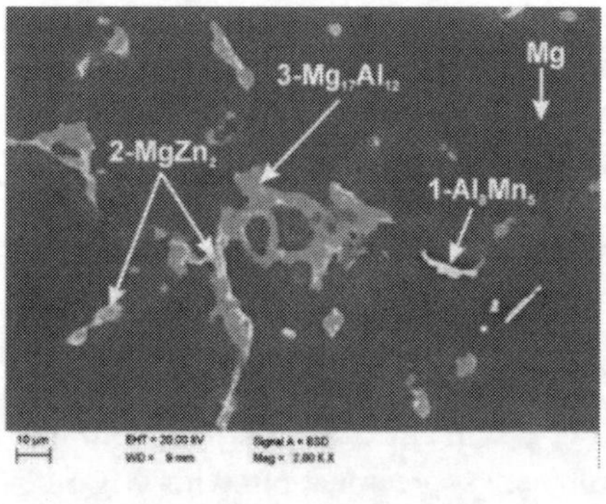

1)

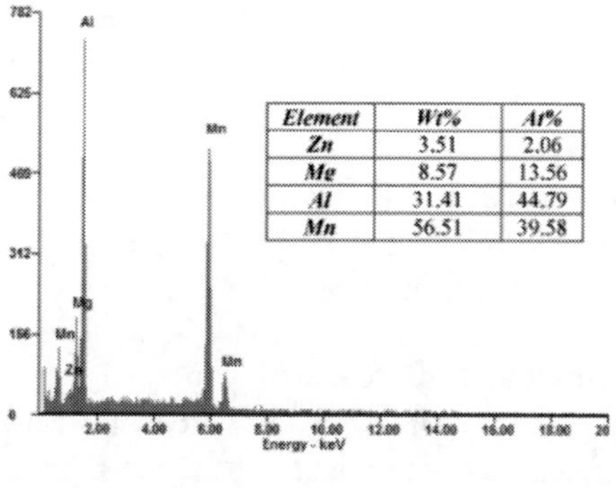

Element	Wt%	At%
Zn	3.51	2.06
Mg	8.57	13.56
Al	31.41	44.79
Mn	56.51	39.58

2) 3)

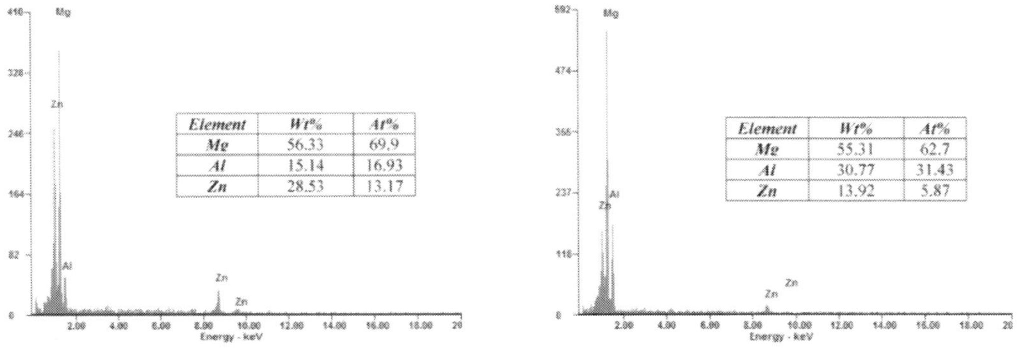

Figure 2. Representative SEM micrographs of an MCMgAl6Zn3 alloy solidify at 2.5°C/s cooling rate with EDS quantitative analysis results at marked points.

3.2. Thermal derivative analysis

The solidification pathways of Mg-Al-Zn alloys have been investigated by cooling curve thermal analysis. Figure 3 and figure 4 shows typical cooling curves and its first derivatives which were used to determine some critical points during solidification. Thermal characteristics of analysed magnesium alloys obtained from the first derivative are summarised in table 2. The solidification of MCMgAl6Zn1 and MCMgAl6Zn3 alloys begins with nucleation of α-Mg dendrites. After the formation of α-Mg dendrites, for MCMgAl6Zn1 alloy phases contains Mg_2Si and phases contain Al and Mn occurs. For MCMgAl6Zn3, after the formation of α-Mg dendrites, only $MgZn_2$ phase occurs. Last stage in the solidification process of investigated magnesium alloys is solidification of eutectic as divorced or partially divorced γ-$Mg_{17}Al_{12}$ in the interdendritic and grain boundary (figure 1, figure 2). It can be seen in figure 3 and figure 4 that the first derivative cooling curves consist three peaks.

Table 2. Thermal characterization of investigated cast magnesium alloys.

Investigated alloy	Cooling rate, (°C/s)	Characteristic points				
		Liquidus, (°C)	Mg_2Si+Al-Mn, (°C)	$MgZn_2$, (°C)	Eutectic, (°C)	Solidus, (°C)
MCMgAl6Zn1	0.6	611.96	529.01	-	430.36	420.72
MCMgAl6Zn1	1.3	619.6	532	-	432.7	408.9
MCMgAl6Zn1	2.5	619.58	535.35	-	429.48	404.76
MCMgAl6Zn3	0.6	607.42	-	393.36	355.68	348.95
MCMgAl6Zn3	1.3	612.38	-	388.45	355.58	344.46
MCMgAl6Zn3	2.5	617.06	-	390.41	354.97	336.56

a) b)

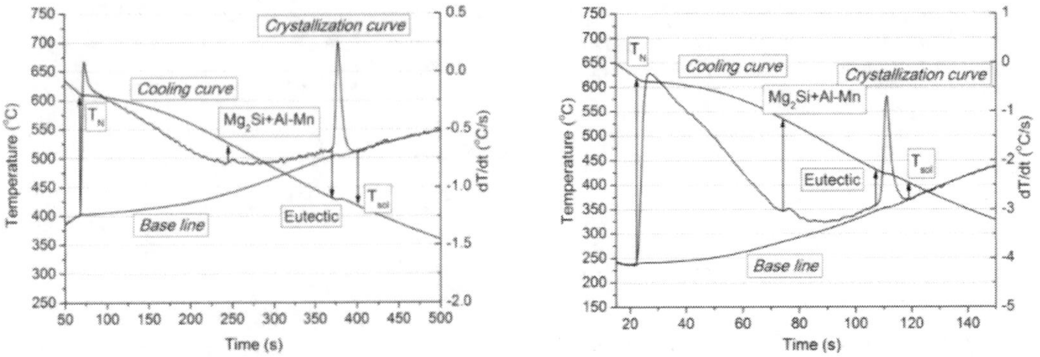

Figure 3. Representative cooling and its first derivative curves and base line of the MCMgAl6Zn1 alloy solidify at a) 0.6°C/s; b) 2.5°C/s.

a) b)

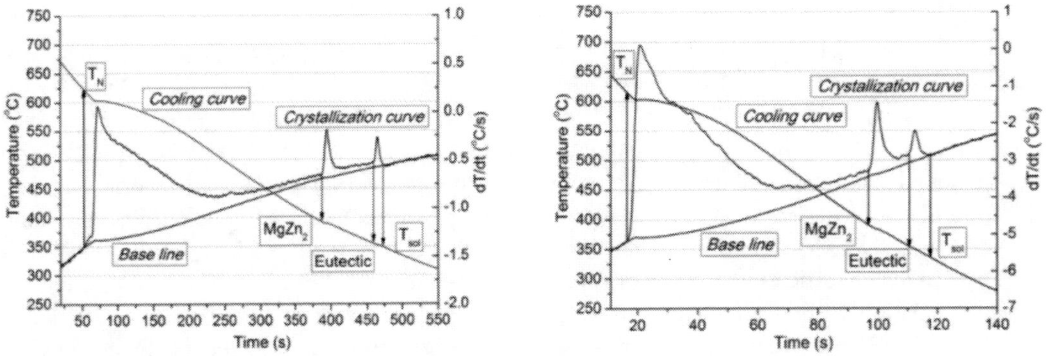

Figure 4. Representative cooling and its first derivative curves and base line of the MCMgAl6Zn3 alloy solidify at a) 0.6°C/s; b) 2.5°C/s

Taking into consideration the first derivative cooling curves, four observable temperature arrests including non-equilibrium nucleation temperature, crystallization of Mg_2Si and Al-Mn or $MgZn_2$ intermetallic, nucleation of the eutectic phase and solidus temperature can be determined. For MCMgAl6Zn1 magnesium alloy, for natural cooling rate, the nucleation temperature of primary magnesium dendrite was recorded at around 612°C. This phase grew on further cooling followed by Al-Mn and Mg_2Si intermetallic precipitation approximately as 530°C. Further cooling caused another sharp and clear positive peak manifested on the first derivative curve at 430°C due to the evolution of latent heat during eutectic transformation. Process finished at solidus temperature at around 420°C where remaining liquid completely transformed to the eutectic phase. It can be inferred that the increase in cooling rate increases the nucleation temperature and decreases the solidus temperature of investigated magnesium alloys. The intermetallic precipitation temperature slightly increases from 529°C to 535°C with increasing cooling rate. The increase in the cooling rate increased the solidification range from about 190°C to 215°C.

Accordingly for MCMgAl6Zn3 magnesium alloy, for natural cooling rate, the nucleation temperature of primary magnesium dendrite was recorded at around 607°C. This phase grew on further cooling followed by $MgZn_2$ intermetallic precipitation approximately as 493°C. Further cooling caused another sharp and clear positive peak manifested on the first derivative curve at 355°C due to the evolution of latent heat during eutectic transformation. Process finished at solidus

IOP Conf. Series: Materials Science and Engineering **95** (2015) 012006 doi:10.1088/1757-899X/95/1/012006

temperature at around 348°C. It can be concluded that the increase in cooling rate increases the nucleation temperature and decreases the solidus temperature of investigated magnesium alloys. Changing the cooling rate does not significantly effect on intermetallic precipitation temperature. The solidification range increases from about 260°C to 280°C with increasing cooling rate. Analysed series of cast magnesium alloys, thermal derivative analysis shows that the increase cooling rate does not effect on eutectic transformation temperature. Based on the results it can be concluded that the increase in Zn content decreases the nucleation temperature, eutectic temperature and solidus temperature of investigated magnesium alloys. In the solidification process dendrites begin to develop to the point where the collide with each other and continuous network become coherent. At this point, the dendrite coherency point, the system behaves less like a liquid and starts to resemble a solid behaviour. On the basis of thermal derivative analysis, the dendrite coherency point is present in the first minimum on the second derivative during formation of the primary magnesium. Figure 5 and figure 6 shows the characteristic points on the second derivative of the examined magnesium alloys.

a) b)

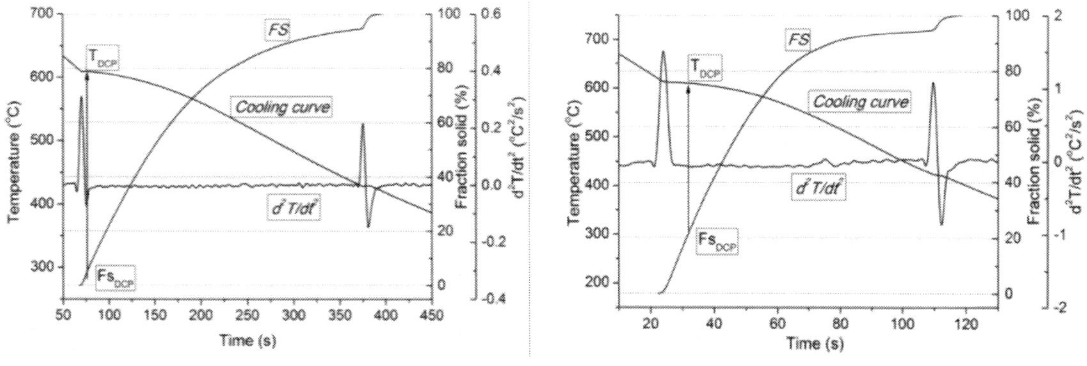

Figure 5. Representative cooling and second derivative curves and associated fraction solid curve of the MCMgAl6Zn1 alloy solidify at a) 0.6°C/s, b) 2.5°C/s.

a) b)

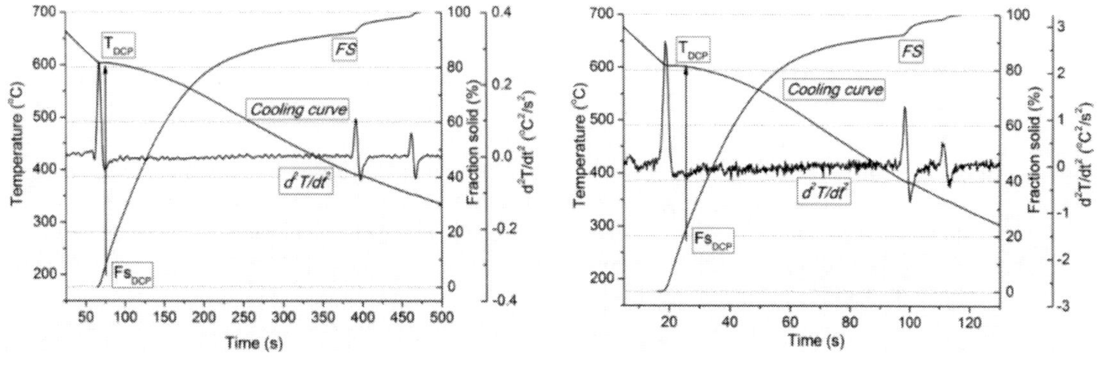

Figure 6. Representative cooling and second derivative curves and associated fraction solid curve of the MCMgAl6Zn3 alloy solidify at a) 0.6°C/s, b) 2.5°C/s.

For the one-thermocouple technique, one thermocouple located in the middle of the solidifying melt has been used to determine the dendrite coherency point (DCP). The minimum value of the second derivative of temperature versus time (d^2T/dt^2) refers to a state that there is a considerable change in thermal conductivity of the alloy. A suitable temperature for the cooling curve (T_{DCP}) is the dendrite coherency temperature. Temperature changes at the dendrite coherency point and solid fraction corresponding to the dendrite coherency (f_{DCP}) as a function of Zn content and cooling rate are presented in table 3. It can be noticed that fraction solid of primary magnesium at the dendrite coherency point increases with the increase zinc content and with the increase in cooling rate. For examined series of magnesium alloys can be observed that increasing cooling rate does not influence on changes in temperature at the dendrite coherency point.

Table 3. Influence of cooling rate and Zn content on temperature at the dendrite coherency point and a fraction solid of investigated magnesium alloys.

Cooling rate [°C/s]	0.6		1.3		2.5	
	DCP, °C	FS, %	DCP, °C	FS, %	DCP, °C	FS, %
MCMgAl6Zn1	609.36	3.41	608.35	6.85	610.01	21.24
MCMgAl6Zn3	603.84	9.11	602.8	15.17	601.97	21.81

4. Conclusions

The dendrite coherency point for magnesium alloys was studied using Thermal Derivative Analyse. It was found that the DCP could be determined by the second derivative curve (using the one thermocouple method). The one thermocouple method simplifies the platforms and lowers the costs of CA-CCA. The solidification pathways of Mg-Al-Zn alloys are obtained. The nucleation temperature of primary magnesium dendrite increases with increasing cooling rate and decreases with increasing of Zn content while solidus temperature decreases with increasing cooling rate and Zn content. For analysed series of magnesium alloys, the solidification range increases with increasing cooling rate.
Knowing the microstructure in the thermal analysis sample, it is possible to have an idea of the microstructure that can be expected in the real part according to its thermal behave and to the mould type.

References

[1] Dobrzański L A, Tański T and Trzaska J 2009 Optimization of heat treatment conditions of magnesium cast alloys *Materials Science Forum* **638-642** pp 1488-1493

[2] Dobrzański L A, Tański T, Malara Sz and Król M 2011 Structure and properties investigation of a magnesium alloy processed by heat treatment and laser surface treatment *Materials Science Forum* **674** pp 11-18

[3] Jafaria H, Idrisa M H, Ourdjinia A and Farahanya S 2013 In situ melting and solidification assessment of AZ91D granules by computer-aided thermal analysis during investment casting process *Materials & Design* **50** pp 181-190

[4] Rad H R B, Idris M H, Kadir M R A and Farahany S 2012 Microstructure analysis and corrosion behavior of biodegradable Mg-Ca implant alloys *Materials & Design* **33** pp 88-97

[5] Krupinski M, Krupinska B, Labisz K, Rdzawski Z and Borek W 2014 Influence of cooling rate on crystallisation kinetics on microstructure of cast zinc alloys *J Therm Anal Calorim* **118(2)** pp 1361-1367

[6] Method and Apparatus for Universal Metallurgical Simulation and Analysis (United States Patent) Patent No : US 7,354,491 B2, Date of Patent: Apr 8 2008

IOP Conf. Series: Materials Science and Engineering **95** (2015) 012007 doi:10.1088/1757-899X/95/1/012007

Durability of cutting inserts coated with ion-plated AlTiN thin films

A Bădănac[1], G Bosoancă[1], V Manole[1], A C Bălan[1], M Popa[1] and V Huşanu[2]
[1] "Gheorghe Asachi" Technical University of Iasi-Romania, Department of Machine Manufacturing Technology, Blvd. Mangeron, No. 59A, 700050, Iasi, Romania
[2] S.C."Rulmenti" S.A., Republicii Street, No.320, 731108, Barlad, Romania

E-mail: badanac_ana@yahoo.com

Abstract. It is known that during the cutting processes, due to the high contact pressures, to the high temperatures, to the relative velocities and shocks between the contact surfaces tool-part can lead to a wear more or less pronounced of the metal carbide cutting inserts. Are known numerous coating processes regarding the increasing the durability of the used cutting inserts, namely by deposition of different materials, having protection role for the cutting inserts locating and clearance surfaces, methods as: vapor chemical deposition at low pressure, pulverization, cathodic arc ion plating, vacuum thermal evaporation and condensation from the vapor phase. Researches carried out by authors in this paper, have followed the deposition of aluminum and titanium materials (AlTiN) in thin layers, on surface of some metal carbide cutting inserts profiled. The depositions of AlTiN materials are new coatings which consists in deposition of the compound solid of AlTiN in the form of vapor in thin layers on different tools or/and the cutting inserts. In the purpose of increasing the cutting inserts durability, they used deposition method by cathodic arc ion plating in vacuum. The authors chose this method due to its advantages, which can be enumerate: the relatively low costs of the equipment, the simplicity in operation, the possibility to be used also to realize researches and industrial installations. As a result of using this method was found a considerable increase of the durability for the metal carbide cutting inserts coated with AlTiN materials, used in the cutting process compared to cutting inserts which were not coated.

1. Introduction

During exploitation of cutting tools, as a result of pressure on the contact surface, of the temperature and relative velocities tool-part and semi-finished-part, active surfaces of cutting tools are subject to wear. This one consists in deterioration the edge of cutting tool during the processing process, deterioration which leads to the fall of processing accuracy and respectively of the surface quality. After [1] achieving to some high quality parts, establishing of some cutting conditions which leads to the increase of processing productivity as well as the rational and efficient use of cutting tools impose the knowledge of their behaviour at wear. Whatever the type and their destination, the cutting tools used in normal regimes limits it wears out only on the settlement face or simultaneously both the settlement face, as well as on the clearance face.

Wear it occurs preponderantly only on one of the active surfaces of cutting tools, or on both surfaces, in the following conditions: wear only on the settlement face occurs in generally in the case of cutting with low speed and low thickness of the splinter, because it increase the specifically

Content from this work may be used under the terms of the Creative Commons Attribution 3.0 licence. Any further distribution of this work must maintain attribution to the author(s) and the title of the work, journal citation and DOI.
Published under licence by IOP Publishing Ltd

mechanical work of friction forces on the settlement face; wear only on the clearance face occurs in generally, for high cutting speed and large splinter thickness because the mechanical work of friction forces on the clearance face is higher; wear on the settlement face and clearance face occurs in medium cutting conditions and is the most common case.

A new modality to increase the durability of cutting tools is represented by the surface coatings with ceramic materials based on titanium compounds, such as: titanium carbide (TiC) and titanium nitride (TiN); titanium-aluminum nitride (AlTiN), titanium carbon-nitride (TiCN). Depositions of thin layers are applied on cutting tools in order to improve the desired properties of the surface, such as corrosion resistance, wear resistance, hardness, or friction. The research problems regarding the production of coatings are one of the most important surface engineering development directions, ensuring the production of coatings with superior properties in the field of applications to the mechanical characteristics and wear resistance [2, 3]. For that surfaces of the cutting tools, which are the most exposed to wear, to be used at maximum capacity these it can harden by several methods, of which are mentioned the coatings processes: physical vapor phase deposition (PVD) and chemical vapor phase deposition (CVD). The coating realized by physical vapor phase deposition or chemical vapor phase deposition, have gained special attention because of their unique physical and chemical properties, for example, extremely high hardness (40-80 GPa), corrosion resistance, excellent resistance to oxidation at high temperatures, as well as high abrasion and erosion resistance [4, 5].

2. Method

In this paper the authors have proposed the deposition on some metal carbide cutting inserts of some thin layers of AlTiN by cathodic arc ion plating method, in order to increase the durability of the cutting inserts and increasing the wear resistance of these. The cathodic arc ion plating method after [6] is the process of deposition of thin layers in which the substrate is subjected to bombardment with ions of a working gas, before deposition and, with ions of material of deposition and of the working gas, during deposition. For AlTiN layers deposition on the cutting inserts was used an equipment type Dreva 400 shown in figure 1 who allowed their introduction together with their fixing device, inside the vacuum enclosure (at 1·10-3-5·10-2 mbar).

The deposition of the AlTiN layer takes place inside the enclosure in which it apply suction for the cutting inserts to be covered (100 pcs.), the rotary table of the machine being performed with the help of device so that, the deposition take place only on the surface which will be in direct contact with the vapor of AlTiN ions of inert gas and the reactant gas. The material to be deposited on the substrate (work piece to be covered - the cutting insert) being in solid state is brought in vapour state as a result of its heating, up to evaporation of the target (AlTiN) and recondensation on the substrate when its temperature is lower than the vapors. The enclosure is cylindrical one, having double walls cooled with water and three arc evaporation sources type AS 65 M mounted along the bathyscaphe walls, used to generate the AlTiN vapors. In the enclosure shaft is mounted a rotary port-tool support equipped with the special devices designed for the settlement of the cutting inserts. The symmetrical positioning to the axis of the cylindrical enclosure makes as the inserts subject to the coating, located in batch, to be in identical deposition conditions for ensure a uniform thickness of the deposition.

Cleaning of the cutting inserts before deposition is carried out also inside the vacuum chamber by spraying with a controlled stream of argon. The AlTiN particles resulting from the evaporation process of the target, on their way to the substrate (the cutting inserts), collide with reagent gas ions (nitrogen) and with the electrons of a glow discharge plasma. Thus, the material particles positively ionized are accelerated in cathodic dark space of the substrate and hitting it with increase energy performs the deposition of some layers with a adherence and structure improved on it.

The AlTiN coating process involves a succession of operations. First, cleaning the plates for 20 min in a tank of benzene to remove grease. Fixing of the cutting inserts on supports in their locations on the machine rotary table. After this, the enclosure is closed and vacuumed for 15-30 minutes, up to 1,1 e-2 mbar. At the same time the Hallow cathod is heated for 20 minutes at 402°C up to 1,2 e-2 mbar. Then, take place Bias voltage for 10 minutes at 425°C up to 9,6 e-3 mbar. Inside of the bathyscaphe enclosure there is an ionization, using argon and nitrogen, for 20 minutes. The coating process begins after removing the covers of the arc evaporation sources for 20 minutes, at a temperature of 350-700°C. This is followed by cooling in two steps, first for 15 minutes at 180°C and the second for 20 minutes at 150°C. To remove the inserts now coated with AlTiN, the bathyscaphe will be repressurized for 20 minutes and its cap will be raised for extraction.

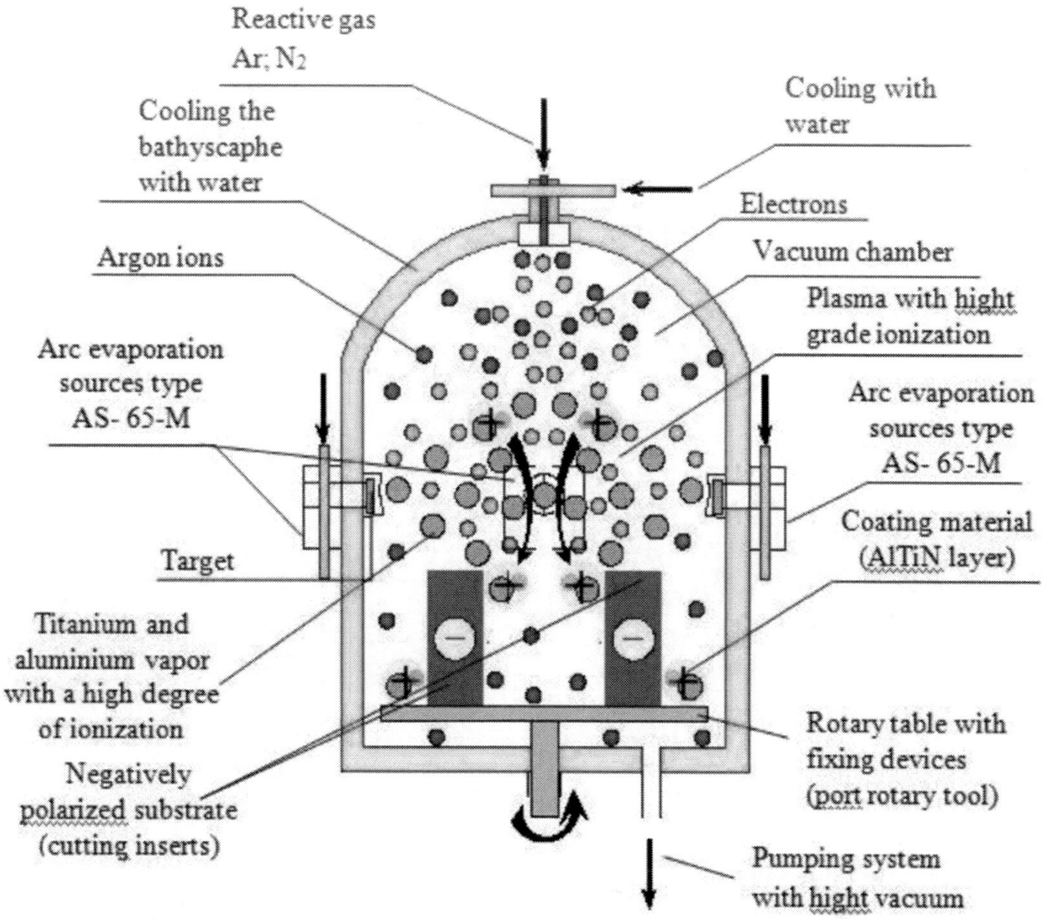

Figure 1. Scheme of the installation used in the application of the cathodic arc ion plating method [7].

3. Results

In order to perform their experimental researches the authors used cutting inserts unprofiled type SPUN 120312. The metal cutting inserts were used to achieve by turning chamfers inside the pieces type roller shown in figure 2, the cutting inserts both coated with AlTiN and uncoated being shown in figure 3, respectively figure 4, and in figure 5 and 6 are shown the cutting inserts after the turning operation, having their wear areas marked.

During the use of these metal carbide inserts type SPUN 120312 in turning process of the roller it was determined their durability, both in case of those uncoated as well as in case of those coated with titanium, at the same time establishing and the number of the machined parts with these cutting inserts, till their wear and loss of quality cutting.

The cutting process by turning was carried out in the same conditions for each case of using the cutting inserts both coated with AlTiN and uncoated. Conditions in which was carried out the cutting process by turning are given in table 1.

Table 1. Conditions in which the turning operation took place.

Cutting depth t[mm]	Cutting speed v[m/min]	RPM n[rot/min]	The working advance f[mm/rot]	The base time t_b[sec]	The volume of chips displaced v[mm^3/min]
0.3	60	1500	0.3	0.5"	185.17

Figure 2. Roller.

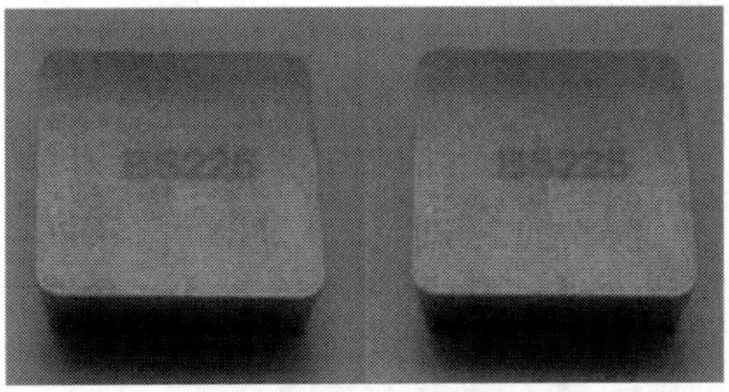

Figure 3. Inserts type SPUN 120312 with coated AlTiN.

Figure 4. Inserts type SPUN 120312 with uncoated AlTiN.

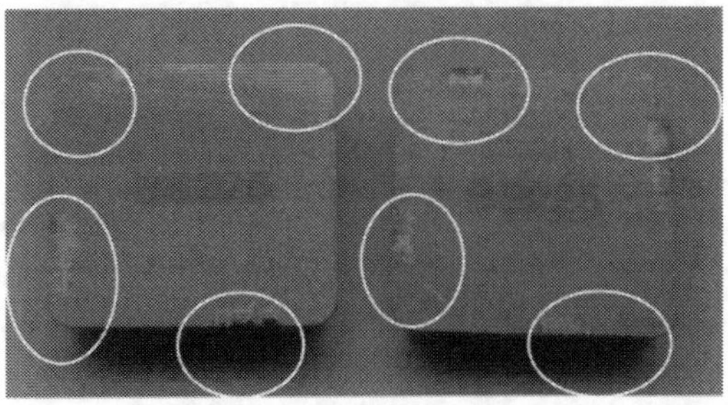

Figure 5. Inserts type SPUN 120312 uncoated with AlTiN and used.

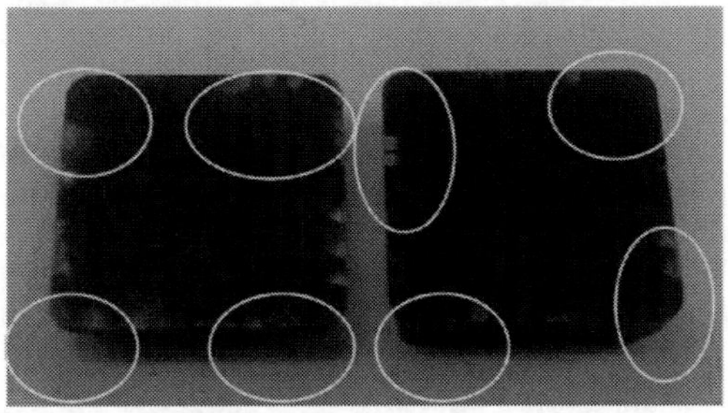

Figure 6. Inserts type SPUN 120312 coated with AlTiN and used.

The values obtained are shown in table 2, respectively table 3, showing a remarkable growth to durability of the cutting inserts undergo the process of thin layer coating with titanium and aluminum by a physical vapor deposition process, for these types of cutting inserts used, in relation to any non-processed by titanium and aluminum deposition. Also in table 4 is shown the total values for durability of the cutting inserts used. The graphical representations on how is the variation in time of the cutting inserts wear, as well as the volume of machined parts, during normal operating time of these, are shown in figure 7 and 8.

Table 2. The obtained time with these cutting inserts during turning operationst

INSERT TYPE	DURABILITY (min)									
SPUN 120312	Edge 1		Edge 2		Edge 3		Edge 4		TOTAL	
	C	U	C	U	C	U	C	U	C	U
	33.4’	0’43”	33’	0’57”	33.57’	0’52”	33’8”	0’55”	33.84’	0’51”

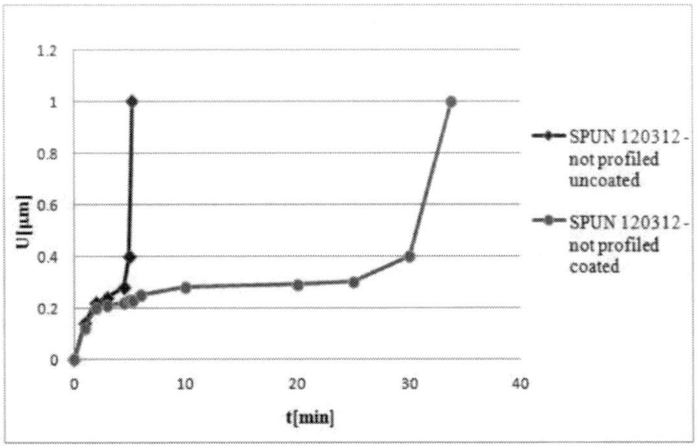

Figure 7. Diagram with variation in time of the wear of the type cutting inserts SPUN 120312 – not profiled uncoated and coated with AlTiN.

Table 3. Pieces number processed with these cutting inserts obtained by turning operation.

INSERT TYPE	DURABILITY (pieces)									
SPUN 120312	Edge 1		Edge 2		Edge 3		Edge 4		TOTAL	
	C	U	C	U	C	U	C	U	C	U
	1010	52	1025	69	1015	63	1010	67	4060	63

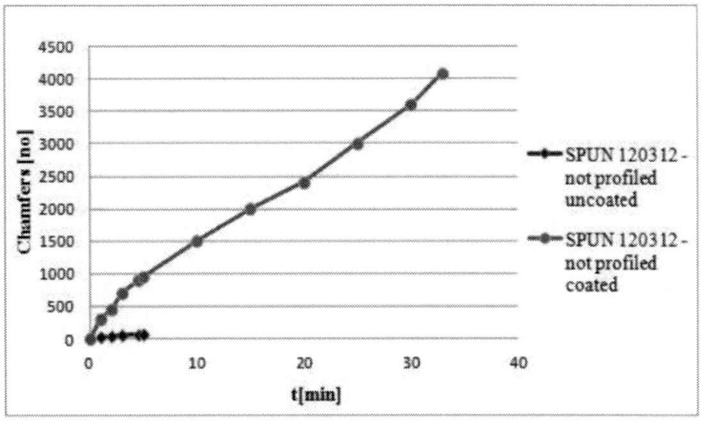

Figure 8. Diagram with the volume of the manufactured parts with the type cutting inserts SPUN 120312 – not profiled uncoated and coated with AlTiN.

Table 4 The durability total values of the cutting inserts used.

NR. CRT.	THE TYPE OF INSERTS		NR. TESTED INSERTS	DURABILITY TESTED [min]	NR. CHAMFERS MADE	SAMPLE TYPE
1	SPUN 120312 not profiled	Uncoated with AlTiN	2	0'51"	63 chamfers	roller
2	SPUN 120312 not profiled	Coated with AlTiN	2	33.84'	4060 chamfers	roller

4. Conclusions

Using in the process of turning of pieces type roller the inserts type SPUN 120312 not profiled, following their coating with thin layer of AlTiN by a physical vapor deposition method, to an increase more than 10 times their durability, with them can be processed more than 10 times pieces until the qualities are lost through cutting wear of the tool.

Regarding the volume of processed parts it is noted a direct proportionality with increasing the time as how the inserts keep their qualities of cutting. The results obtained opens new opportunities for continuing the studies in this domain, future researches following: diversification of the qualitative types of cutting inserts coated with thin layers of AlTiN through a PVD process and the study of their behavior in the processes of turning; realization of coatings and with other types of layers: Ti-Cr, Ti-Cr-Al and the analyze of their influence on the hardness increase and implicitly, the durability of the cutting tools; determination of thickness of deposited layers and determining their optimum values (depending on the layer structure), which allow achieving the highest values of durability.

References

[1] Szalók I 2013 Analysis of drilling surface microgeometry 2nd *Regional Conference Mechatronics in Practice and Education-MECHEDU* pp 111

[2] Dobrzanski L A, Lukaszkowicz K, Zarychta A and Cunha L 2005Corrosion resistance of multilayer coatings deposited by PVD techniques onto the brass substrate *Journal of Materials Processing Technology* **164-165** pp 816-821

[3] Lukaszkowicz K and Dobrzanski L A 2008 Structure and mechanical properties of gradient coatings deposited by PVD technology on to X40CrMoV5-1 steel substrate *Journal of Materials Science* **43** pp 3400-3407

[4] Veprek S, Mannling H D, Karvankova P and Prochazka J 2006 The issue of the reproducibility of deposition of superhard nanocomposites with hardness of $\geq$ 50 GPa *Surface and Coatings Technology* **201** pp 6064-6070

[5] Zou C W, Wang H J, Li M, Yu F J, Liu S C, Guo L P and Fu D J 2010 Characterization and properties of TiN-containing amorphous Ti-Si-N nanocomposite coatings prepared by arc assisted middle frequency magnetron sputtering *Vacuum* **84** pp 817-822

[6] Mateescu G 1998 Straturi subţiri depuse în vid *Tehnologii avansate* (Bucuresti: Dorotea)

[7] Bădănac A, Lupescu O, Manole V and Rusu O T 2014 Researches to improving durability of the cutting inserts coated with titanium thin layers *Academic Journal of Manufacturing Engineering* **12** pp 6-11

Modelling of fracture phenomenon in case of composite materials reinforced with short carbon fibers

R Caliman[1]
[1]"Vasile Alecsandri" University of Bacau, Calea Mărăşeşti, 157, Bacau, 600115, Romania

E-mail: rcaliman@ub.ro

Abstract. The research work presented in this paper describes the composite materials in terms of formation and propagation of cracks using an algorithm that imposes disproportional loads to composite samples. The required parameters that describe the composites fracture demand inputs as: load intensity, geometry features and relative loading direction. In order to obtain reliable results, it should be a good correlation between the model which describes the facture propagation, the composition of the material and the structural homogeneity. The presented study is using a Functionally Graded Material with local homogeneity in fracture area, and a numerical model based on integration of interactions (Mori – Tanaka method). The parameters that describes the fracture behaviour, includes a factor of stress intensity which is important for establish the fracture direction. The model used in simulations is considering a composite sample with rectangular shape and 6 mm thickness. The sample is loaded with predefined stress σ_{ct} (MPa) above and under the fracture line. σ_{ct} represents the critical stress able to lead to fracture propagation. The main objective of this research work it was to generate a numerical model which describes the fracture behaviour of a composite material. The obtained model and its accuracy to describe the fracture behaviour of a composite material is presented in the final part of this paper.

1. Introduction

The composite materials have been used long time before there were known as "composite". Due to permanently growing demands regarding performance and strength of structures in general, especially those for aeronautics and military applications, which require very severe condition during operation, oblige the composite developers to keep pace with [1].

The composite materials are becoming the adopted solution in achieving efficient structures, with applications in all industries. Their implementation in various production fields, as advantageous alternative to classical materials, are coming with some issues arising from their very complex structure and the production possibilities, also insufficiently known behaviour in various applications [2]. Special interest is given, among other things, determining the damages that may occur under loading conditions, their effect on the load bearing structures, as well as their behavior in harsh conditions (large variation of the working parameters). These are only some of the issues raised by researchers in the field, the multitude of published papers demonstrate that the problems are far from resolved [3].

The sustainability of matrix composites can be calculated by using uniaxial and biaxial loading of the composite, studying the wear phenomenon that occurs in these cases. However, this method of loading the composite material offers multiple ways of analyzing its resistance to cracking, because

Content from this work may be used under the terms of the Creative Commons Attribution 3.0 licence. Any further distribution of this work must maintain attribution to the author(s) and the title of the work, journal citation and DOI.
Published under licence by IOP Publishing Ltd

the phenomenon of wear which is characteristic for this kind of material is composed from two important components, wearing of the composite matrix, but mostly from sliding wear and cracking/breaking of fibres [4].

In our days, the optimization of composites structure and properties, such as cracking resistance has become a real need in practice as result of the requirement of making optimizations of complex structures, such as those from aeronautics. To achieve an objective of this kind, there are available specialised software able to deliver very accurate results. The combination between solid modelling and structural analysis, leads to a specially designed system, which can solve engineering problems.

This paper presents a study regarding the modelling process of fracture initiation and propagation in case of polymeric composite materials reinforced with short carbon fibres. In recent times, there are multiple concerns regarding strength and fracture behaviour of composites due to their utilisation in complex structures such as those used in the field of aviation.

2. The calculation algorithm applied to the modelling of crack formation and propagation

Composite materials have been investigated, with regard to crack formation and propagation by using an algorithm which applies disproportional loads, in conjunction with the finite element method. The required parameters in order to describe the material fracture impose the introduction of some intensity factors for load, geometry and load direction [5]. The correspondence between the model which explains the behaviour of crack propagation, the composition and the structural homogeneity, it is done by using numerical algorithm for simulation.

Compared to aluminium based composite materials, already studied and considered "classic", the metal matrix composites reinforced with short carbon fibres, lead to an increase of cracking resistance by lowering the concentration factor in case of matrix loads [6].

Assuming that, at one moment, a symmetrical load will determine a crack generation in the matrix body, simultaneously several parallel lines will appear and cause crack deviation and enlargement, as schematized in figure 1. In this case, the stronger the material oppose, the larger is the cracks deflection.

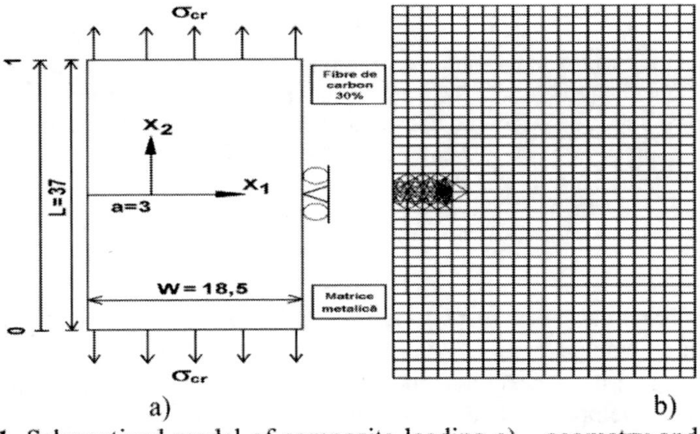

Figure 1. Schematized model of composite loading a) - geometry and loading directions (units and force in mm and N); b) - transposition network model.

On the assumption that the composite behave as a material that has gradual properties, which can be considered to be of FGM (Functionally Graded Materials), but which has local homogeneity in the failure area, known models could be used, based on numerical methods, but are implying a integrating method of inter-reactions [7, 8]. The parameter that describes the behaviour of cracking includes a stress intensity factor which is important in describing the cracking direction on two ways of actions KI and KII.

For the modelling process has been considered a composite surface with rectangular geometry (figure1) and thickness of 6 mm. The model is loaded with a tension σ_{cr} (MPa) above and below the fracture line. The critical load σcr is defined as the value that leads to the crack propagation. Where reference is made to a planar loading state is considered that the in composite material has a zonal homogeneity, and it is treated from the tension point of view as a lamina. The obtained values are presented in table 1.

Table 1. Properties and characteristics of material.

ξ	**E (MPa)**	υ	K_{Ic} **(MPa** $\sqrt{m}$ **)**
0,00	3000	0,35	1,2
0,17	3300	0,34	2,1
0,33	5300	0,33	2,7
0,58	7300	0,31	2,7
0,83	8300	0,30	2,6
1,00	8600	0,29	2,6

where: ξ - defines the point with coordinates x and y from the network; can range between 0.00 and 1.00, as shown in figure 1.

3. Establishing the cracks propagation model and the parameters used in simulation

Based on the crack propagation criterion of Cotterell and Rice, for fragile materials (KII = 0), it will be considered that the crack propagation is made with a gradient whose direction of development is perpendicular to the direction of crack. For modelling is considered a hyperbolic surface, with increasing values for the Young's modulus and for which is applied a fracture pattern, by bending, in three or four-point (figure 2).

The program that allow the description of cracking and fracture propagation, is using FRANC 2D element type (Fracture Analysis Code 2D) and is able to simulate a 2D cracks in similar conditions with the reality. The analysis is based on stepped propagation of the crack. This concept can be extended with good results, for materials that are not homogeneous (maximum 50% reinforced elements).

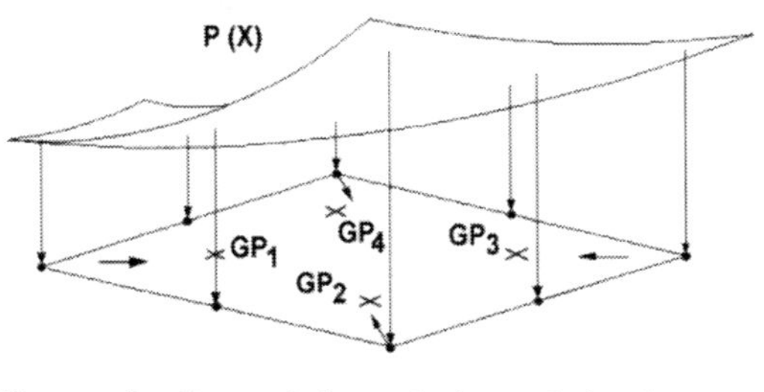

Figure 2. General formulation of the isoparameters based on the use of finite element method.

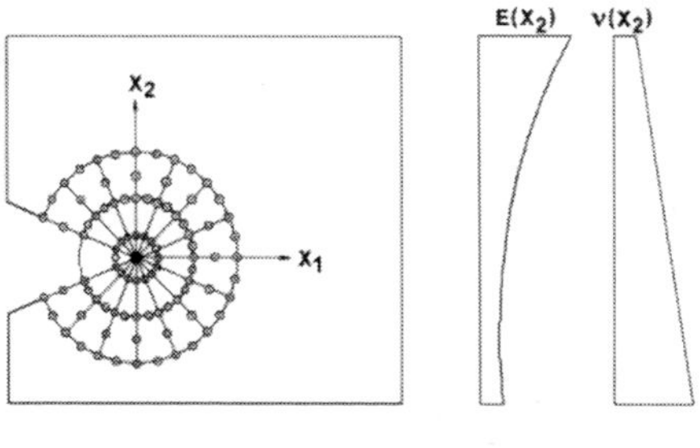

Figure 3. Model development and representation of discontinuous crack propagation in composite materials.

The material properties for Gauss points (PGP) are interpolated with the properties from nodal points (Pi) by equation PGP = Σ Ni·Pi in which N represent a functions ordering.

In order to represent a single node using triangulation with four points and six nodes of type (T6qp), it is necessary to represent a singular area of tension with the origin O (r-1/2), having radial type and circular development (rings), for which the representation of propagation is done by defining circular sectors as it can be shown in figure 3.

The crack propagation geometry is translated into a mesh model, describing the crack propagation in case of a composite material. The numerical values for material properties are defined by a certain area that can be described by a restriction having the following form: $0 \leq \xi \leq 1$.

The Young's modulus and Poisson's coefficient are presented in Cartesian coordinates X_1-X_2. The mesh consists from 818 elements of parallelepiped shape type (Q8), 106 nodes of triangulation (T6) and 12 nodes of triangulation type (T6qp), having a total number of 936 elements of triangulation and 2827 mesh nodes.

The increment constant is Δa = 1mm. The material heterogeneous is has a significant influence on the trajectory of the fracture in case of a composite material. The precise determination of stress intensity factor SIFs is a prerequisite in order to achieve the prediction of initiation and propagation of the fracture. The integration methods of the interactions, taking into account the stress intensity factor are particularly laborious and involves the application of evaluation schemes, using finite element methods. For this activity it is compulsory to call some micromechanics models and simulation calculations of the mechanical properties of materials having gradual properties.

The sequential stages (figure 4) represent a classical procedure of evaluation:

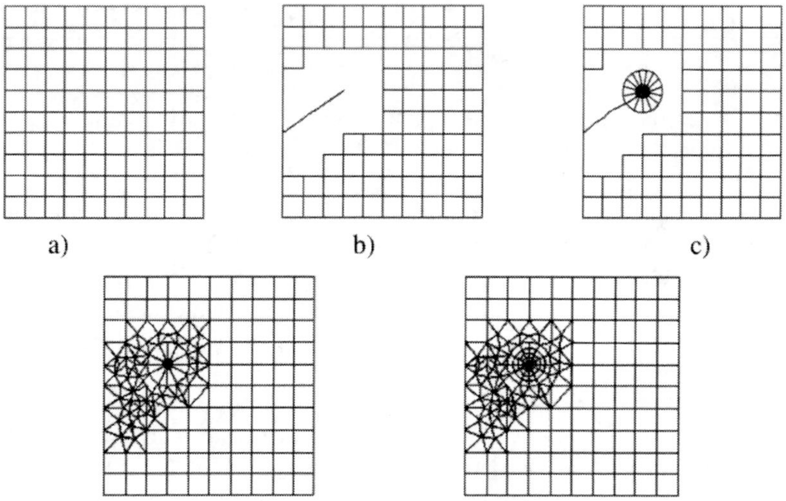

a) b) c)

c) d)

Figure 4. Application of a sequence process for generating a simulation model for crack propagation in the matrix of the reinforced composite with carbon fiber; a - the initial geometry of the composite surface, b - the primary (initial) way of inducing the crack, c - blocking the crack, d - crack development and propagation in the mesh, e - final stage representation of the crack in case of composite materials.

- the surface geometry on which the initial crack will propagate (figure 4a);
- the development of cracking is done by deleting the mesh elements which are associated to the tension zone (figure 4b);
- the development of a triangulation with nodal elements of type (T6qp) around the crack initiation point (figure 4c);
- the triangulation elements are generated by specific algorithm, as is shown in figure 4d;
- the detailed representation of the modified mesh around the crack it is made by increasing the number of elements of triangulation, with representation by jumps in the mesh (figure 4e)

The stress intensity factor can be defined by the following formula:

$$\sigma_{ij}(r,\theta)=\frac{K_I}{\sqrt{2\pi r}}f_{ij}^{I}(\theta)+\frac{K_{II}}{\sqrt{2\pi r}}f_{ij}^{II}(\theta) \quad (1)$$

where σ defines the stress tensor *ij*, K_I and K_{II} represent the stress intensity factor for mode I and II, and $f_{ij}^{II}(\theta)$ the angular positions. The stress intensity factor contains all the parameters required to characterize the behaviour of the composites in case of crack.

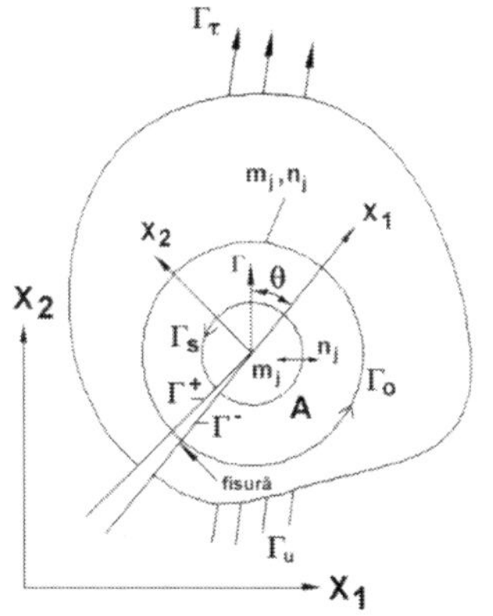

Figure 5. The coordinates and the contours of integration and calculation model.

The auxiliary displacement caused by the existence of tensions can be described by a formula having the following structure:

$$u^{aux} = \frac{K_I^{aux}}{\mu_{tip}}\sqrt{\frac{r}{2\pi g^I(\theta)}} + \frac{K_{II}^{aux}}{\mu_{tip}}\sqrt{\frac{r}{2\pi g^{II}(\theta)}} \tag{2}$$
$$\varepsilon^{aux} = (sim\Delta)\cdot u^{aux}$$

where K_I^{aux} and K_{II}^{aux} are two methods of stress intensity factor to determine the stress, used in fracture of micromechanics.

The formulation by integration of the stress intensity factor can be defined by a integral known as integral *J*.

$$J = \lim \Gamma_{s\to 0}\int \left(W\delta_{1j} - \sigma_{ij}u_{i1}\right)n_j d\Gamma \tag{3}$$

where W is the density of deformation energy, expressed by the expression $W = \sigma_{ij}\varepsilon_{ij}/2$, n_j is the normal vector on Γ_s outline (figure 5), in which the development of integration function is obtained by summing:

$$J = \int_A \left(\sigma_{ij}u_{i1} - W\delta_{1j}\right)q_j dA + \int_A \left(\sigma_{ij}u_{i1} - W\delta_{1j}\right)j_q dA \tag{4}$$

The integral J obtained by extrapolating actual area and the auxiliary area where the crack is developing can be decomposed as $J^s = J + J^{aux} + M$, in which J^{aux} is described by the relation:

$$J = \int_A \left(\sigma_{ij}^{aux}u_{i1}^{aux} - W^{aux}\delta_{1j}\right)q_j dA + \int_A \left(\sigma_{ij}^{aux}u_{i1}^{aux} - \frac{1}{2}\sigma_{ik}^{aux}\varepsilon_{ik}^{aux}\delta_{1j}\right)j_q dA \tag{5}$$

resulting in an integration formula (M) based on a state of non-equilibrium:

$$M = \int_A \left(\sigma_{ij}u_{i1}^{aux} + \sigma_{ij}^{aux}u_{i1} - \sigma_{ik}\varepsilon_{ik}^{aux}\delta_{1j}\right)q_j dA + \int_A \left(\sigma_{ij}u_{i1}^{aux} + \sigma_{ij}^{aux}u_{i1} - \sigma_{ik}\varepsilon_{ik}^{aux}\delta_{1j}\right)j_q dA = M_1 + M_2 \tag{6}$$

The last term of the second integral (M2) in equation (6) can be expressed by the following equality:

$$\left(\sigma_{ik}\varepsilon_{ik}^{aux}\delta_{1j}\right) = \left(\sigma_{ik}\varepsilon_{ik}^{aux}\right)_1 = \left(\sigma_{ij}\varepsilon_{ij}^{aux}\right)_1 = \left(C_{ijk1}\varepsilon_{k1}\varepsilon_{ij}^{aux}\right)_1 = C_{ijk,1}\varepsilon_{k1}\varepsilon_{ij}^{aux} + \sigma_{ij}^{aux}\varepsilon_{ij,1} + \sigma_{ij}\varepsilon_{ij,1}^{aux} \tag{7}$$

IOP Conf. Series: Materials Science and Engineering **95** (2015) 012008 doi:10.1088/1757-899X/95/1/012008

in which $C_{ijk,1}$ represent continuous and derivative functions of auxiliary stresses.

Inserting equation (7) in equation (6) and using the compatibility and equilibrium value for which $\sigma_{ij,j} = 0$ (there are no forces inside the solid) will get another value for M:

$$M = \int_A \left(\sigma_{ij} u_{i1}^{aux} + \sigma_{ij}^{aux} u_{i1} - \sigma_{ik} \varepsilon_{ik}^{aux} \delta_{1j} \right) q_j dA + \int_A \left(\underline{\sigma_{ij}^{aux} u_{i1}} - C_{ij,1} \varepsilon_{k1} \varepsilon_{ij}^{aux} \right) q dA \qquad (8)$$

where the underlined term is a state of non-equilibrium, the introduction of a auxiliary stress, leading to the production of convergent solutions.

4. Conclusions

The fracture analysis is used to assess the durability of a model, having variable geometry defined by the finite elements, when simultaneously or sequentially loads are applied to it. There is possible to run analyses which are linear or nonlinear.

The model presented in this paper can achieve further development as modelling and verification of transient fatigue of composites. This approach can be used to calculate the dynamic stress and the specific effort, thus being able to estimate the wear lifetime of industrial products.

The proposed model offers efficient and flexible tools in order to generate, modify and display the mechanical and thermal properties, necessary for product design, simulation and manufacturing.

The analysis of a model with errors can consume a lot of time and money, and often, errors are not detected even after analysis. The proposed model provides a complete set of mathematical and graphical tools for checking and correcting of the model before its use in simulation.

References

[1] Mori T and Tanaka K 1973 Avarage stress in matrix and average elastic energy of elastic materials with misfitting inclusions *Acta Metallurgica* **21(571)** p 58

[2] Gu P and Asaro R J 1997 Crack deflection in functionally graded materials *International Journal of Solids and Structures* **34 (24)** pp 3085-3098

[3] Wasiluc B and Golos K 2000 Prediction of crack growth direction under plane stress for mixed-mode I and II loading Fatigue and Fracture *Engineering Materials and Structures* **23(5)** pp 381-386

[4] Bueno A and Lambros A J 2002 Investigation of crack growth in functionally graded materials using digital image correlation *Engineering Fracture Mechanics* **69(14-16)** pp 1695- 1711

[5] Becker T L, Cannon R M and Ritchie R O 2001 Finite crack kinking and T-stresses in functionally graded materials *International Journal of Solids and Structures* **38(32-33)** pp 5545- 5563

[6] Chang J H and Chien A J 2002 Evaluation of M-integral for anisotropic elastic media with multiple defects *International Journal of Solids and Structures* **39(9)** pp 2575-2583

[7] Dolbow J and Gosz M 2002 On the computation of mixed-mode stress intensity factors in functionally graded materials *International Journal of Solids and Structures* **39(9)** pp 2557-2574

[8] Khan S M A and Khraisheh M K 2004 A new criterion for mixed mode fracture initiation based on the crack tip plastic core region *International Journal of Olasticity* **20(1)** pp 55-85

IOP Conf. Series: Materials Science and Engineering **95** (2015) 012009 doi:10.1088/1757-899X/95/1/012009

Efficient dielectric fluid approach in electrodischarge finish machining on the material surface roughness of titanium alloy Ti-621/0.8

N V Mihai[1] **and G Brabie**[1]

[1] "Vasile Alecsandri" University of Bacau, Faculty of Engineering, Calea Marasesti 157, 600115, Bacau, Romania

E-mail: niko.vboy@gmail.com

Abstract. This study was developed to reveal the correlation between current intensity and pulse off time with surface roughness of Ti-621/0.8 in fine EDM machining, in a unique manner of dielectric fluid approach. Depth of cut was taken in consideration to be not as high 1 µm due to the fact that good results were taken for more less heights before. In the case of Ti alloy (Ti-621/0.8) bar, EDM machining with low parameters (limited 110 V) is a delicate process, in which reaching SR under 1 µm is a challenge. Cooper and Graphite electrodes were used, die sinker oil dielectric from machine manufacture was used, without thermal modification to workpiece or electrode, just active dielectric compensatory fluid pumping solution with 0.5 MPa.

1. Introduction

The machining of Ti-621/0.8 with conventional machining processes is very difficult because of high toughness and low thermal conductivity. Investigate the effects of cooling time due to *pulse off time* as prime factor, and *peak current* intensity as a second factor, of fine machining without using other cooling options, like cryogenically treatment before or in time of processing [1] which provided smooth surface appearance, the cold effect is provided only from the speed and direction of dielectric fluid in the gap area. Regardless of pulse on time which is one of the main factors in MRR (Material Remove Rate), the surface roughness quality is based on efficiency of recasting time of layers [3].

Controlling the thermal proprieties of the Ti-621/0.8 is capital when we have a process in which cooling is allowed less time than heating up on a regime of finish EDM machining with spark temperatures around melting temperature of Ti alloy.

Optimizing input parameters or other means like dielectric approach, is allowing a larger area of applicability in the industry of Ti-621/0.8.

The influence of the workpiece after solidifying and before machining conditions, were made just for reducing the unwanted deforming lines in the workpiece profile, reducing thermal modifications [2] due to classical machining of fabrication in the pre-process of EDM.

Presented experiments are applicable on the workpiece models that will have a distinct simple geometrical surface to be machined on, and a very good dielectric gap circulation.

Content from this work may be used under the terms of the Creative Commons Attribution 3.0 licence. Any further distribution of this work must maintain attribution to the author(s) and the title of the work, journal citation and DOI.
Published under licence by IOP Publishing Ltd

2. Experimental conditions

The experiments were conducted on a die sinking EDM machine, Model *Fem 110* CNC although, even if kerosene is widely used as dielectric with die sinking machines, at high discharge temperatures, the carbon elements penetrate the electrode surface affecting the normal discharge.

In this study Knuth EDM oil was used, the model provided from the start with the machine.

Figure 1. Dielectric system management of fluid flowing motion lines.

Analyses of dielectric were conducted [5], (using Brookfield DV-e Viscometer) regarding the quality, with temperature around 23 C° and *viscosity* value of 3.07 Centistokes cST or 0.00000307 [m^2/s]. For fine machining 2÷3cST is a range accepted for hard steels.

Many preparations and calibrations of process were made for choosing the dielectric system management figure 1, for analysing materials of workpiece and tool electrode as in table1. Soft abrasive cleaning followed by air pressured cleaning was performed on the circular profile before immersion in dielectric for free of unwanted surface particles.

Table 1. Chemical (%) of workpiece and electrode.

Workpiece Material Titanium Alloy		High Quality brass (Cu)	
C	0.015	Cu	99.5
O	0.106	Zn	0.175
N	0.0052	Cr	< 0.0010
Fe	0.04	Pb	0.0395
Al	5.32	Sn	0.0423
V	-	Fe	0.0837
Ti	Balance	Ni	< 0.0050

The graphite [C] electrode was used only on low energy regimes having no need in dividing the plane surface of 20/20 [mm] preparations like [3]. Divided profile and polished surface High-Quality brass electrode was necessary figure 2 due to depth appearance corrosion of tool electrode and workpiece machined surface.

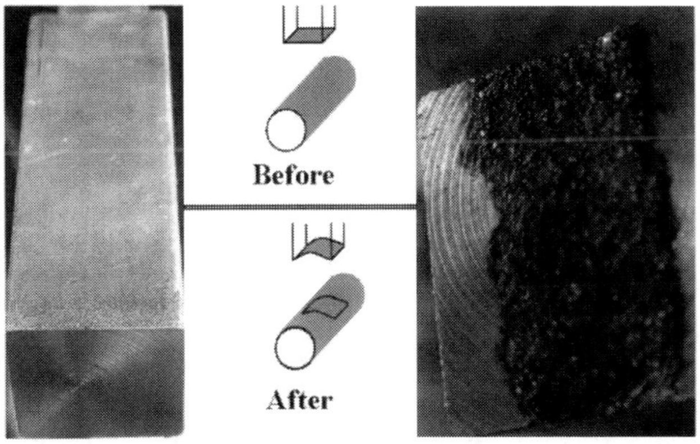

Figure 2. High-Quality brass (Cu) before and after each experiment.

The experimental plan was chosen using Design Expert software. From a total number of 21 machining codes were reduced to 9 experiments by applying the criteria of *pulse off time*13, 10 and 7 μs combined with 6, 10, 14 A *peak current intensity* for the EDM process and input range taken from literature [5].

Parameters that are likely to affect SR and also cause surface changes on the recast layers were identified using a study from the reported results in the literature [1, 4, 3]. Three repetitions for each surface roughness measurement were conducted. The roughness readings were made on Roughness Tester TR-200 (figure 3.) by Namicon Company, with analysing 0.8 μm length of surface each. The surface obtained after each EDM process was photo taken using metallurgical microscope IMM 901 by Metkon Company with the 200x amplification factor shown in figures [3-5].

3. Experimental results

The experimental result are listed in table 2 (lowest SR associated to main parameter current intensity I are *italics*).

Table 2. Experimental results.

Experiment Number.	Peak current intensity I A	Pulse Off Time μs	High-Quality brass tool electrode Cu		Graphite tool electrode C	
			Surface Roughness Ra μm	Average Ra μm	Surface Roughness Ra μm	Average Ra μm
1	6	13	2.302 2.437 2.512	2.417	—	—
2	6	10	1.253 1.72 1.196	1.389(6)	—	—
3	6	7	—	—	0.744 0.871 1.057	*0.890(6)*
4	10	13	2.471 2.542 2.594	2.535(6)	—	—
5	10	10	1.655 1.648	1.638	—	—

IOP Conf. Series: Materials Science and Engineering **95** (2015) 012009 doi:10.1088/1757-899X/95/1/012009

			1.611			
6	10	7	—	—	1.413 1.487 1.523	*1.474(3)*
7	14	13	3.643 3.604 3.578	3.608(3)	—	—
8	14	10	2.208 2.485 2.094	*2.262(3)*	—	—
9	14	7	—	—	2.786 2.805 2.776	2.789

Smallest *pulse off time* is recommended to be used only for finishing stages, very long machining times average 6 hour were recorded for the depth width of 1µm compared to usual timed EDM.

Optical microscopy captions for 200x Ti alloy machined were prepared and chose randomly from machined surface, images are from measurement program of IMM 901 for interpretation by specialists of craters that were chose to be measured, presented as follow:

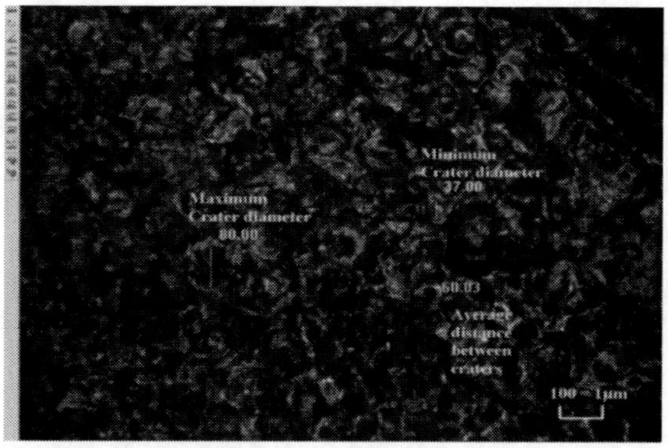

Figure 3. Current of 6 A, pulse off time of 7 µs , pulse on time of 5 µs and C electrode for experiment.

Figure 4. Current of 10 A, pulse off time of 10 µs , pulse on time of 7 µs and Cu electrode for experiment 5.

It is considered almost free of debris, craters 0.37÷0.8 µm were in normal values compared to SR of 0.890(6) µm, with a usual arrangement of craters, average distance between them of 0.6003µm.

The sparks created craters of low depths < craters diameter, which proves a better recast layer.

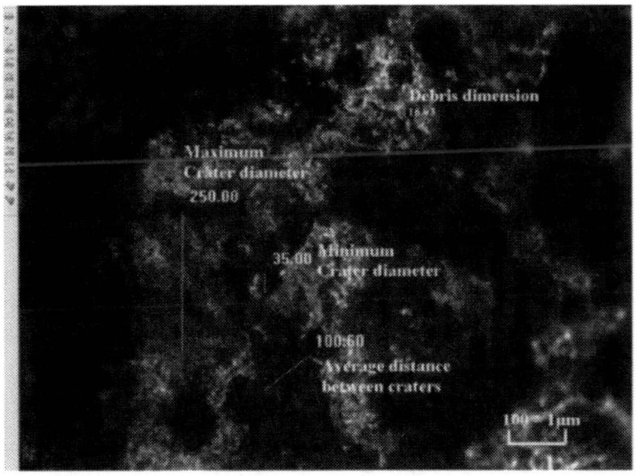

Figure 5. Current of 14 A, pulse off time of 13 μs , pulse on time of 24 μs and Cu electrode for experiment 7.

With major debris diameter found around 0.1603 μm, craters 0.35÷2.5 μm were in normal values compared to SR of 3.608(3) μm with a compact arrangement of craters, average distance between them of 1.006μm.

Cracks were not taken into consideration in measurement of craters due to its atypical forms creation.

Individual contributions on SR are presented for all *peak current intensity* and *pulse off time* input levels in figure 8. Each experiment is being represented as a dot.

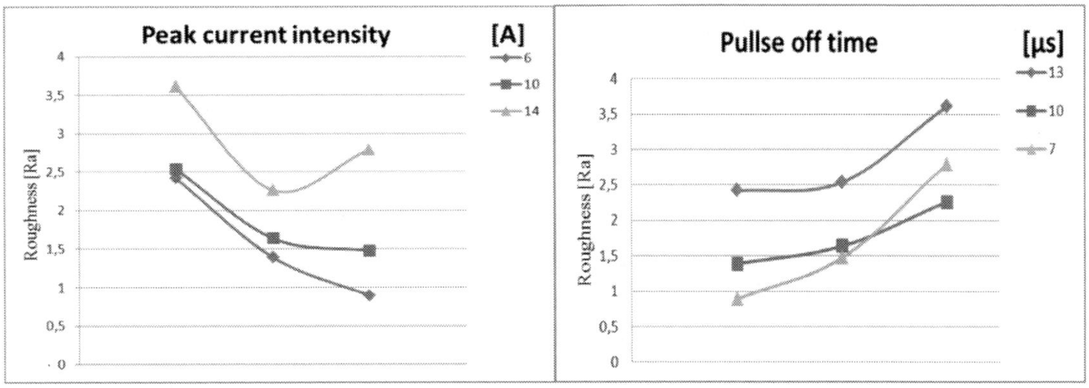

Figure 6. Modifications of SR which depends on Peak current intensity and pulse off time.

Significant decrease in SR was observed in all cases with decrease of energy of EDM, with the presence of low *pulse off time*, in which case the *pulse on time* correlated values were also less than 22÷24 μs to the lowest of 5÷6 μs.

Pulse off time is vital for SR if well combined with dielectric fluid pressure (ex. low fluid pressure for low sparks conditions, conditions associated with *low peak current intensity* or/and low *pulse off time*). The combined influences of input parameters are presented 3D in figure 7 (the lower the cubic pieces are the finest was the machining process).

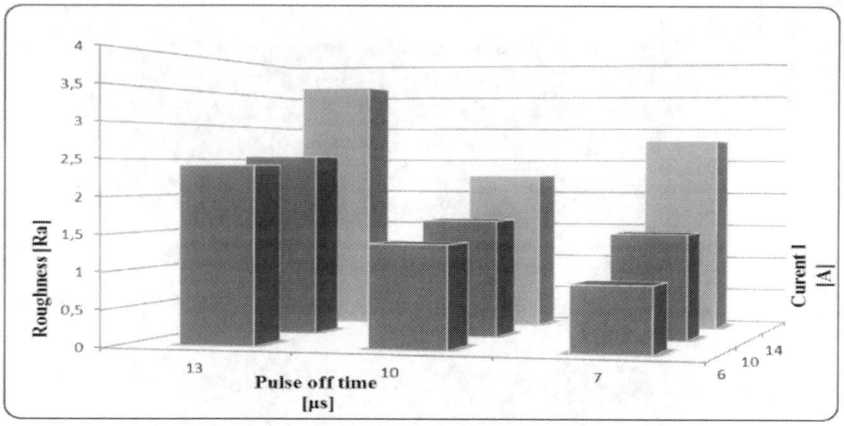

Figure 7. 3D view of SR resulted from parameter of pulse off time and peak current intensity.

4. Conclusions

With the most idealistic environment that consists in: simple geometry Ti-621/0.8 round bar, entire machined surface was dielectric refreshed on each electric contact due to efficient dielectric fluid approach, divided Cu electrode before each experiment, expected results have been reached as follow. Almost free of debris surface roughness was obtained with the pulse off time 7 µs and the pulse on time 5 µs in graphite electrode case.

The graphite electrode used cannot handle well higher peak currents intensity over 10A in fine machining of Titanium alloy (Ti-621/0.8) even if pulse off time used is low, due to his low electrical conductivity compared to High-Quality brass.

Lowering the pulse off time and expecting only to lower SR in the case of graphite is not mandatory as experiment no. 9 was recorded, in which the SR predictive pattern changed dramatically.

In the machining process of finishing with EDM, choosing the 10 µs pulse off time ensures productivity with Cu and C electrodes on Ti-621/0.8 Ti alloy.

The SR predictive pattern on Ti-621/0.8 Ti alloy is applicable only in Cu electrode fine EDM machining.

References

[1] Kumar S, Batish A and Singh R 2015 A mathematical model to predict material removal rate during electric discharge machining of cryogenically treated titanium alloys **229(2)** p 214

[2] Wansheng Z, Zhenlong W and Shichun D 2002 Ultrasonic and electric discharge machining to deep and small hole on titanium alloy *J Mater Process Tech* **101**

[3] Jabbaripour B, Sadeghi MH and Faridvand S 2012 Investigating the effects of EDM parameters on surface integrity, MRR and TWR in machining of Ti–6Al–4V *Mach Sci Technol* **419**

[4] Hascalik A and Caydas U 2007 Electric discharge machining of titanium alloy (Ti-6Al-4V) *Appl Surf Sci* **9007**

[5] Kibria G, Sarkar B R and Pradhan B B 2010 Comparative study of different dielectrics for micro-EDM performance during microhole machining of Ti-6Al-4V alloy *Int J Adv Manuf Tech* **557**

[6] Abdulkareem S, Ali Khan A and Konneh M 2010 Cooling Effect on Electrode and Process Parameters in EDM *Materials and Manufacturing Processes* **462**

Contributions regarding the use of ultrasound capabilities in the reconditioning by metallization of some automotive parts

G Amza[1], D Nitoi[1], C Amza[1], A Dimitrescu[1], Z Apostolescu[1], G Tasca[1] and M Teodorescu[1]
[1] Politechnica University of Bucharest, Splaiul Independenţei no.313, sect.6, Bucharest, Romania

E-mail: amza@camis.pub.ro

Abstract. Paper presents the ultrasonic energy possibility in the reconditioning by powder metallization process. Three separated ultrasonic methods of energy activation in the deposition metal zone are presented. Advantages and disadvantages are studied for the following technological methods: filler material ultrasonic activation, base material ultrasonic activation, reconditioned metal piece activation, both filler and base material ultrasonic activation. Metallization devices design, are presented according to ultrasonic method activation for the ultrasonic field metallization process using metal powder.

1. Introduction

Restoration of automotive parts by reconditioning by powder metallization is the process of restoring the functional size of the piece by deposition of melted particulate filler material mechanically anchored on used surfaces properly prepared, wearing their microroughnesses. When certain energetic conditions are met, appear microweldings formed by microirregularities melting and diffusion [1]. The structure of the deposited layer by powder metallization has the following components (figure 1): particles of filler material 1, flattened due to the impact with the base material (support) 2 and that it adheres to this by wearing the microirregularities of the appropiate prepared surface 4 (line of demarcation). After filling the voids, the filler particles adhere to each other, layer by layer, until the required thickness and the following are happening:

-oxide microparticles 5 formed by oxidation of the filler material, located either at the interface of the base material - the first layer, or between the particles of the various layers deposited 6 (the highest protection is formed between the base material and the first powdered layer), where oxides resulted in surface preparation phase can be found).

-the oxide films reduce adhesion to the base material of the coating, and adhesion to each and other layer.

- a network of microchannels 7 and pores 8 that occur between filler particles deposited on top of another, or between the filler particles and the base material;

-filler particles that were not molten 9, whose adheshion is very weak compared to neighboring particles and lead to poor mechanical properties of metallization [1, 2].

-poor adherence or lack of adherence of filler layer due to weak diffusion process;

Content from this work may be used under the terms of the Creative Commons Attribution 3.0 licence. Any further distribution of this work must maintain attribution to the author(s) and the title of the work, journal citation and DOI.
Published under licence by IOP Publishing Ltd

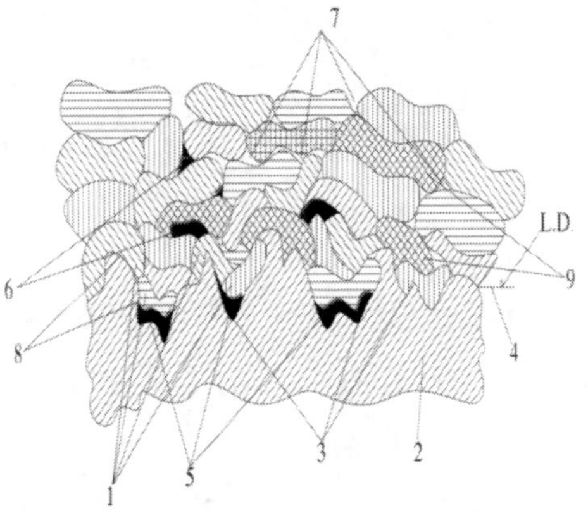

Figure 1. The structure of the deposit layer by powder metallization

1- deposit filler material particles; 2 - base material (suport material); 3 - microirregularities; 4 - line of demarcation (LD); 5 - oxide particles; 7- microchannels; 8 - voids; 9 - unmelted filler material particles

The reconditioning by powder metallization process has several advantages, but also some disadvantages, the most important are [1]:
-low resistance to tensile and bending, to mechanical and thermal shock of the deposited layer and of the resulted joint;
-low resistance of the deposited layers;
-the deposited layers cannot be plastic processed;
-the torque resulting from the metallization cannot be processed threads, grooves and any surfaces that interrupt material fibre of the deposited layer;
-needs post-metallization thermal treatment;
-needs a very good cleaning of the surface to remove any oxide film and impurities to an optimal adherence;
-causes significant emissions of dust and gas and therefore a significant impact on the environment [2, 3].

To prevent and reduce some of these disadvantages, this paper proposes a repairing method – the reconditioning by powder metallization in ultrasonic field, taking into account the main basic effects of ultrasound propagation in different environments. [1, 2, 7].

2. Possibilities of ultrasound use in the reconditioning by powder metallization

Research undertaken on different couples of materials and results obtained by metallization reconditioning [14] process showed that all the changes that occur in the deposited layer, the heat affected zone (if any) and in the support material depend primarily on how to enter ultrasounds in the deposited material, the base material or in both at the same time depending on the geometry of the part to be reconditioned and functional characteristics required [2, 4, 5, 9].

The efficiency of the technological process of powder metallization depends primarily on the behavior of the joint basic material-filler material but also how it connects between atoms marginal homogeneous or the two materials in contact area and near the contact area.

Homogeneous joint formation is the result of technological steps for depositing the filler material over the support material. The most important technological steps in the process of reconditioning by powder metallization are: suitable processing of the used surface that will be covered with filler; cleaning, pickling, degreasing to create optimal conditions of adhesion of filler material to the support material; pre-heating the support material (if appropriate) to reduce the temperature gradient, the actual metallization; ensure optimal conditions of solidification of the deposited material; application of appropriate heat treatment and processing of required dimensions for carrying out the same functional role [1, 2].

Each of these steps is influenced more or less if the ultrasonic energy overlaps conventional energy due to the pursuit phase in ultrasonic field.

The efficiency of the reconditioning by metallization in ultrasonic field depends primarily on the input mode of ultrasonic energy in processing furnace (deposit area). Experiments have shown that ultrasound propagation in the workplace has significant influence over the transfer process of the filler material to the support material, the adhesion process, the process of crystallization and solidification and even on the stress relaxation. All these influences are attributed to three basic phenomena occurring in ultrasound propagation, namely: ultrasonic cavitation, acceleration of diffusion; acceleration of crystallization and solidification.

The most difficult problem that arises in the metallization process in ultrasonic field is related to how the introduce the ultrasonic energy in the loading area of the filler material. In this paper, several variants of introducing ultrasonic energy in the deposition area have been investigated, the best results are obtained by the following: ultrasonic activation of filler material; ultrasonic activation of the reconditioned piece and simultaneous ultrasonic activation of reconditioned piece and filler material [2, 3, 10].

3. Methods proposed for the introduction of ultrasound in the powder metallization

Ultrasound input mode ultrasound in the area of deposition and overlapping ultrasonic energy over conventional energy is essential in obtaining materials with the required properties of the functional role. The introduction of ultrasound in the metallization process must meet the following requirements: ultrasonic system should not be affected by the temperature that develops in the processing area of the filler material, meaning that in the ultrasonic transducer must not be reached the Curie temperature; enabling the phenomenon of cavitation during cleaning and removal of impurities and oxides to create optimal conditions for adhesion of the filler material to the base material; avoiding possible oxidation of filler material droplets on their path to the depositing surface [11],[13]; to accelerate the crystallization and solidification of the deposited material with the base material forming a joint:

- have technological and constructive simplicity;
- enabling economic efficiency.

4. The possibility of introducing ultrasound through the filler material as wire electrode

In this case (figure 2), the filler material as wire electrode is passed through an ultrasonic system [11, 13] that vibrates longitudinally, the tested processes in the paper being [2, 3]:

- oxyfuel flame spray metallization with electrode-wire activated ultrasonic;
- electric arc spray metalization with three wires ultrasonic activated.

In oxyfuel flame powder metallization and ultrasonic activated wire electrode (figure2), the wire electrode 1 made of filler material is driven by the drive rollers 2, passes through acoustic reflector 3, piezoceramic disc 4, acoustic radiant 6, nodal flange 7, amplifier 8, ultrasonic energy concentrator 9 and makes contact with the active element 10 activated ultrasonic, achieving the oxyfuel flame 11 formed by the combustion of fuel gas 12, introduced through the nozzle 13. The wire electrode melts into droplets 14, protective gas entrained introduced by coupling 15 and are projected on the used surface properly prepared very finely pulverized in ultrasonic field created by the active element. Fine spray droplets are entrained by the carrier gas introduced through nozzle 15 and projected onto the used surface adequately prepared of the part 16, resulting filler layer 17 deposited on the surface of the piece. The wire electrode is ultrasonic activated in the active element of ultrasonic energy concentrator, a very fine polarization taking place and controlled projection of the droplets on the used surface [1, 2, 3]. In the case of introducing ultrasound by the wire electrode [8] made from the filler material, the following advantages are obtained: spraying the filler material in the form of very fine droplets with sizes of the order 0.01 .... 10.0 μm; substantial improvement of the transfer of filler material to support material; uniformity of the filler particles that are deposited on the used surface adequately prepared; ultrasonic cavitation occurs that increases the speed of subcooling and cooling and accelerates solidification; prevention of oxidation of the micro particles of filler material that increases the adhesion of the layer deposited on the support material; as a result of the propagation of ultrasound through the environment, accelerating the diffusion of filler material in the base material; degassing and uniformity of pores due to the transmission mode of ultrasound through the filler material; efficiency of deposition is maximized [10].

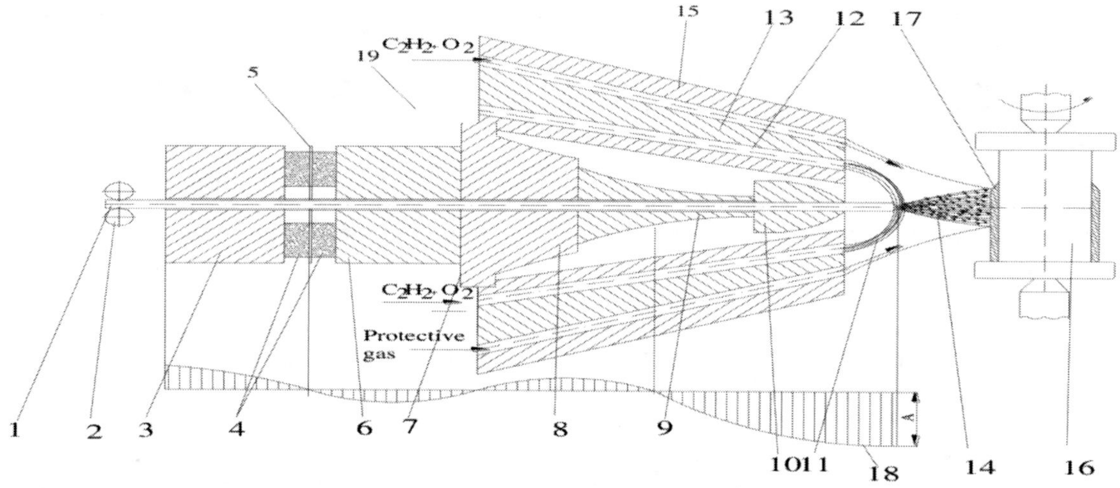

Figure 2. Schematic diagram of the introduction of ultrasound in the metallization process with flame and wire: 1- filler material wire-electrode; 2 - roller drive, 3 - acoustic reflector, 4 piezoceramic discs: 5 - contact electrode, 6 - acoustic radiant, 7 - nodal flange; 8 - amplifier; 9 - ultrasonic energy concentrator; 10 - active electrode; 11 - oxyacetylene flame;12- combustion of fuel gas, 13 - nozzle; 14 - drops of filler; 15 - nozzle; 16 - piece needs to be reconditioned; 17 - deposited layer; 18 - variation diagram for the amplitude of particle velocity along the ultrasonic system.

The disadvantages of this method are related to the construction of the metallization torch and heating the active part vibrating at ultrasonic frequency and creates ultrasonic field. There may also be problems with the transmission of ultrasonic vibration to the wire electrode.

5. Possibility of ultrasound introduction in the part that needs to be reconditioned by powder metallization

In this case, the ultrasonic system is conceived, designed and calculated [11, 13] in such a way that the part to be reconditioned to excite longitudinal ultrasonic waves [7], transverse or on the surface, depending on the geometry of the part and overall dimensions. Possible metallization processes with ultrasonic activation of the parts to be reconditioned are [3, 4, 6, 10]. Metallization with flame and powder and ultrasonic activation of the part to be reconditioned is presented in the figure 4. For metallization with oxyfuel flame and powder with ultrasonic activation of the part to be reconditioned part 1 (figure4.), caught in a drive device 2, selected according to the geometric shape of the surface to be reconditioned, through the active element 3 coupled acoustic with the part to be reconditioned, and are inserted in it longitudinal ultrasonic waves from the ultrasonic system composed of ultrasonic energy concentrator 4, the amplifier 5, the nodal flange 6, through which the system is fixed in the installation to be reconditioned, the acoustic radiant 7, piezoceramic disks 8 and the acoustic reflector 9 mechanical polarized with the polarization screw 10. Through the nozzle 12 is introduced the fuel gas 11 that forms the oxyacetylene flame 13, in which it is brought the powder filler material 16 with the aid of a carrier gas 15 introduced through the central nozzle 14. The powder is melted in form of drops 17, which are projected by the carrier gas to the surface to be reconditioned 18. The ultrasonic system is calculated so that it works under resonance, vibrating according to a diagram of the amplitude variation of the form 19.

When entering the ultrasound directly in the part to be reconditioned by metallization (figure 4), we obtain the following advantages:

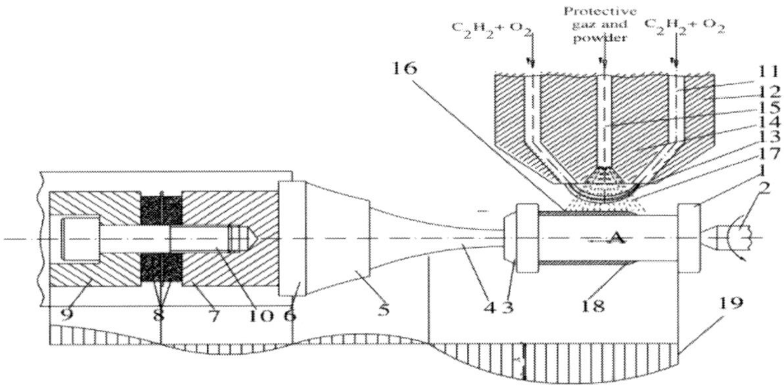

Figure 3. Schematic diagram for metallization with flame and powder with ultrasonic activation of the part to be reconditioned: 1- part to be reconditioned; 2- drive device; 3 - active part;4 - ultrasonic energy concentrator; 5-intermediant element; 6 - nodal flange;7- acoustic radiant element; 8 - piezoceramic discs; 9 - acoustic reflector; 10 - pre-polarization screw; 11- mix C_2H_2 +O_2; 12 - outer nozzle; 13 - oxyfuel flame;14 - central nozzle; 15 - powder carrying gas;16 - powder; 17 - filler material drops; 18- deposited layer; 19 - variation chart of the particle velocity; 20 - along the ultrasonic system.

6. Conclusions

The efficiency of the powder metallization reconditioning depends primarily on the behavior of couples filler and base material and the way how the homogeneous connection is made between atoms of the two materials in fringe of the contact area [12];

The action of ultrasound on the filler particles in the liquid state changes some basic properties such as superficial surface tension in the boundary deposited layer in the liquid state, base layer in solid state; viscosity, solidification temperature; sub-cooling and cooling speed; solidification temperature; degassing and diffusion accelerates; changes the structure; makes dispersion possible of a material in the others etc;

Introduction of ultrasound in the deposition area of the sprayed filler material can be added in several ways, the most effective being: ultrasonic activation of filler material in the form of wire-electrode, ultrasonic activation of the piece to be reconditioned and simultaneous activation of the filler material and the piece to be reconditioned;

In the case of ultrasonic activation of the filler material in the form of wire-electrode, the following advantages are obtained: the powdered filler material in very fine droplets (atomization thereof), a substantial improvement of the transfer process to the base material, increasing the sub-cooling and cooling speed, accelerating the solidification process, a substantial reduction in the oxidation process and increasing the adhesion process; accelerating the diffusion of the filler material to the base material; uniformity of the degassing process and of the distribution of pores etc;

Ultrasonic activation of the piece to be reconditioned by metallization leads to the following advantages: it provides a nearly perfect cleaning of the surface to be loaded, creating the conditions for a very good adhesion; accelerates the diffusion process; ultrasonic cavitation occurs and increases the rate of cooling and solidification of the deposited layer; obtaining a uniform fine grain structure with equiaxed grains; occurs degassing of the deposited layer and obtaining an uniform and controlled porosity; internal thermal stresses disappear and no post-metallization heat treatment is required etc.

References

[1] Dehelean D 1997 *Fusion Welding* (Timisoara: Welding Publishing House)
[2] Amza G 2006 *Ultrasounds, Active applications* (Bucharest: Editura AGIR)
[3] Dumitru G 2009 *Metal part reconditioning* (Bucharest: Bren Publishing House)
[4] Iacobescu G 1999 *Welding Equipments* (Bucharest: Printech Publishing House)
[5] Solomon G 2001 *Theory elements for welding processes* (Bucharest: Bren Publishing House)
[6] Zgura G 2006 *Fusion welding technology* (Didactica si Pedagogica Publishing House)
[7] Amza G, Luchian C, Nitoi D and Dumitrache F 2010 Contributions Regarding Plastic Materials Ultrasonic Welding used in Automotive Industry *MATERIALE PLASTICE* **47** p 90
[8] Kovalev I M and Rybakov A S 1997 The movement of liquid metal in the weld pool in welding in a longitudinal megnetic field *Welding Prod* **24** pp 38-40
[9] Itah Y 2008 Crakking parameter of high strengths steel related to haz cracking *Journal of Japan Welding Society* pp 37-39
[10] John H 2005 *Welded design theory and practice*
[11] Amza Gh, Niţoi D, Amza C, Apostolescu Z and Opran, C 2014 Ultrasonic System Design and Calculation for Polymer Metallization International Conference of technology for Polymeric and Composites Products, *Applied Mechanics and Materials* **760** pp 263-268
[12] Ferber M K and Wereszezak A A 1990 Evolution of stress failure resulting from high – temperature stress – corrosion cracking in a hot isostatically pressed silicon nitride *J. Am. Ceram. Soc.* **78(8)** pp 2129-40
[13] Niţoi D and Taşcă G 2015 Eperimental results Regarding metal plating by welding method in ultrasonic fieldfor high corosed base metal piece *Applied Mechanics and Materials* **760** p 539
[14] Nitoi D F, Dumitru M G, Dumitru B and Mihailescu A 2014 Reconditioning by metal deposition technology. Temperature finite element modelling *Advanced Materials Research* **1029** pp 130-133

Carbon fiber based composites stress analysis. Experimental and computer comparative studies

M Sobek[1], A Baier[1], A Buchacz[1], Ł Grabowski[1] and M Majzner[1]
[1]Silesian University of Technology, The Faculty Of Mechanical Engineering, Institute of Engineering Processes Automation And Integrated Manufacturing Systems, ul. Konarskiego 18a 44-100 Gliwice, Poland

E-mail: michal.sobek@polsl.pl

Abstract. Composite materials used nowadays for the production of composites are the result of advanced research. This allows assuming that they are among the most elaborate tech products of our century. That fact is evidenced by the widespread use of them in the most demanding industries like aerospace and space industry. But the heterogeneous materials and their advantages have been known to mankind in ancient times and they have been used by nature for millions of years. Among the fibers used in the industry most commonly used are nylon, polyester, polypropylene, boron, metal, glass, carbon and aramid. Thanks to their physical properties last three fiber types deserve special attention. High strength to weight ratio allow the use of many industrial solutions. Composites based on carbon and glass fibers are widely used in the automotive. Aramid fibers ideal for the fashion industry where the fabric made from the fibers used to produce the protective clothing. In the paper presented issues of stress analysis of composite materials have been presented. The components of composite materials and principles of composition have been discussed. Particular attention was paid to the epoxy resins and the fabrics made from carbon fibers. The article also includes basic information about strain measurements performed on with a resistance strain gauge method. For the purpose of the laboratory tests a series of carbon - epoxy composite samples were made. For this purpose plain carbon textile was used with a weight of 200 g/mm2 and epoxy resin LG730. During laboratory strain tests described in the paper Tenmex's delta type strain gauge rosettes were used. They were arranged in specific locations on the surface of the samples. Data acquisition preceded using HBM measurement equipment, which included measuring amplifier and measuring head. Data acquisition was performed using the Easy Catman. In order to verify the results of laboratory tests numerical studies were carried out in a computing environment, Siemens PLM NX 9.0.
For this purpose, samples were modeled composite corresponding to real samples. Tests were made for boundary conditions compatible with the laboratory tests boundary conditions.

1. Introduction

Composites were known to mankind for thousands of years. Already in ancient times this technology was used to prepare a composite brick as a durable building material. However, on a large scale composites have come into use relatively recently. Their tendency to manifesting different properties from the constituent materials is so desirable feature that their popularity as engineering materials is growing at a very fast pace. The effect of this phenomenon is the presence of composite materials in

Content from this work may be used under the terms of the Creative Commons Attribution 3.0 licence. Any further distribution of this work must maintain attribution to the author(s) and the title of the work, journal citation and DOI.
Published under licence by IOP Publishing Ltd

every area of daily life. These materials are at the forefront of materials used in the most developed industries, ie. the automotive and aerospace industry, [1, 2, 5].

Composites are divided into different groups depending on the structure and the principle of composition. A specific example of composite is laminates. Reinforcement of such composite form may vary. A composites with a reinforcement made of layer of another material, long fibers, short fibers or dispersed particles can be distinguished (figure 1). A characteristic feature of laminate composites is that as matrix an adhesive binder is used. The purpose of this matrix is to maintain the shape of the composite material, to ensure protection against mechanical damage and transfer the load directly to the reinforcing material. In the case of laminates, the most common matrix material is a resin. In combination with the hardener, which is the initiator of the copolymerization process, the material forms a perfectly checking in manufacturing composites.

Despite their popularity, the composites based on carbon fabrics are unfortunately quite expensive. Carrying out laboratory tests on composites is a long-term and absorbing process. Comparative studies of actual composite samples and their numerical equivalents are intended to verify that this method is likely to replace the actual examination. Comparison of the results will verify the validity and accuracy of this method, [5, 6, 7].

During comparative studies the following course of action was adopted:
-Preparation of the composite samples and determination of measurement points on their surface;
-Construction of a fixing frame;
-Laboratory tests;
-Verification of laboratory tests results by comparing them with the results of simulation studies.

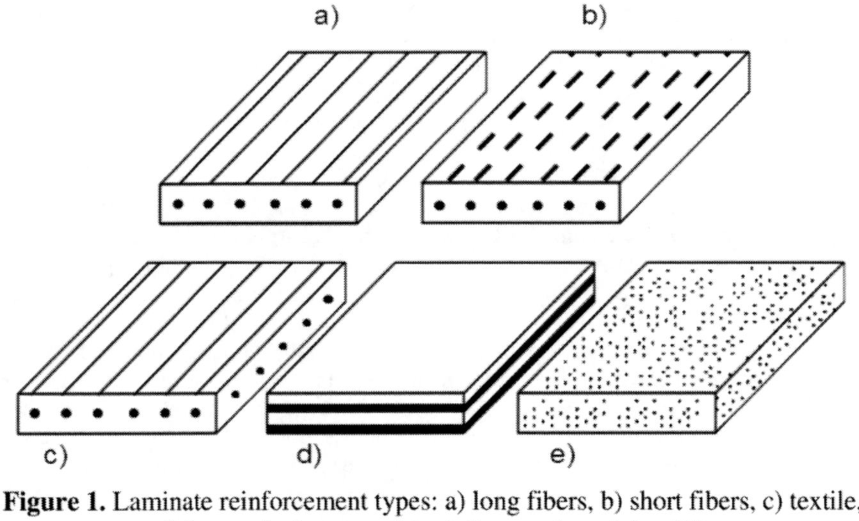

Figure 1. Laminate reinforcement types: a) long fibers, b) short fibers, c) textile, d) layer of other material, e) dispersed particles, [3].

2. Laboratory research part

For the production of the laminate a Toho Tenax HTA40 H13 3K 200TEX carbon cloth was used. I weights 200 [g / m2], and its Young's modulus is equal to 238 [GPa] (figure 2). From available types of strand a plain weave strand was selected, in which the fibers are woven with 90° angle relative to each other. As a composites matrix material an LG 730 epoxy resin was selected, which together with the HG700 initiator creates a very durable composition. The composites were cured at room temperature, which significantly lengthened the cooling process.

Figure 2. Toho Tenax HTA40 H13 3K 200TEX carbon cloth structure.

Figure 3. S42 polyurethane foam.

Figure 4. Tenmex TFr-10/120 delta type strain gauge rosette [11].

Samples were produced to map the area of the vehicles body, and therefore a mold corresponding to the shape had to be produced. The material used to produce the mold was a high density S42 polyurethane foam. It was formed in the cutting process using a three-axis CNC milling machines. Developed surface characterized by a very high accuracy and quality (figure 3), [5, 12].

Laminate was produced in a manual process. The surface obtained in this type of production of the composite is slightly porous, which means that it must be properly prepared before placing strain gauges. For this purpose the surface of the composite in the strain gauges fixing points was polished and cleaned with a special degreaser. For laboratory testing two points on the surface of a laminate were adopted. Additionally, at the intersection of the diagonals of the top surface of the composite an opening for fixing load was made. In laboratory studies a Tenmex TFr-10/120 delta type strain gauge rosettes were used. Figure 4 shows a diagram of the construction of such a system. They consist of three strain gauges spaced at 120°. This system has been attached to the sample using a non-aggressive cyanoacrylate adhesive. The same type of glue is used to attach solder terminal, whose task is to separate strain gauges terminals from each other, [11].

Prepared samples were mounted in a specially constructed frame (figure 5). This solution makes it possible to restrain the test piece in a controlled manner as it is very important to make points, or surfaces of fixing properly defined. This is to avoid measurement errors during simulation studies in this area.

The block diagram of the measuring circuit is shown in figure 6. In order to obtain a reduced stress at the measuring points in the data acquisition software a virtual measurement channel was created. It allowed for immediate calculation of stress by the hypothesis von Misses by reading the stress values for each individual strain gauges of the rosette. Each of the tested composites was loaded by adding metal rings, each with a mass of 66g. These rings were mounted on the M8 hexagonal head screw with suitable nut and a total weight of 158g. Juxtaposition of weights and corresponding forces are shown in table 1. The study was carried out for the double layer and single layer composite, with a number of weights. Strain gauges rosettes were marked with numbers 1 and 2. For each of the measurements graphs of stresses were generated. They contained values for each individual strain gauge and reduced stress calculated according to von Misses hypothesis. Tests results included information on the reduced stress on both strain gauge rosettes.

Table 1. Number of metal rings and corresponding forces.

Number of metal rings	2	3	4	5	7	9	10	11	12	13	14	19
Force [N]	2.9	3.56	4.22	4.88	6.2	7.52	8.18	8.84	9.5	10.16	10.82	14.12

3. Simulation research part

After the completion of laboratory research part a simulation part of examining began. The study was carried out in the Advanced Simulation module of a Siemens PLM NX 9.0 environment. Geometric model of the sample was made in the Modeling module of the same environment.

Figure 5. Mounting frame.

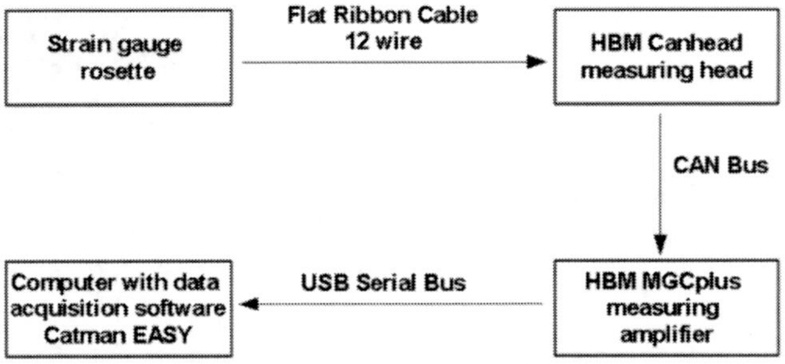

Figure 6. Block diagram of the measuring circuit.

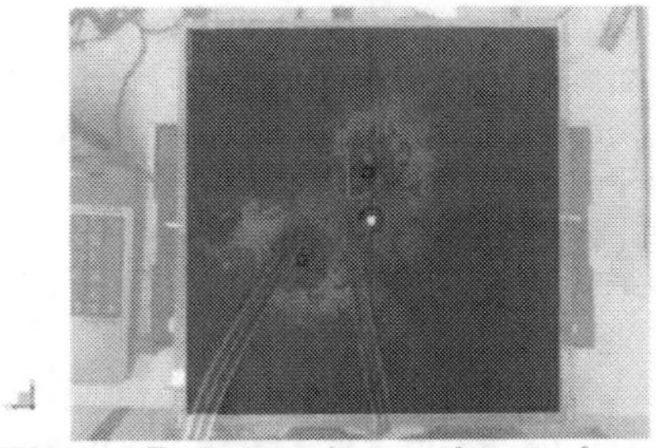

Figure 7. Raster image layer above the CAD model of the test sample.

Figure 8. Fixed faces and force applied on the numeric model of the test sample.

During the modelling process of the simulation sample it was very important to preserve as many of compliance with the actual sample as possible. Therefore, a model used in the modeling process was the same as the one used for the G-code generating for CNC milling machines in the mold production process. This model was also equipped with mounting surfaces corresponding to the areas fixed to the frame and areas of fixing strain gauge rosettes and applying force. In order to accurately designate these areas on a numerical model a picture of the actual sample was applied to the model (figure 7). Test object prepared so was then transferred t

The force was exerted on the surface, which corresponded to the surface covered by the screw with metal rings in the actual model. The force application direction was elected as the opposite to the Z axis. Figure 8 shows the model firmly established and loaded with the forces according to the real model used in the laboratory tests. Red arrows point to the places of force applying. The blue color represents constrained nodes of the finite element mesh.

In individual studies loading force values were set consistent with those used in laboratory studies. The simulation was performed for one- and two-layer composites. The test results are presented in the next section of this paper.

4. Results and discussion

The graphs presented in figures 9 and 10 shows the summary of series of researches which were carried out during the both test. On the chart presented on figure 9 a comparison of the average results of laboratory tests and simulation tests for rosettes No. 1 on the two-layer composite is shown. Figure 10 shows a comparison of the average results of laboratory tests and simulation tests for rosettes number 2 on the two-layer composite.

Results obtained in both types of researches do not differ significantly from each other, especially for smaller values of load. With the increase of loading force discrepancy between the results obtained in the laboratory and simulation researches increases. For the load force of 14.12 N discrepancies reaches about 20%. It is very important to note that in case of maximum force load, which is equal to 14.12 N the shape of the composite sample, has changed. After applying the force the surface transformed from convex to concave. It a drastically changed in the geometry of the sample. Such a change undoubtedly influenced the final indication of sensors.

In case of such gentle sensors like strain gauge rosettes it is very important to fix them properly. The observed differences may be due to both, the laminates structure imperfections and due to the initial stresses caused by the strain gage installation method on a sample. The results of the simulation studies showed a linear stress growth tendency. That tendency leads to a simple conclusion that they were carried out properly in this area. The drastic increase in the discrepancy between the results may be due to exceeding the linear range of applied rosettes. Linear static software simulation solver does not take into account the possibility of such a drastic change of the samples shape during the test. That situation occurred during applying a load of 14.12 N.

Worth noting is an interesting phenomenon, which was observed in both laboratory tests and simulation tests. A sudden increase of strain at a load of 3.56 N is clearly visible on both charts. This situation may be due e.g. to the geometric shape of the sample.

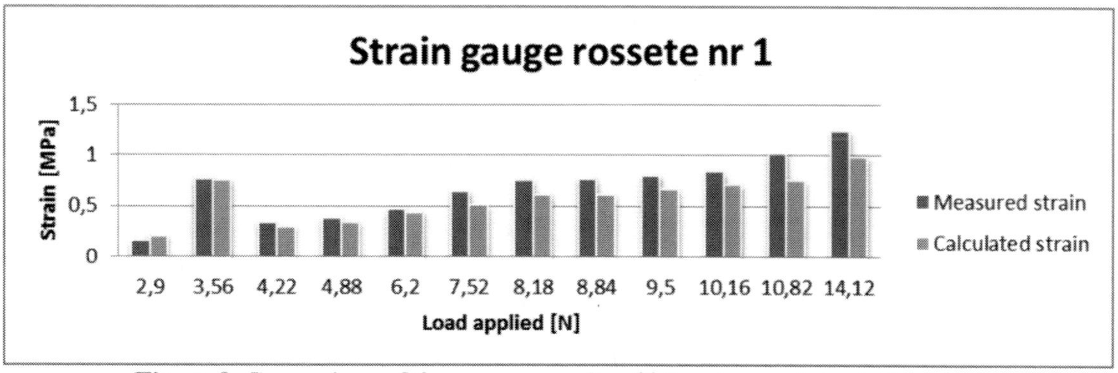

Figure 9. Comparison of the average results of laboratory and simulation tests.

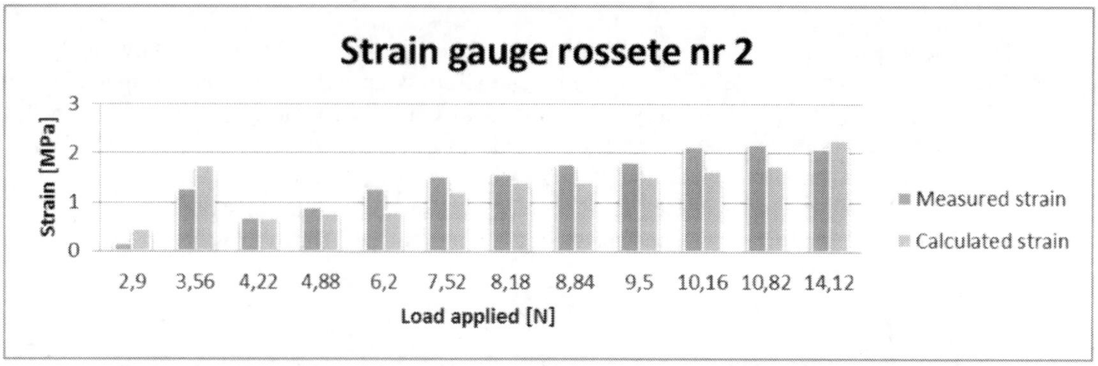

Figure 10. Comparison of the average results of laboratory and simulation tests.

5. Conclusions

After analyzing the results it can certainly be said that, despite some differences in the results the researches were carried out successfully. Simulation test runs much faster allowing to save a lot of time. In addition, no need of producing real samples reduces the necessary financial outlays for research. FEM stress analysis of laminates is a very good substitute for long and expensive laboratory tests.

Acknowledgement

The work was carried out under the project number PBS2/A6/17/2013 realized as a part of the Applied Research Program, funded by the National Research and Development Centre.

References

[1] Chawla K K 1998 Fibrous materials (Cambrigdge: Cambridge University Press)

[2] De Araújo M 2011 Fibrous and composite materials for civil engineering applications - Part 1, 59, (Cambridge: Woodhead Publishing Limited)

[3] Ozimina D, Madej M and Wdowin A 2006 Tworzywa sztuczne i materiały kompozytowe (Kielce: Wydawnictwo Politechniki Świętokrzyskiej)

[4] Dobrzański L A 2002 Podstawy nauki o materiałach i metaloznawstwo, (Warszawa: Wydawnictwo Naukowo Techniczne)

[5] Sobek M 2014 Badanie włóknistych materiałów kompozytowych na bazie tkanin węglowych. Analiza wytrzymałości kompozytowej karoserii samochodu elektrycznego, Gliwice

[6] Baier A, Brajza M, Erm M, Grabowski L, Papaj P and Sobek M 2013 Advanced analysis of the Greenpower race car *International Journal of Materials & Product Technology*

[7] Baier A, Baier M, Dusik D, Sobek M, Papaj P and Grabowski L 2014 Computer aided Process of designing the mechatronic Silesian Greenpower electric car *Advanced Materials Research* **1036** pp 674-679

[8] Baier A, Baier M, Dusik D, Grabowski L, Miera A and Sobek M 2013 Computer aided Process of designing the mechatronic Silesian Greenpower electric car *Selected Engineering Problems* **4**

[9] Niedworok A and Baier A 2015 Numerical Modeling of the Phenomena of Frictional Coupling Between Wheel and Rail to Describe and Verify the Operation of Surface Condition Detector *Solid State Phenomena* **220-221** pp 251-256

[10] Majzner M and Baier A 2014 Computer Aided Design and Analysis of Composite Structure, *Advanced Materials Research* **1036** pp 989-994

[11] http://www.tenmex.pl/

[12] http://www.surfpol.pl/

IOP Conf. Series: Materials Science and Engineering 95 (2015) 012012 doi:10.1088/1757-899X/95/1/012012

Aspects regarding the tribological evaluation of sintered composites obtained from mixture of copper with carbon fibers

R Caliman[1]

[1]"Vasile Alecsandri" University of Bacau, Calea Mărăşeşti, No. 157, Bacau, 600115, Romania

E-mail: rcaliman@ub.ro

Abstract. This paper presents a study of the tribological properties of sintered composite materials made from combination of copper with short carbon fibers. Sintered composite materials are more effective if refer to specific properties per unit volume compared to conventional isotropic materials. Potential advantages of copper - carbon composite materials are: high resistance to breakage and high value ratios strength/density; resistance to high temperatures; low density and high resistance to wear; low or high friction coefficient. The sintered composite materials used in this research work are obtained combining different percentages of copper with short carbon fibres with iron and lead in order to investigate the variation of the friction behaviour. Varying the percentage of copper from 92,2% to 97,6% and the percentage of short carbon fibres from 7,8% to 2,4%, five different composite materials are obtained and tested. Friction tests are carried out, at room temperature, in dry conditions, on a pin-on-disc machine. The friction coefficient was measured using abrasive discs made from steel 4340 having the average hardness of 40 HRC, and sliding velocity of 0,6 m/sec. The main objective of this research work it was to identify a combination of materials with improved friction behaviour. The experimental results revealed that the force applied on the specimen during the tests, is playing a very important role regarding friction coefficient and also the wearing speed. Graphite particles are conveyed consistently inside the net, enhancing scraped spot safety and voltage drop over normal composite material. The static tests demonstrated that this new kind of sintered composite material has preferable electrical execution over previous brush material in the same conditions, and the element tests demonstrated that the temperature climb is not enormous when the brush with this new composite material is exchanged on with huge current and the surface scraped spot is littler amid sliding. The new method of obtained sintered composites, using high-energy and high speed is adequate for production of tribological parts.

1. Introduction

In the last few years, carbon fibres have become an interesting reinforcement for polymeric composites from economically and ecologically points of view. There is increase in the environmental awareness in the world which has aroused an interest in the research and the development of biodegradable and high performance materials. Carbon fibres can be obtained from natural resources as graphite mineral. With the increase of global energy crisis and ecology risk, the unique advantages

Content from this work may be used under the terms of the Creative Commons Attribution 3.0 licence. Any further distribution of this work must maintain attribution to the author(s) and the title of the work, journal citation and DOI.
Published under licence by IOP Publishing Ltd

of carbon fibres such as abundant, non-toxic, non- irritation of the skin, eyes, or respiratory system, non-corrosive property, plant-based fibre reinforced polymer composites have attracted much interest owing to their potential of serving as alternatives reinforcement to the synthetic [1,2].

In all the applications where are used, carbon fibre composites has emerged more efficient and easy to recycle than glass or other kind of synthetic fibres and appeared a good alternative for the following reasons: carbon fibre production has consumed less energy than other fibre and thus lesser pollution emissions, the higher volume fraction of carbon fibres for similar performance has decreased the volume and weight of base synthetic polymer matrix, the lower weight simultaneously with higher volume has improved the fuel efficiency and reduced emission in the use phase [3,4].

Copper alloys are at the moment extensively used in tribological engineering parts, such as bearing and bushings. Due to its properties, the copper alloy is considered as a solid lubricant which has been developed as self-lubricating material under medium or high load conditions. On the other side, carbon fibers have high strength and modulus combination with the good lubrication effect and the wear resistance. It is a good candidate as an additive in the metal matrix composites. Copper–graphite composites are widely known as materials for sliding electrical contacts, due to high thermal and electrical conductivities closed to pure copper and a low friction coefficient caused by a graphite lubricating ability [5–10].

The final parts obtained by sintered metal powders deliver outstanding physical and chemical characteristics, which are influenced by the composition and the structure of the existing phases and the size, shape and distribution of grains. The main characteristic of a material obtained by sintered metal powders is the presence of residual porosity in the structure. Adjusting the remaining porosity of the product, for the same chemical composition allows obtaining parts having desired (and dissimilar) physical and mechanical characteristics.

The stability of sintered composites in a severe environment depends upon its adequate mechanical properties with non-porous and smooth surfaces because these properties are profoundly affected by the action of water [11]. It has also been reported that dissolution or degradation in surface layers may take place in materials that remain in contact with different fluids and some loss of unbound components is suspected may be along with fluid uptake into the basic structure. This fluid uptake in discrete zones of the material may exert unwanted residual pressure thus resulting in softening, degradation or leakage of the materials [12, 13].

The apparent density is one of the important characteristics of composite materials obtained by sintered metal powders and thus of the parts obtained from such materials. For parts having simple geometric form, the apparent density can be determined by weighing and calculating the volume based on the dimensions of the part. For parts with complex shape, the apparent density is determined by hydrostatic weighing. This property is influenced by many factors, including: nature of the metallic powders which are used; dimensions of the powders particle; the required compaction pressure; the amount and type of lubricant used for compaction; the sintering conditions as temperature, time, atmosphere; the chemical composition of the sintered material.

The dimensional variation of parts obtained from composite materials by sintered metal powders is also an important parameter, which must be taken into account. Thus, the application of exact compensation in order to avoid dimensional variations that occur at sintering allows the manufacture of parts without subsequent calibration.

The mechanical properties of composite materials sintered from metal powders are influenced by several factors, among which the main role is played by porosity and the internal shape of the pore, which depend on the characteristics of powders and sintering conditions. The main mechanical properties of composite parts obtained by sintered metal powders are: tensile strength; hardness; bending strength; elongation and fatigue.

For the composites made by sintered metal, in order to obtain better friction behaviour, the most important mechanical property is the hardness. This is a function of binding forces between the particles, the density and the material resistance in the tested area.

IOP Conf. Series: Materials Science and Engineering **95** (2015) 012012 doi:10.1088/1757-899X/95/1/012012

The friction materials sintered from mixtures of metallic and non-metallic powder and used in the manufacture clutches and brakes are tested additionally to determine: the friction coefficient; wear resistance; galling tendency; smooth braking capacity; the heat dissipation capacity.

2. Tribological evaluation of sintered composites from copper-carbon fibers

For applications which requires higher temperature and low friction, and in particular, in the case of high-speed sliding from the electrical contacts, the composite copper-carbon fibres is a suitable material that meets many of the requirements of such applications: high melting temperature, mechanical strength and ductility large enough, high thermal and electrical conductivity, low friction and low speed wear.

Due to a certain degree of immiscibility between copper and carbon fibers, for normal temperatures, it is recommended to improve this feature by using easily fusible metallic binders (as Sn and Pb) or by using carbon fiber coated with various metals, depending on its subsequent use.

By using metal binder it is possible to increase the incipient wetness of the matrix components during the sintering processes. Due to the high temperature from the sliding surface, in critical friction conditions it was observed melting state during the binder phase.

These binders limit the operating range in case of practical applications. In addition the electrical conductivity of copper is very sensitive to impurities. For example, in case of electric brushes, which have to bear high density of current, it is necessary a small electrical resistivity.

Strengthening by processes having high energy and high speed simultaneously, is an alternative and attractive technological process based on the following features: fast processing in order to minimize internal oxidation time; high-speed heating and cooling; local microencapsulation by preferential heating and melting at the interface with carbon-fiber; very dense products can be made with improved mechanical properties by simultaneously forging during unloading.

Figure 1 shows the tool configuration and the mould cavity used in consolidation process. The characteristics of powders and mixtures of the used powders are shown in tables 1 and table 2.

The mould cavity was loaded with approximately 80g of mechanically mixed with copper and carbon-fiber. Thereafter, the powder was pressurized to about 460 MPa, with the purpose of consolidation and bonding of powder particles. After discharge, the pressure was maintained for another 5 minutes, in order to achieve a heat transfer through the pistons, thus obtaining in the end a consolidated product.

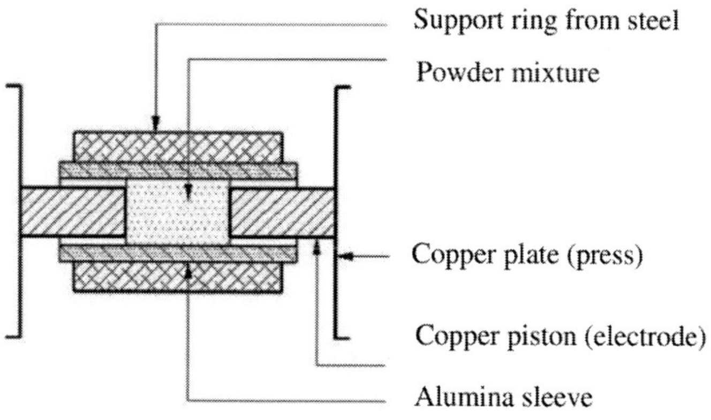

Figure 1. Tool and mould configuration cavity used in strengthening the homopolar generator.

Table 1. Characterization of the copper powder.

No.	Specific feature	UM	Value
1	Dimensions (under 44μm)	mesh	325
	Chemical composition		
2	Cu		99.52
	Fe		0.08
	Pb		0.05
3	Insoluble (SiO_2, Al_2O_3)		0.05

Table 2. Constituents of the consolidated powder mixes.

No.	**Mixture**	**A**	**B**	**C**	**D**	**E**
1	Cu (%)	95.2	95.2	97.6	97.6	97.6
2	Carbon fibre (%)	4.8	4.8	2.4	2.4	2.4

The measurements regarding the relative density are presented in table 3.The relative density is defined as the average of the densities measured relative to the theoretical values for each mixture. It is noted that sample D has the highest relative density from all the analyzed and consolidated samples. specific feature

Table 3. Relative density of the consolidated composite samples.

Sample	A	B	C	D	E
Relative density (g.cm^3)	0.89	0.90	0.89	0.97	0.89

Table 4 presents the superficial hardness of these compressed composites. In the table are included a sample with copper-graphite obtained by powder metallurgy using conventional sintering (11% graphite, 2% Pb, 5% Sn) called control sample, as well a tenacious sample with high percentage of copper, with the purposes of comparison with the analyzed samples. Samples strengthened by high energy high speed method, shows hardness values with 14 ÷ 46% higher than those of conventional control sample.

Table 4. Superficial hardness of the composite samples.

No.	**Sample**	**Specific energy (kJ/kg)**	**Superficial hardness**
1	Martor	-	45.8
2	A	2059	54.8
3	B	2125	52.3
4	C	1650	66.3
5	D	2037	66.8
6	E	2285	66.3
7	F(Cu)	-	75.1
8	B	2125	52.3
9	B	2125	52.3
10	B	2125	52.3
11	Powder Cu	2000	70.5

The electrical conductivity during the process is shown for each sample individually in table 5.

In order to determine the friction coefficient and the wear rate for various mixtures between copper-carbon fiber reinforced through high energy high speed method, was used a testing machine with pin on disc. The discs used were made from 4340 steel with an average hardness of 40 HRC.

Table 5. Changes in conductivity during the consolidation of composite samples.

Sample	Specific energy (kJ/kg)	Electrical conductivity
A	2059	6890
B	2125	6160
C	1650	8500
D	2037	6570
E	2285	5660
Powder Cu	2000	21400

The tests were carried out in air without external lubrication at the room temperature. The sliding speed was of 0.6 m/s. Friction and strain results are presented in table 6. It is noted that all composite samples presents outstanding tribological characteristics. The spectroscopic analysis indicates existence of a interfacial layer (figure 2), which contribute to a substantial decrease of the friction coefficient for all analysed samples from composite, with respect to the samples made from copper.

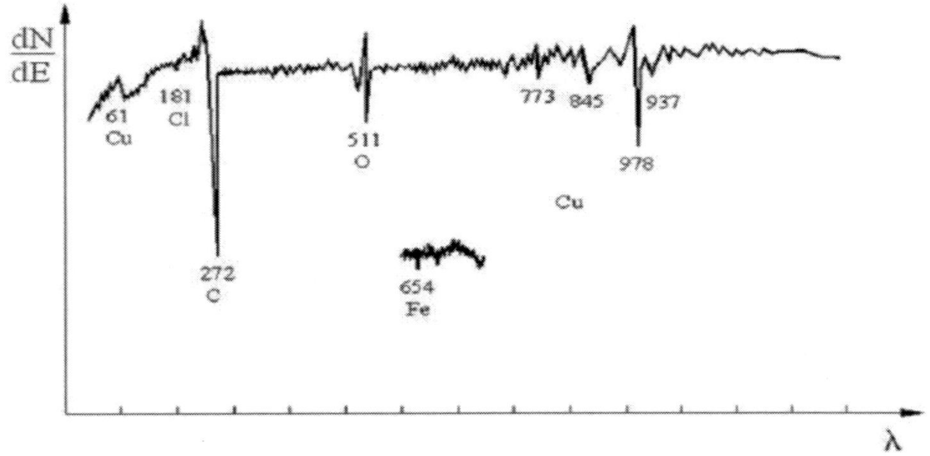

Figure 2. Electron spectrum of a typical film of graphite formed at the sliding interface.

Table 6. Variation of the friction coefficient and composite sample wear.

No.	Load (N)	Diameter of sample (mm)	Load/area (N/mm^2)	Friction coefficient	Wear speed ($mm/Nm \times 10^{-6}$)
1	4.46	3.16	0.57	0.120	0.01
2	4.46	3.30	0.52	0.144	0.07
3	4.46	3.22	0.55	0.097	0.10
4	4.46	3.18	0.56	0.201	4.13
5	4.46	3.23	0.54	0.172	1.51
6	4.46	3.15	0.57	0.172	2.87
7	4.46	3.05	0.61	0.280	58.50
8	4.46	2.66	0.80	0.150	3.21
9	8.92	2.78	1.46	0.207	3.81
10	13.40	2.87	2.07	0.133	8.16

The above presented results demonstrate the fact that the new method of powders consolidation, high energy and high speed is suitable for the production of parts with a good tribologic behaviour, from mixture of copper and carbon fibers.

3. Conclusions

In comparison with pure copper, the wear resistance of the composites obtained by copper and carbon fibers is greatly improved. When the percentage of carbon fibres increases, the friction coefficient simultaneously with wear mass loss decreases but the hardness of the microstructure increases. Increasing of the load and rotating speed, leads to an increase of friction coefficient and wear mass loss, correspondingly.

For pure copper and copper with short carbon fibres composites, the wearing process is adhesive and delaminating wear, respectively. The short carbon fibres improve tribological behaviour of the sintered copper composites by deformation obstruction, clogging the balminess of copper and forming a graphite film on the worn surface. The friction coefficient of copper-carbon composites is considerably less than that of pure copper, the reduction of the friction coefficient observed in an congestion of the carbon fibres at the grain boundaries with resulting in delamination of thinner wear flakes.

References

[1] Cheung Hy Ho Mp, Lau Kt, Cardona F and Hui D 2009 Natural fibre-reinforced composites for bioengineering and environmental engineering applications *Compos Part B Eng* **40** pp 655–63

[2] Wambua P, Ivens J and Verpoest I 2003 Natural fibres: can they replace glass in fibre reinforced plastics *Compos Sci Technol* **63** pp 1259–64

[3] Thomas S and Pothan L 2009 Carbon Fibre Reinforced Polymer Composites. From Macro to Nanoscale: Éd. des Archives Contemporaines

[4] Wötzel K, Wirth and Flake M 1999 Life cycle studies on carbon fibre reinforced components and ABS for automotive parts *Die Angewandte Makromolekulare Chemie* **272** pp 121–7

[5] Kestursatya M, Kim J K and Rohatgi P K 2003 Wear performance of copper–graphite composite and a leaded copper alloy *Mater. Sci. Eng.: A* **339** pp 150–158

[6] Emmer S, Havalda A and Bielek J 1993 Copper–graphite composite-material for application to sliding electrical contacts *J. Phys.* IV **3** pp 1799

[7] Moustafa S F, El-Badry S A, Sanad A M and Kieback B 2002 Friction and wear of copper–graphite composites made with Cu-coated and uncoated graphite powders, *Wear* **253** pp 699–710

[8] Kovacik J , Emmer S, Bielekc J and Lubomır K 2008, Effect of composition on friction coefficient of Cu–graphite composites, *Wear* **265** pp 417–421

[9] Rajkumar K and Aravindan S 2013 Tribological behavior of microwave processed copper–nanographite composites *Tribol. Int.* **57** pp 282–296

[10] Sarmadi H, Kokabi A H and Seyed Reihani S M 2013 Friction and wear performance of copper–graphite surface composites fabricated by friction stir processing (FSP), *Wear* **304** pp 1–12

[11] Domingo C, Arcís R W, Osorio E, Osorio R, Fanovich M A and Rodríguez-Clemente R et al. 2003 Hydrolytic stability of experimental hydroxyapatite-filled dental composite materials *Dent Mater* **19** pp. 478–486

[12] Khan A S, Aziz M S, Paul D, Wong F and Rehman I U 2008 Synthesis and in-vitro analysis of degradative resistance of a novel bioactive composite *J Bionanosci*, **2** pp. 1–12

[13] Gross K A, Berndt C C and Herman H 1998 Amorphous phase formation in plasma-sprayed hydroxyapatite coating *J Biomed Mater Res*, **39** pp. 407–414

IOP Conf. Series: Materials Science and Engineering **95** (2015) 012013 doi:10.1088/1757-899X/95/1/012013

Study on corrosion resistance of high - entropy alloy in medium acid liquid and chemical properties

I Florea[1], **G Buluc**[1], **R M Florea**[1], **V Soare**[2] **and I Carcea**[1]
[1]"Gheorghe Asachi" Technical University, Department of Materials Science and Engineering, Blvd. Mangeron, No. 59A, 700050, Iasi, Romania
[2]National Research & Development Institute for nonferrous and Rare Metals – IMNR 102 Biruinţei Blvd, 077145 Pantelimon, Ilfov, Romania

E-mail: iuliamuscalu@yahoo.com

Abstract. High-entropy alloy is a new alloy which is different from traditional alloys. The high entropy alloys were started in Tsing Hua University of Taiwan since 1995 by Yeh et al. Consisting of a variety of elements, each element occupying a similar compared with other alloy elements to form a high entropy. We could define high entropy alloys as having approximately equal concentrations, made up of a group of 5 to 11 major elements. In general, the content of each element is not more than 35% by weight of the alloy. During the investigation it turned out that this alloy has a high hardness and is also corrosion proof and also strength and good thermal stability. In the experimental area, scientists used different tools, including traditional casting, mechanical alloying, sputtering, splat-quenching to obtain the high entropy alloys with different alloying elements and then to investigate the corresponding microstructures and mechanical, chemical, thermal, and electronic performances. The present study is aimed to investigate the corrosion resistance in a different medium acid and try to put in evidence the mechanical properties. Forasmuch of the wide composition range and the enormous number of alloy systems in high entropy alloys, the mechanical properties of high entropy alloys can vary significantly. In terms of hardness, the most critical factors are: hardness/strength of each composing phase in the alloy, distribution of the composing phases. The corrosion resistance of an high entropy alloy was made in acid liquid such as 10%HNO_3-3%HF, 10%H_2SO_4, 5%HCl and then was investigated, respectively with weight loss experiment. Weight loss test was carried out by put the samples into the acid solution for corrosion. The solution was maintained at a constant room temperature. The liquid formulations used for tests were 3% hydrofluoric acid with 10% nitric acid, 10% sulphuric acid, 5% hydrochloric acid. Weight loss of the samples was measured by electronic scale.

1. Introduction

High entropy alloys have received a special attention in past few years.The first researches conducted on such an alloy were made in 1981 by B.Cantor. In 1995 J.W Yeh began to deepen research on high entropy alloys. Were produced around 40 alloys having the alloying elements Ti, V, Cr, Fe, Co, Ni etc. However the term of high entropy alloy was accepted by science in 2003. High entropy alloys have been defined as consisting of five or more elements, each one having a concentration between 5 and 35 at.% [1].

Content from this work may be used under the terms of the Creative Commons Attribution 3.0 licence. Any further distribution of this work must maintain attribution to the author(s) and the title of the work, journal citation and DOI.
Published under licence by IOP Publishing Ltd

Acid solutions are generally used for the removal of undesirable scale and rust in several industrial processes. Hydrochloric and sulphuric acids are widely used in the pickling processes of metals. Use of inhibitors is one of the most practical methods for protection in the face of corrosion especially in acid solutions to prevent metal dissolution and acid intake. The use of organic compounds containing oxygen, sulphur and nitrogen to reduce corrosion attack on steel has been studied in some details. Sulphur and nitrogen containing compounds are more effective as corrosion inhibitor in sulphuric acid and hydrochloric acid [2].

High entropy alloys subject to corrosion in such environments H_2SO_4 and HCl acid showed a high resistance to corrosion. For example CuCrFeNiMn high entropy alloy immersed in H_2SO_4 at ambient temperature (~25°C) showed that the alloys display a good general corrosion resistance that is mainly influenced by the Cu content and elemental segregation degree. The corrosion resistance degrades when increasing Cu content and elemental segregation degree. The corrosion resistance of NiCoCrFeMnCuC in 10 % HNO_3 - 3 % HF, 10 % H_2SO_4, 5 % HCl and 10 % HF has simple crystal structures with face-centered cubic crystal structure FCC and Quartet and has excellent corrosion resistance in some medium acid liquids [3, 4, 5].

2. Experimental procedures

2.1. Test materials

The type of high entropy alloys manufactured was a five-component alloy of AlCrNiCuMn. Cylindrical alloy ingot (20 cm in diameter and 4 mm thick) were prepared from a compound of pure elements. High purity elemental Al, Cr, Cu, Mn, and Ni were used as gross materials and prepared by induction melting in a medium frequency induction furnace for 8000 Hz. The chemical composition of the alloy given in table 1, is determined using EDAX.

Table 1. Chemical composition (wt.%) of the test material AlCrNiCuMn.

Elements	Al	Cr	Ni	Cu	Mn
High entropy alloy	5,08	6,72	19,09	47.80	21,29

2.2. Microstructural characterization

The metallographic samples of the high entropy alloy AlCrNiCuMn were first mechanically wet ground using a SiC grit paper, then polished with Al2O3 powder of 1 lm diameter, succeed by a chemical atac with NITAL. The mechanically process purpose to put into evidence structural constituents. The development follows the structure via surface attack with chemical reagents, solutions in general. Microstructural characterization was performed by electron microscopy on LMHII VegaTescan equipment using a secondary electron detector at a voltage of 30 kV electron gun.

3. Results and discussions

3.1. Microstructural characterization for high entropy alloy AlCrNiCuMn before immersion

In figure 1 we can observe that the high entropy alloy AlCrNiCuMn has a typical dendritic structure and the dendritic areas are rich in Cr. Cr-rich areas were observed in EDAX analyzes. Figure 2 presents the distribution of elements in the high entropy alloy AlCrNiCuMn. After the analysis of the distribution it is seen that in our high entropy alloy can be formed some different phase like: Al-Cu, Cu-Ni, Al-Ni.

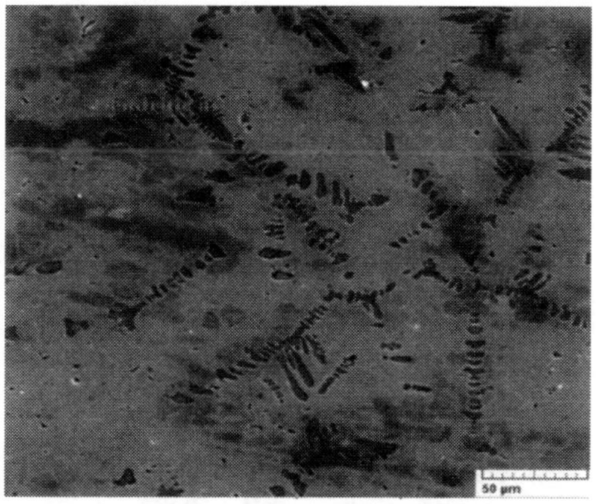

Figure 1. SEM image of high entropy alloy AlCrNiCuMn before immersion.

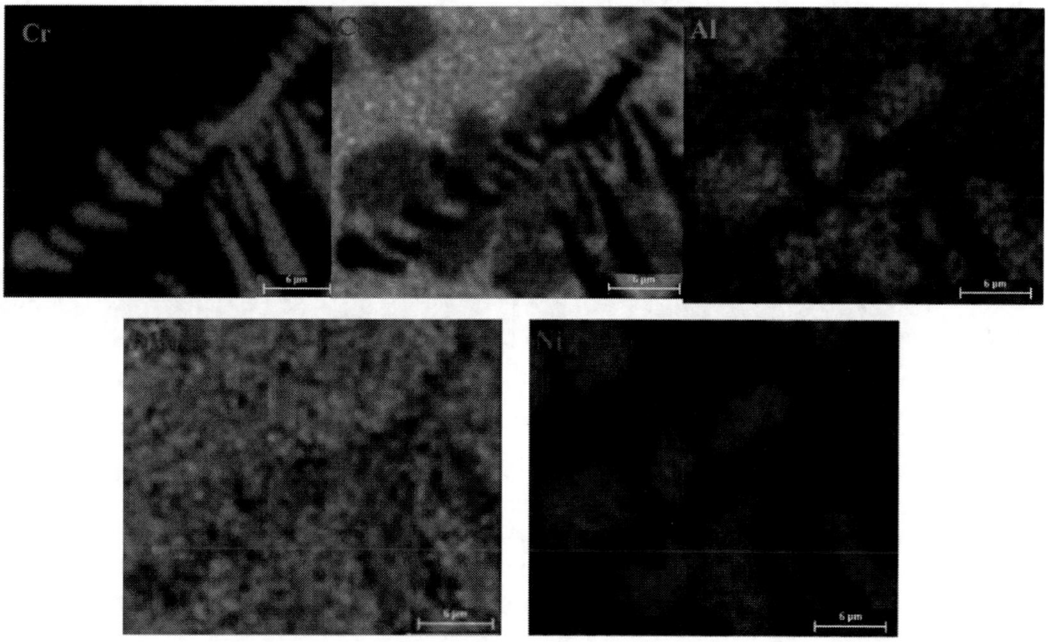

Figure 2. Distribution of the chemical elements in the high entropy alloy AlCrNiCuMn before immersion.

3.2. Microstructural characterization for high entropy alloy AlCrNiCuMn immersed in different acid medium

Another The high entropy alloy AlCrNiCuMn was immersed for 120h in three acid medium namely 3% HF+10% HNO_3, 10% H_2SO_4 and 5%HCl. The samples that were immersed were cute by wire cutting into 8x8x8mm in size. The solution was maintained at a constant room temperature (~25°C). In the figures below are presented the high entropy alloy AlCrNiCuMn immersed in different acid medium. Figure 3 is presented SEM micrography for AlCrNiCuMn immersed in 3% HF+10% HNO_3, Al and Mn have a negative effect but Cr and Ni increase corrosion resistance.

Figure 3. Sem micrography for AlCrNiCuMn immersed in 3% HF+10 %HNO_3.

Figure 4. The Sem micrography for AlCrNiCuMn immersed in 10% H_2SO_4.

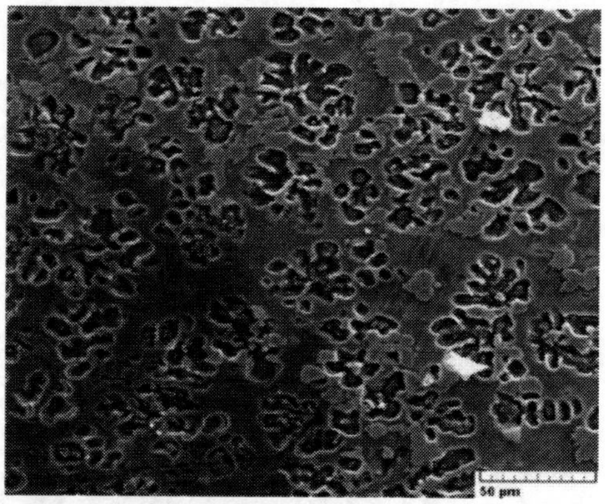

Figure 5. Sem micrography for AlCrNiCuMn immersed in 5%HCl.

After the chemical analysis it could be observed that the alloying elements of high entropy alloy having been subjected to immersion in three different acid medium had a quantitative change. For the sample immersed in 3% HF+10%HNO_3, Al, Cr and Ni alloy undergoes an increase but Mn and Cu suffer an decrease, in the other two cases the samples that was immersed in 10%H_2SO_4 and 5%HCl only Al and Cr increases while the other three alloying elements Mn, Cu and Ni decreases. All these are given in table 2.

Table 2. Chemical composition (wt.%) of the test material AlCrNiCuMn immersed in three different acid medium.

Al	Cr	Ni	Cu	Mn
		3% HF+10 %HNO_3		
6,22	10,05	20,21	42,04	21,45
		10% H_2SO_4		
7,38	8,72	17.37	45,41	21,09
		5%HCl		
5,47	9,59	16,18	48,36	20,37

3.3. Chemical analysis in points

Figure 6 shows the chemical analysis in points for the high entropy alloy AlCrNiCuMn immersed in 3% HF+10 %HNO_3 for 120h. In point A the chemical analysis shows a high quantity of Cu, point B on the figure represents an area rich in Cu and Ni but in point C the predominant alloying element is Cr.

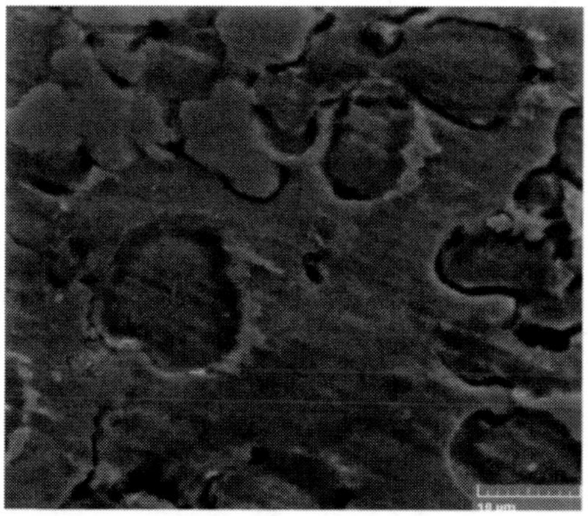

Figure 6. The chemical analysis in points for high entropy alloyAlCrNiCuMn immersed in 3% HF+10 %HNO_3.

Figure 7 shows the chemical analysis in points for the high entropy alloy AlCrNiCuMn immersed in 10% H2SO4 for 120h. In point A the chemical analysis shows a high quantity of Cu,Ni and Mn, point B on the figure represents an area rich in Cu and Ni but in point C the predominant alloying element is Cu and Mn. For the high entropy alloy AlCrNiCuMn immersed in 5%HCl the chemical analysis is identical to the first chemical analysis for the sample immersed in 3% HF+10 %HNO_3 (figure 8).

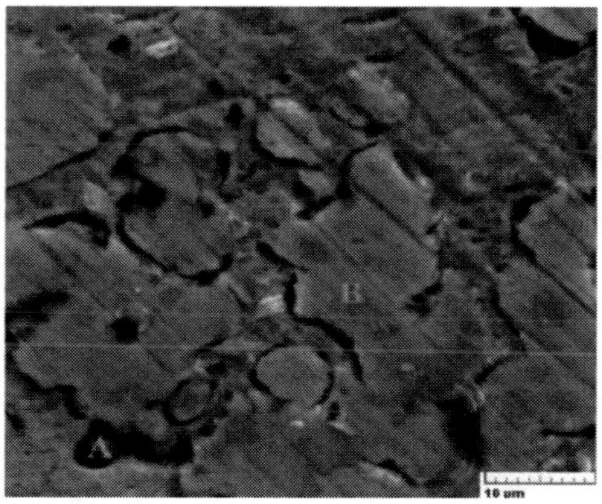

Figure 7. The chemical analysis in points for high entropy alloyAlCrNiCuMn immersed in 10% H2SO4.

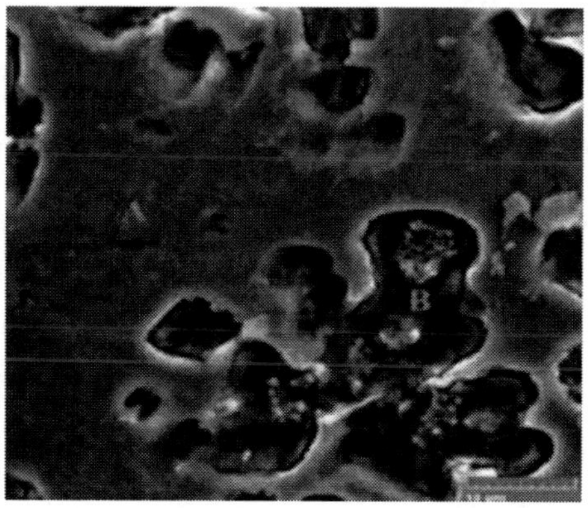

Figure 8. The chemical analysis in points for high entropy alloy AlCrNiCuMn immersed in 5%HCl.

Figures 9, 10, 11 presented some graphs for the weight loss of test material AlCrNiCuMn immersed for 120h in 3% hydrofluoric acid with 10% nitric acid, 10% sulfuric acid and 5% hydrochloric acid. In graphs 9, 10 it can be seen that in the first 48 hours the high entropy alloy AlCrNiCuMn has a bigger weight loss, but for the high entropy alloy AlCrNiCuMn immersed in 5%HCl the weight loss becomes visible in the first 24 hours.

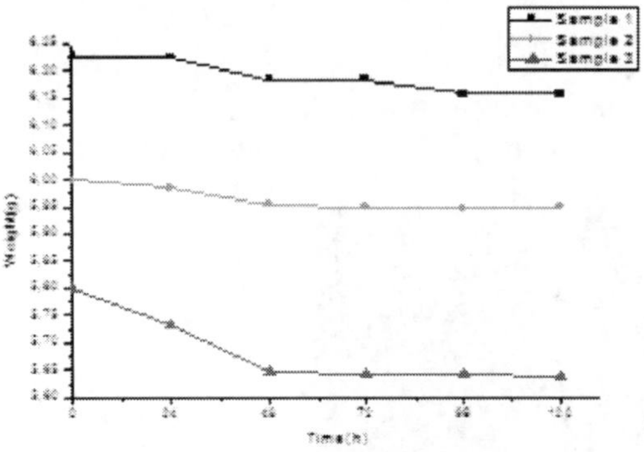

Figure 9. Weight loss for high entropy alloy AlCrNiCuMn immersed in 10%NHO_3-3%HF for 120h.

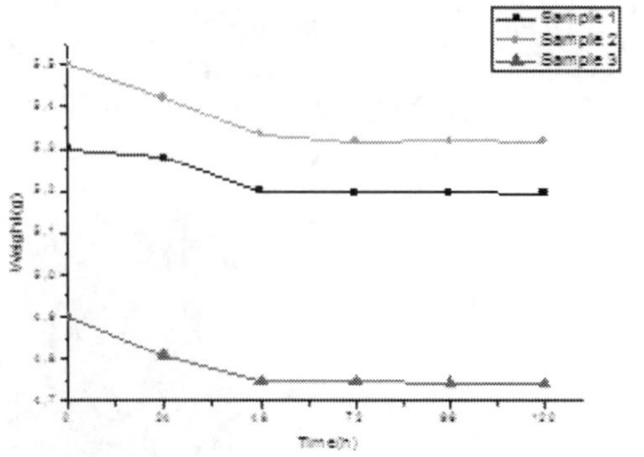

Figure 10. Weight loss for high entropy alloy AlCrNiCuMn immersed in10%H_2SO_4 for 120h.

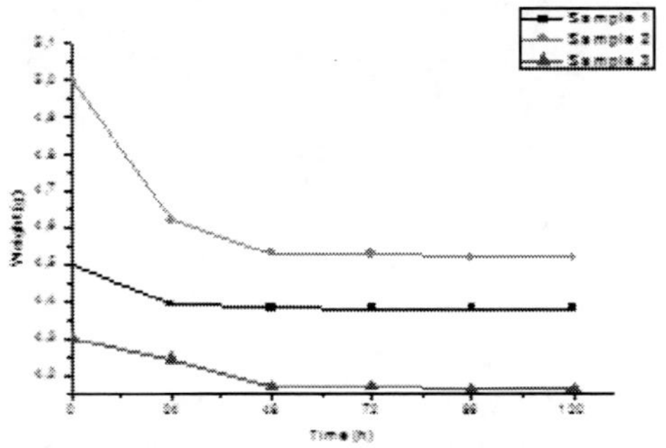

Figure 11. Weight loss for high entropy alloy AlCrNiCuMn immersed in 5% HCl for 120h.

4. Conclusions

Corrosion tests were carried out in three different acid medium for 120h.

Alloying elements from high entropy alloy had suffered various modifications.

AlCrNiCuMn gets good performance of corrosion resistance. Weight loss is only about 4% in 10%HNO_3-3%HF liquid, about 0.4% in 10%H_2SO_4 liquid, about 0.7% in 5%HCl liquid and only about after 120 hours.

Acknowledgments

This work was supported by a grant of the Romanian National Authority for Scientific Research, CNCS-UEFISCDI, project number PN-II-PT-PCCA-2013-4-1048.

This work was supported by the strategic grant POSDRU/159/1.5/S/133652, co-financed by the European Social Fund within the Sectorial Operational Program Human Resources Development 2007- 2013.

References

[1] Ranganathan S 2003 *Curr. Sci.* **85** pp1404-1406
[2] Tang Z, Huang L, He W and Liaw P K 2014 *Entropy* **16** pp 895-911
[3] Mary J, Nagalakshmi R, Rajendran S and Epshipha R 2014 *Eur. Chem. Bull* **3** pp 1031-1035
[4] Chou Y L, Wang Y C, Yeh J W and Shih H C 2010 *Corr. Sci.* **52** pp 3481–3491
[5] Liu Z, Zeng J and Zhan H 2012 *Appl. Mech. Mater.* **117-119** pp 1816-1819

Researches on the behaviour of cellular antiballistic composites based on AlMg-SiC alloys

O Bălţătescu[1], R M Florea[1], I Rusu[1] and I Carcea[1]
[1]Technical University "Gheorghe Asachi" of Iasi-Romania, Department of Materials Science and Engineering, Blvd. Mangeron, No. 59A, 700050, Iasi, Romania

E-mail: oana84rou@yahoo.com

Abstract. The researches presented in this paper refers basically to the impact of a small/medium caliber bullet shot on a light armor built on the base of a AlMg-SiC metallic composite cellular/foam. Thus, we study the antiballistic behavior and protection properties of the armor, based on the effects that occur at the impact zone of the bullet with the composite surface. We performed an antiballistic behavior modeling by means of a finite element analysis, based on a "multi grid" Fast Finite Element (FFE) system. We used for this purpose the DYNA 2D software package. The obtained samples show after the impact the occurrence of concentration / deformation pores effect and intercellular cracks development to the interior of the composite. Those effects, depending on speed, mass and length of the projectile ballistic trajectory, reduce zonal tensions due to the effect of cell walls deformation. It was obtained a good correlation between modeling results and the electron microscope analyse of the impact area. It is worth mentioning that almost all values for impact energy absorbed by the composite armor are in the protection active zone provided by it.

1. Introduction

Structural optimization, properties, impact resistance and ballistic behavior is a real need for application development especially in the military field. In practice occurred requirement to achieve optimization solutions for building and modeling of ballistic structures, or armor needed in areas such as military, naval and aeronautical programs made by dedicated packet type DYNA 2D. These are variants of calculation and solutions while providing a relatively safe with good quality results and descriptive, but not certainties in practical applications. Pair with a type DESIGNER ONE program that can incorporate solid modeling and structural analysis can lead to specific systems, which can quickly resolve issues related to the impact of bodies at high speeds, the energy absorbed and ballistic protection that some materials they can provide direct applications [1].

2. Presentation and general characterization of the modelling process

Till now it was not developed a unified theory that can fully describe the mechanical behavior of composites under ballistic and establish mechanisms dependence of mechanical parameters characterizing the impact:

- dependence of the mechanical properties of some geometrical and physical parameters of request: high speed impact, the time variation of pregnancy by fragmentation, temperature, etc.

Content from this work may be used under the terms of the Creative Commons Attribution 3.0 licence. Any further distribution of this work must maintain attribution to the author(s) and the title of the work, journal citation and DOI.
Published under licence by IOP Publishing Ltd

- dependency on the conditions for obtaining the mechanical properties of the composite; depending on the technology used and the reinforcing elements of the matrix and the heat treatment, etc.

- dependence of the mechanical properties of nature and characteristics of the elements that have made reinforcement matrix cell density, total reinforced material, the nature of the interface and input materials introduced to modify the shape and size of the cells composite.

In the modeling activity is mandatory call micromechanics models and calculations to simulate the mechanical properties of materials with properties gradual. Sequential steps to be taken is in fact a classical procedure for FEA (Finite Element Analysis) evaluation:

- it is the mesh network with a suitable degree of surface geometry that occurs initiation impact (this area generally has the primary form of a grid network name consecrated by Mesh);
- development field area is represented by multiplying the impact of network elements in the mesh polygons representing elements associated clamping zone resulting from impact;
- developing a network with nodal elements around the impact area;
- representation of the composite elements subjected to impact cell were generated by a specific algorithm;
- detailed presentation of modified Mesh network around the impact area is done by increasing the number of dividers, with representation by leaps and play new network tense with a 2D representation may occur as a representation as small rectangles or polygons.

For statistical and comparative analysis in the case of a 9 mm caliber projectile it were generated features obtained from a series of polygon tests. Data are presented in summary in the following tables for which values were selected based on general data already published for 9mm weapons [2, 3].

Table 1. Structural elements of antiballistic armor slightly (caliber projectile used in the study was 9 mm).

Feature ballistic protection for a size 9 x 18mm	The thickness of armor (minimum value)
Carbon steel plate	2 mm
Concrete type B200	4,5 mm
Wall of bricks burned	50 cm
Soft wood	15 cm
Hardwood	12 cm
Layer of earth	15 cm
Snow	50 cm

Table 2. Ballistic characteristics for the analytical study of projectile 9 mm.

Feature caliber 9 mm	Projectile weigh [g]	Speed [m/s]	Energy* [J]	Maximum extension ballistic [mm]	Penetration ballistics gel [mm]
Automatic shooting type	15	300	675	21	410

* The impact energy will be considered to be equal to the energy absorbed by the shield to which is added the residual energy.

To determine the effects of the damage produced by projectile bullet-type small and medium arms, shall be carried out in general a ballistics test. Ballistic test standard is firing a projectile from a distance of 10 m in a target containing a special substance called ballistic gel or paste; range experiments being carried out with different types of weapon [4].

Table 3. Performances range ballistic projectile 9 mm depending on the type of firearm [4].

Ballistic performance		
Firearm-type projectile 9 mm/mass	Speed [m/s]	Energy [J]
Bonded Defense/12g	373	835
Speer Gold Dot JHP/12g	320	614
Speer Gold Dot JHP +P/13	330	702
Federal Hydra-Shok/15g	270	561
US Army Ball FMJ/15g	250	477

As a behaviour mode, the pores of a composite cell to be subjected to the effect of the concentration/deformation and cracking within the composite intercellular for reducing the tension zone by the effect of deformation of the cell walls , thereby reducing the total fracture energy. Neglecting the effects of deformation occurring at a force F it can be considered the total energy absorbed will depend on the number of cells and the structure/geometry composite alveolar cells; but will depend on the relative density and the volume/size of the cell.

Determinations are experimentally assessing fracture and deformation energy as it occurs. The values obtained are dependent on the composite matrix, the shape, cell size, and density of the composite. Ceramic components placed in metal cellular materials have major influences on the matrix plasticity and thermal and mechanical energies of nature that can be absorbed by the matrix. At the same time they influence the composite brittle deformation process, which can be considered as an advantage in terms of load transfer when speaking of the protection can be provided at the time of impact components.

To implement a mathematical model simulating ballistic impact phenomena, we need a network computing discrete set around the two bodies (ie projectile and armor). The mathematical model is transformed into a numerical model equations by approximating the laws governing the movement of bodies and determine the amounts of energy values bodies in motion. This makes the numerical model to work exclusively on a discretized grid computing (mesh), unlike the analytical model which is valid for any field, including the discretized. The transformation is necessary because its structure is made to work with computer programs (numerical models) [5, 6].

3. Experiment and comparison of data obtained with the results of modeling

In the experiment a 9 mm caliber bullet was directed on an armored area-based on composite cellular AlMg-SiC. In order to study the surface deformation armor, composite initial target surface was covered with an epoxy resin stabilization (retain small microparticles that appear at the impact). After impact is obtained as one image shown like in figure 2 and reversed in a FEA modeling in figure 3.

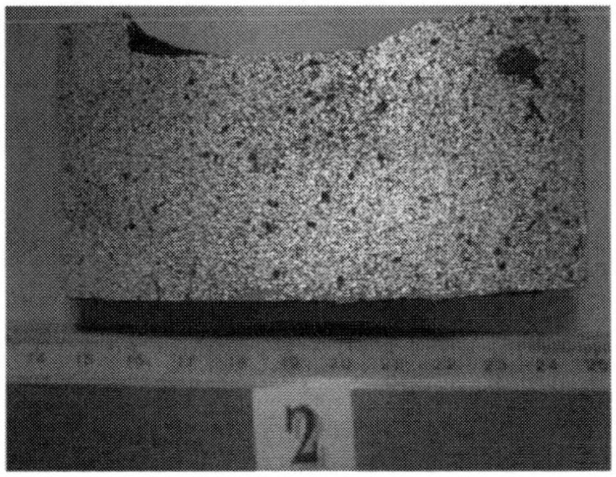

Figure 1. Initial situation of the experiment with the composite and projectile presentation.

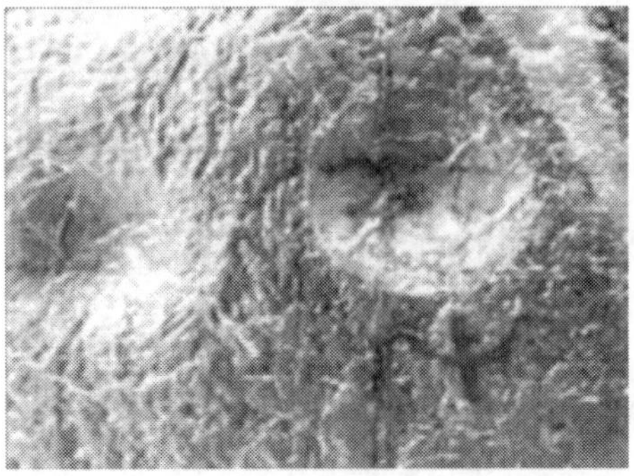

Figure 2. Armor surface deformation that resulted from the impact of projectiles.

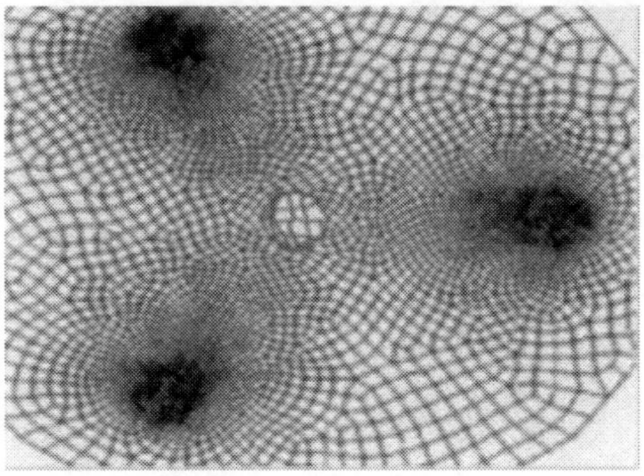

Figure 3. Generating FEA analysis grid with mesh grid representation of the composite surface area of impact projectiles.

For example we have chosen a two-dimensional rendering of the antiballistic experiment for the armor with low section of 20 mm (armor that can be used against small caliber weapons). The pictures show the key steps in ballistic behavior play three distinct ways in which you can assign sequential grid-simplified representation of the armor structure (figure 4).

The figures below show the updated image projectile in three phases with modification and composite structural deformation due to ballistic impact.

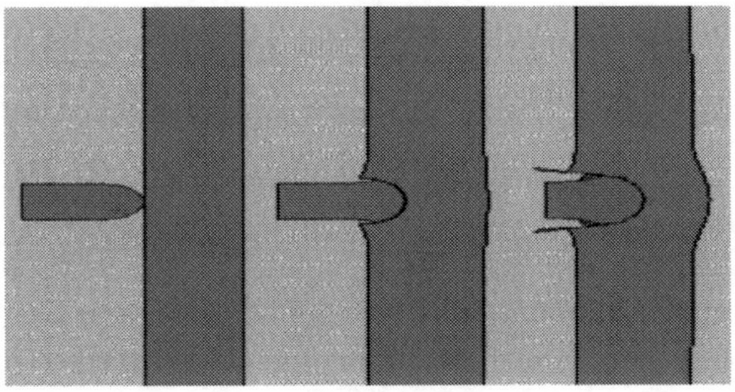

a. b. c.

Figure 4. Application modeling to present sequential volume changes that occur as a result of the projectile impact with the composite surface; a. initial geometry of the assembly projectile- ballistic protection surface, b - representation of quasi-ballistic impact initial presentation of deformation and perforation armor, c - the final state of play time blocking projectile impact and occurrence of major geometric deformation; deformations occur in projectile by its bulging with shortening and deformation of cells that makes up the composite armor structure.

For modeling and numerical simulation parameters were used with minimum values determined experimentally or calculated from the table 4.

Table 4. Maximum values for modeling and numerical simulation.

Characteristics of material	Unit of measurement	Values
Young's modulus - E	N/mm^2	42780
Poisons coefficient - υ	-	0,27
Tensile strength – σr	N/mm^2	125
Density - ρ	kg/m^3	2000
Projectile speed - vp	m/s	300
Projectile mass -mp	kg/s	0,015
Friction coefficient - μ	-	0,22
Average hardness - HB	-	90
Von Mises criterion – maximum value	Pa	4,4x108

On electron microscopy SEM analysis of the impact zone, we can see that a crater formed by deformation, just as a result of laminar flow-plastic composite after ballistic impact (figure 5). We also observed small loose particles composite wall and a major deformation type ductile matrix composite flow especially in the tilt angle under which hit ballistic projectile.

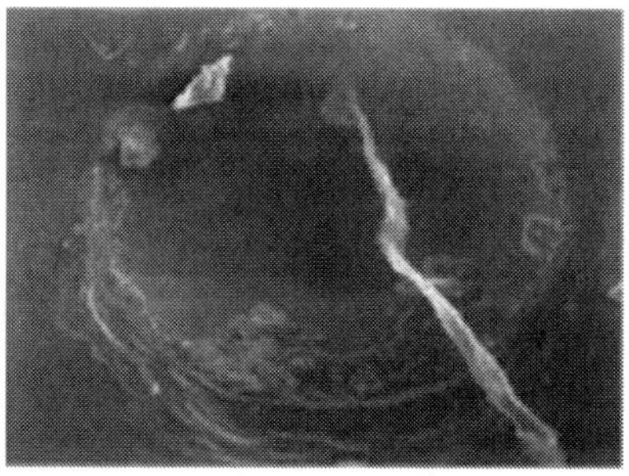

Figure 5. Image of the impact area resulting from the penetration of the tip projectile in composite volume. Small fragments results are deposited in the impact crater.

Figure 6. Laminar flow deformation and destruction of composite walls adjacent at the impact area. High composite compressions due to kinetic energy and armor matrix deformation.

After the projectile impact occurs practically most of the times a major deformity especially the bullet tip. The areas most affected are those little grooves on printed and formed projectile due to existing rifling print a ballistic trajectory correct (a procedural motion to maintain the most accurate trajectory of the bullet). Composite near the impact zone it is suffer important deformation in the volume structure; deformation being a function of speed, projectile mass and length of the ballistic trajectory [7].

Figure 8 shows when it is considered that over 95% of the kinetic energy of the projectile has already been transmitted to the cellular composite armor. Playing network at this time is difficult because the network can no longer represented faithfully take two bodies deformations due to the impact. FEA (Finite Element Analisys) details are very hard to find and even in the FFE (Fast Finite Element) system, but it can be also performed a presentation of the ballistic behavior in FFE network representation. Modeling system generates a detailed representation of volumetric deformation around the projectile in the final stage to unlock the armor. Modeling highlights some of the features of plastic deformation that we presented above electron microscopy analysis which identifies the matrix plasticity around the projectile.

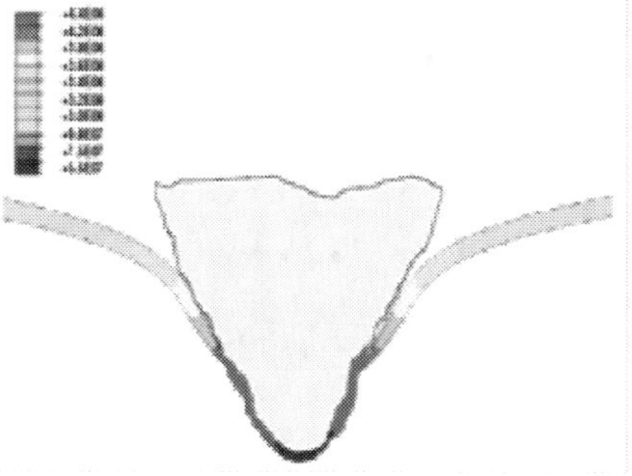

Figure 7. Finite element analysis that highlights of maximum impact stress areas for a projectile with conical tip which penetrates the composite. Maximum tension recorded on the basis of von Mises criterion involves maximum deformation of the composite.

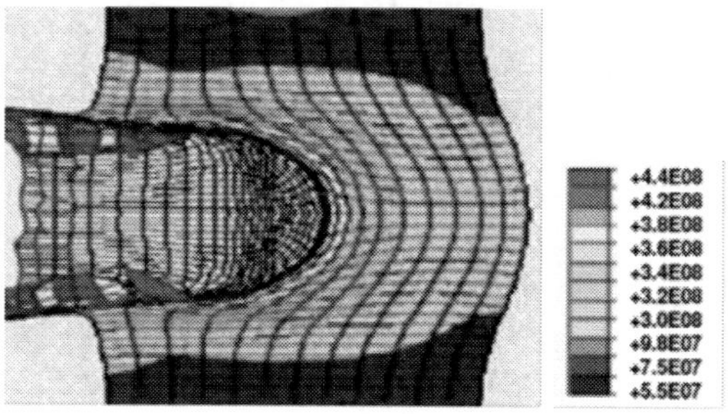

Figure 8. Finite element analysis that highlights the areas that suffered major destruction effects both for surface impact armor and for the projectile that penetrated the armo.r

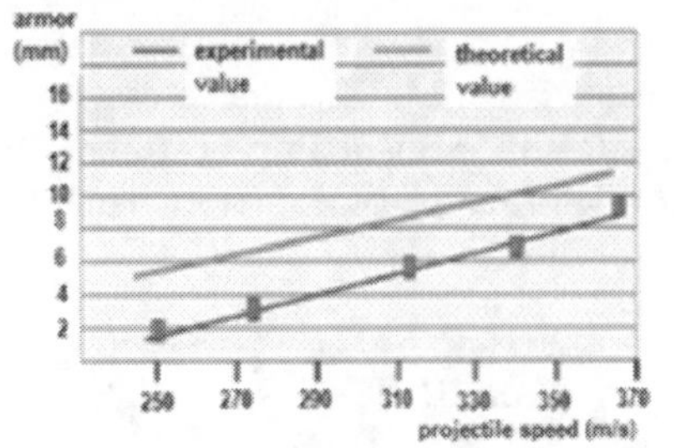

Figure 9. The correlation obtained based on the study for a projectile weighing 15 g and composite armor AlMg15 cell density of 2 kg/dm^3 with 15% SiC particles added.

There is a good modeling correlation made based on finite element analysis, with the results that analyzes the impact zone microscopy. It is noteworthy that almost all values for impact energy absorbed are in the active area of protection provided by the composite armor. From the chart below it can be easily seen that the 20 mm armor made from cellular composite presents sufficient efficiency for a 9 mm projectile (figure 9).

4. Conclusions

The study of this paper is reduced in principle to present only the effect of the impact that it has a small or medium caliber bullet on a lightly armored being constructed of a metal composite cellular/foam from AlMg-SiC. We could thus define based on FEA analysis and electron microscopy, which were the elements of antiballistic behavior, qualities of the cellular composite armor and some effects that occur with the bullet impact with the armor surface.

After the impact in the case of the projectile is practically most often a major deformation and in particular the tip of the bullet. The composite it's suffer important deformation at the volume structure near the impact area, deformation being function of speed, projectile mass and ballistic trajectory length.

There are practical for these type of composites a wide range of applications and combinations to achieve ballistic armor for military vehicles in particular.

References

[1] S Nimmala 2014 *A comparison of DYNA3D, NIKE3D and LS-DY*NA (Oregon State University) http://www.llnl.gov/str/Raboin.html
[2] http://simage1.sportsmansguide.com/adimgs/l/1/186220i_ts.jpg
[3] *MSC Nastran for Accurate, Efficient & Affordable Finite Element Analysys* http://www.mscsoftware.com/Contents/Produtcs/CAE-Tools/MSC-Nastran.aspx
[4] *Bullet Body Armour – Anti-Ballistic* http://www.anti-ballistic.com/shop/
[5] Zohdi T I 2002 Modeling and Simulation of Progressive Penetration of Multilayered Ballistic Fabric Shielding *Comput. Mech.* **29(1)** pp 61-67
[6] Lim C T, Shim V P W and Ng Y H 2003 Finite-element Modeling of the Ballistic Impact of Fabric Armor *Int. J. Impact Eng.* **28** pp 13-31
[7] Grujicic M, Pandurangan B, Zecevic U, Koudela K L, B A and Cheeseman B A 2007 Ballistic Performance of Alumina/ S-2 Glass-Reinforced Polymer-Matrix Composite Hybrid Lightweight Armor Against Projectiles *Multidiscipline Modeling in Materials and Structures* **3** pp 287-312
[8] Wong A K and Connors M L 1971 *A Literature Survey on Correlation of Laboratory Tests and the Ballistic Resistance of Rolled Homogeneous Steel and Aluminum Armors. Technical Report AMMRC SP 72-10* (Watertown MA: Army Materials and Mechanics Research Center)
[9] Mamalis A G et al. 1997 Crashworthy capability of composite material structures *Composite structures* **37** pp 109-134
[10] Jones N 2010 Energy-absorbing effectiveness factor *Int. J. Impact Eng.* **37** pp 754-765
[11] Maine E M A and Ashby M F 2002 Applying the investment methodology for materials (IMM) to aluminium foams *Materials & Design* **23** pp 307-319

IOP Conf. Series: Materials Science and Engineering **95** (2015) 012015 doi:10.1088/1757-899X/95/1/012015

On the role of CFRP reinforcement for wood beams stiffness

A C Ianasi[1]
[1]"Constantin Brancusi" University of Tg-Jiu, Faculty of Engineering, Romania

E-mail: ianasi_c@yahoo.com

Abstract. In recent years, carbon fiber composites have been increasingly used in different ways in reinforcing structural elements. Specifically, the use of composite materials as a reinforcement for wood beams under bending loads requires paying attention to several aspects of the problem such as the number of the composite layers applied on the wood beams. Study consolidation of composites revealed that they are made by bonding fibrous material impregnated with resin on the surface of various elements, to restore or increase the load carrying capacity (bending, cutting, compression or torque) without significant damage of their rigidity. Fibers used in building applications can be fiberglass, aramid or carbon. Items that can be strengthened are concrete, brick, wood, steel and stone, and in terms of structural beams, walls, columns and floors. This paper describes an experimental study which was designed to evaluate the effect of composite material on the stiffness of the wood beams. It proposes a summary of the fundamental principles of analysis of composite materials and the design and use. The type of reinforcement used on the beams is the carbon fiber reinforced polymer (CFRP) sheet and plates and also an epoxy resin for bonding all the elements. Structural epoxy resins remain the primary choice of adhesive to form the bond to fiber-reinforced plastics and are the generally accepted adhesives in bonded CFRP–wood connections. The advantages of using epoxy resin in comparison to common wood-laminating adhesives are their gap-filling qualities and the low clamping pressures that are required to form the bond between carbon fiber plates or sheets and the wood beams. Mechanical tests performed on the reinforced wood beams showed that CFRP materials may produce flexural displacement and lifting increases of the beams. Observations of the experimental load–displacement relationships showed that bending strength increased for wood beams reinforced with CFRP composite plates and sheets compared to those without CFRP reinforcement. The main conclusion of the tests is that the tensioning forces allow beam taking a maximum load for a while, something that is particularly useful when we consider a real construction, so in case of excess lift beam, we have time to take strengthening measures and when is about a catastrophic request (earthquake) the construction remain partially functional. The experiments have shown that the method of increasing resistance of wood constructions with composite materials is good for it. The solution is easy to implement and has low costs.

1. Introduction

Carbon fiber is used in many areas where a combination of high strength and low weight are required. The strongest are approximately five times stronger than steel and considerably lighter. Other useful properties it has are its ability to withstand high temperatures and its inertness. Carbon fiber is a textile consisting mainly of carbon. It is produced by spinning various carbon-based polymers into fibers, treating them to remove most of the other substances, and weaving the resulting material into a fabric. This is usually embedded in plastic — typically epoxy — to form carbon fiber reinforced plastic. The

Content from this work may be used under the terms of the Creative Commons Attribution 3.0 licence. Any further distribution of this work must maintain attribution to the author(s) and the title of the work, journal citation and DOI.
Published under licence by IOP Publishing Ltd

physical properties of composite materials (CFRP) are generally not isotropic in nature, but rather are typically anisotropic (different depending on the direction of the applied force or load). The majority of composite materials use two constituents: a binder or matrix and reinforcement. The reinforcement is stronger and stiffer, forming a sort of backbone, while the matrix keeps the reinforcement in a set place [1]. For instance, the stiffness of a composite panel will often depend upon the orientation of the applied forces and/or moments. Panel stiffness is also dependent on the design of the panel. In contrast, isotropic materials (for example, aluminum or steel), in standard wrought forms, typically have the same stiffness regardless of the directional orientation of the applied forces and/or moments.

2. Experimental results

The main purpose of this paper is to analyze bending phenomenon due to fracture at the wood-composite samples. By using composite materials (CFRP) in reinforcement is expected growth flexural strength and shearing and confinement elements tablets (increased wood resistance).

The type of reinforcement used on the beams is the carbon fiber reinforced polymer sheet SikaWrap 230C with E module of elasticity 230 000 N/mm^2 and traction resistance 4100 N/mm^2, carbon fiber reinforced polymer plates SikaCarbodur S 512 with E module of elasticity E=165000 N/mm^2 and an epoxy resin for bonding all the elements Sikadur 30. The wood part of all 11 beams was formed by beech dry wood which size is equal to 25 by 50 by 500 mm [2]. The beams were reinforced using one carbon fiber plate of thickness equal to 1,5 mm, width equal to 25 mm and the length is equal to 500 cm. The finished dimension of one of the beams is equal to 25 by 51,5 by 500 mm because is a beam stick together with one carbon fiber plate Sika Carbodur S 512 and epoxy resin Sikadur 30 [3,4]. Structural epoxy resins remain the primary choice of adhesive to form the bond to fiber-reinforced plastics and are the generally accepted adhesives in bonded CFRP–wood connections [5]. Advantages of using epoxy resin in comparison to common wood-laminating adhesives are their gap-filling qualities and the low clamping pressures that are required.

Composite material used to strengthen wood samples is the type of carbon fiber sheet carbon fiber plates and an epoxy matrix supplied by Building Velmix Ltd., from Sika S.A Romania. Structural epoxy resins remain the primary choice of adhesive to form the bond to fiber-reinforced plastics and are the generally accepted adhesives in bonded CFRP–wood connections. Advantages of using epoxy resin in comparison to common wood-laminating adhesives are their gap-filling qualities and the low clamping pressures that are required.

To achieve the objective proposed in this paper was presented a particular computer program, Presa.txt, for experimental determination of bending strength for the samples. This was done with Spider 8 data acquisition equipment connected to fixtures and fittings of the samples tested on the universal testing machine. Work equipment and instrumentations used in this case are: universal machine for mechanical tests [6], data acquisition system Spider 8, 12 bit resolution linear WA300 race inductive transducer, force transducer S9 50kN, signal conditioning NEXUS 2692 - A-014, 4391 type piezoelectric accelerometer, IBM ThinkPad R51 notebook. Parameters recorded after the bending tests are: F (kN) - compressive strength of hydraulic press, Ft (kN - transverse compressive strength, CRS (mm) - race piston, Acc (m/s^2) - acceleration of beam vibration (sensor break). The beam is leaning against the head and driven across to the breaking strength recorded maximum cross and displacement (no longer measured axial tensioning force). The results for the un-reinforced beams are reported solely for the purpose of quantitatively evaluating the effectiveness of the interventions through a comparison with the results for strengthened beams. The main purpose is to analyze bending phenomenon due to fracture at the wood-composite samples like we see in the figure number 1 from below. The samples tested were not subjected to lateral instability during loading. The total load on the beam was applied equally at one point equidistant from the reactions (the half length of the beam). We used the bending device of the universal machine for mechanical tests which has the distance between the rollers l = 460 mm.

In figure 2 it can be seen the way of fracture for the reinforced beams with one and two CFRP plates.

Figure 1. Tension failure of an unreinforced beam [2].

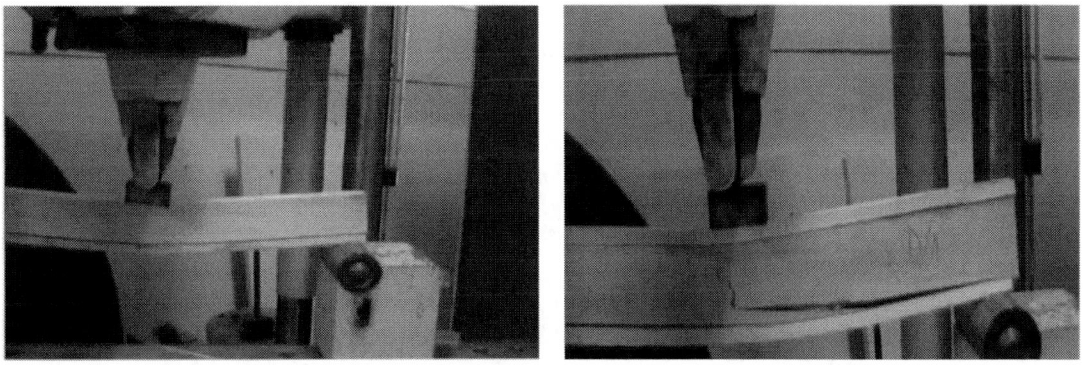

a) b)

Figure 2. Samples of reinforced beams [2]. a) with one carbon fiber plate and one down slide of wood b) with two carbon fiber plates and two up and down slides of wood.

The graphs of these reinforcement are presented in figure 3.

The graphs show that for the un-reinforced beam cracks appear at a low displacement such as 6-7 mm (a) instead of 10 mm (b) for the reinforced beam with one carbon fiber plate and one down slide of wood and 13-14 mm for the reinforced beam with two carbon fiber plates and two up and down slides of wood (c). That it means the CFRP improve the resistance of the beams and permit a growth in flexural properties.

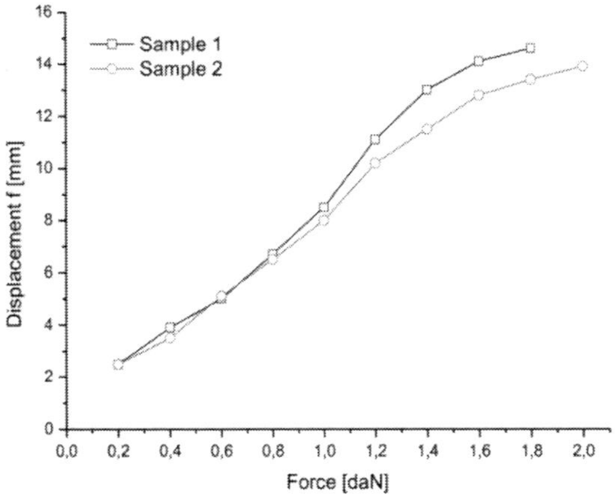

a)

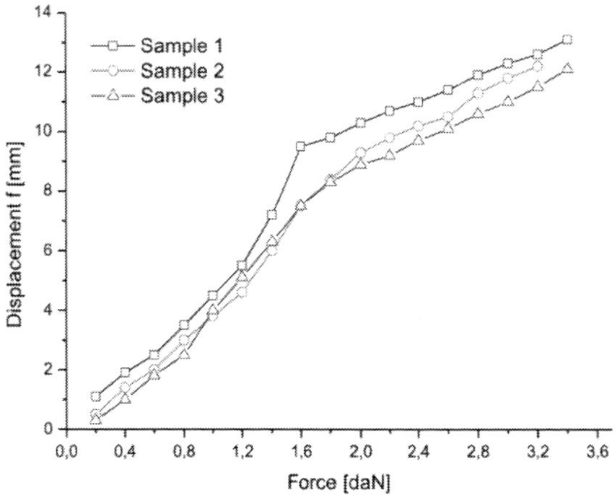

b)

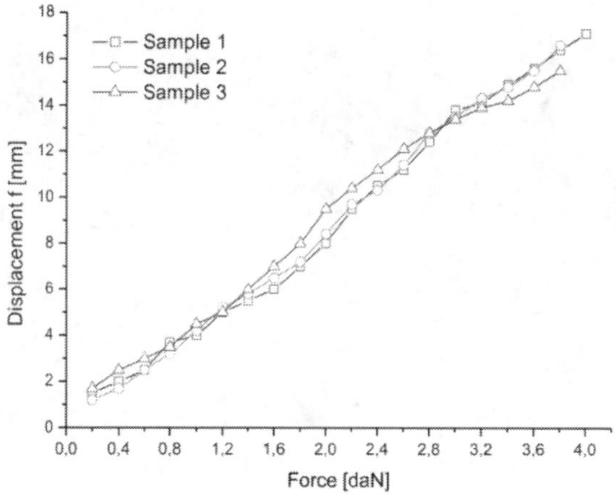

c)

Figure 3. The graphs of failure tensions [2]:
a) the un-reinforced beam; b) the reinforced beam with one carbon fiber plate and one down slide of wood; c) the reinforced beam with two carbon fiber plates and two up and down slides of wood

Studying the graphs we can see that the reinforcement is more efficient for the beam with one carbon fiber plate and one down slide of wood and for the beam with two carbon fiber plates and two up and down slides of wood. The force and displacement growth for these type of beams rising values more bigger than at the un-reiforced beam [7,8]. The un-reinforce beam is broken very quickly at a small load. The other wood beams are broken at a bigger load and have a bigger displacement. The tested wood beams reinforced with carbon fiber sheets are shown in figure 4.

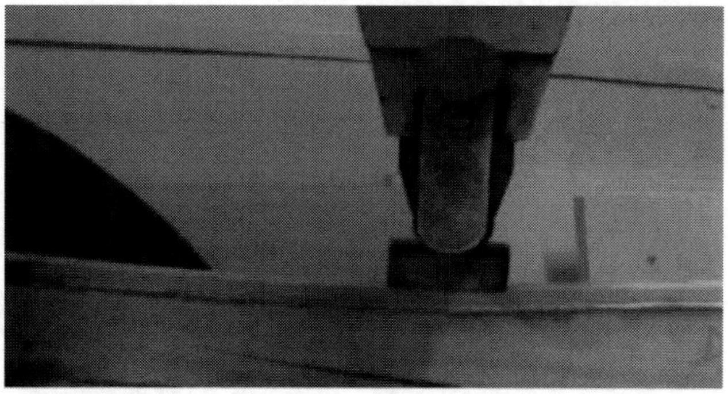

Figure 4. Tension failure of a reinforced beam [2].

The functional dependency graphs for the reinforced wood beam are represented in figure 5 from below.

In the case of the reinforcement with carbon fiber sheets we can observe on the graphs that the results indicate that the behavior of reinforced beams is totally different from that of un-reinforced one. The reinforcement also has changed the mode of failure for the tested beams. The experimental load–displacement relationships shown us that this kind of reinforcement is not so resistant like the carbon fiber reinforcement. Observations of the experimental load–displacement relationships show that flexural strength increased and middle vertical displacement decreased.

Experimental results allowed drawing the follow conclusions:

- the wood beams must be secured to the composite plates, in the mechanical device, to prove the effectiveness of the solution so the type of solidarity was mechanical-link wood beam ends;

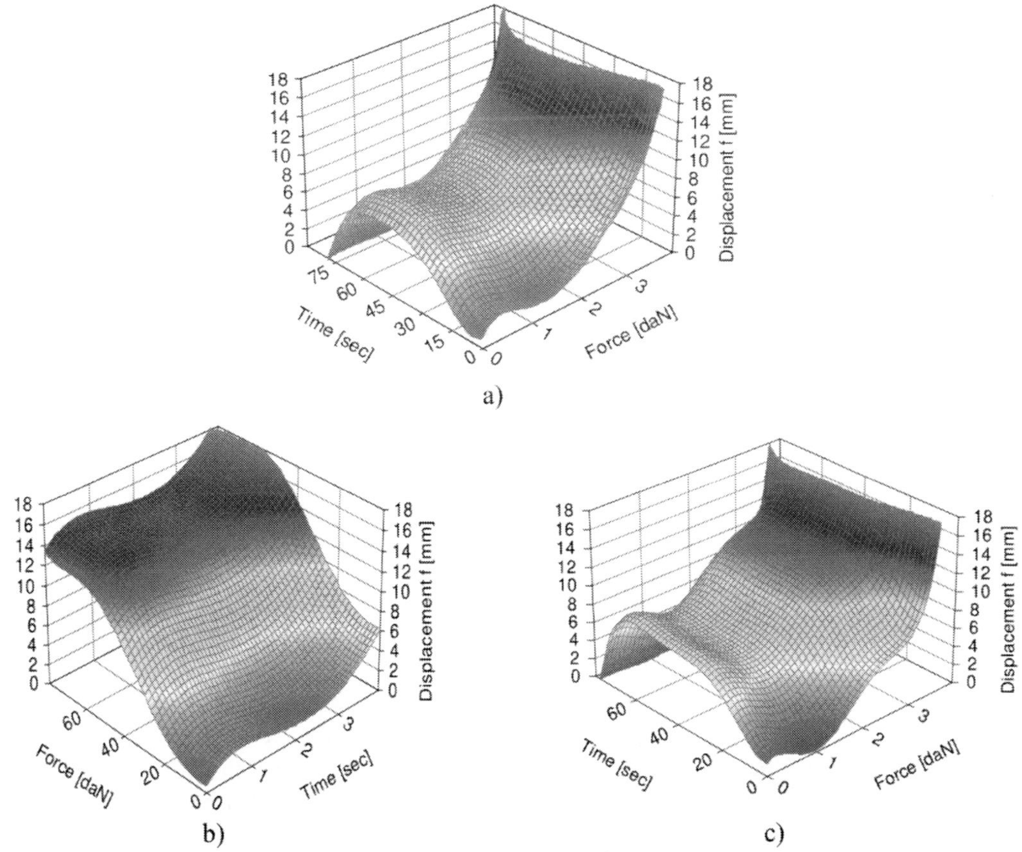

Figure 5. The functional dependency graphs [2]
a) for the un-reinforced wood beam, b) for the reinforced beam with two carbon fiber sheets, c) for the reinforced beam with three carbon fiber sheets.

- initial tension force decreases as the beam is loaded due to local subsidence of wood (in the tensioning device);
- if the mechanical system worked correctly, the lift of the beams increased up to 33 kN, meaning 220% higher than the un-reinforced beam;
- the first cracks in the wood beams appeared at least two times higher than the un-reinforced beam, due to quality wood (beech dry, carefully processed and without tension concentrators in its mass);
- elastic lift of the reinforced beams is significantly influenced by pre-tensioning, most samples having a maximum 8-13 mm flexural displacement that is an improvement over the displacement of the un-reinforced beam.

The maximum load force and the maximum displacement may be, in this case, experimental parameters to quantify the strength beam quality. It should be mentioned that results are easily interpreted in the context of a reference beam (un-reinforced).

By using composite materials in constructions is expected growth flexural strength and shearing, and confinement elements tablets (increased concrete and wood resistance). The reinforcement technology with composite materials offers many advantages over conventional methods, because the composite materials show:

- very high resistance to bending, greater than steel;
- increased resistance and ductility construction without changing the geometry or stiffness;

- consolidation and reinforcement for different materials such as: concrete, wood, steel;
- costs of intervention extremely competitive;
- small weight, high flexibility and availability in various lengths.

Experiments have shown sustainability of the method for wood construction reinforced with composite materials. The use of composites can be applied as a strengthening technique without necessitating the removal of the overhanging portion of the structure. The technique used proved to be easy and fast to execute, even when on in situ parts. In particular, it demonstrated to be very promising in many cases of reinforcement of old, historical structural wood parts. The CFRP pre-tensioning is easy to implement and with low costs. Effectiveness of using composite reinforcements is still modest, requiring further and deeper studies and trying new methods and design alternatives for the tested samples.

3. Conclusions

Following conclusions can highlight: wood is a material with a certain degree of heterogeneity, which makes its mechanical properties vary in a range too wide, so it is especially necessary to improve resistance with composite reinforcement.

As a rigid material with good strength and relatively low cost, we use a composite. Several un-reinforced and reinforced wood beams were tested in order to find their flexural capacity. CFRP materials were conditioned in an environment of 65±5% relative humidity and temperature of 20±2°C as this is the service environment in which CFRP reinforced beams are expected to be used. The results indicate that the behavior of reinforced beams is totally different from that of un-reinforced one. The reinforcement has changed the mode of failure from brittle to ductile and has increased the load-carrying capacity of the beams. Observations of the experimental load–displacement relationships show that flexural strength increased and middle vertical displacement decreased for wood beams reinforced with CFRP composite plates, compared to those without CFRP plates. If in the great strength area of the beam is added composite with greater strength than wood then the wood beam can withstand a bigger force because the maximum strength is bigger too. The presence of carbon plates and sheets causes an interesting increase in stiffness varying from 20,2% to 29,6%, when compared to that of the same wood beams before reinforcement.

References

[1] Information accessed on 19.10.2014,http://www.wisegeek.org/what-are-composite-materials.htm
[2] Ianăşi C 2013 Research on reducing the risk of damage for the resistance elements of wooden building 2nd WSEAS Int. Conf. on Risk Management, Assessment and Mitigation (RIMA '13) (Brasov, Romania) pp 161-164
[3] Information accessed on 21.12.2014, http://www.sika.ro
[4] Li Y F, Yao-Ming Xie Y M and Tsai M J 2009 Enhancement of the flexural performance of retrofitted wood beams using CFRP composite sheets *Construction and Building Materials* **23** pp 411–422
[5] Plevris N and Triantafillou T Creep behavior of FRP-reinforced wood members *J. Struct. Eng. – ASCE* **121(2)** pp 174–86
[6] Rosca V, Ilincioiu D, Marin M and Aştefanei I 2007 *Resistance Testing of Materials* (Craiova: Publishing House TOP Universitaria)
[7] Information accessed on 5.11.2014, http://www.carbonmods.co.uk
[8] Information accessed on 2.9.2014,http://mech.utah.edu/~rusmeeha/labNotes/composites.html

Researches regarding the hydroforming process of aluminum components

V Paunoiu[1], V Teodor[1] and F Susac[1]
[1]Dunarea de Jos University of Galati, Department of Manufacturing Engineering, Galati, 800008, Romania

E-mail: viorel.paunoiu@ugal.ro

Abstract. The lightweighting with aluminum offers the potential for using unconventional fabrication technologies such as sheet hydroforming. Triform process is an unconventional sheet forming technology where the blank sheet is placed under a flexible rubber pad and is deformed under the action of the hydraulic pressure applied from the top of the equipment. The part shape is given by the shape of the die located at the bottom of the equipment. In the paper are presented the experimental results, regarding the influence of the fluid pressure towards the parts properties. The parts quality is evaluated in terms of springback variation. The results could lead to the implementation of the designed procedure for triform die and equipment.

1. Introduction

In the car body constructions the substitution of steel by aluminum presents most interesting lightweighting possibilities [1]. The lightweighting with aluminum is not limited to material substitution, but the different aluminum product forms offer in addition the potential for using unconventional fabrication technologies [1] one of them being sheet hydroforming.

Sheet hydroforming includes a series of manufacturing technologies which have as the main characteristic a liquid pressure medium which acts toward the blank [2, 3]. There are two mainly manufacturing methods: one consists in the action of the liquid pressure toward the blank, the pressure is created by the punch moving into the die filled with the fluid [4], the so called hydromechanical sheet forming (see figure 1) and the other one consists in the action of liquid toward the blank using a flexible rubber and no punch (see figure 2), the so called triform of flexform process [5].

In hydromechanical sheet forming the area of application of fluid pressure may be the punch-blank holder region or the die region (see figure 1) and the mode of application may be the hydrostatic pressure or hydrodynamic pressure.

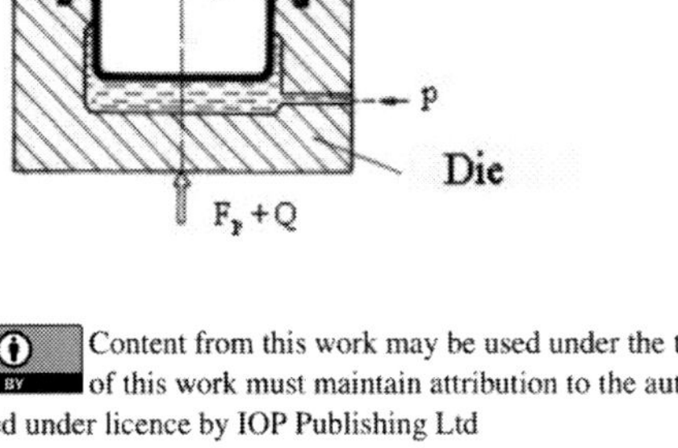

Figure 1. Hydromechanical sheet forming [2].

Content from this work may be used under the terms of the Creative Commons Attribution 3.0 licence. Any further distribution of this work must maintain attribution to the author(s) and the title of the work, journal citation and DOI.
Published under licence by IOP Publishing Ltd

The fluid pressure in this process can be produced by the downward movement of the punch, or be supplied by a hydraulic system, since no rubber diaphragm is used [2]. The tool device in this process is similar to that in a conventional deep dreawing. A rubber seal is used to prevent the flow out of the fluid on the flange. According with figure 1, F_p is the deep drawing force and Q is the restraint force. Since the blank is pushed against the punch by the hydraulic pressure, the punch force is uniformly distributed over the blank, which controls the localized reduction of the thickness in the punch shoulder.

The triform process (see figure 2), is a sheet hydroforming technology where the blank sheet is placed under a flexible rubber pad and is deformed by the action of a hydraulic pressure applied from the top of the equipment.

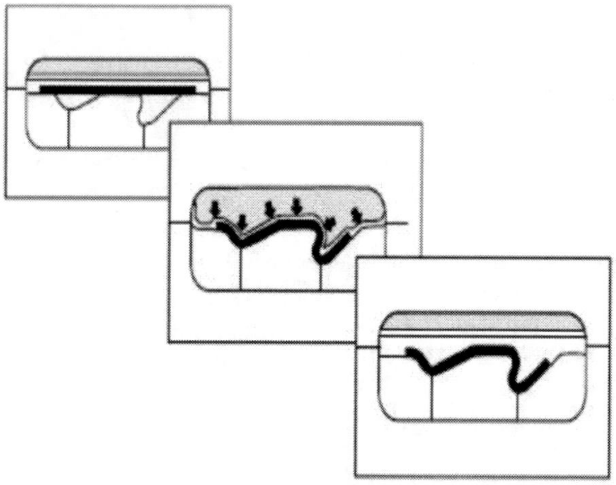

Figure 2. Triform (flexform) process [5].

The rubber is pressurized with hydraulic fluid, and it exerts equal pressure on each point inside the die cavity. The part shape is given by the shape of the die located at the bottom of the equipment [6]. Some advantages of the technology are [5, 7]: reduced tooling costs by 50% to 90% compared to conventional presses because is no need for a mated die to the punch (male die), as the flexible rubber acts as a universal female; due to uniform elongation of the metal, variation in blank to finished part thickness is less than 10% [8, 9]; thanks to the flexible rubber parts are produced with a smooth, uniform surface, typically requiring no additional finishing; tooling and material changes are simple and inexpensive; precise tolerances can be achieved with difficult parts configurations; a wide range of material can be formed such as: high strength steel, carbon steel, aluminum, stainless steel, precious metals, copper and brass [10, 11]. The main drawback of sheet hydroforming is its longer cycle time which causes the process to be costly at high volumes [12, 13]. Another drawback is still lack of knowledge so a lot of researches are being conducted to extend hydroforming's capabilities.

The study of the correlation between the fluid pressure and the part dimensions is a fundamental task to improve the performances of processed parts [14]. In the paper are presented some experimental results applying the triform process to Al alloys for obtaining the above task.

2. Experimental work

A hydraulic press was used for the experimental study of the sheet hydroforming process. The hydraulic system of the press assures a maximum pressure of 25 MPa.

Figure 3 shows the die. The die profile has a radius of 141.5 mm and a diameter of 150 mm. A flexible rubber was used with a thickness of 3 mm. The vulcanized natural rubber used has a Shore A hardness of 55, Tensile strength, Rm=35 MPa at 200C, Elongation at brake 650% at 200C.

For the experimental work, blanks made out of aluminum, Al 99.5, with a thickness of 2 mm and Al alloy series 5000 with 1 mm thickness were used. The blanks diameters were equal with D=150 mm.

According with the dates presented in [3, 7], the material properties are:

- for Al 99.5, 2 mm thickness: density, ρ=2,7 g/cm^3; Elastic module E=70 GPa; Tensile strength, Rm=130 MPa; Yield strength, Rc= 80 MPa; Poisson ratio, ν=0.35.

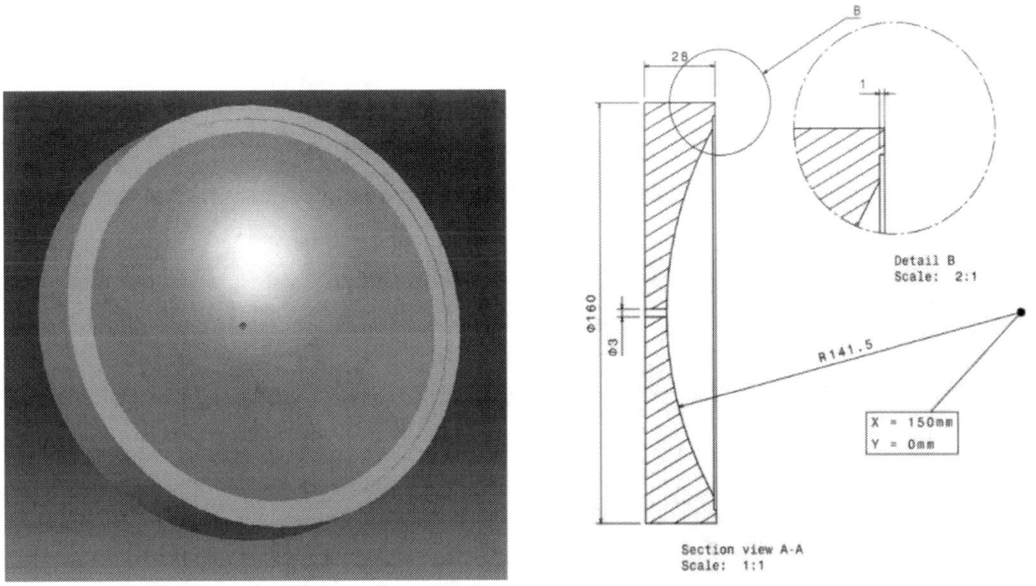

Figure 3. CAD model of the hydroforming die.

- for Al alloy series 5000, 1 mm thickness, the material properties are: ρ=2,7 g/cm^3; Elastic module E=70 GPa; Tensile strength, Rm=320 MPa; Yield strength, Rc=240 MPa; Poisson ratio, ν=0.35.

Four pressures were applied: 20, 40, 60 and 80 Bar. No lubricant was used.

Figure 4. Off-line measurement of part.

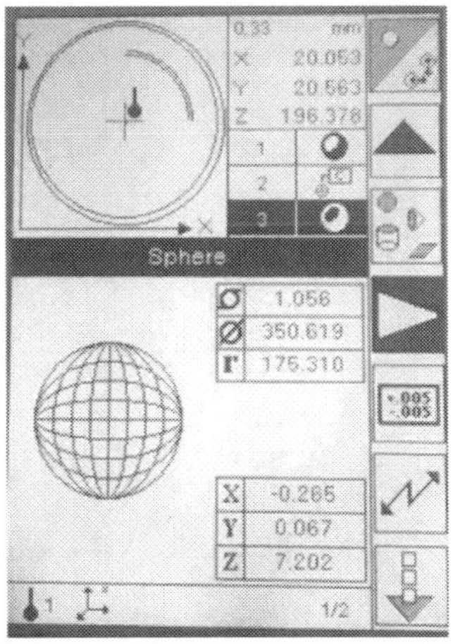

Figure 5. Display informations about part measurement.

The deformed parts were measured on a CM machine, see figure 4.

The machine software gives automatically values of parts diameters and radii, in these cases of simple geometrical forms, see figure 5. The points were measured from the centre of the part to the bottom along the exterior surface.

3. Results and discussions

Figure 6 shows the sample parts obtained by applying the above pressures.

Figure 6. Al Sheet hydroformed parts.

Figure 7 presents the radii variations of the samples for different pressure. It could be seen that the radius of the deformed part decrease with increasing the pressure, after the springback.

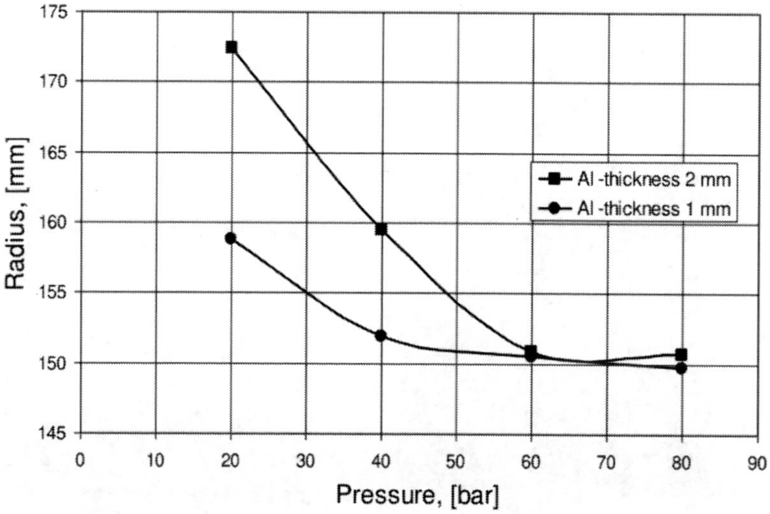

Figure 7. Radius variation function of the applied pressure.

This means that the radius will be more closely to the radius of the die and the part will be more accurate. Starting with the pressure of 60 bar (6 MPa) the part radius is almost the same, insensitive to material characteristics and pressure. We could consider that the only factor which alter the part precision is the flexible rubber characteristics.

Figure 8 presents the samples diameters deviations for different pressures. It could be seen that deformed parts diameters deviations are minimum at 60 bar (6 MPa), for both materials. The deviation was calculated using the relation:

$$Deviation = \frac{d\exp - dt\arg et}{dt\arg et} \cdot 100[\%] \quad (1)$$

where: dexp is the experimental obtained part; dtarget – is the target part diameter, which is equal with 140 mm. At low pressure the diameter variation due to springback is almost constant. By increasing the pressure the part diameter deviation will increase, because the springback will increase. After the pressure reached a certain value, in our case 6 MPa, the diameter variations will decrease. This variation is due to the fact that any lubricant was used. So the adhesion between the blank and the rubber avoid the part to correctly springback.

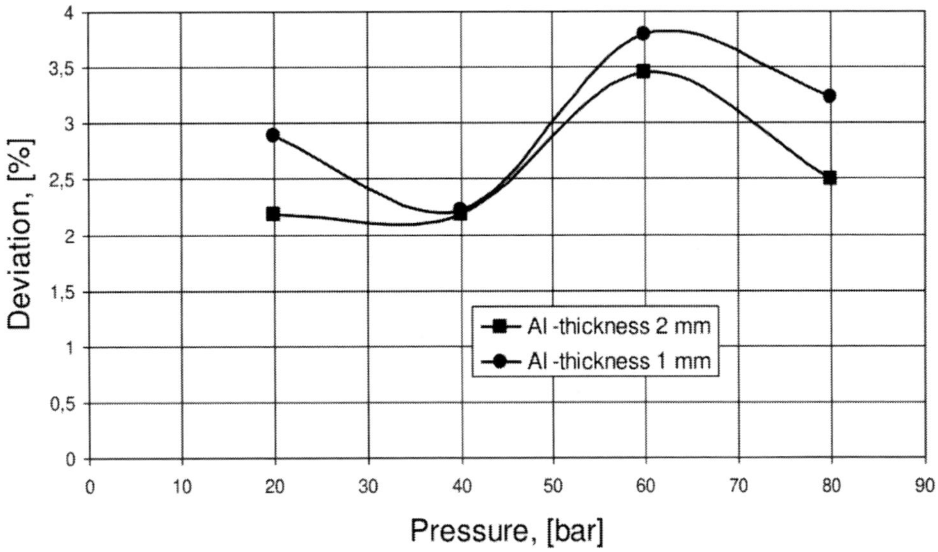

Figure 8. Part diameter deviation function of the applied pressure.

The presence of the solid lubricant is important. Without this, the parts adhere from the rubber. The adherence is greater as the pressure is higher as this influence the part springback.

4. Conclusions

It was concluded that in triform process the pressure has an important role in obtaining high quality parts, but for a certain material it exists a maximum pressure which assures the optimum conditions of deformation. In our case this has the value of 6 MPa. The part dimensions are affected by the presence of the lubricant. This is particularly visible when pressure is high. We are considering that the flexible rubber which assures the uniform pressure toward the blank largely and the lubricant influence the process of deformation. For complex parts we presume that a more flexible and thinner rubber will lead to an improvement of the deformation process. Future experimental studies and numerical ones will follow the above research direction.

References

[1] The Aluminium Automotive Manual, Version 2013 © European Aluminium Association.

[2] Bakhshi-Jooybari M, Gorji A and Elyasi M *Developments in Sheet Hydroforming for Complex Industrial Parts, Metal Forming – Process*, Tools, Design, INTECH, http://dx.doi.org/10.5772/48142

[3] Ramezani M and Ripin Z M 2012 *Rubber-pad forming processes: technology and applications.* (UK: Woodhead Publishing Ltd)

[4] Paunoiu V and Nicoara D 2004 *Sheet Metal Forming Technologies* (Bucharest: University Book Press Bucuresti)

[5] Vollertsen F, Breede R and Lange K 1999 Method for deep drawing with multiple elastomer membranes *CIRP Ann. – Manuf. Tech.* **48(1)** pp 221-226

[6] Information on http://www.triformpress.com; http://industry.avure.com/solutions/flexform-introduction

[7] Ramezani M, Ripin Z M and Ahmad R 2009 Computer aided modelling of friction in rubber-pad forming process *J. of Mater. Process. Technol.* **209(10)** pp 4925-4934

[8] Hama T, Hatakeyama T, Asakawa M, Amino H, Makinouchi A, Fujimoto H and Takuda H 2007 Finite Element Simulation of The Elliptical Cup Deep Drawing Process by Sheet Hydroforming *Finite Elements in Analysis and Design* **43** pp 234 – 246

[9] Lang L, Danckert J and Nielsen K B 2004 Investigation into hydrodynamic deep drawing assisted by radial pressure: Part I. Experimental observations of the forming process of aluminum alloy *J. of Mater. Process. Technol.* **148(1)** pp 119-131

[10] Lang L, Danckert J and Nielsen K B 2005 Investigation into hydrodynamic deep drawing assisted by radial pressure: Part II. Numerical analysis of the drawing mechanism and the process parameters *J. of Mater. Process. Technol.* **166(1)** pp 150-161

[11] Palumbo G, Zhang S H, Tricarico L, Xu C and Zhou L X 2006 Numerical/experimental investigations for enhancing the sheet hydroforming process *Int. J. of Machine Tools and Manuf.* **46** pp 1212-1221

[12] Novotny S and Hein P 2001 Hydroforming of sheet metal pairs from aluminium alloys *J. of Mater. Process. Technol.* **115** pp 65-59

[13] Abedrabbo N, Zampaloni M A and Pourboghrat F 2005 Wrinkling control in Aluminum sheet hydroforming *Int. J. of Mech. Sci.* **(47)3** pp 333-358

[14] Dariani M and Fazli A 2006 Parameter study of the axisymmetric hydromechanical deep drawing process *Part B: J. Eng. Manuf.* **220** pp 1937 – 1944

Investigation of FeNiCrWMn – a new high entropy alloy

G Buluc[1], I Florea[1], O Bălţătescu[1], R M Florea[1] and I Carcea[1]
[1]Technical University "Gheorghe Asachi" of Iasi-Romania, Department of Materials Science and Engineering, Blvd. Mangeron, No. 59A, 700050, Iasi, Romania

E-mail: gheo_u@yahoo.com

Abstract. The term of high entropy alloys started from the analysis of multicomponent alloys, which were produced at an experimental level since 1995 by developing a new concept related to the development of metallic materials. Recent developments in the field of high-entropy alloys have revealed that they have versatile properties like: ductility, toughness, hardness and corrosion resistance [1]. Up until now, it has been demonstrated that the explored this alloys are feasible to be synthesized, processed and analyzed contrary to the misunderstanding based on traditional experiences. Moreover, there are many opportunities in this field for academic studies and industrial applications [1, 2]. As the combinations of composition and process for producing high entropy alloys are numerous and each high entropy alloy has its own microstructure and properties to be identified and understood, the research work is truly limitless. The novelty of these alloys consists of chemical composition. These alloys have been named high entropy alloys due to the atomic scale mixing entropies higher than traditional alloys. In this paper, I will present the microscopy and the mechanical properties of high entropy alloy FeNiCrWMn.

1. Introduction

High entropy alloys are alloys that have in their composition 5 to 11 metallic elements. The percentage of each element in a composition of a high entropy alloys is between 5% and 35%. Mainly in the composition of a high entropy alloys it is found some principal and some secondary metallic elements [2]. They are called principal elements because their molar fraction is higher than 5 % and the secondary their molar fraction below 5%. In this work, we selected Fe, Ni, Cr, W, Mn, to prepare the high entropy alloy FeNiCrWMn and we investigated the mechanical properties of these high entropy alloy.

2. Materials and method

To obtain this alloy we used in chemical composition metallic elements as: Fe, Ni, Cr, W and Mn. As a method of prepare this high entropy alloys we used a medium frequency induction furnace for 8000 Hz which is in the inside of Faculty of Science and Engineering Materials from Iasi.

The alloy obtained was poured into molds made of furan resin-shaped specimens with a diameter of 10 mm.

Content from this work may be used under the terms of the Creative Commons Attribution 3.0 licence. Any further distribution of this work must maintain attribution to the author(s) and the title of the work, journal citation and DOI.
Published under licence by IOP Publishing Ltd

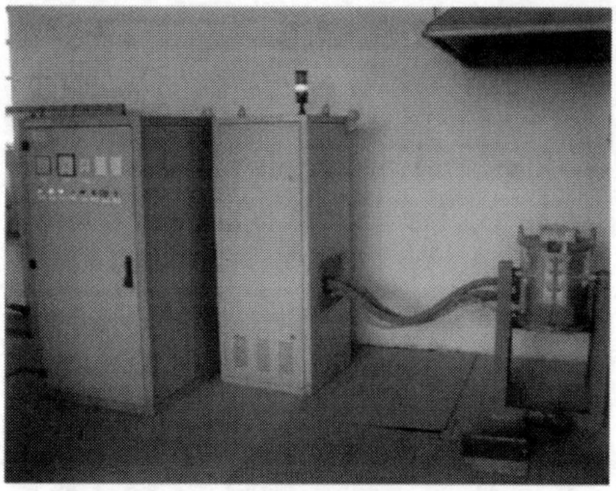

Figure 1. Induction heating furnace of medium frequency 8000 Hz.

Figure 2. Example of specimen for characterization properties of high entropy alloy FeNiCrWMn.

3. Microstructural analysis

For microstructural characterization, the samples were mechanically processed by grinding and chemical attack with NITAL. For the microstructural characterization, we used an electron microscope on LMHII VegaTescan equipment using a secondary electron detector (SE) at a voltage of 30 kV electron gun.

In image 3 we can see the chemical composition for high entropy alloy FeNiCrWMn in 3 points.

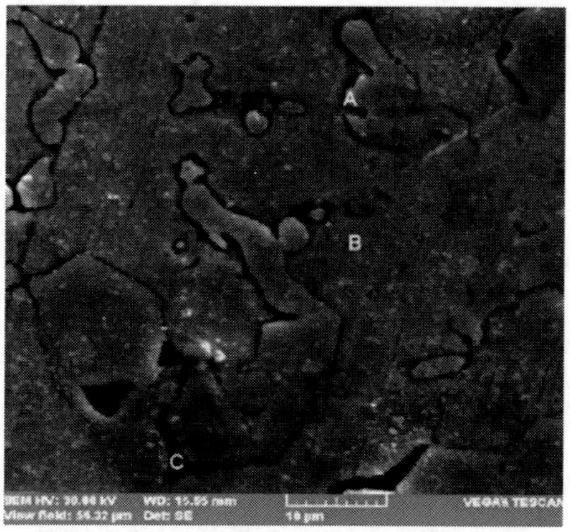

Figure 3. SEM image of a high entropy alloys FeNiCrWMn.

Table 1. Chemical composition (wt.%) and (at%) of the high entropy alloy FeNiCrWMn in point A.

Element	(wt.%)	(at.%)
Chromium	70.40	76.19
Manganese	9.97	10.22
Tungsten	8.44	2.58
Iron	6.07	6.11
Nickel	5.09	4.88
Total	99,94	99,98

Table 2. Chemical composition (wt.%) and (at%) of the high entropy alloy FeNiCrWMn for B point.

Element	(wt.%)	(at.%)
Chromium	13.98	15.56
Manganese	21.97	23.15
Tungsten	4.21	1.32
Iron	18.82	19.51
Nickel	41.00	40.44
Total	99,98	99,98

Table 3. Chemical composition (wt.%) and (at%) of the high entropy alloy FeNiCrWMn for C point.

Element	(wt.%)	(at.%)
Chromium	8.67	9.61
Manganese	32.07	33.63
Tungsten	2.91	0.91
Iron	10.79	11.13
Nickel	45.54	44.70
Total	99,98	99,98

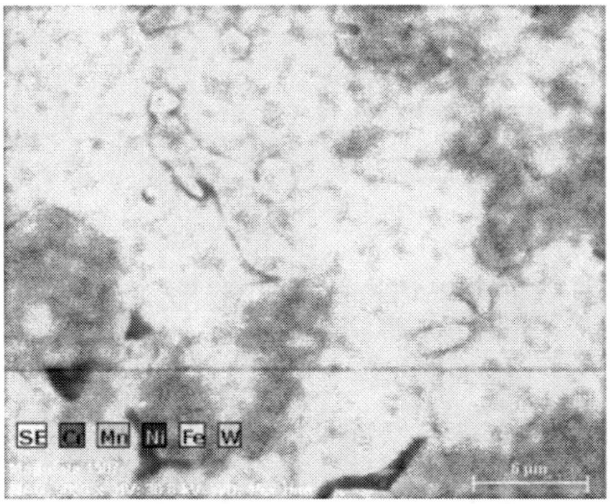

Figure 4. Distribution of chemical elements in high entropy alloy FeNiCrWMn.

We chose to do the chemical composition at various points to observe the degree of mixing of the chemical elements after casting the high entropy alloy FeNiCrWMn.

So, in point 1 we have predominant chemical elements as chromium and manganese, in point 2, manganese and nickel and in point 3 we have manganese and nickel.

In figure 4 we can see the distribution of elements in the high entropy alloy FeNiCrWMn and rich areas in chromium and manganese. Also, from figure 4 it can be seen that tungsten is distributed uniform throughout area the high entropy alloy FeNiCrWMn.

IOP Conf. Series: Materials Science and Engineering **95** (2015) 012017 doi:10.1088/1757-899X/95/1/012017

4. Analysis by optical microscopy

The optical analysis of FeNiCrWMn was performed using a X7P- 6A metallographic microscope and Materials Plus Software was used for the metallographic analysis. The following figures reveal non uniform grain sizes of the FeNiCrWMn alloy. The samples analyzed with Materials Plus Software had a grain size of 300 μm.

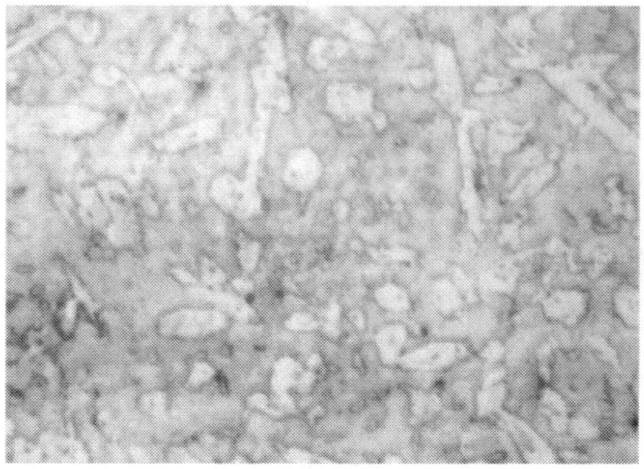

Figure 5. Optical microscopy of high entropy alloy FeNiCrWMn, 150X.

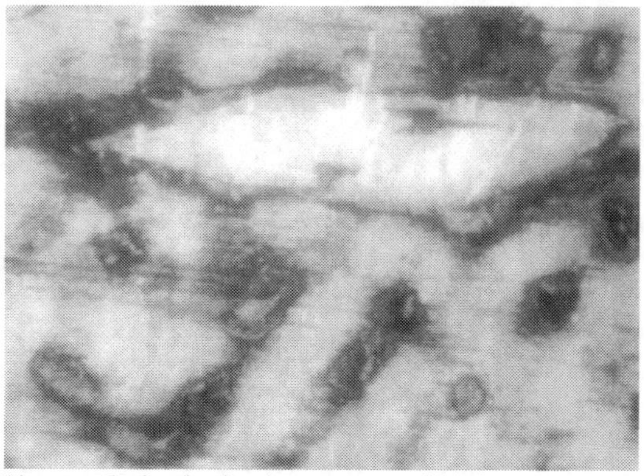

Figure 6. Optical microscopy of high entropy alloy FeNiCrWMn, 500X.

Figure 5 and 6 display micrographs of the surface microstructure of the FeNiCrWMn obtained by optical microscopy. In these pictures it can be observed dendrites with various dimensions and columnar grains.

5. Analysis by optical microscopy

To determinate the results for tensile test high entropy alloys FeNiCrWMn we used the machine Instron 3382.

In figure 7 is presented the characteristic curve of the tensile test. For this analysis we made three samples with the dimension d0 = 4 mm, D = 6 mm, h = 20 mm, L0 = 20 mm. Table 4 presented the results of the tensile test for the high entropy alloy FeNiCrWMn, here we can see that our samples suffered a break at 1,87% for sample 1, for sample 2 the break occurred at 2.03 %, in the case 3 the break occurred exactly like for sample 1.

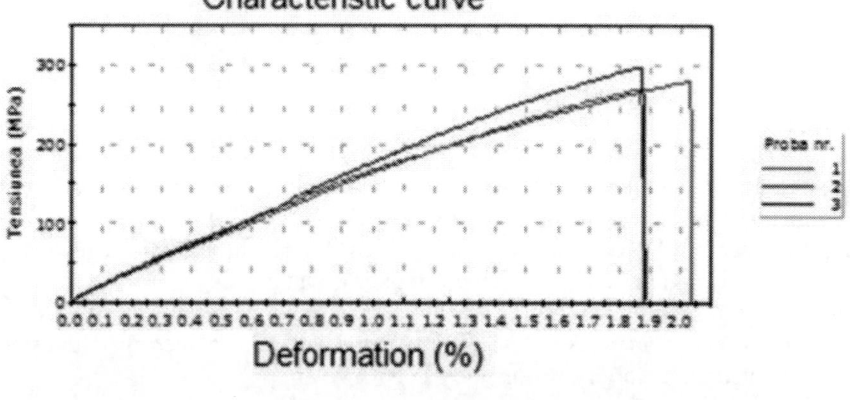

Figure 7. Tensile test of high entropy alloy FeNiCrWMn.

Table 4. Tensile test results for high entropy alloy FeNiCrWMn.

Sample	Maximum Load (N)	Modulus (E modulus)(MPa)	Tensile strain at Break (Standard)(%)	Tensile stress Yield (offset 0,2% MPa)
Sample 1	3,38	15,65	1.87	267.55
Sample 2	3,52	14,69	2.03	274.40
Sample 3	3,75	17,37	1.87	293.35

6. Hardness tests

Hardness is a characteristic of a material who is defined as the resistance to indentation and it is determined by measuring the permanent depth of the indentation. The Rockwell hardness test method as defined in ASTM E-18, is the most commonly used hardness test method. The medium value for high entropy alloy FeNiCrWMn is 38.7 HRC.

7. Conclusions

The high entropy alloy FeNiCrWMn is a new alloy which was successfully obtained and characterized. Results for hardness test show that medium value for high entropy alloy FeNCrWMn is 38. 7 HRC and the tensile stress is 293.359 MPa. Analyzing alloy FeNiCrWMn with the scanning electron microscope and optical microscopy we observed that alloy forms a dentritic structure.

Acknowledgements

This work was supported by the strategic grant POSDRU/159/1.5/S/133652, co-financed by the European Social Fund within the Sectorial Operational Program Human Resources Development 2007 – 2013.

This work was supported by a grant of the Romanian National Authority for Scientific Research, CNCS-UEFISCDI, project number PN-II-PT-PCCA-2013-4-1048.

References

[1] Murty B S, J.W. Yeh J W and Ranganathan S 2014 *High Entropy Alloys* (Elsevier)

[2] Munitz A, Kaufman M J, Chandler J P, Kalaantari H and Abbaschian R 2013 Melt separation phenomena in CoNiCuAlCr high entropy alloy containing silver *Materials Sci. &Eng.* A **560** pp 633–642

[3] Yang J, Zhou Y, Zhang Y et al. 2007 Solid Solution Formation Criteria in the Multi- component Alloys with High Entropy of Mixing *Chinese Mat. Sci. Technol. & Equipment*

[4] Cantor B, Chang I T H, Knight P and Vincent A J B 2004 Microstructural development in equiatomic multicomponent alloys *Mat. Sci. and Eng. a-Structural Materials Properties Microstructure and Processing* pp 213-218

[5] Yeh J W, Chen S K, Lin S J, Gan J Y, Chin T S, Shun T T, Tsau C H and Chang S Y 2004 Nanostructured High-Entropy Alloys with Multiple Principal Elements: Novel Alloy Design Concepts and Outcomes *Advanced Engineering Materials* pp 299-303

IOP Conf. Series: Materials Science and Engineering **95** (2015) 012018 doi:10.1088/1757-899X/95/1/012018

New approach for porous materials obtaining using centrifugal casting

O Bălțătescu[1], M Axinte[1], G Barbu[1] and V Manole[1]
[1]Technical University "Gheorghe Asachi" of Iasi-Romania, Department of Materials Science and Engineering, Blvd. Mangeron, No. 59A, 700050, Iasi, Romania
E-mail: oana84rou@yahoo.com

Abstract. It has been presented different methods for obtaining porous materials, (mainly used for metallic foams) and highlighting a new technology developed in the Faculty of Materials science and engineering, of Iasi. Our technology for obtaining porous materials is called centrifugal casting for porous materials. This technology is included in the method number 8: co-pressing of a metal powder with a leachable powder being in the same time a newer approach in the porous materials field. This technology is currently in the developmental phase. Since now we made experiments on the metallic materials, aluminum alloys. The technology is briefly described in this paper. The obtained parts were used for making samples in order to characterize the properties of the materials. The cellular structure of metallic foams requires special precautions that must be taken in characterization and testing. In this paper we have characterized the samples structurally by its cell topology (open cells, closed cells), relative density, cell size and cell shape and anisotropy. Also it was used scanning electron microscopy (SEM) which is straightforward; the only necessary precaution is that relating to surface preparation.

1. Introduction

Porous materials offer potential for lightweight structures, for energy absorption, and for thermal management; and some of them, at least, are cheap. There are known 9 types of methods used in industry, and presented in literature since now as:

- Bubbling gas through molten alloys [1, 2];
- By stirring a foaming agent into a molten alloy [4];
- Consolidation of a metal powder with a particulate foaming [3];
- Manufacturing a ceramic mold from a wax or polymer-foam precursor [2];
- Vapor phase deposition or electro deposition of metal onto a polymer foam precursor [2];
- The trapping of high-pressure inert gas in pores by powder hot isostatic pressing (HIPing), [1];
- Sintering of hollow spheres [4];
- Co-pressing of a metal powder with a leachable powder [3];
- Dissolution of gas (typically, hydrogen) in a liquid metal under pressure [5, 6].

It has been presented a different method for obtaining porous materials, mainly used for metallic foams and highlighting a new technology called centrifugal casting for porous materials

2. Installation and working method

It has been built a centrifugal casting machine with a vertical axis of rotation and constant speed. It is a machine that required low cost design, production, and maintenance. This type of machine (figure 2) can be used in mass production, in setting at the outset size parts, alloy nature, gravity coefficient,

Content from this work may be used under the terms of the Creative Commons Attribution 3.0 licence. Any further distribution of this work must maintain attribution to the author(s) and the title of the work, journal citation and DOI.
Published under licence by IOP Publishing Ltd

mass form and its construction elements [7].

Usually such machines can be mounted on a conveyor served by several workstations (preparation of molds, pouring the liquid, solidification and cooling of the piece, extracting part).

Figure 1 shows the centrifugal casting machine with a vertical axis of rotation and constant speed. The rotation speed of the centrifugal casting machine is 1390 rot/min.

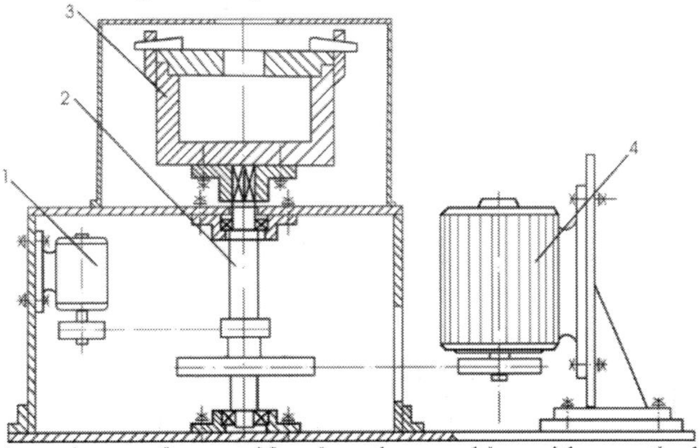

Figure 1. Constructive scheme of a centrifugal casting machine with a vertical axis of rotation and constant speed: 1- tacho; 2- vertical shaft for driving in rotation; 3 - form metal castings (chill); 4 – electric motor [7].

Recorded data for obtaining porous materials are as follows: Initial mass [g]: NaCl – 200; Aluminum – 230; Mass of casted piece (NaCl+Al) [g] – 327; Technological losses [g]: NaCl – 47; Aluminum – 58; Piece mass after dissolving the NaCl [g] – 203 and NaCl dissolved [g]: 124.

It was used food grade salt, grit between (1.6-2.2)mm. The salt was dried at 720°C for 30 minutes.

Figure 2. Centrifugal casting machine with vertical axis of rotation and continuous variation of speed: 1 – push buttons; 2 – electromotor of alternating current, three phase; 3 - mainboard; 4 - protective case; 5 - centrifugal casting machine body.

3. Results and discussions

The characteristics of a foam are best summarized by describing the material from which it is made, its relative density, ρ/ρs (the foam density, ρ, divided by that of the solid material of the cell wall, ρs), and stating whether it has open or closed cells.

$$\rho/\rho_s = 1.66/2.7=0.61 \text{ [g/cm}^3\text{]} \quad (1)$$

where: the foam density ρ=1.66 [g/cm^3] and the solid material of the cell wall density ρs = 2.7 [g/cm^3].

Figure 3 shows casted part and the measured dimensions of our sample which was obtained in the laboratory on the centrifugal casting machine with vertical axis of rotation and continuous variation of speed.

The porous material poured show open pores with uniform distribution due to centrifugal force and vibration that occurs in the machine shaft which presses and forces the particles to occupy the position most favorable to a volume as small pore of them. Knowing that salt particles are actually topology after dissolution voids results, it can be said that by this method can get a good index factor of relative density. This index expresses the uses degree of the material.

In the 4b image is observed salt particles included in the solidified melt casting shape wall. Seeing this figure it can be said that the outside pores are open.

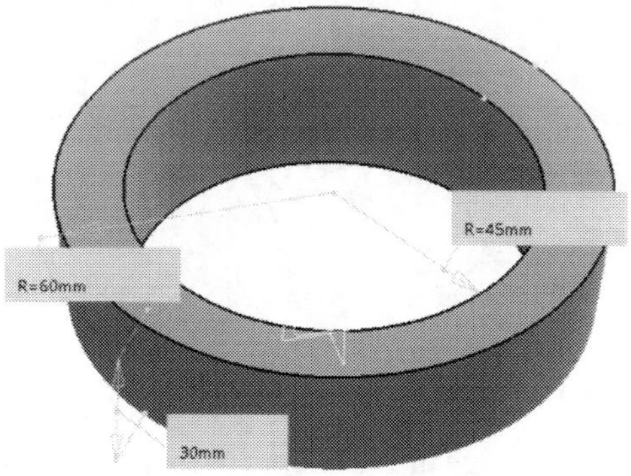

Figure 3. Casted part and the measured dimensions.

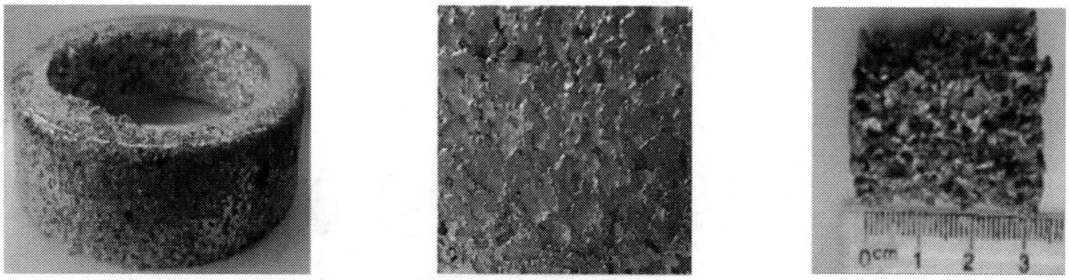

a) b) c)

Figure 4. a) Centrifugal casting piece; b) details on the external surface after the casting; c) detail on the inner surface of the workpiece after the dissolution of the salt.

In the following figures are presented macrographs (figure 5) and micrographs (figure 6) SEM, chemical analysis (figure 7) performed on EDX analyzer probe and scanning electron relief obtained (figure 8).

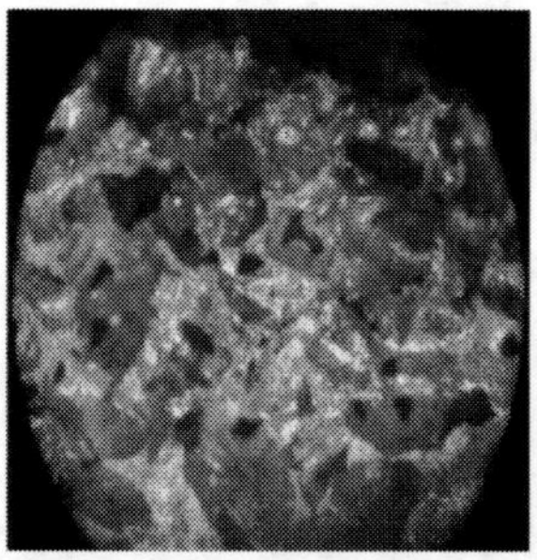

Figure 5. Porous aluminum alloy, macro graphic obtained at a magnification of 10x.

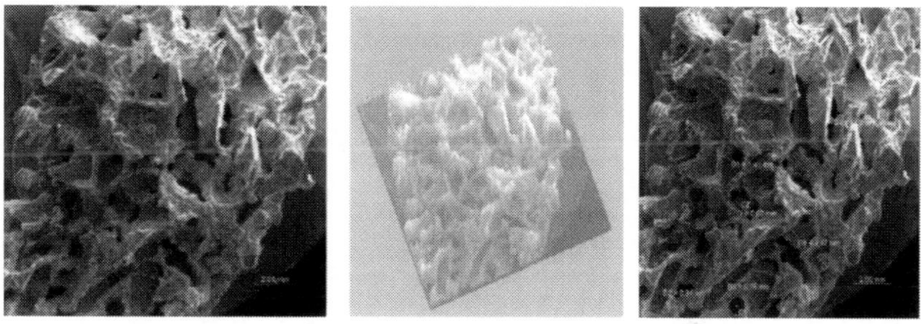

a) b) c)

Figure 6. SEM image of the structure of the porous material obtained by centrifugal casting: a) SEM image of 2mm to increase; b) 3D topography obtained from scanning electron; c) pore size.

Table 1 shows the size and number of the measured pores.

Table 1. The measured pore size.

Value	x [mm]	y [mm]	d [mm]	r [mm]	d [mm]	C [mm]
Obj. count	3	3	5	3	3	3
Summation	14.98	37.63	5.58	1.50	3.00	9.41
Min. value	3.66	9.97	0.75	0.39	0.78	2.45
Max. value	5.71	15.32	1.47	0.62	1.23	3.88
Average value	4.99	12.54	1.12	0.50	1.00	3.14
Std. dev.	0.94	2.19	0.28	0.09	0.19	0.58

It has been structurally analyzed at higher magnification and the results are shown in the following images (figure 7).

The above image is obtained by scanning electrons microscope and capture area of a single pore formed by a salt crystal (subsequently dissolved). It was also observed on the wall surface of the micro porous material inside the pores the appearance of the micro pores. There are two possible causes for these micro pores existence. The first one refers to the small salt crystals adherent to the main crystal surface. The existence of these small crystals is the result of large crystals breakage during operation of dehydration or strong adhesion to the surface of large crystals vibrating sorting operation. The second one can be the appearance of the gas micro bubbles due to the relative humidity of the salt.

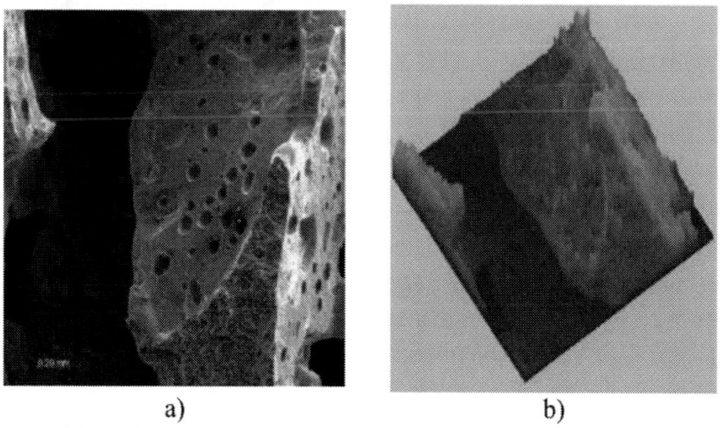

a) b)

Figure 7. Analysis of the microstructure and relief within a macro pores: a) direct image 2D, b) 3D processed image.

Next is presented a preliminary analysis of the micro pores main cavities (figure 8, table 2).

Figure 8. Dimensions between the pore walls.

Table 2. Macropore dimensions on the walls of main pores.

Value	x [mm]	y [mm]	d [mm]	r [mm]	d [mm]	C [mm]	A [mm^2]
Obj. count	22	22	5	22	22	22	22
Summation	81.80	96.11	5.00	1.74	3.47	10.91	0.46
Min. value	0.15	1.57	0.68	0.05	0.10	0.30	0.01
Max. value	7.32	6.56	1.53	0.13	0.26	0.81	0.05
Average value	3.72	4.37	1.00	0.08	0.16	0.50	0.02
Std. dev.	1.94	1.46	0.31	0.02	0.04	0.12	0.01

In table 3 are presented chemical analysis of the sample and chemical compositions made with EDX analyzer.

Table 3. Chemical analysis of the sample.

Element	AN	series	Net	[wt.%]	[norm. wt.%]	[norm. at.%]	Error in [%]
Aluminium	13	K-series	188131	74.46546297	77.79212988	78.47717861	3.760295
Silicon	14	K-series	14413	21.25818296	22.20787012	21.52282139	1.004575
			Sum:	95.72364593	100	100	

4. Conclusions

Compared with previous methods the centrifugal casting presents the following advantages:

- shows a high degree of safety in use as it requires no hazardous substances (gas introduced into the melt);
- preliminary operations are minimized because there are not necessary performs (precursors);
- reduced costs for raw materials and energy;
- installation is easy to maintain and does not require expensive operations and parts;
- it is not necessary to control operating parameters such as pressure, cooling rate;
- this method allows obtaining various pore sizes materials, which are uniformly distributed in the material volume;

- in industrial application the foaming agent may be recovered. After dissolving the salt, brine can be recycled by water evaporation;
- centrifugal casting technology is generally a well-known technology and this new technique can be used for obtaining porous materials just by using existing know how;
- unknown in specialized literature, using centrifugal casting technique is an original method for obtaining porous materials.

References

[1] Davies G J and Zhen S 1983 Metallic foams: their production, properties, and applications *J. Mat. Sci.* **18** pp 1899–1911
[2] Banhart J and Baumeister J 1988 *Production methods for metallic foams*
[3] Sang et al. 1994 *Process for producing shaped slabs of particle stabilized foamed metal* (US)
[4] Shih D S, Evans A G and Wadley H N G 1998 Porous and Cellular Materials for Structural Application Mat. Res. Soc. Proc. 521 (MRS, Warrendale, PA, USA)
[5] Liu X N, Li Y X, Chen X, Liu Y and Fan X L 2010 Foam stability in gas injection foaming process *J. Mater. Sci.* **45** pp 6481–6493
[6] Ashby M F, Evans A G, Fleck N A, Gibson L J, Hutchinson J W and Wadley H N G 2000 *Metal Foams: A Design Guide*
[7] Diaconescu F 2011 *Nonferrous alloys centrifugal casting* (Iasi: Tehnopress)
[8] Nedelcu D, Comăneci R, Chelariu R and Tăbăcaru L 2009 Overview of composite material technology with Si-C particles, reinforcement, Int. J. Modern Manuf. Tech. **1** pp 57-62

Development of infrared staking process for an automotive part

H S Park[1] and T T Nguyen[1]
[1]School of Mechanical Engineering, University of Ulsan, 93 Daehak-ro, Nam-gu, Ulsan, South-Korea, 680-749

E-mail: phosk@ulsan.ac.kr

Abstract. Staking plastic is a new technology that used infrared light as the energy sources in order to assembly the heated plastic parts. In this work, the proposed method coupling between Finite Element Method (FEM), Response Surface Method (RSM), and Genetic Algorithm (GA) is employed to simulate and optimize the staking plastic process. An automotive part, namely Door Trim with polypropylene (PP) material is implemented by using the developed technology. The effects of three key process parameters, such as heating time, cooling time, and air flow rate on the mechanical behavior of the joint has been investigated. The simulation procedure based on DEFORM-3D multi-operation is conducted to integrate sub-process and obtain the numerical results. The optimum values, including heating time, cooling time, and air flow rate are 14 second, 14 second, and 60 (ft^3/h), respectively. The optimizing results indicated that heating time is more contributed to the tensile force, following by cooling time and air flow rate. The correlation between simulation and experimental results indicates the effectiveness of the proposed method.

1. Introduction

The infrared staking process was developed by EXTOL by using halogen lighting in order to procedure the joint between different plastic part [1]. In comparison with the conventional staking technology using heat, hot air, and ultrasonic source, this process provides some advantages, such as higher energy efficiency, product quality, productivity, and better mechanical properties. The process takes place in four basic phases that can be listed as follows (figure 1):

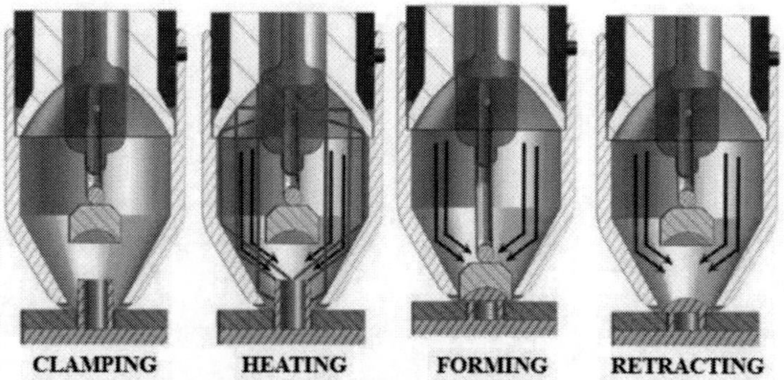

Figure 1. Infrared staking process [1].

Content from this work may be used under the terms of the Creative Commons Attribution 3.0 licence. Any further distribution of this work must maintain attribution to the author(s) and the title of the work, journal citation and DOI.
Published under licence by IOP Publishing Ltd

Clamping: The infrared staking module is positioned over the molded stud or boss to ensure perfect contact with the plastic part. The concentrator directly makes contact with the upper part, clamping it to the lower part and holding them in the proper position. As an added benefit, due to the ideal clamping at the stake points, secondary part clamps are not necessary.

Heating: The energy source used in the infrared staking process is a 12-Volt, 100-Watt, technical-grade of a halogen lamp. The reflector directly transfers the infrared energy from the lamp into a stud around the punch. This column of energy travels downward until it is focused on the full perimeter of the boss, heating it from top to bottom. To make this possible, the surface geometry of both the reflector and the concentrator creates a focal area that is centered on the boss. Ideally, the stud can be heated as quickly and as evenly as possible.

Staking & Cooling: At the end of the heating cycle, the lamp is switched off and the low-impact air cylinder with integrated staking punch is employed which forms over the semi-molten stud. The gold plating and the perfect contact area between the punch and formed part can reduce the temperature of the molten plastic quickly. The cooling air will be flowing into the chamber in order to assist in regulating the punch temperature which prevents the plastic from sticking to it.

Punch Retracts: At the end of the punch time, the punch retracts. The module is retracted from the components, and the cycle is completed. The sub-assembly can be immediately handled. At this point, it is important that the plastic has re-solidified to the point where it can maintain its shape and structure.

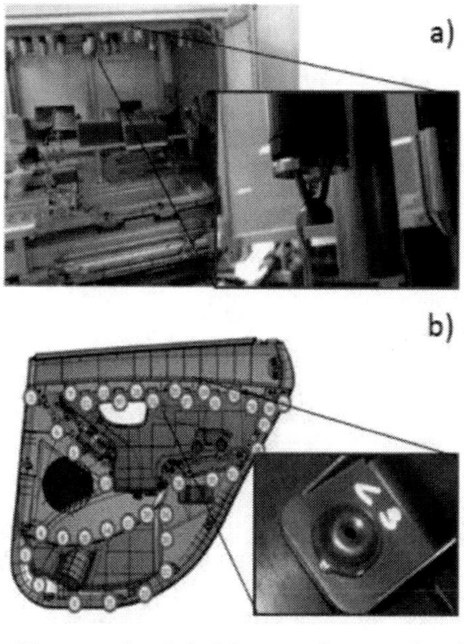

Figure 2. Machine and completed product; a-Developed machine and dies, b-Assembled door trim and completed joint.

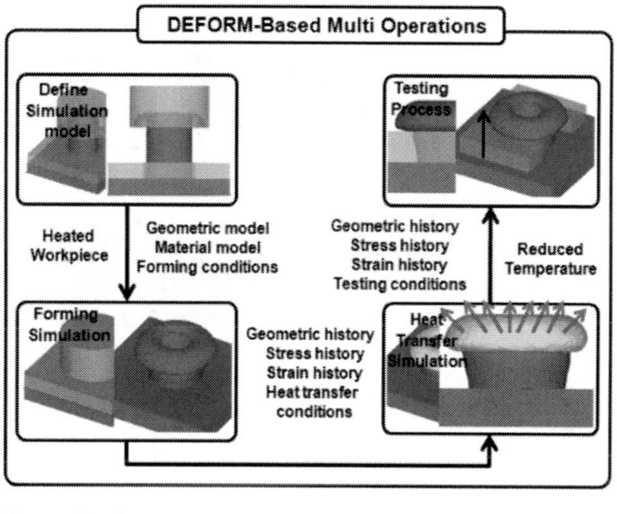

Figure 3. The approach for simulations of the whole process.

In this project, we attempt to develop new machine, which used to assemble the different plastic components of Door Trim. The joint is generated based on PP material and an infrared staking process. The machine, dies, assembled Door Trim, and completed product are shown in figure 2. Recently, this technology is still a lack of the publications. In addition, identification of optimal values and understanding the effects of process parameters on the mechanical behavior after process are necessary. To solve this problem, this paper proposed an approach to simulate and to optimize whole staking process. The key sub-process, including heating, forming, and cooling is integrated into one simulation model. The developed finite element (FE) model considering the data history based on DEFORM-3D multi operation [2] was adopted to conduct the serial numerical experiments. The coupling RSM using Box-Behnken experimental design [3, 4] and GA [5] is used to obtain the optimal values. Finally, the numerical results are compared with physical experiment to validate the effectiveness of the proposed method.

To investigate the material behavior, the sub-process has to integrate into one simulation model, in which the elements have the connection (figure 3). In other words, the results of preceding process are inputs for the next step. The material state is changed by every step in the production process. The behavior of the final part will depend on the results of each step. After sub-process, the material properties and the stress, and the strain state have changed. Therefore the properties of the joints have to be included in a parameterized way. Transferring data between the simulation processes includes geometric, stress, and strain history [6, 7].

Multi-operation in DEFORM-3D was employed to simulate process chains. In the first stage, the simulation model is developed with material model, friction, and heat transfer conditions. After that, the forming process is conducted using boundary conditions. Consequently, these results are transferred to a cooling process considering two kinds of heat transfer, including conduction and convection. Finally, the tensile testing is conducted with the results of the previous stage.

2. Governing equations of forming simulation

2.1. Material model

A material model includes physical and thermal properties of the work piece and tool. Due to the fact that metal cutting and forming operations perform in high temperature and high strain rates, mechanical properties of work piece material are exactly known. Material properties of plastic-PP and die-S45C are shown in table 1.

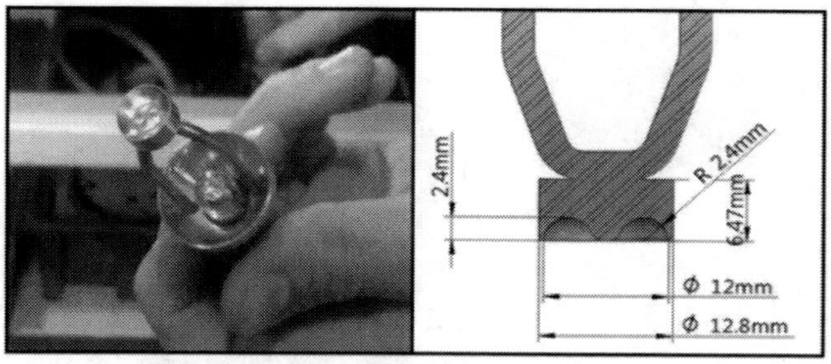

Figure 4. Geometric dimensions of die.

Table 1. Material properties of PP and S45C [8].

Parameters	PP	S45C
Density (g/mm^3)	0.892 x10^{-3}	7.85 x10^{-3}
Young modulus (MPa)	1300	250x10^3
Poison ratio	0.45	0.29
Thermal conductivity(W/mm °C)	0.16x10^{-3}	49.8
Specific heat (J/g °C)	2	0.486

To save the simulation time and increase the accuracy, the tooling was modeled as perfectly rigid, while the work piece was considered as Elasto-plastic properties with isotropic hardening law and following Vonmises Stress yield criterion. FE model of die and PP work piece are consist of tetrahedral elements with four nodes. Flow stress curves of work piece materials must be used especially in the forming analysis. Flow stress curves of PP material change as a function strain, strain rate, and temperature is shown in figure

5 at the strain 5min-1, 0.5min-1, and 0.05min-1. This law was selected due to its ability to follow the true behavior of a material [9]:

$$\bar{\sigma} = (\bar{\varepsilon}, \dot{\bar{\varepsilon}}, T) \tag{1}$$

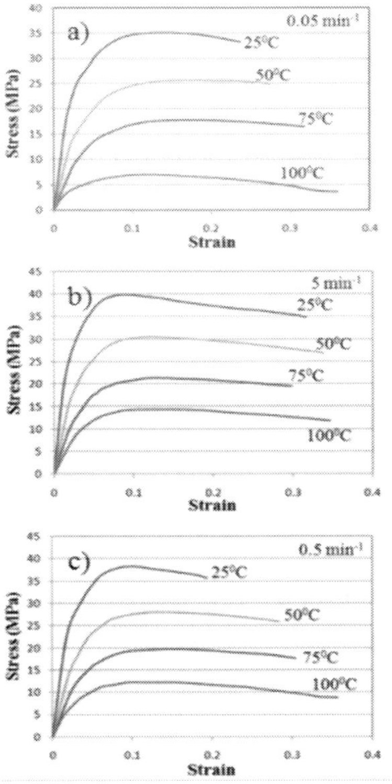

Figure 5. Stress-strain curves of polypropylene; a- at the 0.05 min-1, b-at the 5 min-1; c-at the 0.5 min-1.

According to equation 1, σ is flow stress, $\bar{\varepsilon}$ and $\dot{\bar{\varepsilon}}$, and T are effective plastic strain, effective strain rate, and temperature, respectively.

2.2. Friction model

The friction types allowed are shear and coulomb friction. Shear (sticking); constant shear friction is used mostly for bulk-forming simulations. The frictional force in the constant shear model is defined by:

$$f_s = mk \tag{2}$$

where f_s is the frictional stress, k is the shear yield stress and m is the friction factor. Coulomb friction is used when contact occurs between two elastically deforming objects (could include an elastic-plastic object, if it is deforming elastically) or an elastic object and a rigid object, generally to model sheet-forming processes. The frictional force in the Coulomb law model is defined by:

$$f_s = \mu p \tag{3}$$

where f_s is the frictional stress, p is the interface pressure between two bodies and μ is the friction factor. There must be interfacial pressure between two bodies for frictional force to be present. Due to the complexity of the process, the hybrid friction model was used together both of Coulomb and shear friction models. To obtain the best simulation results, the preliminary simulations were carried out and $\mu = 0.3$ was selected.

2.3. Heat transfer

The initial temperature of the ambient was assumed as 20 °C. The plastic temperature reaches 29 °C (after

8 second), 38 °C (after 14 second) and 43 °C (after 20 second), respectively. The heat losses to the environment from the free surface of the work piece material are determined by the heat flux:

$$Q=h(T-T_0) \tag{4}$$

where $h = 4.5\ W/m^2K$ is the heat transfer coefficient of the work material. T and T_0 are the temperature of the work material and ambient temperature, respectively.

3. Governing equations of heat transfer simulation

Two kinds of heat transfer are parallel conducted: convection and radiation. The plastic temperature is reduced due to conductive heat exchange with the die that decreases the temperature by forced convective heat transfer (figure 6). According to the heat transfer theory and thermal equilibrium relationship by referring to the Fourer Role, the governing equation is established and given as follows:

$$k\left(\frac{\partial^2 T}{\partial x^2}+\frac{\partial^2 T}{\partial y^2}+\frac{\partial^2 T}{\partial z^2}\right)=\rho C\frac{\partial T}{\partial t}+Q \tag{5}$$

where k is the thermal conductivity, ρ is density, C is the specific heat, Q is the latent heat of phase transformation, x, y, z are the coordinates, and t is the time respectively. For the thermal boundary conditions, the adiabatic condition was employed for the others. The heat flow by conduction between the polymer and die [10] can be described as bellow:

$$\dot{Q}=-\frac{kA(\theta_f-\theta_0)}{\sqrt{\pi\alpha t}} \tag{6}$$

where θ_0 is considered at the initial time and the final temperature after cooling is assumed to be at θ_f, α is the thermal diffusivity of the polymer, A denotes the cross sectional area, and t is the cooling time, respectively.

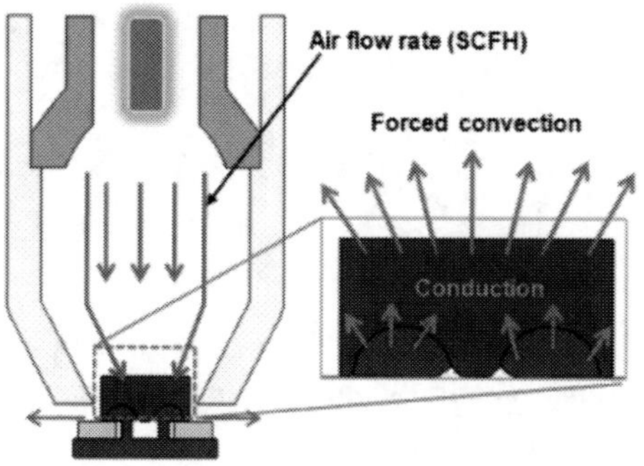

Figure 6. Heat transfer in the cooling process.

4. Optimization process

Response surface methodology is adopted to establish a relationship between process parameters and the performance of objective functions. RSM is a well-known method with higher accuracy and better ease-of use than other popular meta-models, such as radial basis function and kriging model. The second-order RSM model is suitable for modeling the moderate non-linear behavior with few design variables. Prior to the optimization process, relationships between process parameters and objective functions should be created. In this research, the experimental plans are generated using the stipulated conditions using the Box-Behnken experimental designs with 17 runs. Box-Behnken experimental method is one of the effective designs based on multi-dimensional sphere and all the design points lie on a same sphere with at least three or six runs at the center point. The optimization process was resolved based on explicit equations in regression that were obtained through the previous approximation.

Table 2. The parameters the thier levels.

Code	Parameters	Value
A	Heating time (s)	8-14-20
B	Cooling time (s)	8-14-20
C	Air flow rate (ft3/h)	20-40-60

The process parameters, including heating time, cooling time, and air flow rate with three levels are listed in table 2. The sequential numerical simulations are conducted by using boundary conditions to obtain the simulation results that can be used to generate the mathematical model of tensile force (equation (7)).

$$F=630.38+47.65*A+63.32*B+5.53*C+0.2286*A*B-0.0536*B*C-1.6*A^2-2.2*B^2-.041*C^2 \quad (7)$$

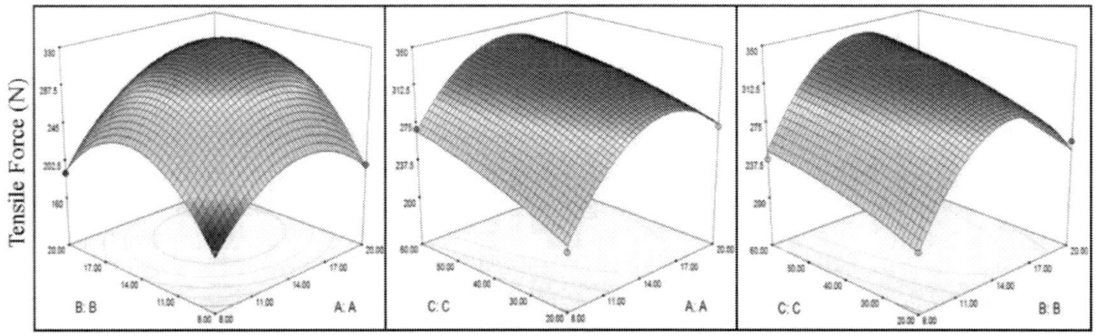

Figure 7. The effects of process parameters on the tensile force.

Figure 7 shows the response surfaces, which show the effect of three key process parameters at the center point in the design space of the tensile force. It can be easily seen from the picture that heating time and cooling time have a large effect on the tensile strength, while volume air flow rate has a little effect on the objective function. The main reasons can be explained as follows. Too low heating time results in poor heat transfer and poor mixing material, thus causing of weak joint. Increasing the heating time increases the heat input to the material, resulting in better fusion, consequently, joint strength increases. However, a further heating time combining with forming pressure decreases the joint strength due to material decomposed and a weak joint is formed. Increasing the cooling time results in reducing more temperature that increases the joint strength. The input energy can be loss due to the diffusion at the contact area. Thus, higher cooling time may lead to decrease the tensile strength. Air flow rate has a significant effect to the tensile force. Increasing the air flow rate leads to reduce temperature and to improve the rigidity of workpiece. Reduction of this factor can lead to the lack of the accuracy of the developed mathematical model.

Process parameters such as heating time, cooling time, and air flow rate have significance and complex effects on the objective function. These process parameters also have contradictory effects on response. To solve the optimization problem, the evolution algorithm, namely Multi-Island GA is adopted. Specific parameters were population size, number of generations, crossover probability, crossover distribution index, and mutation distribution index, with values of 10, 20, 0.9, 20, and 100, respectively. Figure 8 describes the history of the GA-based optimization process.

The optimal parameters, including heating time, cooling time, and air flow rat are 14 second, 14 second, and 60 ft^3/h, respectively. To validate the proposed method, the physical staking process was conducted in HANIL EHWA-Door Trim Manufacturer in South Korea. The error between simulation and experimental results is lower 10% that indicates the effectiveness of the proposed method.

IOP Conf. Series: Materials Science and Engineering **95** (2015) 012019 doi:10.1088/1757-899X/95/1/012019

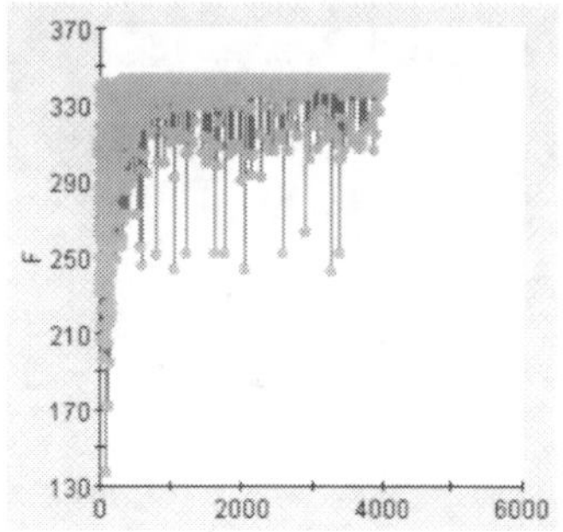

Figure 8. History of GA solving process.

Figure 9. Tensile testing at the company.

Table 3. Optimization values.

Design variables			Tensile Force (N)	
A (s)	B (s)	C (ft3/h)	Simulation	Experiment
14	14	60	348.81	362.8

5. Conclusion

This paper presents a new approach to simulate and optimize infrared staking process. The FE model considering the interaction between sub-process has been developed using DEFORM-3D in order to investigate the effects of process parameters on the mechanical behavior. The maximizing tensile force can be achieved at optimal parameters in 14 second, 14 second, and 60 ft^3/h of heating time, cooling time, and air flow rate, respectively. The optimal results also indicated that heating time is more contribution to tensile strength compared to the others. The proposed method can be used as the effective approach to simulate and optimize the process chains. The effects of forming tool geometry on the mechanical behavior of the joint will be analyzed in future work.

References

[1] Information on http://www.extolinc.com/infrastake

[2] Fluhrer J 2005 *DEFORMTM-3D Version 5.0 User's Manual* (Scientific Forming Technologies Corporation)

[3] Jeff C F and Lan Y 2000 *Experiment Design & Analyses and Parameters Optimization* (Wiley)

[4] Park H S and Nguyen T T 2014 Optimization of injection molding process for car fender in consideration of energy efficiency and product quality *J. Comput. Des. Eng.* **1** pp 266-271

[5] Oujebbour F Z, Habbal A, Ellaia R and Zhao Z 2014 Multi criteria shape design of a sheet contour in stamping *J. Comput. Des. Eng.* **1** pp 187-193

[6] Oudjene M, Ben-Ayed L, Delamézière A and Batoz J L 2009 Shape optimization of clinching tools using the response surface methodology with Moving Least-Square approximation *J. Mater. Process. Technol.* **209** pp 289-296.

[7] Lambiase F and Di Ilio A 2013 Optimization of the clinching tools by means of integrated FE modeling and artificial intelligence techniques Procedia CIRP **12** pp 163-168.

[8] Matweb, Material Property Data.

[9] Zhou Y and Mallick P K 2002 Effects of Temperature and Strain Rate on the Tensile Behavior of Unfilled and Talc-Filled Polypropylene. Part I: Experiments Polymer *Eng. & Sci.* **42** pp 2461-2470.

[10] Carlaw H and Jaeger J 1959 *Conduction of heat in solids* (Oxford at the Clarendon Press, Vol. 2) pp 17-25

IOP Conf. Series: Materials Science and Engineering **95** (2015) 012020 doi:10.1088/1757-899X/95/1/012020

Research concerning optimum cutting parameters according with tool path strategy for finishing procedures

A E Pena[1], **F D Anania1 and M Zapciu**[1]
[1]Politehnica University of Bucharest, Department of Machine Tools and Manufacturing Systems, Spl. Independentei, No. 313, sector 6, 060042, Bucharest, Romania

E-mail: andra.pena@upb.ro

Abstract. Optimization of cutting parameters in NC milling needs to be studied because of its influence on machining time and cost. Today, any CAM software offers many tool path strategies to milling free form geometries. However, the users must have the know-how to choose the strategies according to geometry complexity, cutting tool geometry and its contact on the machined surface. Choosing the right strategy with the right cutting parameter is a rather difficult task to do on the machine tool. In this paper we try to take into account the influence of the toolpath over the surface quality for finishing operation. The main goal is to establish a direct link between machining parameters and toolpath in order to obtain the same surface quality for different trajectories. The first step consist in making a series of experiments for standards toolpaths (which can be found in any CAM software) like one-way, zig-zag, spiral from outside to inside, zig-zag at 45 dgr on a milling center. Based on the results, a correction coefficient for the feed rate was established.

1. Introduction

Generally, the handbook or human experience is used to select convenient machine parameters in manufacturing industry. In process planning of conventional milling, selecting reasonable milling parameters is necessary to satisfy requirements involving machining economics, quality and safety.

The machining parameters in milling operations consists of cutting speed, depth of cut, feed rate and number of passes. These machining parameters significantly impact on the cost, productivity and quality of machining parts. The effective optimizations of these parameters affect dramatically the cost and production time of machined components as well as the quality of final products. [1]

Several researches have also addressed the influence of cutter path on surface roughness, although few studies focus on the impact of tool path strategies on surface texture [2, 3, 4]

The tool path has always been the weak link in the chain –machine tool; cutting tool; CAM software, typically forcing the use of cutting parameters that are on the side of caution rather than productivity. These cautionary tool parameters are detrimental to the tools performance and fate.[5]

One of the main goals in finishing operations is to achieve a very low workpiece surface roughness. However, surface irregularities, which are always present in all machined parts, depend on several factors. In milling operations, surface quality improves at higher cutting speeds. Depth of cut indirectly affects surface quality, since the cutting force, vibration and cutting temperature increase with an increase in the depth of cut. Other factors that influence surface roughness are feed, tool nose radius, tool wear, cutting strategy, the tool's trajectory during cutting, workpiece material,

Content from this work may be used under the terms of the Creative Commons Attribution 3.0 licence. Any further distribution of this work must maintain attribution to the author(s) and the title of the work, journal citation and DOI.
Published under licence by IOP Publishing Ltd

IOP Conf. Series: Materials Science and Engineering **95** (2015) 012020 doi:10.1088/1757-899X/95/1/012020

cooling/lubrication system and the dynamic parameters of machining, such as cutting force, tool deflection, vibration and several thermal phenomena [6].

In general, it is found that surface roughness increases with an increase in the feed rate and depth of cut and a decrease in cutting speed.

The focus of numerous and extensive studies that evaluate surface finish is usually centered on the effect that machining parameters (speed, feed, depth of cut, etc.) have on the surface quality. They do not take into account the influence of the cutting toolpath on the surface quality.

Most of the commercial CAM software today offers several possibilities of algorithms of distributing the tool path in the domain of the designed part. Ones of the commonly used tool path are: One way curves; Zig-zag or raster curves; Helix curves; Spiral curves; Contour curves; Radial curves; Sequential curves.

2. Experimental research

In this paper is presented a way to take into account the different toolpath trajectories as a factor in the calculus of machining parameters. The research is starting with a finishing operation in milling process for basic toolpath trajectories existing in Cimatron E CAM software.

Some machining tests were made and roughness influenced by the type of trajectories was taking into account.

The proposed work investigates the efficiency of the different toolpath strategy for finishing milling.

3. Equipment and materials

The toolpaths evaluated were (figure 1) standard toolpaths from the software CimatronE 11 as follows:

- *One way* - is used climb milling to obtain a good surface quality.
- *Zig-zag*– both climb and conventional milling is used.
- *Helix curves* – the tool trajectory is from outside to inside with corner radius over the trajectories;
- *Radial* – is similar to helix curves from the tool trajectory point of view, but with sharp corner.
- *Zig-zag* at *45 degree* – the tool trajectory is inclined relative to *X* direction with 45 degree (the linear movement is obtained by interpolation of the two axis of the machine (x and y). This trajectory type is used by some mold makers because they obtain a good quality surface.

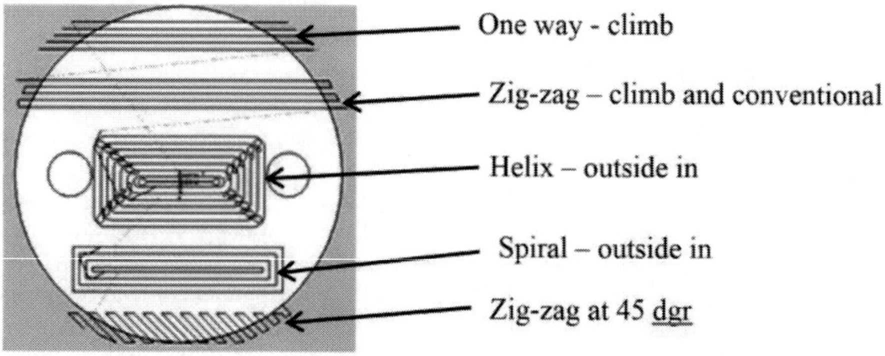

Figure 1. Trajectories types.

The material used was 6061 aluminum, one of the most common alloys of aluminum for general purpose use. This is a precipitation hardening aluminum alloy, containing magnesium and silicon as its major alloying elements. The main mechanical properties are: Density 2.7 [g/cc]; Hardness: Brinell 95; Rockwell A 40; Tensile Yield Strength 276 [MPa]; Modulus of Elasticity 68.9 [GPa]; Poisson's Ratio 0.33; Fatigue Strength 96.5 [MPa]; Fracture Toughness 29 [MPa-m½] Shear Modulus 26 [GPa];

Shear Strength 207 [MPa].

The experiments were made on a 3 axis machining center First MCV300 that has a max. spindle speed of 8000rpm, spindle motor power 7,5kW and a maximum cutting federate of 10000 mm/min.

The cutting tool is an end mill with 12 mm diameter with 3 flute. It is uncoated, solid fine grain, carbide tool from Iscar. All cases investigated were carried out without coolant.

In order to evaluate the surface quality, for the roughness measurement was used a Mitutoyo Surftest SJ210 instrument that has a measuring range (on X axis) of 17.5mm and a measuring speed of 0.25mm/s, 0.50mm/s, 0.75mm/s. The cut off length and the sampling length is 0.08mm, 0.25mm, 0.8mm, 2.5mm (figure 2).

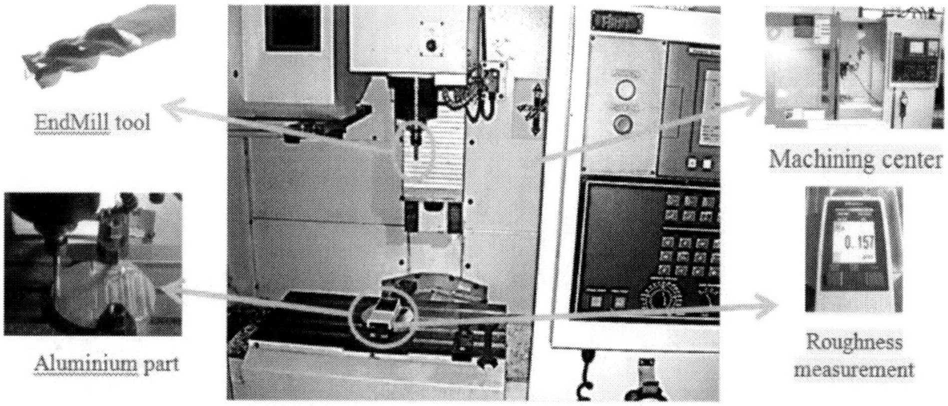

Figure 2. Experimental equipment.

4. Experimental studies and calculus

A number of experimental cutting tests were made in order to identify and quantify the influence of the tool trajectories over the machining parameters.

In the first step the cutting regime for the initial tests was calculated according to the basic formulas for milling (equation (1)).

$$V_c = \frac{\pi \cdot D \cdot n}{100}[\text{m/min}] \quad F = nzf_z\ [\text{mm/min}] \tag{1}$$

Where: D – the tool diameter in mm; n – spindle speed in rpm; F – feed rate in mm/min; z – no. of flutes; f_z – feed/tooth

The cutting regime parameters were kept for all studied trajectories at the same values F=420[mm/min], n=350[rpm], V_c=132 [m/min] and f_z=0.04 [mm/tooth]. The feed/tooth was chosen according to the cutting tool manufacturer's recommendations; and for the preliminary tests the decision was to use the minimum value. The depth of cut was 0.3mm for all tests.

In order to have good results a series of 5 tests were made for each type of tool trajectory. For each test the roughness was measured in 3 points on 2 directions (x and y). Averages were calculated for each direction of measurement (equation (2)).

$$R_{aj}^{x} = \frac{\sum_{i=1}^{n} R_{ai}^{x}}{n} = \ldots R_{aj}^{y} = \frac{\sum_{i=1}^{n} R_{ai}^{y}}{n} \tag{2}$$

where i=1..3 -measurement number;

n- number of measurements (3- for our study case).

j=1..5 -number of test for each type of tool trajectories

R^x_{aj} - Average roughness in x direction for each cutting test for the same trajectories type;

R^x_{ai} – roughness in X direction for each measurement;

R^y_{aj} - Average roughness in Y direction for each cutting test for the same trajectories type;

R^y_{ai} – roughness in Y direction for each measurement;

An average roughness was establish for each cutting test (equation (3))

$$R^t_{aj} = \frac{R^x_{aj} + R^y_{aj}}{2} \tag{3}$$

where j=1..5 -number of test for each type of tool trajectories;

R^t_{aj} –average roughness for each cutting test for the same trajectories;

The measurements results are presented in table 1.

Table 1. Measurements results/

No	Tool trajectory type (t)	R^x_{aj} [μm]					R^y_{aj} [μm]				
	Tests	1	2	3	4	5	1	2	3	4	5
1	One way	0.13	0.10	0.08	0.09	0.12	0.13	0.17	0.12	0.11	0.10
2	Zig-zag	0.10	0.12	0.10	0.11	0.12	0.12	0.15	0.13	0.11	0.13
3	Helix	0.11	0.11	0.11	0.13	0.11	0.12	0.13	0.25	0.37	0.3
4	Spiral	0.11	0.14	0.19	0.11	0.16	0.11	0.32	0.18	0.17	0.15
5	Inclined 45	0.13	0.20	0.22	0.1	0.09	0.33	0.22	0.23	0.14	0.16

In order to have an accuracy results an average roughness was calculated for each tool trajectories type (equation (4)) and the results are in table 2.

$$R^t_a = \frac{\sum_{j}^{m} R^t_{aj}}{m} \tag{4}$$

where j=1..5 -number of each test for each type of tool trajectories;

m - number of test made for a certain tool trajectories (5 in our study case)

R^t_a –reference roughness for each trajectories type;

R^t_{aj} –average roughness for each cutting test for the same trajectories type;

Table 2. Surface quality.

No	Tool trajectory type (t)	R^t_{aj} [μm]					R^t_a [μm]
		1	2	3	4	5	
1	One way	0.13	0.13	0.10	0.10	0.11	0.11
2	Zig-zag	0.11	0.14	0.11	0.11	0.13	0.12
3	Helix	0.12	0.12	0.18	0.25	0.20	0.15
4	Spiral	0.11	0.23	0.19	0.14	0.16	0.16
5	Inclined 45	0.23	0.21	0.23	0.12	0.13	0.18

To calculate the correction factor for machining parameters a constant reference value was established as minimum of R^t_a calculated (equation (5)).

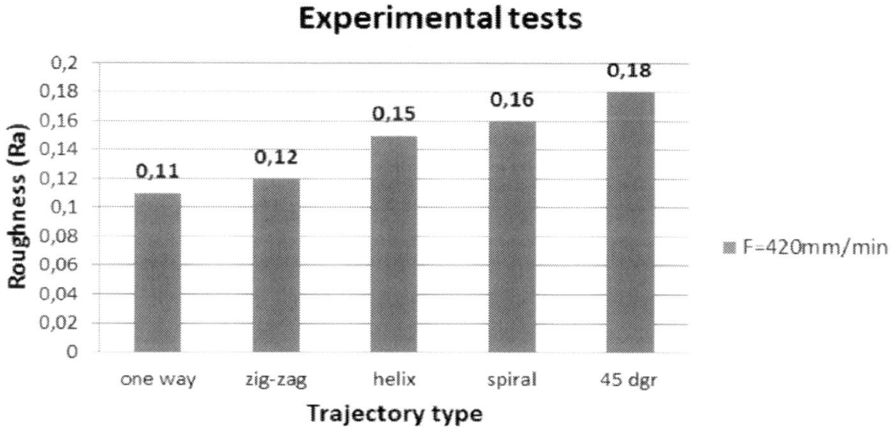

Figure 3. Average roughness(R_a^t) according with trajectory type.

$$R_a^{ref} = \min(R_a^t) \quad \text{where t=1..5} \tag{5}$$

t =tool trajectories with the fallow value: 1-one way; 2 -zig-zag; 3- spiral; 4-helix; 5 – 45 dgr inclined.

5. Experimental studies and calculus

As it can be seen from table 3, the value for roughness differs from one toolpath to another. Based on this data was calculated a correction coefficient that takes into account the tool trajectory. The coefficient was calculated as a rapport between the surface roughness for each trajectory type and the reference. The correction factor was established for each tool trajectories based on formula 6 and the results are presented in table 3.

$$C_{Ra}^t = \frac{R_a^t}{R_a^{ref}} \tag{6}$$

Where t = 1..5 -tool trajectories with the fallow value: 1-one way; 2 -zig-zag; 3- spiral; 4-helix;5 - 45 dgr inclined

R_a^t –reference roughness for each trajectories type;

R_a^{ref} –constant reference value

Table 3. Correction coefficient.

Tool trajectoy	One way	Zig-zag	Helix	Spiral	45 dgr
Correction factor	1	1.15	1.37	1.41	1.60

A formula for the feed rate calculus was established by taking into account the correction factor (equation (7))

$$F = \frac{f_z n z}{c_{Ra}^t} \tag{7}$$

6. Test and validation

In order to validate the correction coefficient another series of tests were made. The validation test was made for maximum feed per tooth supported by the tool (fz = 0.11mm/tooth and 0.3mm cutting depth).

In first step the cutting regime was calculated according with classic formula (equations 1, 2) and in the second step the calculus was made by using correction factor according with value from (table 4).

For each test the average roughness and cutting time was measured on the machining center. The results are presented in table 4 and figure 4 and 5.

Table 4. Validation tests.

	Tool trajectory	One way	Zig-zag	Helix	Spiral	45 dgr
Clasic	Feed rate [mm/min]			1155		
$F = f_z n z$	Cutting time [min]	11	5	5	3	2
	Roughness Ra [µm]	0.30	0.36	0.40	0.50	0.41
Corrected	Feed rate [mm/min]	1155	1004	843	820	722
$F = \frac{f_z n z}{c_{Ra}^t}$	Cutting time [min]	11	6	7	4	3
	Roughness Ra [µm]	0.30	0.31	0.25	0.23	0.22

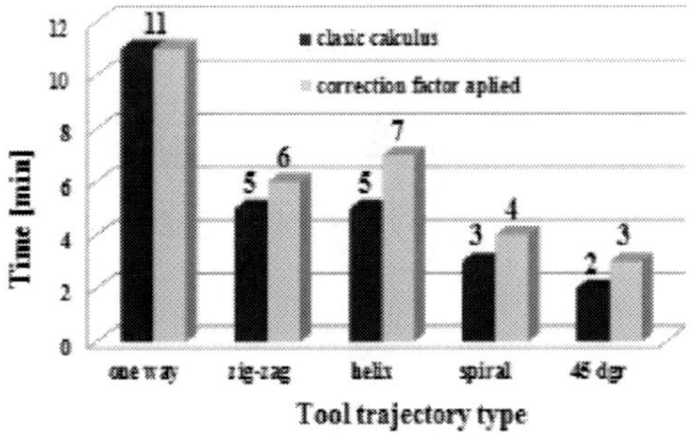

Figure 4. Time variation.

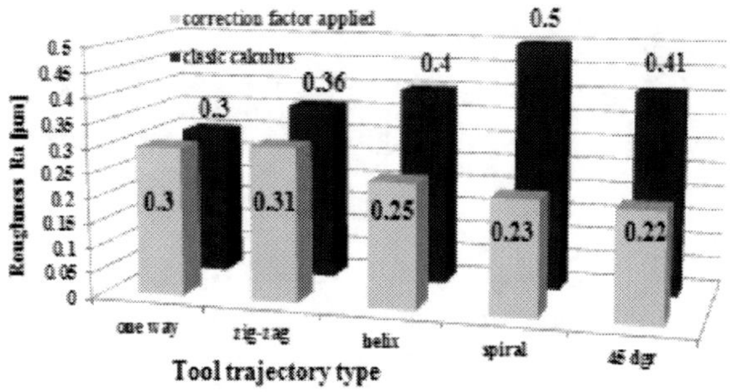

Figure 5. Validation results.

If it is applied this coefficient on the standard formula for calculating the feed rate it can be observed that the feed rate is reduced depending on the cutting tool trajectory in order to obtain similar surface quality for all trajectories.(see equation 7)

It is well known that reducing the feed rates while keeping the spindle speed up lightens the chip load and leads to a nicer surface finish. But lighten the feed rate too much, the tools will start to rub, and tool life will go way down due to the excess heat generated by the rubbing.

In order to evaluate the productivity, during the machining the time was monitored to evaluate the efficiency of applying the coefficient that reduce the feed. As it can be see from table 4 the machining time does not increase significantly although the feed rate is reduced. So we can say that from the point of view of productivity the process is not too much affected, but the surface quality is improved and so the use of this approach is useful.

7. Conclusions

This is the first step for developing an easy to use, fast calculus software for the milling cutting parameters. The difference from the other similar software is that we take into account (besides a number of other factors experimentally or analytically determined) the type of cutting tool trajectories.

Machining parameters are typically adjusted according to the instructions in the tools catalogues and/or handbooks without regard to the roughness requirements and geometrical tolerances of the surface to be machined. Incorrect adjustment of the machining parameters, feed rate and depth of cut lead to tool deflection and consequently reduced surface quality. With increasing feed rate and depth of cut, the tool deflection is increased.

The theory says that the best surface quality is obtained when using climb milling, but we do not always have the possibility to machine in such a manner. Using the method that we propose, to use a coefficient that reduce the feed rate in order to improve the surface quality based on the tool trajectory, but not to affect the productivity, is the right way to approach this problem.

Acknowledgement
The work has been funded by the Sectoral Operational Programme Human Resources Development 2007-2013 of the Ministry of European Funds through the Financial Agreement POSDRU/159/1.5/S/132395.

References
[1] Pena A E, Anania F D and Zapciu M 2014 Research regarding software developing for machining parameters optimization calculus *Applied Mechanics and Materials* **656** pp 192-199
[2] Shajari S, Sadeghi M H and Hassanpour H 2014 The Influence of Tool Path Strategies on Cutting Force and Surface Texture during Ball End Milling of Low Curvature Convex Surfaces *Sci. World J.* **2014**
[3] Toh C K 2004 Surface topography analysis in high speed finish milling inclined hardened steel *Prec. Eng.* **28** pp 386–398
[4] Anania D, Pena A and Mohora C 2014 Implementation of high cutting speed technologies in machine tools for increasing product quality *Proc. Manuf. Systems* Vol. 9 **1** pp 41-46
[5] Mendes de Aguiar M, Diniz A E and Pederiva R 2013 Correlating surface roughness, tool wear and tool vibration in the milling process of hardened steel using long slender tools *Int. J. Machine Tools & Manuf.* **68** pp 1–10
[6] Kuttolamadom M A, Hamzehlouia S and Mears M L 2010 Effect of machining feed on surface roughness in cutting 6061 Aluminum SAE *Int. J. Mater. Manuf.* **3(1)** pp 108-119

IOP Conf. Series: Materials Science and Engineering **95** (2015) 012021 doi:10.1088/1757-899X/95/1/012021

Tool life and cutting speed for the maximum productivity at the drilling of the stainless steel X22CrMoV12-1

A Vlase[1], O Blăjină[1], M Iacob[1] and V Darie[2]
[1]University Politehnica of Bucharest, Production Engineering Department, 313 Splaiul Independenţei, RO-060032, Bucharest, Romania
[2]ees Energyneers Romania S.R.L., 41 Frumoasa Street, RO-010986, Bucharest, Romania
E-mail: aurelvlase@yahoo.com

Abstract. Two addressed issues in the research regarding the cutting machinability, establishing of the optimum cutting processing conditions and the optimum cutting regime, do not yet have sufficient data for solving. For this reason, in the paper it is proposed the optimization of the tool life and the cutting speed at the drilling of a certain stainless steel in terms of the maximum productivity. For this purpose, a nonlinear programming mathematical model to maximize the productivity at the drilling of the steel is developed in the paper. The optimum cutting tool life and the associated cutting tool speed are obtained by solving the numerical mathematical model. Using this proposed model allows increasing the accuracy in the prediction of the productivity for the drilling of a certain stainless steel and getting the optimum tool life and the optimum cutting speed for the maximum productivity. The results presented in this paper can be used in the production activity, in order to increase the productivity of the stainless steels machining. Also new research directions for the specialists in this interested field may come off from this paper.

1. Introduction

The stainless steels are used more and more in various key domains of the technique [1]. The processing of these steels is determined by their specific physical-chemical characteristics and by their technological properties [2, 3]. The research in the last decade regarding the cutting machinability of the stainless steels have highlighted the insufficiency of the data for establishing of the optimum cutting processing conditions and the optimum cutting regime [4, 5]. Therefore the purpose of this paper is the optimization of the tool life and the cutting speed at the drilling of a stainless steel, in terms of the global indicator of the maximum productivity.

For maximize the productivity at the drilling of the steel, a nonlinear programming mathematical model is developed in this paper. The cutting tool life is the optimum solution of this model. The cutting tool speed is obtained on the basis of this value of the tool life. The numerical model, developed for the analyzed case study of the stainless steel X15CrNiSi20-12 (DIN 17243, EN 95), is resolved using the specialized software *WinQSB*. The variation (numerical and graphical) of the productivity depending on the cutting tool life is also presented in this paper, valid only for the studied steel. Thus the results of the analytical solving of the model are confirmed.

Content from this work may be used under the terms of the Creative Commons Attribution 3.0 licence. Any further distribution of this work must maintain attribution to the author(s) and the title of the work, journal citation and DOI.
Published under licence by IOP Publishing Ltd

2. Determination of the productivity of the drilling operation

The time technical norm N_t of a machining operation in a metal cutting process is presented in the specialized technical literature [6] as the function:

$$N_t = t_b + t_{aux} + t_{dt} + t_{do} + t_{on} + \frac{t_{pi}}{n_l} \text{ [min/piece]} \tag{1}$$

where: N_t is time norm on piece (or operation or phase), [min/piece]; t_b – machining time, [min/piece]; t_{aux} – auxiliary time, [min/piece]; t_{dt} – technical maintenance time, [min/ piece]; t_{do} – organizational maintenance time, [min/ piece]; t_{on} – rest and natural necessities time, [min/ piece]; t_{pi} – preparation-closing time on the lot, [min]; n_l – batch of parts that are continuously processed, to the same machine tool.

The machining time t_b for the drilling operation can be written as:

$$t_b = \frac{L}{n \cdot f} \cdot i \text{ [min]} \tag{2}$$

where: L is the length of a hole (including the engagement and exceeding of the drill), [mm]; i – number of holes; f – the cutting feed, [mm/rot].

The rotational speed n is given by the relation:

$$n = \frac{1000 \cdot v}{\pi \cdot D} \text{ [rot/min]} \tag{3}$$

where: D is the diameter of the hole, [mm]; v is the cutting speed, [m/min], given by the Taylor's relation at drilling:

$$v = \frac{C_v \cdot D^{x_v}}{T^m \cdot f^{y_v}} \text{ [m/min]} \tag{4}$$

where: C_v is a constant determined experimentally, according to the couple workpiece material-tool and the cutting conditions; T – cutting tool life, [min]; m – durability exponent of the spiral drill; x_v, y_v – polytropic exponents.

Substituting for v from (4) in (3), and then for n from (3) in (2), the expression of the machining time t_b becomes:

$$t_b = \frac{L \cdot T^m \cdot D^{1-x_v} \cdot i}{318 C_v \cdot f^{1-y_v}} \text{ [min]} \tag{5}$$

The auxiliary time t_{aux} is chosen from normative tables.

The technical maintenance time t_{dt} is calculated by the relation [6]:

$$t_{dt} = \frac{(t_s + t_a) \cdot t_b}{T} \text{ [min/piece]} \tag{6}$$

where: t_s is the consumed time for the tool changing and adjustment remaking, [min]; t_a – the consumed time for the drill sharpening, [min].

The organizational maintenance time t_{do} is given by the relation [6]:

$$t_{on} = \frac{k_2}{100}(t_b + t_{aux}) \text{ [min/piece]} \tag{8}$$

where k_2 is a constant chosen from normative tables.

The obtained expression of the time technical norm for drilling operation is the following:

$$N_t = \frac{L \cdot T^m \cdot D^{1-x_v} \cdot i}{318\, C_v \cdot f^{1-y_v}} \left(1 + \frac{k_1 + k_2}{100}\right) + \frac{L \cdot T^{m-1} \cdot D^{1-x_v} \cdot i}{318\, C_v \cdot f^{1-y_v}} (t_s + t_a) + t_{aux} \left(1 + \frac{k_1 + k_2}{100}\right) + \frac{t_{pi}}{n_l} \tag{9}$$

Finally, the relation of the productivity for the drilling operation can be written as:

$$t_{on} = \frac{k_2}{100}(t_b + t_{aux}) \text{ [min/piece]} \tag{10}$$

3. Mathematical Model to Maximize the Productivity of the Drilling Operation

The optimization mathematical model contains the optimization objective function and several restrictive relations:

$$\begin{cases} \max P & (11) \\ f \le C_f \cdot D^{0,6} \cdot k_s & (12) \\ f^{y_M} \cdot n \cdot v^{z_M} \le \dfrac{9740\eta \cdot P_m}{C_M \cdot D^{x_M} \cdot c} & (13) \\ f^{y_F} \cdot v^{z_F} \le \dfrac{F_{ma}}{D^{x_F} \cdot C_F \cdot c_m} & (14) \\ f^{y_F} \cdot v^{z_F} \le \dfrac{2{,}465 E \cdot I_{\min}}{C_F \cdot D^{x_F} \cdot l_c^2 \cdot c_f} & (15) \\ f_{\min} \le f \le f_{\max} & (16) \\ n_{\min} \le n \le n_{\max} & (17) \end{cases}$$

The objective function (11) of the model is the productivity of the processing operation, P, given by the relation (10), which must be maximized.

The restrictive relation (12) of the cutting feed includes: C_f – a constant which depends on processed material and on precision machining; k_s – a correction coefficient depending on the ratio l/D, where l is the length of the hole, [mm]; D is the hole diameter, [mm].

The restrictive relation (13) of the power consumption of the machining process includes: CM – a constant; x_M, y_M, z_M – polytropic coefficients; c – safety coefficient, c =1.7 in [6].

The restrictive relation (14) of the advance mechanism of the drilling machine includes: F_{ma} – the maximum allowed force of the advance mechanism; C_F – a constant; x_F, y_F, z_F – polytropic coefficients; c_m – safety coefficient, c_m = 1.6 in [6].

The restrictive relation (15) of the buckling resistance of the spiral drill includes: E – modulus of elasticity, [N/mm^2]; I_{min} – minimum moment of inertia, [mm^4]; lc – initial length in console of the spiral drill, [mm]; cf – safety coefficient to buckling, cf = 1.8 in [6].

The restrictive relations (16) and (17) of the drilling machine kinematics require that the two parameters f and n have values in the feed range and, respectively, the rotation range, developed by the drilling machine.

4. Case study for the drilling of the stainless steel X15CrNiSi20-12

The drilling operation of the studied stainless steel X15CrNiSi20-12 (DIN 17243, EN 95) was performed using a machine tool GC0 32 DM3 drilling device and Rp5 high-speed steel spiral drills.

The constants $C_f = 0.031$ and $k_s = 0.9$ were chosen from the specialized literature (table 6.10 and table 6.11 in [6]). The values of the constants and polytropic coefficients for the drilling of the steel X15CrNiSi20-12 were determined based on the experimental measurements: C_F = 2316, x_F = 1.06, y_F = 0.4, z_F = -0.42 in [7]; C_M = 0.72, x_M = 1.84, y_M = 0.34, z_M = -0.35 in [8]; Cv = 4.227, x_v = 0.39, y_v = 0.272, m_v = 0.164 in [9].

The other values of the numeric model are: L = 50 mm; i = 1; D = 16 mm; ts = 0.5 min; ta = 2 min; k1 = 1, k2 = 4 in [6]; taux = 1.48 min; tpi = 15 min; n = 246 rot/min; η = 0.8; Pm = 3.15 kW; Fma = 960 daN; $I_{\min} = 0.043 \cdot D^4 = 2818\,\text{mm}^4$; $E = 2.1 \cdot 10^4\,\text{daN/mm}^2$; lc = 120 mm.

From the relations (12)-(16) it is obtained $f \le 0.24\,\text{mm/rot}$. Because the used drilling machine ensures the feeds: 0.12; 0.20; 0.32; 0.50 mm/rot, it results that f = 0.20 mm/rot.

Finally, the resulted numerical optimization model is a nonlinear programming model without restrictions:

$$\begin{cases} [\max]P = \dfrac{60}{0.684 \cdot T^{0.164} + 1.628 \cdot T^{-0.836} + 1.704} \\ T \ge 0 \end{cases} \tag{18}$$

The objective function P of the above model (18) is a nonlinear function, pozinomial type, of a real variable, T. The model is resolved using the module *Nonlinear Programming* of the specialized software *WinQSB* [10].

For this purpose, first the command *New Problem*, from the main menu *File*, opens the dialog box *NLP Problem Specification*, which is filled as shown in figure 1.

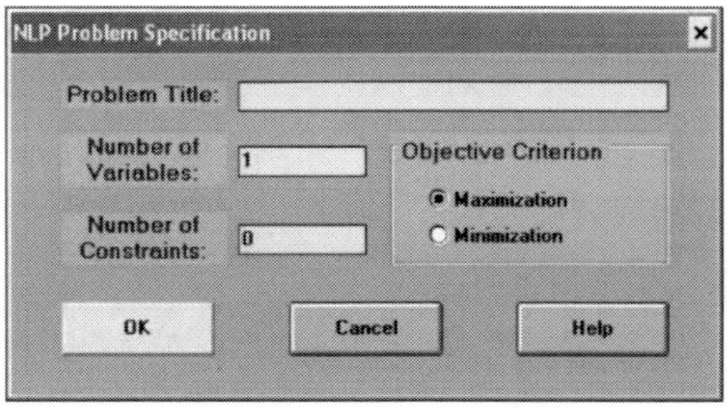

Figure 1. The dialog box NLP Problem Specification.

A click on the *OK* button opens the workspace window *NLP Problem*. The numerical model data are inserted in a table, as shown in figure 2. The command *Solve the problem*, from the main menu *Solve and Analyze*, returns the table *Solution summary* (figure 3), which contains: the optimum value of the tool life, $T_{op} \cong 12$ min; the corresponding maximum value of the drilling productivity, $P_{max} \cong 20$ pieces/h.

NLP Problem

Maximize	60/(0.684*T^0.164+1.628*T^(-0.836)+1.704)
	OBJ / Constraint / Variable Bound
Maximize	60/(0.684*T^0.164+1.628*T^(-0.836)+1.704)
T	>=0, <=M

Figure 2. Numerical model data.

02-05-2015	Decision Variable	Solution Value
1	T	12.1108
Maximized	Objective Function =	20.4358

Figure 3. Solution summary.

The command *Objective Function Analysis*, from the main menu *Solve And Analyze*, returns the tabular analyze (figure 4) and the graphical analyze (figure 5) of the machining productivity *P* depending on the tool life *T*, valid only for the studied steel X15CrNiSi20-12.

Nonlinear Programming

Objective Analysis for NLP Problem -- T

02-05-2015	T	Objective Function
1	6.1100	20.1146
2	7.1100	20.2470
3	8.1100	20.3315
4	9.1100	20.3844
5	10.1100	20.4154
6	11.1100	20.4311
7	12.1100	20.4358
8	13.1100	20.4323
9	14.1100	20.4228
10	15.1100	20.4088
11	16.1100	20.3915
12	17.1100	20.3715
13	18.1100	20.3496

Figure 4. Tabular analysis of the machining productivity depending on the tool life.

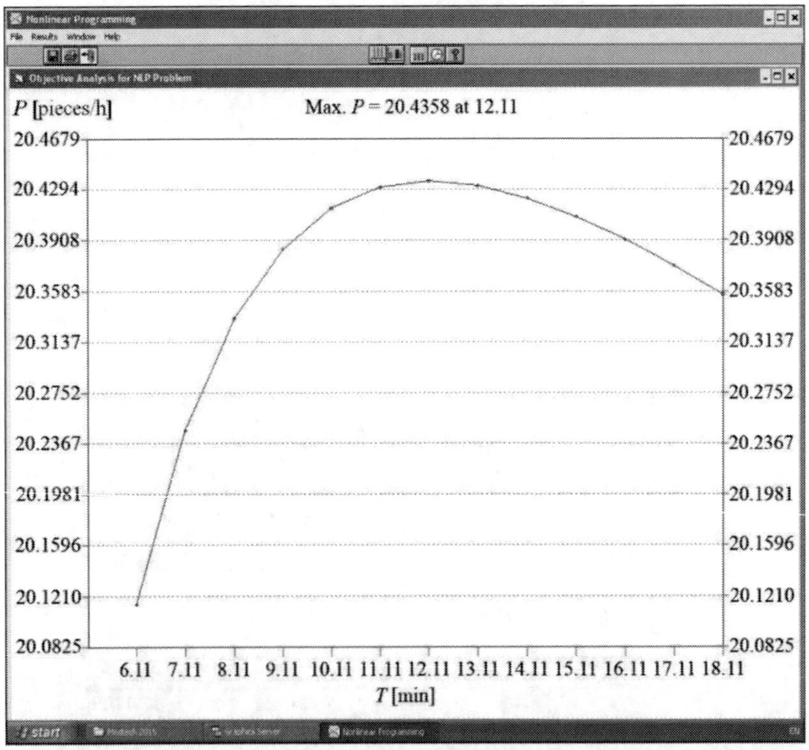

Figure 5. Graphical analysis of the machining productivity depending on the tool life.

The optimum cutting speed, v_{op}, is calculated based on the optimum value of the tool life T_{op}:

$$v_{op} = \frac{C_v \cdot D^{x_v}}{T_{op}^m \cdot f^{y_v}} = \frac{4.227 \cdot 16^{0.39}}{12.11^{0.164} \cdot 0.20^{0.272}} = 12.83 \text{ m/min} \tag{19}$$

The optimum rotational speed, n_{op}, corresponding to the optimum cutting speed, v_{op}, is determined with the relation (3):

$$v_{op} = \frac{C_v \cdot D^{x_v}}{T_{op}^m \cdot f^{y_v}} = \frac{4.227 \cdot 16^{0.39}}{12.11^{0.164} \cdot 0.20^{0.272}} = 12.83 \text{ m/min} \tag{20}$$

5. Conclusions

The maximum productivity of a machining operation is ensured by the optimum cutting tool life, which is derived from the proposed mathematical model in this paper. Using this model allows increasing the accuracy in the prediction of the productivity for the drilling of a certain stainless steel and getting the optimum tool life and the associated optimum cutting speed. The maximum productivity can represent a global indicator for appreciation of the machinability.

The results presented in this paper can be used in the production activity, in order to increase the productivity of the stainless steels machining. Also new research directions for the specialists in this interested field may come off from this paper.

References

[1] Trent E M and Wright P K 2000 *Metal cutting* 4th Edition (Boston: Butterworth Heinemann)
[2] Schnadt R 1991 Austenitischen Stahlfräsen ZWF CIM Vol.86 **6** (München: Verlag) pp 312-315.
[3] Krauss G 2005 *Steels. Processing, Structure and Performance* (Ohio: ASM International, Materials Park)
[4] Cimino T, Schade C and Causton B 1999 High Performance Stainless Steels *Proc. PM2TEC'99* (Vancouver: British, Columbia)
[5] Davis J R 2000 *ASM Specialty Handbook: Stainless Steels* (Ohio: ASM International, Materials Park
[6] Vlase A et al. 2007 Processing Technologies on Drilling Machines (Bucharest: BREN)
[7] Blăjină O, Darie V, Vlase A and Iacob M 2014 Research regarding the relationship of the axial cutting force at drilling of the stainless steel X15CrNiSi20-12 *Proc. Int. Conf. "Innovative Manufacturing Engineering" IManE 2014 Chisinau* (in Applied Mechanics and Materials Vol. 657) pp 3-7.
[8] Vlase A, Blăjină O and Iacob M 2014 Calculus of the relationship of the cutting moment at drilling of the stainless steel X15CrNiSi20-12 *Proc. 7-th Int. Conf. on Adv. Manuf.Tech. ICAMaT 2014*
[9] Vlase A, M. Iacob , Blăjină O and Darie V 2014 Analysis of the relationship of the cutting tool wear at drilling of the stainless steel X15CrNiSi20-12 Proc. *Proc. Int. Conf. "Innovative Manufacturing Engineering" IManE 2014 Chisinau* (in Applied Mechanics and Materials Vol. 657) pp 8-12.
[10] Blăjină O 2011 *Optimum decisions in management with WinQSB 2.0* (Cluj-Napoca: Albastră-Microin-formatica)

The relation between the shop floor energy system and the manufacturing process optimality

N Badea[1], G Frumuşanu[2] and A Epureanu[2]
[1]"Dunărea de Jos" University, Automatic Control & Electrical Engineering Department, Ştiinţei Street 2, 800210 - Galaţi, Romania
[2]"Dunărea de Jos" University, Manufacturing Engineering Department, Domnească Street 111, 800201 - Galaţi, Romania

E-mail: gabriel.frumusanu@ugal.ro

Abstract. Nowadays, the overhead expenses are a significant component of the manufacturing cost. The expenditures occasioned by running the shop floor energy system are one of the most important among the overhead expenses. However, the manufacturing process optimization actually does not take into account these expenses. More than that, the expenses for running the shop floor energy system are not constant, varying after the external environment seasonal conditions, and they are depending on several issues, like the building's energetic exchange with its environment, or the loading degree of the machine tools from the shop floor. This paper approaches the relation between the shop floor energy system expenditures, on one hand, and the manufacturing process optimality, on the other hand. A case study, revealing the influence of the shop floor energy system on the manufacturing cost-optimal solution is also included.

1. Introduction

In today's manufacturing environment, all participants have to fight in order to meet the ever-changing competitive market requirements. To reduce the manufacturing cost, to increase the productivity and to enhance the manufactured products quality, it is highly important to work in optimal conditions. For this reasons, a very large number of researches has already been dedicated to solve the problem of optimizing different types of manufacturing processes. The practical problems' diversity issued many kinds of approaches. The differences between these approaches are mainly regarding the optimization target (the objective function definition), the manipulated variables choice, or the method used for solving the optimization problem.

The manipulated variables are, most frequently, the cutting regime parameters [1, 2, 3], but also the number of passes [4], the driving motor power [5], or the grit size (of abrasive tools) [6]. Despite the fact that the overhead expenses are a significant component of the manufacturing cost, and the expenditures occasioned by running the shop floor energy system are one of their most important elements, we did not find anywhere an approach considering them, when the manufacturing cost is the objective function of the performed optimization.

It should be noticed that, in a country like Romania, where the difference between the monthly average temperatures from January and July is close to 30 °C [7], the expenses for the shop floor climatisation are very significant, with a high variability during the year. Therefore, it easy to presume that they should have an important impact on the manufacturing cost, if considered, also inducing a seasonal fluctuation to it.

Content from this work may be used under the terms of the Creative Commons Attribution 3.0 licence. Any further distribution of this work must maintain attribution to the author(s) and the title of the work, journal citation and DOI.
Published under licence by IOP Publishing Ltd

Regarding the paper structure, the next section is dedicated to problem formulation. The third part suggests a new expression for the objective-function used for performing the manufacturing cost optimization, which takes into account the expenses for shop floor climatisation. The fourth section includes a case study, in order to prove the relevance of the new approach, while the last part of the paper presents the main conclusions.

2. Problem formulation

As it can be noticed in the dedicated literature, the manufacturing process optimization is performed, in practice, by selecting, from the beginning, the machine tool, the cutting tool and the other assets required to perform the considered process. Only the technical aspects are considered in this purpose (e.g. the needed power, the dimensional range etc.), while the subsequent optimization does not extend over this category of elements involved in manufacturing.

The manufacturing process optimization regards, mainly, the cutting regime parameters (the cutting speed, v, the feed rate, s, and the cutting depth, t), as independent variables of the optimization problem. In fact, t is not, effectively, an independent variable, because its value depends on the thickness of the material layer to be detached, and on the dimensional corrections that have to be done in order to achieve the dimensional target. Regarding s, it is matter of course that its maximum possible value is optimal, but this is limited by the required roughness of the machined surface. The only one from the cutting regime parameters, which may be freely varied, is the cutting speed (although even in its case, the instability risk gives restrictions to be considered).

In this paper, we suggest that, if the manufacturing cost is the optimization criterion, then the objective-function should have a new expression, which also consider the expenses of the shop floor energy system. Therefore, some new parameters will become independent variables to be considered for finding the cost-optimal solution of the manufacturing process – for example, the building energy class, the loading degree (in time and space) of the machine tools, or the ratio between the used versus installed power of the machines driving motors.

In other words, the paper targets to reformulate the manufacturing process optimization problem, by analytically emphasizing the overhead expenses for the shop floor climatisation in the expression of the objective-function (meaning the manufacturing cost) and, consequently, by considering new independent variables, relevant for the energy system expenditures.

The paper also aims to assess the relevance of the new approach. Hence, by running a case study, we have determined the impact on the manufacturing optimization of the expenses issued by the shop floor energy system. More precisely, we have studied the variation of the manufacturing minimal cost depending on the exterior environmental conditions and on the new, above-mentioned independent variables.

3. New expression of the objective-function

3.1. Current form of the manufacturing specific cost

By synthesizing the up to date approaches regarding the manufacturing cost calculation, in specific form (referred to the volume of detached material), we may accept, in principle, the following relation:

$$C_s = \sum_{i=i}^{n}\left(\frac{Q_i}{T_i}\right)\cdot\left[1+k+\frac{\tau_{sr}}{T}\right]\cdot\frac{1}{v\cdot s\cdot t}+\frac{\tau_{sr}\cdot c_\tau + c_s}{T\cdot v\cdot s\cdot t}+\frac{c_\tau}{v\cdot s\cdot t}+c_{mater}+c_{se}\cdot c_{energy} \quad (1)$$

In relation (1), the first term gives the specific cost fraction issued by the use of the needed assets. Here Q_i means the value of the ith asset, from the n needed to run a considered process, T_i – the "i^{th}" asset life cycle, k – the ratio between the auxiliary and machining times involved by the process, τ_{sr} – the time for worn tool changing, T – the tool durability, and v, s, t – as we already mentioned, the cutting regime parameters.

The second term reflects the share of the cutting tool cost. Besides the already specified notations, we have c_τ – the wage specific cost (referred to time) and c_s – the tool expenditure between two consecutive tool changes.

The third term refers to the wage cost, the fourth is the specific cost of the detached material (referred to its volume), while the fifth gives the consumed energy cost. Here c_{se} means the specific energy consumption (referred to the volume of detached material) and c_{energy} – the energy price.

3.2. Shop floor microclimate
Running a manufacturing activity requires ensuring a living microclimate inside the space where this activity takes place, no matter of the exterior conditions. We are referring, mainly, to the temperature fluctuations, which in many cases are very significant. This task is covered by the shop floor energy system and it produces costs that are, usually, included in the overhead expenses category.

There are statistic indicators, for example the specific annual consumption, which are giving the annual energy consumption occasioned by covering the various functions of the residence according to the total floor area. This indictor might be also assimilated in the shop floor case. The annual consumption results by multiplying the total floor area with the value of the specific annual consumption. Depending on how global the estimation is, one may distinguish between several cases, which differ in that the evaluation is carried out or not for each function of the residence, separately and in that the evaluation is carried out or not for each form of energy, distinctly.

We should also notice that specific annual consumption of energy for climatisation purpose is highly depending on the building energy class – see table 1, [8].

Table 1. Annual specific consumption, depending on the buildings energy classes.

Energetic class	A	B	C	D	E	F	G
Heating annual specific consumption [kWh/m^2 year]	70	117	173	245	343	500	>500
Cooling annual specific consumption [kWh/m^2 year]	20	50	87	134	198	300	>300

3.3. Specific cost formula when analytically emphasizing the overhead expenses for the shop floor climatisation
We further suggest a modified form of the relation (1) in order to define a new objective-function, which might be used for optimizing the manufacturing process, if considering the expenses for running the shop floor climatisation system.

First of all we expressed the specific cost for climatization as:

$$C_{scl} = \frac{A(1 + k + \tau_{sr}/T)}{n_{mt} \cdot t_c \cdot k_d} \cdot \frac{c_{energy}}{v \cdot s \cdot t} \tag{2}$$

In relation (2), A means the monthly amount of energy needed for heating / cooling (depending on the season and on the energy class of the building hosting the shop floor), n_{mt} – the number of machine tools from the shop floor being in duty, t_c – their scheduled functioning time, in a month (depending on the number of working days and on the shifts number), and k_d – a coefficient reflecting the ratio between the machining effective time and t_c.

Then, we should notice that in the months when heating is needed, the energy consumed in the manufacturing process, finally dissipated as heat, diminishes the amount of energy required for heating the shop floor. At the same time, in the months when heating is no needed, the heat resulted from the manufacturing process becomes an additional load for the shop floor cooling system.

Finally, because the fourth term from the specific cost formula is a constant, while the first has a low variability, we pulled them out from the definition relation of the objective-function, in order to give it the maximum possible relevance.

In conclusion, we are suggesting the following forms of the objective-function, in order to be used for optimizing the manufacturing process, relative to the specific cost criterion:

$$C_s = \frac{\tau_{sr} \cdot c_\tau + c_s}{T \cdot v \cdot s \cdot t} + \frac{c_\tau}{v \cdot s \cdot t} + \frac{A(1 + k + \tau_{sr}/T)}{n_{mt} \cdot t_c \cdot k_d} \cdot \frac{c_{energy}}{v \cdot s \cdot t} \qquad (3')$$

in the months when the shop floor energy balance shows that heating is needed, while

$$C_s = \frac{\tau_{sr} \cdot c_\tau + c_s}{T \cdot v \cdot s \cdot t} + \frac{c_\tau}{v \cdot s \cdot t} + \frac{A(1 + k + \tau_{sr}/T)}{n_{mt} \cdot t_c \cdot k_d} \cdot \frac{c_{energy}}{v \cdot s \cdot t} + 2 \cdot c_{se} \cdot c_{energy} \qquad (3'')$$

in the months when shop floor cooling is required.

We must also specify that relations (2), (3') and (3") are not very specific and are not expected to produce precise results if applied in practice. In reality, in a shop floor there are different types of machine tools, their working times are not uniform etc. However, these formulas enable to perform a relevant case study, while if its results will be considered as significant they could be further customized for solving concrete optimization problems.

4. Case study

We further present a case study regarding manufacturing optimization, if the new objective-function is considered. It was realized by performing a numerical simulation for a hypothetical shop floor, with the help of some original applications developed in MatLab.

The targets of the case study were:

- To reveal the impact of the expenses introduced by the shop floor energy system onto the manufacturing specific cost.
- To assess the influence of the external temperature seasonal fluctuation on the solution of the manufacturing optimization problem.
- To investigate the relevance of the newly considered independent variables (the building energy class, the k_d coefficient, the percentage of using the machine tool installed power).
- To find the sensitivity of the objective-function to some of the most important independent variables (feed rate, cutting depth, degree of effectively using the scheduled functioning time).

The hypothetical shop floor considered has nmt = 25 turning machines, working in a single shift of 8 hours, meaning tc = 14400 minutes in a month, with kd = 0.5. Shop floor surface is of 432 m^2.

The monthly consumption of energy for the shop floor climatisation was calculated on the base of the specific consumption of energy, statistically measured in Galaţi country case [8]. The results, depending on the building energy class, are presented in table 2.

Table 2. The energy for shop floor climatisation, A, depending on building energy class.

	Monthly energy consumption [kWh]					
	Heating					
	A	B	C	D	E	F
Jan	6345	10605	15681	22208	31091	45321
Feb	5410	9043	13371	18936	26511	38646
Mar	4185	6995	10343	14648	20507	29893
Apr	2150	3593	5313	7524	10533	15354
May	415	694	1027	1454	2035	2967
Jun	0	0	0	0	0	0
Jul	0	0	0	0	0	0
Aug	0	0	0	0	0	0
Sep	62	104	154	218	305	445
Oct	2025	3385	5005	7088	9923	14464
Nov	4050	6769	10009	14175	19845	28929
Dec	5597	9356	13833	19591	27427	39981

	Cooling					
Jan	0	0	0	0	0	0
Feb	0	0	0	0	0	0
Mar	0	0	0	0	0	0
Apr	183	459	798	1229	1816	2752
May	1156	2891	5031	7748	11449	17347
Jun	1668	4170	7256	11175	16513	25019
Jul	2018	5046	8779	13522	19980	30273
Aug	2018	5046	8779	13522	19980	30273
Sep	1301	3253	5659	8717	12880	19515
Oct	295	737	1282	1974	2917	4420
Nov	0	0	0	0	0	0
Dec	0	0	0	0	0	0

The specific consumption of energy for running the generic manufacturing process has been determined by using the Curve Fitting Tool from MatLab soft, applied to identify the curve passing by a number of points determined by experimental measurements in the case of a turning process.

In figure 1 it is presented the manufacturing specific cost variation depending on the cutting speed, when the shop floor is of *C* energy class building.

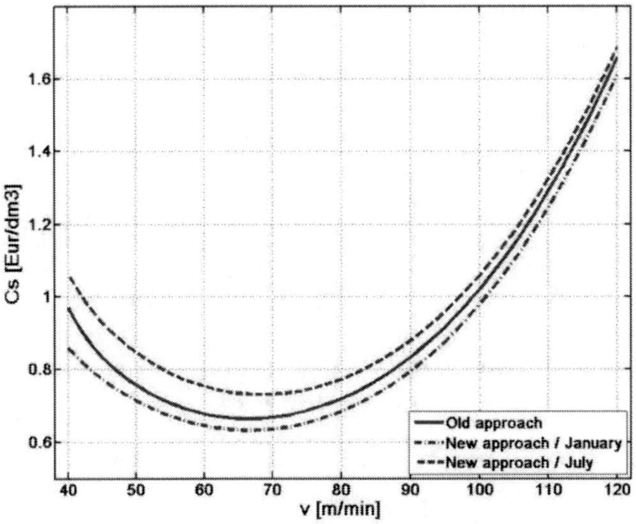

Figure 1. The new approach impact on the specific cost.

The results were obtained by running a MatLab application that uses (3') and (3") formulas, where all the parameters, excepting *v*, were considered constant. The input values were: $s = 0.2$ mm/rot; $t = 3$ mm; $\tau_{sr} = 1$ min; $c_\tau = 0.15$ Euro/min; $c_m = 20$ Euro/dm^3; $k = 1$; $k_d = 0.5$; $c_{energy} = 0.15$ Euro/kWh. The tool durability was calculated, with the well-known Taylor formula, depending on the cutting regime parameters.

The chart from figure 2 presents the minimum manufacturing specific cost and its corresponding cutting speed v_{ec}, on each month of the year, according to the new approach, when the shop floor is of *D* energy class building. The values were determined by running the same application of above, with the same input constants.

The chart from figure 3 illustrates the influence of the shop floor energy class on the minimum specific cost, in the extreme moths (regarding climatisation need), January and July, according to the new approach. The chart from figure 4 shows the influence on the minimum manufacturing specific cost and its corresponding cutting speed of the degree of using the machine-tool installed power (reflected by the cutting regime intensity), in January. By low cutting regime intensity we meant s = 0.1 mm/rot and t = 0.5 mm, by medium – s = 0.2 mm/rot and t = 3 mm, and by high – s = 0.3 mm/rot and t = 7 mm. The chart from figure 5 presents the influence, on the same variables, of the degree of effectively using the scheduled functioning time (k_d coefficient), also in January.

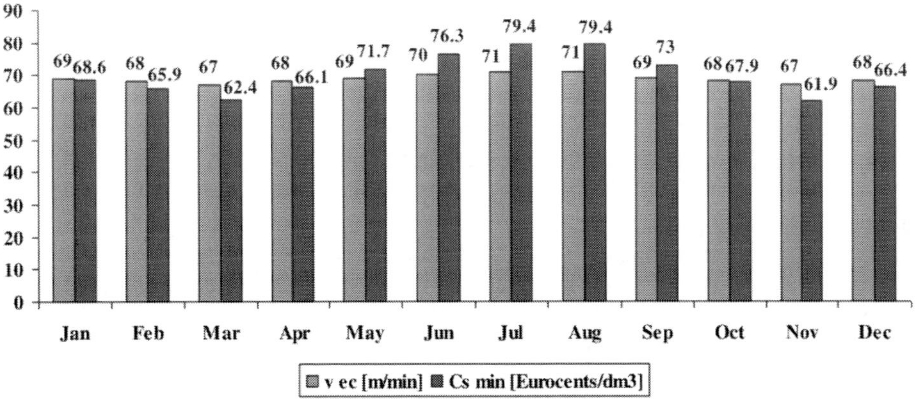

Figure 2. External temperature influence on the solution of the manufacturing optimization problem.

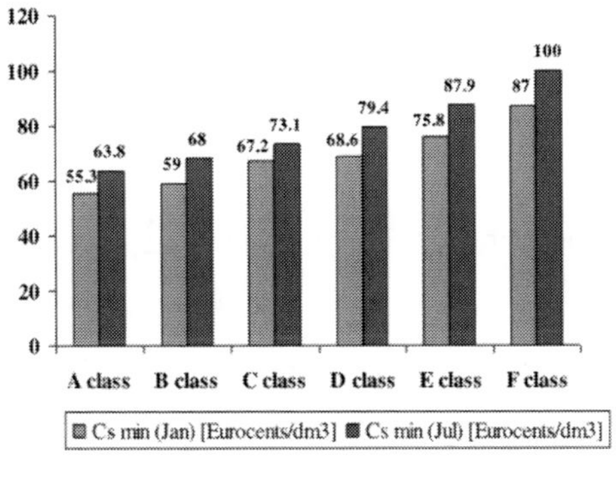

Figure 3. The building energy class influence on the minimum specific cost.

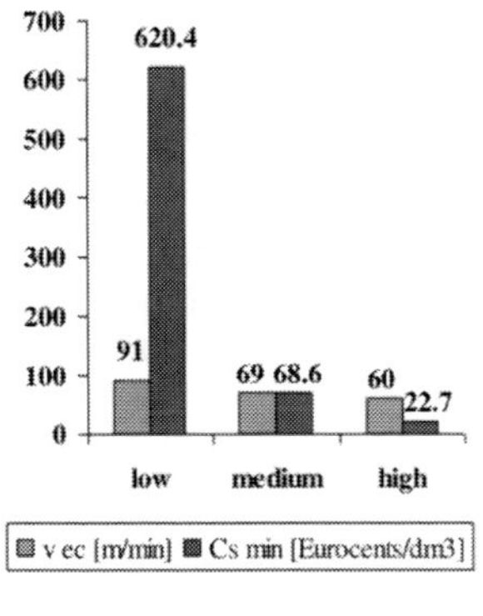

Figure 4. The used/installed power influence

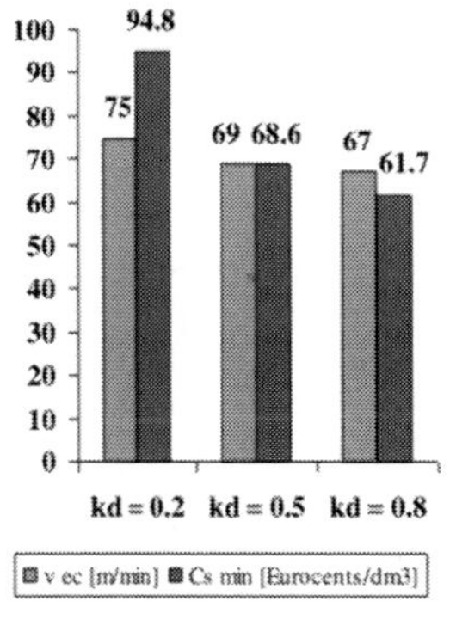

Figure 5. The k_d coefficient value influence

For finding the objective-function C_s sensitivity to the independent variables Iv, defined as:

$$S_{Iv} = \frac{\partial C_s}{\partial Iv} \cdot 100, \ [\%] \quad (4)$$

we firstly gave a variation of ± 25 % to three of the most important independent variables (s, t and k_d) and then we determined C_s relative variation, corresponding to this. Thus, we found S_s = 55.1 % when the maximum shop floor heating is needed (in January) and S_s = 43.7 % when the maximum shop floor cooling is needed (in July). We also found S_t = 47.8 % (in January) versus 46.4 % (in July) and S_{kd} = 16.3 % (in January) versus 9.8 % (in July).

5. Conclusions

In this paper, we have suggested a new expression of the objective-function used for the manufacturing cost minimization, which consider also the expenses of the shop floor energy system. Therefore, some new parameters were considered as independent variables for finding the cost-optimal solution – the building energy class, the loading degree (in time and space) of the machine tools, or the ratio between the used and the installed power of the machines driving motors.

The performed case study entitles us to conclude that:

- The expenses of the shop floor energy system represent an important share of the manufacturing cost and because of that, they have impact on manufacturing process optimality (e.g. on the cost-optimal cutting speed).
- Moreover, the optimality is significantly influenced by the season (the external temperature fluctuation) as well as by the shop floor energy class.
- The new objective-function has a noticeable sensitivity to the feed rate, cutting depth and to the degree of effectively using the scheduled functioning time.

References

[1] Zheng L Y and Ponnambalam S G 2010 Optimization of multipass turning operations using particle swarm optimization *ISMA'10 - 7th Int. Symp. on Mechatronics and its Applications* pp 1-6

[2] Prakasvudhisarn C, Kunnapapdeelert S and Yenradee P 2009 Optimal cutting condition determination for desired surface roughness in end milling *Int. J. Adv. Manuf. Technol.* **41**(5-6): pp 440-451

[3] Zŭperl U, Cŭs F and Gecevska V 2007 Optimization of the characteristic parameters in milling using the PSO evolution technique *J. Mech. Eng.* **6** pp 354-368

[4] Yang W, Guo Y and Liao W 2011 Optimization of multi-pass face milling using a fuzzy particle swarm optimization algorithm *Int. J. Adv. Manuf. Technol.* **54**(1-4) pp 45-57

[5] Oda Y, Mori M, Ogawa K, Nishida S, Fujishima M and Kawamura T 2012 Study of optimal cutting condition for energy efficiency improvement in ball end milling with tool-workpiece inclination *CIRP Ann. Manuf. Technol.* **61** pp 119–122

[6] Chen Z and Li Y 2008 An improved particle swarm algorithm and its application in grinding process optimization *Proc. 27th Chinese Control Conf.* pp 2-5

[7] Information on http://www.meteoromania.ro/anm

[8] Badea N 2014 *Design for Micro-Combined Cooling* (Heating and Power Systems, Springer)

IOP Conf. Series: Materials Science and Engineering **95** (2015) 012023 doi:10.1088/1757-899X/95/1/012023

The influence of milling parameters on the material hardness in the case of magnesium alloy AZ61A

C C Grigoraş[1] and G Brabie[1]
[1]"Vasile Alecsandri" University of Bacau, Faculty of Engineering, Calea Marasesti 157, 600115, Bacau, Romania

E-mail: grigorascosmin_phd@yahoo.com

Abstract. This study was conducted in order to determine the material hardness value due to modification of the milling cutting parameters. The alloy hardness was analysed as a function of different cutting parameters for milling operations (speed, feed and depth of cut), using a face mill. A total number of 17 samples were machined using parameters obtained by combining the input parameters. The total number of combinations is reduced by a pre-optimization, using the DesignExpert software. The cutting process was performed in dry conditions, and it was recorded that dry cutting magnesium-aluminium alloy AZ61A with the used parameters did not lead to chip ignition. The surface hardness was determined based on the Vickers scale (HV), its values ranging from 110.59 [HV] to 121.37 [HV]. The obtained results showed that the feed has a significant contribution in the surface harness modification. The application of the Taguchi method reveals that the material hardness can be improved, together with the manufacturing time, by means of the speed, feed and depth of cut maximization.

1. Introduction

The use of some magnesium alloys [1] as biodegradable orthopedically implants as temporary bone implants is a present topic among bio-engineering research.

In comparison to other implant materials (table 1) magnesium alloys have closer physical and mechanical properties to those of the natural bone [2]. While magnesium and its alloys are among the most interested options [3] low corrosion resistance can be a major step-back.

The low corrosion resistance is consistent to the fact that magnesium and its alloys have a high surface electrochemical activity or high susceptibility to corrosion attack [4]. In the human body, metabolic activities require an amount of 300-400 [mg/day] of magnesium [5], chemically reacting with water according to equation (1) [2].

$$Mg + 2\,H_2O = Mg(OH)_2 + H_2 \quad (1)$$

The hydrogen gas arising from the chemical reaction demands a higher attention when magnesium alloys are machined, if water based cooling agent is used. This issue can be overcome by dry or high speed machining (HSM). However this method, at a critical speed, generates high temperatures that facilitates the adhesion of material in the cutting tool (FBU – flank build-up) [6].

The material hardness in the case of magnesium alloy orthopedically implants is related to the wear resistance mechanism [4]. Research in this field [1, 3, 4, 6] leads to advanced chemical and/or mechanical surface treatments to improve the corrosion rate.

Content from this work may be used under the terms of the Creative Commons Attribution 3.0 licence. Any further distribution of this work must maintain attribution to the author(s) and the title of the work, journal citation and DOI.
Published under licence by IOP Publishing Ltd

Table 1. Summary of the physical and mechanical properties of various implant materials in comparison to natural bone [2].

Properties	Natural bone	Magnesium	Ti alloy	Polylactide
Density [g/cm^3]	1.8-2.1	1.74-2.0	4.4-4.5	1.25-1.29
Compressive yield strength [MPa]	3-20	41-45	110-117	45.5-61.4
Elastic modulus [GPa]	130-180	65-100	758-	3.75

2. Material and experiment setup

The study implies the execution of the following steps:

The AZ61A magnesium alloy was used for the current study in the shape of a round bar. A total number of 17 cylindrical samples, each having 70 [mm] diameter and 30 [mm] thickness, were obtained by cutting the bar. The samples were fixed in a vise and for better grip two groves were made on each sample (figure 1). In order to reduce the risk of chip ignition, dry cutting conditions were used. The chemical composition of the magnesium alloys is given in table 2.

The face milling experiment was conducted on a 3 axes CNC milling machine Knuth Rapmill 700. The cutting tool used was a Sandvik Coromant 490 face mill, having 50 [mm] in diameter and CCMT 0.8 [mm] radius inserts. The materials hardness was measured on the Vickers scale [HV] using the hardness measuring equipment Innovatest Ultramatic 2.

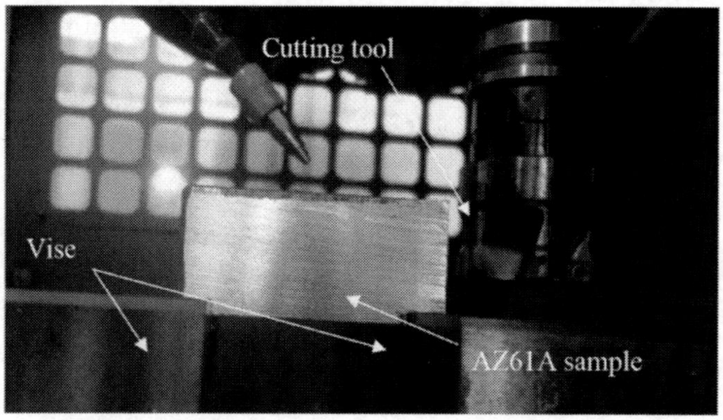

Figure 1. Experimental setup of face milling.

Table 2. Chemical composition of AZ61A magnesium alloy.

Al [%]	Cu [%]	Fe [%]	Mg [%]	Mn [%]	Ni [%]	Si [%]	Zn [%]	Other [%]
5.5-6.5	≤0.1	<0.005	92.3	0.05-0.4	<0.005	≤0.2	0.5-1.5	≤0.1

Given in table 3 are indicated the speed, feed and depth of cut values used in this experiment. The values were chosen so that they are linked one another by the same ratio to ease the pre-optimization and optimization analysis. The total number of samples was calculated by combining all the parameter values (3^3= 27). Thereby having the total number of possible combinations and using DesignExpert software by applying the Box-Behnken model, the number of samples was reduced from 27 to 17. The pre-optimization results given in table 4 where together with the feed, speed and depth of cut are the hardness values, obtained as average of a five point's measurement.

Table 3. Face milling parameters.

Cutting speed [m/min]	Feed [mm/min]	Depth of cut [mm]
200	500	0.15
350	1000	0.25
500	1500	0.35

Table 4. Face milling parameter combinations performed by DesignExpert software and material hardness measurements.

Sample	Cutting speed	Feed	Depth of cut	Hardness
1	500.00	500.00	0.25	113.5
2	500.00	1500.00	0.25	118.5
3	350.00	1000.00	0.25	118.5
4	350.00	500.00	0.15	114.8
5	200.00	1000.00	0.15	114.8
6	200.00	1500.00	0.25	117.2
7	500.00	1000.00	0.35	117.5
8	500.00	1000.00	0.15	117
9	350.00	1500.00	0.15	117.2
10	350.00	1500.00	0.35	116.8
11	350.00	500.00	0.35	114.2
12	350.00	1000.00	0.25	119.7
13	350.00	1000.00	0.25	121.37
14	350.00	1000.00	0.25	118.5
15	350.00	1000.00	0.25	117.5
16	200.00	1000.00	0.35	114.2
17	200.00	500.00	0.25	110.59

3. Results

The initial value for the material hardness was 102 [HV]. After the cutting process the measurements (table 4) shown an increase to a maximum value of 121.37 [HV], meaning 18.9 [%].

The hardness measurements from table 4 were inserted intro DesignExpert software and an ANOVA analysis for response surface was performed. The results given in table 5 show that the model used for this study is significant, with a p-value of 0.0004. The analysis also indicates that the depth of cut is not a major factor in the material hardness modification.

The variation of the material hardness at a constant depth of cut of 0.25 [mm] is given in figure 2. A minimum value of 110.59 [HV] is registred when using a low speed of 200 [m/min] with a feed of 500 [mm/min]. When combining average values for the speed and feed, meaning 350 [m/min] and 1000 [mm/min] the hardness value is 121.37 [HV]. This maximum value indicates an increase of 9.74 [%] in the case of dry face milling of the magnesium alloy AZ61A.

The use of higher speeds and feeds leads to a slight decrease of the material hardness value. When cutting with a speed of 500 [m/min] and a feed of 1500 [mm/min] it is registered a drop of 2.36 [%], corresponding to a value of 118.5 [HV]. It is also noted that the smallest hardness values are concentrated between speeds of 200-275 [m/min] and feeds of 500-750 [mm/min].

An increase of 5.97 [HV] in the magnesium alloy hardness from 110.59 [HV] to 117.2 [HV] corresponds to a variation of the feed from 500 [mm/min] to 1500 [mm/min] at constant a depth of cut of 0.25 [mm] and a speed of 200 [m/min]. For the same values of the feed and depth of cut an increase of 4.40 [%], from 113.5 [HV] to 118.5 [HV], is noted when dry cutting with a speed of 500 [m/min]. The percent in which the material hardness is enhanced is higher for the smaller speeds, but the improvement is in the numerical value that corresponds to the use of higher speed and feeds.

Table 5. ANOVA surface response analysis results performed with DesignExpert.

ANOVA for Response Surface Reduced Quadratic Model Analysis of variance table [Partial sum of squares - Type III]						
Source	Sum of Squares	Df	Mean Square	F Value	p-value Prob > F	
Model	87.53	4	21.88	11.98	0.0004	significant
A-Speed	11.79	1	11.79	6.45	0.0259	
B-Feed	34.49	1	34.49	18.18	0.0010	
A^2	18.35	1	18.35	10.05	0.0081	
B^2	20.62	1	20.62	11.29	0.0057	
Residual	21.92	` 12	1.83			
Lack of Fit	13.13	8	1.64	0.75	0.6648	not significant
Pure Error	8.79	4	2.20			
Cor Total	109.46	16				

The tendency of higher values for the surface hardness can be observed when average and above average speeds and feed are used. From a certain point, near to the maximum values for the input parameters, the hardness value slightly decreases. In this particular case the results shown that the decrease is notable after a speed and feed of over 425 [m/min] respectively 1250 [mm/min].

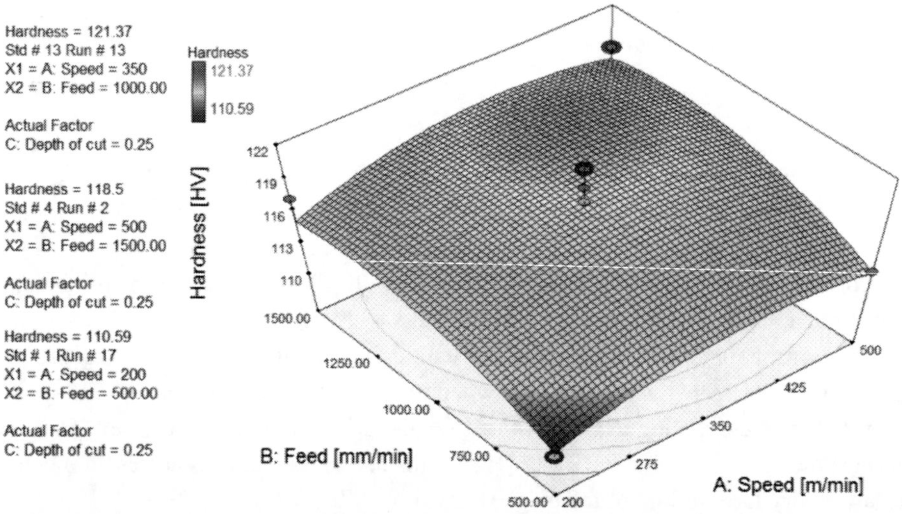

Figure 2. The influence of speed and feed at constant depth of cut on the material hardness.

Significant changes in the material hardness aren't noticed when the depth of cut is modified. In figure 3 (a), (b), (c) and (d) is reported the hardness dependency with respect to the following input parameters: speed of 350 [m/min], feed of 500-1500 [mm/min] and depth of cut of 0.15-0.35 [mm].

The comparison between the results from figure 3 (a) and (c) shows that the depth of cut has a minor impact over the hardness, with a change of only 0.52 [%]. The same insignificant change in the hardness value is noted when comparing the results from figure 3 (b) and (d); a variation of only 1.71 [%], with the change in depth of cut from 0.15 [mm] to 0.35 [mm].

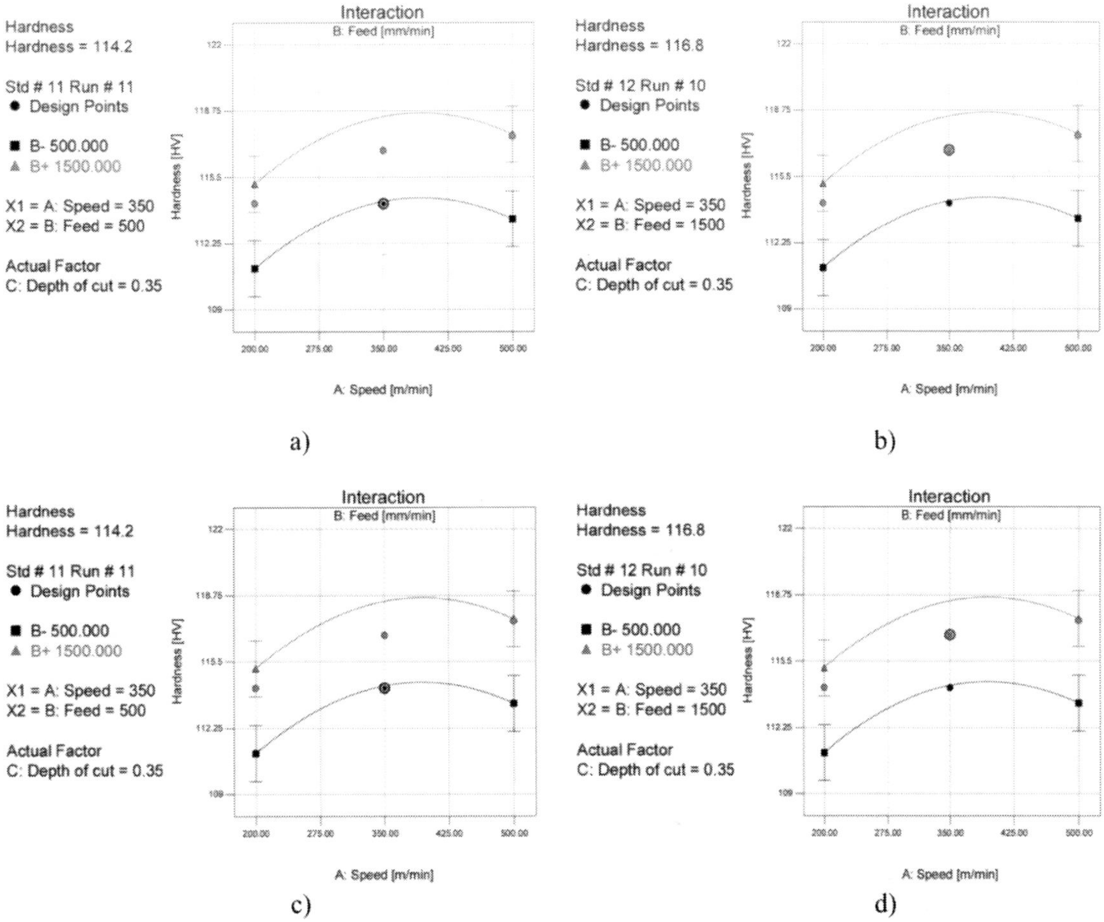

Figure 3. The influence of the depth of cut at constant speed and feed on the material hardness.

As stated above the depth of cut does not have a major role in the material hardness change, when AZ61A magnesium alloy is machined, using the parameters from table 3. However the use of a higher depth of cut ensures a reduced machined time and an increase productivity.

The ANOVA analysis gives the final equation in term of actual factors. It shows the relation between the input parameters and the material hardness. From equation (2) it can be noted that the depth of cut does not have any significant role.

$$HV = 91.426 + 0.072 \cdot Speed + 0.021 \cdot Feed - 9.266E - 5 \cdot Speed^2 - 8.839E - 6 \cdot Feed^2 \quad (2)$$

4. Optimization

In order to enhance the material hardness considering the optimization of the input parameters (feed, speed and depth of cut) the Taguchi method was used. It was taken into account the maximization of the hardness value regarding the intervals for speed and feed in which improvements were noted. By using speeds ranging from 350 [m/min] to 425 [m/min] and feeds from 1000 [mm/min] to 1250 [mm/min] and a maximum depth of cut the optimization predicts a surface hardness of 119.265 [HV]. Having a desirability factor of 0.982 (figure 4) the optimization indicates the use of the following parameter values: speed of 393.5 [m/min], feed of 1233.61 [mm/min] and depth of cut of 0.35 [mm].

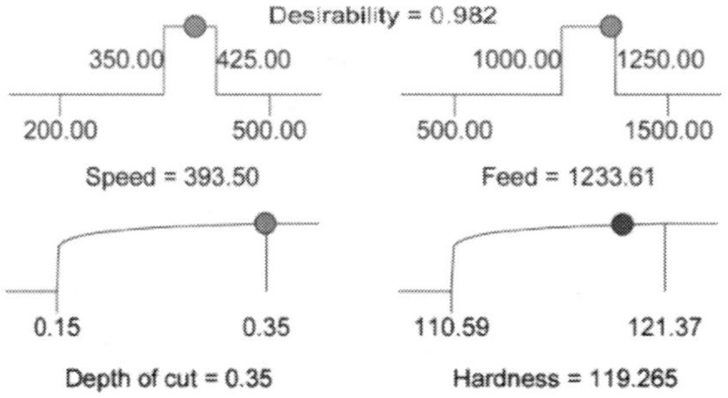

Figure 4. Optimized milling parameters using the Taguchi method.

5. Conclusions

The variation in the material hardness by changing the milling parameters was analyzed in this study.

The pre-optimization indicates that, by reducing the total number of samples from 27 to 17, conclusive results were obtained. The ANOVA analysis points out the relation between the input parameters and material hardness. It was noted that the average and above average speeds and feed offer an increase in the material hardness. Another conclusion was that high speeds and feeds lead to a slight decrease in the hardness value and the depth of cut does not have a significant role.

The optimization, using the Taguchi method, confirms the conclusion drawn from the analysis. The material hardness can be improved by changing the speed and feed until a certain value.

In the case of AZ61A magnesium-alloys the material hardness can be enhanced by using the model resulted in this study.

References

[1] Yang K and Tan L 2013 Control of biodegradation of magnesium (Mg) alloys for medical applications *Corrosion Prevention of Magnesium Alloys* ed G L Song (Woodhead Publishing) pp 509-543

[2] Walter R, Kannan M B, He Y and Sandham A 2013 Effect of surface roughness on the in vitro degradation behaviour of a biodegradable magnesium-based alloy *Applied Surface Science* **279** pp 343-348

[3] Salahshoor M and Guo Y B 2011 Surface integrity of biodegradable Magnesium–Calcium orthopedic implant by burnishing *J. Mech. Behav. Biomed. Mater.* **4** pp 1888-1904

[4] Song G L and Xu Z 2013 5-Surface processing and alloying to improve the corrosion resistance of magnesium (Mg) alloys *Corrosion Prevention of Magnesium Alloys* ed G L Song (Woodhead Publishing) pp 110-134

[5] Hindmarsh J T 1987 *Handbook of Toxicity of Inorganic Compounds* ed H G Seiler and H Sigel (New York: Astrid Sigel Marcel Dekker)

[6] Guo Y B and Salahshoor M 2010 Process mechanics and surface integrity by high-speed dry milling of biodegradable magnesium–calcium implant alloys CIRP Ann. Manuf. Technol. **59** pp 151-154

Influence of the cooling liquid on surface quality characteristics in milling

N C Tampu[1], G Brabie[1], B A Chirita[1], E Herghelegiu[1] and M C Radu[1]
[1]"Vasile Alecsandri" University of Bacau, 157 Calea Marasesti, 600115, Bacau, Romania

E-mail: catalin.tampu@ub.ro

Abstract. Cooling system and cooling liquid characteristics are among the main factors influencing surface quality and tool wear. The aim of this study is to analyse the effect of the cooling liquid, used in different concentrations and at different temperatures, on the quality of the surface layer processed by milling. In order to make this analysis a Minimum Quantity Lubrication (MQL) cooling device is used. Three different volumetric ratios were used to modify the concentration of the cooling fluid (25% water to 75% emulsion, 50% water to 50% emulsion, 75% water to 25% emulsion) and three different temperatures. The studies revealed that surface roughness can be correlated with the variation of the cooling liquid temperature while surface flatness can be correlated to both, cooling liquid temperature and concentration.

1. Introduction

Cooling and lubrication during machining operations are important factors that can determine the surface quality of the processed part. The cooling and lubrication liquids have the tasks: to lower the temperature of both the tool and the part during the machining process, and also to reduce the friction between the surfaces of the tool and the part. They represent also a means of transport for the evacuation of the chips.

A synthetic review of the role of the cutting fluids in machining operations is presented in [1]. The paper presents a model for the formation of fluid mist during machining and discusses its associated health problems. Comparisons are made between dry and wet cutting for different operations, and some strategies for the reduction or elimination of cutting fluids from several processes are presented.

In an effort to address the problems raised by the use of cooling fluids, a new cooling and lubrication method has gained a relatively large interest: minimum quantity lubrication (MQL). This method uses a flow of pressurized air to spray the cooling liquid in the form of fine droplets in the cutting area. An early investigation MQL of performance is presented in [2]. The study revealed, by comparison to the classic method of flood cooling, that although the consumption of lubricant is drastically reduced, the tool wear is comparable, surface roughness is equivalent, there is no significant difference between the cutting forces.

The effects and the mechanisms in (MQL) are examined by the means of an intermittent turning process in [3, 4]. The friction forces between the tool and the newly machined surface was measured. Different types of lubricants were used: synthetic ester and vegetable oil. The study has revealed the existence of a transition temperature, above which the lubricity disappears. In the case of vegetable oil the transition temperature is approximately 200°C.

A comparative study presents in [5] the effect of different cooling conditions in the machining of

Content from this work may be used under the terms of the Creative Commons Attribution 3.0 licence. Any further distribution of this work must maintain attribution to the author(s) and the title of the work, journal citation and DOI.
Published under licence by IOP Publishing Ltd

an aluminum alloy by taking into consideration the cutting forces, surface roughness and tool wear. It was concluded that MQL can replace flood cooling in certain conditions, but a special attention must be drawn to the liquid constituents.

The effect of cutting parameters and cooling technique on surface roughness in milling is experimentally investigated in [6]. The study reveals that surface roughness is influenced by the volumetric concentration of the cutting fluid used in MQL, and that MQL is more effective than flood cooling at higher cutting speeds by decreasing the roughness.

The aim of the present study is to analyze the effect of cooling liquid used in different mixture ratios with water, at different temperatures, on the quality of the surface layer processed by milling, when MQL is deployed.

2. Experimental Procedure

The experiments were conducted on a CNC vertical milling center Rapimill 700 (figure 1). The workpieces made of Al 6061-T6 aluminum alloy were machined with a 50 mm diameter Sandvik CoroMill equipped with 5 uncoated carbide inserts. 6061-T6 is a high strength and high corrosion resistance aluminum alloy with a good workability, which makes it suitable for applications in aircraft and automotive industries, consumer electronics, and numerous others. The mechanical properties are as follows: the modulus of elasticity (Young) of 68.9 GPa, yield strength of 276 MPa, tensile strength 310 MPa, hardness of 107 HV. The chemical composition of the material is presented in table 1.

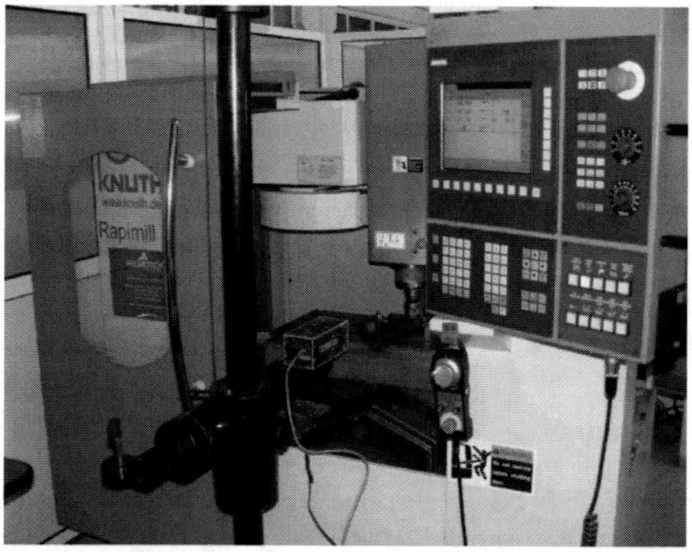

Figure 1. Milling center and infrared thermovision camera.

Table 1. Chemical composition of 6061-T6 aluminum alloy.

Component	Content [%]	Component	Content [%]	Component	Content [%]
Magnesium	0.8-1.2	Silicon	0.4-0.8	Zinc	max. 0.25
Copper	0.15-0.4	Ferrum	max. 0.7	Titanium	max. 0.15
Chromium	0.04-0.35	Manganese	max. 0.15	Aluminum	95.8-98.6

In order to assess the influence of the cooling conditions, cutting parameters were kept constant throughout the experiments, with the following values: the cutting speed vc of 314 m/min corresponding to a spindle speed n of 2000 rpm, the feed rate f of 900 mm/min, and the depth of cut a of 0.15 mm.

The cooling fluid was applied using a SKF MQL system. A synthetic water miscible emulsion TS 30 PT was used for cooling. Three mixture ratios were used for the cooling fluid: 25% water with 75%

emulsion (liquid L1), 50% water with 50% emulsion (liquid L2), and 75% water with 25% emulsion (liquid L3). The cooling fluid was conditioned for three temperatures: 6°C, 12°C, and room temperature (20°C).

The temperatures during machining were monitored using a FLIR® A655 sc thermal vision camera and the Research IR® software. A TR-220 Innovatest® was used for surface roughness (Ra) measurement. The measurements were made parallel to the cutting direction and the results presented in this study represent the averages of three readings for each surface. Hardness was determined using a Ultramatic 2 ultrasonic hardness tester from Innovatest®, while flatness deviation was measured on a Tesa® micro-hite 3D machine.

A run with dry machining at room temperature serves for benchmarking the results of the other tests.

3. Experimental Results and Discussion

Surface quality was examined from the perspective of three parameters: surface roughness R_a, hardness (HV), and surface flatness deviation.

3.1. Surface roughness

Surface roughness affects the functionality of components and it can influence the resistance to corrosion, fatigue resistance, etc. The surface roughness values resulted from the tests are presented in table 2 and graphically illustrated in figure 2. The dotted line represents the value of surface roughness measured for the benchmark part, obtained from milling the aluminum alloy under dry cutting conditions at room temperature. A multiple regression analysis carried in Minitab® software has revealed that the variation of surface roughness cannot be correlated with the concentration of the cooling liquid, but only with the temperature of the liquid (figure 3). Surface roughness values decrease with the increase of cooling liquid temperature. An explanation of this result might be given by the fact that at higher temperatures, the coolant is less viscous so the cooling effect is more effective. A less viscous liquid is capable of a better penetration into the areas where the rake face and the flank face of the tool interact with the material reducing the friction forces.

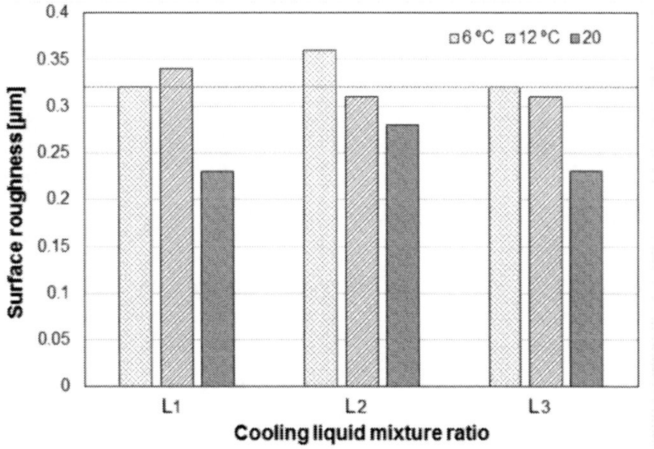

Figure 2. Surface hardness.

3.2. Surface hardness

Surface hardness can improve wear resistance of the parts. The measurements of surface hardness are presented in table 3, while a graphical illustration is in figure 4. The dotted line represents the value of surface hardness that was obtained in the benchmark test. In most of the cases the hardness is over the level of the dotted line. The regression analysis has revealed that a correlation cannot be established between the variation of surface hardness and the two input parameters (the temperature of the cooling liquid and the concentration of the liquid).

Table 2. Surface roughness [µm].

Cooling liquid mixture ratio	Cooling liquid temperature [°C]		
	6	12	20
L1 (25% water, 75% emulsion)	0.32	0.34	0.23
L2 (50% water, 50% emulsion)	0.36	0.31	0.28
L3 (75% water, 25% emulsion)	0.32	0.31	0.23
Benchmark (dry cutting, 20°C)		0.32	

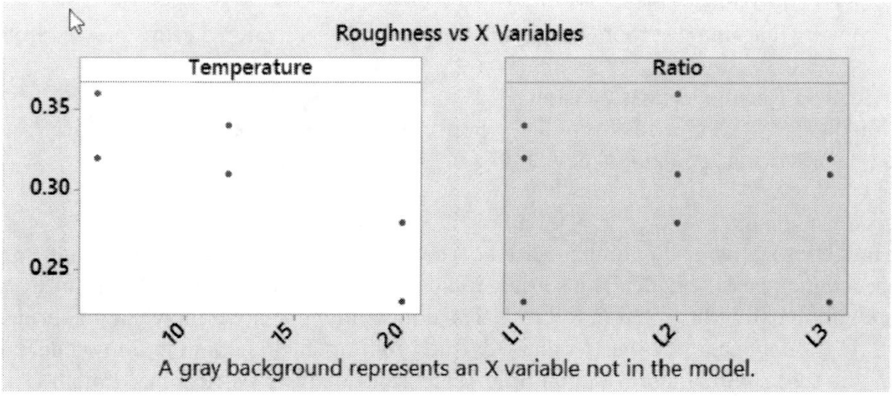

Figure 3. Regression report.

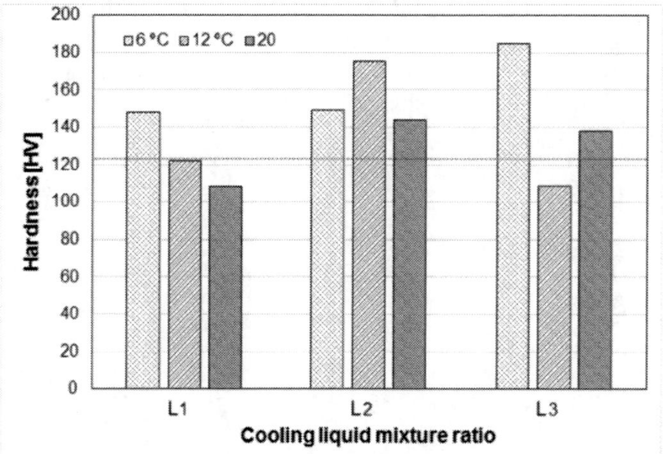

Figure 4. Surface hardness.

Table 3. Surface hardness [HV].

Cooling liquid mixture ratio	Cooling liquid temperature [°C]		
	6	12	20
L1 (25% water, 75% emulsion)	148	122	108
L2 (50% water, 50% emulsion)	149	175	144
L3 (75% water, 25% emulsion)	185	109	138
Benchmark (dry cutting, 20°C)		123	

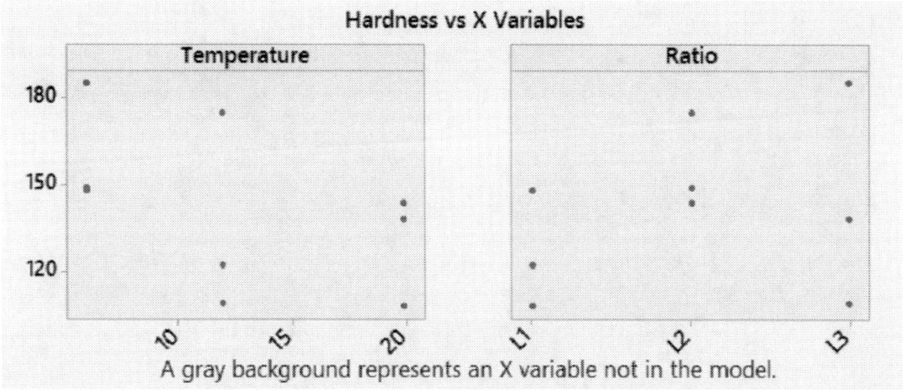

Figure 5. Regression report for surface hardness.

3.3. Flatness deviation

Precision is an important feature for the quality of a part, therefore it should be carefully taken into consideration when cutting conditions are established. The present study has revealed that dry cutting used in the benchmark test produces the largest deviation from the flat shape of the surface. This can result from the fact that during dry cutting, the largest amount of the heat developed in the cutting zone is transferred to the part and the tool, but when MQL is deployed the heat is more efficiently eliminated. This is more obvious especially in the case where the cooling liquid was brought to lower temperatures. The recorded values of deviation are presented in Table 4 and illustrated in figure 6. The regression analysis showed that the deviation values are correlated with both, the cooling liquid temperature and the concentration of the liquid (figure 7).

Table 4. Surface flatness deviation [µm].

Cooling liquid mixture ratio	Cooling liquid temperature [°C]		
	6	12	20
L1 (25% water, 75% emulsion)	2	2	3
L2 (50% water, 50% emulsion)	1	3	5
L3 (75% water, 25% emulsion)	4	1	3
Benchmark (dry cutting, 20°C)		10	

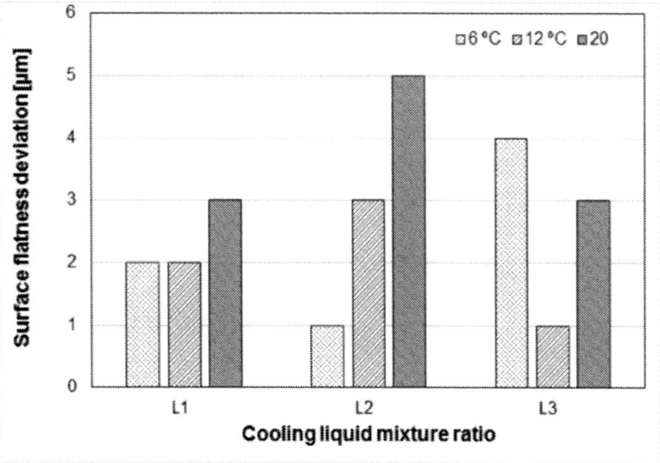

Figure 6. Flatness deviation.

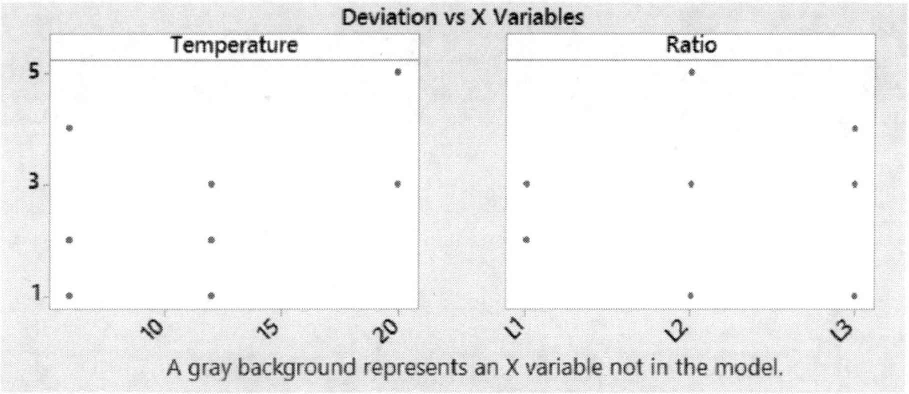

Figure 7. Regression report for the deviation.

4. Conclusions

The present experimental study has analysed the effect of the cooling liquids used in MQL on surface quality resulted from milling a 6061-T6 aluminum alloy. For this, three different concentrations were used for the cooling liquid by modifying the volumetric ratio between the emulsion and the water. Another studied parameter was the temperature of the cooling liquid, case where the liquid was brought to three different temperatures: 6ºC, 12ºC, and 20ºC (room temperature). The results were also compared with the ones resulted from the dry cutting.

The following conclusions can be formulated:

- Surface roughness variation can be correlated with the variation of the cooling liquid temperature. As the temperature of the liquid increases, the surface roughness decreases.
- No correlation was discovered between surface hardness and the concentration of the cooling liquid or its temperature.
- Deviation from surface flatness can be correlated to both, cooling liquid temperature and cooling liquid concentration. The lowest values of deviation were found for the lower liquid temperatures.
- Generally, the quality of the surfaces resulted when milling is done with MQL is superior to dry cutting. This is because MQL assures a better heat dissipation and the reduction of friction between the surfaces of the tool and the processed material.

The study can be continued for different cutting parameters and also for different materials.

References

[1] Adler D P, Hii W W S, Michalek D J and Sutherland J W 2006 Examining the role of cutting fluids in machining and efforts to address associated environmental/health concerns *Mach. Sci. Technol.* **10** pp 23-58

[2] Rahman M, Senthil Kumar A and Manzoor U I S 2001 Evaluation of Minimal Quantities of Lubricant in End Milling *Int. J. Adv. Manuf. Tech.* **18** pp 235-241

[3] Itoigawa F, Takeuchi D, Nakamura T and Childs T H C 2007 Experimental study on lubrication mechanism in MQL intermittent cutting process *Mach. Sci. Technol* **11** pp 355-365

[4] Itoigawa I, Childs T H C, Nakamura T and Belluco W 2006 Effects and mechanisms in minimal quantity lubrication machining of an aluminum alloy *Wear* **260** pp 339-344

[5] Sreejith P S 2008 Machining of 6061 aluminium alloy with MQL, dry and flooded lubricant conditions *Mater. Lett.* **62** pp 276-278

[6] Tosun N and Huseyinoglu M 2010 Effect of MQL on surface roughness in milling of AA7075-T6 *Mater. Manuf. Process* **25** pp 793-798

IOP Conf. Series: Materials Science and Engineering **95** (2015) 012025 doi:10.1088/1757-899X/95/1/012025

Non isothermal drying process optimisation - Drying of clay tiles

M Vasić[1] and Z Radojević[1]
[1]"Institute for Testing of Materials", Bulevar vojvode Mišića 43, 11000 Belgrade, Serbia

E-mail: milos.vasic@institutims.rs

Abstract. In our previous studies we have developed a model for determination of the variable effective diffusivity and identification of the exact transition points between possible drying mechanisms. The next goal was to develop a drying regime which could in advance characterize the real non isothermal process of drying clay tiles. In order to do this four isothermal experiments were recorded. Temperature and humidity were maintained at 350C / 75%; 450C / 70%; 450C / 60% and 500C / 60%; respectively in each experiment. All experimentally collected data were analyzed and the exact transition points between possible drying mechanisms were detected. Characteristic drying period (time) for each isothermal drying mechanism was also detected. The real, non-isothermal drying process was approximated by 5 segments. In each of these segments approximately isothermal drying condition were maintained. Temperature and humidity of the drying air, in the first four segments, was maintained on the same level as in recorded isothermal experiments while in the fifth segment, it were maintained at 700C / 40%. The duration of the first four segments were calculated from the diagrams Deff – t respectively for each experiment. The clay tile in experiment five was dried without cracking using the proposed non isothermal drying regime.

1. Introduction

Drying is very important process in the production of clay tiles. In our previous articles [1-4] calculation methods and computer programs designed for determination of the effective diffusion coefficient were presented. All reported models were created for isothermal conditions. Using those programs and calculation methods we were also able to calculate the variable effective diffusivity, to divide drying curve in segments and to identify the exact transition points between possible drying mechanisms.

Typical curves which represents the dependence of the effective moisture diffusivity as a function of the moisture content or drying time obtained using our calculation method [3] are presented at figure 1. All possible mechanisms of moisture transport and their transition from one to another during the constant and the falling drying period are well identified and are shown at figure 1. Drying segments along with mechanisms that can take place in them are summarized at table 1.

Capillary flow is predominant mechanism within the constant drying period, while in the falling drying period, the evaporation – condensation and vapor diffusion are the predominant mechanisms. Effective diffusivity represents an overall mass transport property of moisture, which includes all possible moisture transport mechanisms that are simultaneously controlling the moisture migration process in a material during drying [3, 4].

The procedure for the determination of time dependent Deff presented in [5] is "nearly" the same as ours [3]. Our calculation procedure was applied to determine the time dependent effective diffusion

Content from this work may be used under the terms of the Creative Commons Attribution 3.0 licence. Any further distribution of this work must maintain attribution to the author(s) and the title of the work, journal citation and DOI.
Published under licence by IOP Publishing Ltd

IOP Conf. Series: Materials Science and Engineering **95** (2015) 012025 doi:10.1088/1757-899X/95/1/012025

coefficient using data taken from reference [5] as well as for the data taken from reference [6]. The estimated effective moisture diffusivities were plotted as moisture ratios and are presented at figure 2. It can be seen that the shape of the curves presented at figure 2 are similar to and comparable with the shape of the curve presented in references [5, 6]. The moisture diffusivity profiles shown at fig 1 are also similar to those reported in the literature [7- 9]. It is also very important that all the transition points of the internal mechanisms registered at figure 1 and explained in table 1, are present an can be registered in the same way on figures 3 and 7 reported in references [5] as well as on figures 1 and 3 reported in references [6], regardless of the fact that the calculation method for the determination of time dependent effective diffusion coefficient were different from the method presented in our study [3].

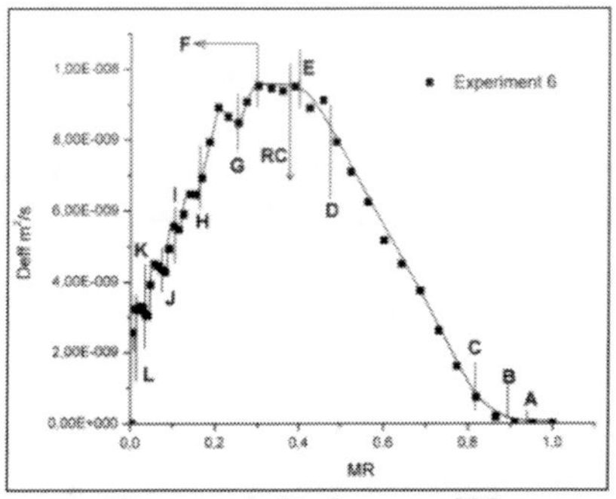

Figure. 3. Reference [3].

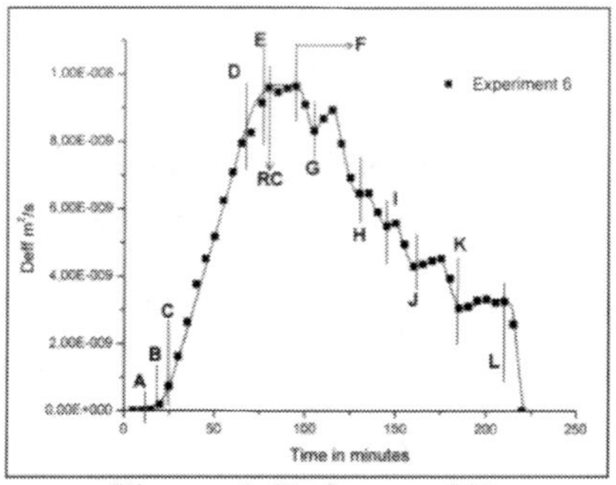

Figure 4. Reference [3].

Figure 1. Typical curves which represents the dependence Deff – MR or Deff – t

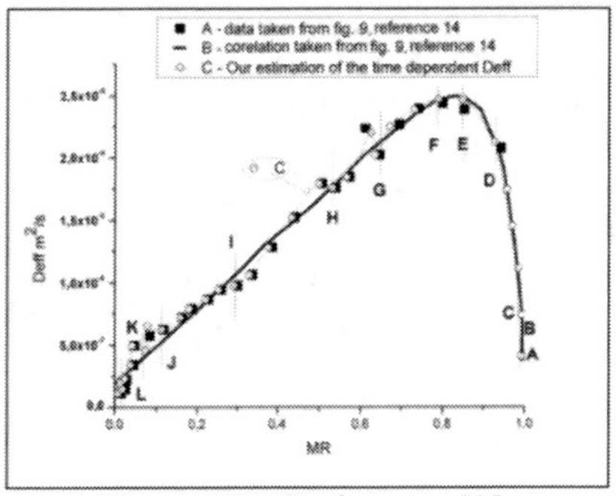

Figure 11. Reference [3].

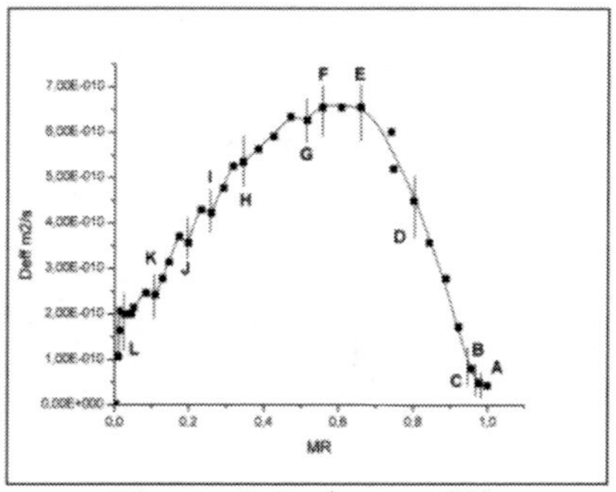

Figure 9. Reference [3].

Figure 2. Estimated effective moisture diffusivities for the data presented in references [5, 6].

Drying is energy intensive process which has a decisive effect on the quality of the clay tiles dried commercially. Escalating energy costs, requirements for eco-friendly and sustainable technologies as well as the rising consumer demand for higher clay tiles quality are the everyday reality in the clay tile production. Industrial draying chambers and drying tunnels are constantly passing the innovation process.

The aim of innovation is to use waste heat from the firing process, shorten the drying time and lower the energy consumption. In order achieve all previously mentioned it is necessary to have a precisely defined drying regime which corresponds with the clay tile raw material. It must be said that there are no widely accepted indicators of innovative performance or common set of indicators to assess the returns on investments. A typical industrial drying regime of clay tiles is presented at figure 3. It can be seen that drying process is not governed isothermally.

The next goal was to create a drying regime which could in advance characterize the real non isothermal process of drying clay tiles. In order to fulfill these objectives the real non isothermal drying regime was obtained by combining five different isothermal drying segments.

IOP Conf. Series: Materials Science and Engineering **95** (2015) 012025 doi:10.1088/1757-899X/95/1/012025

Table 1. Possible drying mechanisms according to reference [3].

Drying segment	Transport of liquid water	Transport of vapor	Additional explanation
A B	Capillary pumping flow through the biggest capillaries	/	0A – initial heating AE - constant period FL - falling period
B C	Capillary pumping flow through macro capillaries hydrodynamic flow	/	
C D	Capillary pumping flow through mezzo capillaries Hydrodynamic flow	/	D – "upper critical" point "funicular state" (continuous threads of moisture are present in the pores) Surface is completely wet Front start to recede into body
D E	Capillary pumping flow (from capillaries in funicular state) Hydrodynamic flow liquid diffusion in the pores	hydrodynamic flow	DE – Partially wet surface
E F	Creeping along the capillary when the liquid is in the funicular state or by the successive evaporation – condensation mechanism between liquid bridges.	hydrodynamic flow (difference in total pressure)	F - "lower critical" point - "pendular state" (continuous threads of moisture are not presented in the pores) - "last" wet patches disappeared from the surface
F G	the successive evaporation – condensation mechanism between liquid bridges of pendular water	hydrodynamic flow (difference in total pressure)	
G H	the successive evaporation – condensation mechanism between liquid bridges of pendular water	Stefan Flux (difference in partial pressure) hydrodynamic flow (difference in total pressure)	
H I	The evaporation – condensation mechanism was still active but is decreasing rapidly	hydrodynamic flow (difference in total pressure)	
I J	/	Molecular diffusion	
J K	/	Transition diffusion	
K L	/	Knudsen diffusion	

2. Materials and Methods

Experiments were carried out with the raw material obtained from the factory „Potisije Kanjiža“. Obtained raw material has passed homogenization process in the factory and was ready for the forming process. Laboratory samples 120x50x14 mm were formed in a laboratory extruder "Hendle" type 4, under a vacuum of 0.8 bar.

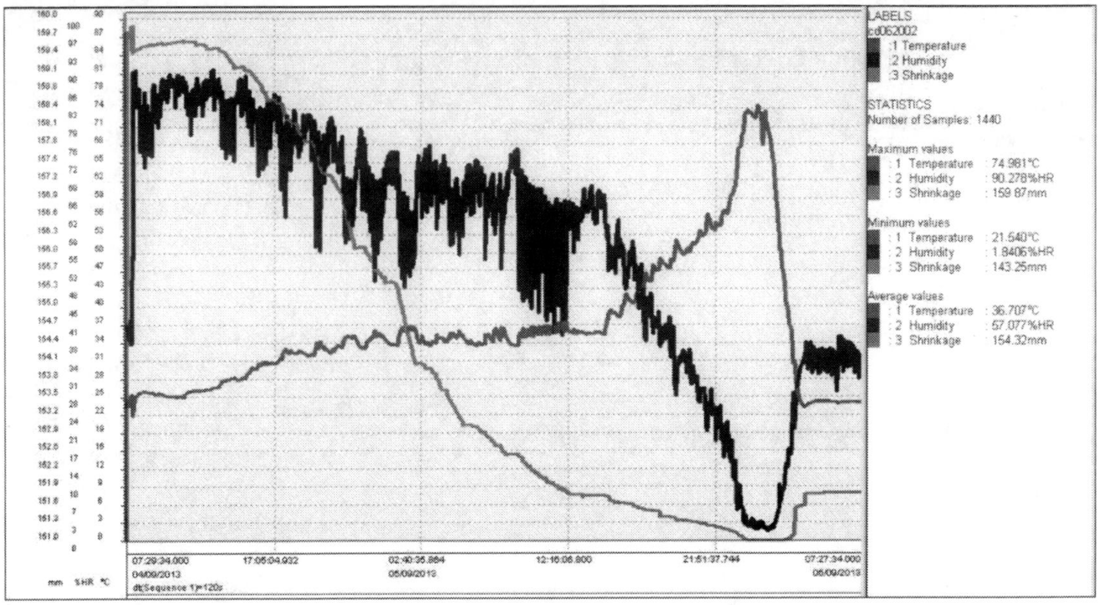

Figure 3. Example of the real non isothermal industrial drying regime.

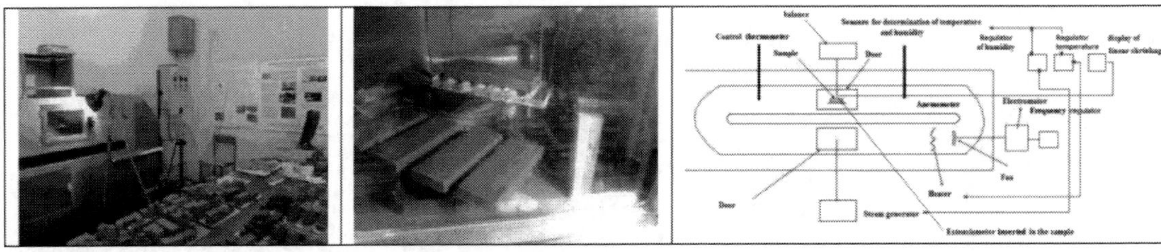

Figure 4. Schematic view of laboratory recirculation dryer.

On prepared clay tiles (samples), in laboratory recirculation dryer, presented at figure 4, under experimental conditions which are presented in table 2, drying kinetic curves were recorded. Technological properties of the raw material during drying and forming process are shown in table 3. In this paper we have used calculation methods previously described in [3, 4].

Table 2. Experimental conditions.

Experiment	Air velocity, W (m/s)	Air temperature, T (0C)	Air humidity, V (%)
1	1	35	75
2	1	45	70
3	1	45	60
4	1	50	60

3. Result and discussion

Drying data obtained, under the experimental conditions presented at table 2 have been simulated using the previously reported variable diffusivity model [3]. Simulated results are presented graphically at figure 5 and figure 6.

The exact transition points between possible drying mechanisms were detected and are presented at figure 5 and figure 6. It can be seen that the effective diffusivity values are increasing from experiment 1 to experiment 4 while the time to reach the characteristic points is respectively decreasing. This is expected because the value of the drying air humidity is decreasing from experiment 1 to experiment 4, while temperature is respectively increasing.

When point E is passed about 90 – 93 % of the shrinkage of the product is over and the clay tile will not crack. Theoretically speaking, when point F is passed and the clay tile is in the funicular state shrinkage of the product is over and the product can not crack. That is the reason why it is very important to know the exact position of the point E. In other words when point E is passed temperature of the drying air has to be increased while simultaneously drying air humidity has to be decreased.

Table 3. Technological properties of the raw material during forming and drying process.

Forming process			Drying process		
The amount of water for forming in (%)		20.94	Bigot's curve	Shrinkage at critical point (%)	5.69
Plasticity according to Feferkorn	Coefficient of plasticity	27.5		Water loss at critical point (%)	7.66
	Plasticity criteria	Good plasticity	Drying sensitivity according to Piltz	Time when first crack is appearing (min)	3.4
				Drying criteria	sensitive

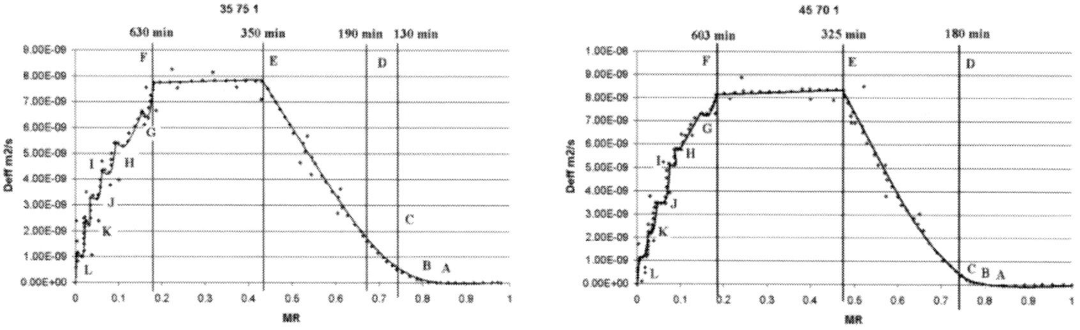

Experiment 1 Experiment 2

Figure 5. Estimated effective moisture diffusivity as a function of moisture ratio for experiments 1 and 2.

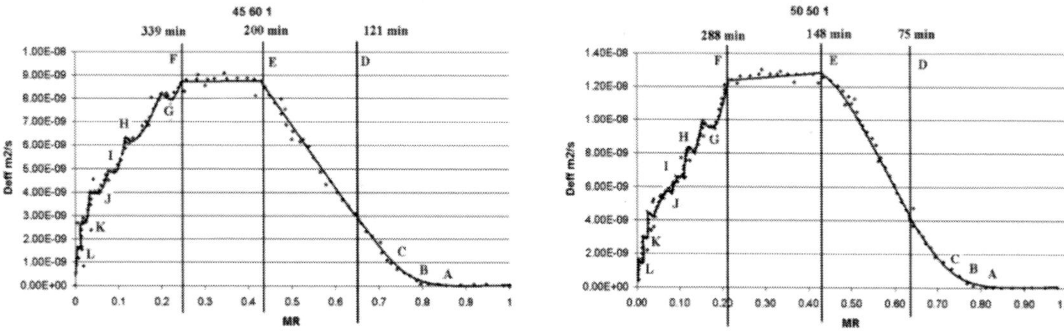

Experiment 3 Experiment 4

Figure 6. Estimated effective moisture diffusivity as a function of moisture ratio for experiments 3 and 4.

The non isothermal drying process was approximated by 5 segments. In each of these segments approximately isothermal drying condition were maintained. In the first four segments temperature and humidity were maintained as in isothermal experiments and in the fifth segment temperature and humidity were kept at 70^0C / 40% respectively.

Duration of the first segment was the same as the duration of the characteristic drying period t_C obtained from the experiment 1 (figure 5), form starting point till point C. Duration of the second segment was the same as the duration of the characteristic drying period t_{CD} obtained from the experiment 2 (figure 5) from point C till point D. Duration of the third segment was the same as the duration of the characteristic drying period t_{DE} obtained from the experiment 3 (figure 6) from point D till point E. Duration of the fourth segment was the same as the duration of the characteristic drying period t_{EF} obtained from the experiment 4 (figure 6) from point E till point F. Duration of the fifth drying segment was one hour. It was obtained experimentally. Previously mentioned drying segments were linked in experiment 5.

The clay tile in experiment five was dried without cracking using the proposed non isothermal drying regime. Dried tile was tested according to the standards EN 1024 in order to see the linearity and planarity coefficient. Results are given in table 4. Tested tile has satisfies regulation of the standard EN 1024 to issues of the linearity and planarity coefficient.

Table 4 Test results according to EN 1024.

Standard SRPS EN 1024		Criteria
Linearity coefficient R (%)	0.14	Allowed deviations of the linearity coefficient: ≤ 1,5% for length >300mm, 2% for length ≤ 300 mm
Planarity coefficient C (%)	0.09	Allowed deviations of the planarity coefficient: ≤ 1,5% for length >300mm, 2% for length ≤ 300 mm

4. Conclusions

Innovation in drying technologies is continuing over the past three decades, although it has not accelerated because of the long life cycles of dryers and relatively unchanged fuel costs over the past decade. The main goal of innovation in clay tile production is to use waste heat from the firing process, shorten the drying time and lower the energy consumption. Replacements of thermo-technical equipments, better control of drying air parameters during convective drying in old tunnel or chamber drying has been achieved over last 10 years. All these measures are worthless if precise drying regime, which is correlated with the nature of the clay raw material, is not used. Determination of the optimal drying regime can occur without heavy investments of time, manpower and funds only when we are able to use a reliable electronic kit to simulate convective drying in the specially constructed laboratory dryer. This implies that a strong base in fundamental knowledge is necessary to set up and characterize the optimal drying regime. This paper represents a pioneer work in which a non-isothermal drying regime was proposed. It was consisted from five isothermal segments. For the first time the duration of the approximately isothermal drying segments were not specified by experience or by methods trial and error. It was detected from the curves Deff – MR. Proposed isothermal drying regime was tested. Clay tile was dried without cracks. Dried tile has satisfies regulation of the standard EN 1024 to issue of linearity and planarity coefficient.

Acknowledgement
Note: This paper was realized under the project ИИИ 45008 which was financed by ministry of education and science of Serbia.

References

[1] Vasić M and Radojević Z 2011 Calculation of the effective diffusion coefficient *Int. J. Modern Manufacturing Technologies* 3 **1** pp 93-98

[2] Vasić M, Radojević Z and Grbavčić Ž 2011 Determination of the effective diffusion coefficiente *Romanian J. Material*, 41 **2** pp 169 – 176

[3] Vasić M, Grbavčić Ž and Radojević Z 2014 Analysis of Moisture Transfer During the Drying of Clay Tiles with Particular Reference to an Estimation of the Time-Dependent Effective Diffusivity *Drying Technology* 32 **7** pp 829-840

[4] Vasić M, Grbavčić Ž and Radojević Z 2014 Determination of the moisture diffusivity coefficient and mathematical modeling of drying *Chem. Eng. and Processing* **75** pp 33-44

[5] Pinto A L and Tobinaga S 2006 Diffusive model with shrinkage in the thin-layer drying of fish muscules *Drying Technology* 24 **4** pp 509-516

[6] Lopez R I I, Espinoza R H, Lozada A P and Alvarado G M A 2012 Analytical model for moisture diffusivity estimation and drying simulation of shrinkable food products *J. Food Eng.* **108** pp 427-435

[7] Chemki S and Zagrouba F 2005 Water diffusion coefficient in clay material from drying data *Desalination* **185** pp 491-498

[8] Batista L M, Cezar R A and Pinto A A L 2007 Diffusive model with variable effective diffusivity considering shrinkage in thin layer drying of chitosan *J. Food Eng.* 81 **1** pp 127-132

[9] Efremov G, Markowski M, Bialobrzewski I and Zielinskam M 2008 Approach to calculation time-dependent moisture diffusivity from thin layered biological materials *Int. Communication in Heat and Mass Transfer* **35** pp 1069-1072

IOP Conf. Series: Materials Science and Engineering **95** (2015) 012026 doi:10.1088/1757-899X/95/1/012026

On methane pyrolysis special applications

D C Toncu[1], G Toncu[2] and S Soleimani[3]
[1]Kazakh-British Technical University, School of Chemical Engineering, 51 Tole-bi Street, 050000 Almaty, Kazakhstan
[2]Ovidius University Constanta, Faculty of Mechanical, Industrial, and Maritime Engineering, 124 Mamaia Blvd., 900527 Constanta, Romania
[3]Politecnico di Milano University, Department of Chemistry, Materials, and Chemical Engineering Giulio Natta, LaBS, 32 Piazza Leonardo da Vinci, Milano, Italia

E-mail: dctoncu@gmail.com

Abstract. Methane pyrolysis represents one of the most important processes in industrial use, with applications rising from the chemical and petrochemical industry, combustion, materials and protective coatings. Despite the intense research, experimental data lack kinetic aspects, and the thermodynamics involved often leads to inaccurate results when applied to various systems. Carrying out a comparative analysis of several available data on methane pyrolysis, the paper aims to study the phenomenon of methane pyrolysis under different environments (combustion and plasma), concluding on the most possible reaction pathways involved in many of its applications. Computer simulation using different database underlines the conclusion, helping to the understanding of methane pyrolysis importance in future technologies.

1. Introduction

The thermal decomposition of methane has formed the study object for decades, but the reaction mechanism is not completely known. Between the two suggested initial reactions in the homogeneous decomposition process, the methyl CH_3 is more probable than methylene CH_2 radical [1]:

$$CH_4 \rightarrow CH_3 + H \quad (1)$$

$$CH_4 \rightarrow CH_2 + H_2 \quad (2)$$

The methyl formation also represents the rate determining step of the methane dissociation. Though small amounts of C_2 hydrocarbons can be found in the equilibrium composition, practically only hydrogen, carbon, and methane are the relevant species up to 1700°C. Experiments also confirmed that hydrogen plays the key role as energy carrier, the C_2 hydrocarbon intermediates and byproducts may lose importance in alternative reaction mechanisms. The activation energy for thermal dissociation of methane was assigned values from 312 to 450 kJ/mol, but under catalyst presence, decreases at 89 … 236 kJ/mol [2].

The initial formation of the methyl radical leads to the uniform deposition on the walls of the reaction vessel, later aimed in special coating processes. The feeding methane is much more stable than methyl radical, and therefore it cannot efficiently decompose into the carbon deposit on the walls, which leads to the conclusion that the chemical specie methyl is predominantly responsible for the carbon deposits. The epitaxial growth process on metal surface results in a highly crystalline carbon form.

Content from this work may be used under the terms of the Creative Commons Attribution 3.0 licence. Any further distribution of this work must maintain attribution to the author(s) and the title of the work, journal citation and DOI.
Published under licence by IOP Publishing Ltd

Further experiments carried out with various diluent gases, such as hydrogen and argon, revealed that methane reactivity is lower at 1100°C than at 1000°C, while there is deposition rate even at zero residence time, which underlined the ability to form pyrocarbon of C_1 species. Furthermore, the deposition rates increased exponentially with residence time, stronger than C_2-hydrocarbons [3].

Even later studies showed that the overall reaction of methane decomposition involves high temperature dehydrogenation [4]:

$$2CH_4 \rightarrow 2C_2H_6 + H_2 \rightarrow C_2H_4 + H_2 \rightarrow C_2H_2 + H_2 \rightarrow 2C + H_2 \tag{3}$$

The difference in experimentally determined activation energies for tubular flow reactors, shock tube reactors, and plasma reactors rises from the heat transfer inside them and the resulting difference in temperature between reaction bulk and reactor walls. The above-mentioned study also revealed that pyrolysis rate of methane on molybdenum carbides in much higher than on tungsten carbides, which leaded many scientists to use molybdenum substrates for carbon coatings. Methane was found more efficient than acetylene for diamond-like-carbon film growth [4].

In most cases, the overall methane decomposition is believed to be a first order reaction, with an order decrease in case of dilution hydrogen environment. The amount of energy required is 216 kJ/mol for CH_4 at 298 K and products at 2000 K [5]:

$$CH_4 \rightarrow C + H_2 \text{ with } \Delta H^o \approx 75 \text{ kJ/mol} \tag{4}$$

The effect of dilution gas on methane pyrolysis has been intensively studied also, conclusions being made for different employed gas.

Despite the detrimental effect of hydrogen dilution on methane conversion and the competition between C_2 formation and carbon deposition, the coke formation decreased and even diamond synthesis was reached. Furthermore, the carbon deposit and the metal wall or substrate may act as catalyst [6]:

$$2CH_4 \leftrightarrow 2CH_3 + H_2 \tag{5}$$

$$2CH_3 + 2s \leftrightarrow 2CH_3 \cdot s \tag{6}$$

$$H_2 + 2s \leftrightarrow 2H \cdot s \tag{7}$$

$$2CH_3 \cdot s + 2H \cdot s \rightarrow 2C + H_2 + 2s \tag{8}$$

During methane combustion, kinetics plays the same vital role. When comparing different mechanisms, it was found that it is feasible to use reduced schemes in engine simulation, though when elementary chemical reactions are based on Arrhenius law, there are discrepancies of temperature profiles between detailed and reduced mechanisms [7]. Kinetic mechanism for two methane combustion mechanisms are given in Table 1, while Arrhenius rate parameters are offered in table 2.

The rate coefficient is expressed as:

$$k = A \cdot T^m \cdot e^{-Ea/RT} \tag{9}$$

The temperature is:

$$T_a = E_a / R \tag{10}$$

Another important aspect concern the missing pollutant or byproduct gases that a first step mechanism cannot predict, since there are no pathways for such compounds.

Another source mentions, for the overall combustion reaction of methane, an Arrhenius coefficient of $8.3*10^{11}$ cm^3/mol s and an activation energy of 30000 cal/mol [8].

Detailed kinetics can be found also in [9], when laminar diffusion flames in vitiated environments are analysed:

IOP Conf. Series: Materials Science and Engineering **95** (2015) 012026 doi:10.1088/1757-899X/95/1/012026

Table 1. Kinetic parameters for various methane combustion mechanisms [7].

Reaction Mechanism	A	m	Ta
Global one-step mechanism			
$CH_4+2O_2 \rightarrow CO_2+2H_2O$	$1.5 \cdot 10^{13}$	0	20000
Four-step mechanism			
$CH_4+0.5O_2 \rightarrow CO+2H_2O$	$4.4*10^{14}$	0	30000
$CH_4+H_2O \rightarrow CO+3H_2$	$3*10^{11}$	0	30000
$H_2+0.5O_2 \leftrightarrow H_2O$	$2.5*10^{19}$	-1	40000
$CO+H_2O \leftrightarrow CO_2+H_2$	$2.75*10^{12}$	0	20000

Table 2. Some Arrhenius rate parameters of some reactions involved in methane oxidation [9].

Reaction	Arrhenius pre-exponential factor, cal/cm^3	Temperature exponent	Activation energy, cal
$H_2 \rightarrow 2H$	$7.15*10^4$	-0.82	103328.3
$H_2O \rightarrow H+OH$	$2.2*10^{13}$	0	105217.9
$CO+OH \rightarrow CO_2+H$	$1.5*10^8$	0	100.5
$CO+O \rightarrow CO_2$	$6.7*10^7$	0	0
$2CH_4+O_2 \rightarrow 2CO+4H_2$	$1.75*10^7$	1	24265

Soot formation involves methane pyrolysis rather than oxidation reactions, especially for temperature interval of 1500 – 1800 K [10].

Some chemical kinetic databases contain information regarding the uncertainty of rate parameters, which form the main input of uncertainty analysis [11].

The present work aims to test various parameters of Arrhenius reaction rates corresponding to several possible pathways involved in methane pyrolysis, either in plasma deposition or soot formation during methane combustion, and their thermodynamic range interval in which such parameters show better match between experimental data and computer simulations. Whenever such parameters allow good correspondence between experiments and simulation, one can conclude that the involved pathways are the right ones involved in reaction mechanisms.

2. Computational simulation with different kinetic parameters

2.1. Methane pyrolysis with hydrogen carrier

The methane pyrolysis in hydrogen plasma was simulated for the geometry of a real chemical vapor deposition existent at IFP CNR Milano: a cylindrical stainless steel reactor with 0.3 m diameter and 0.3 m length. For simplification, only the volume regarding the deposition was simulated, meaning the symmetrical deposition substrate and the gas nozzle with 0.178 mm situated 2 mm above the substrate.

The chemical kinetics used in the *simulation methodology* was simplified as much as possible, in order to allow errors elimination and computational complications in the reaction mechanism. The four gas phase reactions and one surface reaction chosen as model, due to their importance in the process and high occurrence during various deposition techniques, were the ones identified during optical emission spectroscopy diagnosis of plasma gas used in experimental depositions of carbon materials, ranging from crystalline diamond to fullerene and graphite. Acetylene was excluded from the model, since if it resulted, the volume would increase with at least 50%. The kinetic parameters for this case are given in table 3.

Input data comprises methane concentration in hydrogen-methane mixture, velocity of the gas mixture, and, very important, the rate parameters.

Table 3. Kinetic parameters for methane pyrolysis with hydrogen carrier used for simulation.

Reaction	Pre-exponential factor A	Temperature exponent m	Activation energy Ea J/mol
$H_2+\bar{e}\rightarrow 2H+\bar{e}$	1.2E16	0	943685.5 J/mol
$CH_4+H\rightarrow CH_3+H_2$	2.2E4	3	36610 J/mol
$CH_3+H\rightarrow CH_2+H_2$	7.2E14	0	63178 J/mol
$CH_2+H\rightarrow CH+H_2$	4E13	0	0
$CH+H\rightarrow C(s)+H_2$	2.14 mole/cm^3 s	0	30515 J/mol

2.2. Methane combustion

For simplifications, eddy-dissipation model for methane gas combustion with ambient air was simulated in a 0.2 m diameter and 1.8 m length tubular reactor (or cylindrical combustor) using finite-rate chemistry. In order to allow comparison, methane flow rate, equivalence ratio, and flow regime were kept constant. The simulation methodology implies standard k-ε turbulence model and chemical species transport and reaction. The eddy-dissipation turbulence - chemistry computes the rate of reaction under the assumption that chemical kinetics are fast compared to the rate at which reactants are mixed by turbulent fluctuations. Reactions were volumetric, with diffusion energy source, and mixture material methane-air. Calculation was performed assuming that all properties except density are constant, an acceptable assumption for the case of fully turbulent flow, where molecular transport properties play a minor role. Boundary conditions imply velocity air inlet, velocity fuel inlet, and pressure outlet.

Starting with the global one-step reaction mechanism, and assuming complete conversion of the methane to carbon dioxide and water vapor, simulation used several reaction rates, from those existent in the software's database to those found in the literature. The rate-limiting process was assumed to be the turbulent mixing.

The eddy-dissipation reaction model only uses the mixing rate parameters, so the Arrhenius rate are those included in the database, as the turbulent transport plays the major role. However, the rate parameters were changed accordingly, and different output resulted.

3. Results and discussions

3.1. Methane pyrolysis with hydrogen carrier

The kinetic parameters in the presence of major hydrogen gas completely changed the species involved as well as deposited carbon. Hydrogen does not only represent the energy carrier, but also changes the first order elementary reaction into a second order reaction, which involves hydrogen as collision partner.

Methane conversion increases with residence time that much that, by doubling the residence time from 0.027 s to 0.054 s, conversion shifts from about 1.3 % to 99 %. Methane conversion also increases with increasing temperature, due to increased decomposition and transformation through collisions.

Hydrogen yield increases with reactor temperature and residence time, depending also on methane concentration in the feed mixture and conversion. Obviously, hydrogen-consuming reactions are balanced by those generating it.

Surface chemistry is the most dependent on rate parameters, as the presence of carbonaceous material decreasing the activation energy up to about one fifth.

3.2. Methane combustion

Simulations showed that temperature profiles differences between mechanisms decrease with increasing temperature, and especially above 1500 K, and increase with increasing equivalence ratio.

For the same equivalence ratio, detailed mechanisms including more than 4 reaction steps, predict a much lower carbon monoxide emission than reduced to 4-step mechanisms, while the single step mechanism cannot estimate carbon monoxide due to the lack of such a pathway.

For major species carbon dioxide and water, there is a good agreement between mechanisms prediction.

IOP Conf. Series: Materials Science and Engineering **95** (2015) 012026 doi:10.1088/1757-899X/95/1/012026

Taking into account the above-mentioned observations, results are displayed for the single-step reaction mechanism in figures 1-3. The temperature, for the case of constant heat capacity, raised till 3078 K, when applied an initial temperature of 2000 K corresponding to the engine spark (figure 1). In figure 2, methane mass fraction is higher at the nozzle exit, then decreases as the gas dissipates and mixes with combustion air. Carbon dioxide mass fraction is presented in figure 3, the profile corresponding to the reaction consumption with the oxygen present in the combustion air.

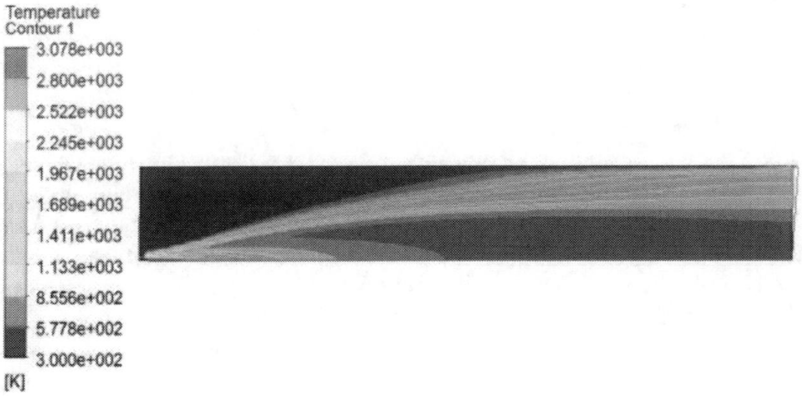

Figure 1. Temperature profile for the CH_4 one-step combustion.

Figure 2. CH_4 mass fraction during one-step mechanism combustion.

Figure 3. CO_2 mass fraction during one-step mechanism combustion.

4. Conclusions

Despite the high scientific and industrial interest in methane pyrolysis, the appropriate kinetic data for many processes still miss. Different rate parameters, experimentally determined in various conditions, such as pressure, gas carrier, and catalyst presence, lead to different estimation results and often mislead conclusions regarding complex phenomena.

In the case of methane pyrolysis for protective coatings, the whole process depends on the accuracy of rate parameters. Plasma environment, metals and their compounds from the reactor material or substrate, carbonaceous material, nature of carrier, all change the rate parameters, giving a completely different reaction mechanism according to real conditions.

In the case of methane combustion and soot formation, differences in kinetic database may reduce significantly, especially when estimating combustion temperature, resulting water and carbon dioxide. For pollutants, from carbon monoxide, nitrogen oxides, to unburnt hydrocarbons, intermediate species and their reaction parameters dramatically change the results. Adjusting the reaction mechanism to the real process becomes more important, and reduced models to one-step reactions often fail to accurately represent the process.

References

[1] Evans E L et al. 1973 Growth of filamentary carbon on metallic surfaces during the pyrolysis of methane and acetone *Carbon* **11** pp 441-445

[2] Wullenkord M 2011 *Determination of Kinetic Parameters of the Thermal Dissociation of Methane* PhD Thesis (German Aerospace Center in Cologne, Germany)

[3] Becker A and Hüttinger K J 1998 Chemistry and kinetics of chemical vapor deposition of pyrocarbon – IV Pyrocarbon deposition from methane in the law temperature regime *Carbon* **36** pp 213-224

[4] Choudhary T V, Aksoylu E and Goodman D W 2003 Nonoxidative Activation of Methane *Cat. Rev.* **45** pp 151-203

[5] Rodat S et al. 2009 Kinetic modelling of methane decomposition in a tubular solar reactor *Chem. Eng. J.* **146** pp 120-127

[6] Rhee B S, Rhee M Y and Hüttinger K J 2001 CVD-kinetics of pyrolytic carbon from methane on the stainless steel 316L at 600oC *CARBON Conf.* (Lexington KY USA)

[7] Ennetta R, Hamdi M and Said R 2008 Comparison of different kinetic mechanisms of methane combustion in an internal combustion engine configuration *Therm. Sci.* **12** pp 43-51

[8] Rouband A et al. 2003 In-cylinder combustion modeling with simplified chemistry *14th Int. Multidimensional Engine Modeling User's Group Meeting* (USA)

[9] Tuovinen H 1994 Modelling of Laminar Diffusion Flames *Vitiated Environments, Fire Safety Science* **4** pp 113-124

[10] Li Q et al. 2012 Experimental study and kinetics modeling of partial oxidation in heavily sooting flames *Chem. Eng. J.* **207-208** pp 235-244

[11] Nagy T and Turányi T 2012 Determination of the uncertainty domain of the Arrhenius parameters needed for the investigation of combustion kinetic models *Reliability Engineering and System Safety* **107** pp 29-34

Graphite pellicles, methods of formation and properties

P Topala[1], L Marin[2], V Besliu[1], P Stoicev[3], A Ojegov[1] and PCosovschii[4]

[1]Alecu Russo Balti State University, Republic of Moldova
[2]National Research and Development Institute for Chemistry and Petrochemistry – ICECHIM, Bucharest, Romania
[3]Technical University of Moldova, Republic of Moldova
[4]State Enterprise "Chisinau Glass Factory", Republic of Moldova

E-mail: pavel.topala@gmail.com

Abstract. The paper presents the results of experimental investigations aimed at the establishing the composition and the functional properties of the graphite pellicles formed on the metal surfaces by the action of plasma in the air media at normal pressure applying electrical discharges in impulse (EDI). It shows that they have the same behavior characteristics as fullerene, avoiding the stick effect between metal surfaces and between metal and liquid glass at temperatures of the order of 400-1200 °C.

1. Introduction

The formation of nano-scale graphite pellicles on the surfaces of pieces made of different metal alloys causes their diffusion in the surface layer accompanied by the formation of carbides with high hardness and, as a result, it increases wear resistance of this layer [1-5]. The formation of graphite films in all cases leads to the surface roughness decrease.

In accordance with the results obtained by the authors [4], the application of films on the surfaces of the work pieces in the kinematic pairs leads to the reduction of the coefficient of friction by at least 3 times.

The experimental tests on the stick effect of the threaded joints has shown that the stick effect caused by the mutual diffusion of the constituent materials of the joint parts is prevented by the presence of graphite films and even when the couple is maintained in the furnace at temperatures within (400-800)°C it has been showed that for the parts made of construction steel the stick effect in the joint is not present [4].

Film formation on the internal surfaces of pipes allows eliminating the stick effect for paraffin stoppers that are a problem for pipelines of oil transporting.

Experimental investigations and trials come to show that deposits are formed more efficiently when the workpiece is connected in the discharge circuit of the current impulse generator as anode, and the formed films can reach up to 7μm thick, they increase the durability of the glass molding form components by at least 2 times, due to the qualities of solid lubricant and anti-refractory properties that they possess [2, 5].

From those mentioned above it follows that the development of an effective technology of graphite deposit formation on the surfaces of the parts from machine building industry would allow to successfully solve a number of problems such as: ensuring the surface refractoriness, obtaining wear resistant surfaces, reducing the coefficient of friction of parts surfaces functioning in the kinematic

Content from this work may be used under the terms of the Creative Commons Attribution 3.0 licence. Any further distribution of this work must maintain attribution to the author(s) and the title of the work, journal citation and DOI.
Published under licence by IOP Publishing Ltd

IOP Conf. Series: Materials Science and Engineering **95** (2015) 012027 doi:10.1088/1757-899X/95/1/012027

couples, buffer film formation at the interface between two surfaces of a couple, partial or total removal of the stick effect between couple parts, etc.

2. Methodology of Experimental Investigations

Experimental investigations on graphite deposition formation on metal surfaces were performed in normal working environment - air. For this purpose electrical discharges in impulse have been applied that interact with electrodes' surfaces in the regime of "cold" electrode spots maintenance to avoid melting, vaporization and removal of material from them. The bars of cylindrical cross-sectional area of (5-7)mm^2 made of technical graphite were used as tool-electrodes. To appropriate to maximum results of interests for production enterprises as workpieces were chosen: nuts and bolts with metric thread M16, plungers applied in molding packaging bottles, component parts of glass molding forms, pipe segments, etc. These pieces were connected in the discharge circuit of the current impulse generator as anode. Materials of which were executed were relative diverse and included group of steels (steel C45, C35), cast iron and bronze alloys. In order to form graphite depositions without melting and vaporization of the machined surface of the workpiece as a source of energy the current impulse generator was applied whose construction and principle of operation is described in [1]. It provides formation current impulses with duration ranging in limits of 10^{-6}-10^{-7} s which corresponds to the life duration of the "cold" electrode spots. Generator offers formation of the current impulses with the following parameters: the energy released in the interstice W_S=0÷4.8J, the energy accumulated in the condenser battery W_C=0÷12J, the charging voltage applied to the condenser battery U_C=(0÷250)V, for its capacity contained within the C=100÷600μF with the step of 100μF. It provides electrical discharge impulses within the interstice values S=0.05÷2.5 mm; discharge frequency f=0÷50Hz. The morphology of the machined surface was studied by means of SEM technique and the chemical composition by means of EDX technique. The adherence of the formed film on the machined surface was achieved by shear tests on a dynamometer type FPZ HECKERT 100. To identify changes in adhesion properties as a result of applying graphite films, it was performed by measuring comparative detachment forces of assemblies made with a strong adhesive - polyurethane - between a set of samples and specimens treated with graphite and without treatment. A set of 3 specimens with graphite treatment and a set of three samples without treatment were supposed to adhesive; the specimens were treated at the end with the graphite film as disclosed in the paper [3]. After adhesive application the specimens overlap an area of 25x25 mm, and after drying of the adhesive they were subjected to the traction.

To measure the wear resistance of graphite films deposited on glass molding plungers the universal microscope of UIM-21 type was used permitting measurement with an error of 1μm as described in [2]. The component parts of glass molding forms (numbered in accordance with the their assembling drawing) were subjected to wear in glassy mass for 44625 cycles (85 hours) for the sample no.19 and for 39900 cycles (75 hours) for the 23 sample. The speed of cycles constitutes v=8.75 cycles/min, the temperature of the glass drop is 1129 °C, the glass of BT-1 type for bottles' molding was used, its chemical composition corresponds to the standard GOST R 52022-2004 SM.

In order to determine the anti-stick properties 6 pairs of bolt joints were selected, of which three pairs of pieces were with deposited on bolt's surface graphite films and another three were kept intact. They were engaged through the action of the same torsion moment. Thus, three pairs were formed for investigations; there were two couples with and without the presence of graphite film in each of them. The studied pairs were placed in the NaCl solution with a concentration of 3% for 24 hours after which they were insert into oven at temperature of 800 °C where were maintained for 10 hours, then the measure of the unscrewing moment of the investigated joints was made in accordance with [4].

3. Results of experimental investigations and their interpretation

The morphology analysis of the surface machined by applying EDI with graphite tool-electrodes showed that physic-chemical changes on the surface does not exceed micrometer sizes. Apart from the initial components of the processed material a considerable amount of carbon (about 80%) in atomic content is attested. The microstructure analysis of the transverse micro section demonstrates that the majority of carbon transferred into the workpiece surface is observed at micrometer depths, which allows concluding that it is possible formation phases of carbides and those of graphite at the interface "metal piece-deposition formed by applying EDI".

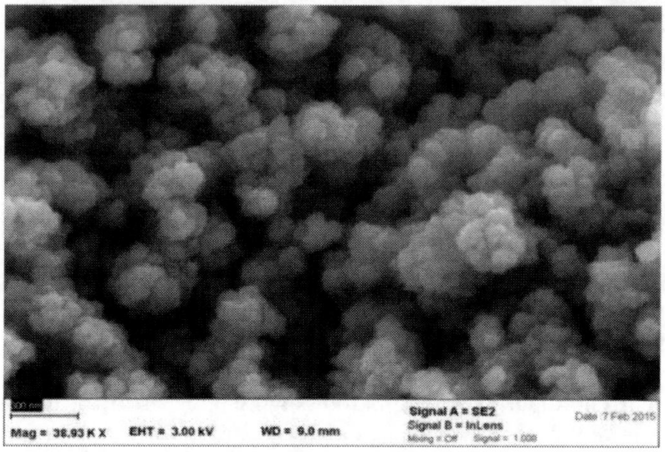

Figure 1. SEM morphology of the graphite film deposited on steel C45 support.

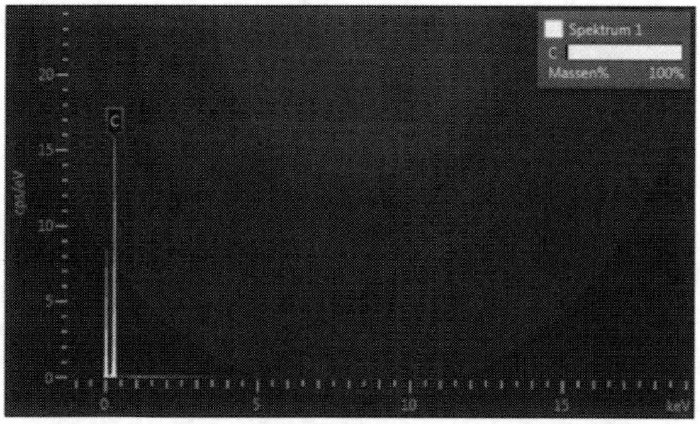

Figure 2. The chemical composition of the formed film determined by the EDX technique.

If you look at those shown in figure 1, we can see that the film is composed of clusters of nano-scale formations. Between these clusters and formations that constitute them the holes (pores) are attested, fact that can explain a set of properties they possess.

As a result of graphite film formation on the piece surface the plunger diameter increases on average by about 14μm beside the initial diameter, i.e. as a result we form the graphite depositions with the respective dimension on the surface of the workpiece as continuous film (figure 2).

Thus, the application of films on the surfaces of the constituent parts of glass molding forms allowed to determine very effective their functionality. It was guaranteed to parts the durability at least 2 times higher compared with the parts coming from factory [2, 5]. This can be explained by the fact that graphite presents a solid state ointment and prevent adherence of the glass to the surface of the workpiece, and respectively its wear by the adherence, and by the fact that the graphite film has the anti-refractory properties and serves as thermal isolator between metal parts and liquid glass surface. The mentioned above is confirmed by the results obtained by the authors of the paper [5], where the plungers were tested under real exploitation conditions, as a result it was established that the plungers of molding forms with formed graphite films on their active surface were worked at 57600 cycles without modifying their initial shape and size. In this regard, the experimental investigations in technological cycle were performed to compare the wear of glass molding forms plungers [2]. The two plungers, the one coated with graphite by EDI and the other unprocessed, were subjected to the test (figure 3).

a)

b)

Figure 3. Overview of finishing pallets: a) with unprocessed surface, b) with graphite film.

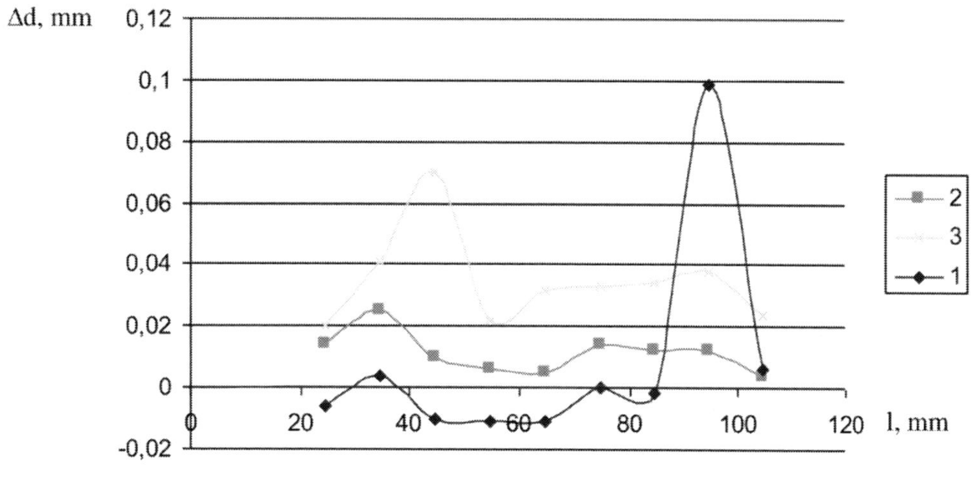

Figure 4. Dependence of the difference of piece diameter on its active length for: 1 – sample no.19 after functioning in technological cycle; 2 – sample no.23 after functioning in technological cycle; 3 – sample no.23 after applying EDI

If we examine the unprocessed by EDI sample (figure 3a) before and after its functioning at technological cycle (figure 4, curve 1), then we can see that in some investigated points its diameter decreases by about 11μm. The sample (figure 3) is studied from three points of view: at the initial stage, then after applying EDI and finally after being subjected to wear. The plunger's diameter increases in the allowable limits due to the graphite depositions on the surface after the piece was processed by DEI. The research of the piece after functioning under the above prescribed regimes showed the decrease in piece diameter by tens of micrometers, as shown in figure 4.

IOP Conf. Series: Materials Science and Engineering **95** (2015) 012027 doi:10.1088/1757-899X/95/1/012027

To confirm or to refuse the adherence on the piece surface of different materials the couples consisted of pieces with graphite films by pasting them surface to surface through the adhesive were tested according to the methodology described in [3, 13].

As a result of traction of assembled by pasting pieces we have observed that untreated specimens have higher values of shear stress than treated one, that constitutes about 85daN/cm^2, mixed-treated specimens have on average value of shear stress of 60daN/cm^2, which tells us that graphite film deposited on one of the specimen reduces the adhesion between the adhesive and the metal surface. The last variant is when both surfaces are coated with the graphite film; shear stress is reduced down to about 50daN/cm^2 that represents a 40% decrease in adhesion beside the untreated specimens. Also, it is noted that untreated specimens' shear takes place within the adhesive, as compared to those treated where shearing takes place in the separation surface of the adhesive with the graphite film. This effect can be explained by the fact that the formed film is porous and, respectively, the adhesive penetrates into its pores.

We may assume that one of the reasons of absorption of component adhesive substances is that the formations of the graphite film present fullerenes and they have the ability to absorb some components from the environment.

The results on solubility of graphite films formed by DEI method on metal surfaces in the interaction with different solvents are presented in [11, 12]. It is also led to the conclusion that the other susceptible chemical species are dissolved after treatment with a suitable solvent. Assuming initially due to electronic microscopy analyzes presented in figure 1, we can state that after the treatment of the metallic surface with graphite films by EDI other chemical formations are also obtained - probably some spatial carbon structures of fullerenes type.

In papers [6-10] it was stated that fullerenes are black, opaque solids, of low hardness and low conductive of heat and electricity. The behavior of samples treated with a specific to fullerenes solvent provides a first indication for assuming, according to which after treatment by EDI with formation of graphite films the same spatial formations composed of atoms of C are also obtained.

In order to broaden the field of application in practice of scientific elaborations and especially of graphite films deposited on metal surfaces by DEI method, a number of attempts on the determination values of screwing-unscrewing moment of threaded joints (bolt-nut) were made by the authors [4]. They were performed measurements on the amount screwing moment in couples "bolt-nut" with the formation of depositions on the frontal surface of the nut and the in the other case on the threaded surface of the bolt (see table 1).

Table 1. The values of the unscrewing moment of the nut coated with graphite M_{un}, N·m.

Number of passes	**Screwing force, kN**			
	0.5	**1**	**1.5**	**2**
2	2	6	9.9	13
3	2.2	5.8	9.5	12.5
4	2.5	4.5	8.2	10.15
5	3	6.1	10.2	14
6	2.9	8	12	16
8	3.5	6.2	11	14.1
10	3.5	5.9	11	13.2

If we compare the results shown in table 1, it can be seen that the screwing moment increases being dependent variable and on the number of passes (number of surface processing by DEI). It would seem that the formed depositions are not beneficial and lead to the increase of the value of screwing moment. Insignificant increase of the screwing moment is also caused by the enlargement of parts forming the couple.

However, at the reverse operation – unscrewing – the stick effect was not attested for parts of joints with graphite films on the active surface maintaining them in ordinary conditions (also in the case when maintaining them at high temperatures or if fitted joints was placed in chemically active environments)

4. Conclusions

As a result of the application of graphite films on metal surfaces by the electrical discharges in impulse method we can say that they have anti-adhesive properties, possess ointment properties, and decrease kinematic couplings corrosion in certain aggressive environments and increase wear resistance of pieces that work at high temperatures by about 10 times;

Behavior of graphite films could be caused by the formation of structures such fullerene synthesis;

Graphite pellicles can be successfully used to eliminate paraffin stoppers formation in pipes at oil and petroleum transportation due to their high anti-adherence properties

References

[1] Besliu V 2008 Research on thermal and thermo-chemical surface treatment of parts by applying electrical discharges in impulse *Summary of PhD thesis Galati* pp 53

[2] Topala P, S Mazuru, Besliu V, Cosovschii P and Ojegov A 2011 Application of EDI in increasing durability of glass moulding forms poansons *Proceedings of The 15th International Conference, Modern Technologies, Quality and Innovation* **II** pp 1093-1096

[3] Topala P, Besliu V and Marin L 2014 Decreasing the adhesion effect of surfaces using graphite pellicle deposition through electric discharges in pulse *Advanced Materials Research* (Switzerland: Trans Tech Publications) **1036** pp 172-177

[4] Topala P, Besliu V and Ojegov A 2011 Application of electrical discharges in impulse with graphite electrode *Mechanical engineering and technosphere of the XXI century* pp 240-245

[5] Topala P, Mazuru S, Besliu V, Cosovschii P and Stoicev P 2010 Increasing the durability of glass moulding forms applying graphite pellicles *Quality and Innovation* pp 635-638

[6] Qin L -C, Zhao X, Hirahara K, Miyamoto Y, Ando Y and Iijima S 2000 Sinteza şi caracterizarea unor nanostructuri cu atomi de carbon *70 Nature* **408** p 50

[7] Hong B H, Joshua Small P, Purewal Meninder S, Mullokandov A, Matthew Sfeir Y, Wang F, Lee J Y, Tony Heinz F, Louis Brus, Kim P and Kwang Kim S 2005 *Applied Physical Sciences* PNAS **102(40)** pp 14155–14158

[8] Moon J M, An K H, Lee Y H, Park Y S, Bae D J and Park G S 2001 Sinteza şi caracterizarea unor nanostructuri cu atomi de carbon 71 *Journal of Physical Chemistry B* **105(24)** pp 5677-5681

[9] Kim K S, Park Y S, An K H, Jeong H J, Kim W S, Choi Y C, Lee S M, Moon J M, Chung D C, Bae D J, Lim S C, Lee Y S and Lee Y H 2000 *Carbon Science* **1(2)** pp 53-59

[10] Ruxanda G, Stancu M, Vizireanu S, Dinescu G and Ciuparu D 2007 Investigation of transport and deposition of species in an arc discharge reactor with carbon electrodes *14th International Conference on Plasma Physics and Applications*

[11] Andrievsky G, Klochkov V K, Bordyuh A B and Dovbeshko G I 2002 Comparative analysis of two aqueous-colloidal solutions of C_{60} fullerene with help of FTIR reflectance and UV–Vis spectroscopy *Chemical Physics Letters* **364** pp 8

[12] Andrievsky G, Klochkov V and Derevyanchenko L 2005 Is the C 60 Fullerene Molecule Toxic?! *Fullerenes Nanotubes, and Carbon Nanostructures* **13(4)** pp 363

[13] Marin L, Topala P, Stoicev P, Besliu V and Ojegov A 2014 Application of graphite micro-films to decrease surface adhesion coefficient *Meridian Ingineresc* Tehnica UTM **1(53)** pp 39-44

A comparative study of morphology and composition on oxide nanopowders elaborated by SPVD

E S Bârcă[1], V Rizea[1], M Abrudeanu[1],A G Plaiasu[1,], B Istrate[2] and C Munteanu[2]
[1]University of Pitesti, Faculty of Mechanics and Technology, Department of Manufacturing and Industrial Management,1st Tg din Vale Str, 110040, Arges, Romania
[2]"Gheorghe Asachi" Technical University of Iasi, Faculty of Mechanics, Materials Science Departament,43 Dimitrie Mangeron Str, 700050,Iasi, Romania

E-mail: eddward_82@yahoo.com

Abstract. The paper proposes a method of obtaining oxide nanopowders in a solar reactor using the solar energy (solar physical vapor deposition). The proposed method is a new method of obtaining nanopowders. Taking into consideration that the properties of the materials change once they get to nanodimensions it is desirable to obtain nanomaterials starting from base materials. Because of the properties that nanomaterials provide, the demand has grown in the last period of time. Also the last few years provided a various number of nanomaterials used in different fields of research. Among the materials that were obtained by SPVD there can be reminded the MoS_2, the MoS_2 doped with Zn, CeO_2 doped with Zr. The CeO_2 is used in applications such as catalysis, gas sensor, fuell cell as well as optical additives. All of the obtained nanopowders were analyzed and characterized by X-Ray diffraction (XRD) and by scanning electron microscopy (SEM) in order to acquire the information about the crystallography and morphology considering the reactions parameters that were used in obtaining the nanopowders. From the X-Ray diffraction can be concluded in the case of the CeO_2 that doping occurred. By controlling the reactions parameters the end result is different in therms of obtained morphology of the nanopowders. All of the above elaborated nanopowders can be deposited in the form of thin films by laser ablation method in order to obtain better properties of the base material.

1. Introduction

Because of the raised interest in nanomaterials different methods for obtaining nano scaled materials have been developed. This paper focuses on obtaining nanopowders by a different, less used method known as SPVD (solar physical vapor deposition) [1]. Obtaining techniques vary and can include chemical or physical vapor deposition, precipitation methods, sol-gel and hydrothermal methods [2-5]. By SPVD there have been obtained different nanopowders including pure ZnO and Bi_2O_3 doped ZnO with better electrical properties than other nanopowders obtained by the same method [6]. Also by the same process there have been obtained Al-doped ZnO nanopowders and it has been noticed that when the amount of Al is smaller the crystallite size is smaller [2].

When switching to nano-dimension materials properties change, making them more appropriate for use in many applications such as electronics, automotive, construction industries and chemical. The interest in materials with nano-dimensions is rapidly increasing because of the large scale of opportunities that are provided by the materials properties. One of the advantages in using the solar

Content from this work may be used under the terms of the Creative Commons Attribution 3.0 licence. Any further distribution of this work must maintain attribution to the author(s) and the title of the work, journal citation and DOI.
Published under licence by IOP Publishing Ltd

furnaces is the fact that the used energy for the reaction is solar energy which has unlimited resources. The solar furnaces uses a solar beam concentrated by mirrors in order to melt down the target which is then vaporized and then reaches a cold support where it remains in order to be collected at the end of the process. In order to have the best end result, studies concerning the orientation of the mirrors of a solar furnace have been carried out by different researchers. Because it is hard to realize a mirror in the needed size, the solution is to use a large number of smaller mirrors that would direct the energy towards the target [7-8]. Manufacturing of different nanopowders by solar physical vapor deposition has shown the possibility of obtaining the desired nanoscale meanwhile controlling the composition and the morphology at the same time.

The elaboration of the nanopowders has been performed in France, Font-Romeu in one of the solar furnace, named Heliotron. The "heliotron" solar furnace is one of the 12 furnaces built in Odeillo-Font Romeu and is made out of mobile plane mirrors which capture the sun light and reflects its radiation toward a parabolic concentrator. The parabolic mirrors focus on a reactor containing a glass flask in which the reaction occurs using solar energy. Inside the glass flask the reaction can take place under an inert gas, the atmosphere being controlled. The advantage of the method is that nano-sized crystals are obtained, the disadvantage being the fact that it needs perfect weather conditions and that it takes several hours to complete the reaction. Due to the properties they have, Al doped ZnO and pure ZnO have been elaborated by this technique. As applications can be reminded the gas sensors, electrodes and piezoelectric devices [9-10].

The paper aims to obtain CeO_2 and ZrO_2 doped CeO_2 nano-powders starting from micro scaled powders. Choosing these powders was made according with the applications they have as fuel cell, gas sensors, free radical scavenger and optical additives [11]. The CeO_2 can also be used to deposit thin films due to the properties and applications the thin layer would have in corrosion prevention, photocatalysis, electrochemical cells, microelectronics, optical devices, thermal coatings and biomaterials. These applications become possible due to the transition to nanoscale. For better results in the chosen application close inspection for surface defects such as oxygen vacancies needs to be carried out. Using the tetravalent Zr as the doping agent for CeO_2 increases the oxygen storage capacity and using the trivalent Zr increases CeO_2 conductivity. Another factor that influences the layers thickness is considered to be the interaction that the Ce atoms have with the substrate [12-13]

2. Experimental procedure

Pure and Zr doped CeO_2 powders with a general formula of $(CeO_2)_x(ZrO_2)_y$ were prepared by solar physical vapor deposition (SPVD). In this formula x/y represents the molar atio of the two powders. The paper presents 2 compositions: $Ce_{0.95}Zr_{0.05}O_{2-x}$ and $Ce_{0.75}Zr_{0.25}O_{2-x}$ obtained by this method.

The SPVD process consists in a material being melted under concentrated solar beam inside a solar reactor. As the name suggests, the power source is the energy from the sun. The process took place in Odeillo-Font Romeu in France in one of the 12 solar furnaces. The solar furnace consists of a plane mirror field named "heliostat" and a parabolic mirror, or concentrator as it can be seen in (figure 1a). The plane mirror captures the sun radiation and directs it to the parabolic concentrator. At the focus of the parabolic concentrator is placed the solar reactor "heliotron" which is composed out of a glass balloon inside which the controlled atmosphere can be maintained. The used target is the powder that we want to obtain in nanoscale, that prior to the process has been prepared in a press of 14.7 to 19.6 kN in order to obtain the initial pills. The pressure inside the balloon depends on the material and on the atmosphere. While the reaction takes place vapor of materials are condensed either on a cold metallic support or on a ceramic filter attached to the glass balloon as presented in (figure 1b). The "heliotron" associates both processes, the collecting on the metallic support and on the filter which increases the quantity of the obtained powder and decreases the condensation process on the balloon walls. The gas flux is regulated as to be in the desired parameters, being higher at low pressures

IOP Conf. Series: Materials Science and Engineering **95** (2015) 012028 doi:10.1088/1757-899X/95/1/012028

Figure 1. (a) Solar furnace in PROMES Laboratory France, Font-Romeu, Odeillo , (b) 2kW solar concentrator with "heliotron" reactor during reaction.

One of the advantages of this method is that there is obtained a direct synthesis of nanopowders. The disadvantage consists in the long period of time necessary in order to complete the process of about 1 to 2 hours in perfect sunny conditions.

All the powders involved in the process were characterized both by X-Ray Diffraction with the X'Pert Pro Mrd diffractometer and by scanning electron microscopy with the Quanta 200 3D electronic microscope.

The X-Ray diffractions were carried out in the 10°-90° 2θ interval, each taken step having the a value of 0.013°(2θ). The link between the diffraction angle θ and the wavelength is given by Bragg's relation:

$n \cdot \lambda = 2 \cdot d \cdot \sin\theta$, where :

n - positive integer

λ - wavelength of the rays

d - the spacing between layers of atoms

θ - the angle between the incident rays and the analyzed surface

The diffractometer works according to Debye-Scherrer principle's, data being transmitted and interpreted by a computer with a dedicated software for this equipment , X'Pert Data Collector, X'Pert High Score Plus and X'Pert Data Viewer, results beeing presented as diffractions. The diffractometer has an X Ray tube with an Cu kα anode, λ=1.54 Å, to which a 45kV tension has been applied at an intensity of 40mA. The used radiation was Cu kα1 with a wavelength of 1.54060 Å and Cu kα2 with a wavelength of 1.54443 Å.

All SEM analysis were carried out in High Vacuum module, at working pressure between 50 - 60 Pa, using the LFD (Large Field Detector). The accelerating voltage of the scanning electron microscope was of 20kV. All of the XRD and SEM analysis were carried out at "Gh. Asachi" Techincal University of Iasi, Romania.

3. Results and discussion

The XRD pattern of the base CeO2 powder is presented (in figure 2.a) and shows the position of the peaks, information about the crystallographic parameters and confirms the system it crystallizes in, cubic with 5.4037x5.4037x5.4037 cell. Also the XRD provides information about the Miller indices and about the FWHM values (in figure 2.b). We see in figure 3 and in figure 4 the XRD pattern for the $Ce_{0.95}Zr_{0.05}O_{2-x}$ and $Ce_{0.75}Zr_{0.25}O_{2-x}$ powders. As can be seen some differences appear when the doping occurs, especially for the nanopowder with a higher quantity of dopant, $Ce_{0.75}Zr_{0.25}O_{2-x}$.

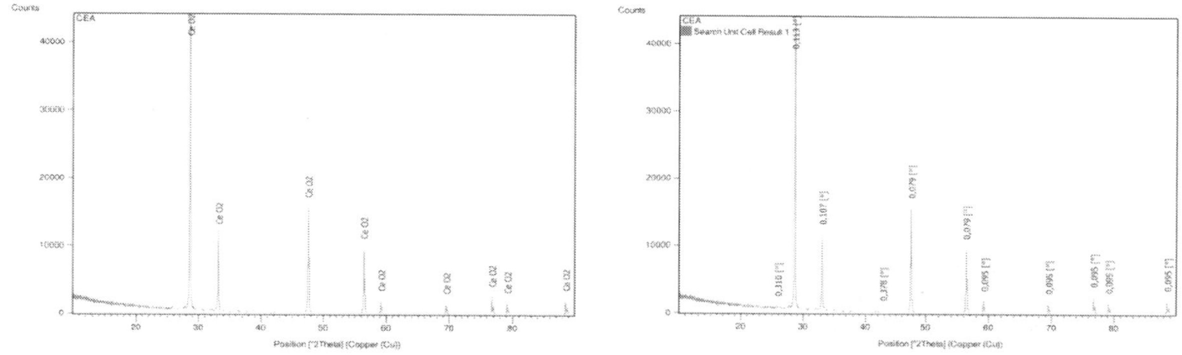

Figure 2. (a) XRD pattern of the CeO_2 powder, (b) FWHM for the CeO_2 powder.

In the next figure (figure 3) we observe the $Ce_{0.95}Zr_{0.05}O_{2-x}$ were the doping occured, being visible that the peaks are slightly different and less higher compared to the analyzed base powder. The system it crystallizes is cubic, with 5.3890x5.3890x5.3890 cell.

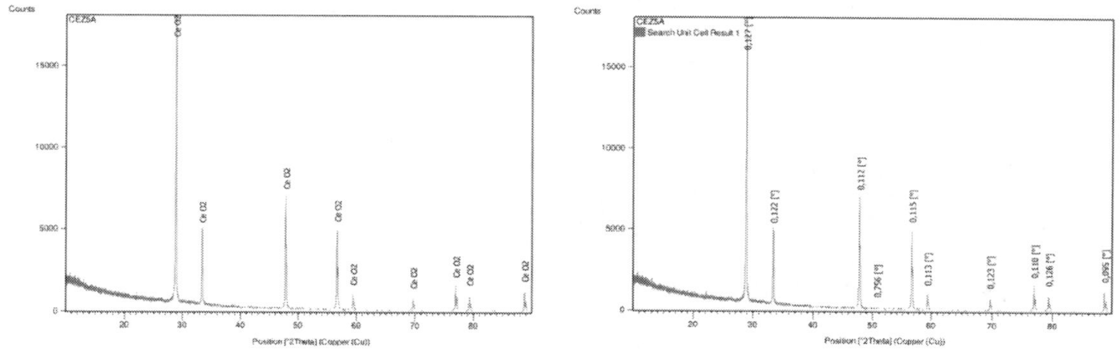

Figure 3. (a) XRD pattern for the $Ce_{0.95}Zr_{0.05}O_{2-x}$, (b) FWHM for the $Ce_{0.95}Zr_{0.05}O_{2-x}$ powder.

The more notable difference can be seen in Figure 4 for the $Ce_{0.75}Zr_{0.25}O_{2-x}$ where the presence of some peaks of Zr appears, the system it crystallizes in is cubic, with 5.4100x5.4100x5.4100 cell.

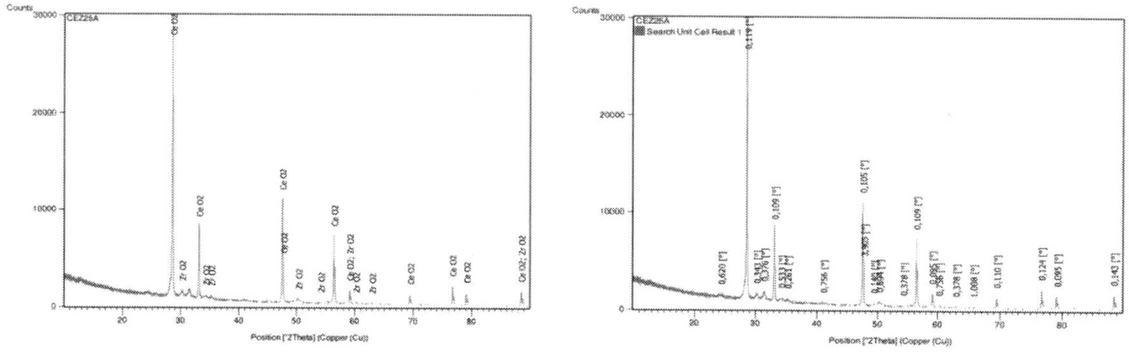

Figure 4. (a) XRD pattern for the $Ce_{0.75}Zr_{0.25}O_{2-x}$, (b) FWHM for the $Ce_{0.75}Zr_{0.25}O_{2-x}$ powder.

For all the CeO_2 powders SEM analysis have been carried out. All of the results are presented in the next 3 figures (figure 5, figure 6 and figure 7). As can be seen after the SPVD process the dimmension of the particle size has smaller values. For both obtained nanopowders $Ce_{0.95}Zr_{0.05}O_{2-x}$ and $Ce_{0.75}Zr_{0.25}O_{2-x}$ the particle size is between 38 and 75nm which proves the fact that the transition to nanoscale occured along with the doping of the powders. In the nanopowders we can observe agglomeration of grains with spherical morphology.

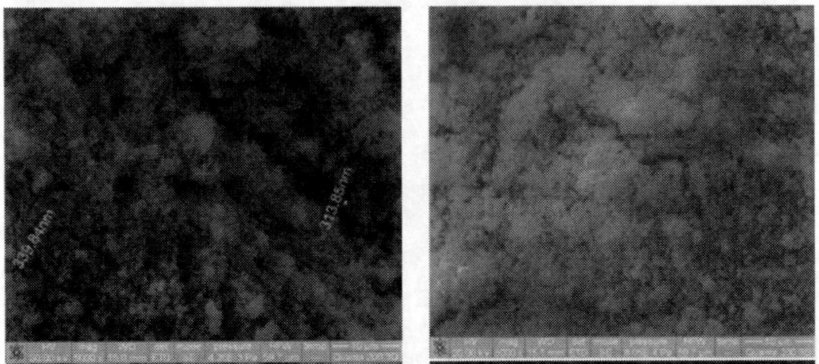

Figure 5. (a) CeO_2 powder (b)CeO_2 after SPVD process.

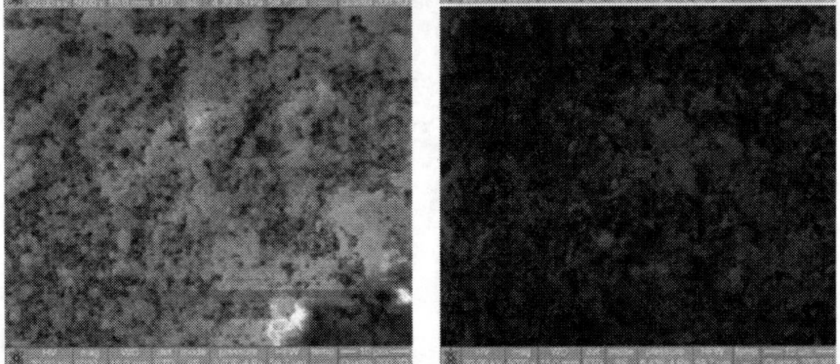

Figure 6. (a) $Ce_{0.95}Zr_{0.05}O_{2-x}$ powder (b) $Ce_{0.95}Zr_{0.05}O_{2-x}$ after SPVD process.

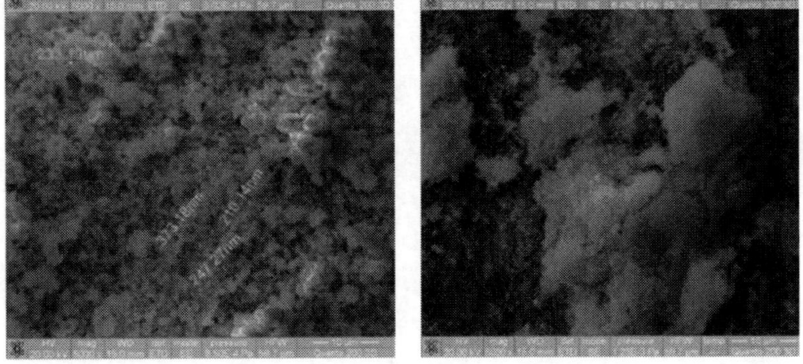

Figure 7. (a) $Ce_{0.75}Zr_{0.25}O_{2-x}$ powder (b) $Ce_{0.75}Zr_{0.25}O_{2-x}$ after SPVD process.

4. Conclusions

All of the solar physical vapor depositions were carried out in a 2kW solar furnace which includes a "heliostat", a concentrator and a solar reactor "heliotron". The solar reactor is positioned in the direct beam of the parabolic mirror thus making the reaction possible. The materials were prepared in the shape of pills in one press in order for the reaction to take place. The nanopowders were synthesized through SPVD from the initial powders with the following composition $Ce_{0.95}Zr_{0.05}O_{2-x}$ and $Ce_{0.75}Zr_{0.25}O_{2-x}$. After the SPVD, $Ce_{0.95}Zr_{0.05}O_{2-x}$ and $Ce_{0.75}Zr_{0.25}O_{2-x}$ nanopowders were obtained with a particle size is between 38 and 75nm which proves the fact that the transition to nanoscale occured along with the doping of the powders. The obtained nanopowders will be used to obtain the targets in order to deposit them as thin films through laser ablation. The study will continue with the analysis of the thin films deposited taking into consideration all the applications that the CeO_2 can be used in, such as corrosion prevention, optical devices, thermal coatings and biomaterials.

Acknowledgements
This work of Bârcă Eduard Sebastian was supported by the strategic grant POSDRU/159/1.5/S/138963 - PERFORM, co-financed by the European Social Fund – Investing in People, within the Sectoral Operational Programme Human Resources Development 2007-2013.
All authors are indebted to SFERA Grant Agreement no. 228296, 7th EUprogram which has provided opportunities to collaborate with researchers from France working with the Odeillo solar facilities on close fields.

References

[1] Guogang T, Jianrong S et al. 2012 Synthesis and characterization of flower like MoS2 nanostructures through CTAB-assisted hydro thermal process, journal homepage: www elsevier com/locate/matlet *Materials Letters* **86** pp 9–12
[2] Kouam J, Ait-Ahcene T, Plaiasu A G et al 2008 Characterization and properties of ZnO based nanopowders prepared by solar physical vapor deposition (SPVD), www elsevier com/locate/solener *Solar Energy* **82** pp 226–238
[3] Townshend A and Jackwerth E 1989 Precipitation of major constituents for trace preconcentration: potential and problems *Pure & App Chem* **61(9)** pp 1643-1656
[4] Hench L L and West J K 1990 The sol-gel process *Chemical Reviews* **90(1)** pp 33–72
[5] Byrappa K, Masahiro Y 2001 *Handbook of Hydrothermal Technology* (Norwich New York: Noyes Publications) Chapter 2: History of Hydrothermal Technology
[6] Ait Ahcene T, Monty C, Kouam J, Thorel A, Petot-Ervas G and Djemel A 2007 Preparation by solar physical vapor deposition (SPVD) and nanostructural study of pure and Bi doped ZnO nanopowders, www elsevier com/locate/jeurceramsoc, *Journal of the European Ceramic Society* **27** pp 3413–3424
[7] Jafrancesco D, Sansoni P, Francini F, Contento G, Cancro C, Privato C, Graditi G, Ferruzzi D, Mercatelli L, Sani E and Fontani D 2014 Mirrors array for a solar furnace: Optical analysis and simulation results, journal homepage: www elsevier com/locate/renene *Renewable Energy* **63** pp 263-271
[8] Plaiasu A G, Abrudeanu M, Dicu M M and Monty C Elaboration and nanostructural study of pure and Al doped ZnO nanopowders *Journal of Optoelectronics and Advanced Materials* **16(9-10)** pp 1116-1120
[9] Plaiasu A G and Anghel D C 2010 Application of functional analysis to establish the process of nanostructured oxide film deposition *Proceedings of the International Conference ModTech* pp 459-462
[10] Soren S, Bessoi M and Purnendu P A rapid microwave initiated polyol synthesis of cerium oxide nanoparticle using different cerium precursors, journal homepage: www elsevier com/locate/ceramint *Ceramics International* available online www sciencedirect com
[11] Castano C E, Matthew J O'Keefe and William G Fahrenholtz 2015 Cerium-based oxide coatings, journal homepage: www elsevier com/locate/cossms *Current Opinion in Solid State and Materials Science* **19** pp 69–76
[12] Parvulescu V I and Tiseanu C 2015 Local structure in CeO2 and CeO2–ZrO2 nanoparticles probed by Eu luminescence, journal homepage: www elsevier com/locate/cattod *Catalysis Today* article in press
[13] Vorokhta M, Matolínová I, Dubau M, Haviar S, Khalakhan I, Ševčíková K, Mori T, Yoshikawa H and Matolín V 2014 HAXPES study of CeOx thin film–silicon oxide interface *Applied Surface Science* **303** pp 46–53

IOP Conf. Series: Materials Science and Engineering **95** (2015) 012029 doi:10.1088/1757-899X/95/1/012029

The shape factor influence on the control process of multiferroic nanoparticle configurations in GHz range

D Ionescu [1] and M Kovaci [2]
[1]"Gh. Asachi" Technical Univ. of Iasi, Faculty of Electronics, Telecommunications and Information Technology, Dept. of Telecommunications, Carol I Blvd, No. 11, 700506, Iasi, Romania
[2] Politehnica University of Timisoara, Faculty of Electronics and Telecommunications, Dept. of Communications, V. Pârvan Blvd, No. 2, 300223, Timisoara, Romania

E-mail: danaity@yahoo.com

Abstract. Materials with piezoelectric properties associated with the ferroic phases generate multiferroic structures at microscopic level. Multiferroic core-shell nanoparticles with different configurations were considered: nanospheres versus nanotubes – spherical (magnetostrictive core / piezoelectric shell), respectively cylindrical (magnetostrictive rod / piezoelectric cylindrical coating) The external sphere diameter / cylinder lengths were of 80 – 100 nm to 360 nm, while the inner sphere / rod diameters were of 20 to 60 nm. The influence of the constituents shape on their control process realized by an applied magnetic field, H_0 (100 – 270 Oe) was investigated. The nanoparticles were simulated using the HFSS 13.0 program, at frequencies in microwave range (16 – 28 GHz), considering the nanostructures with hexaferrites as magnetic phase ($AFe_{12}O_{19}$ M-type hexaferrites, with A an alkali earth metal) and the Bi_2NiMnO_6 perovskite like piezoelectric. The particle configurations response is function of their geometrical shape, modified by the core diameter and the shell thickness. The magnetoelectric (ME) coefficient tensor has been determined using a physical algorithm based on the near-filed values in the particles vicinity, generated by the HFSS, and also the ME voltage coefficient (obtained of tens of mV/cm·Oe). Performance of the two shapes composite nanoparticles to generate a strong ME response was discussed comparatively and the influence of the shape factor was illustrated on graphs. Each composite configuration presents specific advantages in respect with the control process by the H_0 field, considering that the effect of dipolar field generated as response by the electric phase is strongly anisotropic. These features can be elegant illustrated by simulations (non destructive, applied at nanoscale) which help us to decide to the proper nanoparticles shape and the optimal intensity of the applied fields for the desired strength of the ME effect in a defined microscopic area.

1. Introduction

Multiferroic materials with multiple ferroic phases can be obtained using different constituents, in a specific configuration. The core – shell configurations of nanoparticles have demonstrated high values of the magnetoelectric coefficients due to the strong coupling between magnetic and electric phases, mediated by a strain at the certain boundaries between the shell and the core. The shape factor influences consistently the internal processes and consequently the composite properties. The specific quantities control via the shape factor is a method of multiferroic structure optimization which can be easily experimented in practice via simulation. The 3D microscopic model of the composite, realized

Content from this work may be used under the terms of the Creative Commons Attribution 3.0 licence. Any further distribution of this work must maintain attribution to the author(s) and the title of the work, journal citation and DOI.
Published under licence by IOP Publishing Ltd

at crystalline lattice level, can be modified non-destructive in order to improve the structure electromagnetic performances. Based on the models given in literature [4, 5], nanoparticles grouped in core-shell nanospheres versus nanotubes have been simulated in this paper in a static magnetic field, applied for polarization.

Application of the ME structures manufactured in such manner are multiple: voltage tunable multiferroic RF/microwave devices, devices with strong converse ME coupling, used at: tunable bandpass filters, high power phase shifters, magnetic sensors, spintronics, energy harvesters, etc.

2. Theoretical considerations

The free energy of a magnetoelectric material, placed in the electric $\vec{E}$ and magnetic $\vec{H}$ fields, can be written as [1]:

$$F(E,H) = F_0 - P_i^s E_i - M_i^s H_i - \frac{1}{2}\varepsilon_0\varepsilon_{ij}E_iE_j - \frac{1}{2}\mu_0\mu_{ij}H_iH_j - \alpha_{ij}E_iH_j - \\ -\frac{1}{2}\beta_{ijk}E_iH_jH_k - \frac{1}{2}\gamma_{ijk}E_iE_jH_k - \ldots \quad (1)$$

where P^s, respectively M^s is the spontaneous electric, respectively magnetic polarization, $\overline{\overline{\varepsilon}}$ respectively $\overline{\overline{\mu}}$ is the electric permittivity, respectively magnetic permeability tensor, $\overline{\overline{\alpha}}$ is the linear magnetoelectric coupling tensor, β_{ijk}, γ_{ijk} ... are the nonlinear magnetoelectric coefficients. Hence, by differential calculus, one obtains [1]:

$$P_i(E,H) = -\frac{\partial F}{\partial E_i} = P_i^s + \varepsilon_0\varepsilon_{ij}E_j + \alpha_{ij}H_j + \frac{1}{2}\beta_{ijk}H_jH_k + \gamma_{ijk}H_iE_j + \ldots$$

$$M_i(E,H) = -\frac{\partial F}{\partial H_i} = M_i^s + \mu_0\mu_{ij}H_j + \alpha_{ij}E_j + \beta_{ijk}H_jE_i + \frac{1}{2}\gamma_{ijk}E_jE_k + \ldots \quad (2)$$

In multiferroics, the expresions is not so rigorous because $\varepsilon_{ij}(T)$ and $\mu_{ij}(T)$ display field hysteresis. Generally, ferroics are better parameterized in terms of resultant rather than applied fields. In practice, terms of superior order can be neglected, the calculation of α, β and γ coefficients being sufficient for the permittivity and permeability tensors determination.

The following valuable inequality is valuable, which indicates that a large magnetoelectric effect is associated with great values for the permittivity, respectively permeability of the constituents included in the structure [10]:

$$\alpha_{ij}^2 \le \varepsilon_0\mu_0\varepsilon_{ij}\mu_{ij} \quad (3)$$

Consequently, materials where chosen as a double perovskite (electric permittivity $\varepsilon_r \sim 134$) which is magnetoelectric and hexaferrites of M-type, with large magnetic permeabilities ($\mu_r \sim 700\text{-}800$).
Considering the effective values for the susceptibilities, the magnetoelectric voltage coefficient for the polycrystalline samples can be written as:

$$\alpha_{ME} = \frac{dE}{dH} \quad (4)$$

and will be calculated using the data given the HFSS program for our nanocomposite model.

3. Core-shell magnetoelectrics

The magnetoelectric configurations of nanoparticles which we have considered in our study can be described as nanospherical configurations versus nanotubes: spherical core-shell structures, with magnetostrictive core and piezoelectric shell, respectively cylindrical core-shell structures with magnetostrictive rod and piezoelectric cylindrical coating. The external piezoelectric shell is represented by the Bi_2NiMnO_6 double perovskite and the magnetic inner core is a $AFe_{12}O_{19}$ M-type hexaferrites, with A an alkali earth metal: Ca, Sr, Ba.

Characteristics of the crystalline lattice for each phase can be summarized: for the Bi_2NiMnO_6 perovskite: lattice parameters (monoclinic, space group C121): a = 9.4646(4) Å, b = 5.4230(2) Å, c = 9.5431(4) Å, and β = 107.823(2)° [9], [11]; for M hexaferrites: lattice parameters (space group $P6_3mc$): $BaFe_{12}O_{19}$: a = 5.892 Å, c = 23.183 Å; $SrFe_{12}O_{19}$ a = 5.8808 Å, c = 23.2018 Å; $CaFe_{12}O_{19}$: a = 5.8303 Å, c = 22.1652 Å. The M hexaferrites have a RSR*S* crystallographic build up, with 2M molecules per unit cell (figure 1). The structure is hexagonal, represented by a backbone of oxygen anions, the interstices of the close-packed oxygen lattice being occupied by metallic cations, magnetic and nonmagnetic. The cations usually have either four- or six-fold coordination and form tetrahedral, octahedral or trigonal sublattices [12].

The ionic radii of the atomic species included in the considered phases are: cations: Bi^{3+} = 1.17 Å, Ni^{2+} = 0.83 Å, Mn^{2+} = 0.81 Å, Mn^{3+} = 0.72 Å, Fe^{2+} = 0.75 Å, Fe^{3+} = 0.69 Å, Ba^{2+} = 1.49 Å, Ca^{2+} = 1.14 Å, Sr^{2+} = 1.32 Å; anion: O^{2-} = 1.26 Å [8].

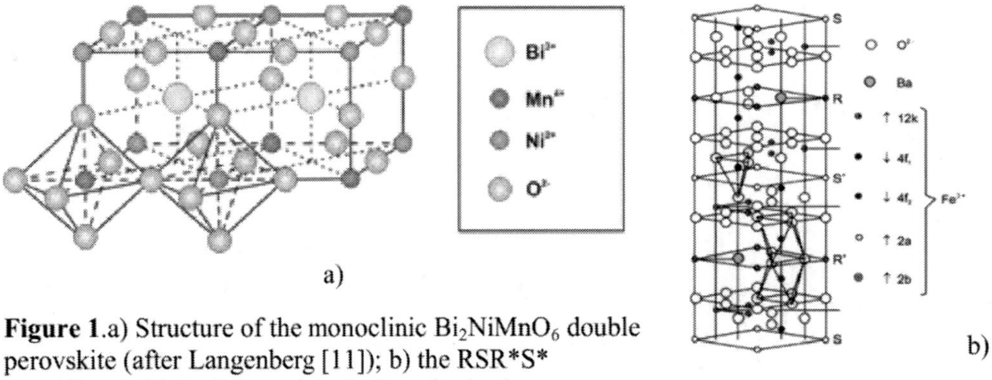

Figure 1.a) Structure of the monoclinic Bi_2NiMnO_6 double perovskite (after Langenberg [11]); b) the RSR*S* crystallographic build up of the M hexaferrite (source Mariño-Castellanos [12]).

The core-shell nanocomposites analysed with our simulation model (figure 2) have an external sphere diameter / cylinder length of 80 – 100 nm to 360 nm, an inner sphere / rod diameter of 20 to 60 nm. Samples are polycrystalline and the intrinsic dimensions depend also on the size of the pores of the membrane. We have estimated that dimensions and relative positions of the nanoconstituents can be controlled in a percent of ± 20 %, based on the data reported in literature. Two stages are indicated in the manufacturing process of the core-shell configurations: synthesizing of the ferrite / ferroelectric nanoparticles and covering the ferrite particle core (sphere or rod) with ferroelelctric particle shell. Chemical methods are prefered, not sintering at high temperatures due to the advantages offered by the chemical low temperature processes [3]. Different methods were reported, such as: the sol-gel process followed by coaxial electrospinning, the template-assisted liquid phase deposition (LPD) method [2], [4].

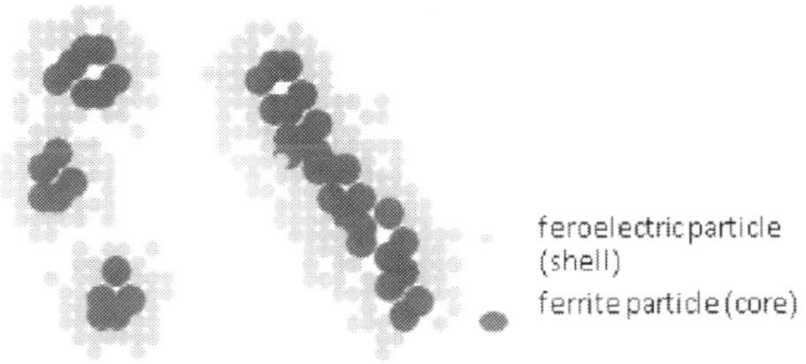

Figure 2. Polycrystalline nanoshereres, respectively nanotubes, with ferroelectric particles in the shell and magnetic particles (ferrite) in the core.

Properties control can be achieved by: modifying the chemical composition of the constituents (the ferrite core in our case), or by varying the ferrite / perovskite ratio by changing the wall thickness of the shell or the core diameter. The shape factor was modified inside the limits predicted by the theory, in order to obtain materials with tunable properties and reproducible characteristics. New possibilities of tuning and control were investigated, using the simulation model of the core-shell ferrite – ferroelectric structures.

At simulation we have to consider the peculiarities offered by the core-shell geometry such as high surface contact between the components and separation of the magnetic particle groups encapsulated in the perovskite shell. The processes at particle interfaces are dependent on the stain generated at interfaces by the magnetostrictive phases, on the degree of coupling with the electric phase and on the interfaces properties: characteristics of the interfaces at the atomic scale, presence of the voids, the curvature of the surface, the intrinsic contact area, etc. [7, 10, 11].

4. Results for the magnetoelectric coefficients

A simulational study was performed in order to determine the influence of the constituents shape on the properties control process of the nanocomposites, realized by an applied magnetic field. Core-shell nanospheres versus nanotubes structures were considered for comparison. Different hexaferrites were taken into account by changing the A alkali earth metal in the $AFe_{12}O_{19}$ combination.

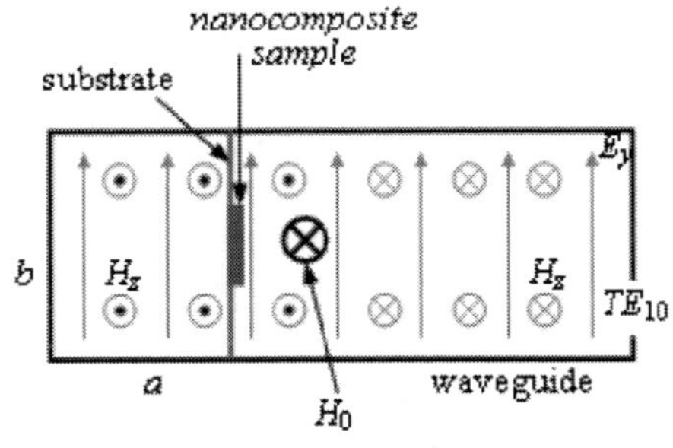

Figure 3. Cross-section of the rectangular waveguide in the inner plane, median at sample level. The propagation mode is TE_{10}.

Samples (4 x 0.8 x 12 mm^3) were simulated inside a rectangular waveguide WR-51 (R180 IEC Standard, K band), excited on TE_{10}, 1mW incident power at the input / output ports (figure 3.). The *S* parameters were determined for each nanoparticle configuration. The testing field has the frequency in the domain of 16 – 28 GHz, above the ferrite resonances. An 100 – 270 Oe external magnetic field was applied parallel with the propagation direction inside the waveguide, for controlling the electromagnetic properties of the nanoparticle samples (magnetic field influences the piezoelectric deformation of the nanocomposite). The response of the core-shell nanoparticle samples is function of the particles geometrical configuration, coordinated by the strength of the magnetoelectric effect inside the material [6, 7, 9].

The components of the magnetoelectric coefficient tensor have been calculated using a physical algorithm based on the near-filed values in the particles vicinity, generated by the HFSS program. The magnetoelectric voltage coefficient (*dE/dH*), of tens of mV/cm·Oe for the considered nanocomposites) has also been computed for each particles configuration.
Results for the magnetoelectric voltage coefficient (dE/dH) were represented in figures 4 – 7.

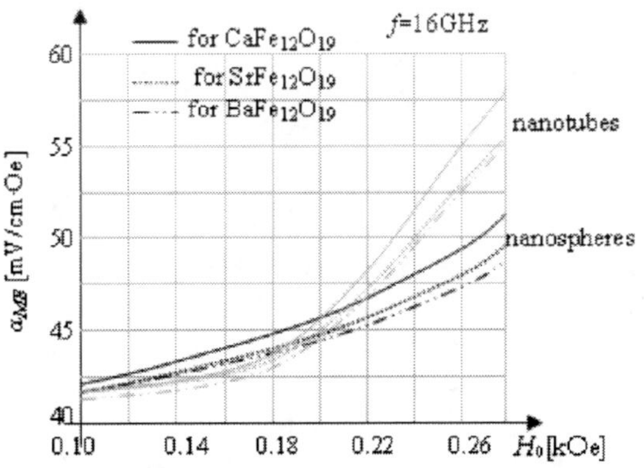

Figure 4. Evolution of the magnetoelectric voltage coefficient (dE/dH) in function of the applied magnetic field H_0, in the composites with nanospheres, respectively nanotubes.

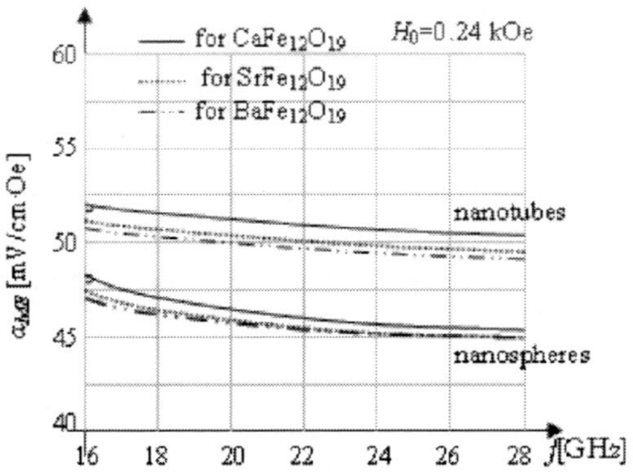

Figure 5. Evolution of the magnetoelectric coefficient in function of frequency, for an applied magnetic field H_0=240 Oe.

The α_{ME} coefficient increases with the applied magnetic field H_0, due to the intensifying of the magnetoelectric effect inside the material samples. Magnetostriction in the magnetic phase is increased, the coupled piezoelectric phase generating a higher polarization. The increasing of the ME effect is almost linear in the case of the nanosphere composites and evolves parabolic at nanotubes. Due to the higher agglomeration of magnetic particles in the rod core inside the cylinder, higher field are necessary to overlap the magnetic interactions between particles and to favour magnetostriction.

The magnetoelectric coefficient decreases slowly with frequency over the ferrimagnetic resonance, as the theory predicts [1], [3], [10], the decreasing being more arched at nanosheres having more accentuate individuality concerning the interactions between constituent particles.

The influence of the shape factor on the magnetoelectric effect inside the material samples was illustrated on graphs in figures 6-7. For an applied magnetic field of 240 kOe, considered optimum for the compromise polarization / energy consumed, the testing frequency was of 16 GHz. The average dimensions of the core-shell particles were varied in the range of 20 to 60 nm for the magnetostrictive inner sphere or rod diameter, respectively of 80 – 200 nm / 80 - 360 nm for the external piezoelectric sphere diameter / cylinder length.

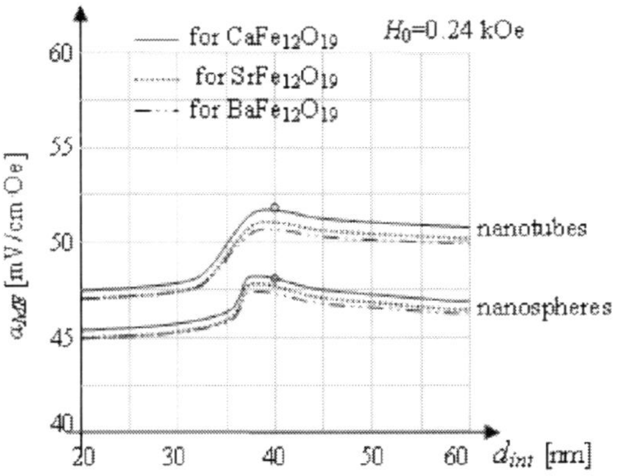

Figure 6. Evolution of the magnetoelectric coefficient in function of the magnetostrictive core diameter d_{int}, for an applied magnetic field H_0 of 240 Oe.

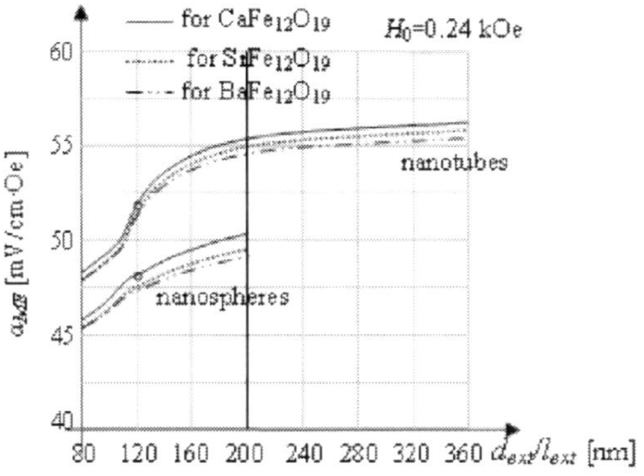

Figure 7. Evolution of the magnetoelectric coefficient in function of the external piezoelectric sphere diameter / cylinder length d_{ext} / l_{ext}, for H_0 = 240 Oe.

Magnetization is directly dependent on the molar fraction of the ferrite phase (controlled by the average diameter of the magnetic core, d_{int}), as so as the electric permittivity increases when the ferroelectric ratios increases because of the enhanced influence of the ferroelectric regions which surround the ferrimagnetic regions [4]. Even if the magnetic permeability of the ferrite increases with the applied H_0, the reduction in the thickness of the shells results in a higher interaction between the magnetic cores [2], [4] the global ME effect being soften, even drastic at low values of d_{ext} / l_{ext}. The same effect of ME softening due to the core interactions apears also when the core dimensions increases over an optimum, which can be determined by simulation methods.

5. Conclusions

Core-shell nanocomposite structures of feroelelctric – ferromagnetic particles have been studied by simulation methods, in order to determine the influence of the shape factor on the strength of the magnetoelectric effect inside the polycrystalline material samples. Each composite configuration presents specific advantages in respect with the control process by the external magnetic field, considering that the effect of the resulting dipolar field generation is strongly anisotropic. Simulations are coming to help us to decide on the proper nanoparticles shape and the optimal values for the applied fields necessary to obtain the desired magnetoelectric coefficient in certain conditions.

In the first instance, one remarks that the close-packed structures like core-shell configurations ensures a high degree of control with low energy consumption (the H_0 polarizing fields are consistently below 1 kOe).

Nanotubes are more easy to be controlled in higher polarization fields, while nanospheres respond almost linear to the magnetic control field. The magnetoelectric coefficient decreases slowly with frequency over the ferrimagnetic resonance, more flexured at nanosheres.

Variation of the geometrical dimensions of the components in the nanocomposite is a good solution to control the magnetoelectric properties of the nanotubes, the dimension influence being higher in their case. The nanospheres response at dimension modification is not so accentuate, but more modulated. The average diameter of the magnetic core controls the magnetization, as so as the ferroelectric ratios influence the electric permittivity, but by increasing the core dimensions over an optimum or by reducing the thickness of the shells, the global ME effect is soften.

Simulation methods impose themselves like powerful tools for analysing the nanomaterial properties at nanoscle, the methods being non-destructive. The results indicate us the proper nanoparticles shape factor and the optimal intensity of the applied fields in order to obtain the desired values of the ME coefficient for a specific material sample.

References

[1] Hajra P, Maiti R and Chakravorty D 2011 Nanostructured Multiferroics *Trans. Ind. Ceram. Soc.* **70(2)** pp 53-64

[2] Raidongia K, Nag A, Sundaresan A and C N.R Rao 2010 Multiferroic and magnetoelectric properties of core-shell $CoFe_2O_4$@$BaTiO_3$ nanocomposites *Applied Physics Letters* 97 062904

[3] Chu I-W 2010 Multiferroic and core-shell nanostructures - Advances in preparations and novel properties, dissertation thesis *City University of New York* p 110

[4] Koo Y S, Song K M, Hur N, Jung J H, Jang T–H, Lee H J, Koo T Y, Jeong Y H, Cho J H and Jo Y H 2009 Strain-induced magnetoelectric coupling in BaTiO3/Fe3O4 core/shell nanoparticles *Appl. Phys. Lett.* 94 032903

[5] Kleemann W and Binek C 2014 Multiferroic and magnetoelectric materials, in: M. Farle, H. Zabel (Eds.) Magnetic nanostructures - Spin dynamics and spin transport *Springer Tracts in Modern Physics* **246** pp 163 - 187

[6] Ionescu D and Ciobanu I B 2008 About the electromagnetic properties of the Fe-Si-B-Nb-Cu soft magnetic composites *Journal of Optoelectronics and Advanced Materials* **10(12)** pp 3305-3310

[7] Ionescu D and Kovaci M 2013 Controlled Properties of the Thin Films of $Ni_{0.5}Zn_{0.5}Fe_2O_4$-$BaTiO_3$ Ferrite-Ferroelectric Nanocomposites - Simulational Study, *Innovative Manufacturing Engineering Book Series: Applied Mechanics and Materials* **37**1 pp 383-388

[8] Atomic Radii Tables: http://home.gwu.edu/~xqiu/links_info/science/atom_radii.htm#ionic

[9] Zhao H, Kimura H, Yao Q, Du Y, Cheng Z and Wang X 2011 New Multiferroic Materials: Bi_2FeMnO_6, in: Dr. Mickaël Lallart (Ed.) *Ferroelectrics - Material Aspects InTech* http://www.intechopen.com/books/ferroelectrics-material-aspects/newmultiferroic-materials-bi2femno6 pp 237-250.

[10] Bichurin M I, Petrov V M and Priya S 2011 Magnetoelectric Multiferroic Composites, in: Dr. Mickaël Lallart (Ed.), *Ferroelectrics – Physical Effect*, (Croatia: InTech) pp 277-302

[11] Langenberg E, Fina I, Ventura J, Noheda B, Varela M and Fontcuberta J 2012 Ferromagnetic and Ferroelectric Bi2NiMnO6 double perovskites *Physical Review* B **86** pp 85-108

[12] Mariño-Castellanos P A, Moreno-Borges A C, García O-M G, Govea-Alcaide E 2011 Structural and magnetic study of the Ti^{4+}-doped barium hexaferrite ceramic samples: Theoretical and experimental results *Physica B: Condensed Matter* **406(17)** pp 3130–3136

IOP Conf. Series: Materials Science and Engineering **95** (2015) 012030 doi:10.1088/1757-899X/95/1/012030

Field control of multiferroic spherical core-shell nanocomposites with applications in microwave range

D Ionescu [1] and M Kovaci [2]
[1] "Gh. Asachi" Technical Univ. of Iasi, Faculty of Electronics, Telecommunications and Information Technology, Dept. of Telecommunications, Carol I Blvd., No. 11, 700506, Iasi, Romania
[2] Politehnica University of Timisoara, Faculty of Electronics and Telecommunications, Dept. of Communications, V. Pârvan Blvd., No. 2, 300223, Timisoara, Romania

E-mail: danaity@yahoo.com

Abstract. Muliferroic spherical nanoparticle configurations consisting of magnetostrictive core encapsulated in piezoelectric shell have been studied. The particles were simulated in a test configuration with help of the HFSS 13.0 (by Ansoft). An external magnetic field H_0 was applied, which determines the apparition of dipolar electric fields of tens of milivolts around nanoparticle structures, used for tuning and control of different microstructures, at molecular level and in spintronics. The nanoparticle compounds are: an $A_2BB'O_6$ double perovskite with large piezoelectric coefficients and a $Me\mathrm{Fe_2O_4}$ spinel or a $AFe_{12}O_{19}$ M-type hexagonal ferrite with large magnetostriction, resulting in nanoparticle configurations with large magnetoeletric (ME) effect. The strength of the ME effect is also increased by the effect of the closed-packed shape of the spherical core-shell configuration. Different magnetic phases (spinel versus hexaferrite) were considered, in order to determine their influence in the interaction process with the applied magnetic field. The diameter of the piezoelectric shell did not exceeded 300 nm, starting from about 50 nm. The components of the magnetoelectric coefficient tensor and the effective electric and magnetic susceptibilities have been determined by simulation in microwave range (2 – 8 GHz for spinel compounds, respectively 12 – 28 GHz for hexaferrite compounds), for different magnetic fields applied for control (0 – 400 Oe), fields depending on particles size. The tensor components depend on the physical properties of the constituents and also on the constituents geometry and relative position. Nanoparticle configurations geometry and the applied H_0 field were modified in order to obtain a convenient variation and control of the compound susceptibilities. Graphs are available for choosing the optimal configuration and parameter values for a specific application.

1. Introduction

Two or more ferroic phases characterizing the same structure determine the multiferroics category of materials and composites, with multiple applications in practice. Among these, magnetoelectric materials contain ferroelectric and ferromagnetic order components, in a nanocomposite structure. Example of such structures can be: core-shell nanoparticles, nanopillars, and nanowires (spherical or cylindrical ferromagnetic core with ferroelectric shell), composite spheres of piezoelectric ceramics or piezopolymers with embedded magnetic nanoparticles, multiferroic nanocomposite films with alternate ferromagnetic / ferroelectric layers, etc.

Content from this work may be used under the terms of the Creative Commons Attribution 3.0 licence. Any further distribution of this work must maintain attribution to the author(s) and the title of the work, journal citation and DOI.
Published under licence by IOP Publishing Ltd

Magnetoelectric materials are used for applications like magnetic field sensors, magnetic data storage, ferromagnetic resonance devices, in magnetic-field-tunable microwave and millimeter wave devices, spintronic devices, where the electric field can control the magnetic effects and also the electric polarizability can be varied with help of an applied magnetic field.

Magnetoelectric composites have better properties if there are synthesized like composed phases, not single phases. The material components must have large magnetic and electric susceptibilities in order to determine a structure with large magnetoelectric coupling [1, 2]. Multilayered structures or core-shell configurations of multiple components present a large magnetoelectric effect due to the magnetostrictive and piezoelectric effects at phase boundaries, mediated by strain. In a magnetic filed, the magnetic phase suffers magnetostriction and the resulted strain is applied to the coupled electric phase, which modifies its electric polarization. Following the idea, the applied fields are used for controlling the structure properties, which can be modulated also by changing the components nature and their geometrical configuration.

2. Material structure

2.1. Core-shell configurations

Nanospheres with core-shell configurations were considered in this paper, presenting large magnetoelectric effect due to a proper choosing of the microscopic constituents. The electric phase consists of the piezoelectric shell, which was chosen like the Bi_2NiMnO_6 double perovskite ($A_2BB'O_6$), while the magnetic phase (ferrite core) is a $MeFe_2O_4$ spinel (*Me* = Mn, Ni) or a $XFe_{12}O_{19}$ M-type hexaferrite (X = alkali earth metal: Ba, Ca, Sr). The core-shell composites with spinel versus hexaferrite were compared in order to determine the magnetic phase influence on the interaction process with the applied magnetic field and the material parameters modifications.

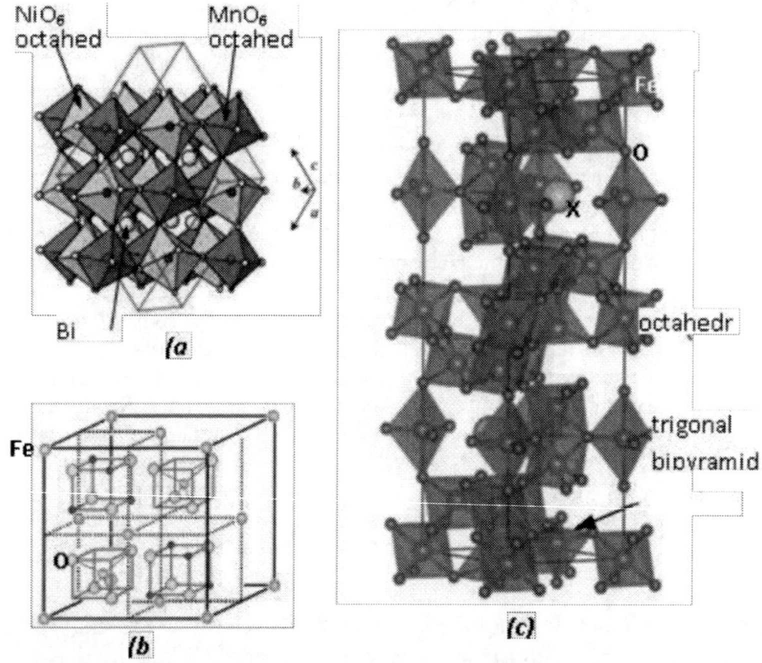

Figure 1. Lattice structure of the electric, respectively magnetic phases:

a) Bi_2NiMnO_6 double perovskite (source: Zhao [10]); b) $MeFe_2O_4$ spinel (source: Ionic Crystals [8]); c) $XFe_{12}O_{19}$ M-type hexaferrite (source:Wang, 2014).

In the Bi_2NiMnO_6 double perovskite (monoclinic, space group C121, figure 1.*a)*), the Bi^{3+} ions are located in the A sites and magnetic Ni^{2+} and Mn^{4+} ions are placed randomly in the B sites (in the three crystallographic sites (M(1), M(2), and M(3), with multiplicities of 2, 4, and 2, respectively) [11]. We have a monoclinic unit cell (space group C2), with a = 9.4646(4) Å, b = 5.4230(2) Å, c = 9.5431(4) Å, and β = 107.823(2)°. The monoclinic unit cells are superimposed. The perovskite itself is a multiferroic due to the combination between B and B' magnetic ions, which give rise to a ferromagnetic coupling.

The magnetic phases: in the spinels (space group Fd3m) the lattice of O^{2-} ions forms tetrahedral and octahedral local symmetry that are referred to as A and B sites, respectively (figure 1.*b)*). The $MnFe_2O_4$ spinel has a = 8.4764 Å, respectively the $NiFe_2O_4$ spinel has a = 8.3378 Å [5]. For hexaferrites (figure 1.*c)*, space group $P6_3mc$), the lattice parameters are: $BaFe_{12}O_{19}$: a = 5.892 Å, c = 23.183 Å; $SrFe_{12}O_{19}$ a = 5.8808 Å, c = 23.2018 Å; $CaFe_{12}O_{19}$: a = 5.8303 Å, c = 22.1652 Å.
The ionic radii of the atomic species included in the considered phases are: cations: Bi^{3+} = 1.17 Å, Ni^{2+} = 0.83 Å, Mn^{2+} = 0.81 Å, Mn^{3+} = 0.72 Å, Fe^{2+} = 0.75 Å, Fe^{3+} = 0.69 Å, Ba^{2+} = 1.49 Å, Ca^{2+} = 1.14 Å, Sr^{2+} = 1.32 Å; anion: O^{2-} = 1.26 Å [3].

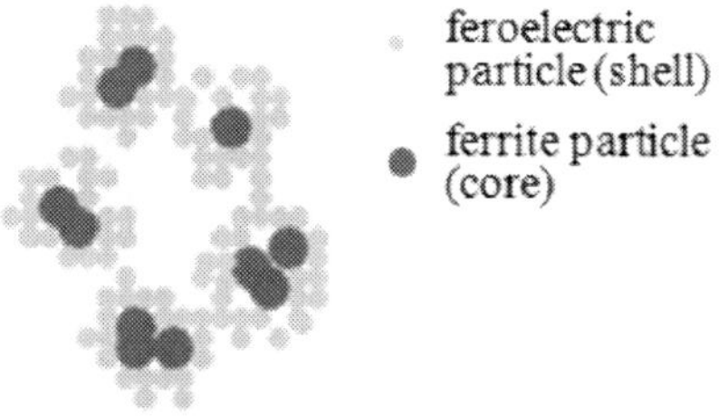

Figure 2. Structure of the core-shell spherical configurations of nanoparticles.

The core-shell nanoparticle configurations were reconstructed using the 3D structural simulation program HFSS 13.0 (by Ansoft) [10], [12], based on the information given in literature. Structures of this type were obtained by different methods, in two steps: synthesis of the core magnetic nanoparticles and covering them with a piezoelectric shell. Examples of different combinations reported in literature: ($BaTiO_3$+$NiFe_2O_4$) obtained by magnetic-field-assisted assembly [1]; ($BaTiO_3$+$CoFe_2O_4$) prepared using a combination of solution processing and high temperature calcinations [7]; ($BaTiO_3$+$CoFe_2O_4$): ferrite nanoparticles prepared through a polyacrylamide gel method, and then coating with ferroelectric using the same gel route [3]; similar - synthesized by combining co-precipitation and organosol methods [6]. Characteristic of the phases are to be observed from there references, in order to define a proper simulational model for the structures. The core-shell configuration has an important role to play in formation of a defined interface between the magnetostrictive and piezoelectric phases, the strain caused by the core and/or shell giving rise to the giant ME coupling. The constituent nanoparticles are polycrystalline (with a random orientation of the particles inside the piezoelectric shell) of ca. 25 - 80 nm for the core magnetic particles, respectively of 20 - 30 nm for the piezoelectric particles in the shell. Diameter of the piezoelectric shell was considered in the range of 60 - 300 nm (Figure 2) and the interdifussion region of e few nanometers

2.2. Theoretical considerations

For the composite structures, the magnetoelectric coefficient tensor $\overline{\overline{\alpha}}$ is given by the formula [2], [9]:

$$D_i = \sum_{j=1}^{3} \varepsilon_{ij} E_j + \sum_{j=1}^{6} e_{ij} S_{uj} + \sum_{j=1}^{3} \alpha_{ij} H_j \qquad i = \overline{1,3} \tag{1}$$

where D = the electric flux density (the response) in multiferroic composite; E_j represents the components of the applied electric field, S_{ij} are the components of the external mechanical stress, H_j are the components of the applied magnetic field vector; ε_{uj} are the permittivity tensor components, e_{uj} are the piezoelectric coefficients and α_{uj} are the components of the magnetoelectric tensor.

The electromagnetic properties of the electric, respectively magnetic phase depend on the electric / magnetic dipole configuration and the lattice structure. An important role is played by the spin-lattice coupling and distinct interactions between spin and dipole [5]. We had to consider that the domain structure is dynamic under the applied fields, meaning that the domain wall thickness and domain width are changeable. For the double perovskite, which is multiferroic, the ferroelectric domain walls present a net magnetization even if the domains are paramagnetic.

The electric, respectively magnetic susceptibilities have been determined using the data given by the HFSS program for the fields inside and outside the material samples:

$$\chi_e(\boldsymbol{E},\boldsymbol{H}) = \frac{\partial P}{\partial E}, \qquad \chi_m(\boldsymbol{E},\boldsymbol{H}) = \frac{\partial M}{\partial H}. \qquad (2)$$

3. Simulation determinations

In the test configuration, the core-shell structures of nanoparticles of different nature were exposed to a magnetic field in the domain 0 – 400 Oe. As a result, a dipolar electric field of tens of milivolts is generated around each core-shell structure, which can be used in practice for tuning and control of different microstructures [4], [10].

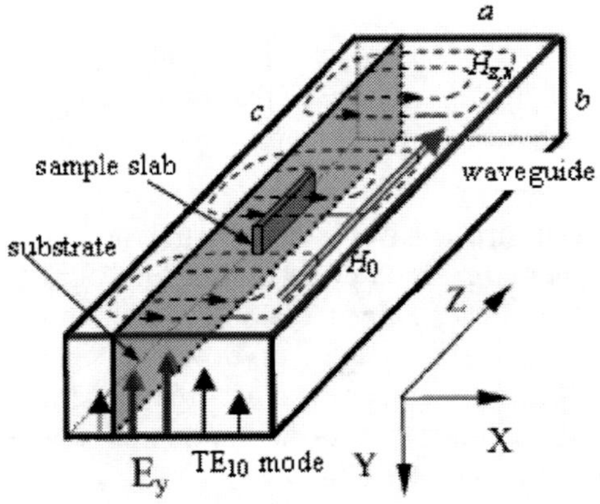

Figure 3. Sample slab inside the rectangular waveguide. The polarization magnetic field H_0 is parallel with the Oz propagation direction.

Samples with dimensions of 4 x 0.8 x 12 mm^3 were placed inside the WR-187 rectangular waveguide (47.5488 x 22.1488 mm^2), excited by TE_{10} mode (figure3.). A static field $\boldsymbol{H_0}$ for polarization was applied on Oz propagation direction, parallel with the long side of the sample. 1mW incident power was applied in the simulation model at each port for determining the S parameters at propagation through the waveguide loaded with sample. The electromagnetic coefficients, electric effective permittivity, respectively magnetic effective permeability of the core-shell polycrystalline samples were calculated from a two-port measurement, using the transmission and reflection coefficients measured with the sample inside in the waveguide. Determinations were made at frequencies above the ferromagnetic resonance of the material samples.

Different magnetic phases (spinel versus hexaferrite) were considered, in order to determine their influence in the interaction process with the applied magnetic field and the capabilities of the modulation of the electric field generated by the nanoparticles, used in applications.

Simulations were performed in microwave range (2 – 8 GHz for spinel compounds, respectively 12 – 28 GHz for hexaferrite compounds). The $\boldsymbol{H_0}$ induced variations in the electric, respectively magnetic susceptibilities are illustrated in the graphs below (figures 4 - 5).

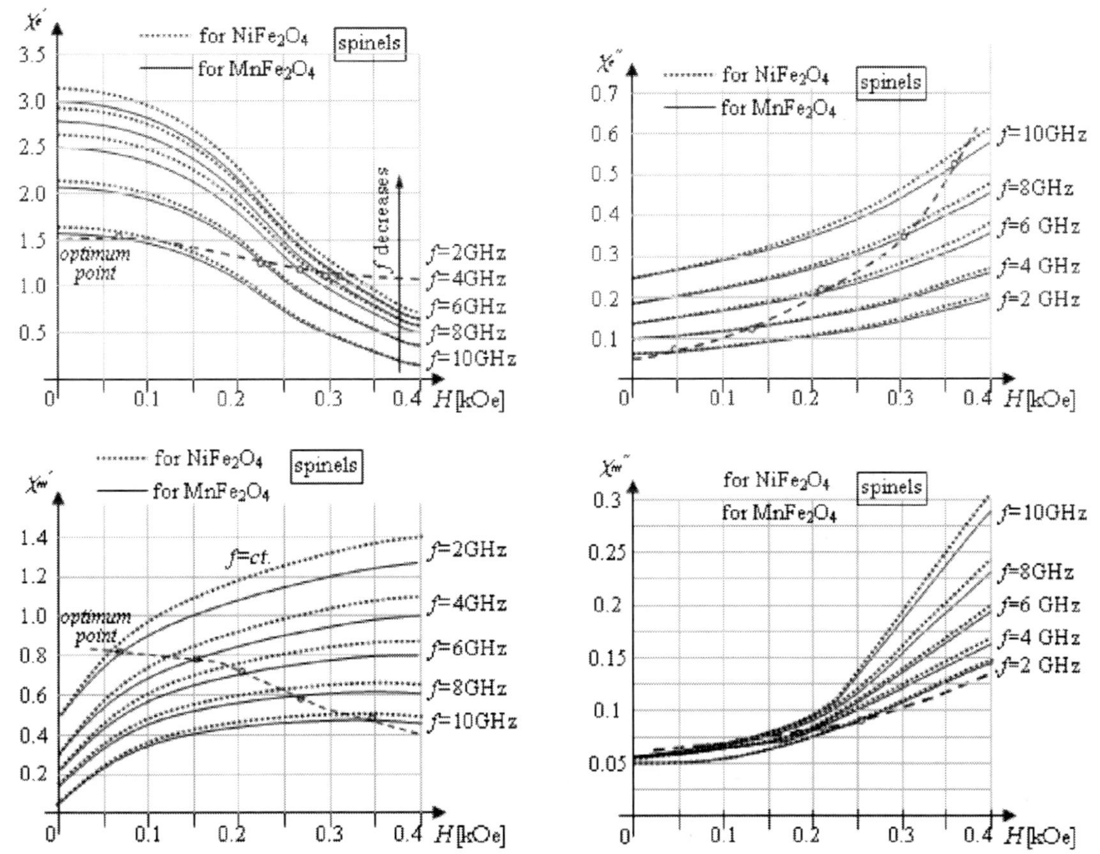

Figure 4. Real and imaginary parts of the electric / magnetic effective susceptibility, χ_e, χ_m, for the considered core-shell nanocomposites with spinels (Bi_2NiMnO_6 + $NiFe_2O_4$; $MnFe_2O_4$). The curve with optimum points is indicated on graphs.

The curve with the optimum points was illustrated on graphs (the dash curve), each point indicating the value of the applied magnetic field, at a given frequency, necessary to obtain the maximum of the electric polarization corresponding to each nanoparticle configuration. The line of the optimum points can be represented on the graphs of all the susceptibility components, in order to obtain a complete characterization of the material in an optimum point.

The susceptibility values depend on the physical properties of the constituents (the piezoelectric coefficients of the ferroelectric, respectively the piezomagnetic coefficients of the ferrimagnetic phase), and also depend on the constituents geometry and relative position. All these quantities can be fine modified in the structural simulation model when the simulations are performed.

The values of the effective electric relative permitivities, respectively magnetic permeabilities have been calculated for the spinel nanocomposites, respectively for the nanocomposites with hexaferrites. Values in the range of 145 – 152 have been obtained for the composites permittivity, higher values for hexaferrites. For the composites permeability, values of about 592-594 were obtained for structures with $NiFe_2O_4$, respectively of about 680-686 for structures with $MnFe_2O_4$. For composites with hexaferrites, $\mu_{r,eff}$ was calculated of around 743-745 for structures with $CaFe_{12}O_{19}$, of around 736-738 for structures with $SrFe_{12}O_{19}$, respectively of around 733-734 for composites with $BaFe_{12}O_{19}$ hexaferrite.

IOP Conf. Series: Materials Science and Engineering **95** (2015) 012030 doi:10.1088/1757-899X/95/1/012030

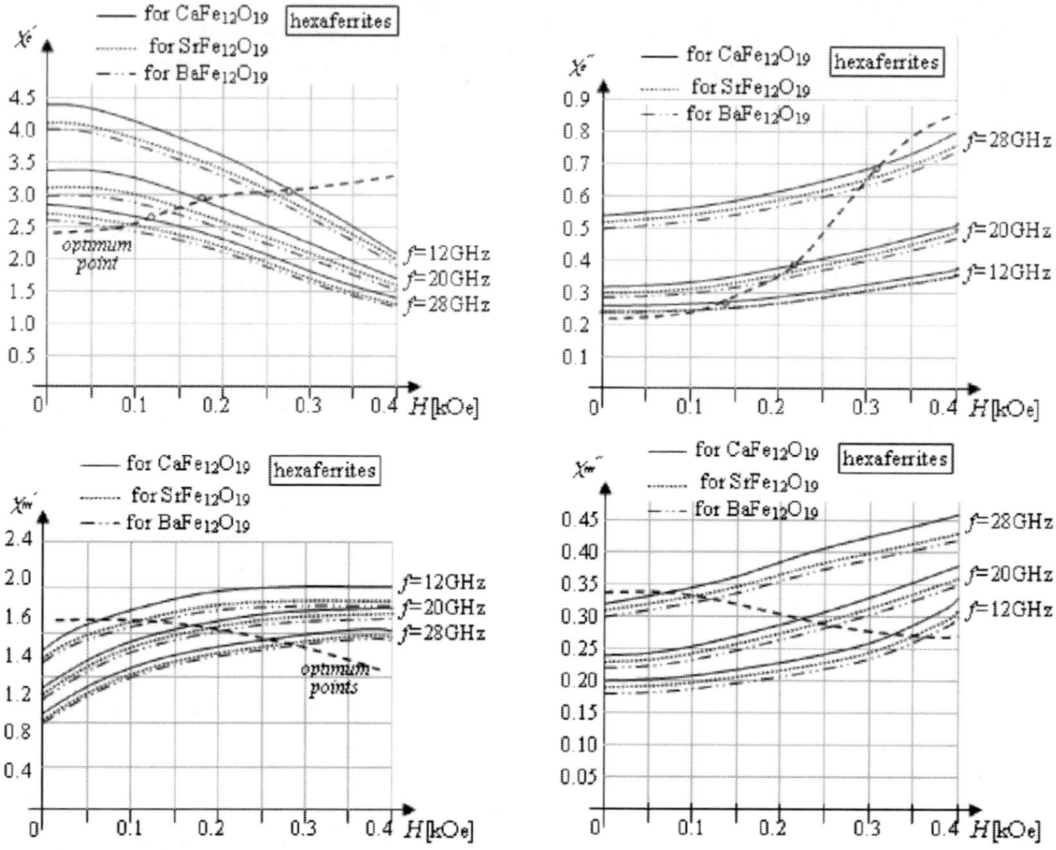

Figure 5. Real and imaginary parts of the electric / magnetic effective susceptibility, χ_e, χ_m, for the considered core-shell nanocomposites with hexaferrites (Bi_2NiMnO_6 + $BaFe_{12}O_{19}$; $CaFe_{12}O_{19}$; $SrFe_{12}O_{19}$) and the curve with optimum points.

4. Conclusions

Core-shell nanocomposites with ferrite – ferroelectric constituents and presenting strong magnetoelectric effect due to close-packed configuration and strain transmission through the certain boundaries have been analyzed in this paper. Magnetic field intensity was modified in order to obtain a convenient variation and control of the electric and magnetic susceptibilities of the nanocomposites. For all compounds, the magnetic susceptibility increases with the applied H_0.

For the composites with spinels, the real part of the electric susceptibility, related to the stored energy within the medium, presents a flexional evolution with the applied field. The curvature modification is a result of the internal processes at sample level (the stain generated by magnetostriction reaches a threshold value and over it the induced electric polarization evolves after a different law with the applied field).

The imaginary parts of the susceptibilities, related to the dissipation of energy within the medium, increases with the applied field and also with frequency. Magnetic losses increases faster, being relative low up to the values of the applied field of about 200 Oe, due to the influence of the resonant energy absorptions which are present over these vales of the field.

For the composites with hexaferrites, curves of evolution of the susceptibility components versus the applied magnetic field are more flat and susceptibilities preset superior values. More magnetic iron ions in the lattice confer a stability of the composite response and lower variations under the applied magnetic fields. Magnetic losses evolve with H_0 not so abruptly, losses being consistent even at lower values of the applied field (this fact indicates energy consumption for magnetic processes at sample level).

Curves with optimum points are illustrated on graphs, in order to recommend the set of punctual parameters which have to be chosen for obtaining the maximum of the electric polarization induced by the applied magnetic field. The simulation method is a non-destructive analysis method, with solutions for choosing the field parameters and geometrical / physical sample configuration.

References

[1] Sreenivasulu G, Popov M, Chavez F A, Hamilton S L, Lehto P R and Srinivasan G 2014 Controlled self-assembly of multiferroic core-shell nanoparticles exhibiting strong magneto-electric effects *Appl Phys Lett* **104** 052901

[2] Hajra P, Maiti R, Chakravorty D 2011 Nanostructured Multiferroics *Trans Ind Ceram Soc 70* **(2)** pp 53-64

[3] Wang W, Yang H, Xian T 2012 Synthesis of $CoFe_2O_4/BaTiO_3$ Core-Shell Composite Nanoparticles Procedia Engineering *Chinese Materials Conference* **27** pp 604–609

[4] Sun N X, Srinivasan G 2012 Voltage Control of Magnetism in Multiferroic Heterostructures and Devices *SPIN 02* 1240004 [46 pages]

[5] Chu I-W 2010 Multiferroic and core-shell nanostructures - Advances in preparations and novel properties *dissertation thesis* (City University of New York)

[6] Du Y, Cheng Z, Dou S X, Attard D J and Wang X 2011 Fabrication, magnetic, and ferroelectric properties of multiferroic BiFeO3 hollow nanoparticles *J Appl Phys* 106 073903

[7] Kanagesan S, Hashim M, Jesurani S, Kalaivani T and Ismail I 2014 Influence of Zn–Nb on the Magnetic Properties of Barium Hexaferrite *J Supercond Nov Magn* **27** pp 811–815

[8] Ionic Crystals http://www tf uni-kiel de/matwis/amat/def_en/index html

[9] Kim H 2013 Magnetoelectric Effect and Magnetodielectric Effect in Magnetic Nanoparticles *dissertation thesis* (Los Angeles: University of California)

[10] Bichurin M and Petrov V 2014 Modeling of Magnetoelectric Effects in Composites *Springer Netherlands, Springer Series in Materials Science* **201** (New York London: Springer Dordrecht Heidelberg)

[11] Shimakawa Y, Azuma M and Ichikawa N 2011 Multiferroic Compounds with Double-Perovskite Structures *Materials* **4** pp 153-168

[12] Ionescu D and Apreotesei G 2013 Novel A Simulational Approach: Electromagnetic Parameters Determinations for Some Ferrite-Ferroelectric Nanocomposites *Rom Rep Phys* **65(4)** pp 1328–1347

IOP Conf. Series: Materials Science and Engineering **95** (2015) 012031 doi:10.1088/1757-899X/95/1/012031

Influence of working parameters on the composite electrodeposition

M M Pasare[1]

[1]University "Constantin Brancusi" of Tg-Jiu, Faculty of Engineering, 30 Eroilor Street, 210135, Tg-Jiu, Romania

E-mail: minodora_pasare@yahoo.com

Abstract. Composite materials are extensively studied, especially in recent years, which have a wide application in industry. Among these composite materials, in this paper we studied about electrolyte composites obtained. The matrix is made up of particles of Ni-P on which have been deposited Si-C hard particles. The amount of phosphorus incorporated into the layer increases with increasing the content of phosphorous acid in the electrolyte. The content of phosphorus incorporated into the coating composition evolves as both solid as well as particle-free solutions. When the content of phosphorous acid in the electrolyte is low and where alloys are micro crystallized, the presence of particulates changes a little the phosphorus content. At the increase of phosphorus content included in the coating is observed that the incorporation of phosphorus is higher for increasing the amount of silicon carbide particles. This matrix study was done according to the amount of phosphorus in the deposited layer. The co-deposited phosphorus modifies the crystallization mechanism so the super saturation nickel network by phosphorus allows the inhibition of the growth of nickel crystallites whose size decreases until the complete amorphization of the solid solution. The development of these deposits was aimed at determining the influence of the variation of phosphorous acid in the electrolyte, and therefore the change in the content of phosphorus included in the layer for the SiC particulate, on the one hand, and, on the other hand, it was intended the influence of heat treatment on the deposition hardness. After the measurements, it is noted that the layer thicknesses are higher for the deposition obtained from electrolyte with small containing of phosphorous acid. The structure of the Ni-P alloy and the composite material Ni-P / SiC, depends on the amount of phosphorus contained in the layer and varies from the crystalline state (the low content of phosphorus) to the amorphous state (the high content of phosphorus). Metallographic investigation and X-ray diffraction on the P20S40 sample structure reveals the existence of nano-crystalline state of the layer, and it is noted that the grains of Ni-P alloy is lower than that incorporated SiC particles that are more evident. Following heat treatment, it was found that the microcrystalline structure of the deposits, consisting of a solid nickel solution supersaturated in phosphorus, is maintained up to 463 K, and at higher temperatures it is producing the precipitation of the Ni_3P compound into the nickel matrix.

Content from this work may be used under the terms of the Creative Commons Attribution 3.0 licence. Any further distribution of this work must maintain attribution to the author(s) and the title of the work, journal citation and DOI.
Published under licence by IOP Publishing Ltd

1. Introduction

The composites of metal/ceramic having a metal matrix reinforced with ceramic particles dispersed in the form of fibers. Making composer electrodeposited metal/hard particles are the result of interactions of various particles/electrolyte, the electrolyte / cathode and particle/cathode. Lately, due to their outstanding properties, were made of composite materials that phosphorus nickel matrix and particles are incorporated in it are diverse (chromium carbide, boron carbide, etc.). Recently were started researches on the introduction of silicon carbide particles in the matrix of nickel-phosphorus [1,2]. Preparation of an electrochemical composite coating requires suspending the particles in an electrolyte. It is necessary above all to ensure the stability of the hard particles in a nickel-phosphorus acidic electrolyte. The incorporation of particulate matter is susceptible to change certain intrinsic characteristics of the deposit; it is first necessary to know the physicochemical characteristics and properties of the deposit when it was obtained from a particle-free electrolyte. Co-deposition yield of solid particles is related to the hydrophobic of the surface charge of the particles and not superficial. The incorporation of hydrophilic particles is more difficult than incorporating hydrophobic particles because it is difficult to remove water around the particles. In the case of silicon carbide, oxidation, leads to the formation of a silicon oxide layer on the surface, where the hydrophilic.

2. Experimental procedure

The development of the samples studied in this paper was made electrolytic, because they have advantageous characteristics. By the electrolytic deposition, made in an electrolytic cell, layers are obtained with a special appearance at the surface and at the same time ensures the atmospheric corrosion protection, to that produced by some chemical and mechanical damage due to friction, abrasion etc. Electrolyte development is a relatively simple process, and the layers thus obtained can be controlled in terms of composition, appearance and mechanical properties. It is a process that takes place at low temperature electrolysis cells have a relatively high life and low cost. Before carrying deposition was performed a media surface preparation. This was necessary because the adhesion between the support and the metal coating material, achieving uniform and homogeneous coating protective structure can be guaranteed only through adequate training support. Surface preparation support included the following:

a) chemically cleaned and degreased surface to be removed traces of fats or oils;

b) they started to surface polishing particles becoming finer to remove existing scratches on the surface;

c) has been supporting the polishing cloth.

Before being introduced into the electrolytic bath, the previously prepared substrate was degreased with acetone. After electro-deposition, the samples were placed in de-mineralized water and subjected to ultrasound for 3 minutes for removal of silicon carbide particles which have not adhered to the substrate. The samples were named PxSy, where X and Y represent the content in g / l of phosphorous acid, respectively content in g / l of particles of silicon carbide (SiC) in the electrolysis bath. Silicon carbide used for the preparation of composite layers had purity of 99.9%, and the particles have size of about 1 μm and the specific surface area of 10.4 m^2/g. The main impurities in silicon carbide were: iron: 0.03%; Aluminum: 0.02%; Vanadium: 0.02%.

Figure 1 shows micrographs of silicon carbide particles and in figure 2 in the powder X-ray diffract gram of silicon carbide appear only characteristic peaks of silicon carbide.

The substrate on which the depositions were conducted is E24 steel with the following composition: C-0.17%; S-0.045%; 1.4% Mn; P-0.045%; N <0.007%; carbon steel (unalloyed); hardness indicated time: 235-255HB. The support of electrolytic copper from electrolysis cell was 99.9% purity, hardness 100 HV, and the thickness of the discs were made of composite layers was 3mm samples, for the samples where phosphorous acid content remained constant and varied the content of silicon carbide particles. These samples were applied a thermal treatment at 463 K.

Samples were named as follows: $P20S40_{TT\text{-}463\ K}$. The development of these deposits was aimed at determining the influence of the variation of phosphorous acid in the electrolyte and therefore the change in the content of phosphorus included in the layer for 40 g/l SiC, on the one hand, and, on the other hand, it was intended influence of heat treatment on hardness deposits. The heat treatment by diffusion phases which it operates, determining structural changes and the increase of adhesion of the coating layer to the sub-layer. Studies in the literature show that the electro-deposition of Ni-P for hard metal heat treatments has a significant influence on the structure and adhesion by removing hydrogen [3,4]. Ni-P alloys are thermodynamically unstable and changes that occur during the thermal stresses are dependent on the phosphorus content. At a temperature of 463 K meets an optimal peak hydrogen evolution, whatever the type of submission. Temperature is falling to exothermic phenomena for that occurring structural change. Because this temperature affects the Ni-P alloy structure and composition of Ni-P / SiC, they will influence the mechanical properties and tribological properties of these materials. The structure of the Ni-P alloy and the composite material thus Ni-P / SiC, depends on the amount of phosphorus contained in the layer varies from the crystalline state (the low content of phosphorus) to the amorphous state (the high content of phosphorus) [5]. Metallographic investigation and X-ray diffraction on the sample structure reveals the existence P20S40 layer of nanocrystalline state (figure 3 and figure 4), and it is noted that the grains of Ni-P alloy is lower than that incorporated SiC particles become more apparent.

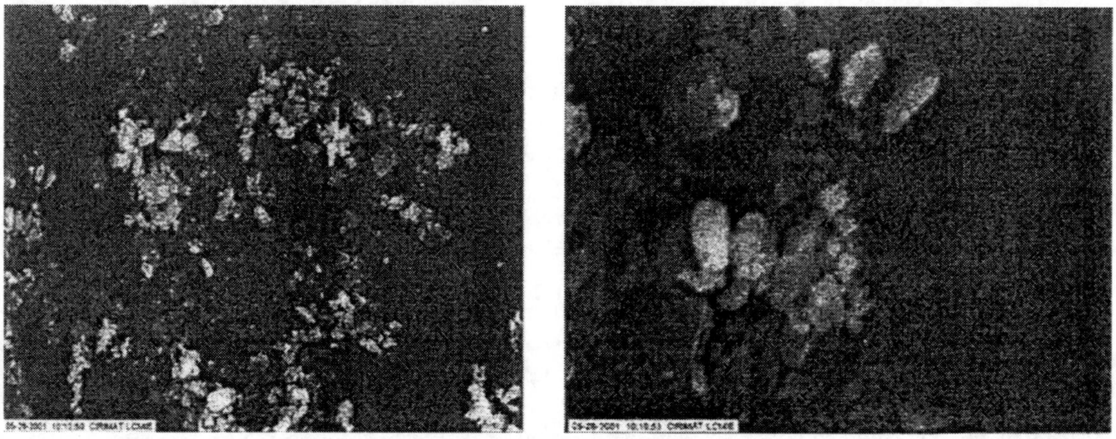

Figure 1. Silicon carbide particle micrographs.

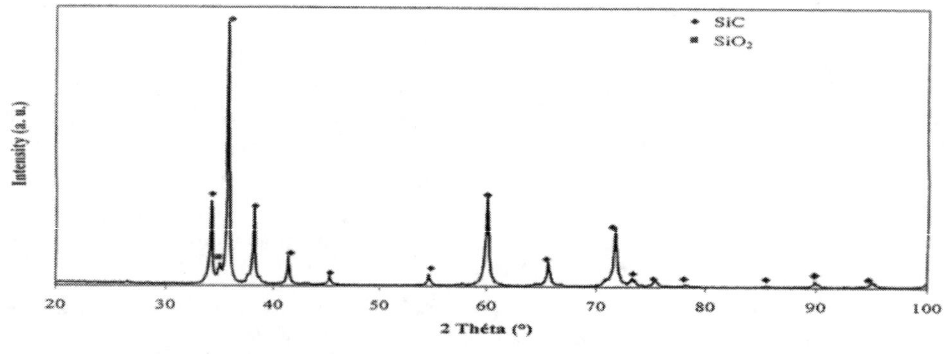

Figure 2. X-ray diffract gram of the powder of silicon carbide.

Qualitative analysis results presented in figure 5 X-ray from below, reveals on the diffraction the presence of two phases, namely: SiC single-phase and majority phase manifested on the diffraction by highest pick. It is noted that this pick is much widened which indicates a very fine grain structure. The position of this maximum corresponds to the Ni structure or to a nickel-based solid solution, in fact a solid solution of Ni-P.

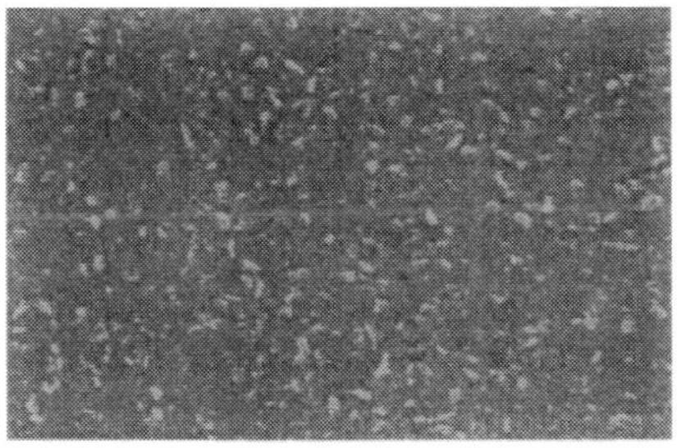

Figure 3. Optical micrograph P20S40 untreated deposit.

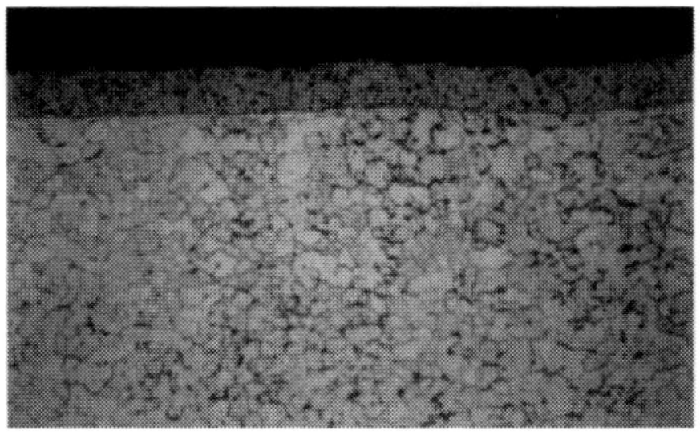

Figure 4. Optical micrograph P20S40 sample treated at 463 K (x150).

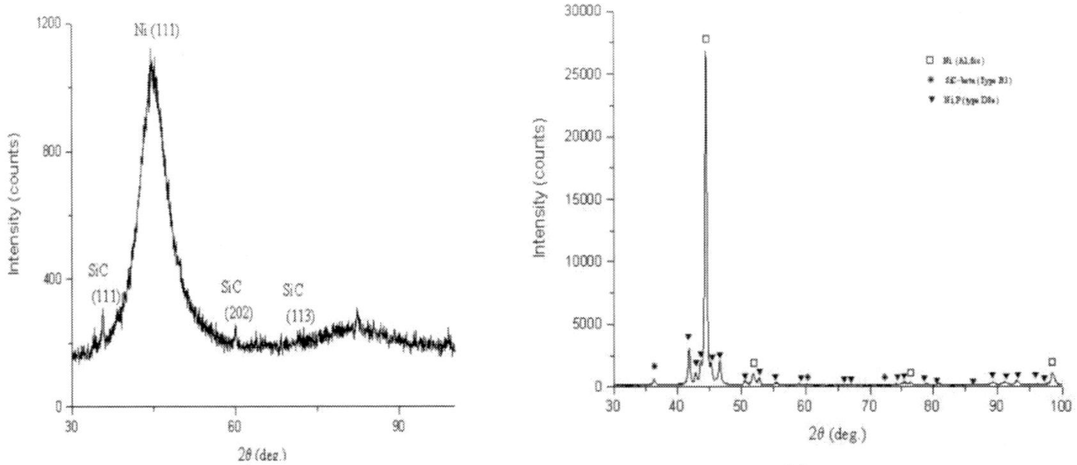

a) b)

Figure 5. X-ray diffraction for P20S40 untreated thermal deposition (a); X-ray diffraction for P20S40 heat-treated at 463 K (b).

The same can be seen in the diffraction figure above for sample P20S40 - heat-treated at 463 K. We meet the same compounds as raw deposit P20S40. As noted before, phosphorous acid and thus phosphorus plays an important role in obtaining the composite material Ni-P / SiC [6]. Variation of phosphorous acid in the electrolyte causes changes in physicochemical and structural layers deposited. Therefore for these samples was measured Vickers micro-hardness by applying loads 15g, 25g, 50g, 100g, 200g, 300g, 500g perpendicular to the surface layer. In the scientific literature specifies that the application of a temperature of 463 K appears strong hydrogen de-sorption, which causes changes in the properties of deposited layers. For this reason the hardness of the deposits was determined annealed at this temperature toughness has been obtained perpendicular to the surface of the film, as well as heat-treated deposits. In Table 1 are presented the results obtained when measuring the hardness and thickness of the coating layers according to the load applied perpendicular to the surface of the film.

Influence of phosphoric acid and the load applied to the layer hardness is shown in the graph of figure number 6.

Table 1. Thickness, load, average hardness deposits.

Sample	Load [g]	Hardness[Hv]	Thikness layer [µm]
P20S40	15	449	24
	25	437	
	50	429	
	100	418	
	200	404	
	300	400	
	500	319	

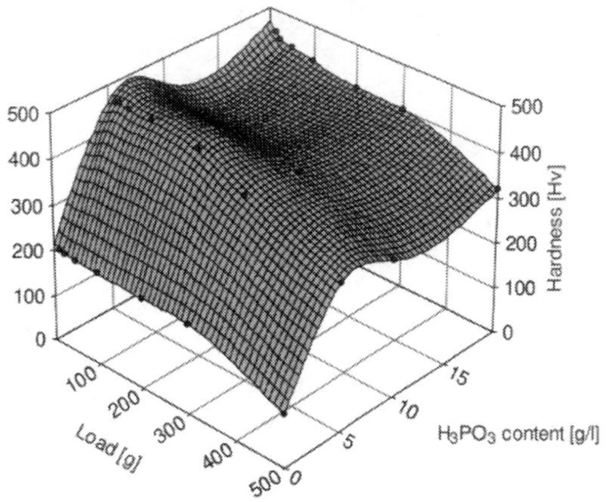

Figure 6. Change in hardness deposits with the load applied to the content of phosphorous acid in the electrolyte.

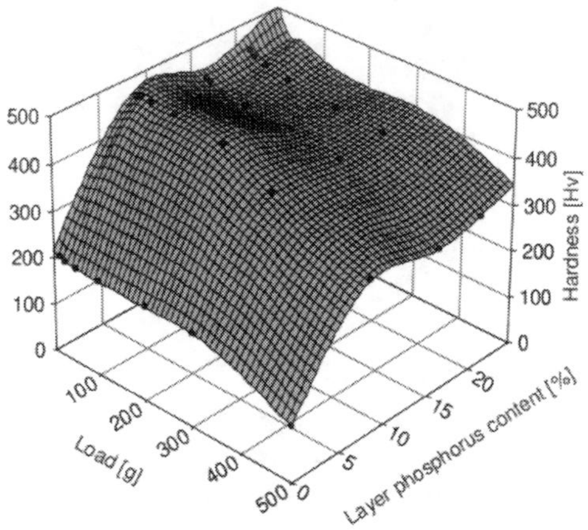

Figure 7. Change in hardness deposits on the load applied and the content of phosphorus incorporated into the layer.

For this series of samples it shows that the micro-hardness measured changes according phosphoric acid composition in three areas: for low content of phosphorous acid in the electrolyte ([H_3PO_3] <5 g/l, the composition domain I), the hardness of the deposits increases with the increase in the content of phosphorous acid; the content of H_3PO_3 between 5-10g/l (the composition domain II), the hardness decreases, then, from [H_3PO_3]=10g/l (the composition domain III) present in higher hardness. It also notes that, as the load applied is bigger the hardness decreases. For loads up to 200g hardness varies not very high (especially in the domain I and II). The load applied 300g hardness begins to decrease and the hardness decreases 500g load increased. Since the variation of the content of H_3PO_3influence the amount of phosphorus incorporated into the layer, the curves were plotted and give change in hardness depending on the content of phosphorus incorporated in the layer as shown in figure number 7 from below:

When we studied the graph in the figure 7 from above we ca see that it has the same allure as the graph of figure 6 given by the hardness variation depending on phosphorous acid in the electrolyte and vary also by the three domains. In the first domain, where the percentage of phosphorus is up to 8.4 at.%, the hardness shows a significant increase; in the second domain (percentage of phosphorus is situated around 16 at.%) we can see a small decrease in hardness which is not greatly influenced by the variation of phosphorus, and over 16% at. the hardness presents a slight increase (domain III). This variation in hardness can be attributed to structural changes arising from the variation of phosphorus content in the layer [7, 8]. When the phosphorus content increases the micro structural changes occur this manifests itself by a gradually structural change of the layer from the amorphous state to the crystallized state. In the area I, the increase in hardness due to the reduction of the crystallite size, and the coating has a crystal structure. In the area where the micro hardness decreases (domain II) where there is a structural disorder coexist micro crystallized phase and nano- crystalline phase.

3. Conclusions

In this paper were studied the conditions of development of deposits by electro-deposition of composite electrolytes containing variable phosphorous acid and constant content of silicon carbide. It was found that by increasing the content of phosphorous acid in the electrolyte between 5% and 20%, we obtained an increase in the amount of phosphorus incorporated into the layer as it evolves as both the solid composition and particle-free solutions. Thermal treatments were carried out at 463 K for phosphor deposits with different content and it was found that the deposition of microcrystalline structure, consisting of a solid solution supersaturated nickel phosphorus, is maintained up to 463 K, and at higher temperatures is produced Ni_3P compound precipitation of nickel matrix. All layers deposited and presented in this paper are characterized by uniform thickness and very good quality of the interface. Metal matrix ensures adherence to the substrate and electrochemical corrosion resistance, anti-friction material is soft phase and hard phases take the mechanical friction, which is a new research direction in the future.

References

[1] Bonino J P, Bruet-Hotellay S, Bories C, Pouderoux P and Rousset A 1997 *Journal Electrochem* pp 33
[2] Bonino J P, Pouderoux P, Rossignol C and Rousset A 1992 *Plat. and Surf. Finish* pp 62
[3] Bonino J P and Vaillant S 2000 *Journal Electrochem* pp 355-362
[4] Pasăre M and Ianăşi C 2013 Another method for determining the surface micro hardness of composite coatings, 13th International Multidisciplinary Scientific GeoConference, doi: 10.5593/SGEM 2013.6. pp 17-22
[5] Srinivasan K N and Thangavelu P 2012 Electroless Deposition of Ni-P Composite Coatings containg Kaolin Particles *Trans. IMF* **90**, pp 105–112
[6] http://labo.univ.poitiers.fr./JE2003/telechargement/JE2003 Information accessed on 18.11.2014,
[7] Low C T and Walsh F C 2013 Multifunctional Nanostructured Metallic Coatings by Electrodeposition, Chapter 10 R.J.K. Wood, (Ed), Multifunctional Materials for Tribological Applications, Pan Stanford
[8] Liu H, Guo R X, Viejo F, Thompson G E and Liu Z 2013 *Trans.IMF* **91** pp 101–112

IOP Conf. Series: Materials Science and Engineering **95** (2015) 012032 doi:10.1088/1757-899X/95/1/012032

Processing of cylindrical hollow parts: piercing vs. extrusion

R Comăneci[1], L Zaharia[1], D Nedelcu[1] and L G Bujoreanu[1]
[1]Technical University "Gheorghe Asachi" Iasi, 700050, Romania

E-mail: comaneci@tuiasi.ro

Abstract. As work principle, piercing (P) and backward extrusion (BE) are similar deforming processes. However, working-load, strain and material flow are quiet different depending on real conditions. Since punch diameter d and workpiece diameter D are the defining parameters for both P and BE, there is a critical d/D ratio separating them as distinctive operations. To have a complete description of P and BE in terms of working load and yield pressure as function of d/D ratio, experimental tests and numerical predictions were conducted. Tridimensional finite element analysis (FEA) for different process designs (i.e. different d/D ratios) revealed distinctive evolutions of the material flow establishing the effective limits of P and BE respectively. Experimental and numerical results can be used in practice to provide estimates of working-load, critical yield pressure and tool dimensions for optimal P and BE in processing of cylindrical hollow parts.

1. Introduction

Although piercing (P) - as preliminary forging operation - and backward extrusion (BE) – as final deforming process - are almost identical deforming processes (at least as work principle) there are however some notable differences in terms of working-load, friction, material flow, stress and strain distribution, and tools which have to be revealed. At first sight, the major differences between P and BE seems to be the extrusion ratio (Ao/Af), maximum working-load (Fmax), and punch diameter (d). From extrusion ratio point of view P is considered as a BE with a smaller reduction in cross-section area. Because of that, the material flow has distinctive evolutions: while in P only a small volume of material flows along the head of the punch, in E the material flows along both of die's wall and punch, the volume of material involved in plastic flow being notable higher (according to the continuity equation). This implies both higher working-loads and friction: in the final stage of the extrusion, the material almost reaches hydrostatic state and the lubricant is ejected. The higher reduction, the bigger working-load and friction. Obviously, P needs smaller punches in diameter and corresponding smaller working-loads. But these are relative variables and they are not related of the workpiece and material itself. For this reason using yield pressure (p/2k) as ratio of normal pressure (p) to the yield stress in shear (2k) becomes more appropriate.

If the yield stress ($\sigma_y = 2k$) evolves with strain (ε) according to the well known relation:

$$\sigma_y = K\epsilon^n \qquad (1)$$

where K is a constant for the given material, n is the work hardening exponent and $\epsilon = \ln\frac{1}{1-\left(\frac{d}{D}\right)^2}$ (d and D are revealed in figure 1), we can easily concluded that yield pressure becomes a function of d/D

Content from this work may be used under the terms of the Creative Commons Attribution 3.0 licence. Any further distribution of this work must maintain attribution to the author(s) and the title of the work, journal citation and DOI.
Published under licence by IOP Publishing Ltd

ratio. Since punch diameter d and workpiece diameter D are the defining parameters for both P and BE, there is a critical d/D ratio (i.e. critical yield pressure) separating them as distinctive operations.

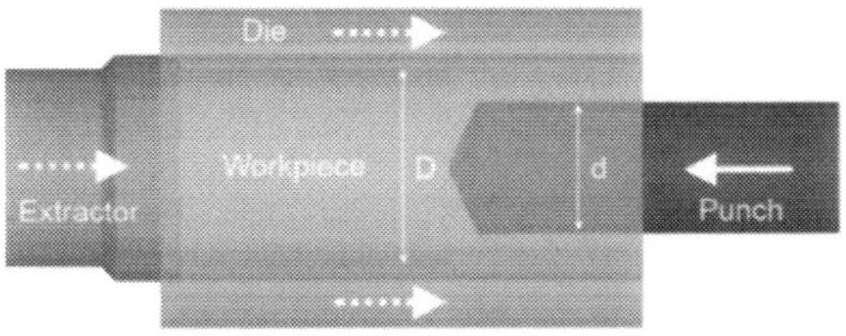

Figure 1. Principle of extrusion/piercing.

Work pressure in extrusion/piercing of hollow tubes starting from either initial solid or hollow billets previously processed with different die/mandrel combinations were calculated using slab [1] or upper bound method [2, 3]. Some theoretical and experimental investigations have been dedicated to friction study for different work conditions [4-7]. It was found that friction can play a positive role and consequently the extrusion with active friction using a mobile container was developed [8, 9]. A major difficulty in piercing is obviously the buckling of punch which has – at the beginning of deformation - one end fixed and other end free, situation which then changes during deformation. The punch stability to buckling being smaller, the length of the punch in piercing should be limited and an appropriate punch holder has to be used. But this is not the single problem in piercing: in practice, the centring of the workpiece becomes essential for a successful operation [10]. Moreover, for a smooth shear surface in piercing, a servo press applying a counter pressure is necessary[11]. Despite P is one of the simplest deforming operations, no much information about it can be found in the literature. Moreover, a criterion separating the two operations (i.e. working-load formula, punch vs. workpiece dimensions) is not clearly stated.

To have a complete description of P and BE in terms of working-load and yield pressure as function of d/D ratio, experimental tests and numerical predictions were conducted in this study. Tridimensional FEA for different process designs (i.e. different d/D ratios) aims to reveal possible distinctive evolutions of the material flow establishing the effective limits of P and BE respectively, setting critical yield pressure and corresponding d/D ratio. Experimental and numerical results can be used in practice to provide estimates of working-load, critical yield pressure and tool dimensions for optimal P and BE in processing of cylindrical hollow parts.

2. Experimental materials and procedures

2.1. Finite element analysis

In the present work, a three dimensional model was considered. Commercial finite element code DEFORM 3D was used to carried out the simulations. A commercial available 99.5% aluminum (AA 1050) was used in this study. The stress-strain relationship of the AA 1050 was experimentally obtained by tensile tests carried out according to ISO 6892-1: 2009, using a universal computer-controlled testing machine (Instron 3382). It was found that flow stress evolves with strain according to the following relation: σ (MPa) = $151.24 \bullet \varepsilon^{0.18}$ which was indicated in DEFORM 3D as constitutive equation (figure 2).

Eight scenarios were simulated, corresponding to the punch diameter of 5, 6, 7, 8, 10, 15, 20, and 25 mm. As can be seen, the increment in punch diameter decreases in the presumed range of piercing (i.e. from 5 to 8 mm) to accurately capture changes in the material flow and confirm a clear trend of its evolution. All simulations were performed at room temperature, with the same speed of punches compared with experimental speed. Also the friction coefficient between inner die walls and cylindrical billets was assumed to be the same as in experiments (0.12) and it has been considered by other authors [12]. The workpiece (ϕ 30 $\times$ 42 mm) was considered a plastic body in whole deformation processes. Poisson's ratio 0.33 and Young's modulus 69 Gpa were assumed. The hardening behavior is considered isotropic and independent of strain rate at room temperature. The die and the punches were modeled with analytical elastic elements. The tolerance, positioning of the workpiece and die, convergence criteria, re-meshing conditions, and boundary conditions were specified before the execution of the simulation processes. The workpiece was discretized in 32000 tetrahedral elements thus forming a sufficiently fine mesh to reveal localized effects. The volume penalty constant which ensures volume constancy of plastic workpiece with minimum loosing volume during re-meshing had to be specified [13].

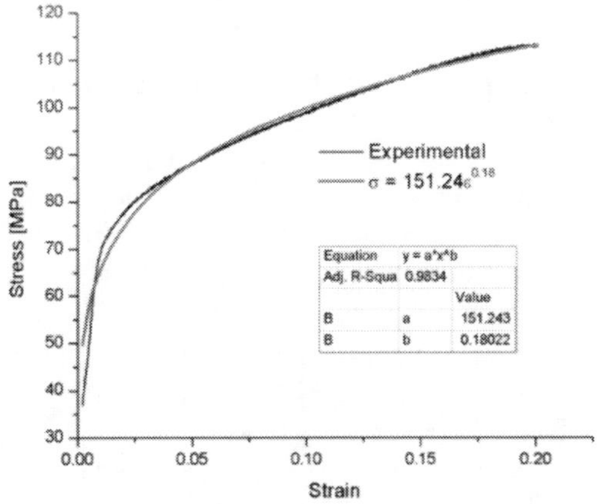

Figure 2. Stress-strain curve of the material used in finite element simulations.

2.2. Experimental piercing/extrusion
After FEA, six punches made from hardened AISI 4140 steel having 5, 8, 10, 15, 20 and 25 mm in diameter were prepared for experimental tests (figure 3).

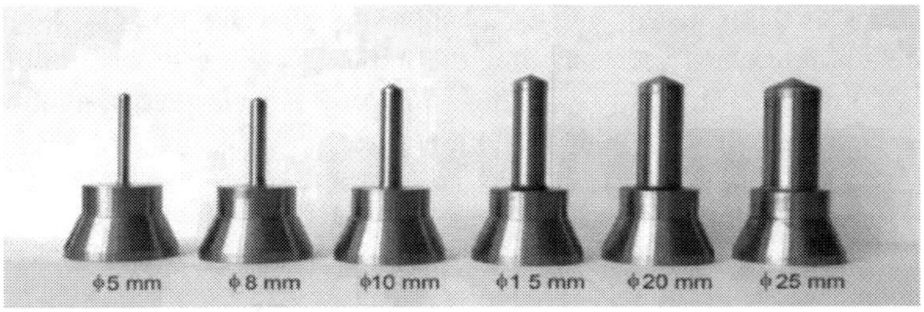

Figure 3. Principle of extrusion/piercing.

Cylindrical specimens of AA 1050 with dimensions of ϕ 30 × 42 mm were machined from as-received material. A subsequently annealing at 523K for 3h was performed before P/BE in order to eliminate strain hardening from previous metal-working operations. Processes were conducted at room temperature with a constant speed of 0.6 mm•s^{-1}, using a hydraulic press of 750 kN and a die with 34 mm in diameter consisting in two half-dies reinforced by two outer circular rings forming a close fit with the die (figure 4). The inner diameter of the die was chosen intentionally larger than that of the workpiece to accurately capture the early stages of material flow.

Figure 4. The P/BE setup used in experimental tests.

All samples and inner walls of the die were lubricated using zinc stearate which ensures a shear friction coefficient of $\mu = 0.12$[14]. The stroke of punches was 38 mm. During P and BE the working-loads were recorded using a National Instruments data acquisition integrated system and a LabVIEW interface. Figure 5 shows the processed specimens.

Figure 5. Experimental specimens processed by P/BE.

3. Results and discussions

3.1. Equations and mathematics

For this study, the material flow becomes the most important because changing in plastic behavior of the material makes the difference between P and BE. As expected, depending on d/D ratio, the material flow had distinctive evolutions, critical ratio being d/D = 0.3.

a) for d/D ≥ 0.3, four stages can be distinguished in material flow as it is suggested by figure 6: flowing along the front head of the punch, upsetting of the material starting around the punch and finishing with full die contact, steady state extrusion, and ending of the extrusion.

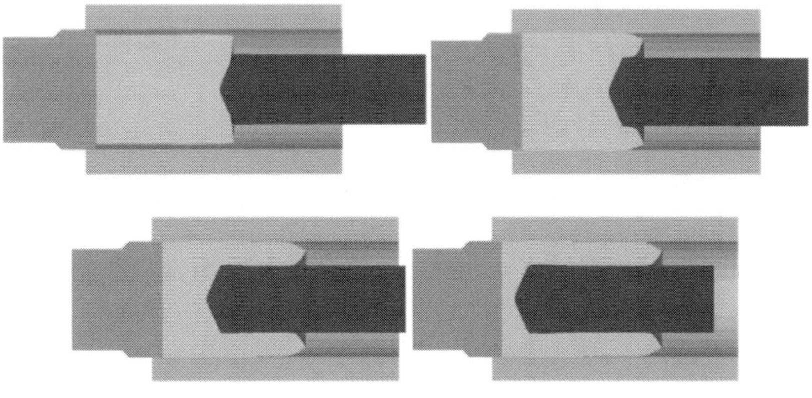

Figure 6. Representative material flow for d/D ≥ 0.3 (d/D = 0.67).

b) for d/D < 0.3, three stages can be distinguished in material flow as it is suggested by figure 7: flowing along the front head of the punch, steady state piercing, and ending of the piercing with a possible upsetting at the bottom of the workpiece (which becomes negligible when d/D ratio decreases).

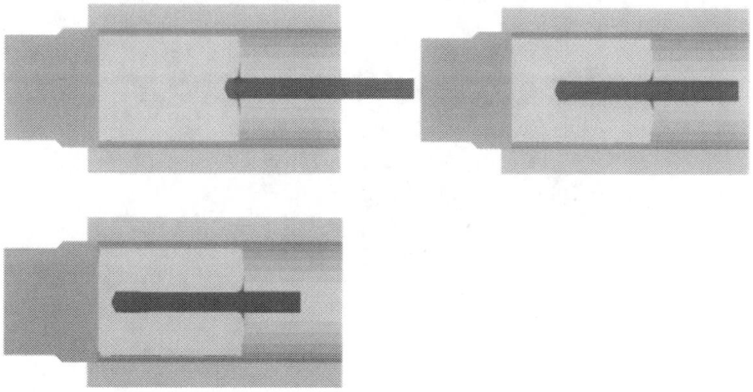

Figure 7. Representative material flow for d/D < 0.3 (d/D = 0.20).

The material flow suggests different ways in changing material shape during deforming processes: with upsetting of the front of workpiece in extrusion and without it in piercing. The experimental results confirmed this hypothesis as can be seen in figure 5.

3.2. Validating of modeling

There are several methods to validate the modeling in simulating deforming processes. Fitting the predicted and experimental working-load is one of them. The level of the maximum load is most interesting from the viewpoint of the die and punch design. The results show that the peak load is reached prior to achieving the steady state and it is higher than the steady state load. This is a normal feature for a strain hardenable material [12]. Figure 8 shows the predicted and experimental load evolution for a middle scenario (i.e. d = 10 mm). The maximum level of working-load and the general evolution are in good agreement with experimental results, confirming the validity of modeling.

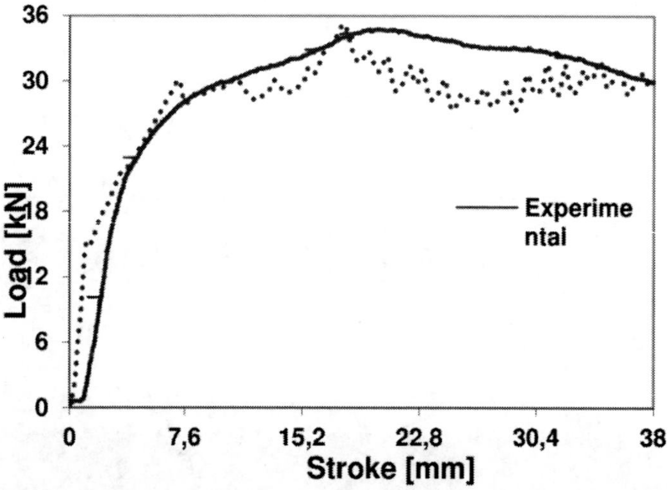

Figure 8. Predicted and simulated working-load for a middle d/D ratio (0.33).

3.3. Piercing vs. extrusion: critical d/D ratio

As shown in both simulated and experimental results (figure 9) the working-load increases with increasing d/D ratio because of higher reduction and friction.

Simulated loads are underestimated because of redundant shear deformation which leads to higher extrusion load. External constraint requires significant internal distortion of the workpiece beyond that strictly necessary for the shape change. This is widely accepted for many deforming processes and especially for drawing15, flow forming16, and extrusion3. For d/D ≥ 0.3 (corresponding to d ≥ 10 mm) the yield pressure remains almost constant having just a slowly decreasing even if working-load increases. According to experimental and FEA results (see figures 5 and 6) this situation clearly corresponds to the extrusion when the material must first fill the container (by upsetting) and then the extrusion can take place. For d/D < 0.3 (corresponding to d < 10 mm), yield pressure suddenly increases with decreasing of d/D ratio even if working-load is getting lower because of smaller reduction. The situation corresponds to piercing without upsetting when the height of the workpiece remains almost unchanged (see figure 5).

The situations can be summarized as follows:

- for d/D ≥ 0.3, Fpiercing > Fupsetting, extrusion takes place after a previous upsetting ;
- for d/D < 0.3, Fpiercing < Fupsetting, piercing takes place without an upsetting.

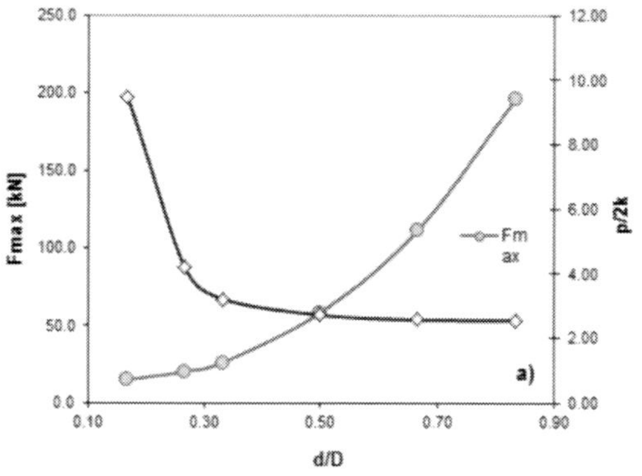

a)

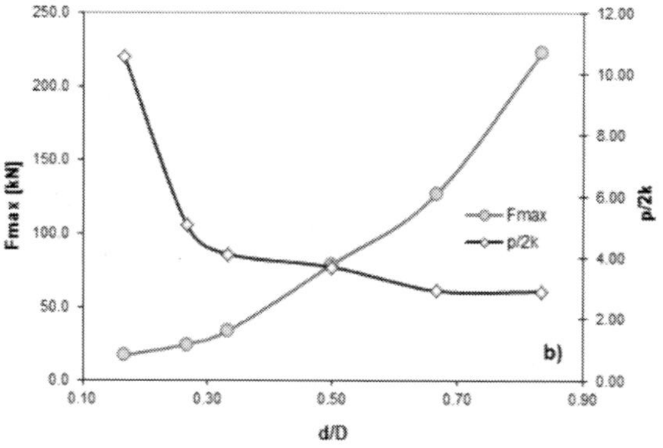

b)

Figure 9. Maximum working-load and yield pressure evolution a) simulation; b) experimental

4. Conclusions

A comparative study between P and BE to separate them as distinctive operations was performed. To have a complete description of P and BE in terms of material flow, working-load and yield pressure as function of d/D ratio, experimental tests and numerical predictions were conducted. Tridimensional FEA confirmed by experiments for different process designs (i.e. different d/D ratios) revealed distinctive evolutions of the material flow establishing the effective limits of P and BE respectively and setting critical d/D ratio as 0.3. It was found that for d/D ≥ 0.3 extrusion takes place after a previous upsetting and for d/D < 0.3 the piercing is that which takes place but without an upsetting. Experimental and numerical results can be used in practice to provide estimates of working-load, critical yield pressure and tool dimensions for optimal P and BE in processing of cylindrical hollow parts.

References

[1] Chitkara N R and Aleem A 2001 2001 Extrusion of axisymetric tubes from hollow and solid circular billets: a generalized slab method of analysis and some experiments *Int. J. Mech. Sci.* **43(7)** pp 1661-1684
[2] Chitkara N R and Aleem A 2001 Axisymetric tube extrusion/piercing using die/mandrel combinations: some experiments and a generalized upper bound analysis *Int. J. Mech. Sci.* **43(7)** pp 1685-1709
[3] Narayanasamy R, Ponalagusamy R, Venkatesan R and Srinivasan P 2006 An upper bound solution to extrusion of circular billet to circular shape through cosine dies *Mater. Design* **27(5)** pp 411-4153
[4] Hsu T C and Huang C C 2003 The friction modeling of different tribological interfaces in extrusion process *J. Mater. Process. Technol.* **140(1-3)** pp 49-53
[5] Flitta I and Sheppard T 2003 Nature of friction in extrusion process and its effect on material flow *Int. J. Mater. Sci. Technol.* **19** pp 837-843
[6] Bakhshi-Jooybari M 2004 A theoretical and experimental study of friction in metal forming by the use of the forward extrusion process *J. Mater. Process. Technol.* **125-126** pp 369-374
[7] Schikorra M, Donati L, Tomesani L and Kleiner M 2007 The role of friction in the extrusion of AA6060 aluminum alloy, process analysis and monitoring *J. Mater. Process. Technol.* **191** pp 288-292
[8] Yuan Li F, Liu S J G and He Z B 2008 Research of metal flow behavior during extrusion with active friction *J. Mater. Eng. Perf.* **17(1)** pp 7-14
[9] Muller K B 2003 Indirect extrusion with active friction (ISA) *Key Eng Mat.* **223-226** pp 323-328
[10] Gordon W A, Van Tyne C J and Moon Y H 2012 Minimizing distortion during extrusions in adaptable dies *Int. J. Mech. Sci.* **62** pp 1–17
[11] Matsumoto R and Utsunomiya H 2013 Cold Piercing of cylindrical aluminum billet with counter punch pressure *Key Eng Mat.* **554-557** pp 613-619
[12] Figueiredo R B, Cetlin P R and Langdon T G 2009 The evolution of damage in perfect-plastic and strain hardening materials processed by equal-channel angular pressing *Mater. Sci. Eng.* **A 518** pp 124–131
[13] Defom 3D v 10.0 2009 User's manual *Scientific Forming Technologies Corporation* (USA)
[14] Chirita C, Comaneci R, Zaharia L and Hanganu A C 2007 Technological aspects concerning severe plastic deformation by equal channel angular pressing *Annals of DAAAM* Katalinic B (ed) pp 139-140
[15] Majzoob G H, Fereshteh Saniee F and Aghili A 2008 An investigation into the effect of redundant shear deformation in bar drawing *J. Mater. Process. Technol.* **201** pp 133–137
[16] Mohebbi M S and Akbarzadeh A 1010 Experimental study and FEM analysis of redundant strains in flow forming of tubes *J. Mater. Process. Technol.* **210** pp 389–395

IOP Conf. Series: Materials Science and Engineering **95** (2015) 012033 doi:10.1088/1757-899X/95/1/012033

Finned double-tube PCM system as a waste heat storage

M H Alhamdo[1], M A Theeb[1] and A S Golam[1]
[1]Mech. Eng. Dept., College of Engineering, University of Al-Mustansirya, Baghdad - IRAQ

E-mail: profmohamedh@yahoo.com

Abstract. In this work, focus is taken on developing a waste heat recovery system for capturing potential of exhaust heat from an air conditioner unit to be reused later. This system has the ability to store heat in phase change material (PCM) and then release it to a discharge water system when required. To achieve this goal, a system of Finned, Water-PCM, Double tube (FWD) has been developed and tested. Different profiles of fins attached to the (FWD) system have been investigated for increasing the thermal conductivity of the PCM. These include using Circular Finned, Water-PCM, Double tube (CFWD) system; Longitudinal Finned, Water-PCM, Double tube (LFWD) system; Spiral Finned, Water-PCM, Double tube (SFWD) system; as well as; Without Fins, Water-PCM, Double tube (WFWD) system. An experimental test rig that attached to an air-conditioner unit has been built to include 32- tubes of the FWD systems for both vertical and horizontal layouts during charging and water discharging processes. Results show a significant performance improvement when using spiral and circular fins during charging process at vertical position. However, longitudinal and without fins showed better performance in horizontal position. Overall, the developed SFWD system in vertical position has been found to exhibit the most effective type due to the fastest PCM melting and solidification. As compared to the WFWD system, the FWD systems have been found to increase the PCM temperature gain of about 15.3% for SFWD system; 8.2% for CFWD; and 4.3% for LFWD system.

1. Introduction

Heat recovery and energy storage systems are used for storing energy during the time when extra energy is available in order to be used afterward. In other words, they are used to correct the disparity between energy supply and energy demand; which results in saving capital cost. Furthermore, this can affect the environmental by reducing pollution. Phase Change Materials (PCM) can be used to store waste heat resulting from industry, furnace, air conditioner, etc.. In spite of great potential of PCM, the practical feasibility of latent heat storage is still limited, mainly due to a rather low thermal conductivity. This low conductivity implies small heat transfer coefficients and, consequently, thermal cycles are slow and not suitable for most of the potential applications.

Medrano et al. 2009 [1] investigated experimentally the heat transfer process during melting and solidification of five small heat exchangers working as latent heat thermal storage systems. Results show that the double pipe heat exchanger with the PCM embedded in a graphite matrix is the one with higher thermal power values. *Joudi et al. 2012* [2] utilized Finite Difference Method for solving transient two-dimensional conduction heat transfer equations with phase change. Results showed that the PCMs in a cylindrical container melt and solidify quicker than the square container. *Lokapure1 et*

Content from this work may be used under the terms of the Creative Commons Attribution 3.0 licence. Any further distribution of this work must maintain attribution to the author(s) and the title of the work, journal citation and DOI.
Published under licence by IOP Publishing Ltd

al. 2012 [3] designed a heat recovery system for air conditioner. The air conditioner condenser was replaced by a special heat exchanger. Results showed not just condenser heat was recovered, but also COP of the air conditioner was enhanced. *Mat et al. 2013* [4] performed a numerical study to investigate the melting process in fined triplex-tube heat exchanger with phase-change material. A two-dimensional numerical model is developed using the Fluent 6.3.26 software program. Around 43.3% reduction in the melting time was achieved as opposed to non-finned tubes. However, no experimental investigation has been conducted in this study. *Chang Liu et al. 2014* [5] accomplished an experimental work to study the phase change heat transfer inside a cylindrical latent heat energy storage that designed with a central finned copper pipe running the length of the cylindrical container. Longitudinal fins were added to the copper pipe to enhance the heat transfer. It was observed that conduction is the dominant heat transfer mechanism during the initial stage of charging, also during the entire solidification process.

The main aim of this work is to design; manufacture and test a new Waste heat Recovery System (WRS) to capture and reuse energy of exhaust hot gas that dissipated from air conditioner condenser. This system consists from a group of PCM double-tubes that arranged in vertical and horizontal layouts with water as a heat transfer fluid for discharging energy.

2. Experimental Work

The test rig is designed and manufactured as shown in figure (1). An air conditioner with a capacity of 24000 BTU/hr, was used. The waste heat exhausted from the condenser was used to supply test section with hot air. Group of extra heaters (12kw) sited inside the duct were used to guarantee that the test section will be supplied with hot air of (70°C) temperature at all times. The water system, which is used to recover waste heat from PCM, consists from 35 W water pumps, PVC connecting pipes and two insulated water tanks with a capacity of 20 Liter each. The second tank is initially empty and drains the water from first tank after passing through the FWD system. Four groups of test tubes (32 tubes in each group) were tested. The effect of various fins shapes of longitudinal, circular, spiral and without fins were investigated for both tube arrangements of vertical and horizontal. The tests were done through condenser hot waste gas charging and water discharging processes. The insulated test section integrates into two parts the perforated duct and the double-tube system. The first part is a square duct which has been manufactured from iron frame enclosed by transparency plastic material (Perspex), of 120 cm length, 30 cm height, and 6 mm thickness. Two of the parallel sides of the duct are perforated; the holes are arranged in a pattern of 5, 4 holes in each line with a total of 32 holes that has a diameter of 10 mm as illustrated and photographed in figure (2). The holes are set evenly in accordance with results obtained in this work for selecting the best distance between the holes as will be explained later.

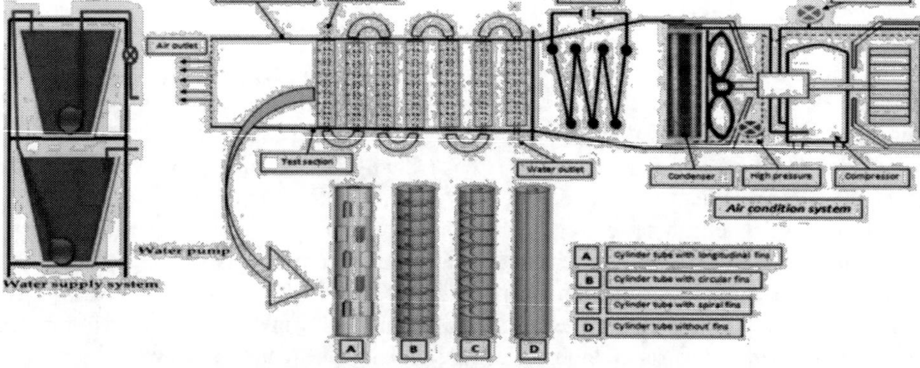

Figure 1. Experimental of the test rig.

The second part is the test tubes with a total number of 32 double-tubes that arranged along with the holes. Each test tube includes an outer glass tube with a diameter of 30 mm, thickness of 1.5 mm and length of 30 cm. The inner copper tube is positioned in the center of the glass tube with a length of 34 cm, inside diameter of 8 mm and thickness of 0.65 mm. The copper tube passes through the duct hole. The space between the glass and the copper tube is filled with PCM (paraffin wax).The outer glass tubes were used to capture waste energy of exhaust hot air from condenser during charging process, while the inner copper tubes were used for discharging process by using water flow. In order to raise the heat transfer process, three different shapes of fins with the same surface area have been welded around the copper tubes as shown in figure 3. All types of fins were made from copper with a thickness of 0.65 mm. To measure water, air, PCM and fins temperatures in the test section, 32 thermocouples of k-type with data logger have been installed in the FWD system. Figure 3 shows selected thermocouples positions of the test tubes

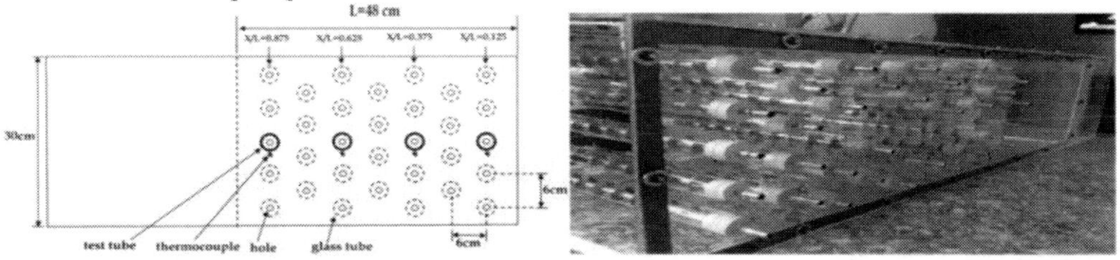

Figure 2. Double–tubes arrangement of the FWD system.

Pure wax produced by major Iraqi oil company (Al-Durra refinery) was selected as a PCM. The thermo-physical properties of the PCM have been measured experimentally and the results are shown in table (1). The thermal conductivity measurements of PCM were based on the ASTM standards. The transient test method is taken primarily from ASTM D5930 and ASTM D5334 standards. To check the reputability of the data, all experiments were performed at least twice (on different days). The error associated with experimental measurements of the temperature and mass flow rate of PCM, hot air; and water were done using the root sum square method. The estimated maximum values of uncertainties for temperature are 1.42% and for mass flow rate is 0.83%.

Table 1. Measured properties of PCM.

Melting point (C)	63.7
Solid Density(kg/m^3)	796
Liquid Density(kg/m^3)	766.5
Viscosity (kg/m.sec)	0.0127

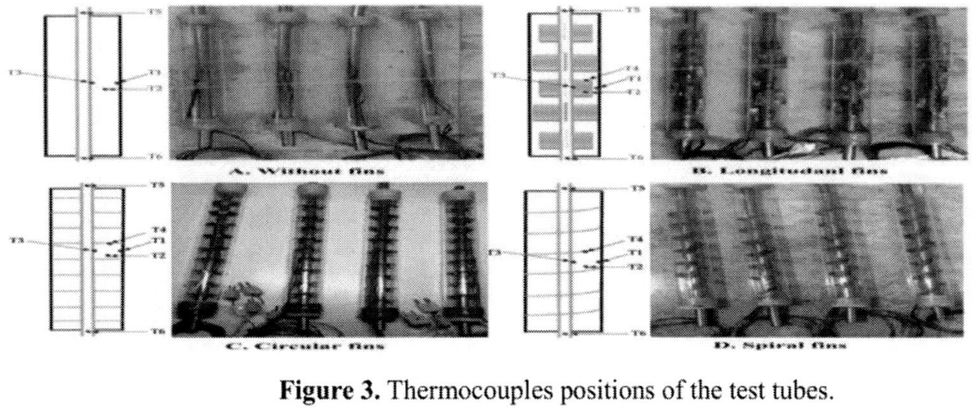

Figure 3. Thermocouples positions of the test tubes.

3. Results

Figure 4 demonstrates comparison between gained temperature in both vertical and horizontal states. It was found that spiral and circular fins had more significant enhancement during charging process at vertical position. However, longitudinal and without fins showed better performance in horizontal position. It is found that about 8.2%, 2.6%, 1.6%, and 1.3% enhancement of gained temperature in vertical state as opposed to the horizontal state for SFWD, CFWD, LFWD, and WFWD respectively.
Figure 5 demonstrate the difference between the temperatures of PCM for each type of WD recovery system. From figure it is seen that PCM temperature is the highest when using spiral fins. As compared to no-fins geometry; the rate of gained temperature increase has been found to be (15.3%, 8.2%and 4.3%) with the use of spiral, circular and longitudinal fins respectively. Overall, spiral geometry fins has been found to exhibit the most effective type due to the fastest the PCM melting and the fastest solidification.

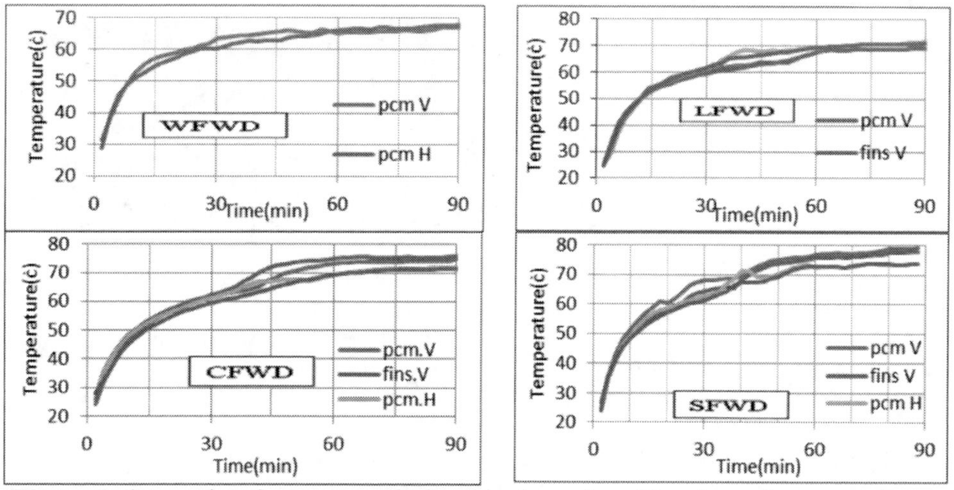

Figure 4. Comparison between vertical and horizontal position thermal response inside WD recovery system.

Figures 6 shows the transient radial temperature distributions during charging process from hot waste gas of the A/C unit through vertical and horizontal CFWD system. It is noted that the fins temperature in vertical position is higher than the PCM temperature .This could be attribute to the melting of PCM close to the lower surface of the circular fins and their separation from the fins.

Figures 7 illustrate the time variation of the temperature during water discharging process. The radial temperature of the PCM at the particular radial position gradually decreases with time until the temporal variation becomes radially consistent after a while. The figures clarify that the temperature decreases particularly in the areas touching the copper tube and fins as a result of good thermal conductivity of the copper. However the rate of discharge of heat varies according to the fins profiles variety. The decreasing rate of the PCM temperature of the WFWD system has been found to be 22%, in the LFWD system was 34%, whereas in CFWD system was 37%, and 32% with SFWD system.

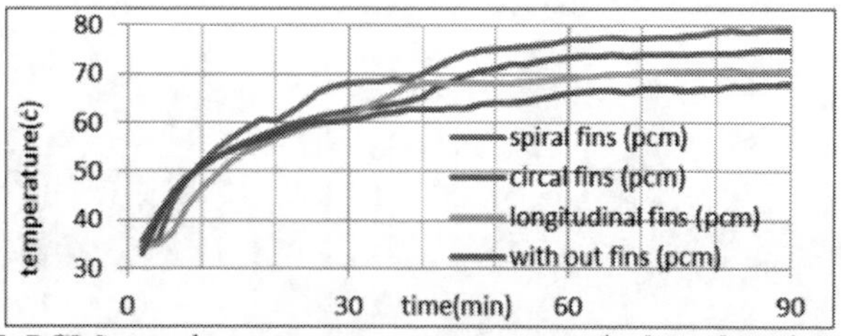

Figure 5. PCM transient temperature response during charging process.

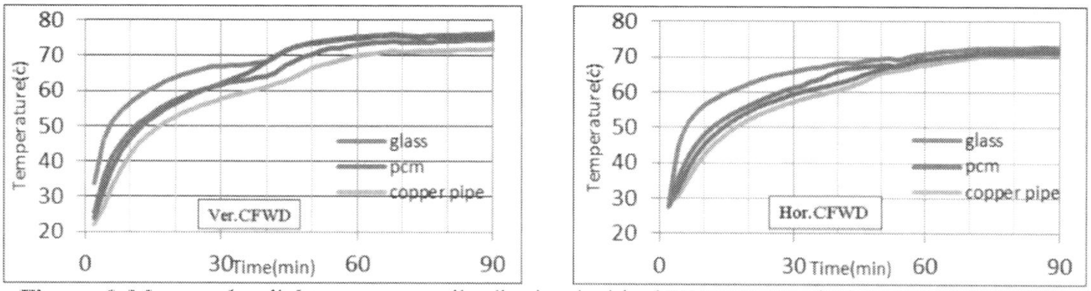

Figure 6. Measured radial temperature distribution inside CFWD system during charging process.

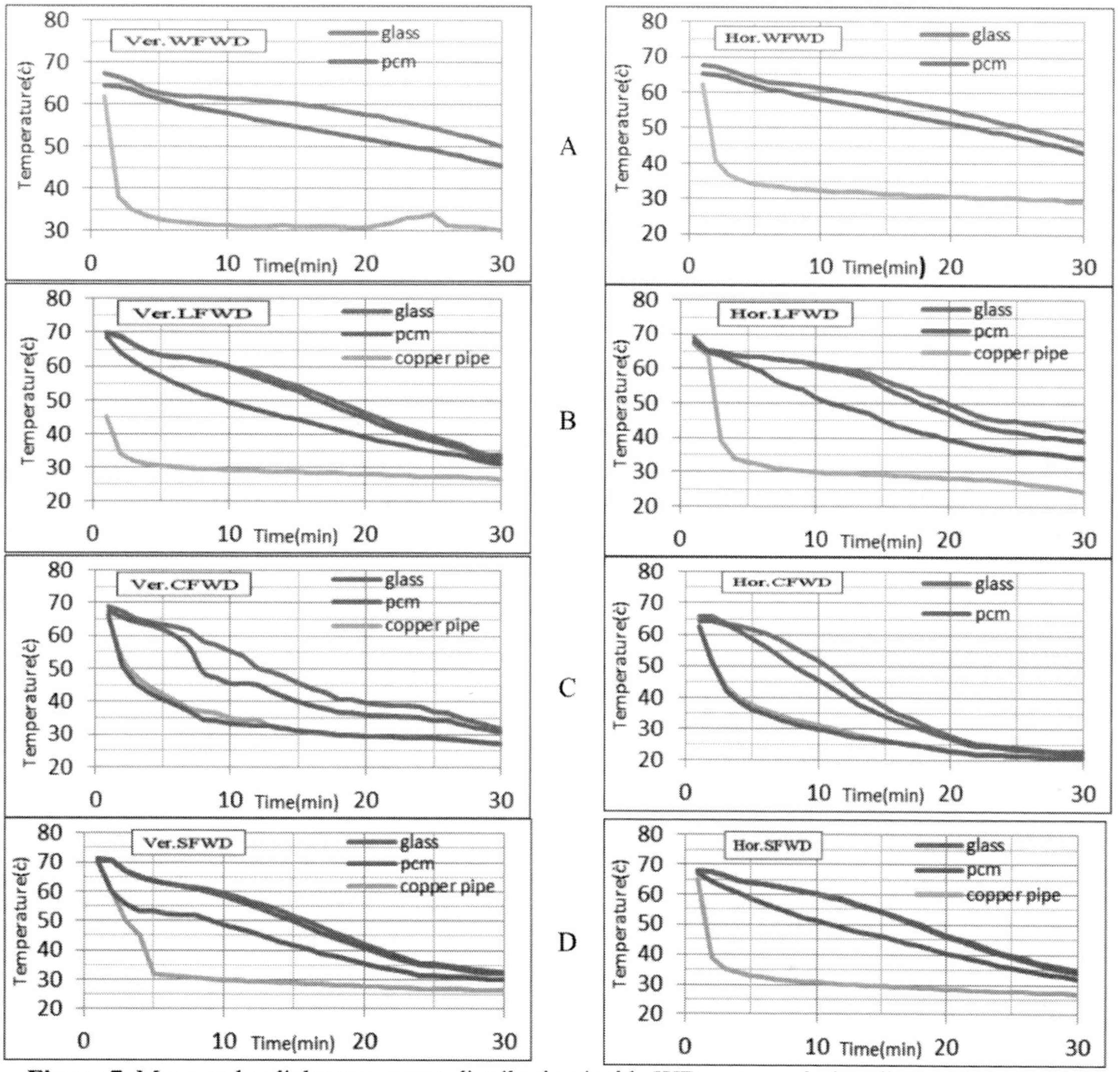

Figure 7. Measured radial temperature distribution inside WD systems during discharging process.

Figure 8 illustrate the water inlet and outlet temperature in both vertical and horizontal positions. It is clear that the curves rapprochement and divergence depends on the gained PCM temperature through the charging process. From results of water discharge system, it is clear that the thermal response of longitudinal and without fins systems showed better performance in horizontal position, whereas, spiral and circular fins system showed better performance in vertical position.

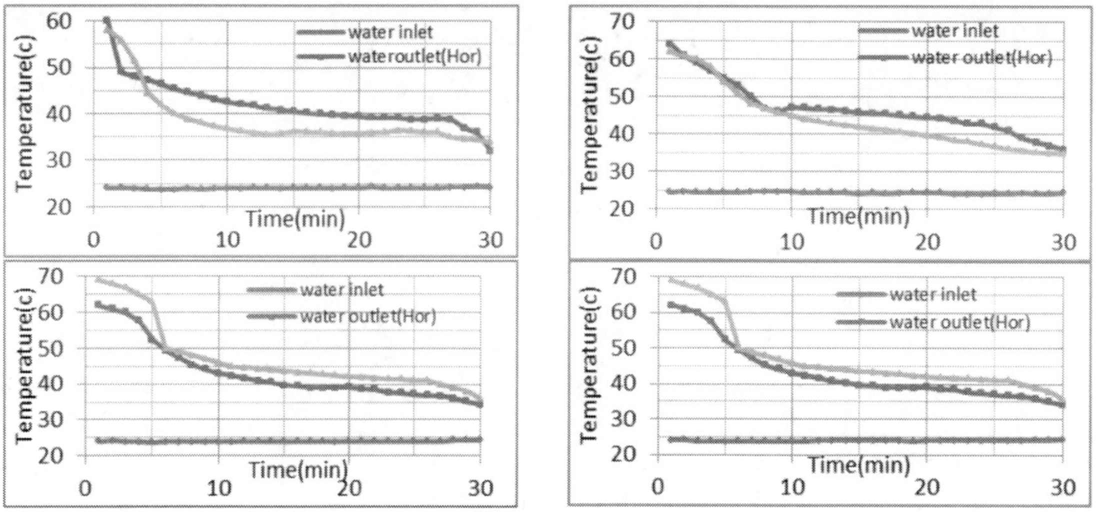

Figure 8. Transient water temperature during discharging process for FWD systems.

4. Conclusions

In general, it is found that using various fins profiles play an important role on enhancing the heat transfer through PCM. As compared to the WFWD system, the FWD systems have been found to increase the PCM temperature gain during charging process of about 15.3% for SFWD system; 8.2% for CFWD system; and 4.3% for LFWD system. During discharging process, when PCM started to lose heat to the water, the decreasing rate of the PCM temperature of the WFWD system has been found to be 22%, in the LFWD system was 34%, whereas in CFWD system was 37%, and 32% with SFWD system.

Results showed different thermal enhancement response between vertical and horizontal arrangement for FWD systems under investigation. Longitudinal and without fins systems showed better performance in horizontal position, whereas, spiral and circular fins systems showed better performance in vertical position. In the horizontal layout, about 1.3% increase in gained PCM temperature has been found with WFWD system and 1.6% in LFWD system as compared to the vertical one. However, CFWD and SFWD systems showed more PCM temperature gain in the vertical layout than horizontal one, of 2.6% and 8.3% respectively. Overall, the developed SFWD system in vertical position has been found to exhibit the most effective type due to the fastest PCM melting and solidification.

References

[1] Medrano M, Yilmaz M O, Nogués M, Martorell I, Roca J and Cabeza L F 2009 Experimental evaluation of commercial heat exchangers for use as PCM thermal storage systems *Journal of Applied Energy* **86** pp 2047–2055

[2] Joud K and Taha A 2012 Simulation of Heat Storage and Heat Regeneration in Phase Change *Material Journal of Engineering* **18** pp 1055-1072

[3] Lokapure R B and Joshi J D 2012 Waste Heat Recovery through Air Conditioning System *International Journal of Engineering Research and Development* **5** pp 87-92

[4] Al-Abidi S A, Sopian K, Sulaiman M Y and Mohammad A T 2013 Enhance heat transfer for PCM melting in triplex tube with internal–external fins *J Energy Conv and Management* **74**

[5] Chang Liu and Groulx D 2014 Experimental study of the phase change heat transfer inside a horizontal cylindrical latent heat energy storage system *Int J of Thermal Sciences* **82** pp 100-110

Design and structural verification of locomotive bogies using combined analytical and experimental methods

I Manea[1], G Popa[2], I Girnita[1] and G Prenta[1]
[1]Softronic LTD, no. 40, Calea Severinului, Craiova, Dolj, 200609, Romania
[2]University Politehnica of Bucharest, Faculty of Transports, Department of Railway Vehicles, Splaiul Independentei nr. 313, sector 6, Bucharest, RO-060042, Romania

E-mail: ion.manea@softronic.ro

Abstract. The paper presents a practical methodology for design and structural verification of the locomotive bogie frames using a modern software package for design, structural verification and validation through combined, analytical and experimental methods. In the initial stage, the bogie geometry is imported from a CAD program into a finite element analysis program, such as Ansys. The analytical model validation is done by experimental modal analysis carried out on a finished bogie frame. The bogie frame own frequencies and own modes by both experimental and analytic methods are determined and the correlation analysis of the two types of models is performed. If the results are unsatisfactory, the structural optimization should be performed. If the results are satisfactory, the qualification procedures follow by static and fatigue tests carried out in a laboratory with international accreditation in the field. This paper presents an application made on bogie frames for the LEMA electric locomotive of 6000 kW.

1. Introduction

T Bogies are complex equipment with vital role in the operation of railway vehicles, having the role of carrying the vehicle body and ensuring the traction and braking forces. Taking into account the bogie importance in the security of railway vehicles on the long run, the nowadays regulations require that the structural strength of the bogies should be validated at various static and dynamic stress configurations that may appear during the life-cycle of the bogie.

Structural strength validation procedures of the bogies used in the European area are regulated by the EN 13749/20011 and EN 15827/2011 standards according to which the validation program shall be established on the basis of: analysis, laboratory static tests, laboratory fatigue tests, track tests. For a new design of the bogie destined to a new application, all four validation stages shall be used, though the fatigue tests can be replaced by other methods that demonstrate the required fatigue life.

The article dealt with procedures followed by Softronic Craiova for the design and structural assessment of the new bogies designed for the electric locomotive with asynchronous motors LEMA 6000kW. The complete design and validation procedure takes into account the relationship with a program for structural validation by tests in a laboratory with international accreditation.

In order to achieve the LEMA bogies, the Softronic procedure follows the following steps:

- 3D design using a CAD dedicated program, such as ProEngineer;
- Structural analysis using a dedicated finite element analysis (FEM) program, such as Ansys;
- Experimental modal analysis (EMA) using dedicated hardware and software, such as Pulse;

Content from this work may be used under the terms of the Creative Commons Attribution 3.0 licence. Any further distribution of this work must maintain attribution to the author(s) and the title of the work, journal citation and DOI.
Published under licence by IOP Publishing Ltd

- Calibration of FEM model by correlation with EMA model, eventually model optimization;
- Analytical assessment of the bogie stress response to the regulated static and dynamic loads, on the bases of the calibrated and optimized FEM model;
- Static and dynamic tests in the international accredited laboratories.
- On track tests;

The paper presents the theoretical background of FEM end EMA, similarities and differences between the two types of analysis, and the need to validate the FEM model through the data obtained by EMA tests. An application of combined analysis carried out on the new design of the LEMA bogie is presented too.

2. Basic principles of applying the combined analysis

From the point of view of experimental modal analysis-EMA, any mechanical system can be modelled by a discrete system consisting of 'n' lumped points with mass 'm_k' joined by elastic elements of 'k_k' rigidity and damping elements 'c_k'. For this system, under the action of a set of external excitation, $\{Q(t)\}$, the response is governed by the following system of equations [3]:

$$[M]\{\ddot{x}(t)\}+[C]\{\dot{x}(t)\}+[K]\{x(t)\}=\{Q(t)\}. \quad (1)$$

The system response to the external excitation, $\{Q(t)\}$, is presented as a sum of 'n' modal contributions due to each separated degree of freedom:

$$\{X(\omega)\}=\sum_{k=1}^{N}\left[\frac{\{\psi^k\}\cdot\{\psi^k\}^T\cdot\{Q(\omega)\}}{a_k(-\mu_k+i(\omega-\nu_k))}+\frac{\{\overline{\psi}^k\}\cdot\{\overline{\psi}^k\}^T\cdot\{Q(\omega)\}}{\overline{a}_k(-\mu_k+i(\omega+\nu_k))}\right], \quad (2)$$

Where:

- $[M]$, $[C]$, $[K]$ - matrix of mass, damping and rigidity;
- $\{\ddot{x}(t)\}$, $\{\dot{x}(t)\}$, $\{x(t)\}$ - column vectors of acceleration, velocity and displacement;
- $\{\psi^k\}$ *and* $\{\overline{\psi}^k\}$ - the 'k' order eigenvector;
- μ_k and ν_k - the 'k' order of damping ratio and eigenfrequency (natural frequency);
- a_k *and* $\overline{a}_k$ - the 'k' order norming constants;
- ω - external excitation frequency.

The essence of experimental modal analysis is to determine the modal parameters, $\{\psi^k\}$, μ_k, ν_k, k=1 ... n, from the experimental measurements made on the equipment brought into controlled vibrating state with simultaneous determination of excitation and response. The controlled vibration state can be achieved by using one of the following low-level energy excitation methods: relaxed step force, sinusoidal excitation, random narrow/wide band excitation or impact force method. The impact force method is very good for modal analysis of bogie frames, and it is widely used, this method being implemented with more advanced modal identification algorithms procedures.

From the point of view of finite element analysis-FEM, the mechanical structure is discretized into finite elements of various types, between which are established mathematical links, equivalent to the physical links between adjacent elements or between them and environment. FEM mathematical model corresponding to physical model is similar to EMA mathematical model, the difference between the two models being that in the finite element analysis it is not taken into account the system damping. FEM mathematical model is governed by the following system:

$$[M]\cdot\{\ddot{x}\}+[K]\cdot\{x\}=\{0\}. \quad (3)$$

where the matrices $[M]$ *and* $[K]$ have the same significations as to the EMA analysis.

Resolving the system leads to the analytical determination of mechanical system in terms of eigenfrequencies and eigenmodes, that is to say, the modal parameters which must to be approximately the same as those determined through experimental modal analysis. The both types of analysis, FEM and EMA, characterize the mechanical system through a set of intrinsic modal parameters that define the system eigenmodes independent of interaction with environment. Through EMA analysis the modal parameters are provided by even the mechanical system itself brought in a controlled vibration state.

The FEM analysis works with an approximate mathematical model, which approximations can be more or less grossly resulting from both geometrical modelling and used material characteristics. Although the geometrical model can be perfect, and motion equations in the mathematical model can be perfect, everything will result in approximately analytical model, because it is known that the material characteristics ever used in modelling features do not coincide with the real material characteristics used in the real structure. In addition, it is well known that any manufacturing process is accompanied by the execution errors more or less grossly.

From the above it follows that a good analytical model cannot be achieved except by validation with data from the real system obtained by experimental modal analysis. Through the correlation analysis of the modal forms obtained from the two models, FEM and EMA results the required information to adjust the analytical model, so that the analytical model best approximates the real system response. This is so-called procedure of analytical model "upgrade", or regeneration of the analytical model. A good analytical model, validated through experimental data can be used with confidence to assess the system response to imposed stresses, as long as they are maintained the hypotheses that formed the basis of grounding the analytical.

3. Application of combined analysis to the LEMA bogie frame

Under Ansys 14.5 was achieved the FEM model of LEMA bogie frame. Figure 1 represents the FEM model and in table 1 are given the mechanical characteristics of materials used in the frame.

The mesh has been structured so to make a network with finite elements of tetrahedron type having a size of 20 mm. The number of used elements is 350791 and the number of nodes is 826173, with the bonded type contacts between the elements.

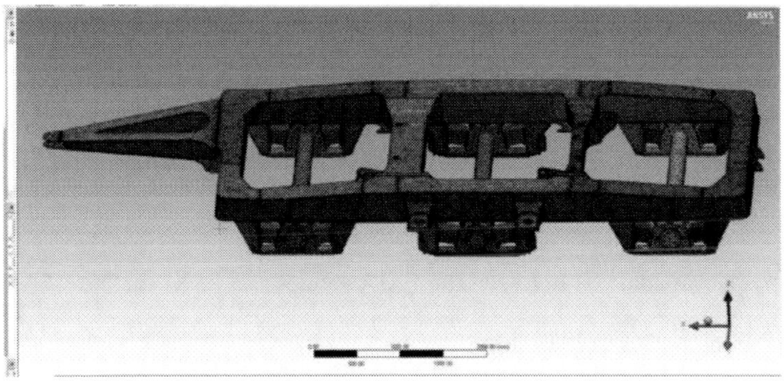

Figure1. FEM model of LEMA bogie frame.

Table 1. Mechanical characteristics of materials used in the LEMA bogie frame.

Young's modulus (elastic modulus)	210 GPa
Yield strength for nonwelded elements	350 MPa
Yield strength for welded elements	325 MPa
Tensile strength	510 MPa

On the analytical model of the bogie frame was run the Modal Analysis module of Ansys, and they was determined the eigenfrequencies and eigenmodes of bogie frame, that are presented in table 2 and figures 3- 6 together with the experimental modal analysis results. The analytical model was exported as a file containing aggregate both geometric model and modal shapes.
On the same bogie frame was applied a test of experimental modal analysis, bogie frame being placed on four springs, as in the normal operating position. For excitation and acceleration measuring response were selected a number of 10 points, as shown in the figure 2. It was used the impact excitation method, for excitation being used an impact hammer type 086D20 of 25 kN

Figure2. Test of experimental modal analysis - location of excitation and measurement points.

The tests execution was performed under PulseLabshop. The modal parameters estimation was performed under PulseReflex-Modal Analysis and was been finished with export of modal shapes and structure geometry. The modal analysis was performed in the frequency range of 0 - 100 Hz, in this domain being determined a number of 7 eigenmodes.

It was carried out the correlation analysis between the EMA model and FEM model using the correlation module PulseReflex-Correlation Analysis, which loads and displays in the same panel the animated modal shapes resulted from EMA and FEM. Correlation analysis provides a way that the analytical models to be verified against test models through graphical comparisons and numerical correlation. Graphical comparisons are performed using:
- Animation: Compare modes using side-by-side, overlaid or difference animations;
- Complexity: Validate mode shape complexity.

Correlation is performed using widely accepted tools to verify the data. These tools include:
- Modal Assurance Criteria (MAC): Evaluate the independence between modes;
- Orthogonality Matrix: Assess the quality of the measured data;
- Cross-orthogonality Matrix: Measure the degree of correlation between two models.

Table 2 presents the result of correlation analysis and in figure 3 to 6 are represented the bogie frame in the first four eigenmodes

Table 2. Correlation analysis between the EMA model and FEM for the LEMA bogie frame.

Mode	FEM Model	EMA Model			Error	MAC	Modal
	Damped Frequency (Hz)	Damped Frequency (Hz)	Damping (%)	Complexity	Damped Frequency (%)	Modal Assurance Criterion	Difference (%)
1	30.73	30.19	0.28	0.0128	1.7925	0.975	1.767
2	37.43	37.95	0.32	0.2374	0.0138	0.910	-1.406
3	48.10	49.08	0.17	0.0015	0.01992	0.968	-2.033
4	60.64	61.86	0.14	0.0014	0.01982	0.901	-2.023
5	66.66	65.66	0.13	0.0031	-0.0152	0.970	1.501

6	81.68	85.30	0.10	0.0048	0.04246	0.900	-4.434
7	88.18	89.92	0.10	0.0013	0.01935	0.902	-1.973

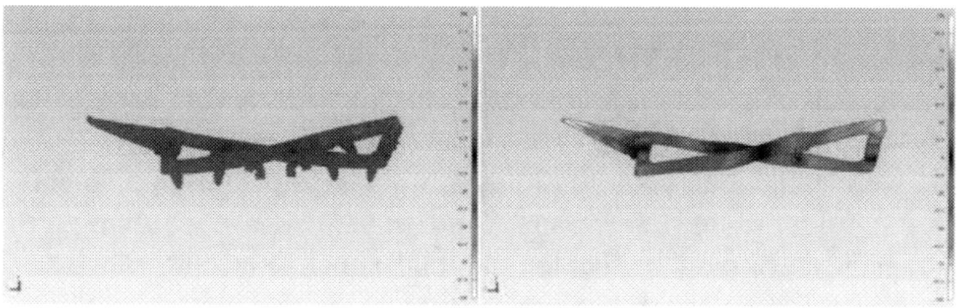

Figure 3. Bogie frame in the 1st eigenmode of 30.19 Hz (FEM left, EMA right).

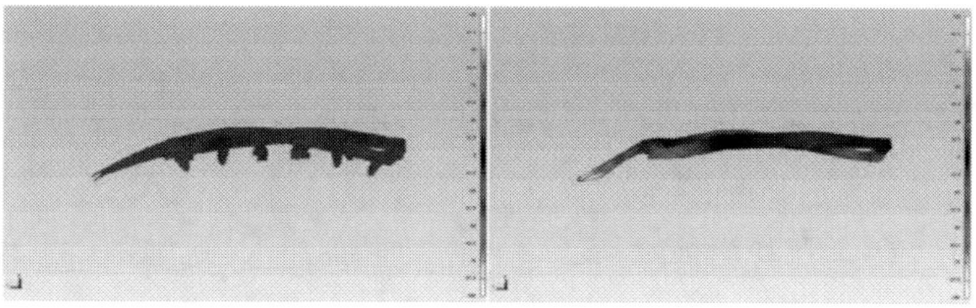

Figure 4. Bogie frame in the 2nd eigenmode of 37.95 Hz (FEM left, EMA right).

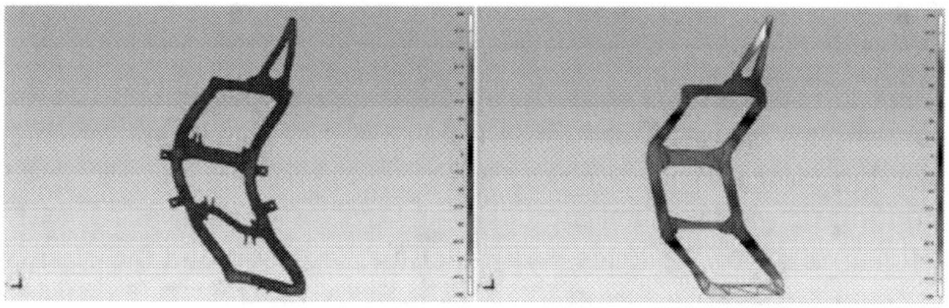

Figure 5. Bogie frame in the 3rd eigenmode of 49.08 Hz (FEM left, EMA right).

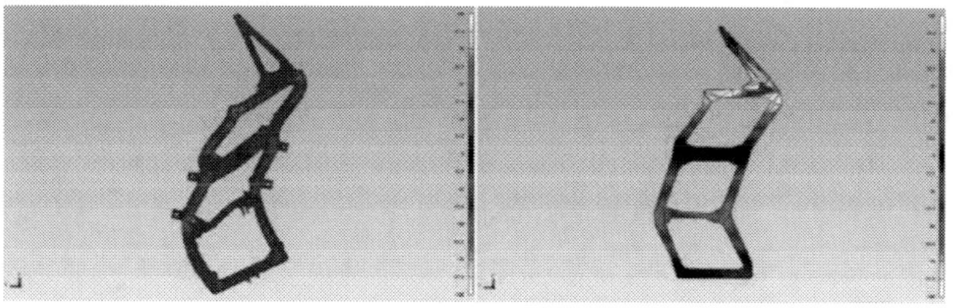

Figure 6. Bogie frame in the 4th eigenmode of 61.86 Hz (FEM left, EMA right).

Analyzing the data in table 2 and figures 3-6, result the following conclusions:

- Modal shapes corresponding to analytical model are highly correlated with experimental modal shapes, taking MAC correlation factor greater than 0.9;
- Global error for assessing the analytical eigenfrequencies fall within 1.7%, compared with the real eigenfrequencies determined by EMA;

- Taking into account the MAC correlation factors and the global error for assessing the analytical eigenfrequencies it can be concluded that the FEM analytical model is correct and can be used for theoretical analysis of bogie frame response to static and dynamic loads applied in accordance with standards EN 13749/20011 and EN 15827/2011.

4. Analytical assessment of the LEMA bogie frame to the imposed loads

For the preliminary check of the bogie frame capability to withstand to exceptional and normal loads applied according to EN 13749/20011, on the above presented analytical model, validated by experimental data, were applied all configurations of static loads imposed by mentioned standard.

From all the test configurations, in the paper is presented the case involving the highest mechanical stresses on the bogie frame, namely the longitudinal shock loads in which a the bogie frame is subjected to a longitudinal force equal to bogie inertia force under an acceleration of 3 g.

The figure 7 shows the scheme of the loads application and the figure 8 shows the distribution on bogie frame surface of the equivalent von Mises stress.

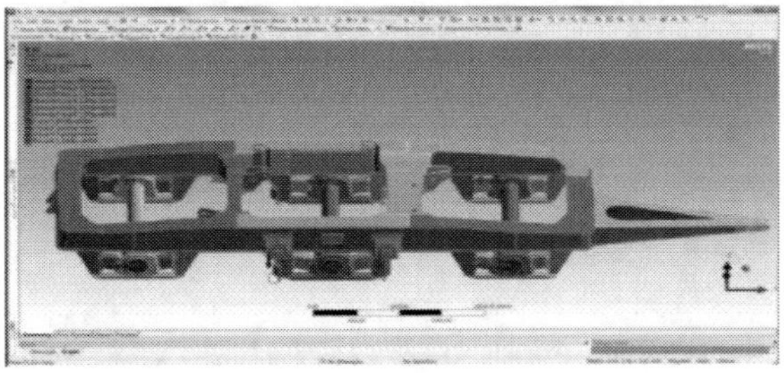

Figure7. Loads for longitudinal shock with 3g acceleration. Scheme of loads application.

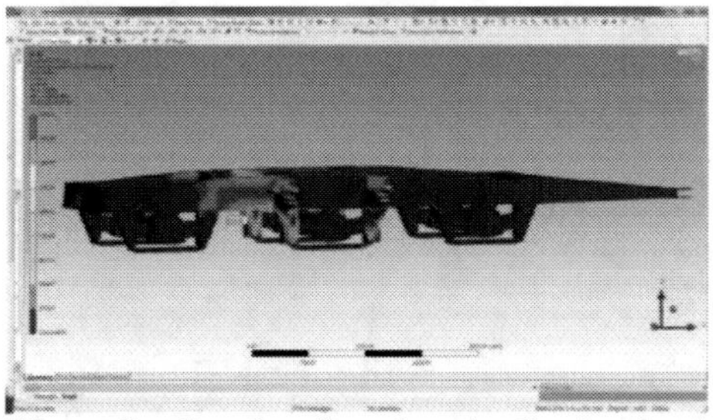

Figure8. Loads for longitudinal shock. Distribution of equivalent Von Mises stress.

In order to assess the bogie frame conformity with the request of loads of longitudinal shock, it is taken into account that the equivalent von Mises stress distribution presents a limited number of areas, reduced in size, the largest tensions having the values of 448.64MPa and 455.15 MPa. These stresses exceed the elastic limit of 350MPa, but are below the tensile strength of 510MPa, given in table 1. Given the exceptional nature of the loads, that solicit the areas beyond the elasticity is very limited, and that the maximum stresses are below the tensile strength of 510MPa, it is concluded that the LEMA bogie frame corresponds to a shock scenario of collision with amplitude acceleration of 3g.

5. Conclusions

A good analytical model developed by FEM, finite element analysis, methods requires validation through experimental data obtained by EMA, experimental modal analysis.

By applying the combined procedures of analytical and experimental analysis, presented in the article, it is concluded that the LEMA bogie frame is correctly designed and sized in terms of withstanding to static and dynamic requirements in accordance with the standards in the field.

Assignation

The work is performed under the contract PN II-PCCA, no. 192/2012.

References

[1] EN 13749 2011 Railway applications - Wheelsets and bogies - Method of specifying the structural requirements of bogie frames
[2] EN 15827 2011 Railway applications - Requirements for bogies and running gears
[3] Ewins D J 2000 Modal Testing: Theory Practice and Application *Research Studies Press Ltd*

An analysis of the performance of an ejector refrigeration cycle working with R134a

F Memet and A Preda[1]
[1]Constanta Maritime University, Faculty of Naval Electro-Mechanics, 104 Mircea cel Batran Street, 900663, Constanta, Romania

E-mail: feizamemet@yahoo.com

Abstract. In the context of recent developments in the field of energy, the aspect related to energy consumption is of great importance for specialists. Many industries rely on refrigeration technologies, a great challenge being expressed by attempts in energy savings in this sector. In this respect, efforts oriented towards efficient industrial refrigeration systems have revealed the necessity of a proper design. The most commonly used method of cooling is based on vapor compression cycles. Compared to vapor compression refrigeration systems, an ejector refrigeration system shows an inferior performance, indicated by the Coefficient of Performance of the cycle, but it is more attractive from energy saving point of view. In this respect, the present study deals with a theoretically analysis of an Ejector Refrigeration System, started with the presentation of the typical ejector design. It is stated that ejector refrigeration is a thermally driven system which requires low grade thermal energy for its working. After a short description of the analyzed system, are given equations for thermal loads and Coefficient of Performance calculation, on First Law basis. The working fluid considered in this research is Freon R134a. The developed study is focused on the effect of generating temperature variation on the Coefficient of Performance (COP) and on the work input to the pump when the cooling effect, the condensation temperature, the evaporation temperature and the reference state temperature are kept constant. Are obtained results in the following conditions: the condensation temperature is $t_c = 33°C$, the evaporation temperature is $t_e = 3°C$, the reference state temperature is $t_0 = 23°C$. The generating temperature varies in the range 82 ÷ 92°C and the cooling effect is 1 kW. Also, are known the isentropic efficiencies of the ejector, which are 0.90, and the isentropic efficiency of the pump, which is 0.75. Calculation will reveal that the Coefficient of Performance is increasing together with the increase of the generating temperature values, the best COP value being 0.178, in the considered range for the mentioned temperature. In the same time, the generating temperature increase leads to the increase of the work input to the pump.

1. Introduction

Refrigeration is a sector which is making possible our modern life. In the framework of global temperature increase, our dependency on refrigeration systems will be stronger in the next decades. Refrigeration systems are welcome not only when we talk about food industry, refrigerators are used in different applications such as ensuring of thermal comfort, medical procedures or industrial sectors. The environmental degradation resulted from refrigeration use comes mainly from generating the electricity needed to run them and from leakage of refrigerants. Expected new technologies, which will solve in a way the actual environmental impact of refrigeration, should be defined by an improved

Content from this work may be used under the terms of the Creative Commons Attribution 3.0 licence. Any further distribution of this work must maintain attribution to the author(s) and the title of the work, journal citation and DOI.
Published under licence by IOP Publishing Ltd

working performance, coming from less combustion of primary energy and mitigation of the ozone damaging and global warming.

Ejector refrigeration systems (ERS) are thermally driven technologies which have been used for cooling applications for many years. These systems are similar to vapor compression refrigeration systems, which are one of the most common types of refrigeration systems; here, instead of the compressor will work an ejector, a generator and a small pump [1].

Ejector refrigeration systems offer advantages of simplicity in construction, installation and maintenance, no moving parts, high reliability and low cost. For this reason, they are more attractive than vapor compression refrigeration systems.

Compared to vapor compression refrigeration systems, ejector refrigeration system offers a lower Coefficient of Performance, but it is more attractive in energy saving [2].

Lately, have been done attempts to extend the penetration of ERS in refrigeration and air conditioning sector by the use of low-grade thermal energy, such as solar energy and waste heat, in the system; thus will be possible to fight against environmental issues, to reduce CO_2 emission from the combustion of fossil fuels [3].

Ejectors are simple pieces of equipment. Their working principle is based on the conversion of internal energy and pressure related flow work contained in the motive fluid stream into kinetic energy [4]. A typical ejector consists of a driving nozzle, a suction nozzle, a mixing section and a diffuser, as seen in figure 1. The performance of an ejector is evaluated by dealing with one dimensional governing equation specific to each component part. The bellow assumptions are considered in this assessment [5]:
- the flow is one-dimensional and steady,
- driving flow and suction flow are considered to be ideal gases,
- the thermo physical properties are constant,
- chock occurs inside the driving nozzle, suction nozzle and mixing section,
-entropy losses during the shock wave or mixing are assessed by the use of different coefficients,
- friction losses along the pipe are neglected.

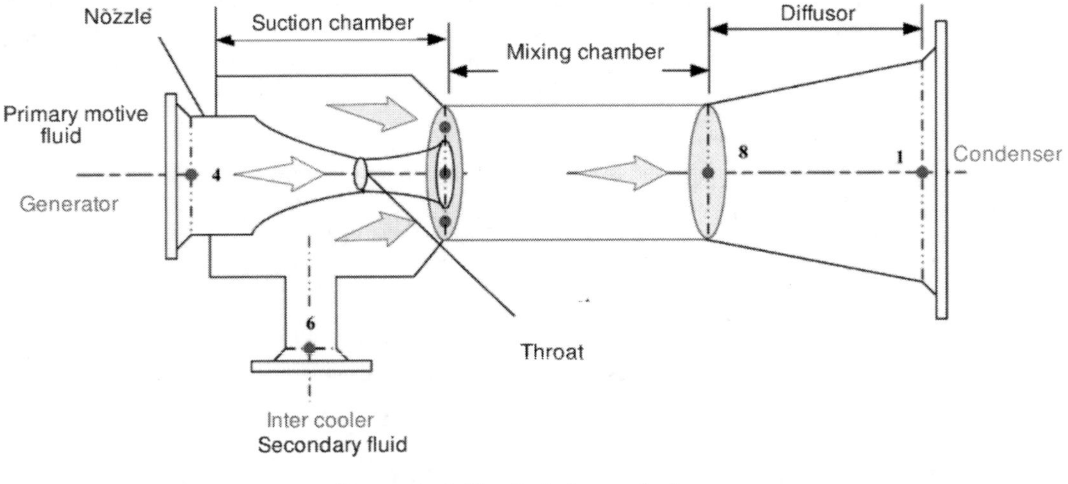

Figure 1. A Typical ejector design.

2. Methods and materials

The ejector is the main component part of the basic ejector refrigeration system. A low grade heat (Q_g) is used in the generator to evaporate the liquid refrigerant having a high pressure. The high pressure refrigerant vapors of state 2, known as primary fluid, enter and expand through the nozzle. Results a reduction in the pressure which enables the entrainment of vapors from evaporator of state 3, known as the secondary fluid. In the mixing chamber occurs the mix of the two fluids, before entering the diffuser section where the flow decelerates and takes place pressure recovery. Then, the mixed fluid flows into the condenser where condenses into liquid by evacuating heat to the environment (Q_c). Resulted liquid of state 5 is divided into two parts: one is pumped back into the generator and the other goes into the evaporator with state 6, after being first expanded through an expansion device. The low pressure liquid refrigerant evaporates in the evaporator producing the cooling effect (Q_o); resulted vapors of state 3 enter in the ejector. The described system is given in figure 2

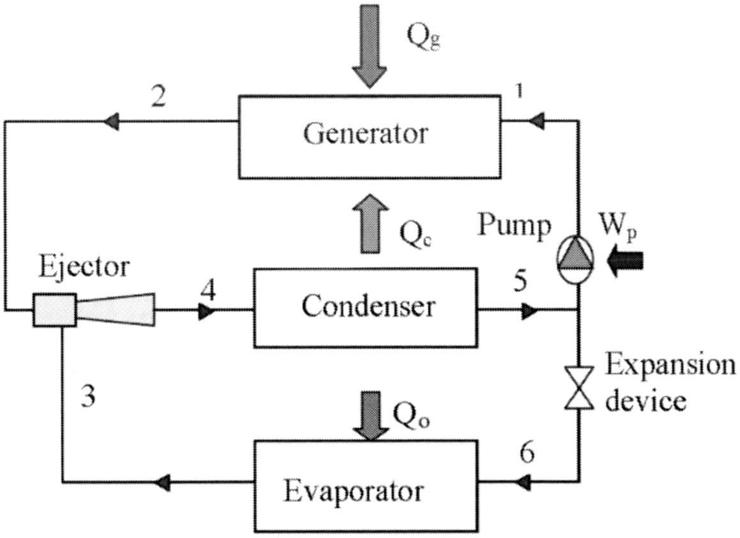

Figure 2. Basic ejector refrigeration cycle

First law analysis offers equations for thermal loads and COP calculation [6]:

heat load in the evaporator:

$$Q_o = m_o(h_3 - h_6) \tag{1}$$

heat load in the generator:

$$Q_g = m_g(h_2 - h_1) \tag{2}$$

heat load in the condenser:

$$Q_c = m_c(h_4 - h_5) \tag{3}$$

pumping work:

$$W_p = m_p(h_1 - h_5) \tag{4}$$

Above, m – mass flow rate (kg/s), h – enthalpy (kJ/kg), Q – heat load (kW), W_p - pumping work (kJ/s)

The efficiency of the ejector refrigeration cycle is assessed by the help of COP (Coefficient of Performance), defined as the ratio between the cooling effect and energy input of the cycle (the sum between the necessary heat input and work input to the pump):

$$COP = \frac{Q_o}{Q_g + W_p} \tag{5}$$

3. Results and discussions

The following results are obtained when ERS is working with R134a, the condensation temperature is t_c = 33°C, the evaporation temperature is t_e = 3°C, the reference state temperature is t_0 = 23°C, the generating temperature is the range 82 ÷ 92°C, the cooling effect is 1 kW, isentropic efficiencies of the ejector are 0. 90 and the isentropic efficiency of the pump is 0.75.

Figures 3 and 4 show the effect of the variation of generating temperature on work input to the pump and COP, when other temperatures are kept constant.

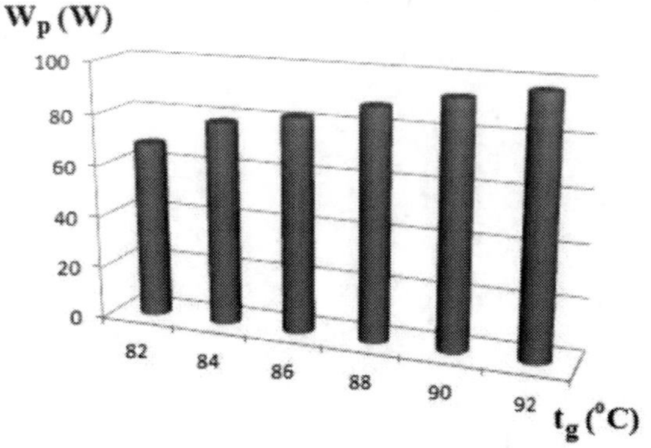

Figure 3. Effect of the generating temperature on work input to the pump

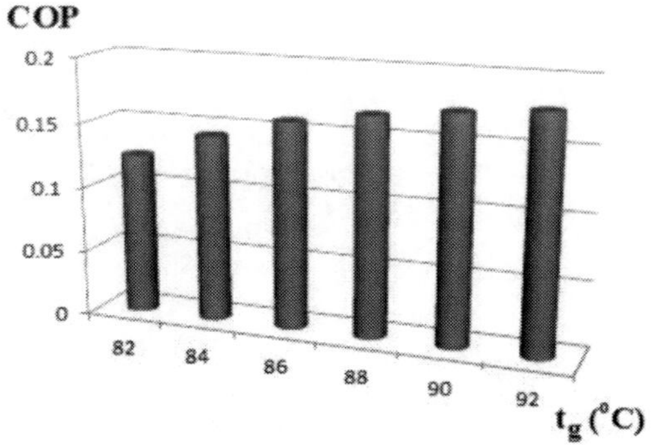

Figure 4. Effect of generating temperature on COP

Together with the increase of the generating temperatures might be seen an increase of work input to the pump and COP. When keeping constant the cooling effect, COP offers best values for high values of the generating temperature; rising this temperature will lead to more consumption in the pump.

4. Conclusions

Despite of the advantages shown by ejector refrigeration systems (ERS), the performance of the basic ERS need to be improved. This paper is an attempt in this respect.

For a specific cooling effect and given condensation and evaporation temperatures, the generating temperature is varying in order to find when best COP is achieved. Thus, was found that:
- first law efficiency (COP) is increasing together with the increase of the generating temperature values;
- in the considered range for the generating temperature (82 ÷ 92°C), best COP value (0.178) is obtained for the highest generating temperature: t_g = 92°C;
- the generating temperature increase leads to the increase of work input to the pump.

References

[1] Aye L, Charters WW S and Rusly E 2001 Combined solar and electric air conditioning *Proc of ISES Solar World Congress* W Y Saman and W E S Charters (Eds) Adelaide, Australia **25-30** pp 517-520

[2] Khajuria R and Singh J 2013 Performance analysis of ejector refrigeration system with environment friendly refrigerant driven by exhaust emission of automobile *Advances in Applied Science Research* **4(5)** pp 232-237

[3] Rahamathullah M R, Palani K, Aridass T, Venkatakrishnan P, Sathiamourthy and Palani S 2013 A review on historical and present developments in ejector systems *International Journal of Engineering Research and Applications* **3(2)** pp 010-034

[4] Elbel S and Hrnjak P 2008 Ejector refrigeration: an overview of historical and present developments with an emphasis on air-conditioning applications *Proc of International*

Refrigeration and Air Conditioning Conference at Purdue (Publisher West Lafayette) pp 14-17
[5] Dang C, Nakamura Y and Hihara E 2012 Study on ejector-vapor compression hybrid air conditioning system using solar energy *Proc of International Refrigeration and Air Conditioning Conference* (Purdue Publisher West Lafayette) pp 16-19
[6] Untea G A, Dobrovicescu Al, Grosu L and Mladin E C 2013 Energy and exergy analysis of an ejector refrigeration system U P B *Sci Bull , Series D* **75(4)** pp 111-126
[7] Aditya J, Agrawal S K, Pachorkar P 2012 Exergy Analysis of the Solar - Driven Ejector Refrigeration System IOSR *Journal of Mechanical and Civil Engineering* **3(3)** pp 30 -36
[8] Faisal S 2014 Performance analysis of ejector enhanced multi evaporator cascade refrigeration system (Department of Mechanical Engineering Thapar University Patiala – 147004)
[9] Bergander M, Butrymowicz D, Karwacki J and Wojciechowski J 2009 Application of two-phase ejector as second stage compressor in refrigeration cycles ExHFT-7 Krakow Poland
[10] Khalil A, Fatouh M and Elgendy E 2011 Ejector design and theoretical study of R134a ejector refrigeration cycle Conception de l'éjecteur et étude théorique d'un cycle frigorifique au R134a à éjecteur *International Journal of Refrigeration* **34(7)** pp 1684–1698
[11] Yapıcı R and Ersoy H 2005 Performance characteristics of the ejector refrigeration system based on the constant area ejector flow model *Energy Conversion and Management* **46(18)** pp 3117-3135
[12] Sun D W 1999 Comparative study of the performance of an ejector refrigeration cycle operating with various refrigerants *Energy Conversion and Management* **40(8)** pp 873-884
[13] Aphornratana S, Chungpaibulpatana S and Srikhirin P 2001 Experimental investigation of an ejector refrigerator: effect of mixing chamber geometry on system performance *International journal of energy research* **25(5)** pp 397-411
[14] Roman R and Hernandez J I 2011 Performance of ejector cooling systems using low ecological impact refrigerants *International Journal of Refrigeration* **34(7)** pp 1707-1716
[15] Selvaraju A and Mani A 2006 Experimental investigation on R134a vapour ejector refrigeration system *International Journal of Refrigeration* **29(7)** pp 1160 -1166
[16] Rusly E, Aye L, Charters W and Ooi A 2005 CFD analysis of ejector in a combined ejector cooling system *International Journal of Refrigeration* **28(7)** pp 1092 -1101
[17] Jia Y and Wenjian C 2012 Area ratio effects to the performance of air - cooled ejector refrigeration cycle with R134a refrigerant *Energy Conversion and Management* **53(1)** pp 240-246

Drag phenomena within a torque converter driven automotive transmission - laminar flow approach

O Alexa[1], M Marinescu[1], Gh Olaru[1], D Costache[1], C O Ilie[1] and V Vinturis[1]
[1]Military Technical Academy, Faculty of Mechatronics and Integrated Systems for Armament, Dept. of Military Automotive and Transportation
Bd. George Cosbuc nr. 39-49, sector 5, Bucharest 050141, Romania

E-mail: alexa.octavian@gmail.com

Abstract. When discussing a torque converter driven, automotive transmission with respect to the vehicle's coasting mode, automotive engineers have to take into account the slip between the converter's propeller and turbine. If the turbine isn't locked to the propellers during coasting process, drag phenomena within the converter's fluid occur and they have to be properly assessed when computing the coasting process dynamics. The best way to make the needed evaluation is to have a separate torque converter and test it on a test bench, if the data provided by the manufacturer, in this respect, weren't available. But there are several issues that could baffle this action. Among them, one could find the lack of information from the manufacturer, missing (bankrupted) manufacturer, classified information, old (out of date) products and so on. An even more challenging situation consists in dealing with a military special vehicle. Actually, the vehicle that would be subjected to the following topic is a military tracked, heavy vehicle (MBT) with a planetary driveline, driven by its engine via a hydraulic torque converter. In the attempt to assess its' coasting dynamic performances, we faced the problem of the reverse rotation of the torque converter that strongly influences the general drag of the vehicle's motion. Hence, this paper tries to provide a method to determine the transmission overall drag considering the torque converter as being its main contributor. The method is based on the experimental research our team has performed in the last several months. Using high-quality software and adjacent mathematics while assuming a certain sort of flow type within the torque converter, we aimed at determining the parameter of interest of the flow. The method can be successfully used for all type of hydrodynamic components of the transmission under the condition of developing the necessary experimental research. As far as the test were concerned, they were the typical ones designed to determine - on experimental basis - the mass inertial moments of the transmission components using the "falling weight" principle.

1. Introduction

Special vehicles require big amounts of money to have their experimental studies developed. Therefore, simulations would be a better second choice whenever comes about assessing a future product's dynamic features. Our institution is being involved in several tests with respect to some special vehicles, aiming at determining their dynamic features. This paper focuses on assessing the coasting parameters of a special tracked vehicle that has a hydro torque converter in its driveline.

Content from this work may be used under the terms of the Creative Commons Attribution 3.0 licence. Any further distribution of this work must maintain attribution to the author(s) and the title of the work, journal citation and DOI.
Published under licence by IOP Publishing Ltd

Figure 1. Experimental rigging to evaluate a vehicle's transmission moments of inertia.

Generally speaking, the dynamic features of the vehicle can be analytically obtained when taking-off [1, 5]. Some of the parameters of the taking-off equations are also used in the coasting equations. We don't hereby intend to promote some vehicle's motion equations, they are already well known [5]. Within both these categories of equations, there is a term that estimates the inertia moments of the moving parts of the vehicle's driveline. When designing the vehicle and manufacturing its transmission's parts, these inertia moments can be easily acquired both from computation or measurements [1,5]. That is impossible for an already built-up vehicle. This is why we tried to figure out a way to compute the global moments of inertia of a vehicle's transmission. The method itself isn't new at all, it is well known among the automotive engineers (figure 1).

The vehicle is missing its tracks on both sides (unlike the one in the picture) so its sprockets, connected to the transmission, are both free. The sprocket on one side of the vehicle hosts a cable, which is coiled around its hub for about 5 to 6 meters long (depending on the crane's height). At the end of the cable, falling down the crane's arm, a weight is hanging. The test itself is quite simple. The sprocket acts like a spool. At the beginning, the weight is uppermost. The driver presses the clutch and depresses the brakes and the weight starts to fall. Knowing the mass of the weight, the spooling radius and the length of the fall, while measuring the falling time, one can find the global moment of inertia of the rotating parts of the driveline. Since the global moment of inertia depends on the gear ratio, the test should be performed for the each vehicle's gear.

2. Particular features of the experiment. Paper's goal.

The experiment is mainly suitable for the vehicles that have mechanical clutches. When disconnecting the clutch from the engine, the transmission parts are completely free to start rotating. No friction, except for the one from gearing, occurs. Moreover, this friction can be kept "under control" when evaluating the moments of inertia, since this is rather easy to assess it.

When a hydro torque converter is involved, beside the above mention sources of friction, another important source comes into being, i.e. the friction of the converter's fluid. The impeller stands still (it is connected to the stopped engine) while the turbine starts to rotate, together with the rest of the transmission parts. Moreover, the angular speed of the turbine is different from gear to gear since the total ratio differs from gear to gear. Eventually, for the same gear, the turbine's angular speed also varies according to the increasing speed of the accelerated weight on its way down.

We'll try to provide a method to determining the global moment of inertia of a transmission that involves a torque converter, taking into account the liquid friction that occurs between the converter's turbine and impeller, after properly processing the experimental data [4].

2.1. Mathematical model of the phenomenon

Figure 2 depicts the diagram of the mathematical model. The first assumption was that the whole transmission's inertia is the same as of a ring that replaces the sprocket's spool with the spooling radius, r. The liquid friction of the torque converter is assumed to act along the ring's one side face, at the same radius, r. The rest of the transmission's friction is considered to be constant, no mater the angular speed or gear ratio and contained by the converter's frictional torque. Hence, the only friction

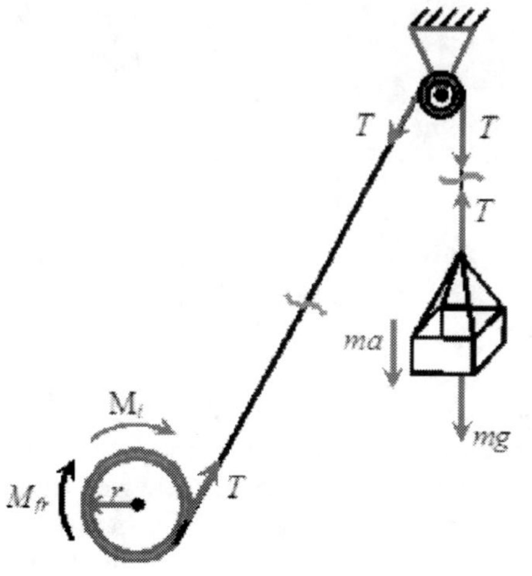

Figure 2. Test's loading diagram.

that varies is the one in the converter, referred as M_{fr}. The moment of inertia M_i brakes the motion while the load mg accelerates it (with m the weight mass).

For the left side of the pulley, the motion equation can be written as $m\frac{dv}{dt} = mg - T$, while for the left side of the pulley, we can write down $Tr = I\frac{d\omega}{dt} + c\omega$, that assumes a laminar (linear dependency) flow within the converter. Adding the cinematic connection, $\frac{dv}{dt} = r\frac{d\omega}{dt}; v = r\omega$, we get the following system:

$$\begin{cases} m\frac{dv}{dt} = mg - T \\ Tr = I\frac{d\omega}{dt} + c\omega \\ \frac{dv}{dt} = r\frac{d\omega}{dt} \\ v = r\omega \end{cases} \qquad (1)$$

In system (1) $a = \frac{dv}{dt}$ is the linear acceleration acting over the right side of the pulley. On the both sides of the pulley T is the cable (thread) tension. The moment of inertia can be written as $M_i = I\frac{d\omega}{dt}$ where I is the global kinetic moment of inertia of the transmission and ω is the angular speed of the flywheel that stands for the whole transmission. Since the flow inside the torque converter was supposed to be laminar, the frictional moment of resistance (drag moment) can be expressed as $M_{fr} = c\omega$, assuming a linear dependence of the drag moment to the angular speed of the equivalent flywheel. In this expression, c is a constant (actually, a drag coefficient). The drag moment consists of a drag force placed at the r radius of the flywheel (a ring-like flywheel, with a single radius r). Eventually, t is the time (independent) variable of the actual process.

Extracting $T = mg - m\frac{dv}{dt}$ from the first equation of system (1) and replacing it in the second one we get $\left(mg - m\frac{dv}{dt}\right)r = I\frac{d\omega}{dt} + c\,\omega$. Yet, taking into account the first cinematic connection it leads to $mgr - mr^2\frac{d\omega}{dt} = I\frac{d\omega}{dt} + c\,\omega$, or grouping the terms according to ω we get:

$$mgr = \left(mr^2 + I\right)\frac{d\omega}{dt} + c\omega \tag{2}$$

This last equation can be further processed, to get some constant terms in front of the angular speed variable and its first derivative: $\frac{d\omega}{dt} = \frac{mgr}{mr^2 + I} - \frac{c}{mr^2 + I}\omega$. Having $\frac{mgr}{mr^2 + I} = b$ and $\frac{c}{mr^2 + I} = a$ it yields to $\frac{d\omega}{dt} = b - a\omega$. Separating the variables, eventually delivers:

$$dt = \frac{d\omega}{b - a\omega} \tag{3}$$

2.2. Integrating the equation of motion

The integration of equation (3) is rather simple. The method of integration involves the separation of the variables. Thus, equation (3) can be written as $\int_0^t dt' = \int_0^\omega \frac{d\omega'}{b - a\omega'}$. For a definite integral it yields to $t\,\Big|_0^t = -\frac{1}{a}\,ln\left|\,-\frac{a}{b}\omega + 1\,\right|\,\Big|_0^\omega$. Since for $\omega = 0,\ ln\,1 = 0$ (limit conditions) then for real, positive numbers, the solution can be written as $t = -\frac{1}{a}ln\left(1 - \frac{a}{b}\omega\right)$, or $-at = ln\left(1 - \frac{a}{b}\omega\right)$. From here, using the exponential function, it turns into $e^{-at} = 1 - \frac{a}{b}\omega$ that eventually provides the solution $\omega = \frac{b}{a}\left(1 - e^{-at}\right)$. Should be appropriate now to replace the previously defined constants, a and b, so we can finally get $\frac{b}{a} = \frac{\frac{mgr}{mr^2 + I}}{\frac{c}{mr^2 + I}} = \frac{mgr}{c}$ and $a = \frac{c}{mr^2 + I}$. Therefore, the final solution of the equation, expressing the angular speed evolution of the sprockets' angular speed when the weight is falling under the given circumstances (among we find laminar flow approach of the converters' frictional torque) will be given by:

IOP Conf. Series: Materials Science and Engineering **95** (2015) 012036 doi:10.1088/1757-899X/95/1/012036

$$\omega = \frac{mgr}{c}\left(1 - e^{-\frac{c}{mr^2 + I}t}\right) \tag{4}$$

Should be noticed that the above equation has, from the experimental data point of view, two unknown parameters, i.e. the global friction constant c (met in the moment of global friction expression as $M_{fr} = c\omega$) and the kinetic moment of inertia I. One way to experimentally find constant c is to horizontally pull the cable of the sprocket at constant speed. Since $M_{fr} = c\omega$, while measuring M_{fr} and knowing the angular speed, the constant will come up, as will be further seen. Eventually, we could compute the kinetic moment of inertia for every gear of the transmission.

3. Experimental data

We have performed several tests with the same falling weight for each gear of the gearbox, except for the 4th gear (a failing transducer prevented in getting reliable results for the last gear [2]; nevertheless, while the remaining three sets of results were good and reliable). Another test has been also performed. It consisted in horizontally pulling the cable from the spool at constant speed while measuring needed force. This test has been performed for each gear and it aimed to experimentally determine the c constant of the friction moment expression $M_{fr} = c\omega$. Since we knew the spooling radius, the pulling speed and force, it was easy to determine the above-mentioned coefficient. The values of the constant c are given in table 1, along with other data.

As an example, Figure 3 depicts the time histories of the measured angular speed for three different tests developed for the third gear of the transmission (due to resolution reasons, only three curves have been plotted). Figure 3 depicts the approximation curves of the experimental data plotted in Figure 3 (after properly processing [4]).

Using the *Curve Fitting* module of the Matlab programming environment [3] and considering the motion as developed according the motion law provided by equation (4), we computed the most accurate global kinetic moment of inertia of the transmission, I, for each researched gear ratio of the transmission. Hence, given equation (4) and its known, contained parameters, the module provided the global kinetic moment of inertia I for each gear. The data are given in table 1.

Analyzing the data in the table one can notice that the constant affecting the angular speed within the drag moment has a rather small distribution of its values around 110...120 [s/kg^2/m]. That could be a proof of the quality of the measurements since it has to keep itself around a certain value [4].

Another way to check for the accuracy of both the measuring procedures and the mathematical model was to check for the terminal velocity (angular speed of the sprocket). Of course, we couldn't measure it for real since the cable was too short, but using the extended time history of the modeled curves (using equation (4)) we have got the results depicted in figure 5.

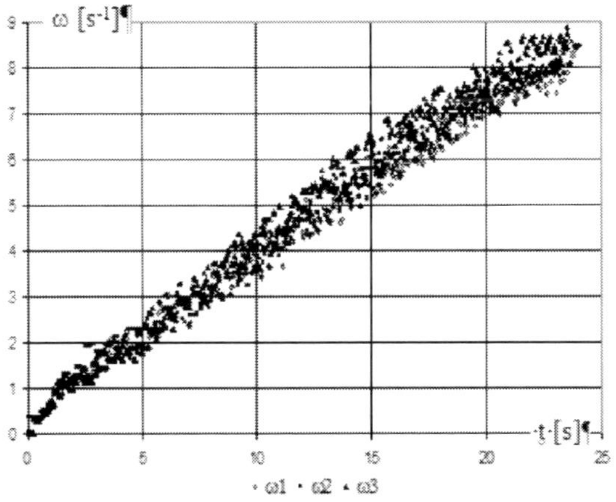

Figure 3. Time histories of the measured angular speed for three different tests developed for the third gear of the transmission.

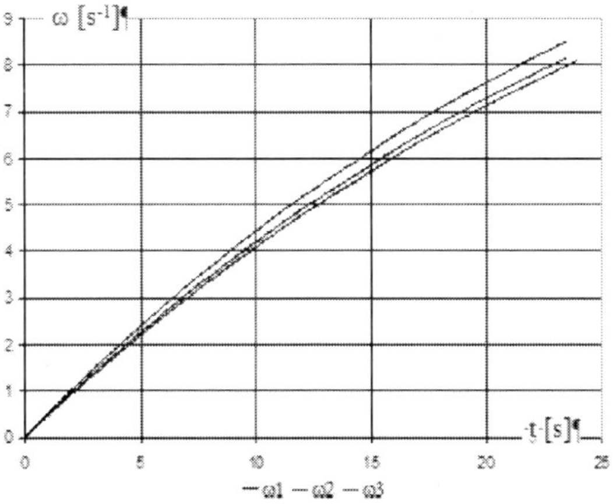

Figure 4. The approximation curves of the experimental data plotted in figure 3.

We overlapped several curves to determine the value of the terminal angular speed of the sprocket wheel. It came out that the value of the angular speed is about 15.8 s-1. It corresponds to a value of about 150 revolutions per minute and it is reached sooner or later, depending on the gear ratio of each gear of the transmission. What is more important is that the vehicle's speed that corresponds to this value is about 15 km/h. This leads to a very important conclusion: the engine brake is extremely effective for this vehicle and, as known, for all the tracked vehicles. It occurs whenever the vehicle's speed exceeds the average values of 12 to 15 km/h. The "Error" column in the table below provides the overlapping error between the approximation curves in figure 4 and the time histories in Figure 3.

Table 1. Measured and computed parameters of the tests.

Article ear	Test no.	m [kg]	r [m]	c [s/kg^2/m]	I [kg·m^2]	Error [%]
I	1	635	0.254	110	8100	7.8
	2			140	7900	6.5
	3			140	8000	5.7
	Average values »			130	8000	6.7
II	1	635	0.254	100	4400	8.5
	2			120	4200	6.8
	3			105	4400	7.6
	Average values »			108	4333	7.6
III	1	635	0.254	100	3300	5.1
	2			100	3200	6.4
	3			100	3000	5.9
	Average values »			100	3167	5.8

4. Conclusions

Tracked vehicles are rarely braked since their propulsion system is featured by a low efficiency. Nevertheless, for this kind of vehicles would be useful to learn what the coasting regime leads to. We tried to experimentally reveal the dragging behavior of the torque converter (within a military vehicle's transmission) and determining the global kinetic moments of the transmission for each gear, hence the coasting regime could be studied.

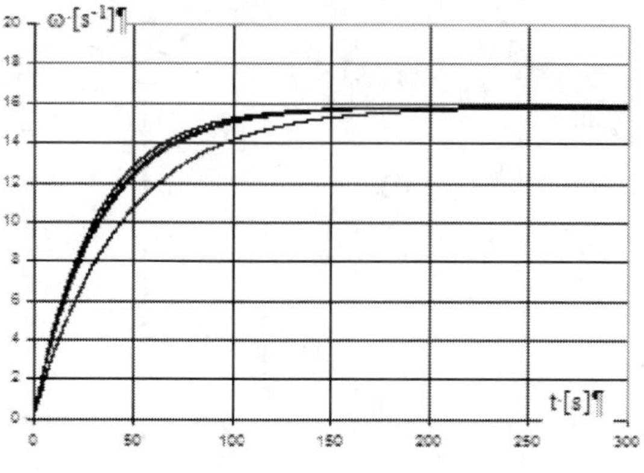

Figure 5. Terminal angular speed of the sprocket wheel

The dragging phenomenon starts whenever the terminal speed is overtaken, and that means about 15 km/h for the actual vehicle. Small fluctuation around this terminal velocity may occur due to the different ratios of the transmission. As a matter of fact, it varies between about 12 km/h in the first gear to abut 15 km/h in the third one.

The mathematical model of the angular speed variation has been confirmed with enough accuracy (in the terms provided by equation (4)). That allowed us to determine the terminal speed.

The model allowed us to compute the global kinetic moments of the transmission. Their values are quite similar to the theoretical ones, validating our theoretical model.

We consider that the model provided in the paper hereby can be further developed using turbulent flow models for the torque converter.

Acknowledgments

This paper has been financially supported within the project entitled "Horizon 2020 - Doctoral and Postdoctoral Studies: Promoting the National Interest through Excellence, Competitiveness and Responsibility in the Field of Romanian Fundamental and Applied Scientific Research", contract number POSDRU/159/1.5/S/140106. This project is co-financed by European Social Fund through Sectoral Operational Programme for Human Resources Development 2007-2013. Investing in people!

References

[1] Alexa O, Ilie C O, Marinescu M, Vinturis V and Truta M 2014 Simulating the torque and angular speed distribution within a heavy vehicle's planetary gearbox *Applied Mechanics and Materials* **659** pp 127-132
[2] Vilau R, Marinescu M, Alexa O, Truta M and Vinturis V 2014 Diagnose method based on spectral analysis of measured parameters *Advanced Materials Research* **1036** pp 535-540
[3] Information on http://mathworks.com
[4] Marinescu M and Ilie C O 2014 Filtering the signal of a measured mechanical parameter 10th *Conference Proceedings* (Bucharest: AGIR Publishing House) pp 85-88
[5] Truţă M, Fieraru O, Vilău R, V Vînturiş V and Marinescu M 2014 Static and dynamic analysis of a planetary gearbox working process *Advanced Materials Research* **837**pp 489-494

Graphical method for profiling hob mill that generate cycloid worms

V Teodor[1], **S Berbinschi**[2], **N Baroiu**[1] **and N Oancea**[1]
[1] Department of Manufacturing Engineering, "Dunărea de Jos" University of Galaţi, România
[2] Mechanical Engineering Department, "Dunărea de Jos" University of Galaţi, România

E-mail: virgil.teodor@ugal.ro

Abstract. The hob mill for generating ordered curls of cycloid surface with non involute profiles may be profiled based on the fundamental theorems of surface enveloping – Olivier – as surface reciprocally enveloping with point like contact. In this paper, is proposed a methodology based on a complementary theorem of the surface enveloping in a graphical expression developed in a graphical design environment – CATIA. The graphical method presented in this paper is developed in two stages: determining of the rack gear model based on the solid model of the surface to be generated, using an original algorithm, following this, based on 3D modelling is determined the solid model of the primary peripheral surface of the hob mill. An application for a cycloid worm is presented – a central screw of helical pumps. In order to prove the quality of method, the analytical and graphical solutions are comparatively presented.

1. Introduction

In order to solve such a problem it is used the second theorem of Olivier – the case of conjugated surfaces, which depends by two independent parameters [1, 3]. It is also possible to use the Gohman theorem [1], when the surface to be generated, in relative motion regarding the peripheral primary surface of hob mill, depends by two independent parameters, usually the angular parameters of the revolution of blank and hob.

They are also knows and applied the fundamentals analytical theorems of surfaces enveloping, based on the method of intermediary surfaces (reference rack gear) [1], as so as the complementary theorems as: the method of "substitutive circles family" [3], the method of "minimum distance" [3], the method of "in-plane generating trajectories" [5] and others.

The development of graphical design environment as AutoCAD, CATIA etc. allows to create methods which use the capabilities of the AutoCAD [8] or CATIA [4] software.

Also, it is highlighted the concerning of the international researchers regarding the issue of profiling generating tools for involute teeth using the capabilities of graphical design environment [4, 9, 10, 11].

In this paper, is presented an application for profiling a hob mill which generates a cycloid worm – the central worm of a helical pump.

An algorithm presented by [4] is proposed. This algorithm made a methodology based on the theorems of helical movement composition and decomposition [2].

Content from this work may be used under the terms of the Creative Commons Attribution 3.0 licence. Any further distribution of this work must maintain attribution to the author(s) and the title of the work, journal citation and DOI.
Published under licence by IOP Publishing Ltd

2. The geometry of rotor with cycloid profile

I In reference system *XYZ*, see figure 1.a, the cycloid profile of the worm is given by equations:

$$X = 2R \cdot sin(\alpha+\varphi) - R \cdot sin(2\varphi+\alpha);\ Y = 2R \cdot cos(\alpha+\varphi) - R \cdot cos(2\varphi+\alpha)\quad (1)$$

where:

R is the base radius equal with the radius of the ruler, R = 10 mm; $\alpha = \pi/4$ - half-angle of the gap between two teeth; R_e =15 mm – external radius of the worm; z = 2 – the rotor teeth number; p_e = 100 mm – helical pitch;

The limits for φ parameter are

$$\varphi_{min} = 0;\ \varphi_{max} = ar\,cos\left[\frac{-R_e^2 + 5R^2}{4R^2}\right].$$

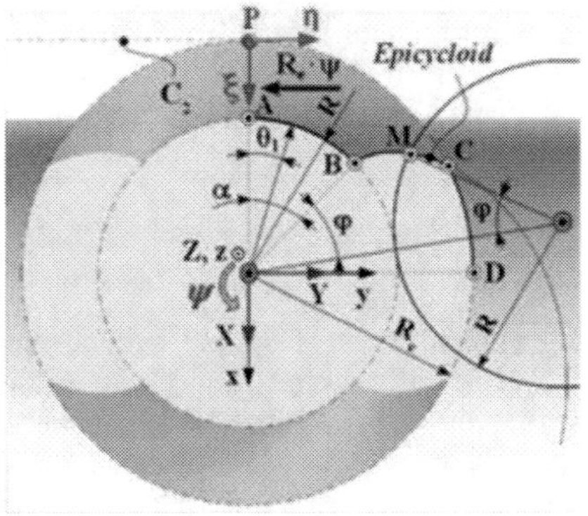

a)

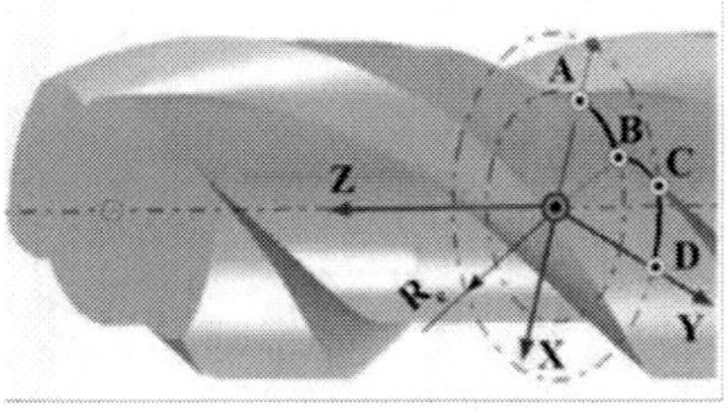

b)

Figure 1. a). Cycloid rotor geometry; b). The 3D model of the cycloid worm.

In figure 1.b, is presented the 3D model of the cycloid worm, for the previously presented dimensions. The frontal profile of the rotor is composed from the $\overset{\frown}{AB}$ arc, with the radius R and the $\overset{\frown}{BC}$ cycloid arc.

3. Generating rack-gear of cycloid worm

If it is accepted that the frontal profile of the cycloid worm is associated with a pair of rolling centrodes: C_1 is a circular centrode with radius R_e associated with the cycloid worm; C_2 - linear centrode associated with the rack-gear, see figure 1.a.

In the movement assemblies:

$$x = \omega_3^T(\psi) \cdot X;\ \xi = x - a;\ a = \| -R_e \quad -R_e \cdot \psi \quad 0 \|^T \quad (2)$$

regarding the global reference system *xyz*, it is determined the family of cycloid profiles, in the reference system associated with the rack-gear, $\xi\eta\varsigma$, see figure 1.a.

In movements (2), with ψ is denoted the value of rotation angle around the Z axe of the axode associated with the cycloid worm.

It is determined the cycloid profiles family and, after this, is determined the Σ profiles families in the reference system of the rack-gear, in principle, in form:

$$\left(\Sigma_{BC}\right)_{\psi} = \begin{vmatrix} \xi = \xi(\varphi, \psi); \\ \eta = \eta(\varphi, \psi); \\ \zeta = 0; \end{vmatrix} \quad \text{and} \quad \left(\Sigma_{AB}\right)_{\psi} = \begin{vmatrix} \xi = \xi(\theta_1, \psi); \\ \eta = \eta(\theta_1, \psi); \\ \zeta = 0. \end{vmatrix} \tag{3}$$

The method of "minimum distance" allow to determine the enveloping of a profile associated with a pair of rolling centrode based on a specific theorem [3, 6]: The enveloping of an in-plane profile associated with a rolling pair of centrodes is the geometric locus of points belong to the profile for which, in various positions of rolling, the distance from the gearing pole is minimum.

In principle, the enveloping of the profiles family (3) is in form:

$$\xi = \xi(\psi); \ \eta = \eta(\psi); \ \zeta = 0, \tag{4}$$

as result of the association between the equations (3) and the specific enveloping condition for minimum distance method [3].

The enveloping of the helical surfaces families Σ_{AB}, Σ_{BC}, Σ_{CD}, represents the flanks of the generating rack-gear. The form of the rack-gear will be used, in following, to determine the peripheral primary surface of the future hob mill.

For problem solving it will be used the method of intermediary surface [1], (the rack-gear) based on the method of "minimum distance" [3].

3.1. Methodology in CATIA

In figures 2, 3 and 4 is shortly presented the method of "minimum distance" for generation of rack-gear conjugated with the crossing section of the cycloid worm, for profile segments profil Σ_{AB}, Σ_{BC}, Σ_{CD}. On the C_1 centrode, associated with the worm's profile, are considered equidistant points which are transferred onto the C_2 centrode of the rack-gear – the points P_1, P_2, …, P_i, representing the relative positions of the gearing pole, P, in the process of rolling between the C_1 and C_2 centrodes.

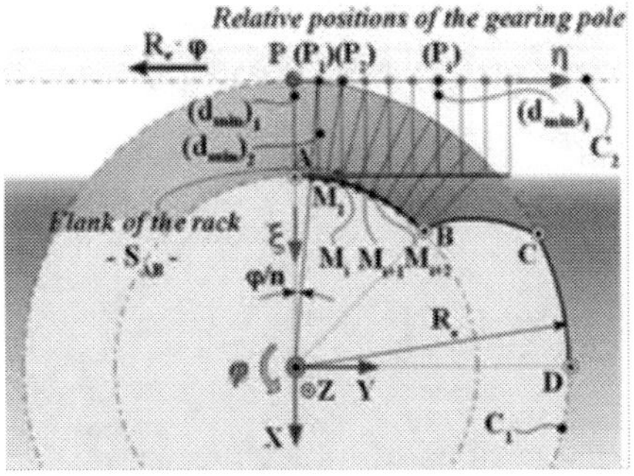

Figure 2. Determination of the rack-gear– the AB flank.

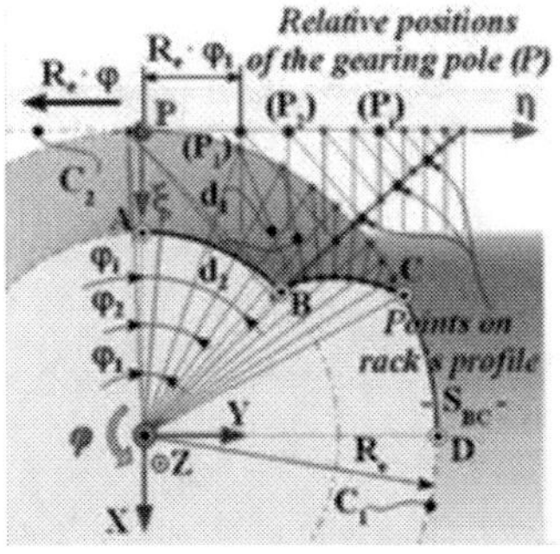

Figure 3. Determination of the rack-gear profile – the BC flank.

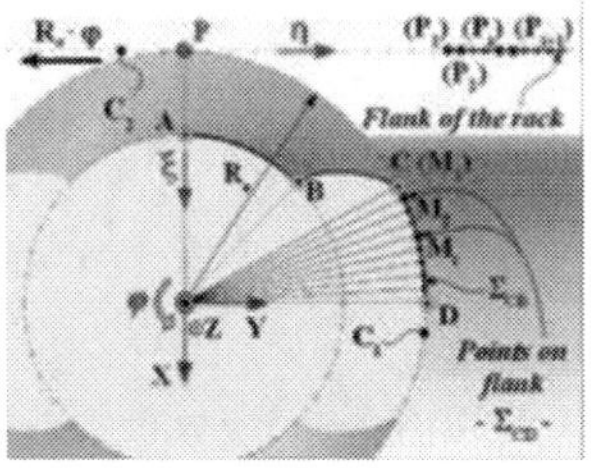

Figure 4. The rack-gear profile flank CD.

The distances measured onto the normals drawn to the Σ_{AB}, Σ_{BC}, Σ_{CD} profiles from the P_i (i=1, 2, …, n) points, by points representing the feets of these normals are the points of the generating rack-gear.

In figure 4, is represented the frontal profile of the cycloid shaft, the crossing section of the rack-gear and the 3D surface of the generating rack – I (I_{AB}, I_{BC}, I_{CD}). The surface of the rack-gear is a cylindrical surface having as directrix the enveloping of the *AB*, *BC* and *CD* profiles and the generatrix parallels with the tangent to the helix with radius R_e of the cycloid worm - $\vec{t}$. We have to notice that for the generating rack-gear form, in figure 6, as so as in following, we not take into account the changes due to the overlapping between the profiles as so as the possible passing curves due to the singular points onto the profile.

4. Helical movement decomposition

Knowing surface *I*, based on the theorem of helical motion decomposition is made an algorithm for determination of the characteristic curve at contact between the surface of the rack-gear and the helical peripheral primary surface of the tool, *S*, with axis $\vec{V}_1$ and parameter p_1.

Based on the specific theorem of the helical movement decomposition, $\vec{V}_1$ p_1, the helical motion is decomposes in two revolution movements with disjoint axis let $\vec{A}$ and $\vec{B}$ these axis. They are defined the $\vec{A}$ and $\vec{B}$ axis, the angular values *α* and *β*, as so as the *a* and *b* distances, measured along the common perpendicular between the axis $\vec{A}$, $\vec{V}_1$ and $\vec{B}$.

The helical motion ($\vec{V}_1$, p_1) is decomposed in a sum of revolution movements with axis $\vec{A}$ parallel with $\vec{V}_1$ and translate along the direction of the cylindrical surface's generatrix, $\vec{t}$.

$$\left(\vec{A}, \alpha = 0\right) + T\left(\vec{t}\right); \ \left(\vec{V}_1, p_1\right). \tag{5}$$

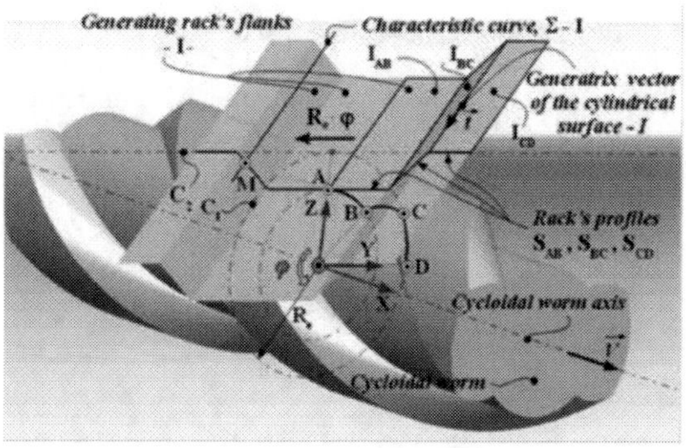

Figure 5. Generating rack-gear - I_{AB}; I_{BC}; I_{CD}; $\vec{t}$ - unitary vector of the cylindrical surface's generatrix; $\vec{V}$ - unitary vector of the cycloid worm's axis.

With $T\left(\vec{t}\right)$ was denoted the translation motion along the generatrix $\vec{t}$ of the cylindrical surface (the *I* rack-gear).

If is denoted with *θ* the angle between $\vec{V}_1$ and $\vec{t}$ then, the position of $\vec{A}$ axis regarding $\vec{V}_1$ is given by the distance: $a = p_1 \cdot tg\theta$, here $\beta = \theta$, and the $\vec{A}$ axis, parallel with $\vec{V}_1$ is positioned at distance *a* from this.

In this way, the *I* surface of rack-gear in the helical motion ($\vec{V}_1$, p_1) is equivalent with the assembly of motion:- rotation around the $\vec{A}$ axis, parallel with $\vec{V}_1$ and the translation along the generatrix of the *I* cylindrical surface, in the direction $\vec{t}$.

The characteristic curve, namely the contact curve between *I* and *S* will depend only by this component of the motion in which the *I* surface will not self-generate. How $\vec{t}$ is the direction of generatrix of *I* surface, the characteristic of rack-gear's surface, in the (5) motion assembly will depend only by the rotation around the $\vec{A}$ axis, previously defined (α=0; $a = p_I \cdot \mathrm{tg}\theta$).

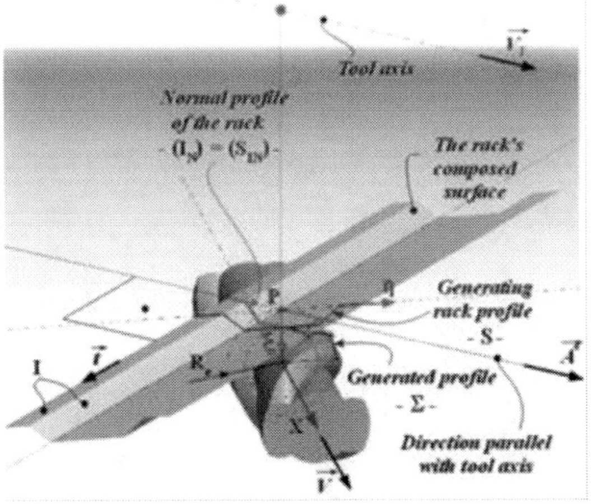

Figure 6. The cycloid worm; generating rack-gear *I*.

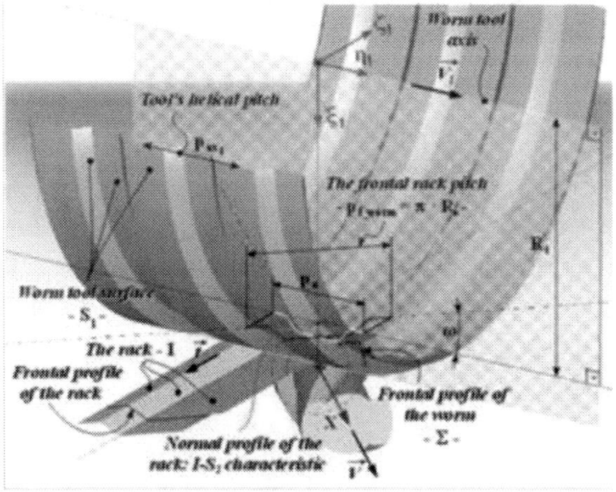

Figure 7. Primary peripheral surface of the hob mill - S_I.

As it is known, the characteristic curve of a surface, *I*, in rotation motion around an axis $\vec{A}$ is the geometric locus of the points belongs to *I* surface, which represent the projection of the rotation axis onto this surface [2].

These reasons allow determining a graphical methodology, developed in CATIA design environment, for profiling the hob mill generating the *I* rack-gear and, as consequence, of the Σ helical surface.

If the intermediary surface *I* is defined, the rack-gear conjugated with the cycloid worm to be generated, see figure 6, through the *P* pole, is drawn a line, $\vec{A}^*$, parallel with the axis of the future hob mill - $\vec{V}_1$. The direction $\vec{A}^*$ make with the $\vec{t}$ generatrix the angle θ, $\theta = \pi / 2$.

It is defined the profile of the generating rack-gear, in a plane normal to the rack-gear generatrix, the normal profile. The normal profile of the rack-gear is positioned in the plane determined by the directions $\vec{A}^*$ and $\vec{V}_1$, parallels, at the distance R_t – the radius of the hob mill (technological value), in the plane normal to the *I* surface's generatrix.

Table 1. Coordinates of the axial profile of hob mill for half-profile.

Crt. no.	η_1 [mm]	ξ_1 [mm]
1	0.0000	-90.0000
2	0.8390	-90.0000
3	1.6770	-90.0000
	...	...
8	8.4970	-88.8680
9	8.9760	-88.3060
10	9.4580	-87.7450
	...	...
16	15.9870	-85.0000
17	16.5670	-85.0000
18	17.1470	-85.0000

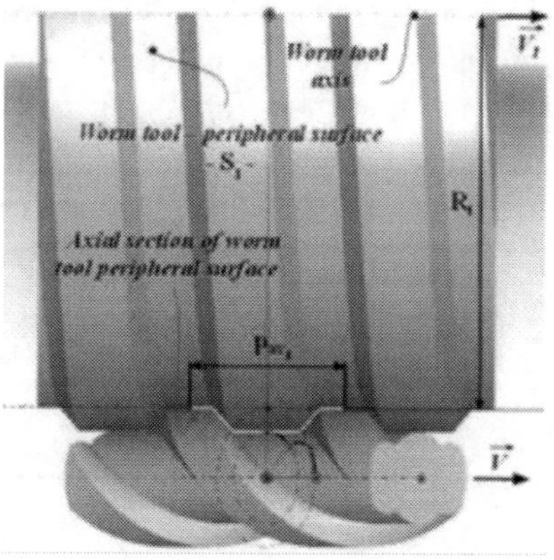

Figure 8. Rack-gear's peripheral surface.

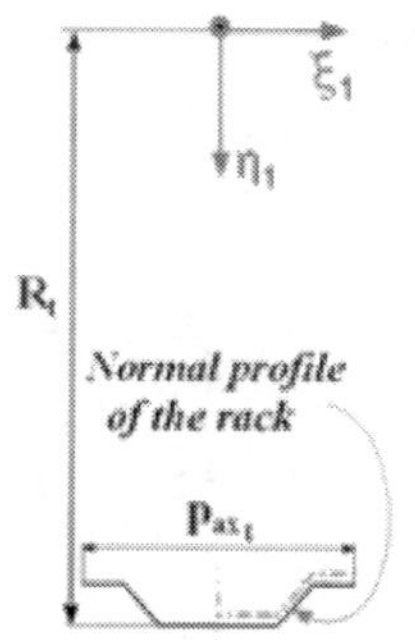

Figure 9. Axial profile.

By the helical motion of the normal profile around the $\vec{V}_1$ axis (the hob mill's axis), as worm with sense contrary to this of the worm to be generated, but with the same inclination on the cylinder with the radius R_t, for the hob mill and R_e, for the cycloid worm, is generated the 3D model of the peripheral primary surface of hob mill, see figure 7, which is a complexes helical surface, with known normal pitch, from the intersection of the 3D model of the generating rack-gear with the plane normal to the $\vec{t}$ direction of its generatrix (the projection of the $\vec{A}^*$ axis on the intermediary surface *I* – the rack-gear).

The angular parameters ω_p and ω_t are the angles presented in figure 7 and the axial section of the helical surface of the hob mill is presented in figure 8. Te intersection of the characteristic curve $\Sigma - I$, see figure 6, with the characteristic I - S_1, figure 8, determines the characteristic point – the contact point between Σ and S_1.

The geometric characteristics: cycloid worm $p_{f\_worm} = \pi \cdot R_e$; normal pitch $p_n = p_{f\_worm} \cdot \sin\omega_p$; $\tan\omega_p = p_{ax_p}/(2\pi R_e)$; rack-gear $p_{n\_rack} = p_n$; hob mill $p_{ax_t} = p_n/\cos\omega_t$; $\tan\omega_t = p_{tx_t}/(2\pi R_t)$.

In table 1 and figure 8, are presented the coordinates and axial profile of the hob mill and its form – the axial section of helical primary peripheral surface which is the peripheral surface of the future tool.

Obviously, this is the primary profile, which may be modified if we considered the issue linked to the presence of singular points and interference trajectories.

5. Conclusions

The graphical method for profiling the hob mill generating of the cycloid worm, based on the complementary theorem of the "minimum distance" was developed in graphical form in CATIA design environment.

The method use an intermediary surface – generating rack-gear conjugated with the surface to be generated.

The proposed method is intuitive and leads to rigorous results of the hob mill's profile. The proposed principle can be applied for other types of teeth (ordered curls of surfaces associated with a centrode).

In this paper, it wasn't approach the issue linked to singular points of the profile, nor does the interference process which always emerges in such a generating process

Acknowledgments
The work has been funded by the Sectoral Operational Program Human Resources Development 2007-2013 of the Ministry of European Funds through the Financial Agreement POSDRU/159/1.5/S/132397.

References
[1] Litvin F L 1984 Theory of Gearing *Reference Publication 1212 NAS* Scientific and Technical Information Division (Washington D C)
[2] Liukshin V S 1968 Theory of Screw Surfaces in Cutting Tool *Machinostroyenie* (Moscow)
[3] Oancea N 2004 Surfaces generation by enwrapping **(I –II)** *Teoreme complementare* (Galaţi: Editura Fundaţiei Universitare „Dunărea de Jos"
[4] Berbinschi S, Teodor V and Oancea N 2013 3D graphical method for profiling gear hob tools *The Int Journ of Adv Manuf Tech* **64(1-4)** pp 291-304
[5] Teodor V G 2010 Contribution to elaboration method for profiling tools which generate by enveloping *Lambert Academic Publishing*
[6] Oancea N and Oancea V G 1997 Geometrical modeling of surface generation through wraping *Journal of Manufacturing Science and Engineering* **119** pp 829-834
[7] Berbinschi S et al. 2011 A 3D Method for Profiling the Shaping Tool for Generation of the Helical Surfaces *The Annals of "Dunărea de Jos" Univ of Galaţi Fasc V Tech in Machine Buildings*
[8] Baicu I and Oancea N 2002 *Profilarea sculelor prin modelare solidă* (Chişinău: Tehnica Info Publishing House)
[9] Skoczylas M J 2007 Modelling the tooth flanks of hobbed gears in the CAD environment *Springer – Verlag London Limited* pp 746 – 751
[10] Dimitriou V and Antoniadis A 2008 CAD – based simulation of the hobbing process for the manufacturing of spur and helical gears *Springer–Verlag London Limited* pp 347–357
[11] Ott O S and Artyukin L L Shaping Involute Profiles by Means of a Disk Tool (London: *Springer–Verlag London Limited*)
[12] Liang Yao, Zhonghe Ye, Jian S Dai and Noiyi Cai 2005 Geometric Analysis and Tooth profiling of the Three-Lobe Helical Rotor of Roots Blower *Journal of Materials Processing Technology* **170** pp 259-267

An engineering model for yield inception in slip-stick elastic contacts

S Spinu[1] **and D Gradinaru**[1]
[1]Department of Mechanics and Technologies, "Stefan cel Mare" University of Suceava, 13th University Street, 720229, Suceava, Romania

E-mail: sergiuspinu@yahoo.com

Abstract. The failure of the mechanical contact due to plastic yielding is generally predicted employing stress analysis coupled with the von Mises yield criterion, which uses the maximum of the second deviatoric stress invariant as a threshold value. This paper aims to establish the relation between the frictional regime and the normal and tangential loading components which lead to yield inception in the slip-stick spherical contact between similarly elastic materials. The Boussinesq and Cerruti fundamental solutions for the elastic half-space are used in a robust semi-analytical method based on the superposition principle applicable in the frame of linear elasticity, and enhanced with an acceleration technique derived from the convolution theorem. A rapid algorithm for accurate computation of elastic stresses induced in subsurface by a known but arbitrary distribution of surface tractions, normal or shear, is advanced. The obtained data is normalized to allow model extension to any elastic constants or contact curvature, and curve fitting is employed to derive simple empirical formulas pertinent to practical engineering applications.

1. Introduction

Stress analysis in the elastic contact between two spheres under combined normal and tangential loading is a fundamental problem in engineering, applicable in the modeling of particle–flow simulations or in study of contact between rough surfaces. While the surface tractions distribution in a normally loaded frictionless contact problem is known from the Hertz theory [1], the tangential contact problem, involving the establishment of a slip annular region surrounding the central stick contact zone, was solved separately by Cattaneo [2] and by Mindlin [3].

The potential contact failure due to plastic yielding is generally predicted using the von Mises yield criterion, which uses the maximum of the second deviatoric stress invariant to locate the plastic yield inception in the elastic body. The problem of assessing the stress field throughout an elastic and isotropic body, loaded by surface shear stresses acting simultaneously and proportional to normal Hertz-type stresses was solved by Hamilton and Goodman [4] for a circular point contact and by Bryant and Keer [5] for an elliptical contact. The latter work advances an explicit solution for the surface and subsurface stresses and displacements in a full slip contact, and indicates that the solution of the slip-stick problem could be obtained by superimposing three full-slip solutions. Due to problem complexity, the study of the influence of the stick region on the stress field is restricted to a single instance. The problem is revisited by Hills and Sackfield [6], who obtained an approximate solution to the stress field induced by the mutual compression of two spheres with dissimilar elastic properties. In

Content from this work may be used under the terms of the Creative Commons Attribution 3.0 licence. Any further distribution of this work must maintain attribution to the author(s) and the title of the work, journal citation and DOI.
Published under licence by IOP Publishing Ltd

the approach advanced in this paper, the normal and tangential contact problems are decoupled, as the influence of shear tractions on the normal displacements is neglected and, consequently, the contact pressure is assumed to conserve a hertzian profile. A numerical solution to the surface stresses arising in the fully coupled slip-stick contact problem was achieved numerically by Spinu and Glovnea [7] and could be used to extend the work reported herein.

Considering the limited information existing in the literature, the goal of this paper is to establish the influence of the frictional regime and of the normal and tangential loading components on plastic yield inception in a slip-stick spherical contact between similarly elastic materials. The study is based on computation of the stress field using a robust semi-analytical method coupled with an acceleration technique based on the convolution theorem.

2. Semi-analytical model

The method applied herein in assessment of subsurface stresses induced in an elastic solid by a known distribution of surface tractions, normal and/or shear, is based on the superposition principle applicable in the frame of linear elasticity, and on the fundamental solutions derived for the elastic half-space by Boussinesq [8] and by Cerruti [9], expressing the stress and displacement field induced in an elastic and isotropic half-space by a normal or tangential concentrated force acting on the half-space boundary. The use of these fundamental solutions is authorized by the so-called half-space approximation employed in the theory of contact mechanics, assuming bodies of arbitrary (yet smooth) surface as semi-infinite solids confined by a plane boundary when computing the corresponding stress and displacement fields. This assumption is reasonable in case of concentrated contacts, when the dimensions of the contact area are small compared to the relevant dimensions of contacting bodies.

The main advantage of this semi-analytical method consists in its ability to handle arbitrary distributions of surface tractions, as long as the effect of the stress gradients can be simulated with adequate accuracy using a piecewise constant digitized counterpart. Stress risers are not to be expected in the contact between bodies assumed smooth in the contact region, and therefore relative errors of less than 1% can be obtained by imposing even coarser grids of 10^5 points in the contact region and its vicinity.

For the contact scenario simulated in this paper, the surface stresses developed during contact evolution are known from the works of Hertz [1] for the normal tractions (i.e. contact pressure), and of Cattaneo [2] and of Mindlin [3] for the shear tractions. It should be noted that the contacting spheres are assumed to have identical elastic constants, which decouples the normal and the tangential contact problems. Consequently, shear tractions developed during application of the tangential load do not affect contact pressure, which preserve a Hertz profile specific to contacts undergoing normal loading only.

The contact model is obtained in a Cartesian coordinate system $x_1x_2x_3$ with the x_1 and x_2 -axes contained in the common plane of contact, i.e. the plane tangent to the contacting surfaces in the initial point of contact. The normal loading consist in a force W generating a contact pressure p, while the tangential T force, acting along direction of $\vec{x}_1$, induces the shear tractions q_1 and q_2. The semi-analytical model can readily handle tangential load components along both directions, but confining the direction of the tangential force to that of $\vec{x}_1$ limits the number of independent input parameters without loss of generality (the coordinate system could be conveniently rotated to accommodate any direction of the tangential force).

Let $\sigma_{ijk}(x_1, x_2, x_3)$ denote the ij stress tensor components induced by the generic surface traction $t_k(x_1, x_2)$ ($t_k \equiv p$ for $k = 3$ and $t_k \equiv q_k$ for $k = 1, 2$) acting on the contact area A_C. By applying the superposition principle in the frame of linear theory of elasticity, the following relation holds:

$$\sigma_{ijk}(x_1,x_2,x_3)=\iint_{A_C} G_{ijk}\left(x_1-x_1',\ x_2-x_2',\ x_3\right)t_k(x_1',x_2')dx_1'\,dx_2' \tag{1}$$

where $G_{ijk}(x_1,x_2,x_3)$, $i,j,k=1,2,3$, denotes the appropriate Green function for the elastic half-space, expressing the ij stress tensor component induced at coordinates (x_1,x_2,x_3) by a unity surface traction acting in origin along direction of $\vec{x}_k$. Closed-form expressions were derived by Boussinesq [8] for $k=3$ and by Cerruti [9] for $k=1,2$. Analytical integration in equation 1 is difficult and leads to lengthy expressions even for the simplest case of Hertz contact, so it won't be performed explicitly in this paper. A robust semi-analytical technique is employed instead, which can be generalized to allow for any known but otherwise arbitrary distribution of surface tractions.

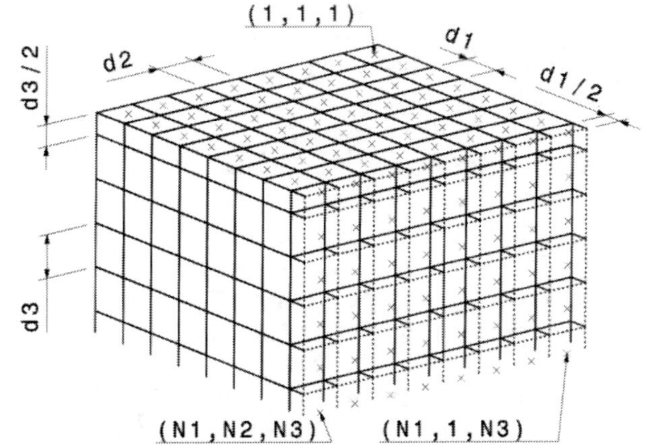

Figure 3. Problem discretization.

In the semi-analytical model of stress analysis, it is convenient to divide the volume domain from the contact proximity into a collection of non-intersecting cuboids using a rectangular three-dimensional mesh having $N_1\times N_2\times N_3$ elementary domains, of side lengths d_1, d_2 and d_3, respectively. In each elementary cuboid, subsurface stresses are assumed constant and equal to the value computed using the coordinates of a representative point, usually the center of the cuboid. Surface stresses are handled in the same manner, but the set of upper peripheral cuboids is chosen so that the plane containing the centers coincides with the half-space boundary, as shown in figure 1. This arrangement guarantees accurate stress estimation on the boundary and immediately under the surface, which is important considering that the maximum von Mises equivalent stress is located on the free surface in high friction regimes.

In this manner, all otherwise continuous problem parameter are substituted by piecewise constant distributions, which brings the advantage of convenient numerical treatment. The stress in every subsurface cuboid is approximated by superposing the individual contributions of each elementary cuboid that touches the half-space boundary. Giving the discrete nature of the semi-analytical model, it is also convenient to introduce a simplified notation, in which continuous coordinates are substituted by integers indexing the position of the referred cuboid in the grid, e.g. the value of any parameter τ in the center of the cuboid (i,j,k), $i=1..N_1$, $j=1..N_2$, $k=1..N_3$, will be denoted by $\tau(i,j,k)$. As the stress is assumed constant within each boundary patch, it can be factored outside the integral operator, yielding the semi-analytical counterpart of equation 1, in which integration with infinity as limits is substituted by a combination of summation and integration over rectangular boundary patches:

$$\sigma_{ijk}(n,p,q)=\sum_{(m,\ell)\in A_C}\left(t_k(m,\ell)\cdot\int_{x_2(\ell)-d_2/2}^{x_2(\ell)+d_2/2}\int_{x_1(m)-d_1/2}^{x_1(m)+d_1/2}G_{ijk}\left(x_1(n)-x_1',x_2(p)-x_2',x_3(q)\right)dx_1'dx_2'\right) \quad (2)$$

where $i,j,k=1,2,3$, $m,n=1..N_1$, $p,\ell=1..N_2$ and $q=1..N_3$. The contribution of each elementary patch loaded by a unity surface traction can be expressed in closed-form, yielding the so-called influence coefficients IC_{ijk}, derived by double integration of the corresponding Green functions G_{ijk} from equation (2). The semi-analytical counterpart of equation (1) can be further expressed as:

$$\sigma_{ijk}(n,p,q)=\sum_{(m,\ell)\in A_C}IC_{ijk}(n-m,p-\ell,q)\cdot t_k(m,\ell) \quad (3)$$

in which the influence coefficient $IC_{ijk}(n-m,p-\ell,q)$ expresses the contribution of a unity surface traction along direction of $\vec{x}_k$, acting on the upper face the peripheral cuboid $(m,\ell,1)$, to the ij stress tensor component in the cuboid (n,p,q). These influence coefficients can be easily obtained in closed-form. The derivation of the influence coefficients for contacts involving thin elastic layers has already been discussed in [10]. It can be seen that equation 3, which is the semi-analytical counterpart of the analytical integration in equation (1), is a discrete double convolution with respect to directions of $\vec{x}_1$ and $\vec{x}_2$. Specialized methods for increasing the algorithmic efficiency of convolution computation can be employed, as discussed in the following section.

3. Improvement of algorithmic efficiency

The double convolution in equation (3) is computationally intensive, having an order of computation of $O(N^2)$ for a grid with N elements, thus constraining dramatically the resolution that can be imposed. The computational complexity can be reduced by employing spectral methods based on the convolution theorem, which transfer the convolution calculation from the space domain (SD) to the frequency domain (FD), where its computational complexity is reduced to an improved order of $O(N\log N)$. The source of this reduction is the convolution theorem [11], which states that convolution in the SD can be computed as an element-by-element product in the FD, in $O(N)$ operations. The operations resulting from applying the direct (FFT) and inverse (IFFT) fast Fourier transform to achieve transfer back and forth between the SD and the FD claim additional computational effort. This technique has already been successfully applied [12] to other types of convolution-related problems in the field of mechanics.

The term in the right-hand side of equation (3) is a two-dimensional discrete convolution with respect to directions of $\vec{x}_1$ and of $\vec{x}_2$, and can be treated efficiently by adapting the DCFFT technique advanced in [13]. The steps of the proposed algorithm are described herein.

Firstly, compute the three-dimensional arrays of influence coefficients. It should be noted that the same operations apply to all terms in equation (3), regardless of the referred traction or stress tensor component, i.e. regardless of the instance of indices $i,j,k=1,2,3$. Therefore, the subscripts denoting the tensor components or the traction will be omitted in this section for brevity, and classical matrix notation will be employed instead, i.e. $IC_{i,j,m}$, $1\le i\le 2N_1$, $1\le j\le 2N_2$, $1\le m\le N_3$, is the (i,j,m) component of any of the 18 three-dimensional arrays of influence coefficients entering equation 3, relating any stress tensor component to any surface traction. For further brevity, as the algorithm steps match along any direction, algorithm description will be limited to the one-dimensional case. The notation of the influence coefficients simplifies accordingly to IC_ℓ, $1\le\ell\le 2N_k$, where k can be any of the two directions $k=1,2$, and N_k is the number of grids in the direction of $\vec{x}_k$. The advanced algorithm can also be applied to any multidimensional configurations [12], when $k>2$. The influence coefficient IC_1 corresponds to the negative greatest distance between two grid nodes taken along the x_k-axis, i.e when the observation mark is indexed with 1 and the source mark with N_k. The situation when the observation and the source points are interchanged corresponds to IC_{2N_k}.

The next step consists in reordering of IC_ℓ terms to avoid the so-called periodicity error related to transfer to and from the FD, and its base is detailed in [14]. The main idea is that, whenever FFT is applied to a series, the latter is assumed periodical, which is not the case in the discussed contact scenario (but could be true for rough contacts problems for which a representative single asperity is considered). A new vector K_ℓ is generated by zero-padding and rearrangement in wrap-around order, i.e. $K_\ell = IC_{N_k+\ell}$ for $1 \le \ell \le N_k$; $K_{N_k+1}=0$ and $K_\ell = IC_{\ell-N_k-1}$ for $N_k+2 \le \ell \le 2N_k$. The same notation convention is applied to the digitized tractions. The generic one-dimensional array of tractions t_ℓ, with $1 \le \ell \le N_k$, $k=1$ or 2, is extended by zero-padding to match the size of the K_ℓ vector: $P_\ell = t_\ell$ for $1 \le \ell \le N_k$ and $P_\ell = 0$ for $N_k+1 \le \ell \le 2N_k$.

In the following step, the K and P vectors are transferred to FD by means of FFT: $\hat{K} = \mathrm{FFT}(K)$, $\hat{P} = \mathrm{FFT}(P)$. The spectral array of elastic stresses if further obtained simply as element-wise product in the FD: $\hat{\sigma}_\ell = \hat{K}_\ell \cdot \hat{P}_\ell$ for $1 \le \ell \le 2N_k$. Its SD counterpart is retrieved by means of inverse Fourier transform, $\sigma = \mathrm{IFFT}(\hat{\sigma})$, and only the $1 \le \ell \le N_k$ terms are kept as algorithm output. This algorithm must be applied for any of the N_3 layers of constant depth, for any of the six stress tensor components and for any of the three surface tractions.

4. Results and discussions

The simulation of the slip-stick elastic contact was conducted by defining a normalized tangential force $\bar{T} = T/(\mu W)$ which acts simultaneously with the initial normal loading. It should be noted that the contact is in a slip-stick regime for $\bar{T} < 1$ and in full sliding for $\bar{T} = 1$.

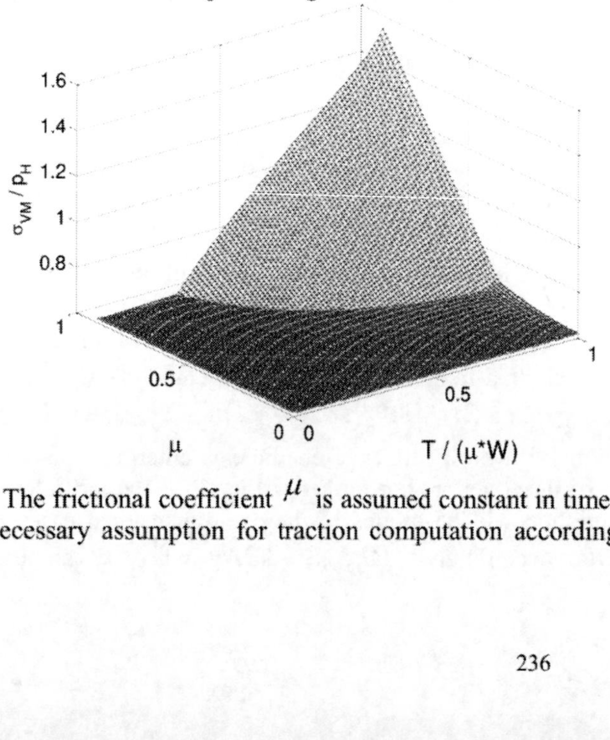

Figure 2. 3D plot of maximum equivalent Von Mises stress.

The frictional coefficient μ is assumed constant in time and uniform on the contact area, which is a necessary assumption for traction computation according to [1, 2, and 3]. This model can handle mapped frictional coefficients as well by employing the contact model advanced in [7] for surface traction assessment, but the number of independent input parameters increases impeding results generalization. Variation of the maximum von Mises equivalent stress σ_{VM} with the frictional coefficient and with the level of the tangential loading is depicted in figure 2. This surface, denoted by $f(\mu,\bar{T})$, was obtained numerically by varying the input parameters with a step of 0.01 for both μ and $\bar{T}$, by computing the corresponding stress state and by extracting the magnitude of the maximum von Mises stress. The data points for specific fixed frictional coefficients were fitted to a polynomial model $f(\bar{T})=\sum_{i=1}^{n} a_i\bar{T}^i$ and the dimensionless best-fit parameters are given in Table 1. It should be noted that, because of the applied normalization involving Hertz contact parameters, these results hold true for any contact load, elastic constants or contact curvature. This model allows rapid estimation of the relation between the load, the frictional regime and the yield inception in the elastic bodies. For a given frictional coefficient, the magnitude of the tangential load that lead to plastic flow can be obtained, according to the von Mises criterion, by solving the equation $f(\bar{T})=Y/p_H$, where Y is the yield strength of the elastic material. Conversely, the maximum allowed frictional coefficient for a given tangential loading level can be found by solving the equation $f(\mu)=Y/p_H$. Alternatively, any combination of frictional coefficient and tangential loading can be checked for consistency with the elastic limit by verifying that the inequality $f(\mu,\bar{T})\le Y/p_H$ is satisfied. This model provides simple, yet effective means of assessing the strength of the mechanical contact subjected to simultaneous normal and tangential loading.

Table 1. Best-fit parameters for assessing the maximum tangential load that can be accommodated for a fixed μ.

μ	a_0	a_1	a_2	a_3	a_4	a_5	a_6	a_7	a_8
0.1	0.62	-0.00306	0.06691	-0.5557	2.362	-5.487	7.091	-4.781	1.311
0.2	0.6201	-0.01374	0.2948	-2.435	10.31	-23.87	30.77	-20.69	5.658
0.3	0.6202	-0.02549	0.5507	-4.553	19.42	-45.42	59.22	-40.32	11.17
0.4	0.6193	0.07853	-1.706	15.2	-66.42	157.8	-206.6	139.4	-37.7
0.5	0.6177	0.2014	-3.548	24.29	-75.88	113.4	-70.7	4.17	8.379
0.6	0.6207	0.00978	-1.661	26.21	-151.8	416.6	-575.4	390	-103.5
0.7	0.628	-0.8206	16.66	-128.2	471.7	-910	961.4	-529.7	119.5
0.8	0.6242	-0.6052	15.62	-146	631.3	-1389	1641	-997.5	245.4
0.9	0.6153	0.2342	0.4083	-44.56	319.4	-876.5	1178	-780.4	204.6

5. Conclusions

This paper reports the derivation of a robust and efficient algorithm for stress analysis in frictional elastic contacts undergoing combined normal and tangential loading that lead to a slip-stick regime. The semi-analytical method employs digitization of a continuous double convolution by assuming piecewise constant distributions of involved parameters. The computation of the resulting discrete double convolution is accelerated by transferring the convolution computation to the spectral domain, where its complexity is reduced to that of an element-wise operation. Although additional treatment of convolution members is required to avoid the periodicity error, the advancement in computational efficiency is remarkable and allows for repeated computation of elastic stresses while varying the input parameters. The relation between the frictional and the loading parameters can be expressed by model fitting the results obtained in this manner, leading to simple, yet effective, equations that can be easily incorporated in engineering models. Moreover, the semi-analytical method advanced herein can be used as a generic tool for stress analysis in the frame of theory of elasticity.

References

[1] Hertz H 1882 Ueber die Berührung fester elastischer Körper, *J. Reine Angew. Math.* **92** pp 156-171.

[2] Cattaneo C 1938 Sul contatto di due corpi elastici: distribuzione locale degli sforzi. I, II, III, *Rend. Accad. Lincei* **27** pp 342-348, 434-436, 474-478.

[3] Mindlin R D 1949 Compliance of elastic bodies in contact *ASME J. Appl. Mech.* **16** pp 259-268

[4] Hamilton G M and Goodman L E 1966 The stress field created by a sliding circular contact *ASME J. Appl. Mech.* **33** pp 371-376

[5] Bryant M D and Keer L M 1982 Rough contact between elastically and geometrically identical curved bodies *ASME J. Appl. Mech.* **49** pp 345-352.

[6] Hills D A and Sackfield A 1987 The stress field induced by normal contact between dissimilar spheres, *ASME J. Appl. Mech.* **54** pp 8–14

[7] Spinu S and Glovnea M 2012 Numerical analysis of fretting contact between dissimilar elastic materials *J. Balk. Tribol. Assoc.* **18** pp 195-206

[8] Boussinesq J 1969 Application des potentiels á l'etude de l'equilibre et du mouvement des solides élastiques *Reed. A. Blanchard* (Paris)

[9] Cerruti V 1882 Ricerche intorno all'equilibrio de' corpi elastici isotropi *Rend. Accad. Lincei* **3** pp 81-122.

[10] Spinu S and Glovnea M 2012 A semi-analytical approach to the contact of thin elastic layers *J. Balk. Tribol. Assoc.* **18** pp 334-345

[11] Press W H, Teukolsky S A, Vetterling W T and Flannery B P 1992 Numerical Recipes in C *The Art of Scientific Computing – Second Edition* (Cambridge: University Press)

[12] Spinu S 2013 A refined numerical solution to the inclusion problem *Mechanika* **19** pp 252-259

[13] Liu S B, Wang Q and Liu G 2000 A versatile method of discrete convolution and FFT (DC-FFT) for contact analyses *Wear* **243** pp 101-111

[14] Liu S B, Wang Q and Liu G 2002 Studying contact stress fields caused by surface tractions with a discrete convolution and fast Fourier transform algorithm *ASME J. Tribol.* **124** pp 36-45

Analysis of heat conduction in a drum brake system of the wheeled armored personnel carriers

A M Puncioiu[1], M Truta[1], I Vedinas[1], M Marinescu[1] and V Vinturis[1]
[1]Military Technical Academy, Faculty of Mechatronics and Integrated Systems for Armament, Dept. of Military Automotive and Transportation, Bd. George Cosbuc, no. 39-49, 5th sector, Bucharest 050141, Romania

E-mail: alin_puncioiu@yahoo.com

Abstract. This paper is an integrated study performed over the Braking System of the Wheeled Armored Personnel Carriers. It mainly aims to analyze the heat transfer process which is present in almost any industrial and natural process. The vehicle drum brake systems can generate extremely high temperatures under high but short duration braking loads or under relatively light but continuous braking. For the proper conduct of the special vehicles mission in rough terrain, we are talking about, on one hand, the importance of the possibility of immobilization and retaining position and, on the other hand, during the braking process, the importance movement stability and reversibility or reversibility, to an encounter with an obstacle. Heat transfer processes influence the performance of the braking system. In the braking phase, kinetic energy transforms into thermal energy resulting in intense heating and high temperature states of analyzed vehicle wheels. In the present work a finite element model for the temperature distribution in a brake drum is developed, by employing commercial finite element software, ANSYS. These structural and thermal FEA models will simulate entire braking event. The heat generated during braking causes distortion which modifies thermo-elastic contact pressure distribution drum-shoe interface. In order to capture the effect of heat, a transient thermal analysis is performed in order to predict the temperature distribution transitional brake components. Drum brakes are checked both mechanical and thermal. These tests aim to establish their sustainability in terms of wear and the variation coefficient of friction between the friction surfaces with increasing temperature. Modeling using simulation programs led eventually to the establishment of actual thermal load of the mechanism of brake components. It was drawn the efficiency characteristic by plotting the coefficient of effectiveness relative to the coefficient of friction shoe-drum. Thus induced thermal loads determine thermo mechanical behavior of the structure of wheels. Study the transfer of heat generated during braking is useful because results can improve and validate existing theory or may lead to the development of a mathematical model to simulate the behavior of the brake system for various tactical and operational situations. Conclusions of this paper are relevant because theoretical data analysis results are validated by experimental research.

1. Introduction

Land vehicles were used in ancient times for the transport of persons and goods, satisfying transportation needs on longer distances, using increasingly higher speeds of increasing loads with weights and volume increased.

Content from this work may be used under the terms of the Creative Commons Attribution 3.0 licence. Any further distribution of this work must maintain attribution to the author(s) and the title of the work, journal citation and DOI.
Published under licence by IOP Publishing Ltd

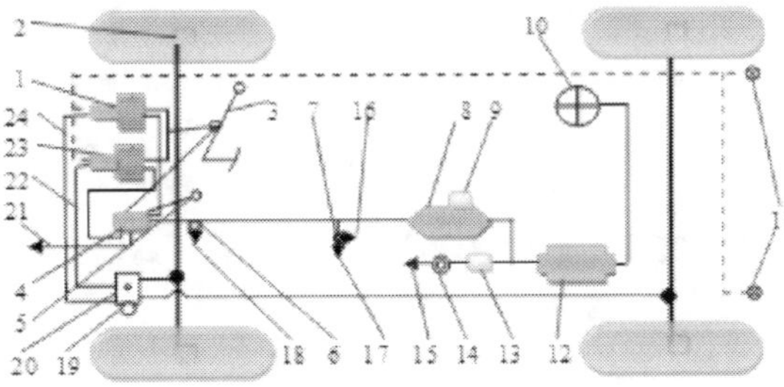

Figure 1. Vehicle's braking system diagram: 1, 23 – amplifying cylinder; 2 – braking cylinder; 3 – brake pedal; 4 – braking regulator; 5 – pressure gauge; 6, 13 – pressure limit valve; 7, 14 – electric valves; 8 – air tank; 9 – safety valve; 10 – compressor; 11 – tail lamp; 12 – pressure regulator; 15 – air pipe for engine stopping; 16 – air pipe of the transfer case's differential locking device; 17 – exhaust valve; 18 – CTIS valve; 19 - warning lamp; 20 – hydraulic warning sensor; 21 – air exhaust pipe; 22, 24 – braking circuits.

Armored personnel carriers are a necessity of the modern battlefield, because they allow movement on undeveloped land at high speeds, proving its usefulness especially in attacks. The high dynamics of warfare involves not only considerable start-up or high speed travel but also the ability to stop on short distances and in safety conditions. We must not forget that it is sometimes necessary to travel on paved roads, in civil traffic conditions, in which case the existence of an effective and reliable braking system is strictly necessary and regulated by the law. Finally, conduct combat actions in the field requires the existence of tight wheel braking mechanisms that do not allow access to indoor contaminants and compromised quality brake pad operation.

The braking system is designed to reduce vehicle velocity faster or even to stop safely, to immobilize the stationary vehicle and to maintain a reasonable speed during downhill. The kinetic energy gained by the vehicle to maintain braking action turns on the one hand in heat (friction), energy which is then dissipated into the environment, and on the other hand is consumed for overcoming rolling and air resistance, which always opposing the movement of the vehicle.The braking system's diagram is given in figure 1. It is a hydraulic braking system assisted in parallel by a pneumatic servomechanism.

In case of a car braking from a higher velocity V_1 to a lower velocity V_2, the braking energy is:

$$E = \frac{m}{2}\left(V_1^2 - V_2^2\right) + \frac{I}{2}\left(\omega_1^2 - \omega_2^2\right)[Nm] \quad (1)$$

Where: m – the vehicle mass, [kg]; V_1 – the velocity at the beginning of braking, [m/s]; V_2 – the velocity at the end of braking, [m/s]; ω_1 – The angular speed of the parts that rotate at the beginning of breaking [1/s]; ω_2 – The angular speed of the parts that rotate at the ending of breaking [1/s].

2. Thermal calculation

Thermal calculation of armored transporter braking system- see figure 2, is based on experimental data relating to real conditions of cooling of the brakes in the braking phase.

Figure 2. The drum brake equipping armored transporter.

IOP Conf. Series: Materials Science and Engineering **95** (2015) 012039 doi:10.1088/1757-899X/95/1/012039

The amount of heat released in a second is determined by:

$$Q = \frac{F_{ft} v_a}{427} = \frac{\mu p_0 \Sigma A v_a}{427} [kcal / s] \tag{2}$$

In which: v is the speed of sliding of the drum on the friction lining ($v_a = (V/3,6)\ (r_t / r_r)$); F_{ft} - the braking force acting on the drum ($F_{ft} = \mu\, p_0 \Sigma A$); ΣA - friction surface; p_0 - average pressure [2].

The heat flux density q_d is determined by the relation:

$$q_d = \frac{Q}{\Sigma A} = \frac{\mu p_0 v_a}{427} \tag{3}$$

It results that, in the case of braking, the heat flux density that is released is proportional to the specific friction power at the drum sliding the friction lining (μp0va).

From the relation:

$$p_0 = \frac{G_a}{\Sigma A} \cdot \frac{r_r}{r_t} \cdot \frac{a_f}{\mu g} \tag{4}$$

Results:

$$\mu p_0 = \frac{G_a}{\Sigma A} \cdot \frac{r_r}{r_t} \cdot \frac{a_f}{g} \tag{5}$$

Considering the relation between v_a and the vehicle velocity V, the specific friction power at the drum sliding the friction lining will be:

$$\mu p_0 v_a = \frac{G_a}{\Sigma A} \cdot \frac{V}{3.6} \cdot \frac{a_f}{g} \tag{6}$$

When comparing vehicles of a certain class, in case of intensive braking to the limit of adhesion (*when* $a_f = g\varphi$) it can be considered that $Va_f = Vg\rho$ is constant. It means that the specific friction power at the drum sliding the friction lining, and therefore, the heat flux density, equation (3) is proportional to the ratio $G_a/\Sigma A$.

It is considered that the ratio $G_a/\Sigma A$ on an average brake shall have the following values:

- cars, $G_a/\Sigma A = 1.5 \div 2.0$ daN/cm^2;
- trucks, buses and trailers, $G_a/\Sigma A = (2.0 \div 4.0)$daN/cm^2.

If the ratio $G_a/\Sigma A > 3$ daN/cm^2 the brakes will be intensive used. This is allowed in vehicles equipped with retarders, which reduce the usage period of the service brake and therefore heating process.

3. Theoretical aspects of non-stationary heat transfer

In non-stationary processes, temperature and heat flow in any point can take variable values in time. The general equation of non-stationary state heat transfer is as follows [3]:

$$\rho c \frac{\partial T}{\partial t} = \frac{\partial}{\partial x}\left(\lambda_x \frac{\partial T}{\partial x}\right) + \frac{\partial}{\partial_y}\left(\lambda_y \frac{\partial T}{\partial y}\right) + \frac{\partial}{\partial z}\left(\lambda_z \frac{\partial T}{\partial z}\right) + Q \tag{7}$$

Hamilton's Principle states that $J = 0$ in relation to the temperature T. Thus, the functional sought is:

$$J(T)=\int_V \frac{1}{2}\left[\lambda_x\left(\frac{\partial T}{\partial x}\right)^2+\lambda_y\left(\frac{\partial T}{\partial y}\right)^2+\lambda_z\left(\frac{\partial T}{\partial z}\right)^2\right]dV+\int_V\left(\rho c\frac{\partial T}{\partial t}-Q\right)TdV+\int_{A_2}\alpha(T-T_2)^2dA+\int_{A_3}qTdA=0 \quad (8)$$

The condition $J=0$ is equivalent formulation to equation (6). It is necessary to be determined the extreme condition for functional $J=0$.

It considers thermal gradients vector $\{T'_{x,y,z}\}$ and thermal characteristics matrix:

$$[\lambda]=\begin{bmatrix}\lambda_x & 0 & 0\\ 0 & \lambda_y & 0\\ 0 & 0 & \lambda_z\end{bmatrix} \quad (9)$$

Record the temperature derivative depending on time $\dot{T}$ and organize T and $\dot{T}$ as vectors with one single component. In this case, functional $J=0$ can be written as a matrix:

$$J(T)=\frac{1}{2}\int_V\{T'\}^T[\lambda]\{T'\}dV+\int_V\{T\}^T(\rho c\{\dot{T}\}-\{\Theta\})dV+\frac{1}{2}\int_{A_2}\alpha(\{T\}-\{T_2\})^T(\{T\}-\{T_2\})dA+\int_{A_3}\{T\}^T\{q\}dA. \quad (10)$$

It is admitted that the calculation of the whole structure is divided into a finite number of subdomains called *finite element* small enough so that each element in the finite temperature function $T=T(x,y,z)$ can be approximated well using interpolation functions $\{\mathbf{T}\}=\sum_i \mathbf{N}_i\mathbf{T}_i=\{\mathbf{N}\}\{\mathbf{T}_i\}$, where $\{N\}$ is the vector interpolation functions $\{\mathbf{N}\}=\{N_1\ \ N_2\ \ N_3\ \ ...\}$, and i is the counter for the number of finite element nodes. For a finite element the tetrahedral relationship $\{\mathbf{N}\}=\{N_1\ \ N_2\ \ N_3\ \ ...\}$ can be written as $\{\mathbf{T}\}=\{\mathrm{N}_1..\mathrm{N}_4\}\{\mathrm{T}_1..\mathrm{T}_4\}^{\mathrm{T}}$.

Interpolation functions N_i are functions of x, y, z, so one can build the temperature derivatives T in relation to x, y, z. It results:

$$\frac{\partial \mathrm{T}}{\partial x}=\sum_i\frac{\partial N_i}{\partial x}T_i=\left\{\frac{\partial N_1}{\partial x}\ \ \frac{\partial N_2}{\partial x}\ \ \frac{\partial N_3}{\partial x}\ \ \frac{\partial N_4}{\partial x}\right\}\begin{Bmatrix}T_1\\T_2\\T_3\\T_4\end{Bmatrix}, \text{ also similar for } \partial\mathrm{T}/\partial\mathrm{y} \text{ and } \partial\mathrm{T}/\partial\mathrm{z} \quad (11)$$

It results:

$$\begin{aligned}J(\mathbf{T})=&\frac{1}{2}\{\mathbf{T}_i\}^T\int_V[\mathbf{B}]^T[\boldsymbol{\lambda}][\mathbf{B}]dV\{\mathbf{T}_i\}+\{\mathbf{T}_i\}^T\int_V\rho c\{\mathbf{N}\}^T\{\mathbf{N}\}dV\{\dot{\mathbf{T}}_i\}-\\&-\{\mathbf{T}_i\}^T\int_V\{\mathbf{N}\}^T\{\mathbf{Q}\}dV+\frac{1}{2}\int_{A_2}\alpha\left(\{\mathbf{T}_i\}^T\{\mathbf{N}\}^T-\{\mathbf{T}_2\}^T\right)\left(\{\mathbf{N}\}\{\mathbf{T}_i\}-\{\mathbf{T}_2\}\right)dA+\\&+\{\mathbf{T}_i\}^T\int_{A_3}\{\mathbf{N}\}^T\{\mathbf{q}\}dA\end{aligned} \quad (12)$$

It requires minimal functional condition in equation (12) with respect to the nodal temperatures $\mathbf{T}_i$ [3]:

$$\delta J = \{T_i\}^T \left(\int_V [B]^T [\lambda][\mathbf{B}] dV \{T_i\} + \int_V \rho c \{\mathbf{N}\}^T \{\mathbf{N}\} dV \{\mathbf{T}_i\} - \right.$$

$$\left. - \int_V \{\mathbf{N}\}^T \{\mathbf{Q}\} dV + \int_{A_2} \alpha \{\mathbf{N}\}^T \left(\{\mathbf{T}_i\}\{\mathbf{N}\} - \{\mathbf{T}_2\} \right) dA + \int_{A_3} \{\mathbf{N}\}^T \{\mathbf{q}\} dA \right) = 0 \tag{13}$$

where $\delta \mathbf{T}_i$ is a small but finite temperature variation to the thermal equilibrium temperature. It follows that it is necessary that value between brackets in equation (13) to be null. It will be obtained:

$$\int_V [\mathbf{B}]^T [\lambda][\mathbf{B}] dV \{T_i\} + \int_V \rho c \{\mathbf{N}\}^T \{\mathbf{N}\} dV \{\dot{\mathbf{T}}\} =$$

$$= \int_V \{\mathbf{N}\}^T \{\mathbf{Q}\} dV - \int_{A_2} \alpha \{\mathbf{N}\}^T \{\mathbf{T}_2\} dA \{\mathbf{T}_i\} + \int_{A_2} \alpha \{\mathbf{N}\}^T \{\mathbf{T}_2\} dA - \int_{A_3} \{\mathbf{N}\}^T \{\mathbf{q}\} dA \tag{14}$$

By grouping the terms it results:

$$\int_V \rho c \{\mathbf{N}\}^T \{\mathbf{N}\} dV \{\dot{\mathbf{T}}\} + \left(\int_V [\mathbf{B}]^T [\lambda][\mathbf{B}] dV + \int_{A_2} \alpha \{\mathbf{N}\}^T \{\mathbf{T}_2\} dA \right) \{\mathbf{T}_i\} =$$

$$= \int_V \{\mathbf{N}\}^T \{\mathbf{Q}\} dV + \int_{A_2} \alpha \{\mathbf{N}\}^T \{\mathbf{T}_2\} dA - \int_{A_3} \{\mathbf{N}\}^T \{\mathbf{q}\} dA \tag{15}$$

The following notations are introduced:

$$\int_V \rho c \{\mathbf{N}\}^T \{\mathbf{N}\} dV = [\mathbf{C}]$$

$$\int_V [\mathbf{B}]^T [\lambda][\mathbf{B}] dV + \int_{A_2} \alpha \{\mathbf{N}\}^T \{\mathbf{N}\} dA = [\mathbf{K}] \tag{16}$$

$$\int_V \{\mathbf{N}\}^T \{\mathbf{Q}\} dV + \int_{A_2} \alpha \{\mathbf{N}\}^T \{\mathbf{T}_2\} dA - \int_{A_3} \{\mathbf{N}\}^T \{\mathbf{q}\} dA = \{\mathbf{R}\}$$

where: $[\mathbf{C}]$ is called heat capacity matrix; $[\mathbf{K}]$ is called thermal rigidity matrix; $\{\mathbf{R}\}$ is called free vector terms.

With this notation, equation (14) becomes

$$[\mathbf{C}]\{\dot{\mathbf{T}}_i\} + [\mathbf{K}]\{\mathbf{T}_i\} = \{\mathbf{R}\} \tag{17}$$

Matrix differential equation (17) is a linear ordinary differential equation system associated with the given element, which means that the heat transfer is linear both in the stationary as well as in dynamic conditions [3, 4]. Following a process of assembly systems it results a linear system of ordinary differential equations associated with the entire structure, whose solution approximates the temperature variation in time and space depending on the external thermal stresses.

4. Modeling of heat transfer in the brake system

The heat generated by the friction depends on the pressure of contact between the friction lining and the drum and also the relative velocity between the parts in contact, the temperature of the component parts that influence, as well as the velocity, the coefficient of friction between the drum and the friction lining. An estimate of the heat that depends on these parameters is difficult to predict. Energy approach to this problem is much simpler, if some simplifying assumptions are adopted:

- work done by friction brake it is fully converted into heat,
- all wheels contribute equally to the work of friction,
- heat flux is distributed on the contact surfaces constantly, being only variable in time,
- heat taken up by the drum is considered to be evenly distributed over the inner surface thereof, the width of contact with the friction lining, so resulting a heat flux equivalent. In reality this amount of heat is transferred to the drum only on the contact surfaces with friction linings that are changing the position due to the rotation of the drum.

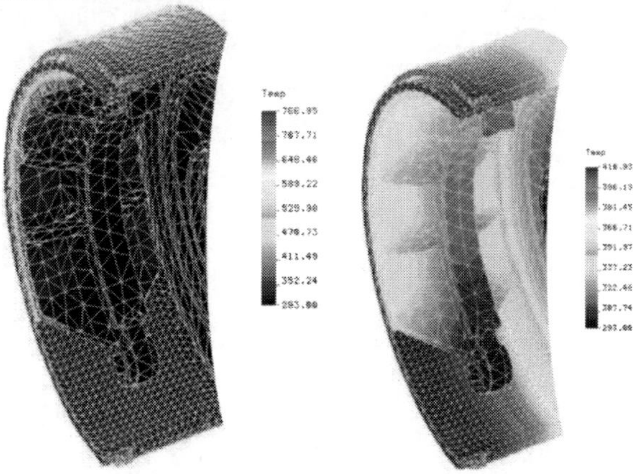

Figure 3. Thermal analysis of drum brake.

5. Conclusions

The modeling of the friction shoe-drum shown in figure 3 allowed simulations of the thermal performance of the two components, in assembly with the metallic element of the shoe. The main conclusions arising from the performance of simulations are:

- the friction lining knows a rapid heating of the surface contact during the period the shoe is applied to the drum. However, once the shoe is withdrawn, the friction lining surface temperature decreases due to the process of convection and transmissibility of the shoe metallic element;
- reducing the temperature of the friction lining surface is enhanced by increased convection because of the high tangential speed of the air driven by the rotation of the drum;
- at the friction lining surface in contact with the rollers is made large increases in temperature during application of the shoe to the drum, followed by intense cooling by heat transfer, so that at the end of a braking cycle the temperature rise does not exceed 40 ° C for an intense braking situation;
- through simulation was determined that by acting intermittent braking the friction lining is kept at an accepted level;
- during braking it is an increase of about 16 ° C ÷ 18 ° C of the drum temperature, field sites being equalized quickly between the range of braking;
- in the shoe was found by simulation the existence of 3 ÷ 8 ° C gradients temperature between the areas studied, the lowest temperatures being recorded in the ribs. This is explained by the local mass increase and by higher surface heat convection adjacent space;
- performed simulations allow identifying the mode of transmission of heat flow, enabling further development of measures to optimize its design of the shoe so as to reduce the operating temperature of friction.

The final conclusion is that, by finite element modeling of the mechanical and thermal simulations allow highlighting the areas localization module with thermal accumulation and intensity of heat transfer processes, both inside the drum and shoe friction lining as and by convection.

Acknowledgments
This paper has been financially supported within the project entitled "*Horizon 2020 - Doctoral and Postdoctoral Studies: Promoting the National Interest through Excellence, Competitiveness and Responsibility in the Field of Romanian Fundamental and Applied Economic Research*", contract number POSDRU/159/1.5/S/140106. This project is co-financed by European Social Fund through Sectorial Operational Programme for Human Resources Development 2007-2013. Investing in people!

References
[1] Limpert R 2011 Brake design and safety *third ed. SAE International* (Warrendale)
[2] Frăţilă Gh and Mărculescu Gh 1986 *The braking systems of motor vehicles* (Bucharest: Technical Publishing House)
[3] Bathe K J 1977 Adina – a finite element program for automatic dynamic incremental nonlinear analysis of temperatures (USA)
[4] Marinescu G and Ivan C 1996 Finite element method Numerical analysis and applications in thermo-elasticity *CIA Bucharest*

Determination of modal attenuation due to external and internal fluids in pipes

M V Predoi[1], C C Petre[2], M Boiangiu[1], A Craifaleanu[1] and O Vasile[1]
[1]University Politehnica of Bucharest, Dept. of Mechanics, Splaiul Independentei 313, 060042, Bucharest, Romania
[2]University Politehnica of Bucharest, Dept. of Strength of Materials, Splaiul Independentei 313, 060042, Bucharest, Romania

E-mail: predoi@cat.mec.pub.ro

Abstract. The ultrasonic nondestructive evaluation of structural integrity of pipes in high risk industries such as chemical or nuclear, represents a domain of highest importance. The inspection of kilometers of piping in rough conditions is a difficult if not an impossible task. Guided ultrasonic waves can propagate however along tens of meters in pipes and bring by the reflected signal, important information concerning the presence of flaws. There are three classes of guided modes in a pipe: longitudinal, torsional and flexural. The longitudinal modes have an axial symmetry of the radial and axial displacements. These waves prove to be most sensitive to partly circumferential flaws. These waves are dispersive, the wave velocity strongly depending on the frequency. The torsional modes are less dispersive, especially the fundamental SH_0 mode. Moreover, the radial displacements are negligible, reducing the interaction with the surrounding fluids and thus reducing the attenuation. The interaction of these waves with axial flaws is more pronounced. The flexural modes are highly dispersive and attenuated. However, if the symmetry of the emitting transducer is not perfect, these modes can propagate in the pipe and their properties must be understood. The presence of fluid inside and in some cases outside the inspected pipe represents a challenging problem of computing the guided modes dispersion curves. The various guided modes velocities and attenuations are determined for several fluids which might be filling and surrounding the common size steel pipes. A dedicated software package developed by our team is used for this purpose. These dispersion curves are used for optimal numerical simulation, using the Finite Elements Method (FEM), in order to verify the attenuation mechanism. The obtained results allow a motivated selection of the least attenuated mode, from the three classes explained before, at a given inspection frequency and for a typical steel pipe, filled with various fluids. The numerical data can be used for comparison with laboratory experiments. The experimental setup is using common ultrasonic transducers in a special geometrical arrangement. The experiments allow the measurement of the numerically predicted modal attenuation. The maximum expected inspection range can be thus determined before the inspection on real industrial piping.

1. Introduction

The inspection of kilometres of piping in chemical or nuclear plants is difficult, time and resources consuming task, even impossible to be done in some locations. Guided ultrasonic waves can propagate along tens of meters in pipes and bring by the reflected signal, important information concerning the

Content from this work may be used under the terms of the Creative Commons Attribution 3.0 licence. Any further distribution of this work must maintain attribution to the author(s) and the title of the work, journal citation and DOI.
Published under licence by IOP Publishing Ltd

presence of flaws. For this reason, guided waves are considered as possible replacements for the classical non-destructive inspection techniques [1, 2].

There are three classes of guided modes in a pipe: longitudinal, torsional and flexural. [3, 4]. The longitudinal modes have an axial symmetry of the radial and axial displacements. These waves prove to be most sensitive to partly circumferential flaws, but are in general dispersive, meaning that the wave velocity depends on the frequency. The torsional modes are less dispersive, especially the fundamental SH0 mode. Moreover, the radial displacements are negligible, reducing the interaction with the surrounding fluids and thus reducing the attenuation. The interaction of these waves with axial flaws is more pronounced. The flexural modes are highly dispersive and attenuated. However, if the symmetry of the emitting transducer is not perfect, these modes will propagate in the pipe and their properties must be understood.

The presence of fluid inside and in some cases outside the inspected pipe represents a challenging problem of computing the guided modes dispersion curves [5, 6, 7]. In the present paper, the various guided modes velocities and attenuations are determined for several fluids which might be filling and surrounding the common size steel pipes. An efficient method, based on the Finite Elements Method (FEM) has been recently developed, with extended capabilities in solving dispersion equations for multi-layered anisotropic tubes [8] A user-friendly dedicated software package developed for this purpose has been also developed [9]. These dispersion curves are used for optimal numerical simulation, using the (FEM), in order to verify the attenuation mechanism [10, 11]. The obtained results allow a motivated selection of the least attenuated mode, from the three classes explained before, at a given inspection frequency and for a typical steel pipe, filled with various fluids. The numerical data can be compared with laboratory experiments. The experimental setup is using common ultrasonic transducers. The experiments will allow the measurement of the numerically predicted modal attenuation. The maximum expected inspection range can be accurately determined before the inspection on real industrial piping.

2. Dispersion Curves

The material used for the industrial piping is homogeneous steel, assumed to have mass density ρ=7800 kg/m^3, Young modulus E=210GPa and Poisson coefficient ν=0.3. The fluid filling the pipe is gasoline or Kerosene, of mass density ρ_f= 800 kg/m^3 and bulk modulus K=1.3GPa. A common type of pipe is considered, of outer diameter D=60 mm and wall thickness h=3mm. The displacement field is defined by radial, tangential and axial displacements respectively [8]:

$$\begin{aligned} u_r(r,\theta,z,t) &= U(r)\cos(m\theta)\exp\left[i(kz-\omega t)\right] \\ u_\theta(r,\theta,z,t) &= V(r)\sin(m\theta)\exp\left[i(kz-\omega t)\right] \\ u_z(r,\theta,z,t) &= W(r)\cos(m\theta)\exp\left[i(kz-\omega t)\right] \end{aligned} \quad (1)$$

in which the modal displacements U(r), V(r), W(r) are shown on Figure1, k is the wavenumber, ω the angular frequency and m = 0, 1, 2, … the modal circumferential order and $i=\sqrt{-1}$.

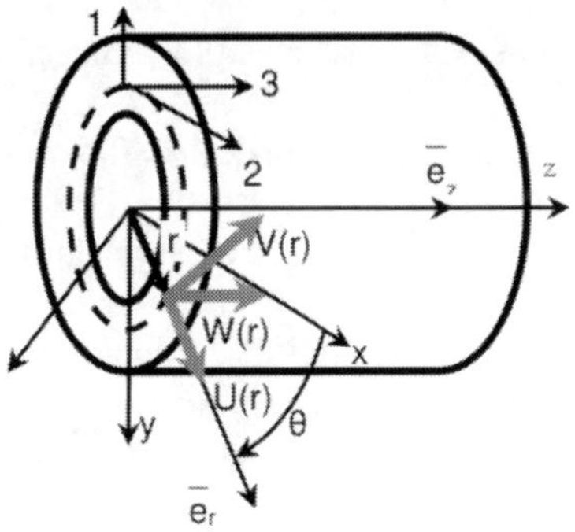

Figure 1. Displacement field on the pipe wall.

The corresponding stress field, leaving aside the propagating factor $\exp\left[i(kz-\omega t)\right]$, is:

$$S_{rr}=\left[C_{11}U'+C_{12}\left(\frac{m}{r}V+\frac{U}{r}\right)+ikC_{12}W\right]\cos(m\theta);\quad S_{\theta\theta}=\left[C_{12}U'+C_{11}\left(\frac{m}{r}V+\frac{U}{r}\right)+ikC_{12}W\right]\cos(m\theta)$$
$$S_{zz}=\left[C_{12}U'+C_{12}\left(\frac{m}{r}V+\frac{U}{r}\right)+ikC_{11}W\right]\cos(m\theta);\quad S_{\theta z}=C_{66}\left[ikV-\frac{m}{r}W\right]\sin(m\theta) \tag{1}$$
$$S_{rz}=C_{66}\left[ikU+W'\right]\cos(m\theta);\quad S_{r\theta}=C_{66}\left[V'-\frac{m}{r}U-\frac{V}{r}\right]\sin(m\theta)$$

The elastic constants C_{11}, C_{12} and C_{66} can be easily obtained [3] from the given material data.

3. Longitudinal axially symmetric L(0,n) waves for a pipe

The differential equations of motion in this case (m=0) can be expressed as [8]:

$$\frac{\partial}{\partial r}\left(rC_{11}U'+C_{12}U+irC_{12}kW\right)-\left(C_{12}U'-irC_{66}kW'\right)-\left(\frac{C_{11}}{r}U+iC_{12}kW\right)-k^2rC_{66}U=-r\rho\omega^2U$$
$$\frac{\partial}{\partial r}\left(rC_{66}W'+irC_{66}kU\right)+irC_{12}kU'+iC_{12}kU-k^2rC_{11}W=-r\rho\omega^2W \tag{2}$$

This set of equations represents the differential form of the dispersion equation. The associated eigenvalue problem can be solved using the algorithm shown in [8] which can be implemented in a commercial available FEM software [12] the L(0,n) "pipe modes" are the modes corresponding to the empty pipe. The first five such modes are shown on figure 2a.

Figure 1. Wavenumbers of the L(0,n) pipe modes (a); Group velocities for the pipe modes (b).

Several practical aspects will be remarked. The shorter wavelengths $\lambda=2\pi/k$ are preferred for their higher sensitivity to smaller defects. This means higher wavenumbers, encountered at higher frequencies. For frequencies above 1 MHz, the shortest wavelength is about 3 mm, for the L(0,1) mode and close to it, for the L(0,2) mode. Another aspect is the group velocity (c_g), representing the velocity of the short signal propagating in the inspected pipe. For this pipe, the group velocities are shown on Figure2b. In the frequency range 1 – 1.8 MHz, the L(0,4) mode will reach first the receiving transducer, followed by the rest of modes L(0,1), L(0,2) and L(0,3). For this reason, the L(0,4) mode would be a good candidate for pipe inspection, if other reasons do not intervene.

4. Longitudinal axially symmetric L(0,n) waves for the fluid filled pipe

It would be very useful if the inspection could take place with pipe filled with the conveyed fluid (kerosene in this case). The fluid inside the pipe is represented by the same equations, but using the bulk modulus ($C_{11}=C_{12}=K$; $C_{66}=0$). This method is faster implementable, but it requires more computing resources, than by using specific fluid wave equations and coupling boundary conditions.
The large number of "fluid modes" representing the guided modes of the fluid inside a rigid pipe is posing a difficult task to the FEM solver. For a reasonable computation time (less than 4 hours on a quad-core I7 – 8 GB RAM computer), the search of eigenvalues was limited to 24 per frequency step (2 kHz in this case).

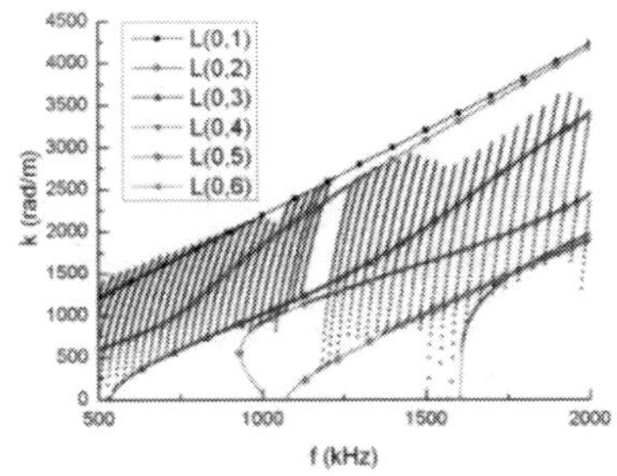

Figure 3. Dispersion curves of the filled pipe.

The dispersion curves are shown on Figure3. It is interesting to notice the large number of fluid modes which are interacting with the "pipe modes", which are superposed as continuous lines for easier comparison. The wavenumbers of the fluid modes increase very fast with frequency (almost vertical lines in the f-k space). The wavenumber of any fluid mode increases rapidly from its cut-off frequency (frequency beyond which the mode can propagate) until reaching a pipe mode. The interaction between the two types of modes can be seen as superposing (along a short frequency range) the fluid modes and the pipe modes, after which the fluid modes continue to rapidly increase in wavenumber (and lower the phase velocity). These fluid modes are very close, so that the pipe modes can be recognized as a succession of short segments.

5. FEM simulation of the wave propagation in a pipe with fluid inside

The signal central frequency of 1.25 MHz is a good compromise between too many propagating modes and a short wavelength. For this reason, a FEM model simulating a pulse propagating at this frequency can determine the most sensitive mode, for an indiscriminate piston-like axial stress applied on one end of the pipe. The pipe is filled with kerosene, keeping the same mechanical parameters. The simulation covers 40 μs of propagation along 200 mm of pipe. The time marching solution required 4 computation hours.

Figure 2. Total displacements after 10 μs (left), 20 μs (middle), 30 μs (right).

The propagating pulse can be seen on Figure4. Being produced by a piston-like excitation, the pulse is made of up to 5 modes, each with different group velocity. For this reason, the highly dispersive behaviour of the ultrasonic pulse is clearly visible. The first pack can be assumed to be L(0,4), having the highest group velocity.

The pressure wave in the fluid inside is propagating as seen on Figure5. The pressure wave velocity in kerosene is 1275 m/s, considerably less than the group velocities in steel. For this reason, the wave front is conical in the fluid. One aspect which is not discussed by other authors is the interference of the conical wave fronts as they reach the axis of the pipe, which can reduce/amplify the pressure wave.

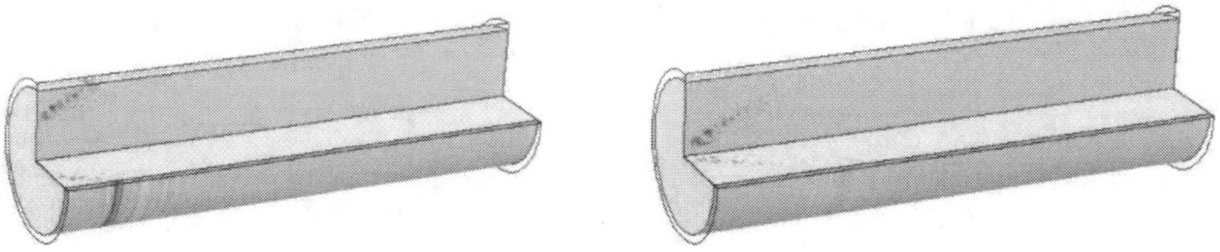

Figure 3. Pressure wave in the fluid filling the pipe, at 10μs (left), 20μs (right).

In this simulation, after 20 μs, the wave front reaching the centre of the pipe, can no longer be followed, due to very low pressure amplitudes.

FEM simulation of the wave propagation in a pipe with fluid outside

The FEM simulation uses the same pipe and input signal. The pipe is empty, but surrounded by fluid (kerosene). The fluid domain is in perfect contact with the pipe wall and non-reflecting condition on the other boundaries, which is a possibility to simulate an infinite fluid domain.

It is known, that in this case, the radial stress on the pipe outer surface, produces pressure waves in the infinite fluid. By this mechanism, wave energy is radiated and the propagating modes are attenuated even if the solid and fluid materials are ideal (lossless).

The total displacements after 10, 20 and respectively 30μs are shown on figure 6. At 10μs, the signal resembles the equivalent one on figure4, but the similarities do not continue in the next two moments, due to different interactions with the fluid.

The pressure wave in the fluid however, follows the expected behaviour (figure 7). The propagation is producing conical wave fronts, having angles depending on the wave's velocities in the two domains. However, since the ultrasonic pulse is made of several modes with different group velocities in the pipe, the wave pattern in the fluid domain has several interfering pressure wave cones, as can be seen on figure 7 (bottom).

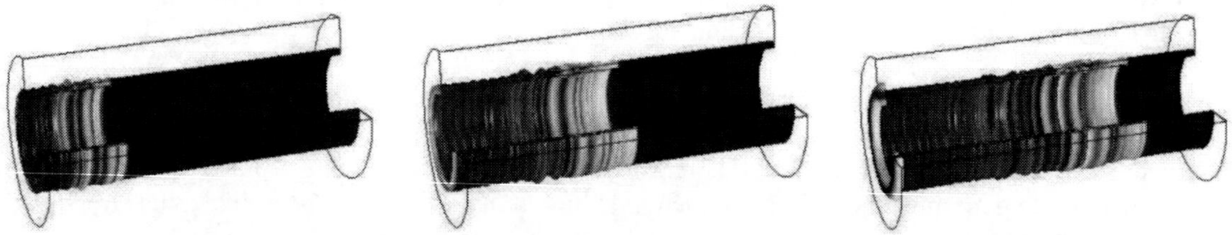

Figure 4. Total displacements after 10μs (left), 20μs (middle), 30μs (right).

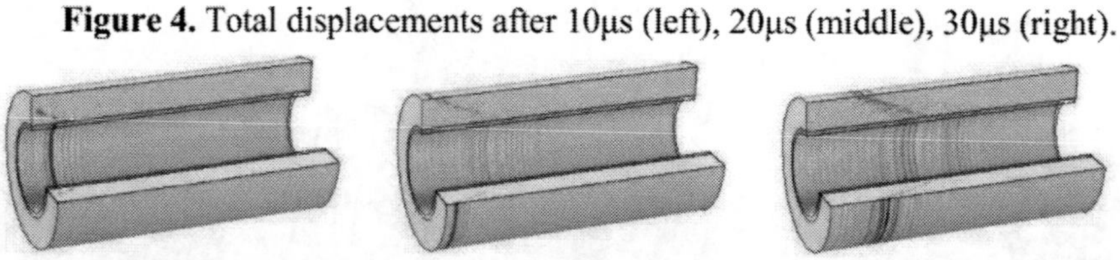

Figure 5. Pressure wave in the fluid surrounding the pipe, at 10μs (left), 20μs (middle), 30μs (right).

These signals can be processed by double FFT in frequency and position, in order to identify the propagating modes.

The experiments require reading signals along a generatrix, following the wave propagation. These signals are proportional to the radial velocity and consequently to the radial displacements at a given frequency. Signal processing provides wavenumbers and modal attenuation. Experimental results confirm the mode propagation and attenuation, and will be developed in next period.

6. Conclusions

The FEM simulations of wave pulse propagation in a kerosene filled and kerosene surrounded pipe have indicated the wave attenuation in both cases. This practical aspect, for the pipe filled with an inviscid fluid, is not mentioned in the literature, because for an infinite duration signal, in an infinite pipe, without material dissipation, the modes should propagate without attenuation. The maximum expected inspection range can be determined with good accuracy, before the inspection on real industrial piping.

Acknowledgments

This research was supported by the UEFISCDI Exploratory Research Project PN-II-ID-PCE-2011-3-0512.

References

[1] Lowe M, Alleyne D and Cawley P 1998 Defect detection in pipes using guided waves *Ultrasonics* **36** pp 147-154
[2] Leger A and Deschamps M (Eds.) 2009 Ultrasonic Wave Propagation in Non Homogeneous Media (Berlin: Springer Proceedings in Physics)
[3] Royer D and Dieulesaint E 2000 Elastic Wave in Solids (Berlin: Springer)
[4] Pavlakovic B 1998 Leaky Guided Ultrasonic Waves in *NDT Ph.D* (Univeristy of London, London)
[5] Vollmann J and Dual J 1997 High-resolution analysis of the complex wave spectrum in a cylindrical shell containing a viscoelastic medium. Part I. Theory and numerical results *J. Acoust. Soc. Am.* **102(2)** pp 896-908
[6] Gsell D, Leutenegger T and Dual J 2003 Modeling three-dimensional elastic wave propagation in circular cylindrical structures using a finite-difference approach *J. Acoust. Soc. Am.* **116(6)** pp 3284-3293
[7] Bartoli I, Marzani A, di Scalea F L and Viola E 2006 Modeling wave propagation in damped waveguides of arbitrary cross-section *Journal of Sound and Vibration* pp 685-707
[8] Predoi M 2014 Guided waves dispersion equations for orthotropic multilayered pipes solved using standard finite elements code *Ultrasonics* **54(7)** pp 1825-1831
[9] Predoi M, Petre C, Vasile O, Craifaleanu A, Boiangiu M and Marinescu S 2014 Tubewave - Interactive software dedicated to guided waves in tubes *Rom. J. Acoust. Vib.* **XI(2)** pp 94-99
[10] Hosten B and Castaings M 2005 Finite elements methods for modeling the guided waves *J. Acoust. Soc. Am.* **117(3)** pp 1108–1113
[11] Marzani A 2008 Time–transient response for ultrasonic guided waves propagating in damped cylinders *Int. J. Solids Struct.* **45** pp 6347-6368
[12] COMSOL Multiphysics User Manual, http://www.comsol.com/comsol-multiphysics

IOP Conf. Series: Materials Science and Engineering **95** (2015) 012041 doi:10.1088/1757-899X/95/1/012041

Static analysis of the hull plate using the finite element method

A Ion[1]

[1]Constanta Maritime University – Romania, Department of Navigation
104, Mircea cel Batran Street, Constanta

E-mail: alina_ion@rocketmail.com

Abstract. This paper aims at presenting the static analysis for two levels of a container ship's construction as follows: the first level is at the girder / hull plate and the second level is conducted at the entire strength hull of the vessel. This article will describe the work for the static analysis of a hull plate. We shall use the software package ANSYS Mechanical 14.5. The program is run on a computer with four Intel Xeon X5260 CPU processors at 3.33 GHz, 32 GB memory installed. In terms of software, the shared memory parallel version of ANSYS refers to running ANSYS across multiple cores on a SMP system. The distributed memory parallel version of ANSYS (Distributed ANSYS) refers to running ANSYS across multiple processors on SMP systems or DMP systems.

1. Introduction

This paper aims at presenting the static analysis for two levels of a container ship's construction as follows: the first level is at the girder / hull plate and the second level is conducted at the entire strength hull of the vessel. This article will describe the work for the static analysis of a hull plate. Static analysis will be conducted for two constructive levels of a container ship: at the panel / plate level of the hull and at the level of the entire ship.

Until the 1970s, the design of ship structures was done with predilection (sometimes even exclusively) according to Rules of the Classification and Shipbuilding Companies (Books). The disadvantages and risks of designing based only on algorithms of the classification companies have revealed , since the 1980s , the need for design and structural analysis of the hull, on rational basis, by widely using the structural theories specific to the mechanics of the continuous media and strength of materials , vibrations and elastic stability, fatigue and fracture mechanics, as well as computers, as tools for implementation of analysis methods, data management, creation of mathematical patterns and numerical calculations.

Because of these facts, the issues of design and resistance calculation of the ship structures always remain important and current, of great interest to ship owners and classification and shipbuilding companies and stuck to the attention of engineers and researchers dealing with design, construction and shipping issues. The computer-aided rational design has as starting point, the goal that the ship must fulfill and takes into account all aspects of performance and structure, achieving precise estimations regarding the behavior of the hull. The designed ships must accomplish their mission in the safest way possible and under economic conditions. Forces (or loads) acting on the hull come either from the interaction with the gravitational field (force of gravity) and the environment (forces created under water and wind pressure) or are generated by the ship in its movement. These loads determine the emergence of solicitations in the hull and in the structures within it [10, 11].

Content from this work may be used under the terms of the Creative Commons Attribution 3.0 licence. Any further distribution of this work must maintain attribution to the author(s) and the title of the work, journal citation and DOI.
Published under licence by IOP Publishing Ltd

The Finite element method (FEM) allows to numerically solve the most diverse problems of solid Mechanics (which also includes the Structural Mechanics of ships), but also from many other areas (Fluids, Thermodynamics, Electromagnetism, etc.). The MEF idea consists in modeling the field studied through a number of elements with finite dimensions, connected to each other in a number of nodal points in which the solution is sought. The operation is called finite element mesh. In structures, the values of linear and, sometimes, angular shifts are determined in nodal points. The Finite element model must simulate as correct as possible the behavior of the real structure which is and remains continue even after deformation and each infinitesimal element inside of it or on its boundary should be in equilibrium under the action of internal and external forces. Static analysis will be conducted for two constructive levels of a container ship: at the level of the panel / plate of the hull and at the level of the entire ship. The material is considered uniform and homogeneous. The pressure is constant on all surfaces where it is applied. Taking advantage of the linearity of the finite element model, a pressure of 1 *Pa* is applied and results can be scaled as needed. The way in which the effects of pressure for different values can be determined, is described in the defining procedure of the analysis and evaluation of results. The operation method for the static analysis of a hull panel is described. The Mechanical software from the ANSYS 14.5 package was used. The program was run on a computer with 4 CPU Intel Xeon X5260 at 3.33 GHz, 32 GB installed memory.

ANSYS Mechanical software is a graphical interface for preprocessing, solver and post-processing mechanical problems. The solver used is Mechanical APDL. The validity of the program results will be verified in the verification book which contains mechanical problems with given solutions. In this case, I will apply this software for static analysis of the effect of wave pressure on the hull. All static analysis will follow the procedures described and applied at this point.

2. General information

The general information concerning the materials properties is presented on the figure 1.

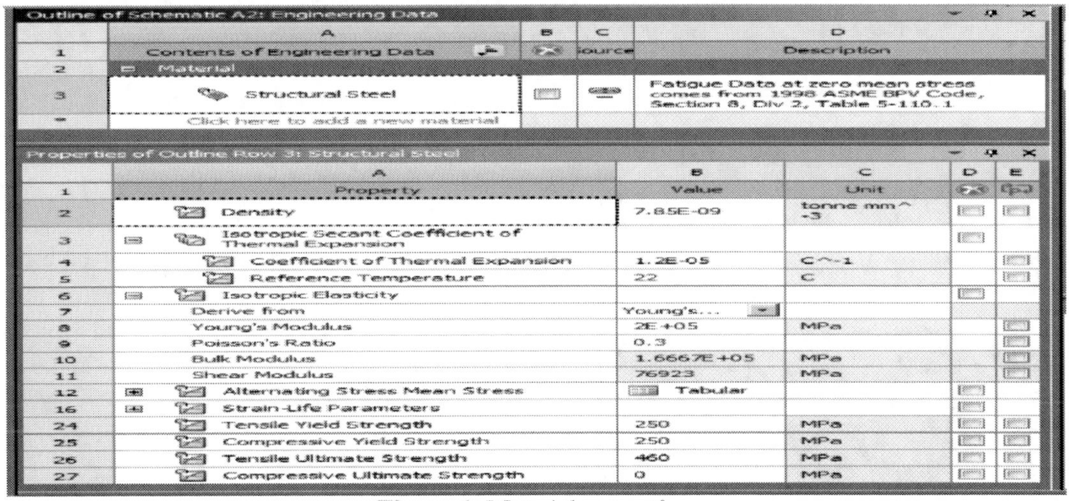

Outline of Schematic A2: Engineering Data

	A	B	C	D
1	Contents of Engineering Data		Source	Description
2	Material			
3	Structural Steel			Fatigue Data at zero mean stress comes from 1998 ASME BPV Code, Section 8, Div 2, Table 5-110.1
*	Click here to add a new material			

Properties of Outline Row 3: Structural Steel

	A	B	C	D	E
1	Property	Value	Unit		
2	Density	7.85E-09	tonne mm^-3		
3	Isotropic Secant Coefficient of Thermal Expansion				
4	Coefficient of Thermal Expansion	1.2E-05	C^-1		
5	Reference Temperature	22	C		
6	Isotropic Elasticity				
7	Derive from	Young's...			
8	Young's Modulus	2E+05	MPa		
9	Poisson's Ratio	0.3			
10	Bulk Modulus	1.6667E+05	MPa		
11	Shear Modulus	76923	MPa		
12	Alternating Stress Mean Stress	Tabular			
16	Strain-Life Parameters				
24	Tensile Yield Strength	250	MPa		
25	Compressive Yield Strength	250	MPa		
26	Tensile Ultimate Strength	460	MPa		
27	Compressive Ultimate Strength	0	MPa		

Figure 1. Material properties.

3. Application of material properties

ANSYS Mechanical software has a large library of linear and nonlinear materials, which optionally can be edited. Furthermore, a material can be created on the basis of properties given as known.

For the static analysis presented below, ordinary steel was used as material and its mechanical properties are presented in figure 1, as shown in ANSYS library according to ASME BPVC standard (American Society of Mechanical Engineers, Boiler and Pressure Vessels Code).

In the static analysis, only the longitudinal elastic modulus (Youngh's Modulus) and the transverse contraction coefficient (Poisson's ratio) are of great interest.

4. Creating geometry

The CAD software from the ANSYS package is Design Modeler. It has most of the capabilities of any CAD software. Therefore, it can be used to create a parametric geometry. The drawback is that it is not used in the design, which makes the preparation of geometry for analysis to be an operation even more complicated than it seems at first sight. There are two methods commonly used in the industry. The first method involves saving geometry in a neutral format and the import in Design Modeler. Through international standards, more neutral extensions have been established: igs, stp, step, x_t. However, geometry still has to be changed, after import, in order to be used in the analysis. The operations of geometry simplification are time consuming and if several variations of the same geometry are running, these operations must be repeated at every new import of geometry. Therefore, for projects involving many tests run on the same geometry with small changes, the procedure of export / import of neutral files is not advantageous. In this case, the graphical interface version can be used; this procedure involves creating geometry in dedicated CAD software and importing geometry in the existing format based on an interface. In this way, geometry changes only in the native CAD program, which requires that the FEA analyst would know the program [6, 7].

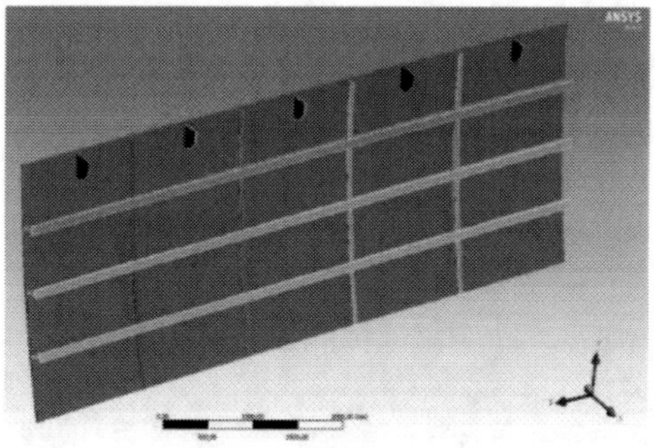

Figure 2. The structure of the steel plate reinforced with three longitudinal and four transverse beams.

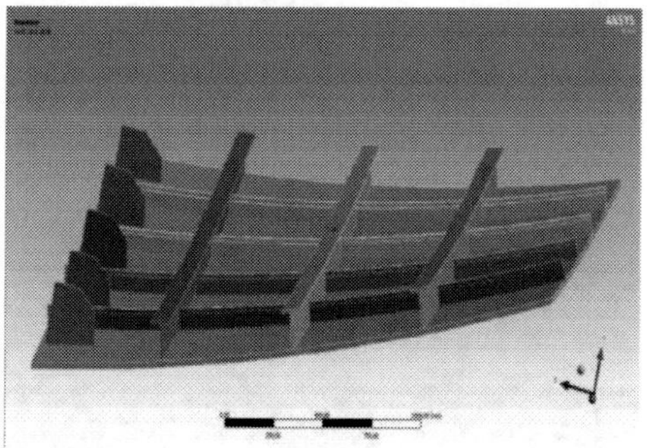

Figure 3. Detail elements of the structure studied- section of longitudinal beams.

In the current issue, which analyzes the hull, geometry was created in a dedicated CAD soft. The hull is steel plate with double curvature (cover), reinforced with longitudinal and transverse bars (figure 2, figure 3). In figure 4 you can see some details of joints and sections of reinforced bars that are welded to the plate throughout their length [4, 5].

Figure 4. Detail elements of the structure studied joining elements between the two types of stiffeners.

5. Achieving the finite element model

At this stage, geometry is replaced by a network of elements. There are several possibilities for the creation of network elements. ANSYS offers several kinds of elements, as shown in a previous chapter. When using the Mechanical software, the element type is selected automatically depending on the settings entered by the user and depending on geometry. Depending on these settings, the type of items can be controlled; for example, by opting to keep the middle edge nodes (*Element Midside Nodes = Kept*), the program will choose a parabolic element. Since it is a solid geometry type, the program will choose a SOLID187type element. The meshing obtained is shown in figure 5 [1, 4, 5].

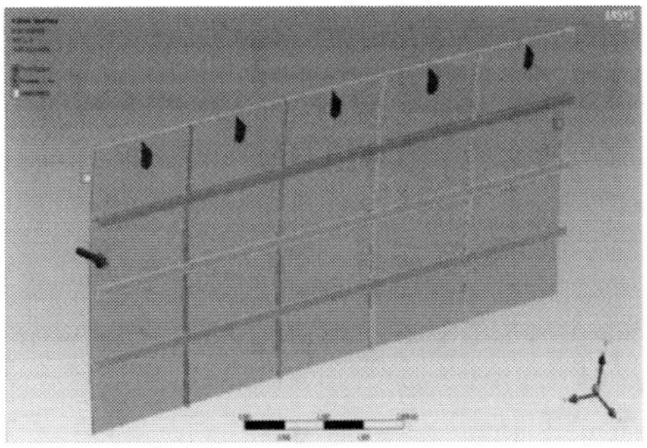

Figure 5. Mesh of the studied structure.

6. Setting the boundary conditions and obtaining the solution

Boundary conditions are affecting directly and essentially the results of the analysis. Although the time required for setting boundary conditions is much lower than that used in meshing, setting boundary conditions is particularly important. Taking into consideration the arrangement of the plate, symmetry conditions for faces at the ends of longitudinal bars were considered; they blocked movements after Z; one side is blocked by all degrees of freedom in order to eliminate the possibility of a mechanism that would lead to a divergent solution; the boundary conditions applied can be seen in figure 6 [1].

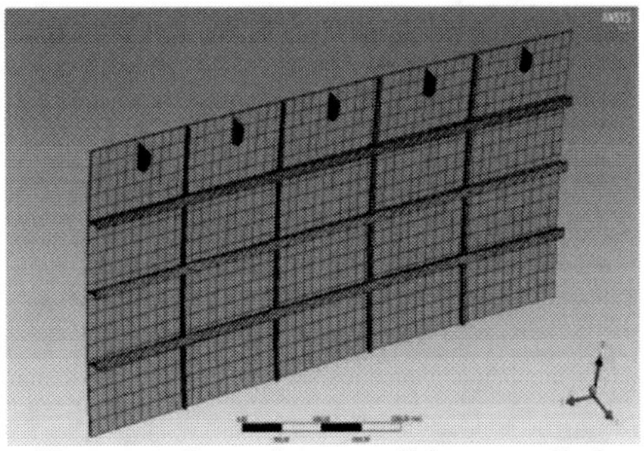

Figure 1. Boundary conditions applied.

For this analysis, it was considered that the plate studied is horizontal and totally submerged in water. Therefore, constant pressure of *1 Pa* is applied to one of the outer faces of the plate as shown in figure 7. Given the linearity of the phenomenon, with results for a load of 1 *Pa*, results for any value of load pressure can be obtained. In other words, if it is desired to determine the state of stress and strain for a load of 1 *MPa,* it is sufficient to multiply by 106 the values obtained for a load of 1 *Pa*. The operation of solving matrix equations resulted from mathematical modeling is started by pressing the *SOLVE* button. The characteristics of how to solve the problem can be changed by the user. For this problem, options recommended by the program were used [1, 4].

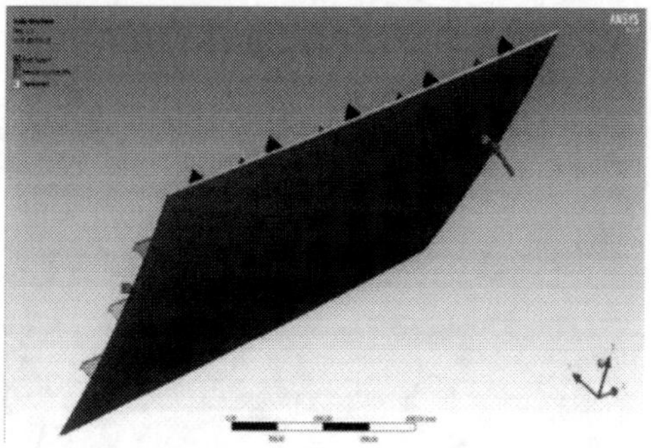

Figure 7. Loading structure with uniform pressure.

7. Viewing the results

The diagram of total displacements and stresses is of great interest in the present static analysis. For the state of stresses, the chart of von Mises equivalent stresses is the most appropriate. Results of the analysis can be seen in the following figures.

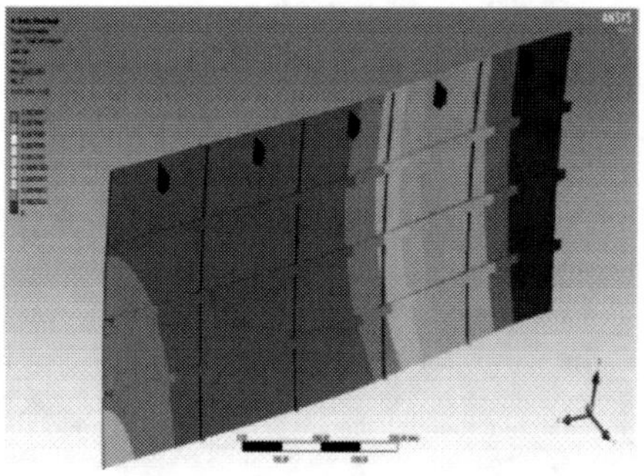

Figure 8. Distribution of total movements

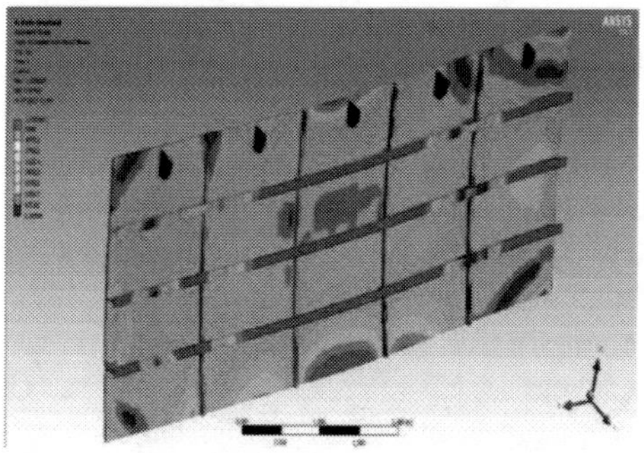

Figure 9. The von Mises equivalent stress status

The maximum equivalent stress must not exceed the yield strength of the material, which is of 250 *MPa*. As it can be seen in the chart, tensions are far below the limit value. If it is desired to determine the charging voltages for a 1*MPa*, which is equivalent to 10 bar, it is enough to multiply the previously obtained values to106. In figure 10 the chart of stresses in the plate, in the case of loading it with a pressure of 1*MPa* is represented. As it can be seen, the values are now expressed in *MPa*, which confirms the linearity of the results. In figure 10 the dark areas are those where the stress exceeds the yield limit, i.e. the material fails [1, 4, 5].

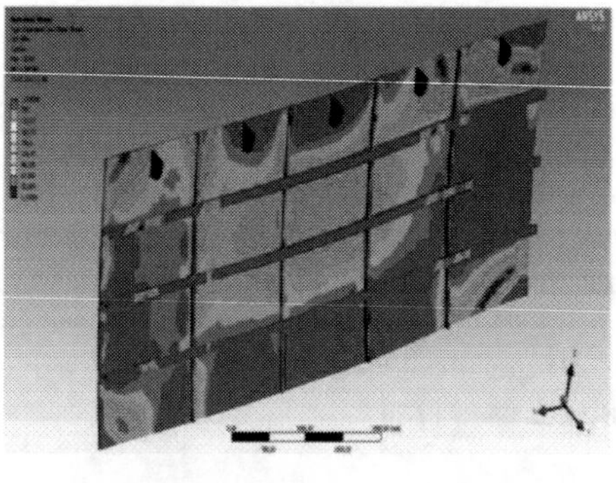

Figure 10. Von Mises equivalent stresses for a 1*Mpa* pressure load

8. Conclusions

For safety of the hull, the loading of 1 *MPa* is dangerous. By using this procedure, the limit pressure to withstand the hull can be determined. In principle, for values of stresses lower than the flow limit, in order to determine the actual safety coefficient, the yield stress is divided to the maximum stress from the diagram. This actual coefficient is then compared to the admissible coefficient. Due to the development of information technology, strong and effective methods of numerical analysis have been developed. In particular, the introduction of the finite element method (FEM) has enabled new approaches to complex problems of structural analysis. By using this method, more accurate resistance calculations and evaluation of criteria for failure of structures can be achieved. FEM also allows the iterative optimization of structural dimensions to fulfill all requirements [1], [9].

Acknowledgments

This article is the result of the project "Increasing quality in marine higher education institutions by improving the teaching syllabus according to International Convention STCW (Standards of Training, Certification and Watch keeping for Seafarers) with Manila amendments". This project in co funded by European Social Fund through The Sectorial Operational Programme for Human Resources Development 2007-2013, coordinated by Constanta Maritime University.

References

[1] Ion A 2014 Research on the calculation of a container ship structural resistance to wave oscillations *PhD Thesis* (Constanta Maritime University)
[2] Ion A, Bocanete P and Ticu R I 2013 The modern design of naval structures *Constanta Maritime University Annals* **20**
[3] Adegeest L J M, Braathen A and Vada T 1998 Evaluation of methods for estimation of extreme nonlinear ship response based on numerical simulations and model tests *Proc. 22nd Symp. on Naval Hydrodynamics* (Washington, DC: National Academy Press) pp 70–84
[4] ANSYS *Mechanical User Guide* Release 14.5, 2012.
[5] ANSYS 1997 *Structural Aanalysis Guide* ANSYS, INC Canonsburg, PA
[6] Avrashi J and Cook R D 1993 New Error Estimation for C Eigenproblems in Finite Element Analysis *Engineering Computations* **10(3)** pp 243-256
[7] Baarholm G S and Moan T 2002 Efficient estimation of extreme long-term stresses by considering a combination of longitudinal bending stresses *J. Marine Science and Technology* **6(3)** pp 122–134
[8] Barkanov E 2001Introduction to finite element method (Institute of Materials and Structures, Riga)
[9] Bathe K J 1982 Finite element procedures in engineering analysis (New Jersey: Prentice-Hall)
[10] Zienkiewicz O C and Taylor R L 2000 *The finite element method* **I** (Butterworth - Heinemann, Oxford)
[11] Zienkiewicz O C and Taylor R L 2005 *The Finite Element Method for Solid and Structural Mechanics* Sixth Edition

IOP Conf. Series: Materials Science and Engineering **95** (2015) 012042 doi:10.1088/1757-899X/95/1/012042

Computer-aided strength analysis of the modernized freight wagon

M Płaczek[1], **A Wróbel** [1] **and A Baier** [1]
[1]Silesian University of Technology, Faculty of Mechanical Engineering, Institute of Engineering Processes Automation and Integrated Manufacturing Systems, Konarskiego St. 18A, 44-100 Gliwice, Poland

E-mail: marek.placzek@polsl.pl

Abstract. In the paper results of computer-aided strength analysis of the modernized freight wagon based on Finite Element Method are presented. CAD model of the considered freight wagon was created and its strength was analysed in agreement with norms described the way of such kind of freight wagons testing. Then, the model of the analysed freight wagon was modernized by adding composite panels covering the inner surface of the vehicle body. Strength analysis was carried out once again and obtained results were juxtaposed. This work was carried out in order to verify the influence of composite panels on the strength of the freight car body and to estimate the possibility of reducing the steel shell thickness of the box in order to reduce weight of the freight wagon.

1. Introduction

Rail transport is a very important part of the modern economy, one of the components determining its dynamic development. Nowadays, numerous studies are conducted, aimed at introducing new technologies and solutions in railway infrastructure, logistics management systems, as well as in the traction vehicles [5, 15-18, 20]. Introduction of modern technology can eliminate or reduce nuisance problems associated with the implementation of any kind of transport or the operation of used technical means [1, 2, 17, 21, 24-26]. This paper contains a report on the part of the research conducted in the research and development project entitled "Analytical and experimental studies and determination of the structural features of components and assemblies in innovative structure of repaired wagons". This project is being implemented under the Programme for Applied Research by Institute of Engineering Processes Automation and Integrated Manufacturing Systems of Silesian University of Technology with consortium partners: DB Schenker Company and Germaz Company. The main objective of the project is to develop technologies to modernize freight wagons for the transport of coal and aggregates, through the use of innovative materials and technologies to repair this type of wagons during periodic repairs [1, 2, 24-26]. Works which have been undertaken within the project are to improve the operating conditions considered types of wagons by increasing their resistance to corrosion and freezes of the freight to the shell of the wagon body in winter, and thus an easier unloading. An additional objective is also to retrofit verification of strength coaches, as well as to estimate the possibility of reducing their weight, while maintaining or increasing the permissible load. Gliwice research centre has a great experience concerning with tasks of modern technical devices [13, 14, 21] as well as analysis and synthesis of mechanical and mechatronic systems [3, 4, 6-12, 19, 22, 23] also supported by computer methods [24].

Content from this work may be used under the terms of the Creative Commons Attribution 3.0 licence. Any further distribution of this work must maintain attribution to the author(s) and the title of the work, journal citation and DOI.
Published under licence by IOP Publishing Ltd

2. The considered freight wagon and its CAD model

In this paper the freight car type 1415 A3 designed for coal and aggregate transport is analysed. This type of freight car is designed for tipping unloading. Examinations of real objects were performed and photographic documentation of damages of freight cars prepared for repair was created. Also employees of repair facility were interviewed about typical defects and operation problems. Recurring damage and their causes were indicated. In case of all considered types of freight cars the main problems occurring during their exploitation are: corrosion of the freight cars shells of body and floors and freezing of the cargo to the sides and floor of the wagons. In case of the type 1415 A3 there is also a problem with damages of the vehicle body shells that occurs as the result of inappropriate methods of unloading (using scoop, excavator, etc.). For the car type 1415 A3 it can be also noticed that it has a very durable wagon supporting frame and frame of the body – there is almost no damages in repaired cars. It means that there may be a possibility to reduce the weight by the use of smaller cross-sections profiles after modernization. The incomplete documentation of the considered freight wagon supplied by consortium member forced the visual inspection of real object and measurements in order to create its CAD model. The main idea of the research project is to use composite panels mounted to the vehicle body shell in order to increase its resistance to corrosion and freezes the freight to the shell of the wagon body in winter. Created CAD models of the original freight wagon and wagon modernized by adding composite panels are presented in figure 1.

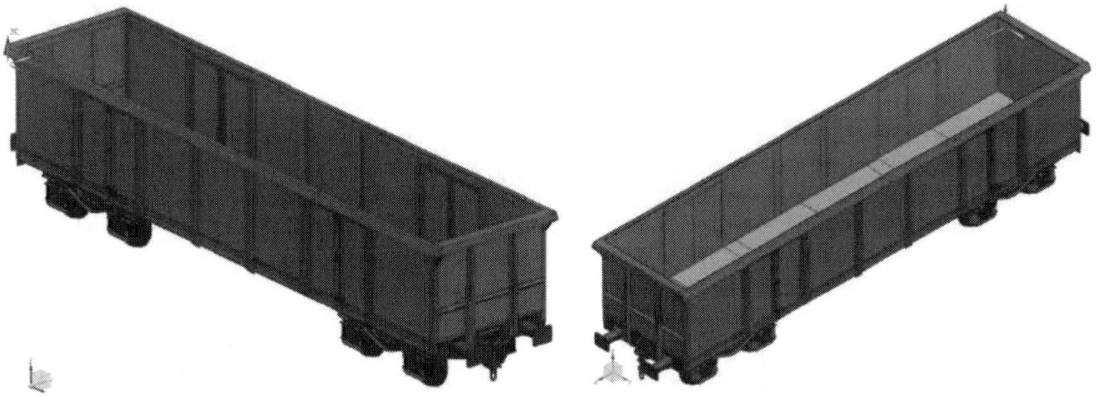

a) b)

Figure 1. The CAD model of the considered freight wagon before (a) and after modernization (b).

The created CAD was verified by juxtaposing the weight of the real object (20600 kg) and the CAD model after the material parameters were applied (19500 kg). The difference in weight is 5.34 %. It should be mentioned that the braking system (air pipes, brake pads, etc.) was not taken into account in the CAD model. It causes the difference in the weight. After this verification it can be assumed that the created CAD model of the real freight car is very precise and can be successfully used in next analysis.

3. Assumptions of the strength analysis and its results

In the first test of the strength analysis the static loading test of the wagon floor was carried out. As a load the maximum permissible mass of freight was assumed, taking into account its uniform distribution on the surface of the floor of the wagon. The analysed model is presented in figure 2.

As the next step of the presented work the strength analysis of the considered freight car type 1415 A3 was carried out. It was realized using Siemens NX 8.5 software and Finite Element Method. In the first step the strength analysis of the freight car's body was carried out. The method of load of the model was assumed in accordance with norm EN 12663-2:2010: Railway applications - Structural requirements of railway vehicle bodies - Part 2: Freight wagons [Required by Directive 2008/57/EC] [27].

Figure 2. The way of the freight wagon forcing during the static loading test of the wagon floor.

In agreement with this norm the structural design and assessment of freight wagon bodies depend on the loads they are subject to and the characteristics of the materials they are manufactured from. Within the scope of this European Standard, it is intended to provide a uniform basis for the structural design and assessment of the vehicle body. The loading requirements for the vehicle body structural design and assessment are based on proven experience supported by the evaluation of experimental data and published information. The aim of this European Standard is to allow the supplier freedom to optimize his design whilst maintaining requisite levels of safety considered for the assessment [27]. The way of verification described by the norm is that the following tests are to be carried out:

Forcing outwards in the horizontal direction at a level of 1.5 m above the floor:
1) force of 100 kN applied at four centre posts of each side wall;
2) force of 40 kN applied at the corner posts of wagons equipped with drop ends.

The significant permanent deformation at the point where the force is applied shall not exceed 1 mm. In addition, the elastic deformation observed during the test shall not result in any encroachment of the loading gauge [27]. In Figure 3a the way of the freight wagon forcing is presented. In agreement of requirement of the norm the CAD model was forced by a system of forces presented in figure 3b. Forces are transferred by rigid beams.

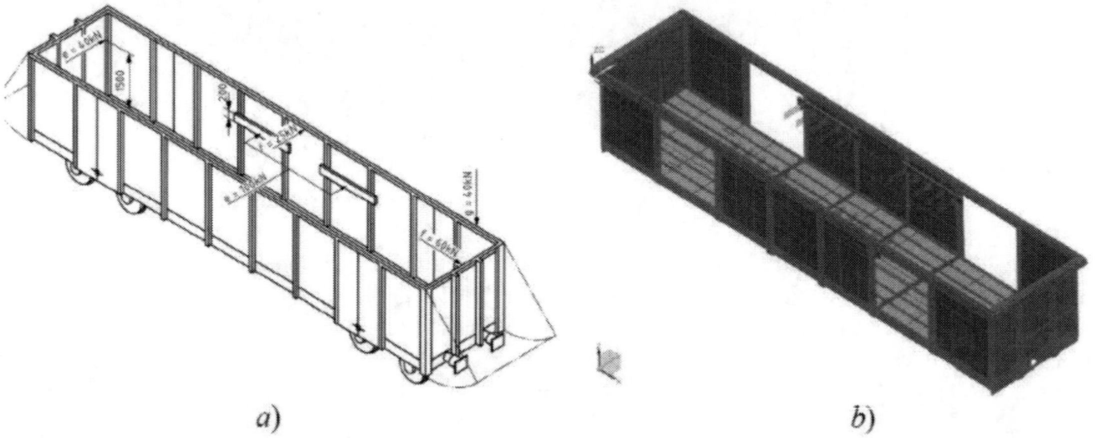

Figure 3. The way of the freight wagon forcing (EN 12663-2:2010) (a) and its implementation on the CAD model (b)

Results of the carried out analysis are presented in figure 4 and 5. Displacements and strains of the model before and after modernization (model with laminate plates) obtained during the static loading test of the wagon floor are presented in figure 4. It can be noticed that for the modernized model of the considered freight wagon values of the maximum strain and displacement are much lower than for the model without laminate plates. The same results were obtained during the static loading test on the four central columns of each wall of the wagon. Those results are presented in figure 5. Result of the carried out analysis is also the possibility to identify elements of the model with maximum or minimum strenuous and identify those which in the future can be upgraded by the thinning of the used steel sections or modernize their structural form. Such modernization can reduce the mass of the wagon and balance the mass growth caused by the adding of composite panels.

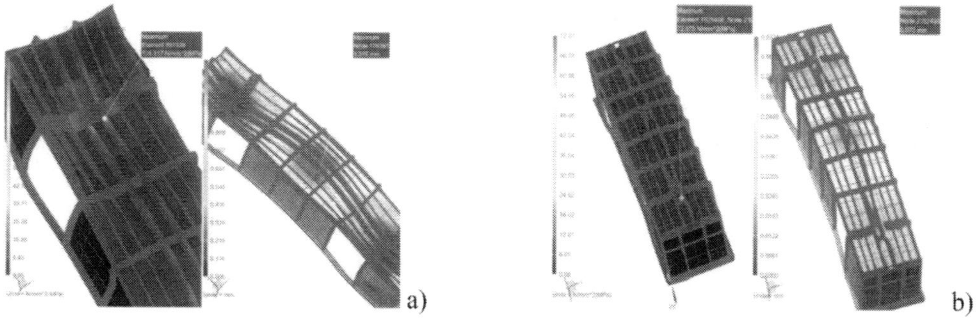

Figure 4. Results (displacements and strains) obtained for the CAD model before (*a*) and after modernization (*b*) obtained during the static loading test of the wagon floor.

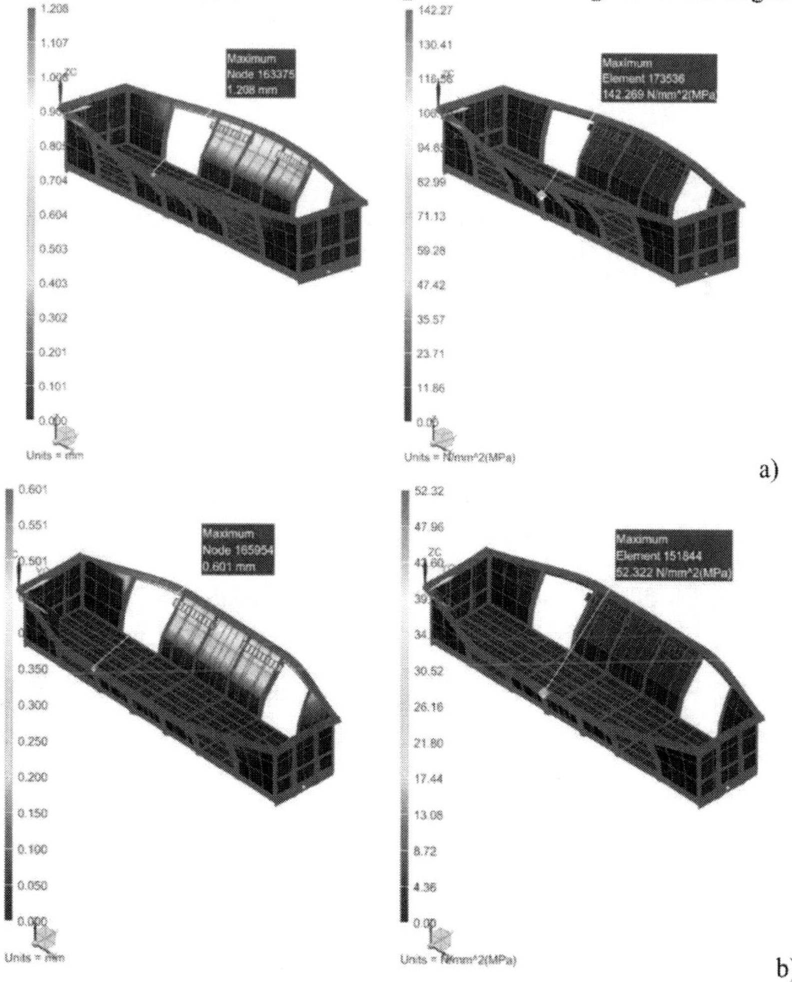

Figure 5. Results (displacements and strains) obtained for the CAD model before (*a*) and after modernization (*b*) during the static loading test on the four central columns of each wall of the wagon.

4. Conclusions

The aim of carried out works that are a part of the research project is to introduce new technologies and solutions in railway traction vehicles. Introduction of modern technology can eliminate or reduce nuisance problems associated with the implementation of this kind of transport. The main nuisance problems that occur during exploitation of the considered freight wagon are corrosion and freezes of the freight to the shell of the wagon body. Those problems should be reduced by adding composite panels mounted to the surface of the wagon's body. The mass of the wagon will increase while adding those new elements. This is why it is necessary to verify the strenuous of wagon's elements and decide if some of them can be replaced by thinner elements or their structural form can be changed. In the paper results of first analysis carried out in order to verify this possibility are presented. It was proved that some elements, for example car body sheet metal plating could be thinner, especially taking into account that laminate panels will be added.

In the future works laboratory stands for testing elements of considered freight wagons will be created and analysis of real objects will be carried out. Results will be juxtaposed with results of strength analysis of virtual models. An attempt will be made to modernize the superstructure of the wagon in order to reduce its weight and balance the additional weight of the composite elements. In the next step the possibility of further reduction of the freight wagon mass will be verify in order to increase permissible load mass of the wagon while maintaining the necessary strength.

Acknowledgments
The work was carried out under the project number PBS2/A6/17/2013 agreement implemented under the Applied Research Program, funded by the National Centre for Research and Development.

References

[1] Baier A et al 2012 Experimental synthesis and analysis of geometric and structural properties of chosen elements of railway wagons (Gliwice: Silesian University of Technology Publishing House)
[2] Baier A and Zolkiewski S 2013 Initial research of epoxy and polyester warp laminates testing on abrasive wear used in car sheathing *Eksploatacja i Niezawodnosc – Maintenance and reliability* **15(1)** pp 37–43
[3] Białas K 2012 Mechanical and electrical elements in reduction of vibrations *J. of Vibroengineering* **14(1)** p 123
[4] Białas K, Buchacz A and Gałęziowski D 2015 Comparison of active and semi-active damping in synthesis of various mechatronic discrete systems *Int. J. Materials and Product Technology* **50 (3/4)** pp 340-355
[5] Bruni S, Vinolas J, Berg M, Polach O and Stichel S 2011 Modelling of suspension components in a rail vehicle dynamics context *Vehicle System Dynamics* **49(7)** pp 1021-1072
[6] Buchacz A and Galeziowski D 2012 Synthesis as a designing of mechatronic vibrating mixed systems *J. of Vibroengineering* **14(2)** pp 553
[7] Buchacz A and Płaczek M 2010 The approximate Galerkin's method in the vibrating mechatronic system's investigation *Proceedings of The 14th International Conference Modern Technologies* pp 147-150
[8] Buchacz A and Płaczek M 2012 The analysis of a composite beam with piezoelectric actuator based on the approximate method *Journal of Vibroengineering* **14(1)** pp 111-116
[9] Buchacz A, Płaczek M and Wróbel A 2013 Control of characteristics of mechatronic systems using piezoelectric materials *Journal of Theoretical and Applied Mechanics* **51** pp 225-234
[10] Buchacz A, Płaczek M and Wróbel A 2014 Modelling and analysis of systems with cylindrical piezoelectric transducers *Mechanika* **20(1)** pp 87-91

[11] Buchacz A, Płaczek M and Wróbel A 2014 Modelling of passive vibration damping using piezoelectric transducers – the mathematical model *Eksploatacja i Niezawodnosc – Maintenance and Reliability* **16(2)** pp 301–306
[12] Dymarek A and Dzitkowski T 2005 Modelling and synthesis of discrete – continuous subsystems of machines with damping *Journal of Materials Processing Technology* **164-165** pp 1317-1326
[13] Gwiazda A, Sękala A, Monica Z and Banaś W 2014 Integrated approach to the designing process of complex technical systems *Advanced Materials Research* **1036(1)** pp 1023-1027
[14] Gwiazda A, Herbuś K, Kost G and Ociepka P 2015 Designing mechatronics equipment based on the example of the Stewart platform *Solid State Phenomena* **220/221** pp 419-422
[15] Hecht M 2009 Wear and energy-saving freight bogie designs with rubber primary springs: principles and experiences *Journal of Rail and Rapid Transit* **223(2)** pp 105-110
[16] Iacob-Mare C and Manescu T S 2013 Study of the freight wagon body through the method of finite elements *Metalurgia 2013* **65(7)** p 13
[17] Jamroziak K, Kosobudzki M and Ptak J 2013 Assessment of the comfort of passenger transport in special purpose vehicles *Eksploatacja i Niezawodnosc – Maintenance and Reliability* **15(1)** pp 25-30
[18] Jönsson P A, Stichel S and Persson I 2008 New simulation model for freight wagons with UIC link suspension *Vehicle System Dynamics* **46** pp 695-704
[19] Klarecki K, Hetmańczyk M and Rabsztyn D 2015 Influence of the selected settings of the controller on the behavior of the hydraulic servo drive *Mechatronics - Ideas for Industrial Application. Advances in Intelligent Systems and Computing* **317** pp 91-100
[20] Kovalev R, Lysikov N, Mikheev G et al 2009 Freight car models and their computer-aided dynamic analysis *Multibody System Dynamics* **22(4)** pp 399-423
[21] Ociepka P, Banaś W, Herbuś K and Kost G 2014 Simulator of the Car for Driving Courses for the People with Mobility Impairments *Advanced Materials Research* **1036(1)** pp 817-822
[22] Płaczek M 2015 Modelling and investigation of a piezo composite actuator application *Int. J. Materials and Product Technology* **50(3/4)** pp 244-258
[23] Wróbel A 2012 Kelvin Voigt's model of single piezoelectric plate *J. of Vibroengineering* **14(2)** p 534
[24] Wróbel A, Płaczek M, Buchacz A and Majzner M 2015 Study of mechanical properties and computer simulation of composite materials reinforced by metal *Int. J. Materials and Product Technology* **50 (3/4)** pp 259-275
[25] Zolkiewski S 2011 Testing composite materials connected in bolt joints *Journal of Vibroengineering* **13(4)** pp 817-822
[26] Zolkiewski S 2015 On force-deflection diagrams of fibre-metal composites connected by means of bolt joints *Int. J. Materials and Product Technology* **50(3/4)** pp 230-243
[27] EN 12663-2:2010: Railway applications - Structural requirements of railway vehicle bodies - Part 2: Freight wagons

IOP Conf. Series: Materials Science and Engineering **95** (2015) 012043 doi:10.1088/1757-899X/95/1/012043

Cold flow simulation of an internal combustion engine with vertical valves using layering approach

G Martinas[1], O S Cupsa[1], L C Stan[2] and A Arsenie[2]
[1] CERONAV, 69 A Pescarilor Str. 900581, Constanta, Romania
[2] Maritime University of Constanta, 104 Mircea cel Batran Str., Constanta, Romania

E-mail: georgemartinas@ceronav.ro

Abstract. Complying with emission requirements and fuel consumption efficiency are the points which drive any development of internal combustion engine. Refinement of the process of combustion and mixture formation, together with in-cylinder flow refinement, is a requirement, valves and piston bowl and intake exhaust port design optimization is essential. In order to reduce the time for design optimization cycle it is used Computational Fluid Dynamics (CFD). Being time consuming and highly costly caring out of experiment using flow bench testing this methods start to become less utilized. Air motion inside the intake manifold is one of the important factors, which govern the engine performance and emission of multi-cylinder diesel engines. Any cold flow study on IC is targeting the process of identifying and improving the fluid flow inside the ports and the combustion chamber. This is only the base for an optimization process targeting to increase the volume of air accessing the combustion space and to increase the turbulence of the air at the end of the compression stage. One of the first conclusions will be that the valve diameter is a fine tradeoff between the need for a bigger diameter involving a greater mass of air filling the cylinder, and the need of a smaller diameter in order to reduce the blind zone. Here there is room for optimization studies. The relative pressure indicates a suction effect coming from the moving piston. The more the shape of the inlet port is smoother and the diameter of the piston is bigger, the aerodynamic resistance of the geometry will be smaller so that the difference of inlet port pressure and the pressure near to piston face will be smaller. Here again there is enough room for more optimization studies.

1. Introduction

During their evolution pistons evolved along with engines. They became lighter and shorter. They use smaller skirts, the cylindrical body of the piston. Aluminum alloys contains more silicon than before pistons being made of that and using this material will reduce expansion by thermal and improves resistance to heat.

Use of different piston crowns or top, the part that is subject to combustion and the part that enters the combustion chamber, is the most advancement way to improve piston design. Most of the older pistons were flat, but many of the bowls on top with today features influence differently the combustion process. Diesel engines use mainly the piston bowl. There is not an ignition phase at diesel engines, so the combustion chamber may be formed by the piston crown itself. The pistons with differently shaped crowns function with direct injection are becoming very popular and are nowadays often used for diesel and gasoline engines, [1].

Content from this work may be used under the terms of the Creative Commons Attribution 3.0 licence. Any further distribution of this work must maintain attribution to the author(s) and the title of the work, journal citation and DOI.
Published under licence by IOP Publishing Ltd

The objective of the present study is the simulation of the CFD functioning on the IC engine with inlet valve, on a manifold with an intake and a piston which uses dynamic mesh. This is the cold flow analysis.

Giving a shape to the airflow in the transient engine cycle with no reactions is a process which is involved by the cold flow analysis. The interaction of the fluid dynamics with the moving geometry of the induction process must be accurately accounted and thus the goal will be achieved: capturing the mixture formation process. This way it can be determined the changing features of the jet of the air flow tumbling into the cylinder with a swirly motion. The process happens when the air enters the intake valves and when it is passed out through the opening and closing exhaust valves. It can also be determined the turbulence production due to swirl and tumble which are generated by the compression and squish, [2].

The information mentioned in the above paragraph is very useful from the combustion and flame propagation point of view, which are influenced by the conditions in the cylinder which is situated at the end of the compression stroke. During the power stroke, the high turbulence levels are facilitated first, by the rapid propagation of the flame and second, by the complete combustion. The right air/fuel ratio of the combustion is ensured by the air flow which is highly turbulent and well mixed. the level of charge stratification can also be assessed by the CFD.

However, reality is not reflected by the flow characteristics which happen during the power and exhaust strokes, because the significant thermodynamic changes which accompany combustion are not included in the cold flow simulations. In terms of validation, even if transparent cylinders and pistons have been used to obtain accurate information about some engine configurations, it is not easy to observe data about PIV (Particle Image Velocimetry) or LDV for cycling engines, due to the analysis of the port flow.

For above mentioned approach, in ANSYS FLUENT 15, can be considered three steps [3] :

1. Mesh the zones properly, after decomposing into the different zones, the geometry. Wishing to apply various strategies regarding mesh motion, to various regions during one simulation, model was divided into various zones.
2. Setup journal is used in setting the engine case in ANSYS FLUENT
3. Perform a transient IC simulation

The cold flow analysis of the combustion chamber is the starting point for the optimization of the geometric configuration of the piston and combustion chamber geometry, [4].

2. CAD and Finite Volume Analysis (FVA) Model of the IC

The goal of this paper is to simulate the aerodynamics of the air induction and end compression stages for a vertical valves IC engine. Therefore the CFD will have as fluid domain the *negative* of the In-Cylinder (the combustion chamber along with its ports) as given below (CAD):

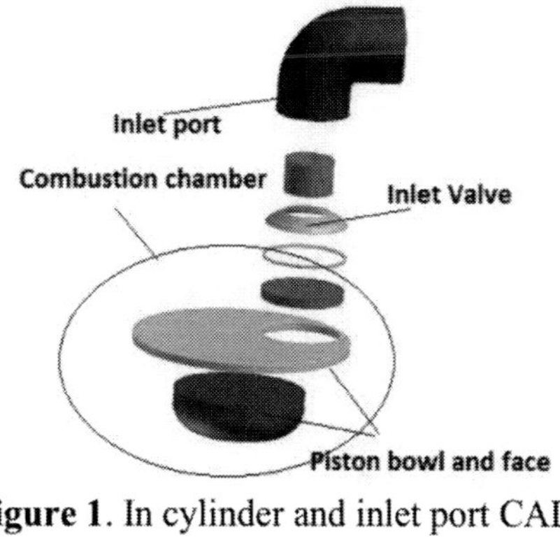

Figure 1. In cylinder and inlet port CAD.

Only the inlet port and valve were modeled for the air admission process.

IOP Conf. Series: Materials Science and Engineering **95** (2015) 012043 doi:10.1088/1757-899X/95/1/012043

Once the piston/valve is moving, the newly created volumes need to be re-meshed using the dynamic meshing function. Three groups of mesh motion methods can be used by ANSYS Fluent. In order to update in the deforming regions the volume mesh considering motion defined at the boundaries: Smoothing Methods; Dynamic Layering; Remeshing Methods. For vertical valves IC the Dynamic layering is involved. [5]
The meshing of the fluid domain has to have defined certain zones with specific mesh requirements, as given in figure2. The mesh requirements are:
1 fluid-piston- any mesh, 2 fluid-ch-lower- layered mesh, 3 fluid-ch-upper -any mesh, 4 fluid-ch-rootname -layered mesh, 5 fluid-rootname-ib -layered mesh, 6 fluid-rootname-vlayer - layer mesh

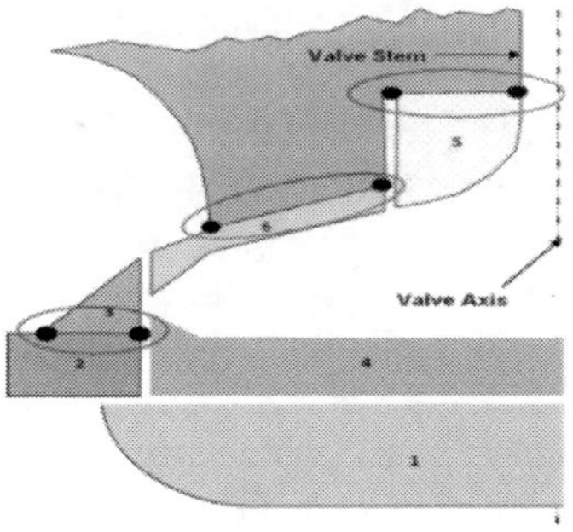

Figure 2. Sketch of Decomposition and Zone Names.

The fluid to be considered in simulation will be air with standard properties. The engine functioning/geometric parameters are: Crank shaft speed 1200 rpm; Crank period 720 deg; Crank radius 25 mm; Connecting rod radius 90 mm, Piston diameter 55 mm.[6]

3. CFA Simulation Results

The results analysis will cover the following crank angles (CA):
CA = 372 (deg) in which the inlet valve is opening;
CA = 445 (deg) in which the piston is moving downwards at halfway of its course;
CA = 535 (deg) where the piston is fully displaced for the air admission phase and inlet valve is closed.

3.1. Results for CA = 372 (deg)

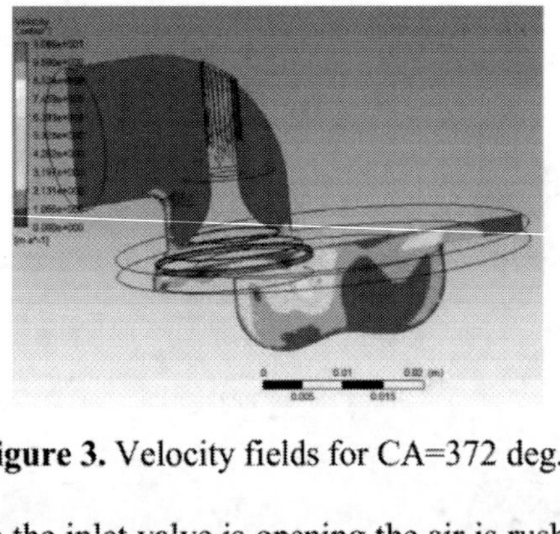

Figure 3. Velocity fields for CA=372 deg.

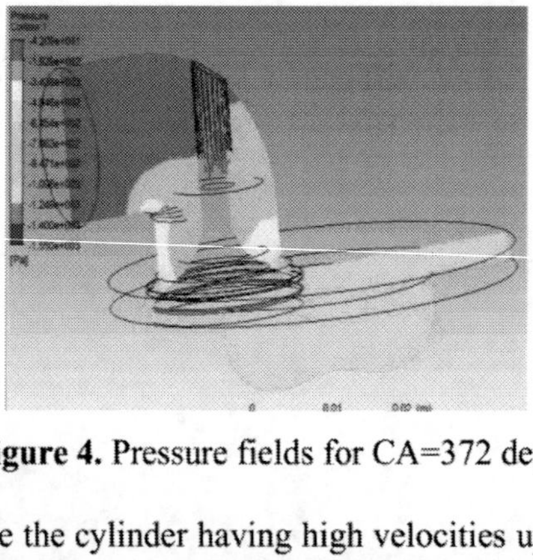

Figure 4. Pressure fields for CA=372 deg.

When the inlet valve is opening the air is rushing inside the cylinder having high velocities up to 38.6 m/s near the valve seat. The air is starting to fill the cylinder with speeds varying from 2.5 to 5 m/s. Immediately under the valve face is detected a *blind zone* where the air velocity is null. One of the first conclusions will be that the valve diameter is a fine tradeoff between the need for a bigger diameter

involving a greater mass of air filling the cylinder, and the need of a smaller diameter in order to reduce the blind zone. Here there is room for optimization studies [7].
The relative pressure indicates a suction effect coming from the moving piston. The more the shape of the inlet port is smoother and the diameter of the piston is bigger, the aerodynamic resistance of the geometry will be smaller so that the difference of inlet port pressure and the pressure near to piston face will be smaller. Here again there is enough room for optimization studies, [8].

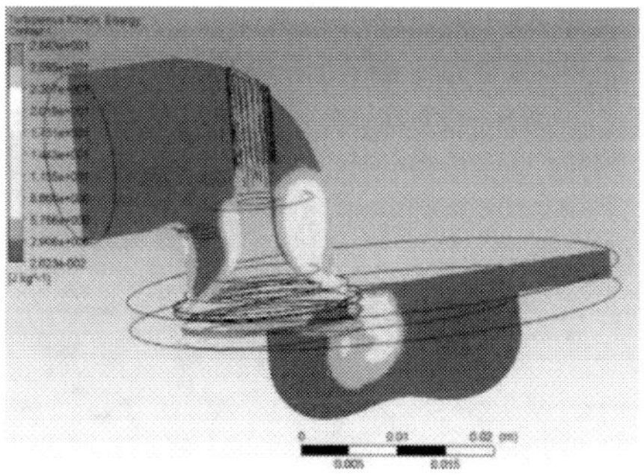

Figure 5. Turbulence kinetic energy for CA=372 deg.

The bigger the turbulence is, the better is for the functioning of the internal combustions engines. That is for the impact of the turbulence upon a better fuel-air mixing leading to finer droplets of fuel which finally is increasing the contact surface between fuel and air and the reaction rate.
The turbulence energy (max. 28.8 J/kg.) is bigger in the valve seat zone and smaller in the blind zone under the valve seat.

3.2. Results for CA = 445 (deg)
As the air is continuing fill the IC space, its velocity continues to increase as well. The maximum velocity becomes 69.1 m/s in the same zone of valve seat and the blind area is decreasing in dimension under the push of the air inside.

The pressure distribution tends to be more uniform across the IC, with peaks on the valve boundary where the air is still pushing upon.

The maximum value of turbulence kinetic energy is relatively decreasing to 106.4 J/kg, but the volume of turbulent regions is increasing instead. Opposed to the inlet valve face the turbulence is minimum as an effect of the blind zone, [9].

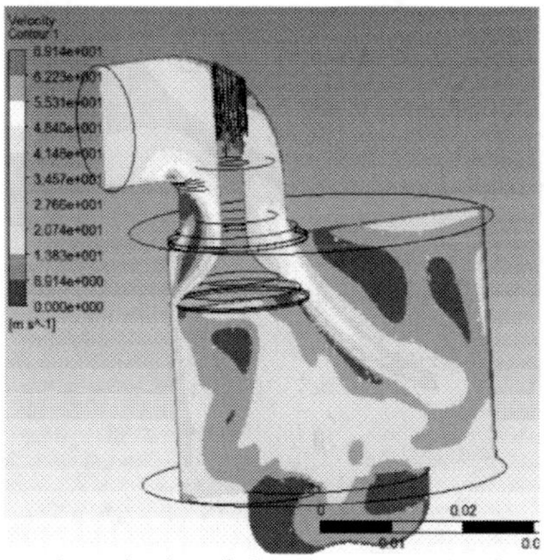

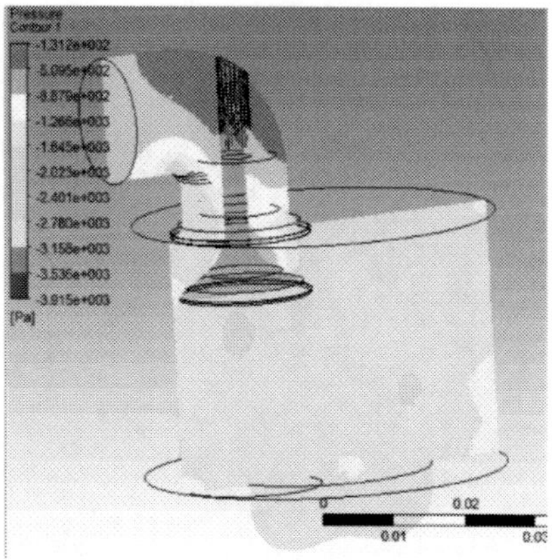

Figure 6. Velocity fields for CA=445 deg. **Figure 7.** Pressure fields for CA=445 deg.

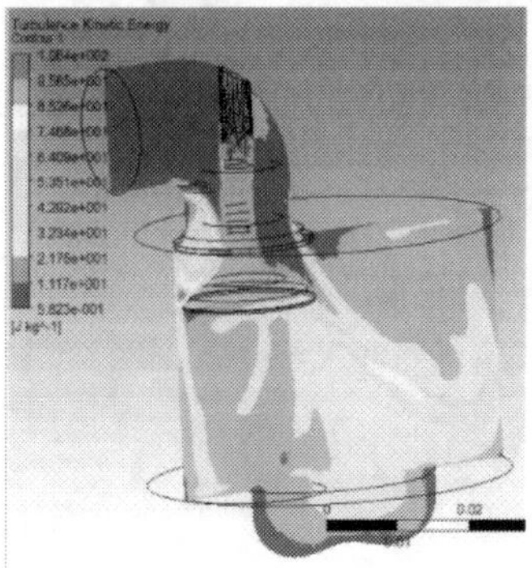

Figure 8. Turbulence kinetic energy for CA=445 deg.

3.3. Results for CA = 535 (deg)

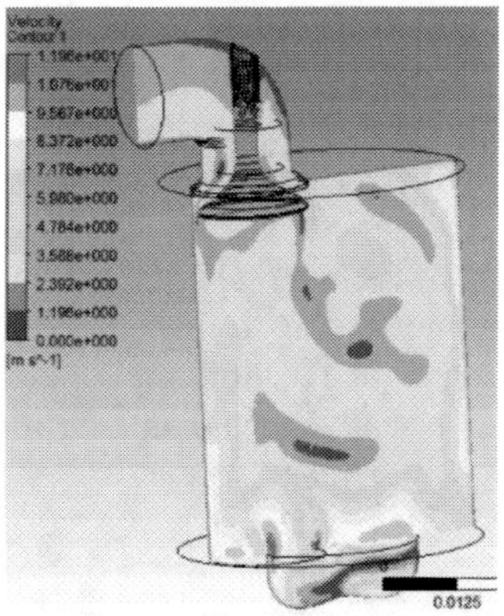

Figure 11. Velocity fields for CA=535 deg.

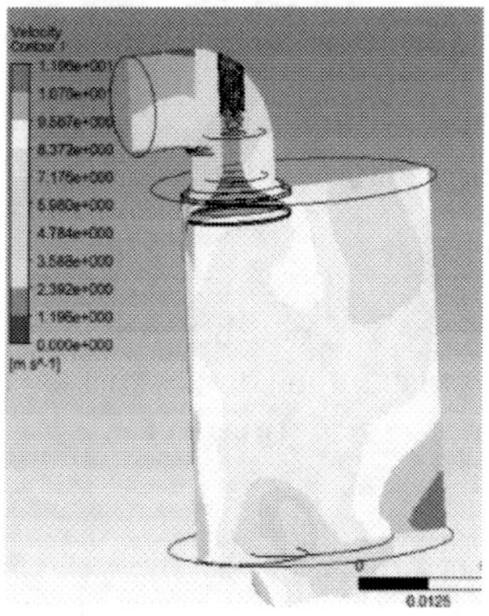

Figure 12. Pressure fields for CA=535 deg.

The inlet valve is closed and no more air is accessing the cylinder. Therefore the velocity field near the valve seat is smaller than before and the maximum velocities are registered near the piston face, with peak values of 11.9 m/s.

The pressure fields tend to redistribute with peaks near the opposed to valve zone of the cylinder.

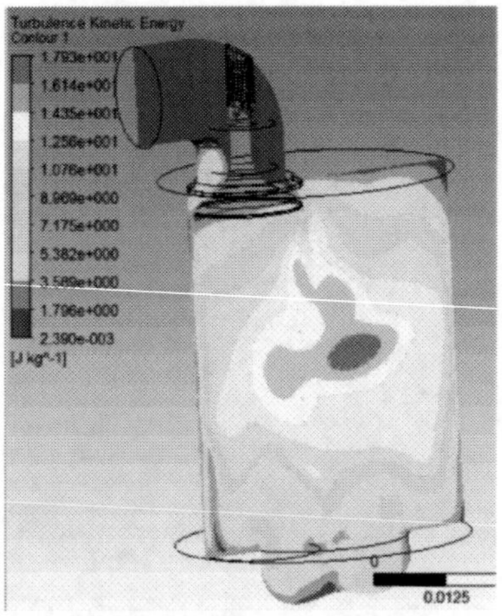

Figure 13. Turbulence kinetic energy for CA=535 deg.

The peak values of the turbulence kinetic energy are diminishing to values of 17.9 J/kg.

4. Conclusions

Air motion inside the intake manifold is one of the important factors, which govern the engine performance and emission of multi-cylinder diesel engines.

Any cold flow study on IC is targeting the process of identifying and improving the fluid flow inside the ports and the combustion chamber. This is only the base for an optimization process targeting to increase the volume of air accessing the combustion space and to increase the turbulence of the air at the end of the compression stage.

One of the first conclusions will be that the valve diameter is a fine tradeoff between the need for a bigger diameter involving a greater mass of air filling the cylinder, and the need of a smaller diameter in order to reduce the blind zone. Here there is room for optimization studies.

The relative pressure indicates a suction effect coming from the moving piston. The more the shape of the inlet port is smoother and the diameter of the piston is bigger, the aerodynamic resistance of the geometry will be smaller so that the difference of inlet port pressure and the pressure near to piston face will be smaller. Here again there is enough room for more optimization studies.

References

[1] Bakar R A and Ismail A R 2008 Computational Visualization and Simulation of Diesel Engines Valve Lift Performance Using CFD Semin, Automotive Focus Group *Malaysia American Journal of Applied Sciences* **5(5)** pp 532-539

[2] Bhagat A R and Jibhakate Y M 2012 Thermal Analysis and Optimization of I.C. Engine Piston Using Finite Element Method *International Journal of Modern Engineering Research* 2(4) pp 2919-2921

[3] ANSYS 15.0 Help Library

[4] Alkidas A C 1989 Performance and emissions achievements with an uncooled heavy duty, single cylinder diesel engine *SAE* **890141**

[5] Alkidas A C 1987 Experiments with an uncooled single cylinder open chamber diesel *SAE Paper* **870020**

[6] Mattarelli E, Rinaldinia C A and Golovitchev V I 2014 2013 CFD-3D analysis of a light duty Dual Fuel (Diesel/Natural Gas) combustion engine *Energy Procedia* **45** pp 929 – 937

[7] Centeno González F O, Mahkamov K, Silva Lora E E, Andrade R V and Jaen R L 2013 Prediction by mathematical modeling of the behavior of an internal combustion engine to be fed with gas from biomass, in comparison to the same engine fueled with gasoline or methane *Original Research Article, Renewable Energy* **60** pp 427-432

[8] Myagkov L L, Mahkamov K, Chainov N D and Makhkamova I 2014 Alternative Fuels and Advanced Vehicle Technologies for Improved Environmental Performance *Advanced and conventional internal combustion engine materials* **11** pp 370-392

[9] Lu G and Li L 2011 Study on Combustion Parameters of Liquefied Petroleum Gas Engine *Energy Procedia* **12** pp 897 – 905

Application of CAD/CAE class systems to aerodynamic analysis of electric race cars

L Grabowski[1], **A Baier** [1], **A Buchacz**[1] , **M Majzner**[1] **and M Sobek**[1]
[1]Silesian University of Technology, Gliwice, Poland, The Faculty Of Mechanical Engineering, Institute Of Engineering Processes Automation And Integrated Manufacturing Systems

E-mail: lukasz.grabowski@polsl.pl

Abstract. Aerodynamics is one of the most important factors which influence on every aspect of a design of a car and car driving parameters. The biggest influence aerodynamics has on design of a shape of a race car body, especially when the main objective of the race is the longest distance driven in period of time, which can not be achieved without low energy consumption and low drag of a car. Designing shape of the vehicle body that must generate the lowest possible drag force, without compromising the other parameters of the drive. In the article entitled „Application of CAD/CAE class systems to aerodynamic analysis of electric race cars" are being presented problems solved by computer analysis of cars aerodynamics and free form modelling. Analysis have been subjected to existing race car of a Silesian Greenpower Race Team. On a basis of results of analysis of existence of Kammback aerodynamic effect innovative car body were modeled. Afterwards aerodynamic analysis were performed to verify existence of aerodynamic effect for innovative shape and to recognize aerodynamics parameters of the shape. Analysis results in the values of coefficients and aerodynamic drag forces. The resulting drag forces *Fx*, drag coefficients $C_x(Cd)$ and aerodynamic factors C_x*A allowed to compare all of the shapes to each other. Pressure distribution, air velocities and streams courses were useful in determining aerodynamic features of analyzed shape. For aerodynamic tests was used Ansys Fluent CFD software. In a paper the ways of surface modeling with usage of Realize Shape module and classic surface modeling were presented. For shapes modeling Siemens NX 9.0 software was used. Obtained results were used to estimation of existing shapes and to make appropriate conclusions.

1. Introduction

In the design process of electric race cars, which main objective of the races is car economy, one of the most important factor affecting achieved result is aerodynamic quality of the car's body shape. The aspiration to create the perfect shape of the vehicle's body, which is characterized by low values of the aerodynamic drag forces, exact the expansion of exploration for innovative and efficient aerodynamic solutions that potentially could influence reduction of aerodynamic drag of vehicles of Silesian Greenpower race team. The aim of research was application of Kammback effect in a race car body, which could directly influence on aerodynamic quality of a shape and thus achieved results of a Greenpower race team. This effect allows to shorten the body in relation to the original shape without worsen aerodynamic properties of shape. Air flows around Kammback shape and creates so-called virtual tail, which is the outline of the shape which has been subtracted from the primary body. It allow to improve model of front part of a car so as to minimize aerodynamic drag. For design of a

Content from this work may be used under the terms of the Creative Commons Attribution 3.0 licence. Any further distribution of this work must maintain attribution to the author(s) and the title of the work, journal citation and DOI.
Published under licence by IOP Publishing Ltd

body Siemens NX 9.0 software were used. Realize Shape, innovative free form modeling tool, were applied. For study of aerodynamics, Ansys Fluent CFD software were used, using the license of Department of Theoretical and Applied Mechanics, located at the Faculty of Mechanical Engineering in Silesian University of Technology. Obtained result of analysis were used to compare with results of previous models of Greenpower Race Team cars [1, 2, 3, 4, 6, 7, 8, 9].

2. Modeling of a SG2014 Shape

Based on the results obtained during analysis of occurrence of the aerodynamic Kammback effect conditions that should be fullfiled for shape for occurrence of this effect were specified. On this basis, the shape of the vehicle of Silesian Greenpower Race Team, SG2014 Bullet was modeled. The dimensions of the vehicle, length limit, were specified in the rules of races organized by Greenpower Education Trust. Length limit is 2800[*mm*]. Compared to the shape of the car's body from previous years, the back part of the body was removed, then the front of the vehicle has been lengthened by this distance, creating more streamline shape, which should improve the aerodynamic quality of the shape [5, 6].

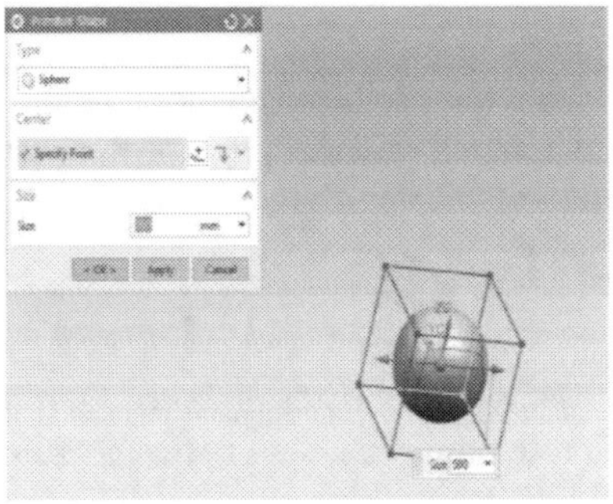

Figure 1. Basic sphere shape.

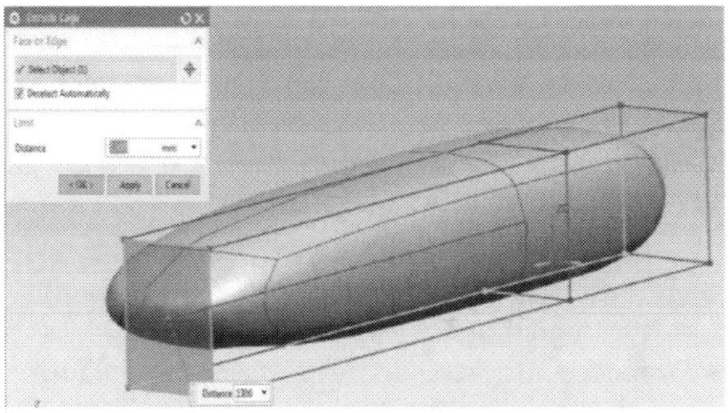

Figure 2. Modifications of basic shape.

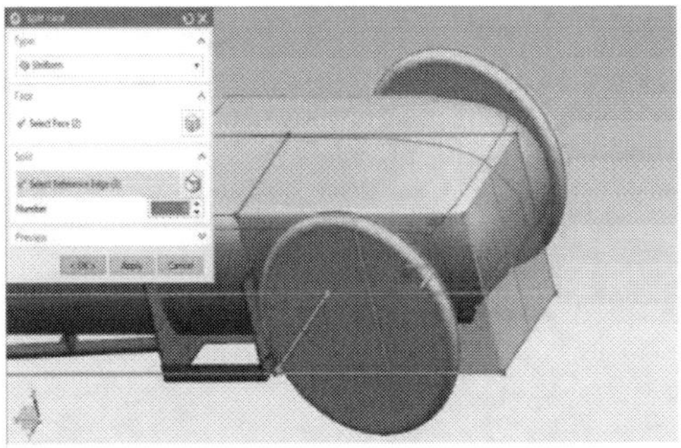

Figure 3. References shapes added.

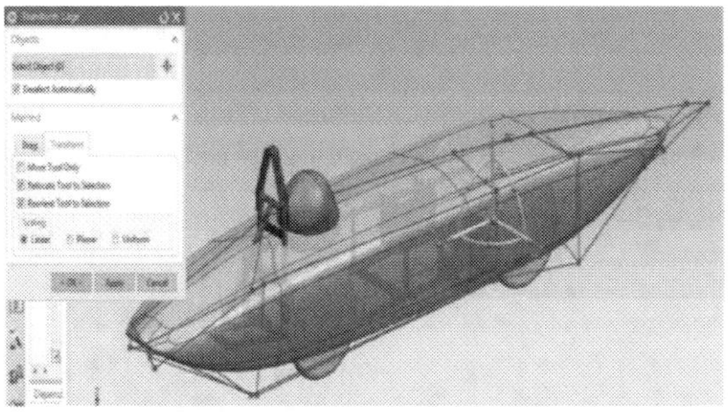

Figure 4. Modeled general shape of car.

The shape was modelled using surface modelling tool, Realize Shape, from Siemens NX program and traditional free form modelling techniques. Modelling using Realize Shape tool is based on modifications of basic shape (pulling / stretching) of the faces of basic body. The modelling process of the SG2014 Bullet car body was started by insertion of the base body, in this case it was a sphere, which was then subjected to modifications. Pre-shape was modelled based on the overall dimensions of the vehicle, and then the vehicle chassis with wheels and model of driver were added. Shape of a body has been adapted to added elements to perfectly describe in the frame and wheels and simultaneously not to increase the frontal area (A), which directly affects the aerodynamic drag.

Driving parameters of the vehicle, including the vehicle turn radius should not be affected. The next step was modelling using traditional techniques of surface modelling, creation of the outline frame using curves of the fairing and then using extrusions modelling of fairing shape were performed. Afterwards, the shape of the fairing was combined with modelled shape of the vehicle body.

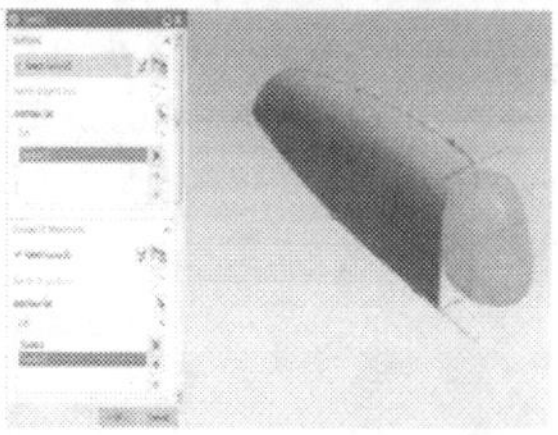

Figure 5. Fairing modeling using traditional free form modeling.

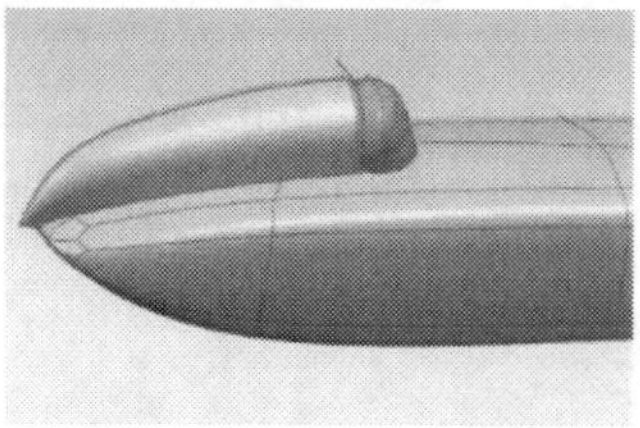

Figure 6. Fairing combined with body.

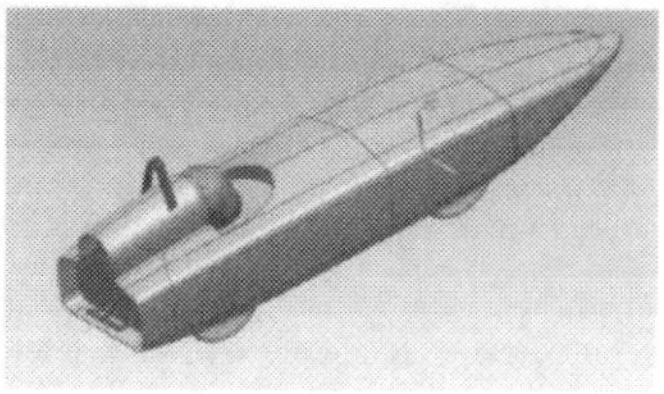

Figure 7. Finished model of SG2014 Bullet.

The next step was to correct the length of the vehicle to fulfill dimensions described in the Rules of Race. Rear part of body were subtracted by the length of 600[*mm*] to create characteristic Kammback effect shape. Afterwards cut-outs in the rear section of the body, necessary to obtain the effect, and hatch for the driver were created. Created shape was modeled as full solid block in order to excision of the vehicle body from the air volume, which is necessary for aerodynamic analysis.
Necessary simplification were made, in order to achieve adequate quality finite element mesh, such as the omission of the holes in wheel neighborhood and the driver's hatch geometry simplification. Modeled shape has been analyzed to determine aerodynamic parameters of the aerodynamic vehicle [1, 2, 3, 4, 5, 6, 7, 9].

3. Aerodynamic analysis of SG2014 Bullet

After modeling of the body of SG2014 Bullet vehicle, analysis to determine the aerodynamic properties of the shape were performed. During analysis, aerodynamic characteristics of Kammback effect of the created shape has also been proven.

The shape of the vehicle was designed in such a way that the air pressure area influencing front part of the body was as small as possible. The shape of a front creates small angle between two surfaces, upper and lower, forming the front part of the body. This should also affect the growth of aerodynamic efficiency of a car. On figure 8 the streamlined shape of the vehicle were showed. Application of innovative rear part of the vehicle which was modeled to use Kammback effect allows to create optimum shape configuration of the front part of a body. It should also have a beneficial influence on the aerodynamic drag values.

Figure 8. SG2014 model prepared for aerodynamic analysis.

For analysis CFD method using Finite Element Method was used. The size of FEM elements, quality and their type are selected with due to the aerodynamic analysis standards for the automotive industry. The dimensions of the virtual wind tunnel were 18000x8000x8000[*mm*]. Tetrahedral Finite Element shape were used for FEM mesh. The maximum size of the initial Finite Element, placed on a wall of a tunnel, was set as 250[*mm*], the minimum size of element was set as 1[*mm*], the decrease of the dimension of the Finite Element with decreasing distance from the walls of the tunnel to the analyzed shape was set as 120%.

In order to increase the accuracy of the aerodynamic analyzes influence zones were applied, in which the density of the finite elements is significantly greater and the Finite Element length has a uniform size. For this type of analysis the size of edge of Finite Element was set as 10[*mm*]. The block of influence, built around the analyzed shape, had size 6000x1000x1000[*mm*]. The next step was the creation of the inflation layers on the surface of the model and on a road. Five layers of elements describing inflation layer were set, growth ratio of 120% of the original size of the size of element [3, 5, 6].

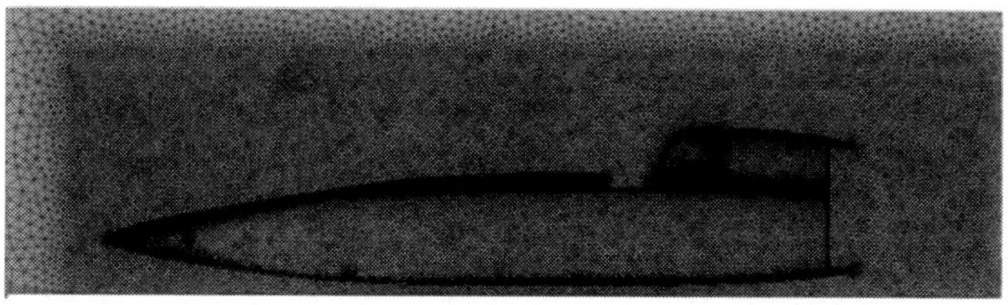

Figure 9. Fragment of Finite Element Method Mesh.

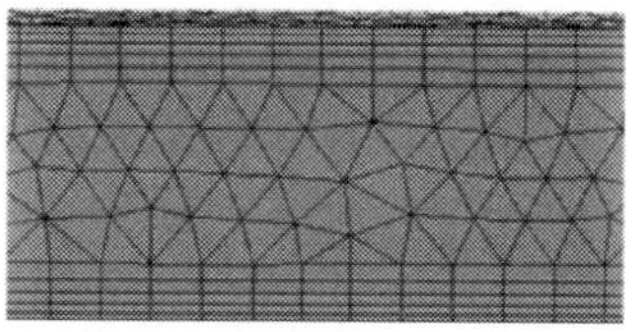

Figure 10. Inflation layer.

Analysis were performed for the following boundary conditions. For virtual aerodynamic tunnel proper properties of air were initialized. Initial air speed was set as 40 $[\frac{m}{s}]$. For analysis, k-omega (k-ω) air turbulent model was used, with Non-Equilibrium Wall Function parameter (responsible for air behavior around the walls). It was chosen on the basis of previous analysis and literature information [3, 4, 5, 6].

Numerical analysis results were obtained, the force of air drag was equal to Fx=9.558211952[*N*]. Drag coefficient was equal to C_x=0.19632. Aerodynamic factor (drag area) was equal C_x*A=0.05399738. Frontal area was equal to A=0.2750478[m^2]. Drag Force component, which was caused by air pressure, was equal to 6.15[*N*], and it is 65% of the total drag force. The force resulting from viscous friction of air is 3.42[*N*], it is approximately 35% of the total aerodynamic drag forces. Calculated results obtained in aerodynamic analysis for the SG2014 Bullet vehicle concept are the best results among all vehicles of Silesian Greenpower Race Team [3, 4, 5].

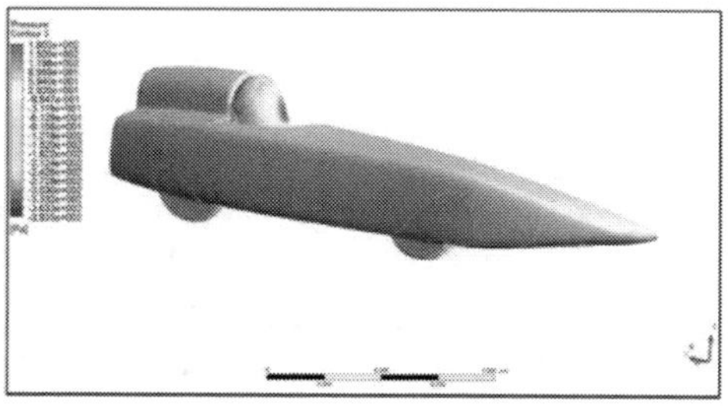

Figure 11. Air pressure distribution on SG2014 Bullet body.

Another part of the test was results analysis of pressure distribution on the vehicle body. Range of a high pressure is marked in red on figure 11. The highest pressure value occur on the driver's helmet, which is result of impact of overlapping air stream along body. In front of the vehicle was a small zone of high pressure, revealing that the shape of the front part of a body causes slight losses caused by high air pressure. At the top of the driver's helmet is a low pressure zone, marked in blue, it leads to the conclusion of occurrence of overlapping streams of air and high air velocity. The area around wheels reveals occurrence of high and low pressure in close proximity, which can cause turbulent flow of air streams around the wheels. Graphical analysis of the results of air pressure distribution on the vehicle body and around shape of a race car reveal similar results.

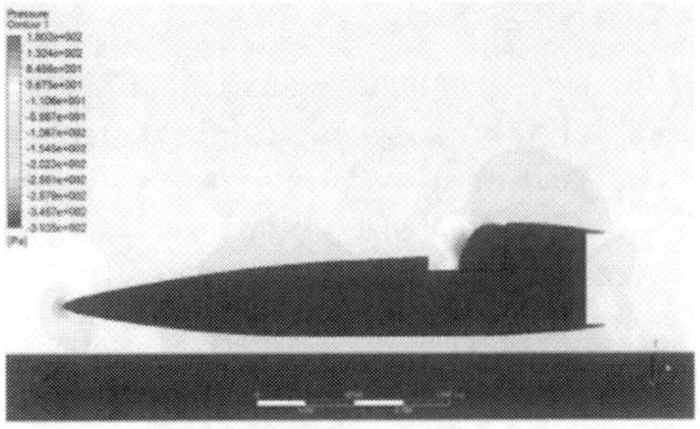

Figure 12. Air pressure distribution around SG2014 Bullet.

Figure 13. Air velocity distribution around SG2014 Bullet.

Analysis of the pressure distribution of air around the body indicates a large area of high pressure on the driver's helmet. This is the main area of the losses resulting from the pressure. At the front of the vehicle high pressure zone occurs on slight area, in comparison to the area on shape of driver's helmet.

Analysis of the results of air velocity showed zone of low air velocity at the rear of the vehicle. In this location turbulences of air could occur. Zone of low pressure behind car creates so-called virtual tail, which is lapped by the air at higher velocities without increasing air drag. This phenomenon is characteristic for Kammback effect shapes. The zone of low velocity also occurs in the driver's cockpit, the air which is reflected from the helmet circulates until outlet from cockpit opening. High-velocity zone is located in the upper part of the driver's helmet; this phenomenon could be caused by overlapping streams of air on helmet's body.

Another element of the study was to analyze the air streams flowing around concept SG2014 Bullet vehicle bodywork. On figure 14 and 15 are shown undisturbed, not turbulent, the course of the streams of air around the body of the vehicle. Streams are not subject to change direction of flow rapidly. Stream of air separated by the front of the vehicle does not stop or not causing turbulences of air. In the side of vehicle stream flows regularly, in a laminar manner.

Behind rear part of a car, to air streams follow so-called virtual tail (virtual removed part) and connect with other streams flowing around vehicle body, which indicates the presence of Kammback effect. Streams are not premature separation from the body, adhere on the entire length of the body. Small air turbulences are visible, around the driver's cockpit; streams are circulating and leaving the cockpit on the side of the driver's helmet (figure 15).

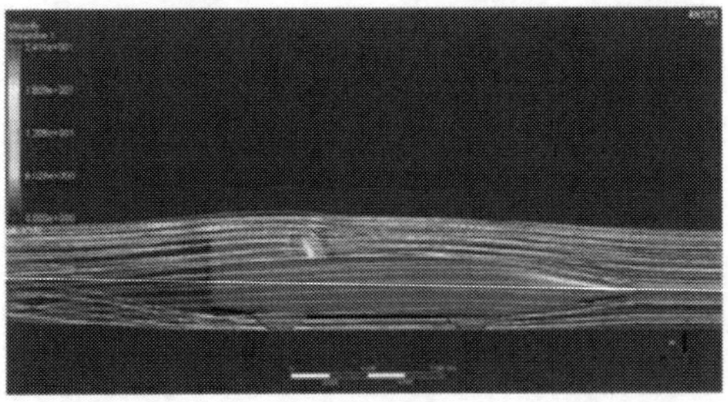

Figure 14. Air stream courses around SG2014 Bullet body.

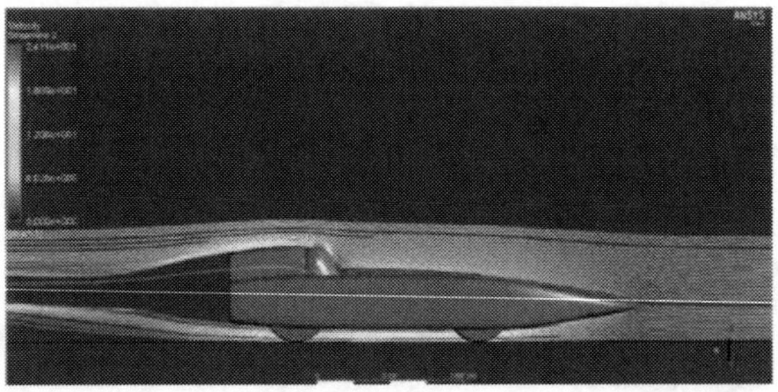

Figure 15. Air stream courses around SG2014 Bullet body.

Air stream create virtual tail of the vehicle, according to the assumptions of Kammback effect theory. Air flowing near the top of the vehicle tends to connect with the streams flowing near the bottom part of a car. In the vicinity of the vehicle body, with exception of the helmet and the wheels vicinity, areas of rapid change of values of velocity of air streams do not occur, which proves the high aerodynamic quality of the shape. Courses of air streams have been analysed in both directions of flow, consistent with the inflow of air and in the opposite direction (figure 16). The purpose of this study was to examine the behaviour of the air within the notch at the rear of the vehicle. Visible are symmetrical, regular air circulation around the notch. This kind of air behaviour is characteristic for shapes applying correctly applying Kammback effect. This is another element proving the existence of the Kammback effect in the concept vehicle SG2014 Bullet.

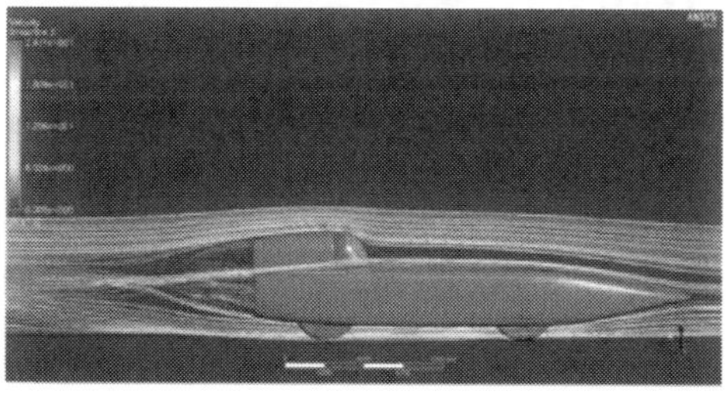

Figure 16. Air stream courses around SG2014 Bullet body.

The analysis results of shape of SG2014 Bullet showed low aerodynamic drag of the vehicle and confirmed the occurrence of Kammback effect around the body.

Table 1 shows a comparison of results of performed aerodynamic analysis on Silesian Greenpower team vehicles. Obtained results of analysis of air flow around the shapes of the bodies of all vehicles revealed the lowest values of air drag of SG2014 Bullet of all the existing Silesian Greenpower race cars. Results allow to claim that the use of the Kammback effect in the shapes of vehicle reduces aerodynamic drag. Usage of Kammback aerodynamic effect in construction of a car could minimalize drag force values resulting in lower energy consumption and creation of more eco-friendly vehicles.

Table 1. Comparison of aerodynamic drag values of Silesian Greenpower race cars.

Lp.	Model	Cx	Frontal area (A)	$Cx*A$	Force	Air velocity
1.	SG2011	0.21	0.275084[m^2]	0.057768	10.22559438[N]	17 [m/s]
2.	SG2012	0.20211	0.286858[m^2]	0.057977	10.26263077[N]	17 [m/s]
3.	SG2013	0.20973	0.2809996[m^2]	0.058934	10.43206284[N]	17 [m/s]
4.	SG2014ver.1	0.19963	0.2736764[m^2]	0.054634	9.670904418[N]	17 [m/s]
5.	**SG2014 Bullet**	**0.19632**	**0.2750478[m^2]**	**0.053997**	**9.558211952[N]**	**17 [m/s]**

Acknowledgments

The work was carried out under the project number PBS2/A6/17/2013 realized as a part of the Applied Research Program, funded by the National Research and Development Centre.

References

[1] Hucho W H 1993 Aerodynamics of Road Vehicles *General Motors Research and Environmental Staff* (Warren, United States)

[2] Orzełowski S 1969 Budowa podwozi i nadwozi samochodowych *WSIP* (Warszawa, Poland)

[3] Piechna J 2000 Podstawy aerodynamiki pojazdów *WKŁ* (Warszawa, Poland)

[4] Von Mises R 1959 *Theory of Flight* (New York, United States: Dover Publications Inc.)

[5] Baier A, Brajza M, Erm M, Grabowski Ł, Papaj P and Sobek M 2013 Advanced analysis of the Greenpower race car *International Journal of Materials & Product Technology*

Accelerated testing of an optimized closing system for automotive fuel tank

A Gligor[1], **S Ilie**[2], **V Nicolae**[2] and **G Mitran**[2]
[1]Kautex - Textron, Stephenson Highway, No. 750, Troy, Michigan, USA
[2]University of Pitești, Târgul din Vale Street, No. 1, 110040, Pitești, Argeș, România

E-mail: adrian.gligor@kautex.textron.com

Abstract. Taking into account the legal prescriptions which are in force and the new regulatory requirements that will be mandatory to implement in the near future regarding testing characteristics of automotive fuel tanks, resulted the necessity to develop a new testing methodology which allows to estimate the behaviour of the closing system of automotive fuel tank over a long period of time (10-15 years). Thus, were designed and conducted accelerated tests under extreme assembling and testing conditions (high values for initial tightening torques, extreme values of temperature and pressure). In this paper are presented two of durability tests which were performed on an optimized closing system of fuel tank: (i) the test of exposure to temperature with cyclical variation and (ii) the test of continuous exposure to elevated temperature. In these experimental tests have been used main components of the closing system manufactured of two materials variants, both based on the polyoxymethylene, material that provides higher mechanical stiffness and strength in a wide temperature range, as well as showing increased resistance to the action of chemical agents and fuels. The tested sample included a total of 16 optimized locking systems, 8 of each of 2 versions of material. Over deploying the experiments were determined various parameters such as: the initial tightening torque, the tightening torque at different time points during measurements, the residual tightening torque, defects occurred in the system components (fissures, cracks, ruptures), the sealing conditions of system at the beginning and at the end of test. Based on obtained data were plotted the time evolution diagrams of considered parameter (the residual tightening torque of the system consisting of locking nut and threaded ring), in different temperature conditions, becoming possible to make pertinent assessments on the choice between the two types of materials. By conducting these tests and interpreting the obtained results, it can be created a clear picture of the capacity of closing system of fuel tank to fulfil the functional requirements following the exposure to values of testing parameters significantly above the values that may appear throughout the entire service life of the vehicle. The proposed accelerated testing method shows the main advantage of simulation in a limited time all the situations which may be encountered in a much longer period of time, namely the service life of the vehicle.

1. Introduction

Today, modern automotive are imposed many requirements related to safety, reduced impact on the environment, resources conservation, competitiveness in the global market etc. The primary responsibility for meeting these requirements lies with the designers; they must improve the existing

Content from this work may be used under the terms of the Creative Commons Attribution 3.0 licence. Any further distribution of this work must maintain attribution to the author(s) and the title of the work, journal citation and DOI.
Published under licence by IOP Publishing Ltd

systems and develop new ones, in accordance with the legal prescriptions more stringent in the field and with user expectations [1,2].
The automotive fuel tank, an integral part of the fuel system, is a complex structure made of plastic, which are imposed a wide range of requirements, because the level of protection afforded by it when an undesirable event occurs (accident, fire etc.) are part of the degree of passive protection offered by the automotive to passengers and other road users. The complexity of this component increased continuously, evolving from a simple container to a complex system for fuel storage and delivery, containing among others the fuel pump and complex closing systems (which provide evaporative emissions reduction).
In order to obtain and validate the optimal solutions of the structure of modern system for fuel storage and supply, it is necessary to know both the legislative requirements related (providing evaporative emissions reduction) and the performance required to be provided by these systems (significant increase of durability and reliability) [3]. In these circumstances, this paper presents the development and implementation of a new testing methodology that allows predicting the behavior of closing system of the fuel tank for a long period of time, by accelerated tests under extreme conditions of assembly and testing.

2. The requirements for fuel tanks and their closing systems

The fuel tanks made of plastics offers major advantages compared to the conventional tanks made of metal. Besides that they can be designed and manufactured in almost any desired shape, using the available space more reasonable compared to those made of metal, they also offer both superior performance for impact and exposure to fire and lighter weight and better resistance to corrosion. The current design also incorporates the fuel delivery module (which includes pump and fuel filter).

The body of this module and most of the components are made of polyoxymethylene (POM), a material that combines superior resistance to fuels and chemicals with good reliability under high temperature, being a good electrical insulator.

In addition, the plastic containers incorporates a valve system that helps to maintain the pressure in the container below the maximum permitted values and, in some cases, incorporates even the automotive carbon canister, to form a complex system for safe storage and delivery of fuel (figure 1).

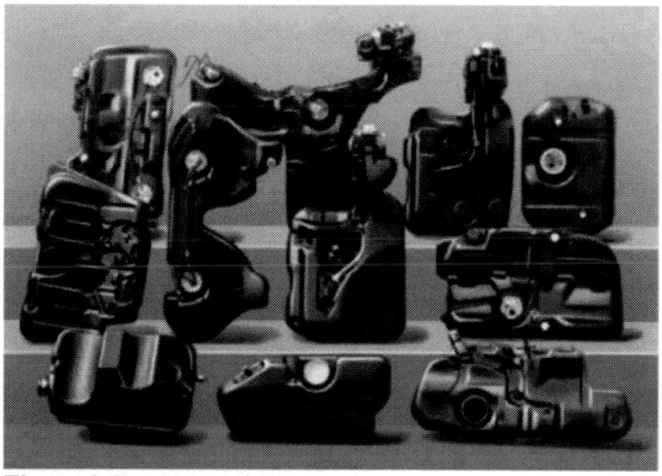

Figure 1. Examples of systems for storage and fuel delivery made of plastic.

The attachment system of the fuel pump to the tank module, called "tank closing system", is a major contributor to the total amount of the evaporative emissions system as a whole. For this reason, the efforts to reduce these emissions had to include as essential element the design and development of new methods and systems, with a much lower permeability, capable of providing a long-term superior sealing and a microscopic hydrocarbon leakage level almost negligible.

To meet the increasingly restrictive regulations on automotive emissions [4], there have been developed many variants, some of the solutions that have provided good results being presented in figure 2 together with the emission level achieved. Out of these, the one shown in figure 2c (KTNA - Kautex-Textron North America) is the one that took place the accelerated tests covered by this research. The basic requirements for these systems are:

- the compression of the gasket must be controlled and maintained between two extreme permitted values (min. 8-10% and max. 25-30%, depending on the material [5]);
- the clamping forces are aligned with each other and pass through, or as close to the center seal;
- the locking system structure and the materials used in the construction of its components, including the gasket seal, must ensure:
- sealing the system throughout the range of temperatures and pressures to which the automotive is designed to run;
- the durability and reliability of the system in accordance with the corporate standards and regulations for all fuels presented in the geographical area where the automotive is sold.

Mason Jar System	**SAE Cam-Lock System**	**KTNA Closing System**
25 ... 40 mg/24 hours	*5 mg/24 hours*	*3 mg/24 hours*

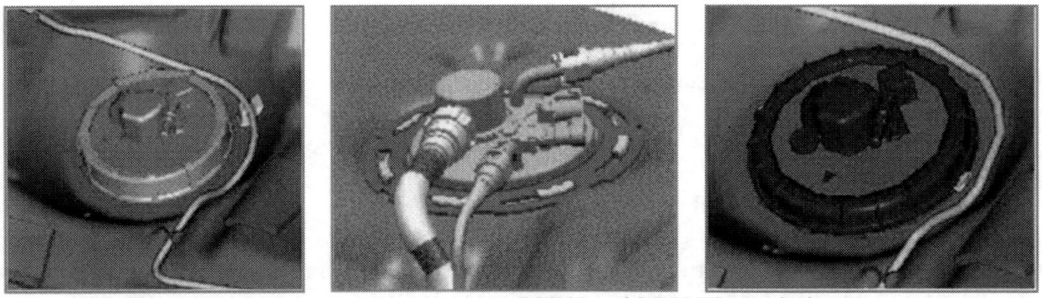

LEV II regulation — PZEV and LEV III regulations

a) b) c)

Figure 2. The closing systems of the fuel tank.

3. Description of KTNA locking system

The KTNA locking system, arose from the need to meet the new requirements stipulated in the regulations regarding emissions [4], contains the following main components (figure 3):

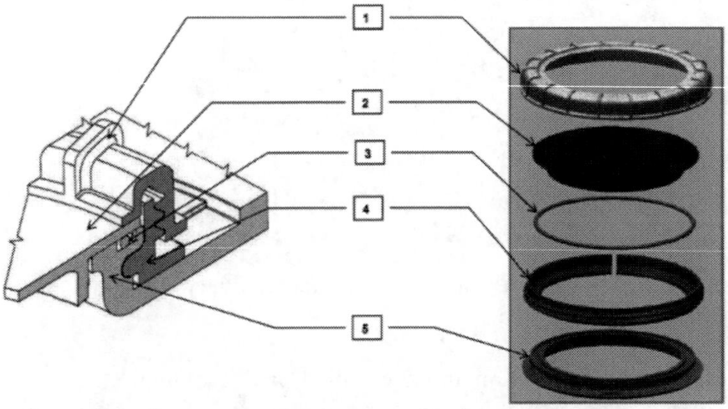

1 - lock nut, made from POM type A material;
2 - fuel pump flange of the fuel delivery unit;
3 - seal gasket, with section "H", made of FKM fluoro-elastomer;
4 - assembled threaded ring, made of POM type A;
5 – tank flange, part of the plastic container body

Figure 3. The KTNA closing system, first solution.

By assembling these components it is created a "sandwich" structure which ensures the compression of the seal gasket. After installing the components, the closure of the fuel tank is performed by turning the lock nut to obtain the prescribed installation torque.

4. The KTNA optimized closing system

After carrying out the initial tests on the system in the initial configuration, it was concluded that it meets the LEV III regulations regarding permeability, but does not provide the requirements regarding sealing, durability and reliability [3].

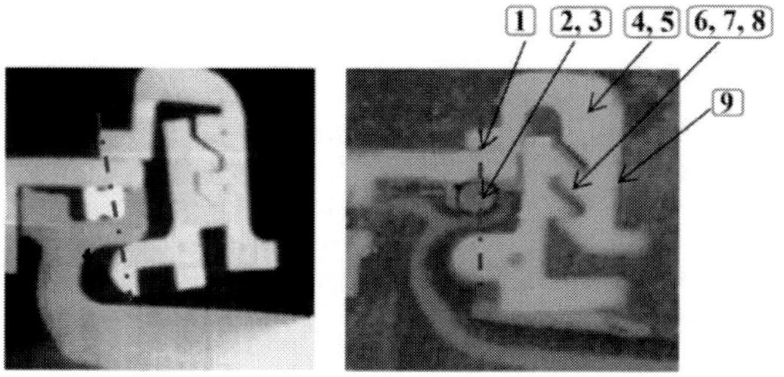

Figure 4. Comparison between the initial and optimized system:
1 - alignment of axial force on the gasket seal;
2 - use of a seal gasket with circular section, replacing the one with section "H";
3 - use of FKM GLT elastomer for manufacturing the gasket seal;
4 - changing the material of the components from POM type A to POM G21CP and POM M25;
5 - redesign of the cross-section of the nut;
6 - changing the thread profile of the screw ring and the lock nut;
7 - change the thread pitch of the screw ring and the lock nut;
8 - change the thread end zone to avoid the tension concentrators;
9 - addition of reinforcement elements to the outside of the lock nut.

5. Durability testing methods applied to the optimized closing system

The accelerated test methodology developed allows estimation of the fuel tank closure system behavior over a long period of time, according to the requirements of the regulations LEV II and LEV III [6]. LEV II Regulation provides that this period it shall be for 10 years, while for compliance with the requirements of the LEV III Regulation it is necessary to ensure the durability of the system for a period of 15 years. Since the tests on periods provided in the regulations is not a realistic option, it became necessary to carry out accelerated tests under extreme conditions of assembly and testing of the system [6]. The goal of testing is to create meaningful and realistic images on the ability of the fuel tank closing system to meet the functional requirements and after being subjected to stress with parameter values well above those that would be encountered throughout the service life of the automotive in all conditions. The test conditions included: *(i) assembling the components of the system at high initial torque*, *(ii) extreme temperature values* and *(iii) extreme pressure values*. During the experiments there are determined various parameters such as *the initial tightening torque, the torque at different points in time during the measurements, the residual tightening torque, the defects occurred (fissures, cracks, breakages), providing leak-free of the system at the beginning and at the end of the test.*

The durability tests included in the test program which will be presented in this paper are: *(I) the test of exposure to temperature with cyclical variation* and *(II) test of continuous exposure to elevated temperature*. In these two tests it was evaluated the behavior of a total of 16 systems optimized for closing the fuel tank, having the gasket seal made of FKM fluoro-elastomer with circular cross section, the lock nut and screw ring being made of POM M25 or POM G21CP for each half of the systems tested. Each of the two tests included a total of 8 tanks assembled, respectively by 4 systems in each category, the torque being measured during tests. The samples that were chosen to conduct the tests were dimensionally controlled and their compliance was validated. The durability criteria chosen to tie the results was *the higher value of the residual tightening torque*, in conditions of fulfillment *the fuel tank sealing requirements*.

5.1. The test of exposure to temperature with cyclical variation

This test consisted of alternating exposure of the 8 container closure systems simultaneously at temperatures between two extremes, namely: - 40 ° C and + 80 ° C. Therefore, the systems have been introduced for 456 hours in two thermal chambers which ensured the temperature conditions mentioned. The two thermal chambers have been placed at a short distance from one another, in order to minimize the time necessary to transfer the containers from one temperature to another. After starting the test, the first measurement of the residual tightening torque was performed at the end of the first 240 hours of testing, the following two measurements are made at 48 hours after each previous measurement.

5.2. The test of continuous exposure to elevated temperature

The total duration of the test was 480 hours and consisted of exposing the tanks at a constant temperature of + 80 ° C, in order to evaluate the maximum effect of creep and relaxation of the material on the component structure and the residual tightening torque. In this test we used a single thermal chamber capable of allowing the conditioning of the 8 systems. After the test, the first torque measurement was performed at the end of the first 240 hours of testing, the following measurements were performed at 48 hours after each previous measurement.

6. Results and discussion

The results of the test of exposure to temperature cycling variation are shown in Figure 5.

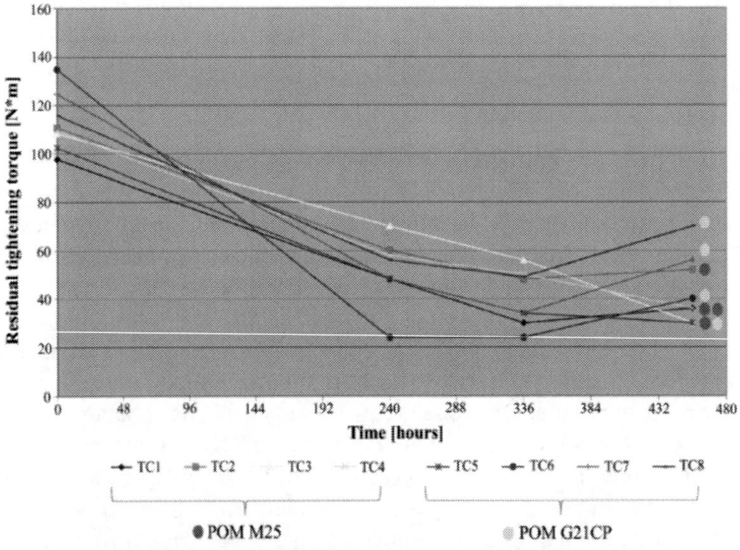

Figure 5. The variation of the residual tightening torque over exposure to cycling variation temperature for 456 hours.

The graphical evolution of the residual tightening torque is influenced by the fact that at each measurement the lock nut has been rotated by about 0.5° - 1.5°, which introduced an additional factor in the variation of torque values. Given that this affected all the 8 systems in a relatively similar manner, the information obtained is representative for the long-term durability assessment of the optimized closing system and for the comparison of the two POM materials analyzed.

The checks for leaks carried out at the end of the tests according to ECE R34 at a pressure of 30 kPa and at three different values of temperature [7] confirmed that the sealing properties were maintained during measurements and the sealing requirements are met successfully in all the 8 cases. Based on these results, it can be concluded that for this test the tanks equipped with closing systems whose components were made of POM G21CP material gave better results compared with those having components made of POM M25.

The results of the continuous exposure at high temperature test are shown in figure 6.

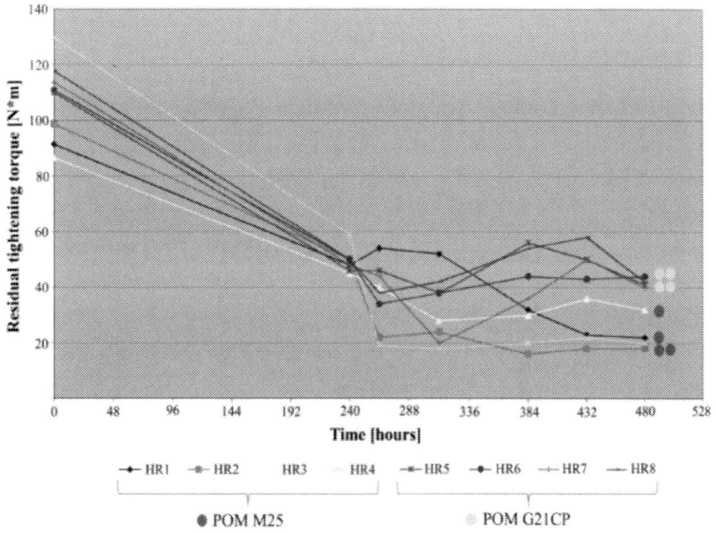

Figure 6. The variation of the residual tightening torque over exposure to high temperature for 480 hours.

In this case, the form of graphs is influenced by the fact that on each measurement the lock nut is rotated by about 0.5° - 1.5°, which introduced an additional factor in the variation of torque values. Based on the results for the residual tightening torque is found that tanks equipped with closing systems with components manufactured of POM G21CP material gave better results compared with those having components made of POM M25.

At the end of the test it was done the fuel tanks leaks check and there were tested and inspected the components of the system to detect the potential areas for cracks and / or stress. It was found that all the 8 tested systems meet the sealing requirements stipulated in ECE R34 [7] without the structure of components show signs of degradation.

7. Conclusions

After closure of a cycle of durability assessment tests of the fuel tank closing system, the results indicated that the systems made of POM G21CP repeatedly showed a residual tightening torque greater than that of the systems made of POM M25. As a result, the option for the POM G21CP material became clear both for the lock nut and for the threaded ring.

By analyzing the results obtained during these tests, it appears a clear picture on the capacity of the fuel tank closing system to fulfill the functional requirements after exposure to the test parameter values well above the values that may occur throughout the entire automotive service life. The main advantage of this method of accelerated testing is that during a relatively short time (480 hours) it is simulated the requests that may be encountered in a longer period of time during the automotive service life (10-15 years).

Acknowledgments
This work of Gabriela MITRAN was supported by the strategic grant POSDRU/159/1.5/S/138963 - PERFORM, co-financed by the European Social Fund – Investing in People, within the Sectoral Operational Programme Human Resources Development 2007-2013.

References
[1] Ilie S 2012 Modelarea Fenomenelor de Impact ale Autovehiculelor (Modeling the Impact Phenomena of Motor Vehicles) (Bucharest: AGIR Publishing House)
[2] Ilie S, Miutescu A, Stoianovici M and Mitran G 2014 Recovery of Precious Metals from Catalytic Converters of Automobiles by Hydrometallurgical Solid-Liquid Extraction Processes *Advanced Materials Research* (Switzerland: Trans Tech Publications) **837** pp 105-109
[3] Gligor A 2012 Enhancements of the Automobile's Passive Safety Performance through the Use of Plastic Materials for the Components of Modern Fuel Tank Systems *PhD Thesis* (University of Pitesti)
[4] Information on http://www.arb.ca.gov/msprog/levprog/leviii/meetings/030210/lev_iii_discussion_paper_2-10.pdf.
[5] Visscher D (ed) 2012 *Dichtomatik O-ring Handbook* Section 4 O-ring sealing elastomers (Dichtomatik North America USA)
[6] Aksal A, Kinoshita S and Gligor A 2010 Torque decay study for a KTNA fuel tank closing system used in a LEV II compliant application *Kautex-Textron Windsor* (Ontario: Canada)
[7] E/ECE/324 // E/ECE/TRANS/505 2003Regulation No. 34, Uniform provisions concerning the approval of vehicles with regard to the prevention of fire risks

Modelling bucket excavation by finite element

O M Pecingina[1]
[1]University "Constantin Brancusi" of Tg-Jiu, Faculty of Engineering, 30 Eroilor Street, 210135, Tg-Jiu, Romania

E-mail: pecingina_olimpia@yahoo.com

Abstract. Changes in geological components of the layers from lignite pits have an impact on the sustainability of the cup path elements and under the action of excavation force appear efforts leading to deformation of the entire assembly. Application of finite element method in the optimization of components leads to economic growth, to increase the reliability and durability of the studied machine parts thus the machine. It is obvious usefulness of knowledge the state of mechanical tensions that the designed piece or the assembly not to break under the action of tensions that must cope during operation. In the course of excavation work on all bucket cutting force components, the first coming into contact with the material being excavated cutting edge. Therefore in the study with finite element analysis is retained only cutting edge. To study the field of stress and strain on the cutting edge will be created geometric patterns for each type of cup this will be subject to static analysis. The geometric design retains the cutting edge shape and on this on the tooth cassette location will apply an areal force on the abutment tooth. The cutting edge real pattern is subjected to finite element study for the worst case of rock cutting by symmetrical and asymmetrical cups whose profile is different. The purpose of this paper is to determine the displacement and tensions field for both profiles considering the maximum force applied on the cutting edge and the depth of the cutting is equal with the width of the cutting edge of the tooth. It will consider the worst case when on the structure will act both the tangential force and radial force on the bucket profile. For determination of stress and strain field on the form design of cutting edge profile will apply maximum force assuming uniform distribution and on the edge surface force will apply a radial force. After geometric patterns discretization on the cutting knives and determining stress field, can be seen that at the rectangular profile appears the "clogging" phenomenon of the cutting edge and at the polygonal profile the point of application remains constant without going inside. From the finite element method done in this paper it can be concluded that the polygonal profiles made of dihedral angles are much more durable and asymmetric cups tend to have uniform tension along the entire perimeter.

1. Introduction

Solid fossil fuels are an important primary source of energy in our country's energy structure which means a share of 39%. As of national primary energy resources lignite is the only domestic primary energy carrier in terms of resources can make significant contributions to consumer demand for electricity. Processes for the extraction of lignite surface excavation technology streaming is achieved using complex excavation-transport-dump high productivity and includes the following technology: selective excavation, transport, deposition and deposition of tailings dumps lignite deposit. The first takes place in continuous flow rotor excavators, working the steps with heights of 25-30 meters.

Content from this work may be used under the terms of the Creative Commons Attribution 3.0 licence. Any further distribution of this work must maintain attribution to the author(s) and the title of the work, journal citation and DOI.
Published under licence by IOP Publishing Ltd

Transport of excavated material is made with the carrier tape with a width of 1000-2250 mm, the maximum speed of 6.15 meters/second, and the transport capacity reaches 12.500 cubic meters per hour. Dump tailings dump is done with plants with flow rates between 2.500 and 12.500 cubic meters per hour discharge arm length is between 65 and 170 meters. These parameters are reflected on the sustainability of the cup shear elements and under the action of excavation efforts leading to deformation occurring whole ensemble. An optimal geometric shape of the cup knives and teeth positively influence the process of excavation and active wear items.

2. Experimental studies

From experimental measurements performed in this work it was observed the dependence of cutting force F_{xm} on the depth of cut h_0 for different careers [1].
It can be observed linear dependence between these two quantities, using our case analysis (static).
The interdependence of these is given in the equation (1) from below:

$$F_{x\max} \cong 3F_{xm} \quad \text{admiting} \quad h_{o\max} = b \tag{1}$$

Were considered two cases of application for the force F: for b = 130mm and for b=150mm.
For the two average forces results the maximum values that we see in equation (2):

$$F_{x\max 1} = 3\cdot(1230\cdot 130) = 479700 \ [\mathrm{N}] \tag{2}$$

$$F_{x\max 2} = 3\cdot(1230\cdot 140) = 516600 \ [\mathrm{N}]$$

We used as force value the maximum force for the Olteţu career because of laboratory analysis it was observed that the cutting force for this area has the highest value field. For determination of stresses and strains on the cutting edge profile constructive form shall apply in maximum force, assuming uniform distribution and edge surface force will fold radial force. To achieve the analysis of the first phase will pass the mesh model, made with tetrahedral elements with 10 nodes, as in the classical cutting edge surfaces are curved surfaces.

In the second phase were considered cutting force and radial force in maintaining solid, as follows:

$$F_x = 86100 \ [\mathrm{N/tooth}]$$

$$F_r = 30135 \ [\mathrm{N/tooth}]$$

Cutting force will be applied on all sides where teeth are located, six in number, and the radial force will be applied only to edge face and teeth: 2, 3, 4, and 5 in clockwise. As a fixed support was considered welded joints of the port ansamblucu knife Cup, which were taken over all degrees of freedom. Models can be treated with an excavator bucket rotor cutters on ERC1499 / 30-7 CALS and modernized as we can see in thr figure number 1 from below [2,3]:

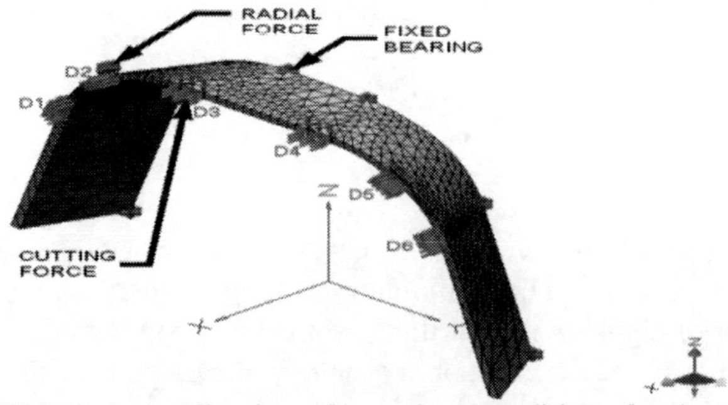

Figure 1. Application of boundary conditions for the classical model.

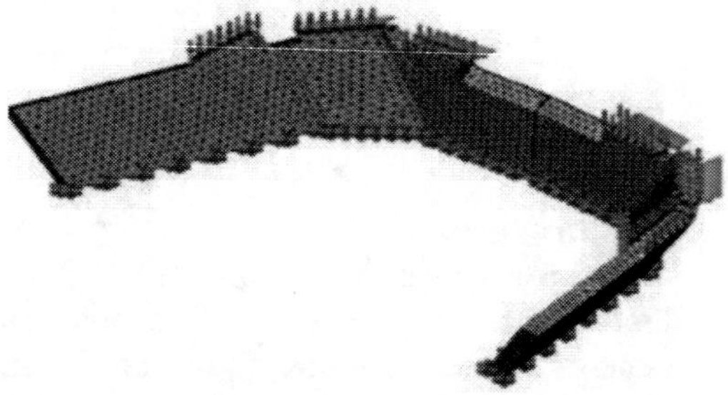

Figure 2. Application of boundary conditions for the modernized cup model.

Because it is considered homogeneous and isotropic medium for the components (tooth-box-knife), we use the steel material features: modulus of elasticity E=2.1x10^5 [MPa], Poisson's ratio =0.29, density ρ=7850[kg/m^3], limit stretch stress $\sigma_l = 448$ [MPa].

Then we analyzed the geometric models considered studying in primary and stress fields in assumptions considered using static analysis. The analysis was performed in order to highlight the differences in stresses and strains for the profiles of the two models, the classical model with curvilinear profile and upgraded model with polygonal profile.

In figure 3 we determined the stress field and deformed configuration under the tangential and radial forces action.

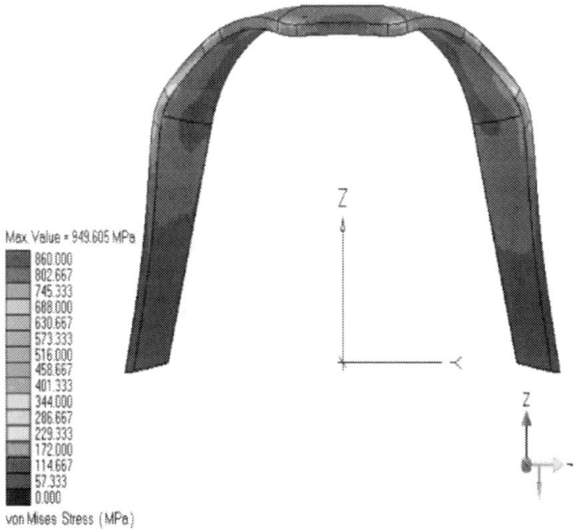

Figure 3. Stress field equivalent to the classical model.

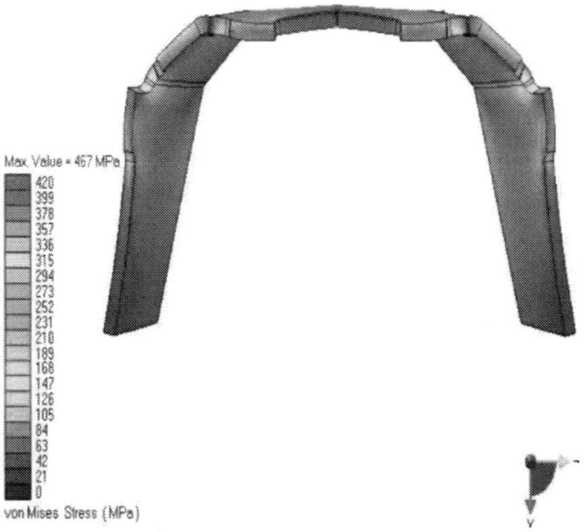

Figure 4. Equivalent stress field for the sheet model.

In figure 4 from above we can se the model of bent sheet metal. It can be observed as equivalent stresses and deformation forces model developed under excavation. It can be seen as equivalent to the classical model stresses are much larger than the model sheet [4]. In reality it was shown that the number of operating hours for polygonal profiles reached approximately 4000 hours of operation and vibrations on excavators decreased.

Also it is observed the importance of dihedral angle made on the central knife, the contour having a maximum radial force due to the action of which cannot deform inward. This can highlight in figure 5 (classic model) and figure 6 (upgraded model) where the resultant displacement under the action of forces is smaller than the classic [5].

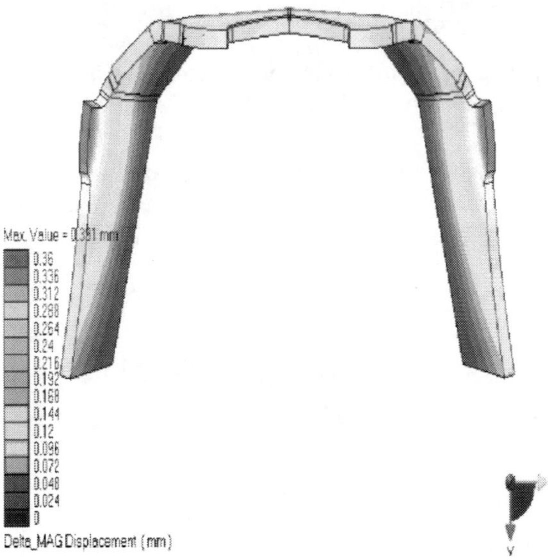

Figure 5. Displacement to modernized model.

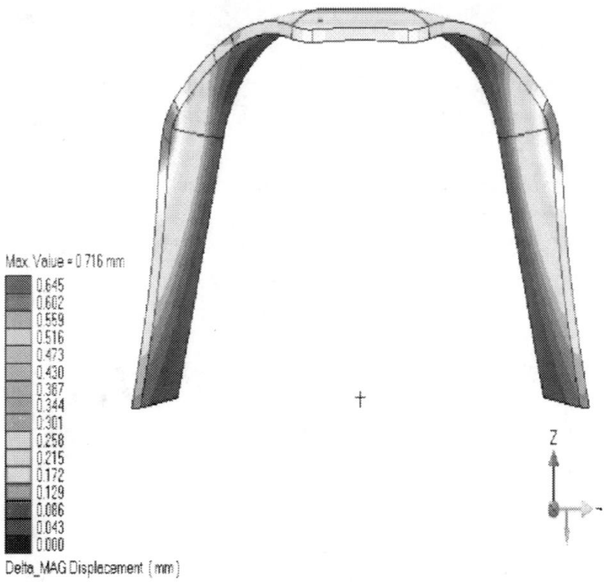

Figure 6. Displacement to the classical model.

Next we studied two models of knives of an asymmetric cups one of them having a profile with a maximum on the knife (an angle a^{0}), forming a dihedral angle and the second having a right profile with the same assumptions calculation and under the action of the same system of forces, changing only the constructive shape of contour cutting edge [6]. Using the same conditions were determined the displacement direction radial force values and equivalent stress as wer can see in the figure number 7 from below:

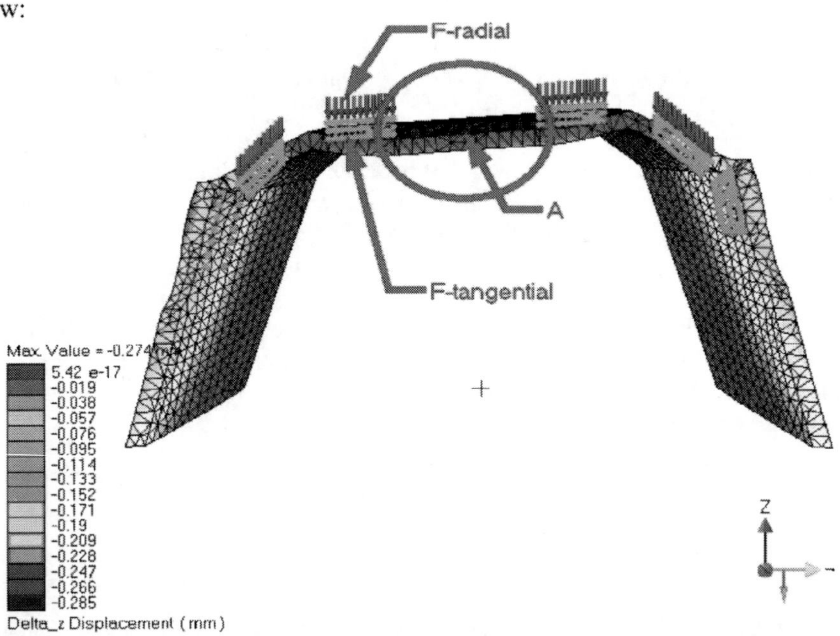

Figure 7. The amount of displacement in the direction of radial force/tangential force on the right profile.

It can be seen that if at the right profile appeared the "clogging" phenomenon of the cutting edge, polygonal profile point "A" remaining constant, without going inside as we see in figure 8.

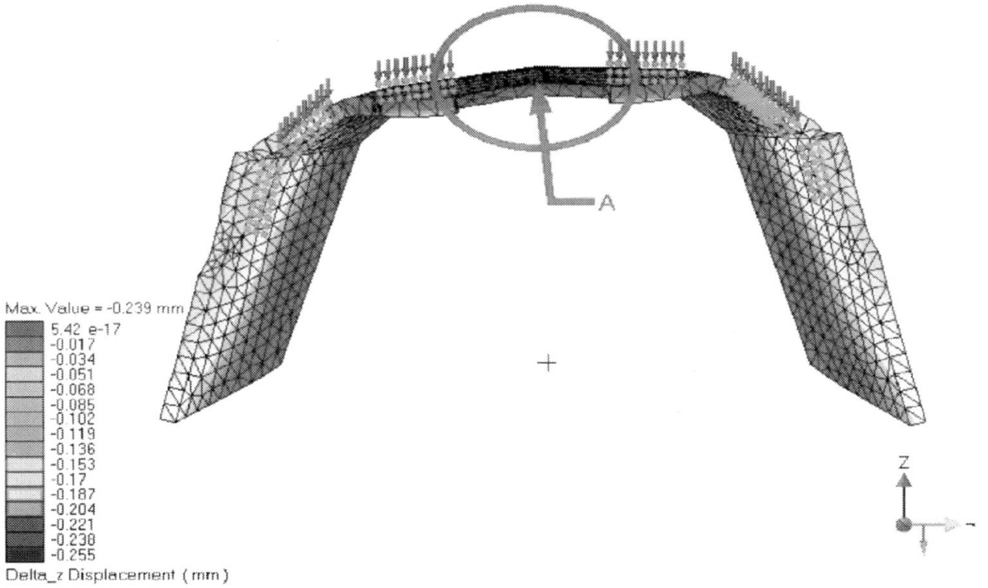

Figure 8. The amount of displacement in the direction of radial force/tangential force on the polygonal profile.

If you follow the figures number 9 and number 10 for the two equivalents stress profiles can be observed that the right profile has maximum points and the tension at the connection planes is much higher [7]. This showed that this type of profile was more resistant than straight profile of the bucket as we can see in figures 9 and 10.

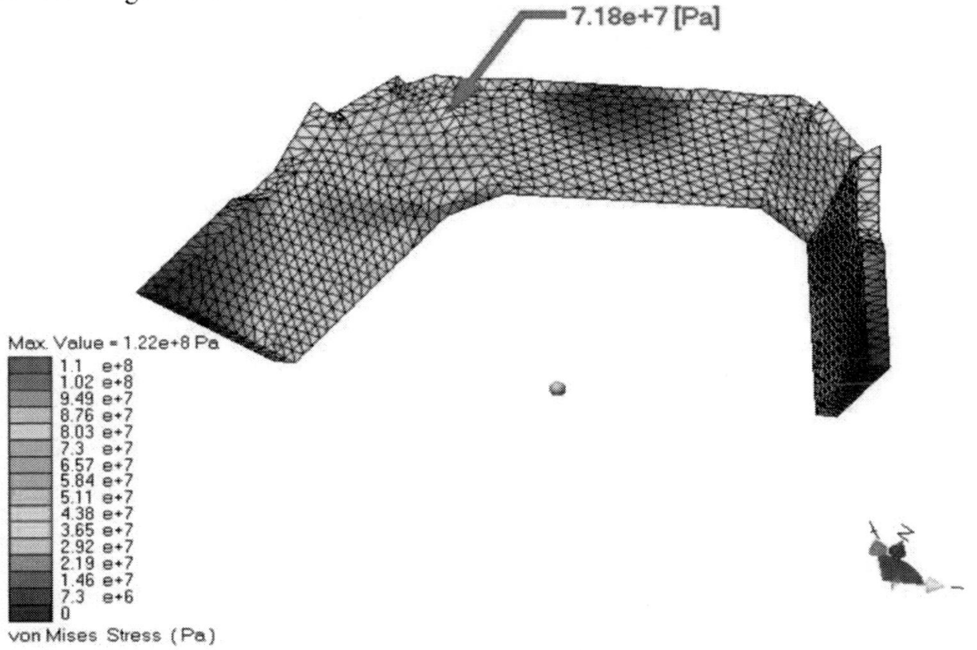

Figure 9. The equivalent stress for the right profile [8].

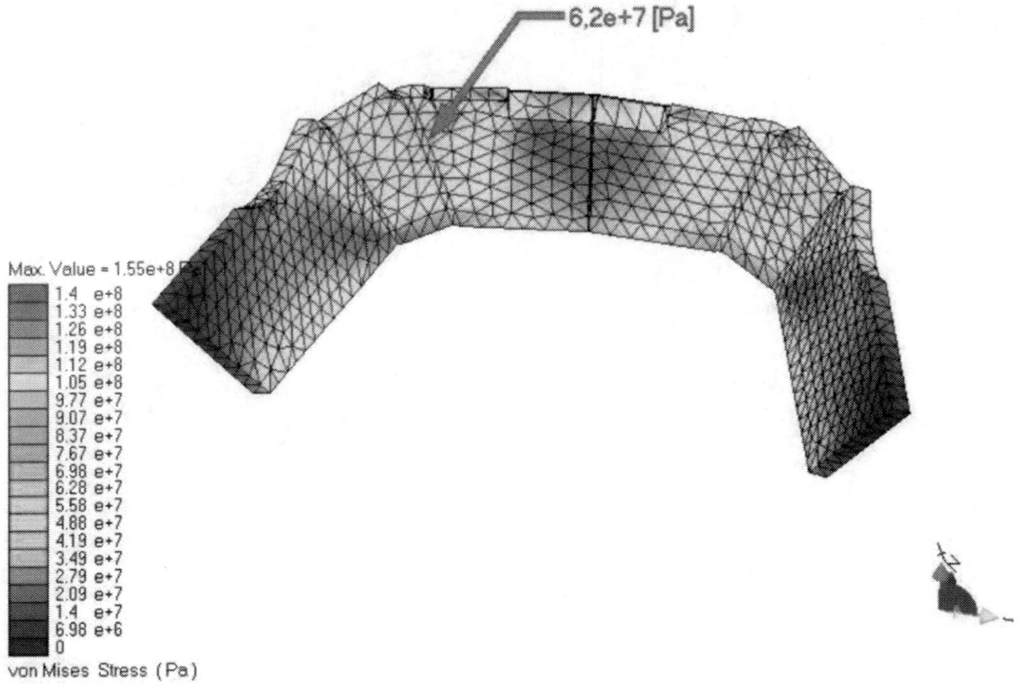

Figure10. Equivalent tensions for the polygonal profile [9,10].

3. Conclusions

From the above analysis it can be concluded that the polygonal profiles made of dihedral angles are much more durable and asymmetric cups tend to have a uniform pressure along the entire perimeter. In practice it was proved that durability at the laminated sheet cups with polygonal profile having a maximum peak is much higher than other models of cutting edges.

References

[1] Păsărin C 1999 *The lignite cutting resistance and the surrounding rocks. Its influence on the cutting body of rotor excavators* (Craiova: Publishing House Sitech)

[2] Blumenfeld M 1995 *Introduction to the finite element method* (ET Publishing House)

[3] Zienkiewiez T 1991 *Finite element method* (Paris: Anfor)

[4] Sontea S 2002 *The technology of materials* (Craiova: Publishing House Spirit Românesc)

[5] Wait P and Mitchell A R 1985 *Finite element analysis and applications* (New York: John Wilwey)

[6] Information accessed on 12.01.2015, http://www.bluparts.ro/dinti-de-cupa

[7] Information accessed on 15.10.2014 on http://www.arhiconoradea.ro/Info/Popa

[8] Information accessed on 15.12.2014, http://issuu.com/masinisiutilaje/docs/mu05_2014

[9] Information accessed on 12.12.2014, http://webbut.unitbv.ro/teze/rezumate/2010/rom

[10] Information accessed on 23.12.2014, http://www.bm-cat.ro/files/resources/193cat-329e-l-ln.pdf

IOP Conf. Series: Materials Science and Engineering **95** (2015) 012047 doi:10.1088/1757-899X/95/1/012047

Analysis of large sheet metal tailored tubes

V M Pomazan[1]
[1]Ovidius University of Constanta, 124 Mamaia, 900527, Constanta, Romania

E-mail: vpomazan@univ-ovidius.ro

Abstract. The present study was triggered by the need to verify and optimize the primary constructive solution, for custom large tubes (section lengths larger than 1000 mm), under the gravity and pressure loads. The cases presented needed to be checked for the reinforcement design. Given the complex tridimensional geometry of the axisymmetric shell structures, the basic shape of the tubes was modelled, with its actual thickness. FEA was used to check the model under static loads and buckling. In order to optimize the weight, an alternative welded reinforcement's grid design was developed and checked for stability. Optimal welding sections along ribs in longitudinal and transversal directions were identified for easier design and further costs reduction.

1. Introduction

Several gaseous media are transported in non-standardized piping, using custom designed large, reinforced steel tubes. Thin-walled tubes have been used extensively to dissipate energy during an impact event [1] and have good behavior when exposed in out-door conditions. The main issues of such structures are the stability of certain segments, as the site topology equation certain tube length and geometry, when big gas volumes needed. These tubes work with gas at design pressures of ± 0.1 bar, temperatures under 100°C, and are built in open air, exposed to wind, rain and snow.

The primary structural calculus of such tubes takes into account the general geometry of the shell-like structure and prompts for holders and props [2]. Custom and empirical reinforcements are designed if section quotient is larger than the material (austenitic steel) thickness.

The weight to strength optimization is constrained by the thin wall and topology of the inlet and outlet to be connected. The sole resource for weight and, ultimately, cost optimization, stands in the stiffeners used along the tube. These are ribs, welded in regular patterns, as rectangular collars, in transversal position and swept ribs, following the median axis of the tube, mounted on each of the four faces at rectangular tubes. The challenge is to design the minimum number of ribs needed to reinforce the tubular structure, in order to reduce the manufacturing and material costs.

2. Rationale

The design of this axial trajectory of the tube must respect the in situ topology and the material mechanical characteristics, for the given loads. The tubes are supposed to be welded and mounted in situ, equipped with supports from extruded profiles, with concrete foundation (figure 1). The width of the square section is 1716 mm; the total length is 17500 mm. The material is chromium-nickel stainless steel SA240 304 and its mechanical properties are presented in table 1.

It was shown that the buckled wave length for a rectangular thin walled tube becomes smaller for the higher pre-buckling stress, if constrained along the boundary [3]. As such, the general design for the

Content from this work may be used under the terms of the Creative Commons Attribution 3.0 licence. Any further distribution of this work must maintain attribution to the author(s) and the title of the work, journal citation and DOI.
Published under licence by IOP Publishing Ltd

stiffeners can follow the rule that their longitudinal frequentation must be smaller than the half of the wavelength (for axial loads) and smaller than the length of the wrinkle (bump shaped) in transversal loads [4].
The present design of the structure followed the primary results of the designer for the reinforcements, correlated with the dimensioning for the supports.
The square sectioned thin wall tube can be considered as a simple supported beam, loaded by its own gravity and standard out-door loads, i.e. distributed forces mainly. The studied tube has horizontal trajectory at both ends and a vertical orientation for the median part, which means that the endings are loaded under bending in two planes while the vertical part must stand mainly for the uniformly distributed axial load from the gravity. Considered as a three dimensional truss, the tube is a statically indeterminate structure.

Table 1. ASME SA240 304 Mechanical Properties [12].

Yield	Tensile	Elongation	Hardness, max	
Min MPa	MPa	Min %	HB	HRC
205	515	40%	201	92

Basic static analysis of the structures indicates the main parameters for welded supports and their location. At large dimensions d and relatively small thickness t (in this case, dimensional factor K= d/t = 270), even though the pressure values are not big, but fluctuant, at some point, safety design considers that stiffeners should be taken into account, to avoid bucking in the most exposed segments [5].
Buckling assessment should take into account the sensitiveness of the critical loading with the imperfections of the shell and assembly [6, 7]. For the horizontal ends, the maximum stress is reached at extremes from the neutral axis and the top and bottom are fully stressed [8].
For thin walls, the area A and the moment I can be expressed as [9]:

$$A = 4dt, I = 2td^3/12 + 2dt[(d/2)]^2 = 2td^3 3/3 \quad (1)$$

The buckling of the slender square tubes combines the local phenomenon with the flexural buckling of the length. It shows as local bumps on the flat sides, without roto-translation and might coexist with flexural behaviour with wrinkling on the length flat faces (figure 2). The stiffeners spacing *s* (figure 2) is the parameter that controls the residual stress distribution between the stiffeners [10].
Flexural buckling can be defined by the coefficient *K*, depending of the aspect ratio width/ thickness and varies with fixture type (equation (2)).

$$K^2=I/A=(2d^3 t/3)/4at\approx d^2/6 \quad (2)$$

The buckling stress is calculated with report to the tangent modulus E_T and the length of the slender strut zone, *L*:

$$\sigma_F=(\pi^2 E_T d^2)/(6L^2) \quad (3)$$

In this case, each wall is considered a flat plate, supported along long edges. Axial stress induces a lateral collapse:

$$\sigma_L=K_L E_T (t/d)^2 \quad (4)$$

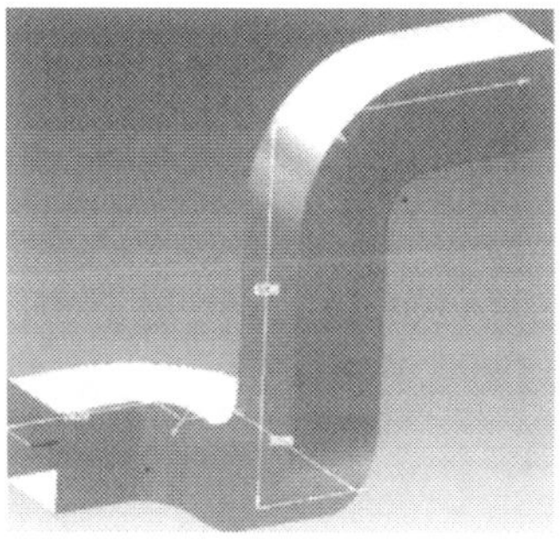

Figure 1. The topology of the thin walled tube

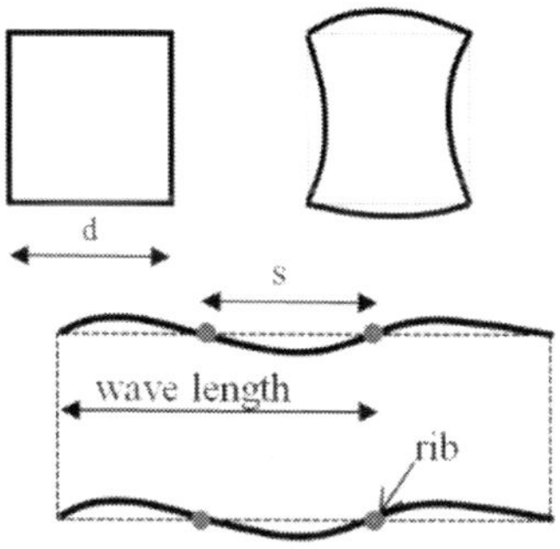

Figure 2. Buckling of square thin walled tube

For design purposes, the local buckling coefficient $K_L = 3.62$, corresponding to a generic Poisson ratio $\nu = 0,3$. This σ_L stress should not exceed the working compressive stress, under the nominal load P:

$$\sigma_W = P/4dt \quad (5)$$

Optimal geometry can be obtained if the buckling, axial stresses equal at most the working stress:

$$\sigma_F = \sigma_W = \sigma_L \rightarrow (\pi^2 E\_T\, d^2)/(6L^2) = P/4dt = 3.62E\_T\,(t/d)^2 \quad (6)$$

The objective function of independent structural variable is referred as the weight W, which is less when the stress in the sectional area A, along the length L, under the load P, is maximum [10]:

$$W = \rho AL = \rho(P/\sigma)L \quad (7)$$

A normalised weight parameter n is defined for depicting the optimum weight range:

$$n = (1/\rho)\ W/L^3 \quad (8)$$

This parameter will be used to control the results of ribs optimal placement and optimal number. Equation 6 is used for optimal dimensioning of the width d and thickness t [11]:

$$d_{opt} = 0.743((PL^3)/E\_T)^{(1/5)},\ t_{opt} = 0.371((P^2 L)/[\![E^2]\!]\_T)^{(1/5)} \quad (9)$$

With these parameters, one can follow Elwi and Kulak recommendation for rational design [12], and assume, further, a stiffener distribution, then check for stability, ensuring that failure should occour rather between the stiffeners than outside. Choosing shell buckling or general buckling for the calculus, one will decide the ribs distribution. At this time, important supplementary aspects as imperfection and residual stress (as a result of welding and assembly on long paths) must be taken into account [13].

3. Method and analysis

3.1. The tube without ribs

The solid model was developed as a swept feature, from the normal cross-section, along the three dimensional path axis. First, parametric CAD INVENTOR ® was used for solid modeling, FEA was performed with Autodesk Simulation Mechanics®, while CATIA® digital prototype was used for double checking the static analysis results. In both models, local constraints were declared on the supporting lines, where the tube will be fixed on the position. Gravity, wind and interior pressure of

0.1 bar load the model. The median surface for shell meshing was generated. Minimum element size (as fraction of average size) is 0.2, while average element size in the mesh is 0.05.

The total mass of the tube without stiffeners is 4984 kg. This value will be used as reference for further study of the optimal stiffeners displacement.

As can be seen in figure 3 (stress and displacement), the critical points stand in the middle of the curved sections of the tube, but the equation univalent stresses remain in the safe zone, under the yield limit. The large displacements present in the flat horizontal faces remain under the material capacity. This steel is designed for vessels under pressure and working at high temperatures. The need for reinforcement is obvious, mainly in the critical zones. The goal is to ensure a safety factor of 4 at least, for the stiffened structure (figure 4).

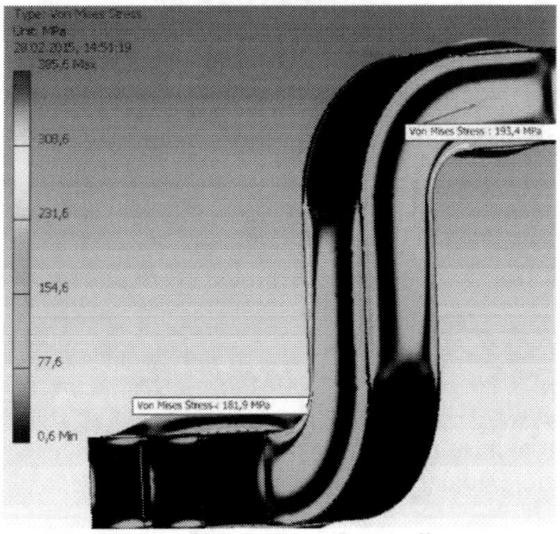

Figure 3. a-Von Misess Stress diagram.

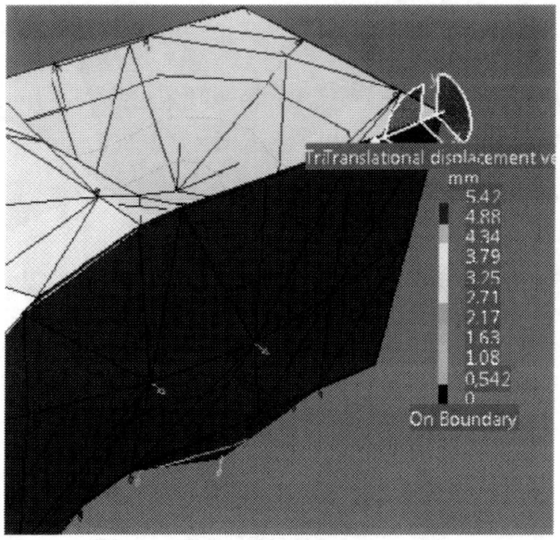

Figure 3. b- Displacement 10x.

3.2. The ribbed tube

The tube was designed with stiffeners that will be modelled as circumferential zones with no radial and torsional displacements. The ribs are separate parts (figure 5), as ribbons of 1716x100x6 mm welded in linear sections, as a circumscribed matrix along the tube. In fact, locally, the tube rectangular cross-section changes, and the flexural buckling coefficient increases with 12%. With stiffeners, as designed initially, the mass reaches 7165 Kg, comprising the welding. The normalized weight parameter increases 43.7%.

A second model was built from the initial tube, with ribs, located with a *s* pitch of 500 mm along length and dividing the sides in three, swept along the tube. Even though modelled for the welding assembly calculus, the stiffeners were declared as formal restrictions in the normal and lateral direction to the surface, and no twisting along the symmetry axis. Otherwise, the complexity of the details would have induced errors at mid-surface shell generation and meshing, altering the results. The stiffeners break the flat surfaces in smaller plates constrained along the edges. The stiffeners alone can improve the bucking behaviour which means that is sufficient to weld the ribs on the shell and not to the web.

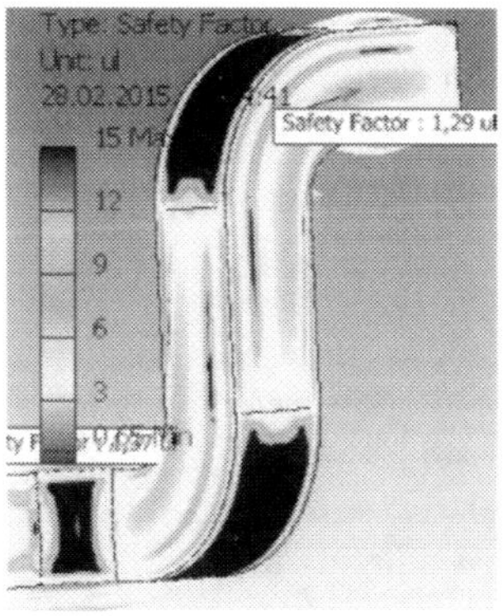

Figure 4. Safety factor for the tube, 10x.

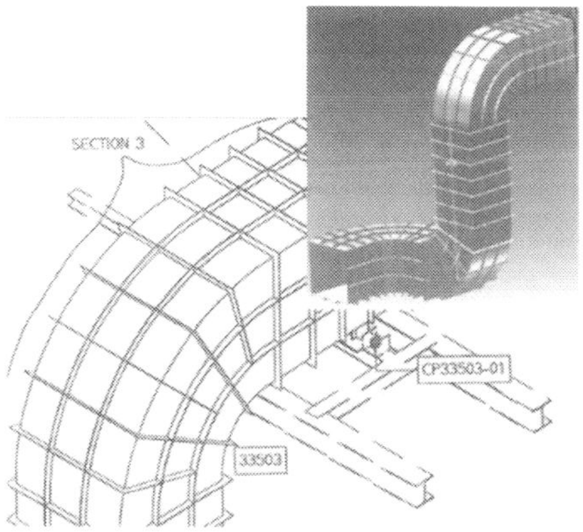

Figure 5. Welded stiffeners on the square section.

Finite element analysis for the ribbed tube, with frictionless contacts defined on the surfaces of the model; reveal the important decrease of the stresses and, practically, the annulation of the static displacements, which barely reach 0.04 mm (figure 6).

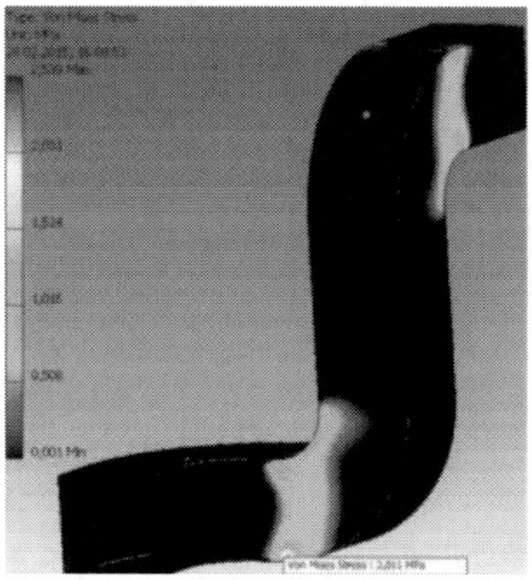

Figure 6. The ribbed rectangular thin walled tube.

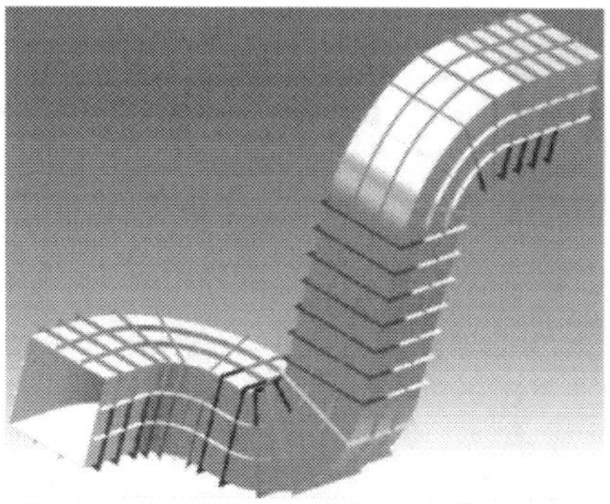

Figure 7. Disposable stiffeners for lighter design.

3.3. The square tube with spaced hoop stiffeners

This initial design was checked for optimization resources. The zones with the lowest stress condition and displacements were considered. The axial stiffener was kept along the tube, due to the uniformity and symmetry of the welding residual stresses and considering the general buckling.

Hoop ribs were reconsidered and spaced at the end and middle of linear zones and discarded on the rest of faces.

As a result, the reinforcements were placed only on the upper plates of the horizontal zones and the laterals of the vertical, where the maximum displacement was identified. The remaining stiffeners are used also for technological reasons, for reliable contact with supporting frame. Each section of the geometry of the tube is supported by a pair of horizontal frame members. The ribs will be used for weldment assembly.

The static analysis indicated a good behaviour and very small displacements, with a small increase in the equivalent stress.

This indicates that the rib design can be changed, for the same pattern along the tube length. The maximum stress is displaced on the bottom flat zones, between the supports, as a result of the load distribution. The stress configuration in the new rib design stays within material potential limits.

The redesign of the stiffeners intended to identify the less stressed zones and to eliminate the disposable ribs. The new design was verified under the same loads, with constraints displacement modified (dark ribs in figure 7).

4. Conclusions

Thin wall slender tubes are prone to buckling deformations under their own mass and designed pressure action. Large deflections of the side walls of the tubes can be prevented with stiffeners. These are ribs welded as a matrix around the perimeter of the tube and along its length. The initial design indicated that the normalized mass parameter n increased up to 43%. Studying the stress distribution and buckling behaviour, a new distribution of the ribs was found (dark zones eliminated in figure 12), that eases the whole structure, reducing the weight parameter n with 25%.

Corresponding stiffeners manufacturing effort decreases with 42%, thus optimizing the objective function. An optimization potential was found in a slender cylindrical tube with high slenderness factor. Given the small values for the stress and displacements, a thickness reduction was applied and checked. The normalized mass parameter n decreased with 25%.

References

[1] Nagel G M and Thambiratnam D P 2006 Dynamic simulation and energy absorption of tapered thin-walled tubes under oblique impact loading *International Journal of Impact Engineering* **32(10)** pp 1595–1620

[2] Rhodes J 2002 Buckling of thin plates and members — an early work on rectangular tubes *Thin-Walled Structures* **40** pp 87–108

[3] Ohkawara T, Kuwako S et al 2006 Analysis of deformation property of the wall surface on rectangular thin walled tube by the bending moment *Developments in Mechanical Structures and Materials* (London: Deeks &Hao, Taylor and Francis) pp 1271-1276

[4] Blake A 2009 Practical Stress Analysis in Engineering Design **II** *CRC Press* (London: John & Wiley) Rondal J, Wurker, K-G. et al. 1992 Structural stability of Hollow sections *Verlag TUV* (Rheinland Koln)

[5] Filipov E, Haseganu E and Smirnov A 2002 Buckling analysis of axially compressed square elastic tubes with weakly supported edges *Technishe Mekanik* **20(1)** pp13-20

[6] Ewi A L and Kulak J T 2006 Large diameter fabricated tubes, in Buckling of thin metal shells Teng T J and Rotter J M (Eds.) (Spon Press) p 344

[7] Shan T, Yuli L and He Y 2013 Effects of geometrical parameters on wrinkling of thin-walled rectangular aluminium alloy wave-guide tubes in rotary-draw bending *Chinese Journal of Aeronautics* **26(1)** pp 242–248

[8] Wierzbicki T and Abramowicz W 2013 On the Crushing Mechanics of Thin-Walled Structures *J. Appl. Mech.* **50(4a)** pp 727-734

[9] Moore I D 1990 Influence of rib stiffeners on the buckling strength of elastically supported tubes *J. Solid Structures* **26(56)** pp 539-547

[10] Rees D W A 2009 *Mechanics of Optimal Structural Design: Minimum Weight Structures* (West Sussex: Wiley and Sons)

[11] Ewi A L and Kulak J T 2006 Large diameter fabricated tubes *Buckling of thin metal shells* Teng T J and Rotter J M (Eds) (Spon Press) pp 344

[12] http://www.oakleysteel.co.uk/pressure-vessel-steel-plate/astm-a516-boiler-plate, accessed 22 february 2015

[13] E/ECE/324 // E/ECE/TRANS/505, Regulation No. 34, Uniform provisions concerning the approval of vehicles with regard to the prevention of fire risks, 2003

A study of fluid flow simulation in convergent–divergent nozzles

I Olaru[1]
[1]"Vasile Alecsandri" University of Bacau, 157 Calea Marasesti, Romania

E-mail: ionelo@ub.ro

Abstract. A study of geometry nozzle is particularly important in optimizing the quantity of coolant in terms to minimize that quantity or coolant can be used for lubrication of machines moving parts. Coolants can be from the simplest (water, air) as well as a mixture of liquid (emulsions) to achieve a good cooling in function of temperature that develops in the area concerned. The study from this paper proposes an analysis and a simulation of flow through the convergent-divergent nozzle type to optimize them and to use a minimum quantity of coolant or lubricant. This study is focus on the analysis of cooling systems for cutting processes on machine tools.

1. Introduction

In industrial applications, coolants are used for different working situations. The cooling fluid is sent in the area to be cooled in particular by means of nozzles. A nozzle is generally a device that increases the velocity of a fluid flow at the expense of this pressure, the flow of fluid through nozzles may be regarded as adiabatic expansion. The nozzles have a special geometry to send a small amount of liquid but to have a good cooling effect [1].

Cooling capacity is a property of liquid cooling/lubrication and is due to evacuation of heat generated in the cutting process, helps to reduce of the temperature of the tool, of the piece and the whole cutting system.

With lubrication fluids are reduced friction forces between the chip and rake face and between the seating face and the machined surface. Cutting fluid is drawn on the chips surfaces with high adhesion forces, forming a film with large surface pressure resistance. Washing effect of cooling liquids, lead to the removal of small chips, metal dust, etc. [2].

Cooling liquids or lubricants used equation a special attention, a greater quantity of liquid consumed can be harmful environment, a larger quantity of fluid can generate high production costs, depending on their chemical composition and reaction with various heated zones of the process may be harmful to the operator [3].

Coolants must have the following properties: does not affect sealing parts of machine tool, to be stable and compatible with the materials used and the type of processing, should not be toxic, and not smoke in contact with cold zone and to be with low flammability [4].

The cutting surfaces lubrication can be achieved under continuous hydrodynamic film, semi-liquid friction and friction limit. In hydrodynamic lubrication regimes, the sliding surfaces are separated by a thick layer of fluid and are sufficiently high to prevent the contact between both surfaces [5].

Content from this work may be used under the terms of the Creative Commons Attribution 3.0 licence. Any further distribution of this work must maintain attribution to the author(s) and the title of the work, journal citation and DOI.
Published under licence by IOP Publishing Ltd

IOP Conf. Series: Materials Science and Engineering **95** (2015) 012048 doi:10.1088/1757-899X/95/1/012048

In industrial applications are used the following types of coolants and lubricants: water, air, oil, water-immiscible oils, fluids emulsifiable in aqueous solutions or organic compounds or inorganic and suspensions or solid lubricant pastes.

The emulsion is a system consisting of two immiscible phases or partially miscible with one phase dispersed in the other state. Water and oil phases can form type of emulsions with water as mainly fluid and oil as a secondary fluid or oil as mainly fluid and water like secondary fluid, however only the main fluid water emulsions can dilute the oil. Appropriate nozzle geometry could use a minimum quantity of coolant that can accurately be sent to the area to be cooled [6].

2. Design of nozzles

For simulations in this paper were considered four types of nozzles: the first type serves as the output surface of the working fluid a series of small diameter holes. In figure 1 was rated nozzle number 1 and is used mainly when cooling fluid with low viscosity. In the case of using a fluid with high viscosity or emulsion can be used nozzle geometries denoted by 2 and 3 in figure 1. The nozzle with number 4 is simple and can be found on many types of machine tools [7].

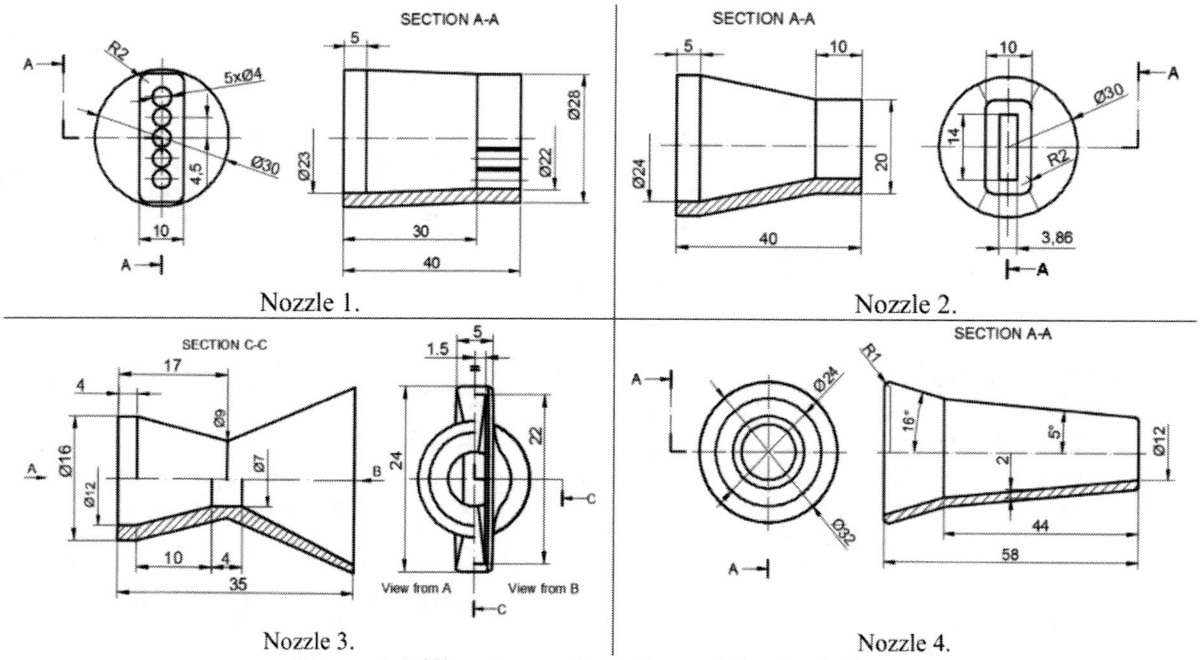

Figure 1. Different type of nozzles used for simulations.

Figure 1 shows the geometric dimensions for each nozzle, nozzle number 3 is a venture type being of convergent-divergent type, nozzle number 1 and 2 only shows a flattening of frontal area to be able to direct the fluid jet as accurately in the area to be cooled and to be lubricated if is necessary.

3. Fluid flow simulations

The simulations are performed using a specialized program with finite element which analysis the flow through the different nozzles with different operating parameters and different geometries of a convergent-divergent nozzle. These simulations can be compared to real working conditions on the machine tool system.

Mathematical model can begin from relation between pressure and velocity of fluid, in the nozzle the distances have small values the forces of gravity can be neglected compared to inertial forces and pressure forces and have a Bernoulli expression for the fluid [8]:

$$\frac{v^2}{2g}+\frac{p}{\gamma}+z=\text{Const.} \tag{1}$$

In what follows we have presented in figure 2 velocity distributions along the nozzle noted down for number 1 [9, 10].

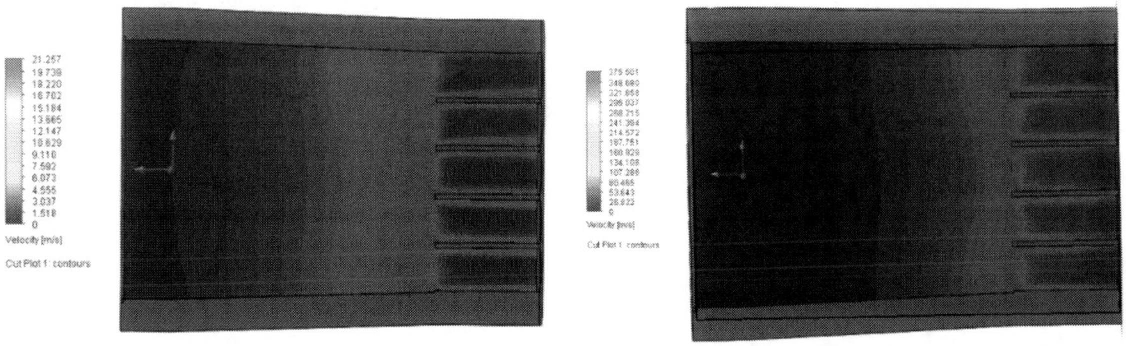

a b

Figure 2. Velocity variation along the nozzle no. 1 for two working fluids: a. water; b. air.

As can be seen in figure 2 the velocities of the working fluid increases in area in which we have holes for guiding the fluid jet. Velocities values are as is normal smaller for water than for air.

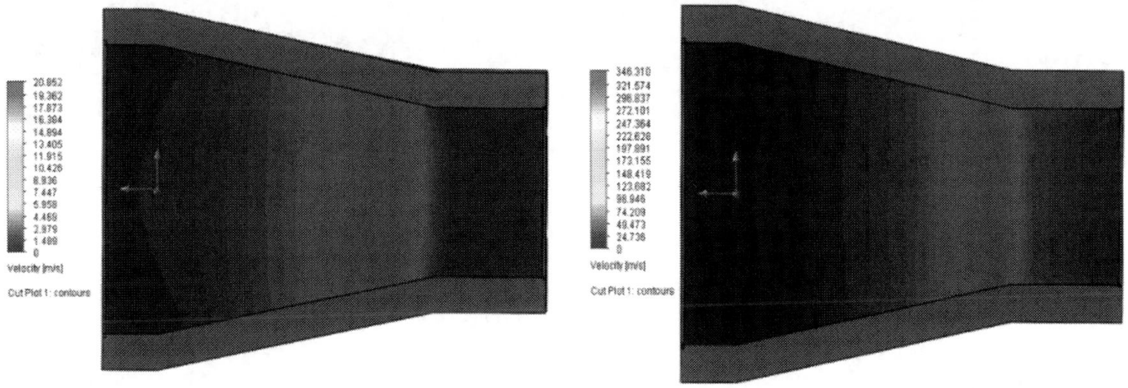

a b

Figure 3. Velocity variation along the nozzle no. 2 for two working fluids:a. water; b. air.

For simulations in this paper were chosen the two fluids due their simplicity, to their low cost and their different properties. Figure 3 presents diagrams velocities for nozzle number 2.

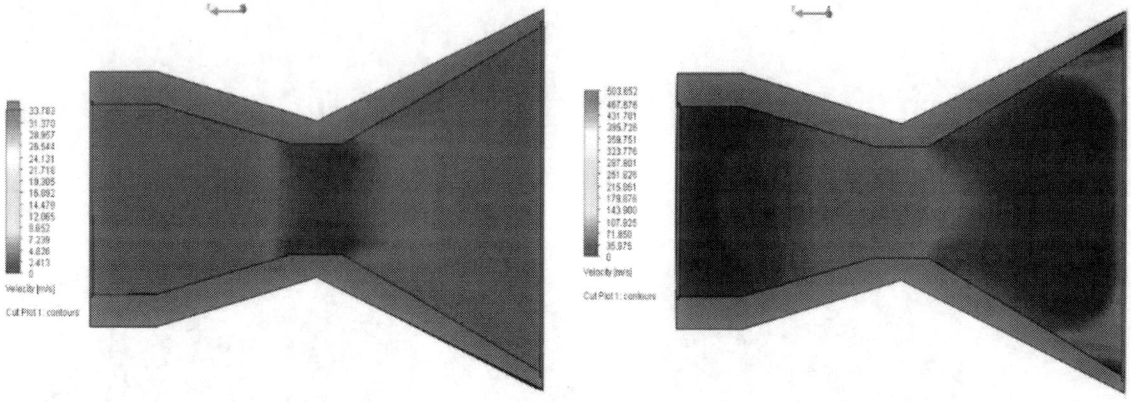

a b

Figure 4. Velocity variation along the nozzle no. 3 for two working fluids:a. water; b. air.

For this type of nozzle (figure 3) been renounced at the holes in the outlet of the fluid, practically in this area is performed only a flattening of the jet of fluid this will record higher values of velocities in this area.

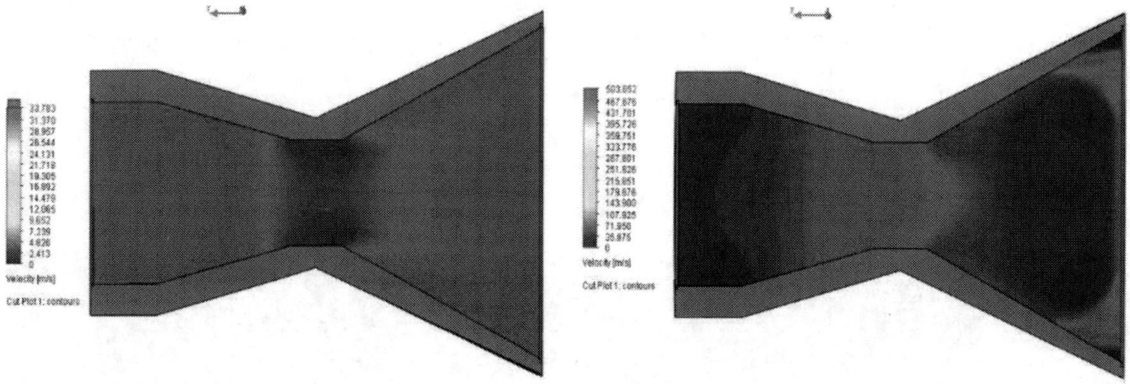

a b

Figure 5. Velocity variation along the nozzle no. 4 for two working fluids: a. water; b. air.

Figures 4 and 5 show a convergent-divergent nozzle and a simply convergent nozzle used for cooling with a wide range of coolants because it is not important the working fluid viscosity.

Simulations were performed using SolidWorks software, initial geometries were established for all four nozzles studied in this paper and all kinds of nozzles have main dimensions similar to be able to analyse better them. As input parameters for simulations we used the two coolants with their physical features, for all types of nozzles have be chosen 3 bar at input pressure, and at the exit of the nozzle was set the atmospheric pressure because the fluid jet is open to the atmosphere.

Figure 6 presents the variation of the velocities charts in usable length of nozzles, for better illustration were grouped by the working fluid used.

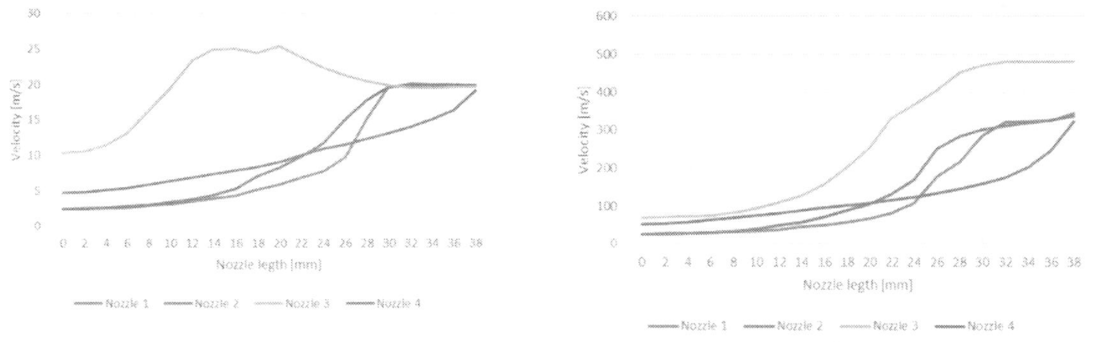

a b

Figure 6. Velocity analysis along the nozzle for two working fluids:a. water; b. air.

The nozzle 3 convergent-divergent type shows a graph a bit different from the other three charts because it has the middle area where the fluid velocity increases with decreasing pressure, which is more pronounced for water. The air does not have the same friction forces with wall, the chart resulted approximately as for other types of nozzles, registering an increase of air velocity in the exit area.

With computer software could simulate the flow inside the chosen nozzle were not taken into account other elements which are not relevant to the study such as: working fluid temperature, convection phenomena of fluid and wall nozzle, conduction phenomenon trough material, was considered a perfect fluid, wall roughness was neglected and load loses of nozzle because it has a relatively small length.

4. Conclusions

After simulations for several nozzle geometries can be concluded from the analysis of flow in the nozzle is very important, following this analysis can optimize the size and shape of the nozzles so that the cooling fluid to be distributed more efficiently in the processed area.

A special importance presents the choice of a cooling fluid depending on the geometry of the nozzle so as to eliminate hydraulic blockages that may arise and can lead to increased fluid pumping power and thus to a higher consumption of lubricant or cooling fluid.

Correct distribution of the fluid in the area leads to a lubricant saving, it no longer wasted, it is recognized that some coolant can be quite expensive, so choosing the right nozzle depending on the physical properties of the lubricant is also important.

Fluid jets are used in many industries; this paper aims to study the jets of fluid that pass through the nozzles particularly used in machine building industry, with those type of nozzles is performed cooling of machining zone. In this sense have been made many specific studies to research how fluid leaves the nozzle and how it wet in the processing area to achieve an equation cooling when is used a small quantity of lubricant.

Study the way in which the fluid flows through the nozzle by computer simulation. In mechanical cutting process is very important to study the cooling fluid jet using minimal quantity of fluid.

This analysis brings more clarification on the cooling fluid flow inside the nozzle and can be analyzing charts speeds, pressure or any analysis on how the vortex flow is formed and the influence of nozzle geometry on fluid distribution in the processed area.

References

[1] Kurnosov N E and Tarnopolskii A V 2007 Air–Liquid Lubricant and Coolant Aerosols in Machining *Russian Engineering Research* **27(10)** pp 689–691

[2] Nedelcu D-I and Olaru I 2013 Study of Cooling Parameters using Fluid Jet in Milling Process Simulation *Applied Mechanics and Materials* **371** (Switzerland: Trans Tech Publications) pp 524-528

[3] Ryzhkin A A, Shuchev K G, Aliev M M and Gusev V V 2008 Dissipative Properties of Lubricant and Coolant Fluid in Cutting and Friction *Russian Engineering Research* **28(12)** pp 1243–1247

[4] Rozati A and Tafti D K 2008 Effect of coolant–mainstream blowing ratio on leading edge film cooling flow and heat transfer – LES investigation *International Journal of Heat and Fluid Flow* **29** pp 857–873

[5] Olaru I 2013 The Fluid Flow Simulation through to a Convergent Nozzle *Journal of Engineering Studies and Research* 19(2) pp 76 - 80

[6] Liu J, Han R, Zhang L and Guo H 2007 Study on lubricating characteristic and tool wear with water vapor as coolant and lubricant in green cutting *Wear* **262** pp 442–452

[7] Olaru I 2013 The Fluid Flow Simulation through to a Venturi Nozzle *Journal of Engineering Studies and Research* 19(1) pp 42 - 46

[8] Florescu I, Florescu D and Olaru I 2003 Fluid mechanics and hydraulic machines *Laboratory Guidance* (in romanian) (Chişinău: Editura Tehnica Info)

[9] Information on SolidWorks: http://www.solidworks.com/sw/resources/solidworks-tutorials.htm. Accessed 19/01/2015

[10] Information on SolidWorks Tutorials: http://www.solidworkstutorials.com. Accessed 23/02/2015

Torsional stress analysis of the squirrel-cage rotor for three phase asynchronous motors

T Axinte[1] and C Nutu[1]
[1]Constanta Maritime University, Faculty of Naval Electro-Mechanics, 104 Mircea cel Batran Street, 900663, Constanta, Romania

E-mail: tibi_axinte@yahoo.com

Abstract. The paper aims to determine the torsional stress occurring in the rotor's shaft of the three phase asynchronous motors for a given rotational speed of the rotor. The three phase asynchronous motor is started using the main starting variants, each variant having its own torque characteristic which produces a certain maximum shear stress. The model uses several hypotheses regarding the geometry of the domain, the dynamic behavior, the friction phenomenon and the mass of the shaft which is disregarded in a model and it is considered in a more accurate model. The maximum shear stress is produced by the shock which occurs in the starting process of the motors. This maximum stress may lead to the failure of the rotor either because it breaks, or because of the fatigue of the material produced by frequent starting of the motor. Information resulted from this study are useful for the selection of the appropriate material, of the according geometry of the design and elaboration of the specifications for the in-service use of the motor.This theoretical study will be followed by the development of more accurate models which will also include experimental studies employed to calibrate the theoretical models and to verify their accuracy.

1. Introduction

The asynchronous electrical machines, also called three phase induction machines are the most frequently used alternative current electrical machines. The term "asynchronous" comes from the fact that the frequency of the stator field is slightly higher than the frequency of the rotor. These asynchronous machines can work both, as a generator and as well as motor. The main advantages of three-phase induction motors are the simplicity of their construction which induces advantageous cost prices and their safety in exploitation [1].

Figure 1. Three phase asynchronous electrical motor (induction motor).

Figure 2. Rotor with two shafts.

Content from this work may be used under the terms of the Creative Commons Attribution 3.0 licence. Any further distribution of this work must maintain attribution to the author(s) and the title of the work, journal citation and DOI.
Published under licence by IOP Publishing Ltd

One main characteristic of the induction motors is that their speed slightly reduces depending on increasing applied loads. That is why the asynchronous motors are generally used where the rotation speed must not vary with the applied load, such as for usual drilling, milling or hard machining mechanical tools, some of the lifting devices etc. [2].

- The main starting variants of the three phase asynchronous motors are:
- direct on-line start variant (DOL);
- the start variant using a star-delta connection, that is extensively used in practical conditions;
- the starting variant using a soft starter which is a modern, electronic alternative to star-delta starting variant for continuous and stepless motor start;
- the starting variant using a frequency inverter, that is the most expensive and modern variant, an important advantage being the fact that it enables a controlled, stepless motor start with rated load-torque.
- Each of the aforementioned starting methods has a certain torque characteristic and a certain relative starting torque determining a distinct torsional stress for each starting version.
- Among the main hypotheses we remind the most important ones, such as: the rotor's shaft has a cylindrical form of diameter D and moment of inertia J, and its steady state rotational speed is n.
- It is also assumed that the motor starts to rotate instantaneously with the angular speed ω. This means that the motor starts to rotate immediately, in this very short time being assumed that no heating due to the friction is produced. Therefore, the entire transmitted kinetic energy is transmitted to the rotor's shaft and transformed in strain energy.
- The maximum torsional stress in the shaft is produced by the shock conditions and they are calculated using two hypotheses. The first one takes into account the mass of the shaft and in the second hypothesis, the mass of the shaft being neglected.

The shock torsion produces high stresses which lead to the occurrence of cracks or even the shock breaking of the shaft. Moreover, the development of the cracks leads to the fatigue of the shafts' material.

2. Theoretical aspects

The following figures are presenting the ensemble induction motor-shaft-flywheel, viewed only from two different angles.

Figure 3. The assembly flywheel – left side – frontal view.

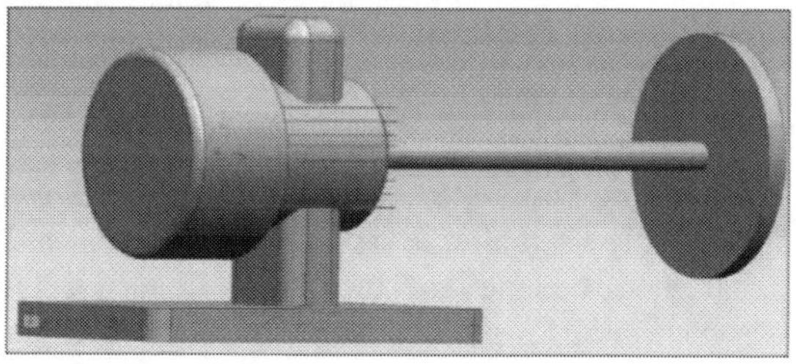

Figure 4. The assembly motor-shaft-flywheel: right motor-shaft-side – back view.

The working assumption used in the paper is that the whole amount of the kinetic energy of the flywheel is transmitted by means of deformation energy of the shaft.

Consequently we shall equal the amount of kinetic energy of the flywheel [3]:

$$W = \frac{1}{2} \cdot J \cdot \omega^2 \tag{1}$$

with the deformation energy of the shaft:

$$W = \int \frac{M_t^2}{2 \cdot G \cdot I_p} dx \quad . \tag{2}$$

In order to express this deformation energy of shaft depending on the shear stress τ_{max}, at the end of the shaft's section, we use the following formulas, thus eliminating the value of torque, M_t from the deformation energy.

$$\tau_{max} = \frac{M_t}{W_p} = \frac{M_t \cdot d}{2 \cdot I_p} \tag{3}$$

The twisting moment is

$$M_t = \frac{2 \cdot \tau_{max} \cdot I_p}{d} \tag{4}$$

and the new form of the energy, [4], is

$$W = \frac{1}{2 \cdot G \cdot I_p} \int_0^l \left(\frac{2 \cdot \tau_{max} \cdot I_p}{d} \right)^2 dx = \frac{2 \cdot \tau_{max}^2 \cdot l}{G \cdot d^2} \tag{5}$$

In our case:

$$I_p = \frac{\pi \cdot d^4}{32} \tag{6}$$

and

$$\frac{\pi \cdot d^2}{4} l = V \quad . \tag{7}$$

It results:

$$W = \frac{2 \cdot \tau_{max} \cdot \dfrac{\pi \cdot d^4}{32} \cdot l}{G \cdot d^2} = \frac{1}{2} \cdot \frac{\tau_{max}}{2 \cdot G} \cdot V = \frac{\tau_{max} \cdot V}{4 \cdot G} \quad . \tag{8}$$

By equating the both, kinetic and deformation energy

$$\frac{1}{2} \cdot J \cdot \omega^2 = \frac{\tau_{max}^2}{4 \cdot G} \cdot V \quad , \tag{9}$$

it results the maximum shear stress τ_{max}:

$$\tau_{max} = \omega \cdot \sqrt{\frac{2 \cdot J \cdot G}{V}} \tag{10}$$

(10),

with the formula of the moment of inertia, J, for a disk-shaped flywheel having the weight N:

$$J = \frac{N \cdot D^2}{8 \cdot g} \quad . \tag{11}$$

3. Results and discussions

Using many values N, of the weight of the disk-shaped flywheel with different diameters, as in the following table:

Table 1. The N and D values.

No	N	D
	[kg]	[cm]
1	10	20
2	15	25
3	20	30

```
>// If we have the following values:
>N = [10, 15,20]; // [kg]
>D = [20, 25, 30]; // [cm]
>g = 981;  // [cm/s^2]
>// Determination of the mass moment of inertia of a disk-shaped flywheel diameter D;
>J = (N*D^2)/(8*981) // [kg*cm*s^2]
  [0.509684,  1.19457,  2.29358]
```

Figure 5. Determination of the mass moment of inertia.

Supposing, in our case, that the diameter of the shaft is d=5 cm having the length, l=3 m, in the blocking moment of the flywheel, the rotation speed was n=150 rpm [5].
The corresponding angular speed is:

$$\omega = \frac{2 \cdot \pi \cdot n}{60} = \frac{2 \cdot \pi \cdot 150}{60} = 5\pi \quad \text{[rad/s]} \tag{12}$$

The volume is:

$$V = \frac{\pi \cdot d^2}{4} l = \frac{\pi \cdot 5^2 \cdot 300}{4} = 1875\pi \quad [\text{cm}^3] \tag{13}$$

The maximal shear stress for this flywheel can be determined:

```
>// Determination of the maximum shear stress of a disk-shaped flywheel diameter D:
>Tmax(1) = ((((2*810000*246.49*J1)/5887.5)^(1/2))/(9.80665)) // [N/cm^2]
  [18.9593,  29.0253,  40.2187]
```

Figure 6. Determination of the maximum shear stress.

The shock-torsion phenomenon for shafts located between the motor and flywheel has been also studied using the finite element method and analysis, based on performant software.

As it can be seen also in the following drawings, the most torsion-stressed element is the shaft, the von Mises stresses at the flywheel, occurring only at the join between the shaft and flywheel.

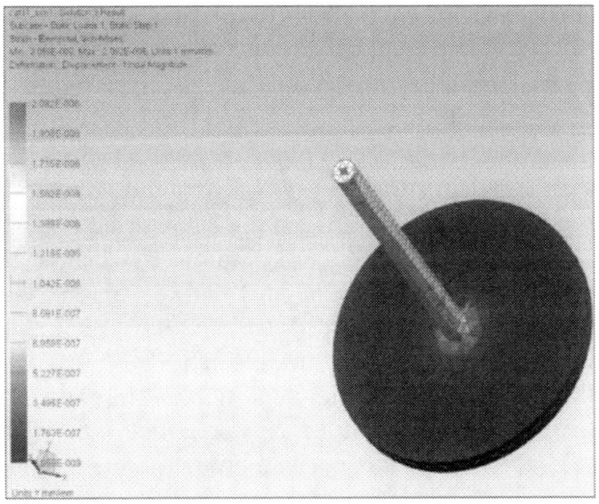

Figure 7. Shaft-flywheel assembly at a reduced torsional stress.

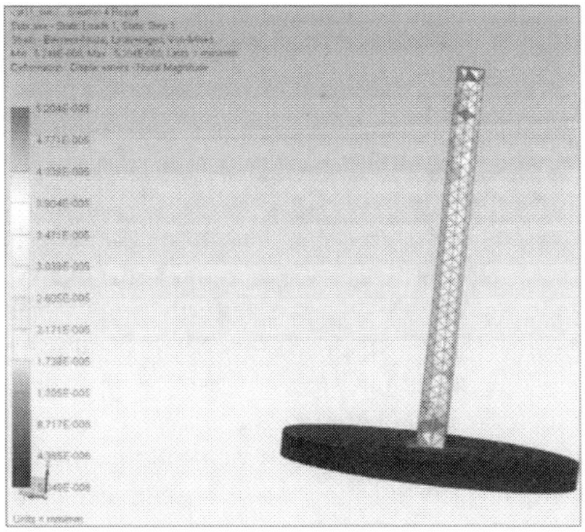

Figure 8. Shaft-flywheel assembly at a higher torsional stress.

Since the flywheel is considered to be as the part of a system that suddenly blocks, then the flywheel suddenly stops, as well.

As it is being shown and can be observed in figure 7 and figure 8, when suddenly halting the flywheel, the intuitive fact that can be remarked, is that the stresses do not occur on the surface of the flywheel, but only in the joint surfaces between the shaft and the flywheel. These stresses are being transmitted, further on, in the shaft body.

The formula used in the present paper for estimation of the maximal shear stress represents a good approximation of the maximal shear stress values occurring in fact, when the shaft-flywheel ensemble suddenly blocks. This simplified formula used is modeling the reality of the torsion phenomenon that occurs in the analyzed ensemble.

Analyzing this approximation and simplified used formula of the maximal shear stress, it results that the main factors determining the values of the maximal shear stresses (τ_{max}) are to be controlled either by modifying the geometry and material of the studied ensemble, or by controlling the initial rotation speed of the ensemble, which is the speed of the motor used to drive the ensemble.

Thus, for certain geometry of the ensemble, this time driven by a rotation speed multiplied by a certain factor "n", in order to obtain about the same value of the maximal shear stress (τ_{max}), the material which has to be used in this second considered alternative, is to have a density which is of about n^2 times lesser than the density of the material initially used in the first alternative.

Thus, the problem of the maximal shear stress occurring in a certain ensemble with different possible geometries becomes either a problem of the discipline of the strength of materials or a problem of the optimization of the geometry of the analyzed ensemble.

In practical engineering the problem of shear stresses is solved by treating these both problems and since the spectrum of the materials used is somehow both limited, at one hand by the materials presently known and on the other hand by their costs, the geometry of the ensemble is the easier part to be controlled by using appropriate software programs.

4. Conclusions

In practical conditions the effects of the shock torsion consist of non-uniform distributed shear stresses occurring on the body of the shaft, figure 7 and figure 8. These maximum shear stresses may produce the deformation and even breakage in some areas of the shaft.

Usually, under the torsional shock conditions the arm between the electrical motor and flywheel is the first which collapses. There are most uncommon cases when the rotor of the electrical motors breaks.

While for the second case, the result is consisting mainly in destroying only the electrical motor by its rotor's collapse, for the first case of arm failing, the entire ensemble (electrical machine, arm and flywheel) could be destroyed, as a result of the breakage.

In order to avoid this, the dimensioning of the shaft should be done always bearing in mind that the stresses occurred are to be minimal, thus calculating them considering the proper safety factors. The paper also provides some initial instruments for the assessment of the stresses in the flywheel-shaft-motor assembly that will be further on used for comparison with the results provided by the models to be developed using the Euler Math Toolbox software and Femap software.

The maximum stresses resulted from the study must be analyzed using the fatigue of materials theoretical background in order to minimize the risks in operating conditions. Moreover, by considering the factors which modify the endurance limit, such as: stress concentrators, size of the part, surface finishing, other miscellaneous influences, there may be conceived products with a longer life in service.

Another concern regards the shocks which may produce large contact stresses that are generating cracks and local high temperatures caused by friction and by the operating conditions. All these joined effects, together with the temperature in steady state conditions may lead to lower values of the material constants and ultimate stresses, the final effect being an early fail of the assembly. The worn supports may lead to vibrations which may also produce shock forces, leading to increased values of the stresses.

To conclude, the paper is an initial analysis which offers some insights regarding the basic phenomena occurring in the electric motor's assembly at this point being approached only from the point of view of the phenomena analytically modeled using the strength of materials theory. In this way, the paper presents a mechanical engineering application in the electrical engineering field.

The research will be continued with more accurate models which will take into account several aspects of the mechanical and electrical phenomena, specific to the electrical machines. The research will also include some experimental studies employed to calibrate the theoretical models and to verify the accuracy of their results. Therefore, it is possible to finally create a hybrid model which includes theoretical (analytical and numerical) aspects as well as experimental studies, [6].

References

[1] *API 684* First Edition1996 An Introduction to Lateral Critical and Train Torsional Analysis and Rotor Balancing
[2] Fitzgerald A E, Kingsley C Jr and Umans S D 2004 *Electric Machinery* **6** (McGraw-Hill)
[3] Buzdugan G 1970 *Rezistenta materialelor* (Bucharest: Tehnica Publishing House)
[4] Voinea R et al. 1989 Introducere in mecanica solidului cu aplicatii in inginerie (Romanian: Academy Publishing House)
[5] Chiasson J 2005 Modeling and High-Performance Control of Electric Machines (online edition: Wiley)
[6] Oanta E, Panait C, Axinte T and Dascalescu A E 2014 Instrumente software originale folosite ca interfețe în cadrul modelelor hibride din ingineria mecanică *AGIR* **4**

IOP Conf. Series: Materials Science and Engineering **95** (2015) 012050 doi:10.1088/1757-899X/95/1/012050

Design and realization of light modules for miniature model passenger coaches

M P Ursu[1], **T Buidos** [1], **G Albert** [2]
[1]University of Oradea, Department of Machine Manufacturing Technology, str.Universităţii nr.1, 410087, Oradea, Romania
[2]Tehnologistic SRL, str.Libertăţii nr.35A, 407035, Apahida, Romania

E-mail: mpursu@uoradea.ro

Abstract. This paper shows the main steps to design and realize an original light module for miniature model passenger carriages, with the possibility to switch on and off individual lights and to operate the lights according to certain patterns, which improves greatly the visual appearance of miniature passenger trains. This continues the previous work of the authors concerning the model railroading and the implications of nonconventional technologies in this field of activity.

1. Introduction

Nowadays, hobbies have become very important human activities, and the fabrication of hobby items requires high-tech designing, engineering and machining, performed by high-skilled and carefully trained operators. The miniature train hobby items, such as locomotives, wagons, railroads and accessories, have become very complex and they are now produced in world renowned factories at astoundingly high quality levels, using cutting-edge technologies, machines and software, both conventional and nonconventional. So, the model railroading hobby has now turned into a very serious and important activity from the point of view of both manufacturers and railway enthusiasts.

Due to their increasingly complex features, the miniature model trains resemble more and more to their real-life counterparts and they behave like them. The new locomotives are fitted with direction-dependent headlights, sounds, conventional or dynamic smoke generators, spark simulators, functional couplers etc, and the passenger carriages are fitted with interior lighting, tail lights, functional doors, sounds and other features that increase their visual and functional appearance. These modern trains and their components can be individually commanded by means of the digital command control (DCC) system [5].

The light modules are very important for the appearance of the passenger coaches (figure 1). First, there were no lights at all, then one or more incandescent bulbs were mounted on the ceiling, electrically connected to the wheel collectors. These lights were permanently on as long there was voltage at the rails, and their intensity depended greatly on the voltage value. So the light increases when the train speeds up, it decreases when the train slows down, and goes off when the train stops, which is not life-like. The modern digital light modules, as the one described in this paper, contain LED's which can be individually switched on and off and their light intensity does not depend on the train speed. These light modules can be commanded by means of the DCC system, but they can also

Content from this work may be used under the terms of the Creative Commons Attribution 3.0 licence. Any further distribution of this work must maintain attribution to the author(s) and the title of the work, journal citation and DOI.
Published under licence by IOP Publishing Ltd

be used on analog railways. However, in order to fully use their capabilities, the DCC system is more suitable.

Figure 1. Passenger carriage, with interior lighting and tail lights.

2. The Shine Plus Maxi light module for passenger carriages

Briefly, the light module (figure 2) consists of an 8-bit ATMEL microcontroller unit [6], which receives, decodes and executes the DCC commands, the adequate peripheral LED drivers and the white LED's, which can be individually commanded – switched on/off, dimmed, flickered etc.

Figure 2. The Shine Plus Maxi light module.

All the components of the light module are surface-mounted-devices (SMD), and they are placed on a double-plated printed circuit board (PCB). Adequate conductive pads are designed for the connections to the wheel current collectors and to the external outputs, such as cockpit lights, tail lights or other additional light modules, individual or not.

As all other DCC devices, the light module has an individual address which is set by the user via the digital command station. Thus, each passenger carriage can be addressed individually, in order to switch on/off all the lights or any of the LED's, tail lights or any other additional lights. Furthermore, if programmed accordingly, the lights can emulate the behavior of good or defective neon lights, can be individually dimmed or flickered and they can be automatically operated following a random pattern. These features increase the life-like appearance of the miniature passenger carriages that roll along the railway layout, as if individual travelers operate the lights of their compartments.

In order to avoid the voltage dropouts caused by imperfect electrical contacts between rails, wheels and current collectors, the light modules are fitted with external buffer electrolytic capacitors, which are connected by the user to the PCB by means of adequate conductive pads.

This light module can be operated on digital or analog railway systems. By adequate programming, the digital operation mode can be partially emulated on analogue systems, but the features of this light module can be fully used on digital railway systems.

3. Design and realization of the Shine Plus Maxi light module

Due to the complexity of the Shine Plus Maxi light module, the design of the electronic diagram and PCB was carried out by means of TVision software, an internal application of Tehnologistic [7] (figure 3, figure 4). The microcontroller was programmed in compliance with the National Model Railroad Association (NMRA) [8] and DCC requirements, and its software can be updated whenever necessary without the need to dismount the module from the passenger carriage. The outputs of the

IOP Conf. Series: Materials Science and Engineering **95** (2015) 012050 doi:10.1088/1757-899X/95/1/012050

microcontroller are connected to the LED drivers that supply the required voltages and currents for the 14 low-current high-brightness white LED's of the light module and for the two additional outputs. All electronic components are SMD type.

The PCB's are double-layered, copper plated, and are manufactured by a partner company according to the TVision project of Tehnologistic. Modern technologies are used, such as photo-etching, galvanizing, tampo printing, computer-aided drilling and cutting etc, so that the resulted PCB's can be taken over by the computer-driven pick-and-place machines that assemble the SMD electronic components according to the design.

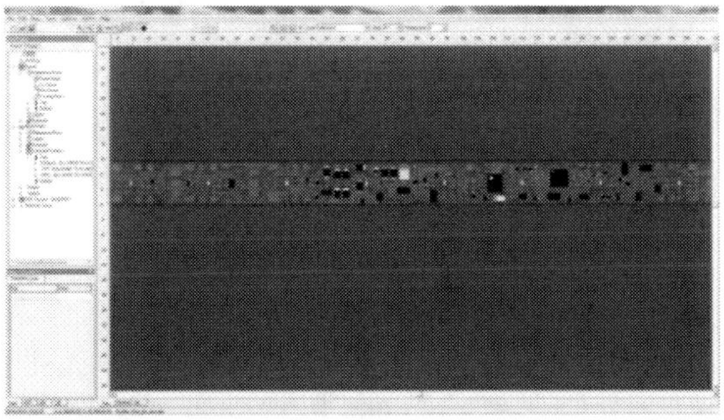

Figure 3. Design of Shine Plus Maxi light module.

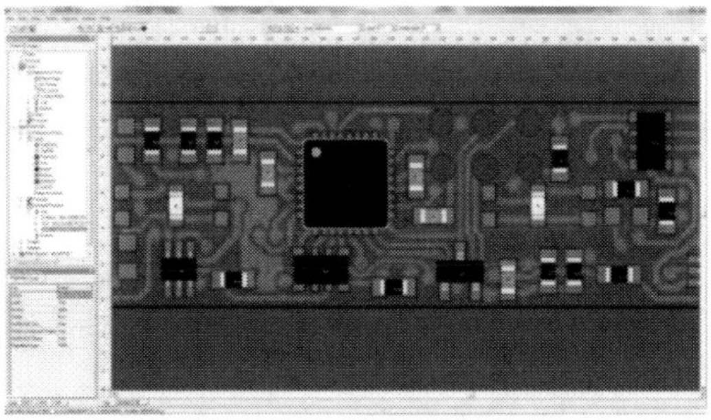

Figure 4. Design of Shine Plus Maxi light module (detail).

According to the project, the solder paste is deposited on the soldering pads of the PCB's by means of a computer-driven Sony SiP850 paste printing machine (figure 5), then the pasted pads are optically checked (figure 6).

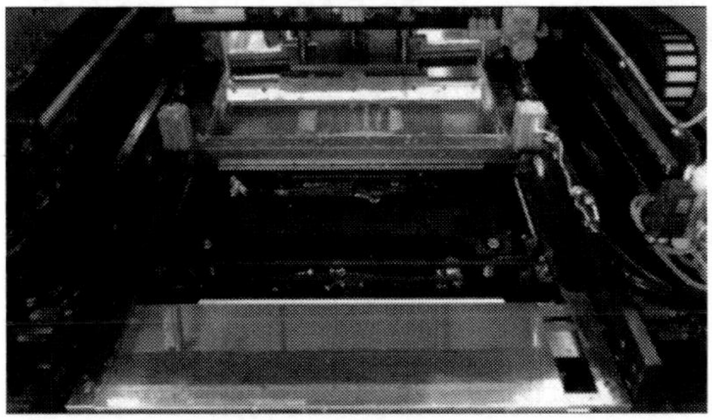

Figure 5. Sony SiP850 machine for solder paste printing.

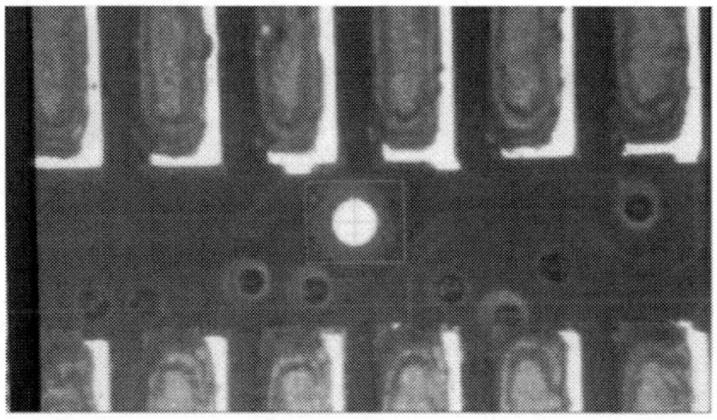

Figure 6. Vision system for optical checking of the pasted pads of PCB.

Figure 7. Siemens Siplace robot, picking up SMD devices from feeders.

Figure 8. Siemens Siplace robot, placing SMD components on pasted PCB's.

Siemens Siplace robots (figure 7) pick up the SMD devices from their feeders and place them on the pasted PCB's (figure 8). Then automated conveyors (figure 9) carry the PCB's to the Rehm Nitro reflow oven (figure 10).

Figure 9. Automated conveyor.

Figure 10. Computer software for the Rehm Nitro reflow oven.

The firmware of the Shine Maxi Plus light modules was developed by means of WinAVR application [9], with Eclipse Integrated Development Environment (IDE) [10] and AVR plug-in [11]. After the completion, the firmware is loaded into the microcontrollers, and the finished light modules are tested by means of a custom self-made tester, according to the pads and functionality of the batch that has been ordered and manufactured (figure 11). The properly functioning modules are then wrapped and sent to the customers.

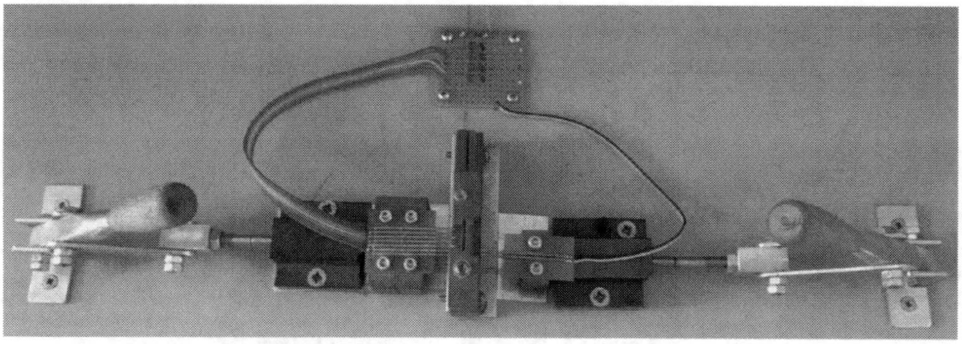

Figure 11. Custom tester for the digital modules.

The design of the Shine Plus Maxi light module features alternate soldering pads for the LED's at 3mm pitch, in order to allow the experienced users to fine-tune the positions of the LED's according to the specific passenger carriage to be illuminated (figure 12). By means of adequate soldering equipment and skills, the LED's can be removed from the factory default positions in order to be re-soldered on the pads of the alternate positions, so that the light module would fit better in the specific passenger carriage.

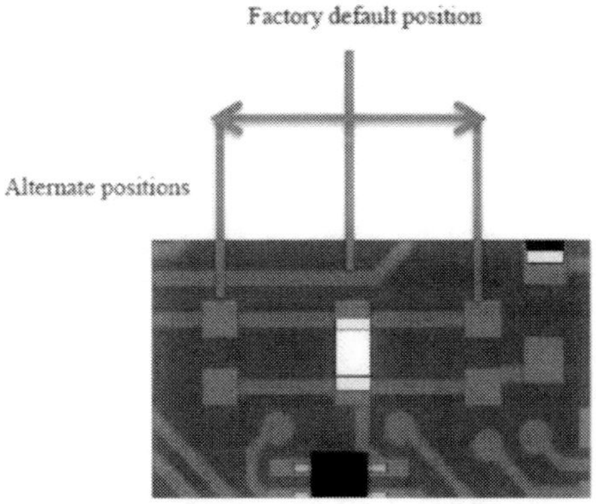

Figure 12. Alternate positions of the LED's on the light module [12].

4. Conclusions

The modern technologies, both conventional and nonconventional, provide quality, accuracy, reliability and reproducibility of the products, so they proved themselves to be very useful for the manufacturers of model railroading items. The railway enthusiasts appreciate greatly the true-to-life miniature models of the real trains, which nowadays can be manufactured at much higher standards than before. The realistic appearance of the model trains is enhanced by additional functional features that can be individually addressed via the DCC system, and one of the most spectacular additions is the interior lighting of miniature passenger carriages. The life-like behavior is achieved by the complex design of the light modules, and the manufacturing requires cutting-edge technologies, machines and software. Further developments of the Shine Maxi Plus light module will be carried out in the future, as the firmware will be periodically updated, and new hardware features will be implemented as the project will be adapted to the requirements of the customers.

References

[1] Ursu M P and Buidoş T 2009 Applications Of Non-Conventional Technologies In Model Railroading *Nonconventional Technologies Review* **1** pp 29-35
[2] Ursu M P, Buidoş T and Iosif D I 2011 Construction Of A Model Railway Building By Means Of Nonconventional Technologies - Case Study *Nonconventional Technologies Review* **3** pp 57-62
[3] Ursu M P, Buidoş T and Albert G 2012 Conventional And Nonconventional Methods In Elaboration Of Digital Decoders For Miniature Trains *Nonconventional Technologies Review* **4**
[4] Ursu M P, Buidoş T and Albert G 2013 Design And Realization Of A Sound Module For A Miniature Model Steam-Locomotive *Advanced Materials Research* **837** pp 500-505
[5] http://en.wikipedia.org/wiki/Digital_Command_Control
[6] http://www.atmel.com/products/microcontrollers/avr/
[7] http://tehnologistic.ro/
[8] http://www.nmra.org/
[9] http://sourceforge.net/projects/winavr/
[10] https://eclipse.org/ide/
[11] http://avr-eclipse.sourceforge.net/wiki/index.php/The_AVR_Eclipse_Plugin
[12] http://www.train-o-matic.com/downloads/manuals/tOm_Shine_Plus_Maxi_Digi_en_v_0.0.9.pdf

Reduction of vibration by using mechatronical subsystem

K Białas[1] **and A Buchacz**[1]
[1]Silesian University of Technology University, Faculty of Mechanical Engineering, Institute of Engineering Processes Automation and Integrated Manufacturing Systems, Konarskiego 18A, 44-100 Gliwice, Poland

E-mail: katarzyna.bialas@polsl.pl

Abstract. The primary aim introduced in this paper is synthesis of mechatronical system understand as planning of this type of systems. Mechatronical system is consisted of fundamental mechanical system and subsystem reducing vibration including electric elements. Fundamental system is received applying reverse task of dynamic (synthesis) and it's including inertial and elastic elements. The subsystem includes electric elements by means moving-coil transducer. The synthesis can also be used to change the already existing systems. Due to the method, introduced in this work, may be performed as early as whilst the designing of future functions. Using this way of designing is support for designers of mechanical systems with active reducing of vibrations.

1. Introduction

During designing a physical object it is needed to take into Investigation many aspects which can influence its process at a later stage. Using the correct modelling of a physical object facilitates the respective optimisation of its project as early as at the design step. At the starting of a design procedure the construction engineer may obtain or take the parameters and texture of the system meeting expected necessities with reference to dynamic characteristics or, in the event of an already existing system – design a system eliminating unexpected phenomena.

There are many procedures of decreasing pulsations. We can split them into passive, semi-active and active approaches. The major dissimilarity between this way of reducing vibration is application passive or active elements to accomplish them. In passive reduction of vibration are used passive elements and their parameters cannot change in time. In this way of reducing vibration energy is disperse or store for some time.

In semi-active way of reducing vibration are used passive elements too. Change between passive and semi-active method is that components in semi-active method can be chance in time. In active way of reducing vibration are used active elements which can chance their parameters in time and energy is equalize by outside supplemental sources. Active elements can be realized by using mechanical, pneumatic, hydraulic, electromagnetic or electro-dynamic elements [1-4].

Shown in this work non-classical approach of designing of mechatronic systems is based on synthesis of these systems. The synthesis is inverse task of dynamics and result of it is receive texture and parameters of system. The task is inverse to analysis because we can obtain system which comply with prerequisite concern the values of poles and zeros or frequency of system. This method of design it is possible to use in existing system and modified its. In case of synthesis of mechatronic system we divide this procedure on two major stages. The first stage is linking with fundamental mechanical

Content from this work may be used under the terms of the Creative Commons Attribution 3.0 licence. Any further distribution of this work must maintain attribution to the author(s) and the title of the work, journal citation and DOI.
Published under licence by IOP Publishing Ltd

passive system build only from inertial and elastic elements. In the second stage we choose way of reducing unwanted vibration and obtain structure and parameters of this subsystem.

The issue presented on this paper concerns introducing reverse task of mechatronic system with electric components as realization of active components to reducing of vibrations. Electric elements are implemented through the moving-coil transducer [1-9].

2. Research problem and method

In an attempt to obtain a system containing active elements, it is necessary to carry out the synthesis of the characteristic function or the identification of the passive system. Thereafter, depending on the texture and parameters of this system and enforcements acting on this system, the structure of the system with active elements and the values of forces generated by these elements are chosen.

To receive structure and parameters of passive fundamental mechanical system which achieve the requirements to deal with spectrum of frequency it is possible to use one of the methods of synthesis. The characteristic function of system may be define by means mobility of slowness (equation (1)-(4)), which can be shown by means of rational functions of the following forms [1,10,11].

- slowness of semidefinite systems:

$$U(s)=H\frac{d_l s^l+d_{l-1}s^{l-2}+\ldots+d_1 s}{c_k s^k+c_{k-1}s^{k-2}+\ldots+c_0} \tag{1}$$

- slowness of restrained systems:

$$U(s)=H\frac{d_l s^l+d_{l-1}s^{l-2}+\ldots+d_0}{c_k s^k+c_{k-1}s^{k-2}+\ldots+c_1 s} \tag{2}$$

where: l – odd degree of numerator, k – degree of denominator, l-k = 1, H – any positive real number;

- mobility of semidefinite systems

$$V(s)=H\frac{c_k s^k+c_{k-1}s^{k-2}+\ldots+c_0}{d_l s^l+d_{l-1}s^{l-2}+\ldots+d_1 s} \tag{3}$$

- mobility of restrained systems

$$V(s)=H\frac{c_k s^k+c_{k-1}s^{k-2}+\ldots+c_1 s}{d_l s^l+d_{l-1}s^{l-2}+\ldots+d_0} \tag{4}$$

where k – odd degree of numerator, l – degree of denominator, k-l = 1, H – any positive real number.

There are two main types of synthesis of passive mechanical systems. One of them is decomposition of a characteristic function into a continued fraction. This characteristic function can be in the form of mobility or slowness (equation (5)). We obtain systems of the cascade structure (semidefinite or restrained) [1, 10, 11].

$$U(s)=\frac{c_1}{s}+m_1 s+\cfrac{1}{\cfrac{s}{c_2}+\cfrac{1}{m_2 s+\ldots+\cfrac{1}{\cfrac{s}{c_n}+\cfrac{1}{m_n s}}}} \tag{5}$$

The next method of synthesis of passive mechanical systems is decomposition of a characteristic function in the form of slowness into partial fractions (equation (6)). As a result of this type of synthesis (decomposition into partial fractions) we obtain systems of the branched structure (semidefinite or restrained) [1, 10, 11].

$$V(s)\frac{1}{H}=\frac{s}{c_1}+\cfrac{1}{m_1 s+\cfrac{1}{\cfrac{s}{c_2}+\cfrac{1}{m_2 s}}+\ldots+\cfrac{1}{\cfrac{s}{c_n}+\cfrac{1}{m_n s}}} \tag{6}$$

After first step of synthesis of mechatronical system one takes structure and parameters of mechanical system – only parameters of inertial and elastic elements. In the second step one determinates external excitations acting on basic system and select elements reducing vibrations. One of the opportunities is active elements. Polar graph of system obtain as result of synthesis with active elements is present in figure 1.

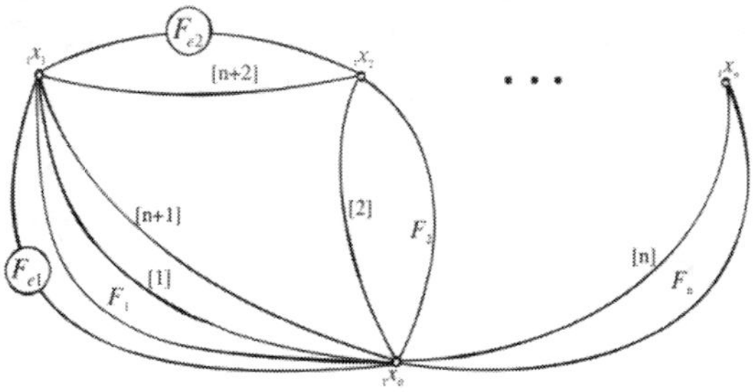

Figure 1. Polar graph of system after synthesis with active elements.

Next step is connected with choosing structure with active element or elements and calculate values of this elements. By solving the system of equations presented in the matrix form (equation (7)), one obtains the values of forces generated by active elements F_{e1}, F_{e2}, …, F_{en}.

$$\begin{bmatrix} K_{1,1}=\dfrac{\partial D(\omega)}{\partial([2],[3]\ldots[n])} & K_{1,2}=-\underset{z}{Sim}(K_{1,1};K_{2,2}) & \cdots & K_{1,n}=-\underset{z}{Sim}(K_{1,n-1};K_{2,n}) \\ K_{2,1}=-\underset{z}{Sim}(K_{11};K_{22}) & K_{2,2}=\dfrac{\partial D(\omega)}{\partial([1],[3]\ldots[n])} & \cdots & K_{2,n}=\underset{z}{Sim}(K_{2,n-1};K_{3,n}) \\ \vdots & \vdots & \vdots & \vdots \\ K_{n,1}=-\underset{z}{Sim}(K_{n-1,1};K_{n,2}) & K_{n,2}=\underset{z}{Sim}(K_{n-1,2};K_{n,3}) & \cdots & K_{n,n}=\dfrac{\partial D(\omega)}{\partial([1],[2]\ldots[n-1])} \end{bmatrix} \cdot \begin{bmatrix} A_1 \\ A_2 \\ \vdots \\ A_n \end{bmatrix} - \begin{bmatrix} F_1 \\ F_2 \\ \vdots \\ F_n \end{bmatrix} = \begin{bmatrix} F_{e1} \\ F_{e2} \\ \vdots \\ F_{en} \end{bmatrix} \tag{7}$$

where

$D(\omega)$ – characteristic equation,

$\dfrac{\partial D(\omega)}{\partial[1]}$ – derivative of structural number in relation to the edge [1],

$\underset{z}{Sim}\left(\dfrac{\partial D(\omega)}{\partial[1]};\dfrac{\partial D(\omega)}{\partial[2]}\right)$ – function of simultaneity of structural number, the inverse image of which contains two oriented edges [1] and [2],

$F_1, F_2, \ldots, F_n$ – excitations,

$F_{e1}, F_{e2}, \ldots, F_{en}$ – excitations generated by active elements.

Possible solution of the physical implementation of active elements is the application of electric elements e.g. the use of a moving-coil transducer (figure 2) [12-15].

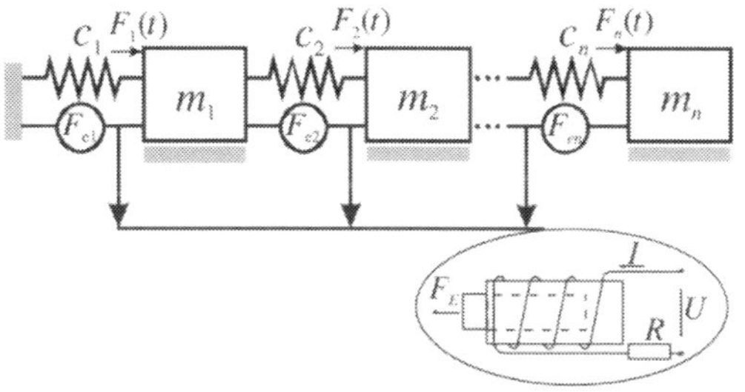

Figure 2. A model of the system with electric elements – moving-coil transducer.

To obtain the constitutive equations of moving-coil transducer is possible to use Faraday's law and Lorentz force law. In compliance with Faraday's law the voltage increment "de" (equation (8)) elementary length "dl" in the direction of the current flow, inducer by the motion of the coil is following:

$$de = v \times Bdl \tag{8}$$

where: v – velocity of the coil, B – magnetic flux density

The complete force of the field acting on an elementary length of the conductor is obtained using equation (equation (9)):

$$df_e = idl \times B \tag{9}$$

where f_e – electromagnetic force, i – current in the conductor

$$de = vBrd\varphi \tag{10}$$

Using an elementary length ($dl = rd\varphi$) the voltage increment from equation (equation (8)) is following (equation (10)):

$$e = 2\pi nrBv = Tv \tag{11}$$

where $T = 2\pi nrB$ – transducer constant

The voltage in the moving-coil transducer in direction of the current is present in equation (11).

$$f_e = i2\pi nrB = Ti \tag{12}$$

The complete power provided to the moving-coil transducer is equal to the sum of the electric energy and the mechanical energy (equation (11), (12)).

Accordingly at any time, there is a balance between electrical energy absorbed by the device and the mechanical energy delivered (and conversely). The moving-coil transducer cannot store energy, and is as a perfect electromechanical converter [16-18].

3. Conclusions

Presented in this work synthesis (reverse task) of mechatronical systems with the active reduction of vibration implemented with electric elements constitute a new approach to the problem of designing such systems. In order to design this system it is necessary to synthesize or identify a passive system and next, depending on the structure of the passive system and on excitations affecting it, select a structure containing active elements from the sequence of structures. In this work is only presents possibility of using electric elements as active reduction of unwanted vibrations. A very important problem is to check the influences between the basic system and the subsystem. This problem will be presented in next works.

Acknowledgement
This work has been conducted as a part of the research project No. PBS2/A6/17/2013 supported by the National Centre for Research and Development.

References

[1] Buchacz A and Świder J 2001 Computer support CAD CAM *Support for construction of systems reducing vibration and machine noise* (WNT Warsaw)
[2] Engel Z and Kowal J 1994 *Vibro-accoustic processes control* (AGH Press: Kraków)
[3] Michałowski S 1994 Active systems in machines construction *Monograph* **171** (Cracow Publication: Cracow University of Technology)
[4] Soong T T and Dargush G F 1999 Passive energy dissipation systems in structural engineering *Chichester* (Wiley: USA)
[5] Buchacz A and Płaczek M 2010 Selection of Parameters of External Electric Circuit for Control of Dynamic Flexibility of a Mechatronic *System Solid State Phenomena* **164** pp 323-326
[6] Buchacz A and Płaczek M 2009 Damping of Mechanical Vibrations Using Piezoelements, Including Influence of Connection Layer's Properties on the Dynamic Characteristic *Solid State Phenomena* **147-149** pp 869-875
[7] Buchacz A and Wróbel A 2010 Computer-Aided Analysis of Piezoelectric Plates *Solid State Phenomena* **164** pp 239-242
[8] Czmochowski J, Moczko P, Odyjas P and Pietrusiak D 2014 Tests of rotary machines vibrations in steady and unsteady states on the basis of large diameter centrifugal fans *Eksploatacja i Niezawodnosc – Maintenance and Reliability* **16(2)** pp 211–216
[9] Jamroziak K and Kosobudzki M 2012 Determining the torsional natural frequency of underframe of off-road vehicle with use of the procedure of operational modal analysis *Journal of Vibroengineering* **14(2)** pp 472-476
[10] Białas K, Buchacz A and Dzitkowski T 2009 Synthesis of vibrating active mechanical systems with dumping in view of polar graphs and structural numbers *Monograph 230* (Silesian University of Technology Press: Gliwice)
[11] Buchacz A 2012 Introduction to synthesis of the torsional vibrating discrete-continuous mechatronic systems by means of the hypergraphs and structural numbers method *Journal of Vibroengineering* 14(2) pp 514-519
[12] Bialas K 2013 Electrical Elements in Reduction of Mechanical Vibrations *Applied Mechanics and Materials* **371** pp 657-661
[13] Bialas K 2012 Mechanical and electrical elements in reduction of vibrations *Journal of Vibroengineering* **14(1)** pp 123-128
[14] Bialas K 2013 Mechanical Subsystem as Implementation of Active Reduction of Vibration *Solid State Phenomena* **198** pp 657-662
[15] Bialas K 2010 Passive and Active Elements in Reduction of Vibrations of Torsional Systems *Solid State Phenomena* **164** pp 260-264
[16] Groszkowski J 1958 *Manufacturing of electric vibrations* (PWT Warsaw)
[17] Onwubolu G C 2005 *Mechatronics Principles and Applications* (Elsevier Butterworth-Heinemann)
[18] Preumont A 2006 *Mechatronics Dynamics of Electromechanical and piezoelectric Systems* (Springer Dordrecht: Netherlands)

IOP Conf. Series: Materials Science and Engineering **95** (2015) 012052 doi:10.1088/1757-899X/95/1/012052

Influence of the excitation parameters of the mechanical subsystem on effectiveness of energy harvesting system

A Buchacz[1], **W Banaś**[1] **and M Płaczek**[1]
[1]Silesian University of Technology University, Faculty of Mechanical Engineering, Institute of Engineering Processes Automation and Integrated Manufacturing Systems, Konarskiego 18A, 44-100 Gliwice, Poland

E-mail: jacky.mucklow@iop.org

Abstract. Piezoelectric transducers are used more and more often in modern technical devices. The wide range of their possible applications is a result of the possibility to use both direct and reverse piezoelectric effect. Nowadays, application of piezoelectric transducers in energy harvesting systems is getting more and more popular. It is caused by the easy way to convert energy of mechanical vibration to the electric voltage using piezoelectric transducers. This paper presents results of influence analysis of the vibrating mechanical subsystem's excitation parameters on the effectiveness of the system designed for energy harvesting. The considered vibrating system is a composite plate with piezoelectric transducer bonded to its surface. Vibrations of the system are excited by means of an actuator with possibility to change the excitation amplitude and frequency. Recovering of electrical energy from mechanical vibrations is possible by using the direct piezoelectric effect – generation of the electric voltage while the transducer is mechanically deformed. In carried out test Macro Fiber Composite (MFC) piezoelectric transducers were used. It was proved that the time that is necessary for switch on the output voltage in analyzed system depends on the frequency of the excitation.

1. Introduction

The development of piezoelectric transducers allows new applications of such kind of materials and improves operation of existing devices. One of those applications is energy harvesting – recovering of the electrical energy from vibrations. Systems designed for such kind of operations cannot generate a huge power but their advantage is that the energy is recovered from phenomena that are unavoidable and normally it would be lost. What is more such kind of energy production allows design systems that work in difficult conditions, where other ways of supplying is difficult or impossible. For example it is impossible to use wires or replace discharged batteries. This is why nowadays a lot of research works are concerned with energy harvesting [1, 4, 12, 13].

In such kind of systems piezoelectric transducers can be easily use for recovering electric energy from mechanical vibrations. In presented work the application of Macro Fiber Composite (MFC) piezoelectric foils in energy harvesting system is presented. MFC transducer is consists of rectangular piezo ceramic rods sandwiched between layers of adhesive, electrodes and polyimide film. Main benefits of the MFC given by the manufacturer are: increased strain actuator efficiency, damage tolerance, environmentally sealed packages, available as elongators and contractors [1, 11].

In this work some results of measurements carried out on laboratory stand are presented. A series of tests with different types of MFC transducers (types P2 and P3) with different field of active area

Content from this work may be used under the terms of the Creative Commons Attribution 3.0 licence. Any further distribution of this work must maintain attribution to the author(s) and the title of the work, journal citation and DOI.
Published under licence by IOP Publishing Ltd

IOP Conf. Series: Materials Science and Engineering **95** (2015) 012052 doi:10.1088/1757-899X/95/1/012052

were carried out. Also the influence of excitation parameters on the system's effectiveness was verified, as well as, the influence of the mechanical subsystem fixing. Influence of the system's fixing (boundary conditions) was also considered. Measured data were acquired using the real time computer CompactRIO produced by National Instruments and analysed using LabVIEW software. Obtained results were juxtaposed and analysed. This work was a first step of the research project that aim is an implementation of the energy harvesting system on real objects and use of the generated electrical energy to supply systems with low energy consumption.

Gliwice research centre has a great experience concerning with tasks of modern technical devices and modern technology [5,7,9,10,16] as well as analysis and synthesis of mechanical and mechatronic systems [2,3,6,8,14] also supported by computer methods [15].

2. Laboratory stand for testing of energy harvesting system

In order to carry out test and measurements the laboratory stand that allows generating of mechanical vibrations with required parameters of excitation was created. It is presented in figure 1. The laboratory stand consists of:

- the composite plate (the mechanical subsystem) with piezoelectric foil bounded to its surface,
- the vibration exciter (in the form of electric motor) with the connecting rod mounted on the eccentric shaft and connected to the laminate,
- the motor speed regulator,
- measuring system consisting of a real-time computer CompactRIO produced by National Instruments with required analogue input card.

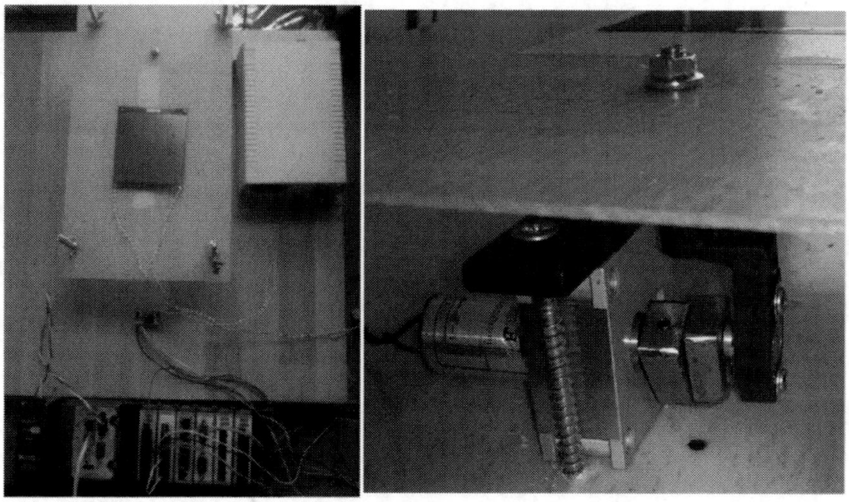

a) b)

Figure 1. The laboratory stand with measurement equipment (*a*) and the system for vibration generation (*b*).

The system for vibration generation was designed in the way that allows adjusting values of vibration frequency of the mechanical subsystem with piezoelectric transducer as well as the amplitude of vibrations. In this paper results obtained in the case of the mechanical subsystem rigid fixing at four points at the corners of the plate are presented while parameters of excitation were changed. Because the CompactRIO real-time computer is a universal tool it was necessary to prepare the corresponding block diagram (computer software) in order to carry out required measurements and data acquisition. The dedicated software was created using LabVIEW.

IOP Conf. Series: Materials Science and Engineering **95** (2015) 012052 doi:10.1088/1757-899X/95/1/012052

In presented tests the Macro Fiber Composite transducer type M8585-P2 was used. It was glued on the surface of the mechanical subsystem and connected to the system for generated electric voltage processing and measuring system.

3. The system for generated electric voltage processing

Energy harvesting systems are usually used as power supply for systems with low power operating conditions, where the use or exchange for new ones standard batteries is difficult or impossible. The electric voltage generated by piezoelectric transducers as a result of vibrations of mechanical subsystems cannot be directly used as power supply. Especially when the vibrations of the system are random. This is why it is necessary to use the for generated electric voltage processing. In presented work the LTC3588 chip that can function as an alternative power source for systems with low power consumption was used. In figure 2 its application diagram and ready to use form are presented.

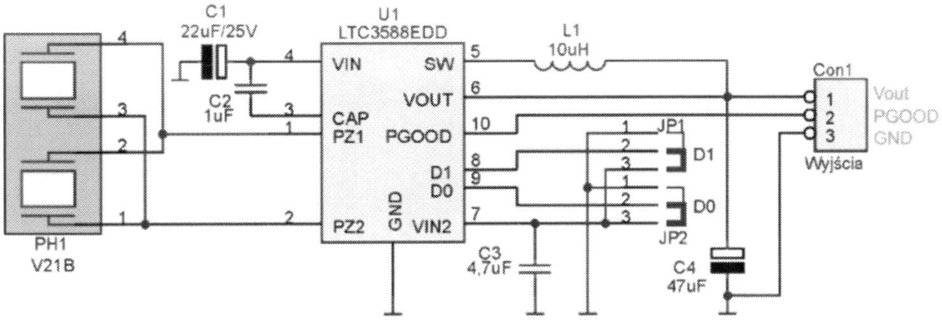

a)

b)

Figure 2. The application diagram of the inverter system (*a*) and ready to use form (*b*).

The converter LTC3588 produced by Linear Technology can be used to generate stabilized voltage of selected value: 1.8V/2.5V/3.3V/3.6V. It has the current capacity up to 100 mA. The system is structurally adapted to supply by the energy derived from unconventional sources, such as, for example, piezoelectric generators, photo- or thermocouples. It has the opportunity to work with a backup power source, for example battery is used when no power from piezoelectric transducer is available. A small number of additional components are needed for the operation of the inverter.

During carried out measurements values of the electric voltage generated directly by the piezoelectric transducer as well as the voltage on output of the inverter and status signal (denoted PGOOD in Figure 2a) were collected. In figure 3 values of the signal measured on the inverter system during loading and unloading of the system are presented. There can be noticed that if the value of the voltage on the system's capacitor reaches a certain value the state of the status signal as well as the output voltage is changed (output voltage is switched on or switched off). Those values are about 5.1 V on the capacitor during the system loading (switching on the output voltage) and 5.6 V during unloading (switching off the output voltage).

IOP Conf. Series: Materials Science and Engineering **95** (2015) 012052 doi:10.1088/1757-899X/95/1/012052

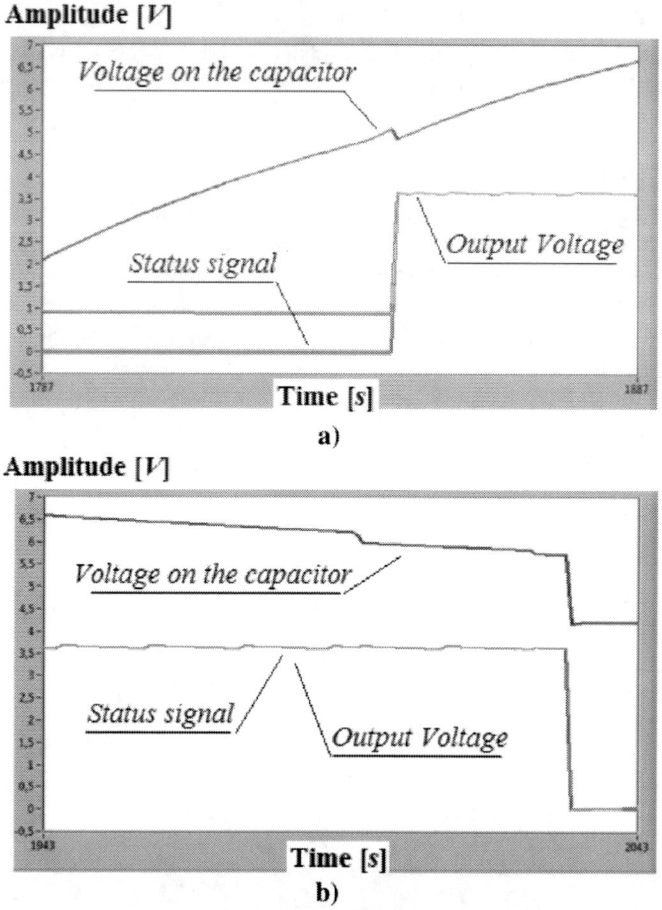

Figure 3. Values of the signals on the inverter system during loading (*a*) and its unloading (*b*).

4. Results of measurements

The main aim of the presented tests was to verify the influence of the parameters of vibrations of the mechanical subsystem on the time that is necessary to generate enough electric energy by the piezoelectric transducer in order to get stabile output voltage on inverter system. In this paper the influence of the vibration frequency is verified. The amplitude of vibrations was the same during all measurements. The frequency was changed by adjusting the rotational speed of the engine that was used for vibration excitation. In Figure 4 results of tests carried out for vibration frequencies 30 Hz, 45 Hz and 180 Hz are presented. On presented graphs measured values of the electric voltage generated by the transducer, voltage on the capacitor of the inverter system, output voltage and status signal are presented. It can be noticed that increase of the voltage on the capacitor has a linear characteristic. Time that is necessary for switch on the output voltage depends on the frequency of the excitation. For the higher frequencies the external equipment can be power supplied by the system faster. For the excitation frequency 30 Hz the time necessary to load the system is about 2000 ms. For the excitation frequency 45 Hz it is about 1100 ms while for the 180 Hz it was only about 950 ms.

5. Conclusions

The aim of this work was to verify influence of the mechanical subsystem vibration parameters on the effectiveness of the system designed for energy harvesting with electric voltage processing based on LTC3588 chip. In this paper influence of the vibration frequency was presented.

The carried out test show that by increasing the frequency of vibration it is possible to reduce the time necessary to switch on the stabilized output voltage that can be used for power supply of any kind of systems with low power operating conditions. The proposed energy harvesting system can be applied in technical means that work in difficult conditions, where other ways of supplying is difficult or impossible. The influence of the other parameters of the mechanical subsystem vibrations will be presented in other publications.

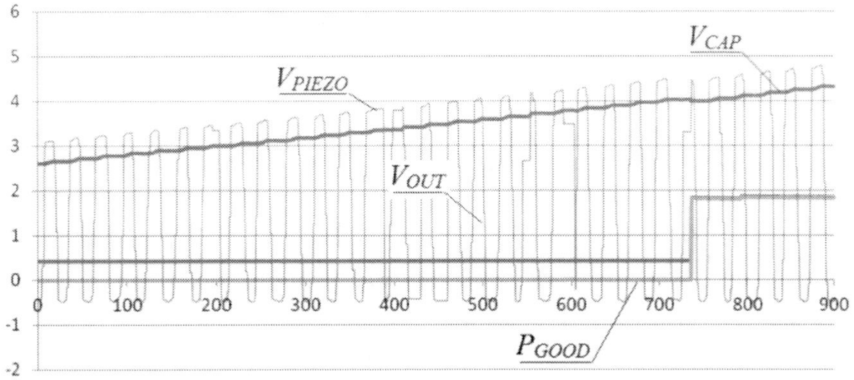

a)

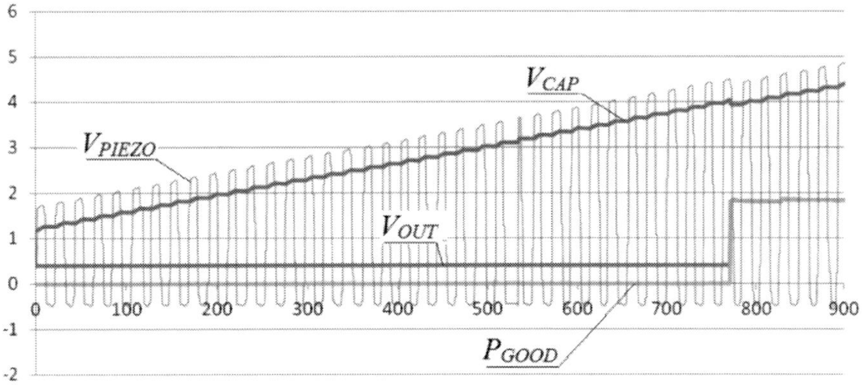

b)

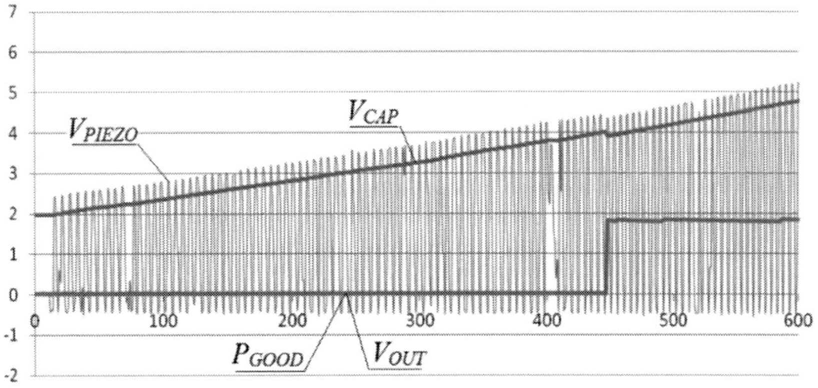

c)

Figure 4. The results of the measurements for mechanical subsystem vibration frequency 30 *Hz* (*a*), 45 *Hz* (*b*) and 180 *Hz* (*c*).

Acknowledgments
The work was carried out under the project number PBS2/A6/17/2013 agreement implemented under the Applied Research Program, funded by the National Centre for Research and Development.

References
[1] Beeby S P, Tudor M J and White N M 2006 Energy harvesting vibration sources for microsystems applications *Measurement Science and Technology* **17** pp 175-195
[2] Białas K, Buchacz A and Gałęziowski D 2015 Comparison of active and semi-active damping in synthesis of various mechatronic discrete systems *Int. J. Materials and Product Technology* **50(3/4)** pp 340-355
[3] Białas K 2012 Mechanical and electrical elements in reduction of vibrations *Journal of Vibroengineering* **14(1)** pp 123-128
[4] Buchacz A, Płaczek M and Wróbel A 2013 Control of characteristics of mechatronic systems using piezoelectric materials *Journal of Theoretical and Applied Mechanics* **51** pp 225-234
[5] Ćwikła G, Grabowik C and Janik W 2014 Case Study of Manufacturing Information Acquisition System (MIAS) in Automated Continuous Production *System Applied Mechanics and Materials* **657** pp 808-812
[6] Dymarek A and Dzitkowski T 2005 Modelling and synthesis of discrete – continuous subsystems of machines with damping *J. of Materials Processing Technology* **164-165** pp 1317-1326
[7] Gwiazda A, Sękala A, Monica Z and Banaś W 2014 Integrated approach to the designing process of complex technical systems *Advanced Materials Research* **1036(1)** pp 1023-1027
[8] Gwiazda A, Herbuś K, Kost G and Ociepka P 2015 Designing mechatronics equipment based on the example of the Stewart platform *Solid State Phenomena* **220/221** pp 419-422
[9] Klarecki K, Hetmańczyk M and Rabsztyn D 2015 Influence of the selected settings of the controller on the behavior of the hydraulic servo drive. Mechatronics - Ideas for Industrial Application *Advances in Intelligent Systems and Computing* **317** pp 91-100
[10] Paprocka I, Kempa W, Grabowik C and Kalinowski K 2014 Sensitivity Analysis of Predictive Scheduling Algorithms *Advanced Materials Research* **1036** pp 921-926
[11] Płaczek M 2015 Modelling and investigation of a piezo composite actuator application *Int. J. Materials and Product Technology* **50(3/4)** pp 244-258
[12] Shashang P 2007 Advances in energy harvesting using low profile piezoelectric transducers *J. Electroceram* **19** pp 165-182
[13] Sodano H and Inman J 2005 Comparison of piezoelectric energy harvesting devices for recharging batteries *Journal of Intelligent Material Systems and Structures* **16(10)** pp 799-807
[14] Wróbel A 2012 Kelvin Voigt's model of single piezoelectric plate *J. of Vibroengineering* **14/2** p 534
[15] Wróbel A, Płaczek M, Buchacz A and Majzner M Study of mechanical properties and computer simulation of composite materials reinforced by metal *Int. J. Materials and Product Technology* **50 (3/4)** pp 259-275
[16] Zolkiewski S 2015 On force-deflection diagrams of fibre-metal composites connected by means of bolt joints *Int. J. Materials and Product Technology* **5(3/4)** pp 230-243

IOP Conf. Series: Materials Science and Engineering **95** (2015) 012053 doi:10.1088/1757-899X/95/1/012053

An overview on internal geared mechanisms with small difference between teeth number

S Macovei (Cîrdei)[1] **and I Doroftei**[1]
[1]"Gheorghe Asachi" Tehnical University of Iasi, Mechanical Engineering, Mechatronics and Robotics Departament, Blv. D. Mangeron, No.63, 700050- Iasi, Romania

E-mail: simina_cost@yahoo.com

Abstract. Internal gears with small difference between teeth number are most commonly encountered in planetary and differential mechanisms (eg. on automatic transmissions for vehicles). One of the advantages of this gear is the high transmission ratio. Their parallel axes make this gear perfect for the cases when a small deviation of the distance between axes is required. Automatic transmission allows internal gear function to high speed. In this, we present a short overview of internal gear with small difference between teeth number.

1. Introduction

Generally, a gear is a mechanism that consists from two or more gears, which send the rotational movement and torque between two shafts, by means of the teeth that are continuously and sequentially in contact [1].

Classification of the parallel axes gears is extensive, but we are interested only on the type of engagement classification [2], presented in figure 1:

- spur gear – consisting in cylindrical shaped gear, with teeth parallel to the axis;
- internal gear- cylindrical type of gear, where the teeth of one of the wheels are inside the circular ring;
- helical gear- the teeth of this gear have a helicoid shape.

a) b) c)

Figure 1. Different types of gears [2]: a) spur gear; b) internal gear; c) helical gear.

The gears are widely use on mechanical transmissions, due their advantages: constant transmission ratio, safety in operation, high durability, high efficiency, reduced size, wide range of powers, speeds

Content from this work may be used under the terms of the Creative Commons Attribution 3.0 licence. Any further distribution of this work must maintain attribution to the author(s) and the title of the work, journal citation and DOI.
Published under licence by IOP Publishing Ltd

and gear rations usage. They are found in industrial applications such as: speed reducers and multipliers, gearboxes, differentials, toys etc. As disadvantages can be mentioned: high accuracy on execution and assembly; complicated technology; noise and vibration during operation.

The planetary gears, that especially use internal gears, are highly found in industrial application. We can classify planetary gears as [3]:

- planetary gears for transmitting power (figure 2a,b);

planetary gears for motion transmission, with a high transmission ration (figure 2 c, d).

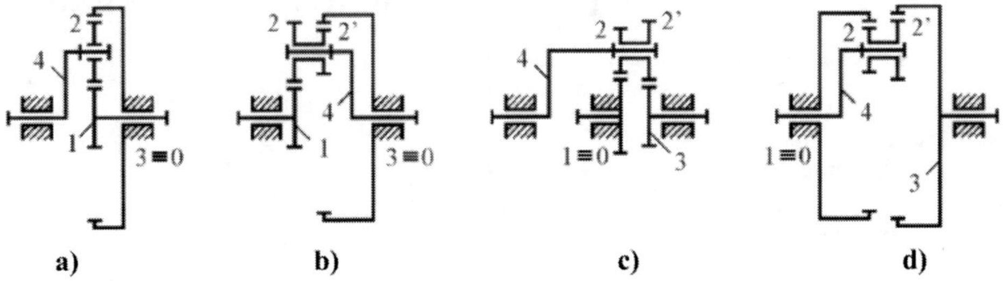

Figure 2. Kinematics of some planetary mechanisms [3]: a), b) for power transmission; c), d) for motion transmission.

Lately research on gears made possible the development of new technologies of mechanical transmissions. Improving gears impose requirements as:

- high load capacity;
- high kinematic precision;
- possibility to transmit motion in sealed environments;
- reduced mass and dimensions.

Challenges encountered by engineers in development of internal gear systems are concerning geometry calculus and damage prevention [4].

One can analyze the internal gear from several points of view: interference phenomenon; contact ratio; transmission ratio. These are aspects that should be very careful analyzed in the case of an internal gear with a small teeth number difference.

We can use functional and kinematic parameters for determining the tooth profile on mechanical transmissions, and also the operation mode of the transmission. Simplifying the calculus of transmission design is a current issue in mechanical engineering.

2. Internal gears

2.1. Interference Phenomenon

Inner cylindrical gears are commonly constructed with straight teeth and consist of two cylindrical wheels, one with external teeth and the other with internal teeth. The directions of rotation for these two wheels are the same. The engagement takes place between a concave profile (of the internal teeth) and a convex one (of the external teeth). This kind of engagement is advantageous in terms of stress for contact teeth. On the internal gear system, the real segment of action is bigger than the external gear's one, this leads to an increase in coverage degree, which is particularly advantageous.

For internal gears with minimum teeth number, the correct engagement occurs only inside of the engagement line, see figure 3 [2]. The connection between the involute base circle to the inner circle requires a special form, so that the teeth tips of the smaller wheel do not intersect the base pinion teeth.

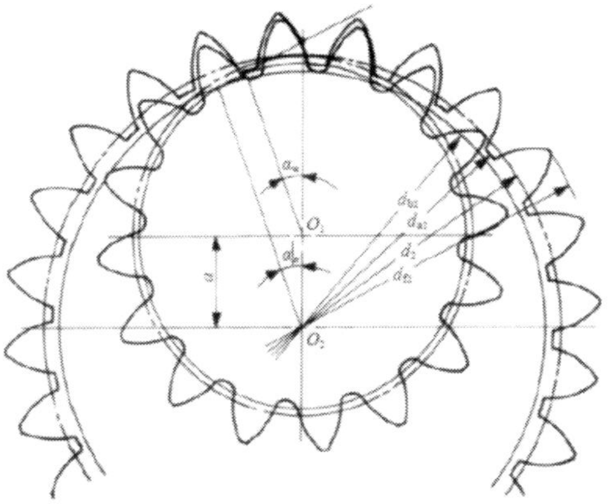

Figure 3. Base condition for correct engagement, with constant transmission ratio [2].

Interference can occur on three different types within internal gears: involute interference, tip interference and fillet interference. Involute interference occurs between the dedendum of the external gear and the addendum of the internal gear, and it is prevalent when the number of teeth of the external gear is small. In other words, the system gearing must begin on the tangent line to the two base circles. This interference can be avoided by respecting the condition [5]:

$$\tan\alpha_{a2} \leq \left(1+\frac{z_1}{z_2}\right)\cdot\tan\alpha_w \qquad (1)$$

where: - α_{a2} is the pressure angle of a tip of the internal gear tooth;

- a_w is the working pressure angle.

The fillet interference appears at the addendum of the external gear and the dedendum of the internal gear and tends to happen when the difference between the numbers o teeth of the two gears is small [2]. In other words this appears when the tips of the teeth of one gear interfere with the fillets at the roots of the teeth of the other gear. To avoid this interference one must calculate very accurate the diameters of the corresponding circles of active profile.

Tip interference appears when the teeth tip of the exterior gear collides with the teeth tip of the interior gear. This type of interference can take place both during engagement and disengagement of gears, see figure 4 [6]. The tip interference can be avoided, in both cases, by satisfying this condition:

$$\theta_p > \left(\frac{z_g}{z_p}\right)\cdot\theta_g \qquad (2)$$

2.2. Contact Ratio

The engagement continuity is verified by the contact ratio. For a spur gear the contact ratio depends only by teeth number, and increases with the increase of this numbers, according to V. Merticaru et al. [5]. The complete contact ratio is determined by approach contact and recess contact . Their expression, established by R. Maiti et al. [6] are:

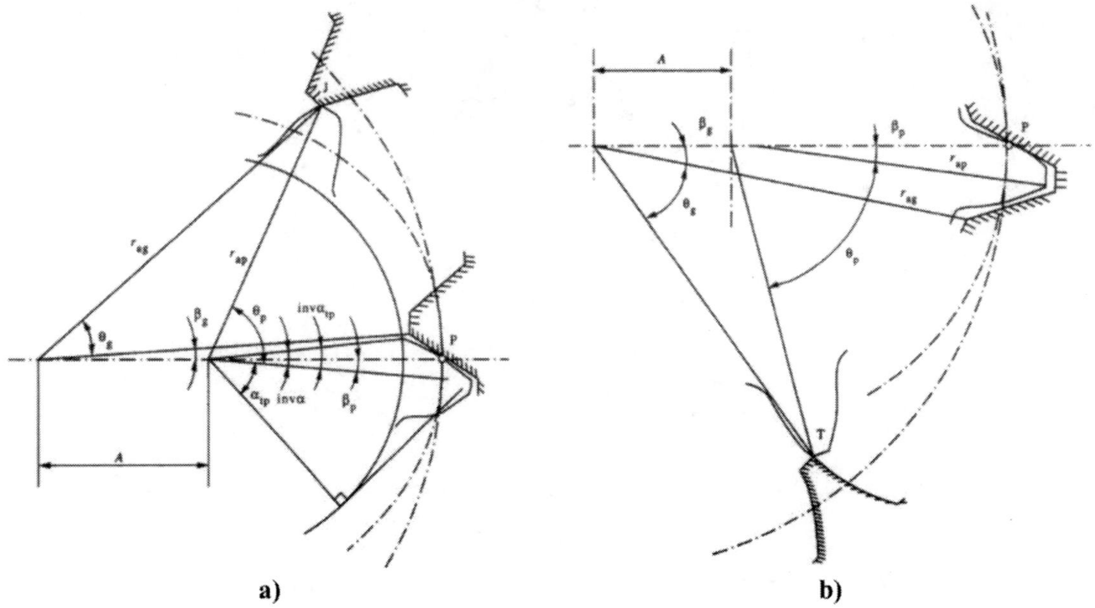

Figure 4. Tip interference [6]: a) during engagement; b) during disengagement.

$$a_c = \sqrt{r_{ap}^2 - r_{bp}^2} - r_p \sin\alpha \tag{3}$$

$$r_c = r_g \sin\alpha - \sqrt{r_{ag}^2 - r_{bg}^2} \tag{4}$$

$$C_r = (a_c + r_c)/(\pi \cdot m \cdot \cos\alpha_0) \tag{5}$$

where r_p and r_g are the working pitch circle radius, r_{ap} and r_{ag} are the tip radius, r_{bp} and r_{bg} are the base circle radius of the pinion and gear respectively, α and α_0 are the working and standard pressure angles respectively, and m is the standard module.

2.3. Transmission Ratio

Transmission ratio for an internal-external gear pair can be calculated as the ratio between the teeth number of the ring gear (z_g) or pinion (z_p) and the difference in the teeth number ($z_g - z_p$) [6]. When the output motion is from the pinion, the transmission ratio is:

$$i = -\frac{z_p}{z_g - z_p} \tag{6}$$

while when the output motion is from the ring gear, we can write:

$$i = \frac{z_g}{z_g - z_p} \quad (7$$

One can observe that the transmission ratio increases as the difference in teeth number decreases, and reaches its maximum when the difference is equal to unity. The goal of many researches is to minimize the difference in teeth number, trough different gear corrections, avoiding tip interference.

IOP Conf. Series: Materials Science and Engineering **95** (2015) 012053 doi:10.1088/1757-899X/95/1/012053

3. New concepts in improving internal gears

An attempt to lower the difference of teeth number, as much as possible, in the internal-external involute gear pair was realized by Maiti, R. and Roy, A.K [6]. This work aimed to accomplish the reducing of tooth difference with avoiding tooth tip interference between involute internal-external gear pair. For this, they had done tooth truncation and center distance modification, keeping the contact ratio within a satisfactory limit. The addendum modification was considerate for avoiding the tip interference, and the results confirmed that maximum contact ratio is achieved if only the addendum of pinion is truncated. The results showed that for 20 and 22.5 involute tooth, the tooth difference cannot be lower than 5. Also, for the harmonic drives, a mathematical relation has been established, for avoiding tip interference when the pinion rim is deflected.

For calculating planetary gear transmission parameters, Muravev developed a scheme that calculates the tooth interference by the means of the Mathematica program [7]. The planetary gear transmission is considered to be used in the reduction gear drive for a worm centrifuge (figure 5 a). For his calculation scheme, he assumed that the known values are number of teeth, displacement coefficient, the tooth modulus and the tooth profile angle. The algorithm incorporates the possibility of evolvent intersection only on the working parts of the meshing teeth. Thus, we can conclude that the interference in meshing can be determined not only by calculation, but also by means of computer-aided design systems. This requires a three-dimensional model for the transmission and the exact specification of the shapes and sizes of the teeth for the wheel.

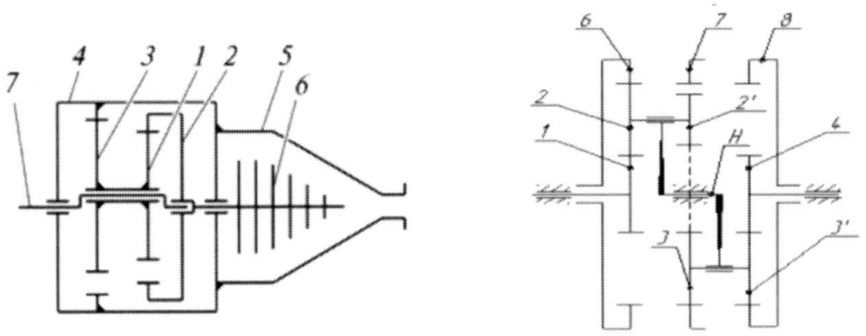

a) **b)**

Figure 5. Kinematic designs :a) reduction gear in a worm centrifuge [7]; b) eight-speed planetary gearbox [8].

A new eight-diameter stepped planetary gear box was proposed by Salakhov, I.I. *et al.* that offer benefits regarding energy consumption and efficiency. The planetary gear box, presented in Figure 5b, implements eight passes of direct gearing and one reverse gear and high efficiency is received due to short kinematic sequences. No interruptions appear in torque delivery, on gear changing, thanks to minimal number of synchronously switching control elements. The minimal number of main links and the less number of control elements resulted into reduced dimensions of the gear box.

Plekhanov, F.I. *et al.* [9] proposed an analytical method, for determining the load capacity of a planetary transmission with internal gear engagement (figure 6 a). The planetary transmissions with small difference in the number of teeth are characterized by large load capacity, large gear ratio and relatively low frictional power losses. The profile of the gear teeth corresponds to a large reduced radius of curvature, therefore is important to calculate the flexural strength of the satellite teeth. From the formula of flexural strength, the authors determine the permissible normal linear load on the satellite tooth. Because on torque application the small gaps between the teeth are shifted, and the load appears simultaneously in several tooth pairs, the authors determined the load capacity of the planetary transmission. Based on their results, it is possible to estimate the load capacity of a planetary transmission and select the number of teeth required for the selected gear ratio.

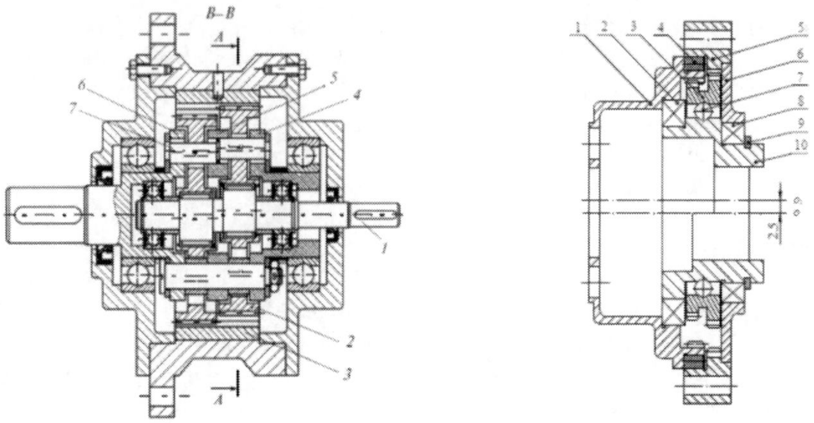

a) **b)**

Figure 6. Assembly diagram: a) gear-roller planetary transmission with internal gear engagement [9]; b) new planetary gear apparatus [10].

A dynamic characteristics analysis was performed by C. Huang *et al.* [10] on a new internal mesh planetary gear with small tooth number difference (figure 6b). The new planetary gear apparatus was analyzed in terms of natural frequencies, vibration modes, dynamic response and acceleration noise. The displacement field and stress field were obtained through static Finit Element simulation. The rotational frequencies of the eccentric shaft were calculated through Finite Element modal analysis and the results showed a lower value. The experimental results confirmed the obtained simulation values.

Shuting Li has presented a theoretical study of the contact problem of internal gears [11]. The calculation model was realized within the limits of the elasticity theory, by considering a pair of elastic bodies, that come in contact one with each other. The assumption made in this model is that deformations are small, the two bodies obey the laws of linear elasticity and the contact surfaces are smooth. The mathematical model can be use for mathematical programming method to find the contact load under the condition of knowing the deformation, and the external forces. Also the Finite Element Model analysis was realized by applying an external torque T on the internal gear and assuming pairs of contact teeth and pairs of contact points. In this paper only FEM calculation results are presented, the experimental tests are still needed to confirm the theoretical results.

4. Conclusions

As general advantages of internal gears we can mention the high transmission ratio for small overall dimensions. It means, they may be very compact for high transmission ratios, which make them candidates for using in industrial manipulators structure. Interference phenomenon; contact ratio; transmission ratio are aspects that should be very careful analyzed in the case of an internal gear with a small teeth number difference. In this paper a short overview on the current existing tendencies into calculating and studying the small teeth number difference gears has been presented.

References

[1] Atanasiu V 2002 *Angrenaje cilindrice evolventice* (Editura Tehnica Stiintifica si Didactica: Iasi)

[2] *Technical Reference from Gear Technical Reference* (Kohara Gear Industry Co)

[3] Petrovic T, Ivanov I and Milosevic M 2009 A new structure of combined gear trains with high transmission ratio *Forsch Ingenieurwes* **73** pp 119-127

[4] Gafitanu M, Cretu S, Racocea C, Coca D, Pavelescu D, Radulescu G, Radauceanu D, Tuleasca C and Vornicu I 1983 *Organe de masini* vol II (Editura Tehnica: Bucuresti)

[5] Merticaru V and Atanasiu V 1996 *Angrenaje cilindrice cu numar mic de dinti* (AC.NMD:Bucuresti)

[6] Maiti R and Roy A K 1996 Minimum tooth difference in internal-external involute gear pair *Mechanical and Machine Theory* **31** pp 475-485

[7] Yu Murav'ev K 2009 Calculating planetary gear transmission parameters for the reduction gears in worm centrifuges *Chemical and Petroleum Engineering* **45** pp 345- 352

[8] Salakhov I I, Voloshko V V, Mavleev I R and Galimyanov I D 2015 Kinematic scheme and design of automatic planetary gear boxes based on a new module *Contemporary Engineering Sciences* **8** pp 1-6

[9] Plekhanov F I and Ovsyannikov A V 2011 Load capacity of planetary transmission with internal gear engagement *Russian Engineering Research* **31** pp 825-827

[10] Huang C, Wang J, Xiao Min Li K and Li J Y 2013 Dynamic characteristics analysis and experimental research on a new type planetary gear apparatus with small tooth number difference *Journal of Mechanical Science and Technology* **27** pp 1233-1244

[11] Li S 2008 Contact problem and numeric method of a planetary drive with small teeth number difference *Mechanism and Machine Theory* **43** pp 1065-1086

Ankle rehabilitation device with two degrees of freedom and compliant joint

C-M Racu (Cazacu)[1] **and I Doroftei**[1]
[1]"Gheorghe Asachi" Technical University of Iasi, Mechanical Engineering, Mechatronics and Robotics Department, Blvd. D. Mangeron, No. 43, 700050 - Iasi, Romania

E-mail: cristina.racu@tuiasi.ro

Abstract. We propose a rehabilitation device that we intend to be low cost and easy to manufacture. The system will ensure functionality but also have a small dimensions and low mass, considering the physiological dimensions of the foot and lower leg. To avoid injure of the ankle joint, this device is equipped with a compliant joint between the motor and mechanical transmission. The torque of this joint is intended to be adjustable, according to the degree of ankle joint damage. To choose the material and the dimensions of this compliant joint, in this paper we perform the first stress simulation. The minimum torque is calculated, while the maximum torque is given by the preliminary chosen actuator.

1. Introduction

Ankle joint is an articulation with a special importance in keeping the balance of an individual, during gate or stance position. According to some studies [1] an important number of people around suffer from sprains of fractures of the ankle every year. In this articulation occur multiple movements, thus a higher rate of injuries. For a normal life it is impetuously that the ankle joint is fully recovered, and the need of recovery exercises emerges. We can divide the devices used for ankle rehabilitation in three categories: parallel mechanism robots, prostheses and powered orthoses and powered exoskeletons. From Yoon, et. al. [2] we find a parallel reconfigurable robot , which has a large range of various exercises. This robot can allow desired ankle and foot motions, four degrees of freedom and it's driven by pneumatic actuators. A soft parallel ankle rehabilitation robot, with four air artificial muscle as actuators, was developed by P.K. Jamwal and his collaborators [3]. This device provides three rotational degrees of freedom to the ankle joint for the necessary range of motion and muscle strengthening exercises. "Rutgers Ankle" [4, 5] is a Steward platform-type haptic interface, which supplies six degrees of freedom. This platform allows patients to exercise at home, true an online database that can be easily accessed by the therapist.

A Lower Limb Exerciser with Intelligent Alloys (Leia) [6] was developed by S. Pittaccio, and it is an orthosis, composed of aluminum structure spanning above and below the ankle joint. It as a pair of motors based on shape memory alloy NiTi, obtained a 40° movement by coiling the wire along a spiraling sequence of pulleys. Anklebot [7], is a wearable therapeutic robot designed by engineers at MIT's Newman Laboratory for Biomechanics and Human Rehabilitation. The subject is seated and wears a modified shoe and a knee brace. The recovery is achieved by moving the ankle along a commanded trajectory, while monitoring static torque-angle relation.

Content from this work may be used under the terms of the Creative Commons Attribution 3.0 licence. Any further distribution of this work must maintain attribution to the author(s) and the title of the work, journal citation and DOI.
Published under licence by IOP Publishing Ltd

Powered exoskeletons are mainly used for learning gait in patients that suffer from a stroke, losing their capability of walking. Although it is used for entire leg (hip, knee, ankle), the exoskeleton is a powerful tool for recovery, especially for the ankle joint, which supports all human body's weight. A lower extremity exoskeleton [8], named BLEEX, was developed aiming to help people for load-carrying and rehabilitation. His unique design allows the user to control the system, while the exoskeleton provides the required strength. It is equipped with a frame that supports different loads, two powered legs and a power unit. ALEX is an active leg exoskeleton [9], used for stroke survivors, the weight of the device is supported by the walker. It has several joints: two joints at hip (abduction-adduction motion, and sagittal plane motion), revolute knee joint and a revolute joint for vertical rotation. Also, it has a parallelogram mechanism for vertical translation, a back support for the subject, pelvic straps and tight brace.

To meet the need for exercise, several mechanisms have been developed for ankle rehabilitation. Traditionally are still used primitive passive mechanisms such as elastic bands, zigzag staircase, and foam roll for balance exercises. However, these mechanisms allow only simple exercises that require patient effort and cannot store information about its progress [10]. Although a variety of systems used in recovery of ankle joint exist [17], we can say that currently we still need a system that fully satisfies the needs of the patient. Therefore, there is a request for easy to use mechatronic systems that facilitate optimal recovery of patients, suffering from various diseases of the ankle. Our system provides a new actuation layer, based on the simple scotch-yoke mechanism, conducting to small dimensions, reduced weight and simple comands.

2. Proposed Ankle Rehabilitation Device

In designing the rehabilitation system, we assumed that the device must provide dorsiflexion and plantar flexion movements and pronation and supination rotations, necessary for a complete recovery of the ankle joint [11]. The kinematics of the new device is presented in figure 1.a, while in figure 1.b the 3D design of the mechanism is presented. This device is based on the simple scotch-yoke mechanism and it consists in a plate connected to a fixed frame through cardanic joints.

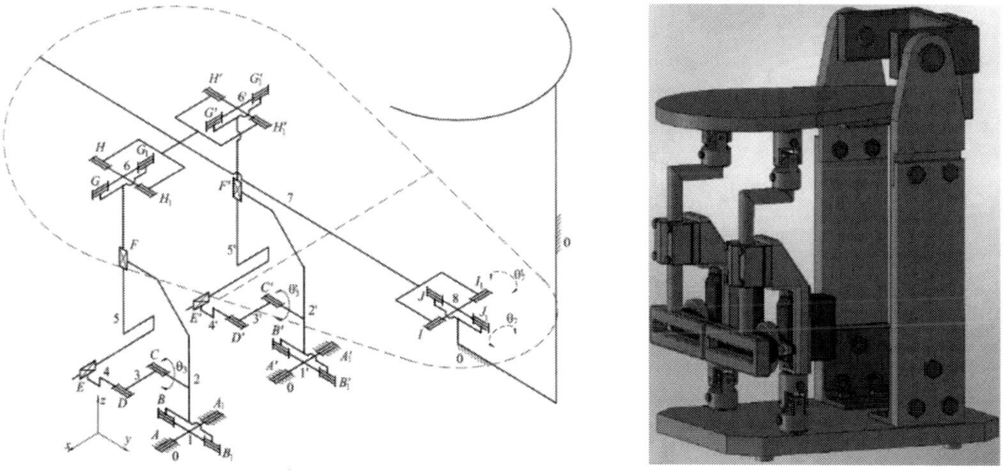

a) b)

Figure 1. The proposed rehabilitation system: a) kinematics; b) 3D design.

The actuation of the system is achieved through servomotors, connected to a lever that actuates the scotch-yoke mechanism. For saving space, we placed the scotch-yoke mechanisms facing each-other. Thus, for plantar flexion-dorsiflexion movement, we will rotate the motors with the same angle, in the same direction. Rotating the motors with the same angle, but different direction will provide us the inversion-eversion movement. The rotational movement is transmitted from the motor shaft through a servo arm (see figure 2.a detail) and passed to a special designed flange. A torsion spring, which is our compliant link, is coupled to the flange and to the rotational arm. By measuring the spring deflection, the joint torque can be measured directly. For that we will need an experimental calibration. It means, we will only need a calibrated spring and an encoder, to measure the rotational movement and the torque.

The spring design was implemented by S. Wang *et al.* [14] in their optimization of an exoskeleton joint, and it consists of two connected Archimedean spirals (see figure 2b). Given the length of the lever arm, we impose a parameter $r_2 = 62\ mm$. The difference between r_1 and r_2 represent the spiral thickness, while b is the spring width. Each of the spirals has an active coil number $n = 1$. The stress distribution and displacement are checked using finite element analysis tool (Solidworks 2014).

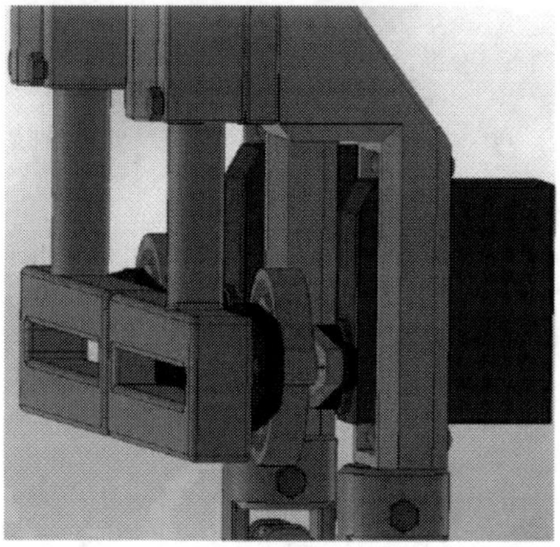

a)

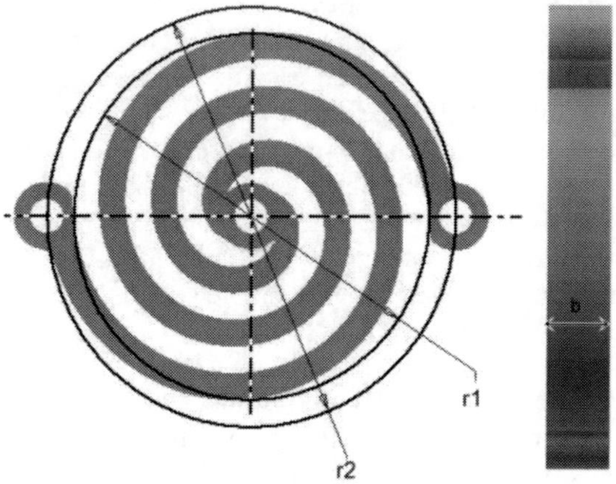

b)

Figure 2. Compliant joint: a) 3D view of the subassembly; b) planar views of the compliant link.

3. Stress Analysis

According to Divecha H M [13], an equation for static forces acting on ankle joint in mid-air horizontal suspension can be determined, offering us the resulting join reaction force of 15 N. Other estimations of forces and moments of lower limb joints offer us a vertical force, for a standing man, of approximately 20 N [14]. Another study, performed by D. Fong et col. [15], consisting in calculating the ground reaction forces through pressure insole data, offer us a range force of 11.79 N for medial-lateral direction and 33.53 N for anterior-posterior direction, for a male subject walking and running. Thus, in our study, we impose an initial force of 10 N, the force that the system must overcome to insure rehabilitation.

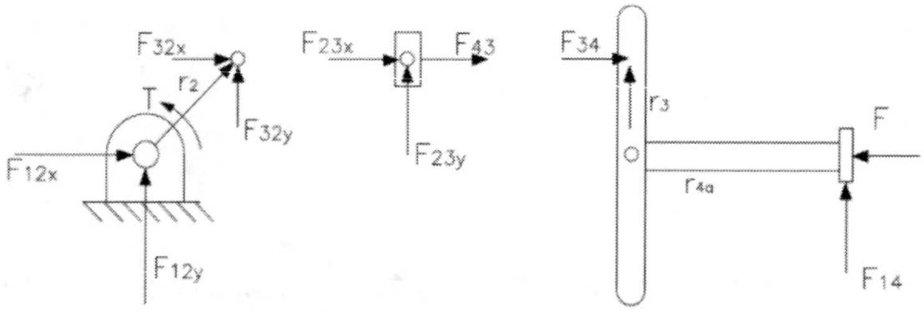

Figure 3. Analyze of the scotch yoke mechanism.

Analyzing the scotch yoke mechanism (figure 3), using the principle of virtual work [16] we find the reaction forces, and the required torque:

$$\begin{bmatrix} F_{34} \\ F_{14} \\ F_{23x} \\ F_{12x} \\ F_{12y} \\ T \end{bmatrix} = \begin{bmatrix} F \\ \frac{r_3}{r_{4a}} \cdot F \\ F \\ 0 \\ F \\ 0 \\ -r_2 \cdot \sin\theta \cdot F \end{bmatrix} \tag{1}$$

For our system r_2 is the length of the lever arm and F is the force exercised by the foot. Thus, in our case $r_2 = 75\ mm$, $\theta = 50°$ and $F = 10\ N$. The minimum torque, necessary to actuate a healthy ankle joint is now:

$$T = -75 \cdot \sin(50) \cdot 10 = 0.575\ N \cdot m \tag{1}$$

The maximum torque provided by a preliminary chosen motor is $T = 2.35\ N \cdot m$. For the Finite Element Analysis we propose two materials to be considered. The first material adopted from program's library is steel AISI 1020 with elastic modulus of 200 000 N/mm^2 and Poisson's ratio of 0.29. The second material is Acrylonitrile butadiene styrene (ABS) plastic with elastic modulus of 2000 N/mm^2 and a Poisson's ratio of 0.394.

Because the spring must always remain in the elastic deformation domain, a Von Mises stress analysis was performed. Spatial state of stress for simulation is given by the relation:

$$\sigma_{ech} = \frac{1}{\sqrt{2}}\sqrt{(\sigma_x - \sigma_y)^2 + (\sigma_x - \sigma_z)^2 + (\sigma_y - \sigma_z)^2 + 6(\tau_{xy}^2 + \tau_{xz}^2 + \tau_{yz}^2)} \tag{1}$$

The spring was deformed by minimum and maximum torque, providing us the total deflection. The boundary conditions imposed to the spring are: fixed extremities, torque applied in the center. For the initial test, we choose following dimensions: spiral thickness $r_2 - r_1 = 4\ mm$ and spring width $b = 10\ mm$. For the minimum value of torque we find the following results:

- alloy steel maximum displacement 0.051 *mm* (figure 4 a);
- ABS maximum displacement 4.662 *mm* (figure 4 b).

Due to the fact that alloy steel displacement was so small, the material choosed for analysis is ABS. This material indentifies a family of engineering thermoplastics, low cost, with a broad range of performance characteristics and easy to machine and fabricate. The most important mechanical properties of ABS are impact resistance and toughness, and it is an ideal material for structural applications when strength and stiffness are required. The material can take complex shapes, through 3D printing, and has good mechanical characteristics, thus making it suitable for our application. Yield strength, or the yield point, is defined as the amount of stress that a material can undergo before moving from elastic deformation into plastic deformation. The ultimate tensile strength for ABS (the limit stress at which the material breaks with sudden release of the stored elastic energy) was found to be around 40 *MPa* [18]. The initial dimensions were too small and the value of ABS displacement was high, thus we modify the dimensions. Our goal is to dimension the spring so we can have from ABS a minimum value, less than 1 *mm* of displacement, at a minimum torque. The first step was to increase the spiral thickness to $r_2 - r_1 = 6\ mm$ and to increase the spring width $b = 20\ mm$.

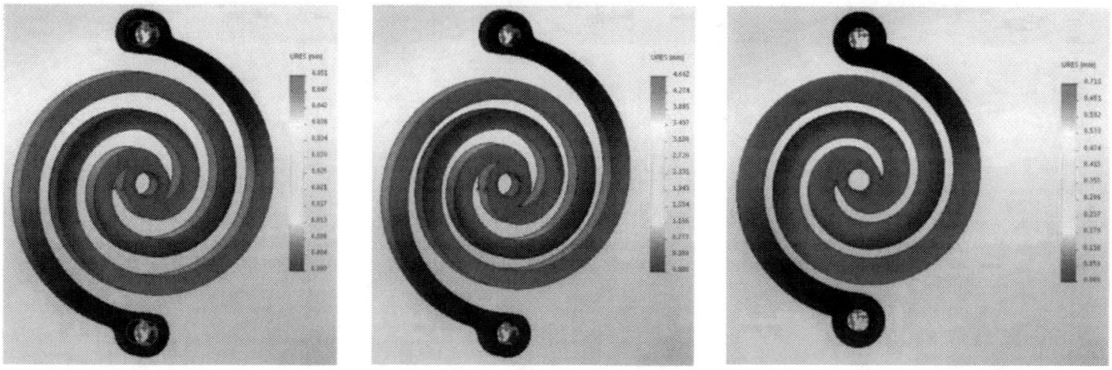

a) b) c)

Figure 4. Displacement at torque 0.575 $N*m$: a) alloy steel 10 mm width ; b) ABS 10 mm width; c) ABS 20 mm width.

By applying both minimum and maximum torque we find the following results:

- minimum torque- maximum displacement 0.711 *mm* (figure 4c);
- maximum torque- maximum displacement 2.81 *mm* .

The maximum value for Von Mises stress at minimum torque was 4.232 *MPa* (figure 5.a) , while at maximum torque, for these dimensions, was found to be 16.9 *MPa* .

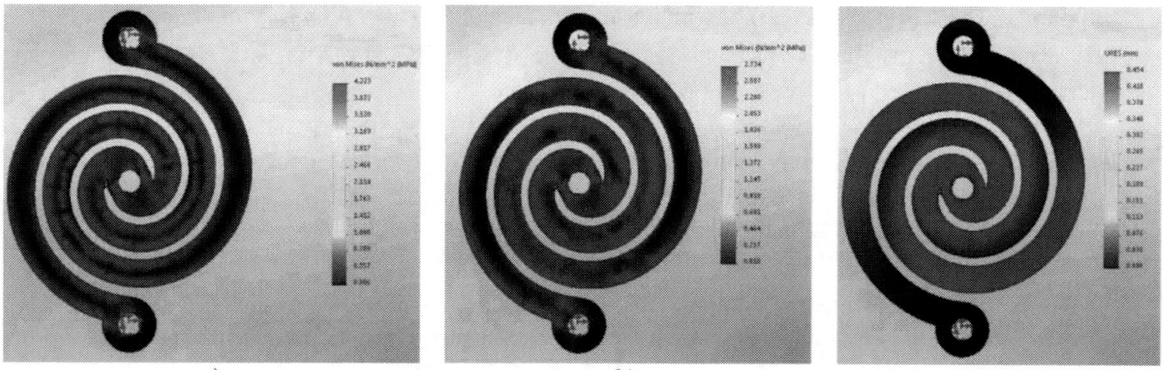

a) b) c)

Figure 5. ABS Von Mises stress and spring displacement under a 0.575 $N*m$ torque: a) stress at 20 mm width; b) stress at 30 mm width; c) displacement at 30 mm width.

Although the displacement found value for minimum torque is small, we propose a further development of the spring, increasing, once again, the spring width. The new values found for minimum and maximum torque are:

- minimum torque- maximum displacement 0.454 mm (figure 5c);
- maximum torque- maximum displacement 1.854 mm .

The Von Mises stress for minimum torque is 2.734 MPa (figure 5b) and for maximum torque is 11.176 MPa, safe within the limits. Further development of the spring will determine angular deflection, and calibration.

4. Conclusions

In this paper we present a novel ankle rehabilitation device with two degrees of freedom, which is using a double spiral spring as a compliant link. The compliance is necessary to avoid ankle joint injure, during rehabilitation. In this paper we performed the first stress simulation of the compliant joint. Optimal width at minimum torque of 0.575 *N*m* and a spiral thickness of 6 *mm* was found to be 30 *mm*. Some angular deflection and calibration will be studied in future work.

Acknowledgement
This work was supported by the strategic grant POSDRU/159/1.5/S/133652, co-financed by the European Social Fund within the Sectorial Operational Program Human Resources Development 2007 – 2013.

References

[1] Waterman B, Owens B, Davey S, Zacchilli M and Belmont Jr P J 2010 The epidemiology of ankle sprains in the United States *The Journal of Bone & Joint Surgery* **92** pp 2279-2284p
[2] Yoon J, Ryu J and Lim K B 2005 A novel reconfigurable ankle rehabilitation robot for various exercises *Robotics and Automation* pp 2290-2295
[3] Jamwal P K, Xie S, Aw K C 2009 Kinematic design optimization of a parallel ankle rehabilitation robot using modified genetic algorithm *Robotics and Autonomous Systems* **57** pp 1018- 1027
[4] Girone M, Burdea G, Bouzit M, Popescu V and Deutsch J E 2001 A Stewart platform-based system for ankle telerehabilitation *Autonomous Robots* **10** pp 203-212
[5] Boisan R F, Lee C S, Deutsch J E, Burdea G C and Lewis J A 2002 Virtual reality-based system for ankle rehabilitation post stroke *Proc.1st Int. Workshop Virtual Reality Rehabilitation* pp 77-86
[6] Pittaccio S and Viscuso S 2011 An EMG- Controlled SMA device for the rehabilitation of the ankle joint in post-acute stroke *Journal of materials engineering and performance* **20** pp 666-670
[7] Lee H, Ho P, Rastgaar M A, Kregbs H I and Hogan N 2011 Multivariable static ankle mechanical impedance with relaxed muscles *Journal of Biomechanics* **44** pp 1901-1908
[8] Kazerooni H, Racine J L, Huang L and Steger R 2005 On the control of the Berkeley lower extremity exoskeleton (BLEEX) *International Conference on Robotics and Automation* pp 4353-4360
[9] Banala S K, Hun Kim S, Agrawal S K and Scholz J P 2009 Robot assisted gait training with active leg exoskeleton (ALEX) *Neural Systems and Rehabilitation Engineering* **1** pp 2-8
[10] Dai J S, Zhao T and Nester C 2004 Sprained ankle physiotherapy based mechanism synthesis and stiffness analysis of a robotic rehabilitation device *Autonomous Robots* **16** pp 207-218
[11] Racu (Cazacu) C M and Doroftei I 2014 Structural and kinematic aspects of a new ankle rehabilitation device *Applied Mechanics and Materials* **658** pp 507-512
[12] Wang S, Meijneke C and van der Kooij H 2013 Modeling,design and optimization of mindwalker series elastic joint *IEEE Int Conf Rehabilitation Robotics*
[13] Dicheva H *Static analysis of the ankle joint* (Cardiff School of Engineering)
[14] Chowdhury S and Kumar N 2013 Estimation of forces and moments of lower limb joints from kinematics data and inertial properties of the body by using inverse dynamics technique *Journal of Rehabilitation Robotics pp* 93-98
[15] Fong D, Hong Y, Yung P, Fung K Y, Lao M and Chan K M 2006 Estimating complete ground reaction forces and ankle joint torques from pressure insole data in walking and running *XXIV ISBS Symposium* (Salzburg: Austria)
[16] Trabia M *Analyze of the scoth yoke mecanism, using the principle of virtual work* Howard R ed (Hughes College of Engineering: Nevada)
[17] Racu (Cazacu) C M and Doroftei I 2014 An overview on ankle rehabilitation devices *Advanced Materials Research* **1036** pp 781-786
[18] In site http://teststandard.com/data_sheets.htm

IOP Conf. Series: Materials Science and Engineering **95** (2015) 012055 doi:10.1088/1757-899X/95/1/012055

Influence of throttling of pneumatic actuators at the positioning accuracy

M P Hetmanczyk[1] **and P Michalski**[1]
[1]Silesian University of Technology, Faculty of Mechanical Engineering, Institute of Engineering Processes Automation and Integrated Manufacturing Systems, Konarskiego 18A, 44-100 Gliwice, Poland

E-mail: mariusz.hetmanczyk@polsl.pl

Abstract. Changing the dynamic motion parameters of pneumatic equipment requires control of pressure and the volumetric flow rate. The article presents a study of the impact of setting control parameters and external disturbances on the positioning accuracy of pneumatic rodless double acting actuator. Testing process was carried out at different values of the parameters (adjusted to the actuator) and external loads. The probe mass was selected within guidelines (a cushioning diagram) so that the operating point was placed on the curve of limiting movement speed of the actuator.

1. Introduction

Machines based on pneumatic drives belong to units with high reliability indexes. Despite many advantages a problem connected with the positioning [1] of linear and rotary actuators still exists. The main cause of problems is among other things the air compressibility.

Furthermore at positioning of the moving parts of pneumatic drives directly affect:

- selected value and fluctuations of a supply pressure;
- pressure drops (directional control valves, throttling valves, pneumatic hose, etc.);
- applied lubrication and operating temperature (affecting the density of the used oil or lubricant);
- air filtration (incorrectly selection of filter causes a penetration of dirt from entering the system and increase frictional resistance);
- characteristics of the load (force: mass or concentrated, fixed or variable).

Additionally the frictional force [4] varies within a change in movement speed of the piston and the pressure in the actuator cylinder. In most cases, the movement speed setting is made by the operator in a manner dependent on the experience.

2. A purpose and characteristics of the test method

The primary objective of presented studies was determination of relationships between the three main parameters that affect the positioning accuracy of pneumatic linear drives (supply pressure, velocity and mass of conveyed element). The researches were carried out according to the plan shown in Table 1. Measurements were carried out in the extreme values of the attenuation coefficient (CF) setting on non-return throttle valves. The border factor was defined as a value resulting in the loss of motion smoothness at the stage of the movement and positioning on the given parameters of the

Content from this work may be used under the terms of the Creative Commons Attribution 3.0 licence. Any further distribution of this work must maintain attribution to the author(s) and the title of the work, journal citation and DOI.
Published under licence by IOP Publishing Ltd

IOP Conf. Series: Materials Science and Engineering **95** (2015) 012055 doi:10.1088/1757-899X/95/1/012055

medium. For the purposes of researches the electropneumatic tests stand consistent with the scheme shown in figure 1 has been developed.

During the measurements, the following parameters were recorded: pressure at the outlet of draining chamber (cushioned chamber), vibration level (based on the effective values of velocity) measured with sensors mounted on the moving elements (connected rigid with the piston) and cylinder cap (at the top side), positioning accuracy (O1D100 position sensor).

Table 1. The scheme and main parameters of the experimental research.

Cushioning factor CF [%]	Value of the supply pressure p [bar]	Active load (where: WM, PM – measurements with and without additional mass)
0 (throttle valve fully opened) 50 border (limiting a movement smoothness)	3 and 6 bars (setting on the pressure reducing valve - air preparation system)	WM – approximate weight of moving parts about 200 grams, PM – additionally mounted steel cylinder with weight of 500 grams (total weight about 700 grams).

The test system was configured with the following components:

- pneumatic linear drive (OSP-P25020000500) with adjustable end cushioning at both sides with connected non-return throttle valves (throttling at the outlet),
- 5/3-way single solenoid valve closed in mid-position (position spring-centered, manual override in both directions), actuation in both directions by solenoids with one winding and pneumatic piloting (P2LAX611EENDDB49 series),
- analog position sensor (O1D100) and noncontact switch (a proximity sensor for piston carrier stopping) connected to modules of relays.

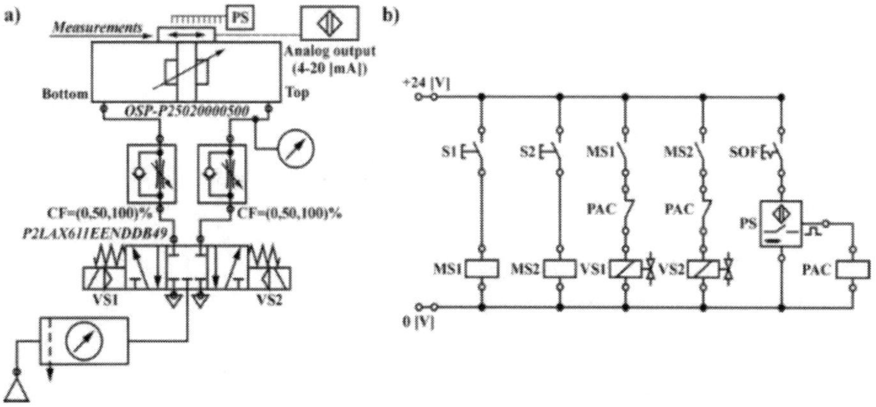

Figure 1. Functional schemes of: a) the pneumatic subsystem, b) the electric subsystem.

The time constant related to the system response is dependent on the value of a delay-time resulting from an activation of relay systems and the induction sensor (figure 2). In order to ensure the measurement reproducibility an automatic interruption has been implemented (in phase of extension movement). The control system provides fully repeatable measurements (power has been switched off when the solenoid actuator came across face of inductive sensor).

Due to an implementation of the inductive sensor applied trigger have form of a metal tag with respectively matched length.

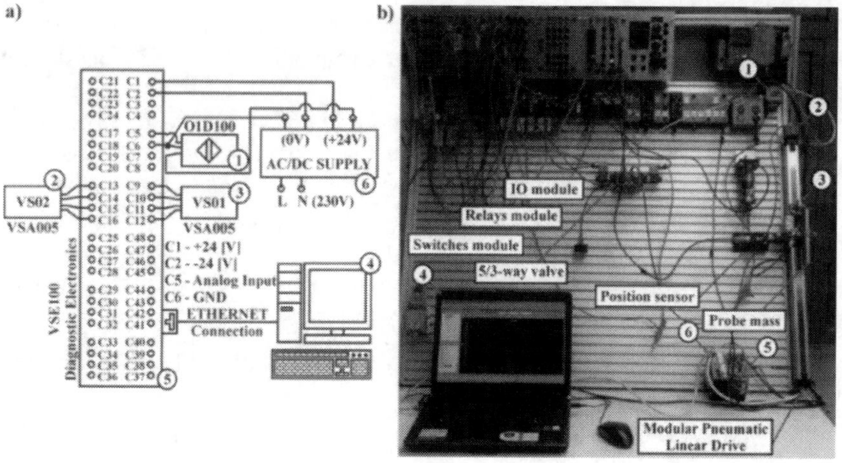

Figure 2. Views of: a) the measurement path configuration, b) the laboratory stand.

3. Results

The basis of the research was a simulation of an emergency stop phase with different settings of parameters. As a reference position of the axis of stopping assumed distance from the sensor to the starting position of the piston equals 225 millimeters. Time used to a determination of the steady state conditions has been determined up to the moment when the peak to peak oscillation does not exceed 0.18 mm/s (or 0.09 mm/s).

Before each cycle (an upward movement of the cylinder) a chamber has been filled with compressed air at a pressure equal to the supply pressure set at regulator valve (compressed air preparation system). The charts (figure 3-8) show the measured values from the sensor mounted on the moving components. Recorded peak velocities of the actuator housing are summarized in table 2.

Oscillations of the piston actuator stem directly from the nature of the control parameter settings. In addition at the accuracy has influence also values:

- supply pressure,
- cushioning factor settings:
- CF = 0% - passage through the stop position with simultaneous exceeding of value that define assumed position (figure 3, 4),
- CF = 50% - stopping in a position close to defined value (figure 5, 6),
- CF = border – settings of the factor at the border values (ie. crawling of moving parts) leads to a slow pressure equalization before and after the non-return throttle valve and stopping before fixed position (figure 7, 8),
- actuator load.

Another negative phenomenon is a piston subsidence under the influence of the compressibility of air trapped in the chamber of the actuator, valve and supply lines.

In the present case a displacement of conveyed mass after stopping was no noticed what results directly from the small value of the load. Differences in reaching the target position are shown in figure 9. The resulting inaccuracies are associated with factors dependent on the time constant of the sensor, an applied weight, orientation of the actuator and the setting of the valves controlling air flow rate.

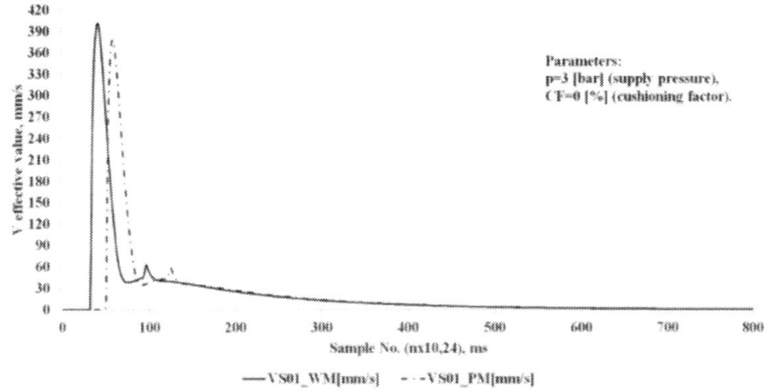

Figure 3. Time characteristics of effective velocity values (p=3 bar, CF=0%) of piston carrier.

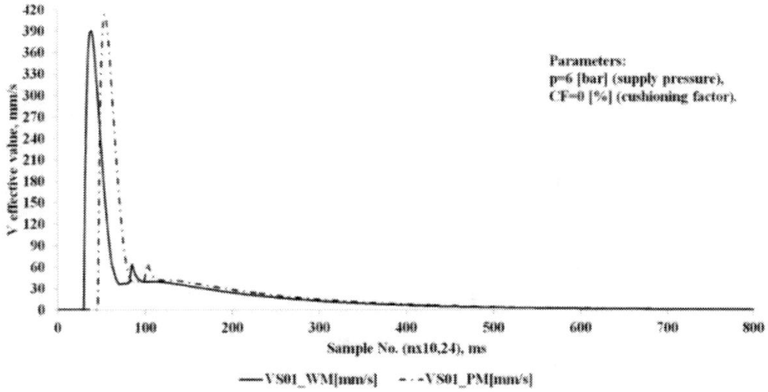

Figure 4. Time characteristics of effective velocity values (p=6 bar, CF=0%) of piston carrier.

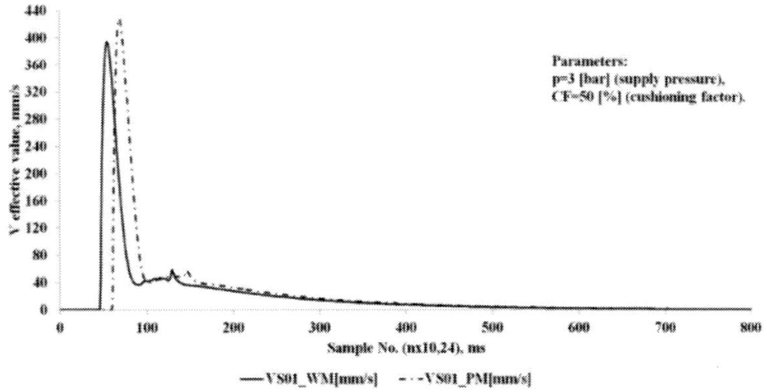

Figure 5. Time characteristics of effective velocity values (p=3 bar, CF=50%) of piston carrier.

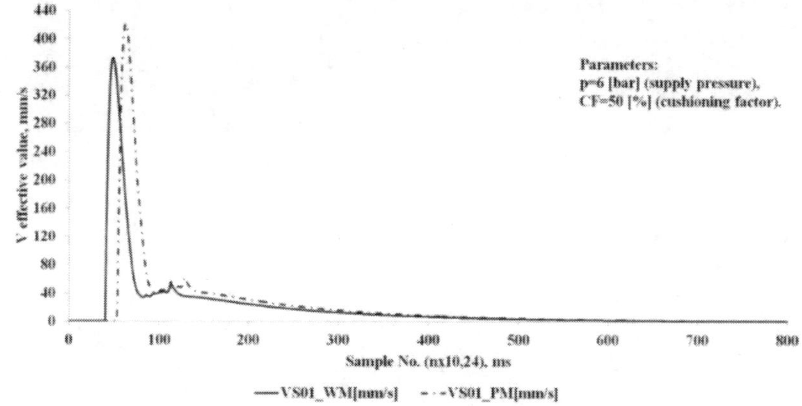

Figure 6. Time characteristics of effective velocity values (p=6 bar, CF=50%) of piston carrier.

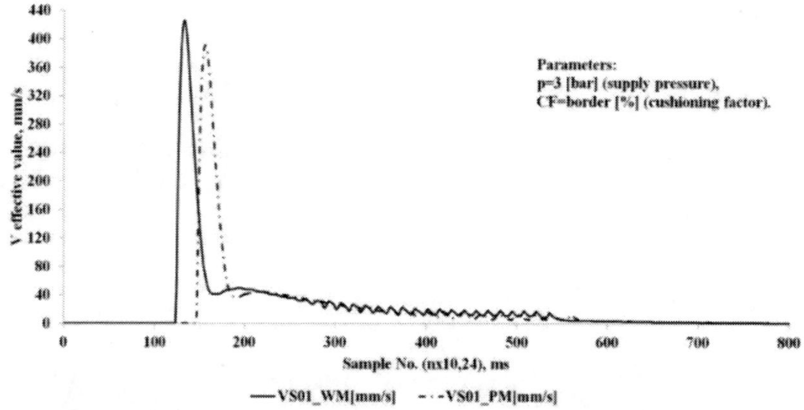

Figure 7. Time characteristics of effective velocity values (p=3bar, CF=border) of piston carrier.

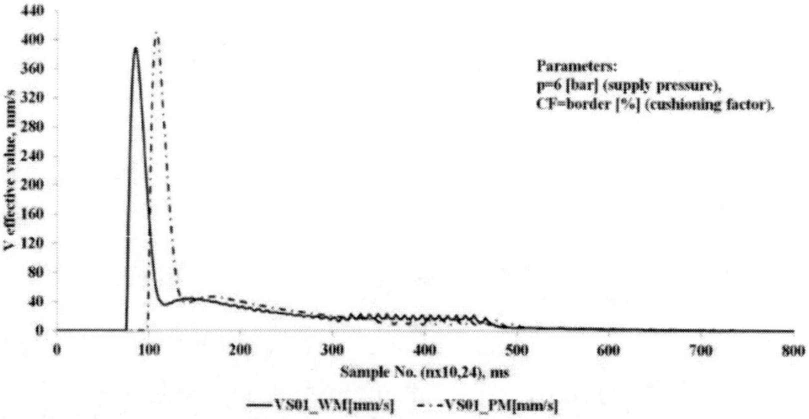

Figure 8. Time characteristics of effective velocity values (p=6bar, CF=border) of piston carrier.

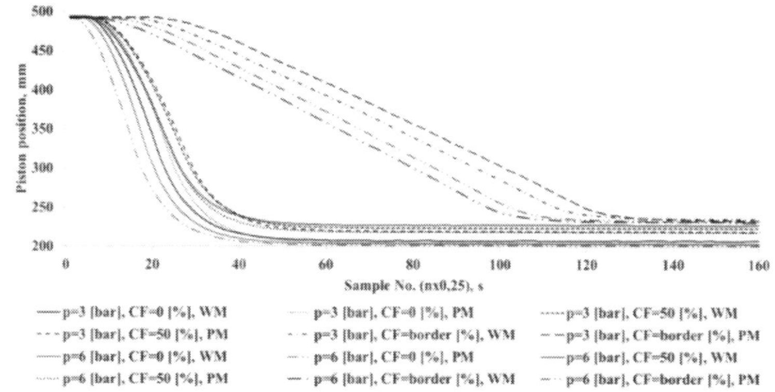

Figure 9. Time characteristics of position courses at different work conditions.

Despite the different effective speed values and operating parameters (mass actuated) attenuate the real time value to the minimum level does not differ from each other in a meaningful way. The only difference is the instantaneous value of the amplitude changes the effective speed. Oscillations around the equilibrium position depend on the time equalization of pressures in fed and discharging chambers. Increasing the degree of throttling leads to rigid suppression characteristics what results in increasing of the vibration energy recorded by the sensor.

Table 2. Summary of effective velocity peak values and cushioning times.

Settings p [bar], CF [%]	Load	V_{peak} [mm/s] VS01	VS02	Time of cushioning T_c [ms] VS01	VS02
p=3, CF=0	WM	401.8	0.3	$12.3\cdot10^{-3}$	$0.2\cdot10^{-3}$
	PM	378.5		$12.7\cdot10^{-3}$	$0.3\cdot10^{-3}$
p=6, CF=0	WM	390.3	0.4	$12.1\cdot10^{-3}$	$0.7\cdot10^{-3}$
	PM	412.8		$11.8\cdot10^{-3}$	$0.6\cdot10^{-3}$
p=3, CF=50	WM	394.1	0.2	$11.7\cdot10^{-3}$	$0.4\cdot10^{-3}$
	PM	428.4		$12\cdot10^{-3}$	$0.3\cdot10^{-3}$
p=6, CF=50	WM	373.1	0.2	$12.1\cdot10^{-3}$	$0.4\cdot10^{-3}$
	PM	421.6	0.3	$12.3\cdot10^{-3}$	$0.2\cdot10^{-3}$
p=3, CF=100	WM	426.2	0.2	$13.3\cdot10^{-3}$	$0.1\cdot10^{-3}$
	PM	391.4	0.3	$13.2\cdot10^{-3}$	$0.2\cdot10^{-3}$
p=6, CF=100	WM	388.5	0.2	$11.9\cdot10^{-3}$	$0.2\cdot10^{-3}$
	PM	410.7		$11.6\cdot10^{-3}$	$2.2\cdot10^{-3}$

Tuning the operational parameters in order to stop in an intermediate position or in Safe Mode requires considerable experience and tools allowing determination and control of actuator dynamics [4].

4. Conclusions

The positioning accuracy of pneumatic cylinders depends on several parameters. Among other things: user-configurable variables (pressure. volumetric flow rate. control and sensor subsystems). External disturbances (pressure fluctuations. improper lubrication or filtration. additional masses).

On the basis of presented researches the impact of a series of sizes on the spool valve positioning accuracy and quality of motion it has been also observed (smaller crosswise sections of distributor induces deterioration caused traffic flow and reducing a value of the border coefficient).

The accuracy of the stop position depends on the load. Speed and direction of movement of the piston. The vector direction of the transported mass. Improvement of the accuracy of positioning may occur through the use of precise pressure regulator placed into the system before or after the divider. In case of industrial equipment (despite of the same location. construction. selection of settings and load value) significant differences in the lifetime of machines still exists [2, 3, 5].

References

[1] Hetmanczyk M P and Michalski P 2014 The self-excitation phenomenon of quasi shielded inductive proximity switches *Advanced Materials Research* **837** pp 405-410

[2] Hetmanczyk M, Swider J and Wszołek G 2014 The systematized data structures oriented towards diagnosis and prediction *Advances in Intelligent Systems and Computing (Recent advances in automation. robotics and measuring techniques* vol 267 ed R Szewczyk C Zieliński M Kaliczyńska (Springer) pp 63-72

[3] Ćwikła G, Grabowik C and Janik W 2014 Case Study of Manufacturing Information Acquisition System (MIAS) Automated Continuous Production System *Applied Mechanics and Materials* **657** pp 808-812

[4] Dymarek A and Dzitkowski T 2013 Reduction vibration of mechanical systems *Applied Mechanics and Materials* **307** pp 257-260

[5] Kalinowski K 2012 Multistage decision making process of multi-criteria production scheduling *Journal of Machine Engineering* **12** pp 20-33

IOP Conf. Series: Materials Science and Engineering **95** (2015) 012056 doi:10.1088/1757-899X/95/1/012056

The qualitative assessment of pneumatic actuators operation in terms of vibration criteria

M P Hetmanczyk[1] and P Michalski[1]
[1]Silesian University of Technology, Faculty of Mechanical Engineering, Institute of Engineering Processes Automation and Integrated Manufacturing Systems, Konarskiego 18A, 44-100 Gliwice, Poland

E-mail: mariusz.hetmanczyk@polsl.pl

Abstract. The work quality of pneumatic actuators can be assessed in terms of multiple criteria. In the case of complex systems with pneumatic actuators retained at end positions (with occurrence of piston impact in cylinder covers) the vibration criteria constitute the most reliable indicators. The paper presents an impact assessment on the operating condition of the rodless pneumatic cylinder regarding to selected vibrational symptoms. On the basis of performed analysis the authors had shown meaningful premises allowing an evaluation of the performance and tuning of end position damping piston movement with usage the most common diagnostic tools (portable vibration analyzers). The presented method is useful in tuning of parameters in industrial conditions.

1. Introduction

Pneumatic actuators have gained great popularity in the industry, thanks to the many advantages among other things cleanliness of work, design simplicity, simple adjustment of durability and high resistance to external factors. Additionally a control of the operational parameters of pneumatic drives does not provide complex problem [1, 3]:

- change of speed (linear or rotational) - throttling the working medium (throttle or non-return throttle valves),
- force or torque modifications - controlling a values of the supply pressure.

Rodless cylinders are often used in design of pneumatic manipulators, linear movement mechanisms and applications with high loads of forces and bending moments. Described drives are equipped with end-position cushioning circuits allowing minimization of piston strokes at lids.

Cushioning at end positions has the following advantages:

- a reduction of internal and external shocks [1] and machine noises,
- a minimization of maintenance activities [2, 3].

2. Measuring station and experiment plan

Primary aims of presented researches are connected with determination a suitability of vibration diagnostic methods for an evaluation of the operation quality of pneumatic systems and minimization of a kinetic energy in a braking phase.

The basis of studies results from a lack of methods allowing tuning of linear pneumatic actuators in order to life increasing, minimization of inertia and vibrations [3, 4] of machines in industrial

Content from this work may be used under the terms of the Creative Commons Attribution 3.0 licence. Any further distribution of this work must maintain attribution to the author(s) and the title of the work, journal citation and DOI.
Published under licence by IOP Publishing Ltd

conditions (with usage of well-known diagnostic methods). In most cases a maintenance service of production plants based on experience without evaluation of measured parameters.

The researches were carried out according to the plan of the experiment shown in the table 1. The selected pressure values have been matched to indicate a normal level in the installation (6 bars) and decrease of pressure (3 bars). For the purposes of researches the electropneumatic tests stand consistent with the scheme shown in figure 1 has been developed.

Table 1. The scheme and main parameters of the experimental research.

Cushioning factor CF [%]	Value of the supply pressure p [bar]	Active load (where: WM, PM – measurements with and without additional mass)
0 (throttle valve fully opened) 50 border (limiting a movement smoothness)	3 and 6 bars (setting on the pressure reducing valve - air preparation system)	WM – approximate weight of moving parts about 200 grams, PM – additionally mounted steel cylinder with weight of 500 grams (total weight about 700 grams).

During the measurements, the following parameters were recorded (figure 2):

- pressure at the outlet of draining chamber (cushioned side) with usage of integrated throttle valve (pneumatic throttling),
- vibration level (effective values of velocity) measured with sensors mounted on moving elements (connected rigid with the piston) and cylinder cap (at the top side).

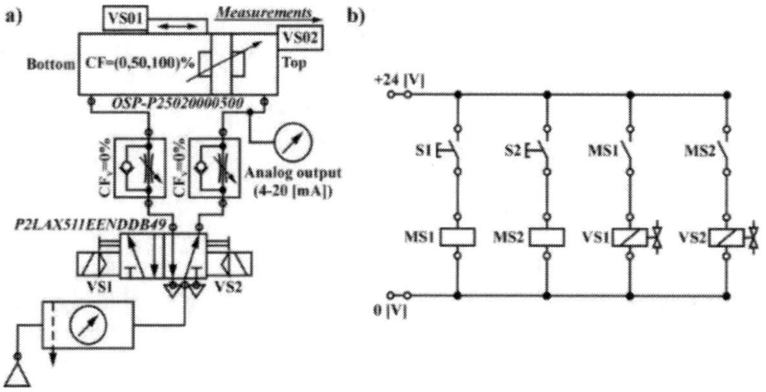

Figure 1. Functional schemes of: a) the pneumatic subsystem, b) the electric subsystem.

The actuator has been inclined with respect to a horizontal direction (figure 2) to determine an effect of a braking weight on the speed and vibrations (in respect of force and torque loads).

The test system has been configured with the following components:

- pneumatic linear drive (OSP-P25020000500) with adjustable end cushioning at both sides with connected non-return throttle valves (throttling at the outlet),
- 5/2-way double solenoid valve actuated by solenoids and pneumatic piloting in both directions, manual overrides (P2LAX511EENDDB49 series),
- analog pressure sensor (PI3054) - monitoring of discharging chamber parameters,
- modules of relays and momentary action switches.

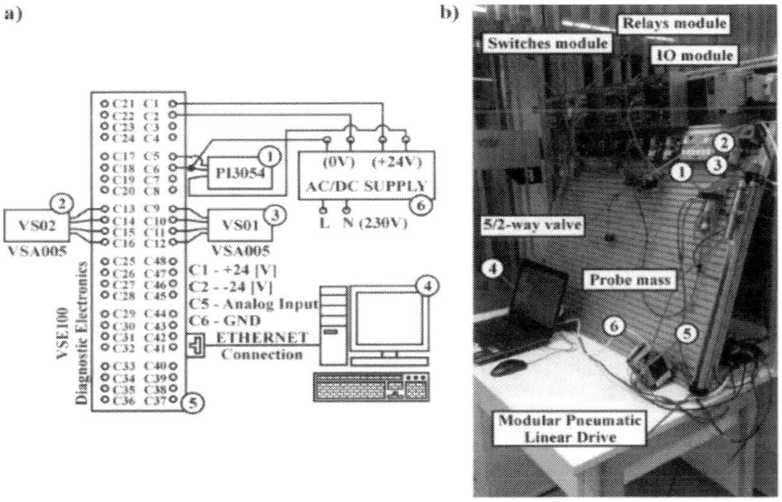

Figure 2. Views of: a) the measurement path configuration, b) the laboratory stand.

The measurements were performed at the maximum speed available at non-throttled flow within the range of selected pressure. The cushioning factor CF relates to throttle valves installed in the rodless cylinder.

3. Results

From the point of view of the operating conditions important vibrational symptoms can be collected from two independent measuring points: a movable table (the VS01 sensor) and actuator covers (the VS02 sensor). Each measuring cycle consisted of data registration from starting position, movement at distance equals 500 millimeters and a velocity descent in the end position (using pneumatic cushioning with different parameter settings).

The impact value on the actuator covers is directly dependent on a speed of the piston and the total weight of moving parts. The additional mass PM (in case of movement in the direction from the bottom to the top) causes a deceleration of the piston. This solution is one of the possibilities used to reduction of the detrimental impact of the piston strokes in cylinder covers.

A small value of the effective velocity (figure 3, 4) results from the use of additional mechanical damping (the plastic liner). Such solution causes faster wear of sealing package, especially in the case of large masses. Throttling of pneumatic medium (CF = 50%) manifested at a slight increase of the effective velocity (figure 5, 6), which is associated with the appearance of the elastic force of the air and increasing pressure on the draining chamber (resulting from throttling).

Setting the value of the damping factor CF = 100% (corresponding to completely close of the throttle valve) causes oscillations (figure 7, 8) around an equilibrium point of the moving part. Higher value of pressure in the chamber enlarges a rigidity of the system, thereby increasing an attenuation value of less velocity and time. It is also visible in decrease of the vibration measured at actuator housing.

IOP Conf. Series: Materials Science and Engineering **95** (2015) 012056 doi:10.1088/1757-899X/95/1/012056

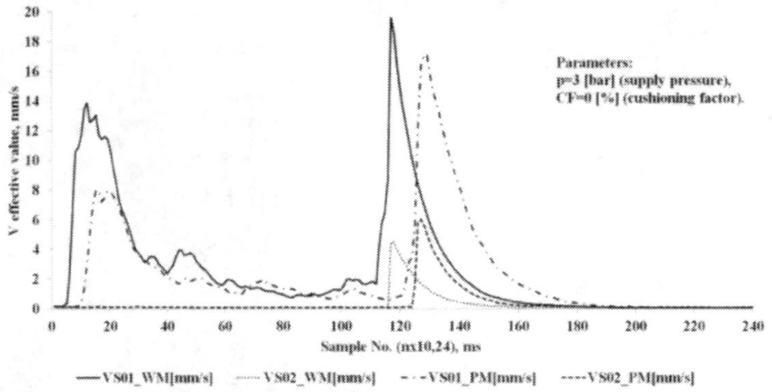

Figure 3. Time characteristics of effective velocity values (p=3 bar, CF=0%).

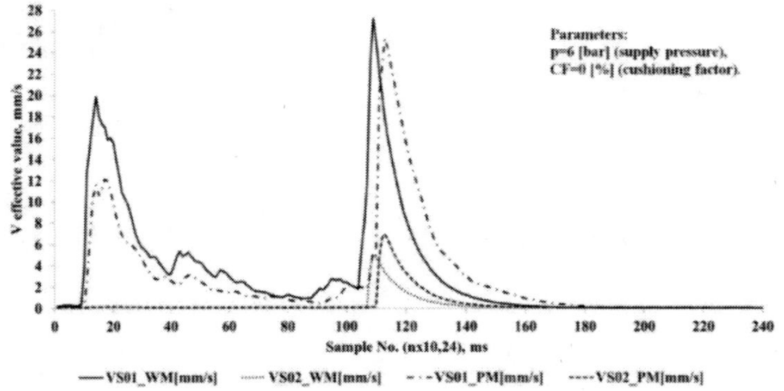

Figure 4. Time characteristics of effective velocity values (p=6 bar, CF=0%).

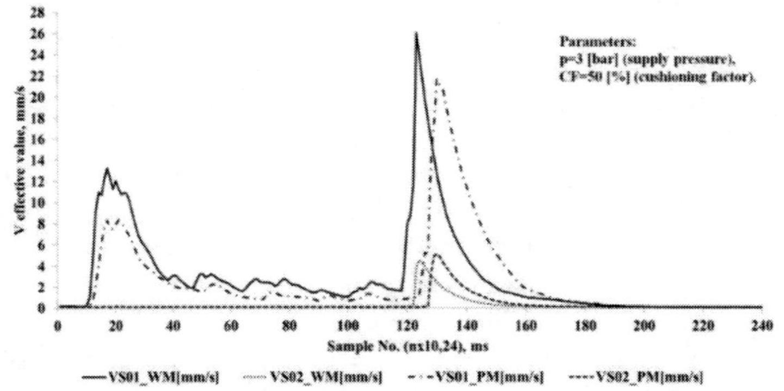

Figure 5. Time characteristics of effective velocity values (p=3 bar, CF=50%).

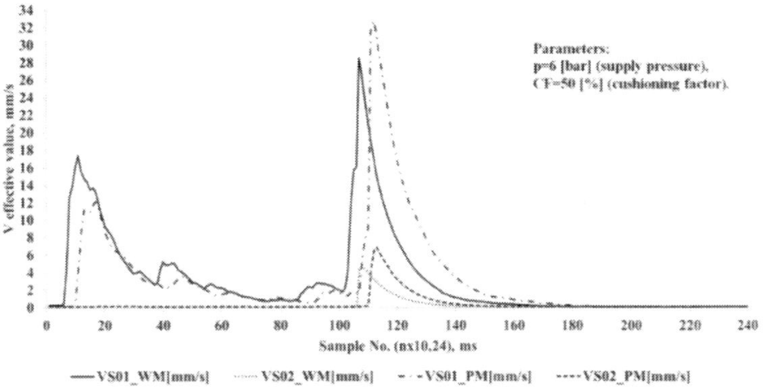

Figure 6. Time characteristics of effective velocity values (p=6 bar, CF=50%).

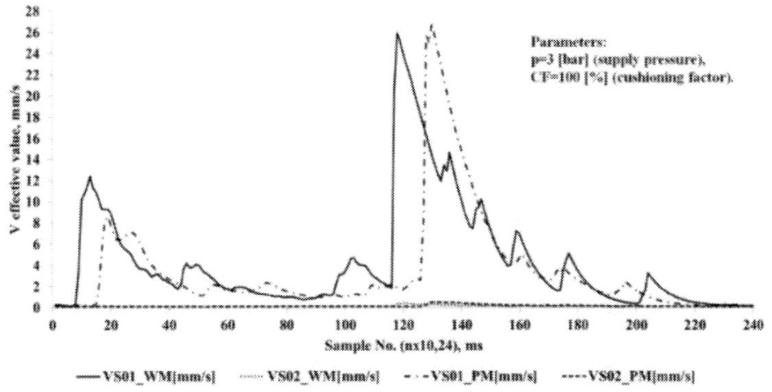

Figure 7. Time characteristics of effective velocity values (p=3 bar, CF=border).

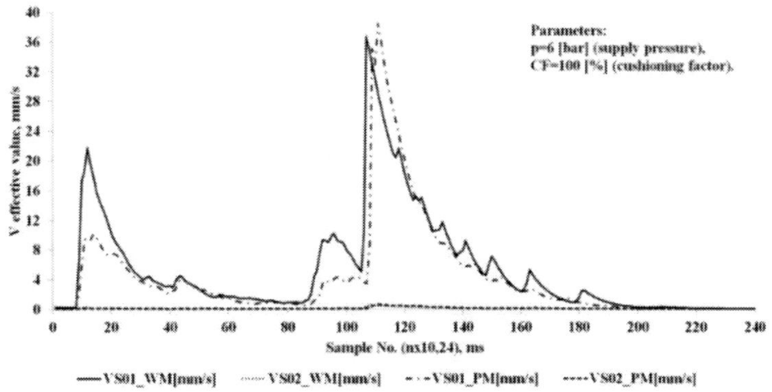

Figure 8. Time characteristics of effective velocity values (p=6 bar, CF=border).

Complete closure of the throttle valve causes positioning errors proportional to a value dependent on the design of damping elements. In presented case a precise adjustment of the supply pressure test system was not used.

The values of pressure (figure 9) are the results of measurements on draining chamber. In the course of measurements values of the damping time was determined (vibration registered on the housing and the piston).

IOP Conf. Series: Materials Science and Engineering **95** (2015) 012056 doi:10.1088/1757-899X/95/1/012056

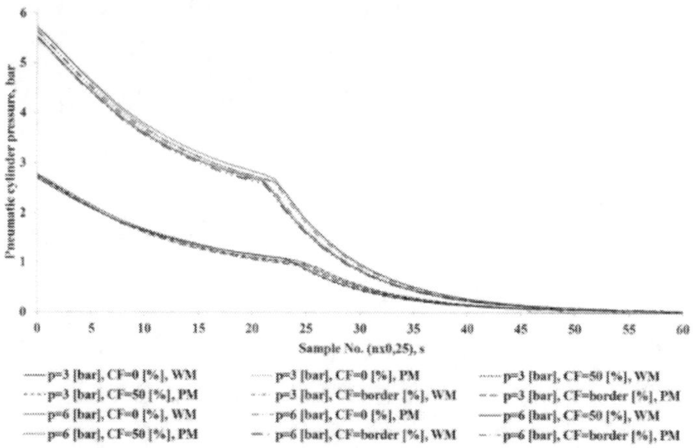

Figure 9. Time characteristics of position courses at different work conditions.

Measurements of effective velocity (table 2) indicate a necessity of settings the upper ranges of CF coefficient (volumetric elasticity coefficient proportional to pressure changes). Vibration measurement enclosure is not meaningful indicator to assess the quality of operation of pneumatic drives. The location of the sensor on the machine frame (rigidly connected with the drive) allowing evaluation of the vibrations propagation on machine parts, but the quality of detention depends on the parameters measured directly on the moving parts. Another important element is an influence of lubrication on the uniformity of a speed of moving parts and actuator subassemblies.

Table 2. Summary of effective velocity peak values and cushioning times.

Settings p [bar], CF [%]	Load	V_{peak} [mm/s] VS01	V_{peak} [mm/s] VS02	Time of cushioning T_c [ms] VS01	Time of cushioning T_c [ms] VS02
p=3, CF=0	WM	19.6	4.5	$0.73 \cdot 10^{-3}$	$0.7 \cdot 10^{-3}$
	PM	17.2	6	$0.74 \cdot 10^{-3}$	$0.69 \cdot 10^{-3}$
p=6, CF=0	WM	27.2	5	$0.71 \cdot 10^{-3}$	$0.51 \cdot 10^{-3}$
	PM	25.3	6.9	$0.98 \cdot 10^{-3}$	$0.73 \cdot 10^{-3}$
p=3, CF=50	WM	26.1	4.5	$0.83 \cdot 10-^{3}$	$0.49 \cdot 10^{-3}$
	PM	21.7	5.1	$0.77 \cdot 10^{-3}$	$0.73 \cdot 10^{-3}$
p=6, CF=50	WM	28.6	4.6	$0.89 \cdot 10^{-3}$	$0.43 \cdot 10^{-3}$
	PM	32.8	6.9	$1.28 \cdot 10^{-3}$	$0.75 \cdot 10^{-3}$
p=3, CF=100	WM	25.9	0.4	$1.26 \cdot 10^{-3}$	$0.69 \cdot 10^{-3}$
	PM	27	0.5	$1.2 \cdot 10^{-3}$	$1.24 \cdot 10^{-3}$
p=6, CF=100	WM	36.8	0.6	$1.3 \cdot 10^{-3}$	$0.5 \cdot 10^{-3}$
	PM	38.5	0.7	$1.9 \cdot 10^{-3}$	$0.66 \cdot 10^{-3}$

4. Conclusions

The positioning accuracy of pneumatic cylinders depends on several parameters, among other things: user-configurable variables (pressure, volumetric flow rate, control and sensor subsystems), external disturbances (pressure fluctuations, improper lubrication or filtration, additional masses).

On the basis of presented researches the impact of a series of sizes on the spool valve positioning accuracy and quality of motion it has been also observed (smaller crosswise sections of distributor induces deterioration caused traffic flow and reducing a value of the border coefficient).

The accuracy of the stop position depends on the load, speed and direction of movement of the piston, the vector direction of the transported mass, Improvement of the accuracy of positioning may occur through the use of precise pressure regulator placed into the system before or after the divider, In case of industrial equipment (despite of the same location, construction, selection of settings and load value) significant differences in the lifetime of machines still exists [2, 3, 5].

References

[1] Dymarek A and Dzitkowski T 2013 The computer aided passive reduction of vibration to desired vibration amplitude *Applied Mechanics and Materials* **430** pp 342-350
[2] Grabowik C, Krenczyk D and Kalinowski K 2012 The hybrid method of knowledge representation in a CAPP knowledge based system *Hybrid Artificial Intelligent Systems* **7209** pp 284-295
[3] Hazem I A, Mohd S B B, Noor Bashi S M and Marhaban M H 2009 A review of pneumatic actuators (Modeling and Control) *Australian Journal of Basic and Applied Sciences* **3** pp 440-454
[4] Michalski P, Hetmanczyk M P and Swider J 2013 Simply-Integrated Method of Judgments of Expert Knowledge Collected in Databases for Objective Computer-Aided Engineering Systems *Hybrid Artificial Intelligent Systems* **8073** pp 262-268

IOP Conf. Series: Materials Science and Engineering **95** (2015) 012057 doi:10.1088/1757-899X/95/1/012057

Comparative analysis of bio fouling microorganisms after treatment with glidarc

M Hnatiuc[1], A Sabau[2], B Hnatiuc[1] and S Ghita[1]

[1]Constanta Maritime University, Mircea cel Batran 104, Constanta, 900663, Romania

E-mail: mhnatiuc@yahoo.com

Abstract. The biofouling of the surfaces immersed in water is one of the most important problems which must be solved in the naval field. This phenomenon is amplified during the harbor operations, when the ships are stationary and there is a high density of microorganisms in the neritic area, developing the biofouling. For this reason, in order to prevent or delay the deposition of the first layers of biofouling, different methods have been used [1]. This paper presents the comparative analysis of microorganisms' behavior obtained by using GlidArc technology for the treatment of the naval metallic surfaces [2, 3]. The main parameters identified for data processing were: the number of microorganisms shared from the point of view of sensibility to the used technology found on the metallic surfaces, the type of the naval paint and the treatment methods. For analysis it was used the epifluorescence microscopy method with. The comparative analysis follows the data processing which has the same input characteristics. Finally it was observed the microorganism's lifetime after the surface treatment.

1. Introduction

Any naval treatment should not affect the sea biodiversity. The biofouling is a well known problem, involving environmental impact. Its consequences are hardly important on naval corrosion topic.
Marine biofouling is typically described by following four stages of ecosystem development: the first two stages, also named "microfouling", are referring to the forming condition of the biofilm, related first to the processes of van der Waals interaction covering the submerged surface with a conditioning film of organic polymers, and then of bacterial adhesion (in the next 24 hours); the third concerns the appearance by the end of the first week of secondary colonizers like spores, micro algae, protozoans, etc.; and the last stage is manifested after around 2-3 weeks by the appearance of tertiary colonizers, referring to "macrofouling".

In this paper is presented a laboratory study on the naval steel samples, using GlidArc as biofouling treatment method. The naval steel was covered with different paints before the experiment. It was studied the microorganism's evolution in time after the GlidArc treatment. The correlation and ANOVA methods were applied to the results in order to identify the efficient methods for treating and cleaning the naval surfaces.

Gliding Arc discharges (GlidArc, see figure 1) provide the most effective treatments among all types of electrical discharges. They are produced between at least two metallic electrodes with divergent shape structure. In our experiments the electrodes were supplied from a voltage power supply (10kV, 50Hz) providing the maximum electrical current value of 50mA. The minimum

Content from this work may be used under the terms of the Creative Commons Attribution 3.0 licence. Any further distribution of this work must maintain attribution to the author(s) and the title of the work, journal citation and DOI.
Published under licence by IOP Publishing Ltd

distance between the electrodes was 2 mm, the maximum distance was 30mm and the height of the electrodes was 80mm. The maximum gas flow amount was 50 Nm^3/min.

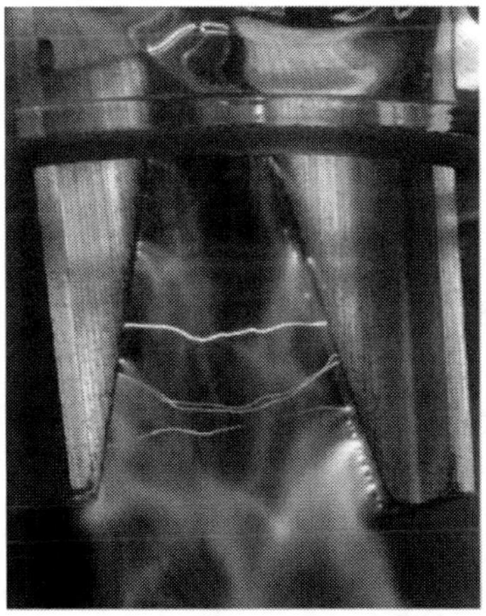

Figure 1. Photo of GlidArc discharges.

2. Methodology

The direct and indirect treatments with plasma produced by GlidArc are applied on the target surface. The direct treatment consists in applying the plasma electrical discharge direct on the metallic surface. The indirect process consists on activating the water with GlidArc discharge then immersing the metallic samples into this solution.

2.1. Experimental

The experiment consists in two parts depending on the types of samples. The process monitoring is achieved using an epifluorescence trinocular microscope, 400-FL type with blue filter of (450-480)nm. For cells quantification and measurement are used the CellC and Image J software. For each experiment, after treatment, the samples are immersed in sea water.

a. The initial experiment was achieved on untreated samples of naval steel, [2]. It is applied the plasma at the distance of 2.5 cm distance lasting of 5 minutes and at the distance of 1.2 cm during 3.5 minutes. One untreated sample was taken as control to compare the microorganisms' evolution. The microorganisms are observed using 40 objective x10 eyepiece magnitude microscope. All the data are recorded at the same moments of time: 30min, 5h, 10h, 24h. With Image J software are achieved 100 measurements/time. The figure 2 depicts the microorganism evolution in time where the green cells represent the living fraction and the red cells represent the dead fraction.

b. The second experiment used three pretreated samples, [3], using different paints during 5 minute. For the cells quantification we used the CellC software and 20 microscopic fields/samples were measured.

The activated water consists in 50mL of distilled water exposed to the non-thermal plasma produced by GlidArc. The time of plasma application for the both methods was 8 minutes.

The adherence period consist in immersion samples in sea water during 6h respectively 24h for biofilm coating. The evolution of microorganisms was studied by harvesting samples at 10, 120, 480 and 1440 minutes after the treatment and processed with epifluorescence microscope. The data results indicate the general microorganisms' evolution and in particular the families as: green algae, cianobacteria and heterotrophic bacteria.

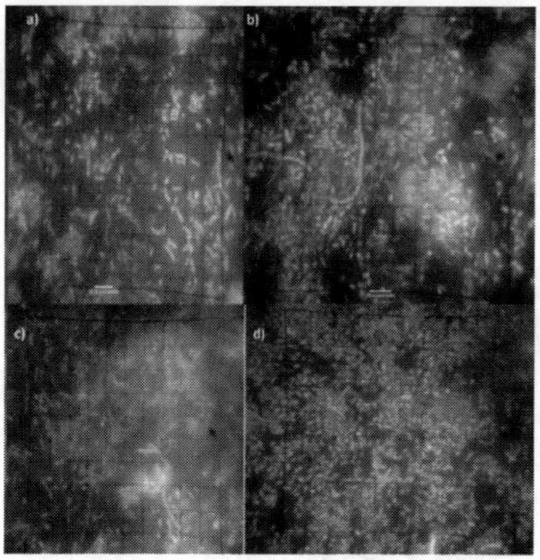

Figure 2. The microorganism behavior in time (a-30 min, b-5 h, c-10h, d-24h).

2.2. Data analysis

As methods to identify the microorganism's behaviors function of GlidArc treatment and the type of surface, the correlation and variance analysis (ANOVA) were used. The correlation describes the behaviors of dead cells in time for the two groups of data and analysis the variance test of significant differences between class means. The ANOVA test is based on comparing two independent estimates of the population variance. ANOVA procedure requires the following assumptions: the observations are independent; the observations in each group come from normal distribution; the population variances in each group are the same [6]. The groups were represented by the samples (red-1, blue-2, and black-3) and the factor by the type of treatment: direct GlidArc (1) or plasma activated water (2). The One-Way ANOVA methods are applied in this case because there is one factor to test the difference between the groups of treatment.

3. Data processing

After treating the naval steel samples, see section 2.1.a, the evolution of dead and alive cells occurred (figure 3). The dead cells on the treated samples have a linear evolution between 5 and 600 min then the evolution of microorganisms are constant up to 24h. The GlidArc is efficient just for a short time period.

The dead cells evolutions reported to the total number of cells in case of 5 min GlidArc exposure are: 76%, 87.3%, 97.5% and 97.4%. It was observed that after 24h the number of dead cells is constant. To observe the GlidArc effect in the both cases it is achieved a correlation between the dead cells. The Pearson Coefficient is 0.989, very closed to 1 so there is a correlation between the responses. In both cases the most of the microorganisms are destroyed until 24h.

The data analysis was continued for the experiments described in section 2.1.b. where the samples were pretreated with different solutions. It was found how the treatment of the samples influences the microorganisms' evolution using the variance analysis intragroup and intergroup. It was identified the microorganisms' evolution, of the first layer of biofouling, for the samples immersed in sea water 6h before the test. Because the microorganisms have not the same evolution, they were split in permeabilized heterotrophic bacteria, cyanobacteria and green algae noted with 5, 6 and 7. It was kept the control on each sample to compare the results noted with 11. The samples are noted 1, 2 and 3 for red, blue, black surface (different types of paints), see figure 4.

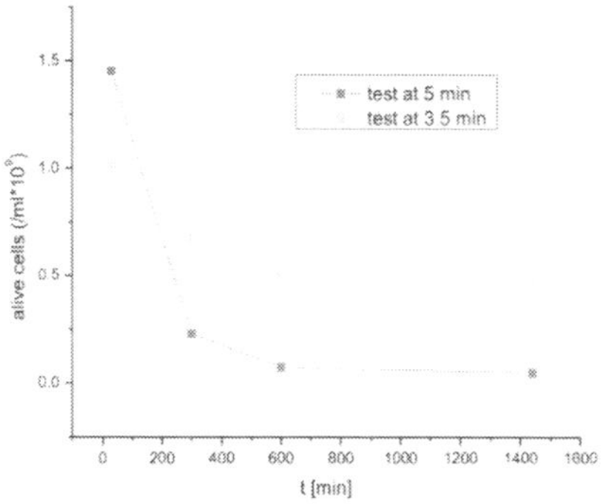

Figure 3. The evolution of the alive cells in two cases of direct GlidAdrc exposure (5 min and 3.5 min).

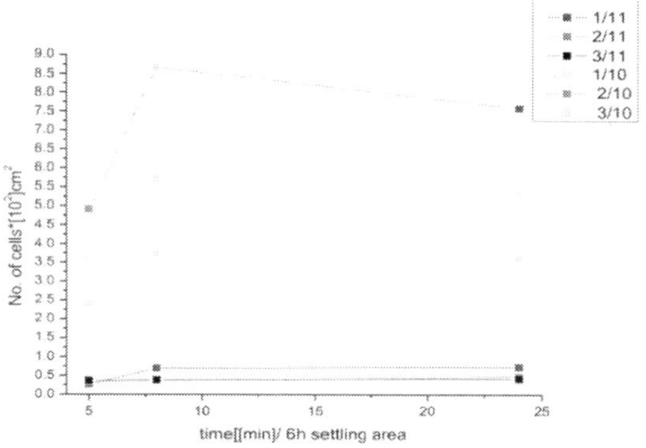

Figure 4. Comparing Dead cells evolution for control (11) and samples 1, 2, 3 treated with direct GlidArc (10); 1/11, 2/11, 3/11 represent the number of samples/the control samples and 1/10, 2/10, 3/10 represent the number of samples /direct GlidArc treatment.

The evolution of microorganism in case of the control sample is constant while in the case of treated samples the number of dead cells is increasing. The number of dead cells for the monolayer biofouling decreases after 8h showing that the effect of GlidArc discharge is finished. The microorganisms were split in three groups in order to observe the treatment effect, see table 1. It was used the notation of the samples type/x, where x can be 5, 6, 7 for the microorganisms group.

Table 1. Microorganisms group correlation function of samples type (direct discharge exposure).

	2/x-3/x	1/x-2/x	1/x-3/x
Pearson Correlation for 5	0.97911	0.99408	0.9954
Pearson Correlation for 6	0.63487	0.98332	0.76481
Pearson Correlation for 7	0.47032	-0.98143	-0.29231

The heterotrophic cells evolution depends on the paint. For red samples the data are negative correlated with the data of blue and black samples. The heterotrophic cells are influenced by both the paint type and the GlidArc treatment. The green algae and heterobacteria are independent by the samples treatment, being autotrotrophic. The cleaning process of biofouling is efficient for the monolayer so the data analysis was continued for the first two layers (see section 2.1b). They were kept the same notations and for the independent factor it was marked the direct GlidArc discharge process with 1 and the tests based on activated water with 2.

Using the same principle as for the analysis described above, it was analyzed the microorganisms time evolution for the two treatment types particularized for three types, see table 2.

Table 2. The correlation of the microorganism's evolution in the case of two treatments (1 and 2).

Samples type/treatments	Pearson corr. 1/1	Pearson corr. 1/2	Pearson corr. 2/1	Pearson corr. 2/2	Pearson corr. 3/1	Pearson corr. 3/2
Pearson corr. 1/1	1	0.35855	0.75042	0.50536	0.80438	0.71148
Pearson corr. 1/2	0.35855	1	0.87698	0.98071	0.0031	0.8327
Pearson corr. 2/1	0.75042	0.87698	1	0.92508	0.46219	0.9745
Pearson corr. 2/2	0.50536	0.98071	0.92508	1	0.09111	0.85442

Pearson corr. 3/1	0.80438	0.0031	0.46219	0.09111	1	0.5529
Pearson corr. 3/2	0.71148	0.8327	0.9745	0.85442	0.5529	1

Note: The columns and rows represent Pearson correlation: ***samples type/treatment type***

The table 2 indicates the correlation results applied between all the data for each sample submitted to the direct treatment and to the plasma activated water. The microorganisms treated with first methods do not have the same evolution in time as the second. The second method is stronger as the first. There is the intergroup correlation but the results are uncorrelated into the intragroup. The data analyses on group of microorganisms (5, 6, 7) for direct treatment and plasma activated water treatment are achieved with ANOVA method [4].

The factor of analysis is the treatment type: direct and indirect. This is applied for the same group of samples (1, 2, 3). The null hypothesis for the analysis is: the treatment for the entire samples group has the same efficiency. The alternative hypothesis is: the effects of treatments are different for each group. The results show that all of them are significantly drawn for normally distributed population $p \geq 0, 05$ (the significance level).

a. Using ANOVA One-Way methods for green algae it was obtained factor value F=2.57861, Prob>F is 0.0629. The factor value F in Distribution Tables for α=0.05 and freedom degree df= (5, 18) is Ftable = 2.7729. In this case the null hypothesis are accepted because F is less than Ftable and p>0.05. There are minor differences between the samples treatment, the Sig Flag = 0 confirming the assumption.

b. In case of cyanobacteria (noted by 6) the ANOVA test applied for the treatment groups in case of each samples shows that F value is 12.70451; Prob>F is 2.20728E-5. Using the F Distribution Tables for the α=0.05 (significance level) and for df = (5, 18) the Ftable = 2.7729. So there is a significant difference between groups because prob computed <0.05 and F computed is bigger than Ftable, showing that the null hypothesis is rejected. There is a difference between the data of the same group and between the different groups. In the case of cyanobacteria the evolution depends on the treatment and the samples type. It was observed that the mean difference between samples intergroup is negative for (3/2, 1 /2) and (2/2, 1 /2), indicating that the red samples have inactivated the cyanobacteria less than the other samples in the both cases of treatment.

c. The permeabilized heterotrophic bacteria results obtained by ANOVA method are F=55.67227 and rob=2.55018E-10, for the same table parameter Ftable for α=0.05 significance degree, df= (5, 18). The F factor is bigger than Ftable and the probability factor is inferior to α. The null hypothesis are rejected and there is a significant difference between heterotrophic bacteria evolution function of the type of treatment and samples pretreatment (figure 5).

The heterotrophic cells inactivation is very sensitive for both types of treatment. There is a big difference behavior between the cells intergroup and intragroup. Consequently these types of cells have unusual evolution compared with the other described above. Difference in response to GlidArc treatment between heterotrophic and autotrophic (cyanobacteria) cells can be attributed to structural differences of the cell wall.

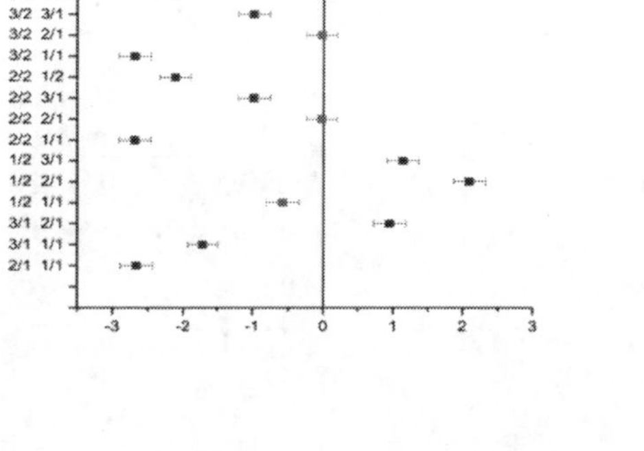

Figure 5. The means of groups, the significant and no significant difference for the permeabilized heterotrophic bacteria evolution.

Prokaryotic cells contain the peptidoglycan layer and this polymer was found in cyanobacteria being considerably thicker than that of most gram-negative bacteria. The unique structure of prokatryotic cell walls makes bacteria vulnerable to decontamination methods.

4. Conclusions

The microorganisms' time evolution has the same results for all the experimental methods in the case of the direct action. For the untreated samples the GlidArc acts in the first period of time, then it loses its influence. The indirect treatment has efficiency in time. The family of microorganisms has different behavior because the green algae and the cyanobateria are autotrophic and the most bacteria are heterotrophic obtaining food from other organisms. The three data analysis described in the last section show that the microorganisms' evolution depends on the treatment type, the type of the paints, the samples roughness and the sea water pretreatments. The results show that the biofouling adherence begins after 8h, but the GlidArc effect has a longer time period. In the case of the biofouling presence prior to the test, the GlidArc action has two tasks: to destroy the biofouling layers and to protect the surface for a long time period.

Acknolegment

This work was conducted within the project UEFISCDI Romania- France bilateral collaboration, TraiBioPlasma, 2015 – 2016, research project 5831/08.10.2014. and the project "Increasing quality in marine higher education institutions by improving the teaching syllabus according to International Convention STCW (Standards of Training, Certification and Watchkeeping for Seafarers) with Manila amendments" POSDRU/156/1.2/G/141534. The second project is co funded by European Social Fund through The Sectorial Operational Programme for Human Resources Development 2007-2013, coordinated by Constanta Maritime University. The content of this scientific article does not reflect the official opinion of the European Union. Responsibility for the information and views expressed in the article lies entirely with the authors.

References

[1] Alkawareek M Y, Gorman S P, Grahamb W G and Gilmore B F 2014 Eradication of marine biofilms by atmospheric pressure non-thermal plasma: A potential approach to control biofouling? *International Biodeterioration and Biodegradation* **86** pp 14 – 18

[2] Hnatiuc B, Ghita S, Sabau A, Hnatiuc M, Dumitrache C L and Wartel M 2014 Treatment with activated water by GlidArc technology of bacteria producing Biofouling *Conference Advanced Topics in Optoelectronics, Microelectronics and Nanotechnologies* (Constanta: Romania)

[3] Hnatiuc B, Sabau A, Ghita S, Hnatiuc M, Dumitrache C L and Pellerin S 2014 Influence of GlidArc treatment on layers formation of biofouling *Conference Advanced Topics in Optoelectronics, Microelectronics and Nanotechnologies* (Constanta: Romania)

[4] Ostertagová E and Ostertag O 2013 Methodology and Application of One-way ANOVA *American Journal of Mechanical Engineering* **1(7)** pp 256-261

IOP Conf. Series: Materials Science and Engineering **95** (2015) 012058 doi:10.1088/1757-899X/95/1/012058

Numerical investigation of forced convection of nano fluid flow in horizontal U-longitudinal finned tube heat exchanger

S M Qasim[1], **A F A Sahar**[1] **and A A Firas**[2]
[1]Al-Mustansirya University College of Engineering, Baghdad, Iraq
[2]Ministry of Oil Gas Filling Company Wassit, Iraq

E-mail: qasim602006@yahoo.com

Abstract. A numerical study has been carried out to investigate the heat transfer by laminar forced convection of nanofluid taking Titania (TiO_2) and Alumina (Al_2O_3) as nanoparticles and the water as based fluid in a three dimensional plain and U-longitudinal finned tube heat exchanger. A Solid WORKS PREMIUM 2012 is used to draw the geometries of plain tube heat exchanger or U-longitudinal copper finned tube heat exchanger. Four U-longitudinal copper fins have 100 cm long, 3.8cm height and 1mm thickness are attached to a straight copper tube of 100 cm length, 2.2 cm inner diameter and 2.39 cm outer diameter. The governing equations which used as continuity, momentum and energy equations under assumptions are utilized to predict the flow field, temperature distribution, and heat transfer of the heat exchanger. The finite volume approach is used to obtain all the computational results using commercial ANSYS Fluent copy package 14.0 with assist of solid works and Gambit software program. The effect of various parameters on the performance of heat exchanger are investigated numerically such as Reynolds' number (ranging from 270 to 1900), volume consternation of nanoparticles (0.2%, 0.4%, 0.6%, 0.8%), type of nanoparticles, and mass flow rate of nanofluid in the hot region of heat exchanger. For 0.8% consternation of nanoparticles, heat transfer has significant enhancement in both nanofluids. It can be found about7.3 % for TiO_2 and about7.5 % for Al_2O_3 compared with the water only as a working fluid.

1. Introduction

The study of improved heat transfer performance is referred to as heat transfer enhancement, augmentation, or intensification. In general, this means an increase in heat transfer coefficient [1].Taborek [2] updated sketchy methods for double pipe heat exchangers especially for longitudinal finned tubes. Calculation methods are presented for plain double pipe units, as well as finned tube units. Equations for the mean temperature difference for units with flow in series-parallel are also given. Mir et al. [3] carried out numerical simulation of studying laminar, forced convection heat transfer in the finned annulus for the case of fully developed incompressible flow. The simulation was corresponding to thermal boundary condition of uniform heat input per unit axial length with peripherally uniform temperature at any cross section. The results showed a good comparison with the literature results. Strand bergandDebendra [4] compared the performance of hydraulic finned–tube heating units utilizingnanofluids with the performance of conventional heat transfer fluid of 60% ethylene glycol and 40% water by using mathematical model. Al_2 O_3and CuO nanoparticles dispersing in 60% EG solution were used. Finned–tube heating performance was enhanced by employing nanofluids. The increase of (11.6%) in finned tube heating output was predicted with 4% of Al_2 O_3and 60% EG. Furthermore, the increase of (8.7%) was predicted with 4% CuO and 60% EG

Content from this work may be used under the terms of the Creative Commons Attribution 3.0 licence. Any further distribution of this work must maintain attribution to the author(s) and the title of the work, journal citation and DOI.
Published under licence by IOP Publishing Ltd

comparing of basefluid. Syed et al. [5] performed numerical simulation of finned annulus in the steady and laminar convection in the thermal entry region, fully developed flow at uniform heat flux. Finite difference based marching procedure was used to compute the numerical solution of the energy equation. The numerical results showed the Nusselt's number has complex depending on the geometric variables like ratio of radii, fin height, and number of fins. The validation of the simulation was performed by comparison with open literature. Present work concerns of studying the performance of U–longitudinal finned tube heat exchanger and comparing it with smooth tube. Then, its performance with nanofluids. This is done after adding nanoparticles to basefluid. The effect of this new technology on estimated reduction in surface area of fins will be examined. Numerical simulation by ANSYS FLUENT 14 package is used to simulate the system and then using laminar flow with and without nanofluids.

2. Mathematical modelling and numerical simulation

Numerical simulation performs analysis of a system involving fluid flow, heat transfer and associated phenomena based on various disciplines of science. These are includes mathematics, computer engineering and physics to provide meaningful modeling of fluid flow. ANSYS FLUENT 14 package has been conducted for performing numerical simulation across the heat exchanger using three–dimensional model. The solution of conservation continuity, momentum and energy equations is used to analyze the flow field inside the heat exchanger. A comparison of heat transfer for smooth, finned tube with or without nanofluids is carried out. A SOLID WORK PREMIUM 2012 is used to draw the geometries of this work which consist smooth tube heat exchanger and U–longitudinal finned tube heat exchanger with inlet and outlet portions. The outside flow is confined by insulating tube having inlet and outlet portions as shown in figure1. Computational domains in the present study are displayed as inlets and outlets of both hot side (water or nanofluids) in the inner tube, and air side in the annuli. Fins are built on the outside surface of the inner tube and the flow on both sides is counter flow.

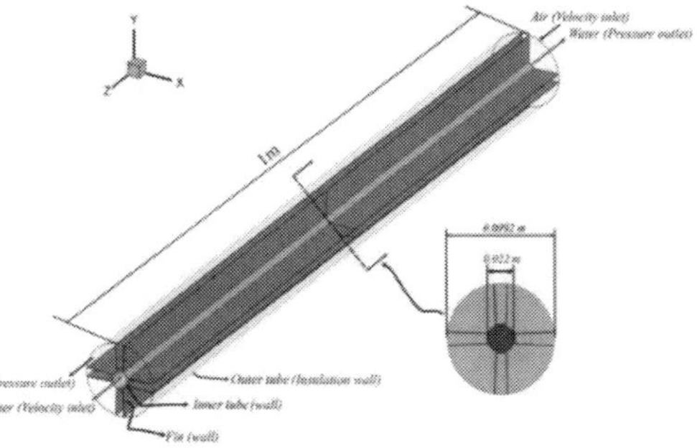

Figure 1. U-longitudinal finned tube heat exchanger.

The assumptions used for water, nanofluid and air during the present study are steady state, Newtonian fluid, incompressible, three dimensional, laminar flow in the inner side (water or nanofluid) and turbulent flow in the annuli side (air).Both fluids are forced convection heat transfer. Buoyancy effect is assumed to be negligible. Radiation heat transfer is not considered. The governing equations for continuity, momentum, and energy of laminar flow in an inner tube are [6]:

Continuity Equation

$$\frac{\partial u}{\partial x}+\frac{\partial v}{\partial y}+\frac{\partial w}{\partial z}=0 \tag{1}$$

Momentum Equations

$$\rho\left(u\frac{\partial u}{\partial x}+ v\frac{\partial u}{\partial y}+ w\frac{\partial u}{\partial z}\right) = -\frac{dP}{dx}+\mu\left(\frac{\partial^2 u}{\partial x^2}+\frac{\partial^2 u}{\partial y^2}+\frac{\partial^2 u}{\partial z^2}\right) \tag{2}$$

$$\rho\left(u\frac{\partial v}{\partial x}+v\frac{\partial v}{\partial y}+w\frac{\partial v}{\partial z}\right)=-\frac{dP}{dy}+\mu\left(\frac{\partial^2 v}{\partial x^2}+\frac{\partial^2 v}{\partial y^2}+\frac{\partial^2 v}{\partial z^2}\right) \quad (3)$$

$$\rho\left(u\frac{\partial w}{\partial x}+v\frac{\partial w}{\partial y}+w\frac{\partial w}{\partial z}\right)=-\frac{dP}{dz}+\mu\left(\frac{\partial^2 w}{\partial x^2}+\frac{\partial^2 w}{\partial y^2}+\frac{\partial^2 w}{\partial z^2}\right) \quad (4)$$

Energy Equation

$$\rho C_p\left(u\frac{\partial T}{\partial x}+v\frac{\partial T}{\partial y}+w\frac{\partial T}{\partial z}\right)=k\left(\frac{\partial^2 T}{\partial x^2}+\frac{\partial^2 T}{\partial y^2}+\frac{\partial^2 T}{\partial z^2}\right) \quad (4)$$

where, (u, v, w) = velocitycomponent,T = temperature, k = thermal conductivity, Cp = specific heat,μ = dynamic viscosity, and ρ = density.

3. Mesh generation

Unstructured mesh is used in the present study to discretize the computational domain into a finite number of control volumes by using the finite–volume scheme. The model was meshed by GAMBIT software [7]. The refinement and generation of mesh system are very crucial to predict the heat transfer in sophisticated geometries. In this study triangular element type is employed for surface mesh and tetrahedron element type is used for three dimensional geometry. This is because it has priority in the sophisticated geometries. Figure 2 shows the mesh of present model. To estimate the performance of the present heat exchanger, some introductory requirements of the physical model are defined adequately as Reynolds number in the inlet was specified for inner and annuli sides during this study. On the other hand the temperature inlet of the inner tube is 60 °C while in annuli is 20°C and specifications of nanofluid particles as in table 1

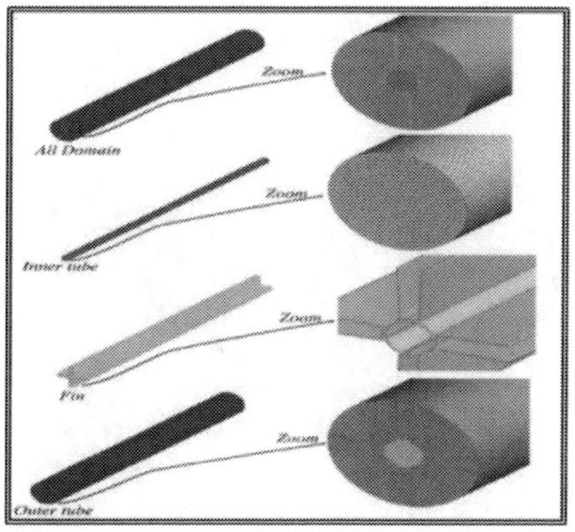

Figure 2. Mesh of present model.

Table 1. Specifications of nanoparticles [9].

Type	D [nm]	$\rho\ [\frac{kg}{m^3}]$	$k\ [\frac{W}{m.°C}]$	$Cp\ [\frac{J}{kg.°C}]$
TiO_2	10-25	4230	8.9	686.2
Al_2O_3	< 80	3970	40	765

The outlet domain is specified as pressure outlet for both sides. No slip boundary condition is specified in the wall of the inner tube. These conditions are used to bound fluid and solid region. It is worth mentioning that convenient numerical control and modeling techniques are very important to step up convergence and stability during the calculation. By adopting control–volume technique, FLUENT shifts the governing equations to algebraic equations that can be solved numerically. The control volume technique involves of integrating the governing equations inside each control volume, yielding discrete equations [8]. Simplifying of geometry which is avoided during this study to get the best degree of accuracy which may be influenced by reducing the resolution as long as the computer ability is capable to achieve the simulation accurately. For this study, an average of 11 million cells is used and it is the maximum number of iterations performed to get the solver terminates. 4000 iterations are needed in this study.

4. Model validation

In order to validate present numerical simulation, a comparison has been carried out with previous numerical results achieved by Hamid Nabati [10]. He had performed a numerical study for predicting the heat transfer in heat exchanger with pin fins by ANSYS FLUENT. Figure 3 shows the temperature contour for both simulations. Good agreement between the present results and reference results [10] which indicates acceptable validation of present simulation.

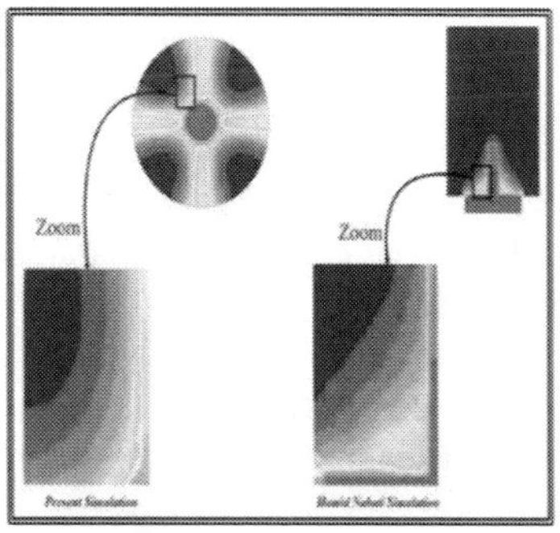

Figure 3. Validation of numerical simulation.

5. Results and discussions

Figures 4 and 5 show both temperature and velocity contours of smooth tube heat exchanger at various water Reynold's numbers and different axial distances.

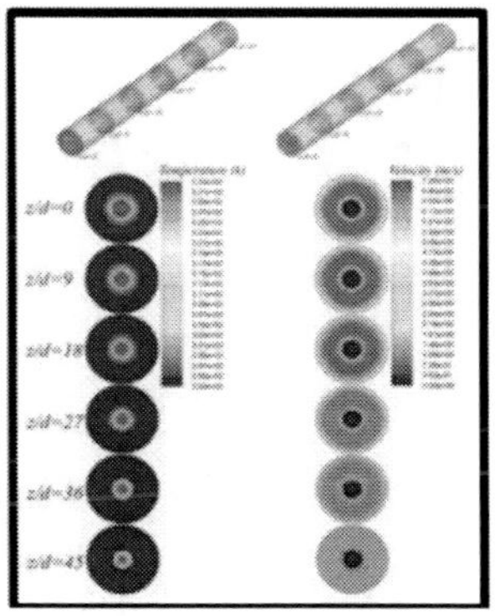

Figure 4. Temperature and velocity contours at Re_h=590, $\dot{m}_c$= 0.05kg/sec for smooth tube.

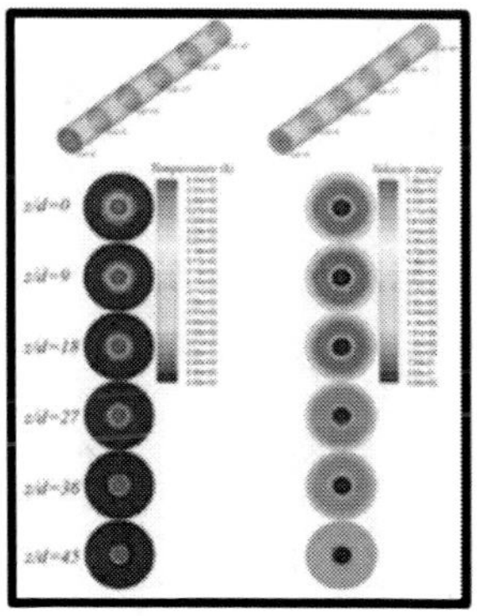

Figure 5. Temperature and velocity contours at Re_h =1900, $\dot{m}_c$ = 0.05 kg/sec for smooth tube.

It can be noted that the maximum air and water temperatures appears at Z/d = 0 while the minimum temperatures appears at Z/d = 45. It can be seen also when observing Z/d = 18, that changing water Reynold's number will not have a significant effect on air side. It seems that the air in annuli has the main reason for this behavior. This is because air thermal conductivity is small velocity contours in these figures show that air inlet velocity distribution at Z/d = 45 are uniform. It will also tend to decrease inside heat exchanger near the walls of inner and outer tubes as shown at Z/d from 36 to 0.

Figures 6 and 7 reveal the temperature and velocity contours of U-longitudinal finned tube heat exchanger for water Reynold's numbers and air mass flow rates at various axial distances ratio.

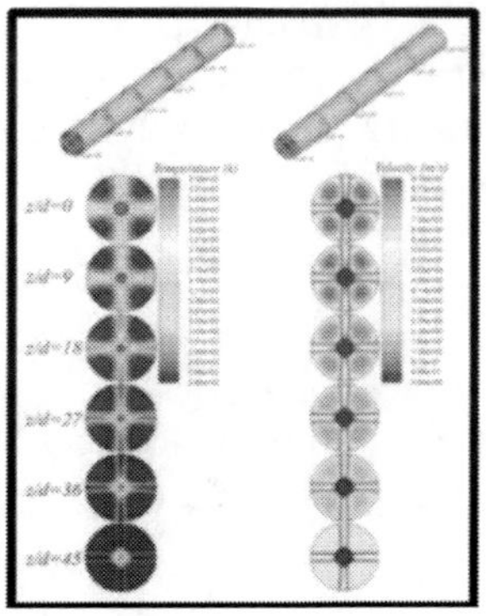

Figure 6. Temperature and velocity contours at Re_h=590, $\dot{\mathbf{m}}_c$= 0.05kg/sec for finned tube.

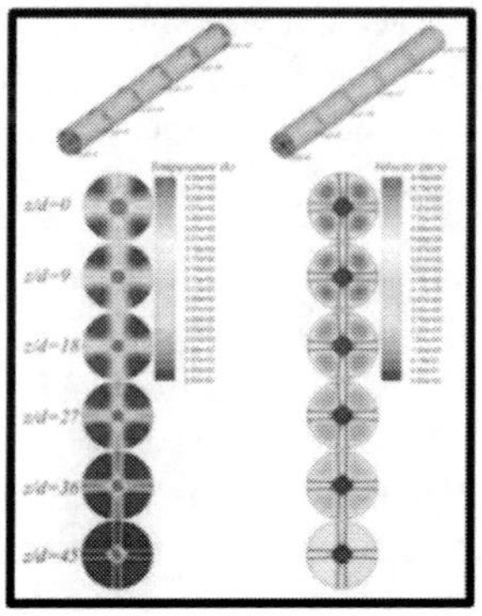

Figure 7. Temperature and velocity contours at Re_h =1900, $\dot{\mathbf{m}}_c$ = 0.05 kg/sec for finned tube.

Results showed a significant heat transfer augmentation in heat exchanger at Z/d from 0 to 45 for air and water sides. The effect of adopting fins on heat transfer enhancement is apparent in both sides. Heat transfer enhancement behavior at Z/d = 27 and m˙_c = 0.05 kg/sec is clear when increasing water Reynolds's number. Velocity contours in these figures reveal that air velocity is uniform at Z/d = 45. However, it increases at the center between every two pairs of fins and U channels at Z/d from 36 to 0.

The effect of wall on decreasing the velocity in air and water sides is apparent in these figures. Also from these figures it can be seen that the distance between the two fins U is important because it clearly give an indication that the velocity between them has been changed and that would effect on the heat transfer in this region. Figures 8and 9 represent the temperature and velocity contours of U-longitudinal finned tube heat exchanger with nanofluids.

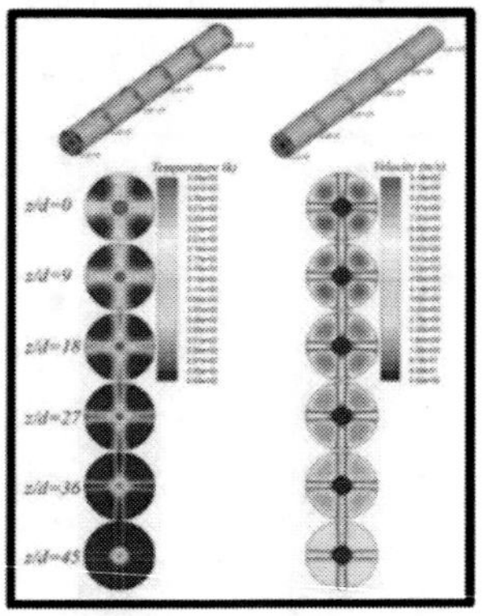

Figure 8. Temperature and velocity contours at Re_h=590, $\dot{\mathbf{m}}_c$= 0.05kg/sec for finned tube with alumina-nanofluid.

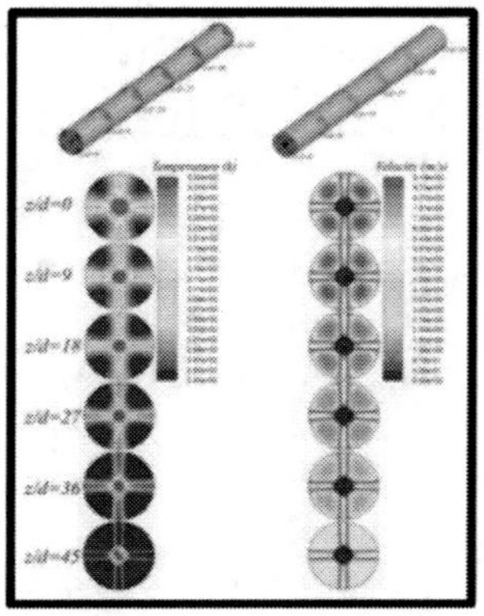

Figure 9. Temperature and velocity contours at Re_h =1900, $\dot{\mathbf{m}}_c$ = 0.05 kg/sec for finned tube with titania-nanofluid.

It can be noted that the heat transfer enhancement in air side appears clearly at Z/d from 0 to 18. But with regards to water side at Z/d from 27 to 45.Thermo physical properties of nanofluid have the main reason of this enhancement .This augmentation in heat transfer behavior can be increased by the increase in thermal conductivity of basefluid. From figures it can be seen that for Al_2O_3nanofluid there is an enhancement but not clearly with respect of TiO_2 but compared with water it can be seen significant enhancement.

Figures 10 and 11 show the variation between Reynolds number and heat transfer coefficient for both of nanofluids ($Al_2O_3$and TiO_2).One of the important advantages of a numerical study is to obtain local heat transfer coefficient values at any point in the test. This observation can lead to improve heat exchanger performance. It can be seen that for Al_2O_3 the increasing in heat transfer about 7.5% and for TiO_2 about 7.3%compared with the water only as a working fluid.

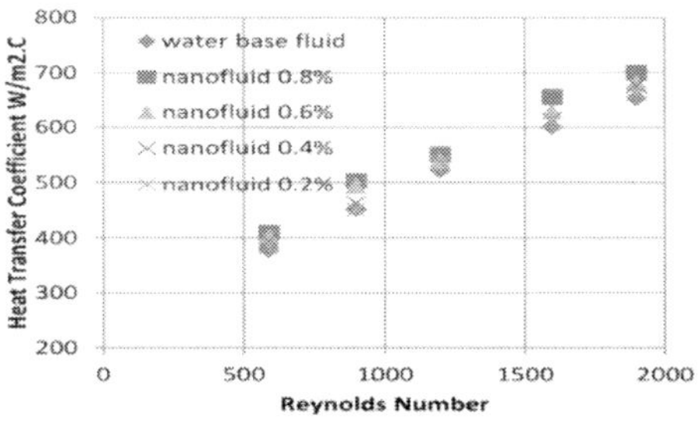

Figure 10. Shows the variation of Reynolds number with heat transfer coefficient for different TiO_2 volume concentration in basefluid.

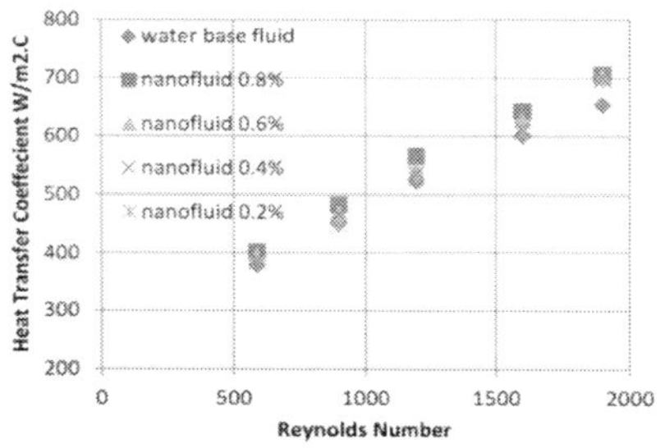

Figure 11. Shows the variation of Reynolds number with heat transfer coefficient for different Al_2O_3 volume concentration in basefluid.

6. Conclusions

Results of numerical simulation showed that adding fins would enhance heat dissipation through heat exchanger. It is noted that heat transfer behavior increases within present model as air mass flow rates and water Reynold's numbers increases. Heat transfer enhancement by adding nanoparticles to basefluid has been found during numerical simulation. Numerical simulation by FLUENT package is successful for predicting both heat transfer and fluid flow in the present heat exchanger. It has been found that enhancement in heat transfer is about 7.5%for $Al_2O_3$and for TiO_2about 7.3% compared with the water only as a working fluid.

References

[1] Bergles A E 1999 *The Imperative to Enhance Heat Transfer* (Department of Mech. Eng., Aeronautical Engineering and Mechanics Rensselaer Polytechnic Institute) pp 13–29

[2] Taborek J 1997 Double-Pipe and Multitube Heat Exchangers with Plain and Longitudinal Finned Tubes *Heat Transfer Eng.* **18 (2)** pp 34–45

[3] Mir N A, Syed K S and Mazhar I 2004 Numerical Solution of Fluid Flow and Heat Transfer in the Finned Double Pipe *Journal of Research (Science)* **15(3)** pp 253-262

[4] Strandberg R and Debandra K D 2010 Finned Tube Performance Evaluation with Nanofluids and Conventional Heat Transfer Fluids *International Journal of Thermal Sciences* **49**

[5] Syed K S, Mazhar I and Mir N A 2007 Convective Heat Transfer in the Thermal Region of Finned Double–pipe *Heat Mass Transfer* **43** pp 449–457

[6] Davidson L 2009 *An Introduction to Turbulence Models* (Department of Thermo and Fluid Dynamics, Chalmers University of Technology, Sweden)

[7] Ansys Fluent User's 2011 Guide Ansys Inc. *South pointe 275 Technology Drive Canonsburg*

[8] Rebay S 1993 Efficient Unstructured Mesh Generation By Means of Delainay Triangulation and Bowyer–Watson Algorithm *Journal of Computational Physics* **106** pp 125-138

[9] Abd ali Abbas F 2014 Augmentation of Heat Transfer by Using Nanotechnology *M.Sc. thesis* (University of Al – Mustansiriya, College of Engineering, Mech. Eng. Dep.)

[10] Nabati H 2008 Optimal Pin Fin Heat Exchanger Surface (Mälardalen University Press) Licentiate Thesis **88**

Mechanical properties of different types of space maintainers

M-A Beldiman[1], I Mârţu[1], B Leiţoiu[2], I Luchian[1], O Lupescu[3] and E S Bârcă[4]
[1] Gr.T.Popa Iaşi University of Medicine and Pharmacy, Faculty of Dental Medicine, 16, Universitatii St, 700033-Iasi, Romania
[2] Gheorghe Asachi Technical University of Iaşi, Faculty of Mechanical Engineering, 59A Mangeron St., 700050-Iasi, Romania
[3] Gheorghe Asachi Technical University of Iaşi, Faculty of Machine Manufacturing and Industrial Management, Department of Machine Manufacturing Technology, Bdul Mangeron, 59A, 700050, Romania
[4] University of Pitesti, Faculty of Mechanics and Technology, Department of Manufacturing and Industrial Management,Tg din Vale Str, 110040, Arges, Romania

E-mail: ionut_luchian@yahoo.com

Abstract. Currently, inside the oral cavity, the dental space maintainers are subjected to forces exerted on them when performing various functions; therefore, it is important to know how each of these mechanisms behave and respond to forces that are applied directly to them. The mechanical properties of the materials used in dentistry are defined by a set of characteristics representing the behaviour of their particular working conditions and it is qualitatively expressed by a number of parameters.The study aimed to determine the pressing force that can be taken by four 4 types of space maintainers frequently used in practice - fixed and removable, applied on four samples realized with human teeth extracted for orthodontic purposes. Static tests were carried out on a machine type short WDW-5 EC with a maximum force of 5 kN and a loading speed of 5 mm/min by a special testing machine, with an innovative appliance; data recording was automatically performed, using a computer with a special program that present the specific diagrams. Experimental determinations included the following aspects: to determine the maximum force that can be supported by each sample, and to observe the deformations. The values obtained indicate that the best option in terms of behavior under the conditions specified is the removable appliance, and the less functional version is the fixed space maintainer using brackets. According to tests conducted, the fracture strength was found to be more important for fixed space maintainers (band and loop, for example) so, in practice is using more frequent these types of space maintainers.

1. Introduction

Currently, into the oral cavity, the dental space maintainers are subjected to forces exerted on them when performing various functions; therefore, it is important to know how each of these mechanisms behave and respond to forces that are applied directly to them. The mechanical properties of the materials used in dentistry are defined by a set of characteristics representing the behaviour of their particular working conditions and it is qualitatively expressed by a number of parameters. The

Content from this work may be used under the terms of the Creative Commons Attribution 3.0 licence. Any further distribution of this work must maintain attribution to the author(s) and the title of the work, journal citation and DOI.
Published under licence by IOP Publishing Ltd

determination of mechanical characteristics/properties uses a special machine, for testing the corresponding goal, aiming the behaviour of samples and records breaking, following the characteristic parameters and analyzing both the way and breaking appearance.

Mechanical tests of strength of materials - tensile or compressive - make it possible to track the behaviour and characteristic parameters diagrams expressing the relationship between the applied forces and specific corresponding linear deformations. [1]. In order to determine the resistance of the materials to dynamic loading, are widely used the bending impact tests, known as the resiliency tests. Under the action of forces, the samples are deforming and moving those points; it has been shown that the values of movements caused by the deformation of the sample are generally small compared to the size of the sample - is allowed so that by deforming, the sample changes its shape and original size, in a relatively small extent. Small deformations are usually elastic deformations, which disappear with the disappearance of the forces applied. Between the applied forces and displacements produced and between internal efforts and deformations of the sample, there is a close relationship expressed by mathematical relations - the elastic movements of these samples are proportional to the forces applied - this observation expresses Robert Hooke's law. Thus, we can write:

$$\Delta = \delta * P \qquad (1)$$

where: Δ is the displacement of any point caused by force P, δ is a coefficient of "influence", which represents the movement of point P considered under a force equal to unity. The magnitude of this coefficient depends on the material, shape and size of the sample, the strength and position of the point considered.

Tensile and compression tests run on special machines that perform axial strain of the sample. From the simplest machines, mechanically operated via a lever system, and till the universal testing machines with many possibilities, there are a variety of systems, from small forces developing for several daN, to those that perform forces as thousands of kN. At any time during the test, it can be register the elongation Δl, corresponding to the axial force N applied to the sample. Than, by plotting the pairs of values obtained during the test, there can be drawn the characteristic curve tensile of the sample. The behaviour of a sample under the action of external forces can be described by the stress-strain curves obtained in tension, compression, torsion, etc. Building a diagram type real tension-real strain, tension grows continuously until rupture, and the actual curve is called tension-deformation [2].

2. Materials and method

The study aimed to determine the pressing force that can be taken by four 4 types of space maintainers frequently used in practice - fixed and removable, applied on four samples realized with human teeth extracted for orthodontic purposes. Static tests were carried out on a machine type short WDW-5 EC with a maximum force of 5 kN and a loading speed of 5 mm/min by a special testing machine, with an innovative appliance; data recording was performed automatically, using a computer with a special program that present the specific diagrams. Experimental determinations included the following aspects: to determine the maximum force that can be supported by each sample, and to observe the deformations, (figure1).

Data recording was performed automatically using a computer which takes the data from each stage of force and forward them to be viewed in the diagrams illustrating the variation curves of the pressing force depending on the strain, respectively, versus time. The samples were tested until the pressing force has reached the maximum value or until the destruction of the sample.

The test samples were performed on extracted human teeth for orthodontic purposes - premolars and third molars (in order to simulate the second temporary molar); extracted teeth were introduced into a mass of acrylic resin, which represent the support for each of the samples. At the level of each of the samples, respectively simulating the edentulous space, by the absence of deciduous tooth, it was applied fixed or removable space maintainer, (figure2, figure 3).

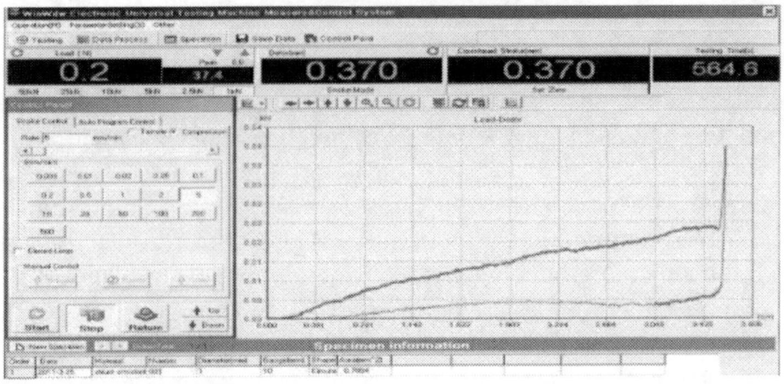

Figure 1. The type of curve recorded for 5 kN force and speed of 5 mm/min.

For the types of fixed space maintainers, fixed orthodontic rings were chosen in a series of rings supplied by Ormco®, and have been adapted on the teeth. Wire loops were made from 1.0 Dentaurum® wires, and welded in a dental laboratory [3].

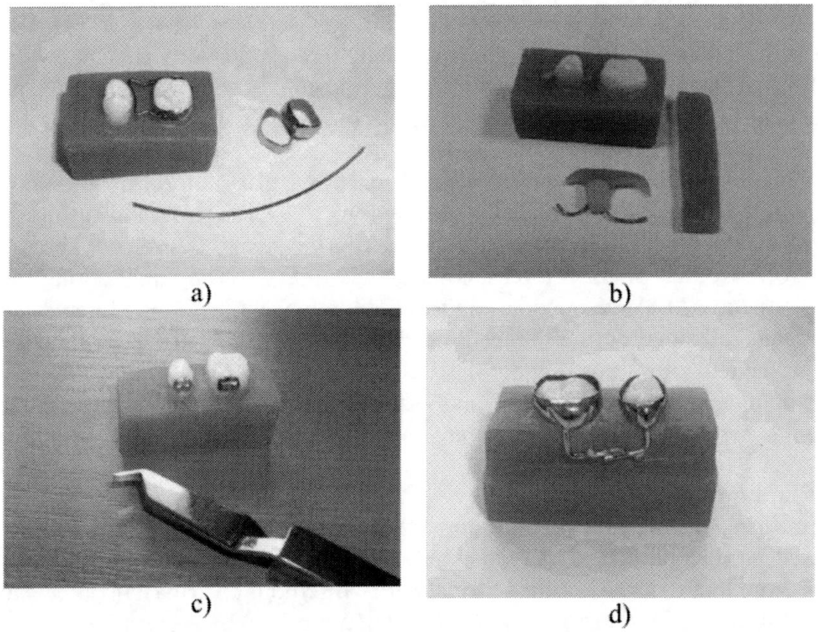

a) b) c) d)

Figure 2. Performing the samples of space maintainers.

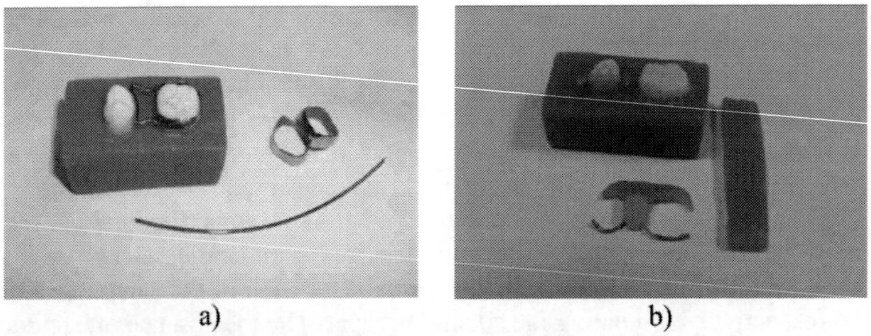

a) b)

Figure 3. The samples (space maintainers) applied on a special support of the testing machine WDW-5 CE.

3. Results and discussions

For each sample, we obtained diagrams of force variations in kN depending on the stain (in mm), versus time (in seconds). So, for first sample, a fixed appliance (brackets), according down force applied, the chart is redrawn to deformation (0.04kN). For the removable space maintainer, the diagram is an ascending one according to force applied (0.42kN), until the fracturing moment; this was the type of space maintainer that had the weakest response to maximum force, (figure 4, figure 5).

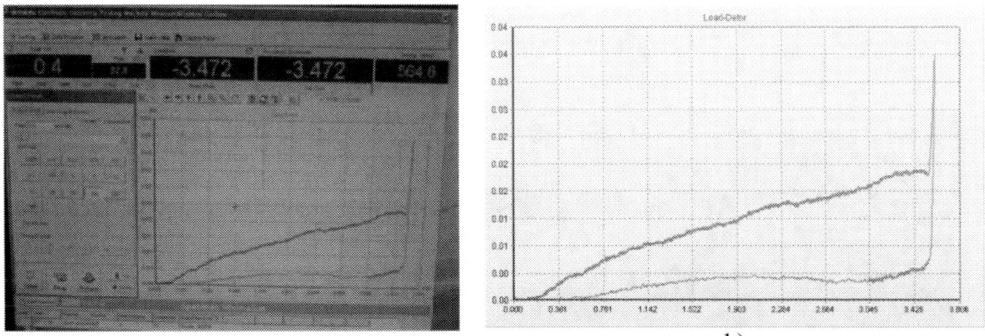

a) b)

Figure 4. Diagrams obtained after application of different forces for the first sample, with fixed orthodontic space maintainer.

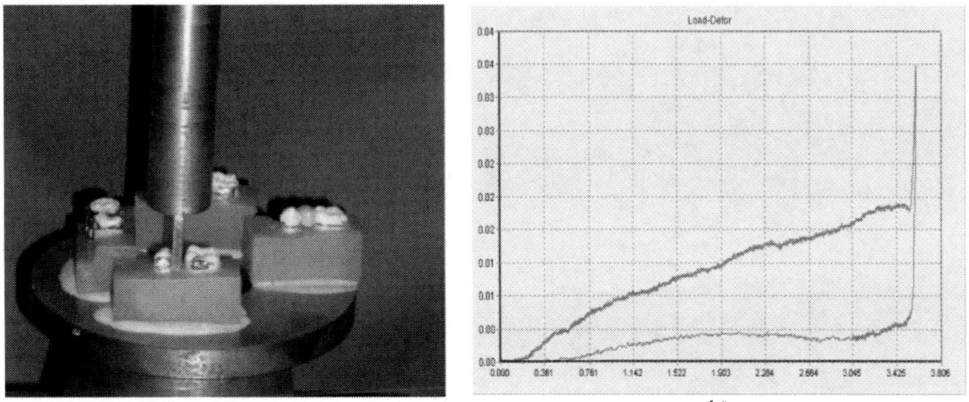

a) b)

Figure 5. The force applied on the first sample and the diagram of force and deformation.

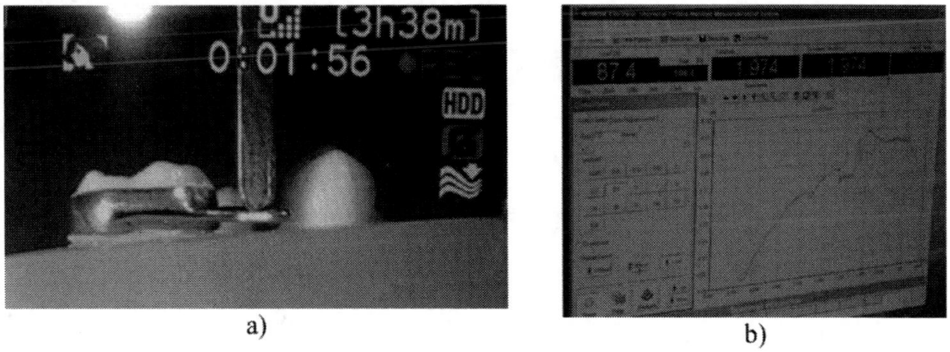

a) b)

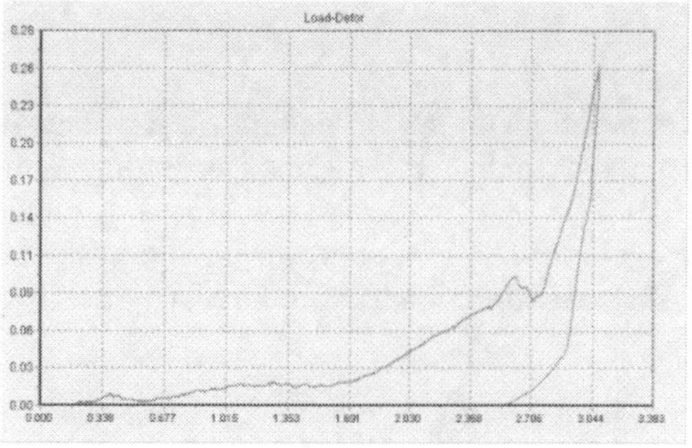

c)

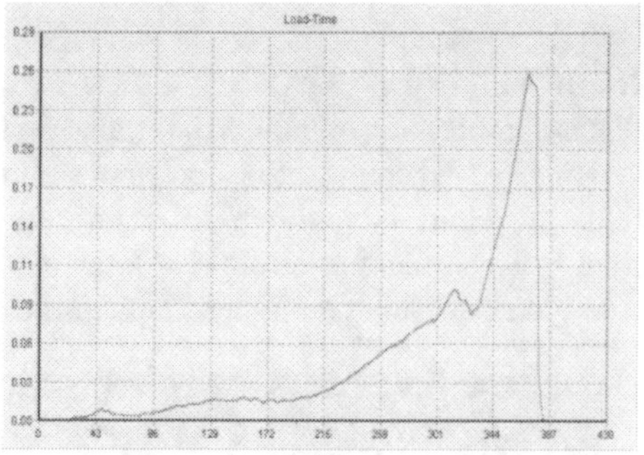

d)

Figure 6. The force applied on the second sample (band and loop) and the diagrams of force and deformation.

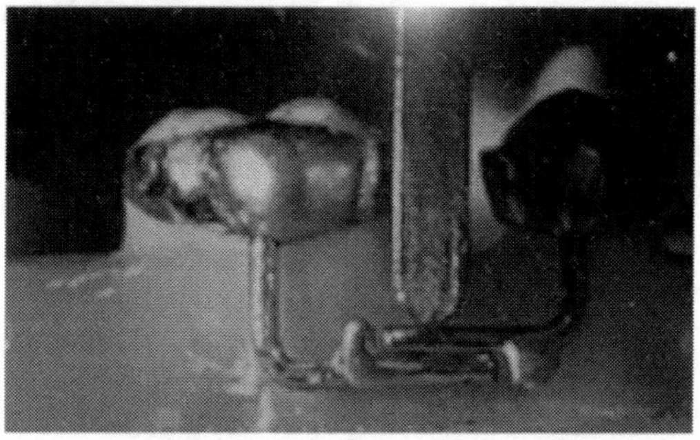

a)

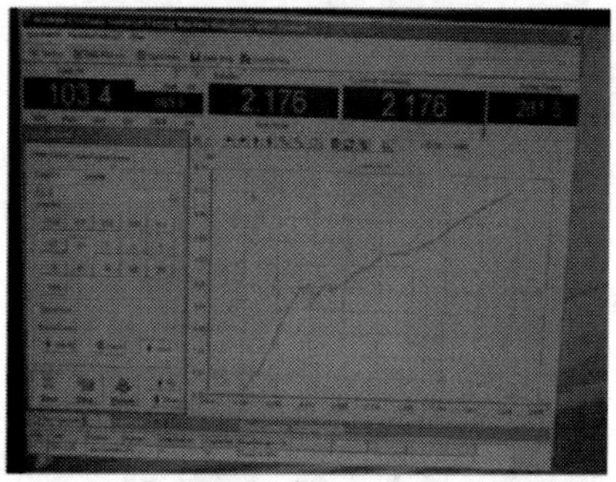

b)

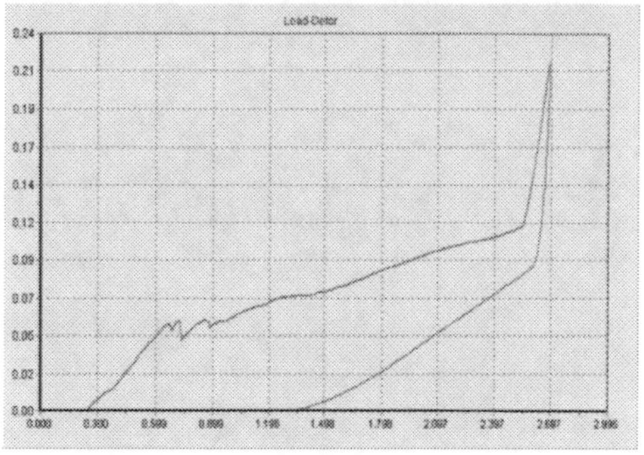

c)

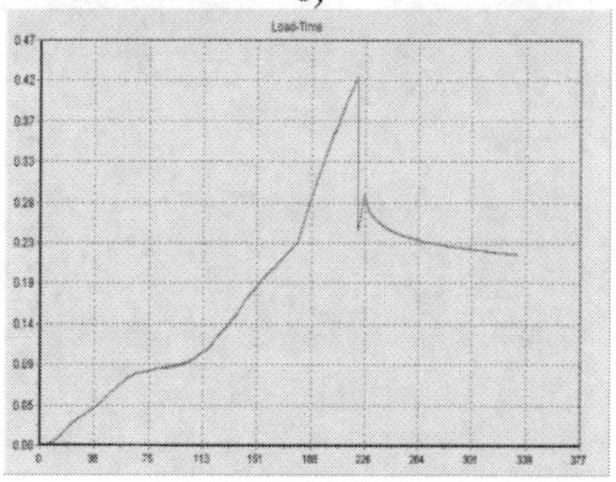

d)

Figure 7. The force applied on the third sample (fixed regaining space orthodontic device) and the diagrams of force and deformation.

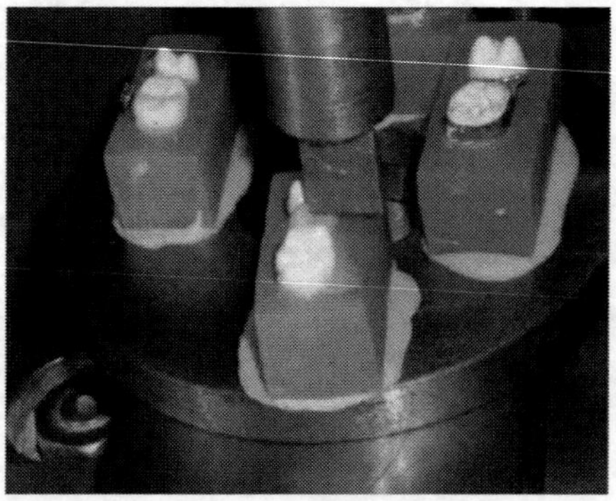

a)

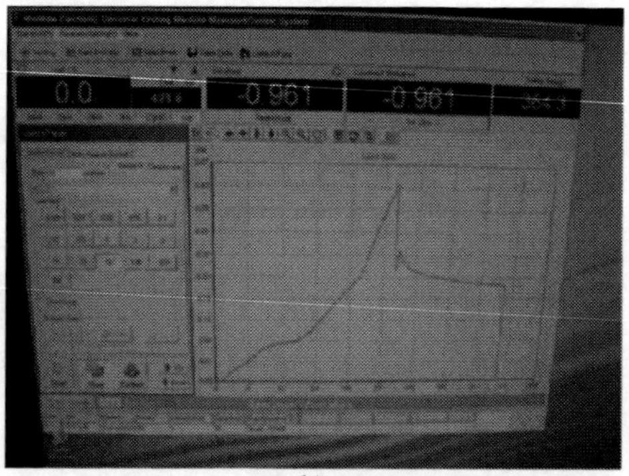

b)

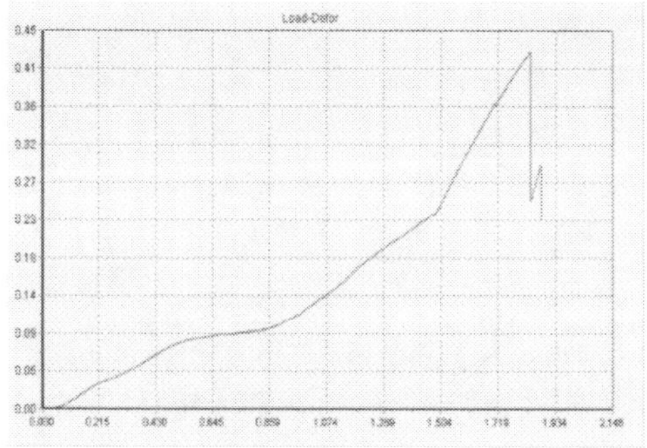

c)

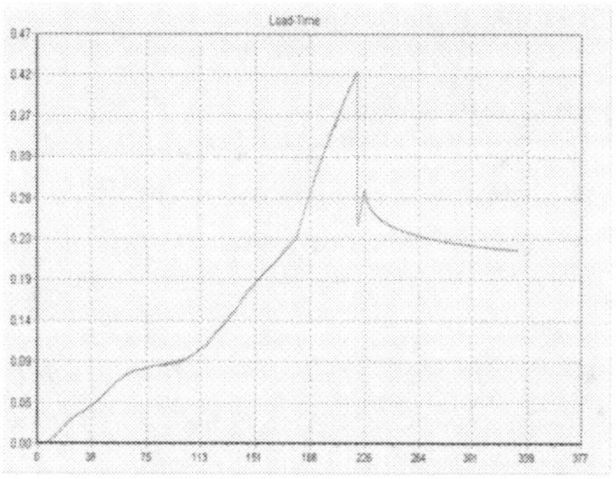

d)

Figure 8. The force applied on the sample with removable space maintainer and the diagrams of force and deformation.

Studies regarding the mechanical behaviour and properties of different types of devices used as space maintainers are rare. There are no scientific studies to evaluate the fracture strength and the effects on space maintainers, both devices made in the laboratory and those prefabricated provided by various manufacturing companies. Krishnan et al. (2004) evaluated in one trial, three types of springs used in the manufacture of orthodontic devices - stainless alloy, titanium beta-timolium, and found that the resistance to fracture differ significantly between the three materials. Although there is, in this study a comparison with the initial evidence, it can be assumed that exist a significantly lower fracture resistance. Babe et al. [4] conducted a study on the fracture strength of orthodontic appliances made in the laboratory, showed the items that were fixed by welding (such as space maintainers - ring and loop, and loop crown cover). The study showed a lower resistance to deformation than the controls, values were not significant.

A recent study showed a greater resistance to bending orthodontic devices using springs, with the following values: Ti - 18% NiCr - 66% and CoCr - over 69%. Testing and fracture resistance appear to be comparable to interlock welding/brazing bands and crowns elements [5].

Further researches are necessary to determine optimal parameters for each device, and to achieve solidarity in particular, for types of space maintainers such as bands/crowns and orthodontic loops, which are the most used fixed appliances, to maintain the edentulous space arrangements until the permanent tooth eruption, (figure7, figure 8).

Table 1. The results indicate the following values for the forces applied.

Sample	1	2	3	4
Loads	0.04 kN	0.26 kN	0.04 kN	0.42 kN

4. Conclusions

The values obtained indicate that the best option in terms of behavior under the conditions stated above is the removable appliance (sample no.4) and the less functional version is the fixed space maintainer using orthodontic brackets (sample no.1). Very similar values are reported for space maintainers consist of orthodontic band and loop, respectively orthodontic rings and double loops. The results of tests carried out are fully consistent with those known from the literature.

According to tests conducted, the fracture strength was found to be more important for fixed space maintainers made of orthodontic band and loop, and orthodontic rings with slider, resulting in more frequent use in practice these types of appliances.

Despite the low resistance, removable space maintainers made from acrylic resins are commonly applied to treat toothless transient (temporary dentition and mixed dentition), because, most often,

IOP Conf. Series: Materials Science and Engineering **95** (2015) 012059 doi:10.1088/1757-899X/95/1/012059

orthodontic appliances included these removable devices, are recommended to treat various malocclusions, common in these age stages.

Acknowledgement

This work of Bârcă Eduard Sebastian was supported by the strategic grant POSDRU/159/1.5/S/138963 - PERFORM, co-financed by the European Social Fund – Investing in People, within the Sectoral Operational Programme Human Resources Development 2007-2013.

References

[1] Atanasiu C 1982 *Încercarea materialelor* **I, II** (Tehnica Publishing House: Bucharest)
[2] Marin C 2006 *Rezistenţa materialelor şi elemente de teoria elasticităţii* (Bibliotheca Publishing House: Târgovişte)
[3] Information on http://www.leone.it, accessed: 1/04/2015, http://www.dentaurum.de, accessed: 1/02/2015, http://www.ormco.com, accessed: 15/02/2015
[4] Babe N, Watanabe I, Liu J and Atsuta M 2004 Mechanical strength of laser-welded cobalt-chromium alloy *J Biomed Mater Res B Appl Biomater* **69** pp 121-124
[5] Quiros O 2006 Bases Biomecánicas y aplicaciones clinicas en Ortodoncia Interceptiva *Amolca* pp 123-126

Considerations regarding the stress relieving in a special product

T Axinte[1] and G Vladu[1]
1Constanta Maritime University, Faculty of Naval Electro-Mechanics,
104 Mircea cel Batran Street, 900663, Constanta, Romania

E-mail: tibi_axinte@yahoo.com

Abstract. The special product has a steel bearing structure located inside the shoulders and it is fitted with a steel cylinder loaded by an internal pressure pi. The value of the maximum internal pressure is high, at limit being equal to the allowable stress (σa = pi), where pi = 3000 MPa. The uniform stresses decrease if there is an external pressure applied on the cylinder. Therefore, it must be produced external pressure in order to reduce the stress σtmax., in other words to achieve a relieving operation. The external diameter of the internal tube is larger than the internal diameter of the external tube. In the normal temperature conditions these two cylinders cannot be assembled. After it is heated up, expanded cylinder is mounted on the internal cylinder and after the quenching, on the contact surface it appears a relieving pressure p. We are interested in computing the parameters of this technological procedure, for the classic way of relieving, namely by heating up the external cylinder. After processing by cutting, radius R2 of the internal tube is greater than δ = 0.05 mm radius R2 of the external tube. After cooling, the two tubes have a common radius R2, which will have an average value between the values they had before the assembling process. This is why the external tube has to suffer a thermal expansion, the internal tube suffering a compression, so that the two displacements summed, measured perpendicular to the contact surface, have a value of δ = 0.05 mm. After the relieving operation, on the contact surface between the two tubes will be produces a contact pressure, p. In the paper we study the effect of the relieving pressure p, onto the stresses if there would be not relieving and the variation of stresses considering relieving of both the internal tube and the external tube. The highest values of the stresses are those on the internal face of the bearing. If calculations are in accordance with the third failure theory it results that the maximum tangential stresses are the highest values. The values of normal and shear stress influence the phenomenon of relieving in the bearing because high internal pressures inside the bearing may break the internal tube leading to collapse of the structure.

1. Introduction

To defend the Romanian coastline, there are various special products such as that one in figure 1 arranged in fixed coast locations. In case of shooting with such product (cannon), the efforts occurred during shooting are taken by the shoulders and the elastic contact located on gun cradle. Gun's shoulder is a bearing, figure 2, which is a tube with thick walls, subjected to an internal pressure p_i during shooting. The shell, figure 3, develops such pressure during shooting both in the gun barrel and the bearing [1].

Content from this work may be used under the terms of the Creative Commons Attribution 3.0 licence. Any further distribution of this work must maintain attribution to the author(s) and the title of the work, journal citation and DOI.
Published under licence by IOP Publishing Ltd

Figure 1. Costal cannon.

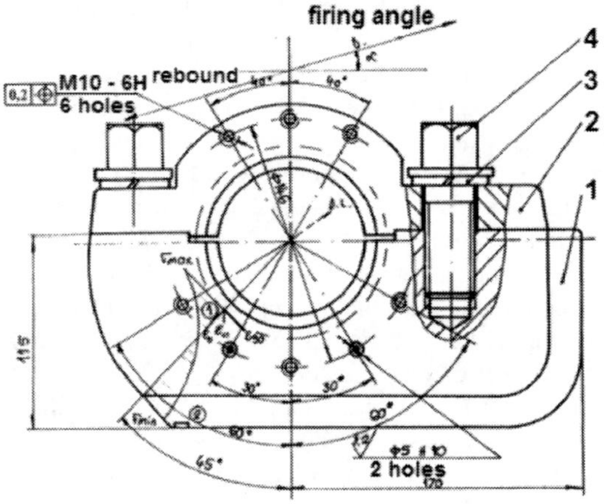

Figure 2. Bearing side view.

Figure 3. Shell.

Theoretical data

If the thick-walled tube is subjected to internal pressure only, then in relations

$$\sigma_r = \frac{R_1^2 p_i - R_2^2 p_e}{R_2^2 - R_1^2} - \frac{(p_i - p_e)R_1^2 R_2^2}{r^2(R_2^2 - R_1^2)} \tag{1}$$

$$\sigma_t = \frac{R_1^2 p_i - R_2^2 p_e}{R_2^2 - R_1^2} + \frac{(p_i - p_e)R_1^2 R_2^2}{r^2(R_2^2 - R_1^2)} \tag{2}$$

is replaced the external pressure $p_e = 0$ and there are deduced the normal stress distribution laws:

$$\sigma_r = \frac{R_1^2 p_i}{R_2^2 - R_1^2}\left(1 - \frac{R_2^2}{r^2}\right) \tag{3}$$

$$\sigma_t = \frac{R_1^2 p_i}{R_2^2 - R_1^2}\left(1 + \frac{R_2^2}{r^2}\right) \tag{4}$$

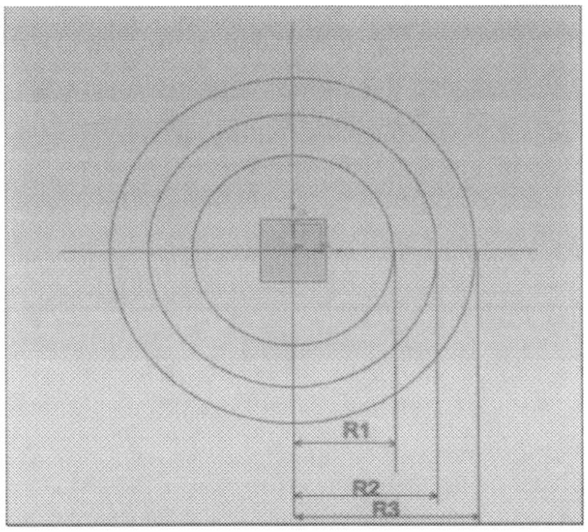

Figure 4. Radii of the bearings.

On the inner surface, where $r = R_1$, the normal stress is σ_{r1}:

$$\sigma_{r1} = -p_i \tag{5}$$

On the outer surface, where $r = R_2$, the normal stress is σ_{r2}:

$$\sigma_{r2} = 0 \tag{6}$$

On the inner surface, where $r = R_1$, the normal stress is σ_{t1}:

$$\sigma_{t1} = \sigma_{t\max} = \frac{\left(R_1^2 + R_2^2\right)p_i}{R_2^2 - R_1^2} \tag{7}$$

On the outer surface, where $r = R_2$, the normal stress is σ_{t2}:

$$\sigma_{t2} = \frac{2p_i R_1^2}{R_2^2 - R_1^2} \tag{8}$$

If there is provided a size after the first stress theory, equation (7) becomes:

$$\sigma_a = \sigma_{t\max} = \frac{\left(R_1^2 + R_2^2\right)p_i}{R_2^2 - R_1^2} \tag{9}$$

But the ratio between the extremes of σ_t is:

$$\frac{\sigma_{t\max}}{\sigma_{t\min}} = \frac{R_1^2 + R_2^2}{2R_1^2} \tag{10}$$

Using the maximum shear stress failure theory there was calculated the maximum tangential stress, located on the contact surface between the cylinders [2]:

$$\tau_{\max} = \frac{\sigma_{t\max} - \sigma_{r1}}{2} = \frac{2p_i R_2^2}{R_2^2 - R_1^2} \tag{11}$$

2. Calculation of normal and tangential stresses

Normal and tangential stresses are calculated using math program EulerMath, figure 4, [3]. For this we consider several values of the internal pressure (pi) since the pressure is variable, while the cylinder sizes are constant.

```
>Pi=[1000, 1075, 1150]; // [MPa]
>//Valori ale razelor:
>R1=100; // [mm]
>R2=200; // [mm]
>R3=300; // [mm]
>// Determinarea tensiunii normale maxime
>((R1^2+R2^2)*Pi)/(R2^2-R1^2)

 [1666.67,  1791.67,  1916.67]

>// Determinarea tensiunii tangentiale maxime
>(2*Pi*R2^2)/(R2^2-R1^2)

 [2666.67,  2866.67,  3066.67]
```

Figure 5. Stresses calculation using the EulerMath software.

The results of the calculi are presented in table 1

Table 1. Values of pressure and stresses.

No	P_i [MPa]	σ_{tmax} [MPa]	τ_{tmax} [MPa]
1	1000	1666.67	2666.67
2	1075	1791.67	2866.67
3	1150	1916.67	3066.67

Stresses variation are drawing (figure 5) with Euler Math program [4]. Normal stress values are represented using the graph in 3D. In this graph, the maximum stress values are shown at the top, and the minimum stress values are shown at the bottom of the graph [4].

The CAD model of the cylinder was developed using Siemens NX. Then, the CAD model was imported in Femap/Nastran where the FEM model was developed: meshing, boundary conditions, loads, material. The results are presented in figures 6 and figure 7.

As is noticed in cylinder 1, von Mises stresses are very high at the edge outwards, figure 6, and the cylinder 2 von Mises stresses are higher inwards, figure 7, [5].

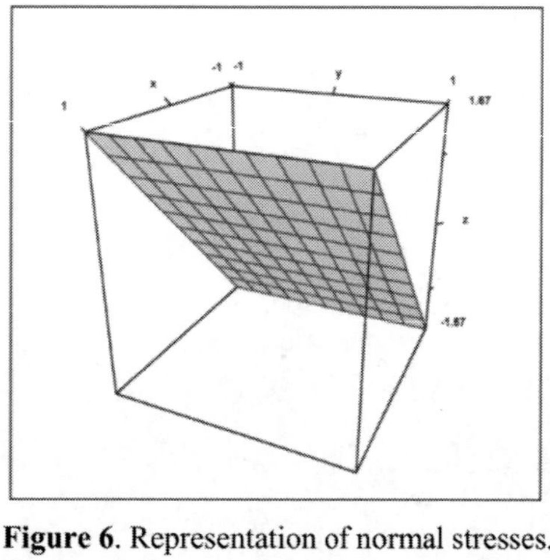

Figure 6. Representation of normal stresses.

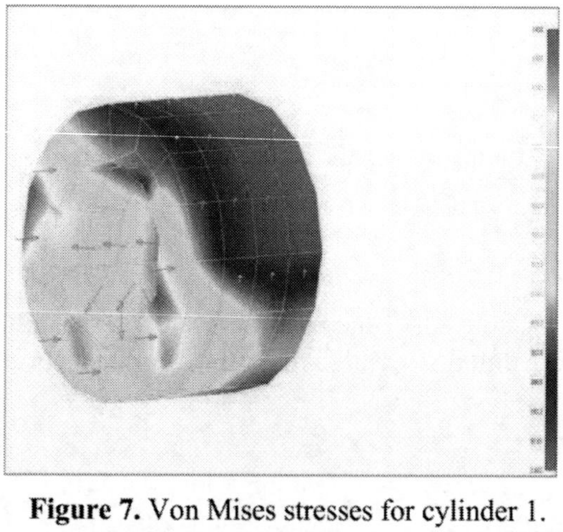

Figure 7. Von Mises stresses for cylinder 1.

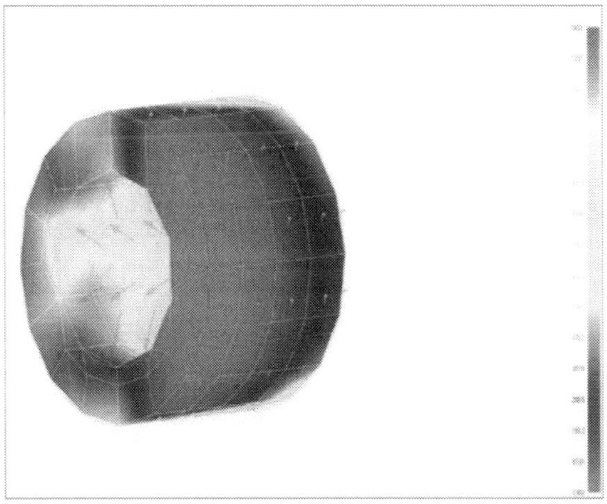

Figure 8. Von Mises stresses for cylinder 2.

3. Conclusions

This study was conducted on a bearing shoulder cannon caliber 130 mm (see figure 2) located in coastal fixed site to determine the normal stresses and tangential camp subjected to strain during drawing.

It demonstrates that for obtaining voltages as low tangential and normal, or must have dimensions of radii smaller cylinders or pressure values as low interior. But because the cylinders are fixed dimensions rays (standard), you need to realize how small pi. Therefore, our product special (cannon) have to work at optimal parameters having the greatest firing cadence.

From normal and tangential stresses calculation is observed that the maximum tangential stress values are higher than the maximum normal stresses. Therefore, the risk to break the bearing is given by the higher values of the stresses.

The values of normal and tangential stresses influences the phenomenon of relieving in the bearing because high internal pressures inside the bearing may break the internal tube leading to the fail of the structure.

References

[1] Vladu G L 2000 Considerations regarding to the robot applications in naval armaments technique *Ph.D. thesis* (Military Technical Academy, Bucharest)

[2] Buzdugan G 1970 *Rezistenta materialelor* (Bucharest: Tehnica Publishing)

[3] Răzvan V 1987 Teoria stabilității (Bucharest: Ştiinţifică şi Enciclopedică Publishing)

[4] www.euler-math-toolbox.de.

[5] www.plm.automation.siemens.com/en_us/

[6] Vladu G L and Craciun C 1995 Consideraţi generale despre stabilitatea gurilor de foc de artilerie navală, montate în structuri fixe pe nave sau amplasamente de coastă *XXVI Sesiune de comunicări ştiinţifice A.T.M* (Bucuresti)

IOP Conf. Series: Materials Science and Engineering **95** (2015) 012061 doi:10.1088/1757-899X/95/1/012061

Electromagnetic pulses bone healing booster

S R Sintea[1], V M Pomazan[2], D Bica[3], D Grebenisan[4] and N Bordea[1]

[1]Constanta Maritime University, Faculty of Naval Electro-Mechanics, 104 Mircea cel Batran Street, 900663, Constanta, Romania
[2]Ovidius University, Faculty of Mechanical Engineering, 124 Mamaia Blvd., 900527, Constanta, Romania
[3]C.M.I Daniela Bica, 10 Industriala, Constanta, Romania
[4]DG Engineering, Canada, 7 Headlands Crescent Witby, L1R1Z9, Ontario, Canada

E-mail: sinteasorin@hotmail.com

Abstract. Posttraumatic bone restoration triggered by the need to assist and stimulate compensatory bone growth in periodontal condition. Recent studies state that specific electromagnetic stimulation can boost the bone restoration, reaching up to 30% decrease in recovery time. Based on the existing data on the electromagnetic parameters, a digital electronic device is proposed for intra oral mounting and bone restoration stimulation in periodontal condition. The electrical signal is applied to an inductive mark that will create and impregnate magnetic field in diseased tissue. The device also monitors the status of the electromagnetic field. Controlled wave forms and pulse frequency signal at programmable intervals are obtained with optimized number of components and miniaturized using surface mounting devices (SMD) circuits and surface mounting technology (SMT), with enhanced protection against abnormal current growth, given the intra-oral environment. The system is powered by an autonomous power supply (battery), to limit the problems caused by powering medical equipment from the main power supply. Currently the device is used in clinical testing, in cycles of six up to twelve months.

Basic principles for the electrical scheme and algorithms for pulse generation, pulse control, electromagnetic field control and automation of current monitoring are presented, together with the friendly user interface, suitable for medical data and patient monitoring.

1. Introduction

In the latest years many people face with periodontal disease. Periodontal condition is favored when patients neglect early symptoms. This condition attacks the mandibular bone. After the bone is affected, the patient must start a very cost expensive treatment. Actually, for this disease the treatment is taking a long time and the used equipment is cost effective. The experimental methods used for healing periodontal disease are based on laser therapy method. Each physician recommends his own method and laser therapy scheme, depending on patient and the disease evolution. Up to know, the laser therapy has fair results in bone healing [1].

From 1996 year a new method for bone restoration process used in medical treatment cabinets replaces laser therapy. This method is based on impregnated oscillatory electromagnetic field in human body in area of affected tissue [2]. This method is also an experimental treatment for healing other human conditions, as it has an encouraging effect on various tissues.

Content from this work may be used under the terms of the Creative Commons Attribution 3.0 licence. Any further distribution of this work must maintain attribution to the author(s) and the title of the work, journal citation and DOI.
Published under licence by IOP Publishing Ltd

The latest medical scientific research institutes have positive results in healing bone conditions by using impregnated oscillatory electromagnetic field [3]. The method is supposed to decrease the healing time of bones.

We use this method in dental domain for healing periodontal disease [4]. We propose an equipment that produces an impregnated oscillatory electromagnetic field, controlled by a microcontroller. This field will be applied in mandibular area of patient. The impregnated oscillatory electromagnetic field will determine a vibration in mandibular bone area that will boost bone healing [5].

2. Rationale

The system is developed based on microcontroller digital solution. The main task of the system is to generate oscillatory electromagnetic field. The field is created using electro-magnetic coil system. The coil is excited by a low voltage oscillatory power module controlled by microcontroller.

The source is powered by a power source module (figure 1). The power module is optimized to ensure optimal current to produce effective impregnated electromagnetic field in human tissue. Also the system monitors the impregnated current into the coil and verifies the coil circuit continuity.

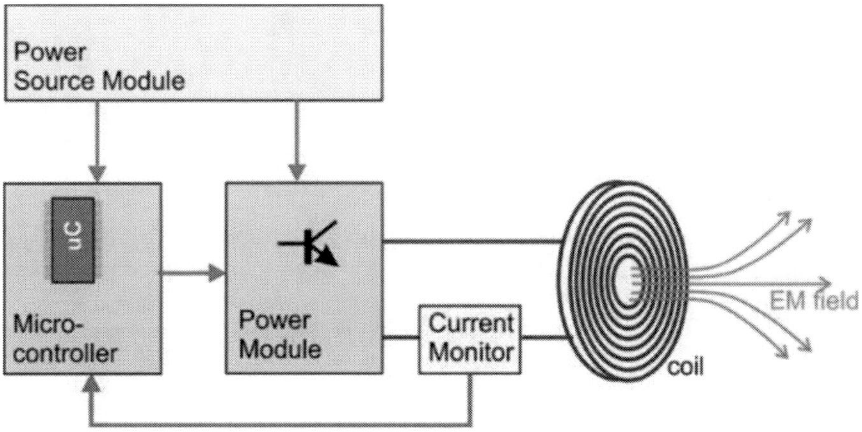

Figure 1. Power source module.

3. The Control

The system is controlled by a microcontroller. For this application we choose Atmel's ATtiny 24 microcontroller. The microcontroller is able to generate digital signals for controlling the power module. The system can generate sinusoidal form pulses simulated using a Pulse-width modulation PWM algorithm (figure 2) or fixed width pulses (figure 3).

Depending by the nature of disease and stage of the disease we can use sinusoidal pulses (figure 2) or train of pulses with different intensity (figure 3) corrected by the width of the pulse of impregnated electromagnetic field.

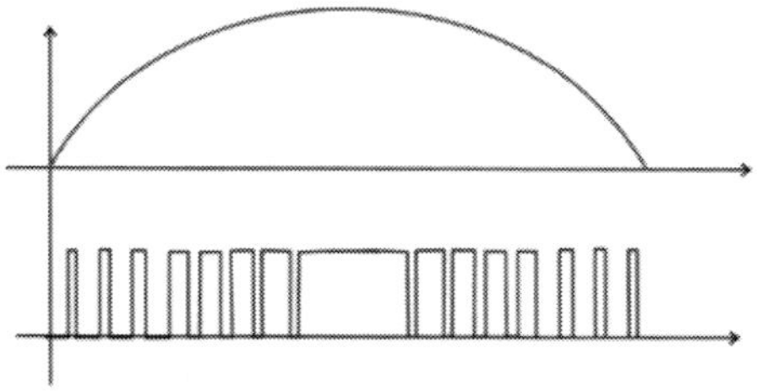

Figure 2. Sinusoidal form pulses.

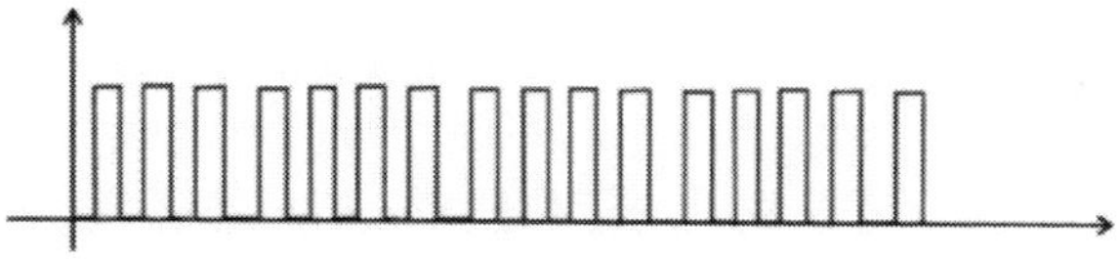

Figure 3. Fixed width pulses.

4. Programming the Unit

4.1. Software Algorithm

The basic software for system control is running in microcontroller. The microcontroller generates the signals based of two algorithms. First one is a basic algorithm used for generating a pulse train with equal width pulses, for generating oscillatory magnetic field. This field is more power full and can be used in short train pulses sequences. Software algorithm is presented in function below, (named GENERATE_0, figure 4).

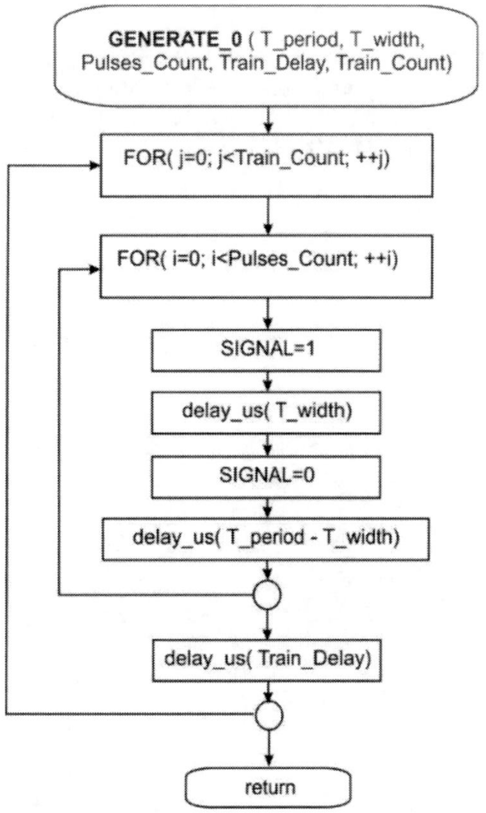

Figure 4. Basic pulse generating algorithm.

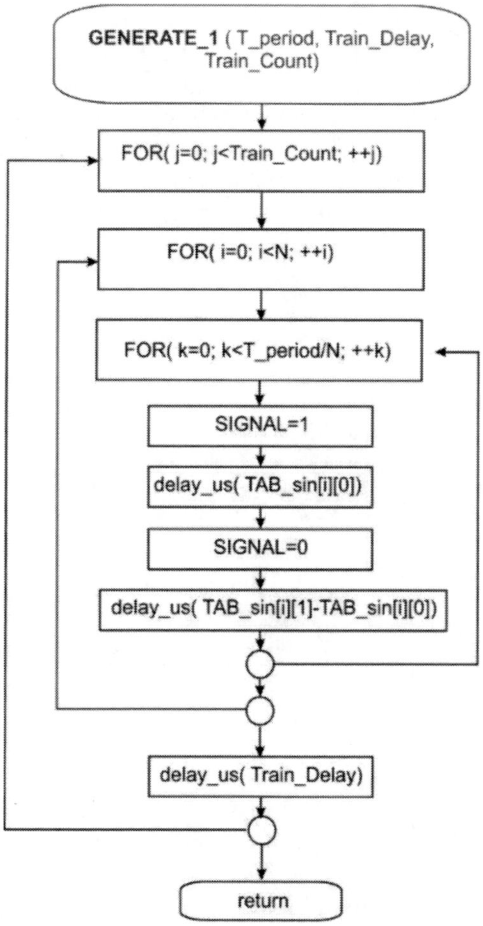

Figure 5. Half sinusoidal signal algorithm.

Parameters of this function are: T_period, represent time period of signal in microseconds, T_width is the time while the signal is active in microseconds, Pulses_Count represents the number of active pulses. This Pulses_Count creates a train of pulses. A sequence of more train of pulses counted by Train_Count is generated with Train_Delay milliseconds pause between them.

The other algorithm used for generating pulses creates an half of sinusoidal signal form based o PWM signal. This generated field is softer and can be used in any application of treatment. Software algorithm is presented in function below in function named GENERATE_1 (figure 5).

The central element of algorithm is table TAB_sin, table which includes integer values for PWM active time of the signal and the period of the PWM signal. These values are calculated so that the output signals to simulate a half of sinusoidal value. Parameters of this function are: T_period, represent time period of signal in milieseconds of half of sinusoidal signal, Pulses_Count represents the number of active pulses. This Pulses_Count creates a train of pulses. A sequence of more train of pulses counted by Train_Count is generated with Train_Delay milliseconds pause between them.

To protect the system by any faults we have a procedure that can be called before applying any signal. This procedure generates a pulse of current through the coil and measure that current. If the current is zero the system is stopped automatically because the coil circuit is interrupted. If the current in coil circuit is greater than a preset value, then the system will be stopped because we have a short circuit.

4.2. The power source check

The microcontroller's software is responsible to check periodically the status of power supply. If power supply generated current is lower than a preset value the magnetic field will be too weak to achieve optimum results. To be able to access the value of power capacity the power line is connected to the microcontroller analog input (figure 6).

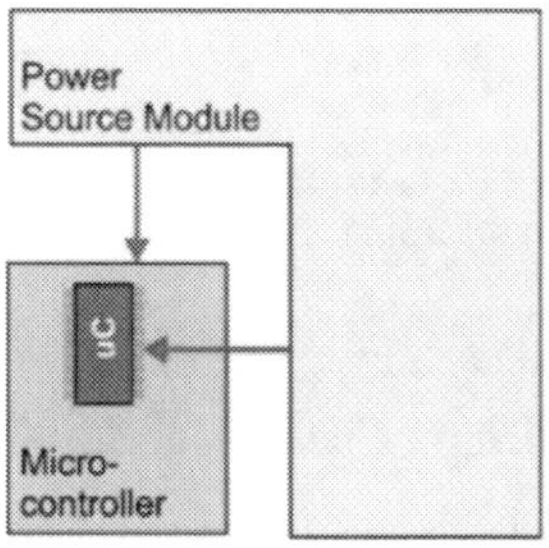

Figure 6. Power check connection.

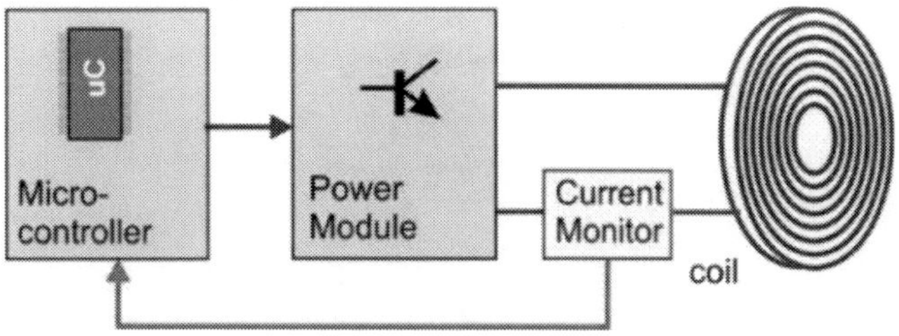

Figure 7. Coil current monitoring.

While generating the electromagnetic pulse, the microcontroller software will check the voltage of power supply. If measured value of power supply voltage will be lower than a preset value, then the controller will stop to activity and will signal error. The software module will return an error message if the value of power supply is lower than a preset value. And also, the function will return a success mesage if power supply voltage is over preset value. In this case the power suppply is able to ensure the coil with enough energy to activate an optimal electromagnetic field.

4.3. The coil current control

The microcontroller's software is also checking the current in the coil.

To be able to measure the current through the coil, the microcontroller will read the current value through the coil (figure 7).

In the first step, before generating any pulse, the software algorithm (figure 8) will send a current pulse through the coil and measure its current value. If the measured value is greater than a presset nominal value, then we have a short circuit in coil electrical circuit. If meassured value is lower than a presset as zero value, the coil will be considered as disconnencted. In both cases the pulses through the coil will be disabled.

4.4. Main program

All modules are integrated into one software program, named main_program. This program controls modules activation and system checking. In according with external configuration switch (figure 9), the software will select one of the following signal generation forms: smooth algorithm or aggressive algorithm.

The main_program is started when microcontroller starts. First function is executed by the main_program program and configures the microcontroller's internal devices and resets the outputs of microcontroller.

The microcontroller's timer interrupts must be tackled in the programm. The software will attach two system events with internal hardware timer:

IOP Conf. Series: Materials Science and Engineering **95** (2015) 012061 doi:10.1088/1757-899X/95/1/012061

- at every one minutes the timer will generate *evPOWER_EVENT* and
- at every 15 seconds the timer will generate *evPULSE_EVENT*.

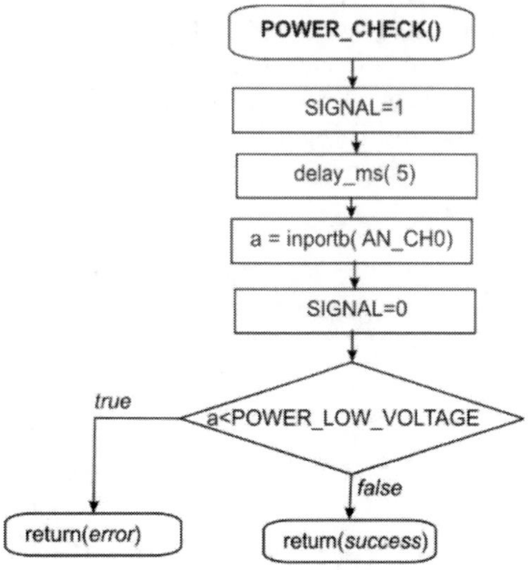

Figure 8a. Power Check Algorithm.

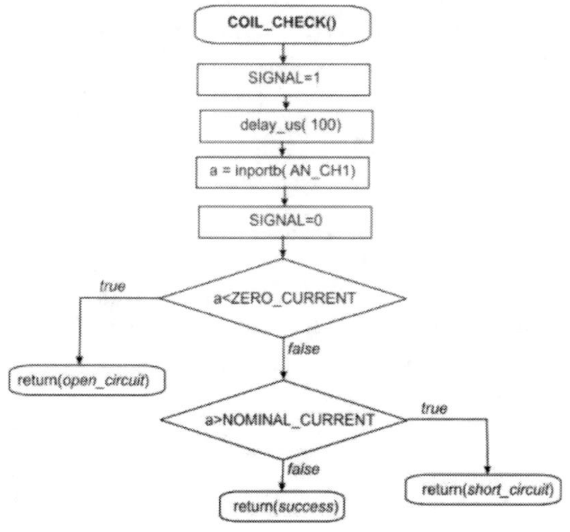

Figure 8b. Coil Check Algorithm.

After programming, the main_programm will opens an infinite loop to check the two timing events.

When *evPOWER_EVENT* event appears the software will check the status of power supply using the *POWER_CHECK* function. If function returns error, then the main loop will be broken and the microntroller stops until the power battery will be replaced. After *evPOWER_EVENT* event is processed the software will rearmed the timer event.

When *evPULSE_EVENT* event appears, the software will re-arme the timer pulse event (*evPULSE_EVENT* event) and test the status of circuit with coil. If we obtain a negative response for the coil circuit test the algorithm will break the loop and the program will be stopped. If coil circuit test algorithm ended successfully, then the status of configuration switch will be checked. If switch is closed the system will generates a train of pulses, using function *GENERATE_0*. In this case the controller will generate an aggressive electromagnetic filed through the coil. If switch is opened the system will generates a train of pulses that simulates a sinusoidal signal, using function *GENERATE_1*. In this case we obtain a smooth signal.

The controller's switch will select between two output types of signals. One type of generated signal is aggressive signal which conduct to aggressive treatment of disease and the other one type of generated signal is a smooth signal which can be used in therapeutic treatment of periodontal disease. All software modules are programmed into microcontroller Atmel's A.Tiny 24 [6]. This microcontroller is part of controller board, which is connected with the coil [7].

5.Conclusions

A pulsed electromagnetic field bone growth stimulator is presented, to offer a dynamic tool to fight with periodontal disease. The controlled wave forms and pulse frequency signal at programmable intervals are easy to be integrated in a light, portable device, easy to be endowed in dental cabinets and easy to manipulate and maintain, safe to be used.

This patient monitoring system enhances the doctor's ability to control the periodontal condition evolution, with strict treatment formulas, visual monitoring and patient history comparison, along with medical parameters monitoring. The research is open for further investigation related with the fine tuning of the magnetic parameters [7] and integration with larger monitoring systems. The algorithm below (figure 9) presents the main_program structure.

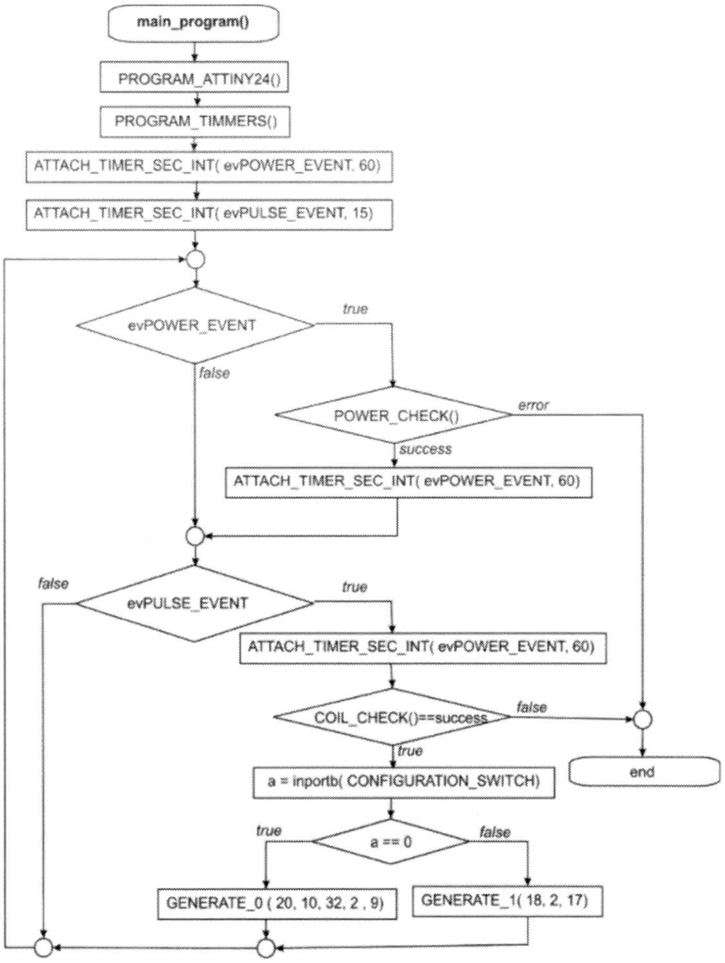

Figure 9. The main algorithm.

References

[1] http://drpawluk.com/updates/magnetics-for-dental-applications/, accessed Feb 27, 2015

[2] Aaron R K, Boyan B D, Ciombor D McK; Schwartz Z, Simon B J 2004 Stimulation of Growth Factor Synthesis by Electric and Electromagnetic Fields *Clinical Orthopaedics & Related Research* **419** pp 30-37

[3] Behrens S B, Deren M E and Monchik K O 2013 A Review of Bone Growth Stimulation for Fracture Treatment *Curr Orthop Pract.* **24(1)** pp 84-91

[4] http://periodontitis-treatment.com/, accessed 27 February 2015

[5] http://www.term.or.kr/journal/view.php?number=484, accessed 27 Feb. 2015

[6] Documentation: Atmel 8 bit AVR Microcontroller with 2k/4kBytes In-System Programmable Flash, ATtiny 24, Atmel, 2010

[7] PEMF Stimulation Therapy Method and System, US Patent 5,743,844, Apr 28, 1998

Contributions regarding calculus and design of an ultrasonic system used in plasma metallization

G Amza[1]

[1]Politechnica University of Bucharest, IMST Faculty, Splaiul Independenţei No. 313, sect.6, Bucharest, Romania

E-mail: amza@camis.pub.ro

Abstract. Paper presents the calculus elements and construction for the ultrasonic system used in reconditioning process by metallization. A series of necessary elements used in ultrasonic system dimensioning are given and an ultrasonic system used in reconditioning process by plasma and grain metallization are presented. Also, a calculated ultrasonic system is modelled to work in resonance regime at the frequency f = 22Khz. Stress map inside material, internal energy variation, lost energy variation on length and volume unit are presented.

1. Introduction

Efficiency of spray technology [3,4,5] depends primarily on the ultrasonic field of the ultrasonic energy input mode of the load area of the filler material. Experiments have shown that ultrasound propagation in the workplace has significant influence over the process of transfer of the filler material to support material, on the accession process, the process of crystallization and solidification and even the process of tempering [1,2,8]. All these influences are attributed to three basic phenomena occurring ultrasound propagation namely ultrasonic cavitation; accelerate the diffusion and acceleration of crystallization and solidification. The most difficult problem that arises is linked to the means of introducing the ultrasonic energy into the load.

This paper presents the experimental results obtained for spray metallization piece reconditioned ultrasound activation using plasma metal spraying and powder [3,9]. The metallization plasma jet and ultrasonic activation powder reconditioned piece (figure 1), part 1, is placed on the active element 2. Vibrating ultrasonic frequency is related to ultrasonic energy concentrator amplifier 3 and 4. Because the ultrasonic energy does not dissipate throughout the system ultrasonic is fixed to the drive device 6 through the nodal flange 5. The ultrasonic transducer [9] is composed of sound reflector 9, the ceramic piezoceramic disc 8 and the radiant sound 7, assembled by mechanical bias means of a screw 10. Plasma jet 11 is formed between the tungsten electrode 12 and the nozzle 13, cooled by liquid fluid 14, the arc formed and axially and radiating constrained by a plasma gas 15 together with the plasma gas and the filler material is introduced in the form of powder 16, who arrived in the plasma jet turns into drops 17, which are designed to be reconditioned surface, forming deposited layer 18. The ultrasonic shall be compiled and designed as the active element particle vibration amplitude and have longitudinal vibration and radial vibration.

Content from this work may be used under the terms of the Creative Commons Attribution 3.0 licence. Any further distribution of this work must maintain attribution to the author(s) and the title of the work, journal citation and DOI.
Published under licence by IOP Publishing Ltd

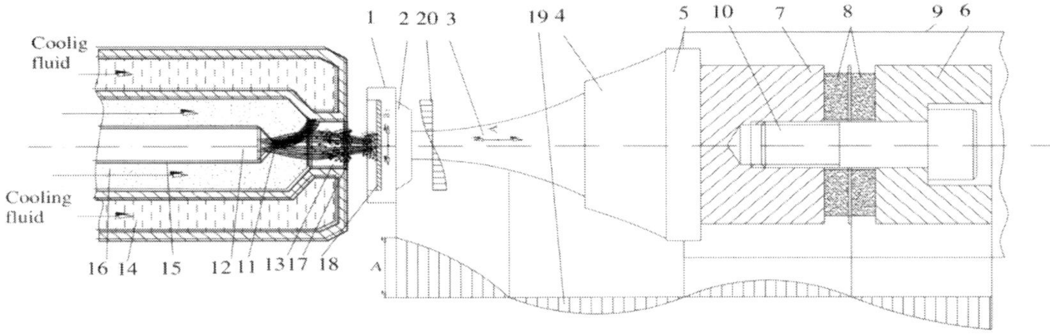

Figure 1. Schematic diagram for metal plasma and dust with ultrasonic activation reconditioned piece: 1-piece refurbished; 2 active element; 3- concentrator of ultrasonic energy; 4-intermediate element; 5- nodal flange; 6- clamping device; 7- radiant sound; 8- ceramic piezo discs; 9- acoustic reflector; prepolarization screw 10; 11 plasma jet; 12- tungsten electrode; 13- nozzle; 14 water-cooled nozzle; 15-nozzle for plasma gas and dust; 16- powder filler; 17 drops of filler; 18-deposited layer; 19- diagram particle velocity amplitude variation along the ultrasonic system; 20 - diagram of radial oscillation amplitude variation.

Table 1. Initial data necessary ultrasonic dimensioning system components.

Initial data necessary dimensioning	U.M.	Value	Reflector diameter d_R	mm	48.37
Average permeability ε_0	F/m	$8.854.10^{-12}$	Ceramic piezo disc diameter	mm	35.45
Resonance frequency f_0	Hz	$2.2.10^4$	Acoustic radiant diameter d_r	mm	56.75
Resonance angular frequency ω_0	Hz	$4.4.\pi.10^4$	Radiation area of radiant A_r	cm^2	28.26
Maximum a tip amplitude of the active part A	µm	107	Aria acoustic reflector A_R	cm^2	28.26
Electrical input power P_{in}	W	1000	Ceramic piezo disc area	cm^2	28.26
Minim acoustic intensity I_u	W/cm^2	9.6	Active element radius r_r	cm	3.0
Performance acoustic-mechanical η_{am}	-	0.75	Electromagnetic conversion coefficient n_p	N/V	5.23
Electromechanical coupling factor k_{em}	-	0.65	Acoustic impedance of the reflector Z_R	Kg/s	$5.248.10^4$
Performance electro acoustic η_{ea}	-	0.92	Acoustic impedance piezo disc Z_p	Kg/s	$6.432.10^4$
Ceramic piezo material density ρ	Kg/m^3	$7.6.10^3$	Acoustic impedance acoustic radiant Z_r	Kg/s	$9.637.10^4$
Relative permittivity ε_{rp}	F/m	2600	Effective Voltage U_p	V	$3.062.10^3$
Young's modulus Y_p	N/m^2	$7.4.10^{10}$	Active element capacitance C_p	nF	2.45
The angle losing σ_p	deg	0.687472	Ideal resonance acoustic power	W	$3.654.10^3$
Piezoelectric constant k_p	m/V	665.10^{-12}	Sound power at resonance by considering losses	W	$2.06.10^3$

IOP Conf. Series: Materials Science and Engineering **95** (2015) 012062 doi:10.1088/1757-899X/95/1/012062

The calculation resulted ultrasonic system dimensions shown in figure 2.

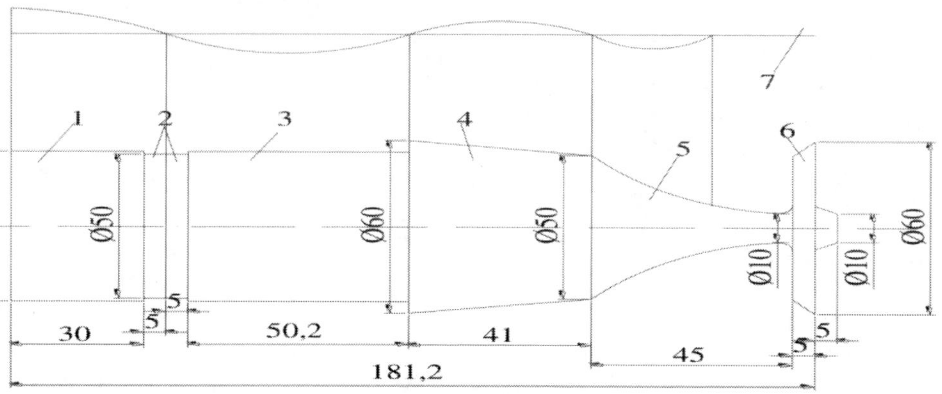

Figure 2. The ultrasonic refurbishing used by metallization in ultrasonic field SUS 01-RM model: 1 - acoustic reflector; 2 - ceramic piezo disc pads (PZT4); 3 - radiant sound; 4 -acoustic amp; 5 - energy concentrator ultrasound; 6 - ultrasonic tool; 7 - diagram of oscillation amplitude variation along the ultrasonic system.

2. Experimental

For ultrasonic system modeling and simulation was done with specialized software that enables knowledge KREEL ENGINEERING state parameters and calling for energy concentrator ultrasonic and ultrasonic system wide: the gain; node and antinode location, size amplitude curves of variation of losses and states of intense stress concentrator, block, respectively ultrasonic transducer ultrasonic. Introducing data into specialized software were obtained following diagrams and dimensions of ultrasonic energy concentrator, figures 3 and 4, [8].

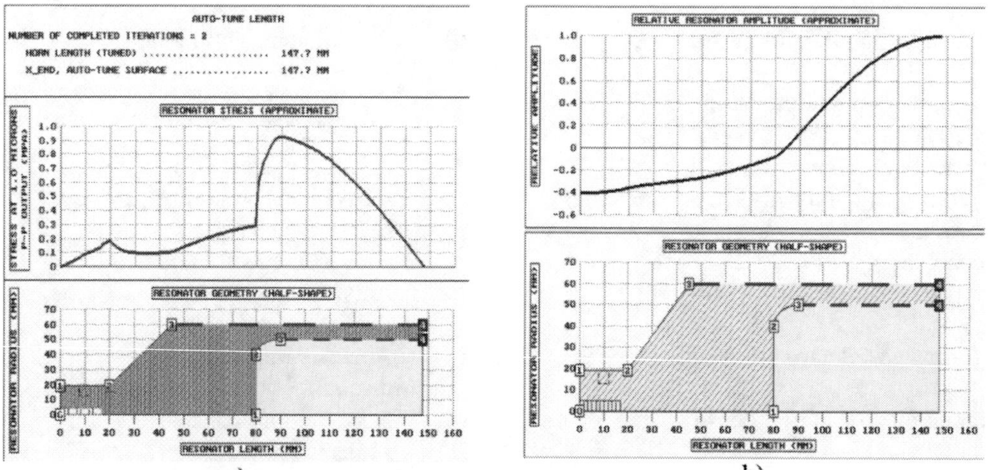

a) b)

Figure 3. Diagram of efforts in ultrasonic system structure (a) Relative amplitude variation diagram tool work (b).

The simulation was performed of a design and dimensional ultrasonic system operating at f = 22 kHz was used in research and obtained the following results:

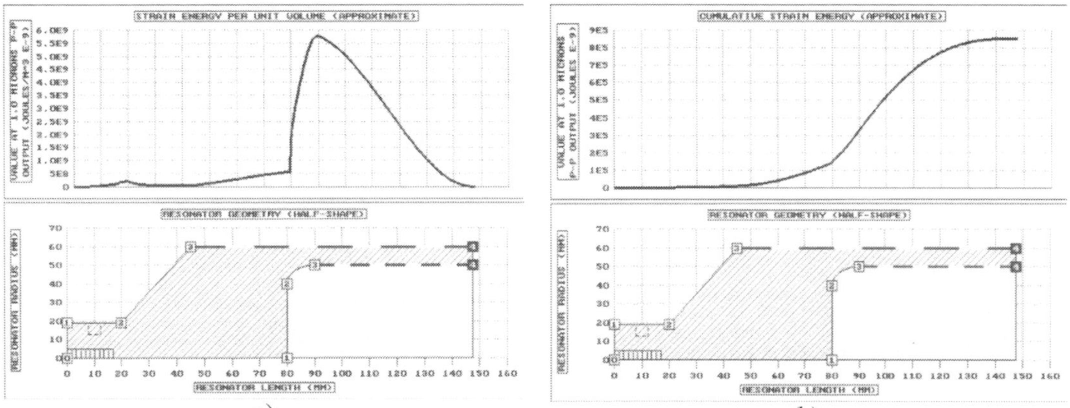

a) b)

Figure 4. Mechanical loading energy per unit volume (a) Specific loss per unit volume (b).

- Elements dimensional ultrasonic system (figure 5);

23e	ø Expo	173.96	3.00			Titan Grade 2
22e	ø Straight	115.95	13.00	0.00		Titan Grade 2
21e	ø Straight	99.95	13.00	45.00		Titan Grade 2
20e	ø Straight	99.75	12.60	-90.00		Titan Grade 2
19e	ø Straight	99.75	13.00	-45.00		TiGr2
18e	ø Straight	99.25	14.00	0.00		TiGr2
17e	ø Arc	67.89	14.00		0.50	TiGr2
16e	ø Straight	67.39	14.50	-90.00		TiGr2
15e	ø Straight	67.39	17.00	0.00		TiGr2
14e	ø Straight	65.39	17.00	90.00		TiGr2
13e	ø Arc	65.39	14.50		0.50	TiGr2
12e	ø Straight	64.89	14.00	0.00		TiGr2

Figure 5. Elements dimensional ultrasonic system at 700 kHz

- Form and dimensional of ultrasonic system with L = 181.2 mm (figure 6);

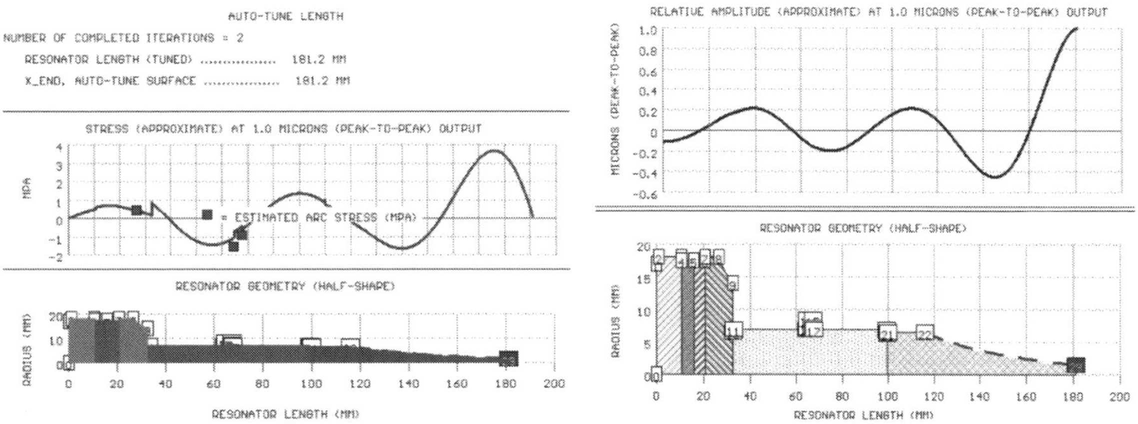

a) b)

Figure 6. Form and dimensions resulting from ultrasonic system auto calibration program (a) Graph of variation of ultrasonic oscillation amplitude (b).

- Characteristic elements of ultrasonic energy concentrator (table 2);

Table 2. Elements characterization tool the results of the simulation program

No.	Elements of characterization	Value
1	Length working instrument	181.23 mm
2	Auto calibration tool length	48.2 mm
3	Frequency	22000 Hz
4	Amplifier factor	6.005 S
5	Maximum effort (1 μm output)	3.64 MPa la 166.0 mm
6	Coordinate oscillation node 1	16.0 mm
7	Coordinate oscillation node 5	60.3 mm
8	Internal effort(1 μm output)	1.05E-4 Joules
9	Power dissipation 1 μm output)	1.6 E-4 watt
10	Form factor Q	26688

- Variation of state of internal application and of internal energy per length (figure 7);

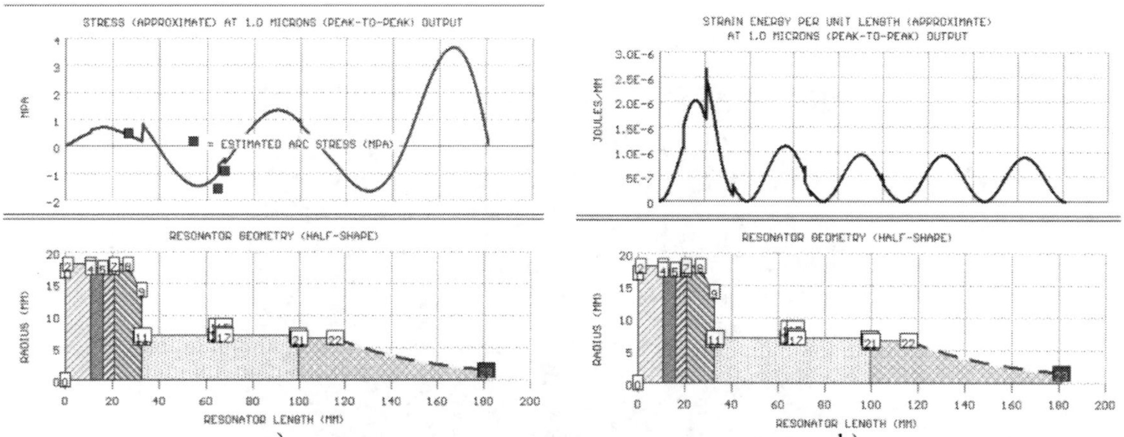

a) b)

Figure 7. The graph of variation of state of internal application of the material ultrasonic tool (a); The graph of the variation of internal energy per length (max = 142.92 MPa at x = 181.2 mm) (b).

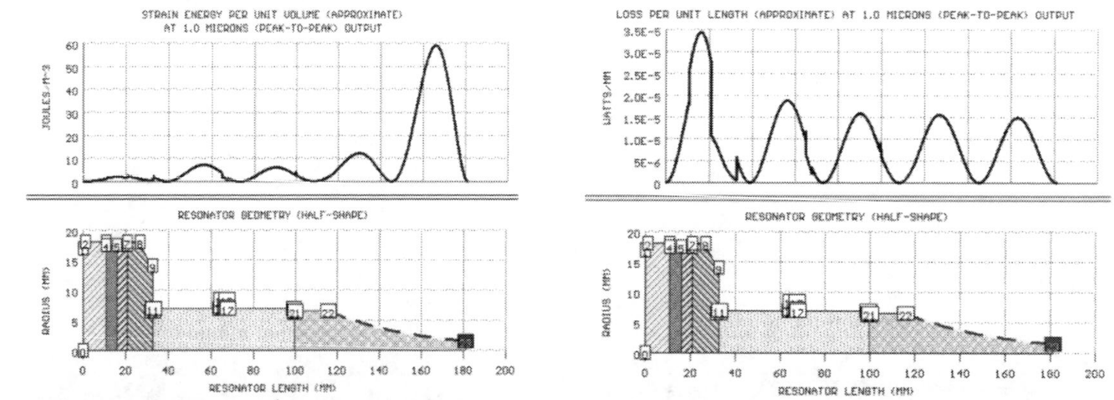

a) b)

Figure 8. The graphic variation of internal energy by volum (2676 Joule/m3 la x =181,2 mm) (a); The graph of the variation of losses per unit length (13.77 x10-5 W / mm at x = 16 mm) (b).

IOP Conf. Series: Materials Science and Engineering **95** (2015) 012062 doi:10.1088/1757-899X/95/1/012062

The graph in figure 7b reveals five maximum internal energy variation along transformer amplitude maximum corresponding nodal areas 6.4. 10-6.Joule / mm in the first nodal point xnod1 = 16.0 mm and 35.10-6 Joule / mm in the fifth nodal area xnod = 60.3 mm. These areas are susceptible to heat for a long working life. The graph in figure 7b reveals five maximum internal energy variation along transformer amplitude maximum corresponding nodal areas 6.4. 10-6.Joule / mm in the first nodal point xnod1 = 16.0 mm and 35.10-6 Joule / mm in the fifth nodal area xnod = 60.3 mm. These areas are susceptible to heat for a long working life. The graph of figure 8b reveals two peaks of variation in the loss per unit length along the concentrator ultrasonic energy maximum corresponding to the nodal areas, 13,77.10-5 W / mm at the first nodal point xnod1 = 16.0 mm and 59, 04.10-5W / mm, the fifth nodal point xnod5 = 106.5 mm. The above areas are susceptible to heat for a long working life.

3. Conclusions

The ultrasonic system is the most important any plant reconditioning by metallization coating ultrasonic field because the efficiency of deposition of filler material in the loading depends primarily on its correct functioning;

Elements of calculation for ultrasonic transducer are: the condition mechanical resonance; electromechanical resonance condition; acoustic power emitted at resonance; acoustic power at low frequencies; frequency characteristic of acoustic power in near resonance; electrical impedance of the transducer; efficiency of acoustic electro transducer; sensitivity electromechanical transducer resonance regime; receptor sensitivity at low frequency; specific characteristic frequency reception sensitivity and currents at resonance;

Identify methods each section vibrating ultrasonic system is necessary and possible the design of by finite element analysis after which the plot diagram of particle velocity amplitude variation along the ultrasonic system and determine the type of ultrasonic oscillations;.

The correct operation of the an ultrasonic system is necessary to know the following: ultrasonic oscillation amplitude variation in any section; state of application of the material in any section; variation state of internal application of the material; according to length variation of internal energy; losses per unit length variation; variation losses per unit volume and characteristics of ultrasonic energy concentrator.

References

[1] Dehelean D 1997 *Fusion Welding* (Welding Publishing House: Timisoara)

[2] Amza G 2006 *Ultrasounds, Active applications* (AGIR Publisher: Bucharest)

[3] Dumitru G 2009 *Metal part reconditioning* (Bren Publishing House: Bucharest)

[4] Iacobescu G 1999 *Welding Equipments* (Printech Publishing House: Bucharest)

[5] Niţoi D and Taşcă G 2015 Eperimental results regarding metal plating by welding method in ultrasonic fieldfor high corosed base metal piece *Applied Mechanics and Materials* **760** pp 539

[6] AmzaG 2013 Research on enviromental impact assesment of flame oxyacetilene welding processes *Metalurgija* **52(4)** pp 457-460

[7] Amza G, Luchian C, Niţoi D and Dumitrache F 2010 Contributions Regarding Plastic Materials Ultrasonic Welding used in Automotive Industry *Materiale Plastice* **47** p 90

[8] Kovalev I M and Rybakov A S 1997 The movement of liquid metal in the weld pool in welding in a longitudinal megnetic field *Welding Prod.* **24** pp 38-40

[9] Amza Gh, Niţoi D, Amza C, Apostolescu Z and Opran C 2015 Ultrasonic System Design and Calculation for Polymer Metallization *Applied Mechanics and Materials* **760** pp 263-268

IOP Conf. Series: Materials Science and Engineering **95** (2015) 012063 doi:10.1088/1757-899X/95/1/012063

A point of view on Otto cycle approach specific for an undergraduate thermodynamics course in CMU

F Memet[1] and A Preda[1]
[1]Constanta Maritime University, Faculty of Naval Electro-Mechanics, 104 Mircea cel Batran Street, 900663, Constanta, Romania

E-mail: feizamemet@yahoo.com

Abstract. This paper refers to the description of the way in which can be presented to future marine engineers the analyzis of the performance of an Otto cycle, in a manner which is beyond the classic approach of the course of thermodynamics in Constanta Maritime University. The conventional course of thermodynamics is dealing with the topic of performance analysis of the cycle of the internal combustion engine with isochoric combustion for the situation in which the working medium is treated as such a perfect gas. This type of approach is viable only when are considered relatively small temperature differences. But this is the situation when specific heats are seen as constant. Instead, the practical experience has shown that small temperature differences are not viable, resulting the need for variable specific heat evaluation. The presentation bellow is available for the adiabatic exponent written as a liniar function depending on temperature. In the section of this paper dedicated to methods and materials, the situation in which the specific heat is taken as constant is not neglected, additionaly being given the algorithm for variable specific heat.For the both cases it is given the way in which it is assessed the work output. The calculus is based on the cycle shown in temperature- entropy diagram, in which are also indicated the irreversible adiabatic compression and expansion. The experience achieved after understanding this theory will allow to future professionals to deal successfully with the design practice of internal combustion engines.

1. Introduction

Maritime Education and Training is the path towards a successful career in shipping industry, which despite of the present global economic crises, is asking for highly qualified marine engineers in the context of the world's fleet growth [1].

Developments in marine engineering, found in new and future ships, offer major benefits in terms of fuel consumption, maintenance, carrying capacity, environmental behavior, etc. All these aspects must be reflected in the education and training of future marine engineers [2].

In Constanta Maritime University (CMU), an important educational outcome of its engineering programme is the capacity of graduates to design a component or a system. Taking into account that population of the world is increasing and the natural resources are depleting, future marine engineers will face the energy efficiency challenge. In this context, future graduates will deal with the problem related to the increase of the efficiency of power systems.

Thermodynamics is an important component of mechanical engineering curricula, as well as of marine engineering curricula. In Constanta Maritime University, future marine engineers, encounter the Otto cycle during Thermodynamics 1 course, delivered in their second year of study. Because

Content from this work may be used under the terms of the Creative Commons Attribution 3.0 licence. Any further distribution of this work must maintain attribution to the author(s) and the title of the work, journal citation and DOI.
Published under licence by IOP Publishing Ltd

IOP Conf. Series: Materials Science and Engineering **95** (2015) 012063 doi:10.1088/1757-899X/95/1/012063

large temperature differences exist in the practical cycles, the constant specific heat assumption is not valid, therefore this topic should be extended when it is delivered to students. The efficiency evaluation of the cycle of the internal combustion engine with isochoric combustion, considering the heat transfer and also constant and not constant specific heats of the gas, by the involvement of the finite- time thermodynamics, leads to a better understanding of the modern aspect of this discipline.

The aim of these algorithms presented bellow is to provide to future marine engineers guidance for the design of practical Otto engines, by choosing the optimal situation for the desired use.

2. Methods and Materials

The air standard Otto cycle is shown on temperature – entropy coordinates in figure 1. The closed thermodynamic cycle 1-2_t-3-4_t consists of four internally reversible processes: reversible adiabatic compression (1-2_t), constant volume heat addition (2_t-3), reversible adiabatic expansion (3-4_t) and constant volume heat rejection (4_t-1). In the mentioned figure, it is possible to be seen the processes 1-2 and 3-4, which are the irreversible adiabatic processes – in which internal irreversibility in the real compression and expansion processes are considered.

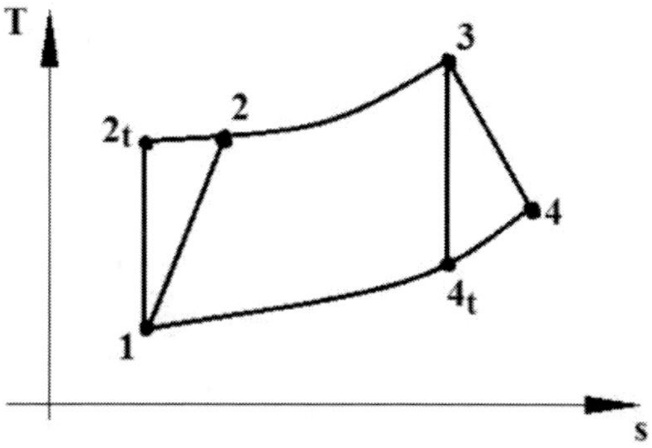

Figure 1. (T-s) diagram for the air standard Otto cycle.

The Otto cycle analysis, which is developed on the assumption that the working fluid behaves as an ideal gas with constant specific heat, is viable for relatively small temperature differences; but in the practice of thermal machines, the temperature differences are important, making useful the examination of the variable specific heat case [3].

In this study, dealing with the performance of the irreversible Otto cycle, it will be used Finite Time Thermodynamics, which is a tool able to reveal the optimal time path of any cyclic process with friction and heat leakage and to offer a clear image of how irreversibility influences the performance of thermal processes [4].

The performance of the analysed air standard Otto cycle is evaluated with:

$$\eta = \frac{W_{out}}{Q_{in}} \qquad (1)$$

meaning the ratio between the work output and the heat added to the working fluid.

2.1. Constant specific heat case

When working with an ideal gas for which are known its initial conditions, the maximum temperature and minimum volume, it is easy to find the other states, the heat input, heat rejected and the work output in the analyzed cycle, by the use of ideal gas laws and the first principle of thermodynamics [5].

For the isentropic compression process (1-2_t) and isentropic expansion process (3-4_t) are written the following equations:

$$\frac{T_{2_t}}{T_1} = r^{k-1} \tag{2}$$

$$\frac{T_{4_t}}{T_3} = r^{k-1} \tag{3}$$

where:

r – the compression ratio for the engine ($r = V_1 / V_2$),
V_1 – the total volume of the cylinder,
V_2 – cylinder clearance volume,
k –ratio of the specific heats ($r = c_p / c_v$),

The investigation is focused on reversible and irreversible processes, thus it is requested the definition of the compression and expansion efficiencies, which describe the internal irreversibilities of the adiabatic processes [6]):

The investigation is focused on reversible and irreversible processes, thus it is requested the definition of the compression and expansion efficiencies, which describe the internal irreversibilities of the adiabatic processes [6]):

$$\eta_c = \frac{T_{2_t} - T_1}{T_2 - T_1} \tag{4}$$

$$\eta_c = \frac{T_{2_t} - T_1}{T_2 - T_1} \tag{5}$$

The total energy of the fuel (Q_{fuel}), the heat added to the working fluid (Q_{in}), the heat leakage (Q_{ht}) and the work output (W_{out}) are evaluated with [7]:

$$Q_{fuel} = \eta_{com} m_f Q_{LHV} \tag{6}$$

$$Q_{in} = Q_{fuel} - Q_{ht} \tag{7}$$

$$Q_{ht} = m_t B(T_2 + T_3) \tag{8}$$

where:

m_f – fuel amount,
m_t – air-fuel amount inducted into the cycle,
η_{com} – combustion efficiency,
B – constant,
Q_{LHV} – the lower heating value of the fuel.

$$m_t = m_f \left[1 + \frac{1}{\left(m_{air} / m_f \right)_s \Theta} \right] \tag{9}$$

Above, Φ is the equivalence ratio and “s” refers to stoichiometric conditions.

$$Q_{in} = \frac{R_{air} m_t}{k-1}(T_3 - T_2) \tag{10}$$

$$W_{out} = \frac{R_{air} m_t}{k-1}(T_1 - T_2 + T_3 - T_4) \tag{11}$$

where R_{air} – characteristic gas constant of air.

2.2. Variable specific heat case

Having in view the existence of large temperature difference in the practical cycles, the specific heat with constant pressure and the specific heat with constant volume depend on temperature, resulting that the adiabatic exponent "k", depends on temperature also.

A reversible adiabatic process, between states i and j, for which k is variable, can be divided into infinitesimally small processes, each of them with constant k. For any of these processes, when occur infinitesimally small changes in temperature (dt) and in volume (dV) of the working fluid, the equation for reversible adiabatic process with variable k is written as [8]:

$$TV^{k-1} = (T + dT)(V + dV)^{k-1} \tag{12}$$

According to Ebrahimi [9]:

$$k = k_0 - k_1 T \tag{13}$$

Above, T is absolute temperature and k_0, k_1 are constants. In this respect, from equation (12) results:

$$Q_{in} = m_t \int_{T_2}^{T_3} c_v \, dT = m_t \int_{T_2}^{T_3} \left(\frac{R_{air}}{k_0 - k_1 T - 1} \right) dT = \frac{m_{air} R_{air}}{k_1} \left[1 + \frac{\Theta}{(m_{air}/m_f)_s} \right] \ln \left(\frac{k_0 - k_1 T_2 - 1}{k_0 - k_1 T_3 - 1} \right) \tag{14}$$

$$Q_{out} = m_t \int_{T_1}^{T_4} c_v \, dT = m_t \int_{T_1}^{T_4} \left(\frac{R_{air}}{k_0 - k_1 T - 1} \right) dT = \frac{m_{air} R_{air}}{k_1} \left[1 + \frac{\Theta}{(m_{air}/m_f)_s} \right] \ln \left(\frac{k_0 - k_1 T_1 - 1}{k_0 - k_1 T_4 - 1} \right) \tag{15}$$

$$W_{out} = \frac{m_{air} R_{air}}{k_1} \left[1 + \frac{\Theta}{(m_{air}/m_f)_s} \right] \ln \left[\frac{(k_0 - k_1 T_2 - 1)(k_0 - k_1 T_4 - 1)}{(k_0 - k_1 T_3 - 1)(k_0 - k_1 T_1 - 1)} \right] \tag{16}$$

3. Results and discussion

An assessment aiming the performance evaluation of the Otto cycle in the case in which internal irreversibility in the real compression and expansion processes are considered was described in terms of thermal efficiency.

In the presented calculation procedure was considered the case of constant specific heats, which is a theoretical situation and also the case of variable specific heats, which corresponds with reality since in practice temperature differences are important.

For both presented cases are given formulas leading to the calculation of the thermal efficiency of the cycle of the internal combustion engine with isochoric combustion (Otto).

The irreversibility of the two adiabatic processes of the cycle was not ignored and were given formulas of the specific efficiencies. For the first case in discussion, T_{2t} can be found if are known r, η_c, η_E and T_1 by the use of relation (2). By introducing T_{2t} in equation (4), results T_2. The temperature at the end of combustion is calculated by replacing equation (7) in equation (10).

T_{4t} results from equation (3); when it is introduced in equation (5), it is possible to get T_4.With T_1, T_2, T_3 and T_4 evaluated, are used equations (11) and finally (1), to have a theoretical basis of the cycle performance analysis.

For the second case in discussion, T_{2t} can be calculated if known r, ηc, ηE and T1 are from equation (14). Then, T_{2t} is introduced in equation (4), to be found T_2.

The temperature at the end of the combustion is calculated using relations number (7) and (15), while the temperature and the end of the theoretical expansion results from formula (14). When it is introduced in equation (5), it is possible to get temperature at the end of real expansion. Being known temperatures in the characteristic points of the actual cycle, by using equations (17) and (1), it is possible to complete the proposed assessment.

4. Conclusions

An efficient marine engineering program, such is the one developed in CMU, provides to future marine engineers the ability to deal with engineering design. The short review of the standard cycle specific to the internal combustion engine with isochoric combustion (Otto), is the foundation of the theoretical exposure in discussion.

In the context of new technologies penetration in the maritime sector and of the need for highly qualified marine engineers, thermodynamics course in CMU was updated. Such an aspect is demonstrated through this article. The classical approach related to the performance of Otto cycles was based on the influence of the constant specific heat. This situation is available only when considering small temperature differences; in practice, because of the large temperature differences, the above mentioned should be reconsidered.

As a result, when delivered to the students, this topic was extended to the analysis of the effect of the specific heats depending on temperature on the efficiency of the irreversible cycle of the internal combustion engine with isochoric combustion, in order to offer to future professionals a tool for optimal design and working conditions assessment specific to above mentioned cycles.

References

[1] Baylon A M, Ma V E and Santos R 2011 The challenges in Philippine Maritime Education and Training *International Journal of Innovative Interdisciplinary Research* **1** pp 34-43

[2] Saharuddin A H, Sulaiman O O, Kader A S A and Wan Nick W B 2011 Marine Engineering as future career in Malaysia *Business Management Dynamics* **1(2)** pp 01-10

[3] Anwar Beg O, Rashidi M M and Mehr N F 2014 Comparative thermodynamic study of air standard cycles with heat transfer and variable specific heats of the working fluid *Int. J. Appl. Math. and Mech.* **10(2)** pp 41-60

[4] Mehta H B and Bharti O S 2009 Performance analysis of an irreversible Otto cycle using Finite Time Thermodynamics *Proc. of the World Congress on Engineering* **II** p 5

[5] Huleihil M and Mazor G 2012 Irreversible performance characteristics of air standard Otto cycle with polytropic processes *Applied Mechanical Engineering* **1(3)** p 6

[6] Ust Y, Sahin B and Safa A 2011 The effects of cycle temperature and cycle pressure ratios on the performance of an irreversible Otto cycle *Acta Physica Polonica A* **120** pp 412-416

[7] Ebrahimi R 2010 Theoretical study of combustion efficiency in an Otto engine *Journal of American Science* **6(2)** pp 113-116

Location of an intermediate hub for port activities

Ş Burciu[1], C Ştefănică[1], E Roşca[1], V Dragu[1] and F Ruscă[1]
[1]Politehnica University of Bucharest, Spl. Independenţei, 313, Bucharest, Romania
Department of Transport traffic and logistics, Faculty of Transport

E-mail: stefanburciu@yahoo.com

Abstract. An intermediate hub might increase the accessibility level of ports but also hinterland and so it can be considered more than a facility with a transhipment role. These hubs might lead to the development of other transport services and enhance their role in gathering and covering economic centres within hinterlands and also getting the part of logistic facility for the ports, with effects on port utilization and its connectivity to global economy. A new location for a hub terminal leads to reduced transport distances within hinterland, with decreased transport costs and external effects, so with gains in people's life quality. Because the production and distribution systems are relatively fixed on short and medium term and the location decisions are strategic and on long term, the logistic chains activities location models have to consider the uncertainties regarding the possible future situations. In most models, production costs are considered equal, the location problem reducing itself to a problem that aims to minimize the total transport costs, meaning the transport problem. The main objective of the paper is to locate a hub terminal that links the producers of cereals that are going to be exported by naval transportation with the Romanian fluvial-maritime ports (Galaţi, Brăila). GIS environment can be used to integrate and analyse a great amount of data and has the ability of using functions as location - allocation models necessary both to private and public sector, being able to determine the optimal location for services like factories, warehouses, logistic platforms and other public services.

1. Introduction

In order to distinguish the need for facility location actions one must foresee the material flows that they would generate, the "inputs" and the "exits" brought by the new locations, so, the products that would be transported (raw materials and products). The essential role of transportation is to reduce distance and allow people and economic agents to put aside space. In a continuous space, in theory, all relations are possible but, in reality, defining the transport system takes into account different restrictions like time, transport cost and space itself that might be ameliorated or overcame by certain conditions and in favour of space continuity (law cost and split-able infrastructures, high quality services etc.).

Literature regarding economic activities location has evolved from classical, descriptive modelling to mathematical and probabilistically modelling. In the 18th century, Turgot established the fundaments of commercial location theory and Von Thunen [1], at the beginning of the 19th century, explained that "the optimal location for agricultural activities is represented by the points where the land rent is maximum". In 1909, Weber [2] has developed the first industrial location model that has the objective of minimizing the transport costs and that was updated later by authors like Hoover, Lösh and more recent, based on rectangular distances, by Huriot and Smith [3]. August Losch and

Content from this work may be used under the terms of the Creative Commons Attribution 3.0 licence. Any further distribution of this work must maintain attribution to the author(s) and the title of the work, journal citation and DOI.
Published under licence by IOP Publishing Ltd

Walter Christaller, based the central point theory [4] that supposes that a consumer will always search to buy a product or a service by travelling the shortest possible distance. Owen et al. [5] has grouped network location models (discrete space) in three major categories: covering problems, where serving a client depends on the distance from him to the production facility that he is allocated to, distance lower than a pre-established limit, satisfying all needs with minimum total costs; Church and ReVelle [6], based on the covering concept, underlie the maximum covering problem (Maximum Covering Location Problem - MCLP), trying to maximize the served demand with a limited number of new production centres; the p-central problem that aims establishing p production centres and allocating each client to the closest of them so that the maximum distance between demand points and production centres would by minimum; the p-median problem [7], where locating the p production centres must be realised so that all demands are covered and the total costs are minimum, in the following hypothesis: the number of facilities is apriority known, the costs of establishing new facilities are the same no matter the potential location and the facilities have no capacity restrictions. The Uncapacitated Facility Location Problem (UFLP) is derived from the p-median problem, by endogenous determination of the number of facilities to be located and was followed by the Capacitated Facility Location Problem (CFLP) [8].

Recent development of the location-allocation models took into account integrating these models within the geographical information systems (GIS) by generating GIS data and by structuring the objective functions of the different models like software modules in GIS [9]. By this integration, decidents might limit the search area of new locations based on well fundament decisions.

2. The location-allocation model

The location-allocation models select optimal locations from a set of possible locations in the most efficient manner and allocate demand for a certain service to these facilities, depending on the demand distribution. The main objective of the location models is assuring the covering of all demand within a specific area for a certain length of the route, predefined, within a standard response time, also predefined.

Location-allocation models have three base components:

- the distribution of the demand points and of the raw material supply points;
- candidate production centres` position;
- distances/travel times matrix from the points of demand/raw materials to consumption.

The study within the paper integrates GIS models and geo-spatial analyse based on location-allocation models with the main objectives:

- evaluation of the existing locations of the silos and their territory covering;
- necessity of locating new silos in order to reduce costs and realise a better coverage of the territory;
- examining the way that these models are sensitive to the change of location for different points on the network.

The minimum impedance problem of the ArcGIS software tries to find new locations so that the weighted costs sum between production and consumption points would be minimum, with certain restrictions. The first set of restrictions says that the demands outside the impedance limit will be considered unallocated and the second set imply that the demands outside the impedance limit will be allocated to the closest centre [11]. This type of problem is similar to the p-median problem of Hakimi but further restriction regarding the capacity of the silos has to be taken into consideration.

The p-median problem consists in finding the locations of p centres on a network so that the total cost is minimized, supposing the situation where the cost of covering the demands of node i is given by the product of the demand of node i and the distance from node i to warehouse/silo j [7]. It can be formalised like an integer linear programming problem:

$$v(P) = \min \sum_{i=1}^{n} \sum_{j=1}^{n} h_i d_{ij} x_{ij} \qquad (1)$$

with the restrictions assuring that every j node is allocated to just one i node that has to be a median and that determine the exact number of medians to be located. Church and Reveille [6] introduced the maximum covering condition (MCLP – Maximum Covering Location Problem) that puts aside the condition of serving all clients and implies a limited number of new facilities, with the formulation:

$$\max \sum_{i \in I} h_i y_i \tag{2}$$

and with restrictions regarding the fact that i client's demand is covered if there is at least one facility situated at most the service distance and that the total number of facilities is limited.

For the case study the area chosen was the one of Brăila and Galați counties as major cereals flows pass through this area to be shipped abroad. The existing silos of the two ports have capacities of 25000 tones, the one in Brăila and 31000 tones, the one in Galați. The silos taken into consideration were the ones with capacity exceeding 10000 tonnes in order to be able to realise a complete train to the destination (table 1) [12,13]

Table 1. Silos in the two counties analysed.

Silos	County	Capacity [t]
Vădeni	Brăila	14000
Traian	Brăila	60000
Ianca	Brăila	29000
Făurei	Brăila	10000
Dudești	Brăila	44100
Jirlău	Brăila	10300
Tecuci	Galați	10000
Independența	Galați	59000
Ivești	Galați	13000

3. Data handling and modelling

The data set in GIS format has been realised by grouping the data-bases used in network formalisation in a structure of network geo-database type within ArcCatalog. The attributes of the links and nodes of the network were taken into consideration, length and capacity. Then, the layers with the necessary data were built in order to analyse the different location-allocation models (figure 1). From the multitude of models offered by ArcGIS, the Maximize Capacitated Coverage model was chosen as the facilities are located such that as many demand points as possible are allocated to solution facilities within the impedance cut-off; additionally, the weighted demand allocated to a facility can't exceed the facility's capacity.

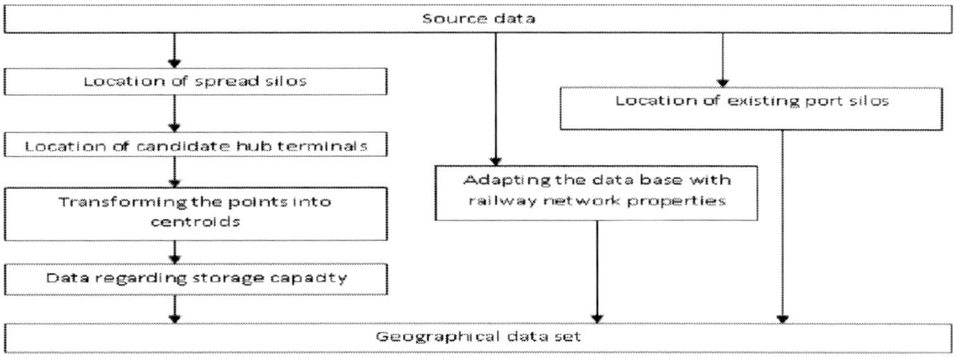

Figure 1. The process of data integration within GIS environment.

Three hypothesis were taken into account, the first were the demand had to be covered by the existing silos in the three ports, were the lack of capacity played a major role, the second considering establishing two new silos of higher capacities in the two ports and the third were only one silo had to be put in place, with a capacity necessary to cover all demand (figure 2).

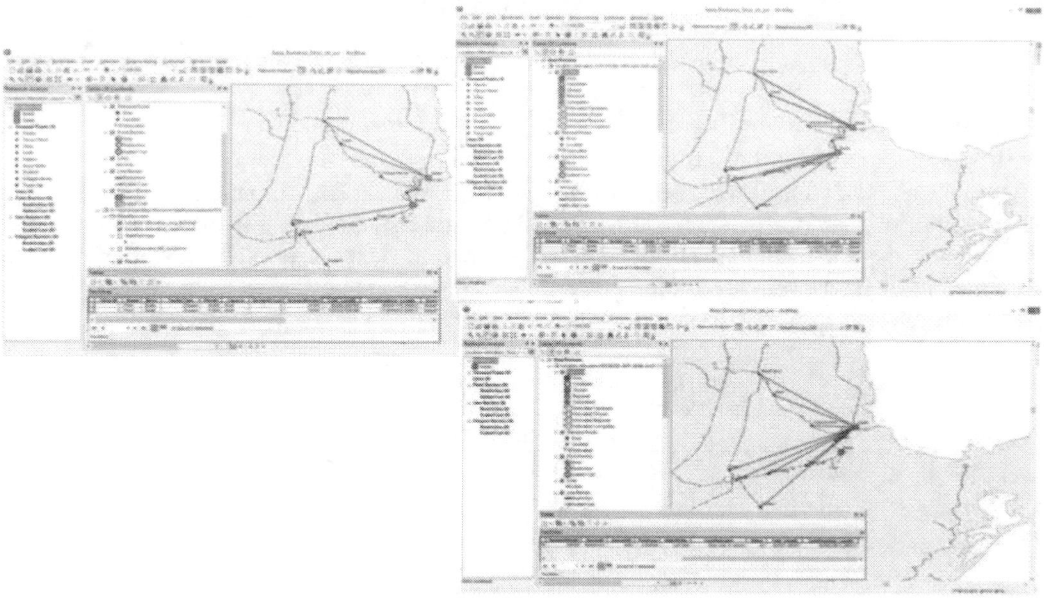

Figure 2. Result of the simulation of the three hypotheses.

4. Results and discussions

After running the three models the logistic parameters regarding the transport process were determined (table 2).

Table 2. Transport parameters in the three hypotheses

Transport parameters / Considered model	Total length [km]	Total weighted length [t·km]
Existing port silos with limited capacity	233835	10658480409
Two new port silos with the necessary capacity	463911	11323218123
One new silo with the necessary capacity (Galați)	637021	16209229914

Among them [10]:
- total length covered by the transport means:

$$D_{\hat{i}} = \sum_{p=1}^{\theta}\sum_{w_i=1}^{\Omega_i} \frac{Q_p^{(w_i)}}{q_{o/p}^{(w_i)}} \times \frac{1}{l_{o/p}^{(w_i)}} \times d_{\hat{i}/p}^{(w_i)} \qquad (1)$$

- total weighted length:

$$P_u = \sum_{p=1}^{\theta}\sum_{w_i=1}^{\Omega_i} v_p^{(w_i)} \times d_{\hat{i}/p}^{(w_i)} \times q_{o/p}^{(w_i)} \times \frac{l_{o/p}^{(w_i)}+1}{2} \qquad (1)$$

where: $D_{\hat{i}}$ is the loaded trip for all transport means, during θ time interval,

IOP Conf. Series: Materials Science and Engineering **95** (2015) 012064 doi:10.1088/1757-899X/95/1/012064

$Q_p^{(w_t)}$ – transport flow for the covered area with $w_t\left(w_t = \overline{1,\Omega_t}\right)$type transport vehicles in a $p\left(p = \overline{1,\theta}\right)$ period,

$q_{o/p}^{(w_t)}$ – average size of a delivery,

$l_{0/p}^{(w_t)}$ – average number of deliveries in a trip,

$d_{i/p}^{(w_t)}$ – length of the average loaded trip.

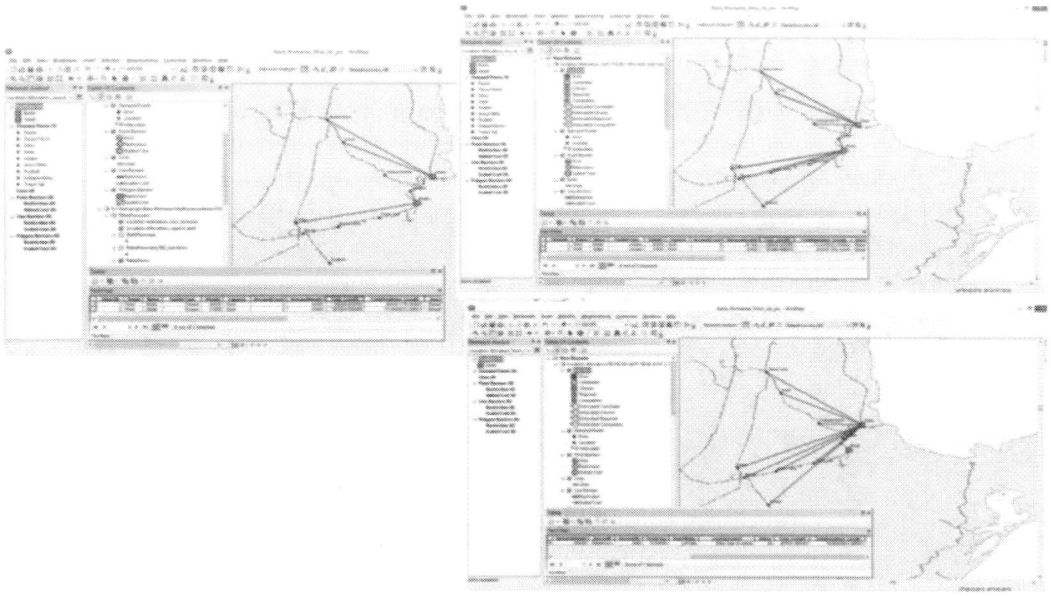

Figure 3. Result of the simulation of the three hypotheses.

The existing situation is not able to cover all demand at the right time, bringing up major waiting times and so, relevant costs increase. In the second hypothesis the demand could be covered but with major strategically decisions necessary as establishing two facilities is a hard thing to do and not feasible as the Port of Galaţi has an on-going project for building an intermodal terminal. In the third hypothesis, even though the total length increased by 34%, the total weighted length, the one bringing incomes to the transport operators, increased by 43%.

5. Conclusions

In order to discern the need for location/relocation actions we must know the transport flows that would be generated and that would eventually become transport processes together with the set of points regarding production areas and possible locations. Location decisions are critical in realising the logistic chains and for their operation. Misplaced facilities might lead to costs increase and low service quality no matter stock, transport or information politics. As the number of located facilities increases they become closer to users and determine transport cost diminishing with higher operational costs. As location decisions are strategic and long term decisions, the facility location models must take into account uncertainties regarding possible future situations. One might see that most models consider equal production/storage costs, the location problem becoming a problem with the objective of minimising the total transport costs, the so called transport problem, with a solution that might be perceived as the determination of the set of market and production areas [11].

Optimization of the multitude of the logistic activities supposes accepting lower performances for some activities so that the whole logistic chain efficiency would increase. By comparing the transport indicators of the models even though the total length increased by 34%, the total weighted length, the one bringing incomes to the transport operators, increased by 43% and so, the facilities were able to be used almost at their full capacity, bringing up benefits to the port and to the storage and handling operators within. GIS might be used to integrate and analyse a great amount of data and has the capacity of using functions like the location-allocation models, so necessary to the private and public sector by being able to determine optimal locations for private and public services.

Acknowledgement
The work has been funded by the Sectoral Operational Programme Human Resources Development 2007-2013 of the Ministry of European Funds through the Financial Agreement POSDRU/159/1.5/S/132397

References

[1] Asami Y and Isard W 1989 Imperfect information, uncertainty and optimal sampling in location theory: An initial reexamination of Hotelling, Weber and Von Thunen *Journal of Regional Science* **29** pp 507-521
[2] Cooper L and Katz I N 1981 The Weber problem revisited *Computers and Mathematics with Applications* **7** pp 225-234
[3] Huriot J M, Smith T E and Thisse J F 1989 Minimum-cost distances in spatial analysis *Geographical Analysis* **21** pp 239-315
[4] Kuby M 1989 A location-allocation model of Lösch's central place theory: testing on a uniform lattice network *Geographical Analysis* **21** pp 316-338
[5] Owen T W, Carlson T N and Gillies R R 1998 An assessment of satellite remotely-sensed land cover parameters in quantitatively describing the climatic effect of urbanization *International Journal of Remote Sensing* **19** pp 1663-1681
[6] Church R and ReVelle C S 1974 The maximal covering location problem *Papers of the Regional Science Association* **32** pp 101-118
[7] Farahani R Z, Hekmatfar M, Arabani A B and Nikbakhsh E 2013 Hub location problems: A review of models, classification, solution techniques, and applications *Computers & Industrial Engineering* **64** pp 1096–1109
[8] Drezner Z 1995 *Facility Location A Survey of Application and Methods* (New York: Springer)
[9] Yen B C 1987 Reliability of Hydraulic Structures Possessing Random Loading and Resistance Duckstein L and Plate E J (Eds.) *Engineering Reliability and Risk in Water Resources* pp 95-113
[10] Raicu Ş 2007 *Sisteme de transport* (Bucharest: AGIR)
[11] Portugal L S, Morgado A V and Júnior 2011 Location of cargo terminals in metropolitan areas of developing countries: the Brazilian case *Journal of Transport Geography* **19** pp 900–910
[12] Information on http://resources.arcgis.com/en/Tutorials/, accessed: 24/11/2014
[13] Information on http://www.madr.ro/docs/depozite/galati.pdf, accessed: 15/10/2014
[14] Information on http://www.madr.ro/docs/depozite/2014/braila.pdf,accessed: 22/10/2014

IOP Conf. Series: Materials Science and Engineering **95** (2015) 012065 doi:10.1088/1757-899X/95/1/012065

Testing market informational efficiency of Constanta port operators

E Roşca[1], **M Popa**[1], **F Ruscă**[1] **and Ş Burciu**[1]
[1]Politehnica University of Bucharest, Spl. Independenţei, 313, Bucharest, Romania
Department of Transport traffic and logistics, Faculty of Transport

E-mail: eugen.rosca@gmail.com

Abstract. The Romanian capital market is still an emergent one. Following the mass-privatization process and the private investments, three of the most important handling and storage companies acting in Constantza Port (OIL Terminal, Comvex and SOCEP) are listed on Romanian Stock Exchange. The paper investigates their evolution on the market, identifying the expected rate of return and the components of the shares risk (specific and systematic). Also, the price evolution could be analyzed through the informational efficiency which instantly reflects the price relevance. The Jarque-Bera normality test regarding the shares return rate distribution and the Fama test for the informational efficiency are completed for each company. The market price model is taken into consideration for price forecasting, computing the return rate auto-correlations. The results are subject of interpretation considering additional managerial and financial information of the companies' activity.

1. Introduction

The Romanian capital market has been developed following the Mass Privatization Program, through which a part of the public companies shares had been freely distributed. After years of regulations, the secondary securities market is represented by Bucharest Stock Exchange (BSE) and the Romanian OTC Market (Rasdaq), which are now unified on a single trade platform. The company shares listed on BSE are traded according to their dimension, the number of shares, the atomicity of owners and the financial results. Among them, there are six companies acting in transportation, handling and storage industry, two of them being located in Constantza Port.

Oil Terminal S.A. (OIL) is connected with the oil industry, considered a very dynamic field. SOCEP Constantza (SOCP) is one of the greatest port operators. It has a container terminal and one of general wares, especially for chemicals, metallic products and cereals.

Based on the market prices evolution between August (2014) - February (2015), the present paper focuses on computing the investment performance of the two companies' shares, testing the market price model and the informational efficiency.

2. Formatting the title, authors and affiliations

The shares return rate is a measure of the growth in wealth, resulting from the investment in the company's equity. It is computed over a specific time interval, according to the investor trading horizon:

Content from this work may be used under the terms of the Creative Commons Attribution 3.0 licence. Any further distribution of this work must maintain attribution to the author(s) and the title of the work, journal citation and DOI.
Published under licence by IOP Publishing Ltd

$$R_t = \frac{P_t - P_{t-1}}{P_{t-1}} + \frac{D_t}{P_{t-1}} \qquad (1)$$

While this takes care of all the explicit payments, there are other benefits that may derive from holding a stock, including the right to vote on corporate governance, tax treatment, rights offerings, and many other things.

These extra benefits are typically reflected in the price fluctuation of the shares.

The first term in the right member of equation (1) denotes the capital gains/losses resulting from the price evolution, while the second term represents the dividend yield. Usually, the return rate on small intervals (day, month) is computed without taking into consideration the dividend yield, because the dividend repartition is followed by a long period before payment. Also, equation (1) does not consider any tax payment.

Stocks return rate may be riskier or more volatile. The associated risk with a share is denoted by the standard deviation of the return rate:

$$\sigma^2 = \frac{1}{T-1}\sum_{t=1}^{T}[R_t - E(R_t)]^2 \qquad (2)$$

where:

T is the number of observed intervals

$E(R_t)$ - the expected return rate.

Assuming a normal distribution for the return rate, a confidence interval of it could be determined. Testing normality should be done using the Jarque-Bera test:

$$JB = \frac{T}{6}\left[s^2 + \frac{(k-3)^2}{4}\right] \qquad (3)$$

where s is the distribution skewness (asymmetry factor) and k its kurtosis (flatness factor). The JB statistics has a χ^2 distribution with two degrees of freedom.

The influence of the market on the share price could be analyzed through the market model that assumes a linear regression:

$$R_t = \alpha + \beta R_{Mt} + \varepsilon_t \qquad (4)$$

where:

R_t is the share return rate at the moment t

R_{Mt} - the market return rate.

In equation (4), α denotes the influence of the steady factors upon the return rate of the shares, excepting the market contribution. β represents the volatility factor, expressing the sensitivity of the return to the market evolution. The residuals ε_t quantify the stochastic influence of the random factors. Parameters α and β are estimated using the least squares method and additional statistic tests should be made to determine the validity of the market model. The coefficient of determination is used to measure the goodness of fit. The market return rate is computed using the appropriate market index:

$$R_{Mt} = \frac{I_t - I_{t-1}}{I_{t-1}} \qquad (5)$$

where I_t is the market index. Therefore, the shares risk is:

$$\sigma^2(R_t) = \beta^2\sigma^2(R_{Mt}) + \sigma^2(\varepsilon_t) \quad (6)$$

The first term of the shares risk represents the systemic component, due to the market, while the second term is the specific component of the risk.

Securities price evolution could also be analyzed through the market efficiency perspective. Informational efficiency represents the capacity of securities prices to reflect instantly and fully all relevant available information regarding them. Thus no excess return is possible on such market. According to Fama [1], there are three levels of informational efficiency:

- *weak form* characterized by the instantly and fully reflection of all information concerning the history of securities prices in the current prices
- *semi-strong form* characterized by the integration of all public available information related to securities in their prices
- *strong form* supposing the instantly and full reflection of all information related to securities (public and private) in their prices.

The absence of short term auto-correlation in the return rate sample is translated into a high probability of market weak efficiency [2].

A random walk stochastic process for securities prices evolution is than assumed:

$$\ln(P_t) = \ln(P_{t-1}) + \varepsilon_t \quad (7)$$

where ε_t is the white noise. Since $\varepsilon_t = \Delta \ln(P_t) \cong \frac{P_t - P_{t-1}}{P_{t-1}} = R_t$, the weak efficiency supposes that R_t are not correlated.

Some empirical tests on Romanian capital market, completed by Preutu *et al.* [3], Dragotă & Mitrică [4], and Dragotă *et al.* [5] led to different conclusion with respect to the weak form of efficiency.

3. Results and discussions

Securities daily return rate distribution of the two companies is depicted in figure 1. The main descriptive statistics are shown in table 1.

Table 1. Securities return rate statistics.

	SOCP	OIL
Mean	0.122%	-0.146%
Standard deviation	2.245%	1.408%
Kurtosis	0.347	-2.012
Skewness	1.56	10.67
Jarque-Bera test value	9.16	268.88

The Jarque-Bera normality test proves a normal distribution for SOCP return rate, with a 99% level of significance, while OIL fails the test. The OIL return rate frequency (figure 1b) shows a very sharp shape and also right-side asymmetry.

The market price model for securities is tested taking into account the appropriate BET market index in equation (4). The graphic correlations between shares return rate and market indices are depicted in figure 2. Regression statistical analysis is shown in table 2.

IOP Conf. Series: Materials Science and Engineering **95** (2015) 012065 doi:10.1088/1757-899X/95/1/012065

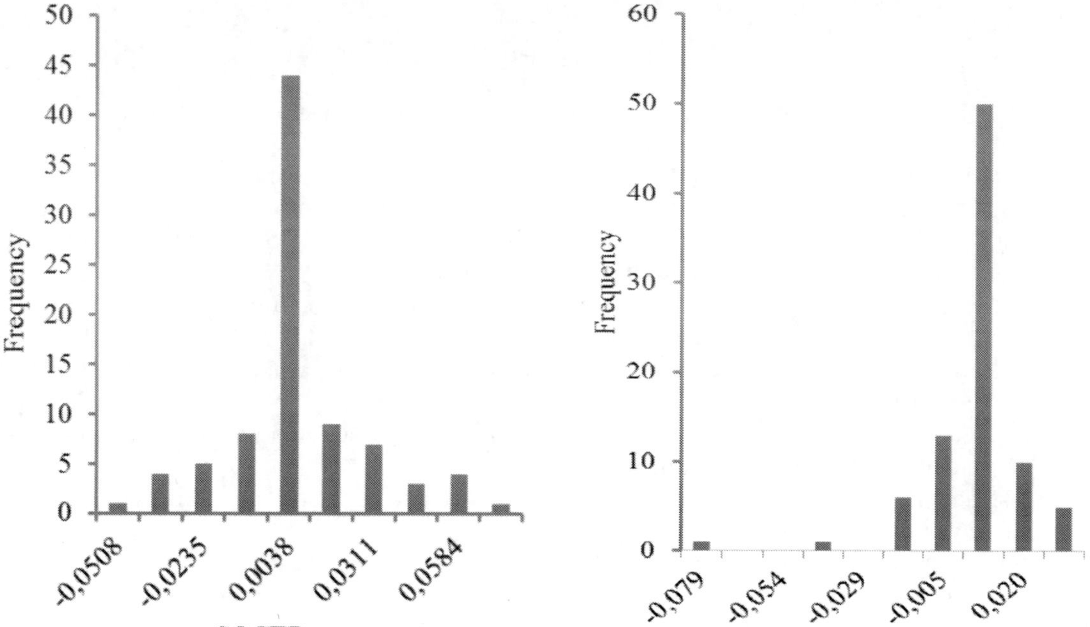

a. SOCEP return rate b. OIL return rate

Figure 1. Securities return rate.

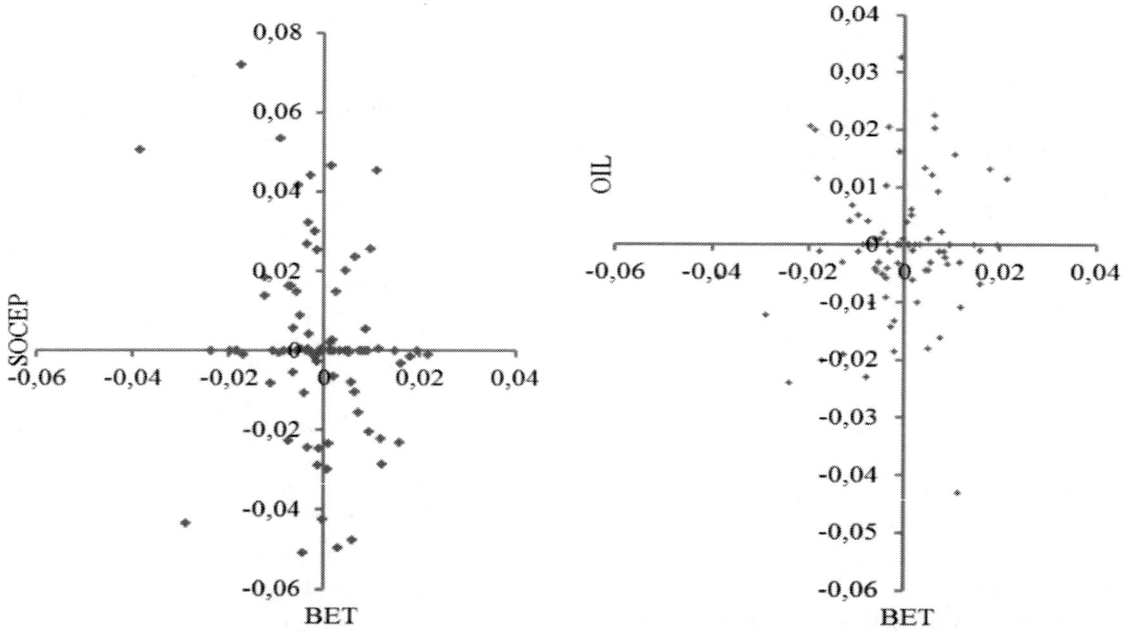

Figure 2. Market index (BET) – securities return rate correlations.

The plots show for the three securities a wide spread of the points in all the quadrants. Applying a Student t-test for coefficients α and β one can conclude that the null hypothesis H$_0$: $\hat{\alpha}=0$ and H$_0$: $\hat{\beta}=0$ could be accepted for both OIL and SOCP. The multiple determination coefficients R^2 have small values in each case. At most 3.9% for SOCP and 1.3% for OIL return rate can be explained by the market evolution. Thus, the return rates evolution of the shares have no linear connection to the market general index and should be subject of other considerations than market generally performance. The arbitrage models could be useful in such a situation. Fama and French [6] proposed to use market equity, book-to-market ratio, price-earnings ratio or indebtedness level as correlation variables.

Table 2. Regression analysis.

	SOCP – BET	OIL – BET
Coeff. α	0.00076	-0.0013
Coeff. β	-0.425	0.155
Coeff. α standard error	0.0024	0.0015
Coeff. β standard error	0.229	0.146
α t-stat	0.317	-0.853
β t-stat	-1.852	1.063
Multiple determination coefficient R^2	0.039	0.013

OIL has a quite null value of the price-earnings ratio (PER), considering the dividend for 2013 and the last price of the same year. This value could be a consequence of the stockholders expectations. The company is owned by two significant investors and 16% of shares are owned by minor stack holders.

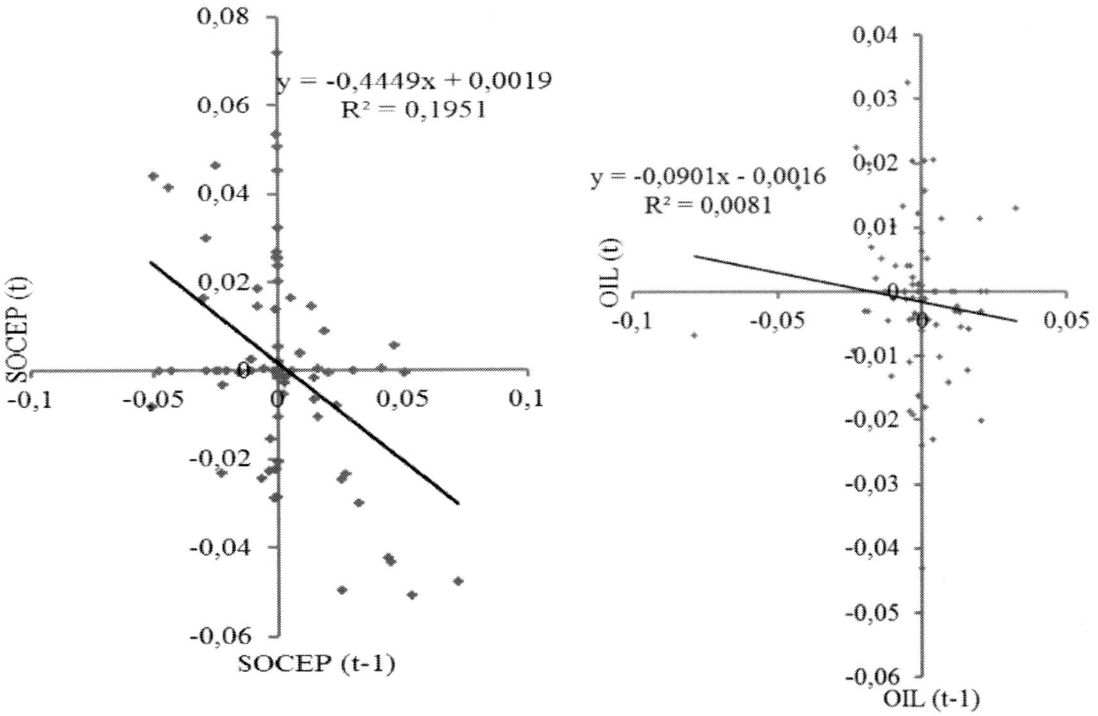

Figure 3. Return rate auto-correlations.

The SOCP's PER value is 19.75. There are two major investors in SOCP, and 33 % of shares are owned by minor investors. The dividend policy is also steady, providing a 1% dividend rate which is quite smaller and reflects the major investor's policy to capitalize the gains.
Figure 3 shows the auto-correlation of the return rates with a gap of one day. There is no obvious correlation between return rates for OIL, because the points (R_{t-1}, R_t) are approximately uniformly distributed in all four quadrants of the graphs and the correlation coefficient is 0.09. One can state that there is no auto-correlation between the return rate values on short terms. Hence, the weak informational efficiency form is assumed for OIL. The correlation coefficient for SOCP is 0.44 and the weak form of the informational efficiency is rejected. Thus, peculiar information could be used by the major investors to obtain benefits.

4. Conclusions

Constantza Port is a strategic point for the European core-transport network, and its importance is outlined by the political evolution around the Black Sea. Despite the importance of the Constantza Port for the national economy and particularly for the transport industry, the presence of handling, storage and transport companies on the Romanian capital market is quite feeble and the dependency to the whole market evolution (market models) is not satisfied. The evolution of the shares prices depends to a small extent on the market return rate. The arbitrage models are more appropriate to quantify the prices evolution of the shares. The random walk process for the return rates is satisfied and the weak efficiency of the market might be assumed for OIL Company. Otherwise, for SOCP Company the weak informational efficiency is rejected. Further analysis should take into consideration the equity structure, dividend policy and stakeholders expectations for explaining the market evolution of securities.

Acknowledgement

The work has been funded by the Sectoral Operational Programme Human Resources Development 2007-2013 of the Ministry of European Funds through the Financial Agreement POSDRU/159/1.5/S/132397.

References

[1] Fama E 1970 Efficient Capital Market: A Review of Theory and Empirical Work *Journal of Finance* **5** pp 34-105
[2] Fama E 1965 The Behaviour of Stock Market Prices *Journal of Business* **1** pp 15-34
[3] Preutu A, Călinescu A, Burlacu M and Macovei D 1998 Teoria eficienţei pieţelor financiare Stancu I (Eds.) *Articole fundamentale în teoria financiară* (Bucharest: DOFIN)
[4] Dragotă V and Mitrică E 2001 Romanian Capital Market – Testing Efficiency *28th European Working Group for Financial Modelling* (Vilnius)
[5] Dragotă V, Stoian A M, Pele D T, Mitrica E and Bensafta M K 2009 The Development of the Romanian Capital Market: Evidences on Information Efficiency *Romanian Journal of Economic Forecasting* **10** pp 147-160
[6] Fama E and French K 1992 The Cross-Section of Expected Stock Returns *Journal of Finance* **XLVII(2)** pp 427-465

Appraisal and regulation of the ship energy performance

N Badea[1], A Epureanu[2], G V Badea[3] and G Frumuşanu[2]

[1]"Dunărea de Jos" University, Automatic Control & Electrical Engineering Department, Ştiinţei Street 2, 800210 - Galaţi, Romania
[2]"Dunărea de Jos" University, Manufacturing Engineering Department, Domnească Street 111, 800201 - Galaţi, Romania
[3]Aspen Institute, Bucharest Business Park, Entrance A, 3rd Floor 1A, Bucuresti-Ploiesti National Road, Bucharest, Romania

E-mail: nicolae.badea@ugal.ro

Abstract. The ship may be viewed as a living environment associated with two industrial environments, one corresponding to the transport industry and other one to the processing, services, or other specific type of industry developed aboard. Each environment has its own energy system and changes energy with the other two. Nowadays, the appraisal and regulation of the ship energy performance is based on the Energy Efficiency Design Index (EEDI). Its definition covers the three mentioned systems, without distinction between them. This paper addresses the assessment and regulation of the ship energy performance, bearing in mind that, by far, the main purpose is to increase the level of performance by selecting, from the available measures of performance improvement, those that are the most effective. The paper highlights the EEDI shortcomings, explaining that they appear mainly due the fact that this index covers a couple of energy systems that are far too different (though these energy systems are intimately interpenetrated).

1. Introduction

The shipping sector has emitted 1,015 million tonnes of CO2 in 2007, corresponding to 3.3% of the global emissions. Of this, international shipping has emitted 870 million tonnes or about 2.7% of the global total in 2007 [1]. The propulsion power and energy conversion on-board installation – namely the ship's energy conversion system – is the main contributor to the overall emission footprint, cost effectiveness, and energy performance of a vessel.

Anticipated climate changes on the international agenda require reducing the global CO2 emissions in order to achieve stabilization at 2 °C of mean global temperature increase. The existing and impending stringent environmental legislations could be very strong drivers for growing the sustainability regarding the environmental footprint and energy performance of the vessels.

Energy policy is among the fastest, highest impacting, and most cost-effective ways of reducing greenhouse gases (GHG) emissions. Over the years, the European Union has introduced a number of directives, regulations [2, 3], and initiatives to encourage and support Member States, regional authorities, individuals and so on to increase the energy performance in the different sectors, including buildings, transport, and products.

In the figure 1, the energy forms and energy demands on board ships are suggested [4]. Energy systems of the ship must be autonomous and it includes the entire chain of energy production,

Content from this work may be used under the terms of the Creative Commons Attribution 3.0 licence. Any further distribution of this work must maintain attribution to the author(s) and the title of the work, journal citation and DOI.
Published under licence by IOP Publishing Ltd

IOP Conf. Series: Materials Science and Engineering **95** (2015) 012066 doi:10.1088/1757-899X/95/1/012066

distribution and use. Energy production is necessary for moving vessel and for meet other needs that are required shipboard (electricity, heat and cooling) depending on the destination of the ship. Prime mover or main engine converts the chemical energy of the fuel as primary energy source to kinetic energy for ship moving. The propulsion power is result of the kinetic energy which is dissipated to the sea, overcoming the ship's resistance. Some ships have on main engine shaft, one electrical generator called shaft generator which supply electricity on board ship. Moreover, to satisfy all electricity need, all ships have auxiliary engines which drives electrical generators.

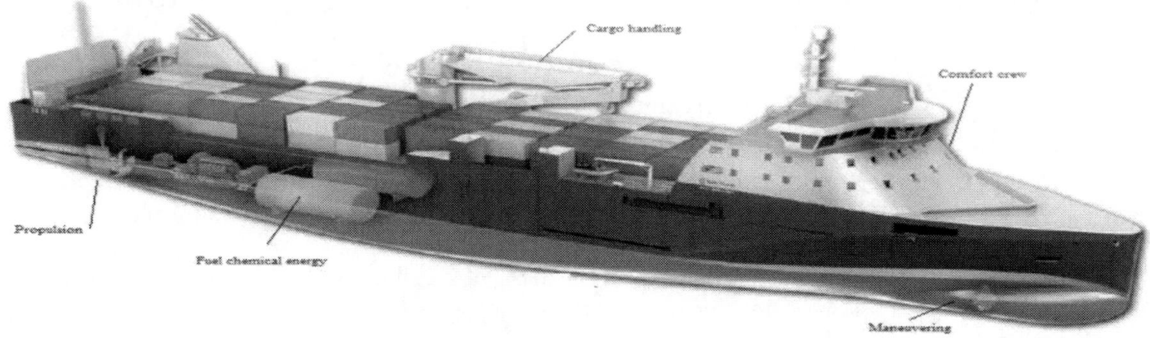

Figure 1. Energy forms and demands on board ships [4].

Besides electricity and mechanical power for propulsion, the heat is required for ships. It is produced by ship boiler resulting from combustion (including the exhaust gases from the main propulsion), nuclear reaction, or thermal effect of electric current, and is transferred to the water. The water is heated, evaporated, and converted into saturated steam and if the continuous heating process, the steam produced is over-heated steam. The steam consumption on board depends on the type and class of ship. In naval practice, the ship boilers fall into two categories: fuel boiler, which use heat energy produced by burning a quantity of oil, and exhaust-gas boiler, using heat energy accumulated by the exhaust gas resulted from the main engine. In the first case, fuel boiler only works when the ship is stationary, whereas during the voyage the ship has a sufficient amount of steam to supply all consumers on board. For ships which have main boiler, donkey boiler operates continuously since these vessels do not have other sources of steam.

On the other hand, two terms – namely the 'energy savings' and 'energy efficiency' - are used (often with little precision or accuracy) to express the performance in the area of energy policy. The definitions, provided in the Energy Efficiency Directive (EED), establish a clear relation between the two terms, as follows:

- Article 2.4 defines *energy efficiency* as "the ratio of output of performance, service, goods, or energy to input of energy ".
- Article 2.5 on the other hand defines *energy savings* as "an amount of saved energy determined by measuring and/or estimating consumption before and after implementation of an energy efficiency improvement measure, whilst ensuring normalization for external conditions that affect energy consumption". Specifically, energy saving is defined as the result of energy efficiency improvements. Savings are measured as the difference in energy consumption before and after the efficiency improvements.

As conclusion, the ship it may view as a living environment associated with two industrial environments, one corresponding to the transport industry and other one to the processing, services, or other specific type of industry developed aboard. Each environment has own energy system and changes energy with the other two.

The purpose of this paper is to highlight the shortcomings of both the EEDI definition and the regulations regarding the implementation of this index, as well as to propose a new approach through which these shortcomings could be diminished. As a result, the next section presents the gaps of the current appraisal and regulation system, and how they affect the qualitative attributes of the system, namely the relevance, sensitivity, precision, sustainability, and coherence. Further, the third section presents a new approach of the appraisal and regulation system, which aims to diminish the shortcomings. The numerical example given in the fourth section highlights the benefits of the new approach. Finally, the conclusions are summarised in the fifth section.

2. Criticism of the current appraisal and regulation system

The current system [5, 6] is based on the following features:

a) *Energy Efficiency Design Index* (EEDI), which quantifies the amount of carbon dioxide that a ship emits in relation to the goods transported. The actual EEDI of a vessel is called the "attained EEDI" and is calculated based on guidelines published by IMO. The result must be below the limit ("required EEDI") prescribed in MARPOL.
b) *International Energy Efficiency Certificate* (IEEC), which is to be issued at the vessel's initial survey, and provides that the EEDI has been verified (for applicable vessels) and the SEEMP is on board.
c) *Ship Energy Efficiency Management Plan* (SEEMP), having as main purpose to lay down a mechanism for a company and/or a ship to improve the energy efficiency of a ship's operation. All ships must have a SEEMP on board before the issuance of the first IEEC. The SEEMP is not subject to approval or verification by the Administration.
d) *Energy Management Systems* (ISO 50001: 2011), which is an ISO standard that establishes a framework for a company's energy management system. It follows along similar lines to *Environmental Management Systems* (ISO 14001).

The EEDI index is a fundamental element for the actual system. This index measures a certain attribute of the ship, namely its level of aggressiveness over the environment, measured through the quantity of CO_2 emitted. The index is designed so that, by regulating the maximum value allowed, the aggression is limited.

The attained EEDI is calculated on the base of the technical design parameters and standard conditions of operation, and provides specific figure for an individual ship design, expressed in grams of CO_2 per ship's capacity-mile. That can be illustrated by its definition formula, for instance those given by-Germanischer Lloyd [7]:

$$Attained\ EEDI = CO_2\,emissions/\,Transport\ work = \frac{\sum P \cdot SFC \cdot C_F}{f_i \cdot Capacity \cdot V_{ref} \cdot f_w} \tag{1}$$

where: P- power, C_F - conversion between fuel and CO_2, SFC - specific fuel consumption, f_i - capacity factor, f_w - non-dimensional coefficient indicating the decrease of speed in representative sea conditions of wave height, frequency and wind speed, V_{ref} – reference speed, and Capacity - a consistent set of 75 % main engine power and the corresponding speed when the ship is carrying its full capacity in calm weather.

The numerator of this fraction includes the amount emissions of main engine, auxiliary engine and shaft generator/motor, minus emission savings from innovative technologies.

The required EEDI is the maximum allowable value of the attained EEDI and is given by the following equation:

$$\text{Attained EEDI} \leq \text{Required EEDI} = (1\text{-X}/100) * \text{baseline value} \tag{2}$$

where X is the reduction factor of the required EEDI, compared to the EEDI baseline value.

The baseline values shall be calculated as follows:

$$\text{Baseline value} = a \cdot b^{-c} \tag{3}$$

Where: *a*, *b*, and *c* are the parameters given for the different ship types. The baseline value is defined by a diagram which is mathematically defined.

The shortcomings of the current appraisal and regulation system are presented below:

a) The name of the index is inappropriate as energy efficiency is a variable defined as a ratio between output of performance, service, goods, or energy to input of energy. Though it has the same naming, the EEDI is referring to the ratio between the quantity of CO_2 emitted and the transport work. The measurement unit for the EEDI is grams of CO_2/tonne nm. More to it, the actual name defies intuition as the increase in the ship's efficiency is highlighted by the decrease in the index value.

b) The definition of the index is inappropriate for the following reasons:

- The quantity of CO_2 that forms the numerator of the ratio is referring to both the energy used for ship propulsion and all the onboard users of energy, regardless of their energy consuming activity, while the denominator of the ratio contains only a certain measure that evaluates the transport activity, namely the transport work (defined as multiplication between the ship's capacity and speed). Indeed, the energy used for propulsion and the transport work are directly correlated, and consequently they can define an attribute. But the other consumers are referring to other activities that are not correlated to the transport work, for instance, loading and unloading the ship is correlated to the ships capacity, crew accommodation is correlated with time, regardless of the fact that the ship is moving or not, transported goods maintenance is correlated with the quantity and type of good and with the travel time. As a result, according to the actual definition, the numerator and denominator of the index are insufficiently correlated. This is affecting the *relevance* of the index.
- The components forming the numerator have far too different values so that some of them are insignificant to the others. For instance, the power of the main engine is dozens of times higher than the sum of all the other. Thus, the modification of certain components leads to very small modifications of the index value. This is affecting the *sensitivity* of the index.
- Moreover, it appears that the effect of the component modifications has comparable values to the effect of the appreciation errors of the variables value. This is affecting the *precision* of the index (namely the precision with which the index is measuring the effect of the modifications).

c) The regulation regarding the index baseline is inappropriate for the following reasons:

- The baseline is referring to the level of CO_2 emissions of the ship and does not take into account the fact that, even though the EEDI is under the baseline level, the global emission can surpass the actual level through the increase in the number of ships under exploitation. This is affecting the *sustainability* of the index.

The baseline for ships is not in agreement with the baseline for the similar terrestrial constructions, sustaining the same type of activity (transport, residential, or industrial). This is affecting the *coherence* of the index.

3. The proposed approach

From our perspective, the appraisal should first of all pinpoint the effect of the measures to reduce the CO_2 emissions, therefore helping the target groups to compare the possible solutions and highlight the most efficient. The target groups of the regulations are the designers / builders, owners and operators of ships. The index should sustain the procedures that the target groups applied for identification and application of the most efficient solutions.

Moreover, the appraisal should be coherent with the methodology applied in the other fields where the CO_2 emissions are considered the attribute reflecting the level of aggression over the environment.

On the other hand, the index should be regarded as any measurement instrument, intended for the evaluation (through measurement or calculation) of certain variables (in this case, the variable is the specific quantity of CO_2 emitted as a consequence of an activity). For this reason it has the same qualitative attributes as any measurement instrument and particularly, the relevance of the measured value, sensitivity to small variations of the variable value, precision of repeated measurement of the same value, sustainability in time (linked to the fact that the index leads to the same conclusion, though the measurement conditions have evolved in time), and finally coherence with other similar instruments (respectively, the indexes used in similar fields).

The regulations should oblige the target groups to modify the adopted solutions, both in design and in exploitation of the ships, in order for the value of the index to decrease.

To materialize this vision we propose the following:

Firstly, the name of the index should be *Energy Impact Design Index (EIDI)* as the index points out the *unwanted environmental impact* of the used energy for developing a certain useful activity and not the quantity of energy used (as the name Energy Efficiency Design Index suggests). This way, the intuition is not "aggressed" as for a good use of energy, the "Energy Efficiency" is high and the "Energy Impact" is low.

Secondly, we propose that both the index and the baseline to be *vectors* with three components called *naval* component, *residential* component, and *industrial* component. Thus, the index will have the naval EEDI component, residential EEDI component and industrial EEDI component. Similarly, the baseline will have the naval baseline component, residential baseline component and industrial baseline component. Each component, either of the index or of the baseline, is defined by a calculation formula and through standard conditions in which the values of the variables from the calculation formula are evaluated.

In what the *calculation formula* is concerned, the components of the index would therefore be defined as follows:

- Naval EIDI is the ratio between CO_2 amount and the transport work (the last being expressed as multiplication between capacity and speed).
- Residential EIDI is the ratio between CO_2 amount and the residential work (the last being expressed as multiplication between total floor area and time).
- Industrial EIDI is the ratio between CO_2 amount and the industrial work (the last being expressed specifically for the onboard industrial activity undergone).

In what the *standard conditions* are concerned, the following key ideas should be considered:

- The three energy systems exchange energy between them. Nonetheless, the standard conditions will correspond to the situation in which there is no exchange of energy. This decision has a double motivation. The first motivation is that the energy exchange is not permanent but only occasional, for limited term. For each of the three energy systems there are situations in which they have to function independently. Sizing the system components is achieved through design and is based on the standard condition of independent functioning. The second motivation refers to the fact that the index characterises the energy system by itself and not by the occasions in which it functions.
- The components of the baseline are defined separately from one another. The naval baseline is defined according to the current methodology, while the industrial and residential baselines are defined according to the methodologies applied to similar terrestrial constructions.

Thirdly, we propose that the actual regulations should foresee the obligation to periodically weighting the current baselines with the percentage of increase in global volumes of the shipping transport activities so that the quantity of CO_2 emitted into the atmosphere is decreased at a global level, even though the volume of this activity increases.

Numerical example

This example is for Anchor Handling Tug Supply vessels (AHTS) [8] that is capable of operations in the harshest environments. AHTS vessels mainly perform anchor handling duties and towage of offshore drilling units and floating production units. As AHTS vessels generally have free deck area, and some of them have tanks below deck, they can also carry cargo and bulk

a) Acoording to the current system, the index EEDI have only one component:

$EEDI=((P_{me}*CF_{me}*SCF_{me})+(P_{ae}*CF_{me}*SCF_{me}))/DWT*Vref=\underline{21.43\ gCO_2/\ t\ nm}$ for:

Capacity = 700+4,580 t; Speed =17 knots; MCR_{me} = 6,000kW; P_{me}= 0,75*(MCR_{me}) =4,500kW; P_{ae}=(0,025*MCR_{me})+250kW = 400kW; Vref =17 knots; CF_{me} = 3.206 gCO_2/g-fuel;

CF_{ae} = 3.206 gCO_2/g-fuel; SFC_{me} = 190 g/kWh; SFC_{ae} = 215 g/kWh.
As in other examples, different structures of energy system lead to the modification of the EEDI value with just a few percent.
b) According to the proposed system, the index EIDI have the following three components:
1. $EIDI_{naval}$ = (P_{me} *SFC_{me} * CF_{me})/DWT*Vref = 19.47 gCO_2/ t nm
2. EIDI $_{accomodation}$ = CO_2/surface = 63.19 $kgCO_2/m^2$year, for: P = 394.6kW; Time = 2,190h; Energy = 864,182kWh; Surface = 9,429m^2; EPC = 91.65 kWh/m^2 corresponding to the B energetic class; CO_2 (calculated by using [9]) is 595,898 kg/year.
3. EIDI $_{industrial}$ = (P_{me} *SFC_{me} * CF_{me})/Brake Power =4500*3.206*190/19,000 = 144.27 gCO_2/kWh
By its components, the proposed index highlights areas where are real possibilities to improve the energy performance.

4. Conclusions

A strategy for ship energy efficiency is welcome, but regulation only monitors fuel consumption instead of directly reducing it, and covers only CO2 and not air pollutants like SO2 or NOx.
A simple analogy with eco-design directive or energy performance of buildings indicates lack of ecological labelling of ships. All ship type is possible in the future, to use the cleaner fuel (gas fuel or hydrogen fuel) and is expected to achieve the green ship by using efficiency of renewable technologies. The use of renewable energy technologies in ship is not a usual practice yet. Even if the market can offer adequate technology, neither clears mandatory actions, nor provisions stimulating the increase of renewable energy use in especially hydrogen technology for new ships. The path towards a greener marine future and sustainable shipping involves the more widespread use of wind, solar and electric power to help propel and power vessels of all sizes.

Finally, the current appraisal and regulation system is a non-prescriptive, performance-based mechanism that leaves the choice of technologies to use in a specific ship design to the industry. As long as the required level of energy impact is attained, the ship designers and builders are free to use the most cost-efficient solutions for the ship to comply with the regulations.

References

[1] Buhaug, Corbett and J J Second IMO GHG 2009 study *International Maritime Organization* (London)
[2] Directive 2012/27/EU of the European Parliament and of the Council of 25 October 2012 on energy efficiency pp 1-56
[3] Directive 2010/31/EU of the European Parliament and of the Council of 19 May 2010 on the energy performance of buildings OJ L 153 pp 13-35
[4] http://www.rolls-royce.com/marine/products/
[5] Bazari Z 2011 Norway -Assessment Of Imo Mandated Energy Efficiency *Measures For International Shipping* (London UK and Tore Longva: Lloyd's Register) DNV Oslo
[6] http://www.imo.org/OurWork/Environment/PollutionPrevention/AirPollution/Pages/Technical-and-Operational-Measures.aspx
[7] Germanischer Lloyd SE Rules for Classification and Construction VI *Additional Rules and Guidelines 13 Energy Efficiency* (Hamburg: Germanischer Lloyd SE)
[8] http://www.slideshare.net/bourbonoffshore/bourbons-fleet
[9] http://www.epa.gov/cleanenergy/energy-resources/calculator.html#results

Maritime vessel obsolescence, life cycle cost and design service life

O Dinu[1] **and A M Ilie**[1]
[1]Department Transport, Traffic and Logistics, Faculty of Transport, University Politehnica of Bucharest Spl. Independenţei, No.313, Bucharest, 060042, Romania

E-mail: aoana.dinu@upb.ro

Abstract. Maritime vessels have long service life and great costs of building, manning, operating, maintaining and repairing throughout their life. Major actions are needed to repair, renovate, sometime built or even replace those scrapped when technology or demand changes determine obsolescence. It is regarded as a concern throughout vessel's entire life cycle and reflects changes in expectation regarding performances in functioning, safety and environmental effects. While service live may differ from physical lives, expectations about physical lives is the main factors that determines design service life. Performance and failure are illustrated conceptually and represented in a simplified form considering the evolution of vessels parameters during its service life. In the proposed methodology an accumulated vessel lifecycle cost is analyzed and obsolescence is characterized from ship's design, performances, maintenance and management parameters point of view. Romanian ports feeding Black Sea are investigated in order to provide comprehensive information on: number and types of vessels, transport capacity and life cycle length. Recommendations are to be made in order to insure a best practice in lifecycle management in order to reduce costs.

1. Introduction

Maritime shipping has seen several major technical innovations aiming to improve the performance of ships or their access to port facilities [1]. Most important of them are: size, speed, specialization of ships and automation technologies. Ship design has significantly improved to steel, aluminum and composite materials hulls therefore the hulls of today's ships are the result of considerable efforts to minimize energy consumption, construction costs and improve safety.

Failure occurs if performance falls below levels that decision makers evaluate to be unacceptable, maritime vessels tend to be ineffective or is very likely to become so in the near future or maintaining and operating costs become considerably high [2].

There is no general consensus on how types of ship structural failures can be classified. Various methods including load type, stress type, degradation type, crack sizes and others have been used. Some studies grouped failure modes according to crack sizes - two levels of crack severity, namely, nuisance and fracture cracks were used in the classification [3]. Other identified five failure categories considering as a criterion longitudinal strength of a hull girder [4]. Vessels failure modes can also be categorized according to the severity of consequences resulting from failures (failure modes were

Content from this work may be used under the terms of the Creative Commons Attribution 3.0 licence. Any further distribution of this work must maintain attribution to the author(s) and the title of the work, journal citation and DOI.
Published under licence by IOP Publishing Ltd

classified into catastrophic, end of serviceability, serviceability limiting, non-limiting and nuisance failure modes) [5]. Ship structure failures are grouped into three types: primary, secondary and tertiary, the most important being the primary behavior, associated with the ship as a whole [6] or according to the type of load that induces the failure mode and consequently two categories, dynamic and static loading were used in the classification [7].

2. Vessel lifecycle cost analysis

The goal of life cycle analysis, therefore, is a holistic understanding of the long-term economic, social and environmental effects of design, construction, operation and maintenance and disposal of a vessel. This understanding is used for efficient management of the system. In a life cycle analysis, all the short-term and long-term costs (financial, physical, service, environmental), benefits and risks involved in operating the structural system are assessed, evaluated and used for optimal decision making [8, 9].

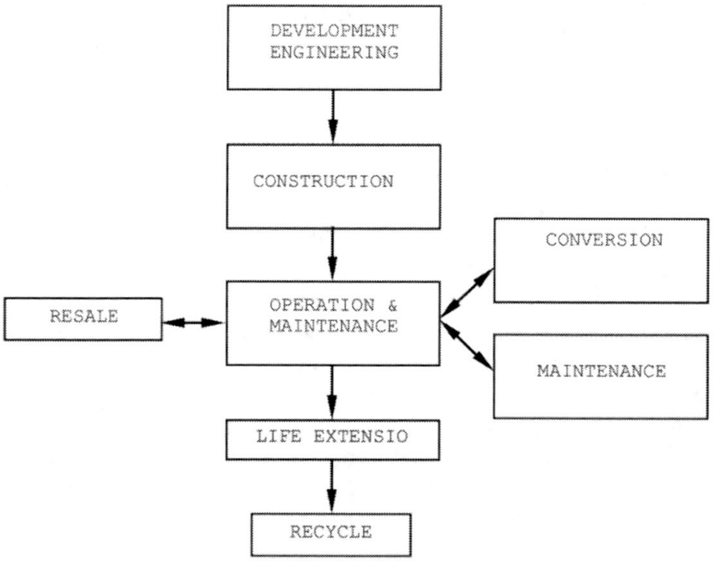

Figure 1. Life Cycle of a Vessel System (Source [8]).

Three of the phases presented in figure 1 are considered major in a vessel life cycle, as follows:

Regarding construction, ship structural systems in particular have traditionally been constructed from steel. Highly stiffened thin steel plates are commonly used to achieve minimum weight structure at optimal cost. The life cycle of a structure depends on the quality of construction, which is affected by the type of welding technique, type of electrode, nature of the surface, qualification and experience of the welders and inspection of the welds.

Modern ship operation is a big and sophisticated process and in order to insure appropriate performance levels maintenance is needed. Larger repair work or rebuilding may occur as part of maintenance, vessel operator is the one that choses how often and the level of maintenance during lifetime. Data shows that approximately 10% of the steel amount is added during vessel's lifetime. It is estimated that half of the materials from the construction phase is changed over a ship lifetime. For each docking 50% of the area below the water line is painted with primer and antifouling.

Recycle takes place after 25 – 30 years of service or when repairs cannot be financial justified. Normally the ship owner sells the ship to a ship scrap yard for demolition. At the yard all the steel and some of the equipment is reused or sold in the secondhand market. Once a ship has completed is commercial life, it will be brought to a scrapping yard to be disassembled and recycled.

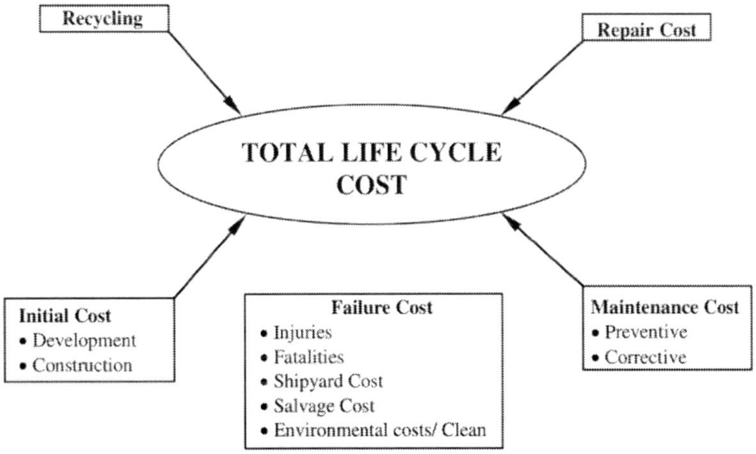

Figure 2. Costs involved in life cycle analysis.

Life-cycle cost is the expected net cost over the lifetime of the structure. Initial cost and all subsequent expected costs of significance, as well as disposal costs, are included in economic life cycle cost. For a vessel, the total economic life cycle cost is illustrated in figure 2, is given by equation (1) and the time value of money needs to be considered in evaluating.

$$C_T = C_O + C_M + C_F + C_D \quad (1)$$

where C_T represents total life cycle cost;
CO - initial cost;
CM - maintenance cost (this could include inspection, repair, layup, conversion and modification and resale costs);
CF - failure cost and
CD - disposal cost (this could include resale cost).

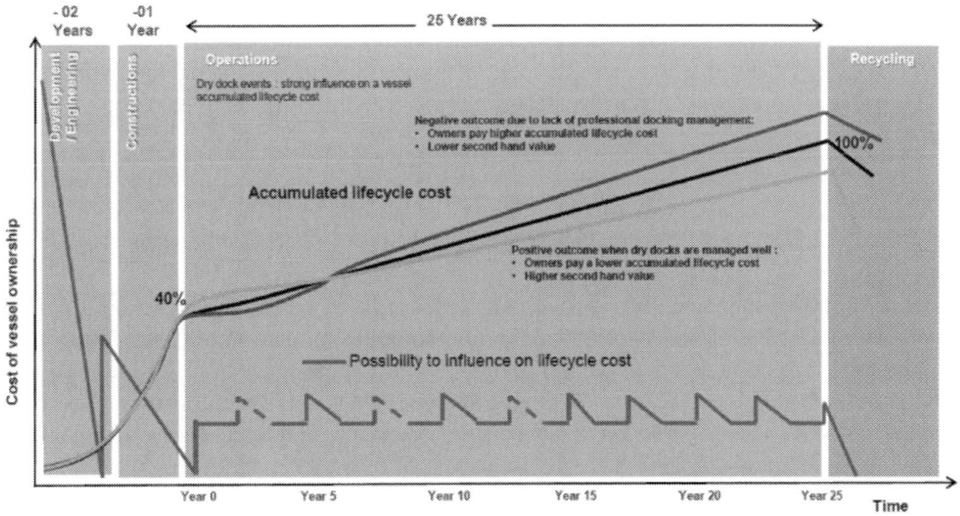

Figure 3. Vessel lifecycle cost analysis (Source [10]).

Since designers and operators of ship structural systems cannot see into the future, then all the above components of life cycle cost are uncertain. Therefore, probability-based techniques should be used in the costing process. The costing process should be based on analysis, data and experience.
In figure 3 are shown the opportunities to influence ship life cycle costs during the ship's life span. The red curve shows that first of all design and then construction determines the future costs. During operation the dry dock maintenance works also enable measures to raise efficiency. The operation costs consist of fuel costs, administration costs, crew costs, insurances, spares and maintenance costs.

3. Obsolescence and design service life

Obsolescence appears when changes in requirements or anticipations considering the utility of the vessel are registered. In most cases, vessels continue to function but at levels below standards [2]. One of the defining characteristics of a ship is the long service life, of many decades [11].

There is a large number of obstacles faced trying to predict vessels design life including uncertainties on climate and variables influencing deterioration, our limited knowledge of mechanisms of ships' degradation, lack of sufficient information and inherent complexities of the problem.

In mathematical terms, vessels performance may be represented in equation (2):

$$Performance = P(S_j, D_j, t) \tag{2}$$

where S_j represents supply vector of services j that the vessel provides to various groups (e.g., users, owners);

D_j = demand vector of services j that the vessel provides to various groups (e.g., users, owners);

t = time, measured from the vessels completion of construction and start of commissioning.

In general, the supply of services, S_j, is characterized or predicted as a function (equation (3)) of design and operational characteristics of the facility and its management

$$S_j = S(X_j) \tag{3}$$

where X_j represents a vector of descriptive and functional characteristics i of the facility (load type, stress type, degradation type, crack sizes, steel modulus of elasticity)

If performance falls below levels that decision makers evaluate to be unacceptable, maritime vessels tend to be ineffective or is very likely to become so in the near future or maintaining and operating costs become too high. Mathematically, failure can be shown as

$$P(t) < P^T \tag{4}$$

In equation (4), P^T represents minimum acceptable performance.

Planers and vessel designers generally seek to assure that effectiveness, reliability and cots are balanced to achieve „optimum" performance during the design lifetime as in equation (5):

$$t \leq T^D, P(t) \geq P^F \tag{5}$$

where T^D represents design service life. „Optimum" in this context generally means reliable effective service at lowest possible cost. In practice, however, it may imply lowest construction cost, tolerance of what some users will view as reliable service and other compromises.

Figure 4 illustrates conceptually and in simplified form the progression of a vessel's performance during its service life. Performance immediately following the completion of construction, at commissioning, is typically less than the design ideal (optimum performance). The new constructed vessel will continue to deliver the performance at a reasonably steady level for some years (barring catastrophe and with proper operations and normal maintenance) and then, a slow decline inevitably begins, owing to wear and aging - eventually performance falls to a level judged to be the minimum acceptable.

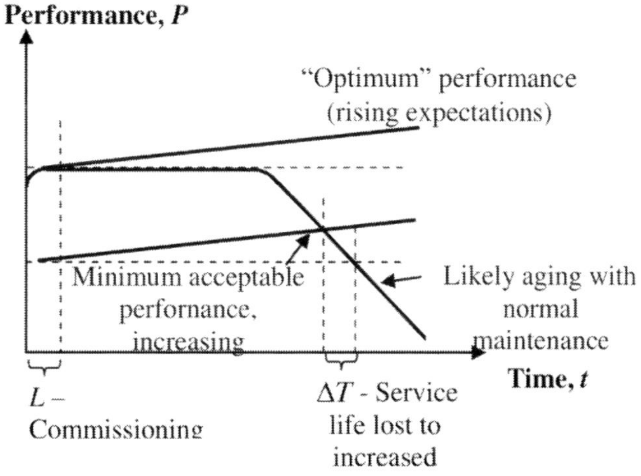

Figure 4. General representation of performance (Source [2]).

More typically, peoples' expectations change over time as a result of newer facilities development, introduction of new products and materials and increased experience, leading to the reach of minimum acceptable performance levels much sooner than they would otherwise

4. Investigation of Romanian ports feeding Black Sea

Globally, on 2013 year level, maritime shipping industry is serviced by more than 100.000 commercial vessels of more than 100 tons falling into four broad categories [1]: passenger vessels (subdivided into two categories: passenger ferries, where people are carried across relatively short bodies of water and cruise ships, where passengers are taken on vacation trips of various durations, usually over several days), bulk carriers (differentiated into liquid bulk and dry bulk vessels), general cargo and Roll on-Roll off (RORO) vessels.

For the same year, in Romanian ports 71.079 ships entered, 64.631 of them interior navigation vessels, representing 90,92% and 6.448 of them maritime vessels, representing 9,08%. Total number of ships that exit Romanian ports was 72.505, 66.236 of them interior navigation vessels, representing 91,35% and 6.269 of them maritime vessels, representing 8,65% [12]. Regarding most important Romanian Ports feeding Black Sea, Constanta Port, from all registered maritime vessel number, it attracted 4764, meaning 75% of Romanian maritime traffic.

Table 1. Maritime vessels type distribution in Constantza Port (Source [13]).

Types of vessels	2008	2009	2010	2011	2012	2013
Bulk carriers	415	386	419	401	439	533
Container ships	1201	694	523	577	651	579
General cargo	2881	2748	3145	2879	2692	2525
Oil tankers	957	724	647	632	673	636
Others	415	356	411	341	550	492
All vessel types	5869	4908	5145	4830	5005	4764

As presented in table 2, in 2013, the average age (per ship) was highest for general-cargo ships (25 years), followed by other types (22,6 years), oil tankers (16,7 years), container ships (10,8 years) and dry-bulk carriers (9,9 years). Following the surge of new buildings in the dry-bulk segment, almost half of the dry-bulk dead weight tonnage is only 4 years old or younger, overtaking for the first time container ships as the youngest vessel category.

Table 2. *Maritime* vessels age distribution in 2013 (Adapted from [14]).

Types of vessels		Age [years]					
		0-4	5-9	10-14	15-19	+20	Average
Bulk carriers	[%] of total vessel number	41	10	9	16	24	11,77
	Average size [DWT]	80772	65854	60514	75693	47053	
Container ships	[%] of total vessel number	21	23	15	25	17	12,83
	Average size [DWT]	56530	41481	28210	22545	13619	
General cargo	[%] of total vessel number	11	12	5	8	63	25,38
	Average size [DWT]	6396	4194	5808	4342	3102	
Oil tankers	[%] of total vessel number	24	14	7	12	43	18,69
	Average size [DWT]	64176	59987	74818	37046	6404	
Others	[%] of total vessel number	20	15	9	11	45	20,19
	Average size [DWT]	5112	5269	4909	4265	4224	
All types	[%] of total vessel number	20	14	8	11	46	20,21
	Average size [DWT]	35193	22382	25060	23249	6856	

More, 20% of all seagoing merchant ships were younger than 5 years, representing 40% of the world's deadweight tonnage. New container ships are on average three times the size of those built 20 or more years ago, and only 5 per cent of the container ship tonnage is older than 20 years. Oil tankers, too, tend to be replaced relatively early; only 4 % of the existing oil-tanker tonnage was built more than 20 years ago. As a reflection of most recent ships being larger than older ones, the global average age per ship shows an age of 20,3 years.

5. Conclusions

Most of the time, service life differs from physical lives therefore expectations about physical lives determines, for the most part, the designed service life that is typically recommended to be 20 – 30 years. Vessels design service life and expectations of service life are based mainly on experience and testing. Considering data from Table 1 and Table 2 most common type of maritime vessel in Constanta Port is the general cargo vessel, 53% from total traffic and in the same time average age for this specific type of vessel is of 25,38 years, the biggest registered value. As number of vessels, is followed by oil tanker, 13,35% average age for this specific type of vessel is of 18,69 years. Concluding, almost tree thirds of maritime vessels entering and exiting Constanta Port are at their life cycle limit, most of them reaching the recommended designed service life of 25 years.

References

[1] Rodrigue J P 2013 The Geography of Transport Systems (Rutledge Publishing)
[2] Lemer A 1996 Infrastructure Obsolescence and Design Service Life *Journal of Infrastructure Systems* pp 153-161
[3] Stambaugh K A and Wood W A 1990 Ship Fracture Mechanisms Investigation *Report SSC 337* Parts **1, 2**
[4] Daidola J C and Basar N S 1980 Probabilistic Structural Analysis of Ship Hull Longitudinal Strength *SSC 301*
[5] Ayyub B M and all 1989 Estimation of the Structural Service Life of Boats *Naval Engineers Journal American Society of Naval Engineers* **101(3)** pp 156-166
[6] Mansour A E & all 1997 Structural Safety of Ships *Trans SNAME* **105** pp 61-98
[7] Ma K-T and Bea R G 1995 A Repair Management System for Fatigue Crack in Ships *Society of Naval architects and Marine Engineers* (Transactions) pp 343-369

[8] Ayyub B M & all 2000 Risk-based Life Cycle Management of Ship Structures *Report SSC 416* pp 149 -167
[9] Pawlik T 2011 Ship life cycle assessment and management *Gesellschaft für angewandten Umweltschutz und Sicherheit im Seeverkehr GmbH* (Bremen)
[10] Srinath M 2014 Efficient Ship Operations – through vessel lifecycle management *Wilhelmsen Ship Management*
[11] Ayyub B M and White G J 1990 Structural Life Expectancy of Marine Vessels *Marine Structures* **3(4)** pp 301-317
[12] Cazan L A 2013 Autoritatea Navală Română *Raport privind Incidentele Navale*
[13] Information on http://www.portofconstantza.com, Constantza Port Anual Report 2013.
[14] Duval Y and Pawlik T 2013 Review of Maritime Transport *UNCTAD/RMT/2013 United Nations Publication* pp 35-50

IOP Conf. Series: Materials Science and Engineering **95** (2015) 012068 doi:10.1088/1757-899X/95/1/012068

Manufacturing and design of the offshore structure Froude scale model related to basin restrictions

I C Scurtu[1]
[1]Constanta Maritime University, Mircea cel Batran Street 1, 900663, Constanta, Romania

E-mail: scurtucristian@yahoo.com

Abstract. Manufacturing steps for a modern three - column semi-submersible structure are delivered using CFD/CAE software and actual Froude scaled model testing. The three- column offshore is part of the Wind Float Project already realized as prototype for wind energy extraction in water depths more than 40 meters, and the actual model will not consider the wind turbine. The model will have heave plates for a smaller heave motion in order to compare it with the case without heave plates. The heave plates will be part of the Froude scale model.. Using a smaller model will determine a smaller heave motion and this will affect predictions of the vertical movement of the three- column offshore structure in real sea. The Froude criterion is used for the time, speed and acceleration scale. The scale model is manufactured from steel and fiberglass and all parts are subjected to software analysis in order to get the smallest stress in connections inside the model. The model mass was restricted by scale dimensions and also the vertical position of centre gravity will be considered during the manufacturing and design process of the Froude scale offshore structure. All conditions must converge in model manufacturing and design in order to get the best results to compare with real sea states and heave motion data.

1. Introduction

The ship model basin was built between1960 and 1965 and is currently the only basin equipped with Cusson appliances and equipment that can meet the requirements of a research basin. This basin gives the possibility for applied research to be done both on marine vessels, and on floating structures, so as to provide the naval field with new development concepts. This basin offers the University of Galati the opportunity to realize applied research not only for industry partners, but also for the young Romanian researchers in their doctoral program, whose potential will be exploited further on .The University of Galati in partnership with other institutions of research, development and innovation at European level can easily carry out their naval projects.

Experimental tests are a way of studying the behaviour of ships and semi-submersible offshore structures. Scaled models contribute to knowledge improvement in the field of naval hydrodynamics, thus, accepting the challenge to find the most modern design solutions.

Software evaluation of different hydrodynamic phenomena has developed a lot lately having better and better results. However, control and validation can only be done with experimental research, in a complex analysis, taking into account variables whose quantification is not easy to estimate. The possibility to study the behaviour of semi-submersible structures on the waves at the "Dunărea de Jos" University model requires adequate and well-trained personnel. Comprehensive studies are needed to

Content from this work may be used under the terms of the Creative Commons Attribution 3.0 licence. Any further distribution of this work must maintain attribution to the author(s) and the title of the work, journal citation and DOI.
Published under licence by IOP Publishing Ltd

develop appropriate CFX simulations for the studied case. In Romania, studies validity can be checked only by using equipment and specialists from the "Dunărea de Jos" University.

The size of the basin, as well as the measurement equipment requirements will influence the model studied, that is the reason why only by using the similarity criteria can we validate numerical solutions from CFX solver. The experimental investigation of the performance of any model must take into account the possible interferences in the basin study and differences from not operating in open sea. The basin has a width of 4 m, a depth of 3 m and a length of 40 m and its measuring equipment can support a model with a maximum weight of 200 kg, that is why a Froude scaled model will face restrictions.

2. Scale model size calculation according to Froude criterion

The scale is limited by the measurement equipment limits for amplitude, and the maximum weight allowed. Due to the wide range of heave and pitch measurement the main restriction is the model's weight. Based on the Froude criterion, $Fr_m = Fr_n$, where Fr_m is for the model and Fr_n is for the natural scale structure, we will determine the scale by intersecting functions of the model mass according to scale $f(x)=3974000/x^3$ and the constant function $f(x)=200$. The calculated result is 26.44 and this corresponds to a model weight of 200 kg and this is the maximum value allowed by the measurement equipment of the University of Galati. We will choose a high value for the scale, 1:30, corresponding to a mass of 147 kg. All sizes and scales are presented in table 1. This scale is achieved by a geometric similarity between structure and designed model and all the other dimensions are proper to the test.

Table 1. Sizes and scales table.

Size \ Structure	Cylinder	Cylinder 1:30	Semi-submersible	Semi-submersible Model 1:30
Column diameter [m]	10	0.33	10	0.30
Column height [m]	20	0.66	20	0.66
Heave plate diameter [m]	20	0.66	20	0.66
Draft [m]	16	0.53	16	0.53
Weight[t]	1256	0.046	3974	0.147
VCG [m]	8.5	0.28	8.5	0.28

Using Froude model we will establish the relation between model and the real semi-submersible structure. We will use the length scale as follows:

$$k_L = {}^{L_n}/_{L_m} = 30 \tag{1}$$

The speeds scale is determined from $Fr_m = Fr_n$ and this is also $\frac{v_n^2}{gL_n} = \frac{v_m^2}{gL_m}$.

And the result is (for our values):

$$\frac{v_n^2}{v_m^2} = \frac{L_n}{L_m} = k_L = 5.47 \tag{2}$$

Time scale will be determined from $k_t = k_l/k_v$ and the result is (This value can be used for periods):

$$k_t = {}^{t_n}/_{t_m} = \sqrt{k_L} = 5{,}47 \tag{3}$$

The frequency scale is:

$$k_f = {}^{f_n}/_{f_m} = {}^{1}/_{k_t} = {}^{1}/_{5,47} = 0{,}182. \quad (4)$$

The mass scale is:

$$k_\Delta = {}^{\Delta_n}/_{\Delta_m} = {k_L}^3 = 30^3 \quad (5)$$

The acceleration scale is:

$$k_a = {}^{a_n}/_{a_m} = 1 \quad (6)$$

2.1. Pre-analyzed model design

The semi-submersible structure with 3 columns materials will be chosen based on the rigorous analysis presented in Ansys CFX. A static structural analysis will be performed in each model case to get the best guidance in the implementation stages. Modern model manufacturing industry is based on steel models, fiberglass models, wooden models and plastic models. A model manufacture is a difficult job when the model has to respect some rules because it has to be geometrically similar to the original analyzed structure. We will present a successful way to pre-analyze (figure 1) the model in a coupled field analysis.

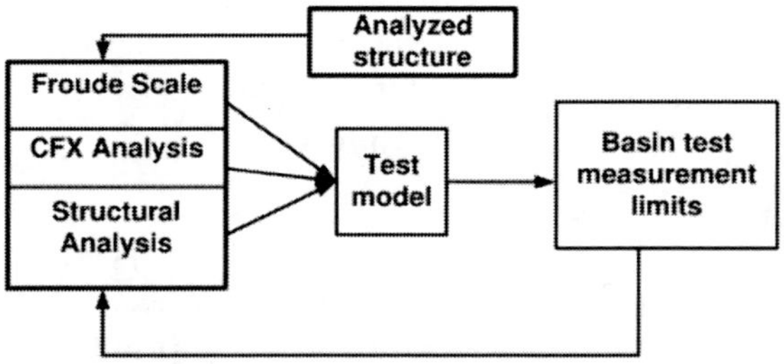

Figure 1. Pre-analyzed model design.

The pre-analyzed model in a coupled field analysis can use software and cost- free tools to evaluate the scale, the structure behavior in waves, and the dimensions required for any structural analysis. At any scale a Fluid-Structure interaction case with CFX and Static structural will be studied before building the model.

3. CFD/CAE numerical investigation

In accordance with the test plan, experimental tests will be carried out on semi-submersible model for wave directions 0, 60 and 90 degrees. To estimate the forces occurring on the 3D model of the semi-submersible with three columns we will use advanced Volume of Fraction simulation in ANSYS CFX. CFX software is able to simulate wave action on model from any angle with a parametric simulation where wave angle is the input parameter for analysis [1]. CFX analysis is able to determine model forces, moments and displacements in regular waves. CFX-Pre module can model input parameters of the analysis as required [2]. Based on the results of CFX-Post module we can view and measure the results for static mechanical analysis on the scale model. It will have to withstand forces and moments during experimental tests regardless of the direction of the wave. Graphical results are presented for simulated wave action on semi-submersible model (figure 2, CFX-Post results). At each step, we will evaluate values of forces and torque (figure 3) looking for the maximum values for ANSYS structural analysis.

Figure 2. CFX-Post results regular wave action simulation.

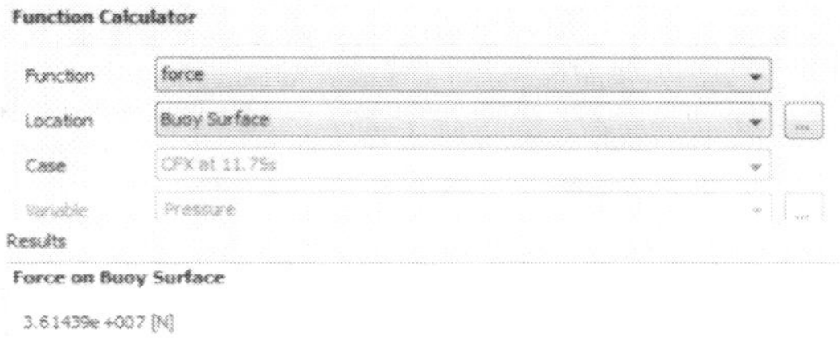

Figure 3. CFX-Post semisubmersible resulting forces/moments on the scaled model.

The study will be conducted to assess the vertical oscillation, pitch and depreciation amortization thanks to the plates with diameters of 1.5 to 2 times larger than the diameter of the column. The model is made of two different materials: steel and fiberglass. The connection is made by M8 stainless steel bolts in order to obtain a rigid model. Fiberglass columns and stainless steel reinforcements will be studied with 3D software from ANSYS Structural Static (figure 4.) with the gravity forces and buoyancy forces simulated in ANSYS CFX.

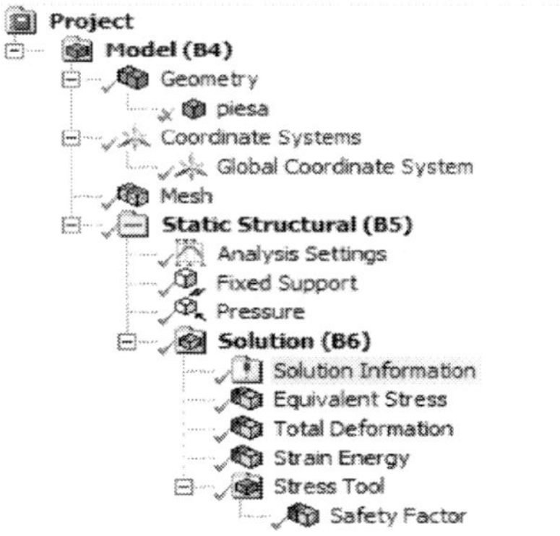

Figure 4. Pre-analyzed model design.

ANSYS structural static analysis for the scale model involves assessing the forces and pressures that occurs in basin testing. The CAD model should be tested on the maximum simulated load. For this analysis we will consider a fixed cylinder while the other two will have an applied pressure equal to the maximum hydrostatic pressure. The evaluation results will be included in the model pre-analysis until the size of the model will be able to support these loads. For this steps we use: mesh (figure 5), total deformation (figure 6), stress values (figure 7) and safety factor (figure 8).

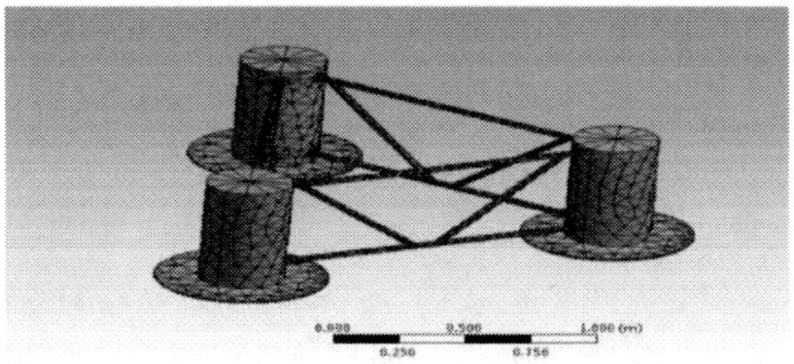

Figure 5. Creating in ANSYS Mesh model studied.

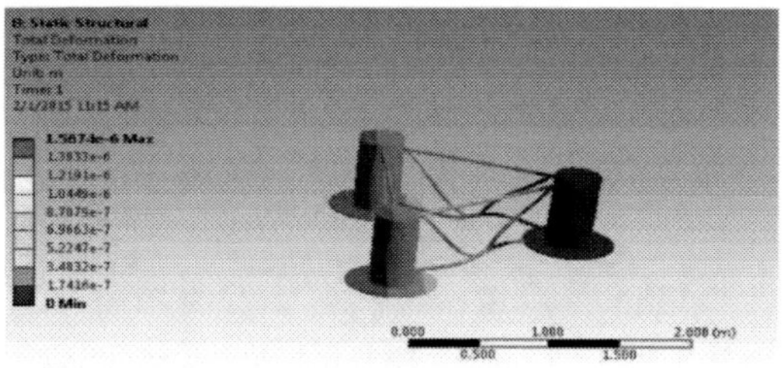

Figure 6. Total deformation model studied with final values generated.

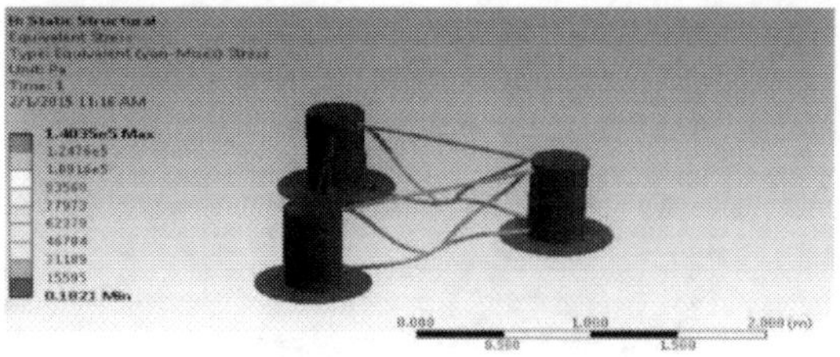

Figure 7. Stress values in the studied model structure WindFloat final values generated.

Figure 8. Safety factor at the request analyzed to study with the final model generated.

Using this method of analysis software we are able to estimate the minimum diameter of the steel bars and the minimum fiberglass wall thickness for the vertical columns. Results indicated for the presented basin that bars with diameter of 30 mm and columns with 5 layers of fiberglass are enough to withstand the stress given by the basin tests.

4. Equations and mathematics

The model mass is limited to basin restrictions and the mass is decided also for easy handling during the test. To build the semi-submersible model to an appropriate scale according to the project requirements and basin restrictions, we will use approximately 40 m^2 of fiber glass and about 12 m of round steel pipe. This means that the final weight of the model developed will be about 30 kg. Fiber glass is a material used in boatbuilding, automotive and implementation of highly complex parts such as ship models and semi-submersible offshore structures. The most important properties [4] of glass fiber are: high tensile strength, heat resistance, chemical resistance, non-toxicity and non degrading bacteria and fungi. Fiber glass can be of several types and has variable percentage composition: SiO2, Al_2O_3, Na_2O, CaO, MgO, B_2O_2, FeO_3 and PbO. These fibers come from suppliers in the form of rolls of material of different widths and are processed with the resin to fill the mold forms with various patterns. The glass fiber acquires a tensile strength of 60 MPa and an elongation at break of 1%.

The round steel 30mmX1.5 mm pipe is made of 304 L which gives it the property to be stainless steel because it has the composition of 18% chromium, 8% nickel, 0.03% carbon, 1% manganese, 0.045% phosphorus, and 0.03% sulfur. These pipes have a low carbon content which reduces the problems of precipitation of the bath of molten metal during welding. Therefore, this material lends itself well to the weld when using stainless steel electrodes.

Dimensional requirements are established according to the Froude scale. Model data for fiberglass wall thickness and stainless steel thickness are used in simulation of structural strength. Dimensions based on a schematic process design will lead to a well prepared model for basin test according to restrictions involved.

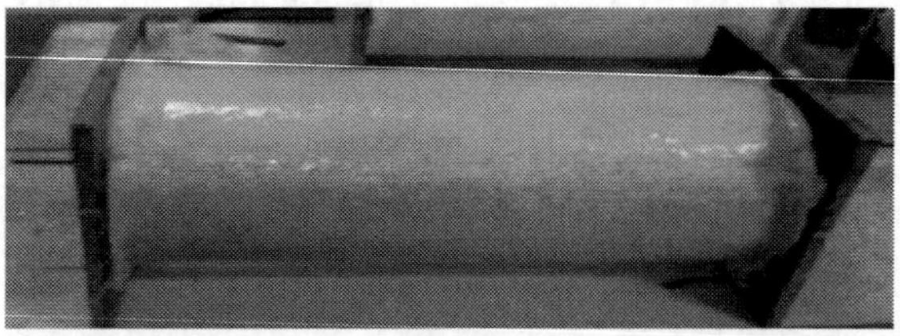

Figure 9. Cylindrical floats casting mold.

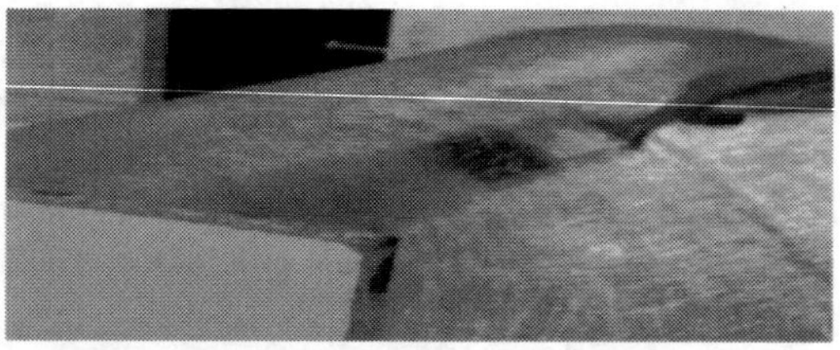

Figure 10. Bonding the layers of fiber glass.

During construction, each layer of fiberglass is placed in a new direction to create a strong material after hardening. The connections between fiberglass and steel are based on screws in the pre-assembly model (figure 11). During the model construction process dimensions are set to the precise value obtained from Froude scale (figure 12).

Figure 11. Model pre-assembly fiber and steel.

Figure 12. Final assembly of the model.

The semi-submersible model is build after CFD/CAE pre-analysis [5] and according to requirements of Hull Basin from University of Galati. It was already tested in the basin to assess heave and pitch motion in regular waves to determine RAO functions. To facilitate experimentation in the tank, the measuring equipment is mounted above center of gravity of the model deck surface with a wooden white deck. The wooden deck is used for placing measurement equipment above the centre of gravity on board the model during tests, and also for initial stability test.

The model for the WindFloat[6] semisubmersible was assembled by CirusPlast Constanta in December 2014, and the tests in the Hull Basin, in January 2015, were successful.

5. Conclusions

A concept for manufacturing models for three - column semi-submersible emerged from presented CFD/CAE software. The CFD analysis for wave action on the model, and for real scale structure is done with Ansys CFX. The three- column semi - submersible is analyzed with and without heave plates through CFX solver. The force and torque calculator will indicate the force acting on the Froude model and the values can be used in further static structural analysis. The model was tested successfully in the basin of the "Dunărea de Jos" University of Galati. All basin restrictions and technical suggestions will be modified according to the design in the pre-analysis. The response in regular waves depends on the model dimensions. A smaller model will determine a smaller output, and this will affect sensor readings for movements of model in basin tests. Through the use of ANSYS we will predict, before testing, if the scale model manufactured from steel and fiberglass will pass basin tests without technical difficulties. The result will be a rigid structure, able to face waves in basin tests, from all directions, according to the parametric simulation from CFX.

References

[1] Stern F, Wilson R V, Coleman HW 2001 Comprehensive approach to verification and validation of CFD simulations—part 1: methodology and procedures *J Fluid Eng* **123** pp 793–802

[2] Hänninen S K, Mikkola T and Matusiak J 2012 On the numerical accuracy of the wave load distribution on a ship advancing in short and steep waves *Journal of Marine Science and Technology* **17(2)** pp 125-138

[3] Khalid S 2013 Three-dimensional numerical simulation of a vertical axis tidal turbine using the two-way fluid structure interaction approach *Journal of Zhejiang University - Science A: Applied Physics & Engineering* **14(8)**

[4] Dumitrache R 2012 Calculation of the resistance of a fiber reinforced composite *Journal of Marine Technology and Environment* **II** pp 67-78

[5] Sharma R, Kim T W, Sha O P and Misra S C 2009 Semisubmersible design faces challenges. *Offshore Marine Technology* pp 22 - 30

[6] Wind Float project: www.principlepowerinc.com/products/windfloat.html

IOP Conf. Series: Materials Science and Engineering **95** (2015) 012069 doi:10.1088/1757-899X/95/1/012069

Optimizing ships' behaviour when sailing in following seas

O C Acomi[1], A Boştină[2], A L Boştină[3] and N Acomi[3]

[1]Robert Gordon University, Aberdeen Business School, Oil and Gas MBA Candidate, Garthdee House, Garthdee Road, Aberdeen, AB10 7QB, Scotland, UK
[2]Marine Department, Quai d'Arenc, No. 4, 13235, Marseille, France
[3]Constanta Maritime University, Navigation Department, Mircea cel Bătrân Street, No. 104, 900663, Constanta, Romania

E-mail: ovidiu.acomi@gmail.com

Abstract. Ships are built for sailing and transporting cargo on seas and oceans in weather conditions that are not friendly all the time. Even if the weather forecast is transmitted to vessels, the way of acting is a matter of officers' judgement, based on their knowledge and experience. The subject of this paper is to analyse the behaviour of a port container vessel in different weather conditions. The method consists in using a specially developed software which takes into account the main particulars, the actual stability and the dynamic characteristics of the individual ship in the real voyage conditions, in order to obtain the Dangerous Zone Diagram. In order to avoid surf-riding and broaching the master should reduce speed to less than the critical speed. The results of the study are presented in a diagram that allow us to analyse the situation and to determine the ways for avoiding dangerous conditions by changing the course or the speed. The study should be regarded as a supporting tool during the decision making process.

1. Introduction

The ship's behaviour certainly depends on the type of ship and the weather conditions. Nevertheless, we have to consider that having one type of vessel sailing on the same weather characteristics, the vessel will behave differently, subject to particular stability situation. The International Maritime Organization addressed this situation by developing an instrument that could be used by the captain of the vessels navigating in severe water conditions, in order to decide about actions to be taken to avoid dangerous situations.

The International Maritime Organization, through Maritime Safety Committee, has synthetized and approved this instrument, published as Guidance to the Master for Avoiding Dangerous Situations in Adverse Weather and Sea Conditions [1, 2]. These guidelines define the severe weather conditions and the situations that might occur while vessels operate under different conditions. There are studies addressing these situations through designing against parametric instability in following seas using analytical predictions [3, 4, 5].

The concept of following and quartering seas refers to the situation when the direction of the wave relative to the course of the ship, takes values between 0° and 45° from the stern of the ship. The dangerous situations that could occur when the vessels are sailing in severe weather conditions are:

- Surf-riding and broaching;

Content from this work may be used under the terms of the Creative Commons Attribution 3.0 licence. Any further distribution of this work must maintain attribution to the author(s) and the title of the work, journal citation and DOI.
Published under licence by IOP Publishing Ltd

- Reduction of intact stability caused by riding on the wave crest at mid-ship;
- Synchronous rolling motion;
- Parametric rolling motion or
- A combination of dangerous phenomena, which may lead to vessel's capsizing.

The reaction of a ship to dangerous situations depends on the actual stability parameters, hull geometry, ship size and ship speed. This implies that the vulnerability to dangerous responses, including capsizing, and its probability of occurrence in a particular sea state may differ on a case to case basis.

2. Case study

When the vessel sails in following seas its behaviour will be much different than in head seas, existing the possibility for the above mentioned dangers to occur. The case studies a model ship of a Panamax port-container vessel sailing in severe following and quartering seas. According to the Guidance to the Master for Avoiding Dangerous Situations in Adverse Weather and Sea Conditions [1], there are two critical conditions of encountered waves:

The first critical condition is (equation (1)), when the component of the ship speed in the wave direction approaches the speed of the waves, the ship can be accelerated to reach surf-riding and broaching. There is a marginal zone (equation 2)) where a large surge motion may occur.

$$V_{critical} = 1.8 \cdot Lpp^{1/2} [kn] \quad (1)$$

$$V_{m\,arg\,inal} = (1.4...1.8) \cdot Lpp^{1/2} [kn] \quad (2)$$

where:V – the speed of the vessel [kn]

Lpp – length between perpendiculars [m]

The second critical condition is related to the component of the ship speed in the wave direction that approaches the group velocity of the wave system; in this situation the ship may be overtaken by large waves, which may be twice the height of the observed significant wave height.

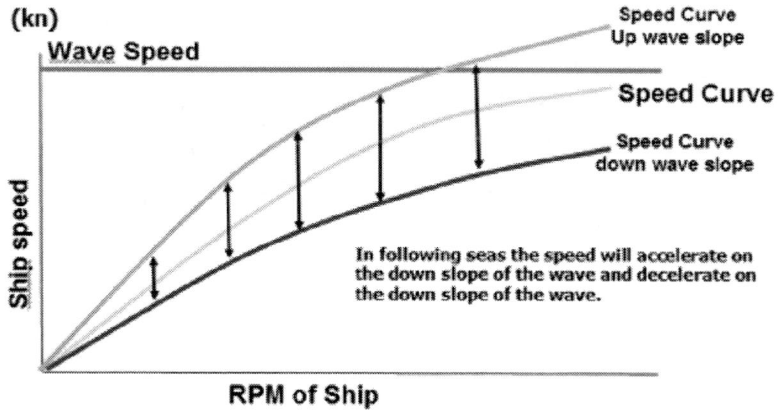

Figure 1. Critical Conditions [1].

IOP Conf. Series: Materials Science and Engineering **95** (2015) 012069 doi:10.1088/1757-899X/95/1/012069

When the speed of the vessel approaches the speed of the waves, the ship can be accelerated to reach surf riding. In that situation the ship will experience a surge motion and may lose steerage and thereby lose control of the ship. Figure 2 presents the stages for the ship model's speed to approach the constant value of wave speed.

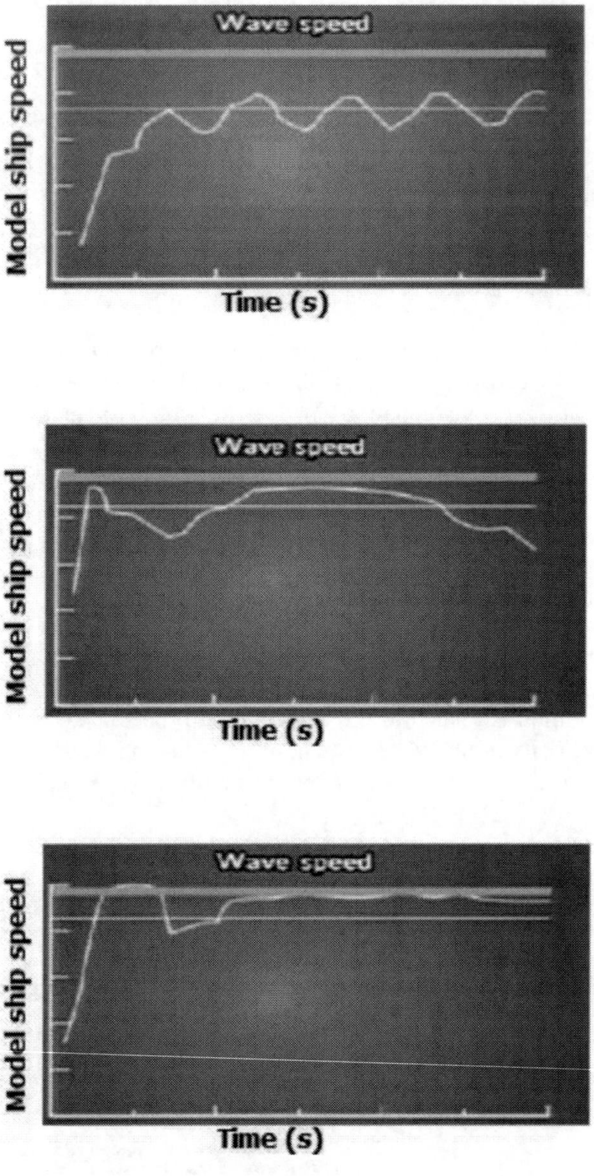

Figure 2. Model ship speed / Wave speed.

To avoid surf-riding and broaching the Captain of the vessel have to reduce speed to less than the critical speed as resulted from equation (1). The critical speed causing surf-riding and broaching are determined upon the length of the vessel and are presented in the table 1.

Table 1. Critical speed [kn] related to Ship's length [m].

Ship length [m]	10	20	50	70	100	150	200	300
Critical speed [kn]	6	8	13	15	18	22	25	31
Marginal speed [kn]	5	6	10	12	14	17	20	24

The figure 3 was elaborated by Maritime Safety Committee [1] to provide the officers of the vessel assistance on their decision on navigation, to avoid dangerous phenomena.

This study considered a Panamax containership travelling at 18 knots in stern quartering sea with a wave angle of 30 deg. The ratio of ship speed to length is calculated (equation (3)):

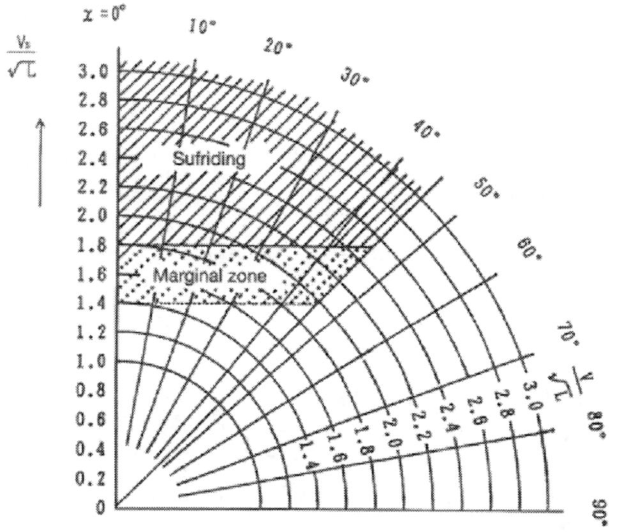

Figure 3. Critical Conditions [1].

$$\text{The\_ratio} = \frac{V_S}{L_{1/2}} = \frac{18[\text{kn}]}{284.7^{1/2}[\text{m}]} \approx 1.1 \qquad (3)$$

From the diagram (figure 4) it can be noticed that the speed is not critical, and there is no need for future actions to be taken.

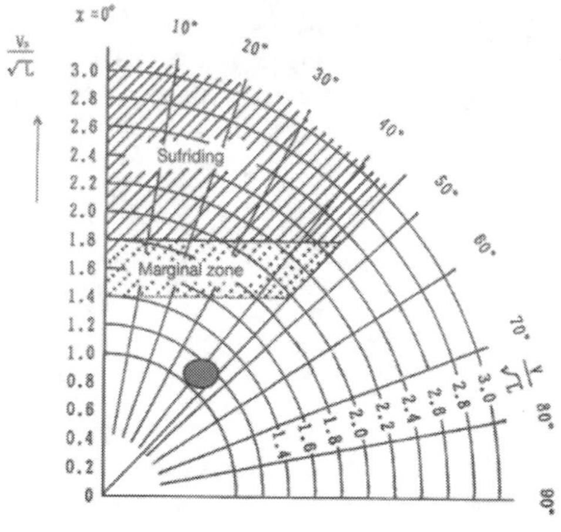

Figure 4. Dangerous zone relative to surf-riding for Panamax containership.

However, in case of large surge accelerations and surge motions, precautions must be taken as regards speed and heading. It should be noted the group velocity of the wave system is less than the velocity of the dominant waves speed.

Dangerous situations occur when a ship is sailing in rough seas that involve irregular waves with sudden serial high wave attacking the ship regularly from aft. It is seen that the sea becomes more regular in following seas, and high waves emerge continuously.

This means that when a ship is sailing in the same direction and at the same speed as the high wave group, the ship may be constantly hit by severe high waves. This phenomena is called Dangerous Encounter Wave Grouping. Figure 5 presents the wave data observed from a fixed point (a), and Wave data observed from a travelling point in following and quartering seas (b).

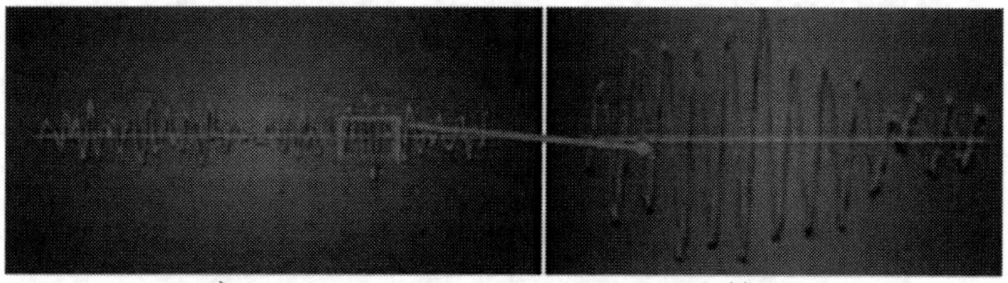

a) b)

Figure 5. Sailing in rough seas involving irregular waves. a) wave data observed from a fixed point, b) wave data observed from a travelling point in following and quartering seas.

The International Maritime Organization through Maritime Safety Committee, and Revised Guidance to the Master for Avoiding Dangerous Situations in Adverse Weather and Sea Conditions [2], suggests to avoid the dangerous zone as defined by the relationship between the ship speed and the wave period within following and quartering sea. The ratio is defined by equation (4) and should be placed in the range from 0.8 to 2.0.

$$\text{The\_ratio} = \frac{V}{T} = 1.5 \quad (4)$$

where T – the period of waves [s]

3. Results

The study was conducted for a Panamax containership travelling at 16 knots in stern quartering seas with a wave angle of 30 deg. The considered weather conditions: the height of the swell Hs = 10 [m] and the period of the waves of approximately 15 sec.

It resulted that in order to avoid dangerous phenomena, it is important that the ship's officer to adjust the speed of the vessel or to change the course. The calculated ratio of the ship's speed to the waves' period:

$$\text{The\_ratio} = \frac{V_S}{T} = \frac{16[\text{kn}]}{15[\text{s}]} \approx 1.1 \qquad (5)$$

From the diagram shown in figure 6 it is seen that the speed of the vessel is in the lower part of the dangerous zone, and, the speed should be slightly reduced or the course should be altered towards the beam sea.

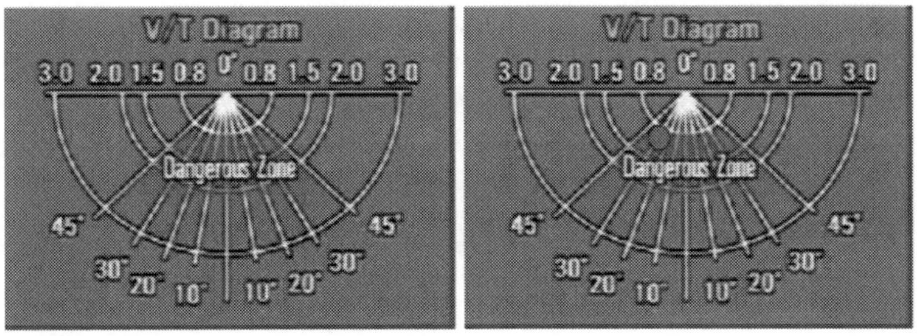

Figure 6. Critical Conditions [1].

4. Conclusions

Synchronous and Parametric Rolling highly depend on the encountered wave period and the natural roll period of the ship. Synchronous Rolling occurs when the encountered wave period is nearly equal to the natural roll period frequency and Parametric Rolling occurs when the encountered wave period is nearly equal to the half of the natural roll period frequency. In order to avoid these two phenomena each situation has to be analysed individually. Considering the results of this study, that surf-riding and broaching may occur even at lower speeds, the Captain have to adjust the speed of the vessel or to change its course in case of large surge accelerations.

References

[1] Maritime Safety Committee 1995 MSC/Circ.707 *Guidance to the master for avoiding dangerous situations in adverse weather and sea conditions* IMO

[2] Maritime Safety Committee 2007 MSC/Circ. 1228 *Revised guidance to the master for avoiding dangerous situations in adverse weather and sea conditions* IMO

[3] Spyrou K J 2000 Designing against parametric instability in following seas *Ocean Engineering* **27** pp 625–653

[4] Maki A et al 2010 Analytical formulae for predicting the surf-riding threshold for a ship in following seas *Journal of Marine Science and Technology* **15** pp 218-229

[5] Maki A et al 2014 Analytical methods to predict the surf-riding threshold and the wave-blocking threshold in astern seas *Journal of Marine Science and Tech nology* **19** pp 415-424

The rigid bi-functional sail, new concept concerning the reduction of the drag of ships

I Țicu[1], I Popa[2] and M Ristea[3]
[1]Constanta Maritime University, Faculty of Navigation and Naval Transport, 104 Mircea cel Batran Street 1, 900663, Constanta, Romania
[2,3]"Mircea cel Batran" Naval Academy, Faculty of Navigation and Naval Management, Fulgerului Street 1, 900218, Constanta, Romania

E-mail: ionela.ticu@yahoo.com

Abstract. The policy of the European Union in the energy field, for the period to follow until 2020, is based on three fundamental objectives: sustainability, competitiveness and safety in energy supply. The "Energy – Climate Changes" program sets out a number of objectives for the EU for the year 2020, known as the "20-20-20 objectives", namely: the reduction of greenhouse gas emissions by at least 20% from the level of those of 1990, a 20% increase in the share of renewable energy sources out of the total energy consumption as well as a target of 10% biofuels in the transports energy consumption. In this context, in order to produce or save a part of the propulsive power produced by the main propulsion machinery, by burning fossil fuels, we suggest the equipping of vessels designed for maritime transport with a bi-functional rigid sail. We consider that this device may have both the role of trapping wind energy and the role of acting as a deflector for reducing the resistance of the vessel's proceeding through the water by conveniently using the bow air current, as a result of the vessel's heading through the water with significant advantage in reducing the energy consumption for propulsion insurance.

1. Introduction

Starting from the idea that, before the advent of the industrial revolution, ships sailing on the ocean used to transport goods worldwide and reach speeds up to 16 knots, without polluting, in what follows we are going to resume the application of this type of energy which at that time it was used unrestrictedly and cost free. We also take into consideration the fact that the wind energy which seemed to be sufficient at that time can be also used to some extent nowadays for operating the power of the main propulsion machinery involving significant fossil fuel savings and gas combustion reductions.

Although over the years, transport vessels have strayed from the ecological path and begun to use air currents more and more rarely, until this method was used only in racing and by small pleasure yachts, we hold the idea that nowadays there is a need to modernize certain types of ships, so that the renewable energy be stored in the vessel's outer marine environment at a certain moment.

Before the advent of the industrial revolution, the most popular form of energy used for ship's propulsion was the wind. In this respect, nowadays ship builders have promoted creative solutions to trap the wind energy even when ships are equipped with modern internal combustion machinery. Thus, the present solutions mostly used are the following [1]:

Content from this work may be used under the terms of the Creative Commons Attribution 3.0 licence. Any further distribution of this work must maintain attribution to the author(s) and the title of the work, journal citation and DOI.
Published under licence by IOP Publishing Ltd

Unfortunatelly for the fitting of merchant ships this solution has the disadvantage of beeing dependent on the type of vessel and the enviromental factors, as well as the direction and force of wind.

Figure 1. Ship equipped with classic sails.

Figure 2. Ship equipped with rigid sails.

Hybrid propulsion is based on the opportunity of trapping wind energy of the certain moment replacing thus a part of the energy produced by conventional methods but it also presents the disadvantage that in normal conditions of underway the helical surface of the vessel grows and consequently the headway resistance increases as a result of air friction.

Figure 3. Ship equipped with 4 Fletner rotors.

Figure 4. Ship equipped with a kite type sail.

Although considered a constructive and cost efficient solution Fletner rotors propulsion involves the turning of propusion cylinders which can pose difficult problems for the crew on deck and this in turn could enter into contact with these moving masses. Also the rotors lengths implies the rising of the ship's center of gravity with negative values upon the ship's stability or with the necesity of a supplimentary ballast for lowering the center of gravity.

The use of the kite type sails represents an actual element of trapping wind energy, though this solution involves high instalation costs which can be hardly supported during ship service and it can be used only with high wind speeds with important energy savings. It also presents the disadvantage of introducing great drifting forces which may lead to the ship's plying off course.

The new concept of bi-functional rigid sail which is the aim of our present paper developed as a result of our thesis entitled "Research on the use of alternative sources of energy to ship's propulsion" represents a new solution for conventional energy saving on board ships and by case for trapping wind energy with the purpose of using it as a means of propulsion energy.

2. Some considerations regarding the increase in transport speeds of present maritime vessels

The increase in the share of freight transport by sea, particularly of the specialized cargo transport has lately determined a growth in the ships service speed. Thus, current trends show that the service speed of ferryboats or container ships often exceeds 22 Kt, this trend also applying to tanker vessels with service speeds of 16 Kt. In this context, the velocity of air coming from the fore part of the ship, that is the Vsc, grows high enough as a result of the vessel's heading through the water and in order to be possible to use the wind energy trapped from the environment at a given moment there must be a wind from the stern, Vrv, which is strong enough, so that the true wind velocity can be achieved, that is the Vva, from athwartships (see figure 5).

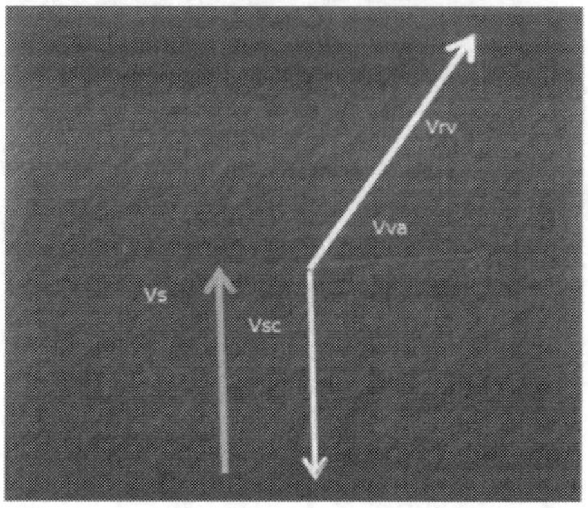

Figure 5. The true wind direction (Vva)
Vs = Service ship velocity
Vsc = Water current velocity
Vrv = Real wind velocity
Vva = True wind velocity

Therefore, only under such circumstances it is possible to trap the wind energy when the true wind Vva can produce an active bearing stress in each trapping device. For example, the kite type sail under figure 4, which has recently appeared on "the unconventional energy market" gives excellent results on condition that the wind speed be 4-5 knots higher than the ship's speed that is, 25 Kt for a liner ship which would be regarded as a very high speed, would cause swells in the sea and the meteorological conditions would become unfavorable for trapping wind energy [2].

2.1. Equipping shipping vessels with bi-functional rigid sails

Given the abovementioned conditions of the increase in transport speed, we have developed an innovative device called bi-functional rigid sail (see figure 6), which has two important roles with the purpose to replace fossil fuel consumption, as demonstrated by experiments on modelsand by simulation. Thus, this device is installed in the bow and can have two functions: role of deflector when the velocity of air comes mostly from the bow and of rigid sail when the direction of the real wind is favorable.

The components of the bi-functional rigid sail are (figure 6): 1 – the rigid sail; 2 – the sail's axis; 3 – the symmetric profile of the sail; 4 – the sail joint which allows tilting as deflector; 5 – the sail chock for allowing the sail's orientation into the wind for trapping wind energy.

The bi-functional rigid sail has two degrees of freedom and allows two-plan rotation. According to the two declared roles, that is, of properly conducting the air current in conformity with the direction of the true wind and capturing the wind energy respectively when the sea wind speed (stern wind) is greater than the speed of the vessel underway.

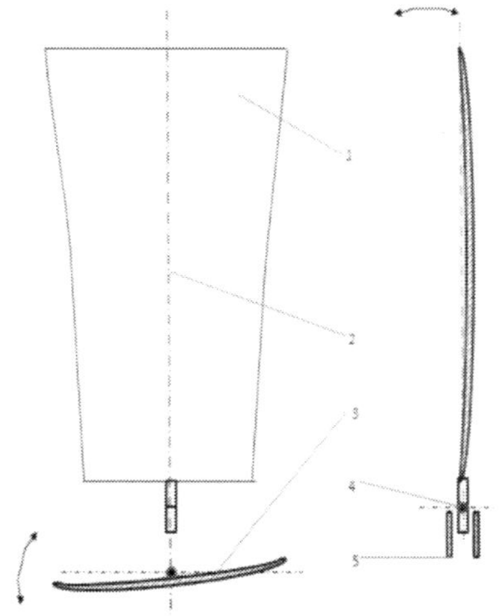

Figure 6. Bi-functional rigid sail.

2.2. Analysis of the Bi-functional rigid sail

To demonstrate the solution suggested, Ansys Workbench 14.0 software was used, in order to reveal the lowering of the ship's heading resistance by using the bi-functional rigid sail. In this regard, air flows, as compressible fluid, have been integrated around the ship's quickworks [3].

The simulated ship trials based on the use of finite element method [4], searched the optimal angle of inclination of the bi-functional rigid sail, and also its location in relation to the bow (see figure 7). In order to carry out these trials, a ship model has been produced at a scale of 1:10 where the sail had, in relation to the ship's bow, three successive locating aft points. The model under consideration was derived from a transport ship, preforming handling operations and whose architecture fitted the installation of bi-functional rigid sail, and having the following characteristics: Width: 10.768 m; Length: 29.599 m; Height: 6.993 m, Total area: 552.911 m^2; Volume: 2228.823 m^3.

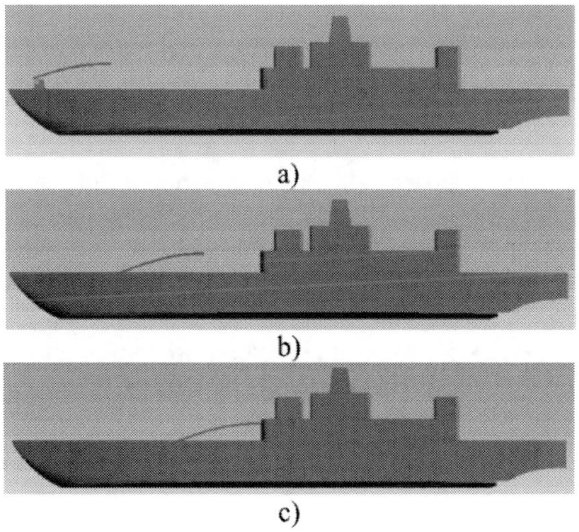

Figure 7. Placing the bi-functional sail type with deflective role
a) placing the deflector in the fore part of the ship
b) placing the deflector in the middle of the ship's fore part
c) placing the deflector in front of the ship's superstructure

Checking by simulation, the values of the ship's heading resistance due to air currents, significant variations thereof were observed, determined for the ship equipped with bi-functional rigid sail as compared to the conventional ship type [5]. Thus, it was found that the highest heading resistance occurred in b) at a 30° angle of loll, when the heading resistance, decreased by 35 % (figure 8).

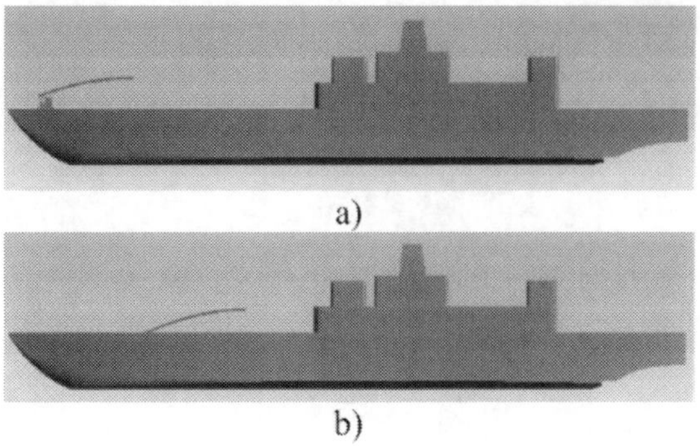

Figure 8. Specific variations of speeds
a) distribution of speeds
b) deflector placed near the superstructure, tilted at an angle of 30°

3. Results

The results of the comparative study connected to the two scenarios considered to equipping the ship, with and without bi-functional rigid sail highlight major variations in thrust (see figure 9). It appears that these increases are not linear, maximum values being recorded for cases where the wind speeds 10 to 17.5 m/s when the heading resistance increases for the unequipped ship with 332% and 330% respectively (see figure 10).

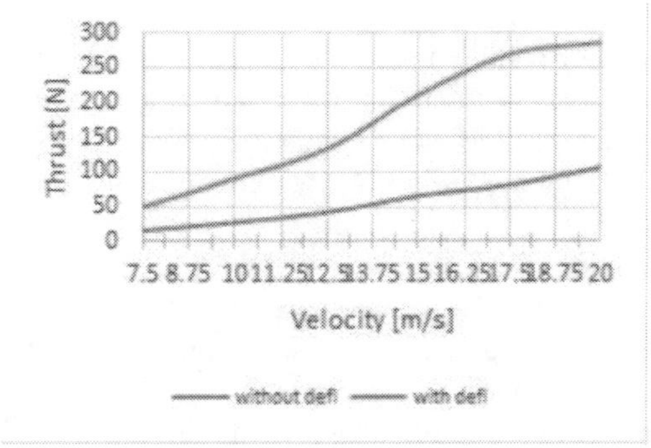

Figure 9. Thrust variations for the two cases considered.

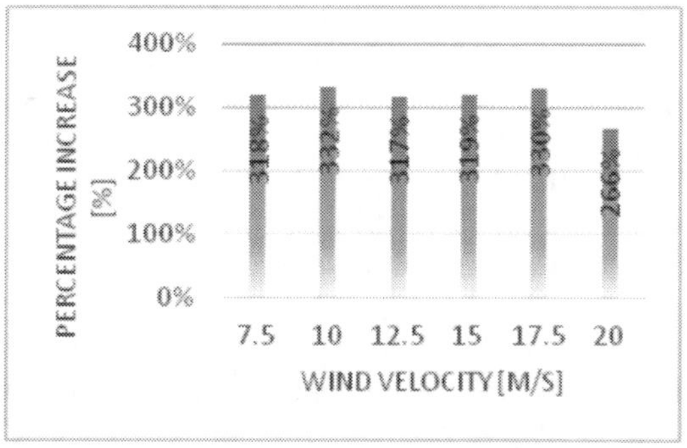

Figure 10. Comparison of thrust values set for the ship equipped with/without deflector.

4. Conclusions

Where the configuration of the ship allows the placement on board of such a bi-functional rigid sail and technical difficulties represented by its development and placement can be overcome so that using this solution would not affect the safe operation of the ship, this innovative technique could be a viable alternative to increase the energy efficiency of the entire energy complex represented by the ship and implicitly of the effective trapping wind energy from the marine environment.

Therefore the analysis of the doctoral theisis "Research on the use of alternative sources of energy to ship's propulsion" [5] leads to the folowing advantages concerning the equipment of vessels with this type of sail, as folows:

- diminishing the headway resistace as a result of the convenient swing of the air curent produced by the vessel's headway;
- trapping the stern wind for the active propulsion and discharge of vessel's reactive propulsion with important savings of fossil fuels;
- increasing the crew's working on deck level of comfort by a convenient deviation of the air curent produced by the vessel's underway speed;

- constructive simplification and increased fiability without involving important technological solutions;
- reduced maintenace and simple usage;
- the posibility of working optimization regarding the direction and force of wind at a certain moment;
- a low price for equipping a vessel with such a type of sail.

This article is the result of the project "Increasing quality in marine higher education institutions by improving the teaching syllabus according to International Convention STCW (Standards of Training, Certification and Watchkeeping for Seafarers) with Manila amendments". This project in co funded by European Social Fund through The Sectorial Operational Programme for Human Resources Development 2007-2013, coordinated by Constanta Maritime University.

The content of this scientific article does not reflect the official opinion of the European Union. Responsibility for the information and views expressed in the article lies entirely with the authors.

References

[1] AEA Energy & Environment 2008 Greenhouse gas emissions from shipping: Trends, projections and abatement potential (London: AEA)
[2] Alvik S, Eide M, Endresen Q, Hoffmann P and Longva T 2010 Pathway to low carbon shipping abatement potential towards 2030
[3] Hoffmann A, Klaus–Chang T S, Siddiqui S and Papadakis M 1996 Fundamental Equations of Fluid Mechanics (Wichita) *Engineering Education SystemTM* pp 67208-1078
[4] Oprişa-Stănescu P D 1997 The "ROFEM" Finite Elements Analysis Package– Structure and Operating Mode *"Politehnica" from Timisoara* **42(56)**
[5] Ţicu I 2014 Research on the use of alternative sources of energy to ship's propulsion (PhD Thesis: Constanta Maritime University)

The use of ECDIS equipment to achieve an optimum value for energy efficiency operation index

N Acomi[1], O C Acomi[2] and C Stanca[1]

[1]Constanta Maritime University, Navigation Department, Mircea cel Bătrân Street, No. 104, 900663, Constanta, Romania
[2]Robert Gordon University, Aberdeen Business School, Oil and Gas MBA Candidate, Garthdee House, Garthdee Road, Aberdeen, AB10 7QB, Scotland, UK

E-mail: nicoleta.acomi@cmu-edu.eu

Abstract. To reduce air pollution produced by ships, the International Maritime Organization has developed a set of technical, operational and management measures. The subject of our research addresses the operational measures for minimizing CO_2 air emissions and the way how the emission value could be influenced by external factors regardless of ship-owners' will. This study aims to analyse the air emissions for a loaded voyage leg performed by an oil tanker. The formula that allows us to calculate the predicted Energy Efficiency Operational Index involves the estimation of distance and fuel consumption, while the quantity of cargo is known. The electronic chart display and information system, ECDIS Simulation Software, will be used for adjusting the passage plan in real time, given the predicted severe environmental conditions. The distance will be determined using ECDIS, while the prediction of the fuel consumption will consider the sea trial and the vessel experience records. That way it will be possible to compare the estimated EEOI value in the case of great circle navigation in adverse weather condition with the estimated EEOI value for weather navigation.

1. Introduction

The studies related to the marine pollution show that it takes years from the proposal till the moment of ratification [1], even if the problem is recognized worldwide and thoroughly investigated [2]. After several years of debate, the concept of Energy Efficiency Operational Index (EEOI) was introduced, representing the amount of emissions related to vessel's performance [3, 4]. Considering the main purpose of the vessel, to carry goods for different distances, this recommended instrument for determination of air emissions is more a qualitative than a quantitative measure.

The main parameters that influence the value of EEOI being the quantity of cargo, fuel consumption and distance, the paper emphasizes the importance of using electronic navigational charts and Electronic Chart Display and Information System - ECDIS Simulator functions to choose the optimum route from the environmental point of view.

2. Concept of EEOI

In order to assess the level of CO_2 emissions resulted from marine traffic, the International Maritime Organization created and recommend the use of EEOI indicator, as resulted from Guidelines for voluntary use of the ship energy efficiency operational indicator - MEPC.1/Circ.684 (2009) [5, 6]. The main factors that affect its value are the voyage distance, comprising of various voyage legs, in

Content from this work may be used under the terms of the Creative Commons Attribution 3.0 licence. Any further distribution of this work must maintain attribution to the author(s) and the title of the work, journal citation and DOI.
Published under licence by IOP Publishing Ltd

nautical miles, the quantity of cargo in tonnes and the amount of marine fuel consumed for transporting the above-mentioned cargo, in tonnes. The EEOI formula expresses the CO2 emitted per unit of transport work (equation (1)):

$$\text{EEOI} = \frac{MCO_2}{\text{Transport\_Work}} \quad (1)$$

where

- the quantity of CO2 emitted is calculated based on fuel type and quantity, consumed per voyage, as recorded in Engine Log Book,
- the transport work represents the quantity of cargo in tonnes multiplied by total distance of the voyage in nautical miles.

For a single complete voyage where the vessel carries a certain quantity of cargo, the ship-owner uses the formula equation (2), considering different types of fuel (j), as required by navigation area and types of consumers: main engine, boilers.

$$\text{EEOI} = \frac{\sum_j FC_j \times C_F}{m_{c\arg o} x D} \quad (2)$$

To calculate the Energy Efficiency Operational Index for multiple voyages (i – the number of voyage), the International Maritime Organization recommends the use of Average EEOI, formula Eq. 3, by introducing the sum of cargo as per Bill of Lading, carried for each voyage recorded in Bridge Log Book.

$$\text{AverageEEOI} = \frac{\sum_i \sum_j \left(FC_{ij} \times C_{Fj}\right)}{\sum_i m_{c\arg o,i} x D_i} \quad (3)$$

where:

- FC – fuel consumption [tonnes],
- m – quantity of cargo [tonnes],
- D – distance [nautical miles]
- C_F – conversion factor corresponding to fuel type as per IMO [6].

A smaller value of EEOI as calculated by the use of equation (2) or equation (3) represents a more energy efficient ship.

A typical voyage comprises of a succession of voyage legs, performed by the vessel from the end of the previous voyage, when the vessel departed the discharge port facility, to the next end of voyage. Such voyage legs include primarily the ballast voyage from the previous discharge port to the loading port and the loaded voyage from the loading port to the next discharge port. Each of these include periods of port stays, anchor and manoeuver, which do not incur distance sailed, but add to the total fuel consumption. The paper aims to emphasize the influence of the distance sailed comparatively for a voyage performed on the shortest distance in adverse weather, compared to a longer distance adjusted to align to the weather navigation principles.

The first distance was calculated using the initial passage plan of the voyage, created before starting the voyage and the second passage plan was adapted as the marine environmental conditions required. For the second passage plan, the speed was determined considering the initial ETA of the vessel – estimated time of arrival. The type of marine fuel has not been changed during the voyage and the same applies to the quantity of cargo carried out.

2.1. Description of the Voyage

The analysed scenario represents the legs of the voyage performed by the ships model of an Oil tanker 82078 TDW (figure 1), equipped with Diesel Engine 1x9847kW, having the following characteristics Length 228.0 [m], Breadth 32.2 [m], between:

- loading port, Angra Dos Reis, Brazil, and
- port of discharge, Cape Town, South Africa

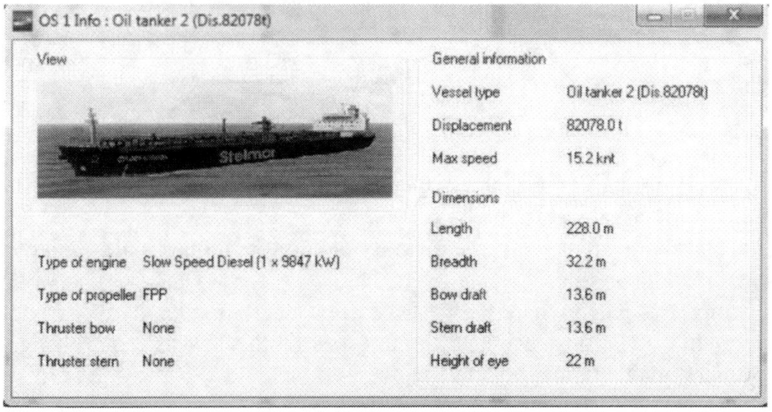

Figure 1. Oil Tanker characteristics.

2.2. Method

ECDIS represents an excellent software developed for the safety of navigation. During ECDIS training period, the students are required to create routes, conduct and monitor navigation using different ships' models. One function of the ECDIS Simulator allows the instructor to track record the own ship's manoeuvre and the ships' evolution relative to different marine environmental conditions (figure 2), by selecting the parameters from a drop down menu. For this study, the analysed parameters were (figure 3): current direction, current speed, wind direction, wind speed, wave amplitude and sway.

The initial passage plan has been created using ECDIS Simulator. Using the same software, it was possible to create the marine environment condition depending on instructor's option: Weather type Force 5, Wind speed 19.knots, Wind direction 080°, Wave height 2.5 m, Visibility 10 nautical miles, Current direction 220°, Current speed 7 knots.

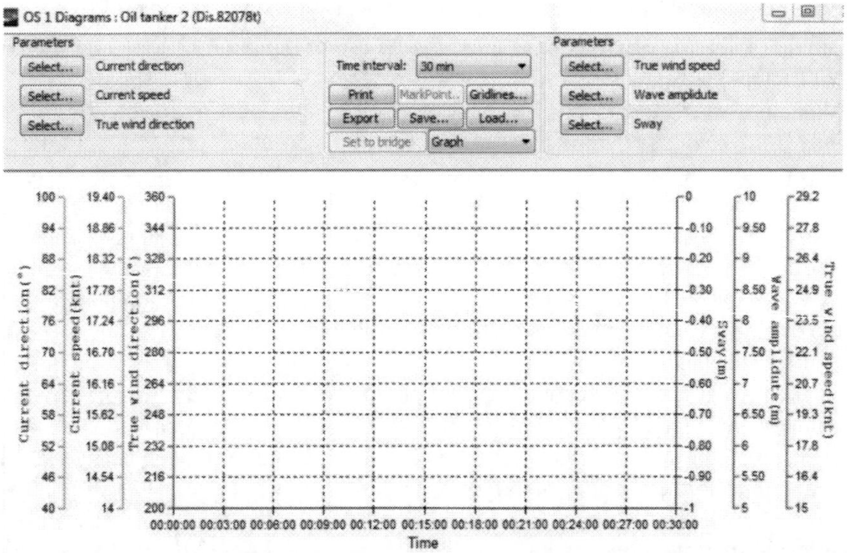

Figure 2. Environmental conditions.

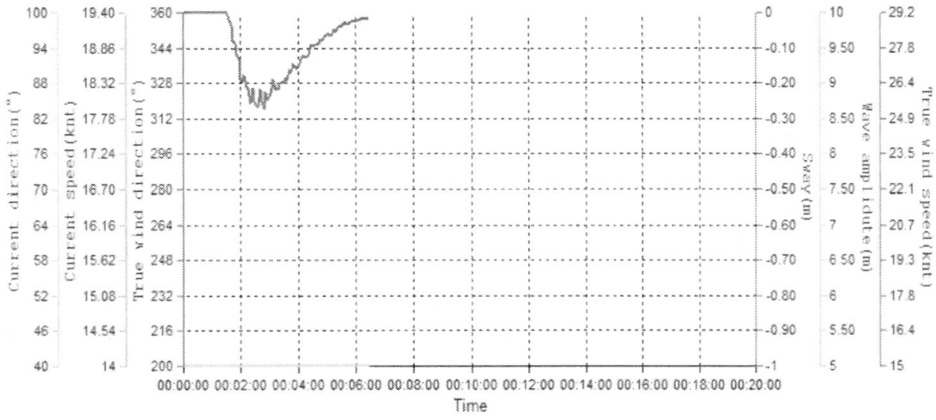

Figure 3. Ships parameters relative to different marine environmental conditions.

3. Comparative analysis

Considering the fuel consumption for the two situations, using the direct and the alternative route it is emphasized the variation of the EEOI, taking into account the environmental conditions.
The Oil tanker carried 75,000mt of cargo during the considered voyage leg. The initial distance calculated for normal environmental conditions was 3,327.6 nautical miles representing the first scenario, red line on the figure 4. Due to severe environmental conditions predicted, the vessel considered a second scenario as an alternative route of 3,389.6 nautical miles, represented with the blue line (figure 4).

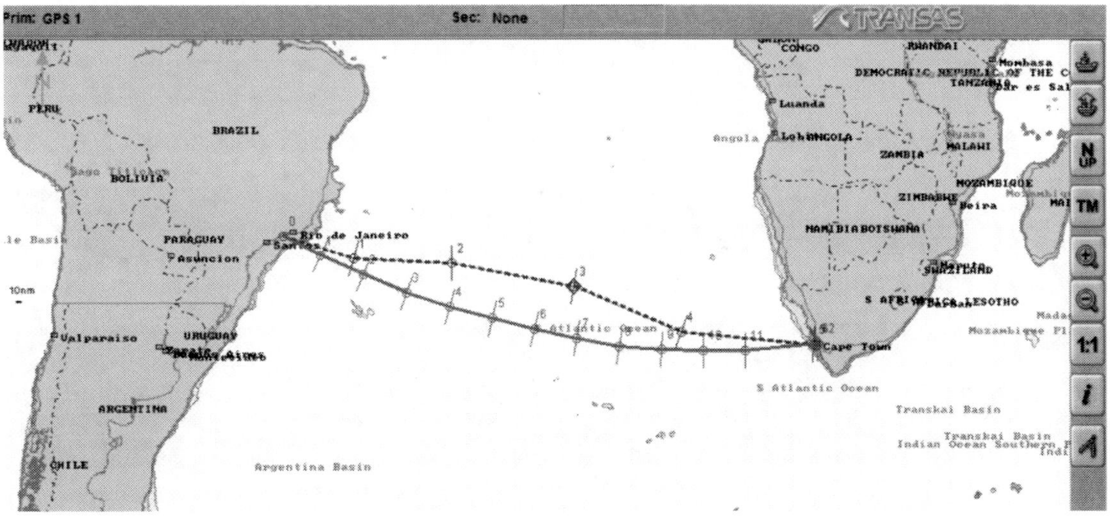

Figure 4. Ship's voyage, 1st scenario (red), 2nd scenario (blue).

The first situation considered a speed of 12.5knots to meet the estimated time of arrival (ETA), with an average consumption of 40mt/day, while the second situation met the same ETA requirement on a longer distance, with a different average speed of 12.73knots and a fuel consumption of 37.7mt/day. Both consumptions were considered according to the vessel's fuel consumption experience records and the ship trials for the particular speed and weather conditions.

IOP Conf. Series: Materials Science and Engineering **95** (2015) 012071 doi:10.1088/1757-899X/95/1/012071

A commercial software calculates both the EEOI and the average EEOI based on the following entries: type of fuel, distances and the quantity of fuel. The free software calculator developed by Totem Plus Company is used for both scenarios: 1st scenario (figure 5) and 2nd scenario (figure 6).
For different types of fuel, there are different carbon content and, consequently, different correction factor [1]:

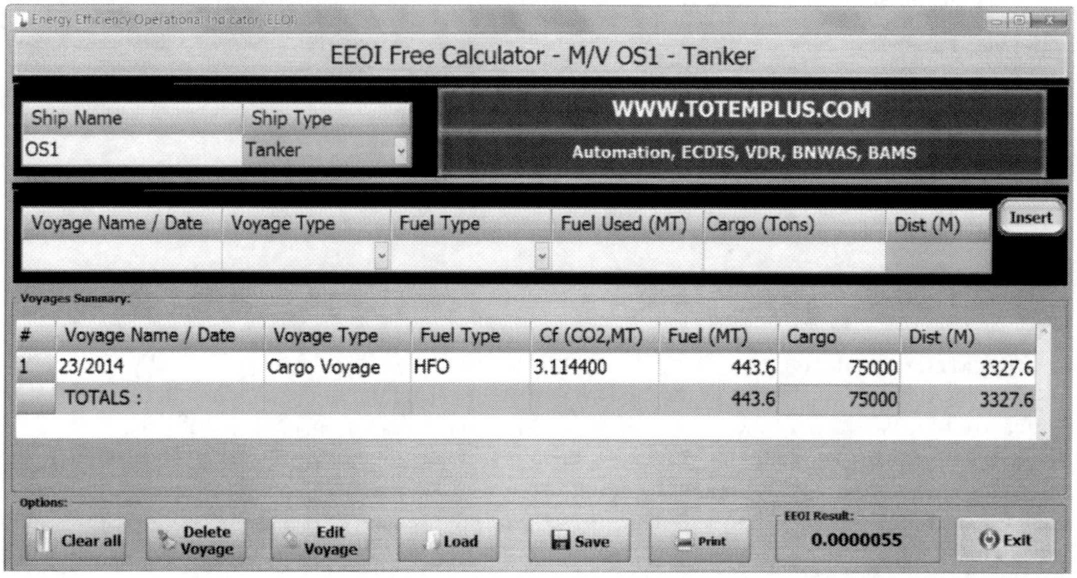

Figure 5. The interface of the EEOI Software calculator, 1st scenario (Source: TotemPlus, 2013).

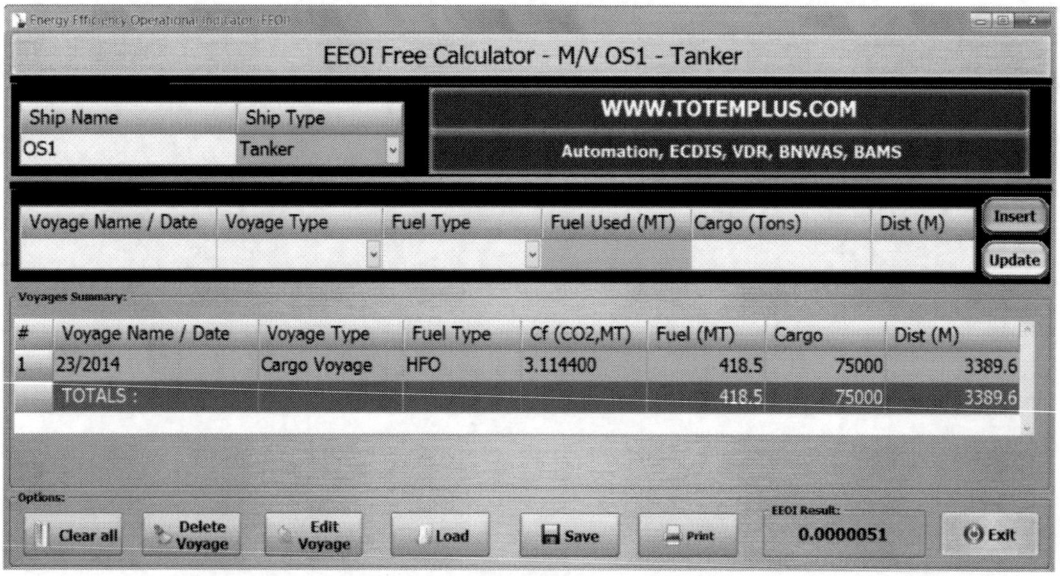

Figure 6. The interface of the EEOI Software calculator, 2nd scenario (Source: TotemPlus, 2013).

Table 1. The Carbon content per fuel type.

Type of fuel	Reference	Carbon Content	C_F (t-CO_2/t-Fuel)
Diesel / Gas Oil	ISO8217Grades DMX through DMC	0.875	3.206000
Light Fuel Oil (LFO)	ISO8217Grades RMA through RMD	0.86	3.151040
Heavy Fuel Oil (HFO)	ISO8217Grades RME through RMK	0.85	3.114400

4. Conclusions

The results of this study reveal the importance of using ECDIS software functions for achieving relevant values for Energy efficiency operational index and CO2 emissions and they represent a measure of changing EEOI related to meteorological condition. Comparing the values obtained for EEOI by applying the free commercial software for the voyage, having the same departure and arrival ports but using two different routes, can be noticed a difference of the EEOI values per voyage, of about 7.3%. This difference between the two values of the EEOI resulted from choosing an alternative route due to severe environmental conditions. The variation of distances was 62 nautical miles longer for the second scenario, but due to environmental condition encountered during the first route, even if the total distance was shorter, the total fuel consumption would have exceeded the total consumption of using the alternative route. The results show that the weather conditions encountered by vessels could greatly influence the value of the EEOI, and a good weather prediction, followed of accurate weather navigation while considering the fuel consumption experience records could reduce the EEOI and, consequently the CO_2 emissions.

Acknowledgements

The acknowledgement for the Totem Plus Company for their kindness in allowing us to use the EEOI Calculator Free Software.

References

[1] Anechitoae C 2009 The protection of community "design" right (drawing and design) within the domain of naval activities *Towards the sustainable marine technology and transportation* Gören Ö, Okan B, Karakaş Ş (Eds) **III** pp 1043-1048

[2] Ghiţă S 2007 *Ecology Course* Nautica Ed Romania

[3] IMO 2009 MEPC 59/INF.10 Second IMO GHG Study

[4] IMO 2005 MEPC.1/Circ.471.CO2 Interim guidelines for voluntary ship CO2 emission indexing for use in trials

[5] IMO 2012 MEPC 59/24/Add.1 ANNEX 19 Guidance for the development of a ship energy efficiency management plan SEEMP

[6] IMO 2009 MEPC.1/Circ.684 Guidelines for voluntary use of the ship energy efficiency operational indicator EEOI

Comparative analysis of different loading conditions on large container ships from the perspective of the stability requirement

C Stanca[1], N Acomi[1], C Ancuta[2] and S Georgescu[1]

[1]Constanta Maritime University, Navigation Department, Mircea cel Bătrân Street, No. 104, 900663, Constanta, Romania

E-mail: costel.stanca@gmail.com

Abstract. Container ships carry cargoes that are considered light from the weight point of view, compared to their volumetric capacity. This fact makes the still water vertical bending moment to be in hogging condition. Thus, the double bottom structure is permanent subject to compressive load. With the enlargement of container ships to the Post Panamax vessels, the breadth to depth ratio tends to be increased comparative to those of Panamax container ships that present restriction related to maximum breadth of the ship.The current studies on new build models reveal the impossibility for Panamax container ships to comply with the minimum metacentric height value of stability without loading ballast water in the double bottom tanks. In contrast, the Post-Panamax container ships, as resulted from metacentric height calculation, have adequate stability even if the ballast water is not loaded in the double bottom tanks. This analysis was conducted considering two partially loaded port-container vessels. Given the minimization of ballast quantities, the frequency with which the still water vertical bending moment reaches close to the allowable value increases.This study aims to analyse the ships' behaviour in partially loaded conditions and carrying ballast water in the double bottom tanks. By calculating the metacentric height that influences the stability of the partially loaded port container vessels, this study will emphasize the critical level of loading condition which triggers the uptake of ballast water in the double bottom tanks, due to metacentric height variation.

1. Introduction

The Post Panamax vessels fitted with double bottom structure are required to gain greater flexibility in loading more containers while fulfilling the stability and cross stability requirements. Thus, various loading scenarios that were not possible on Panamax vessels, now are being applied to Post Panamax ships that can stack lighter weight containers inside the holds [1, 2]. On the other hand, the breadth of the Post Panamax ship increased, along with the increase of lateral loads due to water pressure, while the weight of containers even under normal circumstances is reduced, to balance these loads. Therefore, prompt that appears in the double bottom structure occurs in the double bottom and tends to increase as far as weight of loaded containers in the storage decreases [3].

An empty loading section, along with the adjacent sections, is one of the loading scenarios (particularly in terms of the tensions that arise due to inhomogeneity load) required by designers of port container ship structure. The MOL Comfort ship accident resulted in various loading scenarios to be investigated and, after analysing the results, answered to the query whether applications on the

Content from this work may be used under the terms of the Creative Commons Attribution 3.0 licence. Any further distribution of this work must maintain attribution to the author(s) and the title of the work, journal citation and DOI.
Published under licence by IOP Publishing Ltd

double bottom structure are close to or exceeding in value those of the one-bay empty loading conditions. Investigations are carried out for 21 loading scenarios [1] as set forth in the table below.

Table 1. One bay empty condition.

	One-bay empty	All loaded compartments/roster
Storage loaded containers	30.0 Ton/FEU (design load)	0 la 30.0 Ton/FEU (7 scenarios at 5 Ton interval)
On caps loaded containers	130.0 Ton/stack (Design load for 40" containers)	Design load (130 Ton/Stack) 2/3 Design load (87 Ton/Stack) 1/3 Design Load (43 Ton / Stack)

The results are shown in the graph below where on the horizontal axis is represented the container loads in the hold, while the transverse stress occurring in the bottom shell plates are represented on the vertical axis. Maximum transverse stress for each load condition is shown by the three lines for the three container loading scenarios on the hatch cover. (at the design value, at 1/3 of design value, and at 2/3 of design value).

For comparison, under one bay empty condition, the maximum transverse stress present in the bottom shell plates is between 85 and 90 N /mm^2 and is represented by the red zone.

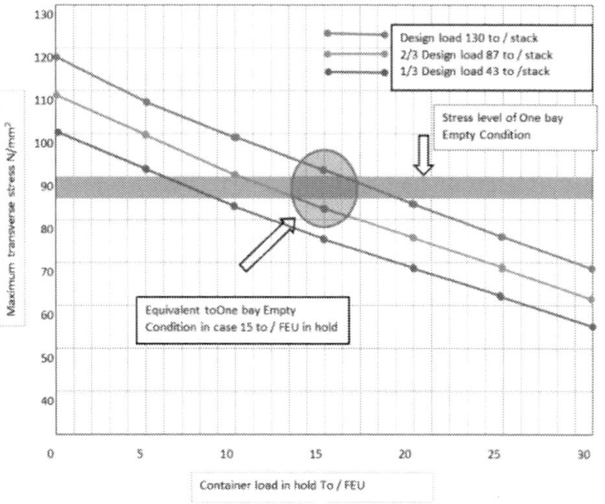

Figure 1. Comparison of transverse stresses from bottom floor tables for load conditions [1].

As emphasized in figure 1, the value of transverse stress is equal to that of one-bay empty condition in the area where the line cuts the red zone.

For example, when the container load on the hatch cover is between 1/3 and 2/3 of design load and the load inside the hold is about 15 tons/FEU, the transverse stress is equal to that one, due to the one-bay empty load condition.

The result reinforces the idea that Post Panamax ships, while winning in terms of load capacity, might meet conditions where the transverse stress from the bottom floor tables equal those from one-bay empty condition even when loading scenario is very similar to that from current operating real conditions.

2. Loads acting on the structure of the double-bottom structure of port container ships

Generally, port container ships are regularly subject to hogging that induce stress in both the deck and the bottom shell plates area throughout the ship's operation. It is a feature of the port container ships with a capacity greater than 10,000 TEU to have the engine room and the accommodation located in the ships' semi-aft.

The bottom sea pressure, container load, ballast water and the fuel from the double bottom fuel tanks are factors acting on the structure of the double bottom structure.

The upward load due to bottom sea pressure is a major load, as well as the lateral load acting on the structure of the double bottom, because the weight of goods is relatively small in relation to load due to the bottom sea pressure. Bottom sea pressure is determined by the hydrostatic pressure corresponding to the draft and the pressure induced by the waves.

This pressure is compensated to some extent when the double bottom tanks are ballasted or the fuel tanks are filled, because the loads induced by their weight is opposite to that due to bottom sea pressure (to a greater extent in the case of ballast, due to higher specific gravity).

Compressive loads due to side water pressure acts upon the side shell of the ship in a transverse direction.

We can structure the three loads acting on the double-bottom as shown below:

1. Compression loads due bending moments
2. Vertical loads due to water pressure
3. Transverse loads compression due to lateral pressure water

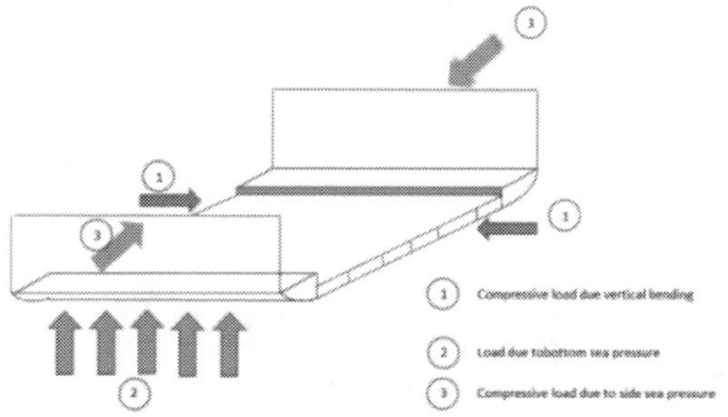

Figure 2. Acting forces on the double-bottom space [4].

Loads due to bending moment induce longitudinal compressive stress in the bottom shell plates and the transverse induce transverse compressive stress. On the other hand, the upward load due to bottom sea pressure causes convex deformation in the structure of the double bottom structure, consisting of bottom shell plate and longitudinal, girder and floor. The result is that the compressive stress in the double-bottom structure area arises in the central area; deformation structure is maximized near partial bulkhead in the longitudinal direction.

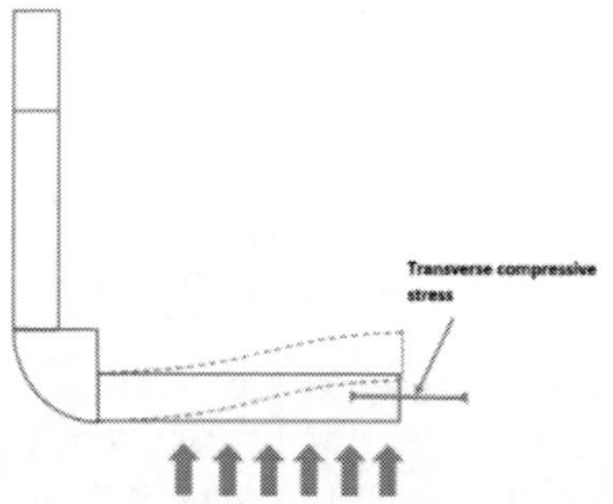

Figure 3. Transverse Compressive stress [4].

In conclusion, the combination of two forces at the middle area (centreline) and near partial bulkhead overlap in the longitudinal direction 1 and 2, and transverse, 2 and 3.

Increasing of loading capacity of Post Panamax port container ships was given by removing the width restriction (32m) so that the ratio B/D (width/height) of the ship increased accordingly, especially for the ships over 8,000 TEU. However, this coefficient is directly related to ship stability, specifically to the GM metacentric height (G_oM considering free surface).

The reference literature related to Post Panamax container vessels [4] calculated the B/D values for port container ship generations and values of GM in case the ship is or not ballasted in double bottom tanks.

3. Case studies

This study aims at analysing the ships' behaviour in different conditions loaded/unloaded and carrying ballast water in the double bottom tanks. By calculating the metacentric height that influences the port container vessel stability for vessels carrying containers in a limited number, different by maximum allowed for that type, this study will emphasize the level of loading capacity where the Post Panamax vessels require to take ballast water in the double bottom tanks, due to metacentric height variation.

The study concerns two vessels, Panamax and Post-Panamax, for which the Draught and the metacentric height was calculated using the loading program for port-container ships, MACS3 developed by Interschalt, figure 5. The program was applied for different situations of loading ballast water in the double bottom tanks.

The first vessel is a Panamax with the following characteristics:

Length overall: 207.46 m
Breath overall: 29.80 m
Draught (max): 11.40 m
Capacity: 2.474 TEU

seacos MACS3 by Interschalt v. NET 1.1 Mon Feb 23 2015 18:47
CONTAINER VESSEL (DEMO)
ArrivalAbidjan.mxml

Voyage No.: 838W
ETD: 9/1/1994 1:00:00 PM

Cargo Overview

Loading Condition total						
	Weight tons	LCG m.fr.AP.	VCG m.a.BL.	TCG m.fr.CL.	Fs m*t	Cont TEU
Constants	53.0	29.25	11.13	0.00		
Tanks	3192.7	87.53	4.29	0.92	386	
Container	22397.9	101.85	14.24	-0.10		1986
Deadweight	25643.6	99.92	12.99	0.03	386	
Dead Load	0.0	0.00	0.00	0.00		
Light Ship	10590.0	81.28	11.60	-0.07		
Displacement	36233.6	94.47	12.59	0.00		
Deadw. Reserve	8760.0	SUMMER Freeboard				

Figure 4. MACS3 programme developed by Interschalt [5].

Results 1. There were calculated two situations: for 1729.4 m^3 of ballast water in double bottom space and without ballast water. The results are presented in table 2.

Table 2. The results for Panamax vessel.

Scenario	A1	A2
Water ballast in double bottom tanks	1729.4 to ballast	0.0 to ballast
Draught Mean	9.69 m	9.33 m
GM'(corrected)	0.989/0.400 mlimit	0.376 m/0.400 m limit

The second vessel is a Post Panamax with the following characteristics:
Length overall: 366.08 m
Breath overall: 48.20 m
Draught (max): 15.50 m
Capacity: 12.552 TEU

Results 2. For the second vessel, there were loaded different quantities of ballast water in double ballast tanks: first scenario 517.9 m^3 of ballast water in double bottom space and the second scenario 352.9 m^3 of ballast water. This quantity cannot be unloaded due to the size of the vessel, and represents the minimum amount remaining distributed on the 18 ballast water tanks. The results are presented in table 3.

Table 3. The results for Post Panamax vessel.

Scenario	A1	A2
Water ballast in double bottom tanks	517.9 to ballast	352.9 to ballast
Draught Mean	13.71 m	13.71 m
GM'(corrected)	2.4049/0.802 limit	2.398 m/0.802 m limit

4. Conclussions

The results of the two scenarios emphasized the variation of Panamax and Post Panamax vessels' metacentric heights in different conditions carrying ballast water in the double bottom tanks. The values show the small variation of metacentric height 0.069 m, while varying the amount of ballast water in the double bottom tanks.

References

[1] Sumi Yet al 2013 Interim Report of Committee on Large Container Ship Safety *Committee on Large Container Ship Safety* (Japan)

[2] Oka M, Ogawa Y and K. Takagi 2011 A Fatigue Design for Large Container Ship Taking Long-Term Environmental Condition Into Account *J. Structures, Safety and Reliability* **2** pp 229-237

[3] Ogawa Y 2010 Study on the structural safety of large vessels (National Maritime Research Institute: Japan)

[4] Class N K 2014 Investigation Report on Structural Safety of Large Container Ships (Japan)

[5] Interschalt 2014 IS MACS 3 Loading Computer technical data sheet

IOP Conf. Series: Materials Science and Engineering **95** (2015) 012073 doi:10.1088/1757-899X/95/1/012073

The finite element method for calculating the marine structural design

A Ion[1] and I Ticu[2]
[1] "Constanta Maritime University" – Romania, Department of Navigation
104 Mircea cel Batran Street, Constanta, 900663
[2] "Constanta Maritime University" – Romania, Department of Fundamental Sciences and Humanities 104 Mircea cel Batran Street, Constanta, 900663

E-mail: alina_ion@rocketmail.com

Abstract. The aim of this paper is to optimally design and dimension marine structures in order for them to fulfil both functional and safety requirements. A master level of structural mechanics is vital in order to check tests and analysis and to develop new structures. This study can improve the calculation and estimation of the effects of hydrodynamics and of other loads; movements, strains and internal forces in fixed and floating platforms and ships. The finite element method (FEM) ensures basic understanding of the finite element model as applied on static cases including beam and plate elements, experience with static analysis of marine structures like platforms and ships, along with the basic understanding of dynamic response of systems with one degree of freedom and simple continuous beams, and also how analysis models can be established for real structures by the use of generalized coordinates and superposition.

1. Introduction

The Finite element method (FEM) allows to numerically solve the most diverse problems of solid Mechanics (which also includes the Structural Mechanics of ships), but also from many other areas (Fluids, Thermodynamics, Electromagnetism, etc.). The FEM idea consists in modeling the field studied, through a number of elements with finite dimensions, connected to each other in a number of nodal points in which the solution is sought. The operation is called finite element mesh. In structures, the values of linear and, sometimes, angular displacements are determined in nodal points. The Finite element model must simulate as correct as possible the behavior of the real structure which is and remains continuous even after deformation and each infinitesimal element inside of it or on its boundary should be in equilibrium under the action of internal and external forces. Through the very idea of FEM, elements in which the actual structure is divided, are satisfying the continuity condition at nodes, and in the preparation process of the FEM equations system, the nodes balance is also achieved. Based on them, movements, specific strains and stresses at any point of any finite element are determined.

In principle, the finite element method has the following main steps: meshing the structure into finite elements; choosing the form functions; establishing the finite element equation and its specific matrices; assembling the finite elements and implementing boundary conditions; solving the system of equations; determining the stress and strain state in each finite element [9].

Content from this work may be used under the terms of the Creative Commons Attribution 3.0 licence. Any further distribution of this work must maintain attribution to the author(s) and the title of the work, journal citation and DOI.
Published under licence by IOP Publishing Ltd

2. Structure mesh into finite elements

The structure mesh into finite elements represents the selection of the nodal points in which it is desired to obtain the finished solution and to set the affiliation of finite elements to the selected set of nodal points. This operation requires not only experience, but also some knowledge regarding the response of structures with certain geometries to various types of external actions (forces, imposed displacements, temperature variations). The structure mesh must, therefore, take into account geometry and how to load it, links, symmetry, possible proximity with infinite media, concentrations of power etc. Depending on the geometry of the meshed structure, elements can be mono-dimensional, two-dimensional or three-dimensional. Since most software dedicated to the finite element analysis has limited modeling capabilities, it is recommended to create geometry in dedicated CAD programs. Meshing is actually the process by which geometry representing the field with infinite degrees of freedom is replaced by a network with a finite number of elements to simulate as closely as possible the given geometry. The loading type and how it applies determine a specific behavior of the structure and, thus require a certain manner to mesh it. In points where external forces are applied, links or required movements, as well as in those where stiffness changes or characteristics and / or material changes occur, nodes are necessarily provided. The density of the nodes network determines the size of finite elements. Reducing these dimensions, thus increasing the number of nodes generally leads to increasing the accuracy of the calculation. In determining the network nodes, the following factors should be considered: achieving the optimum between the accuracy pursued, the computer performances and the analysis cost. The excessive increase in the number of nodes may not be the ideal solution. The tests demonstrated that there was a limited number of elements that if surpassed, the solution did not improve, in fact it could even get worse. ANSYS program has applications in almost all fields of engineering: structural, mechanical, electrical, electromagnetic, electronic, thermal, fluid and biomedical. Usually, solving in ANSYS includes: building the finite element model; applying geometric and mechanical boundary conditions (loads) and obtaining the solution; viewing results. Building the finite element model consumes the most time in an analysis. The element type determines the types of degrees of freedom, the element shape and the analysis size. The finite element model can be created in two ways: by modeling and direct engendering. By modeling, the analyst means creating geometry, which then the program automatically fills with nodes and elements; the model can be controlled by the user at the level of size and shape of the elements. The direct engendering means manually creating the finite element model; the user creates each node, and elements are after defined based on the nodes created. The analysis type is determined depending on the load conditions and on the response of the structure. Once the results determined, they can be viewed by using a post- processor [4].

3. Solutions and experimental tests on the dynamic analysis of a hull through FEM

Vibrations are dynamic phenomena, most often unwanted, manifested in almost all fields. In principle, the mechanical vibration occurs after an exchange of energy between two components of a system: one that develops kinetic energy and another that stores potential energy. The energy is transferred from one component to another, thereby generating swinging motion around a position considered of equilibrium. If the movement stops after a while, it is said that the vibration is damped, i.e. the system loses energy through various forms: friction, viscous or structural damping. Being an undesirable phenomenon, it occurs most often accidentally or it is uncontrolled. It may happen that after the removal of a system from the balance, the disturbance would cease. That is the case of free vibrations. But if disturbances tend to repeat or last a longer period of time, it is said that vibrations are maintained. The characteristics of disturbances transmitted to the system are generating deterministic or random vibrations [8].

3.1. Frequencies and natural vibration forms of a hull

For the modal analysis of the hull, a finite element model is created, according to the study. Since the vessel is floating in a continuous medium, in the modal analysis, the vessel does not require additional links, i.e. the modal analysis will be of free-free type. For such an analysis, the first six vibration modes come from the rigid solid movement; they can be easily identified by frequencies that are very small; theoretically, frequencies of rigid solid movements are void, but due to errors inherent in the solver, they will have values of MHz order. In order to estimate the influence of different stiffeners on frequencies and vibration modes, three constructive variants of hull will be analyzed: without stiffeners, with transverse stiffeners and with transverse and longitudinal stiffeners. For each constructive type studied, frequencies and associated modes were extracted. In the study of wave action, it is not needed to study modes with frequencies higher than 5 Hz, but as good engineering practice, between 100 and 200 modes are extracted; this requirement arises because of multiple excitation sources that may arise in the operation of a ship [5].

3.2. Idealized hull without stiffeners

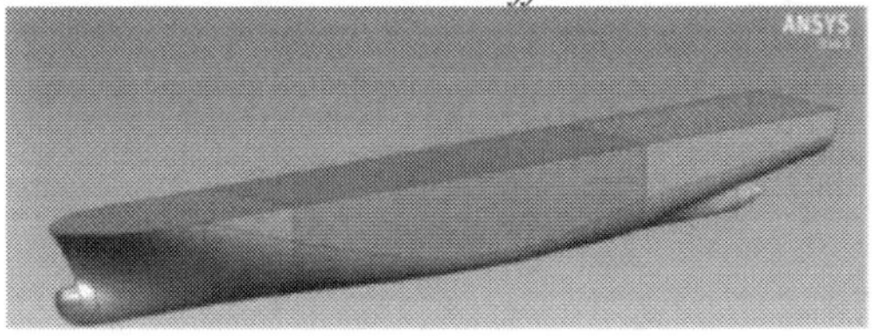

Figure 1. CAD model of the idealized hull without stiffeners – isometric view from outside.

Figure 2.CAD model of the idealized hull without stiffeners- isometric view with transparency.

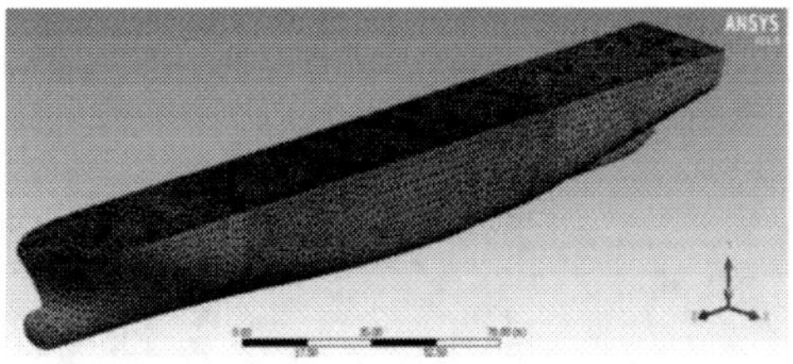

Figure 3. Mesh preparation and the finite element model for the hull without stiffeners.

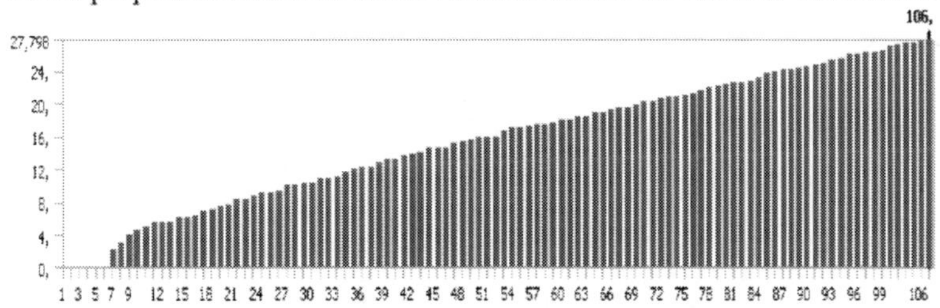

Figure 4. The first 100 natural frequencies of the idealized hull without stiffeners.

Figure 4 shows the first vibration mode of the hull without stiffeners corresponding to the minimum frequency of 2.142 Hz [1].

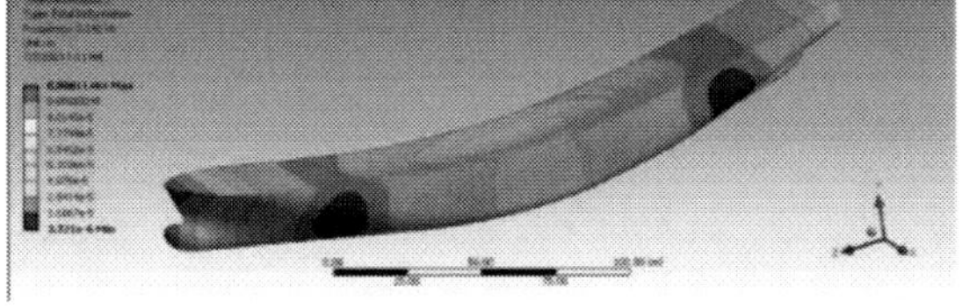

Figure 5. The vibration mode of the hull with-out stiffeners at the min frequency of 2.142 *Hz*.

Figure 6. Vibration mode of the hull without stiffeners – frequency of 6.1305 *Hz*.

3.3. Hull with transverse stiffeners

For the hull with transverse stiffeners, the mesh preparation and the final model can be seen in figures 7a and b.

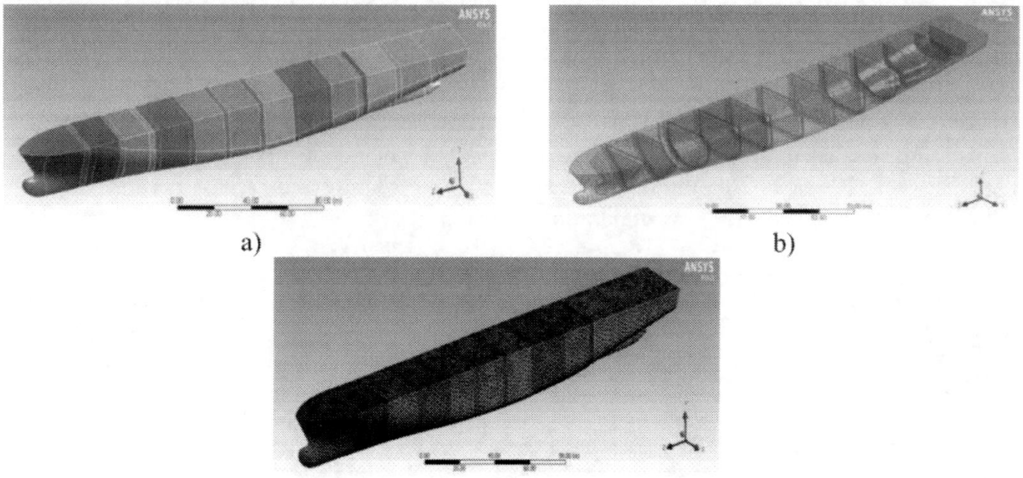

a) b)

c)

Figure 7. Mesh preparation (a, b) and the finite element model (c) for the hull with transverse stiffeners: stiffening structure.

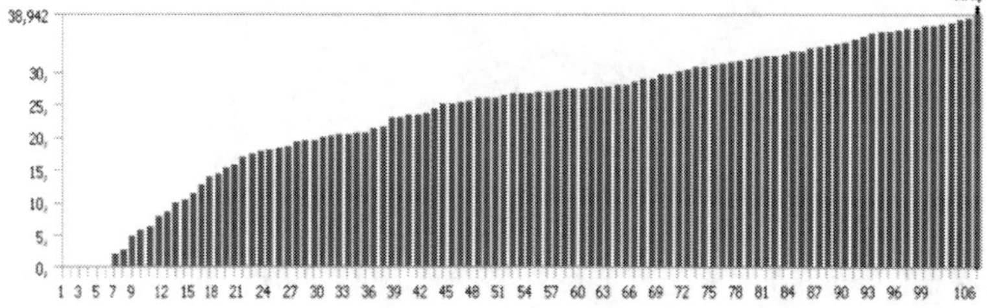

Figure 8. The first 100 natural frequencies of the hull transversely stiffened.

The frequencies of the first 100vibration modes of the hull transversely stiffened are shown in figure 8. Figure 9 and 10 show the first form of vibration of the hull transversely stiffened corresponding to the minimum frequency of 1.965 Hz and 11.209 Hz at the frequency of 11.209 Hz and the frequency of 11.209 Hz [2].

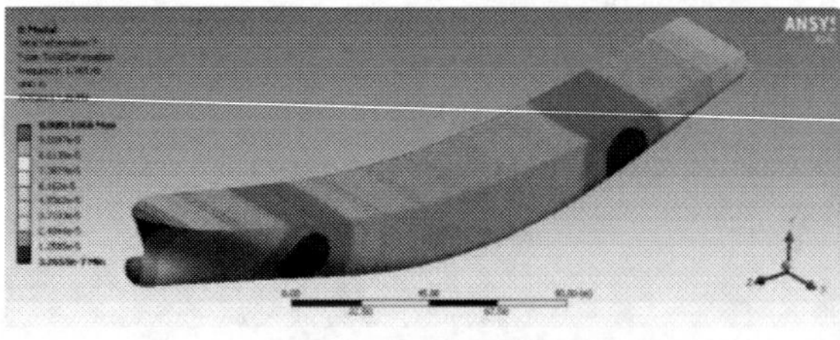

Figure 9. The vibration form of the hull transversely stiffened at the minimum frequency 1.965*Hz*.

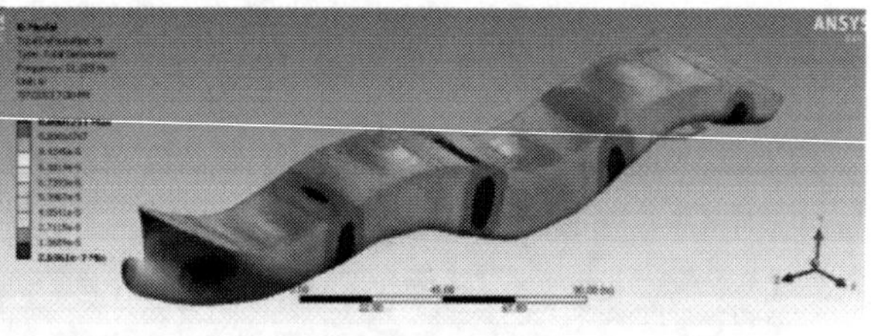

Figure 10. Vibration mode of the hull transversely stiffened at frequency 11.209 *Hz*.

3.4. Hull with transverse and longitudinal stiffeners

For the hull with transverse and longitudinal stiffeners, the mesh preparation and the final finite element model can be seen in figures 11a and b, and figure 12.

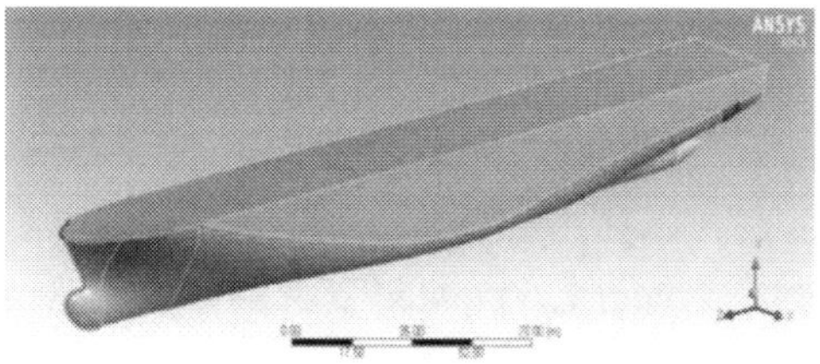

a)

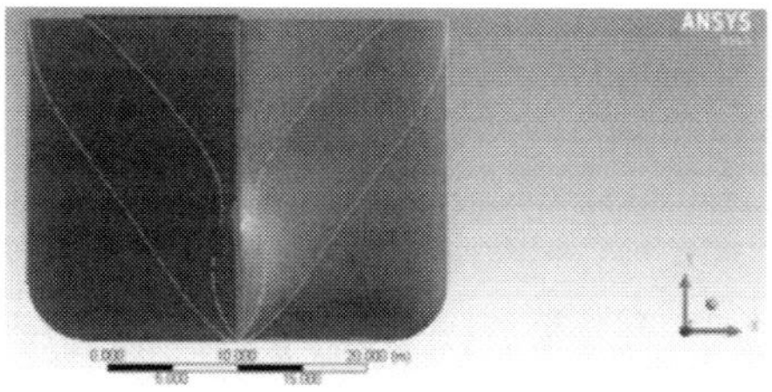

b)

Figure 11. Virtual topologies created to assimilate small surfaces (a, b).

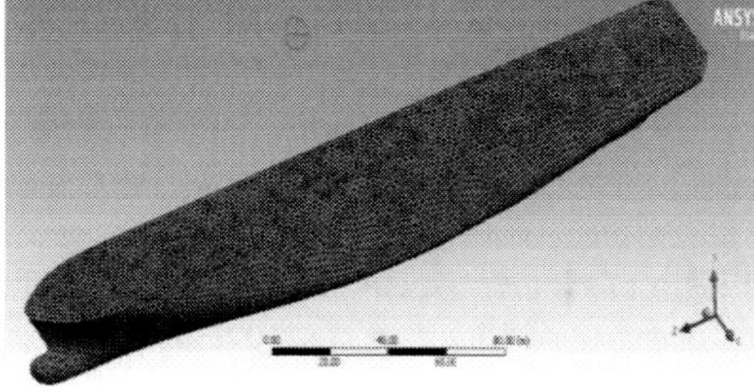

Figure 12. Mesh preparation and the finite element model for the hull transversely and longitudinally stiffened: finite element model.

The frequencies of the first 100 vibration modes of the hull transversely stiffened are shown in figure 13, figures 14 and 15 presents the first form of vibration of the hull transversely and longitudinally stiffened at the minimum frequency of 2.057 Hz and at the frequency of 12.435 Hz.

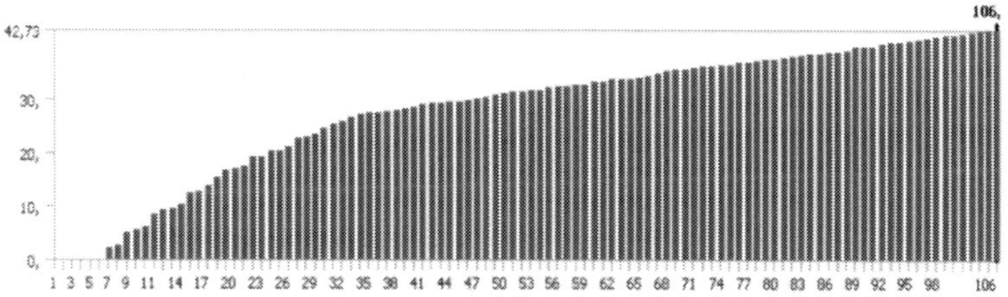

Figure 13. The first 100 natural frequencies of the hull transversely and longitudinally stiffened.

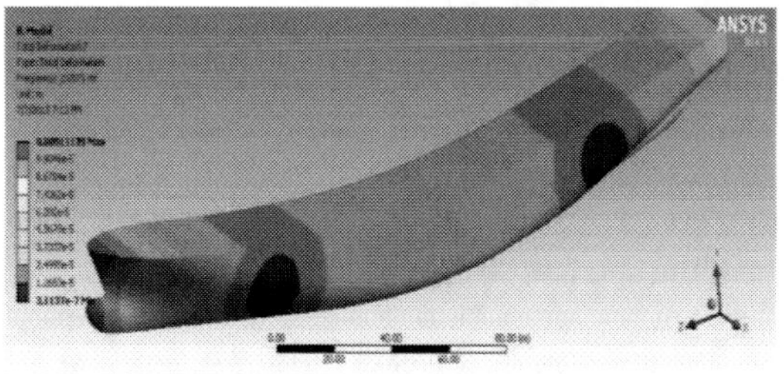

Figure 14. Form of vibration of the hull transversely and longitudinally stiffened at the minimum frequency of 2.057 *Hz*.

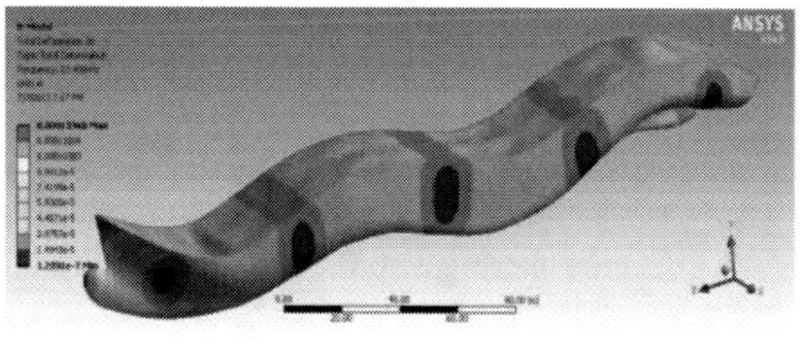

Figure 15. Form of vibration of the hull transversely and longitudinally stiffened at a frequency of 12.435 *Hz*.

It is noted that, for all three constructive configurations, the frequencies of the first vibration modes are higher than the frequencies of waves. The results in tables 1, 2, 3 show that, with the stiffening of the structure, frequencies of the homologous vibration modes are growing.

4. Conclusions

Data from these plots are rather qualitative than quantitative; even though the modal forms are represented as displacements, they are not real, being associated only to the mode of deformation of the structure at the appropriate frequency. A modal form of a structure is unique, but vibration amplitudes after this modal form are not unique, being directly proportional to the energy of the excitation source, in our case the waves [5].

The estimation of external forces induced by waves and stresses occurring in the structure of the ship due to them is a very complex issue and its solution requires knowledge of several areas, such as the probability theory, the theory of random processes, the statistical theory of extreme values, the analysis of experimental data, naval hydrodynamics, numerical analysis. Since the estimation of wave characteristics, the deduction of the spectral density function from experimental data is equivalent to the total characterization of the sea surface, the knowledge of wave spectra for different situations and geographical areas is particularly important for the structural analysis of ships' behavior at real sea, the prediction of characteristic extreme values is necessary. Static analyzes performed for three different structural models of a container port hull (ideal without stiffeners, with transverse stiffeners and transverse and longitudinal stiffeners) submitted to loads, come from the pressure distribution in calm water, allow to determine the influence of stiffeners and loading manner on the stress state.

The introduction of the finite element method (FEM) has enabled new approaches to complex problems of structural analysis. By using this method, more accurate resistance calculations and evaluation of criteria for failure of structures can be achieved. FEM also allows the iterative optimization of structural dimensions to fulfill all requirements [7].

This article is the result of the project "Increasing quality in marine higher education institutions by improving the teaching syllabus according to International Convention STCW (Standards of Training, Certification and Watchkeeping for Seafarers) with Manila amendments". This project in co funded by European Social Fund through The Sectorial Operational Programme for Human Resources Development 2007-2013, coordinated by Constanta Maritime University. The content of this scientific article does not reflect the official opinion of the European Union. Responsibility for the information and views expressed in the article lies entirely with the authors.

References

[1] Ion A 2014 Research on the calculation of a container ship structural resistance to wave oscillations PhD Thesis(Constanta Maritime University)

[2] Ion A, Bocanete P and Ticu R I 2013 The modern design of naval structures *Constanta Maritime University Annals* **20**

[3] Adegeest L J M, Braathen A and Vada T 1998 Evaluation of methods for estimation of extreme nonlinear ship response based on numerical simulations and model tests *Proc. 22nd Symp. on Naval Hydrodynamics* (Washington DC: National Academy Press) pp 70–84

[4] ANSYS Mechanical User Guide Release 14.5, 2012.

[5] Avrashi J and Cook R D 1993 New Error Estimation for C Eigenproblems in Finite Element Analysis *Engineering Computations* **10(3)** pp 243-256

[6] Baarholm G S and Moan T 2002 Efficient estimation of extreme long-term stresses by considering a combination of longitudinal bending stresses *J. Marine Science and Technology* **6(3)** pp 122–134

[7] Bathe K J 1982 *Finite element procedures in engineering analysis* (New Jersey: Prentice-Hall)

[8] Beards C F 1995 *Engineering vibration analysis with application to control systems* (London: Arrowsmith Ltd)

[9] Zienkiewicz O C and Taylor R L 2005 *The Finite Element Method for Solid and Structural Mechanics* Sixth Edition

IOP Conf. Series: Materials Science and Engineering **95** (2015) 012074 doi:10.1088/1757-899X/95/1/012074

Risk assessment techniques with applicability in marine engineering

E Rudenko[1], F V Panaitescu[2] and M Panaitescu[2]

[1]Odessa National Maritime University, 34, Mechnikov Street, Odessa, 65029, Ukraine
[2]Constanta Maritime University, Faculty of Naval Electro-Mechanics, 104 Mircea cel Batran Street, 900663, Constanta, Romania

E-mail: magnit_od@ua.fm

Abstract. Nowadays risk management is a carefully planned process. The task of risk management is organically woven into the general problem of increasing the efficiency of business. Passive attitude to risk and awareness of its existence are replaced by active management techniques. Risk assessment is one of the most important stages of risk management, since for risk management it is necessary first to analyze and evaluate risk. There are many definitions of this notion but in general case risk assessment refers to the systematic process of identifying the factors and types of risk and their quantitative assessment, i.e. risk analysis methodology combines mutually complementary quantitative and qualitative approaches. Purpose of the work: In this paper we will consider as risk assessment technique Fault Tree analysis (FTA). The objectives are: understand purpose of FTA, understand and apply rules of Boolean algebra, analyse a simple system using FTA, FTA advantages and disadvantages. Research and methodology: The main purpose is to help identify potential causes of system failures before the failures actually occur. We can evaluate the probability of the Top event.The steps of this analize are: the system's examination from Top to Down, the use of symbols to represent events, the use of mathematical tools for critical areas, the use of Fault tree logic diagrams to identify the cause of the Top event. Results: In the finally of study it will be obtained: critical areas, Fault tree logical diagrams and the probability of the Top event. These results can be used for the risk assessment analyses.

1. Introduction

Any danger is realized, bringing damage, as a result of any cause or multiple causes, so preventing hazards or protection from them is only possible at revealing the causes. Between dangers realized and causes there is a causal relationship: the danger is a consequence of a cause which, in turn, is a consequence of another cause, and so on. Thus, the causes and dangers form hierarchical, chain structure or systems. Graphic representation of these dependencies resembles a branching tree, so in the literature analyzing safety of objects is used term "fault tree".

Fault tree underlies the logical-probabilistic model of cause-effect relationships of system failures with failures of its elements and other events (impacts). In the analysis of a fault, fault tree consists of sequences and combinations of faults and disturbances, and thus it is a multilevel graphological structure of casual relationships received as a result of tracking hazardous situations in the reverse order to find their possible causes [1].

Content from this work may be used under the terms of the Creative Commons Attribution 3.0 licence. Any further distribution of this work must maintain attribution to the author(s) and the title of the work, journal citation and DOI.
Published under licence by IOP Publishing Ltd

The objectives of FTA are [2]:

- identification of the cause or combination of causes leading to the final (top) event;
- identification the causes which can affecting the reliability, safety, efficiency and other performance characteristics of the analyzed system;
- identification of conflicting requirements or specifications that may affect the reliability, safety and other performance characteristics;
- determination of whether system characteristics meet the established requirements;
- identification of common cause failures (common events).

The steps for construction a *Fault Tree logic diagrams* are: 1) description of the system, when boundaries and data needs are determined (description of the processes, characteristics of the materials and equipments, description of the maintenance procedure, etc.); 2) the identification of the hazard (site inspections, check-lists, hazard and operability studies); 3) the construction of the Fault Tree (events and gates) [1, 2, 4]; 4) the qualitative analyse of the structure of the Fault Tree (using the laws of Boolean algebra); 5) the quantitative evaluation of the Fault Tree (the calculus of the Top event frequency or probability [1, 3, 4].

2. Basic components of Fault Tree Analysis

2.1. The diagrams of Fault Tree.

Fault Tree diagrams consist of gates (AND, OR) and events. These are connected with lines.
Clasical FTA use different shapes for the events (*rectangle* –for the negative event; this shape is located at the Top of the tree and it is the only symbol that will have a logic gate and input events below it; *circle* – for the base event; this shape is located on the bottom tiers of the tree (figure 1).

a) b) c) d)

Figure 1. Basic components of Fault Tree.
a) AND gate; b) OR gate; c) rectangle (the negative event); d) circle (the base event).

2.2. Case studies of Fault Tree Analysis

Case study 1

A necessary condition of human death from electrical is his or her body inclusion in electric circuit with sufficient for death amperage. Therefore, occuring the accident (Top event "A") needs simultaneous existence of three conditions (figure 2).

Event "B" – a potentially high voltage on the electrical installation.

Event "C" means the appearance of a man on a conductive surface connected to ground.

Event "D" – a human body touching of electrical installation.

In turn, the event "B" may be the result of either of two events – prerequisites "E" and "F", where "E" – lowering the resistance of insulation of live parts and the event "F" – live parts touching of electrical installation.

Event "C" is also due to two assumptions: "F" – joining a man on a conductive surface, and "Z" – a human body touching of grounded elements of the room.

Event "D" is the result of the appearence of one of three prerequisites: "I" – the need for repair, "J" – the need for maintenance and "K" – intended use of the electrical installation, or normal operation of the installation.

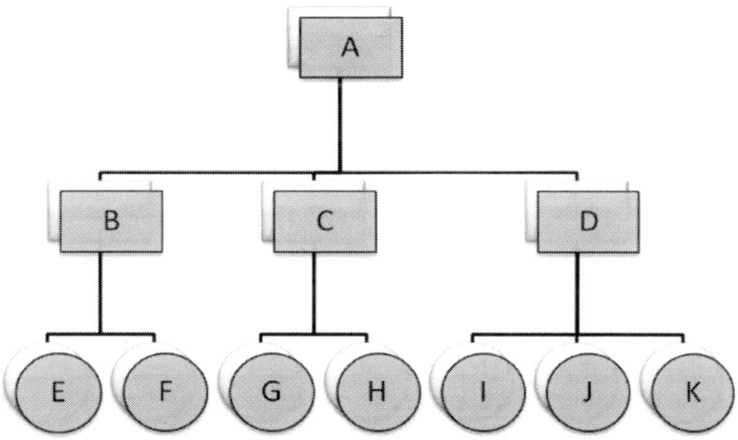

Figure 2. Fault tree for electrical injury.

Analytically the expression equation (1) for condition of realization of the accident (Top event "A") takes the form:

$$P(A) = P(B) \times P(C) \times P(D) = [P(E) + P(F)] \times [P(G) + P(H)] \times [P(I) + P(J) + P(K)] \quad (1)$$

Note. On fault trees the failure probabilities tend to be small with very small error term [1], [4]. For this, we can make an assumption (equation (2)) for specialy events:

$$P\ (X \text{ or } Y) \approx P(X) + P(Y),\ P\ (X \cap Y) \approx 0. \quad (2)$$

Case study 2

Unloading operation is one of the most intensive technological operations for tanker. It requires a lot of energy and leads to high load of SPP (ship power plant) comparable to energy consumption during vessel motion.

In case of failure of the above technical facilities performance degradation of cargo-pumping system occurs, that leads to an increase in the discharge time and downtime of the tanker, at a cost to the shipowner in the tens of thousands of dollars per day.

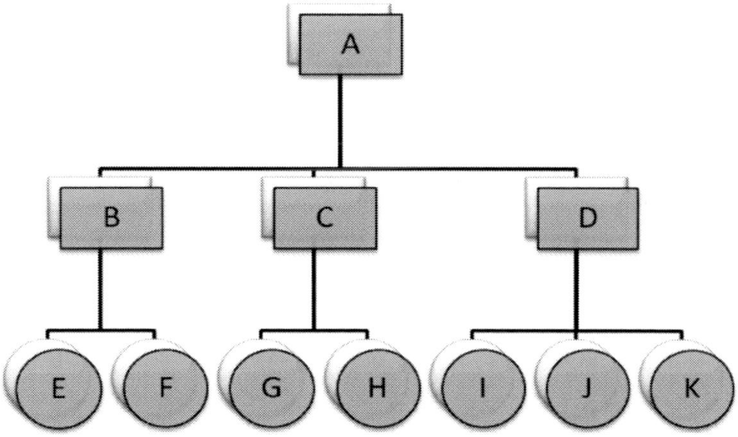

Figure 3. Fault tree for cargo-pumping system failure.

IOP Conf. Series: Materials Science and Engineering **95** (2015) 012074 doi:10.1088/1757-899X/95/1/012074

Figure 3 shows the developed model (fault tree) needed for solving the problem of finding the frequency of tanker cargo-pumping system failure during unloading. Values of the probabilities can be taken from the practice of exploitation of this and similar equipment.

Event "A" (Top event) – failure of cargo-pumping system, that leads to an increase of tanker unloading time.

For the cargo-pumping system failure occurring of one of the following three events is sufficient:

- Event "B" – reducing the cargo-pumping system supply because of stopping one of the cargo pumps.
- Event "C" – reducing the cargo-pumping system supply because of failure one of the auxiliary boilers.
- Event "D" – ship de-energization.

For the event "B" to occur one of the following three events is sufficient:

- Event "E" – cargo pump № 1 stopping.
- Event "F" – cargo pump № 2 stopping.
- Event "G" – cargo pump № 3 stopping.

For the event "C" to occur one of the following two events is sufficient:

- Event "H" – auxiliary boiler № 1 failure.
- Event "I" – auxiliary boiler № 2 failure.

For the event "D" to occur the following three events must occur simultaneously:

- Event "J" – diesel-generator № 1 failure.
- Event "K" – diesel-generator № 2 failure.
- Event "L" – diesel-generator № 3 failure.

For the event "E" to occur one of the following two events is sufficient:

- Event "M" – cargo pump № 1 failure.
- Event "N" – cargo pump № 1 drive failure.

For the event "F" to occur one of the following two events is sufficient:

- Event "O" – cargo pump № 2 failure.
- Event "Q" – cargo pump № 2 drive failure.

For the event "G" to occur one of the following two events is sufficient:

- Event "R" – cargo pump № 3 failure.
- Event "S" – cargo pump № 3 drive failure.

Analytically the expression (equation (3)) for condition of realization of the accident (Top event "A") takes the form [5]:

$$P(A) = P(B) + P(C) + P(D) = \{[(P(M) + P(N)] + [(P(O) + P(Q)] + [(P(R) + P(S)]\} + [P(H) + P(I)] + [P(I) \text{ x } P(J) \text{ x } P(K)]. \quad (1)$$

Advantages of FTA [1, 2, 3, 6]:

- easy to read and understand;
- can construct many combinations of failures;
- can manage the actions to minimize the risk;
- can manage the data for the risk assessment process.

Disadvantages of FTA:

- FTA involves thorough details for designing a system;
- significant training and experience is necessary to use this technique properly;
- in calculating the probability of an event in FTA there is uncertainty regarding the basic events; there is the possibility of errors when they are omitted causal links.

3. Conclusions

Fault Tree analysis is a technique that allows the existing situation analysis: analyse current needs, identification of key issues in their context, arrangement of issues according to the cause-effect relationship in the form of charts/tree, analyse quantitatively the systems and possible causes and ways of failure occurrence, analyse in the each stage of operation how can you determine the most signifiant failures can occur and which are the connections between causes, failures and importance of the different ways of fault tree analysis.
In the finally of study it will be obtained:
1) critical areas: in the case study 1-the accident (top event "A"), in the case study 2- failure of cargo-pumping system,
2) fault tree logical diagrams- Fault tree for electrical injury and Fault tree for cargo-pumping system failure and the probability of the Top event. On fault trees the failure probabilities tend to be small with very small error term and for this we must make an assumption for specialy events (e.g. joining a man on a conductive surface, a human body touching of grounded elements of the room).
3) the developed models (fault trees) needed for solving the problem of finding a necessary condition of human death from electrical domain for body inclusion in electric circuit with sufficient for death amperage or finding the frequency of tanker cargo-pumping system failure during unloading.
4) The analitical expressions for the failure probabilities which we have obtained can be used for the risk assessment analyses.

Acknowledgements
This article is a result of the project "Increasing quality in marine higher education institutions by improving the teaching syllabus according to International Convention STCW (Standards of Training, Certification and Watchkeeping for Seafarers) with Manila amendments" POSDRU/156/1.2/G/141534. This project in co funded by European Social Fund through The Sectorial Operational Programme for Human Resources Development 2007-2013, coordinated by Constanta Maritime University. The content of this scientific article does not reflect the official opinion of the European Union. Responsibility for the information and views expressed in the article lies entirely with the authors.

References
[1] Ericson C 1999 Fault Tree Analysis. A History *Proceedings of the 17th International System Safety Conference* **1(9)** pp 150-163
[2] Vesely W et al. 1981 *Fault Tree Handbook* NUREG-0492 Nuclear Regulatory Commission pp 33-54
[3] American Institute of Chemical Engineers 1989 Guidelines for Chemical Process Quantitative Risk Analysis pp 548-740
[4] http://www.academia.edu/6621646/M3_4c_Fault_Tree. Accesed:09/03/2015
[5] http://en.wikipedia.org/wiki/Fault_tree_analysis.Accesed: 26/02/2015
[6] http://psam12.org/proceedings/paper/paper_581_1.pdf. Accesed: 01/12/2014

IOP Conf. Series: Materials Science and Engineering **95** (2015) 012075 doi:10.1088/1757-899X/95/1/012075

An approach to the implementation of ISO 14000 at a shipping company

F V Panaitescu[1] **and M Panaitescu**[1]

[1] Constanta Maritime University, Faculty of Naval Electro-Mechanics, 104 Mircea cel Batran Street, 900663, Constanta, Romania

E-mail: viopanaitescu@yahoo.ro

Abstract. We present here a successful implementation of the Environmental Management System (EMS) in a shipping company. The EMS of the each company must be revised and expanded and new analytical requirements included in it. This new EMS concept is available to everyone on board and must be studied by every seafarer. All environmental requirements to be strictly followed all over the world. The main points of new EMS are: new open reporting system, environmental tag system (ETS) which introduce non re-usable numbered seals, pollution prevention equipment (OWS, OCM, Incinerator and Sewage Treatment Plant), the environmental defect reports system – extraordinary E/R operations and leakages log book.

1. Introduction

The implementation of ISO 14000 quality management system onboard ships is not required now by IMO (International Maritime Organization). However, in the last years, more and more shipping companies decides to introduce this certification, as a prove of the Company reliability. The respect for environment give to the ship transport company more credibility and class, more international recognition, being the mark of his power and competence.

On the other side, in the event that the company is found violate repeatedly the environmental rules imposed through MARPOL (Marine Pollution Convention), IMO can impose to the shipping company to introduce an ISO 14000 system. In a period of time of 1-2 Years, the Company remains under probation and need to respect strict environmental requirements. This can include extern audit from the Authority (generally, inspectors from the Class Register or Marine Coast Guard).

2. Methodology of the ISO 14000 implementation

The methodology proposed here includes specific points of application. This points constitutes the base of a minimum measures; they can be completed with others (depending also of the ship type: oil tanker, passenger, Ro-Ro, Chemical, Bulk Carrier, etc.). Also each point needs to support improvement in the implementation period and after. This improvement need to come from the crew members, especially from management level officers: Master and Chief Mate from the bridge team, first engineer and Chief engineer from the Engine team part.

The implementation of ISO 14000 quality management system in Company starts with a new and upgraded/expanded Environmental Management System (EMS). This EMS Manual need to be available on board for every seafarer and the seafarer is obliged to read and know this Manual.

Content from this work may be used under the terms of the Creative Commons Attribution 3.0 licence. Any further distribution of this work must maintain attribution to the author(s) and the title of the work, journal citation and DOI.
Published under licence by IOP Publishing Ltd

2.1. Open Reporting System (ORS)

This system allow every seafarer on board to report anonymously (or signed), any environmental non-compliances they are witness. Also this is an obligation, and failure to use and can be considered as legal violation. Masters must inform ship crew about ORS. Report can be done on paper form, introduced in the special locked box onboard ship, e-mail, tax free telephone or fax. The keys of the ORS box is kept at the office (no one on board do not have the key), papers are collected by the company representatives when inspect the ship. The Company will take no actions against those who use ORS, except in cases of intentional false reporting.

2.2. Environmental Tag System (ETS)

Consists in non re-usable numbered seals (figure 1.) that prevent unauthorized use of piping systems or valves. The seals are received by Master, who keep the Master's Spare Seals Inventory and maintains replacement seals in a secure location. Also the Master verifies if tags do have duplicate numbers.

Figure 1. Non re-usable seals.

The Chief Engineer:
- Open and keep up to date the Engine Room Seal Log and records the following:
1. All flanges and valves fitted with a seal (figure 2, figure 3);
2. Every time a seal is broken: the location of seal, date and number of seal removed, reason for removal, personnel involved; date, number of the new seal fitted, personnel involved;
3. Records in Oil Record Book (ORB) seal numbers of Oily Water Separator (OWS) and Oil Content Meter (OCM);
4. Numbers all flanges with seals starting from discharge of pump / equipment and ending at overboard valve;
5. Marks flanges and valves fitted with seals at copies of ship's piping plans;
6. Attaches the copies of piping plans at E/R Seal Log.

Also EMS will specify what flanges and pipes need to be sealed.

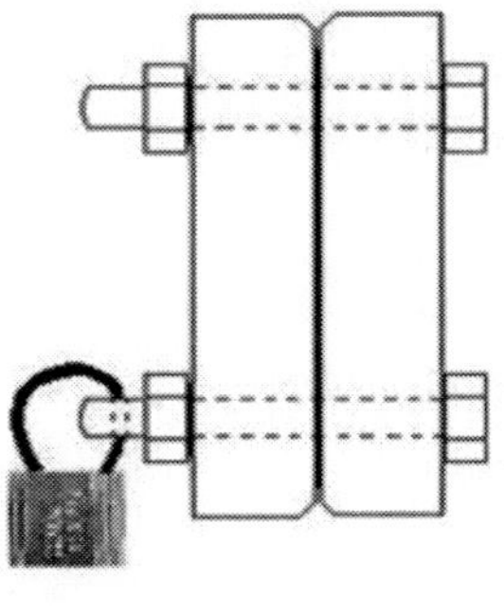

Figure 2. Seal on a flange.

Figure 3. Seal on a valve.

Will be accepted following reasons to break a seal: normal (planned) maintenance works, repair works (unplanned), discharge/testing of OWS, Register Class request.

2.3. Pollution prevention equipment

This equipment is from now very important for the ship and must be treated in consequence. Pollution prevention equipment is OWS including OCM, Incinerator and Sewage Treatment Plant. Always these equipments will be kept operational and properly maintained.

If any failure of this equipment: this will be noted in the Engine Log Book, in the Extraordinary Engine Room (E/R) Operations and in the Leakages Log Book. Second, an Environmental Defect Report will be completed. Also entries will be made in the ORB for OWS & OCM and for Incinerator.

All critical spares of these three environments will be on board, according to List of Critical Spares Related to Pollution Prevention Equipment (the Office recommendation is to double that stocks).

The Officers will give special attention for the testing schedule of OWS, OCM, and Incinerator: will be weekly operational tests, recorded in the ORB, including date/hour place and person in charge. Tests will be as follow:

The OWS: tested for at least one hour with outboard discharge valve closed; clean water should come out of the MARPOL (IMO) line,

The OCM: tested depending on the maker, either through a function that simulates a value higher than 15 ppm (e.g. Fellow Kogyo – Focas 1500C, 1800) or by the use of an object or liquid that activates optical sensor (e.g. Deckma – OMD-21, 2005),

Incinerator: no additional test required, when under routine operation at the testing date. (When the vessel is within port limits or areas where emission restrictions apply, test will be postponed).

Chief Engineer is from now the only person who will order to start the OWS; prior this, they will instructs personnel giving himself appropriate training.

Also instructions for OWS Operation will be posted near OWS and in the Engine Control Room (ECR): Operating Instructions for OWS and MARPOL Annex I & OWS Operation Policy.

Person who operates OWS, OCM and Incinerator need to fully review and understands posted operating instructions, and also understands environmental policies and the civil and criminal consequences in case of non-compliance.

2.4. Tank Soundings

Tank sounding give to the auditor and also to the Office precious information about proper application of ISO 14000 onboard.

The engineering officer in charge, records and sign the soundings daily. A Log must be completed for retained quantity in every designed Tank; the Tank Sounding Log must be handwritten and signed by the Chief Engineer. Entries of this log must be in correspondence with ORB entries.

2.5. Flexible hoses inventory

Every flexible hose on board shall be numbered (by fixing a numbered metal tag). The flexible hose inventory need to specify their storage location, diameter, made material, length of hose and usage. Flexible hoses inventory has two parts: one for deck department and one for engine department; the inventory is the responsibility of both Chief Officer and Chief Engineer.

2.6. Fleet Engineering Survey and Environmental Reporting

The Fleet Engineering Survey documents give to all employees the responsibility for their actions. These forms are completed by every engineering officer after signing-on onboard vessel:

-Declaration of Environmental Commitment;

-Declaration of Environmental Commitment – During Embarkation;

-Declaration of Environmental Commitment – Prior Signing off.

The Environmental Reporting Documents help the Environmental Department to upgrade the survey system and to recover problems appeared during implementation of the ISO 14000 standard.
They are :

- Environmental Weekly Report: contains dates of weekly testing of the OWS, OCM and Incinerator, the retained on board quantity of bilge water and sludge, and the maximum quantity of bilge water and sludge.
- Environmental Monthly Report: concerns all environmental operations / events of the last month. Include a Checklist.
- Environmental 4-Monthly Report: about the compliance with specific requirements of the EMS – Checklist included

Other items also are (in brief):

- Handover Notes Of Master, Chief Engineer And 2nd Engineer; - Crew Member Appraisals; - Ballast Water Management Plan; - Garbage Management Plan; - Sewage (Black Water) discharges; bilge water discharge; sludge discharge; - Air pollution – SOx control: Fuel Oil Management Plans; NOx control; - Vessel General Permit (VGP); - Shipboard Oil Pollution Emergency Plan (SOPEP); - Non Tank Vessel Response Plan (NT-VRP); - Liabilities of ship and shore personnel.

3. Methodology

Besides the upgrade of ship equipments required by the ISO 14000, also training of the personnel must be done. The training is conceived in two parts: first, a prior embarkation training and after, during voyage, periodically lectures and assessments.

All personnel is trained and evaluated on environmental issues prior embarkation at the Crewing Agency Office. Deck officers will receive a specific training about use of all pollution prevention equipment for deck operations. Engineer officers receive a upgraded training for operation and maintenance of all pollution prevention equipment (OWS, OCM and Incinerator) and the relevant environmental procedures and policies, as well as the operation of the Data Logger (where applicable). After embarkation personnel is trained and evaluated regularly on environmental onboard issues.

Also vendors, technicians and other non-crew members need to be Informed for company environmental policies when onboard; they must sign the Environmental Familiarization Report for the Non Crew Form.

All personnel receive first a short 4 hour to 6 hour lecture familiarization with the EMS requirements; this lecture use PowerPoint Presentation and video material. After familiarization, they complete three CBTs program. Each program has a training part and a assessment part. Assessment results are recorded for every employee. Minimum requirements are imposed: 60% for ordinary seaman and 70% for officers.

First CBT program used is an ISO14001 Environmental Management made by Seagull Marine Software Developer. The software includes four lessons: Introduction, The Earth and the Environment, Environmental threats and Protection of the environment. The CBT ends with an Assessment: 16 questions, each with 3 forgiven responses. The CBT return a Training report wich include: trainee name, date completed, total time used, completed lecture for each lesson and total, assessment results (figure 4).

The Second CBT used is made by TOTAL QUALITY CONSULTANTS Software Training. Also the CBT include a first lecture part and end with an 24 questions assessment. The software offer full information about the trainee results and is a valuable document for Office Environmental Department and also for external auditor (figure 5). You can see in the Attestation (figure 5) the eleven chapters included in the training session.

Third training CBT is online, created by SeaLearn; user can connect to the learning session through internet connection from home, using a username/password. Different modules are available in the Environmental Protection Series:

- MARPOL 73/78 Annex I - Prevention of Pollution by Oil: About MARPOL, Waste Oil Management - Oily Water Separators, Keeping Records in the Oil Record Book, Special Requirements for Tankers, Shipboard Oil Pollution Emergency Plan (SOPEP), Best Practices;

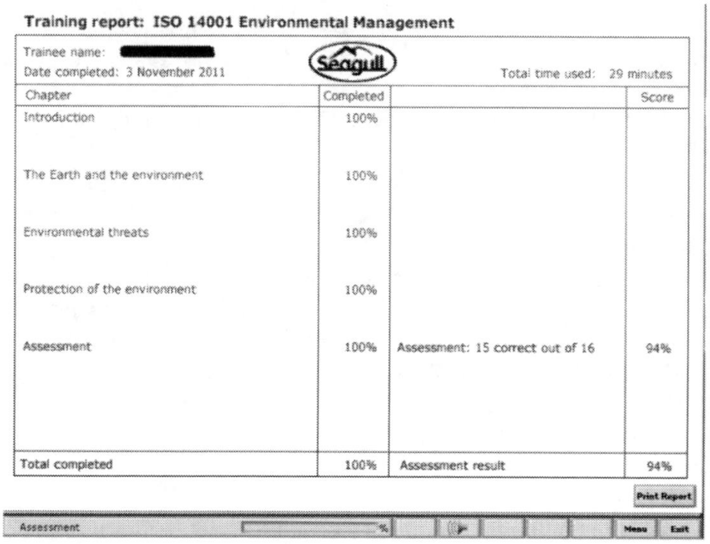

Training report: ISO 14001 Environmental Management

Trainee name: [redacted]
Date completed: 3 November 2011
Total time used: 29 minutes

Chapter	Completed		Score
Introduction	100%		
The Earth and the environment	100%		
Environmental threats	100%		
Protection of the environment	100%		
Assessment	100%	Assessment: 15 correct out of 16	94%
Total completed	100%	Assessment result	94%

Figure 4. Training report.

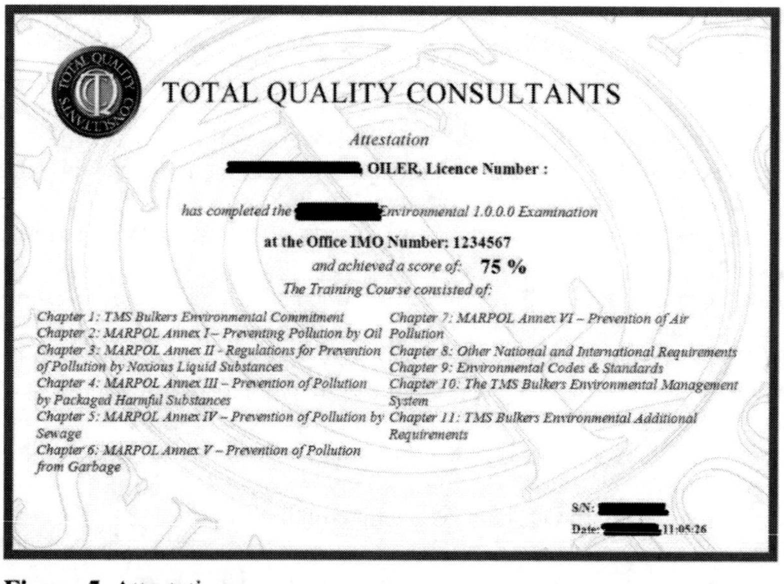

TOTAL QUALITY CONSULTANTS

Attestation

[redacted], **OILER, Licence Number :**

has completed the [redacted] *Environmental 1.0.0.0 Examination*

at the Office IMO Number: 1234567

and achieved a score of: **75 %**

The Training Course consisted of:

Chapter 1: TMS Bulkers Environmental Commitment
Chapter 2: MARPOL Annex I – Preventing Pollution by Oil
Chapter 3: MARPOL Annex II - Regulations for Prevention of Pollution by Noxious Liquid Substances
Chapter 4: MARPOL Annex III – Prevention of Pollution by Packaged Harmful Substances
Chapter 5: MARPOL Annex IV – Prevention of Pollution by Sewage
Chapter 6: MARPOL Annex V – Prevention of Pollution from Garbage
Chapter 7: MARPOL Annex VI – Prevention of Air Pollution
Chapter 8: Other National and International Requirements
Chapter 9: Environmental Codes & Standards
Chapter 10: The TMS Bulkers Environmental Management System
Chapter 11: TMS Bulkers Environmental Additional Requirements

S/N: [redacted]
Date: [redacted] **11:05:26**

Figure 5. Attestation.

- MARPOL 73/78 Annex IV - Prevention of Pollution by Sewage from Ships;
- MARPOL 73/78 Annex V - Prevention of Pollution by Garbage from Ships;
- Vessel General Permit (VGP).

Test start with a lecture session and finish with an Assessment. Assessment results for every seafarer can be accessed online; the program generates a LEARNING PROGRESS REPORT (figure 6), a LEARNING ASSESSMENT HISTORY REPORT and issue a CERTIFICATE.

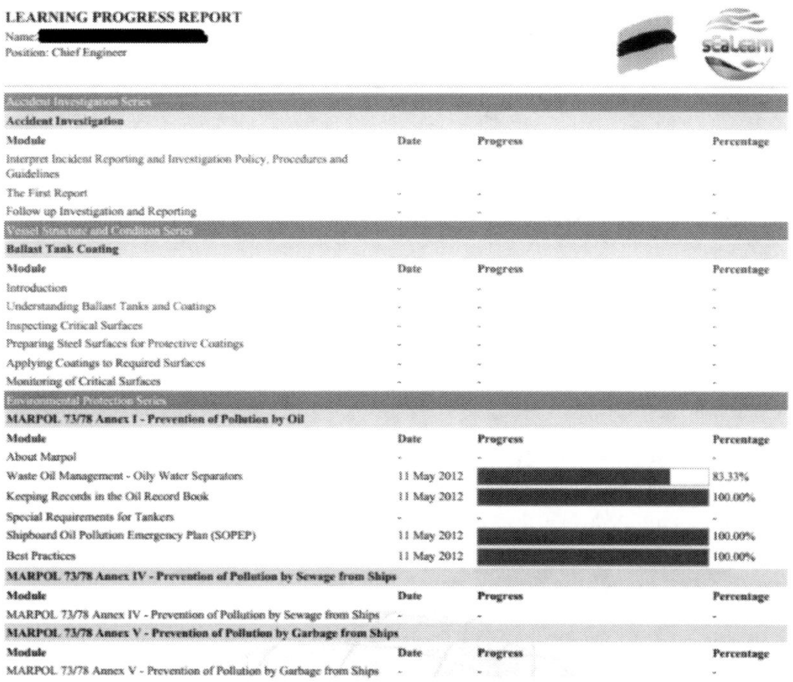

LEARNING PROGRESS REPORT
Name: [redacted]
Position: Chief Engineer

Accident Investigation Series			
Accident Investigation			
Module	**Date**	**Progress**	**Percentage**
Interpret Incident Reporting and Investigation Policy, Procedures and Guidelines	-	-	-
The First Report	-	-	-
Follow up Investigation and Reporting	-	-	-
Vessel Structure and Condition Series			
Ballast Tank Coating			
Module	**Date**	**Progress**	**Percentage**
Introduction	-	-	-
Understanding Ballast Tanks and Coatings	-	-	-
Inspecting Critical Surfaces	-	-	-
Preparing Steel Surfaces for Protective Coatings	-	-	-
Applying Coatings to Required Surfaces	-	-	-
Monitoring of Critical Surfaces	-	-	-
Environmental Protection Series			
MARPOL 73/78 Annex I - Prevention of Pollution by Oil			
Module	**Date**	**Progress**	**Percentage**
About Marpol	-	-	-
Waste Oil Management - Oily Water Separators	11 May 2012		83.33%
Keeping Records in the Oil Record Book	11 May 2012		100.00%
Special Requirements for Tankers	-	-	-
Shipboard Oil Pollution Emergency Plan (SOPEP)	11 May 2012		100.00%
Best Practices	11 May 2012		100.00%
MARPOL 73/78 Annex IV - Prevention of Pollution by Sewage from Ships			
Module	**Date**	**Progress**	**Percentage**
MARPOL 73/78 Annex IV - Prevention of Pollution by Sewage from Ships	-	-	-
MARPOL 73/78 Annex V - Prevention of Pollution by Garbage from Ships			
Module	**Date**	**Progress**	**Percentage**
MARPOL 73/78 Annex V - Prevention of Pollution by Garbage from Ships	-	-	-

Figure 6. Learning progress report.

4. Conclusions

Training of the crew personnel is a key point for a successful implementation of a ISO14000 Environmental Management onboard ships. Prior embarkation lectures/assessments and periodical training during voyage, is the best practice for maintaining a high level preparation. Use of CBT (Computer Based Training) off and online, was founded to offer the most efficient and cheap solution.

Acknowledgements

This article is a result of the project "Increasing quality in marine higher education institutions by improving the teaching syllabus according to International Convention STCW (Standards of Training, Certification and Watchkeeping for Seafarers) with Manila amendments" POSDRU/156/1.2/G/141534. This project in co funded by European Social Fund through The Sectorial Operational Programme for Human Resources Development 2007-2013, coordinated by Constanta Maritime University.

The content of this scientific article does not reflect the official opinion of the European Union.Responsibility for the information and views expressed in the article lies entirely with the authors.

References

[1] Christopoulou V and Margelis G 2012 *EMS implementation* (Athens)
[2] Panaitescu F V and Panaitescu M 2006 *Using Deck and Engine Room Simulators in the practical training of the Maritime University of Constanta graduate students and officers* (Performantica Publishing)
[3] Jurian M, Panaitescu V, Panaitescu M and Dima S 2005 The importance of educational factor to assure the safe and security on the sea *3rd Balcan Region Conference on Engineering Education* (Sibiu)

IOP Conf. Series: Materials Science and Engineering **95** (2015) 012076 doi:10.1088/1757-899X/95/1/012076

Analysis for the amplitude oscillatory movements of the ship in response to the incidence wave

M G Chiţu[1], R Zăgan[2] and E Manea[3]

[1] Ovidius University of Constanta, Blvd. Mamaia, No. 124, 900527, Constanta, Romania
[2] Maritime University of Constanta, Mircea cel Bătrân Street, No. 104, Constanta, Romania
[3] Constanta Shipyard - Greece Representative Office, Romania

E-mail: amgchitu@gmail.com

Abstract. Event of major accident navigation near offshore drilling rigs remains unacceptably high, known as the complications arising from the problematic of the general motions of the ship sailing under real sea. Dynamic positioning system is an effective instrument used on board of the ships operating in the extraction of oil and gas in the continental shelf of the seas and oceans, being essential that the personnel on board of the vessel can maintain position and operating point or imposed on a route with high precision. By the adoption of a strict safety in terms of handling and positioning of the vessel in the vicinity of the drilling platform, the risk of accidents can be reduced to a minimum. Possibilities in anticipation amplitudes of the oscillatory movements of the ships navigating in real sea, is a challenge for naval architects and OCTOPUS software is a tool used increasingly more in this respect, complementing navigational facilities offered by dynamic positioning systems.
This paper presents a study on the amplitudes of the oscillations categories of supply vessels in severe hydro meteorological conditions of navigation. The study provides information on the RAO (Response Amplitude Operator) response operator of the ship, for the amplitude of the roll movements, in some incident wave systems, interpreted using the energy spectrum Jonswap and whose characteristics are known (significant height of the wave, wave period, pulsation of the wave). Ship responses are analyzed according to different positioning of the ship in relation to the wave front (incident angle ranging from 10 to 10 degree from 0 to 180), highlighting the value of the ship roll motion amplitude. For the study, was used, as a tool for modeling and simulation, the features offered by OCTOPUS software that allows the study of the computerized behavior of the ship on the waves, in the real conditions of navigation. Program library was used for both the vessel itself and navigation modeling environment, for regular waves as well for the irregular waves which was modeled using Jonswap energy spectrum.

1. Introduction

Supply vessels do as their name implies; they are used to transport some materials, equipments and personnel to the offshore oil drilling platforms and of them to the land, as follows [1]:

Content from this work may be used under the terms of the Creative Commons Attribution 3.0 licence. Any further distribution of this work must maintain attribution to the author(s) and the title of the work, journal citation and DOI.
Published under licence by IOP Publishing Ltd

-goods transported to the work platform is intended for drilling platforms activities and consists of: solid cargo - drill pipes, equipments (containerized or not), anchors, chains etc.; bulk cargo - fuel, drinking water, pulverized cement, drilling mud, brine etc.
-goods transported from the platform to the shore are intended for recycling or scrapping and consists of: outworn or out of service equipments; waste oil and sanitation, various used chemicals.

Additionally, supply vessels can carry specialized personnel during crew exchanges. Also they are used for: supply platforms and other offshore installations; platforms towing and mooring; underwater construction and installation; repair and maintenance of underwater installations; mounting of submarine cables; assistance for various offshore activities etc. In general, these vessels are of medium size, with lengths ranging from 60-70 meters to 120-150 meters and are built considering certain characteristics and specific facilities. The design of supply vessels respect that they are often dedicated to operations in a particular area. A typical aspect of the design of a supply vessel is the forward location of the deck house with the large open space aft. Exposed aft deck is considered as working deck and therefore need to be more extensive and free of obstacles. According to [1], loads due to the cargo (and use for the strength calculus of the deck) are at least 1.5 tons per square meter plus 80% of design pressure of the wave. Usually, the total load taken into account is 5-10 tons per square meter. As offshore support vessels must operate even in rough weather and sea, the bow forms are optimized to have good water resistance and seakeeping both into calm water and high waves. Exposed bow deck extends over at least 2-3 superstructure storey above the main deck and forecastle of the modern ships are protected to avoid taking water on the open deck. Since these vessels operate in rough sea and near offshore oil drilling platforms (risk of collision and crash) they are provided with elements designed to increase the body strength. Offshore support vessels have good manoeuvrability thanks to powerful propulsion equipments and systems. Usually, propulsion is with nozzles propellers or azimuth thrusters. The ship steering system must be able to bring the rudder from 35 degrees port to 30 degrees starboard (and vice versa) in a maximum of 20 seconds at top speed headway ship. Modern ships are equipped with aft and bow thruster which must be able to keep fixed point positioning the ship and keeping this position (or by using anchor systems or dynamic positioning systems). Study on ship movements done while navigating on the sea waves, is of particular importance, especially for ships in the fleet serving the offshore oil and gas industry, and justified interests of authors of this paper for the topic. Among the issues that concern naval architects and engineers regarding study of the general movements of the ship, is noted [2, 3]:
- deterioration or loss of buoyancy, which may lead to sinking (unless tightness conditions are not ensured) as a consequence of deck flooding at the entrance of a board, bow or aft of the ship in wave (in terms of harsh regimes of roll and/or pitch);
- loss of transverse stability, which may lead to the overthrow of the ship caused by: a dangerous inclinations on board determined by ship gyration in terms of a rough sea; dangerous board slopes caused by displacement of the wrong mooring transported cargoes; hard roll oscillations caused by wave action at resonance;
- deterioration due to the ship steer caused by: decrease of maneuverability under unfavorable hydrodynamic wave action on the rudder; loss of stability on the way through the phenomenon of drift generated by unfavorable actions of the wind and sea currents when, for various reasons, the steering system can no longer function;
- reducing the march qualities due to: worsening of propulsion system function, throughout the entire driveline (propellers is out of the water, blade empty revolve and, when reinforce to the water, they are overburdened and cause shocks in the system); vessel speed reduction due to drag increase and malfunction of propellers in rough sea conditions; reducing autonomy of the ship by increasing the energy consumption;
- loss the nautical quality of small amplitude oscillations, on the strong impact between the hull and waves, with the consequences: the emergence of seasickness (accompanied by discomfort of passengers and crew on board); occurrence of vibration due to the effects of slamming (hitting the bow

on wave) whipping (hitting stern on wave) and slams (repeated shocks); worsening the function of equipments and systems onboard caused by dynamic actions of water on deck;
- damage of elasto-plastic behavior of hull structures due to: general and additional local loads; weakening of general strength by fatigue; the motions and accelerations on the work deck; the vertical relative motions between the waves and the deck edges (deck wetness); the effect of shipped water in pipes transported on the open deck of supply vessels.

2. Theoretical considerations

The subject of ship movements on the waves modeled as rigid body with six freedom degrees, has been addressed in the literature by various authors. Those who have the attention of authors are: Bhattacharya (1978) "Dinamics of marine vehicles" [4]; Chakrabarti (1987) "Hydrodynamics of offshore structures" [5]; Edward (1989) "Principles of naval architecture" [6]; Fossen (1994) "Guidance and Control of Ocean Vehicles" [7] and (1995) "Nonlinear modeling of marine vehicles in 6 degree of freedom" [8]; Domnişoru (2001) "Ship Dynamics. Oscillations and Vibrations of the Ship Hull" [9].

The general equation of motion of the ship, in a vector form, referred to the moving reference system is well known and is:

$$M\dot{v}_r + C(v_r)v_r + D(v_r)v_r + g(\eta) = \tau_M + \tau \qquad (1)$$

where:

$v_r = v - v_C$, relative velocity vector; $\dot{v}_r$, relative acceleration vector;

$v = [v_1^T \quad v_1^T]^T = [u \quad v \quad w \quad p \quad q \quad r]^T$, overall velocity vector of the ship;

$v_c = [v_{2c}^T \quad v_{2c}^T]^T = [u_c \quad v_c \quad w_c \quad 0 \quad 0 \quad 0]^T$, overall velocity vector of marine currents;

$\eta = [\eta_1^T \quad \eta_2^T]^T = [x \quad y \quad z \quad \varphi \quad \theta \quad \psi]^T$, position vector of an arbitrary point belonging to the vessel;

$g(\eta)$, generalized vector of restored forces and moments;

τ, generalized vector of the disruptive forces and moments due to operations control and manoeuvre executed with the propulsion and steering systems;

τ_T, generalized vector of the disruptive forces and moments generated by the currents and waves on the wetted surface of the hull respectively by the action of the wind on the sail area of the vessel;

M, the matrix of the ship mass inertia and additional water masses;

$C(v_r)$, complementary matrix for the ship motions and additional water masses;

$D(v_r)$, the damping matrix determined by the dynamic action of the water on the ship surface, caused by the moving of the vessel.

The general equation of motion of the ship, in a vector form, referred to the fixed reference system is also well known:

$$M_\eta(\eta)\ddot{\eta} - C_\eta(v_r\eta)\dot{\eta} + D_\eta(v_r\eta)\dot{\eta} + g_\eta(\eta) = \tau_{M\eta}(\eta) + \tau_\eta(\eta) \qquad (2)$$

3. Results and Discussions

For the case study was chosen a platform supply vessel (often abbreviated as PSV). The ship has a hull without bulbous in bow and aft in mirror. The principal dimensions are: length m, width m, draft m; buoyancy 12537 cubic meters. Service speed, knots (maximum speed knots). Propulsion is made by two main azimuth thrusters aft mounted, plus other three azimuth thrusters, two on sides and one retractable on bow, with power 880 kW each. The vessel is equipped with a diesel-electric propulsion system with four generator units and two main engines Wärtsilä 6L34DF type with nominal power

P=2610 kW at 720 rotation per minute. The study was conducted for hydrometeorological conditions of the North Atlantic.

For the case study developed in this paper, the evolution of the most probable extreme amplitudes of roll oscillations, obtained for speeds between 0 and 15.6 knots and encountering angles between 0 and 360, is illustrated in figure1. This summative graph was obtained by superimposing detailed schedules for each rate included in the study (from 0 knots - standing ship up to the maximum speed of the ship, 15.6 knots) finding, obviously, increased range of roll motion with increasing travel speed of the ship.

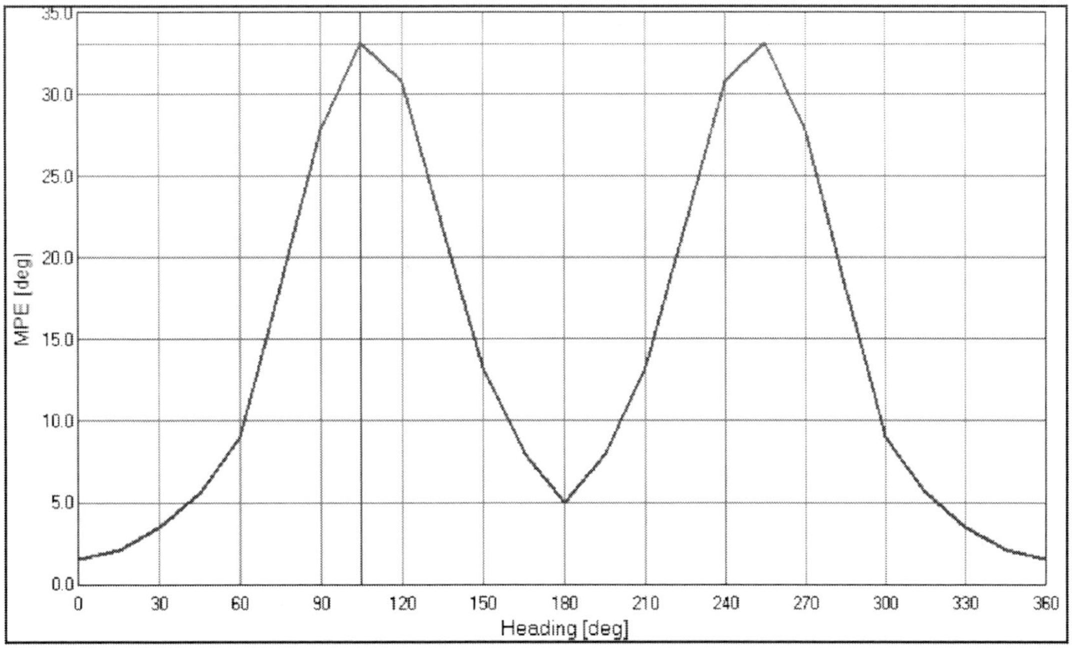

Figure1. Envelope of roll motion most probable amplitude (speeds 0; 3.9; 7.8; 11.7; 15.6 knots).

In figure 2 was exemplified the results for three speeds of the ship (5; 9.5 and 15 knots). Significant wave heights analyzed varies from 1 meter to 13.6 meters and are indicated in the figure legend.

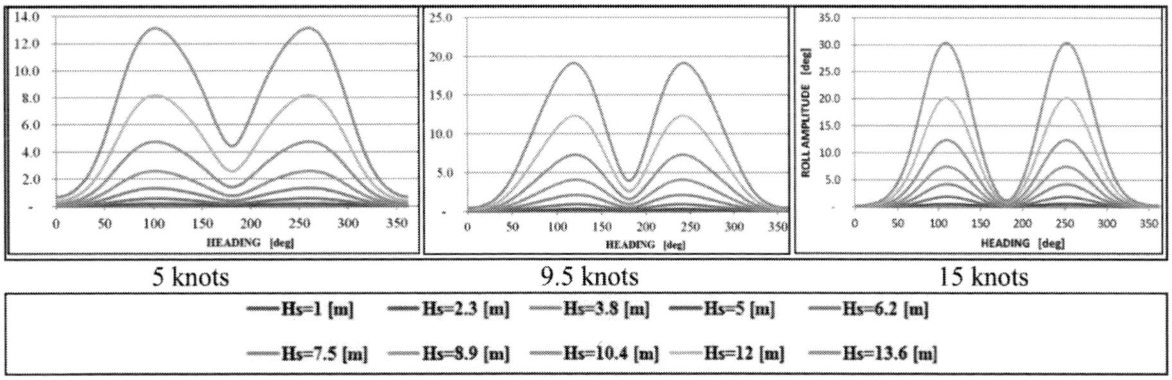

Figure 2. Roll motion most probable amplitude (speeds 5; 9.5; 15 knots).

For the analyzed navigation conditions, it is was found that the maximum values of the amplitudes of the roll motion (most dangerous for the stability of the ship) are reached for encountering angle of

105 degrees (or 255 degrees) and are, in relation to the speed of the ship, those indicated in Table 1. The maximum values for roll motion amplitudes illustrated in table 1 are not recommended for navigation and must be avoided. The absolute maximum value of 30.1 degree was found for significant wave heights of 13.6 meters and for ship speed of 15.6 knots.

Table 1. Roll amplitude in respect with ship speed.

Ship speed [knots]	5	9.5	15
Roll amplitude [deg]	13	18.5	30.1

The study of this paper show more results than those illustrated in figure 2 and table 1, being able to specify the roll amplitude for any vessel speed between 0 and 15.6 knots and encountering angles of the wave between 0 and 360 degrees with a sampling step of 1 degree. In Figure3 is exemplified aspect of the response amplitude spectra (RAO) for the roll for motion, for the five speeds of the ship considered in the study (0; 3.9; 7.8; 11.7 respectively 15.6 knots) and calculated for significant height of waves, Hs = 2.1 meters and for zero crossing period, Tz = 4.42 sec.

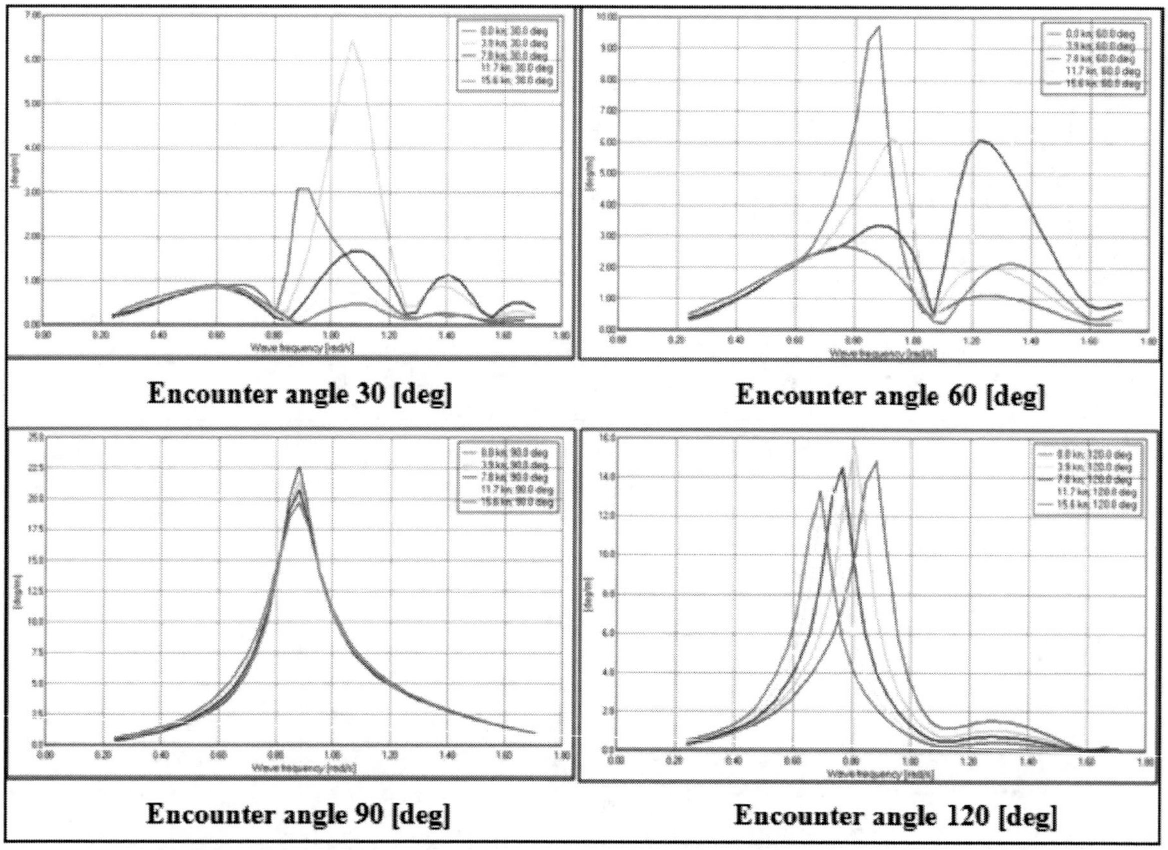

Figure 3. Roll motion RAO in respect with ship speed and encounter angle.

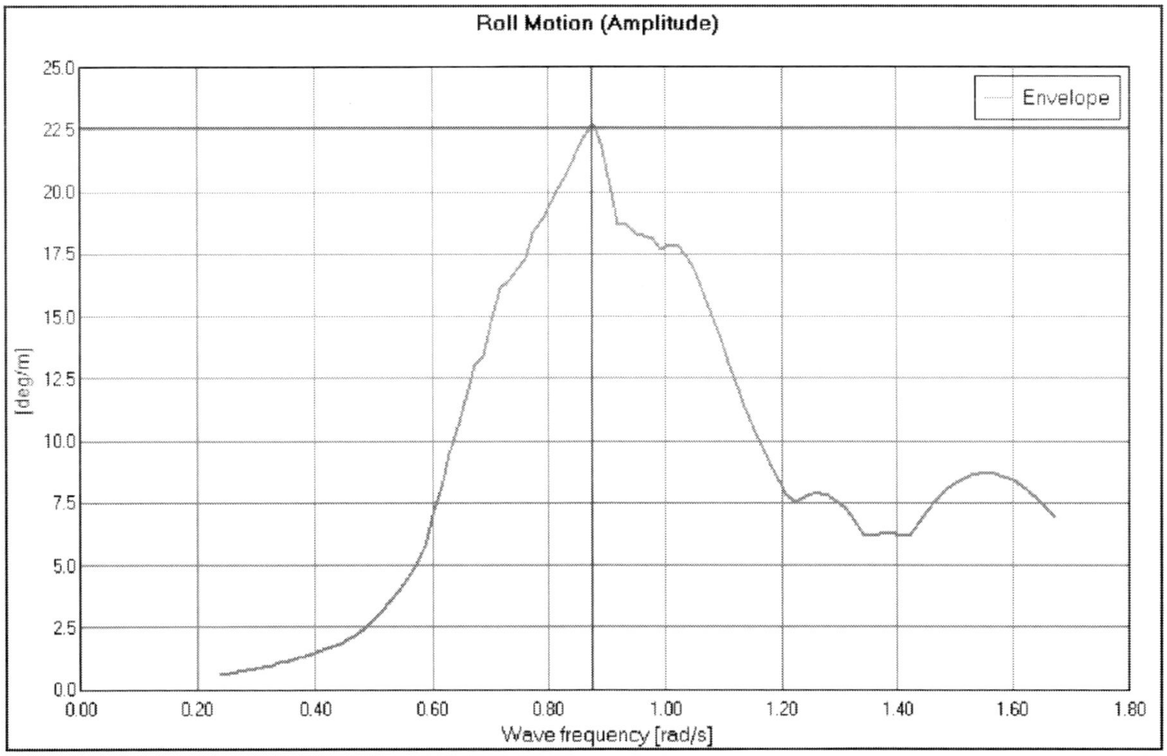

Figure 4. Envelope of roll motion RAO in respect with wave frequency.

4. Conclusions

World-leading shipping companies have used OCTOPUS onboard ships since 2003 for route planning and speed optimization, heading and fuel consumption in every weather condition. OCTOPUS software combines wave measurements, weather forecasts, and navigation data like speed, course, and the voyage plan, with ship characteristics, loading conditions and motion sensor measurements. [10]. One aim of this paper is to demonstrate the usefulness of OCTOPUS software as real-time analysis tool for ship movements (considered as rigid body with six freedom degrees) under various navigation conditions. OCTOPUS software is a powerful tool used in navigation to handling the ship in waves, together with other devices and facilities like dynamic positioning systems heave or roll compensator devices. OCTOPUS-onboard are use in container, LNG, offshore and many others shipping activities; the students of maritime universities should be familiar with it.

The case study presented in this paper aims is to highlight the usefulness of this program: for research studies in the field of complex interactions between body and navigation environment; for navigation as a tool for the efficient operation of vessels with functional features (work near drilling platforms, underwater work assistance, anchor handling, oversized pieces transport by sea etc.); for future navigation officers proper training.

The given limitations on enlargement of the paper resulted in partial presentation of the obtained results of the study. Those who may be interested to access the full case study results must contact authors.

5. References

[1] Ionaş O 2014 Technical ships *Galaţi University Press*

[2] Chiţu M G 2003 Contributions to the Ship Oscillations on the Waves Ph.D. Thesis (Politehnica University of Bucharest)

[3] Journée J M J and Massie W W 2001 *Offshore Hydromecahanics* **1** (Delft University of Technology)
[4] Bhattacharyya R 1978 *Dynamics of marine vehicles* (NewYork: John Wiley & Sons)
[5] Chakrabarti S K 1987 *Hydrodynamics of offshore structures* (Southampton: Computational Mechanics Publications)
[6] Edward V L 1994 Principles of Naval Architecture *SNAME* (New Jersey) **1, 2, 3**
[7] Fossen T 1994 *Guidance and Control of Ocean Vehicles* (University of Trondhaim Norway: John Wiley&Sons)
[8] Fossen T and Fjellstad O E 1995 Nonlinear modelling of marine vehicles in 6 degree of freedom *Journal of mathematical modeling of systems*
[9] Domnişoru L 2001 Oscillations and Vibrations of the Ship Hull *Ship Dynamics* (Bucharest Romania: Tehnica Publishing House)
[10] *** 2010 OCTOPUS Office 6 User Manual, Amarcon B V.
[11] Chiţu M G and Zăgan R 2009 Comparative study of dynamic nautical features of turning computer asist and sea trial *International Journal of Modern Manufacturing Technologies* **I(1)** pp 21-24
[12] Maier V 1997 General loads in modern naval architecture (Bucharest Romania: Tehnica Publishing House) **1, 2, 3**
[13] Voinea R 1989 *Introduction to Solid Mechanics with Engineering Applications* (Bucharest Romania: Academia Publishing House)

IOP Conf. Series: Materials Science and Engineering **95** (2015) 012077 doi:10.1088/1757-899X/95/1/012077

On computing special functions in marine engineering

E Constantinescu[1] **and M Bogdan**[1]

[1] Constanta Maritime University, Faculty of Naval Electro-Mechanics, 104 Mircea cel Batran Street, 900663, Constanta, Romania

E-mail: mirelcon@imc.ro

Abstract. Important modeling applications in marine engineering conduct us to a special class of solutions for difficult differential equations with variable coefficients. In order to be able to solve and implement such models (in wave theory, in acoustics, in hydrodynamics, in electromagnetic waves, but also in many other engineering fields), it is necessary to compute so called special functions: Bessel functions, modified Bessel functions, spherical Bessel functions, Hankel functions. The aim of this paper is to develop numerical solutions in Matlab for the above mentioned special functions. Taking into account the main properties for Bessel and modified Bessel functions, we shortly present analytically solutions (where possible) in the form of series. Especially it is studied the behavior of these special functions using Matlab facilities: numerical solutions and plotting. Finally, it will be compared the behavior of the special functions and point out other directions for investigating properties of Bessel and spherical Bessel functions. The asymptotic forms of Bessel functions and modified Bessel functions allow determination of important properties of these functions. The modified Bessel functions tend to look more like decaying and growing exponentials.

1. Introduction

Solutions of many engineering and mathematical physics problems are connected with so called special functions. One of these functions, Bessel and modified Bessel functions, are especially important for a lot of models from acoustics, wave propagation, heat conduction theory, aeronautics, hydrodynamics, mechanics of solids, elasticity, atomic and nuclear physics. Some complicated models in marine engineering related with wave propagation theory, vibration, hydrodynamics, heat conduction or elasticity could be solved using Bessel functions.

Differential equations with variable coefficients cannot be solved using the same techniques like in differential equations with constant coefficients (by algebraic methods obtaining solutions as elementary functions). In such cases, it is searched for solutions in form of infinite series (method of Frobenius, method of Picard). The Bessel's equations are of this type. In engineering, the Bessel's equations are always to be expected when partial differential equations are used in the study of configurations possessing cylindrical symmetry. We also encountered Bessel functions when we solve Laplace equation in cylindrical coordinates.

In mathematics, Bessel functions, denoted as $J_n(x)$, are canonical solutions Bessel's differential equation [1]:

Content from this work may be used under the terms of the Creative Commons Attribution 3.0 licence. Any further distribution of this work must maintain attribution to the author(s) and the title of the work, journal citation and DOI.
Published under licence by IOP Publishing Ltd

$$x^2\frac{d^2y}{dx^2}+x\frac{dy}{dx}+\left(x^2-n^2\right)y=0 \tag{1}$$

where n is called the order of the Bessel function and represents a real or complex number. The most used case is for integer n. For example, the solutions of Laplace's equation in cylindrical coordinates are Bessel functions of integer order, often called cylinder functions. Other very important case is for n half-integer for spherical Bessel functions as solutions of Helmholtz equation transformed in spherical coordinates. It's important to observe that simply replace x with λx will obtain a more general parameter form of Bessel's equations [1]:

$$x^2\frac{d^2y}{dx^2}+x\frac{dy}{dx}+\left(\lambda^2x^2-n^2\right)y=0 \tag{2}$$

In practical situations, when modeling real systems, it's difficult to recognize the form of the equation and then develop Bessel functions as solutions.

2. Bessel functions

Using the method of Frobenius, it could be searched Bessel's differential equation solution in the form $y(x)=\sum_{m=0}^{\infty}a_m x^{m+r}$; after substitution in equation (1), by identification, the indicial equation becomes $(r+n)(r-n)=0$, so $r_1=n$, $r_2=-n$. In the first case, it can be obtained the solution [2]

$$J_n(x)=\sum_{k=0}^{\infty}(-1)^k\cdot\left(\frac{x}{2}\right)^{n+2k}\cdot\frac{1}{k!\Gamma(n+k+1)} \quad n\geq 0, \tag{3}$$

called Bessel function of the first kind, order n (this series converges for all values of x). Let's mention that $\Gamma(n)$ is the Gamma function (the generalized factorial function).

Taking into account Gamma function properties, if n are positive integer or zero, then $\Gamma(n+k+1)=(n+k)!$ and the formula for Bessel function of order n becomes [3]:

$$J_n(x)=\sum_{k=0}^{\infty}(-1)^k\cdot\left(\frac{x}{2}\right)^{n+2k}\cdot\frac{1}{k!(n+k)!} \quad n=0,1,2,... \tag{4}$$

The behavior of Bessel functions could be seen in the figure 1 plotted using Matlab:

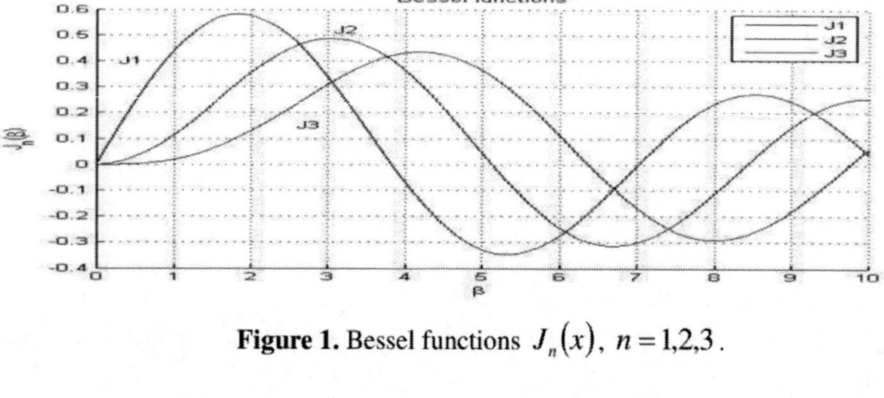

Figure 1. Bessel functions $J_n(x)$, $n=1,2,3$.

It can be observed that the functions are bounded at 0. These graphs look like the graphs of the functions sine or cosine that decompose proportionally to $\frac{1}{\sqrt{x}}$, which could be explained if it is taken into account the asymptotic forms for positive n of Bessel functions [5]. $J_0(x)$, $J_1(x)$ and $J_2(x)$ have no complex roots, but each of them has an infinite number of distinct roots which separate each other.

In the second case, by replacing n with $-n$, can be found the relation $J_{-n}(x)=\sum_{k=0}^{\infty}(-1)^k\cdot\left(\frac{x}{2}\right)^{-n+2k}\cdot\frac{1}{k!\Gamma(-n+k+1)}$. The last function is called Bessel function of negative order n and let observe that also this series converges for all values of x.

The most used series are for n = 0, 1, 2. According to [2], it can be mentioned interesting properties of these functions:

$$\frac{d}{dx}J_0(x)=-J_1(x),\ \frac{d}{dx}\{xJ_1(x)\}=xJ_0(x) \tag{5}$$

For n not an integer, Bessel functions J_n and J_{-n} are linearly independent; therefore, $y(x)=c_0J_n(x)+c_1J_{-n}(x)$ will represent the general solution of Bessel's differential equation.

Bessel functions J_n and J_{-n} are important for engineering applications for $n=\frac{1}{2}$ and $n=-\frac{1}{2}$. It is not difficult to obtain $J_{\frac{1}{2}}(x)=\sqrt{\frac{2}{\pi x}}\sin x$ and $J_{-\frac{1}{2}}(x)=\sqrt{\frac{2}{\pi x}}\cos x$. Using the plots (figure 2) of these two functions, it is easy to observe that one is bounded at 0 and the other not. These means that the functions $J_{\frac{1}{2}}(x)$ and $J_{-\frac{1}{2}}(x)$ are linear independent.

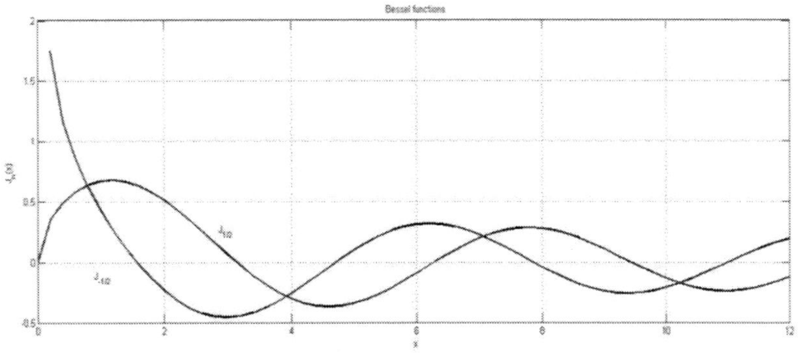

Figure 2. Bessel functions $J_n(x)$, $n=-1/2, 1/2$.

In the case $n\in\mathrm{Z}$, using the infinite series representations, it can be obtained the relation: $J_{-n}(x)=(-1)^nJ_n(x)$ [3]. It is necessary to mention that Gamma function exist only for positive arguments. Because the functions J_n and J_{-n} are not linearly independent, it's necessary to find another second linearly independent solution for Bessel's differential equation.

The following recurrence formulas can be proven:

$$xJ_n'(x)=nJ_n(x)-xJ_{n+1}(x),\ xJ_n'(x)=-nJ_n(x)+xJ_{n-1}(x),\ 2\frac{n}{x}J_n(x)=J_{n+1}(x)+J_{n-1}(x).$$

The function [1]:

$$Y_n(x) = \frac{\cos(n\pi)J_n(x) - J_{-n}(x)}{\sin n\pi}, \tag{6}$$

it is known as Bessel functions of the second kind. It's easy to observe that in origin the functions are infinite. Obviously, $J_n(x)$ and $Y_n(x)$ are linearly independent if n is not an integer. The general solution of Bessel's differential equation is written as: $y(x) = AJ_n(x) + BY_n(x)$, where usually the constants are determined from boundary conditions of the equation.

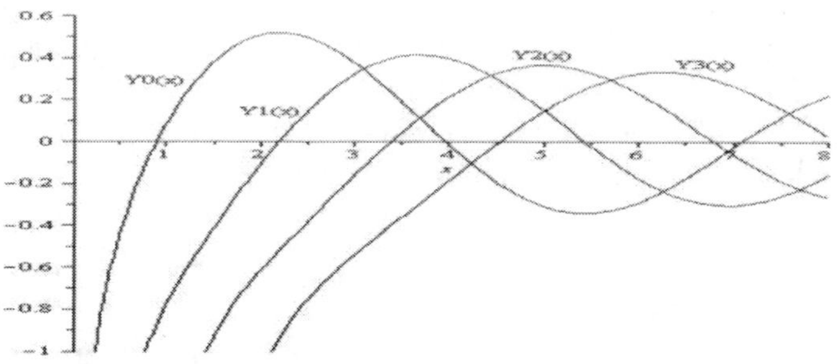

Figure 3. Bessel functions $Y_n(x)$, $n = 0,1,2,3$.

It is easy to verify that the Bessel function of second kind satisfy the relation: $Y_{-m}(x) = (-1)^m Y_m(x)$.

The examination of the surface $z = J_n(x)$ show the behavior of these function when x and n varies continuously. (figure 4, for $0 \le n \le 10$, $0 \le x \le 10$).

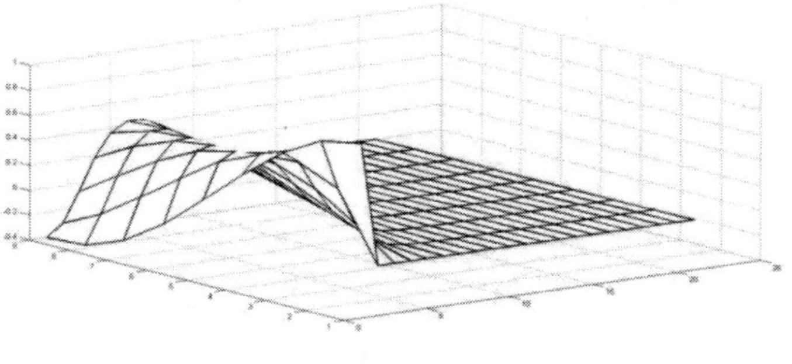

Figure 4. Surface of $z = J_n(x)$, $0 \le n \le 10$, $0 \le x \le 10$.

For $n = 0$, the function $J_0(x)$ equals 1 when x equals 0.

If $n > 0$, $J_n'(0) = 0$; the tangent line at origin is horizontally.

If $n < 1$, $J_n'(0) = \infty$; the tangent line at the origin is vertically.

For $n = 1$, $J_1'(0) = \frac{1}{2}$; tangent line is oblique.

IOP Conf. Series: Materials Science and Engineering **95** (2015) 012077 doi:10.1088/1757-899X/95/1/012077

3. Spherical Bessel functions

Let consider the equation $x^2 \frac{d^2 y}{dx^2} + 2x\frac{dy}{dx} + \left[k^2x^2 - l(l+1)\right]y = 0$. The general solution of this equation is $y = \frac{A}{\sqrt{x}} J_{l+\frac{1}{2}}(kx) + \frac{B}{\sqrt{x}} Y_{l+\frac{1}{2}}(kx)$, which can be written as $y = A \cdot \sqrt{\frac{2k}{\pi}} \cdot \sqrt{\frac{\pi}{2x}} \cdot J_{l+\frac{1}{2}}(kx) + B \cdot \sqrt{\frac{2k}{\pi}} \cdot \sqrt{\frac{\pi}{2x}} \cdot Y_{l+\frac{1}{2}}(kx)$, where $j_l(x) = \sqrt{\frac{\pi}{2x}} \cdot J_{l+\frac{1}{2}}(x)$ and $y_l(x) = \sqrt{\frac{\pi}{2x}} \cdot Y_{l+\frac{1}{2}}(x)$ are the spherical Bessel functions. The spherical Bessel function of most interest are those of integer and half integer order. Important engineering applications of these functions are diffusion problems (especially Helmholtz equation) and asymptotic analysis. According to [3], these functions can be written in terms of elementary functions, for n a non-negative integer:

$$j_n(x) = (-1)^n x^n \left(\frac{1}{x}\frac{d}{dx}\right)^n \left(\frac{\sin x}{x}\right), \qquad y_n(x) = -(-1)^n x^n \left(\frac{1}{x}\frac{d}{dx}\right)^n \left(\frac{\cos x}{x}\right),$$

$$h_n^{(1)}(x) = -i(-1)^n x^n \left(\frac{1}{x}\frac{d}{dx}\right)^n \left(\frac{e^{ix}}{x}\right), \quad h_n^{(2)}(x) = i(-1)^n x^n \left(\frac{1}{x}\frac{d}{dx}\right)^n \left(\frac{e^{-ix}}{x}\right).$$

We shall plot the graph of spherical Bessel functions for $n \in \{1,2,3\}$.

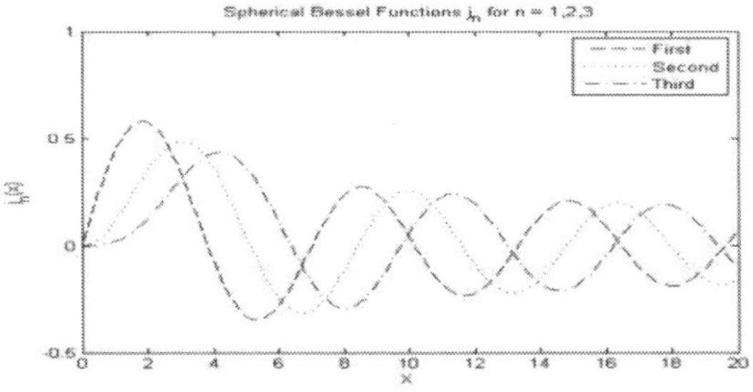

Figure 5. Spherical Bessel functions $j_n(x)$, $n = 1,2,3$.

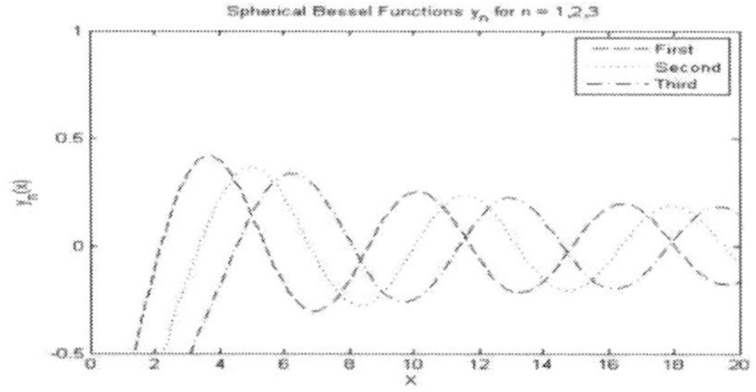

Figure 6. Spherical Bessel functions $y_n(x)$, $n = 1,2,3$.

4. Application

Bessel functions are connected with other special functions such as: Gamma, Hypergeometric, Struve function. Bessel-real and Bessel-imaginary have applications in physics, such as the resistance of conductors of alternating current.

Bessel functions are important in problems of wave propagation, static potentials, studying electromagnetic waves in a cylindrical waveguide, heat conduction in a cylindrical object.

Solving problems, hardly find Bessel equation in canonical form (1). When it is possible, it is useful to know the reduction of the equation to the form (1).

To simplify writing the solution of presented examples, it could be used $AJ_n(z) + BY_n(x) = Z_n(z)$

Table 1. Examples of equations which can be reduced to Bessel equation.

Name	Equation	Solution
	$\frac{d^2y}{dx^2}+\frac{1}{x}\frac{dy}{dx}+\left(k^2-\frac{n^2}{x^2}\right)y=0$	$Z_n(kx)$
oscillations of a heavy wire suspended from one end	$\frac{d^2\varphi}{dx^2}+\frac{1}{x}\frac{d\varphi}{dx}+\frac{\omega^2}{g}\frac{\varphi}{x}=0$	$Z_n\left(2\omega\sqrt{\frac{x}{g}}\right)$
perfect circular motion of a membrane stretched	$\frac{d^2z}{d\rho^2}+\frac{1}{\rho}\frac{dz}{dx}+\frac{1}{\rho^2}\frac{\partial^2 z}{\partial\rho^2}=k^2\frac{\partial^2 z}{\partial t^2}$	$e^{j\omega t}\cos n\varphi J_n(k\omega\varphi)$
electromagnetic oscillation of a cavity in the form of rotating drum	$\frac{d^2U}{dz^2}+\frac{1}{\rho}\frac{d}{d\rho}\left(\rho\frac{\partial U}{\partial\rho}\right)+\frac{1}{\rho^2}\frac{\partial^2 U}{\partial\rho^2}+k^2U=0$	$U=J_n(a\rho)_{\sin}^{\cos}n\varphi_{\sin}^{\cos}qze^{jk\frac{t}{\sqrt{\mu\varepsilon}}}$
wave modulated frequency spectrum	$u(t)=U\sin\left(\Omega_0 t+\beta\int_0^t\sigma(t)dt\right)$	$u(t)=U\sum_{n=-\infty}^{\infty}J_n(\xi)\sin(\Omega_0+n\alpha)t$

5. Conclusions

It have been briefly analyzed, analytical and numerical, the most known Bessel functions and also other modified Bessel functions. Using Matlab facilities seems to be much easier to investigate the functional behavior of the Bessel functions. Since the low order integer Bessel functions are so much used in various applications, it was mainly numerically treated the Bessel functions, spherical Bessel functions for order n = 0, 1, 2, 3. Obviously, the ordinary Bessel functions are oscillating as functions of a real argument. On the other hand, the modified Bessel functions tend to look more like decaying and growing exponentials.

It's also interesting to compare the behavior of low order integer Bessel functions and modified Bessel functions for $x \to 0$ and for $x \to \infty$. These results are used in evaluating boundary conditions for boundary value problems. Using Matlab, the asymptotic forms of Bessel functions and modified Bessel functions allow determination of important properties of these functions.

Taking into account the large number of engineering problems and also mathematical physics problems connected with Bessel functions, it is sure that using Matlab (as numerical computing environment and programming language) could improve not only computing Bessel functions and modified Bessel functions, but also studying properties of these special functions.

References

[1] Abramowitz M and Stegun I A 1985 Handbook of Mathematical Functions (National Bureau of Standards) *Applied Mathematics Series* **55** (Dover Publications)
[2] Karris S T 2004 Numerical Analysis using Matlab 2nd Edition *Orchard Publications*
[3] Watson G N 1996 A Treatise on the Theory of Bessel Functions *Cambridge University Press*
[4] Korenev B G 2002 Bessel Functions and their Applications *Series Analytical Methods and Special Functions* **1**
[5] Arfken G B and Weber H J 2005 Mathematical Methods for Physicists, 6th edition *Academic Press*

Ship's behaviour during hurricane Sandy near the USA coasts. Simulation results

B Chiotoroiu[1], **N Grosan**[1] **and L Soare**[1]

[1] Constanta Maritime University, str. Mircea cel Batrin, 900663 Constanta, Romania

E-mail: b.c.chiotoroiu@gmail.com, v.n_grosan@yahoo.com and loredanaasoare@yahoo.com

Abstract. The aim of this study is to analyze the impact of the stormy weather during hurricane Sandy on an oil tank using the navigation simulator. Meteorological and waves maps from forecast models are used, together with relevant information from the meteorological warnings. The simulation sessions were performed on the navigation simulator from the Constanta Maritime University and allowed us the selection of specific parameters for the ship and the environment in order to observe the ship's behavior in heavy sea conditions. Simulation results are important due to the unexpected environmental conditions and the ship position: very close to the hurricane centre when the storm began to change its track and to transform into an extra tropical cyclone.

1. Introduction

The simulations of the ship route and behaviour were performed on a deck simulator. Specific parameters for the ship and real meteorological data, recorded during the cyclone passage, were used during the simulation. The aim was the reconstruction of a real situation – an oil tank drift on October 29-30 2012 and the analysis of the vessel's movements. The simulation results are important from the safety point of view; they could contribute to a better understanding of ship movements in heavy seas.

Tropical cyclones are no longer a threat to life and maritime activities when they move to higher latitudes. Some could however evolve to fast-moving extra tropical cyclones that produce rainfall, large waves and strong winds. The modeling of this transformation is currently considered as one of the most challenging forecast problems [1]. That is the case of hurricane Sandy, the most notable storm of the 2012 season [2]. It was different from other hurricanes in the area because of the lack of an eye, the track, the large horizontal extension and the transformation into an extra-tropical cyclone. Sandy developed from a category 1 hurricane in Jamaica to a category 3 in eastern Cuba on 23 October 2012, before quickly weakening to a category 1 two days later. It started then to grow in size and moved NE-ward due to a middle - to - upper-level trough (figure1). It regained hurricane strength on October 27. Its structure became unusual, with a large radius of maximum winds, over 100 Nm, the strongest winds located in the western semicircle of the cyclone, the formation of a warm front in the NE quadrant and of another weak stationary boundary to the NW of the center, serving to enhance the convection and strong winds there [3]. Most of the cyclones in the North Atlantic move eastward, which is typical for the extra tropical transition of a hurricane [1]. However Sandy deepened on October 29, SE of New Jersey and began to curve to the west. This track change was due to an upstream trough and its cold air mass. An anomalously high pressure over Atlantic Canada and

Content from this work may be used under the terms of the Creative Commons Attribution 3.0 licence. Any further distribution of this work must maintain attribution to the author(s) and the title of the work, journal citation and DOI.
Published under licence by IOP Publishing Ltd

IOP Conf. Series: Materials Science and Engineering **95** (2015) 012078 doi:10.1088/1757-899X/95/1/012078

Greenland extended from 40° to 60° of latitude. The cyclone movement into the mid-latitudes was to the left of the subtropical ridge over the Atlantic and downstream of the mid-latitude trough located over the USA; the ridge was situated between the eastward propagating trough to the west and a cut-off low to the east of it. The cyclone intensified as cold continental air encircled the warm core vortex; the intensification occurred in response to shallow low-level convergence below 850 hPa (850 hPa charts are not shown in this paper). The cyclone weakened and slowed over the continent, while following a northward movement.

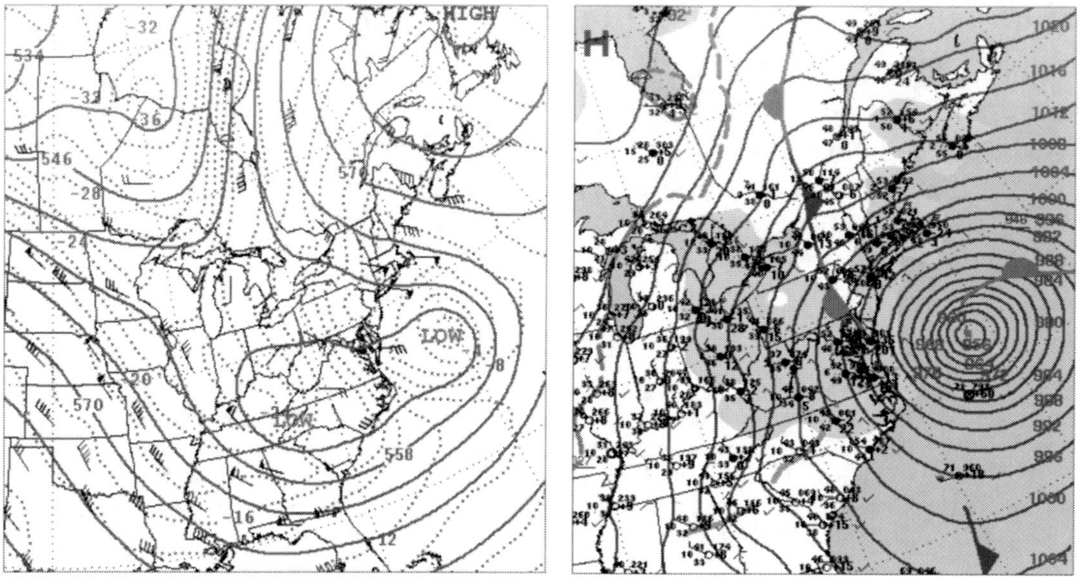

Figure 1. NCEP Reanalysis maps (500 hPa geopotential - left, sea level pressure - right).

1.1. Cyclone's track and ship's route planning

The meteorological forecast of Sandy's lowest central pressure (an indicator of hurricane's force) in the ECMWF analysis, of 947 hPa, was very similar to 946 hPa estimated by the NHC [5, 3]. Comparisons of several forecast models highlighted that the ECMWF model was one of the first to show the NW-ward turn of Sandy, even while most of the rest of the guidance showed the cyclone staying offshore of the East Coast [3, 6, 7]. The turn to the left of the cyclone was sharper for the higher model resolutions [5]. The cyclone's turn and extratropical transition and intensification represented a high risk for the ships in the New York port area.

The oil tank we considered in this study arrived in the port of New York and started the fuel oil discharging operations on October 27 2012. Its main characteristics were: a double hull oil tanker classified as: 1A1 (E), oil carrier, a deadweight of 158000 MT, LOA 274,4 M, LBP 264 M.

The same day at 21H00 UTC a hurricane warning was sent by the Meteorological Office in Washington to all the ships in the area: "hurricane center located near 30.2°N and 75.2°W [...]. Present movements toward the NE or 25 degrees at 11 kts. Estimated minimum central pressure 961 mb, maximum sustained winds 65 kts with gusts to 80 kts" [8]. On October 28, 10H22 UTC, Sandy was: "north of area near 31.9°N/ 73.3°W with 960 mb, maximum sustained winds 65 kt, gusts 80 kt [...] and 270 Nm NW quadrants with seas to 36 FT (significant waves)" [8]. Because of the high risk due to the approaching cyclone, the Captain of the Port of New York and New Jersey [...] "in addition the COTP hurricane severe weather plan" required on October 28: "commercial deep draft vessels greater than 500 gross tons are not authorized to remain in port alongside a pier after 6 p.m. today; all the vessels must be out of Bay Ridge, Stapleton and Gravesend Bay Anchorage Grounds by 6 p.m. today" [9].

That is why the oil tank left the port in the afternoon and was, on October 28/29 under way to north along the USA coasts.

1.2. Simulations of the ship's route and behaviou
The deck simulator from Constanta Maritime University allowed us the selection of a specific area and of a type of vessel with identical characteristics to those mentioned above. For a more accurate simulation of different wind forces and sea states, we made several successive tests for different wind speeds and directions and corresponding waves' heights and directions. The input data (wind conditions and sea states) are real meteorological records, done all the 6 or 12 hours and they were extracted manually from the meteorological telegrams [10] and from the weather bulletins and warnings received on board ship.

The simulator's functionalities are however still limited, especially when it comes to the dynamic of the air masses [11]. The present situation was complicated by the fact that the wind changed its direction frequently near the centre of the cyclone.

As a result of the simulation sessions we obtain a progress curve of the ship for a period of several hours which is very similar to the real ship's route from the ECDIS maps (recorded on ship board but not shown in this paper). The progress curve in figure 2, starts on October 29 when the ship tried to move away from the coast. The ship went adrift starting with 29.10.2012, 13H00, because of the very rough sea. The storm waves were pushing the ship to the shore so the oil tank had to change its course starting with October 29, 1H p.m. and to head to the east. The ship began therefore to move away from the initial course in a ESE direction but very quickly afterwards it was gradually pushed to SW and then W by the large waves (figure 2).

Three days later, after hurricane Sandy landfall, the ship headed back to the port of New York in order to complete the unloading operations. figure 2 (left) shows the approximate position of the ship on October 29, under its way to a place of refuge.

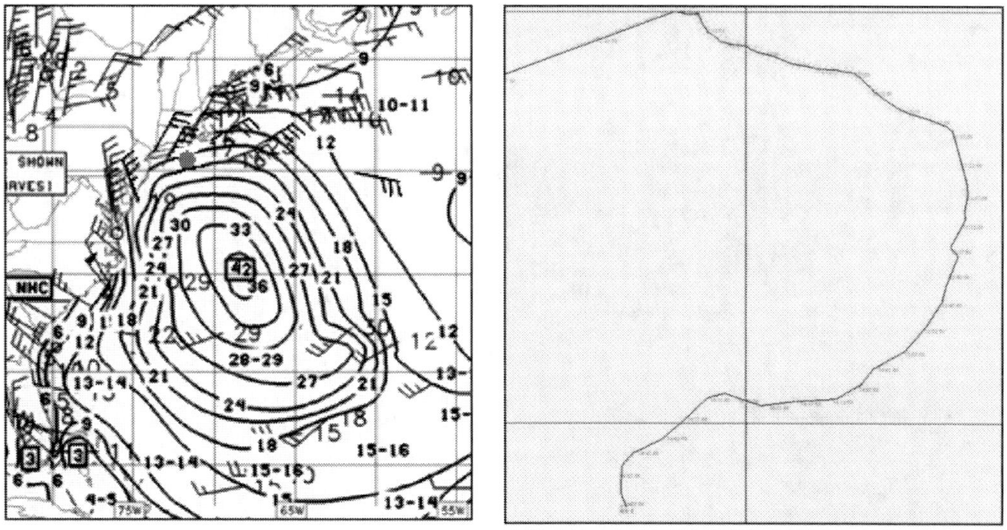

Figure 2. Wind speed, significant waves and ship position on 29.10.2012, 13H00, noaa.gov. – left; ship's course on the simulator – right.

Variations of hydrodynamics and ship parameters' for periods of 1 and 5 minutes of simulation are shown in figures 3-6. The values on the diagram in figure 4 were recorded at the end of the simulation. Diagrams in figures 3 and 4 show that:

-the pitching and rolling moments (lines 3 and 4), record high values and rapid changes within the negative and positive limits. The high values are normal, considering the heavy pitching and rolling (roll in figs. 5 and 6), [12]. Ship oscillatory movements around the longitudinal and transversal axis (that define the roll and pitch) are violent; the ship has a continuously motion to port and starboard sides, as well as raising and lowering movements of bow and stern;

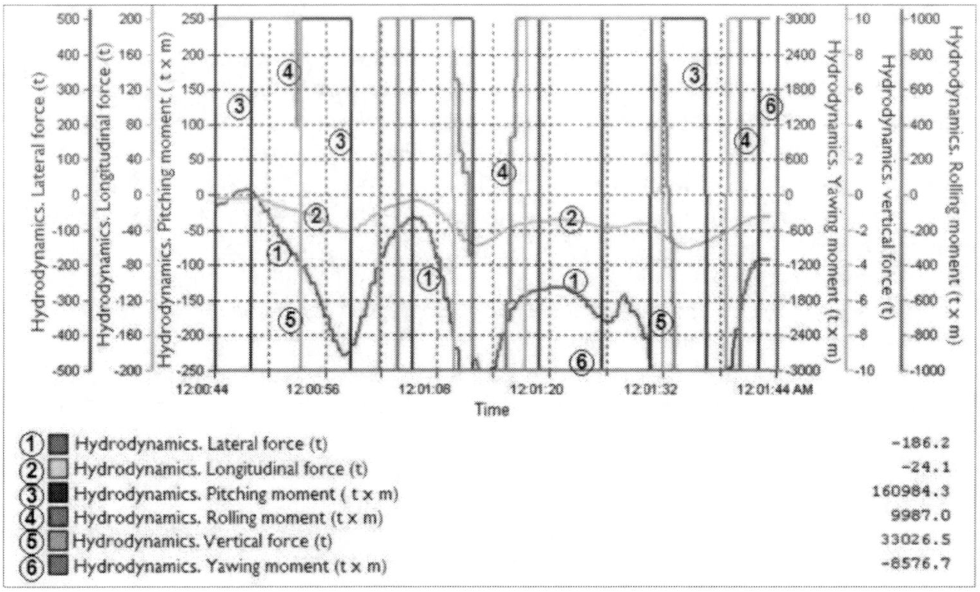

Figure 3. Evolution of hydrodynamic selected parameters for 1 minute of simulation.

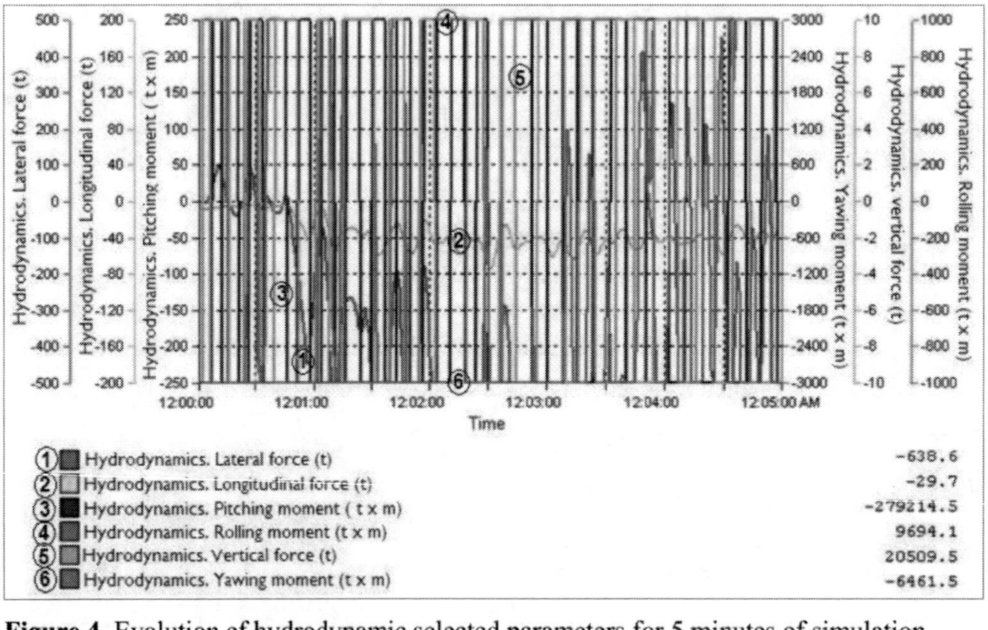

Figure 4. Evolution of hydrodynamic selected parameters for 5 minutes of simulation.

Figures 5 and 6 show that:

-the roll angle (line 1) has very high values that largely exceed the limits set for the simulation sessions; the ship has fast oscillatory movements to the port and starboard sides, along its fore-and-aft axis; during this heavy roll, values change rapidly from port to starboard and vice versa. At 12.00.16 the ship have a list to the port side and the next 6 seconds the value exceeds the limit set for the

starboard. The diagrams also show that for short periods of time the ship remains 3 degrees list to port side, then the heeling angle becomes greater on the same side, followed by an immediate inclination to the starboard side. This is a violent roll; the vessel loses its equilibrium position because of strong external forces such as wind force and wave impact force.
-the speed over the ground (line 3) has an initial value of 4 knots, with slight variations. During the 90 minutes simulation session, ship speed over the ground decreased, last values being equal to 1 knot or to zero, even if the main engine running is on "slow ahead" and "half ahead".
-the yaw angle (line 4) is defined as the angle formed by the fore-and-aft line with the ship's heading line. The movement occurs around a vertical axis passing through the vessel's centre of gravity and is different from the movement of lateral deviation that occur around a vertical axis passing through the vessel's centre of gyration.

The variation of the angle of deviation of the extremities is within the maximum limits, to port and starboard sides; this indicates that the extremities of the vessel are in constant motion due to frequent changes of wind direction. As a consequence of the heavy roll, the drafts fore and aft change (any inclination to starboard increases the draft this side and decreases the draft in the opposite side i.e. the port side). Therefore, the increasing draft determines an increase of the hydrostatic thrust forces at the starboard side, larger than the hydrostatic forces on the port side; an additional hydrostatic thrust force appears, that will push the bow to the starboard side and the stern to the port side. Ship's behaviour will be the same if the ship is inclined to the port side: due to the additional hydrostatic thrust force, the bow is pushed to the starboard side and the stern to the port side.

Bow and stern are continuously drift away and the ship keeps the course only for very short periods: 2 minutes at the starboard side, between 12.01.47 and 12.03.47; then the bow turns to the starboard side where it stays another 2 minutes (from 12.03.47 to 12.05.47), figure 6.

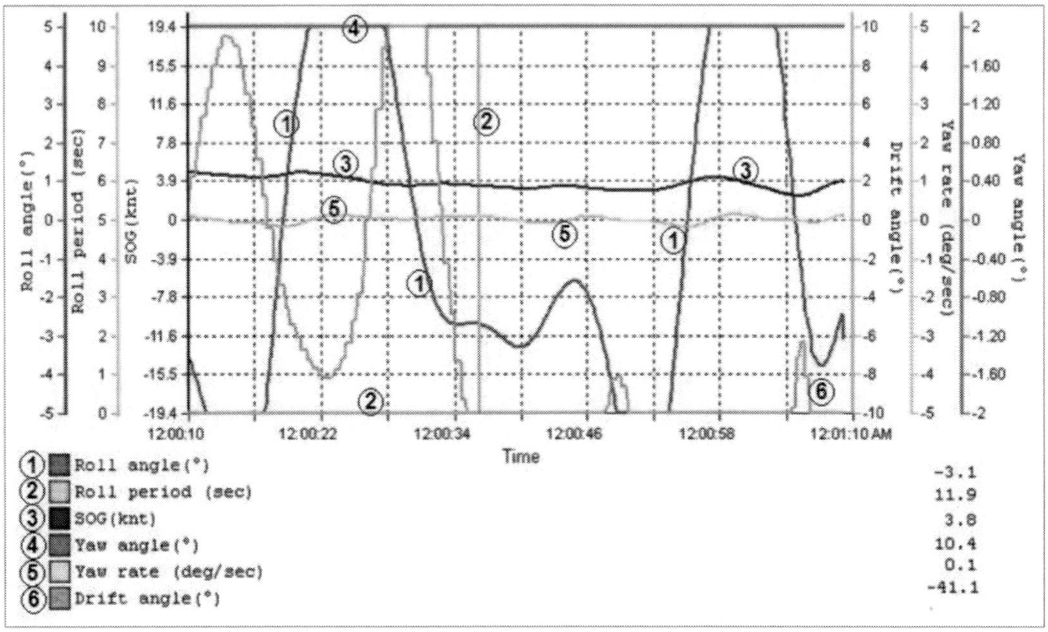

Figure 5. Evolution of ship's selected parameters for 1 minute of simulation.

2. Conclusions

Simulations of a ship's progress curve and ship's behaviour in heavy seas were conducted on a TRANSAS NaviTrainer 5000. The purpose is a better understanding of ship movements in heavy seas. The simulator allowed us the use of real meteorological input data, the selection of a certain navigation area and of a certain type of ship, and thus restoring a real situation.

The selected ship's and hydrodynamic parameters are the best indicators for ship behaviour in extreme weather conditions. The 1 and 5 minutes periods were chosen because they provide a more detailed representation and a clearer reading of the results. A longer period for 6 parameters that vary continuously would make the results interpretation impossible. The results show dangerous pitching and rolling moments, this aspect being very important from the safety point of view.

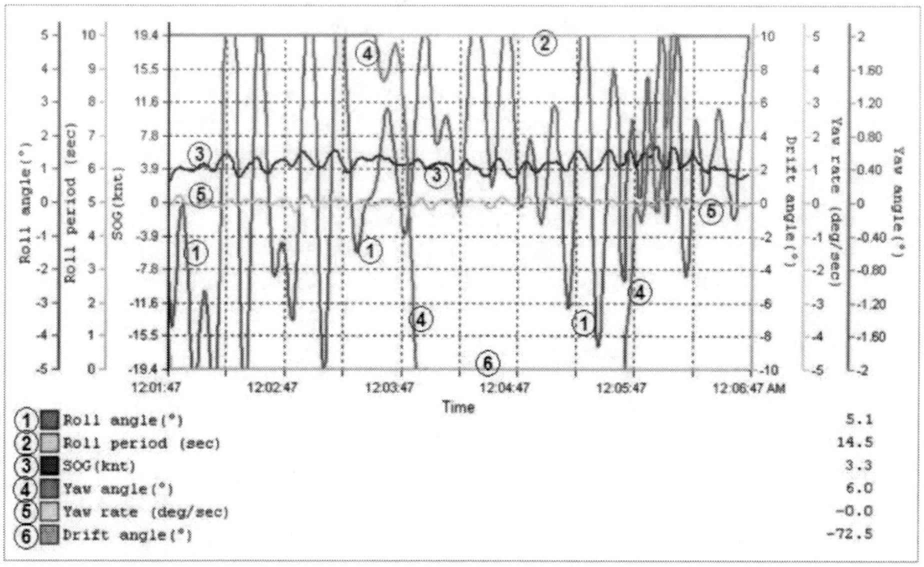

Figure 6. Evolution of ship's selected parameters for 5 minutes.

References

[1] Jones S C, Harra P, Abraham J, Bosart F L, Bowyer P J, Evans J L, Hanley D E, Hanstrum B N, Hart R E, Lalaurette F, Sinclair M S, Smith R K and Thorncroft C 2003 The Extratropical Transition of Tropical Cyclones: Forecast Challenges *Weather and Forecasting* **18** pp 1052-1092

[2] Hurricane/Post-tropical cyclone Sandy 2012 NOAA Service Assessment http://www.nws.noaa.gov/os/assessments/pdfs/Sandy13.pdf.

[3] Blake E S, Kimberlain T B, Berg R J, Cangialosi J P and Beven J L 2014 Tropical Cyclone Report- Hurricane Sandy http://www.nhc.noaa.gov/data/tcr/AL182012_Sandy.pdf.

[4] *** http://www.hpc.ncep.noaa.gov/dailywxmap/index_20121029.html

[5] Magnusson L, Bidlot J R, Lang S T K, Thorpe A, Wedi N and Yamaguchi M 2014 Evaluation of Medium-Range Forecasts for Hurricane Sandy *Mon. Wea. Rev.* **142** pp 1962-1981

[6] Gall R, McCarren D and Toepfer F 2013 Deterministic versus Ensemble Forecasting. The case from Sandy *Focus* **3(2)**

[7] Johnson S K 2012 Why European forecasters saw Sandy's path first *Arstechnica* arstechnica.com/science/2012.

[8] Information on www.weather.gmdss.org.

[9] Information on www.uscgnews.com/go/doc/4007/1591335/Port-of/New-York-and-New-Jersey-set-port-condition.

[10] www.ogimet.com

[11] Transas NaviTrainer 5000 2014 *Technical book*

[12] D.J. House 2007 Ship Handling *Theory and practice Amsterdam* (London: Elsevier Butterworth-Heinemann)

IOP Conf. Series: Materials Science and Engineering **95** (2015) 012079 doi:10.1088/1757-899X/95/1/012079

The influence of the waves in the areas near the port

F V Panaitescu[1], M Panaitescu[1] and I A Anton[1]

[1] Constanta Maritime University, Faculty of Naval Electro-Mechanics, 104 Mircea cel Batran Street, 900663, Constanta, Romania

E-mail: viopanaitescu@yahoo.ro

Abstract. The paper investigates the influence of the waves in the areas of Constanta port and the North of Mamaia resort sandy coast. The port has an asymmetric design that protects them from the sea. This study analyses the port' influence on the waves, the wave driven current and the sediment transport patterns. The harbour is on a sandy coast consisting of sand with d_{50}=0.70 mm (where d50 is diameter of sand particle) and s_g=0.75 m in the entire area (where s_g is significant wave height). Irregular, directional waves move towards the coast from 270_0 (West direction). Waves move over a water level corresponding to MWL=1m. The wave characteristics at the offshore boundary are H_{m0}=1.61 m (where H_{m0} is average wave height) and T_p=10.00 s (where T_p is the wave period). We are looking at an area that spans 10 000 m in the long-shore direction and 2500 m in the cross-shore direction. It is used the wave simulation with MIKE 21 PMS, and the hydrodynamic simulation with MIKE 21 HD. Irregular, directional waves are added at the offshore boundary of the wave model. The frequency spectrum is of JONSWAP type, but around the mean direction of wave propagation (270^0) a $\cos^n$ distribution of wave energy is used. At the other open boundaries symmetry boundary conditions are used. The wave simulation takes wave breaking into the consideration. The bed roughness is not considered. In this study, it is of interest to examine how the presence of the harbour influences the waves, the wave-driven current and the sediment transport patterns. When simulating of the wave driven currents with MIKE 21 HD we use constant values of velocity and Manning number all over the model area. The time slots are set to five seconds and the simulation runs for ten hours to ensure stable conditions. Finally the data are visualized flow, level and influence over the area of the coast.

1. Introduction

One of the most important elements in sediment transport and coastal zones are inshore currents and waves that they generate.

Most coastal areas dissipate power comes from wind generated waves and dispersion and transport of pollutants in coastal areas, brackish and sediment transport, erosion of coast lines and the formation of sand bars are some of the results from this dissipate energy in the coastal zones [1, 2]. The waves and the currents they generate are therefore the primary factors in the transport and deposit of sediments. Dislodges the material swells and suspended above the sea bed, and captures this material and the currents they carry along the beaches. The main parameters that characterize the waves are wave height (H_m) and period (T_p). They depend on the speed and duration of the winds and waves that generate the distance which waves are generated. Types of surfing at the beach and their seasonal variation are known as the wave climate. Deep-water waves are long and have a sinusoidal shape. However, as the waves come into shallow water propagation speed and degrease in length and the

Content from this work may be used under the terms of the Creative Commons Attribution 3.0 licence. Any further distribution of this work must maintain attribution to the author(s) and the title of the work, journal citation and DOI.
Published under licence by IOP Publishing Ltd

height there of and the sharp peaks of their growing up so the waves will consists of sharp ridges separated by flat areas. This transformation of the waves starts from the depths at which the waves "feel the seabed ". The corresponding depth at which to start the process is about half the length of the wave. The entries into shallow water waves are subject to refraction through ridges tend to become parallel to the contours of the seabed [3, 6, 8]. At the same time, the process of diffraction wave causes a transfer of energy, along the crest of waves, high waves from small ones. For coastal areas, with straight parallel strokes, the refraction decreases the angle between the direction of propagation of the wave and the normal to the line of the coast, and diffraction is causing a scattering of energy along the ridge. Accordingly, changes in the height and direction of waves along the coast are depending on the period of waves, their direction and contour of the seabed.

Coastal zones are dynamic environments (figure1) that change consistently under the action of natural factors, the ratio between the processes of erosion and accretion causing shoreline configuration for various time intervals [5, 6]. The waves in this area contribute directly to the processes of erosion, acting via mechanical abrasion processes in areas of beach and surf, by displacing the particles of sediment it carries in the offshore areas.

Another important phenomenon that manifests itself in coastal areas is the action of longitudinal currents carrying significant amounts of sediment along the shoreline, changing the look and feel of the beach. In the event of storms or extreme, large volumes of sediment are brought about in a state of suspension and beach areas will be subject to such a process of degradation much more intensely.

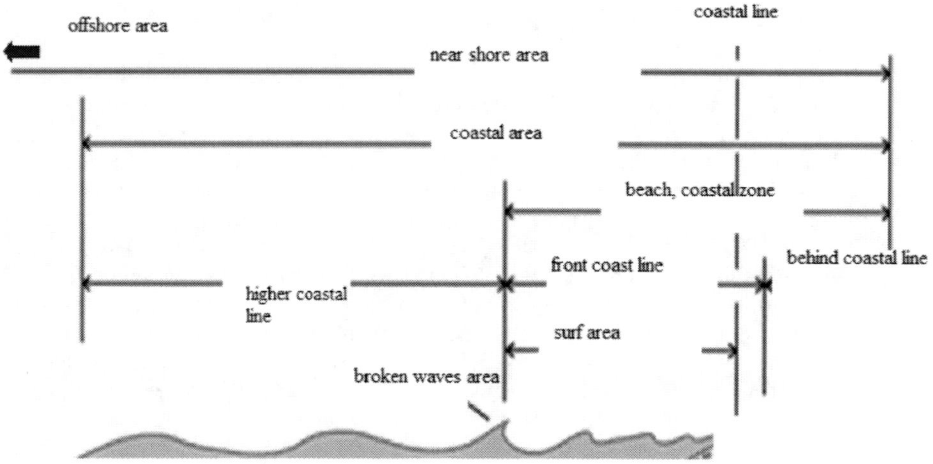

Figure 1. The general structure of coastal zone [9].

1.1. Establish a study area coordinates

The bathymetry of area in study. The paper investigates the influence of the waves in the areas of Constanta port and the North of Mamaia resort sandy coast. The port has an asymmetric design that protects them from the sea. This study analyses the port' influence on the waves, the wave driven current and the sediment transport patterns. The harbour is on a sandy coast consisting of sand with d_{50}=700.10^{-3} [m] and s_g=750.10^{-3}[m] in the entire area. Irregular, directional waves move towards the coast from 270^0 (West direction). Waves move over a water level corresponding to MWL=1[m]. The wave characteristics at the offshore boundary are H_{m0} =1.61 [m] and T_p=10.00 [s]. We are looking at an area that spans 8 000...10 000 [m] in the long-shore direction and 2 500....2800 [m] in the cross-shore direction. It is used the wave simulation with MIKE 21 PMS, and the hydrodynamic simulation with MIKE 21 HD. First, establish a study area coordinates. The next step is choosing the design of the system data of the study; start Mesh generator (Mike 0) and import boundary data *(Import boundary)* with files in xyz, *(import scatter data)* and obtain the geometry of the land (figure 2). After this, insert the nodes and point properties and choose the study polygon. Export mesh, introduce the number of nodes and the simulation time and generate the flow model.

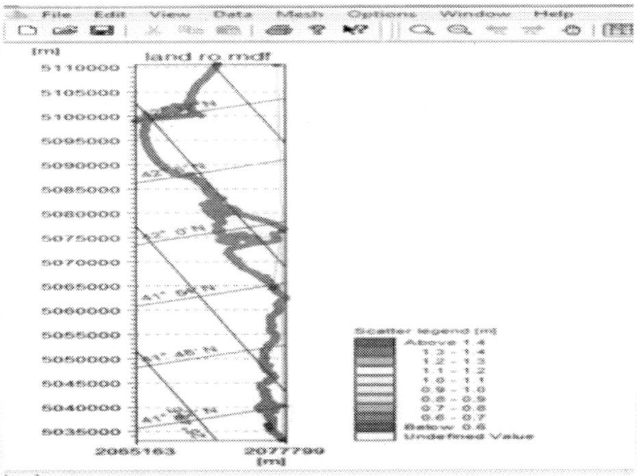

Figure 2. The geometry of the study.

Wind set-up. The wind speed is constant, [35 m/s] and wind direction is 270^0 (West direction) (figure 3). The value of wind friction coefficient is 2526.10^{-6}.

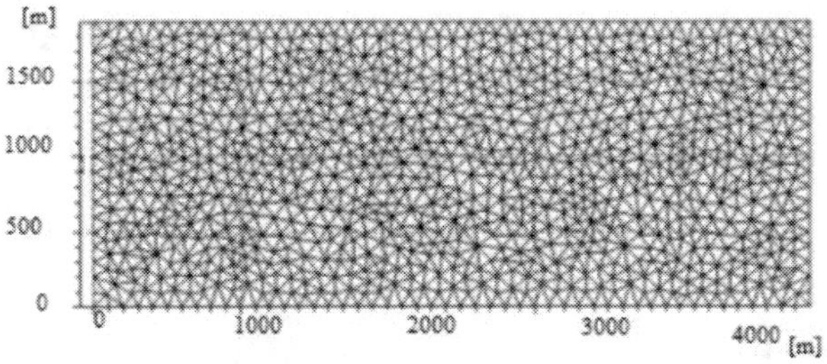

Figure 3. Wind set-up.

The configuration of domain study. The domain of problem is presented in the figure 4:

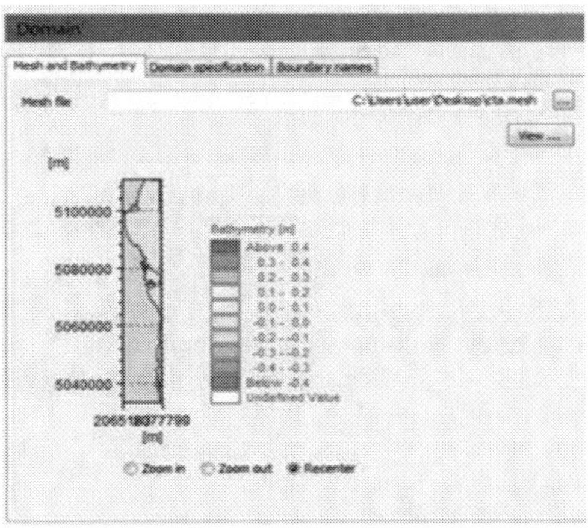

Figure 4. The configuration of domain study.

1.2. The equations of flow

Wind conditions. The driving force due to this wind is calculated equation (1) [1]:

$$c_w = \frac{\rho_{air}}{\rho_{water}} W^2 \tag{1}$$

where cw is the coefficient of wind friction, ρ is the density and W is the wind velocity.

The Wind Friction. The flow is characterized by a flow resistance defined on Chezy number between the limits: Cf = 98.10-4(rocks) and Cf = 3.10-3(fine sand) [2].

We consider Courant number equation (2) [3, 10]:

$$C_R = c\frac{\Delta t}{\Delta x} \tag{2}$$

where: c is the celerity, Δt the time step and Δx the grid spacing. The celerity for tidal wave is calculated equation (3) [3, 10]:

$$c = \sqrt{gh} \tag{3}$$

where: g is gravity and h is the water depth. The maximum value which we can correctly use depend on our bathymetry [1, 2, 3, 4].

Eddy viscosity. In the momentum equations we must consider momentum fluxes due to turbulence, vertical integration and subgrid scale fluctuations. The formulation of the eddy viscosity [3, 4, 10] in the equations is implemented in two ways: flux formulation equation (4) and velocity formulation equation (5):

$$\frac{\partial}{\partial x}\left\{E\frac{\partial P}{\partial y}\right\}_{x.momentum} \tag{4}$$

where P - the flux in the Ox-direction and E-the eddy viscosity coefficient;

$$\frac{\partial}{\partial x}\left\{h\cdot E\frac{\partial u}{\partial x}\right\}+\frac{\partial}{\partial y}\left\{h\cdot E\frac{\partial u}{\partial y}\right\} \tag{5}$$

where u-the velocity in the Ox-direction and h-the water depth.

The eddy viscosity coefficient E can be specified on the Smagorinsky concept [8, 9] equation (6):

$$E = C_s^{\,2}\Delta^2\sqrt{\left(\frac{\partial U}{\partial x}\right)^2+\frac{1}{2}\left(\frac{\partial U}{\partial y}+\frac{\partial V}{\partial x}\right)+\left(\frac{\partial V}{\partial y}\right)} \tag{6}$$

where U,V are depth-averaged velocity components in the Ox- and Oy-direction, Δ is the grid spacing [5] and Cs = 0.25....1.0 (we choose 0.28). We can optimized the study with the Smagorinsky formulas for the shear stresses (equation (7)) [9, 10]:

$$\frac{\partial}{\partial x}\left(hE\frac{\partial U}{\partial x}\right)+\frac{\partial}{\partial y}\left[\frac{1}{2}hE\left(\frac{\partial U}{\partial y}+\frac{\partial V}{\partial x}\right)\right] \tag{7}$$

which is in agreement with technical recommendations [4, 5, 8, 10] (the Smagorinsky formulation, the Smagorinsky factor Cs).

2. Results

We studied the hydrodynamics of the waves in the areas and obtained wind forcing, spectral waves, waves breaking, significant wave height, peak period, mean wave direction. The final results are presented in the table and graphical template (figure 5a, b).

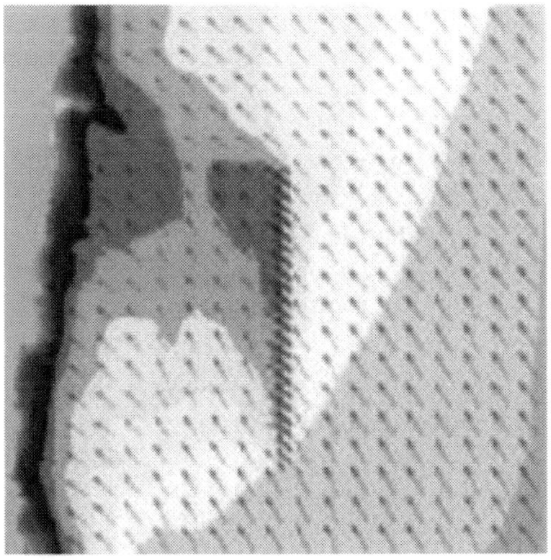

Geographic View | Output specification | Integral wave items

Parameters	Total	Wind Sea
Significant wave height	—	1.61
Maximum wave height	—	2.3
Peak wave period	—	10.00
Wave period T01	—	3.4
Wave period T02	—	4.5
Wave period Tm10	—	10.00
Peak wave direction	—	270
Mean wave direction	—	west
Directional standard deviation	—	☐
Wave velocity components	—	u,v
Radiation stresses	—	☐

Figure 5. The evolution of waves.

3. Conclusions

The wave simulation takes wave breaking into the consideration. The bed roughness is not considered. In this study, it is of interest to examine how the presence of the harbour influences the waves, the wave-driven current and the sediment transport patterns. It can be seen that the results provided by the JONSWAP MIKE are generally acceptable, but becomes less accurate as the bathymetry gets more complicated. Larger data errors model is due to a patchy bathymetry and some goals in data interpolation [6, 9]. The simulation runs for ten hours to ensure stable conditions.

The obtained results have highlighted the patchy field of three dimensional currents in the shallow area, showing the existence of important gradients on vertical direction, determine both speed differences and bathymetric gradients that exist in flow direction at different levels, in particular in the areas of characteristic coastal slope breaking South, rocky bottomed in the sector South of the Capul Tuzla to Vama Veche. Finally the data are visualized flow, level and influence over the area of the coast.

Acknowledgements

This article is a result of the project "Increasing quality in marine higher education institutions by improving the teaching syllabus according to International Convention STCW (Standards of Training, Certification and Watchkeeping for Seafarers) with Manila amendments" POSDRU/156/1.2/G/141534. This project in co funded by European Social Fund through The Sectorial Operational Programme for Human Resources Development 2007-2013, coordinated by Constanta Maritime University.

References

[1] Aupoix B E 1984 Viscosity Subgrid Scale Models for Homogeneous Turbulence *Macroscopic Modelling of Turbulent Flow* Lecture Notes in Physics

[2] Kofoed-Hansen H, Slot P, Sorensen O R and Fuchs J 2001 Combined numerical and physical modelling of seiching in exposed new marina *In proc. 27th Coastal Eng.* pp 3600-3614

[3] Leonard A 1974 Energy Cascades in Large-Eddy Simulations of Turbulent Fluid Flows *Advances in Geophysics* **18** pp 237-247

[4] Lilly D K 1966 On the Application of the Eddy Viscosity Concept in the Inertial Subrange of Turbulence *NCAR Manuscript* (National Center for Atmospheric Research) **123**

[5] Madsen P A, Rugbjerg M and Warren I R 1988 Subgrid Modelling in Depth Integrated Flows *Coastal Engineering Conference* pp 505-511

[6] Panaitescu F V, Deleanu D and Panaitescu M 2013 Processing of the data obtained with the Advanced Doppler Velocymeter *Advanced Materials Research* **836** p 360

[7] Rodi W 1980 Turbulence Models and Their Application in Hydraulics - A State of the Art Review *Special IAHR Publication*

[8] Sorensen O and Sorensen L S 2001 Boussinesq type modelling using an unstructured finite element technique *Proc. 27th Coastal eng. Conf.* pp 190-202

[9] Sorensen O, Schaffer H A and Sorensen L S 2004 Boussinesq type modelling using an unstructured finite element technique *Coastal Eng.* **50** pp 181-198

[10] Wang J D 1990 Numerical Modelling of Bay Circulation the Sea *Ocean Engineering Science* **9(B)** pp 1033-1067

[11] *** Information on http://www.dhigroup.com/upload/publications/mike21/Hashim_2010.pdf

The influence of transport network vulnerability for maritime ports

F Ruscă[1], E Roşca[1], A Rusca[1], M Roşca[1] and S Burciu[1]

[1] Politehnica University of Bucharest, Faculty of Transport, 313 Splaiul Independentei, 060042 Bucharest, Romania

E-mail: florin_rusca@yahoo.com

Abstract. The concepts of reliability and vulnerability are quite important in assessing the ability of transport networks from land to provide continuity in operation tacking in consideration the relation with seaports. Transport infrastructure modernizing and extension according to land use and sustainable development requirements still represents a challenging issue among policy makers, regional/local communities and scientists. The interest for the researches reliability and vulnerability for transport networks who connect maritime ports with interior city's is generate by the natural disasters, the terrorist acts, unconventional war, etc.. Transport network modelling enable the development of mathematical models used in evaluating the reliability and vulnerability. Nodes or link disruption could have an important impact over transport network users. In our paper we investigates the Romanian road network vulnerability related to Danube crossing versus maritime ports (Constanţa and Mangalia) and its mitigation by improving network topology. We use Visum software to promote a methodology to assess road transport network vulnerability versus Romanian seaports and we propose some solutions to reduce probability of road transport network to fail.

1. Introduction

Seaports are important nodes in the supply chains, joining high capacity transport modes: rail, auto and maritime transport. Transport infrastructure modernizing and extension according to land and maritime use, respectively sustainable development requirements still represents a challenging issue among policy makers, regional/local communities and scientists. Due to the importance of the transport networks for logistics production the reliability and vulnerability of transport networks are key factors in network management, prioritization of investments, maintenance and repair, emergency planning, evaluation of regional disparities and collateral effects [1]. The current literature defines reliability and vulnerability as two related, but different concepts. Transport network reliability has mainly a system engineering approach. Reliability is perceived as an expression of the probability of links functioning and thus as the stability of the service provided. Bell and Iida [2] analyze the network reliability as connectivity and travel time reliability. Nicholson et al. [3] introduce capacity reliability and flow decrement as measures for reliability. Husdal [4] states that a reliable network exhibits a high degree of operability as expressed by its serviceability, accessibility and non-variability under any circumstances, due to the presence of redundancy, robustness and resilience in the network. Chen &all [5] have a series of study about transport network capacity reliability. Mathematically it is defined as being equal to the probability that the maximum capacity of the network to be greater than or equal to the demand when capacity values of arcs may have some random variations. While

Content from this work may be used under the terms of the Creative Commons Attribution 3.0 licence. Any further distribution of this work must maintain attribution to the author(s) and the title of the work, journal citation and DOI.
Published under licence by IOP Publishing Ltd

probability is a major concern in transport network reliability, the consequences of links failure is the main focus of vulnerability [1]. Berdica [6] define in her paper a new concept: vulnerability of transport networks. This new network attribute is defined as "*a susceptibility to incidents that can result in considerable reduction in network serviceability, namely the operability of the link/route or network during a given period*". Also Taylor and D'Este [7] have made some research about how accessibility of nodes is correlated to notion of vulnerability .

The interest for the researches reliability and vulnerability for transport networks who connect maritime ports with interior city's is generate by the natural disasters, the terrorist acts, unconventional war, etc.. Also it is a EU demand to connect seaports to TEN-T network. Romania has 2 Core Network Corridors crossing its country. One is The Orient/East-Med Corridor who connects the German ports Bremen, Hamburg and Rostock via Czech Republic and Slovakia, with a branch through Austria, further via Hungary to the Romanian port of Constanta [8]. So it is important to determinate influence of vulnerability for maritime ports of land transport network.

2. Method

A method to evaluate transport network vulnerability is proposed by Jenelius et al. [9] who use, as a measure of reduced performance of the network, the increase in the generalized cost of travel (time, distance, money) for the users. When a link k is closed, the network may be divided into several disconnected parts, so that a number of trips from origin i are not able to reach the destination j. Thus results an unsatisfied demand.

$$u_{ij}^{(k)} = \begin{cases} \varphi_{ij} & \text{if } c_{ij}^{(k)} = \infty \\ 0 & \text{if } c_{ij}^{(k)} \neq \infty \end{cases}, \tag{1}$$

where φ_{ij} represents the travel demand from node *i* to node *j* and $c_{ij}^{(k)}$ is the generalized cost of travel from node *i* to *j* when link *k* is closed.

Therefore, there is a dichotomy of the link importance according to travel cost increasing and unsatisfied demand into the network. If the link k belongs to the set of non-cut links, the importance of the link k for the whole network is

$$\Omega(k) = \frac{\sum_i \sum_{j \neq i} \varphi_{ij}(c_{ij}^{(k)} - c_{ij}^{(0)})}{\sum_i \sum_{j \neq i} \varphi_{ij} c_{ij}^{(0)}} \tag{2}$$

where is the generalized cost of travel from node i to node j in the undamaged network.

The importance regarding the unsatisfied demand of a link *k* is

$$\Omega_{uns}(k) = \frac{\sum_i \sum_{j \neq i} u_{ij}^{(k)}}{\sum_i \sum_{j \neq i} \varphi_{ij}} \tag{3}$$

In addition, the link disruption is translated into nodes exposure. The demand weighted exposure of node i is the maximum value over all non-cut links:

$$\Phi(i) = \max_{k \in L^{n-c}} \frac{\sum_{j \neq i} \varphi_{ij}(c_{ij}^{(k)} - c_{ij}^{(0)})}{\sum_{j \neq i} \varphi_{ij} c_{ij}^{(0)}}. \tag{4}$$

The exposure regarding the unsatisfied demand for the node i is

IOP Conf. Series: Materials Science and Engineering **95** (2015) 012080 doi:10.1088/1757-899X/95/1/012080

$$\Phi(i) = \max_{k} \frac{\sum_{j \neq i} u_{ij}^{(k)}}{\sum_{j \neq i} \varphi_{ij}}. \tag{5}$$

3. Experimental procedure

The variation of the travel time of the users due to the inoperability of the critical road links influences their perception on network performance. The identification of the threshold values above which the travel times lead to the journey refuse is one of the main concerns of engineers in the transport sector [10]. The following methodology assesses the transport network vulnerability according to the travel time.

Due to the complexity of modelling the land transport network which normally have thousands of nodes and links, it is necessary to use a specialized software (e.g. VISUM, TransCad) with user's friendly graphical interface and allowing accurate network representation [10].

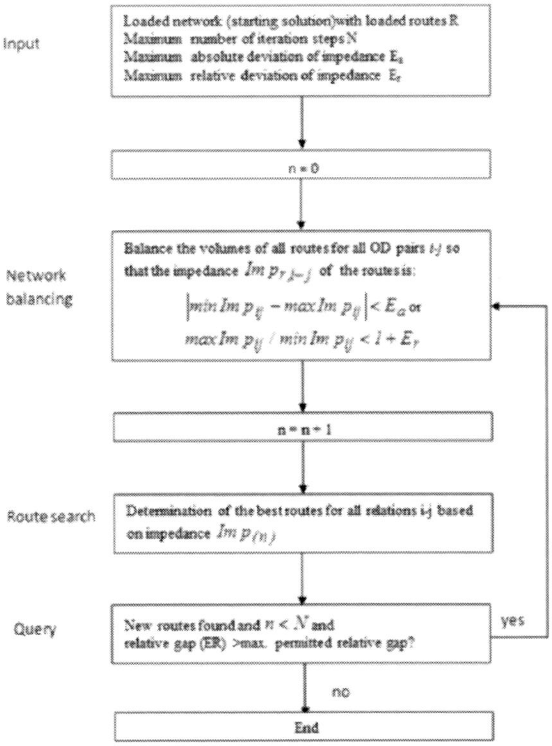

Figure 1. Equilibrium Assignament Algorithm [11].

In our model the generalized cost of travel from node *i* to node *j* is calculate as travel time. To determinate the initial value for travel time in model is used *Equilibrium Assignment Algorithm* (figure 1). This procedure distributes demand according to Wardrop's first principle: "*Every individual road user chooses his route in such a way that his trip takes the same time on all alternative routes and that switching routes would only increase personal journey time.*" The state of equilibrium is reached by multi-successive iteration based on an incremental assignment as a starting solution. In the inner iteration step, two routes of a relation are brought into a state of equilibrium by shifting vehicles. The outer iteration step checks if new routes with lower impedance can be found as a result of the current network state [11].

Travel times for motorised transport are determined by the saturation of links and intersections which result from the traffic volume and the capacity of these network objects. Also is necessary to know travel time at free traffic flow determined from the link length and the free-flow speed v_0. The impedance of a route results from the sum of the link impedances of the route. It is assumed that impedance is equal to the current travel time and that the current travel time is calculated using a travel time volume delay functions. Because the procedure only terminates when all routes of any Origin-Destination relation are in the equilibrium state, the procedure provides realistic results

In table 1 are presented relation used to calculate travel time on a single network link:

Table 1 Travel time volume-delay functions (after [12]).

No.	Author	Year	Function
1.	Irwin&all	1961	$T_a(\varphi_a, C_a) = \begin{cases} t_a^0 + a_a\varphi_a & \varphi_a < C_a \\ t_a^0 + \beta_a\varphi_a + (\alpha_a - \beta_a)C_a & \varphi_a \geq C_a \end{cases}$
2.	Smock	1962	$T_a(\varphi_a, C_a) = t_a^0 \exp\left(\frac{\varphi_a}{C_a}\right)$
3.	Mosher (1)	1963	$T_a(\varphi_a, C_a) = t_a^0 + \ln(C_a) - \ln(C_a - \varphi_a)$
4.	Mosher (2)	1963	$T_a(\varphi_a, C_a) = \alpha_a - \frac{C_a(t_a^0 - \alpha_a)}{\varphi_a - C_a}$
5.	Soltman	1965	$T_a(\varphi_a, C_a) = t_a^0 2^{\frac{\varphi_a}{C_a}}$
6.	Davidson	1966	$T_a(\varphi_a, C_a) = t_a^0\left(1 - \alpha_a \frac{\varphi_a / C_a}{1 - \varphi_a / C_a}\right)$
7.	Akcelik	1991	$T_a(\varphi_a, C_a) = \begin{cases} t_a^0\left(1 + a_a \frac{\varphi_a/C_a}{1 - \varphi_a/C_a}\right) & \varphi_a < \rho C_a \\ t_a^0\left(1 + a_a \frac{\rho}{1-\rho}\right) + \frac{\alpha_a}{(1-\rho)^2}\left(\frac{\varphi_a}{C_a} - \rho\right) & \varphi_a \geq \rho C_a \end{cases}$

where t_a^0 is travel time at free traffic flow, T_a travel time for traffic volume φ_a on link a, C_a is link capacity and calibration parameters. In our model we use a BPR function [13]

$$T_a = t_a^0[1 + (\frac{\varphi_a}{C_a})^4] \tag{6}$$

4. Results and discussion

In our study case we investigate the Romanian road network vulnerability related to Danube crossing versus maritime ports (Constanţa and Mangalia). Using Visum software, we implement a methodology to assess road transport network vulnerability versus Romanian seaports. In our study case are studied some routes between Romanian Seaports (Constanta Port and Mangalia Port) and road border points (Albita, Bors, Nadlac, Calafat, Giurgiu, Moravita, Siret and Vama Veche).

The density and the poor connectivity of the road transport network are responsible for its vulnerability to structural, natural and traffic factors. The Danube Bridges represent critical

IOP Conf. Series: Materials Science and Engineering **95** (2015) 012080 doi:10.1088/1757-899X/95/1/012080

infrastructure elements. The disruption of their functionality generates increasing of travel times and induces network vulnerability.

The Romanian road network has a length of 82,386 km of which 15,396 km of national roads (6,188 Km of European roads), and 529 km highways. The modelled network used in the case study has a total length of 17,530 Km including national roads, highways and major county roads (figure 2). The network is separated by the Danube River in two sub-networks connected by two bridges (Cernavodă Bridges-in reality are two bridges connected by a land link on Balta Ialomitei Island, Giurgeni Bridge). The initial traffic flow on road transport network is presented in figure 3 and is obtained using Equilibrium Assignment Procedure [10].

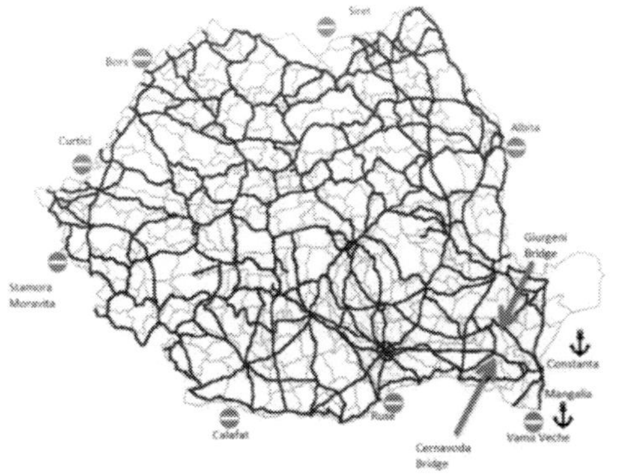

Figure 2. Road transport network with interest points.

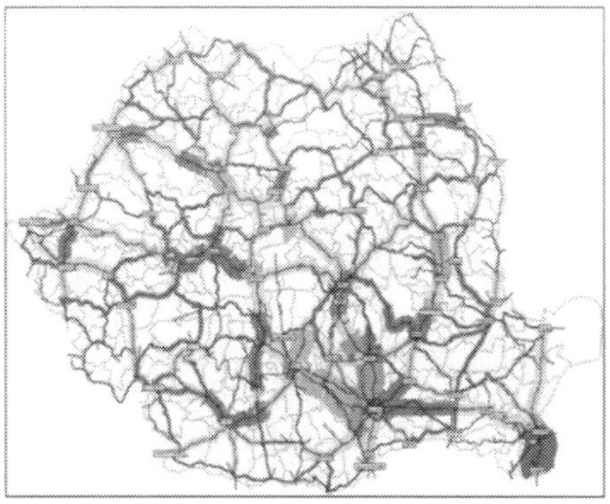

Figure 3. The initial road traffic flow.

In our study case is taken in consideration for the generalized cost of travel the duration of travel time on loaded road network and is examinate the situation when Cernavoda Bridges (figure 4) or Giurgeni Bridge (figure 5) are closed.

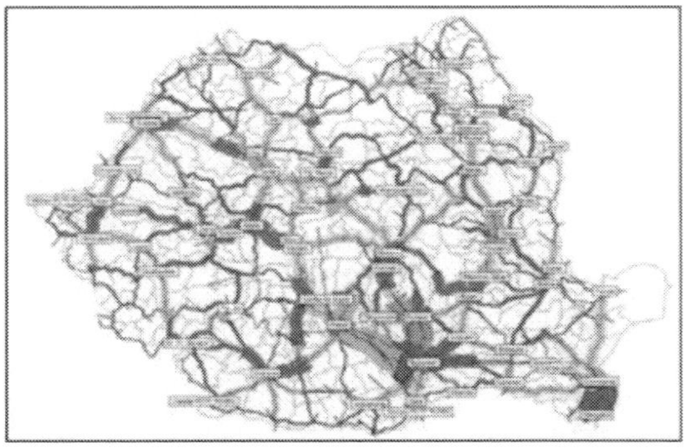

Figure 4. Road traffic flow –Cernavodă Bridges closed.

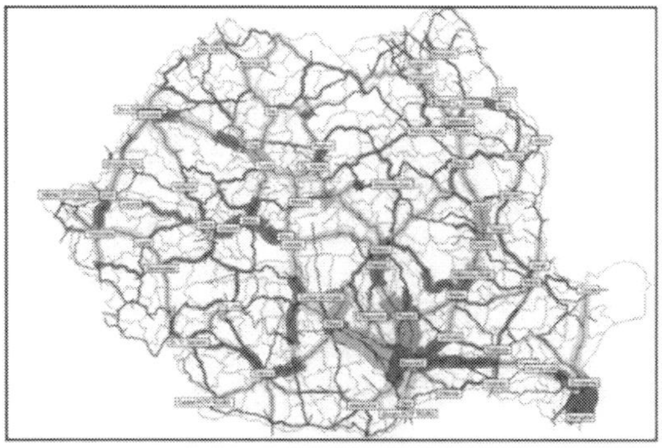

Figure 5. Road traffic flow –Giurgeni Bridge closed.

Using equation (4) is calculate the exposure regarding the unsatisfied demand for the Romanian seaports (Constanta and Mangalia) for road border points induced by Danube Bridges. The result are presented in figure 6.

The fail of the present Danube crossing points generates important increasing in the vulnerability of the Romanian seaports versus road crossing border. The link containing the Cernavoda bridges is an important link of the network. Its fail results for the demand weighted exposure parameter big value for relation with Calafat, Moravita and Nadlac. When Giurgeni bridge is closed the routes from Romanian seaports to road border points (Albita, Bors, Nadlac, Calafat, Giurgiu, Moravita, Siret and Vama Veche) are not affected. Also in both case the route to Vama Veche is not affected.

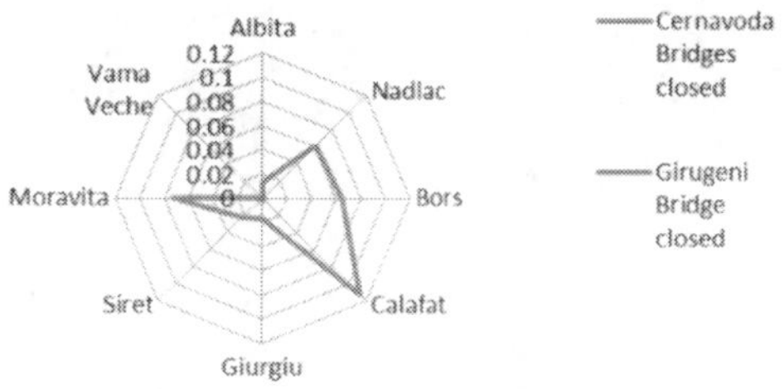

Figure 6. The demand weighted exposure.

5. Conclusions

Events like natural disasters, the terrorist acts, unconventional war generates from researcher a particular interest to evaluate transport network reliability and vulnerability. The developed procedure leads to some result which highlights the importance of Danube Bridge for Romanian seaports. If Cernavoda Bridges are closed for routes from maritime ports to road border points Calafat and Moravita we obtain a big value for the demand weighted exposure parameter. So it highlights the role of Danube Bridges in connection of maritime ports with European Inland Network. The opening of a new bridge near Brăila will also increase accessibility in the region and will reduce the vulnerability of the network versus Danube Bridges: Cernavoda and Giurgeni.

Acknowledgement

The work has been funded by the Sectoral Operational Programme Human Resources Development 2007-2013 of the Ministry of European Funds through the Financial Agreement POSDRU/159/1.5/S/132397

References

[1] Raicu S, Rosca E, Popa M, Dragu V, Costescu D and Rusca F 2007 Transport network vulnerability *CEEX Proceedings* Brasov

[2] Bell M and Iida Y 1997 Network Reliability, in Transportation Network Analysis Wiley&Sons Chichester pp 179-192

[3] Nicholson A, Schmöcker J, Bell M and Iida Y 2003 Assessing Transport Reliability: Malevolence and User Knowledge, in The Network Reliability of Transport, eds. M. Bell and Y. Iida *Elsevier Science, Kidlington* pp 1-22

[4] Husdal J 2004 Reliability and vulnerability, *Journal of the Norwegian Institute for Transport Economics* **2** pp 28-30

[5] Wang J, Ehrgott M and Chen A 2014 A bi-objective user equilibrium model of travel time reliability in a road network *Transportation Research Part B* **66** pp 4-15

[6] Berdica K 2002 An introduction to road vulnerability: what has been done, is done and should be done *Transport Policy* **9** pp 117-127

[7] Taylor M and D'Este G 2003 Concepts of network vulnerability and applications to the identification of critical elements of transport infrastructure, in The Network Reliability of Transport eds. M. Bell and Y. Iida *Elsevier Science, Kidlington, G.* pp 23-44

[8] Information on http://ec.europa.eu/transport/themes/infrastructure/ten-t-guidelines/doc/ten-t-country-fiches/merged-files/ro.pdf.

[9] Jenelius E, Petersen T and Mattson L 2006 Importance and exposure in road network vulnerability analysis *Transportation Research A* **40(7)** pp 537-560.

[10] Popa M, Rosca E and Rusca F 2012 On transport network reliability, Transport problems **7(3)** pp 127-134.

[11] VISUM 10.0 2007 User Manual-Volume 1 PTV Traffic Mobility Logistics 2.136-2.140.

[12] Tu H 2008 Monitoring travel time reliability on highways PhD. thesis T.U. Delft.

[13] Bureau of Public Roads 1964 *Traffic Assignment Manual* U.S. Dept. of Commerce Urban Planning Division, Washington D.C.

IOP Conf. Series: Materials Science and Engineering **95** (2015) 012081 doi:10.1088/1757-899X/95/1/012081

Control of gaseous pollution via the leaves of non-edible trees

S J B Al-Maliky[1]

[1]Environmental Engineering/ AlMustansiriya University- Iraq

E-mail: salambash2000@yahoo.com

Abstract. The accelerated increase of the use of various transportation means, industrial machinery and other power consuming technologies has led to tremendous degradation of outdoor air quality all around the world. Green solution was tested here as an innovative gas control mean via non edible Myrtus communis green leaves as natural sorption media. Statistical analyses were applied in order to examine the correlation between various parameters of this study. The tests of gas records around the tree that was targeted by a gas stream of 5 KW power generators have demonstrated an excellent gas control role of the green leaves, with average efficiencies of about 75% and 82% for the removal of Nitrogen Dioxide and Carbon Monoxide, respectively. An interesting finding of this research was that the sorption role of green leaves has promoted their sizes and Chlorophyll Content Index.

1. Introduction

The control of gaseous pollutants; such as Nitrogen, Carbon and Sulpher Oxides, that are originated from various commercial and industrial uses of fossil fuels as power sources was a big concern for many scientists, researchers, and institutions for decades (CEC, 2005). Although many effective solutions were suggested and implemented in order to onsite emission reduction of these gases from stationary sources, such as air-fuel control and selective non-catalytic reduction (SNCR), the problem is still an issue for the moving sources, in addition to the technical and cost concerns that are related to these mentioned solutions [9]. From another side, huge efforts and resources were dedicated towards the search of green alternatives for both power sources and pollution control means, as a crucial step for the implementation of global sustainability. Ground heat and aerobic digestion of solid wastes in addition to solar, wind and tide and web sources have demonstrated excellent alternatives for fossil fuels [3,4]. In parallel to that, many researchers have examined the performance of roots and leaves of indoor plants like dwarf date palm, areca palm and Boston fem, for the control of harmful gas contaminations via absorption process that is followed by converting these pollutants to food that is beneficial for the growth of these plants in turn [7]. This reflects the multi role of these plants, as they act as gas control means in addition to being natural filters for the possible surrounding dust, and noise dumpers while supplying fresh Oxygen via photosynthesis process, nevertheless acting as friendly decoration and stress relieving for the inhabitants. The major possible restriction of the use of plants for the control of outdoor gaseous emissions was there durability against high temperatures that are related to various power sources/consumers., hence it is essential to search for proper kinds of plants that are either naturally durable such as Myrtus Communis or artificially made to be so [5]. Myrtus communis was examined against high temperatures and proved good tolerance for it leaves that emitted mono terpenes; such as α-pinene as a result showed no negative impact on its food metabolism [2].

Content from this work may be used under the terms of the Creative Commons Attribution 3.0 licence. Any further distribution of this work must maintain attribution to the author(s) and the title of the work, journal citation and DOI.
Published under licence by IOP Publishing Ltd

1.1. Myrtus communis
It is a highly deficiency lenient evergreen tree with dense flora that has so ancient history of being part of many traditional celebrations and religious rituals for the Mediterranean regions and Middle East nations, as these region`s mostly sunny and humid weather was found to be the perfect for blooming and cultivation of this tree. Its aromatic leaves were finely creased to be used for healing (antiseptic and anti-fever drugs) [6]. Strong emission of isoprenoids was observed by Affek and Yakir [1], whom reported the storage of monoterpenes in firmly impenetrable structures inside he leaf, detached from photosynthesis. monoterpene emission stimulus was also observed by Loreto et. al.,[8] in the leaves of Quercus ilex that are recovering from ozone stress. The isoprenoids could have been serving as a quenching agent for reactive oxygen species that were formed due to almost all stressful circumstances. Also, Myrtus communis and Viburnum tinus were found resistable and no indications were detected on any of the plants subsequent to 35 days of exposure to 30 ppb ozone [10].

Myrtus communis tree shall be tested here as a green widely available and low cost controller for the outdoor gas pollution originated from 5 KW power generator.

1.2. Setup and methodology
Small household 5 KW, MAC-AFRIC Petrol power generator was adopted to operate 3-5 hours per day as the experimental gas pollutant`s source that was set not less than 25cm apart from middle aged Myrtus communis bush that almost surrounds the generator. Concentrations of Nitrogen dioxide and Carbon Monoxide were recorded via MX6 iBrid portable multi gas detector, for six months; three times per week, just before and behind the tree in order to determine the role of it as a green gas sorbent. The possible impacts of this setup on some life indicators of Myrtus communis tree were tested by selecting samples of leaves from various locations of the tree in order to record their sizes and chlorophyll content (via CCM-200 plus portable device) prior and after the application of gas stream. The experimental data were statistically analyzed in order to assess the degree of possible correlation between gas reduction and impacts on the tree leaves.

2. Results and discussions
The average of three weekly tests was calculated as a representative of that week to be drawn versus time for each of Carbon monoxide and Nitrogen dioxide gas concentrations just before and behind the Myrtus communis tree, as illustrated in figure 1 and figure 2. The gradual contamination increase of gas emission from the generator as time passes after the start of this study that is shown in these figures for both gases under study was due to the intentional irregular maintenance in order to account for the highest possible contamination environment. The first 7-10 weeks of operation were observed to have, somehow, fluctuated outlet concentrations for both gases on the other side of the bush in a sign of unstable sorption performance of the tree leaves and this may be attributed to the adapting process against sudden gas streams, after which, leaves tended to show more stable performance as illustrated in the low and semi constant gas concentrations behind the tree.

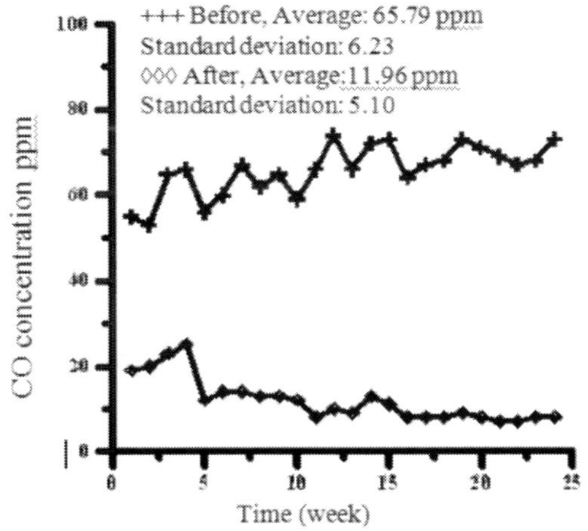

Figure 1. Role of Myrtus communis in the control of CO gas.

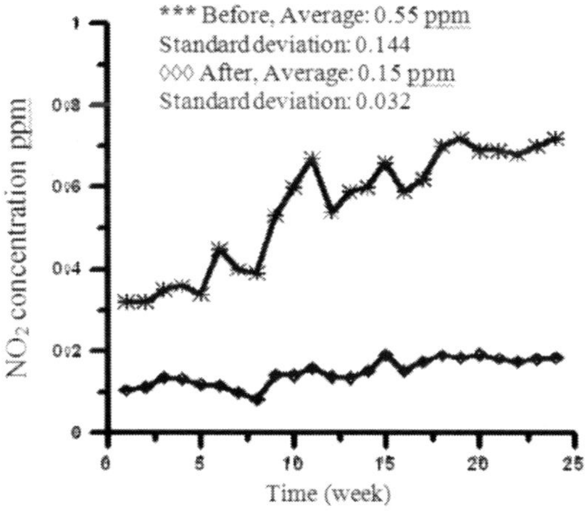

Figure 2. Role of Myrtus communis in the control of NO2 gas.

The previous indications were more clarified by figure 3 and 4, that show low and unstable efficiencies for the first 7 weeks (less than 60s% and 70s% for Nitrogen monoxide and Carbon dioxide gases respectively), however, with an increasing trend towards higher values. That increase rate tended to slow down after the tenth week to show more semi flat form for the rest of study period, in an interpretation of continuous stable control performance of the Myrtus communis tree against such harmful gases with mean efficiencies of about 75% and 82% for Nitrogen monoxide and Carbon dioxide gases respectively.

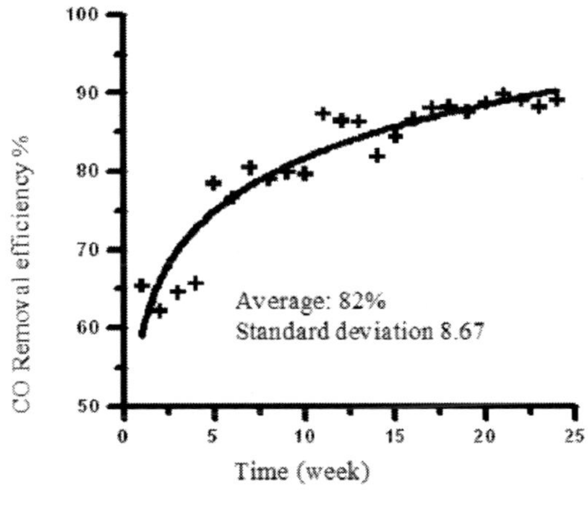

Figure 3. Removal efficiency of Myrtus communis for CO gas.

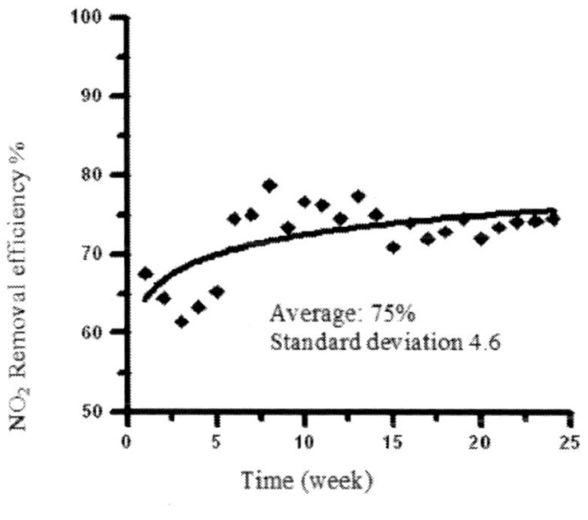

Figure 4. Removal efficiency of Myrtus communis for NO2 gas.

As regarding the reactions of the tree health indicator with the contaminated gas stream, Chlorophyll content index CCI was measured for various leaves in order to determine its trend as a result of the sorption of the tested gases, as illustrated in figure 5 that infer a slight decrease from CCI initial value of 7.5 to values in the range of 5, has occurred during the first four weeks, followed by sharp upsurge rate during the next few weeks that led to double its original value at the tenth week. The CCI has tended to almost keep that value for the rest of study period in a remarkable indication for the stable gain of gas sorption process for the benefit of photosynthesis process that is exchange ably related to the chlorophyll content. In consecutive action, the leaf sizes of randomly selected samples from all around the gas attacked bush has demonstrated slow gradual promotion for almost all tested ones for the first 16 weeks to reach the size of about 200 mm^2 as compared to their initial size of 120 mm^2, and almost keep that size or little higher for the rest of period of study as shown in figure 6, which, in addition to the promotion in CCI are strong interpretation for the energetic metabolism process for the conversion of absorbed food into the required chemical energy for plant`s activities, as a result of the gas sorption process.

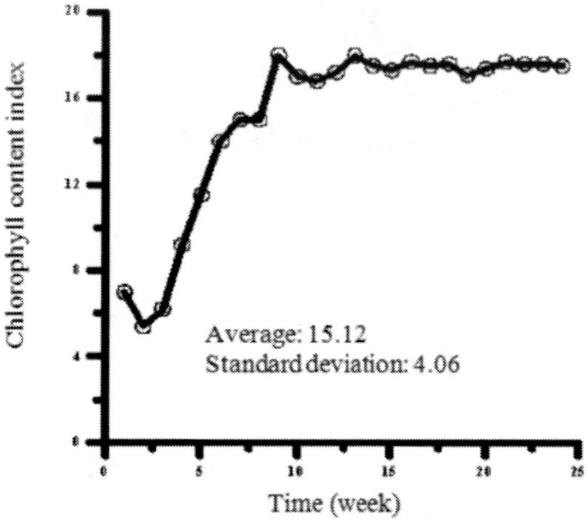

Figure 5. Myrtus communis tree CCI vs. Time.

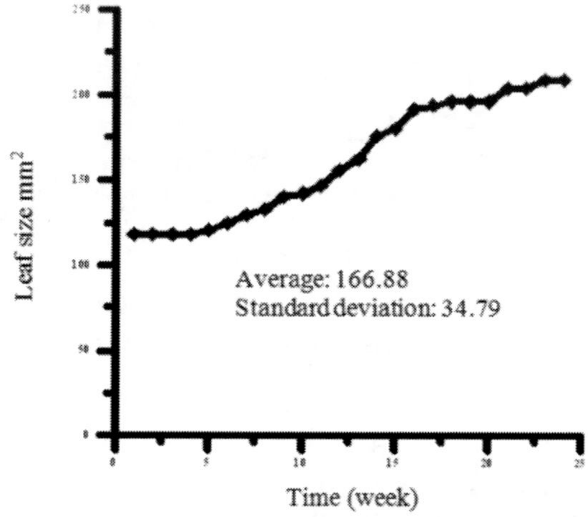

Figure 6. Average Myrtus communis leaf size vs. Time.

2.1. Statistical analysis

The previous experimental findings were statistically examined to determine the correlation levels between each pair of the various indicators and parameters of this study; contamination gases, Leaf size, and CCI, and to determine the closest mathematical relations that describe the interrelations, the results of which are tabulated in table 1. The high positive correlation coefficients between each pair of the tested parameters remark a possible causal relation or dependency relations between the sorption of Nitrogen dioxide and Carbon monoxide (as independents) from one side and the promotion of CCI and leaf size (as dependents) from the other one, even though, correlation coefficient (r) does not proof, certainly, that causal, however it boosts further searches.

Table 1. Statistical analysis between various parameters of the study.

parameter1 (P1)	parameter 2 (P2)	Analysis		
		Predicted relation	Coefficient of determination R2	Correlation Coefficient (r)
CO removed ppm	CCI	$P_2 = -41.38 + 1.858 * P_1 - 0.0146 * P_1^2$	0.798	0.84
NO2 removed ppm		$P_2 = -11.78 + 214.02 * P_1 - 387.91 * P_1^2$	0.94	0.90
CO removed ppm	Average leaf size mm^2	$P_2 = e^{(0.02\ P1) * 53.625}$	0.673	0.80
NO2 removed ppm		$P_2 = e^{(2.62 * P1) * 85.031}$	0.781	0.86

That anticipation of the existence of causal relations between the tree`s health indicators and the sorption of contamination gases was vastly demonstrated by the strong positive correlation between each pair of data in the scatter plots illustrated figure7 to figure10, which are close to the predicted mathematical relations listed in table 1 with their considerably high coefficients of determination (R^2).

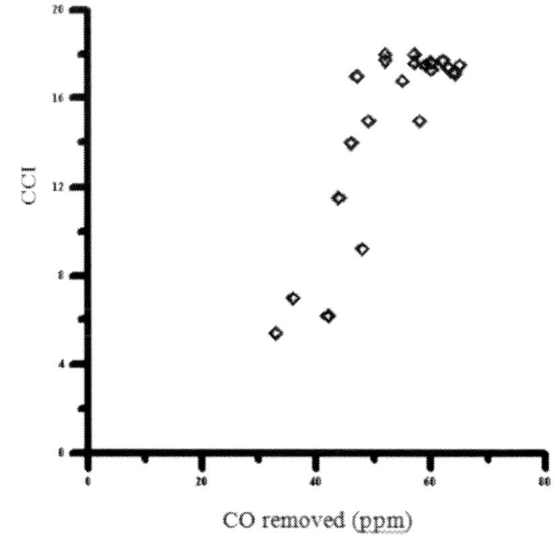

Figure 7. Scatter plot of CCI vs. CO removed.

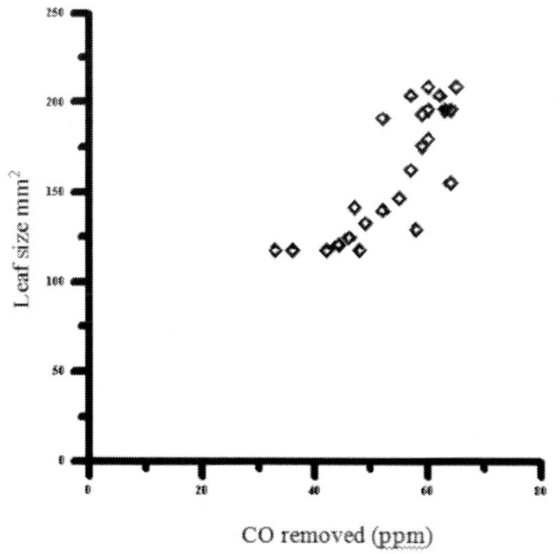

Figure 8. Scatter plot of average leaf size vs. CO removed.

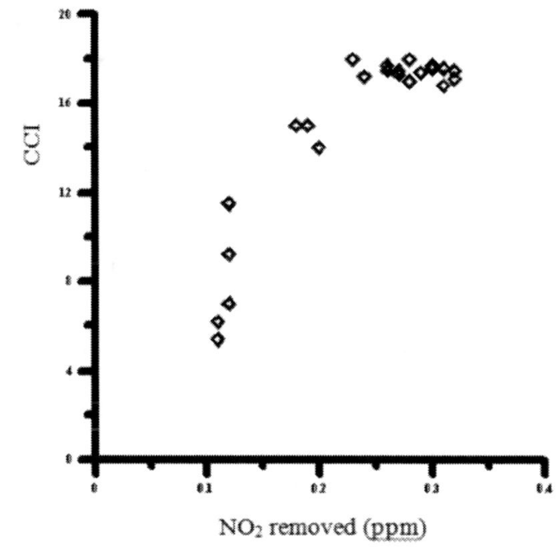

Figure 9. Scatter plot of CCI vs. NO2 removed.

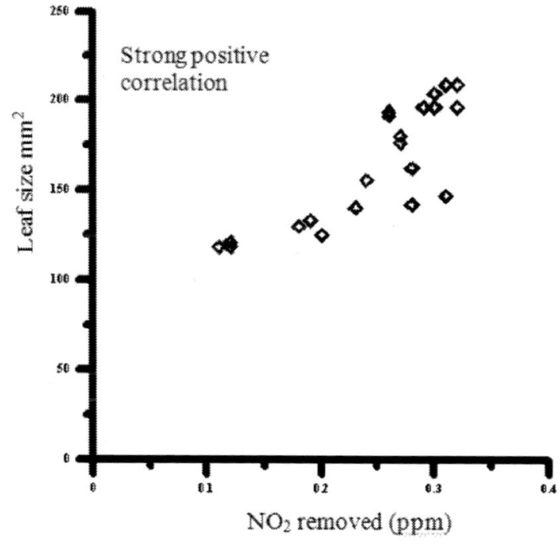

Figure 10. Scatter plot of average leaf size vs. NO2 removed.

3. Conclusions

The green simple and low cost control of harmful outdoor gaseous emissions of various power producing machines and technologies was studied via the adoption of one of the innovative solutions for this significant consequence of the huge industrialization era, by means of the non-edible Myrtus Communis tree which is a well-known plant in various parts of the world especially the Middle east and Mediterranean sea region. This may be considered as a great step towards global sustainability. The results of 24 weeks test of the mutual effects of gaseous emissions (Nitrogen dioxide and Carbon monoxide gases were selected as representatives of these emissions) as the sorbates and the health parameters of the tree (sorbent), has demonstrated significant performance in the sorption of both gases (average efficiencies of 73% and 82% for NO2 and CO respectively). In addition to the fact that almost no or little harmful effects were observed on the tree leaves, due to exposure to the contaminated gas stream, the analyzed health indications of these leaves (Chlorophyll content and leaf size) has demonstrated remarkable enhancement by the end of tests period, in a notation of well adaption and beneficial relation between all parameters. The statistical analysis of these data has made a strong support to the previous conclusions as noticed by the high positive correlation coefficients and scatter plots between various pairs of data, in a reference to a highly possible causal relations.

References

[1] Affek H P and Yakir D 2003 Natural abundance carbon isotope composition of isoprene reflects incomplete coupling between isoprene synthesis and photosynthetic carbon flow. *Plant Physiology* **131** pp 1727-1736

[2] Alessio G A, De Lillis M, Fanelli M, Pinelli P and Loreto F 2004 Direct and indirect impacts of fire on isoprenoid emissions from Mediterranean vegetation. *Functional Ecology* **18** pp 357-364

[3] AlMaliky S B 2010 Feasibility of Anaerobic Digestion of Flotation Skim and Its Potential as Renewable Energy Source *Journal of Sustainable Development* **3(3)**

[4] AlMaliky S B 2011 Investigation of the Readiness of Ground Soils for the Installation of Ground Heat Exchange Systems in Baghdad City. *Journal of Geography and Geology* **3(1)** doi:10.5539/jgg.v3n1p200

[5] Bita C E and Gerats T 2013 Plant tolerance to high temperature in a changing environment: scientific fundamentals and production of heat stress-tolerant crops. Front Plant Science **4** doi:10.3389/fpls.2013.00273

[6] Çakir A 2004 Essential oil and fatty acid composition of the fruits of Hippophae rhamnoides L. (Sea Buckthorn) and Myrtus communis L. from Turkey. *Biochemical Systematics and Ecology* **32** pp 809–816

[7] Kobayashi K D, Kaufman A J, Griffis J and McConnell J 2007 Using houseplants to clean indoor air. Cooperative Extension Service, College of Tropical Agriculture and Human Resources, University of Hawaii at Manoa. Ornamentals and Flowers, OF-39

[8] Loreto F, Pinelli P, Manes F and Kollist H 2004 Impact of ozone on monoterpene emissions and evidences for an isoprene-like antioxidant action of monoterpenes emitted by Quercus ilex (L.) leaves. *Tree Physiology* **24** pp 361–367

[9] Moretti A L and Jones C S 2012 Advanced Emissions Control Technologies for Coal-Fired Power Plants. Technical Paper BR-1886, Babcock & Wilcox Power Generation Group, Inc.Barberton, Ohio, U.S.A. Presented at:Power-Gen Asia, Bangkok, Thailand

[10] Orendovici T, Skelly J M, Ferdinand J A, Savage J E, Sanz M J and Smith G C 2003 Response of native plants of northeastern United States and southern Spain to ozone exposures; determining exposure/response relationships. *Environmental Pollution* **125** pp 31–4

Numerical methods for time-domain and frequency-domain analysis: applications in engineering

R D Tamas[1]
[1] Constanta Maritime University, Str. Mircea cel Batran nr.104, 900663 Constanta

E-mail: razvan.tamas@cmu-edu.eu

Abstract. Numerical methods are widely used for modeling different physical phenomena in engineering, especially when an analytic approach is not possible. Time-domain or frequency-domain type variations are generally investigated, depending on the nature of the process under consideration. Some methods originate from mechanics, although most of their applications belong to other fields, such as electromagnetism. Conversely, other methods were firstly developed for electromagnetism, but their field of application was extended to other fields. This paper presents some results that we have obtained by using a general purpose method for solving linear equations, i.e., the method of moments (MoM), and a time-domain method derived for electromagnetism, i.e., the Transmission Line Matrix method (TLM).

1. Introduction

Characteristic equations modeling different phenomena in engineering do not generally lead to analytic solutions. Numerical methods are therefore widely used in order to compute solutions to most equations with linear operators e.g., differential and integral equations.

Most phenomena are time-dependent, so they can be modeled as functions of time. A time-domain approach would better model a transient behavior; however, when an oscillatory or repetitive process is investigated, a frequency-domain representation might be preferred. This actually is the case of many problems to be solved in mechanics, electromagnetism, heat transfer, etc. A frequency-domain approach saves time provided that the response of a system is needed for a given frequency of oscillation and/or for a small number of harmonics. Nevertheless, a time-domain analysis with a pulsed excitation would provide thorough information on the oscillatory response for a myriad of frequencies by means of an inverse Fourier transform, and that despite of a long computing time. In this paper, we present and compare two numerical methods primarily employed in electromagnetism, but potentially applicable to other fields, as similar equations are to be solved.

Although some of our examples belong to the field of electromagnetism, they are far from being typical applications and we show therefore the high flexibility and the potential of application of the two methods.

The Method of Moments (MoM) [1, 2] is a general purpose method for solving equations with a linear operator applied to a function to be found. It is based on an approximate expansion of the unknown function, by using known basis functions. Basically, applications of the MoM in electromagnetism deal with functions of a coordinate (e.g., current distributions); though, we show that time-domain equations can also be accurately solved, and further a frequency-domain approach is absolutely feasible.

Content from this work may be used under the terms of the Creative Commons Attribution 3.0 licence. Any further distribution of this work must maintain attribution to the author(s) and the title of the work, journal citation and DOI.
Published under licence by IOP Publishing Ltd

The Transmission Line Matrix method (TLM) [3,4] is a more specific method that could solely be applied to wave-type differential equations, as that method is inspired from the way in which a pulse propagates within a transmission line mesh. It is therefore more appropriated to the time-domain analysis of the propagation of electromagnetic waves, sound, heat, etc. An insight of how the two methods could be applied to other fields of engineering is eventually provided.

2. Method of moments

The general form of an equation to be solved with the MoM is:

$$L_z\,[I(z')] = h(z) \tag{1}$$

with L_z a linear operator. In (1), L_z and $h(z)$ are known, and $I(z')$ is an unknown distribution.

The method of moments technique is based on a finite expansion of $I(z')$, by using of a function basis, $\Phi_n(z')|_{n\in\mathbf{N}}$. The resulting approximate form of (1) can be written as:

$$I(z') \cong \sum_{n=1}^{N} \Im_n \Phi_n(z') \tag{2}$$

where $\Im_n$ are weights to be calculated. Since the operator is linear, the equation to be solved becomes:

$$\sum_{n=1}^{N} \Im_n L_z[\Phi_n(z')] = h(z) \tag{3}$$

By sampling the z variable, equation (3) can be put in the following discrete form:

$$\sum_{n=1}^{N} \Im_n L_z[\Phi_n(z')]_{z=z_m} = h(z_m) \tag{4}$$

with m=1,2,…M and further in the following matrix form:

$$[\mathbf{A}]\,[\boldsymbol{\Im}]=[\mathbf{B}]. \tag{5}$$

The resolution of the equation actually reduces to a matrix inversion leading to the weights $\Im_n$.

3. Transmission line matrix method

The transmission line matrix (TLM) method [3,4] has originally been developed to model propagation phenomena, based on the analogy between the space and time variation of the characteristic parameters of the process under consideration, and the voltage and/or current variation within a 2D or 3D transmission line mesh, respectively. Examples of 2D and 3D nodes are given in figures 1 and 2, respectively [5].

The resolution of the equations of propagation is performed iteratively, including pulse scattering in each node of the mesh, propagation between two nodes, and boundary conditions. As an example, when characteristic figures are modeled as incident and reflected voltages in the mesh, scattering and propagation can be expressed as

$$_{k+1}[V]^r = [S]\cdot_k[V]^i \tag{6}$$

and

$$_{k+1}[V]^i = [C]\cdot_{k+1}[V]^r, \tag{7}$$

where $[S]$ is the scattering matrix, $[C]$ is the propagation matrix, and k denotes the time step.

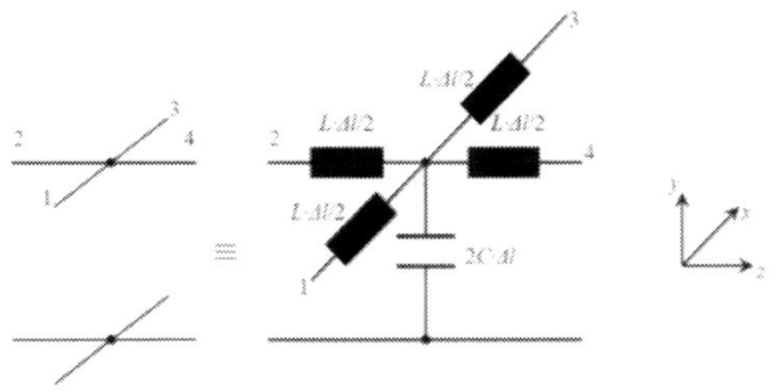

Figure 1. Example of 2D node and the corresponding lumped-element model.

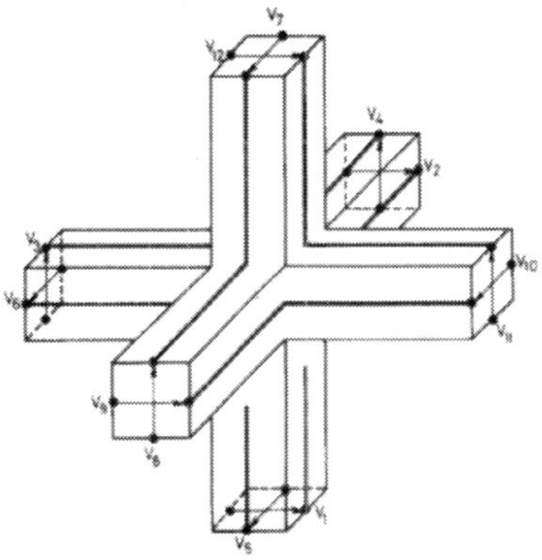

Figure 2. Example of 3D node.

3.1. Example of application of the MoM in kinematics: decomposition of a complex displacement

Let us consider a mobile platform moving along the x-axis with the instantaneous acceleration $a_x(t)$. The relationship between displacement and acceleration is given by a second order differential equation,

$$\frac{d^2x(t)}{dt^2} = a_x(t) \tag{8}$$

Equation (8) is one with a linear operator, namely:

$$L\{I\} = \frac{d^2}{dt^2} \tag{9}$$

where the displacement $x(t)$ is still unknown.

In order to compute the unknown distribution $x(t)$, an alternative to the double integration of equation (8) would be to expand the movement into a finite series by means of appropriately chosen basis functions, and to apply the method of moments, in order to calculate the coefficients of the series.

In that case the finite series is:

$$x(t) \cong \sum_{n=1}^{N_b} A_n \cdot \varphi_n(t) \tag{10}$$

where A_n are the unknown weights, $\varphi_n(t)$ are the basis functions, and N_b is the total number of basic functions.

In order to assess the stresses to which the mobile platform is subjected and to assess the effects of movement on the crew comfort, the basis function should be chosen such as to correspond either to different orders of accelerated movements, or to oscillations of different frequencies. In practice, a model with accelerations of order of no more than three (throttle acceleration) can be used. The most important parameter for assessing the crew comfort is the second order acceleration (speed of acceleration), the third order acceleration being particularly important for space crafts. Thus, the model includes three basis time-domain polynomial functions.

The basic functions of order more than 4 are oscillations of different frequencies. As for a Fourier series, half of them will be sine functions and half cosine functions.

We used the point matching technique, that is, the test functions used for discretization are delayed Dirac pulses [6,7].

For validation purposes, we measured the instantaneous acceleration onboard a small boat. We used the method of moments with N=107 basis functions, in order to extract the x-coordinate. Then, we calculated the second derivative of the x-coordinate resulting from the method of moment approach, and we compared it to the measured, instantaneous acceleration [6, 7]. Both results are given on figure 3.

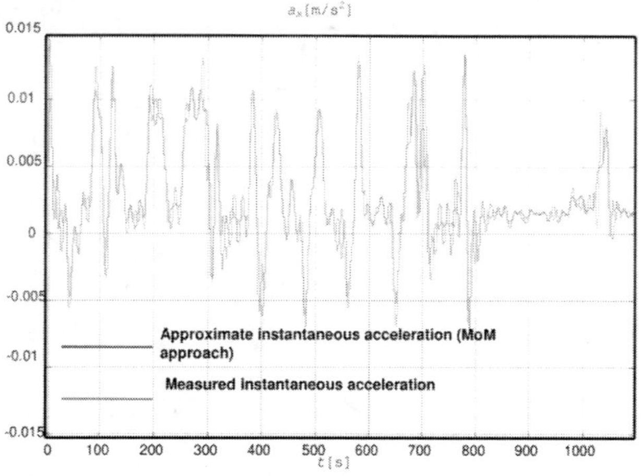

Figure 3. Approximate, instantaneous acceleration resulting from the MoM approach versus measured, instantaneous acceleration.

3.2. Example of application of the TLM method: characterization of a short dipole antenna
Typical applications of the TLM method in electromagnetism include antenna characterization in the time-domain, and particularly evaluation of the antenna response to a pulsed excitation [8]. Figure 4 gives the waveform of the normalized, electric field radiated by a short dipole antenna; simulated and analytical results are given. A good agreement can be noted, although the TLM approach is based on some approximations.

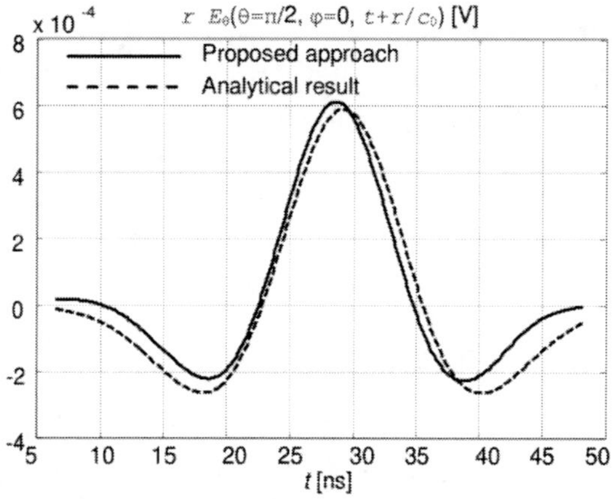

Figure 4. Response of a short dipole to a Gauss pulse

3.3. Example of joint application of the MoM and TLM method: characterization of ultra-wide band antennas
Let us consider an antenna with a pulsed excitation. In practice, the impulse response of the antenna must be calculated when excitation and radiated field waveforms are known. Referring to the notations in figure 5, the equation to be solved is

$$V_g(t) * h_t(\hat{\mathbf{r}}, t) = r\, E(\mathbf{r}, t + r / c_0) \tag{11}$$

where $V_g(t)$ is the antenna input voltage, $h_t(t)$ is the impulse response, and $E(\mathbf{r}, t+r/c_0)$ is the electric far-field. Both the potential signature i.e., the right-hand side member of (11), and the impulse response depend on the direction alone.

The unknown impulse response can be approximated as follows:

$$h_t(\hat{\mathbf{r}}, t) \cong h_{t,app}(\hat{\mathbf{r}}, t) = \sum_{n=1}^{N} h_n \Phi_n(t) \tag{12}$$

where $\{\Phi_n|_{n=1,N}\}$ are the basis functions. By discretizing the time variable t, the coefficients h_n can be computed by solving a linear system of equations that can be written in matrix form as follows:

$$[\boldsymbol{V}]\,[\boldsymbol{h}] = [\boldsymbol{E}] \tag{13}$$

where $[\boldsymbol{h}]$ is the vector of the weighting coefficients h_n, $[\boldsymbol{E}]$ is the vector of samples of the potential signature, and $[\boldsymbol{V}]$ is a matrix whose elements are $V_{mn}=(V_g*\Phi_n)(t_m)$, $m, n=1\ldots N$.
Let t_1 be the duration of the input signal and t_2 the duration of the potential signature. Prior to calculate the coefficients h_n time-shifted Gauss pulses were used as basic functions:

$$\Phi_n(t) = \exp\left[-\left(\frac{b \cdot t - (2n-1)t_1/2}{2T}\right)^2\right] p_n(t, t_1) \tag{14}$$

with $n=1, 2, \ldots N$, $b = Nt_1/(t_2-t_1)$, and $p_n(t, t_1)=1$ if $t \in [(n-1)t_1/b,\ nt_1/b]$ and 0, otherwise.

As an example, we consider a rectangular notched monopole antenna on a square ground plane (figure 6). The excitation is a voltage pulse proportional to the first derivative of a Gauss function. The maximum of the spectrum magnitude of the input waveform occurs around 100MHz. We used the TLM method to compute the time-domain near-field on a closed surface including the antenna, and then the electric far field $E_\theta(\mathrm{r}, t)$ on the direction ($\theta=\pi/2$, $\varphi=0$) was calculated by a time-domain near-field to far-field transformation. Subsequently, the approximate impulse response $h_{t,app}$ was computed (figure 7) by deconvolution using the time-domain approach that we propose.

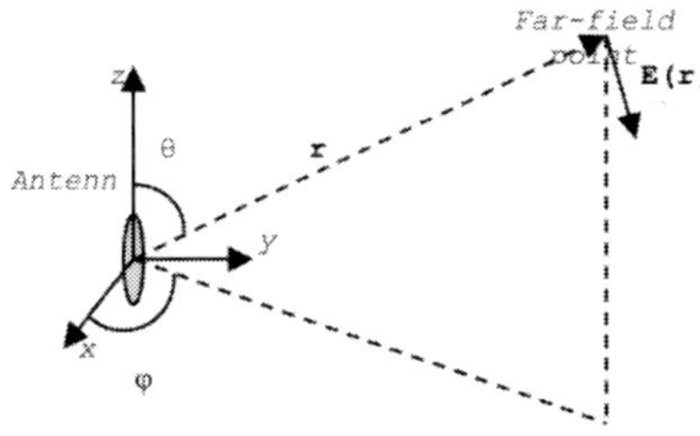

Figure 5. Electric far-field of an antenna: notations.

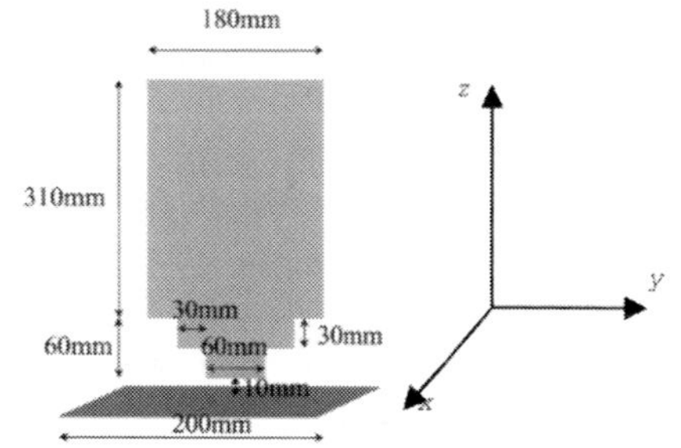

Figure 6. Rectangular notched monopole antenna

The accuracy of the method is evaluated by reconstructing the potential signature from the resulting approximate impulse response and from the excitation, i.e., $h_{t,app}(t)*V_{g0}(t)$. This result is compared to the original potential signature, computed by using the TLM method and a time-domain, near-field to far-field transformation (figure 8).

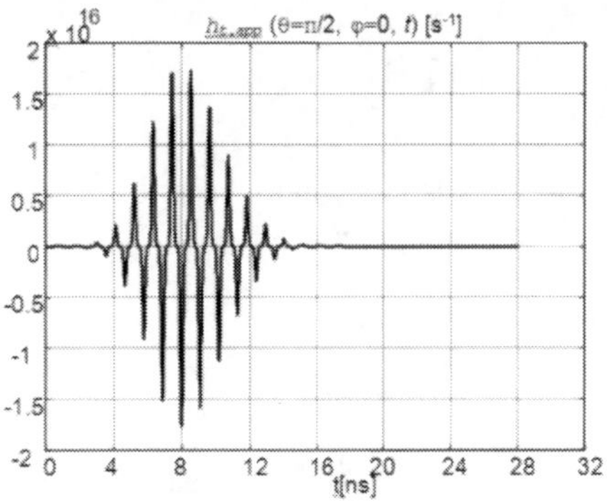

Figure 7. Approximate impulse response.

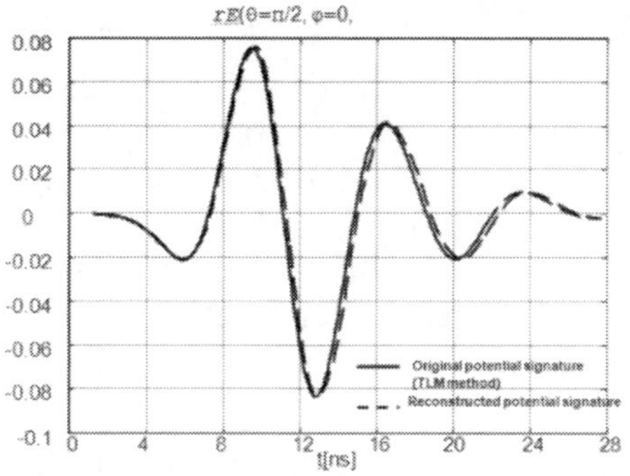

Figure 8. Potential signature for an excitation proportional to the first derivative of a Gauss function.

4. Conclusions

In our first example, we compared the instantaneous, measured acceleration provided by an inertial navigation system to the approximate, instantaneous acceleration resulting from the second derivative of the approximate coordinate, calculated with the MoM. Then, we compared the impulse response of a short dipole computed by using the TLM method to the analytic result.

Finally, we showed how an approximate form of the impulse response of an ultra-wide band antenna can be found, by using the method of moments. We compared the time-domain response of an antenna, issued from simulation by using the TLM method, for the first derivative of a Gauss function as an excitation, to the response calculated by convolving the approximate impulse response with that excitation waveform.

As a conclusion, the MoM and TLM approaches yield accurate results for every one of the applications we have presented; nevertheless, the TLM method can be applied for modeling other propagation phenomena (sound, heat, etc.), and the MoM can also be used to solve other linear operator equations, in order to model electrical circuits, wave propagation, current distribution on high-frequency devices etc.

References

[1] Harrington R F 1992 Origin and development of the method of moments for field computation, in E.K. Miller et al. *Computational Electromagnetics*, Frequency-Domain Method of Moments IEEE Press pp 43-47
[2] Cheng A H D and Cheng D T 2005 Heritage and early history of the boundary element method *Engineering Analysis with Boundary Elements* **29** pp 268-302
[3] Hoefer W 1985 The Transmission-Line Matrix Method - Theory and Applications, IEEE Trans. *Microwave Theory and Tech.* **10** pp 882-893
[4] Hoefer W 1991 Huygens and the Computer - A Powerful Alliance in Numerical Electromagnetics, *Proceedings of the IEEE* **11** pp 1459-1471
[5] Johns P B 1987 A Symmetrical Condensed Node for the TLM Method, *IEEE Transactions on Microwave Theory and Tech.* **4** pp 370-377
[6] Dumitrascu A, Tamas R D, Dragulinescu M and Caruntu G 2014 A method of moments approach for decomposition into elementary movements of displacement data from an inertial navigation system *Proc. 21st Saint Petersburg* pp 380-383
[7] Dumitrascu A, Tamas R D and Dragulinescu M 2014 On the accuracy of the method of moments approach for decomposition into elementary movements of displacement data from an inertial navigation *Advanced Materials Research* **1036** pp 981-986
[8] Tamas R D 2010 Antenna Theory: Traditional versus Modern Approach *Nautica* (Constanta)

IOP Conf. Series: Materials Science and Engineering **95** (2015) 012083 doi:10.1088/1757-899X/95/1/012083

Model of decision system for ^{13}C Isotope Separation column

M L Boca[1]
[1]University "1 Decembrie 1918" of Alba Iulia-Romania, Faculty of Science and Engineering, Blvd. N. Iorga, No. 11-13, 515900, Alba Iulia, Romania

E-mail: loredana_boca1@yahoo.com

Abstract. This paper presents the model of a decisional system for ^{13}C Isotope Separation column, which is used to detect mission critical situation. The start model was a model of one distributed control system of critical situations that may arise in the operation of the distillation column. The research work it is proposed a model of decision system which implement a temperature sensor inside of liquid nitrogen level in the condenser. The condenser is a part of column where take place the cryogenic process using nitrogen liquid. The work temperature is very low about -192°C, and because the temperature can grow or go down more than 2 degrees is a very critical location inside the column. In this way the column has a deeply monitor and supervised and it take a decision in a proper time when the temperature is grow up or getting down and became a critical situation. For monitor and supervised it was used MatLAB SimuLink. The model, the decision system gives a signal to one sensor when something is wrong in the condenser which is the most critical place of the isotopic column. In this way it creates an alarm that something is getting wrong in the isotopic column.

1. Introduction

In the last 15 years, the availability of increasingly powerful computers at a decreasing cost, on one part, and advances in operations research science that have resulted in improved algorithms, on the other part, have given the possibility to solve larger and more complicated problems than ever before [1]. The need of an optimisation-based decision support system, which can also be viewed as an application-specific delivery system for an industrial process [1], arises when one technology can be used to support highly repetitive decisions.

The level of continue increasing of complexity in industrial processes causes a corresponding growth in the complexity of control algorithms. Similarly, the structure of controlled technological objects and process control requirements is affecting the control of the system and its software architecture and system– wide quality requirements.

The column, in simplified scheme, separates the carbon isotopes based on the cryogenic distillation of pure carbon monoxide, which is fed at a constant flow rate as a gas through the feeding system.

Most of the systems are simultaneously control and real-time systems, since their proper functioning is directly dependent on the satisfaction of certain time restrictions. For time restrictions should be adopt a way of planning tasks.

In time with technological progress, monitoring systems for remote control and distributed control systems has proven their value through some contribution, to increasing the effectiveness and efficiency of the activities of the industry processes.

Content from this work may be used under the terms of the Creative Commons Attribution 3.0 licence. Any further distribution of this work must maintain attribution to the author(s) and the title of the work, journal citation and DOI.
Published under licence by IOP Publishing Ltd

2. Model of ^{13}C isotope separation column

The column separates the carbon isotopes based on the cryogenic distillation of pure carbon monoxide, which is fed at a constant flow rate as a gas through the feeding system [2].

Take in consideration the very-low operating temperature, an efficient thermal isolation vacuum jacket is necessary. Since the "elementary separation ratio" [2] is very close to unity in order to raise the (^{13}C) isotope concentration up to a desired level, a permanent counter current of the liquid-gaseous phases of the carbon monoxide is created by the main elements of the equipment: the boiler in the bottom-side of the column and the condenser in the top-side.

The gaseous carbon monoxide upstream (from boiler and from the feeding system) condenses on the "coldwall" of the condenser, cooled with liquid nitrogen by atmospheric pressure and falls, in small drops, downstream to the electrically heated boiler [3].

The (^{13}C) isotope, slightly heavier than the predominant (^{12}C) isotope, accumulates in liquid phase and will be extracted as end product at the bottom of the column, while the (^{12}C) component accumulates in vapours phase and will be extracted as waste at the top of the column. The column operates with two zones: the stripping zone, from the feeding point to the top of the column and the enriching (rectifying) zone, in its lower part [3].

The characteristics of the carbon cryogenic separation column, presented, needs a necessity and modern control strategies.

The counter current cryogenic distillation column related is a highly complex plant, nonlinear, multivariable, with many time constants that overcomplicate the control solutions. The control system is to maintain the column operation parameters constant, by eliminating the effects of disturbances while keeping the overall system stable, despite the uncertainties that may arise [4].

The main parameters that need to take in use for monitoring and control are:

- The liquid nitrogen level in the condenser of the column. The drop of the liquid nitrogen level below a critical value would lead to the impossibility of efficiently condensing the vapours upstream and thus would compromise the entire separation process.
- The electrical power supplied in the boiler part. High variations of the power supplied would affect the separation by modifying the upward gaseous stream.
- The vacuum pressure of the column. Variations in the vacuum pressure bring about the loss of the efficient thermal isolation and cause the increment of the inner column temperature.

It is very important to know the material and energetic column equation to establish the mathematic model of the separation process [3].

The material equation for the separation column is [8]:

$$\frac{d(N \cdot x)}{dt} = F_a x_a - F_v y - F_b x \tag{1}$$

Where: N is the number of moles of liquid from the column, and Fa, Fv, Fb are the molar flow of feeding at the top and base of the column.

Energy balance equation for separation column will determine the enthalpy of the liquid and vapour in the system and is [3]:

$$\frac{d(N\lambda)}{dt} = F_a\lambda_a - F_v\lambda_v - F_B\lambda - a_p + Q_{ag} \tag{2}$$

Because the ^{13}C isotope separation column is a critical process I try to make a model of one decision system which will be able to monitor the column and make one decision when some critical situation appear.

In these simulations, a wireless communication network is used, to transmit data. It was taken as study case, two critical situations that may arise due to increasing pressure parameter:

a) increasing the pressure parameter inside the column;

b) increasing the pressure parameter in the vacuum cover of the column .

In the first case it is necessary the action of an electro-valve which will open pipe connecting the column and a vacuum container, resulting in pressure drop in the column to the reference value. To automatically adjust the electro-valve's position and implicit the flow rate of pipe on which it is acting, it is used a PI controller. In the second case it is necessary the consecutive start-up and in parallel of two vacuum pumps.

For proper operation of the distillation column is required thermal insulation from the outdoor environment. Thermal insulation is achieved by ensuring a high vacuum values in column casing. On the occurrence of an alarm that signals the increase of pressure in casing of column, first of all is turned on high vacuum pump.

If pressure in casing does not drop for a specific time, in parallel with the first, the second pump is turned on.

The transfer function of the electro-valve is obtained using the relation:

$$H_{EV}(s)=\frac{10}{(0.005s+1)(0.11s+1)}=\frac{10}{0.00055s^2+0.115s+1} \quad (3)$$

To automatically adjust the position of electro-valve and the flow pipeline which it acts, it is use PI regulator (Kessler method) which transfer unction is:

$$H_{PI}(s)=0.22(1+\frac{1}{0.11s}),\ \text{unde}\ K_P=0.22,\ \text{iar}\ T_I=0.11\text{sec.} \quad (4)$$

In the figure 1 it is shows the simulation scheme of distributed control system for critical situations afferent to column

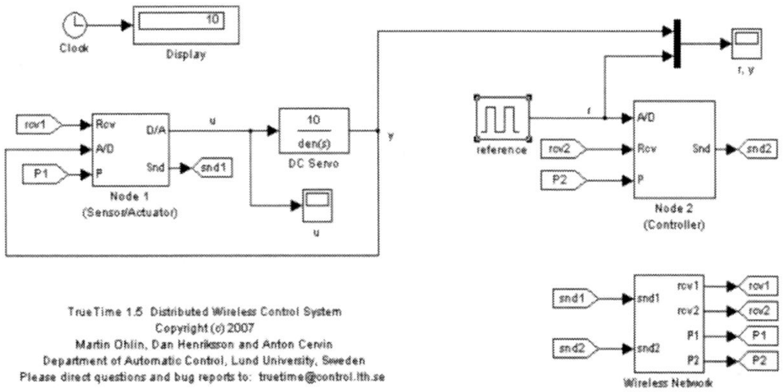

Figure1. Simulation scheme of distributed control system for critical situations afferent to column.

3. Generalised Model of critical situation-tolerant Control Systems

In the community control, the most used technique to describe a control system is based on logical function blocks. This approach is also supported by most generic and custom design tools for control processes. Each function block can have several inputs, performs some function, and produces a number of outputs.

Several function blocks can be combined in different topologies, each function block can be decomposed into more primitive blocks.

This model is used for theoretical system analysis and to design and implement on control processes. Moreover, this model-based approach becomes more and more popular for dealing with all kinds of industrial processes.

There are further benefits of using function blocks. When designing distributed systems, the innermost function blocks represent basic units of allocation to processing elements. A composite function block, on the other hand, can be executed in distributed mode. Furthermore, on the top (system or context) level and entire control system – distributed or not- can be considered as a single function block.

A generalised model of a control system needs to take critical situation tolerance and real-time issues into account that are typically not integrated in the traditional development methods for control applications. The model considered has been designed for easy implementability on diverse general distributed architectures [4, 5].

Furthermore this model is use for control the liquid nitrogen level in the condenser of separation column. Its basic idea is to combine control functions (composed of function blocks) with a monitoring function and reconfiguration capabilities, and to include the notion of system resources.

The monitoring function supervised the behaviour of input and output signals that control the nitrogen level in the condenser which is critical for column functionality. For this it used the model Decisional Critical Situations System (DCSS).

A DCSS combines both the algorithmic part of a control application and all resources needed to perform the task given. The algorithmic part is represented by a set of function blocks [6]. In addition, a DCSS encloses also the critical situation tolerance and temporal features of the encapsulated blocks [7].

In the context level, an entire control system is considered as a single global DCSS that contains all hardware resources and program code. Its objectives are expressed generally (e.g., "control one process").

This DCSS can be decomposed into a set of more primitive DCSS that perform a number of subtasks of the primary objective.

Instead of running the control application code, the global DCSS is managing its sub-blocks. Each sub-block owns a subset of resources and a part of program code allocated to the superior block. On this level, the objectives are more specific (e.g., "maintain the temperature of a mixture"). This decomposition can be applied recursively, i.e., DCSS can be decomposed hierarchically.

On the lowest level of the hierarchy, application code of control functions is executed. Since a specific objective can be accomplished in different ways, it is not necessary that all sub-blocks are active at a specific point in time. Selecting alternative function blocks and/or alternative DCSS also provides for both critical situation tolerance and timeliness [7].

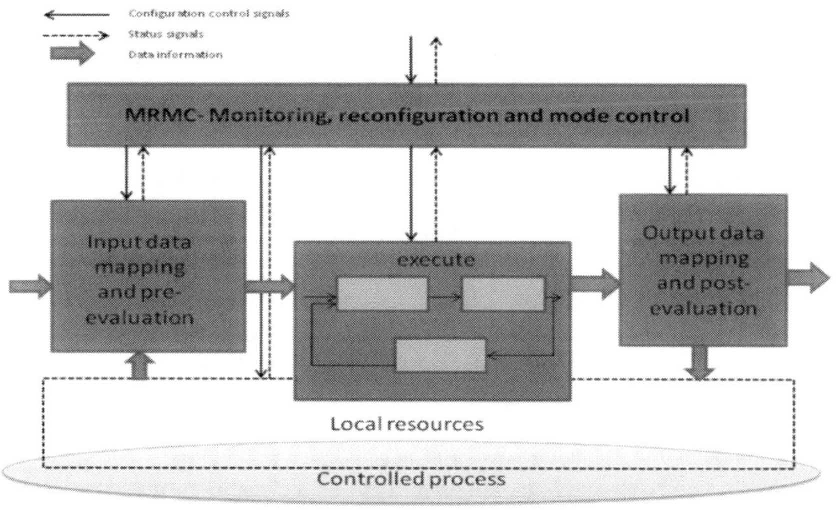

Figure 2. Structures of a DCSS path [7].

The internal configuration of a DCSS at a specific instant is selected by configuration control signals provided by higher levels in the DCSS hierarchy, or by the operator on the top-most level. Each configuration also inherently performed. In the case of hard real-time tasks these parameters are firm. For soft real-time tasks, some of its can be given in ranges, or by some fuzzy rules. This allows for better flexibility in cases that only limited resources are available or of transient overloads.

If there are not enough local resources available to perform a selected configuration status information is generated to the upper levels in the DCSS hierarchy and, eventually, to the operator. Consequently, at an upper level, alternative scenarios can be chosen, the general objective can be degraded, or the system can be shut down in a controlled manner. If a DCSS can handle hardware, software, and timing critical situations locally, a higher level of hierarchy is not notified.

The internal structure of DCSS path is shown in figure 2. In general, any DCSS acquires some input information, processes and produces outputs. The inputs are read either from sensors in local resources, from operators, or from other DCSS [7, 9].

Traditionally, sensors, actuators, and other resources are not considered to be parts of function blocks, as in the case for DCSS. For instance, a DCSS measures temperature instead of just reading a sensor. Since the corresponding sensor is part of the DCSS local resources, in case of a critical situation the temperature may be determined in an alternative way, if possible without the need to notify the superior DCSS.

A conversion function (e.g., from an acquired raw integer to a physically correct floating-point temperature value) may be defined within the DCSS, together with bounds and/or other plausibility control. For an alternative sensor, a different function may apply. Similarly, DCSS is influenced the plant through changing some physical properties (e.g., by heating the liquid in a tank, or by increasing/decreasing fuel flow). This approach allows for a higher degree of flexibility and for better critical situation tolerance [8, 10].

4. Description of DCSS

A complete DCSS is composing of:

Local resources. Any DCSS is associated with a set of resources: processing capabilities, input/output interfaces, sensors and actuators, etc. Some of the resources are available exclusively to a single DCSS, and some can be shared among several DCSS. For example, more than one DCSS can be executed on a single processing element, or sensor data can be read by several DCSS.

However, an actuator will most likely be associated with one single DCSS, only. A set of resources available can change during local or global reconfiguration of a system. The resources allocated to a specific DCSS are available in all its sub-blocks as well. There can be no resources locally assigned to a sub-block, which are not part of superior block, too. Therefore, the DCSS on the context level owns all resources of the control system [7].

Monitoring, reconfiguration and mode control (MRMC). The module manages the DCSS. Based on external configuration control signals, the MRMC allocates and configures available resources and other components of the DCSS to perform a requested task. According to the configuration selected, the programs performing the control algorithm are invoked.

The MRMC also monitors all associated resources and traces any malfunctions in them. If a critical situation is detected by MRMC during the execution of the executive DCSS, the MRMC firstly, tries to deal with it locally by using redundant or alternative components, by reconfiguring the local sub-blocks, etc. If a critical situation cannot be dealt with locally, the MRMC immediately notifies the superior entity by means of appropriate status information.

Input data mapping and pre-evaluation. This module receives all input data signals needed by the DCSS to perform a given task. The inputs originate either from the local resources of from external information paths. It maps external data values into the data domain used by the DCSS. For example, it can map integer values produced by an A/D converter into floating-point quantity that represents the actual control parameter [7, 9].

The module can check the plausibility of input data and inform the MRMC if anomalies are detected. It also serves as a multiplexer if alternative information sources for some input exist.

For instance, several redundant sensors may by allocate to the DCSS. After start-up, one of them serves as actual data input. If this sensor fails, the MRMC can transparently switch to the next available one without the need to restart the primary control function. In some cases, if no alternative for some input is available, the MRMC can activate an auxiliary module that will provide the missing information by an approximation based on other sensorial inputs. This information is then delivered to the same data input as if it came from the sensor. Not all input data needs mapping and probably not every signal is tested.

Executive. The executive performs the DCSS control functions according to the configuration parameters set by the MRMC. On non-leaf levels of the block hierarchy, the executive consists of several subordinate DCSS [8].

The MRMC controls which subordinate DCSS should be activated, and how they interact with each other. On the elementary level, the executive executes program code of control functions on the processing element, some dedicated hardware component, etc., which are part of the local resources.
In either case, the executive provides also some information on the MRMC if required. As noted before, sometimes the MRMC should know if subordinate DCSS cannot execute their tasks within required deadlines, or if subordinate DCSS do not have the capability to perform the tasks at all.

5. Conclusion and future work

In this paper it is propose a model DCSS use to deeply monitor the ^{13}C isotope separation column the most important part of column being the liquid nitrogen level in the condenser. In this condenser the temperature has to be deeply monitoring because it can't have fluctuations more the 2 up or down degrades.

The DCSS has rolled to make a decision in the column when some critical situation appears at condenser level.

In the future it is wanted to implement this model inside the condenser of the ^{13}C isotope separation column, more then that to apply this model inside of vacuum column.

References

[1] Vânătoru M 2001 Automatic conduction of industrial processes *Control Engineering Series* **1** (Universitaria Publishing House: Craiova, Romania)

[2] Boca M L, Secara M and Abrudean M 2013 Methods for control of some critical situations afferent of a ^{13}C isotope separation column *International Journal of Modern Manufacturing Technologies* **IV** pp 21-26
[3] Axente D, Abrudean M and Bâldea A 1994 15N, 18O, 10B, 13C isotopes separation by isotopic exchange Ed. (Science Book House: Romanian)
[4] Ciortea E M 2013 Analytical interpretation to optimize transport systems *Advanced Materials Research* **027** pp. 881-885
[5] Boca M L and Secara M 2014 Analogy between "mission critical" detection in distributed systems and ^{13}C Isotope Separation Column *Proc. of SPIE* **9258**
[6] Muntean I, Unguresan M L, Savu A and Abrudean M 2012 Modeling and composition estimation of an isotopic distillation column, *Automation Quality and Testing Robotics (AQTR)* pp 109-113
[7] Matjaz C and Wolfgang A H 2003 Architectural support for predictability in hard real-time systems *Control Engineering Practice* pp 51-59
[8] Hermann K, Damm A, Koza Ch, Mulazzani M, Schwabl W, Senft R and Zainlinger Ch 1989 Distributed Fault-tolerance real time systems *IEEE Micro* **9** pp 25-40
[9] Matjaz C, Domen V and Wolfgang A H 2008 *Distributed Embedded Control System* Springer
[10] Pop C I and Dulf E H 2010 Control Strategies of the ^{13}C Cryogenic Separation Column *CEAI* **12** pp 36-43

IOP Conf. Series: Materials Science and Engineering **95** (2015) 012084 doi:10.1088/1757-899X/95/1/012084

Integration of the virtual 3D model of a control system with the virtual controller

K Herbuś[1] and P Ociepka[1]

[1] The Silesian University of Technology, The Faculty of Mechanical Engineering, The Institute of Engineering Processes Automation and Integrated Manufacturing Systems, Konarskiego 18A, 44-100 Gliwice, Poland

E-mail: krzysztof.herbus@polsl.pl

Abstract. Nowadays the design process includes simulation analysis of different components of a constructed object. It involves the need for integration of different virtual object to simulate the whole investigated technical system. The paper presents the issues related to the integration of a virtual 3D model of a chosen control system of with a virtual controller. The goal of integration is to verify the operation of an adopted object of in accordance with the established control program. The object of the simulation work is the drive system of a tunneling machine for trenchless work. In the first stage of work was created an interactive visualization of functioning of the 3D virtual model of a tunneling machine. For this purpose, the software of the VR (Virtual Reality) class was applied. In the elaborated interactive application were created adequate procedures allowing controlling the drive system of a translatory motion, a rotary motion and the drive system of a manipulator. Additionally was created the procedure of turning on and off the output crushing head, mounted on the last element of the manipulator. In the elaborated interactive application have been established procedures for receiving input data from external software, on the basis of the dynamic data exchange (DDE), which allow controlling actuators of particular control systems of the considered machine. In the next stage of work, the program on a virtual driver, in the ladder diagram (LD) language, was created. The control program was developed on the basis of the adopted work cycle of the tunneling machine. The element integrating the virtual model of the tunneling machine for trenchless work with the virtual controller is the application written in a high level language (Visual Basic). In the developed application was created procedures responsible for collecting data from the running, in a simulation mode, virtual controller and transferring them to the interactive application, in which is verified the operation of the adopted research object. The carried out work allowed foot the integration of the virtual model of the control system of the tunneling machine with the virtual controller, enabling the verification of its operation.

1. Introduction

The dynamic growth of simulation capabilities, offered by modern engineering software of the CAE class, action allows conducting a simulation of a virtual 3D model of the given control system [7,12-14,16,18]. In the mentioned class of software the process of operation simulation is always preceded by a process of creating the concept, using different methods of designing and modeling [1,2,6,8,10-11,15,17]. While the integral part accompanying the verification process of operation of a system is the process associated with creating the control application using the software for programming the

Content from this work may be used under the terms of the Creative Commons Attribution 3.0 licence. Any further distribution of this work must maintain attribution to the author(s) and the title of the work, journal citation and DOI.
Published under licence by IOP Publishing Ltd

PLC controllers, by means of which it could be, in a rational way, verified the correctness of a signal flow, with respect to the proposed drive system. The solution consisting in verification the operation of the given control system, in the form of the 3D model basing on control signals from the virtual controller, requires the use of an integrator [4,5,9]. The integrator has the task receiving output signals from the software of a virtual controller and transferring them to the system with the 3D model of the analyzed system, where on their basis is performed the simulation of its operation. The presented procedure is of particular importance in the case of designing specialized machines, manufactured in the regime of the piece production, which is especially important in the case of specific operating conditions of this device. The machine, of this type, is the tunneling machine for trenchless work. The mentioned machine allows drilling a tunnel in accordance with the micro-tunneling technology, without the necessity of digging up roadways, pavements or other elements of the urban infrastructure. Wherein micro-tunneling consists in drilling a horizontal tunnel between the two previously realized chambers: start one and closing one, using a tunneling machine (TM - tunneling machine). Due to the use in this technology machines with their own independent drive, the diameters of drilled tunnels are in the range from 1000 to 4000 [mm], wherein due to technological development the upper value of the diameter range is constantly increasing. The presented micro-tunneling technology is used in: realization of sewage and water pipelines, construction of pipelines for energetic media (gas, oil, etc.), installing casing pipes on other conduits (gas piping, telecommunication cables, heating conduits, etc.), drilling multi-conduits tunnels [3].

The paper presents the conceptual design of a tunneling machine for trenchless work, in relation to which was conducted the verification process of previously adopted operation cycle. The main subsystems of the conceptual design of the tunneling machine for trenchless work (figure 1) are:

- The outer jacket, which is the base element in relation to the other components of the tunneling machine;
- The set of four hydraulic cylinders of a linear motion, which allows for the movement of the whole tunneling machine. The cylinders, included in that set, are mounted within the outer jacket of a tunneling machine and influence on a support of the tunneling machine;
- The support of the tunneling machine, which props on the tubing system;
- The set of four hydraulic cylinders of a rotary motion, which drives the mining jacket;
- The mining jacket, which is the base element in the mining process of a ground;
- The system of the manipulator together with crushing head, which is used for output crushing;
- The system of power transmission from the engines of a rotary motion on the mining jacket;
- The system of output transporting.

In the phase of conceptual work it was assumed that in order to drill a tunnel, the tunneling machine should realize the following tasks:

- Running the system of output transporting;
- Running the mining jacket by the set of hydraulic cylinder of a rotary motion using the power transmission system;
- Running the drive system of the manipulator together with the crushing head;
- Running the system of the set of hydraulic cylinders of a translatory motion.

Figure 1. Conceptual design of the tunnelling machine for trenchless work.

1.1. Preparing the model of the tunneling machine for operation simulation

Preparation of the 3D model of the control system of the tunneling machine for trenchless work for motion simulation was realized in software of the VR (Virtual Reality) class, EONStudio. In the first stage of the work was created an interactive visualization of operation of the 3D model of the tunneling machine. In the elaborated interactive application were created procedures allowing controlling the translatory motion system, rotary motion one and the drive system of the manipulator. It was also created the procedure enabling turning on and off the output crushing head, mounted on the last element of the manipulator. Actuators in particular control sub-systems (figure 2) are: four hydraulic cylinders of linear motion (drive system of translatory motion of the whole tunneling machine), four hydraulic motors of a rotary motion (drive system of the cutting ring), five hydraulic cylinders of a linear motion (drive system of the manipulator), and one hydraulic motor of a rotary motion (drive system of the output crushing head). Due to the scope of conducted researches, concerning the verification of operation system of the tunneling machine for trenchless work, in the 3D model of the tunneling machine, from elements of control systems, it was mapped drives and power transmission systems.

The mentioned drives operate, in the created interactive simulation, in accordance with the accepted characteristics of their operation (figure 2), where in:

- piston1+, piston1-, piston2+, piston2-, piston3+, piston3-, piston4+, piston4-, piston5+, piston5-, means the procedure related with the characteristic of moving out and in of the piston rod of the i-th linear drive, where i=1,…,5;
- shaft1_on_off, shaft2_on_off, shaft3_on_off, shaft4_on_off, shaft5_on_off, means the procedure related with the characteristic of turning on and off of a rotary motion of a shaft of a rotary motor with the assumed rotational speed;

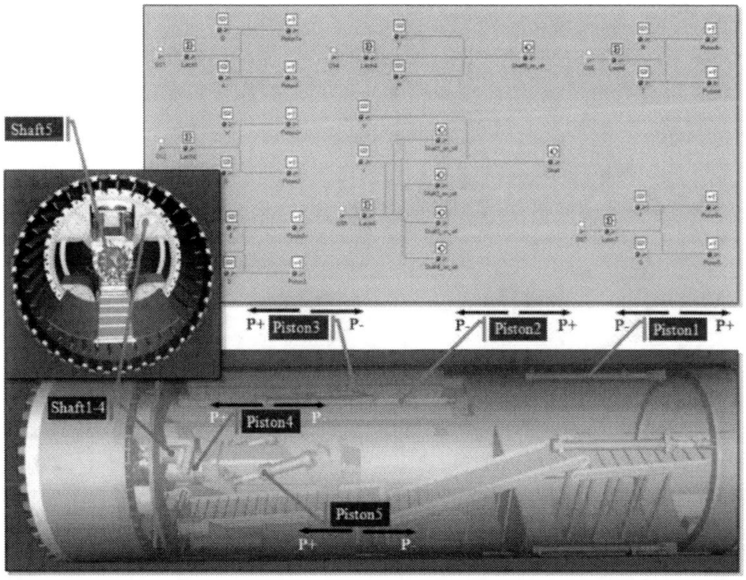

Figure 2. Model prepared for motion simulation in the EONStudio program.

In the adopted model were also created interactive components performing the functions of virtual distribution valves, logically mapping the function of the real two-position distribution valve (latch1, latch2, latch3, latch4, latch5, latch6, latch7). The described model could be manually controlled using the keyboard buttons (Q, A, W, S, E, D, R, F, T, G, Y, H, J, U), or using the integration with the virtual controller. In the created interactive application it has been established procedures for receiving input data from an external application, on the basis of dynamic data exchange (DDE), and assigning them to the components OS1, OS2, OS3, OS4, OS5, OS6, OS7, what allows controlling actuators of particular control systems of the considered machine. It is assumed that the control is based on a digital two-state signal of 0 or 1 (true or false).

1.2. Creating the program controlling the system in a virtual controller

The stage of the work, associated with the creation of the control program, was carried out in the CoDeSys program. The control program for the virtual controller is elaborated on the basis of the Ladder Diagram (LD) language. The control program (figure 3) was developed on the basis of the adopted operation cycle of the tunneling machine, and namely, the first must be turned on the hydraulic motors of a rotary motion which, by the power transmission system should start the mining ring. Then should be activated actuators of a linear motion moving the whole tunneling machine and, at the same time, should be activated work of the manipulator and output crushing head. The whole machine is moving, in one cycle, by a distance equal to the width of the tubing, used to reinforce the drilled part of the tunnel and to prevent the phenomenon of ground sliding.

In the created control program it was assumed that all actuators of a linear motion are in the start position, in which piston rods are in its extreme, rear position. The created visualization component allows applying excitations on the particular drives realizing the movement of the whole tunneling machine, by transferring the signal to the elements performing the role of virtual two-position valves.

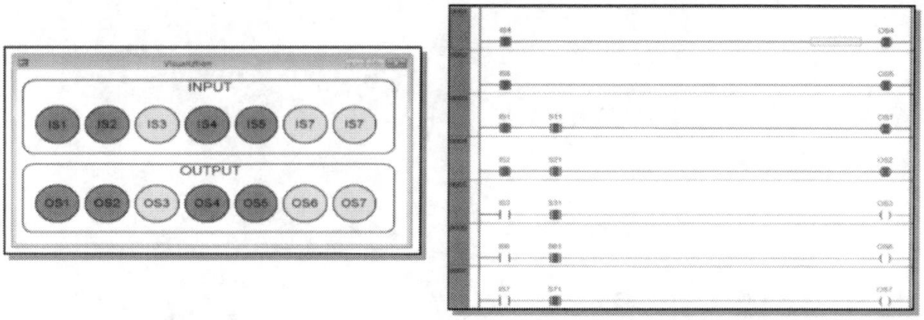

Figure 3. Control program created in the CoDeSys software.

1.3. Integration of the VR system with the virtual controller

The element integrating the virtual model of the tunneling machine for trenchless work with the virtual controller is the application (figure 4) written in a high level language (Visual Basic). In the developed application were created procedures responsible for receiving data from the controller, operating in the simulation mode, where the results of their actions are presented in the fields marked with the names of the signals: OS1, OS2, OS3, OS4, OS5, OS6, OS7. Then they are transferred to the interactive application, in which is verified the operation of adopted object of researches.

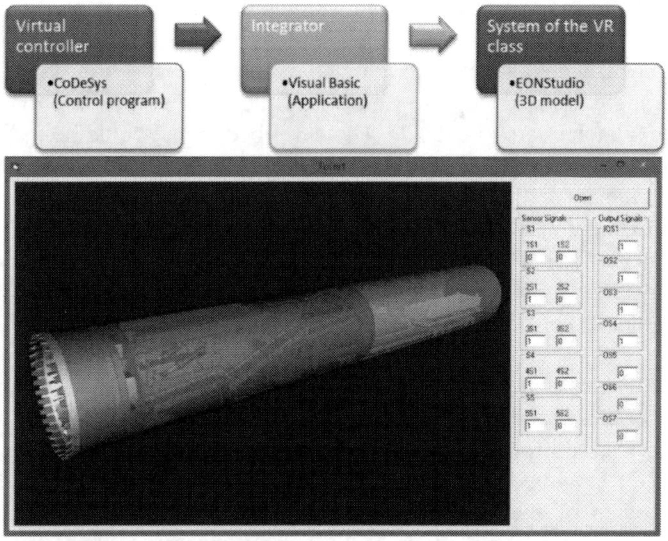

Figure 4. Scheme of the signals flow between the virtual controller and the system of the VR class.

While iSJ signals, where i=1,…,5, j=1,2, reflect the states of virtual sensors indicating the position of the piston of cylinders in their extreme positions.

2. Conclusions

The important problem, related with the integration process of a model of the control system with the virtual controller, is correct mapping of operation of the analyzed system, on which allows software of the VR class, EONStudio. This mapping must include the operation characteristics of the particular drives of the control system.

The conducted works enabled the integration of the virtual model of the control system of the tunneling machine with the virtual controller, what allows verifying its operation.

The virtual analysis of the systems, basing on the integration of the virtual 3D model with the virtual controller, allows step by step eliminating errors without fear of material losses.

Acknowledgements
The work is realized within the project titled: "Modular automated production stand with instrumentation for non-invasive confirmation of product quality" funded by The National Centre for Research and Development, agreement No UOD-DEM-1-495/001.

References

[1] Świder J and Herbuś K 2006 Zastosowanie funkcyjnych obiektów elementarnych do wspomagania modelowania maszyn zorientowanego na analizę ruchu, Monografia, Wydawnictwo Politechniki Śląskiej
[2] Świder J, Ociepka P and Herbuś K 2005 The application of object methods for functional features representation *Achievements in Mechanical and Materials Engineering* Gliwice-Wisła pp 655 – 658
[3] Kolonko A, Madryas C, Szot A and Wysocki A 2006 Mikrotunelowanie, Dolnośląskie Wydawnictwo Edukacyjne, Wrocław
[4] Banaś W, Herbuś K, Kost G, Nierychlok A, Ociepka P and Reclik D 2014 Simulation of the Stewart platform carried out using the Siemens NX and NI LabVIEW programs *Advanced Materials Research* **837** pp 537-542
[5] Herbuś K, Kost G, Reclik D and Świder J 2014 Integration of a virtual 3D model of a robot manipulator with its tangible model (phantom) *Advanced Materials Research* **837** pp 582
[6] Ociepka P, Herbuś K and Gwiazda A 2014 Application of the CBR method for adding the design process of module manipulators *Advanced Materials Research* **1036** pp 1011-1016
[7] Ociepka P, Banaś W, Herbuś K, Kost G 2014 Simulator of the Car for Driving Courses for the People with Mobility Impairments *Advanced Materials Research* **1036** pp 817-822
[8] Herbuś K, Ociepka P and Gwiazda A 2014 Application of functional features to the description of technical means conception *Advanced Materials Research* **1036** pp 1001-1004
[9] Herbuś K, Ociepka P and Gwiazda A Conception of the integration of the virtual robot model with the control system *Advanced Materials Research* **1036** pp 732-736
[10] Ociepka P, Herbuś K and Gwiazda A 2014 Application of the method basing on engineering knowledge and experience for adding the hexapod design process *Advanced Materials Research* **1036** pp 1005-1010
[11] Gwiazda A, Herbuś K, Kost G and Ociepka P 2015 Designing mechatronics equipment based on the example of the Stewart platform *Solid State Phenomena* **220/221** pp 419-422
[12] Gwiazda A, Herbuś K, Kost G and Ociepka P 2015 Motion analysis of mechatronic equipment considering the example of the Stewart platform *Solid State Phenomena* **220/221** pp 479-484
[13] Banaś W, Gwiazda A, Herbuś K, Kost G, Ociepka P and Reclik D 2015 Analysis of the dynamic properties of the mechatronic integrator of control procedures of the vehicle driven by persons with disabilities *Solid State Phenomena* **220/221** pp 3-8
[14] Gwiazda A, Sękala A, Banaś W and Monica Z 2014 Protection of hydraulic systems against dynamic loads using multi-valve approach *Ad. Mat. Res.* **1036** pp 547-552
[15] Gwiazda A, Sękala A, Monica Z and Banaś W 2014 Integrated approach to the designing process of complex technical systems *Ad. Mat. Res.* **1036** pp 1023-1027
[16] Gwiazda A 2014 Design of the roof support with strait-line mechanism *Ad. Mat. Res.* **1036** pp 553-558
[17] Gwiazda A 2014 System of designing complex technical means using fuzzy analysis *App. Mech. and Mat.* **474** pp 147-152
[18] Płaczek M 2014 Computer Aided Analysis of Systems with MFC Actuators *Advanced Materials Research* **1036** pp 711-714

Implementation of the safety components base on industrial networks

P Michalski[1] **and M P Hetmańczyk**[1]

[1] Silesian University of Technology, Faculty of Mechanical Engineering, Institute of Engineering Processes Automation and Integrated Manufacturing Systems, Konarskiego St. 18A, 44-100 Gliwice, Poland

E-mail: piotr.michalski@polsl.pl

Abstract. Capabilities of modern automation systems, especially in the field of industrial networks applications, encouraging them to favor during the process of developing the automation project [2,3,5]. Extensive network topologies and simplicity and clear structure causing displacement of the classic wiring. The article presents: project workflow for creating laboratory stand based on industrial network and base on it implementing the safety system. The individual steps of the process are described. In addition, the paper presents the main concept for controlling the safety functions of the system.

1. Introduction

A perfect example is the comparison of classical topology and wiring distribution automation equipment in relation to a modern industrial network topology of the AS-interface. Topology simplification is not the only benefits that we get through the selection of industrial networks. Built control systems must often provide the required level of safety. Thus, the designer of the automation system is required to enable the handling of so-called safe I/O modules, safety light curtains, emergency stop buttons, pull-wire switches, etc. This article describes the application possibilities of industrial network such as AS-interface in the implementation of safety at work technology. The authors presents the process of creating a simple application, which includes: the position limit switches, actuators, electro-pneumatic control units, modules of analogue inputs and outputs, push buttons modules, the frequency converter and its controlled asynchronous 3-phase motor. Figure 1 shows the test application designed to simulate a simple industrial system. Cyclic operation is controlled actuators in accordance with the adopted histogram, the work of an asynchronous motor with different speeds and possible pressure drop in the pneumatic system. It is possible to choose the manual control mode and fully automatic. Once launched, the authors present the basic functionality of the implementation process, on the existing structure of hardware, security components. The sample safety program, security restarting the system after the breach of security, in a development environment ASIMON, has been written.

2. General description

To simulate a simple industrial process [1,6], the electro-pneumatic laboratory system of the drink production line has been assembled. The main control program consists of a series of sequences written in sequential function chart (SFC). The whole process from activation by choosing glasses,

Content from this work may be used under the terms of the Creative Commons Attribution 3.0 licence. Any further distribution of this work must maintain attribution to the author(s) and the title of the work, journal citation and DOI.
Published under licence by IOP Publishing Ltd

pour a drink, provide the glasses on the tray can be divided into certain stages. Each of them is assigned to individual steps in the main program in the SFC with the same name. Beneath it lies for each step of the program written in functional block diagram (FBD), which carries out the tasks assigned to the stage. All variables, both parts programmed in SFC and FBD, are in the main program.The program contains three sub-programs, also written in FBD. To clarify the whole process steps simple figures has been created. The first, choosing the glass, is shown on figure 2.

.

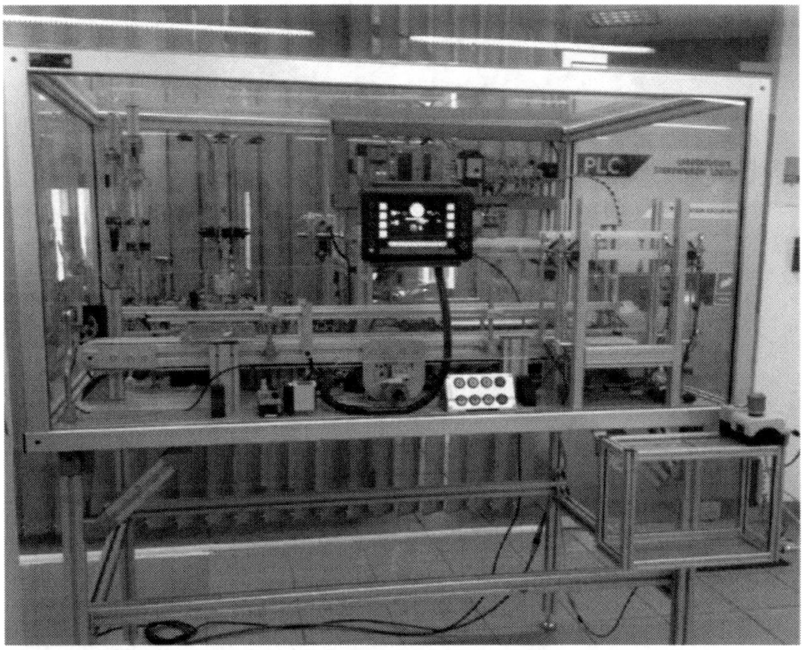

Figure 1. Laboratory application designed to simulate drink production line.

Whole process starts from magazine of glass (first subsystem). In the project it was program challenge to create the matrix of variable of glasses positions and adequate algorithm for searching the first available glass. Authors uses electro-pneumatic system based on suspended transport system with vacuum griper.

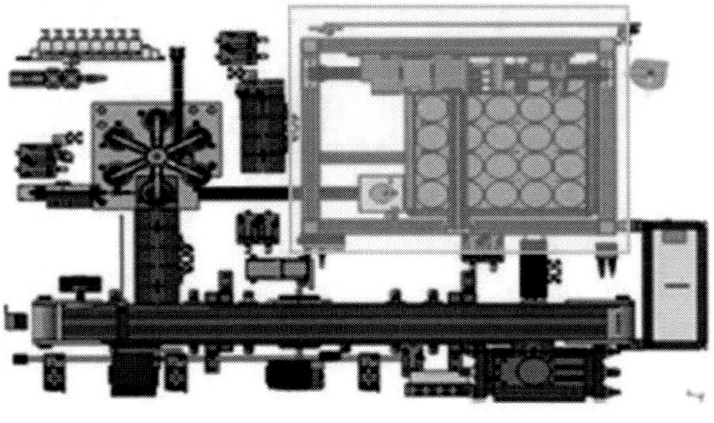

Figure 2. Drink production line – subsystem of glass matrix.

System is ready to start the next step after it download the glass and give it to the transport tray. Then the glass is transported to the drink distribution area. As soon as the glass is transported to the drink distribution area pneumatic cylinder with vacuum griper should base the glass under the rotary distributor. Six drinks are available to choose. Algorithm of control allow to prepare mixtures of maximum two liquids. Proportions are freely set via the HMI panel. Figure 3 shows the next subsystem – six liquids rotary distributor. Figure 4 shows the photo of real implementation.

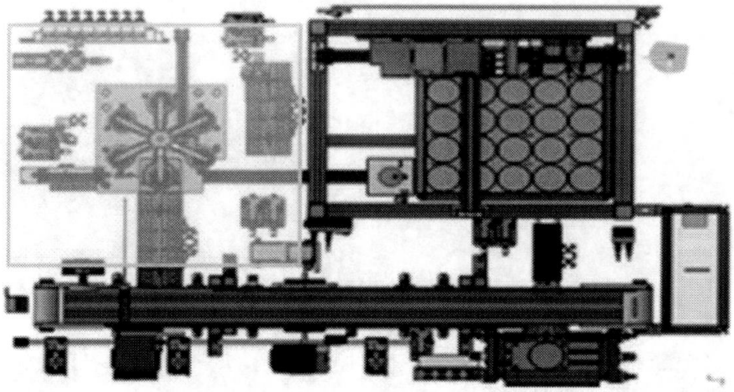

Figure 3. Drink production line – subsystem of rotary distributor.

Figure 4. Drink production line – real photo of subsystem of rotary distributor.

When this subsystem confirm that the mixture is ready then the next conveyor starts working. This part consist of barrel conveyor system and one long range pneumatic cylinder with griper. Glass filled of liquid is gently pushed in to direction of belt conveyor system. As soon as the optical sensor sense the glass pneumatic cylinder goes back, and the belt conveyor system starts his movements. Laser diffuse reflection sensor determines the distance on which the glass should stop traveling. This subsystem is also equipped with condition monitoring of the bearings of the drive motor. For this task the VE1001 sensor of ifm electronic have been chosen. If of some circumstances the vibration of this laboratory set will increase and reach the unacceptable level of RMS then this information will effect in changing the safety input state [4]. Glass stack on the path of belt conveyor will also be trade as non-safety situation (stopping the glass movements in area between start point and end point of travel is equal to error situation). In normal situation the glass on this subsystem should reach the end point as soon as possible, without any unnecessary stopping. As it is shown on the figure 5 – on the end of belt conveyor system the last pneumatic cylinder responsible for glass movements is located. When the glass stops on the end point on the conveyor, the vertical gates will be opened (goes up) and the arm of pneumatic cylinder will push the glass to final position – glass collection point. The view of this subsystem – real photo shows figure 6.

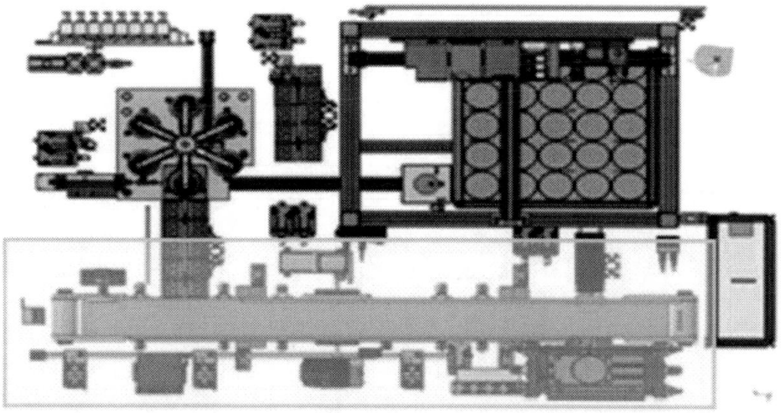

Figure 5. Subsystem of belt conveyor with distance and vibration monitoring.

Figure 6. Drink production line – real photo of subsystem of belt conveyor.

2.1. Implementation of safety functions

The functional task of the AS-interface safety monitor is to continuously specify the state(s) of the OSSD in accordance with the configuration specified based on the states of the configured devices and to activate or deactivate the assigned safe switching outputs or safe actuators (the main concept is shown on figure 7). The ASIMON software automatically arranges the devices in the respective windows during configuration.

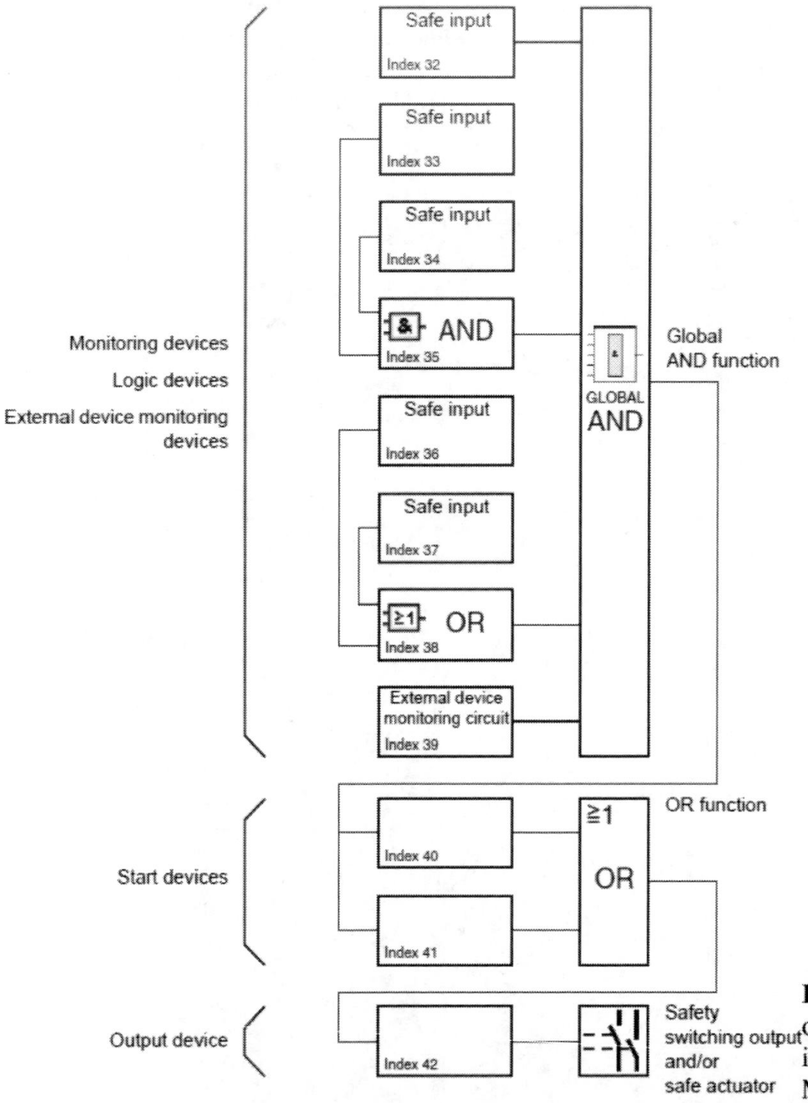

Figure 7. Main structure of the safety management in AS-interface Safety Monitor.

Each device can take on two states: ON state (switched on, logical "1"). This state means that the device has agreed to validate the circuit, i.e. to activate the safe switching outputs. Depending on the device type, various conditions must first be met. OFF state (switched off, logical "0"). This state means that the device has not agreed to validate the circuit, i.e. it results in the switching off of the safe switching outputs. In the first step of the evaluation, the states of all monitoring, logic and EDM devices are linked to one another by means of a global logic AND function, i.e. only when all configured monitoring, logic and EDM devices have the ON state is the result of the AND function equal to ON. In principle, the device states are evaluated in the same way as in an electrical safety circuit in which all safety switch elements are connected in series and validation is possible only when all contacts are closed.

In the second step, the start devices which determine the startup behaviour of the OSSD are evaluated. A start device enters the ON state when the result of the global AND function from the first step of the evaluation is equal to ON and when the respective start condition is fulfilled. With regard to the start condition, the start devices have a lock. The start condition must therefore only be fulfilled once. A start device is reset (OFF state) when the result of the global AND function from the first step of the evaluation returns the OFF state. The states of the start devices used are linked to one another with an OR function, i.e. only one of the start devices needs to be in the ON state in order for the internal validation of the circuit to occur.

In the third step, the output device is then analyzed. If the circuit has been internally validated (result of the OR function from the second step of the evaluation is equal to ON), the output device switches the message and safe switching outputs of the OSSD in accordance with its function characteristics and time behaviour, i.e. the relays trip and the switching contacts close or the safe AS-interface output is set.

3. Conclusions

This approach simplifies the maximum number of components in the way of industrial automation from sensor to PLC. Instead of: multi-core cables, intermediate boxes, connectors, terminals, hundreds of cable descriptions, automation systems integrator chooses increasingly clear industrial network bus. With the adoption of the proposed safety monitor, to prepare for the process automation project development, it is possible to implement a safety system based on AS-interface industrial network, as well as a significant acceleration of the final run of the application.

References

[1] Fraden J 2004 Handbook of Modern Sensors: Physics, Designs, and Applications New York: *Springer Verlag*

[2] Gwiazda A 2007 Multi-criterion Analysis Technique in a Process of Quality Management, Journal of Achievements in Materials and Manufacturing Engineering **25(1)** pp 75-78

[3] Gwiazda A 2008 Technical Project Management Using Quality Methods *Journal of Achievements in Materials and Manufacturing Engineering* **31(2)** pp 739 – 746

[4] Michalski P and Hetmanczyk M P Simply-integrated method of judgments of expert knowledge collected in databases for objective computer-aided engineering systems *Lecture Notes in Computer Science* **8073** 0302-9743

[5] Krenczyk D, Kalinowski K Grabowik C 2012 Integration Production Planning and Scheduling Systems for Determination of Transitional Phases in Repetitive Production, in: E. Corchado, V. Snasel, A. Abraham et al. (eds.), *Hybrid Artificial Intelligent Systems II Book Series*: Lecture Notes in Computer Science **7209** pp 274-283

[6] Kalinowski K, Krenczyk D and Grabowik C 2013 Predictive - reactive strategy for real time scheduling of manufacturing systems *Applied Mechanics and Materials* **307** pp 470-473

IOP Conf. Series: Materials Science and Engineering **95** (2015) 012086 doi:10.1088/1757-899X/95/1/012086

Technological process supervising using vision systems cooperating with the LabVIEW vision builder

P Hryniewicz[1], W Banaś[1], A Gwiazda[1], K Foit[1], A Sękala[1] and G Kost[1]

[1] Silesian[1]Silesian University of Technology, Faculty of Mechanical Engineering, Institute of Engineering Processes Automation and Integrated Manufacturing Systems, Konarskiego 18A, 44-100 Gliwice, Poland

E-mail: przemyslaw.hryniewicz@polsl.pl

Abstract. One of the most important tasks in the production process is to supervise its proper functioning. Lack of required supervision over the production process can lead to incorrect manufacturing of the final element, through the production line downtime and hence to financial losses. The worst result is the damage of the equipment involved in the manufacturing process. Engineers supervise the production flow correctness use the great range of sensors supporting the supervising of a manufacturing element. Vision systems are one of sensors families. In recent years, thanks to the accelerated development of electronics as well as the easier access to electronic products and attractive prices, they become the cheap and universal type of sensors. These sensors detect practically all objects, regardless of their shape or even the state of matter. The only problem is considered with transparent or mirror objects, detected from the wrong angle. Integrating the vision system with the LabVIEW Vision and the LabVIEW Vision Builder it is possible to determine not only at what position is the given element but also to set its reorientation relative to any point in an analyzed space. The paper presents an example of automated inspection. The paper presents an example of automated inspection of the manufacturing process in a production workcell using the vision supervising system. The aim of the work is to elaborate the vision system that could integrate different applications and devices used in different production systems to control the manufacturing process.

1. Introduction

Modern companies increasingly use computer systems allowing integrate many areas of their business. In this regard, it is possible to distinguished, among others, advanced systems of the CAE class in which are realized complex projects of technical means [1,2]. It should also be mentioned the systems designated for specialized areas of technical applications [3,4] and even the inspection of material features of manufactured parts [5-7]. Another direction of development are mechatronic systems which combining both mechanical and electrical components together with controls systems, basing on electronics and computer techniques [8,9]. In the latter range more and more important become vision systems.

The vision systems are increasingly used in industry and automation due to lower and lower manufacturing costs of their components and lighting systems. With the rapid development of technology of light-sensitive matrices, used in vision systems, their price and the resolution are more and more satisfactory. By increasing the resolution of photosensitive matrices the accuracy of vision

Content from this work may be used under the terms of the Creative Commons Attribution 3.0 licence. Any further distribution of this work must maintain attribution to the author(s) and the title of the work, journal citation and DOI.
Published under licence by IOP Publishing Ltd

systems is also increasing. The vision systems enable full control of the production process and the final quality control of manufactured products. Unfortunately, vision systems also have limits resulting from the need to integrate video systems with the real-time computers of high computational capabilities and hence also expensive ones. Another problem in machine vision systems is the lighting of a work scene. Wrongly determined could cause erroneous measurement results. One of the basic processes in which are used vision systems is the inspection of manufacturing operations realized in workcells. In the LabVIEW Vision Builder it is possible to realize the automatic inspection which allows initially preparing the program for controlling the entire production process.

2. Creating the application for object recognition

The work in vision systems required preparation of the environment in which these systems should realize the tasks. It is necessary to appropriately select the workplace exposure because it is one of the most important criteria for correct work of vision systems. The less lighted work scene the greater shadows fall on the elements. Hence it is harder to get an accurate picture of the analyzed object, as well as of the whole work scene. If, however, the lighting would be too intensive one could achieve reflections which can cause erroneous work of the system.The first step in the process of automated inspection is choosing and setting the camera parameters due to which one can get the image of a work scene (figure 1). It should be stated that the better is given camera the more data should be processed (what is considered having higher resolution). Whereas the weaker are the parameters of a camera, the more probable is situation of obtaining erroneous results.

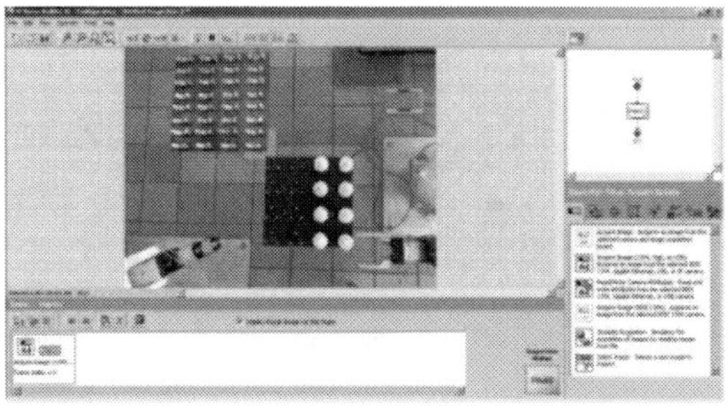

Figure 1. Work scene seen by the camera.

In order to improve the viewing area, after connecting the camera, some parameters such as brightness, contrast, exposure, etc. have to be set. The obtained image of the camera is in the RGB color palette and hence the amount of data is large enough that it could cause long work of the whole vision process. In the next step the color image in RGB color palette should be changed into monochromatic one in order to reduce the amount of processed data (figure 2). For this purpose, in the Vision Builder program, should be used the option Vision Assistance in the tool Enhance Images.

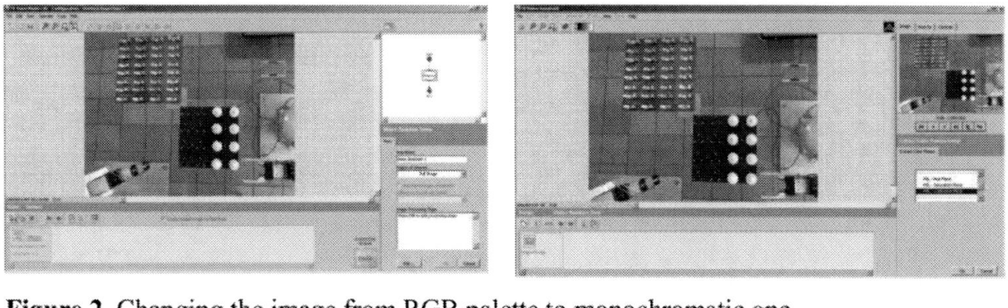

Figure 2. Changing the image from RGB palette to monochromatic one.

The function Vision Assistance allows selecting the range of a camera view to be processed (Region of Interest) according two options:

- Constant – chosen area of the camera view is constant during the entire operation of the vision system
- Full Image – the work area during image processing is equal to the area of the camera view.

Selecting the option Region of Interest, it is possible to use the program Vision Assistance by editing the option in the function Image Processing Steps In the tool Processing Functions: Color there is the option Extract Color Planes. To improve the monochrome image (Luminance Plane function), it is possible to change settings in the tool Processing Functions: Image (figure 3). In this function it is possible to set brightness, contrast and gamma in order to improve the image seen by the camera. Incorrect settings of these parameters may result in inaccurate measurements and object recognition.

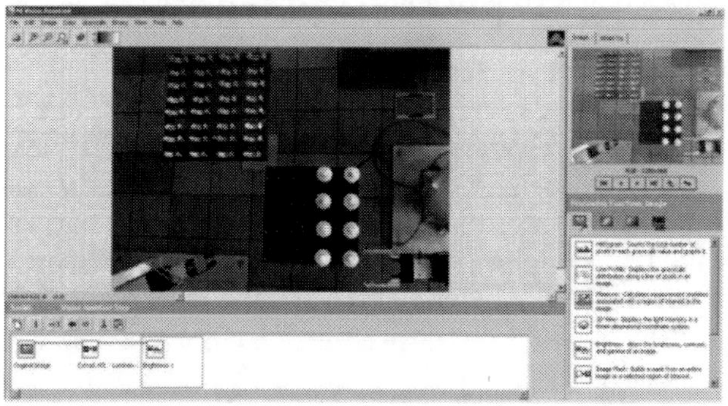

Figure 3. Image improving using Brightness function.

When the image is improved it is possible to realize the next step of automated inspection and namely to create the template of an object (figure 4). The template could be created using the option Match Pattern. This is an option that allows locating regions of an image and also defining the templates. The option Match Pattern let to omit problems with image noise, blur, and uniform lighting changes. It bases only on 2D intensity and edge information.

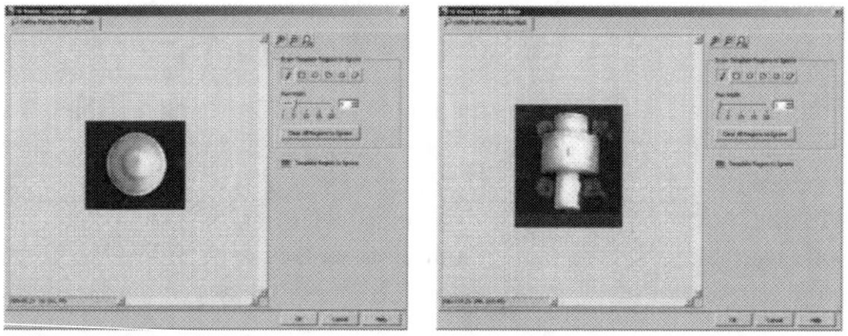

Figure 4. Template determining in the program Vision Builder.

In the tool Select Region Template, it is possible to select the area in which is located the object being analyzed. After selecting the analyzed area, considered with the object, one should use the option Define Pattern Matching. In this option the image of an object is contoured using the tool Draw Template Regions. The area, which is indicated with the red color will not be considered as a part of the template. After obtaining the area consistent with the desired pattern it is possible to continue the process of the application creation. At this stage one could set the other specific parameters for the template. It is also possible to modify the parameters set previously. Among setting parameters one could distinguish:

- Number of Matches to Find which determines the number of items we want to see in the picture,
- Minimum Score which allows specifying the minimal compliance of an element in the image, relative to the standard: from 0 - no mapping (no match) till 1000 - exact matching to the template (perfect match).
- Search Level Specifies how thoroughly the step searches for matches.
- Search for Rotated Patterns which allows searching the work area for rotated and reflected images of the original template.

3. Utilization of the created application

Vision systems could be used in a wide range of technical applications. They replace, inter alia, applications basing on sensors measuring different indirect signals [10,11]. Vision systems consist of a camera, lens, and computer processing data in real time as well as equipment illuminating the work scene. Because of realized tasks, and related with them needs, the vision systems could be divided into the following types:

- vision sensor (soft sensor) - integrated unit, having in one housing both the camera, the computational processor as well as the system illuminating the workspace. Due to the low resolution (up to 640 - 680 pixels) this type of sensor is used in less complex tasks (reading bar code, simple measurements, etc.),
- intelligent camera - camera with high image parameters (up to 1200 to 1600 pixels), integrated with a computer and software allowing to perform difficult tasks. It is the most commonly used type of video systems in the industry,
- camera with the PC-class computer - the most optimal type of vision systems allowing for quick modification of the entire system (including cameras changing), depending on the needs of the production supervision process as well as for rapid task modification by changing the software of the real-time computer.

For practical use of the presented applications one must specify the scope of its work. To accomplish this it is needed, in the tab Limits, set the minimal and maximal number of templates to be found that are used in the process of automatic inspection (figure 5). Using this method we could analyze and visualize the process of automatic inspection [12,13].

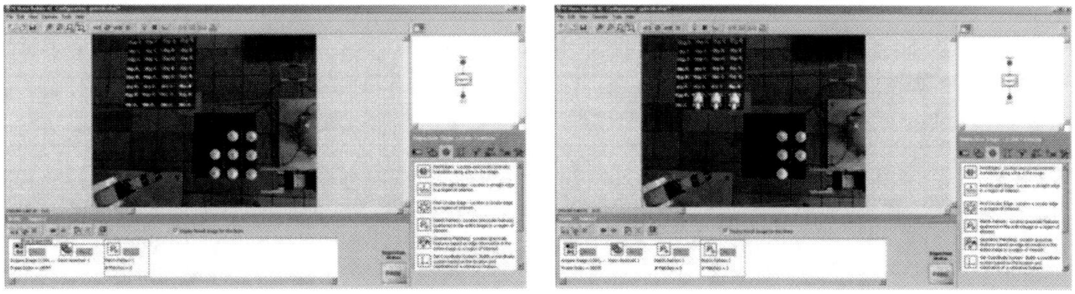

Figure 5. Identification of images consistent with templates.

As seen in figure 6, regardless of the number of stored templates and depending on the settings of correctness of lighting exposure and processing of the received image, the application allows correctly recognizing many static and moving elements that are in the workspace of the camera. It is determined by the parameters of a compliance with the template, image position and finally with the rotation, relative to the original template.

In figure 7 is seen the code in the LabVIEW program that has been obtained by exporting data from program LabVIEW Vision Builder, where was prepared the application. To obtain the code in the LabVIEW program one must use tools form program migration to the LabVIEW graphical user interface. The graphical interface allows to analyzing all relations defined in the program and easy modifying particular functions and procedures. The developed application is used to test functioning of a robotized manufacturing workcell.

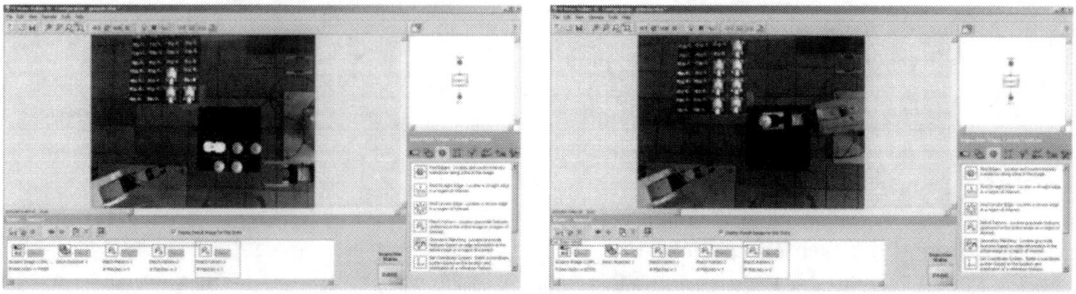

Figure 6. Searching elements that are consistent with stored template.

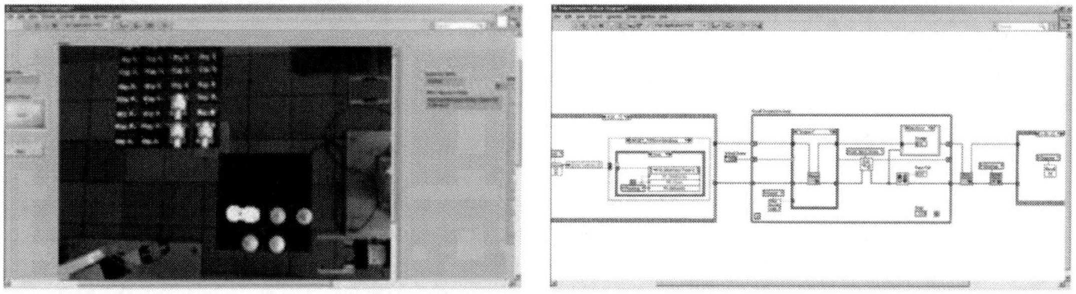

Figure 7. Migrating the program code from Vision Builder to the LabVIEW program.

The purpose of these researches is to develop automated procedures for controlling the robot working in a workcell by equipping it with an external vision system. It is also planned to connect the vision system with other sensors systems [14,15]. To realize the automatic control of a production process the vision control system should be integrated with other system what provides correct data acquisition [16,17]. The most appropriate are proximity sensors and ultrasonic ones. However, with the increase in the accuracy of the technology they are one of the most important branches of non-contact sensors used in automation. They should allow creating independent production planning system [18] that could operate in automatic cycles. It is also possible to use them in independent, autonomous production systems utilizing artificial intelligence [19-21].

4. Conclusions

In the article is presented the integrative approach to the problem of vision systems creating basing on the LabVIEW program. The use of this software facilitates the production monitoring process. Information on the amount of semi-finished products and their positions allow improving the supervising process. It should be, however, taken into account the quality of lighting of a work scene as the factor influencing the monitoring process. Unfortunately, current technology of vision systems could not be fully independent to control industrial processes because of the possibility of erroneous results. Because of this they are utilized with other types of sensors like lidars or sonars.

Acknowledgements

The work is realized within the project titled: "Modular automated production stand with instrumentation for non-invasive confirmation of product quality" funded by The National Centre for Research and Development, agreement No UOD-DEM-1-495/001.

References

[1] Gwiazda A, Herbuś K, Kost G and Ociepka P 2015 Designing mechatronics equipment based on the example of the Stewart platform *Sol. St. Ph.* **220-221** pp 419-422
[2] Grabowik C, Ćwikła G and Janik W 2014 The new approach to design features identification *App. Mech. and Mat* **657** pp 750-754
[3] Dymarek A and Dzitkowski T 2013 Reduction vibration of mechanical systems *App. Mech. and Mat.* **307** pp 257-260
[4] Gwiazda A, Sękala A, Monica Z and Banaś W 2014 Integrated approach to the designing process of complex technical systems *Adv. Mat. Res.* **1036** pp 1023-1027
[5] Ćwiek J, Łabanowski J, Topolska S and Sozańska M 2012 Determination of failure causes of a steam turbine casting *Sol. St. Ph.* **183** pp 37-42
[6] Topolska S and Łabanowski J 2015 Duplex stainless steels toughness impact investigations *J. Mat. inTehn.* **5** pp 114-119
[7] Łabanowski J, Świerczyńska A and Topolska S 2014 Effect of microstructure on mechanical properties and corrosion resistance of 2205 duplex stainless steel *Pol. Mar. Res.* **21(4)** 84 pp 108-112
[8] Hetmańczyk M 2015 The prediction oriented analysis of mechatronic machine structures in terms of the signal stream flow *Sol. St. Ph.* **220-221** pp 423-428
[9] Hetmańczyk M 2013 The multilevel prognosis system based on matrices and digraphs methods *Sol. St. Ph.* **199** pp 79-84
[10] Płaczek M 2015 Modelling and investigation of a piezo composite actuator application *Int. J. Mat. and Prod. Tech.* **50(3/4)** pp 244-258
[11] Buchacz A, Płaczek M and Wróbel A 2014 Modelling of passive vibration damping using piezoelectric transducers – the mathematical model *Maint. and Rel.* **16(2)** pp 301–306
[12] Foit K 2007 Introduction to use virtual reality visualisations in the exploitation and virtual testing of machines *J. Achiev. Mater. Manuf. Eng.* **25(2)** pp 57-60
[13] Foit K 2014 Mixed reality as a tool supporting programming of the robot *Adv. Mat. Res.* **1036** pp 737-742
[14] Wróbel A 2012 Kelvin Voigt's model of single piezoelectric plate *J. of Vibr.* **14(2)** pp 534-537
[15] Wróbel A, Płaczek M, Buchacz A and Majzner M 2015 Study of mechanical properties and computer simulation of composite materials reinforced by metal *Int. J. Mat. and Pr. Tech.* **50(3/4)** pp 259
[16] Ćwikła G 2014 The methodology of development of the Manufacturing Information Acquisition System (MIAS) for production management *App. Mech. and Mat.* **474** pp 27-32
[17] Ćwikła G Methods of manufacturing data acquisition for production management – a review. *Adv. Mat. Res.* **837** pp 618-623
[18] Krenczyk D and Olender M 2014 Production planning and control using advanced simulation systems *Int. J. of Mod. Man. Tech.* **VI(2)** pp 38-43
[19] Banaś W and Sękala A 2014 Concepts of flexible production line, on the example of robotic cell *Adv. Mat. Res.* **1036** pp 749-754
[20] Sękala A, Gwiazda A and Banaś V 2014 Agent-based systems approach for robotic workcell integration *Adv. Mat. Res.* **1036** pp 721-725
[21] Gwiazda A, Sękala A, Monica Z and Banaś W 2014 Integrated approach to the designing process of complex technical systems *Adv. Mat. Res.* **1036** pp 1023-1027

Stand for testing the electrical race car engine

M Baier[1], J Franiasz[1], P Mierzwa[2] and D Wylenzek[2]

[1] Silesian University of Technology, Faculty of Mechanical Engineering, 18A Konarskiego Street, 44-100, Gliwice, Poland
[2] Silesian University of Technology, Faculty of Automatic Control, Electronics and Computer Science, 16 Akademicka Street, 44-100, Gliwice, Poland

E-mail: maja.baier@gmail.com

Abstract. An engine test stand created especially for research of electrical race car is described in the paper. The car is an aim of Silesian Greenpower project whose participants build and test electrical vehicles to take part in international races in Great Britain. The engine test stand is used to test and measure the characteristics of vehicles and their engines. It has been designed particularly to test the electric cars engineered by students of Silesian Greenpower project. The article contains a description how the test stand works and shows its versatility in many areas. The paper presents both construction of the test stand, control system and sample results of conducted research. The engine test stand was designed and modified using PLM Siemens NX 8.5. The construction of the test stand is highly modular, which means it can be used both for testing the vehicle itself or for tests without the vehicle. The test stand has its own wheel, motor, powertrain and braking system with second engine. Such solution enables verifying various concepts without changing the construction of the vehicle. The control system and measurement system are realized by enabling National Instruments product myRIO (RIO – Reconfigurable Input/Output). This controller in combination with powerful LabVIEW environment performs as an advanced tool to control torque and speed simultaneously. It is crucial as far as the test stand is equipped in two motors – the one being tested and the braking one. The feedback loop is realized by an optical encoder cooperating with the rotor mounted on the wheel. The results of tests are shown live on the screen both as a chart and as single values. After performing several tests there is a report generated. The engine test stand is widely used during process of the Silesian Greenpower vehicle design. Its versatility enables powertrain testing, wheels and tires tests, thermal analysis and more.

1. Introduction

One of the crucial elements of the vehicle being studies is a DC motor. All participants of the race are equipped in identical 240W motors and 12V batteries. In order to choose the motor with optimal parameters, several of them are being thoroughly tested on the engine test stand designed and built by participants. The research lead to excluding the worse motors from further use [1].

The paper covers both description of the control system of the engine test stand and results of the research. The study was conducted on the engine test stand designed for the measurement of the whole drive system. It was designed to keep conditions as close as possible to those on the racetrack. Versatility of the test stand allows to verify only motor with powertrain or the whole car, depending on needs. The principle of operation of the measurement is based on the behavior of kinetic energy. Test stand shaft's moment of inertia I rotating at a predetermined angular velocity ω, corresponds to the

Content from this work may be used under the terms of the Creative Commons Attribution 3.0 licence. Any further distribution of this work must maintain attribution to the author(s) and the title of the work, journal citation and DOI.
Published under licence by IOP Publishing Ltd

vehicle of mass m and moving with velocity v. In addition, the test stand is equipped with loads and measurement system.

2. The test stand control system

The aim of the test stand is to provide racing conditions in laboratory. Actual components tested on the stand are particularly: motor, batteries, high beam current, motor controller, measurement system and many transducers. The elements participating in the motion simulation of the vehicle include mechanical and electronic components cooperating with the control and measurement software. These components allow to simulate racing conditions properly, which is mainly to provide calculated load. The load may be put manually or it may be calculated basing on advanced mathematical models describing vehicle motion in particular conditions [2].

Application used on the test stand was written in the LabVIEW programming environment. Measurement environment enable creating measurement-control software for complex and fast-changing processes, which includes controlling of test stand elements.

Software works closely with controllers manufactured by National Instruments. The test stand performance is controlled by myRIO-1900 controller. A multitude of analog and digital inputs and outputs allows to provide highly advanced test stand (figure 1.). In the figure 1. MyRIO is closing the feedback loop by allowing for interpretation and determination of measurement signals for main engine control and brake engine control. Brake engine is of the same kind as the tested engine but they are rotating in opposite directions. Brake engine operates in regenerative mode and thanks to a special device we can regulate the amount of energy dissipated and thus the braking torque.

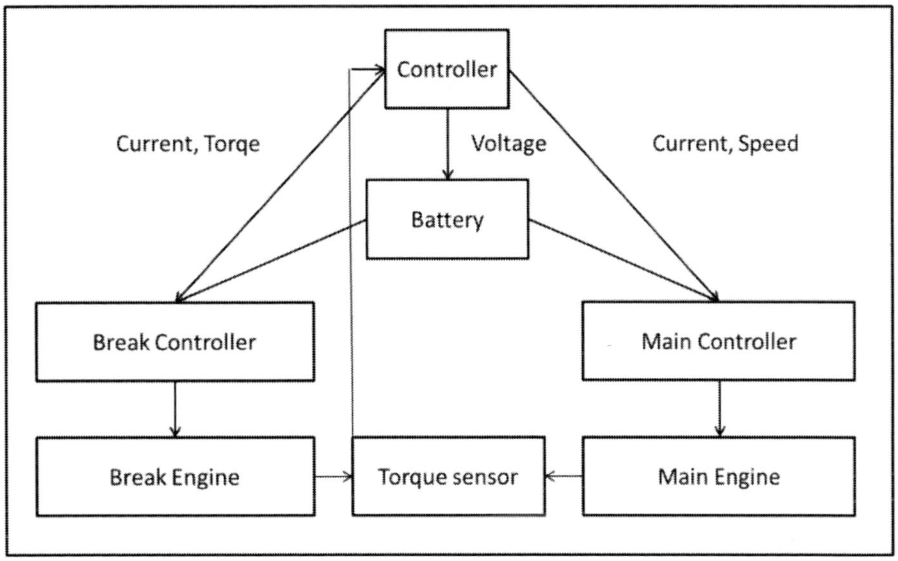

Figure 1. Schematic diagram of the test stand.

The test stand control system (figure 2) provides the regulation of any of the given values. Controlling process works in three modes: constant current, constant speed, constant braking torque. The control system is based on PID controller implemented on the FPGA of myRIO controller. Depending on chosen mode controller settings and control signals are selected. The control system in myRIO controller has known parameters but includes a closed autonomous control system - Kelly controller [3]. Knowing the mathematical model of myRIO and input and output signals, there is a possibility to estimate the parameters of the Kelly controller.

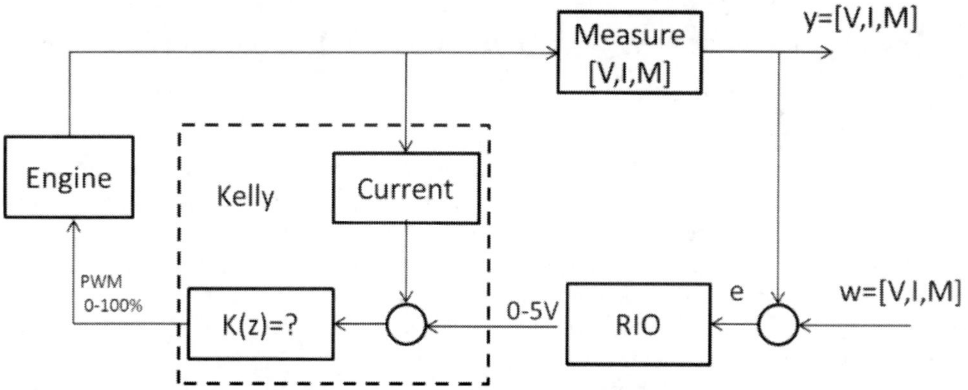

Figure 2. The engine test stand control system.

An additional advantage of the test stand is the ability to carry out automated tests which consist of a sequence controlling test stand elements, often performed repeatedly. The test procedure and its parameters are set in the sequence file. This approach allows to perform the exact same test for various motors [4].

LabVIEW implementation makes it easy to create a user control panel. The user is able to control the test stand and also observe measured values diagrams during a test. Three applications are working on the test stand. FPGA of myRIO controller is programmed and its real-time system. The application that combines everything as a whole is shown in figure 3.

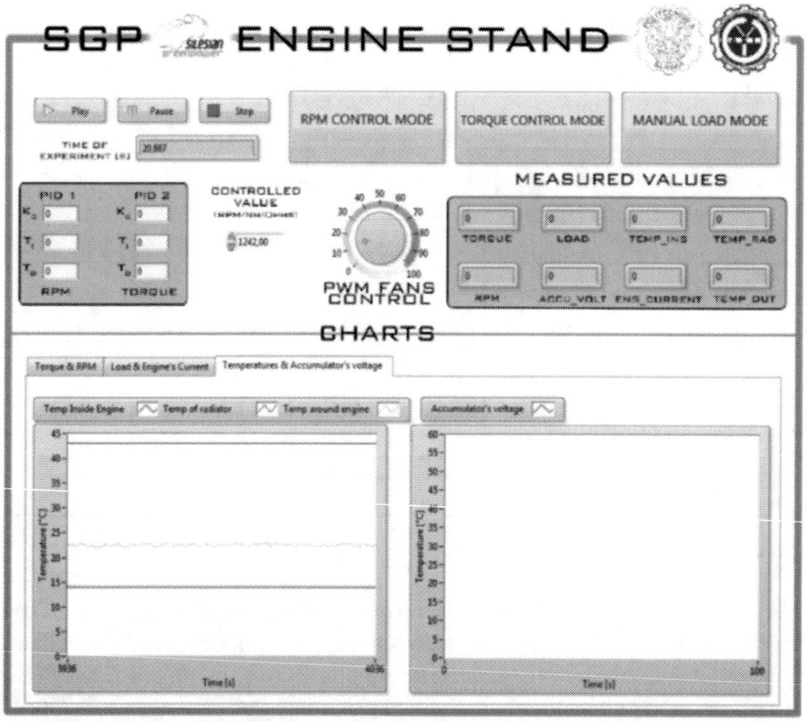

Figure 3. User control panel.

3. Research of the motors

The stand is the real object consisting of the same motor and powertrain that are present in the vehicle. Research conducted on the test stand focused mainly on the parameters which are crucial in terms of the race, i.e. speed, torque and energy consumption. In order to maintain comparable conditions during all tests, several parameters were always equal: current, initial battery voltage supply, ambient temperature, level of air pressure in the wheel, downforce of the wheel to the shaft, gear ratio and measurement time [5].

During the study the following phases of motor performance can be defined: starting and accelerating to full speed, maintaining the maximum speed, free run and stop. To determine the value of the energy consumed by individual motors, battery voltage and current consumption were measured for each of the above phases. The graphs of relative voltage and current are presented in figure 4. And figure 5. Motors have been numbered in order to ensure unambiguous identification.

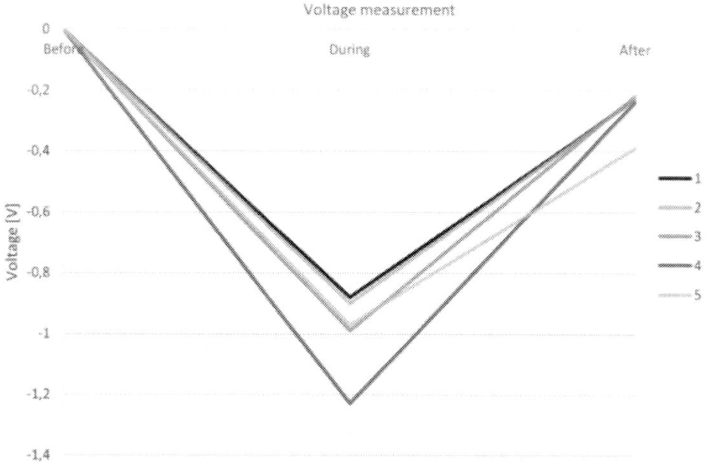

Figure 4. Voltage-time graph for 5 motors.

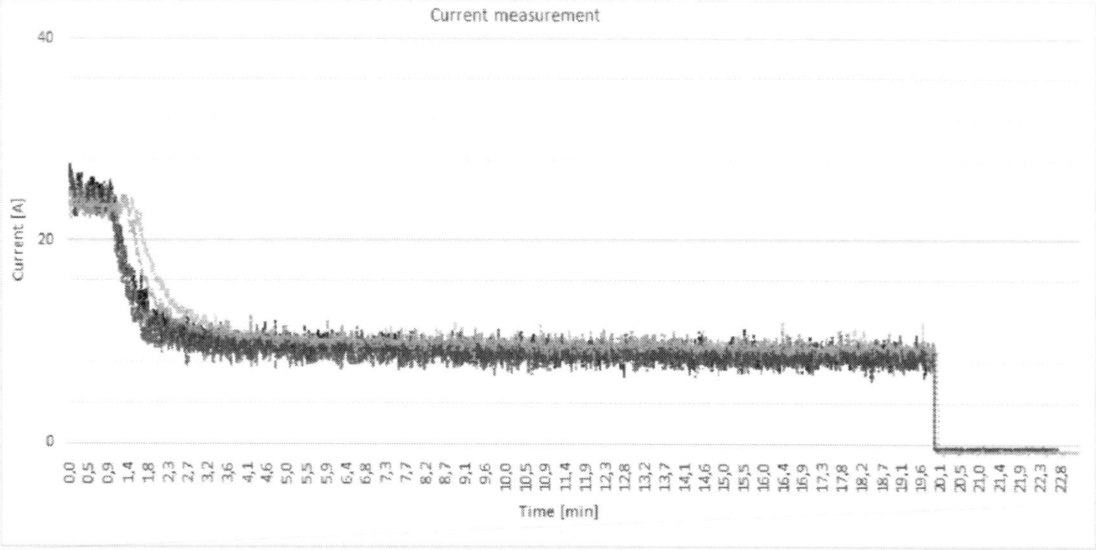

Figure 5. Current-time graph for 5 motors.

The obtained data and the transformations of the simple mathematical relations allowed to obtain value of the energy (table 1). The maximum speed of individual motors were significantly different to each other. In order to analyse them further they were converted to rotational speed as shown in the following table. Basing on the speed gained in time and known motor parameters, average torque was calculated during the first phase of work - acceleration.

Table 1. Selected parameters of the motors.

Parameter	Unit	Motor 1	Motor 2	Motor 3	Motor 4	Motor 5
Energy consumption	Wh	144,55	147,64	138,48	141,34	264,81
Max. rotational speed	RPM	1900,6	2018,2	1934,1	1833,3	2102,4
Average torque	Nm	1,930	1,556	1,945	1,737	1,585

Usefulness of the motors was determined by comparing them to each other and by confronting them with nominal values specified by the manufacturer (figure 6.).

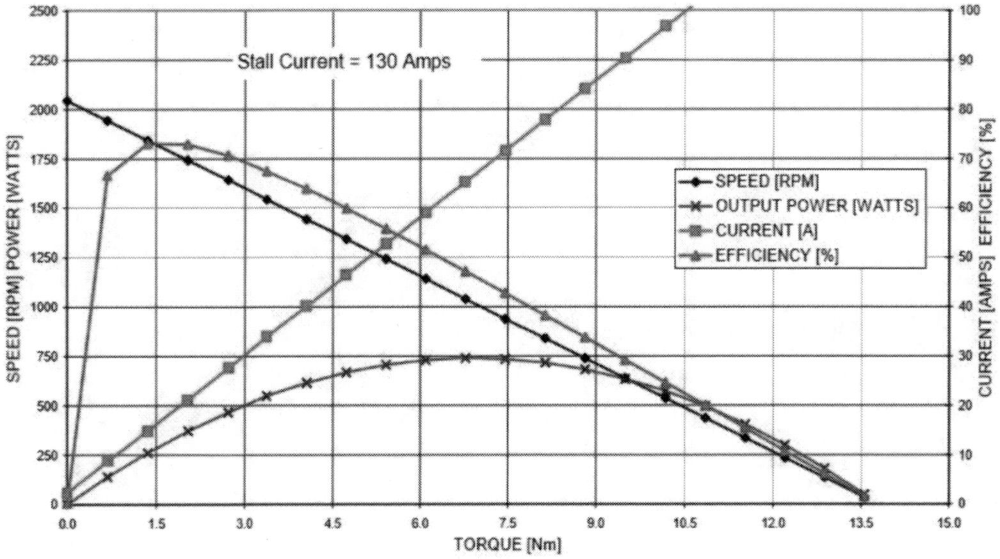

Figure 6. Nominal parameters of the motor provided by FRACMO.

The graph above shows that the motor maximum speed is slightly over 2000 RPM. The motor 5 exceeds this value going faster even being under load. However, this motor uses the greatest amount of energy and additionally, generated comparably low torque. In terms of these facts motor 5 was labelled as not suitable to be used during the race. Motor 3 optimally meets the established criteria - it has the lowest energy consumption, high torque and the speed is satisfactory.

4. Conclusions

The synergy between automation and mechanics enabled creating the test stand (figure 7.) which occurred to be a powerful tool in determining parameters of DC motors. The test stand control system allows to set values of tests adjusted to current needs of the user thanks to user-friendly application. Everything is gathered in complex device – myRIO controller – which makes the control and measurement systems highly effective. The combination of tested motor and braking motor is the optimal solution to depict race conditions in laboratory comparably accurately. The motors research presents how crucial it is to choose the motor which is characterized by the most optimal parameters. Such operations lead to improved performance of the vehicle on the race track and have priceless meaning during the contest.

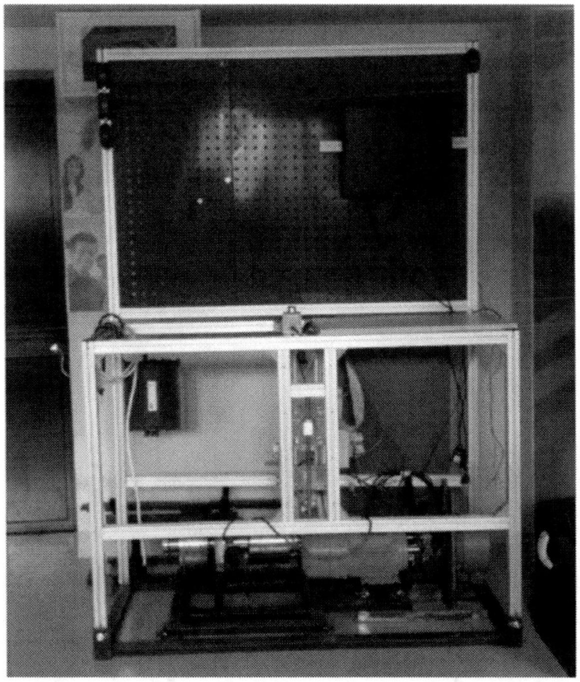

Figure 7. Stand for testing the electrical race car engine – front view.

References

[1] Baier A, Baier M, Sobek M, Grabowski L, Dusik D and Papaj P 2014 Computer aided process of designing the mechatronic Silesian Greenpower electric car *Advanced Materials Research* **1036**

[2] Burt C, Piao X, Gaudi F, Busch B and Taufik N 2006 Stand for testing the electrical race car engine, *ITRC Report* 06-004

[3] DC motor controller Kelly KDS user manual, available at http://kellycontroller.com/mot/downloads/KellyKDSUserManual.pdf, accessed: 1.02.2015

[4] Dziewior M, Mierzwa P, Wylenżek D and Żuk M 2014 Stanowisko do badania regulatora silnika DC dla elektrycznego bolidu Silesian Greenpower. W: Automatyzacja procesów dyskretnych, Gliwice: Wydawnictwo Pracowni Komputerowej Jacka Skalmierskiego pp 281-288

[5] Niedworok A, Baier A and Orzech L 2012 The concept of transducer for detection of trackway surface condition basing on frictional coupling of wheel with a rail and its experimental verification, *Journal of Vibroengineering* **14**

A new method for the accuracy evaluation of a manufactured piece

E V Oniga[1] **and M Cardei**[1]

[1] Department of "Terrestrial Measurements and Cadastre", Faculty of Hydrotechnical Engineering, Geodesy and Environmental Engineering,„Gheorghe Asachi" Technical University of Iasi, Dimitrie Mangeron Boulevard, no.67, 700050, Iasi, Romania

E-mail: ersilia.oniga@tuiasi.ro

Abstract. To evaluate the accuracy of a manufactured piece, it must be measured and compared with a reference model, namely the designed 3D model, based on geometrical elements. In this paper a new method for the precision evaluation of a manufactured piece is proposed, which implies the creation of the piece digital 3D model based on digital images and its transformation into a 3D mesh surface. The differences between the two models, the designed model and the new created one, are calculated using the Hausdorff distance. The aim of this research is to determine the differences between two 3D models, especially CAD models, with high precision, in a completely automated way. To obtain the results, a small piece has been photographed with a digital camera, that was calibrated using a 3D calibration object, a target consisting of a number of 42 points, 36 placed in the corners of 9 wood cubes with different heights and 6 of them placed at the middle of the distance between the cubes, on a board. This target was previously tested, the tests showing that using this calibration target instead of a 2D calibration grid, the precision of the final 3D model is improved with approximatly 50%. The 3D model of the manufactured piece was created using two methods. First, based on digital images, a point cloud was automatically generated and after the filtering process, the remaining points were interpolated, obtaining the piece 3D model as a mesh surface. Second, the piece 3D model was created using also the digital images, based on its characteristic points, resulting a CAD model, that was transformed into a mesh surface. Finally, the two 3D models were compared with the designed model, using the CloudCompare software, thus resulting the imperfections of the manufactured piece. The proposed method highlights the differences between the two models using a color palette, offering at the same time a global comparison.

1. Introduction

In recent years non-metric digital cameras have known a great development, being used in a wide range of applications, such as: preservation of cultural heritage, the shoreline dynamics study, the wave movement for costal protection, soil erosion, industry, etc.

Taking into account that a manufactured piece may have imperfections, in order to determine their size, the piece must be measured and compared with a reference model, namely the designed 3D model, based on geometrical elements. This measuring process require a high precision device, either a micrometer or a CMM (Coordinate Measuring Machine), is semi-automatic and time consuming and

Content from this work may be used under the terms of the Creative Commons Attribution 3.0 licence. Any further distribution of this work must maintain attribution to the author(s) and the title of the work, journal citation and DOI.
Published under licence by IOP Publishing Ltd

the comparison process is done manually. On the other hand, if we want to highlight the missing parts of the manufactured piece, we have to do it visually.

So, a new method to evaluate the accuracy of a manufactured piece and the representation completeness, in a completely automatic way, is necessary. Knowing that most of the designed 3D models are CAD models, we suggested to transform them into 3D mesh surfaces, so they can be used in the comparison process. The differences between the two models represented as mesh surfaces, the manufactured piece 3D model created based on digital images and the designed model, are now calculated using the Hausdorff distance, considering as reference model the designed model.

The camera calibration process is necessary to obtain metric information from two-dimensional (2D) images of the three-dimensional (3D) world. The purpose of camera calibration is to describe the projection model that relates both coordinate systems, and to identify intrinsic camera parameters so it can be used as a measurement device. Such parameters are usually calculated from a calibration target that contains easily and accurately detectable features in the captured image [1].

The motivation of this paper is to create a precise 3D model of a manufactured piece by close-range photogrammetry, which will be used to evaluate the accuracy of the manufactured piece and the representation completeness, in a completely automatic way.

2. Material and methods

Equipments. For this case study a small manufactured piese was chosen and for the images acquisition a Canon PowerShot SX120 IS digital camera (10 Mega pixel), equipped with a 5.744 mm by 4.308 mm image sensor was used. The digital image has a resolution of 2816 x 2112 pixels.

The 3D calibration target. Calibration using 3D calibration objects yields very efficient results, although the calibration elements must be accurate and require an elaborate configuration [1].

In the mentioned context, a target consisting of a number of 42 points, 36 of them being placed in the corners of 9 wood cubes with different heights and 6 of them on a board, was used as a 3D calibration object for the calibration model. This target was then attached to a room wall in order to make image observations. A coordinate measuring machine (CMM), produced by Aberlink, with an uncertainty within the working space of 2 μm, was used to place the target in the world coordinate system. The 42 control points coordinates were measured in the *(X, Y, Z)* coordinate system, corresponding to the world coordinate system, with the XOY plane, the board plane and the OZ axis perpendicular to the board plane [2].

The Hausdorff distance. Hausdorff Distance - named after Felix Hausdorff, is the most famous metric for comparing two mesh surfaces, providing a global comparison [3].

Hausdorff symmetric distance $d_S(S,S')$, is defined as follow [4]:

$$ds(S,S')=\max[d(S,S'), d(S',S)] \quad (1)$$

The symmetrical distance offers a more accurate measurement of the differences between two surfaces, because the one-side distance can lead to an underestimation of the distance values between the two surfaces [4].

The distance accompanied by its sign was introduced in the "CloudCompare" software for a independent evaluation of the areas that belong to the first surface and are situated inside or outside the space, relative to the second surface [5,6].

3. Results and discussions

The Canon PowerShot SX120 IS digital camera calibration. Having a 3D object, one image is enough to estimate the camera parameters through the calibration process, but for this experiment we used 3 images, taken from 3 different positions, the camera parameters being calculated as an average.

The control points are often circular, because they are easy to make and accurate to measure in subpixel precision from digital images.

For the following experiments the camera calibration toolbox for Matlab (version 3.0) implementing the Heikkila and Silven's method was used [7,8]. This Matlab toolbox is available at www.ee.oulu.fi/~jth/calibr/ and utilizes a new bias correction procedure for circular control points and a nonrecursive method for reversing the distortion model. The intrinsic parameters such as focal distance (f), optical center point (u_0, v_0), correction of radial distortion (k_1, k_2), correction of decentering distortion (p_1, p_2) and the image scale factor s_u, for the three images taken with the Canon PowerShot SX120 IS digital camera, are presented in Table 1.

Table 1. Physical Canon PowerShot SX120 IS digital camera parameters.

Image	s_u	$f\,[mm]$	$u_0\,[pixels]$	$v_0\,[pixels]$	$k_1\,[mm^{-3}]$	$k_2\,[mm^{-4}]$	$p_1\,[mm^{-4}]$	$p_2\,[mm^{-4}]$
1	1.0025	5.9898	1435.7557	1063.5828	8.97139	-3.45656	-5.88285	-4.01607
2	1.0029	5.9856	1432.6982	1067.9399	9.13208	-3.53836	-6.88640	-4.42121
3	1.0027	5.9932	1444.6962	1080.9282	9.16506	-3.63519	-6.17076	-5.17371
Average	-	**5.9895**	**1437.7167**	**1070.8170**	**9.08951**	**-3.54337**	**-6.31334**	**-4.53700**

The radial and decentering distortion profiles for the Canon PowerShot SX120 IS digital camera, computed for the 3 images, are presented in figure 1. To model the radial distortion Δr the odd-order polynomial $\Delta r = k_1 r^3 + k_2 r^5$ was used, where r is the radial distance. The decentering distortion $P(r)$ was graphically represented in a manner similar to the radial distorsion, using the function $P(r) = \sqrt{p_1^2 + p_2^2} \cdot r^2$.

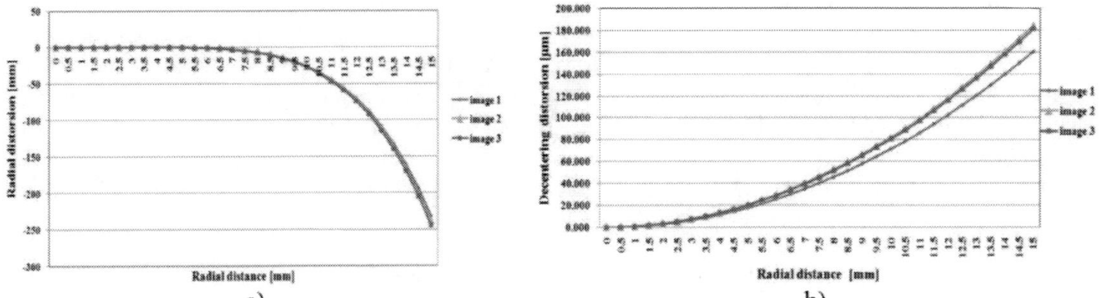

a) b)

Figure 1. (a) Radial and (b) decentering distortion profiles for the Canon PowerShot SX120 IS digital camera, computed for 3 images.

The designed piece 3D model creation. The designed piece 3D model was created based on its dimensions using the Autodesk 3ds Max software (figure 2a) and was converted into a mesh surface by exporting the model into *.stl file format (figure 2b).

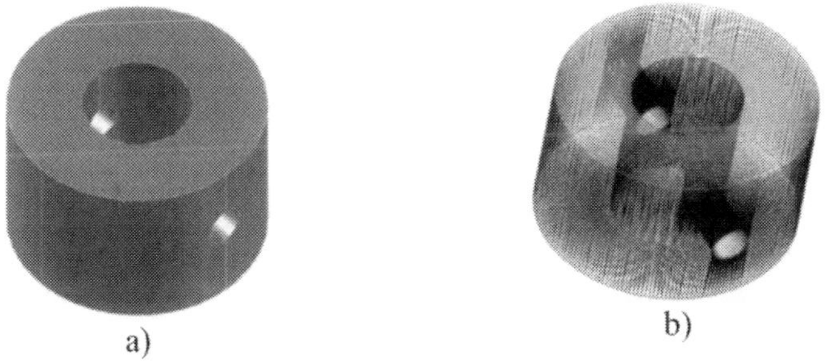

Figure 2. The designed piece 3D model (a) CAD model, (b) mesh surface.

The manufactured piece 3D model creation as point cloud. To create the 3D model of the manufactured piece, the piece was placed on a 2D grid of squares and photographed all around with the 6 mm minimum focal length of the Canon PowerShot SX120 IS digital camera, 22 images being taken from 22 different camera positions distributed circularly around the object. Using the "3DF Zephyr Pro" software [9], a dense point cloud (figure 3a) and a mesh surface (figure 3b) were automatically generated. In order to scale the model, four control distances were used, representing the sides of the squares situated in the immediate vicinity of the piece, the RMS (Root Mean Square) being 0.144 mm and the scale factor 2.23. Then, using the local coordinates of six control points representing the grid intersections, the piece 3D model was brought in a local system with the XOY plane in the grid plane, the RMS being 0.193 mm. The point cloud and the mesh surface were imported into „MeshLab" software where, through the filtering process, only the points, respectively the mesh corresponding to the piece surface were extracted.

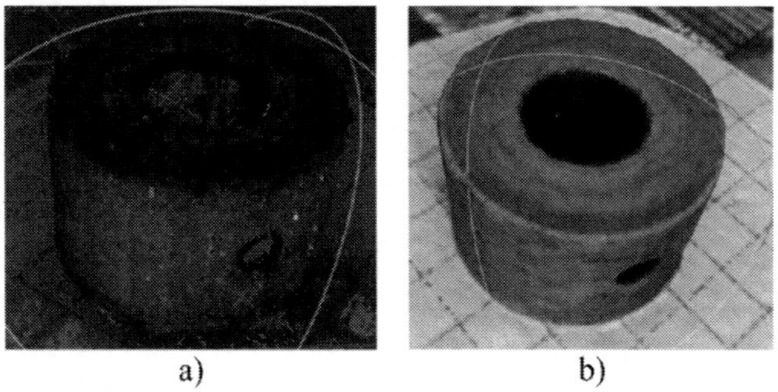

Figure 3. The designed piece 3D model, created in the "3DF Zephyr Pro" software (a) point cloud, (b) mesh surface.

The manufactured piece 3D model creation based on its characteristic points (CAD model). To create the 3D model of the manufactured piece based on its characteristic points, the same images as in the previous case were used.

The characteristic points of the manufactured piece and also the corners of the 2D grid, were manually marked on each image individually. By processing the data, using the *bundle adjustment algorihm* implemented into *"PhotoModeler Scanner 2012"* [10], the 3D coordinates of all referenced points were computed, their measurement precision was estimated and also the exterior orientation parameters for each camera position were calculated. Thus, were computed the 3D coordinates of a number of 127 points, of which a total of 83 points represent the 2D grid corners (figure 4a). Knowing that the 2D grid size is 10 mm, the object was scaled using a 70 mm line whose ends were clearly identified in the images, the scale factor being 8.46.

IOP Conf. Series: Materials Science and Engineering **95** (2015) 012088 doi:10.1088/1757-899X/95/1/012088

The manufactured piece 3D model (figure 4b) was created based on its characteristic points, i.e the points defining the top and inferior bases, the top hole limit and the limits of the small holes on the sides, using the *"PhotoModeler Scanner"* software specific functions, as following: point-line-surface.

For this case study, all the image coordinates errors were less than 5 pixels tolerance suggested by "PhotoModeler Scanner". The overall residual of the project was 0.847 pixels, less than the recommended of 5 pixels. The total error for determining the world coordinates, range between 0.018 mm and 0.186 mm. The angles between the projection rays range between 140.0765 ÷ 890.9712, the recommended angle being of 900, with an average of 830.2070.

Comparing the buildings under study by the proposed method. To determine the accuracy of the manufactured piece by the proposed method, the 3D models created as a point cloud and the CAD model, must be compared with a reference model considered for this case study the designed 3D model. In order to compare the 3D models, they must be brought in coincidence. For this case study the models were aligned using 3 common points.

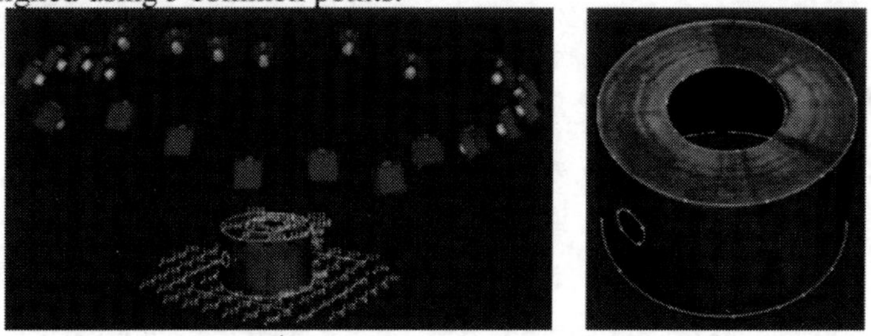

a) b)

Figure 4. The manufactured piece 3D model, created in the "PhotoModeler Scanner" software, (a) the resulted elements of the bundle adjustment, scaling and rotation processes and (b) detail.

After comparing the manufactured piece 3D model, created as a point cloud, with the designed 3D model of the piece, the following differences were obtained: maximum positive of *0.58 mm*, maximum negative of *-1.42 mm*, mean difference of *0.23 mm* and the standard deviation of *0.32 mm*, as can be seen in figure 5. Analyzing the color palette we can easily obtain the value for the Hausdorff distance, for example some poins situated on the top of the 3D model are colored in blue tone, corresponding to a value of about *-1.40 mm* on the scale. On the other hand, the points around the upper hole are colored in yellow and orange, corresponding to values between *0* and *0.33 mm* and the rest of the top points are colored in green, corresponding to a value of around *-0.73 mm*, so we can say that there is a small dent on the top of the manufactured piece. For the side of the piece, the differences are in range of *-0.30 mm ÷ 0* and *0 ÷ 0.30 mm*.

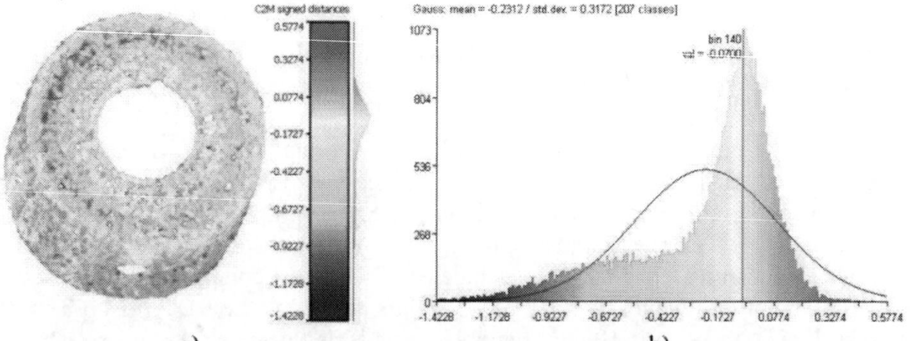

a) b)

Figure 5. (a) The differences between the manufactured piece 3D model, created as a point cloud and the reference model, (b) the differences distribution histogram.

We can compare the manufactured piece 3D model, created based on its characteristic points and represented as a mesh with the designed 3D model, represented also as a mesh, but the Hausdorff distances are calculated only in the vertices of triangles (figure 6a). In order to obtain a better accuracy, a sampling of each triangle surface is necessary. Thus, for each triangle which belongs to the mesh surface, 100 sample points were generated and after computing the distances between each point and the designed 3D model mesh (figure 6b) [11,12], the following differences were obtained: maximum positive of *0.19 mm*, maximum negative of *-0.37 mm*, mean difference of *0.09 mm* and the standard deviation of *0.09 mm*.

Analyzing the color palette we can easily say that the differences for the side of the 3D model are around *-0.37 mm* (blue color) for a small upper part, between -0.16 mm and -0.02 mm (green color) for the most of the side and between -0.02 mm and *+0.19 mm* (yellow to red) for a small bottom part. The same differences can also be observed for the top of the 3D model.

If we want to highlight the missing parts of the manufactured piece, we assume that our 3D model doesn't have the side holes. Second, we create sample points for each triangle of the mesh surface and we calculate the Hausdorff distance for each point. From figure 6c we can see that the points located in the right of the holes are colored in red, blue and yellow, showing the maximum differences between the two models, while the rest of the points are colored in green, the differences being close to *0*. We can conclude that from our 3D model the holes are missing.

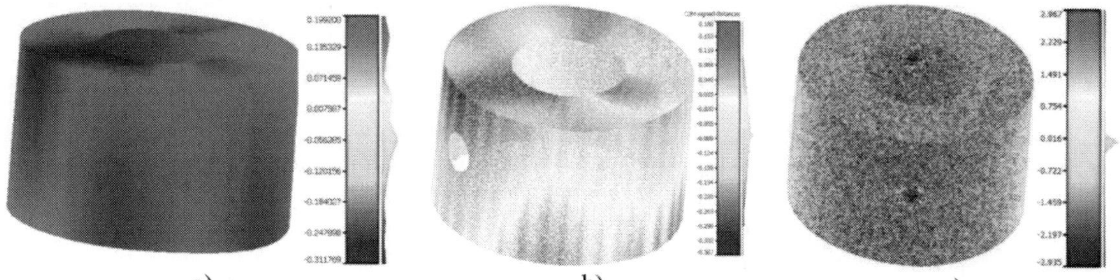

a) b) c)

Figure 6. The differences between the piece 3D model, created based on its characteristic points and the reference model (a) mesh 3D model, (b) sample points of each triangle of the mesh, (c) sample points of the mesh without the holes on the side.

4. Conclusions

Before any camera system is used for photogrammetric projects, the intrinsic parameters of the camera: focal length (f), optical center point (u_0,v_0), correction of radial distortion (k_1, k_2), correction of tangential distortion (p_1, p_2) and the medium image scale factor s_u, as well as the extrinsic parameters (r_{ij}, X_0, Y_0, Z_0) should be determined in the calibration process.

The accuracy in 3D reconstruction of an object is directly proportional to the accuracy of determining the scale factor, so it is recommended to use a 2D grid to be photographed simultaneously with the object.

When creating a 3D model as a point cloud by using the Structure from Motion algorithm, the result is obtained with high accuracy, in a completely automated way and in a very short time. By comparing it with the design 3D model, by the proposed method, the imperfections are measured and highlighted using a color palette. For a correct estimation of the imperfections size, we have to take into acount the RMS of the 3D model.

When creating the 3D model as point-line-surface (CAD model), we lose much of the representation fidelity, i.e small holes, big scratches, etc. and when comparing it with the design 3D model, we obtain only dimensional differences.

The proposed method for 3D models comparison, offers a comprehensive assessment of the differences between two 3D models by using a color palette, the user being able to quickly and accurately identify the errors of the 3D model that was compared with a reference model considered with no errors. It also illustrates the missing elements of the 3D model that was evaluated by generating sample points for each triangle surfaces which are components of the mesh surface, or by changing the reference model with the compared model.

The disadvantage in using this method for 3D models comparison is that we can compare only the exterior of the manufactured piece, because the images are taken only for the exterior of the piece.

References

[1] Zhang Z 2000 A flexible new technique for camera calibration *IEEE Transactions on Pattern Analysis and Machine Intelligence* **22** pp 1330-1334

[2] Oniga V E and Diac M 2013 Metric and non-metric cameras calibration for the improvement of real-time monitoring process results *Environmental Engineering and Management Journal* **12** 4 pp 719-726

[3] Oniga V E and C. Chirilă 2013 Hausdorff distance for the differences calculation between 3D surfaces *RevCAD* **15** pp 193-202

[4] Aspert N, Santa-Cruz D and Ebrahimi T 2002 Mesh: measuring errors between surfaces using the Hausdorff distance. *In Proc. of the IEEE International Conference in Multimedia and Expo (ICME)* **1** pp 705-708

[5] Cignoni P, Rocchini C and Scopigno R 1998 Metro: Measuring error on simplified surfaces *Computer Graphics Forum* **17(2)** pp 167–174

[6] *** Information on http://www.danielgm.net/cc/documentation.html

[7] Heikkila J and Silven O 1997 A four-step camera calibration procedure with implicit image correction *Proceedings of the 1997 Conference on Computer Vision and Pattern* **1106**

[8] J. Heikkila 2000 Geometric camera calibration using circular control points *IEEE Transactions on Pattern Analysis and Machine Intelligence* **22** pp 1066-1077

[9] *** Information on http://www.3dflow.net

[10] *** Information on http://photomodeler.com

[11] Roy M, Foufou S and Truchetet F 2004 Mesh comparison using attribute deviation metric *International Journal of Image and Graphics (IJIG)* **4(1)** pp 127-140

[12] Oniga V E 2014 A new method for buildings 3D models comparison *Journal of Geodesy and Cadastre RevCAD* **17** pp 60-67

IOP Conf. Series: Materials Science and Engineering **95** (2015) 012089 doi:10.1088/1757-899X/95/1/012089

Designing a system for measuring the flow of material transported on belts using ultrasonic sensors

N M Mihuţ[1]

[1] University "Constantin Brancusi" of Tg-Jiu, Faculty of Engineering, 30 Eroilor Street, 210135, Tg-Jiu, Romania

E-mail: nicoleta_simionescu@yahoo.com

Abstract. Excavation tailings (scraping) and extracting the useful (lignite) in surface mine pits in Mining Basin Oltenia is achieved with technological lines of excavation - transport - dump of high productivity. A correlation of working capacity of the main components of technological lines (motor rotor, high capacity transport, car dumps) is necessary for economic reasons on electricity consumption. To achieve experience in the process was chosen excavator SRS 1400 from South Jilt career in the CET Turceni. The question of coal excavated volume has a great importance in the mine pits. At the excavation is desired a density estimate for each machine production tracking, cost estimation and tracking product unit profitability of each band on various sections zones. Permanent display size excavated volume snapshots in the excavator's cabin permits to track tape loading, eliminating unproductive times and information management to determine profitability. Another important requirement is closing the loop of the machine drive system of an excavator for a uniform deposition of carbon on the strip, thus achieving automatic control of the loading belt. Such equipment is important for the system dispatching in surface mine pits. Through a system of three ultrasound transducers to determine the smart instant of coal excavated section which, coupled with the tape speed, integrated over time will determine the amount of excavated coal. The basis of the system developed is a device for determining the volume and quantity of coal excavated acting on the march and optimize the system speed excavator working order. The device is designed primarily following the careers of lignite production: rotor excavators, rubber conveyor belts and dump facilities. Newly developed system aims to achieve the following determines: the optimum energy excavation depending on the nature of excavated material - lignite, shale, clay, etc., economic times to use the excavator bucket teeth rotor, energy optical regime to rubber belt conveyors, eliminate damage to the plant by conveyor belt breakage detection or tread and eliminating time and energy losses by limiting reproductive idle, monitoring the amount of coal excavated, control power consumption. Systems in general and particularly automated systems cannot be designed without taking into account their degree of effectiveness, compliance with minimum consumption of time, energy and materials, insofar as they are requested and used all the resources it has, at minimum cost production, etc. For this reason any matter of calculation, design, analysis and operation of transport systems continuously subordinate requirements optimality.

1. Introduction

The question of coal excavated volume per unit time is important for tracking production per machine, per product unit cost estimation and tracking profitability on various sections of each band schedules

Content from this work may be used under the terms of the Creative Commons Attribution 3.0 licence. Any further distribution of this work must maintain attribution to the author(s) and the title of the work, journal citation and DOI.
Published under licence by IOP Publishing Ltd

[1,2]. Permanent display size excavated volume snapshots in the excavator cabin can track the tape loading, eliminating unproductive times and information management to determine profitability. The mechanism is installed so as to read information directly on the conveyor belt and perform such coal volume measurement using a computer system [3,4]. figure 1 below is an example of such a measurement mechanism.

Figure 1. Equipment for measuring the volume of coal belt.

Such equipment, important for the system dispatching in surface mine pits, can be achieved using three ultrasonic sensors that measure carbon-sensor distances in three points. The schematic diagram of the apparatus for measuring the flow rate of the carbon / sterile is shown in figure 2.

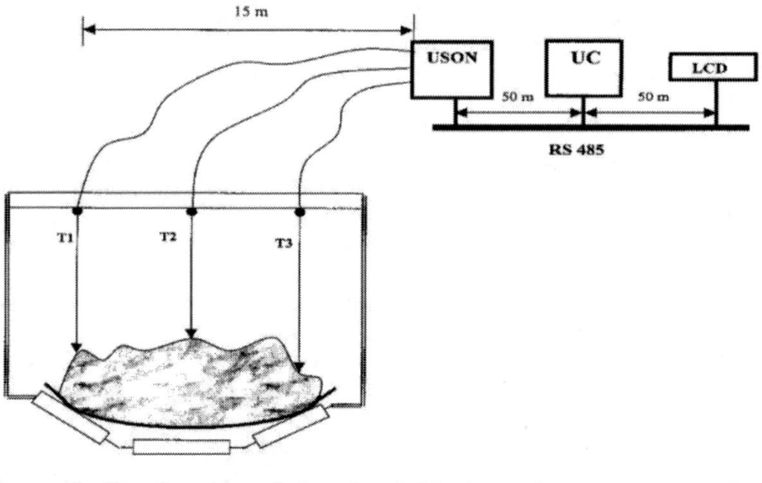

Figure 2. The location of functional blocks and sensors measuring device.

In this figure shows the sensors T1, T2, T3 installed on the conveyor belt that transmit information gathered in the computer system USON-UC-LCD which processes them and calculates the volume of coal/sterile excavated. Section instantaneous current flowing material on tape, coupled with integrated belt speed while determining the amount of coal excavated [5]. The software package developed (called PP USON) to determine the flow of material transported large capacity tape is written in C^{++} assembly language for microcontroller 80C552 type [6,7]. The computer calls for useful data packet by USON and not vice versa. Once the application executed, the computer waits for 50 milliseconds order fulfillment. If USON does not respond within this time frame, there are three possibilities:

a) wireless communication interrupted;
b) communication wire cage;
c) lack of supply voltage USON;

With 10 such applications useful data packet and honored by the computer running, the computer decides that USON is not responding and has no reason to do so. This situation is recorded in a file called serial error 'ComErr.dat'. With the variable giving the sum of the volume calculated by the computer relative to a time of x seconds, it would therefore be excavated material flow. This flow is transmitted on the serial interface COM2 in the excavator cabin with tape condition. Recall that there in the excavator cabin is a display unit features a 2 x 16 character LCD, 80C552 microcontroller and a serial interface functionally compatible with the PC. The computer is not equipped with any video card or graphics display device, outside the mounted LCD display on the front panel where excavated material flow along the strip state. Such unit display is shown in figure 3 below.

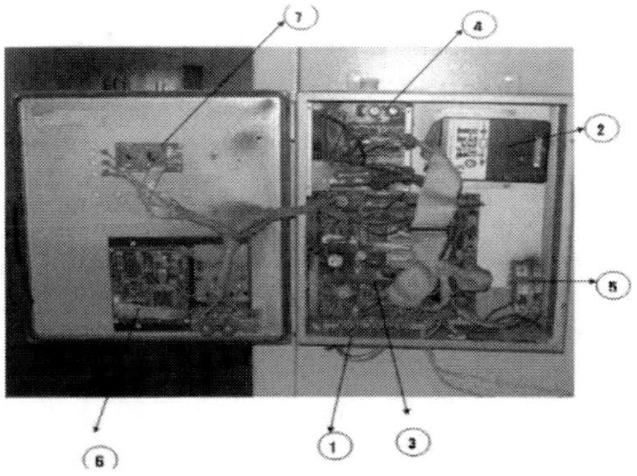

Figure 3. Equipment for real-time data acquisition: 1-PC; 2-HDD; 3-Motherboard; 4-Source (12V on output); 5-Circuit; 6-Unit power microcontroller; 7-LCD 2X16 characters.

The device type is for scanning response USON three ultrasonic sensors and inductive encoder revolutions. All values thus obtained are digital type. The four values are dynamic and depend upon the band (both the rate and the thickness of the web of coal excavated in a time t). This equipment also has a Digital Display local distances are displayed in the three sensors blanket strip coal and condition (functional / non-functional). The data thus acquired by USON are sent to UC series. This is actually an industrial computer equipped with a fast microprocessor interprets the data, managing to answer three ultrasonic sensors correlate with dynamic belt speed to a value of "Volume coal digger". This dynamic value is displayed on the two systems functionally independent (that is display equipment data). All calculations massive volumes that are running are allowed UC's because it has a fast microprocessor suitable processing speed required [8]. Besides being recorded on HDD dynamic behavior of excavator, referring only to the volume of excavated the UC contacts on a serial data bus with the USON to provide a data stream according to the scope of the equipment. Communication rate USON - UC is very fast and can detect any of the equipment listed lack other to eliminate possible defects interconnection.

From a functional perspective, the software consists of the following distinct parts:

- Serial communications that have implemented some data structure corresponding to the descriptor issued by the microcontroller package on request. Serial communication is between local equipment USON and the machine on the PC structure and between the PC and display unit from the excavator cabin;
- Interpolation algorithms for finding cross-section of the band (relatively instantaneous "photo" of belt profile - coal layer) which is shown in figure 4 from below;
- Decision algorithms on the data that is stored on the hard disk.

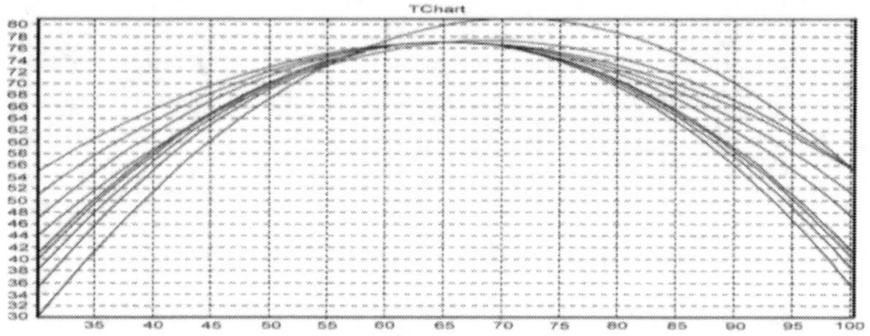

Figure 4. Relatively instantaneous "photo" of belt profile - coal layer.

2. Experimental data

To achieve experience in the process was chosen excavator SRs 1400-1430/7 Jilt South career within Turceni Electrical Plant. The time of the entry corresponding to the type excavator SRC 03 1400-1430/7, from Jilt South career within is 14.02.2013-19.04.2014, when there were corresponding average flow values approximately 700-800 m3 / h. Table 1 presents the results of measurements corresponding time periods studied.

Table 1. The evolution of production capacity in operation on 14.02.2013.

Year	Month	Day	Hour	Belt condition	Volume of excavated coal
2013	2	14	13: 05:45	START	7.38515587133861E-0002
2013	2	14	13:12:54	START	0.00000000000000E+0000
2013	2	14	13:12:55	STOP	0.00000000000000E+0000
2013	2	14	13:14:18	START	0.00000000000000E+0000
2013	2	14	13:14:18	STOP	0.00000000000000E+0000
2013	2	14	14:27:23	START	0.00000000000000E+0000
2013	2	14	14:28:08	STOP	0.00000000000000E+0000
2013	2	14	14:28:34	START	0.00000000000000E+0000
2013	2	14	14:29:04	STOP	0.00000000000000E+0000
2013	2	14	14:29:40	OK	8.25616892870748E+0001

Figure 5 highlights the operation of the machine during excavating of coal layers with small thickness (g <1m) on 14/02/2013.

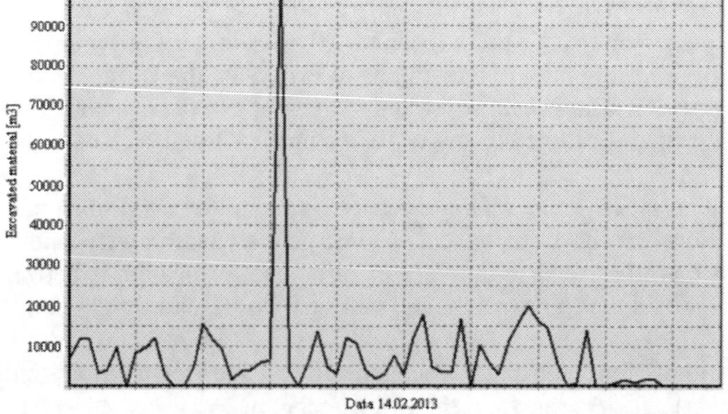

Figure 5. Graph functioning machine for coal layer thickness g <1m.

In table 2 from above were revealed the parameters of the excavation regime on 14.02.2013 which were obtained in a usually day of work with an 8 hours program.

Table 2. The parameters of the excavation regime on 14.02.2013.

Data	Hour	T (s)	ΔT (s)	V (m^3)	Q (m^3/h)
14.02.2013	13: 05:45	51687	32	262,967	509,1535
14.02.2013	13:12:54	51745	33	271,3927	922,7167
14.02.2013	13:12:55	51777	30	246,7251	898,2012
14.02.2013	13:14:18	51810	27	222,2733	985,7439
14.02.2013	13:14:18	51840	30	247,0731	1097,841
14.02.2013	14:27:23	51867	31	255,3229	987,7055
14.02.2013	14:28:08	51897	29	238,5969	956,9799
14.02.2013	14:28:34	51928	28	230,6965	1021,877
14.02.2013	14:29:04	51957	58	477,9813	1057,265
14.02.2013	14:29:40	51985	45	370,4693	512,3883

The data in the table above are decoded and stored in the computer's RAM. It appeals to a procedure that calculates the cross-sectional area of coal tape only if the quantum rotation frequency while giving the conveyor belt drum is less than a predetermined value, given that the first band starts hard and achieve speed rated in a negligible time. The algorithm for obtaining the cross-sectional area based on a numerical method of Lagrange interpolation in series. By applying the Lagrange interpolation method, the procedure returns to square meters cross-sectional area of the strip occupied by charcoal. In the event that one of the three sensors returns a large distance, close to within 2% of the distance bottom band - rod sensors, it is considered that the band at that point is slacking / sterile. In a particular case, all three sensors T1, T2, T3, may have very similar responses than 2% of the distance mentioned above, where the cross-sectional area at that point of time tends strongly to zero, no need laborious calculation. The latter is just a particular case, but quite common in practice after measurements and it is imperative to introduce the general algorithm [9]. Using the data obtained from the measurements was performed in the graphical representation of figure 6 corresponds to the change in the flow of material , depending on the time and the amount of excavated material:

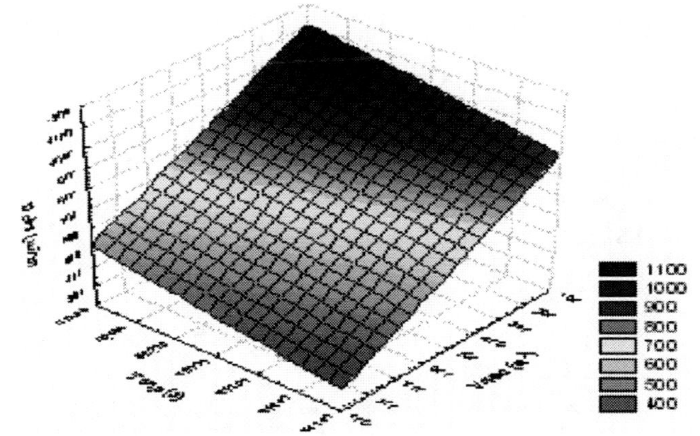

Figure 6. Material flow variation versus time and the amount of material excavated [7].

The corresponding characteristic equation of material flow as a function of volume and time $Q = f(t,V)$ is of the form presented in equation (1) from below:

$$Q = -7759,9958 + 0,1701 \cdot t - 0,7467 \cdot V \tag{1}$$

Following the evolution of the parameters in the above figure it is observed that the material flow varies between 400-1100 m3/h while the average flow is 700-800 m3/h. These data are important for estimating the volume of production in certain times and for determining the number of machines and the number of shifts to be used to obtain the estimated production volume.

3. Conclusions

Increased automation of transport continuous belt facilities is subject to equipping them with a system of sensors and transducers for online tracking of the production process. Continuous transmission control unit must do the following operations:

- Automatic control of state band,
- Order long ball towards the band, band adjustment effort,
- Measuring the amount of material transported,
- Avoid agglomeration of material,
- Removal of metal bodies,
- Avoiding spillage of material,
- Cleaning tape, etc.

State control involves the control strip longitudinal fracture, complete or partial transverse band, metal inserts control, skid control, etc. Exiting alignment produce material spread on the lower branch, causing premature wear, stiffs lower roller train which causes additional energy consumption. Optimization of the constructive parameters of the installation for continuous transport belt requires the use of a database drawn up on-line, necessary for controlling the dynamic operation thereof in order to improve the technological parameters of the transport conveyor. Solution using an ultrasonic sensor device solves the problem of knowledge coal excavated volume per unit time, essential for tracking production per machine, per unit product cost estimation and tracking profitability on various sections of each band schedules. Permanent display size excavated volume snapshots in the excavator cabin permits tracking tape loading, eliminating unproductive times and information management to determine profitability. Transportation costs greatly decrease with increasing flow rate, especially when the carrier has great length, load it with the rated load of the drive motor which requires a good power factor naturally occurring, thus making it an economy of energy.

References

[1] Manolea Gh 2004 Automated electromechanical drive *Universitaria Publishing* Craiova
[2] Manolea Gh 2004 The quality of energy in electrical drives with asynchronous motors and static converter *Acta Electrotehnica* **45(3)** pp 45-47
[3] Ancău M and Nistor L 1996 Numerical optimization techniques in computer-aided design *Technical Publishing Bucuresti* pp 86-92
[4] Babutia I 1985 Automatic management processes *Facla Publishing* Timişoara pp.63
[5] Magyari G T 1990 Mining machinery *Technical Publishing House* pp 33-37
[6] Batatorescu A 2003 Linear optimization methods *Universităţii Publishing House* pp 75-81
[7] Mihut N M 2004 Research on increasing transport capacity conveyor belt by improving teaching - receiving material transfer point. Research Contract Nr. 306 C from 20.07.2004 between University "Constantin Brâncuşi" of Tg-Jiu and Uzina de Reparaţii LTD, Tg-Jiu
[8] *** Information accessed on 18.11.2014, www.sensorsportal.com/HTML/SENSORS/
[9] *** Information accessed 12.11.2014, on www.ipp.md/public/biblioteca/22/ro/

IOP Conf. Series: Materials Science and Engineering **95** (2015) 012090 doi:10.1088/1757-899X/95/1/012090

Analysis of possibility of applying the PVDF foil in industrial vibration sensors

A Wróbel[1]

[1] Institute of Engineering Processes Automation and Integrated Manufacturing Systems Faculty of Mechanical Engineering, Silesian University of Technology, Konarskiego 18a Street, 44-100 Gliwice, Poland

E-mail: andrzej.wrobel@polsl.pl

Abstract. There are many machines using the piezoelectric effects. Systems with smart materials are often used because they have high potential applications for example transducers can be applied to receive required characteristic of projected system. Every engineer and designer know how important it is properly mathematical model and method of the analysis. Also it is important to consider all parameters of analyzed system for example glue layer between elements. Geometrical and material parameters has a significant impact on the characteristics of the all system's components because the omission of the influence of one of them results in inaccuracy in the analysis of the system. In article the modeling and testing of vibrating systems with piezoelectric ceramic materials transducers used as actuators and vibration dampers. The method of analysis of the vibrating sensor systems will be presented, mathematical model, and characteristics, to determine the influence of the system's properties on these characteristics. Main scientific point of the project is to analyze and demonstrate possibility of applying new construction with the PVDF foil or any other belonging to a group of smart materials in industrial sensors. Currently, the vibration level sensors are used by practically all manufacturers of piezoelectric ceramic plates to generate and detect the vibration of the fork.

1. Introduction

What is the piezoelectricity? By the force action on some crystals you can make electricity flow through them. It is simple piezoelectric effect presented on figure 1a. In other way if you pass electricity through the same crystals they change geometry (figure 1b). Piezoelectricity was found in 1880 by a French physicists and now it has a lot of apps. Direct piezoeffect it is generation of an electric potential when stress is given to the piezoelectric material, also other way the reverse piezoeffect it is production of deplacement when an electric is fixed. In this paper analysis of possibility of using the PVDF foil in industrial vibration sensors will is presented.

In the existing vibration sensors direct and reverse piezoelectric effects are presented. In those level sensors piezoelectric plates are used as actuators and as vibration dampers with the external electric circuit. Usually the plates are fixed on the surface of a mechanical subsystem or only press by mechanical elements to generate vibrations or also to control amplitude of vibrations. Electric potencial is produced by external box and applied plates. A passive electric circuit is adjoined to transducer's clamps. Piezoelectric topic is often described by the authors of the publication [8-11].

Content from this work may be used under the terms of the Creative Commons Attribution 3.0 licence. Any further distribution of this work must maintain attribution to the author(s) and the title of the work, journal citation and DOI.
Published under licence by IOP Publishing Ltd

There are two basic applications of this idea. What is more there are many commercial applications of this idea presented for example in [6, 11].

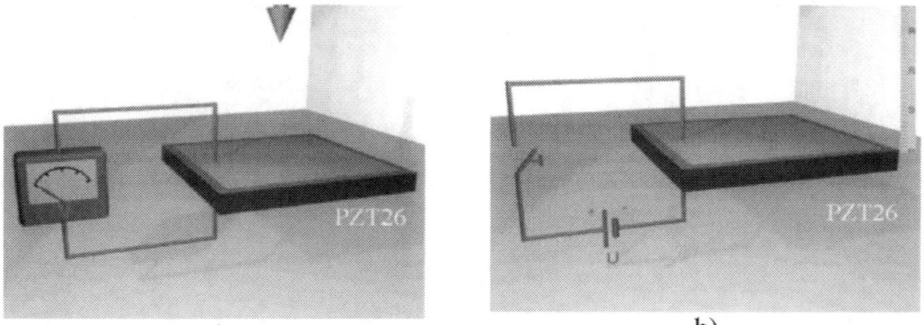

a) b)

Figure 1. Direct (a) and converse (b) piezoelectric effect.

Mechatronic systems with piezoelectric sensors or actuators are widely used because piezoelectric transducers can be easy controled in order to obtain required dynamic characteristic of designed system [2-7]. A very convenient method of analysis of mechatronic systems are non-classical methods presented in publications [1-3]. The model under examination with the description of input and output parameters is shown in figure 2.

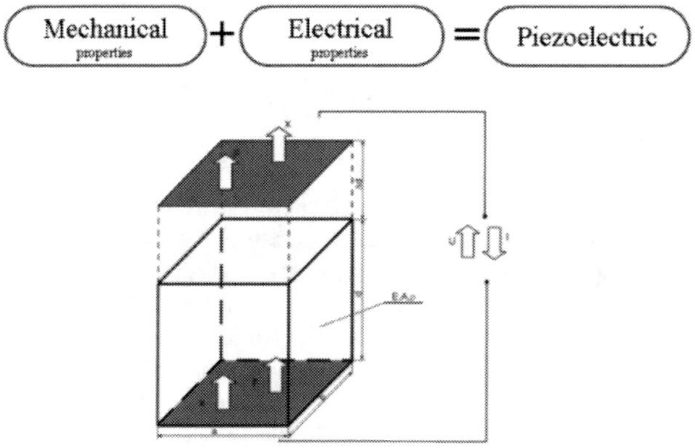

Figure 2. Model under examination.

This article is a theoretical consideration of possibility of applying the PVDF in industrial sensors. The article does not contain mathematical calculations and computer simulations of new construction, which will be the subject of future studies. The discussed subject is important due to increasing number of applications, both simple and reverse piezoelectric phenomena in various modern technical devices. The process of modeling of technical devices with piezoelectric is complex and requires large amounts of time because of the complexity of the phenomena occurring in these systems. The correct interpretation of the system by mathematical model during the design phase is fundamental condition for proper operation of designed system.

2. Vibration sensor unit and proposition of changes

Currently proposed detectors (presented in figure 3) have many important defects using the piezoelectric ceramic elements consists of two pairs of plates, and receiver of two or three plates broadcasting connected as a stack. An alternating voltage fed into the transmitting plate, the thickness changes proportionally to the applied voltage.

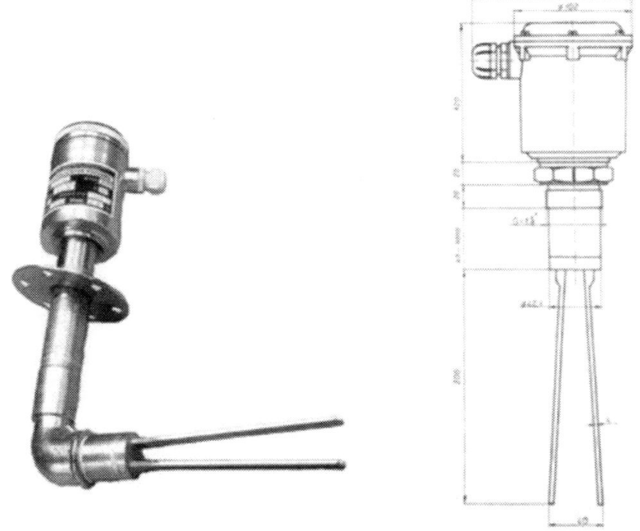

Figure 3. Construction of currently used level detectors.

Changes in thickness of the plate lead the vibration of the mechanical element, so-called fork. If the fork is not buried by the material deformation of the full tiles broadcasting is transmitted to receiver plate. As a result of crushing and stretching of the plates receiver the cladding is formed on the potential difference proportional to the force. The value of this voltage is processed by an electronic circuit. In the case of backfilling forks, tile receiving cease to be crushed and stretched. At the same time is not generated the potential for cladding tiles. New construction of level sensor is composed exclusively of a single beam which is fixed to a piezoelectric layer, it was presented in figure 4.

Figure 4. Proposition of new construction of level detector.

Utilitarian the scientific purpose of the project is to analyse and demonstrate possibility of applying the PVDF foil or any other belonging to a group of smart materials in industrial level sensors. Currently, tey are used by practically all manufacturers of piezoelectric ceramic plates to generate and detect the vibration of the fork. This approach is associated with a number of drawbacks, including:

- The need to use several piezoelectric, with the increase number of components is increased defect sensors. The plates are made of brittle materials which also causes crashes and damage,
- After laying the tiles in the so-called piezoelectric stack, use the calibration screw perform tuning fork resonance frequency and the frequency of an electronic circuit that generates vibrations.
- Sensors are produced with two forks resonance, using these in moist materials is often the case that the material remains between the forks and at the same time causes a measurement error, and

sometimes the need for intervention conservator. Given the fact that currently produced sensors should be characterized by high accuracy and reliability.

- Vibration sensors are performe as complex construction, two separate forks welded to the so-called membrane. On the other side of the membrane a stack of ceramic innovation proposed by the present design allows for a considerable simplification of the device is mounted.

The main reason to take consideration of this subject are researches done by many years on Faculty of Mechanical Engineering, Silesian University of Technology conducted by author and other. The proposed research problem is therefore a continuation and extension of previous work carried out by the applicants whose aim is to develop new mathematical models and universal methods of analysis and synthesis of complex mechatronic systems.

3. Mathematical model of new design of level sensor

The aim of future work will be to develop the most exact mathematical model and geometrical new sensor industry. In programm CAx it is possible to conduct simulations and the natural frequency of the sensor. The analysis will also be a mathematical model that will confirm the accuracy of the computer simulation done.

The different ways of restraint systems investigated, and the impact of geometric parameters and plastic on their characteristics will be presented.

Mechanical system will be described in the equation taking into account the weight of the continuous distribution. The transmitter will be presented in while using appropriate electrical schematics replacement and the corresponding equation. In the case of continuous models, both the mechanical subsystem and describe the piezoelectric transducer, having a continuous distribution of their mass. In addition, developing various mathematical models will refine them, starting with the assumption of ideal attaching the transducer to the surface of the mechanical subsystem, to take account of the phenomena occurring in the adhesive connecting the system components.

In the future strings of mathematical models of industrial sensor models with varying degrees of detail mapping real circuits will be developed. This approach is the result of inability to use the exact method for the analysis of mechatronic systems, which is why using approximate methods, determines the characteristics of the individual components and shall examine the effects of the changes in the parameters of plastic, geometrical and electromechanical compared to and evaluate the mathematical models developed in terms of the impact of the level of detail of the complexity and time-consuming calculations performed and the results obtained. In terms of the adopted criteria, indicate the most appropriate place to attach a piezoelectric foil of the developed models.

The main idea of future work will be to designate dynamic characteristics of a mechatronic system with piezoelectric transducer used as an actuator or passive vibration damper. It is a cantilever beam which has a rectangular constant cross-section, length, width and thickness. Young's modulus of the beam is denoted. Piezoelectric of length l and it is fix to the beam's surface within the distance. The transducer is bonded by a glue layer of thickness hp and Kirchhoff's modulus G. The glue layer has homogeneous properties in overall length. The system under consideration in both cases (with piezoelectric actuator) is presented in figure 5.

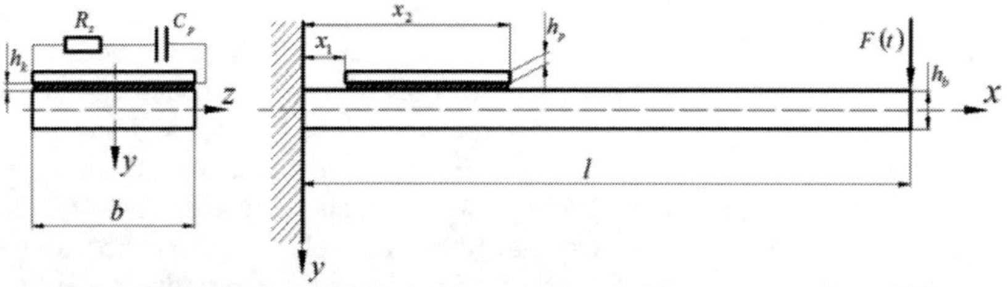

Figure 5. Model of new sensor construction.

In order to analyse vibration of the systems following assumptions were made:

- material of which the system is made is subjected to Hooke's law,
- the system has a continuous, linear mass distribution,
- the system's vibration are harmonic,
- planes of sections that are perpendicular to the axis of the beam remain flat during deformation of the beam – an analysis is based on the Bernoulli's hypothesis of flat sections,
- displacements are small compared with the dimensions of the system.

4. Conclusions

Piezoelectricity has found a lot of applications, for example, it is used in vibration level sensors.

Main point of this article was to demonstrate possibility of applying the PVDF foil or any other belonging to a group of smart materials in industrial sensors. Currently, the vibration level sensors are used by practically all manufacturers of piezoelectric ceramic plates to generate and detect the vibration of the fork. This approach is associated with a number of drawbacks. The new design of presented sensor eliminates the problems of the currently used sensors. For example in new type of sensors the need of use ceramic piezoelectric stacks will be eliminated, it will cause a longer service life of sensors and their reliability. By use just one vibrating element it will be eliminate locking material into forks. This article is only an attempt to deliberations on a new design of level sensors. In future work mathematical model will be calculated and author will try to create a prototype. This article is only an attempt to deliberations on a new design of level sensors. In future work mathematical model will be calculated and author will try to create a prototype.

Acknowledgments

The work was carried out under the project number PBS2/A6/17/2013 agreement implemented under the Applied Research Program, funded by the National Centre for Research and Development.

References

[1] Białas K 2008 Graphs and structural numbers in analysis and synthesis of mechanical systems *Journal of Achievements in Materials and Manufacturing Engineering* **29** pp 151-154
[2] Buchacz A 1997 Computer aided of synthesis and analysis of machine's sub-assembly by graph and structural number's methods *Silesian University of Technology Press*
[3] Buchacz A and Wróbel A 2008 Homogenization of mechatronical systems in modeling piezoelectric layer *Machine Dynamics Problems 25* **32(3)** pp 7-11
[4] Buchacz A and Wróbel A 2009 Modelling of multilayer transducer by means of the graph and structural numbers method *80th Annual Meeting of the International Association of Applied Mathematics and Mechanics* **9(1)** pp 393–394
[5] Buchacz A and Wróbel A 2010 Modelling and study of piezoelectric effect influence on the characteristics of mechatronic system University of Technology*: Publisher Silesian*
[6] Buchacz A and Płaczek M 2010 The approximate Galerkin's method in the vibrating mechatronic system's investigation *Proceedings of The 14th International Conference Modern Technologies, Quality and Innovation ModTech* pp 147-150
[7] Dymarek A and Dzitkowski T 2005 Modelling and synthesis of discrete – continuous subsystems of machines with damping *Journal of Materials Processing Technology* **164-165** pp 1317-1326
[8] Gwiazda A, Herbuś K, Kost G and Ociepka P 2015 Motion analysis of mechatronic equipment considering the example of the Stewart platform *Solid State Phenomena* **220-221** pp 479-484
[9] Jamroziak K and Bocian M 2010 Identification of pierced materials characteristics in the aspect of selected degenerated models *Journal of KONES* Powertrain and Transport **17(1)** pp 169-176
[10] Mason W P 1948 Electromechanical Transducers and Wale Filters. Van Nostrand
[11] Płaczek M 2014 Computer Aided Analysis of Systems with MFC Actuators *Advanced Materials Research* **1036** pp 711-714

IOP Conf. Series: Materials Science and Engineering **95** (2015) 012091 doi:10.1088/1757-899X/95/1/012091

Complementary methods for nondestructive testing of composite materials reinforced with carbon woven fibers

R Steigmann[1,2], N Iftimie[1], R Sturm[3], P Vizureanu[4] and A Savin[1]

[1] National Institute of R&D for Technical Physics, 47 D.Mangeron Blvd., 700050, Iasi, Romania
[2] University Alexandru Ioan Cuza, 11 Carol Blvd, 700006, Iasi, Romania
[3] University of Ljubljana, Kongresni trg 12, 1000 Ljubljana, Slovenia
[4] Technical University Gh.Asachi Iasi, 67 D.Mangeron Blvd., 700050, Iasi, Romania

E-mail: steigmann@phys-iasi.ro

Abstract. This paper presents complementary methods used in nondestructive evaluation (NDE) of composite materials reinforced with carbon woven fibers as two electromagnetic methods using sensor with orthogonal coils and sensor with metamaterials lens as well as ultrasound phased array method and Fiber Bragg gratings embedded instead of a carbon fiber for better health monitoring. The samples were impacted with low energy in order to study delamination influence. The electromagnetic behavior of composite was simulated by finite-difference time-domain (FDTD) software, showing a very good concordance with electromagnetic nondestructive evaluation tests.

1. Introduction

Carbon fiber reinforced plastics composites have been started to be used at large scale in aerostructures and automotive industry. Majority of structures have been manufactured using carbon-fiber-reinforced polymers (CFRP), which are thermoset pre-impregnated carbon fibers in the form of unidirectional tape or woven fabric. But, in order to reduce the costs, to increase the production rates and avoid the impact on environment, it requires new manufacturing processes and materials, as for carbon-fiber-reinforced thermoplastics (CFRTP). Despite the fact that thermoplastic composites are more expensive, they have been preferred, due the advantages of thermoplastic matrix as recyclability, aesthetically finishing, high impact resistance, chemical resistant, hard crystalline or rubbery surface option, eco-friendly manufacturing, making the entire process costless [1]. Woven carbon fiber is most suitable for applications requiring a high strength-to-weight ratio. Polyphenylene sulfide (PPS) has excellent properties [2], framing into the advantages described above. Woven carbon fibers/PPS laminates are characterized by reduced damages but are susceptible to impacts with low energies, leading to delaminations, desbonding of carbon fibers and/or matrix cracking [3]. The aerospace industry has the requirement for the highest quality control and product release specifications [4]. The raw material standard, prepreg materials, require mechanical property testing (e.g. interlaminar strength and tensile strength) from specimens. Final parts also require NDE [5] and structural health monitoring (SHM) [6]. Alternative techniques based on phenomenological changes on the composite materials were developed during the past several decades, to measure damage due to impacts: acoustic emissions [7], infrared imaging [8], electrical resistance [9] or non-contact techniques such as digital

Content from this work may be used under the terms of the Creative Commons Attribution 3.0 licence. Any further distribution of this work must maintain attribution to the author(s) and the title of the work, journal citation and DOI.
Published under licence by IOP Publishing Ltd

image correlation [10] and X-ray tomography [11], but these methods have their limitations, being complementary. Nowadays, NDE methods are developed for post-damage inspection and assessment as thermography [12], ultrasonic C-scan [13], and eddy current [14] offering quantitative results. Other techniques, such as piezoelectric sensor [15], optical fiber [16], can be used for on-line health monitoring of CFRP structures, embedding external sensors or additional fiber input in CFRPs.

This paper presents complementary methods used in NDE of composite materials reinforced with carbon woven fibers as two electromagnetic methods using sensor with orthogonal coils [17] and sensor with metamaterials lens [18] as well as ultrasound phased array method [19] and Fiber Bragg gratings embedded instead of a carbon fiber [20]. The studied samples were impacted with low energy in order to determine the damage inside the composite and provide more advanced imaging system and image of impacted area reconstruction. The electromagnetic behavior of composite was simulated by finite-difference time-domain (FDTD) software, the samples being CAD designed following textiles features, showing a very good concordance with electromagnetic NDE (eNDE) tests.

2. Studied samples

Plates of CFRTP made by Tencate, the Netherlands [2] having the dimensions 150 mm×100 mm×4.2 mm have been taken into study. The composite plate contains 12 layers of carbon fibers, 5 Harness satin woven type as reinforcement and the matrix is PPS –CETEX. The carbon fibers volume ratio is 0.5±0.1. The carbon fibers are T300JB type. This composite has been used in the construction of Airbus and Boeing aircrafts. In figure 1 are presented the studied samples and the layout of the 5 Harness satin woven. The samples were impacted with energies of 2 J, 4 J, 6 J, 8 J, 10 J, and 12 J, using equipment FRACTOVIS PLUS 9350-CEAST-Instron USA with a hemispherical bumper head having 20 mm diameter and 2.045 kg weight, in order to induce delaminations.

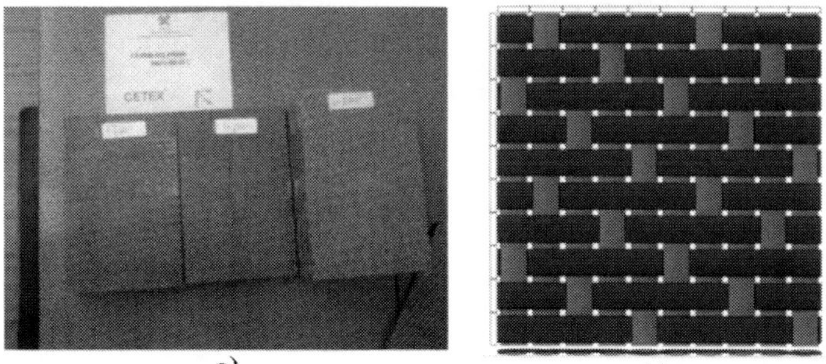

a) b)

Figure 1. Studied samples: a) composite plates; b) 5 H satin layout.

Under impact, the CFRPT suffer delamination that is usually accompanied by a dent. The dent causes a reduction in the spacing between fibers in the thickness direction and this causes an increase in fiber contact leading to decrease of electrical resistance in the thickness direction. This situation modifies the electrical conductivity local both in the plane of the fibers and perpendicularly on fibers so that eNDE methods can be applied.

2.1. Ultrasound Phased array

The US measurements were effectuated with Phasor XS-General Electric, which can work both in conventional mode (pulse-echo or emission-reception) as well as in phased array mode, classic or top view. A linear phased array type with 32 elements and pitch of 0.5 mm, working at 5MHz central frequency. The Phasor XS equipment allows the selection of the domain for the angular scan in terms of refracted angles in the examined material after that its properties were input in the program. Usually, phased array sensor is fixed on a wedge (figure 2) [19].

Figure 2. Phased array sensor.

2.2. Fiber Bragg gratings

The principal problem of CFRTP is the desbonding of fibers near subsurface during the impact leading to exfoliation of underskin that cannot be "seen" with US method. Therefore, CFRTPs with FBG sensors embedded were monitoring in order to emphasize their behavior at impact [20].

The optical block diagram of measuring method is show in figure 3.

It consist of a broadband light source (LED) that illuminates the optical input line and a reference cell after passing a 2 by 2 coupler. The reflectance cell consists of a double FBGs housed in an athermal package and reflects two wavelength peaks (around 1529 nm and 1571 nm). Together with the reflected signals coming from the optical input, the reflected spectrum will be analyzed after the 2 by 2 coupler using an optical spectrum analyzer.

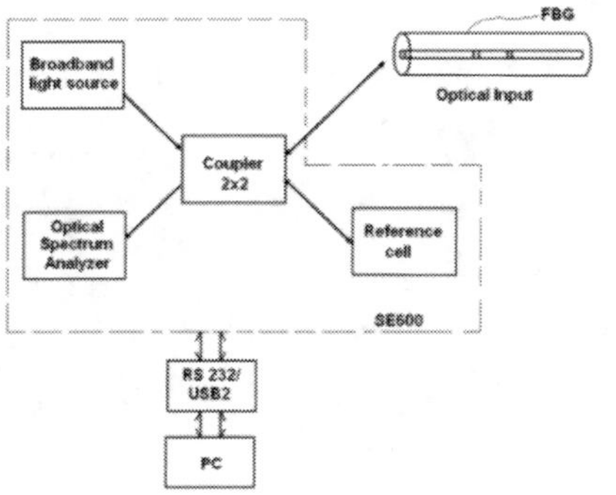

Figure 3. Optical block diagram.

2.3. Electromagnetic sensors

Sensor with orthogonal coils. First of eNDE method of the studied samples, a focused eddy current sensor with orthogonal coils [21] has been used. The sensor is absolute send receiver type. The emission part is winded inside a high frequency ferrite cup core with 2.5 mm inner diameter, 3.5 mm outer diameter and 2.8 mm height. The reception coil is orthogonal on the emission one (figure 4a). The surface is XY raster scanned with 1mm step in both directions and give information about the amplitude and the phase of the signal. The wave front of the field created by the emission coil in material can be considered quasispherical.

Sensor with metamaterials lens. Second electromagnetic sensor involved in characterization of studied samples is a sensor that contains a metamaterial lens for detection (figure 4b). The lens is constructed from two conical Swiss rolls (CSR) with the characteristics and functioning mode given in [18], with large basis face to face. The rectangular frame used for the generation of TMz polarized electromagnetic field has 20 mm x 60 mm dimensions from 1.2 mm diameter Cu wire.

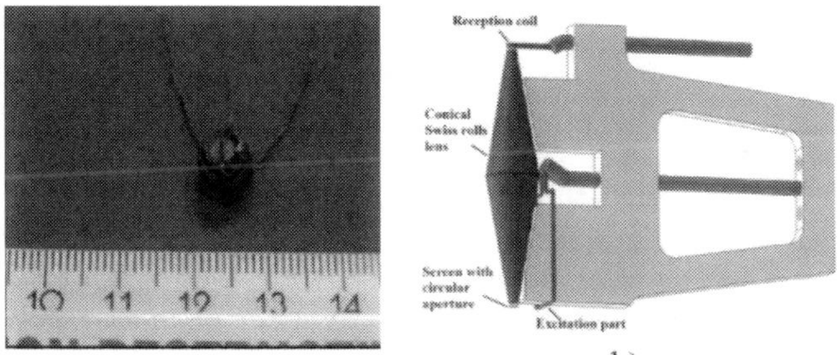

a) b)

Figure 4. Electromagnetic sensors: a) with orthogonal coils b) with metamaterial lens.

The electromagnetic sensor (either the one with orthogonal coils as well as the one with metamaterials lens) is coupled to a Network/Spectrum/Impedance Analyzer, 4395A type – Agilent USA. The sample is fixed on a XY motorized stage type Newmark USA. The command, the displacing and the acquisition of the amplitude and the phase of the signal received by the sensor are made by PC. The equipment Agilent 4395A is coupled with PC through a parallel interface type IEEE 488.2 and the XY motorized stage through serial interface type RS232 (figure 5). The commands and the data acquisition are carried out through a code developed in Matlab 2011b.

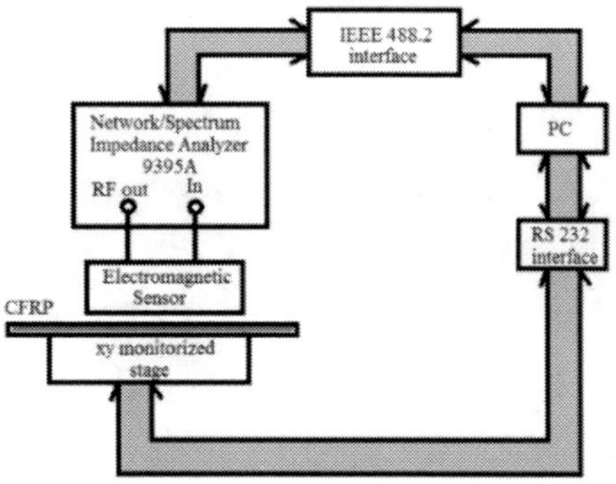

Figure 5. The experimental set-up for electromagnetic methods - block diagram.

3. Experimental results

A region of 60x60 mm from the sample has been scanned in both directions with 1 mm step, using the electromagnetic sensors described above. For the sensor with orthogonal coils, the working frequency was 2.2 MHz and the lift-off 1 mm. For the sensor with metamaterials lens, the frequency was 478 MHz and lift-off 0.5 mm. The obtained electromagnetic images are presented in figure 6a and b, for the sample impacted with 8 J energy. This energy is the threshold from which delaminations start to appear in this type of composite.

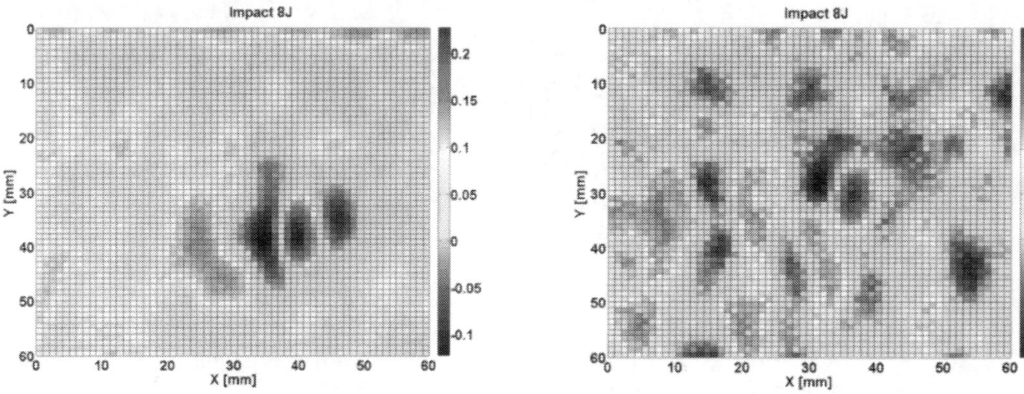

a) b)

Figure 6. Electromagnetic image: a) obtained with the sensor with orthogonal coils; b) obtained with sensor with metamaterials lens.

It can be shown that the metamaterials lens allow the increasing of spatial resolution, the layout of the woven being emphasized. This is possible due to the appearance of the evanescent modes between the carbon fibers at the excitation with TMz polarized wave [18].

One cell of carbon fiber woven has been designed in TexGen -Textile Geometric modeler software (figure 7a) and exported as CAD format in order to be used in Finite Difference Time Domain software XFDTD Remcom USA. The result of the simulation is presented in figure 7b.

It is shown the apparition of evanescent modes on the edge of the carbon fibers tow, allowing their manipulation using the sensor with metamaterials lens [18].

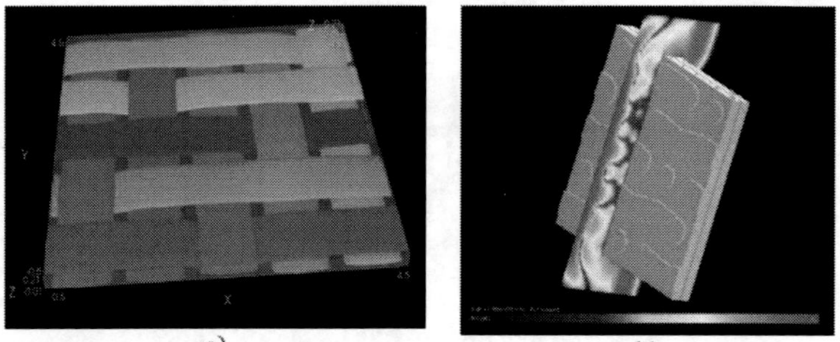

a) b)

Figure 7. One cell carbon woven simulation: a) TexGen; b) XFDTD.

In figure 8 is presented the image obtained at the scanning of the sample with ultrasound phased array sensor. The equipment operate in TOP VIEW mode, the scanning being made with the help of an encoder, in order to have accurate measurement of the distance.

It can be observed that US scan overestimate the area of delaminated surface. The under estimation of delaminated area using electromagnetic methods is due, to the modification of transverse electrical conductivity in the region of the impact, where the material suffers a plastic deformation [14].

In one of the sample taken into study, single tower grating having core diameter 6 μm and fiber diameter (with cladding made of ORMOCER) of 125 μm has been embedded. Centre wavelength has been 1535 nm with strain sensitivity $7.8x10^{-7}\mu\varepsilon^{-1}$ and temperature sensitivity $6.5x10^{-6}K^{-1}$. During the experiments, the temperature has been maintained constant at $19\pm1^{0}C$. The sample was subjected to a series of static flexural load with increment of 0.5 N under INSTRON 8801 testing machine. The FBG was connected at a stand –alone system for monitoring one optical line type Spectral Eye 600 coupled with a notebook via RS232-USB2 connector. The data processing has been made in Matlab 2011b, the temperature correction being made, the equipment indicating the temperature of fiber grating, too. The measurements were effectuated on sample loaded with 528.2 N force and respectively unloaded, the experimental results representing the average of 10 measurements. Two parameters were recorded simultaneously, the deformation of the sample and the reading from optical sensor. The results are presented in figure 9. The modifications in the position of diffraction peak are very small so that the image has been zoomed in. The shifts of the reflected wavelength showed a steady and consistent deformation from 0 N to 528.2 N.

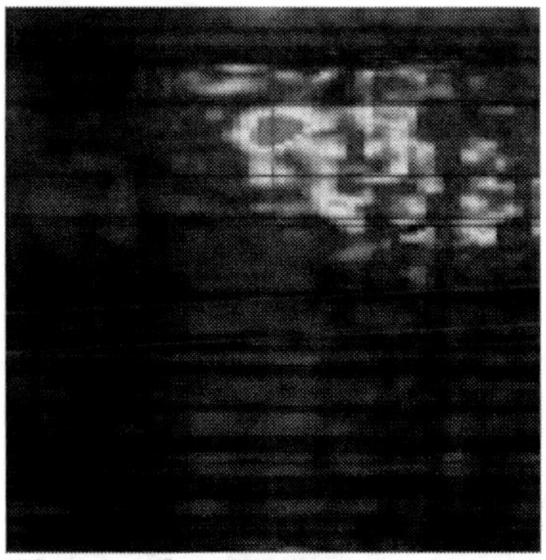

Figure 8. Ultrasound image using phased array sensor and TOP VIEW mode.

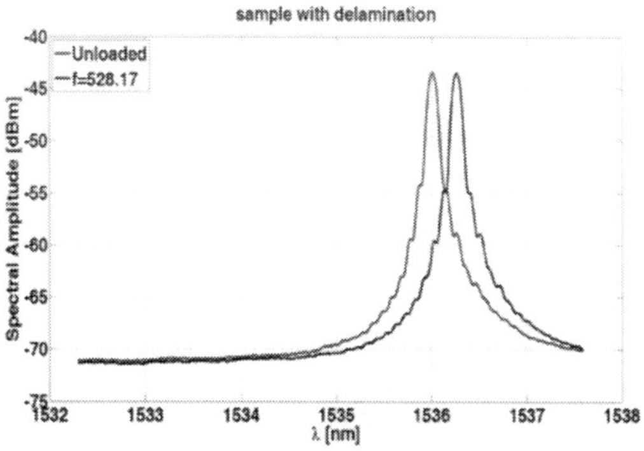

Figure 9. The response of FBGs embedded in sample delaminated under 8 J impact, for loading-unloading, four points bend flexural type.

4. Conclusions

CFRTP were impacted with low energies in order to obtain delaminations. They were tested with complementary methods to estimate the influence of delamination over the composite behavior. Due to impact, local modification of the conductivity appear. Only delaminations were emphasized, not breaking of fibers. It has been found that at the scanning with electromagnetic sensors, an improving of the spatial resolution can be obtained if metamaterials lens is used, due to manipulation of evanescent waves that appear between carbon fibers, fact emphasized also by numerical simulation using FDTD software. The evaluation with ultrasound phased array sensor shown an over estimation of delaminated area. Replacing a carbon fiber with a FBG grating has allowed determination of the sensitivity at plastic deformation of material due to the apparition of delamination. These methods are suitable for SHM of final parts made from these composites.

Acknowledgments

This paper is supported by Romanian Ministry of National Education under project PN-II-ID-PCE-2012-4-0437, Nucleus Program, Contract PN 09 43-01-04 and by the strategic grant POSDRU/159/1.5/S/137750, cofinanced by the ESF within the Sectorial Operational Program Human Resources Development 2007 – 2013.

References

[1] Vielle B and Taleb L 2011 About the influence of temperature and matrix ductibility on the behavior of carbon woven-ply PPS or epoxy laminates *Compos. Sci. Technol.* **71** pp 998-1007
[2] *** Information on http://www.tencate.com
[3] Kaw A K 2006 Mechanics of Composite Materials 2nd ed. New York: *Taylor and Francis*
[4] Standard AS9100/2009. Quality Systems - Aerospace - Model for Quality Assurance in Design, Development, Production, Installation and Servicing

[5] Elmarakbi A, Mollón V, Bonhomme J, Viña J and Argüelles A 2013 Advanced Composite Materials for Automotive Applications: Structural Integrity and Crashworthiness, *J. Wiley&Sons Ltd*
[6] Boller C 2000 Next generation structural health monitoring and its integration into aircraft design *International Journal of Systems Science 31.11* pp 1333-1349
[7] Adden S, Pfleiderer K, Solodov I, Horst P and Busse G 2008 Characterization of stiffness degradation caused by fatigue damage in textile composites using circumferential plate acoustic waves *Compos. Sci. Technol.* **68(7–8)** pp 1616–23
[8] Ahsan M, Han X, Islam S and Newaz G Fatigue damage detection in graphite/epoxy composites using sonic infrared imaging technique *Compos. Sci.Technol.* **64(5)** pp 657–66
[9] Todoroki A, Shimazu Y and Mizitani Y 2012 Electrical resistance reduction of laminated Carbon Fiber Reinforced Polymer by dent made by indentation without cracking *JSME* **6(12)** pp 1042-1052
[10] Dattoma V and Giancane S 2013 Evaluation of energy of fatigue damage into GFRC through digital image correlation and thermography *Compos Part B: Eng* **47** pp 283–9
[11] Schilling P J, Karedla B R, Tatiparthi A K, Verges M A and Herrington P D 2005 X-ray computed microtomography of internal damage in fiber reinforced polymer matrix composites *Compos. Sci. Technol.* **65(14)** pp 2071-2078
[12] Hung Y Y, Chen Y S, Ng S P, Liu L, Huang Y H, Luk B L, Ip R W L, Wu C M L and Chung P S 2009 Review and comparison of shearography and active thermography for nondestructive evaluation *Materials Science and Engineering: R: Reports* **64(5–6)** pp 73–112
[13] Kaczmarek H and Maison S 1994 Comparative ultrasonic analysis of damage in CFRP under static indentation and low-velocity impact *Compos. Sci. Technol.* **51** pp 11–26
[14] Grimberg R, Savin A, Steigmann R, Bruma A and Barsanescu P 2009 Ultrasound and eddy current data fusion for evaluation of carbon-epoxy composite delaminations *INSIGHT 51* pp 1-25
[15] Zhao X, Gao H, Zhang G, Ayhan B, Yan F, Kwan C and Rose J L 2007 Active health monitoring of an aircraft wing with embedded piezoelectric sensor/actuator network: I. Defect detection, localization and growth monitoring *Smart materials and structures* **16(4)** pp 1208
[16] Takeda S, Obabe Y and Takeda N 2002 Delamination detection in CFRP laminates with embedded small-diameter fiber Bragg grating sensors *Composites Part A* **33** pp 971–80
[17] Grimberg R, Savin A and Rotundu C R 2001 Eddy current microscopy applied to graphite-epoxy composite *Sensors and Actuators A-Physical* **91(2-1)** pp 73-75
[18] Grimberg R and Tian G Y 2012 High Frequency Electromagnetic Non-destructive Evaluation for High Spatial Resolution using Metamaterial *Proc. of the Royal Society A* **468** 2146 pp 3080-3099
[19] Grimberg R, Savin A, Steigmann R, Iftimie N and Bruma A 2008 Using of ultrasound phased array for non-destructive evaluation of carbon epoxy composites *Proceedings of the 8th International Conference on Technology and Quality for Sustained Development* pp 285-290
[20] Kinet D, Mégret P and Goossen K W 2014 Fiber Bragg Grating Sensors toward Structural Health Monitoring in Composite Materials: Challenges and Solutions *Sensors* **14(4)** pp 7394-7419
[21] Grimberg R, Savin A, Radu E and Mihalache O 2000 Nondestructive evaluation of the severity of discontinuities in flat conductive materials using the eddy current sensor with orthogonal coils *IEEE Trans on Mag.* **36(1)** pp 299-300

IOP Conf. Series: Materials Science and Engineering **95** (2015) 012092 doi:10.1088/1757-899X/95/1/012092

The laboratory station for tyres grip testing on different surfaces

K Kalinowski[1], C Grabowik[1], W Janik[1], G Ćwikła[1] and M Skowera[1]

[1] Silesian University of Technology, Faculty of Mechanical Engineering, Institute of Engineering Processes Automation and Integrated Manufacturing Systems Konarskiego 18A Str., 44-100 Gliwice, Poland

E-mail: krzysztof.kalinowski@polsl.pl

Abstract. The paper presents the conception of the device for tyre grip testing in the laboratory conditions. The main purpose is to provide a device working in confined spaces, which enables rapid changes of the tested samples of the road surfaces. Among the key assumptions the minimization of the device dimensions and the relative ease of transportation and mobility – the ability to quick assemble and disassemble were also assumed. The main components of the projected workstation includes: the replaceable platform for mounting samples of a road surface, the roller conveyor, the drive of the platform, the wheel mounting assembly and the axial force measuring system. At the design the station a morphological structure method has been used, particular elements have been optimized individually.

1. Introduction

Nowadays in the European Union countries, four types of devices (vehicles) measuring grip factor (traction, friction between tyre and road surface) are used, which differs depending on the method of measurement of traction of tested wheel rolling

- Devices with full wheel lock – e.g. SRT-3 (Skid Resistance Tester). This type of a device carry out the grip measurement of the tested wheel/surface combination in the conditions of the total blockade.
- Devices with constant skid of the wheel – wheel being tested is braked hydraulically and moves relative to the surface with a fixed, known slip, e.g. Griptester used in Norway.
- Devices with variable wheel skid – this group of devices measures friction under conditions of variable skid of the tested wheel. The tested wheel is braked, but not with fixed value of force. The value of the wheel slip is known and may vary, so that these devices can obtain a full characterization of the coefficient of friction. An example of such a device is Norsemeter OSCAR used in Norway.
- Devices with measured wheel positioned angularly corresponding to the direction of movement, allowing measuring not only the longitudinal grip strength, but also cross-grip. The friction coefficient is determined by comparing the lateral force acting on the measured wheel with the static load on the wheel. The measured wheel is not aligned along the direction of movement, it is inclined at a fixed angle, which causes lateral force acting on the wheel, e.g. SCRIM used in Germany.

Content from this work may be used under the terms of the Creative Commons Attribution 3.0 licence. Any further distribution of this work must maintain attribution to the author(s) and the title of the work, journal citation and DOI.
Published under licence by IOP Publishing Ltd

IOP Conf. Series: Materials Science and Engineering **95** (2015) 012092 doi:10.1088/1757-899X/95/1/012092

All of these devices perform measurements on the real road surface, but there is a need for a device capable of measuring these parameters in a laboratory conditions. During the literature review no device suitable for use in the laboratory environment were found. Therefore, it was decided to adapt conception of one of the types of equipment used in road conditions (traction force measuring device via measuring deflected wheel) to the laboratory conditions.

2. Main assumptions

Aim of this study was to develop a device (laboratory station) for measuring the grip of tires on different pavement surfaces. According to the structural assumptions the device should perform measurements in an enclosed room (hall, laboratory). The measurement of the interaction parameters of the pavement surface with the tire is very important because it allows the development of both tires providing better traction, as well as the application of new types of pavement to improve road safety [1].

Working in confined spaces also requires compliance with standards for acceptable noise and emission of substances harmful to health. In the construction of the device the existing standards in the field of tyre grip measuring were taken into account, among others PN-ISO 23671, PN93/C94300/009 and PN-88/C-94300/030.

The laboratory station for tyres grip testing on different surfaces should allow to assess traction of different types of tires on various types of pavement surfaces, so its design should allow easy and quick exchange of tires and pavement samples. Due to the fact that each tire has a different adhesion, weight and other parameters, it cannot be accurately specified what load can occur during the test, the structure of the device should withstand correspondingly higher loads to prevent damage [2].

Another aim was to determine the minimum and maximum of the size of the tire, and thus main dimensions of the device. It was found that the device is capable of carrying out the research of tires for passenger cars in the range from 185/30 X10 X23 to 315/75. The device has been modelled in Autodesk Inventor.

2.1. Design of the laboratory station for measuring the grip of tires on different surfaces

During the construction of such a device, it is advisable to rely on systematic methodology that ensures obtaining the expected results A method of morphological structures have been used to develop the project of the test station, the whole structure have been divided into separate segments which can be individually optimized later Appropriate analysis have been also carried out, using advanced CAD and CAE software (Autodesk Inventor)

During the development of the concept of a device for testing grip of tires on different surfaces main design criteria and assumptions have been formulated:

- minimum dimensions of the device,
- the possibility of assembly and disassembly of the device indoors,
- avoiding excessive noise and releasing of harmful chemicals and fumes during the test,
- the tested wheel is freely turning, it is tilted at an angle relative to the direction of the linear movement performed by the test pavement surface,
- the device must allow changing of the pavement sample surface as easy as possible to be able to run simulations on different surfaces,
- the device should allow mounting of the widest range of the rims used in typical cars.

2.1.1. The concept of the device

The tested wheel is fixed in the holder, mounted on a shaft with appropriate bearing, which is positioned angularly to the moving sample of the pavement. The forces on the bearings of the shaft are dependent on the grip of tested tire. The basic elements of the measuring device are (figure 1): the linear drive unit moving the pavement sample (actuator 1), the pavement sample frame (2) with supporting rollers and its frame, the tested wheel support with force measuring unit (3), and the tested wheel mount (4).

IOP Conf. Series: Materials Science and Engineering **95** (2015) 012092 doi:10.1088/1757-899X/95/1/012092

Rectangular pavement sample is mounted in the frame made of angle bars, allowing easy exchange of the sample. The pavement sample surface must be made in the same way for all types of pavements and tires, so that all tests of tyres have been carried out under similar conditions.

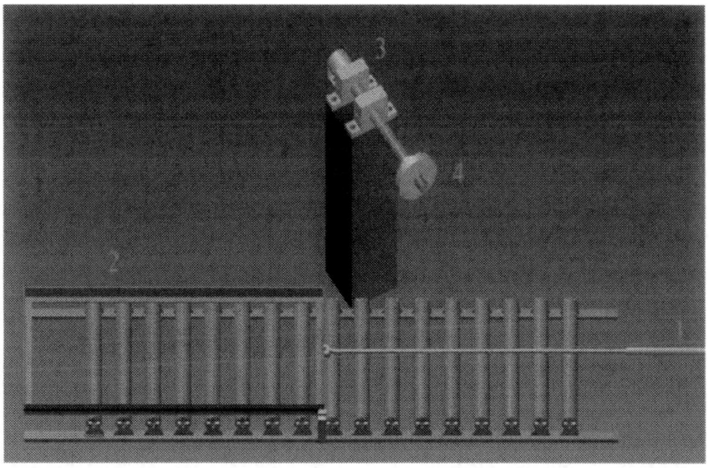

Figure 1. Simplified model of the laboratory station for tyres grip testing.

Pavements samples should be made in such a way as to reflect the specific characteristics of the specific surface, the surface must be flat and positioned evenly, relative to the tested tyre. The minimum dimensions of the pavement sample (and its frame) sufficient for testing are 625 x 1500 mm. The pavement sample is placed in the steel frame and supported by rigid superstructure in order to avoid deformations resulting from the pressure of the tested wheel.

The device support frame have load-bearing rollers mounted, which are necessary to reduce the friction force between the sample surface frame and the main support frame structure, which could distort the results of measurements of friction. Rollers used in the design (RT1-80, manufactured by Europa Systems) come from the belt conveyors used in mining (figure 2), they are characterized by a very simple construction and, above all, very high load carrying capacity, ease of installation and reliability. The rollers will be mounted in a support structure with a length of 3000 mm, sufficient for purpose of testing friction between tyres and pavement surface.

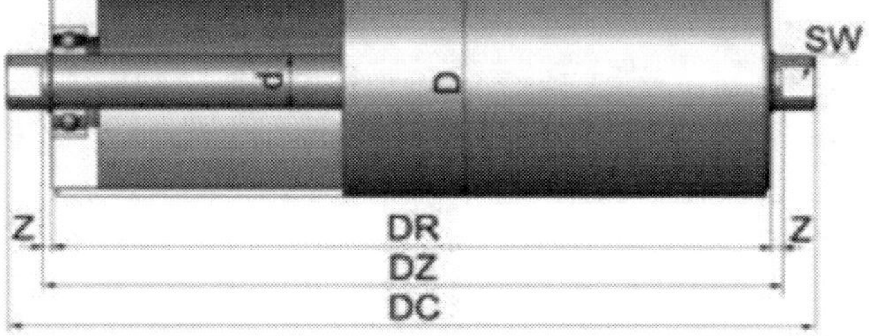

Figure 2. Half section of the proposed roller.

2.1.2. Linear drive unit moving the pavement sample frame

The double-acting hydraulic actuator has been selected as an element driving frame with pavement sample. Hydraulic drive will allow adjusting the speed by using a flow regulator. Additional advantage of the hydraulic system is the ability to carry very heavy loads and overload, which in determining the unknown coefficient of friction can be important. In addition, the use of a hydraulic drive advantage is the lack of additional structural elements, like clutches and gears. The power supply of hydraulic system requires the use of noisy pump which should be installed outside of the laboratory. The hydraulic cylinder used in the design should have a pitch ~ 1,5 m, dependent on the length of the carrier path of the sample and the sample pavement frame length.

2.1.3. Tested wheel support unit

This module is a crucial part of the device, because it must ensure the main shaft positioning and bearing with expected accuracy, in the same time allowing measuring of axial and radial forces on bearings, coming from interaction between tested tyre and pavement sample surface. Main shaft is also part, on which tested wheel mount is fixed, carrying load of whole wheel and forces of tyre-pavement contact. The set of sensors measuring force and other parameters must also be installed in the tested wheel support unit, allowing measurements of parameters required in order to calculate friction force.
Bearings must withstand variable loads that stems from variable pressure of tested wheel on a moving pavement sample. These loads also depend on the size of the tested wheel. The proposed bearing setting consists of two radial ball bearings and one axial bearing, mounted in the appropriate holders. In the construction of a bearings, angular contact bearing have been abandoned, due to the unbalanced load of bearing which in turn could lead to abnormal pressure on the force sensor measuring the pressure, and erroneous readings.

It is also necessary to ensure the possibility of change the distance from the whole tested wheel support unit to the pavement sample in order to adjust device to the given diameter of the tested wheel. Whole tested wheel support unit will be installed on support frame, allowing adjusting of the position (height).

The main shaft of the device should allow the carrying of loads occurring in the system and has been designed using advanced CAD and CAE (Finite Elements Method - FEM) software (figure 4 and figure 5). The same software was used for parameter selection and verification of bearings and their fittings, and other components.

At the end of the main shaft, tested wheel mount is installed, so the shaft must be appropriately shaped, allowing interfacing with the family of other components. The total length of the shaft equals 830 mm.

The sensor set installed in the tested wheel support unit is intended to provide the ability to measure key test station parameters, such as the axial and diagonal forces in bearings, wheel rotational speed and linear speed of the pavement sample frame.

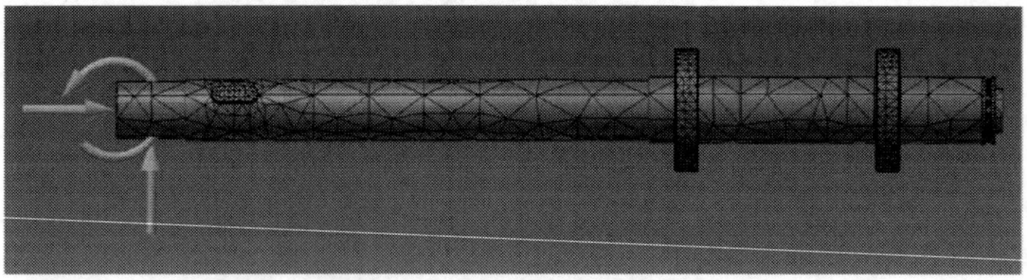

Figure 4. The main shaft FEM analysis – the load and FEM mesh.

The pressure (force) values (axial and radial forces) will be measured by the sensors, e.g. Tecsis power converter F12131310436 (measuring range 0…1 kN, measurement accuracy 0,1%, measurement time 2 ms, signal output: voltage 0...10V) [7].

Another value measured to calculate the kinetic friction of the tire on pavement sample is the speed of movement of pavement samples, so it is necessary to use the linear position sensor, allowing calculation of the speed of the pavement sample. Among the available sensors, it was decided to use magnetic linear sensor, e.g. ETMA 1 (Resolution: 0,1 mm (0,025 mm if by reading each front, zero

IOP Conf. Series: Materials Science and Engineering **95** (2015) 012092 doi:10.1088/1757-899X/95/1/012092

pulse every 5 mm, repeatibility +/- 0,025 mm, output type – line driver/push pull, max speed 4 m/s, magnetic tape EBM L = 0,5 m – 32 m) [8]. This sensor have been chosen because of its high accuracy and, above all, resistance to vibration, much higher than rotary-pulse sensors.

The magnetic linear encoder is composed of diamagnetic sensor and a three-layer magnetic tape which can be fixed on pavement sample frame, allowing direct measurement of pavement sample speed.

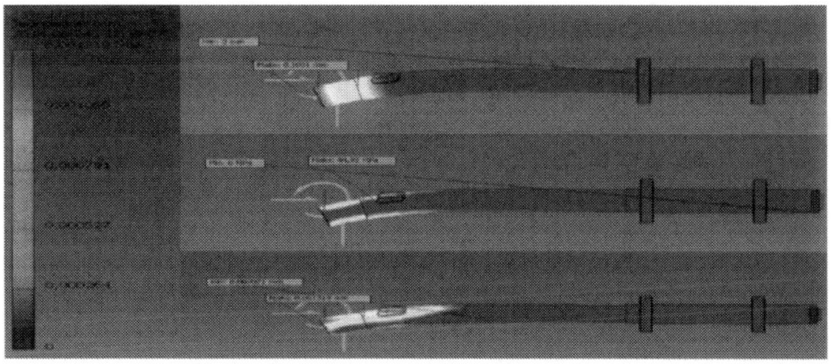

Figure 5. Examples of main shaft FEM analysis: the total displacement in all axes, equivalent stress (Von Mises), the tension relative to the XX axis.

2.1.4. Tested wheel mount

A wheel (tyre with rim) is mounted on the main shaft, set angularly relative to the direction of movement of the pavement sample. It is important that tested wheel mount should assure the lack of slipping between the tested wheel and shaft, and its operation at a constant angle. In addition, mounting and shaft, which is mounted on the wheel must be strong enough to move the load acting on them. In addition, tested wheel mount and shaft, on which this mount is fixed, must be strong enough to endure the load acting on them.

Tested wheel mount will be installed on the main shaft extension.

Interface allowing fixing the tested wheel to the shaft must be made in such a way as to allow mounting of wheel rims of different diameters of the bore, so that there will be no need to provide different adapters for each tyre/rim size. The solution to this problem would be fixing the rim using the central clamp on the cone.

Examples of such mountings used in wheel balancing machines, are depicted in the figure 6.

Figure 6. Two versions of the wheel mount: with truncated centering cone, and with centering disc.

3. Conclusions

Within the work carried out, device measuring the friction and adhesion of the tire in different conditions in the enclosed laboratory room have been designed, based on the concept of the design of one of the devices intended to conduct such tests in the real road conditions (group of devices with measure parameters of wheel positioned angularly corresponding to the direction of movement). The morphological structure method has been used to design the test station, the whole structure is divided into separate segments, which have been individually optimized. The dimensions of the device were chosen in such a way as to allow testing of all the typical sizes of tyres used for passenger cars. In order to test tyres with different pavements, it was decided to design the frame that allows installation of samples of different pavement surfaces. Pavement sample frame moves along main support frame with rollers. To move the sample it was decided to use a double acting hydraulic actuator that provides force sufficient to move the sample and to overcome the friction of the tyre. The last important part of the laboratory station is a tested wheel support unit with sensors, allowing measurement of forces and speed, necessary to calculate friction between tyre and pavement surface.

Strength calculations, support reactions and stress analysis had been carried out in Autodesk Inventor, and indicated the proper selection of the construction parameters.

References

[1] Carcaterra A and Roveri N 2013 Tire grip identification based on strain information: Theory and simulations *Mechanical Systems and Signal Processing* **41(1–2)** pp 564-580
[2] Beer M and Fisher C 2013 Stress-In-Motion (SIM) system for capturing tri-axial tyre–road interaction in the contact patch *Measurement, Elsevier* **46(7)** pp 2155-2173
[3] Vaiana R, Capiluppi G F, Gallelli V, Iuele T and Minani V 2012 Pavement Surface Performances Evolution: an Experimental Application *Procedia - Social and Behavioral Sciences* **53** pp 1149–1160
[4] Gwiazda A 2014 System of designing complex technical means using fuzzy analysis *Applied Mechanics and Materials* **474** pp 147-152
[5] Gwiazda A, Sękala A, Monica Z and Banaś W 2014 Integrated approach to the designing process of complex technical systems *Advanced Materials Research* **1036** pp 1023-1027
[6] M. Płaczek 2014 Computer Aided Analysis of Systems with MFC Actuators *Advanced Materials Research* **1036** pp 711-714
[7] *** Tecsis power converter F12131310436, http://www.tme.eu/en/details/f12131-1-10/pressure-sensors/tecsis/f12131310436/ - access
[8] *** ETMA 1, Magnetic linear sensor, http://motion.nl/historie/products/eltra/etma.pdf
[9] *** Wheel balancer DSP7705, http://www.hunter.com/balancer/dsp7705/specs.cfm

IOP Conf. Series: Materials Science and Engineering **95** (2015) 012093 doi:10.1088/1757-899X/95/1/012093

Analysis hierarchical model for discrete event systems

E M Ciortea[1]

[1] University "1 Decembrie 1918" of Alba Iulia-Romania, Department of Science and Engineering, Blvd. N. Iorga, No. 11-13, 515900, Alba Iulia, Romania

E-mail: ciortea31mihaela@yahoo.com

Abstract. This paper presents the hierarchical model based on discrete event network for robotic systems. Based on the hierarchical approach, Petri network is analysed as a network of the highest conceptual level and the lowest level of local control. For modelling and control of complex robotic systems using extended Petri nets.

Such a system is structured, controlled and analysed in this paper by using Visual Object Net ++ package that is relatively simple and easy to use, and the results are shown as representations easy to interpret. The hierarchical structure of the robotic system is implemented on computers analysed using specialized programs.

Implementation of hierarchical model discrete event systems, as a real-time operating system on a computer network connected via a serial bus is possible, where each computer is dedicated to local and Petri model of a subsystem global robotic system. Since Petri models are simplified to apply general computers, analysis, modelling, complex manufacturing systems control can be achieved using Petri nets. Discrete event systems is a pragmatic tool for modelling industrial systems. For system modelling using Petri nets because we have our system where discrete event. To highlight the auxiliary time Petri model using transport stream divided into hierarchical levels and sections are analysed successively. Proposed robotic system simulation using timed Petri, offers the opportunity to view the robotic time. Application of goods or robotic and transmission times obtained by measuring spot is obtained graphics showing the average time for transport activity, using the parameters sets of finished products.

1. Introduction

When using Petri nets, events are associated with transitions. Activities are associated to the firing of transitions and to the marking of places which represents the state of the system. In addition to its graphic representation differentiating events and states, Petri nets allows the modelling of true parallelism and the possibility of progressive modelling by using stepwise refinements or modular composition. Libraries of well-tested subnets allow components reusability leading to significant reductions in the modelling effort. The possibility of progressive modelling is absolutely necessary for flexible manufacturing systems because they are usually large and complex systems. The refinement mechanism allows the building of hierarchically structured net models which leads to the implementation of hierarchical and distributed control [3].

Formally, a Petri net has two types of nodes, called places and transitions. A place is represented by a circle and a transition by a bar. The places and transitions are connected by arcs. The number of places and transitions are finite and not zero. An arc is connected directly from one place to a

Content from this work may be used under the terms of the Creative Commons Attribution 3.0 licence. Any further distribution of this work must maintain attribution to the author(s) and the title of the work, journal citation and DOI.
Published under licence by IOP Publishing Ltd

transition or a transition to a place. In other words a Petri net is a bipartite graph, i.e. places and transitions alternate on a path made up of consecutive arcs. [3]

An ordinary Petri net is represented by the 5-tuple $G=\{P,T,F,W,M_0\}$ such that:

$P=\{p_1,p_2,...,p_n\}$ is a finite, not empty, set of places;

$T=\{t_1,t_2,...,t_m\}$ is a finite, not empty, set of transitions;

$F\subseteq(P\times T)\cup(T\times P)$ is the set of arcs;

$P\cap T=\phi$, i.e. the sets P and T are disjointed;

$W:F\rightarrow\{1,2,3,...\}$ is the weighting function arcs;

$M_0:P\rightarrow\{0,1,2,3\}$is the initial marking.

Presentation formalization is:

1. Crowds T and P are disjoint, $P\cap T=\phi$
2. To ensure the definition, crowds T and P satisfy the condition $P\cup T=\phi$
3. Define the weighting function can extend the set of all ordered pairs of $(P\times T)\cup(T\times P)$, $W:(P\times T)\cup(T\times P)\rightarrow\{0,1,2,3,...\}$, pairs that are not in the set F, function value W is 0 and those pairs are not graphically represented. As the crowd F corresponding to non-zero pairs that share only this plot a graph.
4. The network capacity is infinite, because the definition of Petri network believes that all network positions may contain any element.
5. They follow the notation of topology which specifies that a general Petri net structure $N=\{P,T,F,W\}$, that is without any specification related to the marking is denoted by N .
6. A Petri net with initial marking M_0 is denoted (N,M_0).
7. A Petri net with some marking M is denoted (N,M).

In modelling system conditions and transitions positions represent events. Event have a number of entry and exit positions for the event. Items in position to be understood as logical values true condition associated with the position.

2. Description of the process

The analysis is an advantage for industrial systems, so using production cell consists of 6 components: 2 conveyors, 1 robot arm, 1 elevator, machine tool, a crane transport.

The system analysed in its initial form, the classic is relatively simple, but hierarchical system leads to a complex system, and graphical representations which can be obtained after the simulation provides useful information for any system.

The whole system forms a closed circuit. The production cycle is continuous and dynamic. Robot with two arms is to take the elevator blank, submit it to the machine tool and the second arm picks a work piece and deposited in the warehouse. The elevator rotating charged with material from the conveyor belt and positioned to the robotic arm.

Conveying and storage are powered by electric motor that can be turned on or off depending on the program of monitoring and control. A photoelectric cell mounted at the end of each band serves as a flow tracking.

Elevator rotary role performs vertical and rotational movement of the table depending on the robot arm. Robot arm is unable to work vertically. Electromagnets, attachments are not rotating and can not work at varying angles. The advantage is that there is placed an analogue potentiometer which indicates the angle of rotation of the table. Boolean switches are activated when the weight reaches the upper or lower position.

The robot consists of extendable arms placed at different levels and pivot rotation. The position of each arm must be withdrawn when the robot rotates or perform the duties of an arm, that processing or

IOP Conf. Series: Materials Science and Engineering **95** (2015) 012093 doi:10.1088/1757-899X/95/1/012093

submission blank piece. Robot arm extensions and rotation angles are indicated by means of potentiometers. In the processing, because it is a different placement level of the robot arm, there are two loading positions of the blank on each machine tool and a work piece discharge from the machine tool. Adjust the positions corresponding to switches the signal. The role of the crane is to provide a continuous supply of semi-finished goods storage conveyor without involving other additional external component. The set of sensors of the crane comprises switches that are activated if the retainer is positioned above the belt and a potentiometer that determines the height of the grip.

3. Modelling with hierarchical Petri Net

To be able to express security requirements relating to physical devices, the model must be designed to contain a description of the working environment to adapt their behaviour to situations that may occur (not provided).

Planning complex processes face the problem of describing the workload characteristics and their interactions in a concise and easy to understand. One of the basic strategies for achieving this function is a hierarchical approach.

Within the complex processes by analysing the characteristics of the system according to the blank transported considering the general case. A particular case where it can be studied in future work in the system is to be used for the transport of hot blanks, that is, in the steel rolling system. [4]

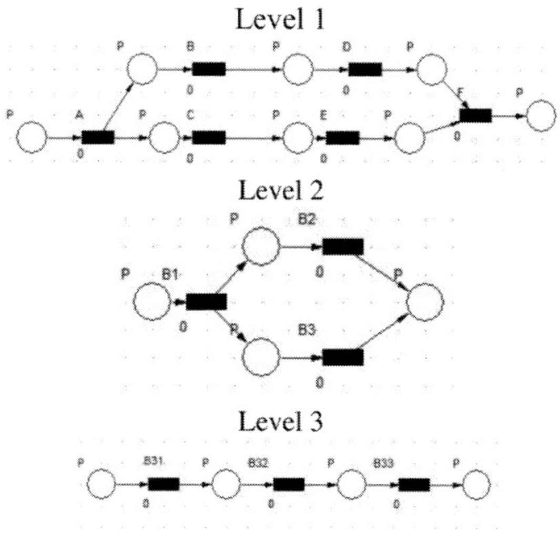

Figure 1. The hierarchical levels with Petri Nets [4].

In level 1 Petri Net is presented to the highest level network, a hierarchical transition and is used for level 2 Petri network. At level 2 transition Petri Net is a lower level hierarchical Petri Net. This mechanism can be used to achieve hierarchical decomposition of a complex system.

Under this system could be divided into different levels of detail from top to bottom. The concept of hierarchical structure information can be used effectively. The upper level offers a higher level of abstraction and detail a complete plan without extensive and complex hierarchical system. At the lowest level it offers a high level of detail on Petri Nets modelling form and is working at the complex tasks of system. [4]

The concept of transition hierarchical modularity provides features programming technique using Petri net based systems. Many repetitive technology components can be found and adapted to complex systems. These units can be combined repetitive work in "mock" and referenced a "hierarchical transition". This "transition hierarchical" network can be specified as a number system based on repetitive components in the system. The modularity can be adopted for the benefit schemes involving repeat modules workplaces.

IOP Conf. Series: Materials Science and Engineering **95** (2015) 012093 doi:10.1088/1757-899X/95/1/012093

4. System modelling

Figure 2 presents the classical structure modelled with Petri nets, from which depart in order to analyse hierarchical system. The classic is the representation of Petri Nets model with factors indicated in the first chapter.

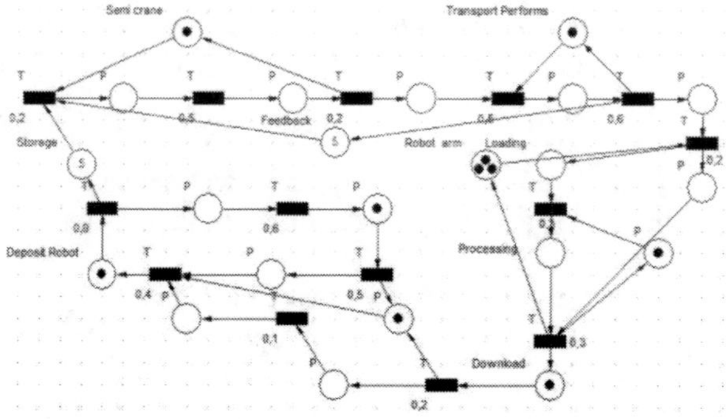

Figure 2. Classical structure modelled with Petri nets.

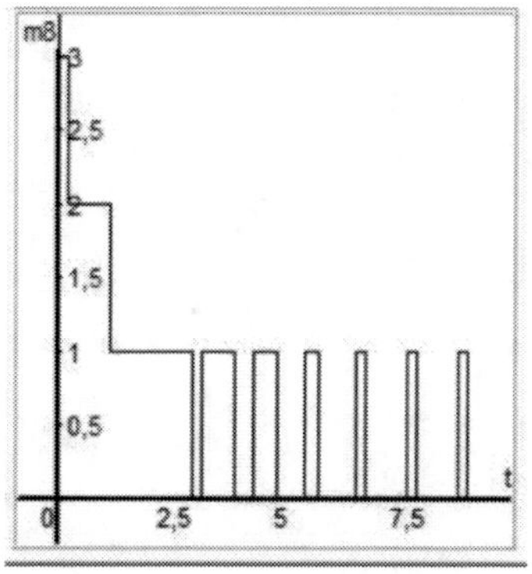

Figure 3. Increased flow after initial processing.

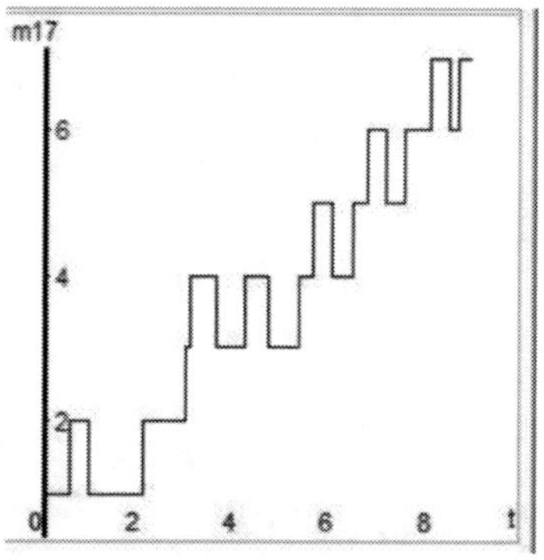

Figure 4. Transport blank in the classical system.

Shaped structure of the system under analysis is provided below and is presented analysis system.

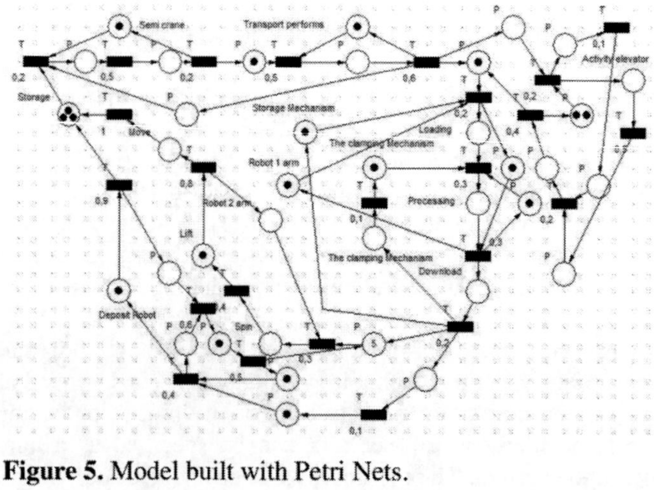

Figure 5. Model built with Petri Nets.

The modelling used to describe a manufacturing system consisting of a set of sorting or selection of material and then using the appropriate conveyor belt six work stations or components which interfere with the flexible robot arm with a role of charging the material to be processed and after processing requires the use of robot arm for depositing. Ready processed material is transported in mobile storage where the process continues, but not shown in the model. Real time processing times are selected from the specifications of processes and transport times were selected after changing the location of the equipment system to optimize transport times, but without interfering calling strings.

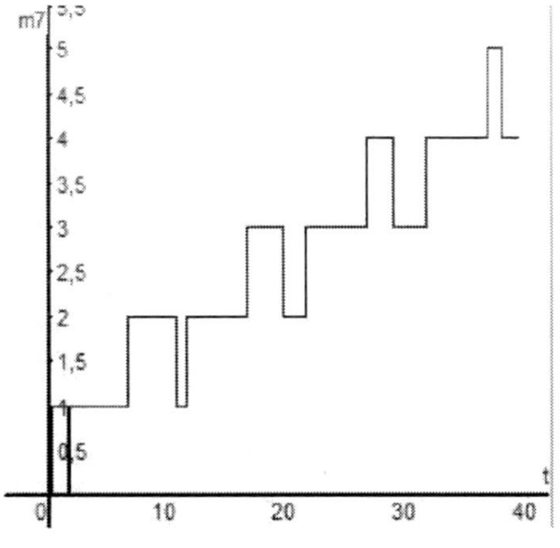

Figure 6. Graphical representation of the average for the first conveyor transport.

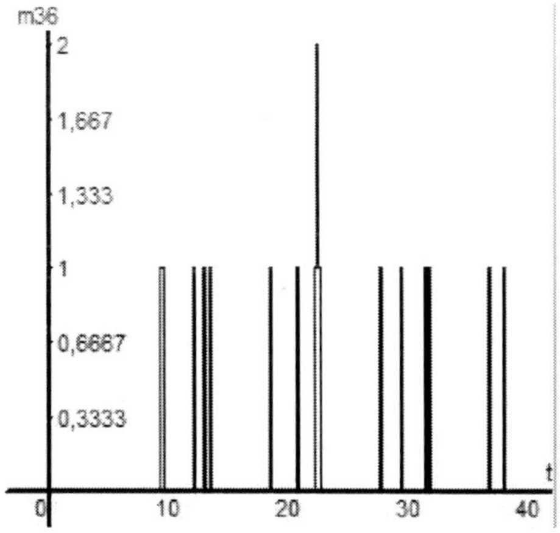

Figure 8. Graphical representation of average flux through the semi finished products elevator.

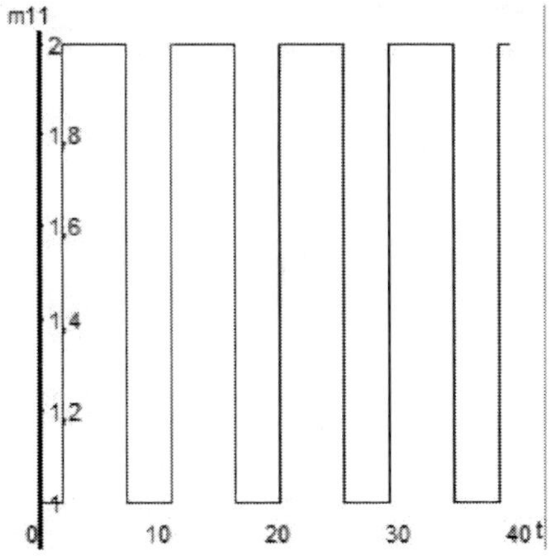

Figure 7. Graphical representation of the average occupancy of the machine.

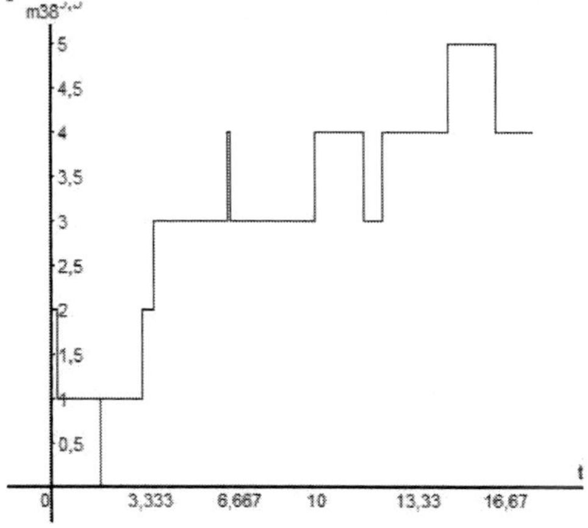

Figure 9. Graphical representation the system formonitoring the first robot arm.

For the analysis of variance monitoring and control of production flow are considered only nodes and transitions without extensions. Monitoring and control system was developed and analysed step by step, as described in the literature. The model of cooperation is described abstractly in sync aspect cell components.

The bottom is made up of the following steps:

- Analysis of reusable models on communication,
- Communication between devices, identified and modelled as Petri nets,
- Complete model is built by composing courts by merging communication places.

The description has led process control with Petri nets.

Figure 5 shows the basic relationship between neighbourhoods. The entry and exit restrictions are modelled. According to the requirement of mutually exclusive, should be run to ensure a good flow in order to avoid errors in the manufacturing system.

Robot arms are held separately to increase the degree of parallelism in the production cell. To avoid collisions shared regions are set to be used exclusively. In this study, are problems that disappear in a constructive way so the robot arms and crane are allowed to move only when the arms are retracted and the crane is raised.

For analysis, figure 6, figure 7, figure 8, figure 9 are considered three types of communication model regions in the order of input, output, control and release.

5. Conclusions

Analysis with Petri nets in this type of hierarchical control system is applied after each step. Analysis of a successful model at a given hierarchical level is considered a prerequisite for the next stage of modelling. Thus, faults and design errors can be detected early.

To date models based on Petri nets to control cell production, analysed the hierarchical levels have developed valuable qualitative properties, which can be used successfully for production systems.

The paper has not been described Petri net model, but used to delimit restrictions and management analysis tools.

References

[1] Boca M L, Secara L and Abrudean M 2012 Methods for control of some critical situations afferent of a 13C isotope separation column *International Journal of Modern Manufacturing Technologies* **IV(1)** pp 21-26

[2] Heiner M, Deussen P and Spranger J 1999A Case Study in Design and Verification of Manufacturing System Control Software with Hierarchical Petri Nets *The International Journal of Advanced Manufacturing Technology* Springer-Verlag London Limited

[3] Yasuda G 2010 Petri Net Model Based Implementation of Hierarchical and Distributed Control for Discrete Event Robotic Manufacturing Cells,Nagasaki Institute of Applied Science *J. Software Engineering & Applications* **3** pp 436-445

[4] Sawhney A 1997 Petri net based simulation of construction schedules *Proceedings of the 1997 Winter Simulation Conference*

[5] Stecke K E 1986 A hierarchical approach to solving machine grouping and loading problems of flexible manufacturing systems *European Journal of Operational Research* **24** pp 369-378

[6] Sethi S P, Yan H, Zhang H, and Q. Zhang Optimal and Hierarchical Controls in Dynamic Stochastic Manufacturing Systems: A Survey, *Institute of Applied Mathematics Academia Sinica*, Beijing,China

[7] Păstrăvanu O, Matcovschi M, Mahulea C 2002 Aplicaţii ale reţelelor Petri în studierea Sistemelor cu evenimente discrete *Publisher Gh. Asachi*, ISBN 973-8292-86-7

IOP Conf. Series: Materials Science and Engineering **95** (2015) 012094 doi:10.1088/1757-899X/95/1/012094

Time-dependent solution for the manufacturing line with unreliable machine and batched arrivals

W M Kempa[1], I Paprocka[2], C Grabowik[2] and K Kalinowski[1]

[1]Silesian University of Technology, Faculty of Applied Mathematics, Institute of Mathematics, 23 Kaszubska Str., 44-100 Gliwice, Poland

[2] Silesian University of Technology, Faculty of Mechanical Engineering, Institute of Engineering Processes Automation and Integrated Manufacturing Systems, Konarskiego 18A, 44-100 Gliwice, Poland

E-mail: wojciech.kempa@polsl.pl

Abstract. Time-dependent queue-size distribution in a finite-buffer manufacturing line with unreliable machine is investigated. Successive jobs arrive in batches (groups) with sizes being generally distributed random variables, and are being processed individually with exponential service times. Applying the approach based on the memory less property of exponential distribution and the total probability law, a system of integral equations for the transient queue-size distribution conditioned by the initial level of buffer saturation is derived. The solution of the corresponding system written for Laplace transforms is found via linear-algebraic approach.

1. Introduction

The in-depth analysis of the operation of a manufacturing line with an unreliable service station at different traffic scenarios requires the knowledge about the behaviour of key stochastic characteristics of the system, like e.g. the queue-size distribution of waiting jobs. It is worth noting that in the literature analytical results obtained for queues with server breakdowns are dedicated mainly to the stationary state of the system (see e.g. [1] and [8]). In the article we study the time-dependent queue-size distribution in a single-machine manufacturing line with failures, modelled by a finite-buffer queuing system with server breakdowns (see e.g. [2-6] for transient queue-size distribution in different queuing models). A typical input stream of individual jobs is generalized by considering batched arrivals occurring according to a compound Poisson process. Jobs are processed individually with exponentially distributed service time. The evolution of the system can be observed during successive working periods followed by repair times. If the arriving job finds the machine busy, joins the buffer queue or, if the buffer is saturated, is being lost. Applying the approach based on the memory less property of exponential distribution and the formula of total probability we derive a system of integral equations for the queue-size distribution conditioned by the initial state of the buffer queue. The solution of the corresponding system written for LTs (=Laplace transforms) of conditional distributions we obtain by using the linear-algebraic approach.

Content from this work may be used under the terms of the Creative Commons Attribution 3.0 licence. Any further distribution of this work must maintain attribution to the author(s) and the title of the work, journal citation and DOI.
Published under licence by IOP Publishing Ltd

2. Mathematical Description of the Model

Let us take into consideration a finite-buffer queuing model with the input stream of jobs described by a compound Poisson process with intensity λ. Sizes of successive batches are independent random variables, and the batch consists of exactly k jobs with probability p_k, where $\sum_{k=1}^{\infty} p_k = 1$.We assume thatjobs are being processed individually, with mean μ^{-1} of exponential processing time. The maximal number of jobs simultaneously present in the system is bounded by a non-random value N,i.e. we have $N-1$ places in the buffer queue and one place in service station. If the size of the arriving batch of jobs exceeds the number of free places in the buffer, the buffer is being saturated and the remaining jobs are removed (so, we accept the so called partial batch acceptance strategy (PBAS)). According to the jobs waiting for service in the buffer queue, the FIFO discipline is used. We assume that the system may contain a number of jobs before the opening at time $t = 0$, and that the machine is working at this time. After exponentially distributed working period, with mean γ^{-1}, a failure occurs and the repair period is assumed to be generally distributed with a DF (=distribution function) $F(\cdot)$.

At the completion epoch of the repair time the machine renews the processing normally (next working time begins) and so on. It is allowed for a machine failure to occur only when the machine is busy with a processing of a job. Successive interarrival, processing, working and repair times are assumed to be independent random variables.

3. Integral Equations for Transient Conditional Queue-Size Distribution

Introduce the following notation for the transient queue-size distribution of jobs, conditioned by the initial level of buffer saturation:

$$Q_n(t,m) = P\{X(t) = m \mid X(0) = n\}, \quad 0 \le m,n \le N, \ t > 0, \tag{1}$$

where $X(t)$ denotes the number of jobs present in the system at time t. Let us consider, firstly, the case in which the buffer queue is empty before the opening of the system. The following integral equation is then true:

$$Q_0(t,m) = \lambda \int_0^t e^{-\lambda x}\left[\sum_{k=1}^{N-1} p_k Q_k(t-x,m) + Q_N(t-x,m)\sum_{k=N}^{\infty} p_k\right] + \delta_{m,0}e^{-\lambda t}, \tag{2}$$

where the notation $\delta_{i,j}$ stands for the Kronecker delta function. Indeed, if the first arrival occurs at time $x < t$ and the batch size equals $k \le N-1$, then the system renews the operation at time x with k jobs present and the queue size is measured after time $t-x$ (the first summand on the right side of equation (2)). If the buffer is being saturated simultaneously with the first arrival at time $x < t$, then the system renews the operation with N jobs present (second summand). If the first batch of jobs arrives after t (third summand on the right side of equation (2)), then the only possibility is $m = 0$.

To describe precisely the behavior of the system being non empty before the opening, we need some auxiliary notation. Introduce the following random events:

- E_1 – the first arrival occurs before t and precedes the first departure and failure occurrences;
- E_2 – the first departure occurs before t and precedes the first arrival and failure occurrences;
- E_3 – the first failure occurs before t and precedes the first arrival and departure epochs;
- E_4 – the first arrival, departure and failure occur after time t.

Evidently, random events E_1, E_2, E_3 and E_4 are separable in pairs and moreover

$$Q_n(t,m) = \sum_{i=1}^{4} Q_n^{(i)}(t,m) \stackrel{\text{def}}{=} \sum_{i=1}^{4} P\{(X(t) = m) \cap E_i \mid X(0) = n\}. \tag{3}$$

Assume now that the system starts working at $t = 0$ with $1 \le n \le N-1$ jobs present in the buffer queue. Due to memoryless property of exponential distributions of interarrival, service and machine working times, moments of arrival, departure and failure occurrences are Markov epochs in the evolution of the system. Basing on this conclusion we can write the following integral equations:

$$Q_n^{(1)}(t,m) = \lambda \int_0^t e^{-(\lambda+\mu+\gamma)x}\left[\sum_{k=1}^{N-n-1} p_k Q_{n+k}(t-x,m) + Q_N(t-x,m)\sum_{k=N-n}^{\infty} p_k\right] dx; \tag{4}$$

$$Q_n^{(2)}(t,m) = \mu \int_0^t e^{-(\lambda+\mu+\gamma)x} Q_{n-1}(t-x,m)dx; \tag{5}$$

$$Q_n^{(3)}(t,m) = \gamma \int_{x=0}^t e^{-(\lambda+\mu+\gamma)x}\left\{\int_{y=0}^{t-x}\left[\sum_{k=0}^{N-n-1}\frac{(\lambda y)^k}{k!}e^{-\lambda y}\sum_{j=k}^{N-n-1} p_j^{k*} Q_{n+j}(t-x-y,m) + Q_N(t-x-y,m)\left(\sum_{k=N-n}^{\infty}\frac{(\lambda y)^k}{k!}e^{-\lambda y} + \sum_{k=0}^{N-n-1}\frac{(\lambda y)^k}{k!}e^{-\lambda y}\sum_{j=N-n}^{\infty} p_j^{k*}\right)\right] dF(y) + \int_{y=t-x}^{\infty}\left[I\{n \le m \le N-1\}\sum_{k=0}^{m-n}\frac{[\lambda(t-x)]^k}{k!}e^{-\lambda(t-x)}p_{m-n}^{k*} + \delta_{m,N}\left(\sum_{k=N-n}^{\infty}\frac{[\lambda(t-x)]^k}{k!} + \sum_{k=0}^{N-n-1}\sum_{j=N-n}^{\infty}\frac{[\lambda(t-x)]^j}{j!}p_j^{k*}\right)e^{-\lambda(t-x)}\right] dF(y)\right\} dx; \tag{6}$$

$$Q_n^{(4)}(t,m) = \delta_{m,n} e^{-(\lambda+\mu+\gamma)t}, \tag{7}$$

where $I\{A\}$ and p_j^{i*} denote the indicator of a random event A and the jth term of the i-fold convolution of the sequence (p_k) with itself, respectively.

Finally, if the buffer is saturated at the opening, we obtain similarly

$$Q_N^{(1)}(t,m) = \lambda \int_0^t e^{-(\lambda+\mu+\gamma)x} Q_N(t-x,m)dx; \tag{8}$$

$$Q_N^{(2)}(t,m) = \mu \int_0^t e^{-(\lambda+\mu+\gamma)x} Q_{N-1}(t-x,m)dx; \tag{9}$$

$$Q_N^{(3)}(t,m) = \gamma \int_{x=0}^t e^{-(\lambda+\mu+\gamma)x}\left[\int_{y=0}^{t-x} Q_N(t-x-y,m)dF(y) + \delta_{m,N}\bar{F}(t-x)\right] dx; \tag{10}$$

$$Q_N^{(4)}(t,m) = \delta_{m,N} e^{-(\lambda+\mu+\gamma)t}, \tag{11}$$

where $\bar{F}(t) = 1 - F(t)$.

Introduce now the LT (=Laplace transform) of the functional $Q_n(t,m)$ on the argument t in the following way:

$$q_n(s,m) = \int_0^{\infty} e^{-st} Q_n(t,m)\, dt, \quad 0 \le n \le N, \ Re(s) > 0. \tag{12}$$

Now, from the representations equation (2)-equation (11), we get

$$q_0(s,m) = \frac{\lambda}{\lambda+s}\left[\sum_{k=1}^{N-1} p_k q_k(s,m) + q_N(s,m)\sum_{k=N}^{\infty} p_k\right] + \frac{\delta_{m,0}}{\lambda+s}; \tag{13}$$

$$q_n(s,m) = b(s)\{\lambda[\sum_{k=1}^{N-n-1} p_k q_{n+k}(s,m) + q_N(s,m)\sum_{k=N-n}^{\infty} p_k] + \mu q_{n-1}(s,m) + \gamma[\sum_{k=0}^{N-n-1} a_k(s)\sum_{j=k}^{N-n-1} p_j^{k*} q_{n+j}(s,m) + q_N(s,m)(\sum_{k=N-n}^{\infty} a_k(s) + \sum_{k=0}^{N-n-1} a_k(s)\sum_{j=N-n}^{\infty} p_j^{k*}) + I\{n \le m \le N-1\}\sum_{k=0}^{m-n} p_{m-n}^{k*}\beta_k(s) + \delta_{m,N}(\sum_{k=N-n}^{\infty}\beta_k(s) + \sum_{k=0}^{N-n-1}\beta_k(s)\sum_{j=N-n}^{\infty} p_j^{k*})] + \delta_{m,n}\}, \quad 1 \le n \le N-1 \tag{14}$$

and

$$q_N(s,m)\left[1-b(s)\left(\lambda+\gamma f(s)\right)\right]=b(s)\left[\mu q_{N-1}(s,m)+\delta_{m,N}(\gamma s^{-1}(1-f(s))+1)\right], \quad (15)$$

where we utilize the following notation:

$$a_k(s)=\int_0^\infty e^{-(\lambda+s)y}\frac{(\lambda y)^k}{k!}dF(y); \quad (16)$$

$$b(s)=(\lambda+\mu+\gamma+s)^{-1} \quad (17)$$

$$\beta_k(s)=\int_0^\infty e^{-(\lambda+s)y}\frac{(\lambda y)^k}{k!}\bar{F}(y)dy. \quad (18)$$

To rewrite the equations of the system equation (14) in a specific form, let us define the following functional sequences:

$$\alpha_0(s)=\mu b(s), \alpha_1(s)=\gamma a_0(s)b(s), \alpha_{k+1}(s)=b(s)\left[\lambda p_k+\gamma\sum_{i=0}^{k}p_k^{i*}a_i(s)\right], k\geq 1; \quad (19)$$

$$\varphi_n(s,m)=-b(s)[q_N(s,m)A_n(s,m)+B_n(s,m)], \quad (20)$$

where

$$A_n(s,m)=\lambda\sum_{k=N-n}^{\infty}p_k+\gamma\left(\sum_{k=N-n}^{\infty}a_k(s)+\sum_{k=0}^{N-n-1}a_k(s)\sum_{j=N-n}^{\infty}p_j^{k*}\right) \quad (21)$$

and

$$B_n(s,m)=$$
$$\gamma\left[I\{n\leq m\leq N-1\}\sum_{k=0}^{m-n}p_{m-n}^{k*}\beta_k(s)+\delta_{m,N}\left(\sum_{k=N-n}^{\infty}\beta_k(s)+\sum_{k=0}^{N-n-1}\beta_k(s)\sum_{j=N-n}^{\infty}p_j^{k*}\right)\right]+$$
$$\delta_{m,n}. \quad (22)$$

Applying now equation (19)-equation (22), we can transform the system equation (14) to the following one:

$$q_n(s,m)=\sum_{k=-1}^{N-n-1}\alpha_{k+1}(s)q_{n+k}(s,m)-\varphi_n(s,m), \quad 1\leq n\leq N-1. \quad (23)$$

4. Compact-Form Solution for the LT of the Queue-Size Distribution

Let us use in equation (13), equation (15) and equation (23) the following substitutions:

$$d_n(s,m)=q_{N-n}(s,m), \quad \psi_n(s,m)=\varphi_{N-n}(s), \quad 0\leq n\leq N. \quad (24)$$

Now, the system of equation (13), equation (15) and equation (23) can be rewritten as follows:

$$d_0(s,m)\left[1-b(s)\left(\lambda+\gamma f(s)\right)\right]=b(s)\left[\mu d_1(s,m)+\delta_{m,N}(\gamma s^{-1}(1-f(s))+1)\right]; \quad (25)$$

$$\sum_{k=-1}^{n-1}\alpha_{k+1}(s)d_{n-k}(s,m)-d_n(s,m)=\psi_n(s,m), \quad 1\leq n\leq N-1 \quad (26)$$

and

$$d_N(s,m)=\frac{\lambda}{\lambda+s}\left[\sum_{k=1}^{N-1}p_{N-k}d_k(s,m)+d_0(s,m)\sum_{k=N}^{\infty}p_k\right]+\frac{\delta_{m,0}}{\lambda+s}. \quad (27)$$

In [7] one can find the following lemma:

4.1. Lemma 1

Let $(\alpha_k),\ k \geq 0$, and $(\psi_k),\ k \geq 1$, be two given sequences, where additionally the condition $\alpha_0 \neq 0$ is assumed. Define the following infinite-sized system of linear equations with unknowns $y_1, y_2, \ldots$:

$$\sum_{k=-1}^{n-1} \alpha_{k+1} y_{n-k} - y_n = \psi_n, \qquad n \geq 1. \tag{28}$$

Successive components of arbitrary solution of (28) can be represented in the form

$$y_n = CR_n + \sum_{k=1}^{n} R_{n-k}\psi_k, \qquad n \geq 1, \tag{29}$$

where $C$is a constant and the sequence $(R_k),\ k \geq 0$, is defined recursively by means oftwo given sequences $(\alpha_k),\ k \geq 0$, and $(\psi_k),\ k \geq 1$, in the following way:

$$R_0 = 0,\ R_1 = \alpha_0^{-1},\ R_{k+1} = R_1\big(R_k - \sum_{i=0}^{k} \alpha_{i+1} R_{k-i}\big),\ k \geq 1. \tag{30}$$

Let us note that, since equation (26) has the same form as equation (28), then we can write

$$d_n(s,m) = C(s,m)R_n(s) + \sum_{k=1}^{n} R_{n-k}(s)\psi_k(s,m), \qquad n \geq 1, \tag{31}$$

where $(R_k(s))$ is given by equation (30) with $(\alpha_k(s))$ given by equation (19). The remaining to do is to find the representation for $d_0(s,m)$ (occurring also in the definition of $\psi_k(s,m)$, see equation (20) and $C(s,m)$. Substituting $n = 1$ in equation (31) we obtain $d_1(s,m) = C(s,m)\alpha_0^{-1}(s)$. Comparing the right side of this representation with equation (25), after eliminating $d_1(s,m)$, we get $C(s,m)$ in a function of $d_0(s,m)$, namely

$$C(s,m) = \alpha_0(s)[\Gamma(s)d_0(s,m) + \Delta(s,m)], \text{ where} \tag{32}$$

$$\Gamma(s) = \big(\mu b(s)\big)^{-1}[1 - b(s)(\lambda + \gamma f(s))], \Delta(s,m) = -\mu^{-1}\delta_{m,N}(\gamma s^{-1}(1 - f(s)) + 1). \tag{33}$$

Now, similarly, substituting $n = N$ into equation (31) and comparing the right side to the right side of equation (27), applying equation (20), equation (24) and equation (33), we eliminate $d_0(s,m)$ explicitly as follows:

$$d_0(s,m) = q_N(s,m) = [H_1(s,m) + H_2(s,m)][H_3(s,m) + H_4(s,m)]^{-1}, \text{ where} \tag{34}$$

$$H_1(s,m) = \alpha_0(s)\Delta(s,m)R_N(s) - b(s)\sum_{k=1}^{N} R_{N-k}(s)B_{N-k}(s,m) - \frac{\delta_{m,0}}{\lambda+s}; \tag{35}$$

$$H_2(s,m) = \frac{\lambda}{\lambda+s}\left[\sum_{k=1}^{N-1} p_{N-k}\big(b(s)\sum_{i=1}^{k} R_{k-i}(s)B_{N-i}(s,m) - \alpha_0(s)\Delta(s,m)R_k(s)\big)\right]; \tag{36}$$

$$H_3(s,m) = \frac{\lambda}{\lambda+s}\left[\sum_{k=1}^{N-1} p_{N-k}\big(\alpha_0(s)\Gamma(s)R_k(s) - b(s)A_{N-i}(s,m)\sum_{i=1}^{k} R_{k-i}(s)\big)\right]; \tag{37}$$

$$H_4(s,m) = \frac{\lambda}{\lambda+s}\sum_{k=N}^{\infty} p_k - \alpha_0(s)\Gamma(s)R_N(s) + b(s)\sum_{k=1}^{N} R_{N-k}(s)A_{N-k}(s,m). \tag{38}$$

Collecting now the formulae equation (20), equation (24), equation (31)-equation (34), we can state the following:

4.2.Theorem 1

The representation for the LT$q_n(s,m)$ of conditional queue-size distribution in the considered finite-buffer manufacturing model with unreliable server is following:

$$q_n(s,m) = \left(\alpha_0(s)\Gamma(s)R_{N-n}(s) - b(s)\sum_{k=1}^{N-n} R_{N-n-k}(s)A_{N-k}(s,m)\right)q_N(s,m) - \alpha_0(s)\Delta(s,m)R_{N-n}(s) + b(s)\sum_{k=1}^{N-n} R_{N-n-k}(s)B_{N-k}(s,m), \quad 1 \le n \le N-1, \tag{39}$$

wherethe formulae for $\boldsymbol{b(s)}, \boldsymbol{\alpha_0(s)}, \boldsymbol{A_k(s,m)}, \boldsymbol{B_k(s,m)}, \boldsymbol{R_k(s)}, \boldsymbol{\Gamma(s)}$,$\boldsymbol{\Delta(s,m)}$and $\boldsymbol{q_N(s,m)}$are found in equation (17), equation (19), equation (21), equation (22), equation (30), equation (33) and equation (34), respectively.

The result obtained in Theorem 1 may be used in the analysis of the effect of frequently changing traffic parameters of the production line (e.g. the intensity of job arrivals) on the process of accumulation of jobs in the buffer, and on their waiting times. Such an information makes it possible to build a preventive algorithmfor redirecting jobs to other machines in the case of risk of buffer overflow or too long delay.

5.Conclusions

In the paper transient queue-size distribution in a finite-buffer model of manufacturing line with unreliable machine and batched job arrivals is studied. Using the approach based on the memoryless property of exponential distributions of interarrival, processing and failure-free times and the total probability law, a system of integral equations for time-dependent queue-size distribution conditioned by the number of jobs present in the system initially is built. The compact-form solution of the corresponding system written for Laplace transforms is derived via linear-algebraic approach.

Acknowledgement

The project was financed with subsidies from the National Science Centre in Poland, granted by virtue of the decision number DEC-2012/07/B/ST6/01201.

References

[1] Govil M K and Fu M C 1999 Queueing theory in manufacturing: A survey *J. Manuf. Syst.* **18** pp 214-240

[2] Kempa W M 2010 The transient analysis of the queue-length distribution in the batch arrival system with N-policy, multiple vacations and setup times *Applications of mathematics in engineering and economics* **1293** pp 235-242

[3] Kempa W M 2012 On transient queue-size distribution in the batch arrival system with the N-policy and setup times *Math. Commun.* **17** pp 285-302

[4] Kempa W M 2013 A direct approach to transient queue-size distribution in a finite-buffer queue with AQM *Appl. Math. Inf. Sci.* **7** pp 909-915

[5] Kempa W M 2014 On transient queue-size distribution in the batch-arrivals system with a single vacation policy *Kybernetika* **50** pp 126-141

[6] Kempa W M, Paprocka I, Kalinowski K and Grabowik C 2014 On transient queue-size distribution in a single-machine production system with breakdowns *Advanced Materials Research* **1036** pp. 505-510

[7] Korolyuk V S 1975 Boundary-value problems for compound Poisson processes *Naukova Dumka* Kiev

[8] Papadopoulos H T and Heavey C 1996 Queueing theory in manufacturing systems analysis and design: A classification of models for production and transfer lines *Eur. J. Oper. Res.* **92** pp 1-27

Designing of robotic production lines using CAx software

A Wróbel[1] and P Langer[1]
[1] Silesian University of Technology, Faculty of Mechanical Engineering,
Institute of Engineering Processes Automation and Integrated Manufacturing Systems,
Konarskiego 18A, 44-100 Gliwice, Poland

E-mail: andrzej.wrobel@polsl.pl

Abstract. Present market conditions causes that modern control systems of robotized manufacturing cells should be characterized by the much greater degree of flexibility, self-organization and, above all, adaptability to emerging outer excitations. The phenomenon of information distribution is one of the most important features of modern control systems. In the paper is presented the approach, based on application of multi-agent systems, for supporting the operation of robotized manufacturing cells. The aim of this approach is to obtain the flexible response to outer excitations and preventing situations that might cause the delay of the production process. The presented paper includes description of the concept of an informatics system designed for controlling the work of production systems, including work cells. Such systems could operate independently if it would be equipped with the self-organization mechanism. It is possible in the case of the proposed multi-agent system. The implementation of the presented concept will follow the present analysis of the described concept. The advantage of the proposed concept is its hierarchical depiction that allows integrating different utilized informatics tools in one complex system. It allows preparing the final computer program.

1. Introduction

Delmia is a virtual environment which allows to design fully robotic stations of production line and simulate it work. Program developed by Dassault is based on Catia files which causes that it is able to import any geometry modelled using Catia. Primary functionality of Delmia is graphically offline robot programming. This program belongs to Computer Aided software type and is used by robotics engineers all around the world.

Basic types of files used by Delmia:

- .cgr type - contains only basic information's about geometry, does not contain any additional information, requires little space in RAM memory;
- CATpart - contains information about geometry and stages of its creation or link to geometry included in other files, allows to storage additional information like frames of interest;
- CATproduct - contains links to one or a few geometries contains in other files, allows to define additional information like frames of interest and kinematic properties;
- process - main type of files generated by Delmia V5, contains location of all parts in station, robots programs and points used during offline programming.

One of the greatest problems during using CAx programs in work on big projects are hardware limitations. Model of robotic station can includes many elements what requires high computing power

Content from this work may be used under the terms of the Creative Commons Attribution 3.0 licence. Any further distribution of this work must maintain attribution to the author(s) and the title of the work, journal citation and DOI.
Published under licence by IOP Publishing Ltd

and provides slowdown the simulation. To avoid this problem stations should be built from the as great as possible number ".cgr" elements. Additionally most elements can be modelled only by their external geometry, interior can be replaced by empty space or solid. Element which extremely borne the computer and should be mandatory replaced by simpler geometry is thead. Engineers must find compromise between accurate models and labour productivity of the program. Because of that some element like a cables are modelled in a very simplified way and it is possible that simulation of their movement will not be consistent with reality. It is very important to sign that one of the worst problems is a difference between virtual model and a real devices or production hall geometry because of using out of date information. Most of program errors is caused by bad communicate between programmer and a customer or between programmer and designer responsible for tool and external devices design. To avoid this errors programmers should use files included in one compact database which has to be constantly updated by all the users and customer.

Positioning parts into virtual robotic cell, definition their properties and creating robots programs are based on a few types of points. Types are divided into two groups: frames of interest and points stored in process file.

Frames of interest is a group of points which is used to specify properties of models and contains three basic types:

- base – type used during positioning parts or devices in cell;
- designe – allows to define kinematic properties;
- tool – defines tool centre point of appliance;

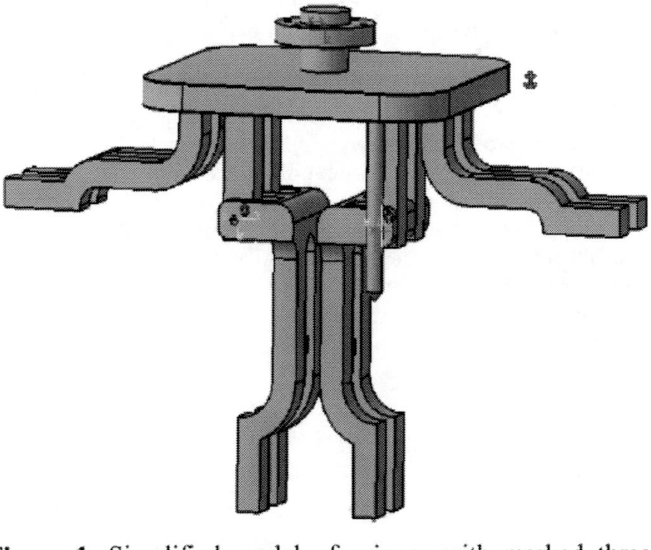

Figure 1. Simplified model of gripper with marked three types of points: blue-design points, red-tool, violet-base point.

As shown on the picture 1 blue points on moveable tool parts are connected with points on fixed one. Based on that points programmer set constrains and axes of rotation. Violet one should be used for positional the gripper on model of robot during building a virtual cell. Red frame determines tool canter point of the gripper. Its position was determined by method of positioning manipulated part on the gripper.

Points which is stored in process file is called manufacturing. This type determines point which are parts of program and define following arrangements of robot. Manufacturing points are called tags and collected in tag groups. Programs can use many different tag groups and tags can be used more than once. Tag groups make easier to organize points because of their position in space or program membership. Each program contains additional information about tags and divides them on via and process points. Process points defines point in space where operation should be executed of where should it starts and ends. Via points define tool path between process points.

Robotic engineers use most often two of Delmias modules: Device Building and Device Task Definition. First one allows to specify kinematic properties of moveable models, second is used for building cells with prepared elements, robot programing and simulation. As in most of CAx programs, basic element of program Delmia V5 is tool called tree. Tree contains names of all elements of the simulation and relationships between them. It allows to switch between modules by double click on the name of process or resource and modify some properties of resources. Delmia V5 gives a possibility to modify all joint ranges values and joint speeds values in device model so as make it in line with reality even if the robot or other machine has been modified by the customer (figure 2).

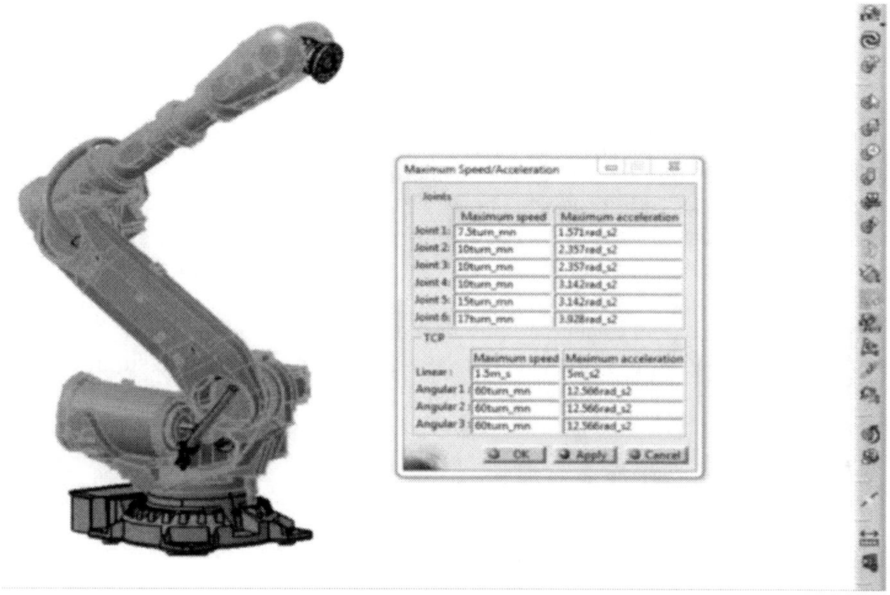

Figure 2. Modification kinematical properties of robot model.

Basic term used during robot programming is tool canter point. This point defined by frame of interest type tool determines place of tool which is the last point robot. Correct definition a TCP significantly simplifies improving tasks in factory. Tool canter point can be defined on physical elements of tool like an electrode or on immaterial parts of geometry like an axis or circle canter. It is very important to notice that movement of tool in critical parts of task should be realized along one of the axis defined by TCP. This fact is very helpful during taking detail from magazine or mounting it into manufactured item.

2. Programming in Delmia V5

Robot programming in Delmia V5 is based on conception of using four profiles:

- tool profile,
- accuracy profile,
- motion profile,
- object profile.

It is recommended to use many different profiles of the same type during robot programming if task requires it. Profiles allow easy patching tasks in factory and correct functionality of programs.

Tool profile contains information about TCP and should be created every time when robot changes its tool or other part of tool becomes the most important during operation. Accuracy profile determines approximation of the path in specified points (figure 3). Selection high level of accuracy may result robot motion slowdown. It should be emphasized that there is no reason to know exact robot path in many parts of the task, when process does not require it and there is no risk of clash.

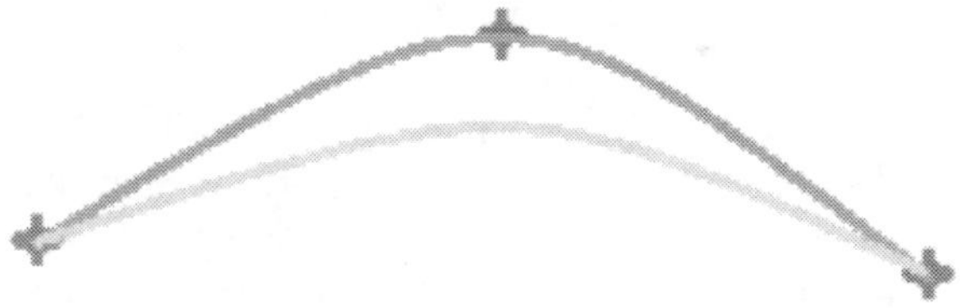

Figure 3. Green line represents path with high accuracy of the middle point, orange with low value of accuracy.

Motion profile determines speeds of robot drives during movement. Object profile contains information about local reference system used during points creation. Object profiles should be created for tasks related with external devices like magazines or manufactured products.

For example during mounting some elements into manufactured product, robot should take out part from storage and put it down into product. There is a need to use two different object profiles because process is related with two different objects: manufactured product and external magazine. There is need to know that all elements in real factory has positional inaccuracy of different value. Because of that we should use two different object profile during the task. Movement in different parts of task should be realized with different accuracy and speed. When robot moves near some elements of its scene movement should be slower and accurate because of possibility of clash, in the rest of task we should avoid low speeds and accurate movement to obtain the shortest cycle time. Positioning mounted part during taking it out from storage and mounting it into product can be positional by different parts of a tool, so it is justified to use two tool profiles. Depends on settings Delmia V5 can make possible selection config of the robot during reaching following points, interpolation type and allows to add support external devices. Robot path can be saved in program using points in space or following orientation robot arms saved as known joints values. This feature can be especially helpful if is necessary to reach positions near singularity of the robot to determine its configuration. Singularity is a term describing situation when point in the workspace can be reached on infinitely many ways. It's possible when axes of two rotary joints lying on the same line and there is an infinite number of combinations their relative positions. Other application of this functionality is predefined position programming. For example position which has no connection with robot scene like transport position or home position. Both of theme should has no connection with environment of the robot.

3. Communication signals and determination of collisions

Communication between robots and other parts of the station is realized by signals. Delmia V5 allows to add signals during robot programming and simulate work of PLC controllers. This feature allows to simulate simultaneous work all of station parts using tool named Multi Resource Simulation. For accuracy definition work of all controllers managing robots movement programmer should use additional module named Real Resource Simulator. RRS allows also designation accuracy cycle time and make sure that there will be no collision. It is necessary because of little differences in work between controllers responsible for determination of robot movement developed by different company. Without RRS Delmia gives an approximated path and cycle time and this approximation depends on values set in accuracy profiles.

Collision can be divided into two groups:

- collisions between robots or their tools and unmovable elements called static collisions,

• possible collisions between robots during work inclusive collisions provided by unexpected events like delay or stop one of the robots in station. For detection and prevention fist type programmer can use tool clash which can search and indicate time and position in space when clearance between marked falls below specified level. Collisions between robots have to be detected by programmer during program development. Prevention can be realized for two ways:

• by path planning,
• using signals as an interference zone.

Path planning is method programing in such way that robots have to cover greater distance but their paths do not intersect. Interference zone can be understood like a flag or resource. Two robots compete for a flag when enter into hazard zone. If first one occupy interference zone program of the second is stopped to time release the resource. It means that interference zone has no relation with space only a parts of robots programs. Interference zone as other signals should be sent only when all information about state of process are known, especially robots positions. For this reason signals should be sent only when robot stops in point specified with full accuracy. Each stop of the robot results in a lengthening cycle time, so sometimes it is better to send signal much earlier instead add additional stop. It means that one interference zone submitted in space can be different sizes and in different places depending on the robot which it use. Robot programing consist of creation programs for normal work of robots, service tasks enabling tool maintenance or assessment of damage after collision and tasks enabling tool change.

Tool maintenance consist of improving tool geometry or replacing some tool parts.

Service tasks are programs which consist of achievement points on measuring element by a key part of tool. If after start-up service task tool centre point do not achieve established point staff knows that damages are serious and further work of station is impossible until the repair.

Programs can also contain command "call task" (figure 4) which runs other task. This functionality can be used for create main programs consisting of small parts. Each part store information about path in some part of manufactured product. When product has a few types and the greater part of the treatment do not change, it is possible to divide path for main and secondary parts. In that case main program can call main and type paths in desired order.

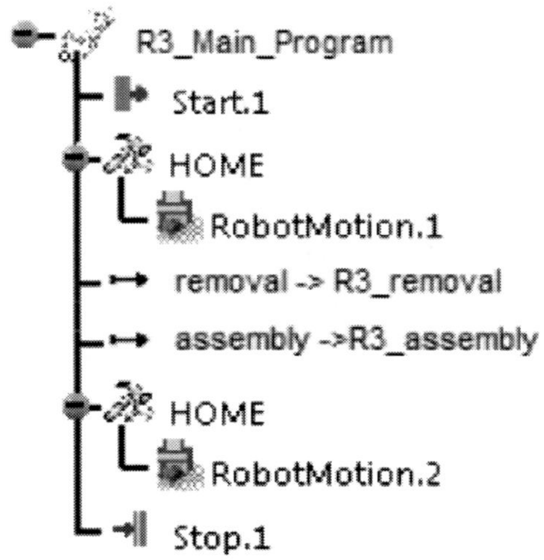

Figure 4. Main program consisting two "call task" command.

Robot programming using Delmia V5 allows to save a lot of time and many because it is not required to stop production for long time. Delmia allows creation programs even when factory is still being build. One of the greatest difficulties during during work using CAx software is need for good project management. Project lider has to ensure that data on which programmers work are always up-to-date and direct the work of the team.

4. Conclusions

In this article possibilities what is offered by the use of modern CAx software in the design of manufacturing processes was shown. It is also shown as by two programs NX and DELMIA simulation and offline programming of robots can be created. The result of this article - design and simulation of robotic work station on the production line was done. In future articles authors will show presented software tools for modeling car production line and they will try to apply simulation in real industry plant.

Acknowledgments

The work was carried out under the project number PBS2/A6/17/2013 agreement implemented under the Applied Research Program, funded by the National Centre for Research and Development.

References

[1] Zdanowicz R 2012 Basics of of robotics *Silesian Technical University Publishing House*
[2] Olszewski M, Barczyk J and others 1985 Manipulators and industrial robots *Automatic handling machines Scientific and Technical Publishing*
[3] Morecki A and Knapczyk 1999 Basics of robotics. The theory and elements of manipulators and robots *WNT Warszawa*
[4] Buchacz A Computer aided of synthesis and analysis of machine's sub-assembly by graph and structural number's methods (Silesian University of Technology Press: Gliwice)
[5] Buchacz A and Wróbel A 2008 Homogenization of mechatronical systems in modeling piezoelectric layer *Machine Dynamics Problems 25* **32(3)** pp 7-11
[6] Buchacz A and Wróbel A 2009 Modelling of multilayer transducer by means of the graph and structural numbers method *80th Annual Meeting of the International Association of Applied Mathematics and Mechanics* **9** pp 393–394
[7] Buchacz A and Płaczek M 2010 The approximate Galerkin's method in the vibrating mechatronic system's investigation *Quality and Innovation* pp 147-150
[8] Dymarek A and Dzitkowski T 2005 Modelling and synthesis of discrete – continuous subsystems of machines with damping *Journal of Materials Processing Technology* **164-165** pp 1317-1326.
[9] Gwiazda A, Herbuś K, Kost G and Ociepka P 2015 Motion analysis of mechatronic equipment considering the example of the Stewart platform *Solid State Phenomena* **220-221** pp 479-484
[10] Jamroziak K and Bocian M 2010 Identification of pierced materials characteristics in the aspect of selected degenerated models *Journal of KONES* **17** pp 169-176
[11] Kost G and Zdanowicz R 2005 Modeling of manufacturing systems and robot motions *Journal of Materials Processing Technology*
[12] Foit K, Kost G and Reclik D 2011 Automatic programming and generation of collision-free paths for the Mitsubishi Movemaster RV-M1 robot *Journal of Achievements in Materials and Manufacturing Engineering* pp 57-65

Analysis of the Hexapod Work Space using integration of a CAD/CAE system and the LabVIEW software

K Herbuś [1] and P Ociepka [1]
[1] The Silesian University of Technology, The Faculty of Mechanical Engineering, The Institute of Engineering Processes Automation and Integrated Manufacturing Systems, Konarskiego 18A, 44-100 Gliwice, Poland

E-mail: krzysztof.herbus@polsl.pl

Abstract. The paper presents the problems related to the integration of a CAD/CAE system with the LabVIEW software. The purpose of the integration is to determine the workspace of a hexapod model basing on a mathematical model describing it motion. In the first stage of the work concerning the integration task the 3D model to simulate movements of a hexapod was elaborated. This phase of the work was done in the "Motion Simulation" module of the CAD/CAE/CAM Siemens NX system. The first step was to define the components of the 3D model in the form of "links". Individual links were defined according to the nature of the hexapod elements action. In the model prepared for movement simulation were created links corresponding to such elements as: electric actuator, top plate, bottom plate, ball-and-socket joint, toggle joint Phillips. Then were defined the constraints of the "joint" type (e.g.: revolute joint, slider joint, spherical joint) between the created component of the "link" type, so that the computer simulation corresponds to the operation of a real hexapod. The next stage of work included implementing the mathematical model describing the functioning of a hexapod in the LabVIEW software. At this stage, particular attention was paid to determining procedures for integrating the virtual 3D hexapod model with the results of calculations performed in the LabVIEW. The results relate to specific values of the jump of electric actuators depending on the position of the car on the hexapod. The use of integration made it possible to determine the safe operating space of a stationary hexapod taking into consideration the security of a person in the driving simulator designed for the disabled.

1. Introduction

The subject of conducted studies isthe stationary hexapod. It is the main executive system of a driving simulator for people with lower limb disabilities, (figure 1). The presented simulator was considered as the result of the research projectand was the subjectof many investigations conducted by theresearch team [5,8].On the hexapod is mounted the fiat panda car that has been mechanically modified. Modifications included the moving of mechanical systems, which are not required for correct functioning of a driving simulator (e.g.: transmission gearbox, suspension). The described system is connected to the control cabinet, which is responsible for the correct functioning of the hexapod and for cooperation with a virtual environment in which the car avatar is moving. Information from the virtual environment and from control elements of the modified car (gas pedal, brake one, clutch one, lights controls and speed indicator) are forwarded as inputs to the control cabinet. In this cabinet, basing on acquired information obtained from stored mathematical model, decisions are made

Content from this work may be used under the terms of the Creative Commons Attribution 3.0 licence. Any further distribution of this work must maintain attribution to the author(s) and the title of the work, journal citation and DOI.
Published under licence by IOP Publishing Ltd

concerning the mode of "behave" of the hexapod. It should be noted that the method of driving the modified vehicle does not differ from that of driving a car moving in the real traffic. This is due to the fact that all control elements (gas pedal, brake one, clutch one, lights controls and speed indicator) from the point of view of the driver look and function identically as in an unmodified car.

Figure 1. Virtual model and the real system of the simulator for car driving learning for disabled persons.

The whole simulator completes the virtual world image projection system, which consists of four multimedia projectors and four large screens. Three of them are connected to each other and allow realizing the projection of the driver's front and sides view, and rear screen allows to ride "over the shoulder" and control traffic by rear mirror mounted inside the cabin of the car.

The aim of investigations was to determine the maximal values of movement for each axis that could be performed by the frame with attached car body using the approach basing on the integration of the model prepared for driving simulation, created in the system Siemens NX program, and the control application created in the Lab VIEW program. In connection with realizing works leading to the best possible tuning of work of the real hexapod system, in the adopted environment of the simulator designed for driving lessons for people with movement dysfunctions, the important problem is determining the movement "margin" which could be realized by the hexapod without danger of collision of the car with the elements of the work environment of the simulator. The elements of the work environment of the simulator, with which may collided the car body, mounted on a hexapod, are predominantly components of the projection system of virtual environment. In contrast, the hexapod itself is exposed to the possibility of a collision between its drives and the pocket cavity of a foundation that has been made due to insufficient height of the room in which is installed the simulator system.

2. Preparation of the hexapod model for motion simulation

The issues related to the designing and modeling are the basis for the creation of virtual models of technical systems [1,2,3,4,6,7,9,10,11,12,13,14]. Preparation of the presented 3D model of the hexapodwas realized on the basis of the 3D model in the "Motion Simulation" module in the software Siemens NX 8.5 of the CAD class (figure 2). In the first stage of works the components of the"link" typewere defined, among which it could be distinguished: platform_base_link (1), cross_link (2), motor_link (3), actuator_link (4), and body_link (5).

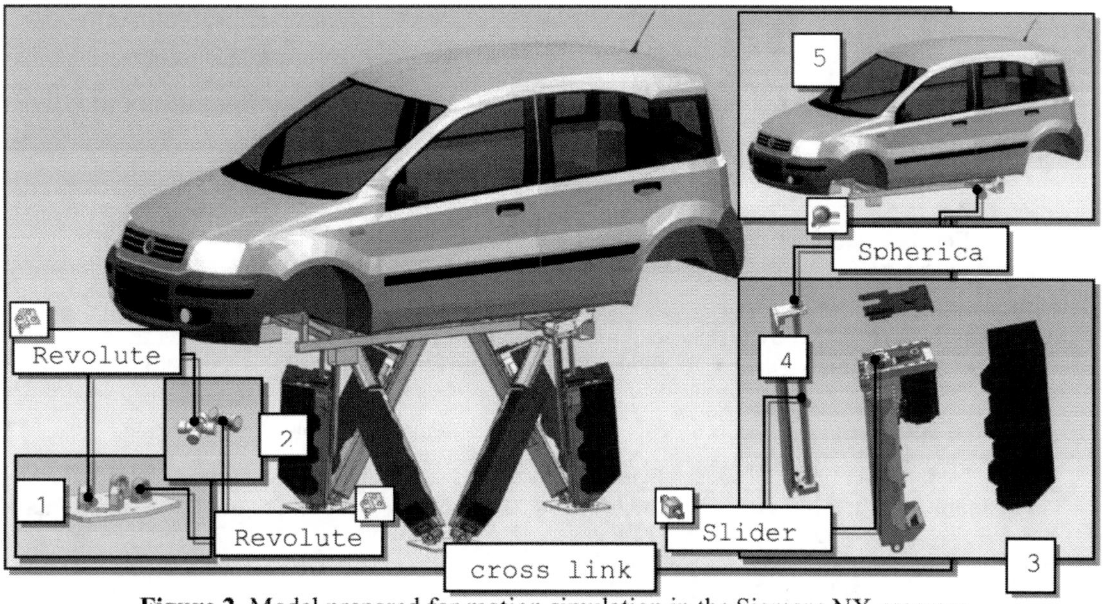

Figure 2. Model prepared for motion simulation in the Siemens NX program.

Due to the complex solid model of the analyzed system the particular components of the "link" type contains many elements of the "solid body" type. It should be noted that the elements being includedin the given component of the "link" type do not change their relative position during the motion simulation. Accordingly, the component of the "link" type is treated, by the simulation environment, as a single object, but its properties depend on the properties of its elements (e.g.: the weight of a component of the "link" type depends on the material properties assigned to particular elements of the "solid body"type). For mapping the hexapodoperation in a virtual environment between the components of the "link" type was established the components of the"joint" type.Among them it could be distinguished: revoluta joint -between the platform_base_link and the cross_link; revoluta joint - between the cross_link and the motor_link; slider joint - between the motor_link and the actuator_link and spherical joint - between the actuator_link and the body_link.

3. Control application created in the LabVIEW program

The developed control application was created in the LabVIEW program. The input parameters for the described application are the characteristic points ofthehexapod,with which is described the initial position (neutral) of the hexapod (figure 3). By neutral position of the hexapod is understood such position in which its all drives reach the displacement value equal to half of the maximal extension of the drive screw. On the basis of the neutral position is determined the length between the corresponding characteristic points representing the linear dimensions of the hexapod legs in the given position. In order to verify the correctness of work of the adopted mathematical model, describing the relationship between the position of the frame with a car and lengths of particular leg of thehexapod, it was used the virtual measurement tool, in relation to the 3D model of the hexapod in the Siemens NX software.

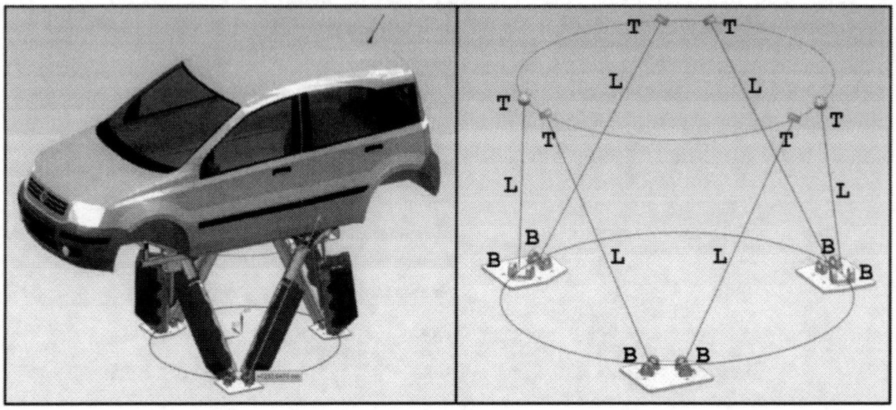

Figure 3. Adopted, simplified model used for determining the length of hexapod legs.

The leg length L_i ($i = 1,\ldots,6$) of the hexapod, in the neutral position,are determinedon the basis of coordinates valuesof points T_i ($i = 1,\ldots,6$) and B_i ($i = 1,\ldots,6$) (table 1).

Table 1. Values of coordinates of pointsin the neutral position of the hexapod.

Coordinates of points of the hexapod located in the upper and lowerring					
T1; B1 [mm]	**T2; B2 [mm]**	**T3; B3 [mm]**	**T4; B4 [mm]**	**T5; B5 [mm]**	**T6; B6 [mm]**
X = 582; 77	X = 697; 756	X = 114; 679	X = -114; -679	X = -697; -756	X = -582; -77
Y = -469; -829	Y = -271; 347	Y = 739; 481	Y = 739; 481	Y = -271; 347	Y = -496; -829
Z = 1125; 37	Z = 1125; 37	Z = 1125; 37	Z = 1125; 37	Z = 1125; 37	Z = 1125; 37

In figure 4 is presented the view of the application using which it is possible to determine the values of the lengths of all hexapod legs depending on the position of the coordinate system.

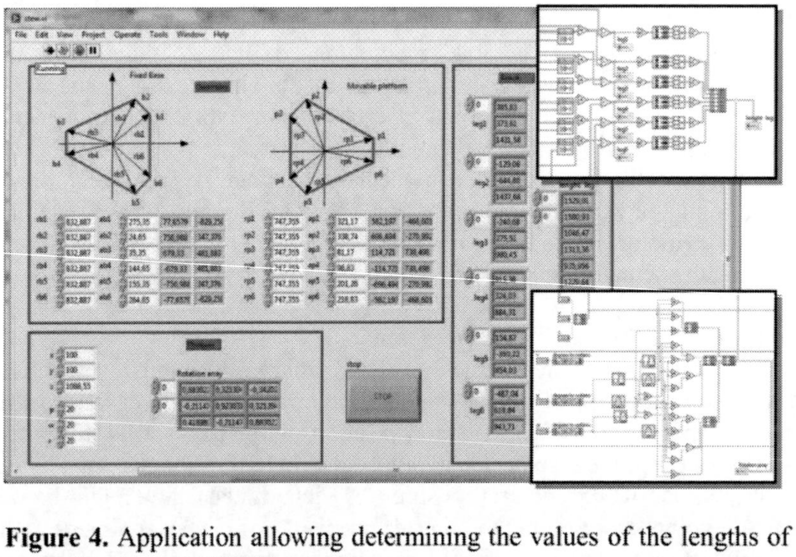

Figure 4. Application allowing determining the values of the lengths of all hexapod legs created in the LabVIEW program.

IOP Conf. Series: Materials Science and Engineering **95** (2015) 012096 doi:10.1088/1757-899X/95/1/012096

4. Analysis of hexapod displacements using an integrative approach

The process of movements analyzing of the hexapod, in the adopted simulator environment,consists of: changing the position and orientation of the coordinate system, closely connected with thehexapod frame; determining the length of the legs of the hexapod in thegiven position using the matrix of moving system transformation, associated with the frame, to the stationary system, connected to the hexapod base; determining the difference between the length of the legs in the position *i* and *i-1*, which is the value of displacement of the drive screw in the CAD model; storing the values to the integrator in the form of an Excel file; analyzing the positions of the car in the adopted CAD environment of Siemens NX basing on data from the integrator.

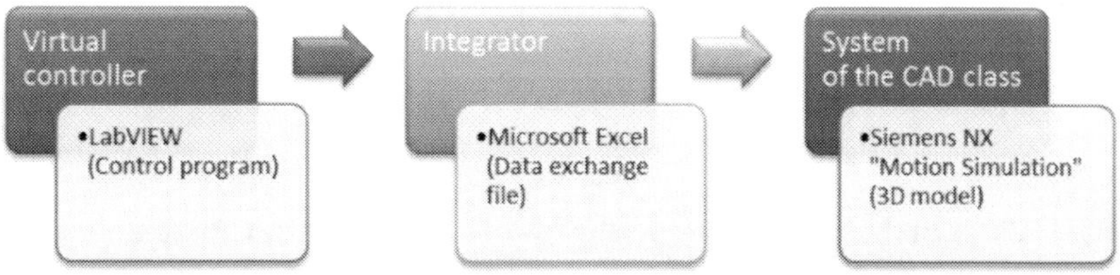

Figure 5. Scheme of the integration of the control application with the virtual 3D model.

The obtained values of displacement of the frame, with the mounted car body,in the particular axis of the system are shown in figure 6.

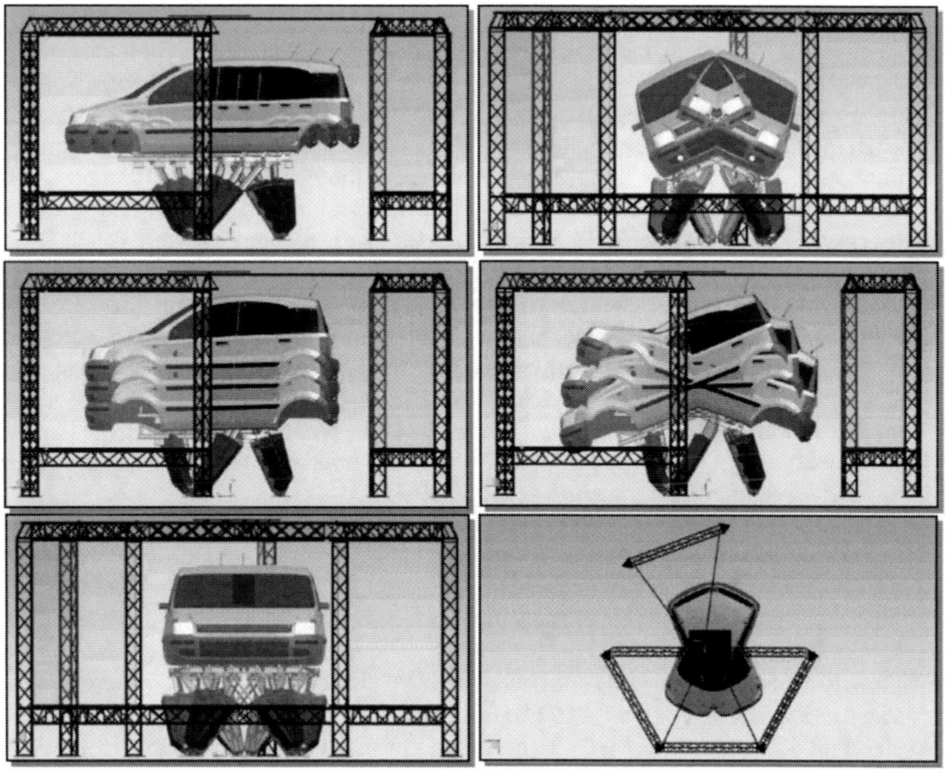

Figure 6. Locations of the hexapod, with the mounted car body, in the simulator work environment.

5. Conclusions

An important problem, related to the integration process of the control system model with the control application, is correct representation of operation of the analyzed system, whatis possible usingthe Siemens NXsoftware of the CAD class.The carried out works allowed the integration of the virtual model of the control system of thehexapodwith the control application created in the LabVIEW program, what allowedfor determination the displacements of the hexapod in the particular axes of the coordinate system.

Virtual analysesof systems, basing on the integration of the 3D model with the control application, allow determiningthe"margin" of movement concerning the analyzed hexapod.

Acknowledgements

The work is realized within the project titled: "Modular automated production stand with instrumentation for non-invasive confirmation of product quality" funded by The National Centre for Research and Development, agreement No UOD-DEM-1-495/001.

References

[1] Świder J and Herbuś K 2006 Zastosowanie funkcyjnych obiektów elementarnych do wspomagania modelowania maszyn zorientowanego na analizę ruchu *Monografia* (Wydawnictwo Politechniki Śląskiej Gliwice: Polish)

[2] Świder J, Ociepka P and Herbuś K 2005 The application of object methods for functional features representation *Achievements in Mechanical and Materials Engineering* pp 655 – 658.

[3] Herbuś K, Kost G, Reclik D and Świder J 2014 Integration of a virtual 3D model of a robot manipulator with its tangible model (phantom) *Advanced Materials Research* **837** pp 582-587

[4] Świder J, Baier A, Ociepka P and Herbuś K 2005 Zastosowanie metod obiektowych w procesie projektowo – konstrukcyjno – wytwórczym, Inżynieria Maszyn, *Komputerowe wspomagania w inżynierii maszyn i materiałów* **Rok 10** pp 61 – 83

[5] Ociepka P, Banaś W, Herbuś K and Kost G 2014 Simulator of the Car for Driving Courses for the People with Mobility Impairments *Advanced Materials Research* **1036** pp 817-822

[6] Janik W, Grabowik C and Ćwikła G 2014 The practical approach to freeform shape elements reverse engineering *Applied Mechanics and Materials* **657** pp 755-759

[7] Herbuś K, Ociepka and Gwiazda A 2014 Conception of the integration of the virtual robot model with the control system *Advanced Materials Research* **1036** pp 732-736

[8] Dymarek A, Dzitkowski T, Herbuś K, Kost G and Ociepka P 2013 The simulator for teaching how to drive a car for people with disabilities *Solid State Phenomena* **198** pp 59-64

[9] Świder J and Herbuś K 2005 The idea of modelling – aided computer method oriented on the motion *analysis Intelligent Manufacturing Systems* pp 236-241

[10] Grabowik C, Ćwikła G. and Janik W 2014 The new approach to design features identification *Applied Mechanics and Materials* **657** pp 750-754

[11] Świder J and Hetmańczyk M P 2009 Hardware and software integration of mechatronic systems for an example measurement path for temperature sensors *Solid State Phenomena* **147-149** pp 676-681

[12] Dzitkowski T and Dymarek A 2013 Active synthesis of machine drive systems *Applied Mechanics and Materials* **430** pp 178-183

[13] Dzitkowski T and Dymarek A 2013 Active synthesis of discrete systems as a tool for stabilisation vibration *Applied Mechanics and Materials* **307** pp 295-298

[14] Świder J and Hetmańczyk M P 2010 Adaptation of the expert system in diagnosis of the connection of the PLC user interface system and a field level *Solid State Phenomena* **164** pp 201-206

The role of multi-agent systems in improving performance of manufacturing robotized cells

A Sękala[1], G Ćwikła[1] and G Kost[1]
[1] Silesian University of Technology, Faculty of Mechanical Engineering,
Institute of Engineering Processes Automation and Integrated Manufacturing Systems,
Konarskiego 18A, 44-100 Gliwice, Poland

E-mail: agnieszka.sekala@polsl.pl

Abstract. Present market conditions causes that modern control systems of robotized manufacturing cells should be characterized by the much greater degree of flexibility, self-organization and, above all, adaptability to emerging outer excitations. The phenomenon of information distribution is one of the most important features of modern control systems. In the paper is presented the approach, based on application of multi-agent systems, for supporting the operation of robotized manufacturing cells. The aim of this approach is to obtain the flexible response to outer excitations and preventing situations that might cause the delay of the production process. The presented paper includes description of the concept of an informatics system designed for controlling the work of production systems, including work cells. Such systems could operate independently if it would be equipped with the self-organization mechanism. It is possible in the case of the proposed multi-agent system. The implementation of the presented concept will follow the present analysis of the described concept. The advantage of the proposed concept is its hierarchical depiction that allows integrating different utilized informatics tools in one complex system. It allows preparing the final computer program.

1. Introduction

The increasing requirements of customers, shortening of the product life cycle, balancing between the production costs, the quality and timeliness of orders, the production of a wide range of products in batches of a low size and necessity for immediate response to the emerging demand are just some of the factors with which have to deal companies operating in the variables market conditions at the beginning of the XXI century. This involves, among others, the increasing complexity of information systems used in enterprises. Therefore, it becomes necessary to search for new ways of development, without which it is impossible to gain the sustainable competitive advantage. Increasingly decreases the odds of success of companies that do not operate comprehensively and rely only on the development of certain areas of their business. For example if they focus only on reducing production costs or controlling the profitability index (like in the case of General Motors). The task of complex data analysis could be accomplished by investing in machines that could work, round the clock, without supervision. It is only possible with the use of industrial automatics systems like industrial robots. These systems require also the automatic maintenance sub-systems that could ensure its maintenance-free operation.

Content from this work may be used under the terms of the Creative Commons Attribution 3.0 licence. Any further distribution of this work must maintain attribution to the author(s) and the title of the work, journal citation and DOI.
Published under licence by IOP Publishing Ltd

In the age of advanced technologies, robotization has become an integral part of development of any production system. The increase of the physical work costs causes that the initial investments outlays on the purchase of robots are rapidly reimbursing. In the case of two-shift work they are reimbursing after about 2-4 years, while in the case of three-shift work after 1-3 years [1, 3-5]. The increase of availability and the dynamic development of industrial robots caused that they are implemented in almost every industry branch. Robots are increasingly replacing human workers in monotonous works or works requiring enormous precision that is not attainable for a man. Automation and robotization of technological/production processes affects, in a very significant way, all aspects of the enterprise business activity. Nowadays, the most robotized industry branches, in quantitative terms, are: the automotive industry, the electronic industry, the production of metal goods and the production of rubber and plastics goods [1, 5-11].

2. System architecture

The main objective of the implementation of multi-agent systems (MAS) is to create the virtual environment, which enables direct flow of data and information from the level of production management (strategic level) the level of a robotized manufacturing work cell (operational level). The information exchange, between these levels, is necessary to take appropriate action in response to changing conditions in the work cell and in the market as well as to allow flexible managing the operations realized in the work cell.

The data flow system, in an enterprise, is organized hierarchically what is shown in the figure 1. The information flows from the strategic level (management level) to the operational level (workcell). Basing on this description the problem of data flow between the presented levels/layers could be modeled using an agent system, with generally defined objectives. The MAS structure is formulated on the bases of precise defining the information flow in the enterprise information system. It is realized in the process of decomposition of the system into smaller agent-based subsystems, which are, in some sense, independent one from another.

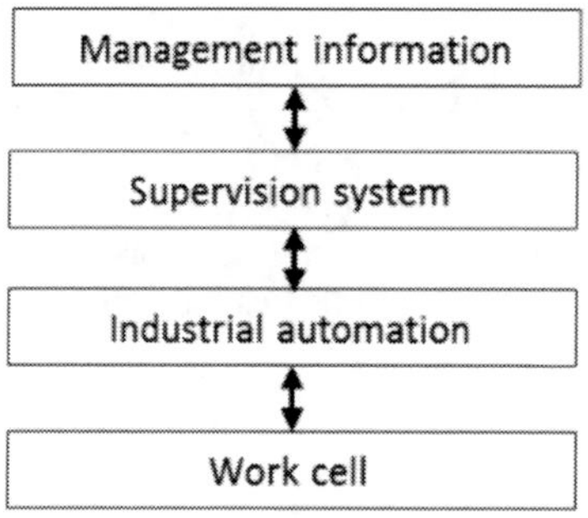

Figure 1. Hierarchical control architecture (based on [12]).

The complexity of the processes, taking place in the robotized work cell, causes that during the operation of a work cell it should be fulfilled, by the control system, many, often contradictory, requirements such as: enabling the access to data in real time, utilization of different communication media, or providing the uniform model for data presentation. Moreover, the decentralization of systems causes that the important problem is the distribution of control functions to its particular components. Hence, without the proper integration of various components, consisting robotized manufacturing workcells, it is very difficult to create the efficiently working system that realizes the tasks concerning the selection the appropriate technical and organizational solutions. The application of MAS systems allows generalizing complex problems, that is, to realize the problems that are abstract ones.

IOP Conf. Series: Materials Science and Engineering **95** (2015) 012097 doi:10.1088/1757-899X/95/1/012097

The agent, in the literature, is determined as an autonomous unit, embedded in an environment that acquires data from this environment through sensors and, basing on these data and the internal reasoning process, performs actions on that environment. The ability of an agent to act autonomously, also in the group, means the ability to participate in the high-level interactions such as, for example, cooperation [13-15]. The possibility of cooperation with other agents offers great opportunities associated with the implementation of multi-agent systems for the efficient functioning of robotized manufacturing workcells. The proposed approach is the extension of the concept presented in [2]. It assumes the use of the mechanisms of self-organization that allows organizing the independent information channels and providing different information channels what increases the efficiency of the information system.

The system consists of a user interface, a multidimensional database, and agents cooperating with each other. In the structure of the proposed system (figur 2) are distinguished four decentralized and partially independent units (agents) focusing themselves only on the key responsibilities. Each agent contains detailed information concerning the tasks assigned to them. Furthermore, each of the agents is equipped with a decision-making module. Between all groups of agents is established cooperation, which is considered one of the most attractive features of the multi-agent approach.

The analysis of the presented architecture the system allows distinguishing in it several basic units creating the multi-agent system. These include:

- **Coordinator agent** – its task is to coordinate and monitor of the whole manufacturing workcell.
- **Task order agent** – its task is to manage the sequence and allocation of realized tasks, in accordance with the current situation in the workcell.
- **Resources agent** – agent which is responsible for proper utilization of available resources, which are: technological tooling of a robot, machine tools, tools, buffers, auxiliary equipment, transport equipment and reorientation stands. Moreover, the agent controls the flow of elements manufactured using mentioned resources.
- **Robots agent coordinator** – it is responsible for selection and suitable work of a particular robot. In the structure of the robots coordinator, which plays a supervising part, it could be distinguished n agents responsible for planning the trajectory of an industrial robot (trajectory planner agents).
- **Trajectory planner agent** - its task is to plan the optimal trajectory of a manipulator movement, in a changeable environment of a robot scene, according determined algorithms basing on the information gathered using e.g. a vision system, force sensors as well as RFID (*Radio-frequency identification*) chips. After completion of analysis the returnable information is transferred to a robot, which generates, basing on suitable algorithms, the new trajectory of a manipulator movement.

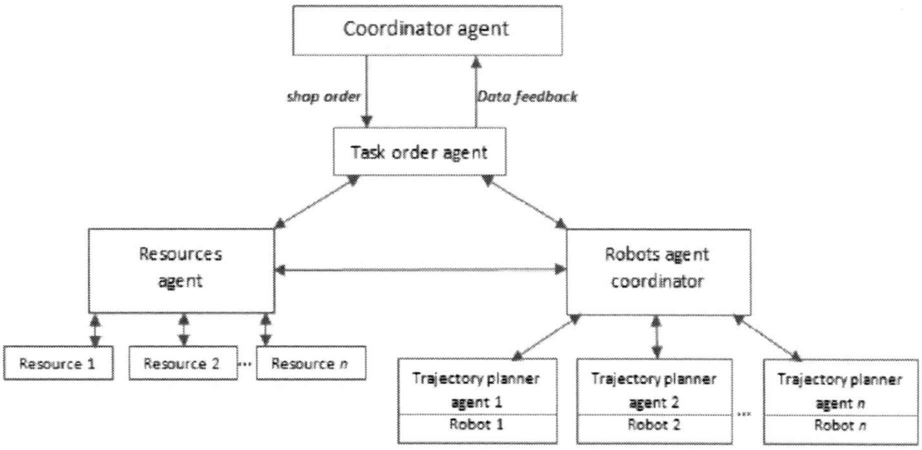

Figure 2. The proposed architecture of the system.

At the strategic level of the proposed system are proposed two main agents responsible for supervising the work of the controlled work cell (figure 3). The task order agent is responsible for dynamical scheduling the production system, in real time, to obtain the highest possible production efficiency. The coordinator agent is responsible for solving resources conflicts, including the time conflicts of tasks realized in the supervised work cell.

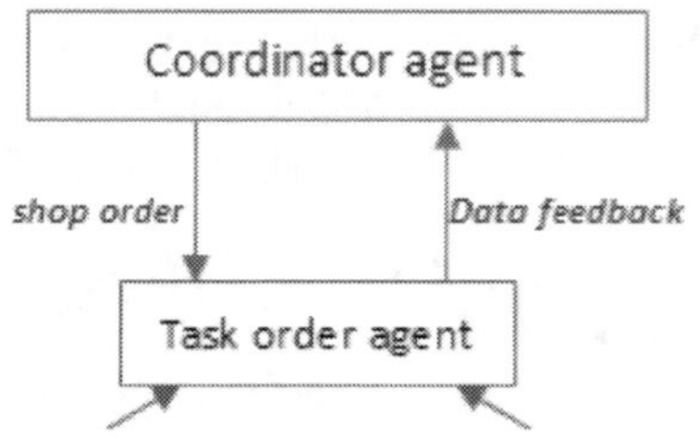

Figure 3. Strategic layer of the proposed system.

At the resources domain the agent structure is organized according the resources structure (figure 4). There are assumed five main resources: physical ones, financial ones, human ones, information ones and time. The most important are agents responsible for physical resources and time. The physical resources agent is supervising all agents responsible for components of a wok cell (machine tools, conveyors, auxiliary devices, industrial robots, etc.).

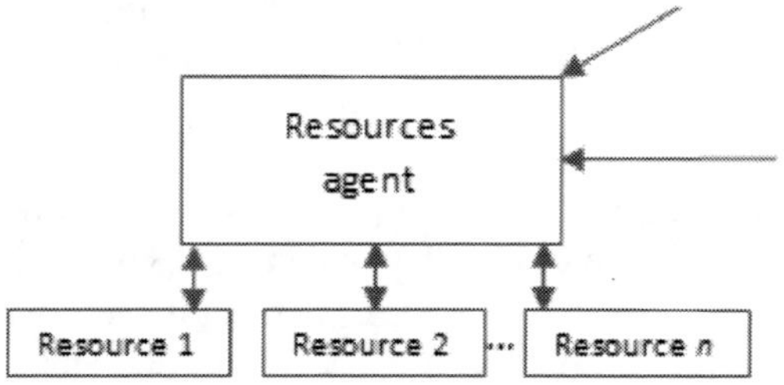

Figure 4. Structure of resources agents.

The schedules and task concerning industrial robots are clustered in the other group of independent agents (figure 5). These agents are responsible for planning the operation of industrial robots in the supervised manufacturing system (work cell) according to master production plan.

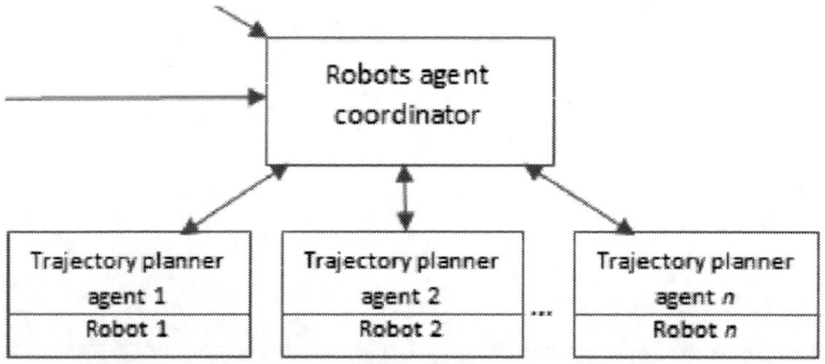

Figure 5. Structure of agents responsible for industrial robots.

Agent, which plans the trajectory, is one of the new proposals in the presented system. It is designated for managing the work cell operations. This agent structure is based on the idea of an intelligent agent [14]. Basing on the inner learning module it could modify its own base of knowledge. In this manner the trajectory planner agent could adapt to changing work conditions, to changing work environments and to changing work requirements. The adaptation process is dependent on the learning module efficiency. The structure of the trajectory planner agent is shown in figure 6.

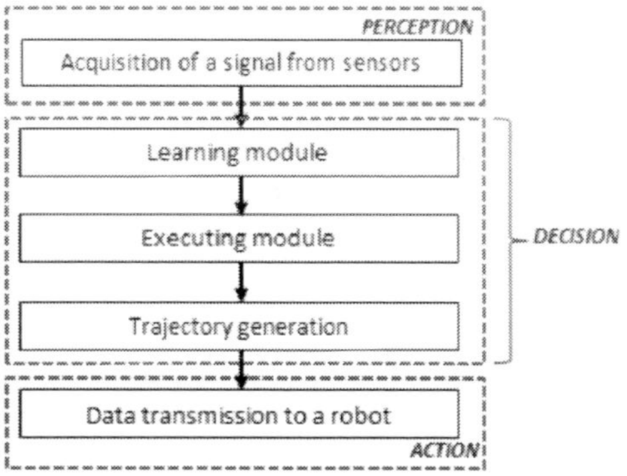

Figure 6. Structure of the trajectory planner agent.

The vision system may consist of one or multiple cameras, cooperating with each other, and could be placed, for example, directly on the robot arm, in dependency on the nature of operations that should be performed in a workcell. The purpose of vision systems is to identify the position and angular orientation of elements which should be e.g. gripped by the robot arm. The use of force sensors allows, in dependency, for identification the force or torque applied to the gripper. Whereas, in relation to RFID chips the information stored on a data carrier, attached to the product or transport pallet, could provide to the system the information about the type and characteristics of the object.

The proposed concept of integration could be used both in relation to one workcell, as well as to several robotized workcells. In the second case, each workcell is treated as a separate agent, so called workcell agent. Between workcell agents is programmable established communication and cooperation. While the agent structure of a single workcell does not change.

3. Conclusions

The solutions integrating an enterprise should ensure the efficient flow of information between all its components. From one side there is the flow of information for manufacturing workcells controlling, while to the other side is transmitted information on the realization of ordered tasks, about material flow, about problems encountered and about the system state. The proposed in the work the possibility of application the multi-agent techniques for supporting the operation of a robotized manufacturing workcell is designed to facilitate the flow of information within the robotized workcell production. The use of agent technologies could also significantly speed up the flow of information within the robotized manufacturing workcells.

The presented concept includes solutions that should be implemented in the prepared, specialized MAS system. One of the requirements is considered with the need for utilization the already existed and licensed software. The second is considered with the need for integration of different data format as it is generated by different computer programs. These requirements should be realized by utilization of common databases that are intended to be linked with particular supervising agents. The existing computer program, using special program interfaces, could exchange data by these databases through mentioned supervising agents. These elements could be determined as the main qualitative advantages of the proposed concept of the MAS architecture.

Acknowledgements
The work is realized within the project titled: "Modular automated production stand with instrumentation for non-invasive confirmation of product quality" funded by The National Centre for Research and Development, agreement No UOD-DEM-1-495/001.

References
[1] Information on http://www.ifr.org/.
[2] Sękala A, Gwizada A and Banaś W 2014 Agent-based systems approach for robotic workcell integration *Advanced Materials Research* **1036** pp 721-725
[3] Banaś W and Sękala A 2014 Concepts of flexible production line, on the example of robotic cell *Advanced Materials Research* **1036** pp 749-754
[4] Gwiazda A, Sękala A and Monica Z 2014 Integrated approach to the designing process of complex technical systems *Advanced Material Research* **1036** pp 1023-1027
[5] Ćwikła G, Sękala A and Woźniak M 2014 The expert system supporting design of the Manufacturing Information Acquisition System (MIAS) for production management, *Advanced Materials Research* **1036** pp 852-857
[6] Gwiazda A 2010 System of designing complex technical means using fuzzy analysis *Applied Mechanics and Materials* **474** pp 147-152
[7] Foit K 2014 Mixed reality as a tool supporting programming of the robot, *Advanced Materials Research* **1036** pp 737-742
[8] Dymarek A and Dzitkowski T 2013 Active synthesis of discrete systems as a tool for reduction vibration *Solid State Phenomena* **198** pp 427-432
[9] Białas K and Buchacz A 2014 The influence of changing the parameters of electrical components implementing the active reduction of vibration *Applied Mechanics and Materials* **657** pp 614-618
[10] Ociepka P and Świder J 2004 Object-oriented system for computer aiding of the machines conceptual design process *Journal of Materials Processing Technology* **157-158** pp 221-227
[11] Kalinowski K, Grabowik C, Paprocka I and Kempa W 2014 Interaction of the decision maker in the process of production scheduling *Advanced Materials Research* **1036** pp 830-833
[12] Ferreira P, Reyes V and Mestre J 2013 A web-based integration procedure for the development of reconfigurable robotic work-cells *Int. Journal of Advanced Robotic Systems* **10** pp 295-304
[13] Jennings B Multi-agent architecture for distributed-in load control and overload protection (http://cordis.europa.eu/infowin/acts/analysys/products/thematic/agents/ch3/mariner.htm)
[14] Wooldridge M and Jennings N R 1995 Intelligent agents: theory and practice *Knowledge Engineering Review* **10/2** pp 115-152
[15] Deen S M (ed.) 2003 *Agent based manufacturing* (Springer Verlag: Berlin)

IOP Conf. Series: Materials Science and Engineering **95** (2015) 012098 doi:10.1088/1757-899X/95/1/012098

Object positioning in storages of robotized workcells using LabVIEW Vision

P Hryniewicz[1], **W Banaś**[1], **A Sękala**[1], **A Gwiazda**[1], **K Foit**[1] **and G Kost**[1]
[1] Silesian University of Technology, Faculty of Mechanical Engineering,
Institute of Engineering Processes Automation and Integrated Manufacturing Systems,
Konarskiego 18A, 44-100 Gliwice, Poland

E-mail: przemyslaw.hryniewicz@polsl.pl

Abstract. During the manufacturing process, each performed task is previously developed and adapted to the conditions and the possibilities of the manufacturing plant. The production process is supervised by a team of specialists because any downtime causes great loss of time and hence financial loss. Sensors used in industry for tracking and supervision various stages of a production process make it much easier to maintain it continuous. One of groups of sensors used in industrial applications are non-contact sensors. This group includes: light barriers, optical sensors, rangefinders, vision systems, and ultrasonic sensors. Through to the rapid development of electronics the vision systems were widespread as the most flexible type of non-contact sensors. These systems consist of cameras, devices for data acquisition, devices for data analysis and specialized software. Vision systems work well as sensors that control the production process itself as well as the sensors that control the product quality level. The LabVIEW program as well as the LabVIEW Vision and LabVIEW Builder represent the application that enables program the informatics system intended to process and product quality control. The paper presents elaborated application for positioning elements in a robotized workcell. Basing on geometric parameters of manipulated object or on the basis of previously developed graphical pattern it is possible to determine the position of particular manipulated elements. This application could work in an automatic mode and in real time cooperating with the robot control system. It allows making the workcell functioning more autonomous.

1. Introduction

The rapid development of electronics and the increasing demand for faster and faster processing the data on surrounding environment significantly accelerated the development of methods for image processing and analysis being a part of vision systems. These systems are already widely used in almost all fields of industry and science. These systems have also been used to create devices used in police and military applications. Modern manufacturing enterprises use different CAE application that are added with artificial intelligence methods [1, 2, 3]. These applications utilize vision systems beginning from delivering and tracking raw materials on production lines [4], through their orientation of robotized manufacturing work cells [5], ending in checking the quality of the products and the protection of the entire production process [6,7,8]. It helps to supervise the production system and scheduling it in real time [9, 10].

Content from this work may be used under the terms of the Creative Commons Attribution 3.0 licence. Any further distribution of this work must maintain attribution to the author(s) and the title of the work, journal citation and DOI.
Published under licence by IOP Publishing Ltd

With the rapid development of microprocessors the computing power increases allowing for faster and more accurate (the use of more precise calculation method) processing and analyzing digital images. Also the cost of vision systems is getting smaller. In the paper is presented a tool for image acquiring, processing and analyzing of National Instruments. It is the Vision Builder environment.
The main parameters to be taken into account using vision systems are:

- Resolution,
- field of view,
- working distance,
- sensor size,
- depth of field,
- image,
- pixel,
- pixel resolution.

Depending on the resolution, it is possible to identify smaller or larger parts of the image (the higher resolution the more accurate measurements [11] and also increasing quantity of received image data to be processed). Field of view is the area seen by the camera. Working distance is the distance between the scene (investigated object) and the lens of the camera. Knowing the field of view, working distance and sensor size, it is possible to determine the focal length. Depth of field is the ability to obtain a sharp image of the analyzed object, and maintain its sharpness throughout the entire process of image analysis and processing. Reliability of vision systems also depends on two elements: light (stronger light does not always mean better image) and the inclination angle of the camera relative to the FOV. The ideal position of the camera is, when it is aligned perpendicular to the plane of the viewing area. Camera inclinations from the angle of 90 degrees could lead to errors associated with the change of perspective.

Lighting is one of the most important factors [12] of the vision systems. With well-matched lighting for carried out image analysis in the vision systems it is possible to obtain more precise results and the analysis itself could be faster. Bad lighting of a scene could lead to erroneous results as well as to measurement-control errors. Insufficient lighting of a scene could lead to blurring the images of a scene, while too strong lighting could introduce distortions associated with the reflection of light from the smooth, drooling elements.

2. Vision Builder application

The first step in programming using the Vision Builder tool from National Instruments is choosing the camera that is connected to the given computer (desktop one or real-time one). After obtaining the test image, checking the camera settings and scene lighting one should convert the image from the camera to a monochrome one (figure 1) to obtain a smaller amount of data to be processed (one pixel of a monochrome image is represented in the binary system as 2^8, which gives 256 points in a grayscale; the color image in the RGB palette contains three color components and therefore the number of processes data is three times larger).

To obtain a monochromatic image one should use the tools Vision Assistant, in which is determined the area of a scene containing the analyzed object, or a full view of the camera. Using the function Extract Color Planes it is possible to get a monochrome image. To sharpen the image and to get details of the analyzed image, it is needed to adjust the contrast and saturation of a scene with light (figure 2). Using the Vision Assistant one should choose the function Brightness to make appropriate adjustments of contrast, brightness and color saturation which, in order to extract the elements being the object of the analysis.

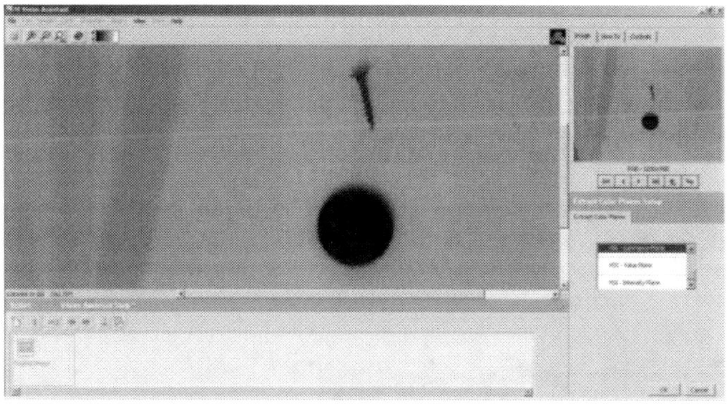

Figure 1. Obtaining of a monochromatic image.

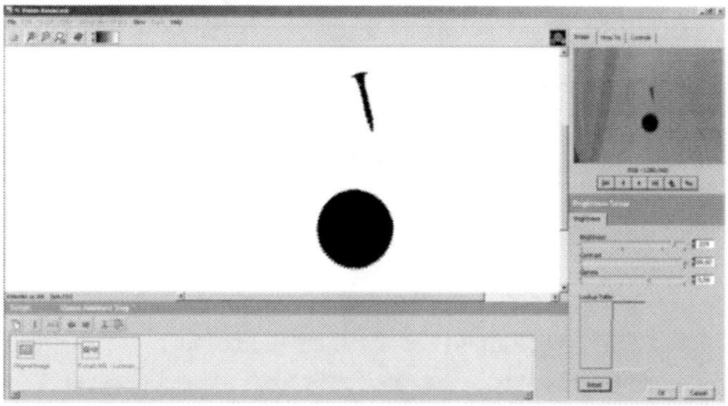

Figure 2. Background contrast setting in relation to analyzed elements.

To obtain better image quality, for further processing, in Vision Assistant should be utilized the tool Table Lookup. Depending on the characteristic of the element analyzed in the work scene one could use one of functions of this tool. In this case it is the function Reverse (figure 3).

After all the operations related to improvement of the displayed image needed for further processing, during the elements identifying process, on could close Vision Assistant [13, 14].The next step, to search the work scene, is to determine matching elements in the form of templates (patterns). These patterns are obtained using the tool Geometric Matching in the Template Editor (figure 4). During a pattern modeling one should remember to mark the possible smallest area of a work scene, which contains the analyzed element, to obtain the possible fast processing of data by the program.

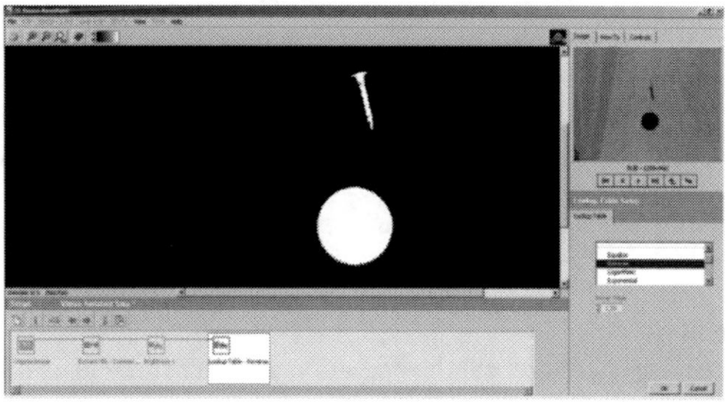

Figure 3. Sequent contrast improving and colors reversing.

IOP Conf. Series: Materials Science and Engineering **95** (2015) 012098 doi:10.1088/1757-899X/95/1/012098

With the functions: Edge Threshold, Edge Filter Size, Minimum Lenght, Row Search Step Size and Column Search Step Size it is possible to set the area which should be seen as a template. The function Draw Regions to Ignore allows omitting places, which deform the template image. After the template setting one should go to the next step, it means to the Template Editor – Specify Match Offset. The function Specify Match Offset allows determining the reference point of the template in relation to the camera view (figure 5). Having determined this function it is possible to find the desired element and determine its location in the working scene.

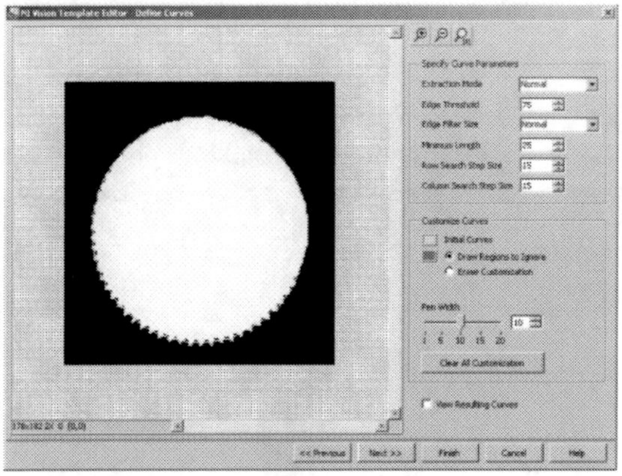

Figure 4. Determining the template basing on the element recorded by the camera.

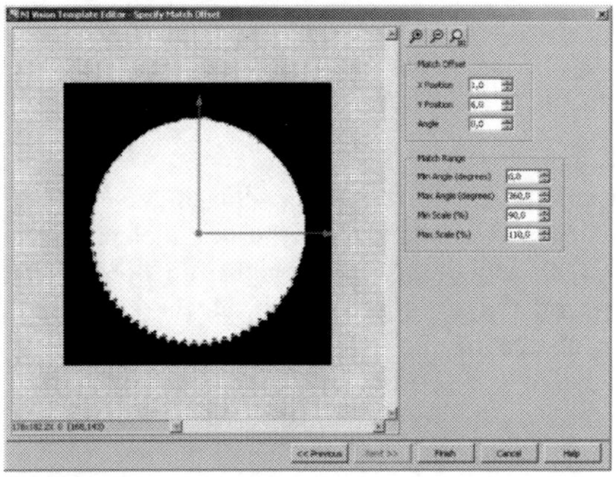

Figure 5. Setting the reference point for the analyzed element.

After closing the Template Editor, it is possible go to the function Geometric Matching Setup, where it is possible to set the parameters characteristic for the analyzed task. In the tab Main, one could set the name of the process of the template searching and the Region of Interest, where this template should be localized. In the tab Template it is possible to improve the template, which was previously determined or store it again. In the tab Curve it is possible to set parameters considering the template detection process, i.e. Extraction Mode, Edge Filter Size, etc.

3. Testing the application

After setting all options one should repeat the operation as many times as is need to store all templates being localized in the analyzed work scene. Is presented an example of the application functioning

[15]. It was used to localize two different templates in the work scene. In figure 6 (upper image) is presented the process of locating the single elements matching to two stored templates. Each of the following elements has its own characteristics by which it is possible to determine the position and the element mapping level in relation to the template.

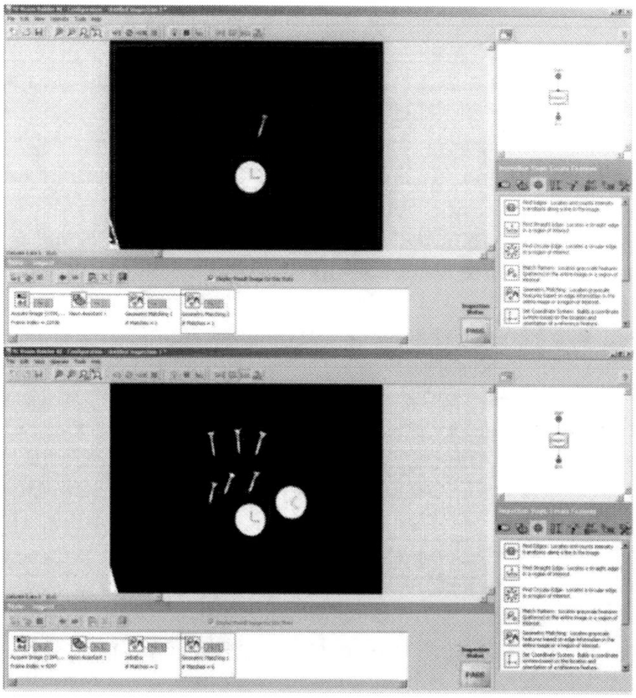

Figure 6. Elements localized in the work scene basing on the previously generated templates.

In figure 6 (lower image) is presented the localities process for the case of higher number of analyzed elements [16, 17]. With the function Scaling, using the base dimensions of template elements, it is possible to determine the distance from the camera to the element and the rotation angle in relation to the template.

4. Conclusions

Vision systems could not yet be used as the only tool for work cell supervision because of the difficulty of data acquiring and processing. Despite this inconvenience the vision systems should be treated as future tool for control systems of businesses projects. Realizing investigations considered with such automatic system one should solve both problems considered with work of particular system components and with data flow within the entire system.

Development of these systems is directed to business systems creation that could control all business production including planning operations or automatic data acquisition. Such integration seems to be the new challenge in this field of systems engineering.

Acknowledgements

The work is realized within the project titled: "Modular automated production stand with instrumentation for non-invasive confirmation of product quality" funded by The National Centre for Research and Development, agreement No UOD-DEM-1-495/001.

References

[1] Gwiazda A, Herbuś K, Kost G and Ociepka P 2015 Designing mechatronics equipment based on the example of the Stewart platform *Sol. St. Ph.* **220-221** pp 419-422
[2] Grabowik C, Ćwikła G and Janik W 2014 The new approach to design features identification *App. Mech. and Mat.* **657** pp 750-754
[3] Dzitkowski T and Dymarek A 2014 Active reduction of identified machine drive system vibrations in the form of multi-stage gear units *Mechanika* **20/2** pp 183-189
[4] Hetmańczyk M P 2015 The Prediction Oriented Analysis of Mechatronic Machine Structures in Terms of the Signal Stream Flow *Solid State Phenomena* **220-221** pp 423-428
[5] Hetmańczyk M 2013 The multilevel prognosis system based on matrices and digraphs methods [in:] Mechatronic systems and materials *Solid State Phenomena* **199** pp 79-84
[6] Ćwiek J, Łabanowski J, Topolska S and Sozańska M 2012 Determination of failure causes of a steam turbine casting *Sol. St. Ph* **183** pp 37-42
[7] Topolska S and Łabanowski J 2015 Duplex stainless steels toughness impact investigations *J. Mat. in Tehn. / J. Mat. and Techn.* **5** pp 114-119
[8] Łabanowski J, Świerczyńska A and Topolska S 2014 Effect of microstructure on mechanical properties and corrosion resistance of 2205 duplex stainless steel *Pol. Mar. Res.* **21/4** pp 108-112
[9] Paprocka I, Kempa W, Kalinowski K and Grabowik C 2014 On Pareto optimal solution for production and maintenance jobs scheduling problem in a job shop and flow shop with an immune algorithm *Adv. Mat. Res.* **1036** pp 875-880
[10] Paprocka I and Kalinowski K 2014 Pareto optimality of production schedules in the stage of populations selection of the MOIA immune algorithm *App. Mech. and Mat.* **657** pp 869-873
[11] Hetmanczyk M P and Michalski P 2014 The self-excitation phenomenon of quasi shielded inductive proximity switches *Adv. Mat. Res.* **837** pp 405-410
[12] Ćwikła G 2014 The methodology of development of the Manufacturing Information Acquisition System (MIAS) for production management *Applied Mechanics and Materials* **474** pp 27-32
[13] Krenczyk D and Olender M 2014 Production planning and control using advanced simulation systems *Int. J. of Mod. Man. Tech.* **VI(2)** pp 38-43
[14] Ćwikła G 2014 Methods of manufacturing data acquisition for production management – a review *Adv. Mat. Res.* **837** pp 618-623
[15] Banaś W and Sękala A. 2014 Concepts of flexible production line, on the example of robotic cell *Adv. Mat. Res.* **1036** pp 749-754
[16] Sękala A, Gwiazda A and Banaś W 2014 Agent-based systems approach for robotic workcell integration *Adv. Mat. Res.* **1036** pp 721-725
[17] Gwiazda A, Sękala A, Monica Z and Banaś W 2014 Integrated approach to the designing process of complex technical systems *Adv. Mat. Res.* **1036** pp 1023-1027

The modular design of robotic workcells in a flexible production line

W Banas[1], A Sekala[1], A Gwiazda[1], K Foit[1], P Hryniewicz and G Kost[1]
[1]Silesian University of Technology, The Faculty of Mechanical Engineering, Institute Of Engineering Processes Automation And Integrated Manufacturing Systems, ul. Konarskiego 18a 44-100 Gliwice, Poland

E-mail: waclaw.banas@polsl.pl

Abstract. In the case of large-scale and mass production lines often the same model of an industrial robot is used in various places of the line and is intended to various task. However, the replacement of one industrial robot to another is a long lasting and arduous process. It requires stopping all the production line and sometimes even dismantling the whole workcell. Such situations are not frequent in production lines that are not flexible. They are related the most often with the failure on an industrial robot. However, during the designing of a flexible production line the ability to replace any robot, which is unrestricted, fast and trouble-free, greatly increase the flexibility level of such line. It could be realized by modular design of the proposed production line. In this way it could be possible to change any elements of such production system. But this approach needs to apply the specialized informatics system.
This paper presents the obtained design of several versions of the same production workcell. Each, succeeding version of the designed production workcell contains more and more modular elements. Thereby it would be presented the evolution of a workcell design beginning from the typical design and ending with the fully modular one. One of tools needed to realize this task is the elaboration of a base of modules and typical joint and mounting elements that could be utilised in the described designing process. It is also presented the guidance information about the designing and programming processes useful at each stage of analysed process.

1. Introduction

Because of years we have robotics and now we cooperate with the company on increasing the flexibility of their cell, it was necessary to present opportunities to increase the flexibility of the cell. To this end, we chose a sample cell and we modelled successive stages of increasing flexibility. Many manufacturers offer elements to increase the flexibility of the cell figure 1. However sometimes industrial robotics development is very slow because of cost a robot. The first step is the application of industrial robots to the most dangerous tasks [1, 2]. In this case, the person is made the pieces together, and a robot welding them together. After working robot leaves the safety position and the man pulls made the item and put the next one [3, 4]. Such work is dangerous and ineffective, although the quality of the weld is better but not much increase efficiency [10, 11].

Content from this work may be used under the terms of the Creative Commons Attribution 3.0 licence. Any further distribution of this work must maintain attribution to the author(s) and the title of the work, journal citation and DOI.
Published under licence by IOP Publishing Ltd

Figure 1. Quick change systems (http://www.schunk.com).

2. Successive stages of increasing flexibility robotics workspace

The next step is to the old workcell, new equipment apply. Sometimes it happens that a robot from another manufacturer is cheaper, and then workcell looks like in figure 2. Such workcell can work without major problems for many years [12]. The cost of maintaining such a workcell is high, because periodic repair, you have to call two different teams. False economy is expensive to maintain, so you should buy works from the same manufacturer [13, 15]

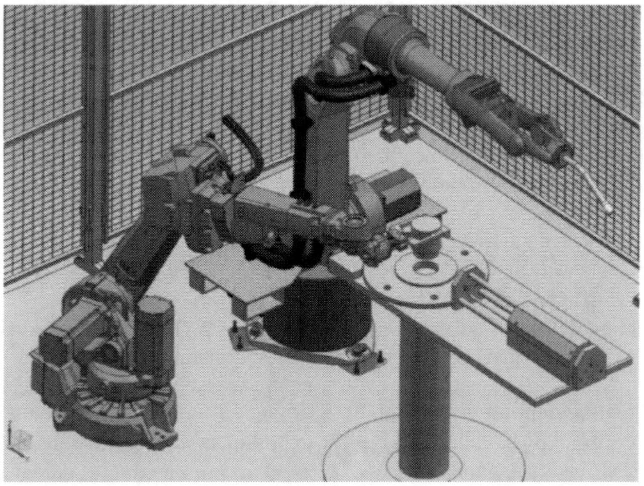

Figure 2. Workcell composed with two different robots.

In this case, it is easier to configure communication between robots and periodic repairs are cheaper. Such workcell is easier to maintain, both robots have the same components and the same software. So it is easier to reprogram the entire workcell. All robots can be loaded onto the same programs and just the right is running. It is easier to train an employee to operate an industrial robot. When welding robot is damaged, you can replace it with a second robot, until repair. If this type of robots are many in our factory, you can have one additional pre-loaded with all programs. This robot robot can replace any downtime will be shorter then, and lose less time and less money.

This workcell not flexible, changing the workpiece in this cell is connected with its complete reconstruction. If, instead of clamps is permanently mounted plinth, we will use the universal table, and clamp mount on it, it is easy to change the clamp and do something else. Production is more flexible in that workcell [5, 6]. This table can be made outside workcell then downtime of the production line is shorter. Note if workcell is busy at the same time may be made to the two elements. In this case, it is necessary reconfiguration workcell that robots do not disturb each other. When one robot is welding element, the second robot is composing the second element. [7, 8, 9].

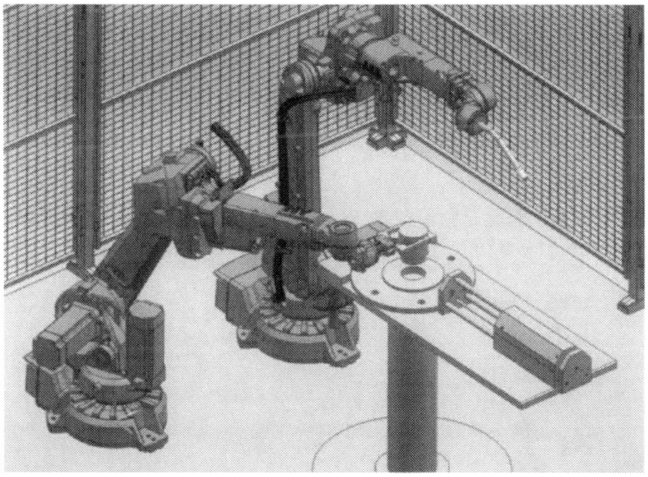

Figure 3. No workcell inflexible comprised of two identical robots.

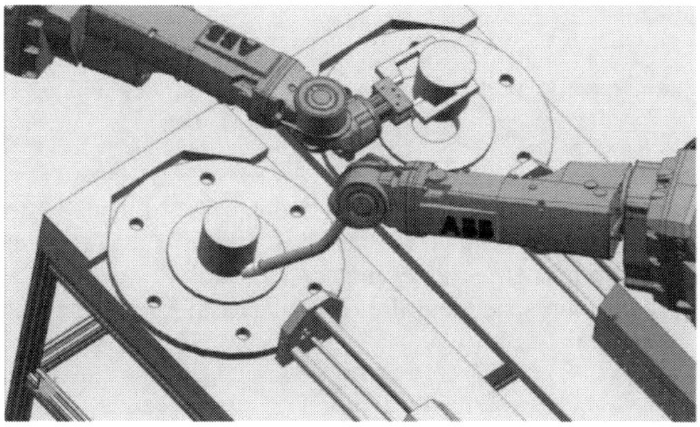

Figure 4. Two robots working in the same time

Replacement of a specialized clamp, the constructed specifically for the element and workcell, the modular unit that can be easily modified increases the flexibility of such a cell. Introducing modularity remember the exact positioning of the robot. If the new table is not fixed to the ground at exactly the same place as the old one, set the frame again. You can help with a special tool but also on the table should be points for easy setting frames. The robot operator should know how to set the frame.
If the developer builds a virtual cell should be strictly defined frames, each fixed elements of the cell. If he draws an existing workcell to reprogram robots must know the robot frame.

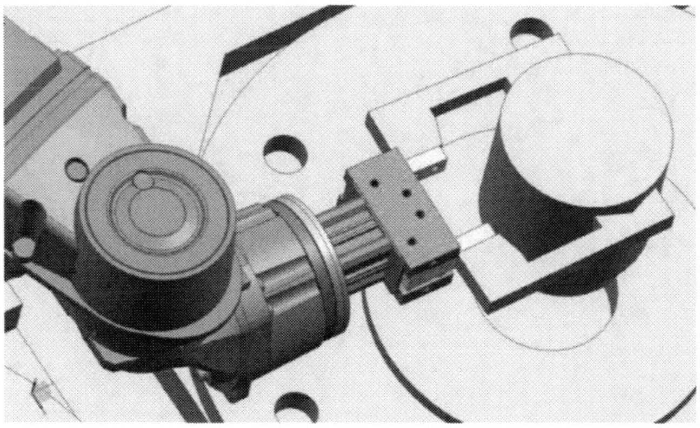

Figure 5. Robot with adapter

When made in workcell element changes need to replace the tool fixed to the robot wrist. For easy tool change and to be able to use the tool for different types of robots it is necessary to make the adapter. This adapter will not only facilitate the exchange of tools, but also made properly extend life of the robot wrist and tools.

Such an adapter is shown in figure 5 is composed of two parts, one fixed permanently to the robot wrist and the other attached to the tool. Both parts are typically connected by screws. Use an adapter also makes it easy assembly and disassembly tool.

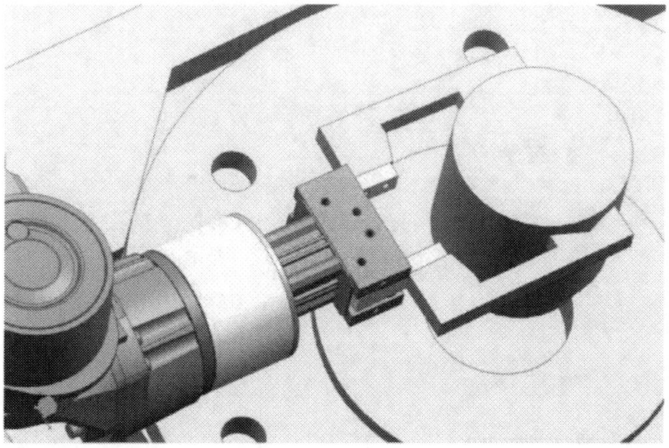

Figure 6. Robot with quick change systems

Flexible workcell is also flexible update tool attached to the robot wrist. Many manufacturers figure 1 produce such a quick change systems. For example, as shown in figure 6 quick change systems Schunk. But keep in mind that you need to make the robot grippers parts store.
Each gripper used in workcell should have your own space and the robot needs to know where the gripper is.

3. A fully modular cell
The last method presented in this article is a division workcell on modules. Currently, robot and all non-moving parts workcell are mounted directly to the ground. When we want to move a robot or another part of the attachment we have to destroy the old and make new ones. It is a work of long-term and very dangerous. After a few such moving on our floor can be completely destroyed and we can not fix properly robot in a new place.

Use of special pallets attached to them, robots or parts is a good idea. If we can build workcell as puzzle taking ready modules [13, 14]. They can be freely configured by building his own cell. This facilitates both workcell design and its implementation. Modules can be prepared on the outside and the time and cost of installation will be incomparably lower.

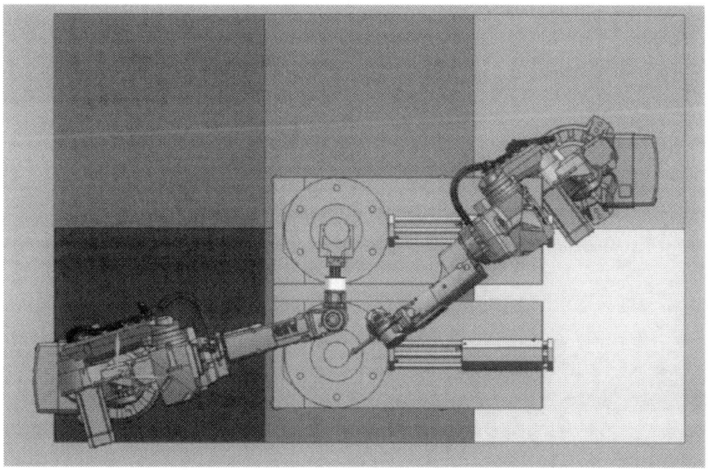

Figure 7. A fully modular cell.

4. Conclusions

Only factories producing the same item without any modification for many years may have not flexible production lines. Development of the forces of production line flexibility and forces you to increase the range of products. It is important that cell was less downtime as the most, and it was easy to reconfiguring the variable production.

Previously designed robotic production lines are composed of individually designed production cells. It happens that each robotic cell is designed by another person. At the end, the entire robotic cell is submitted as non-matching blocks by another person. Often supervisor must match the adjacent elements robotic cell.

This method causes each individual cell is not reproducible. The components are designed specifically for the cell and cannot be used anywhere else. If it was possible to unify even some elements of the cell would have reduced costs and easier repair. Using the same model in several cell robot also reduce cell maintenance costs.

If the design stage in a few places will use those same elements in several places reduce the cost of generation and number of spare parts. In addition, parts are designed more flexibly which increases the flexibility of the cell.

Acknowledgements

The work is realized within the project titled: "Modular automated production stand with instrumentation for non-invasive confirmation of product quality" funded by The National Centre for Research and Development, agreement No UOD-DEM-1-495.

References

[1] Hetmanczyk M P 2015 The prediction oriented analysis of mechatronic machine structuresin terms of the signal stream flow *Solid State Phenomena*, **220-221** pp 423-428

[2] Hetmanczyk M P and Michalski P 2014 The self-excitation phenomenon of quasi shielded inductive proximity_switches *Advanced Materials Research* **837** pp 405-410

[3] Herbuś K, Kost G, Reclik D and Świder J 2014 Integration of a virtual 3D model of a robot_manipulator_with its tangible model (phantom) *Advanced_Materials Research* **837** pp 582-587

[4] Gwiazda A, Herbuś K, Kost G and Ociepka P 2015 Designing mechatronics equipment based on_the_example of the Stewart platform *Solid State Phenomena* **220/221** pp 419-422

[5] Gwiazda A, Herbuś K, Kost G and Ociepka P 2015 Motion analysis of mechatronic equipment considering the example of the Stewart platform *Solid State Phenomena* **220/221** pp 479-484

[6] Ociepka P, Herbuś K and Gwiazda A 2014 Application of the CBR method for adding the design process of module manipulators *Advanced Materials Research* **1036** pp 1011-1016
[7] Herbuś K, Ociepka P and Gwiazda A 2014 Application of functional features to the description of technical means conception *Advanced Materials Research* **1036** pp 1001-1004
[8] Herbuś K, Ociepka P and Gwiazda A 2014 Conception of the integration of the virtual robot model with the control system *Advanced Materials Research* **1036** pp 732-736
[9] Ociepka P, Herbuś K and Gwiazda A 2014 Application of the method basing on engineering knowledge and experience for adding the hexapod design process *Advanced Materials Research* **1036** pp 1005-1010
[10] Paprocka W, Kempa M, Kalinowski K and Grabowik C 2014 A production scheduling model with maintenance *Advanced Materials Research* **1036** pp 85-890.
[11] Paprocka W, Kempa M, Kalinowski K and GrabowikC 2014 On Pareto Optimal Solution for Production and Maintenance Jobs Scheduling Problem in a Job Shop and Flow Shop with an Immune Algorithm *Advanced Materials Research* **1036** pp 875-880
[12] Paprocka W, Kempa M ,Grabowik C and Kalinowski K 2014 Sensitivity Analysis of Predictive Scheduling Algorithms *Advanced Materials Research* **1036** pp 921-926
[13] Paprocka W and Kalinowski K 2014 Pareto Optimality of Production Schedules in the Stage of Populations Selection of the MOIA Immune Algorithm *Applied Mechanics and Materials* **657** pp 869-873
[14] Sękala A, Banaś W and Gwiazda A 2014 Agent-based systems approach for robotic workcell integration *Advanced Materials Research* **1036** pp. 721-725
[15] Sękala A and Dobrzańska-Danikiewicz A 2015 Possibilities of application of agent-based systems to support functioning of e-manufacturing environment *Mechatronics systems and materials VI. [Ed:] A.V. Valiulis. Solid State Phenomena* **220/221** pp 781-784

IOP Conf. Series: Materials Science and Engineering **95** (2015) 012100 doi:10.1088/1757-899X/95/1/012100

Application of CBR method for adding the process of cutting tools and parameters selection

P Ociepka[1] **and K Herbus**[1]
[1]Silesian University of Technology, The Faculty Of Mechanical Engineering, Institute Of Engineering Processes Automation And Integrated Manufacturing Systems, ul. Konarskiego 18a 44-100 Gliwice, Poland

E-mail: piotr.ociepka@polsl.pl

Abstract. Modern enterprises must face with the dynamically changing market demand what influences the designing process. It is possible by linking computer tools with information gathered by experienced designers teams. The article describes the method basing on engineering knowledge and experience to adding the process of tools selection and cutting parameters determination for a turning operation. The method, proposed by the authors, is based on the CBR (*Case Based Reasoning*) method. CBR is a method of problem solving that involves searching for an analogy (similarity) between the current task to be solved, and the earlier cases that properly described, are stored in a computer memory. This article presents an algorithm and a formalized description of the developed method. It was discussed the range of its utilization, as well as it was illustrated the method of its functioning on the example of the tools and cutting parameters selection with respect to the turning process.

1. Introduction

Market trends related to the diverse and rapidly changing demand, forcing producers to significantly accelerate the design and manufacture of new products. Time for the realization of designing, constructing and production preparation processes is getting shorter. This determines the need for powerful informatics tools that will be able to significantly speed up these processes while maintaining profitability and the proper quality and durability of the product.

In the process of designing and constructing are applied advanced and specialized tools of the CAD/CAE class, which significantly accelerate these processes [1-10]. In the process of production preparation and in the manufacturing process are used flexible manufacturing systems which, in combination with CNC machine tools and computer programs of the CAD/CAM class, enable quick and efficient adaptation to new products.

The important step in the preparation process of the products manufacturing technology is the selection of appropriate tools and machining parameters, in relation to the realized technological tasks. In the case of a machining process the selection of tools and suitable cutting parameters is particularly important and greatly affects the quality and efficiency of the realized process. In the case of a mass production the properly selected tools and cutting parameters often determine the profitability of a conducted task. The process of selection of tools and cutting parameters requires from technologist having vast knowledge and experience concerning the machining process. Engineering experience is particularly important in this process, which is the result of practical utilization of possessed

Content from this work may be used under the terms of the Creative Commons Attribution 3.0 licence. Any further distribution of this work must maintain attribution to the author(s) and the title of the work, journal citation and DOI.
Published under licence by IOP Publishing Ltd

knowledge. So it is reasonable to search for effective computer methods adding storage and reuse of the acquired experience.

2. CBR method

Case Based Reasoning (CBR) is a method of problem solving basing on experience. It consists of looking for an analogy (similarity) between the currently discussed task and previously elaborated tasks, which have been recorded in the "base of cases" of the CBR system. The software that aids problems solving basing on this method should be equipped with the mechanism for quick databases searching and efficient methods for determining the similarity between the current task, and cases stored in the base. The main advantage of the CBR method is the improvement of the process of knowledge and experience acquiring [11, 12, 13]. This process runs in parallel with the solution being created, when the knowledge about this solution is the most complete (e.g. at the moment of task realization or completion). The next advantage of this method is fact, that the new solution (new case) is immediately available during solving the following tasks and could be applied or adapted to solve them. There is here a very simple, and yet effective learning process, what at continuous use of such a system gives a possibility to gather a very large stock of knowledge and experience in a certain field.

The knowledge and experience, stored in the base of the CBR system, could be also applied to the other type of an inference module, e.g. in an advisory system. Another advantage of this method is the ability to store negative cases, i.e. solutions that have been proven to be incorrect. This lets to avoid making the same mistakes in the future. The process of problems solving using the CBR method could be presented as a cycle of operations, which in the literature is called the loop of four R (Retrieval, Reuse, Revision, Retainment) [14,15,16,17] (figure 1).

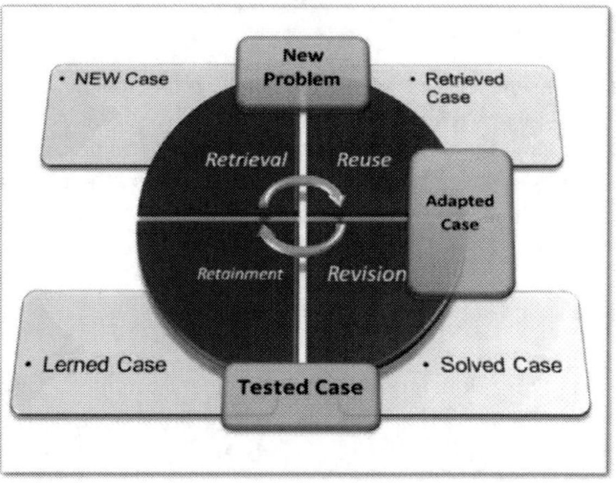

Figure 1. R^4 model of the CBR cycle.

3. Formalized description of the elaborated method

The computer system, created on the basis of the proposed method, works as follows. Given is the description of the task, *T_case*, in which a user specifies the attributes and weights describing the technological task.

$$T\_case = \{ WA_1, w_1, WA_2, w_2, \ .... \ WA_j, w_j \}. \qquad (1)$$

where: WA_j – the value of the j – th attribute describing the technological task,
w_j – the weight of the j – th attribute describing the technological task.

In the CBR database is stored the set of cases $Case_s$, which includes technological solutions (used tools and machining parameters) implemented in the past.

$$Cases = \{ case^1 [T\_case_1, T_1, P_1, H_1, M_1], \ldots case^i [T\_case_i, T_i, P_i, H_i, M_i] \}. \quad (2)$$

where: $case^i [T\_case_1, T_1, P_1, H_1, M_1]$ – the *i*- th technological case; T_case_i – description of the *i*-th technological case; T_i – the tools description for the *i*- th technological case; P_i – parameters description for the *i*- th technological case; H_i – fixing method description for the *i*- th technological case; M_i – machine tool description for the *i*- th technological case.

Table 1. Register of attributes and exemplar values in relations to the data sets T_1, P_1, H_1, M_1.

Data set	Attribute	Exemplar values
	Tip form	R, S, C, D, V
	Tip size	R (06), S(12), C(25)
	Nose radius	04, 08, 1.2, 1.6
	Clearance angle Kr	75°, 90 °, 135 °
	Type	Positive, negative
T_i	For the type of a material	P, M, K, N, S, H
	Tool body type	A90 °, B75 °, D45°
	Type of tip fixing	C, D, M, W, P, S
	Tip version	R, N, L
	Tip marking	SNMG 120408-R2
	Cutting speed	V_c=…
	Cutting depth	a_p=…
P_i	Rotational speed	N =…
	Feed	f_n=…
H_i	Fixing type	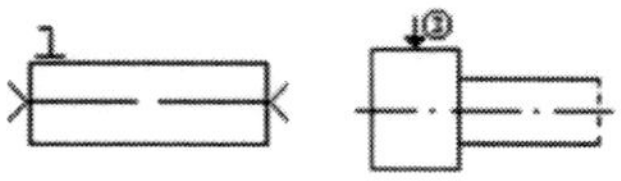
	Machine tool type	CNC lathe
M_i	Machine tool model	AVIAturn 35SM
	Individual machine tool	AVIAturn 35SM – WT1

Table 2. Attributes and exemplar values with respect to the *T_case* task description.

Data set	Attribute	Exemplar values
	Machining type	(R, M, F)
	Feed type	transverse, longitudinal, mixed
	Operation type	(outer/inner)
	Outer diameter	(min/-/max)
	Inner diameter	(min/-/max)
T_case	Allowance value	(min/max)
	Material	P, M, K, N, S, H
	Surface roughness	Ra=1.25
	Accuracy class	IT 7
	Pass	R, L, BB_L, BB_R
	Production	(piece, series, mass)
	Goal function	Output, costs

On the basis of values of particular attributes and weights, recorded in the description *T_case*, the CBR calculation mechanism, according to the formulas (3) and (4), determines the degree of similarity between the task description *T_case*, and cases recorded in the database [11,15].

$$Sim\,(T\_case,\ case^i)) = 1 - Dist\,(T\_case,\ case^i). \tag{3}$$

$$Dist\,(T\_case\ ,\ case^i) = \left(\frac{1}{k}\cdot\sum_{j=1}^{k} w_j^2 \cdot \left[T\_case_j - case_j^i\right]^2\right)^{\frac{1}{2}} \tag{4}$$

where: $case^i_j$ – the value of the *j*- th attribute in the *i*- th case; *k* – the number of corresponding themselves attributes; w_j – the weight coefficient of the *j*- th attribute in the analyzed case.

Then is generated the set of cases with the desired degree of similarity $Cases^s$.

$$Cases^s=\{case^1\ [T\_case_1,\ T_1,\ P_1,\ H_1,\ M_1],....\ case^k\ [T\_case_k,\ T_k,\ P_k,\ H_k,\ M_k]\}. \tag{5}$$

In the next step, the system sorts the selected solutions according to the degree of similarity and creates the ordered set of cases $Cases^s_u$.

$$Cases^s_u = <\ case^1\ [T\_case_1,\ T_1,\ P_1,\ H_1,\ M_1],....\ case^k\ [T\_case_k,\ T_k,\ P_k,\ H_k,\ M_k]\ >. \tag{6}$$

The ordered set $Cases^s_u$ = <....> is the set of solutions concerning the currently analyzed technological task.

4. Functioning of the elaborated method and its integration with the CAM system

In the figure 2 is shown the algorithm of the system basing on the proposed method concerning the manufacturing process. In the first step the manufacturing process (a set of technological operations) is decomposed into the so-called elementary technological tasks, for which must be selected cutting tools and parameters. Next, for particular "elementary technological tasks" is created the task description *T_case*. In the next step, using the elaborated system, analogous cases realized in the past are looked for. On their basis the tools and machining parameters, for the current task, are selected. Depending on the similarity degree it is possible to adapt these parameters to the currently concerning task.

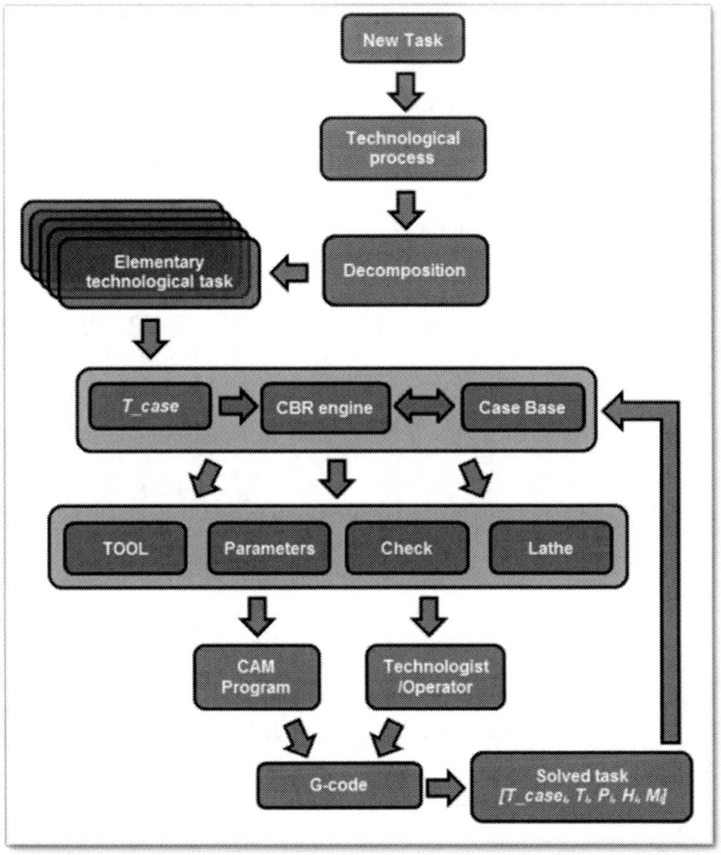

Figure 2. Algorithm of the system functioning.

In the next step, the data are entered into the CAM system or they could be directly applied by machine tool operator. In the CAM program is generated the code on a CNC machine tool. The realized technological process is the new case, which is stored in the CBR base. Information about tools and cutting parameters used in this process could be reused in following, analogous cases in the future.

5. Conclusions

In the paper is discussed the possibility of using the artificial intelligence method for adding the process of cutting tools and parameters selection. Application of the elaborated method greatly simplifies and speeds up works concerning the manufacturing preparation of a product. It also facilitates quick gaining of knowledge and experience by young technologists. It eliminates errors resulting from the lack of experience of a technologist in the case of elaborating new technological processes for new products.

Acknowledgements
The work is realized within the project titled: "Modular automated production stand with instrumentation for non-invasive confirmation of product quality" funded by The National Centre for Research and Development, agreement No UOD-DEM-1-495/001.

References
[1] Banaś W, Herbuś K, Kost G, Nierychlok A, Ociepka P and Reclik D 2014 Simulation of the Stewart platform carried out using the Siemens NX and NI LabVIEW programs *Advanced Materials Research* **837** pp 537-542
[2] Ociepka P, Banaś W, Herbuś K and Kost G 2014 Simulator of the Car for Driving Courses for the People with Mobility Impairments *Advanced Materials Research* **1036** pp 817-822
[3] Herbuś K, Ociepka P and Gwiazda A 2014 Application of functional features to the description of technical means conception *Advanced Materials Research* **1036** pp 1001-1004
[4] Herbuś K, Ociepka P and Gwiazda A 2014 Conception of the integration of the virtual robot model with the control system *Advanced Materials Research* **1036** pp 732-736
[5] Gwiazda A, Herbuś K, Kost G and Ociepka P 2015 Designing mechatronics equipment based on the example of the Stewart platform *Solid State Phenomena* **220/221** pp 419-422
[6] Gwiazda A 2014 System of designing complex technical means using fuzzy analysis *App. Mech. and Mat.* **474** pp 147-152
[7] Gwiazda A, Herbuś K, Kost G and Ociepka P 2015 Motion analysis of mechatronic equipment considering the example of the Stewart platform *Solid State Phenomena* **220/221** pp 479-484
[8] Banaś W, Gwiazda A, Herbuś K, Kost G, Ociepka P and Reclik D 2015 Analysis of the dynamic properties of the mechatronic integrator of control procedures of the vehicle driven by persons with disabilities *Solid State Phenomena* **220/221** pp 3-8
[9] Gwiazda A, Sękala A, Monica Z and Banaś W 2014 Integrated approach to the designing process of complex technical systems *Advanced Materials Research* **1036** pp 1023-1027
[10] Gwiazda A 2014 Design of the roof support with strait-line mechanism *Advanced Materials Research* **1036** pp 553-558
[11] Ociepka P and Świder J 2004 Object-oriented system for computer aiding of the machines conceptual design process *Journal of Materials Processing Technology* **157-158** pp 221-227
[12] Ociepka P, Herbuś K and Gwiazda A 2014 Application of the CBR method for adding the design process of module manipulators *Advanced Materials Research* **1036** pp 1011-1016
[13] Ociepka P, Herbuś K and Gwiazda A 2014 Application of the method basing on engineering knowledge and experience for adding the hexapod design process *Advanced Materials Research* **1036** pp 1005-1010
[14] Ociepka P and Grabowik C 2008 Zastosowanie metody CBR do wspomagania procesu projektowo-konstrukcyjnego i przygotowania technologii, *Mechanik* **8-9** pp 728-733 (in Polish)
[15] Ociepka P and Świder J 2007 Współczesne systemy doradcze do wspomagania projektowania maszyn, *Monografia* (Gliwice: Wydawnictwo Politechniki Śląskiej) (in Polish)
[16] Świder J, Baier A, Ociepka P and Herbuś K (2005) Zastosowanie metod obiektowych w procesie projektowo – konstrukcyjno – wytwórczym, Inżynieria Maszyn, Komputerowe wspomaganie w inżynierii maszyn i materiałów, *Rok* **10**, Zeszyt 1 – 2 pp 61-83 (in Polish)

IOP Conf. Series: Materials Science and Engineering **95** (2015) 012101 doi:10.1088/1757-899X/95/1/012101

Strength analysis of parallel robot components in PLM Siemens NX 8.5 program

P Ociepka[1] **and K Herbus**[1]
[1]Silesian University of Technology, The Faculty Of Mechanical Engineering, Institute Of Engineering Processes Automation And Integrated Manufacturing Systems, ul. Konarskiego 18a 44-100 Gliwice, Poland

E-mail: piotr.ociepka@polsl.pl

Abstract. This article presents a series of numerical analyses in order to identify the states of stress in elements, which arise during the operation of the mechanism. The object of the research was parallel robot, which is the basis for the prototype of a driving simulator. To conduct the dynamic analysis was used the *Motion Simulation* module and the *RecurDyn* solver. In this module were created the joints which occur in the mechanism of a parallel robot. Next dynamic analyzes were performed to determine the maximal forces that will applied to the analyzed elements. It was also analyzed the platform motion during the simulation a collision of a car with a wall. In the next step, basing on the results obtained in the dynamic analysis, were performed the strength analyzes in the *Advanced Simulation* module. For calculation the *NX Nastran* solver was used.

1. Introduction

The dynamically changing demand on a market forces manufacturers to significantly accelerate designing and production of new products. It results that the engineers, who implement these processes, must use powerful informatics tools that are able to significantly speed up these processes while maintaining profitability, the proper quality and durability of the product. Modern programs of the CAD/CAE class [1-10] have great potential for carrying out advanced stress analysis, fatigue analysis or kinematic and dynamic ones. They allow simulating and analyzing many phenomena which occur during the operation of the mechanism. This makes possible to carry out a number of tests regarding the behavior of the virtual object in the virtual environment. Due to these analyses, it is possible to quickly perform a preliminary study of the considered object, what in turn allows limiting the scope, time and costs of researches conducted on real objects.

In the presented article is described the series of strength tests carried out for elements of the parallel robot, which is the basis of a driving simulator, created for people with disabilities (figure 1) [11,12,13,14]. These analyzes were performed in the NX 8.5 program. The aim of the analysis was to determine the maximal stresses and displacements that occur in elements of the parallel robot during operation of the simulator. For this purpose, during realization the dynamic analysis it was conducted, in the virtual model of the platform, the simulation of collision of a car with a wall. The simulation of platform operation was carried out in the *Motion Simulation* module. In the next step, basing on the results of the dynamic analysis, were performed strength analyzes in the *Advanced Simulation* module.

Content from this work may be used under the terms of the Creative Commons Attribution 3.0 licence. Any further distribution of this work must maintain attribution to the author(s) and the title of the work, journal citation and DOI.
Published under licence by IOP Publishing Ltd

Figure 1. Photo of the car simulator.

2. Identification of the trajectory and motion parameters of the parallel robot

In order to determine the maximal loads, which occur in elements of the parallel robot, was carried out the identification of the trajectory and motion parameters of the platform during the simulation of a car colliding with a wall. It is the most dynamic event that is simulated in the simulator. Therefore, the load of platform elements is then the highest. To determine the trajectory of the platform motion was used the fast camera Mega speed HHC x4. Two films were made which record, in two planes, the behavior of the selected points of the platform in the simulator space.

Processing and analysis of recorded films allowed establishing the approximate motion trajectory of the platform. In order to determine the motion parameters of the platform, during collision (velocity and acceleration), the measurements were realized using the Xsens Mtw Kit6 sensors. The analysis of recorded results allowed identifying the velocity and acceleration of the platform frame during the recorded movement. In figure 2 is shown the process of identification the trajectory and motion parameters of the simulator during the collision simulation.

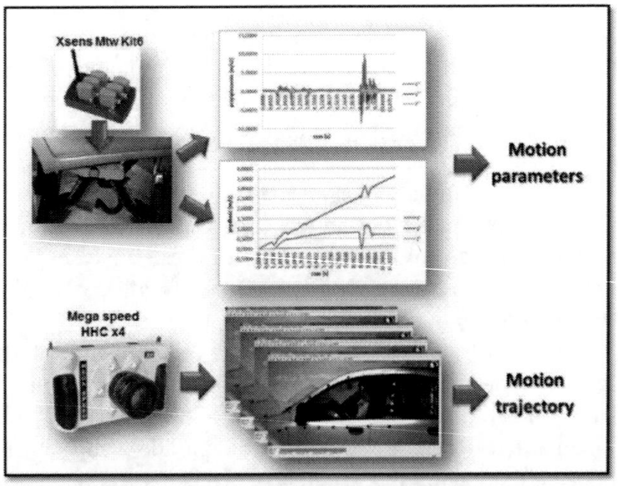

Figure 2. Identification of the motion parameters and trajectory during collision simulation.

The data, obtained in this way, were the basis for the dynamic analysis considered to the whole tested parallel robot.

IOP Conf. Series: Materials Science and Engineering **95** (2015) 012101 doi:10.1088/1757-899X/95/1/012101

3. Dynamic analysis in the *Motion Simulation* module

In order to conduct the dynamic analysis it was created the solid model of the parallel robot in the environment of the NX 8.5 software. In the *Modeling* module were created particular elements of the platform, and in the *Assembly* module was created the assembly of the model. The dynamic analysis was carried out in the *Motion Simulation* module. For calculation, were used the RecurDyn solver. The first step in preparing the model for the dynamic analysis was identifying the main components (links) involved in the conducted simulation of platform movement. It should be noted that one component may consists of many structural elements of the platform. The condition is that the elements, grouped in a single component (left), did not change the relative positions to each other during the simulation of platform movement. The prepared model of the whole system of the platform consists of seven components (links):

- **Base** - which consists of a foundation and attached to it bases with eyes to which the platform is attached;
- **Cross** - the lower joint component, which allows for the proper connection of the parallel robot leg with the base;
- **Engine** - which contains casing of an engine with all elements of a frame that allow for correct fitting of the engine;
- **Screw** - which consists of a screw with a frame and an end of a ball-and-socket joint;
- **Frame** - which consists of a frame, to which is fixed a car body and 6 slots of ball-and-socket joints;
- **Dr_sht** - a drive shaft, which ensures the transfer of power, from the engine to the nut of the screw, by means of a transmissions with a toothed belt;
- **Nut** - nut that allows transferring a torque on the screw of a screw transmission.

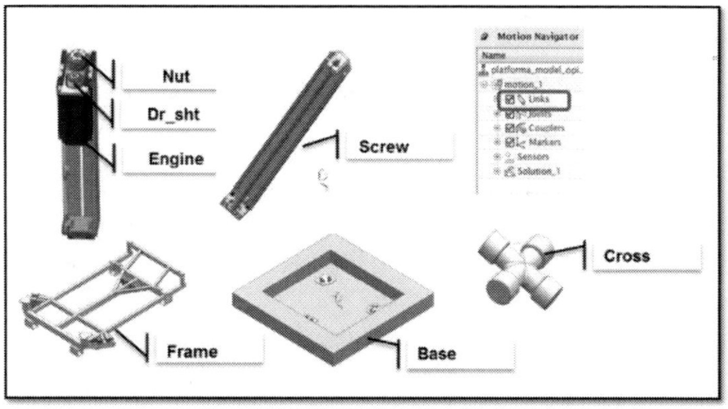

Figure 3. Components created to conduct the dynamic analysis.

The next step of the model developing was determination of joints (figure 4) between the particular components of the model to reconstruct the operation of the real object with the virtual model. For the implementation of the motion simulations the following types of joints were created [11,12]:

- ***Base_fix*** - joint of the *fixed* type that removes six degrees of freedom of the **Base** component,
- ***Ki_Base*** - joint of the *revolute joint* type allowing only for rotation of the **Cross** component around the chosen axis,
- ***Ki_Mi*** - joint of the *revolute joint* type allowing only for rotation of the **Engine** component around the chosen axis,
- ***Mi_Wni*** - joint of the *revolute joint* type allowing only for rotation of the **Dr_Sht** component around the chosen axis,
- ***Mi_Ni*** - joint of the *revolute joint* type allowing only for rotation of the **Nut** component around the chosen axis,

- ***Screw_i*** - joint of the *screw* type creating a screw transmission between the components **Nut** and **Screw**,
- ***Sfr_i*** - joint of the *spherical type* determining the possible motion between the components **Screw** and **Frame**,
- ***Gear_i*** - joint of the *gear* type determining the method of motion transmission from the **Dr_sht** component on the **Nut** component,
- ***Slider_i*** - joint of the *slider* type determining linear motion between trucks mounted on the **Nut** component and guides mounted on the **Screw** component.

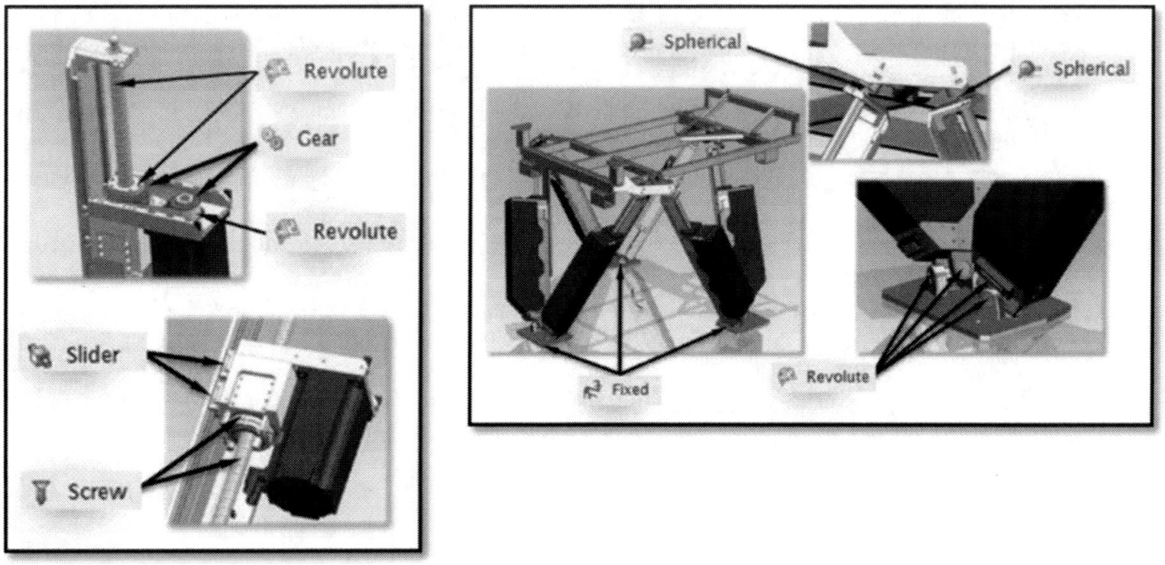

Figure 4. Joints needed for obtaining the proper operation of the model.

In the next step were determined the motion characteristic reflecting the movement trajectory and motion parameters identified during the measurements of the real object. The dynamic analysis of the platform movement was carried out in order to determine the maximal forces that are applied to the components of the platform (forces actions in particular joints).

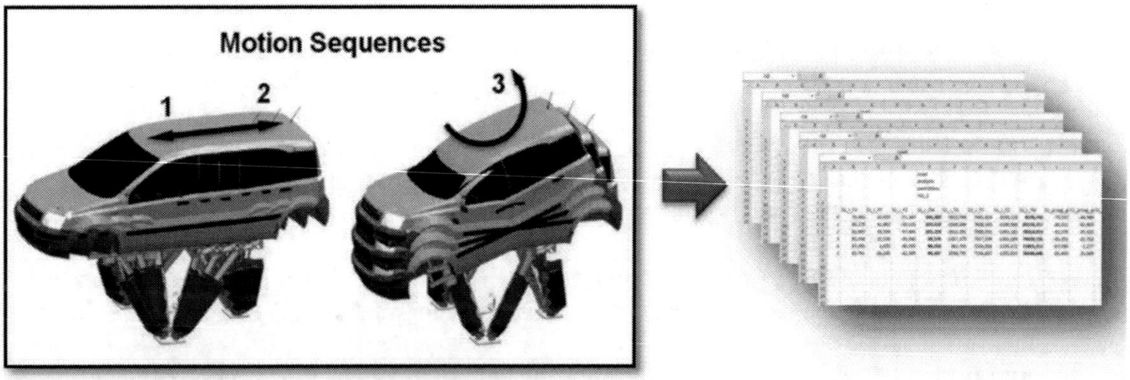

Figure 5. Results of the dynamic analysis.

In order to analyze the obtained results they were exported to the Microsoft Excel program, where they were processed. First was selected the leg of the platform, in relation to which the loads, during the simulation, were the highest. And then was selected this step of the dynamic analysis during which was observed the maximal loads of the analyzed components of the platform. In the figure 5 are shown the simulation results of the analyzed movement of the platform. The results, obtained in this way, were then the basis for conducting the strength analysis with the FEM method.

4. Strength analysis in the *Advanced Simulation* module

The last stage of the realized research cycle was carrying out the strength analysis with the FEM method. This analysis was conducted in the *Advanced Simulation* module using the NX Nastran solver. It was analyzed this leg of the parallel robot in which occurred the highest loads. The model discretization, into finite elements model (TETRA 10) was done. It was defined material properties, introduced contacts between mating elements and defined appropriate joints (geometric constraints) in the model. The next step was to download the results of the dynamic analysis, in the form of forces acting on elements of the platform leg, and activating them in the FEM model. The program automatically introduced these forces (values, direction and senses) in the place where constraints were defined during realized dynamic analysis. Using such prepared FEM model were performed static strength analyses. The obtained results, in the form of contours of stresses and displacements, are shown in figure 6. It was found that the most enforced element is the ball-and-socket joint connecting the platform leg with the frame, to which is attached the vehicle body. The stresses occurring there (about 140 MPa) are not high what indicates that it is possible to increase the value of generated accelerations during the collision simulation. Further analyses showed that increasing the acceleration value of three times do not cause the risk of the joint damage. However, the generation of such acceleration during the motion simulation could be uncomfortable and dangerous for a simulator user.

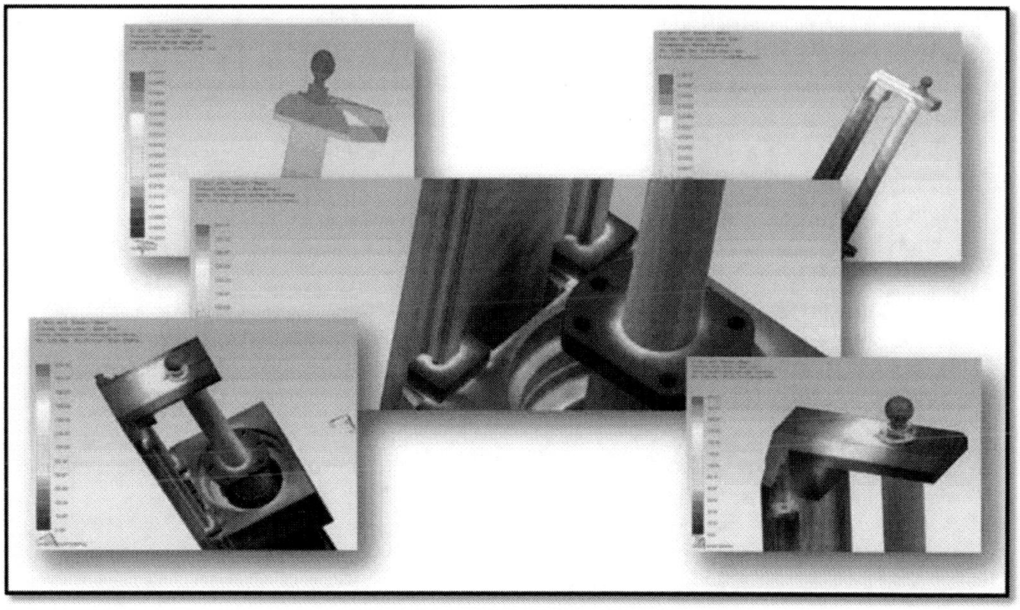

Figure 6. Graphical representation of the results of realized strength analysis.

5. Conclusions

The use of advanced programs of the CAD/CAE class, during the design-constructional process, significantly speeds up creating a new product. The investigations let to determine the behavior of selected elements of the platform during operation of the virtual mechanism. The obtained results allow determining the motion parameters that could be obtained at the

maximal effort of the used components. The functioning simulator of car driving for the disabled persons could be seen on the website
https://www.youtube.com/watch?v=D9yFqFPpWi4.

Acknowledgements
The work is realized within the project funded by The National Centre for Research and Development, agreement No UOD-DEM-1-495/001.

References

[1] Ociepka P and Świder J 2004 Object-oriented system for computer aiding of the machines conceptual design process *Journal of Materials Processing Technology* **157-158** pp 221-227

[2] Ociepka P, Herbuś K and Gwiazda A 2014 Application of the CBR method for adding the design process of module manipulators *Advanced Materials Research* **1036** pp 1011-1016

[3] Herbuś K, Ociepka P and Gwiazda A 2014 Application of functional features to the description of technical means conception *Advanced Materials Research* **1036** pp 1001-1004

[4] Ociepka P and Grabowik C 2008 Zastosowanie metody CBR do wspomagania procesu projektowo-konstrukcyjnego i przygotowania technologii, *Mechanik* **8-9** pp 728-733 (in Polish)

[5] Gwiazda A 2014 System of designing complex technical means using fuzzy analysis *App. Mech. and Mat.* **474** pp 147-152

[6] Gwiazda A, Sękala A, Monica Z and Banaś W 2014 Integrated approach to the designing process of complex technical systems *Advanced Materials Research* **1036** pp 1023-1027

[7] Gwiazda A 2014 Design of the roof support with strait-line mechanism *Advanced Materials Research* **1036** pp 553-558

[8] Płaczek M 2014 Computer Aided Analysis of Systems with MFC Actuators *Advanced Materials Research* **1036** pp 711-714

[9] Ociepka P and Świder J 2007 Współczesne systemy doradcze do wspomagania projektowania maszyn *Monografia* (Gliwice: Wydawnictwo Politechniki Śląskiej) (in Polish)

[10] Banaś W, Herbuś K, Kost G, Nierychlok A, Ociepka P and Reclik D 2014 Simulation of the Stewart platform carried out using the Siemens NX and NI LabVIEW programs *Advanced Materials Research* **837** pp 537-542

[11] Ociepka P, Herbuś K and Gwiazda A 2014 Application of the method basing on engineering knowledge and experience for adding the hexapod design process *Advanced Materials Research* **1036** pp 1005-1010

[12] Gwiazda A, Herbuś K, Kost G and Ociepka P 2015 Designing mechatronics equipment based on the example of the Stewart platform *Solid State Phenomena* **220/221** pp 419-422

[13] Gwiazda A, Herbuś K, Kost G and Ociepka P 2015 Motion analysis of mechatronic equipment considering the example of the Stewart platform *Solid State Phenomena* **220/221** pp 479-484

[14] Banaś W, Gwiazda A, Herbuś K, Kost G, Ociepka P and Reclik D 2015 Analysis of the dynamic properties of the mechatronic integrator of control procedures of the vehicle driven by persons with disabilities *Solid State Phenomena* **220/221** pp 3-8

Optimization of the production process using virtual model of a workspace

Z Monica[1]

[1]Silesian University of Technology, The Faculty Of Mechanical Engineering, Institute Of Engineering Processes Automation And Integrated Manufacturing Systems, ul. Konarskiego 18a 44-100 Gliwice, Poland

E-mail: zbigniew.monica@polsl.pl

Abstract. Optimization of the production process is an element of the design cycle consisting of: problem definition, modelling, simulation, optimization and implementation. Without the use of simulation techniques, the only thing which could be achieved is larger or smaller improvement of the process, not the optimization (i.e., the best result it is possible to get for the conditions under which the process works). Optimization is generally management actions that are ultimately bring savings in time, resources, and raw materials and improve the performance of a specific process. It does not matter whether it is a service or manufacturing process. Optimizing the savings generated by improving and increasing the efficiency of the processes. Optimization consists primarily of organizational activities that require very little investment, or rely solely on the changing organization of work. Modern companies operating in a market economy shows a significant increase in interest in modern methods of production management and services. This trend is due to the high competitiveness among companies that want to achieve success are forced to continually modify the ways to manage and flexible response to changing demand. Modern methods of production management, not only imply a stable position of the company in the sector, but also influence the improvement of health and safety within the company and contribute to the implementation of more efficient rules for standardization work in the company. This is why in the paper is presented the application of such developed environment like Siemens NX to create the virtual model of a production system and to simulate as well as optimize its work. The analyzed system is the robotized workcell consisting of: machine tools, industrial robots, conveyors, auxiliary equipment and buffers. In the program could be defined the control program realizing the main task in the virtual workcell. It is possible, using this tool, to optimize both the object trajectory and the cooperation process.

1. Introduction

Optimization process is a sequence of action that leads to selection of the best solution from a set of available alternatives. Taking into account a wide range of problems considered with production management one could define the optimization process as determination the values of parameters that characterize the given production process to adjust its course without violation previously specified constraints. From this point of view the process optimization is related with the lean production approach. The last is directed to iterative elimination of all process elements that violate its main requirements. In this iterative process could be distinguished three sub-processes [1] of elimination (in their Japanese forms):

Content from this work may be used under the terms of the Creative Commons Attribution 3.0 licence. Any further distribution of this work must maintain attribution to the author(s) and the title of the work, journal citation and DOI.
Published under licence by IOP Publishing Ltd

- *muda* (waste elimination);
- *mura* (unevenness elimination);
- *muri* (overburden elimination).

The *muda* process is considered with elimination the seven wastes determined with the acronym TIMWOOD. Particular letters mean the next wastes: T- wastes in transportation (product damage), I - wastes in inventory (not used material that do not bring a profit), M - wastes in motion (not products damage), W - wastes considered with waiting, O - wastes considered with over-processing (product should fulfil customer requirements an do not exceed them), O - wastes considered with over-production (production of larger number of products than is required by the customer) and finally D - wastes caused by defects (considered with products and process and required additional work). The *mura* process is considered with establishing a smooth production cycle. For large production system it is realized using the JiT philosophy. It leads to higher productivity at lower storage capacity. Finally the *muri* process is directed to standardization of parts and processes. By simplification it allows increasing the productivity and quality levels. In figure 1 is presented an overview of an optimization process including Japanese approach to lean production. Concluding it should be stated that lean production is centered on making only actions that add value and on reducing everything else. The value adding action, from the perspective of the customer for whom the product or service is designed, is any action that the customer could be able to determine as necessary and valuable.

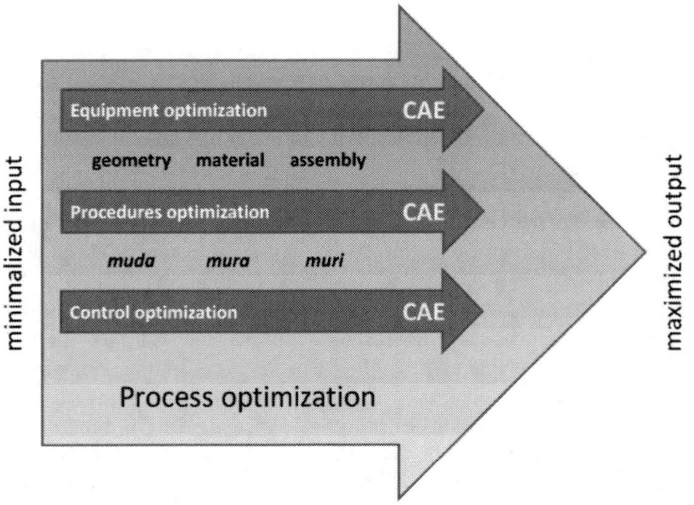

Figure 1. Scheme of process optimization.

In the figure 1 are presented additional elements considered with the optimization process. For the system of a production process it was distinguished three main elements: equipment created this system, procedures linking equipment to realize a production flow and control system supervising functioning of two previous elements.

During the optimization process should be taken into account three designing parameters: geometry, material and assembly method. The geometry parameter is consider with determining the form of process elements using a system of the CAD/CAE class [2,3,4]. This process is added with more specifies analysis like dynamic one or strength one [5,6,7]. The second factor is considered with material optimization. This is particularly important for some types of manufacturing systems [8,9,10]. At least the last factor is considered with assembly method and proper assembly parameters that influence the characteristic of a product.

2. The life-cycle approach in the optimizing process

The second problem, related with process optimization issue, is the life-cycle of a technical mean [11]. Optimization criteria must include requirements considered with all possible stages of this cycle. It must be stated that the life-cycle is a complex scheme that presents sequence of different forms of a product "life", what is shown in figure 2. Three main periods could be distinguished in the product life-cycle: construction creation phase, material "existence" one and liquidation one.

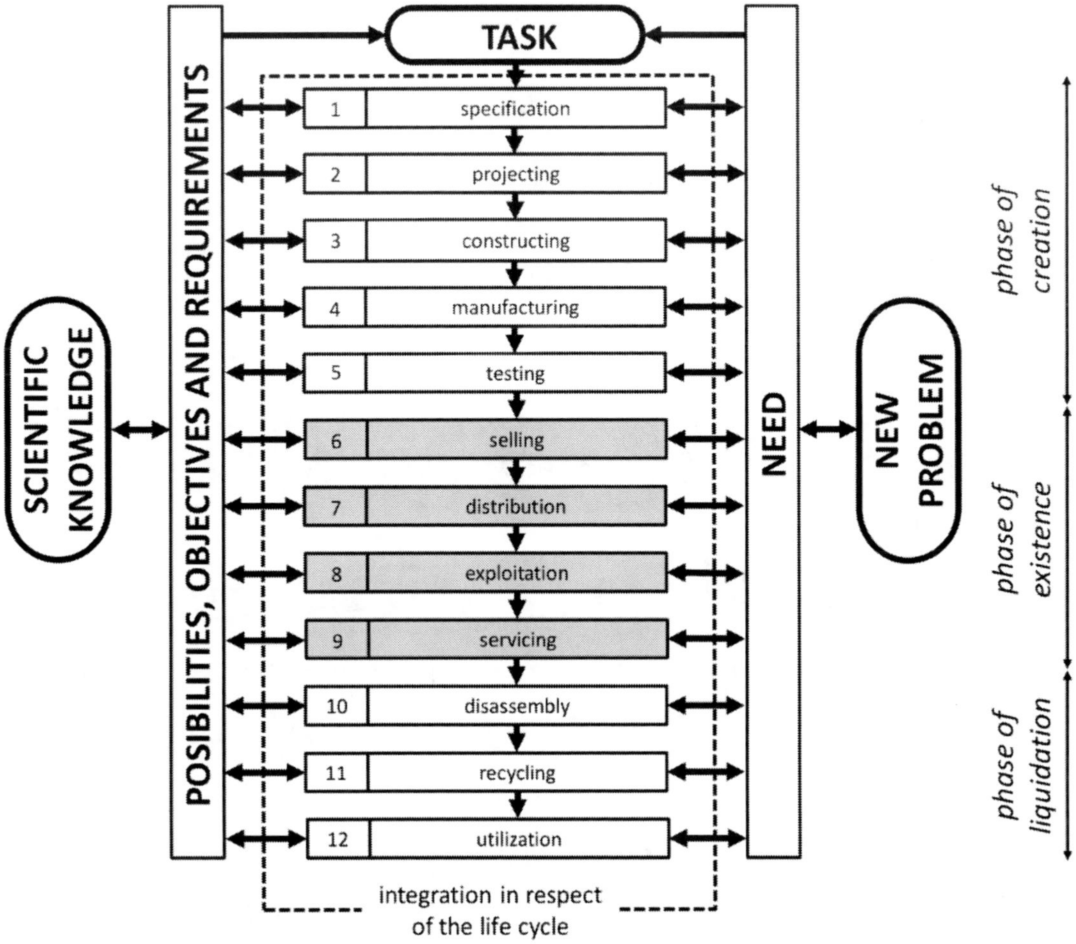

Figure 2. Scheme of the life-cycle of a technical mean [11].

One of the most susceptible phase of the whole technical mean life-cycle is the creation phase. Mistakes made at this phase are the most serious and costs of their fixing are the highest. So it is very important to conduct that phase carefully and, on the other hand, quickly according to a market demand. The other phases do not influence the costs of the whole technical mean life-cycle so much.

Therefore taking into account costs considered with product elaboration it should be stated that computer tools allow decreasing them in different stages of the technical mean life-cycle. One of techniques, applied for time and costs decreasing, is simulation. Modern computer systems of the CAE class allow designing complex technical means [12] and conduct complicated engineering analysis including simulations tests.

3. Simulation of a virtual workcell as the tool for process optimization

The elaborated problem was considered with designing a virtual production system, in the form of a workcell, to analyze the problem of production system optimization. The scheme of the analyzed system is presented in figure 3. The system consists of a lathe, a milling machine and a robot. The robot is also responsible for controlling the production process in this model workcell. Optimization process is conducted using outer computer on which optimizing procedures are implemented. In figure 4 is presented the virtual model of a lathe. This model is equipped with a virtual controller allowing executing programs written in the G-code.

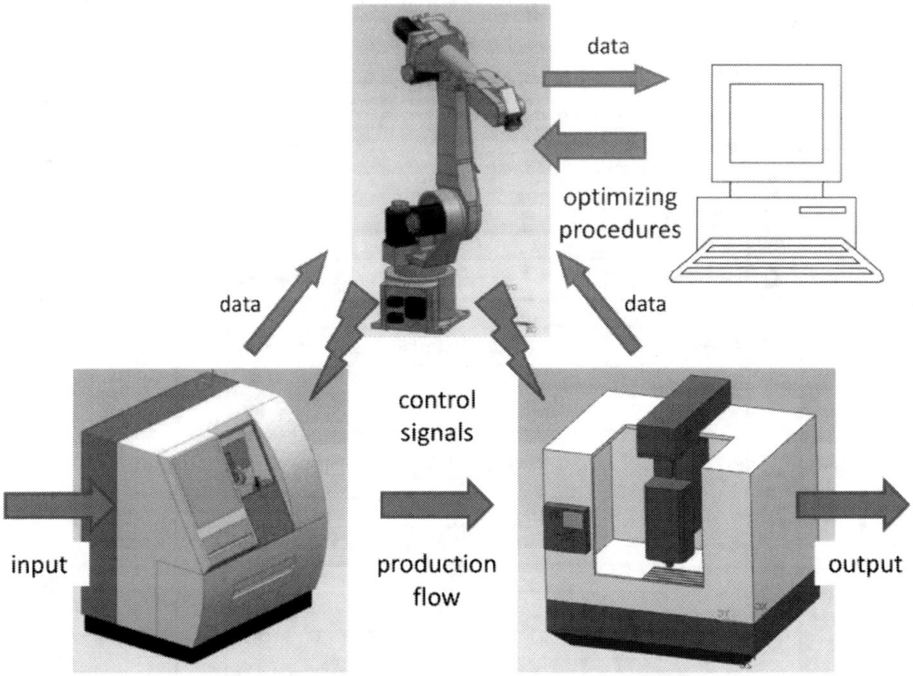

Figure 3. Scheme of the analyzed production system.

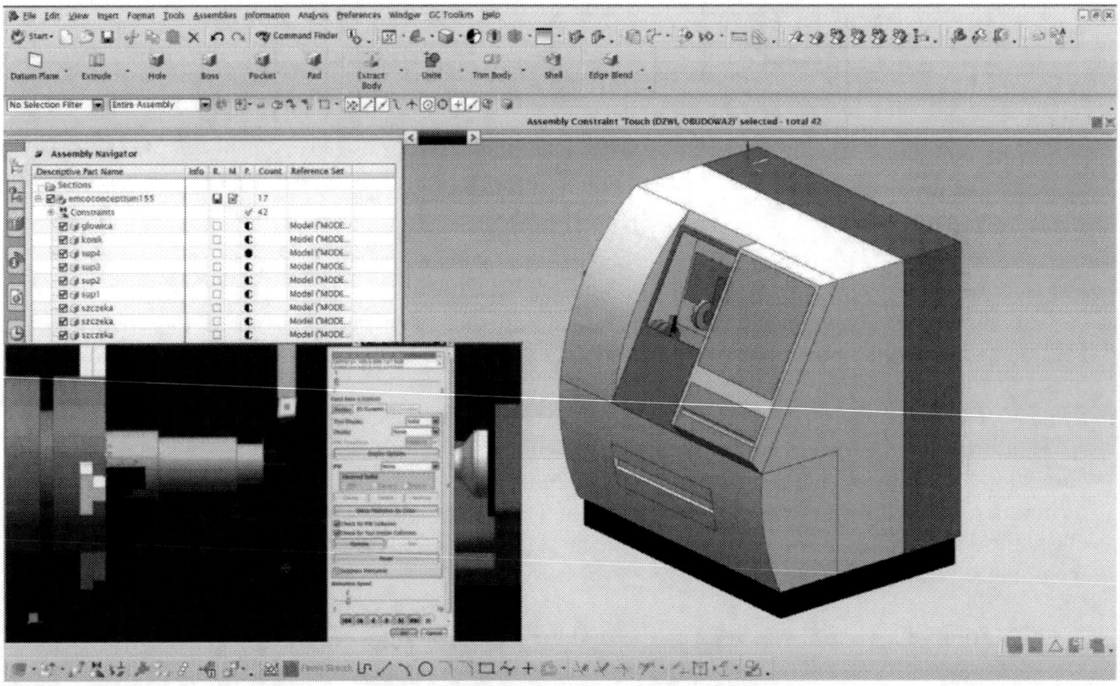

Figure 4. Virtual model of a lathe.

In a similar way is developed the virtual model of a milling machine. It also allows the implementation of procedures stored in the G-code. The sequence of transport operations, realized in the presented virtual manufacturing system, is supervised by the robot controller. The purpose of this research is to develop a protocol of cooperation of indicated equipment and of data exchange to make possible the optimizing process of the working workcell on the basis of past experience. Nowadays was accomplished tests considered with cooperation of mentioned devices.

Figure 5. Virtual model of a milling machine.

The whole workcell could be controlled using supervising program that rum programs controlling particular devices. It is possible to use virtual sensors which allows simulating a real flow of control signals in a real control system.

4. Conclusions

The presented approach of production system optimizing using a virtual environment of a production system is only the transitional stage to the fully automated production system that could optimize itself. To realize this task one should applied visualization systems [13,14] that could help to supervising the system work. It is also needed to implement algorithm of artificial intelligence to speed up the calculation processes and to work around the problem of the global optimum.

The element of future research should be also the problem of linking and analysis of data from external sensors [15,16] for the implementation in procedures performed in virtual reality. Thus, it is possibility to make more real the operating conditions of virtual devices. These devices may be any real controllers or and sensors of any type. [17,18]. It should be also noted that the important role, in further work, will play control programs basing on algorithms of artificial intelligence [19,20]. They allow creating the environment of production systems, as intelligent and adaptive.

References

[1] Womack J, Jones D and Roos D 1990 The Machine that changed the world: the story of lean production (New York: Harper Collins)

[2] Herbuś K, Monica Z and Ociepka P 2008 Zastosowanie obiektów elementarnych w procesie projektowania i przygotowania wytwarzania *Górnictwo Odkrywkowe* **4-5** pp 104-109

[3] Herbuś K, Ociepka P and Gwiazda A 2014 Conception of the integration of the virtual robot model with the control system *Adv. Mat. Res.* **1036** pp 732-736

[4] Banaś W, Gwiazda A, Herbuś K, Kost G, Ociepka P and Reclik D 2015 Analysis of the dynamic properties of the mechatronic integrator of control procedures of the vehicle driven by persons with disabilities *Solid State Phenomena* **220-221** pp 3-8

[5] Dzitkowski T and DymarekA 2014 Active reduction of identified machine drive system vibrations in the form of multi-stage gear units *Mechanika* **20/2** pp 183-189

[6] Gwiazda A 2014 Virtual analysis of the mining support under loads of the roof *Adv. Mat. Res.* **837** pp 393-398

[7] Gwiazda A, Herbuś K, Kost G and Ociepka P 2015 Designing mechatronics equipment based on the example of the Stewart platform *Solid State Phenomena* **220-221** pp 419-422

[8] Ćwiek J, Łabanowski J, Topolska S and Sozańska M 2012 Determination of failure causes of a steam turbine casting *Solid State Phenomena* **183** pp 37-42

[9] Łabanowski J, Świerczyńska and Topolska S 2014 Effect of microstructure on mechanical properties and corrosion resistance of 2205 duplex stainless steel *Pol. Mar. Res.* **21/4(84)** pp 108-112

[10] Gwiazda A, Sękala A, Banaś W and Monica Z 2014 Protection of hydraulic systems against dynamic loads using multi-valve approach *Advanced Materials Research* **1036** pp 547-552

[11] Gwiazda A 2013 Designing of technical means using the integration methodological approach. GIG Publishing (Katowice:Polish).

[12] Gwiazda A, Sękala A, Monica Z and Banaś W 2014 Integrated approach to the designing process of complex technical systems *Adv. Mat. Res.* **1036** pp 1023-1027

[13] Foit K 2007 Introduction to use virtual reality visualizations in the exploitation and virtual testing of machines *J. Achiev. Mater. Manuf. Eng.* **25/2** pp 57-60

[14] Foit K 2014 Mixed reality as a tool supporting programming of the robot *Adv. Mat. Res.* **1036** pp 737-742

[15] Płaczek M 2015 Modelling and investigation of a piezo composite actuator application *Int. J. Mat. and Prod. Tech.* **50-3/4** pp 244-258

[16] Płaczek M 2014 Modelling of electric circuit's type influence on dynamic flexibility of system with passive vibration damping *Adv. Mat. Res.* **1036** pp 511-516

[17] Hetmanczyk M and Michalski M 2014 The self-excitation phenomenon of quasi shielded inductive proximity switches *Adv. Mat. Res.* **837** pp 405-410

[18] Michalski P and Hetmańczyk M 2014 The self-excitation phenomenon of quasi shielded inductive proximity switches *Adv. Mat. Res.* **837** pp 405-410

[19] Hetmanczyk M P 2015 The Prediction Oriented Analysis of Mechatronic Machine Structures in Terms of the Signal Stream Flow *Solid State Phenomena* **220-221** pp. 423-428

[20] Płaczek M 2014 Modelling of electric circuit's type influence on dynamic flexibility of system with passive vibration damping *Adv. Mat. Res.* **1036** pp 511-516

Construction typification as the tool for optimizing the functioning of a robotized manufacturing system

A Gwiazda[1], W Banas[1], A Sekala[1], K Foit[1], P Hryniewicz[1] and G Kost[1]
[1]Silesian University of Technology, The Faculty Of Mechanical Engineering, Institute Of Engineering Processes Automation And Integrated Manufacturing Systems, ul. Konarskiego 18a 44-100 Gliwice, Poland

E-mail: aleksander.gwiazda@polsl.pl

Abstract. Process of workcell designing is limited by different constructional requirements. They are related to technological parameters of manufactured element, to specifications of purchased elements of a workcell and to technical characteristics of a workcell scene. This shows the complexity of the design-constructional process itself. The results of such approach are individually designed workcell suitable to the specific location and specific production cycle. Changing this parameters one must rebuild the whole configuration of a workcell. Taking into consideration this it is important to elaborate the base of typical elements of a robot kinematic chain that could be used as the tool for building Virtual modelling of kinematic chains of industrial robots requires several preparatory phase. Firstly, it is important to create a database element, which will be models of industrial robot arms. These models could be described as functional primitives that represent elements between components of the kinematic pairs and structural members of industrial robots. A database with following elements is created: the base kinematic pairs, the base robot structural elements, the base of the robot work scenes. The first of these databases includes kinematic pairs being the key component of the manipulator actuator modules. Accordingly, as mentioned previously, it includes the first stage rotary pair of fifth stage. This type of kinematic pairs was chosen due to the fact that it occurs most frequently in the structures of industrial robots. Second base consists of structural robot elements therefore it allows for the conversion of schematic structures of kinematic chains in the structural elements of the arm of industrial robots. It contains, inter alia, the structural elements such as base, stiff members - simple or angular units. They allow converting recorded schematic three-dimensional elements. Last database is a database of scenes. It includes elements of both simple and complex: simple models of technological equipment, conveyors models, models of the obstacles and like that. Using these elements it could be formed various production spaces (robotized workcells), in which it is possible to virtually track the operation of an industrial robot arm modelled in the system.

1. Introduction

Typification is one of the stages of the normalization process. This process causes reducing costs, enabling mass production, functioning together devices from different manufacturers and replacement of worn parts. It facilitates also making commercial contracts. The whole normalization process includes: classification, typification and unification. Classification consists of searching similar constructions or constructional features and creating groups of similar objects. The example of classification is the group technology - manufacturing similar products in one production batch [1,2,3].

Content from this work may be used under the terms of the Creative Commons Attribution 3.0 licence. Any further distribution of this work must maintain attribution to the author(s) and the title of the work, journal citation and DOI.
Published under licence by IOP Publishing Ltd

Typification is a process of limiting variety of constructions or their features (e.g. dimensions). This is realized by selecting a number of constructions or their features that however allow meeting demand in its whole range. The example of typification is the system of standardized dimensions (e.g. diameters of screw joints). Generally for small features the distance (in a similarity space) between typified objects is much smaller than for large ones (e.g. Renard sequence). As the results of typification one could obtain the series of types. Unification is the use the same elements and assemblies in different technical means. It concerns also features of produced elements. Generally speaking unification is then when only one element or assembly of the considered type is used for manufacturing. The example of unification is utilization of the same engine in different car models or one type of file format (e.g. Step). As it was stated unification is also considered with unification of construction features. For example in the Polish car CWS T1 (figure 1) was use the only type of screw joints (M10x1.5) and the sparking plug thread M18x1,5. So using one double-ended spanner it was possible to take the whole engine and car to pieces [4].

Figure 1. Virtual model of a replica of the CWS T1 Torpedo [5].

The presented discussion of normalization techniques in machine building shows its undeniable advantages. It is important to consider this approach in being elaborated and applied methodologies of technical means designing. Modern engineering informatics environments in the form of the system of the CAD/CAE/CAM [6, 7] class enable and facilitate such approach not only in designing of a particular technical means but also of complex technical systems. Particularly in the case of the last the possible profit not could be underestimated [8, 9].

2. Typification as the designing criterion

Typification process, as leading to types of series creation, should be included into methodology of designing, as it was mentioned previously. In the classical methodology [10] the designing process is treated as general, heuristics approach to the problem of new technical solutions searching. It is needed to systematize that process to show main directions of new technical solutions development to obtain an optimized construction. One approach to this issue could be pointed two groups of sub-processes leading to typified construction [11]. These two groups of sub-processes intermingled creating 9 optimizing requirements.

The first group relates to designing of typified constructional forms (figure 2). It is possible to determine three fundamental forms of a construction [12]. In other words, according to this assumption, construction is the result of synergetic assembly of three components called constructional forms. The first one is the geometric form. It describes construction as regards geometric relations and dimensions (scale). The typification process at the geometric form designing relate to simplification of variety of shapes and of dimensions of technical means. Sometimes the form is unified and dimensions create s series of types. The second constructional form relates to used material so it describes the micro- and nano-structure of technical means. Material typification is frequently replaced with material unification. In the designing process one drives towards eliminating material which are expensive, almost unworkable, not functional and not eco-friendly [13, 14, 15].

This phase includes also determining typical material dimensions like a hardening temperature. Finally the third process of constructional form designing and typifying is considered with assembling machine tools. Typification, at this phase, drives for limiting the variety of assembly procedures and dimensions (the example of an assembly dimension is the screw tightening torque [17].

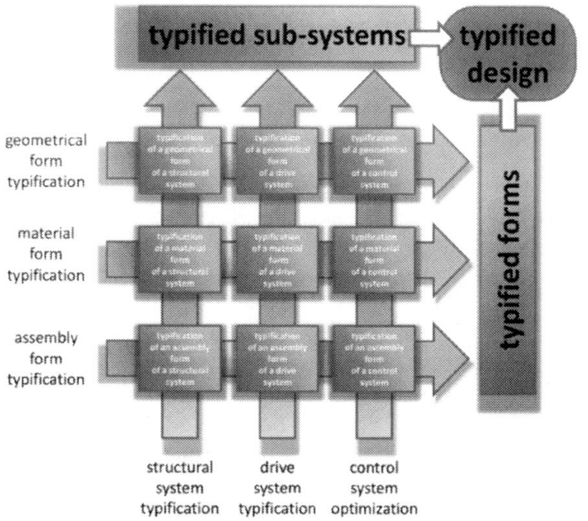

Figure 2. Typification processes in a designing methodology plan.

The second group of typification operations is considered with previously determined sub-systems of a technical mean [11]. The first one concerns to the structural sub-system. It includes functional elements of a technical mean except the two next systems. During designing process of the structural subsystem of a technical mean three typification operations are conducted. They are related with typification of the structural subsystem under conditions concerning the three constructional forms (typification of the geometrical form of the structural subsystem, typification of the material form and typification of the assembly form). These typification operations lead to the typified structural sub-system as a technical solution for this one engineering task.

Realization of all three processes of designing designated sub-systems of a technical means allows conducting nine typification operations. As the result one could obtain both the typified constructional forms and sub-systems of a technical mean. The typification operation must include also mutual typification of particular processes (their influence on each other). The typified design is the result of mutual optimization of typified sub-systems and typified constructional forms.

3. Typification of industrial robot constructions and work scenes

One of areas of application the typification approach in the design process is elaborating constructions of industrial robots. Analyses show that there is a huge variety of designs of industrial robots (figure 3). In this situation each workcell has to be designed to the characteristics of an industrial robot and any exchange of a robot enforces its reconfiguration.

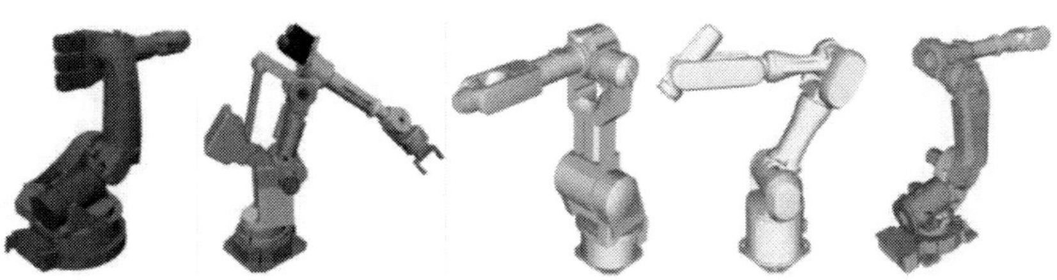

Figure 3. Exemplar designs of industrial robots [16].

In the conducted project it was assumed that it should be changed the philosophy of design. The challenge was to design robots to the previously designed workcell. Below is presented an exemplar workcell which should be robotized (figure 4). The workcell has been designed using typified machine tools and conveyors.

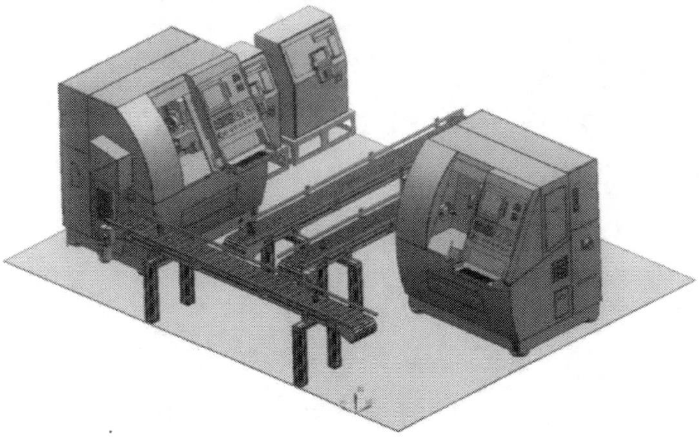

Figure 4. Exemplar workcell.

The realization of such design process was based on utilization previously designed series of types of components of an industrial robot. It was selected next groups of typified elements: robot bases (with linear and rotational drives), robot drives (linear and rotational), robot arm parts (a series of types) and gripers. Using this components it is possible to assembly any robot configuration of any work space (of course in the limited range). The robots resulting from the assembly process are presented, in the simplified form, in the figure 5. The simplified form is utilized for the control procedure to analyze collisions and accessibility of a robot arm.

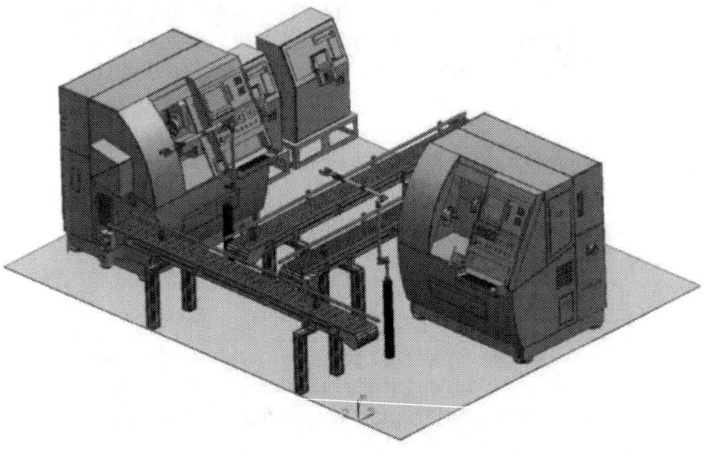

Figure 5. Exemplar workcell with two typified robots (in the simplified form).

Utilization of typified components allows designing robots taking into consideration the requirements of existing workcells, which should be robotized or in which should be changed the robotized unit.

4. Conclusions

The presented analysis and practical example of typification process utilization allows stating that this approach is valuable and profitable and could be computer aided. One of identified problems is considered with data acquisition that is considered with widespread production computer systems [17].

The second problem is considered with realization specific optimization problems e.g. concerning the optimal vibrations characteristic [18,19,20].

The next investigations will be considered with determining rules allowing creating the minimal families of construction at required level of dimensional variety. The modern CAD/CAE environments let to elaborated such typified construction in shorter time and to test them in a virtual environment what causes decreasing of designing costs.

Acknowledgements
The work is realized within the project titled: "Modular automated production stand with instrumentation for non-invasive confirmation of product quality" funded by The National Centre for Research and Development, agreement No UOD-DEM-1-495/001.

References

[1] Paprocka I, Kempa W, Kalinowski K and Grabowik C 2014 On Pareto optimal solution for production and maintenance jobs scheduling problem in a job shop and flow shop with an immune algorithm *Adv. Mat. Res.* **1036** pp 875-880
[2] Paprocka I and Kalinowski K 2014 Pareto optimality of production schedules in the stage of populations selection of the MOIA immune algorithm *App. Mech. and Mat.* **657** pp 869-873
[3] Krenczyk D and OlenderM 2014 Production planning and control using advanced simulation system *Int. J. of Modern Manufacturing Technologies* **VI/2** pp 38-43
[4] Information on http://www.polskie-auta.pl/stara/cws3.htm (Polish)
[5] Information on http://cwst1.pl/index.php?t=8 (Polish)
[6] Ociepka P, Herbuś K and Gwiazda A 2014 Application of the CBR method for adding the design process of module manipulators *Adv. Mat. Res.* **1036** pp 1011-1016
[7] Grabowik C, Ćwikła G and Janik W 2014 The New Approach to Design Features Identification *App. Mech. and Mat.* **657** pp 750-754
[8] Herbuś K, Ociepka P and Gwiazda A Application of functional features to the description of technical means conception *Adv. Mat. Res.* **1036** pp 1001-1004
[9] Chwastyk P and Kolosowski M 2009 Features method for CAD/CAM system integration with estimation of cost production *Annals of DAAAM* pp 1005-1007
[10] Pahl G and Beitz W 1997 Science of constructing. Methods and applications (Berlin: Springer Verlag) (in German)
[11] Gwiazda A 2013 Designing of technical means using the integration methodological approach (Katowice: GIG Publishing) (in Polish)
[12] Dietrych J 1978 System and construction (Warszawa: WNT) (in Polish)
[13] Ćwiek J, Łabanowski J, Topolska S and Sozańska M 2012 Determination of failure causes of a steam turbine casting *Sol. St. Ph.* **183** pp 37-42
[14] Topolska S and Łabanowski J 2015 Duplex stainless steels toughness impact investigations *J. Mat. in Tehn. / J. Mat. and Techn.* **5** pp 114-119
[15] Łabanowski J, Świerczyńska A and Topolska S 2014 Effect of microstructure on mechanical properties and corrosion resistance of 2205 duplex stainless steel *Pol. Mar. Res.* **21/4(84)** pp 108-112
[16] Information on http://www.3dcadbrowser.com/
[17] Ćwikła G 2014 Methods of manufacturing data acquisition for production management – a review *Adv. Mat. Res.* **837** pp 618-623
[18] Dzitkowski T and Dymarek A 2014 Active reduction of identified machine drive system vibrations in the form of multi-stage gear units *Mechanika* **20/2** pp 183-189
[19] Hetmanczyk M P 2015 The prediction oriented analysis of mechatronic machine structures in terms of the signal stream flow *Sol. St. Ph.* **220-221** pp 423-428
[20] Hetmanczyk M P and Michalski P The self-excitation phenomenon of quasi shielded inductive proximity switches *Adv. Mat. Res.* **837** pp 405-410

IOP Conf. Series: Materials Science and Engineering **95** (2015) 012104 doi:10.1088/1757-899X/95/1/012104

Modular industrial robots as the tool of process automation in robotized manufacturing cells

A Gwiazda[1], W Banas[1], A Sekala[1], K Foit[1], P Hryniewicz[1] and G Kost[1]
[1]Silesian University of Technology, The Faculty Of Mechanical Engineering, Institute Of Engineering Processes Automation And Integrated Manufacturing Systems, ul. Konarskiego 18a 44-100 Gliwice, Poland

E-mail: aleksander.gwiazda@polsl.pl

Abstract. Recently the number of designed modular machine was increased. The term modular machine is used to denote different types of machinery, equipment and production lines, which are created using modular elements. Modular could be both mechanic elements, and drives, as well as control systems. This method of machine design is more and more popular because it allows obtaining flexible and relatively cheap solutions. So it is worth to develop the concept of modularity in next areas of application. The advantages of modular solutions are: simplification of the structure, standardization of components, and faster assembly process of the complete machine Additional advantages, which is particularly important for manufacturers, are shorter manufacturing times, longer production series and reduced manufacturing costs. Modular designing is also the challenge for designers and the need for a new approach to the design process, to the starting process and to the exploitation process. The purpose for many manufacturers is the standardization of the components used for creating the finished products. This purpose could be realized by the application of standard modules which could be combined together in different ways to create the desired particular construction as much as possible in accordance with the order. This solution is for the producer more favorable than the construction of a large machine whose configuration must be matched to each individual order. In the ideal case each module has its own control system and the full functionality of the modular machine is obtained due to the mutual cooperation of all modules. Such a solution also requires the modular components which create the modular machine are equipped with interfaces compatible one with another to facilitate their communication. The individual components of the machine could be designed, manufactured and used independently and production management task could be divided into subtasks. They could be also outsourced to an independent manufacturer. Standardization and run of the entire modular machine should be easier if standardized are individual modules. The advantages of modular design, in addition to those mentioned above, there are many more.

1. Introduction

Modular design is nowadays one of the most popular approaches to the process of complex technical means designing. In the designing philosophy there are two opposed approaches: complex product design and modular products design [1]. It is possible to describe the complex design as designing the whole product beginning from product requirements. All components are design anew. This approach is considered with high costs but the resulted product is almost ideally adapted to customer requirements. Modular communication interfaces should help to reduce the amount of wiring, which is

Content from this work may be used under the terms of the Creative Commons Attribution 3.0 licence. Any further distribution of this work must maintain attribution to the author(s) and the title of the work, journal citation and DOI.
Published under licence by IOP Publishing Ltd

IOP Conf. Series: Materials Science and Engineering 95 (2015) 012104 doi:10.1088/1757-899X/95/1/012104

of significance for subsequent servicing equipment [2,3]. Also, in the event that any modifications machine changes its individual modules can be implemented in less time than would be the case, the machine created a classic. The paper presents an example of application the modular approach to the process of industrial robot design to make the process of robotized workcell design and exploitation more flexible.

On the other hand modular design allows utilizing in the designing and production processes the already elaborated and verified modules. This causes that the designing process is shorter and cheaper (sometimes much cheaper) but it is less adapted to the customer requirements. This approach to the designing process could also change the philosophy of a production system [4,5,6]. In the case of Hyundai shipyard the modular design (figure 1) allow changing over the production regime from piece system to series one. It causes that the backlog of orders reaches twenty years.

Figure 1. A ship module in the Hyundai shipyard [7].

Below are presented typical modules utilized in a design of military ships (figure 2). To the presented one should add such modules as: engines, transmission systems, a hull parts (like in figure 1), equipment of particular compartments etc. This approach allows, in the shipyard industry, offering a wider range of types and variants of ships at lower designing and manufacturing costs.

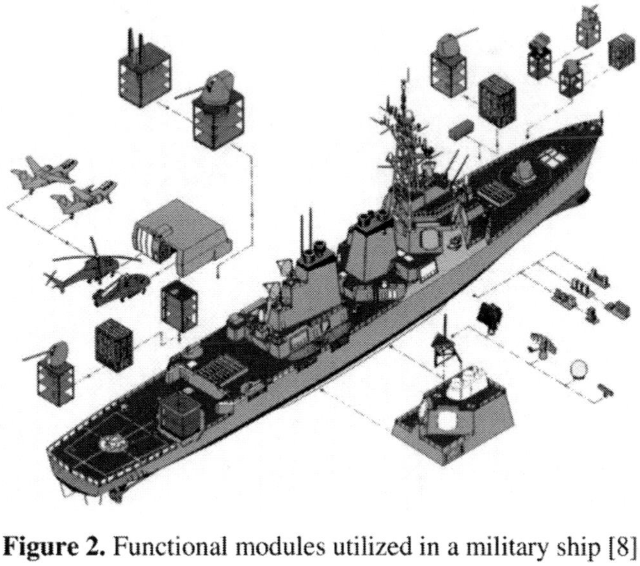

Figure 2. Functional modules utilized in a military ship [8].

The presented shortened survey lets to conclude that modular design characterizes of many advantages that facilitate the design process and could bring a lot of economic benefits. Modular design could be summarized in two statements [9]: modular design creates options and modular design evolves as the options are pursued and exercised. According the first statement one should point out that modular design allows managing the whole design process more precisely, enables parallel work (concurrent engineering) and is tolerant of uncertainty. Referring to the second statement it is needed to state that modular designs should evolve through the process of module evolution. The exchange of a module could modernize a whole product.

2. Modularity in a design philosophy

The design process is frequently described from the perspective of chosen design methodology as a specific design philosophy. To analysis the modular approach to a design methodology could be used the integrative thinking in a design process [10]. In this approach are determined two groups of processes: the processes related with designing a particular constructional form and the processes related with designing a particular sub-system of a technical mean. Among the first group of processes it is possible to distinguish three passages considered with determining: the geometrical form (geometrical relations and dimensions), the material form (material types and parameters) and the assembly form (assembly method and parameters).

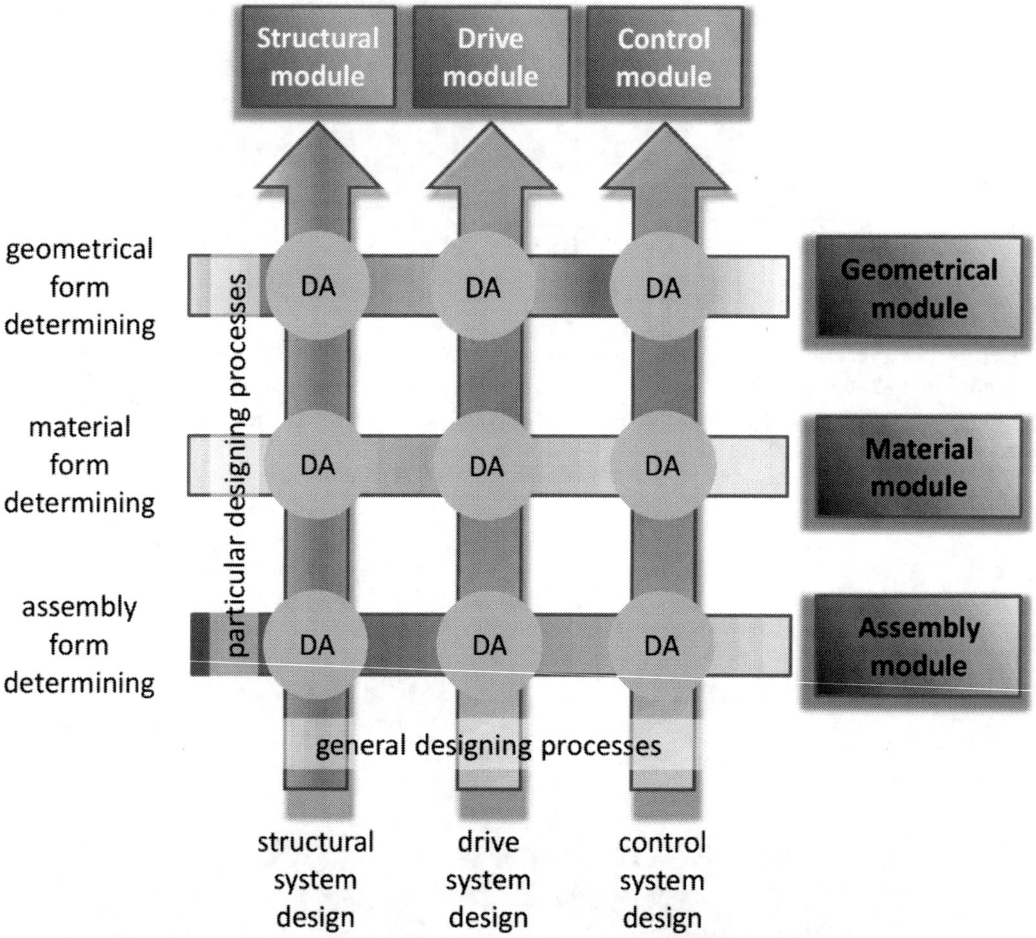

Figure 3. Designing methods of module creation (DA - design action).

The second group of processes concerns the methods of determining: the structural system (housing and other functional sub-systems) the drive system (drive type and characteristic) and the control system (type, characteristic and parameters).

The mentioned processes lead to different types of modules, used in designed technical means. In the result of the first group of processes one could obtain: geometric modules (of the same geometry but of different type: structural, drive or control), material module (the same material use in different functions and the same technological process) [11, 12, 13] and assembly module (the same assembly process) [14]. These modules are frequently used as components of complex modules of the types: structural, drive (like series of types of engines for one technical mean) and control (like PLC controllers).

3. Modularity in industrial robots design

One of the areas, where could be used the modular approach in designing is industrial robotics. In this area could be utilized two types of modules: structural and drives (assuming utilization of a standard control system), which are presented in figure 4. The structural elements are considered with the manipulator arm elements and create a series of types (particularly in length). The drive modules are divided into translatory motion modules and rotary motion modules.

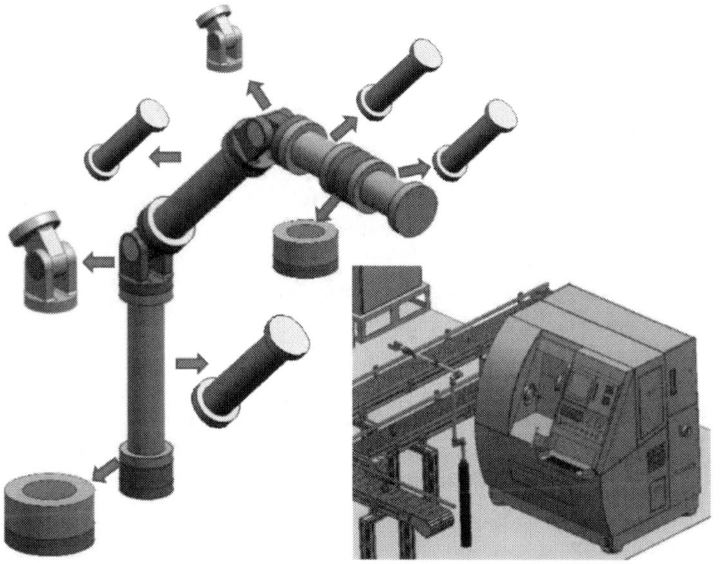

Figure 4. Schematics representation of some proposed modules.

The presented modules are elaborated in the PLM Siemen NX program. They are stored in a database and could be used by the special application (figure 5). It allows choosing particular modules and creating (assembling) the model of a manipulator in accordance to the previously designed or existing scene of robot work (e.g. workcell). The application has the mechanism of attributing proper joints to particular modules. It should be noted that in figures are presents the resulting manipulators in a schematic for due to limitation considered with computer capacity. In the right, bottom corner of the figure 4 is presented an exemplar manipulator in a workcell.

The presented application, basing on modular approach in designing, allows creating any manipulators models, basing on the closed set of modules. Moreover it is possible not only to configure the kinematic chain but also to differ in its dimensions. In the case of a large scale production this type of manipulator should be cheaper than traditional solutions.

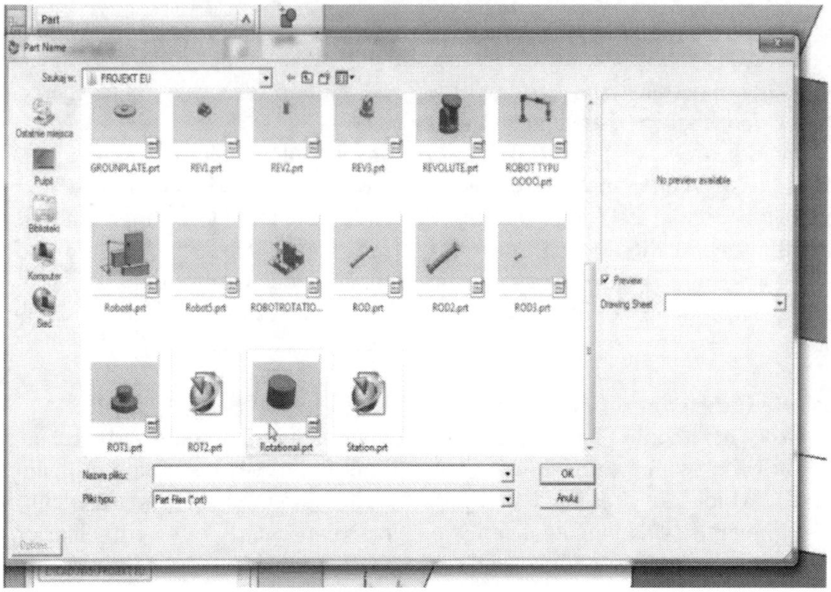

Figure 5. Control panel for the modular robot creation.

4. Conclusions

The presented approach of modular design could be develop due to increasing possibilities of the CAE environment allowing conducting complex designing processes [15,16]. It is also considered with utilization of different tools of artificial intelligence [17, 18, 19]. Taking into account production data acquisition [20] in real time it is possible to create automatic designing systems for adding modular components creation. Next investigations in this area will be considered with elaboration the proposal of a designing system for workcell equipment elaboration.

The result of the presented work is the computer system that allows modelling robots, taking into account the requirements of an existing workcell. When the placement of particular machine tools and other workcell equipment is imposed from above the one of possible solution of robotization of this manufacturing system is to design a modular industrial robot. The elaborated program allows mating particular modules creating the open kinematic chain of the designed robot. Using joints, defined in the program, it is possible to link particular modules to obtain the working model of a robot. In the case of linear modules it is possible to check its different lengths (dimensional modules). Such elaborated model could be placed in the working scene (workcell) to estimate its functioning. It decreases the time needed to design such a robot in a traditional way.

Acknowledgements

The work is realized within the project titled: "Modular automated production stand with instrumentation for non-invasive confirmation of product quality" funded by The National Centre for Research and Development, agreement No UOD-DEM-1-495/001.

References

[1] Baldwin C Y and Clark K B 2000 Design Rules: The power of modularity, (Boston: MIT)

[2] Hetmanczyk M 2013 The reliability model of AC-asynchronous drive based on the multilevel prognosis system supported by matrices and digraphs methods *Sol. St. Ph.* **199** pp 85-90

[3] Hetmańczyk M and Michalski P 2014 The aid of a mistake proofing with the use of mechatronic systems according to the Poka-Yoke methodology *Adv. Mat. Res.* **837** pp 399-404

[4] Krenczyk D and Olender M 2014 Production planning and control using advanced simulation systems *Int. J. of Modern Manufacturing Technologies* **VI/2** pp 38-43
[5] Paprocka W, Kempa M, Grabowik C and Kalinowski K 2014 Sensitivity analysis of predictive scheduling algorithms *Adv. Mat. Res.* **1036** pp 921-926
[6] Paprocka W, Kempa M, Kalinowski K and Grabowik C 2014 A production scheduling model with maintenance *Adv. Mat. Res.* **1036** pp 885-890
[7] Information on http://gcaptain.com/wp-content/uploads/2011/02/110213-QEC-Govan-unit-move-036_Med.png
[8] Information on http://www.aocinc.net/capabilities/open-systems-architecture
[9] Baldwin C Y and Clark K B 2002 The Option Value of Modularity in Design (Harvard: Business School)
[10] Gwiazda A 2013 Designing of technical means using the integration methodological approach (Katowice: GIG Publishing) (in Polish)
[11] Ćwiek J, Łabanowski J, Topolska S and Sozańska M 2012 Determination of failure causes of a steam turbine casting *Sol. St. Ph.* **183** pp 37-42
[12] Topolska S and Łabanowski J 2015 Duplex stainless steels toughness impact investigations *J. Mat. in Tehn. / J. Mat. and Techn.* **5** pp 114-119
[13] Łabanowski J, Świerczyńska A and Topolska S 2014 Effect of microstructure on mechanical properties and corrosion resistance of 2205 duplex stainless steel *Pol. Mar. Res.* **21/4(84)** pp 108-112
[14] Dymarek A and Dzitkowski T 2013 Reduction vibration of mechanical systems *Applied Mechanics and Materials* **307** pp 257-260
[15] Chwastyk P and Kolosowski M 2012 Integration CAD/CAPP/CAM systems in design process of innovative products *Annals of DAAAM* **23/1** pp 397-400
[16] Grabowik C and Janik W 2013 The concrete casting matrixes inserts design preparation based on the master models *Adv. Mat. Res.* **702** pp 259-262
[17] Grabowik C, Ćwikła G and Janik W 2014 The New Approach to Design Features Identification *App. Mech. and Mat.* **657** pp 750-754
[18] Ociepka P, Herbuś K and Gwiazda A 2014 Application of the method basing on engineering knowledge and experience for adding the hexapod design process *Advanced Materials Research* **1036** pp 1005-1010
[19] Herbuś K, Ociepka P and Gwiazda A 2014 Application of functional features to the description of technical means conception *Advanced Materials Research* **1036** pp 1001-1004
[20] Ćwikła G 2014 Methods of manufacturing data acquisition for production management – a review *Adv. Mat. Res.* **837** pp 618-623

IOP Conf. Series: Materials Science and Engineering **95** (2015) 012105 doi:10.1088/1757-899X/95/1/012105

Determination of the robot location in a workcell of a flexible production line

W Banas[1], **A Sekala**[1], **A Gwiazda**[1], **K Foit**[1], **P Hryniewicz**[1] **and G Kost**[1]

[1]Silesian University of Technology, The Faculty Of Mechanical Engineering, Institute Of Engineering Processes Automation And Integrated Manufacturing Systems, ul. Konarskiego 18a 44-100 Gliwice, Poland

E-mail: waclaw.banas@polsl.pl

Abstract. Location of components of a manufacturing cell is apparently an easy task but even during the constructing of a manufacturing cell, in which is planned a production of one, simple component it is necessary, among others, to check access to all required points. The robot in a manufacturing cell must handle both machine tools located in a manufacturing cell and parts store (input and output one). It handles also transport equipment and auxiliary stands. Sometimes, during the design phase, the changes of robot location are necessary due to the limitation of access to its required working positions. Often succeeding changes of a manufacturing cell configuration are realized. They occur at the stages of visualization and simulation of robot program functioning. In special cases, it is even necessary to replace the planned robot with a robot of greater range or of a different configuration type. This article presents and describes the parameters and components which should be taken into consideration during designing robotised manufacturing cells. The main idea bases on application of advanced engineering programs to adding the designing process. Using this approach it could be possible to present the designing process of an exemplar flexible manufacturing cell intended to manufacture two similar components. The proposed model of such designed manufacturing cell could be easily extended to the manufacturing cell model in which it is possible to produce components belonging the one technological group of chosen similarity level. In particular, during the design process, one should take into consideration components which limit the ability of robot foundation. It is also important to show the method of determining the best location of robot foundation. The presented design method could also support the designing process of other robotised manufacturing cells.

1. Introduction

Robotic manufacturing lines are increasingly replacing traditional workplace [11,12] In the case of the concept of ergonomics. If the platform is built with the principles of ergonomics worker does not get tired, it has easy access to all the necessary parts at his workplace. Workplace built without these principles are impractical and difficult to work worker can cause accidents. [15,16]

So it is with the robot to place the robot in the correct position will provide security jobs, providing access to all necessary positions [13, 14]. Now, the slot configuration is based on the experience of the designer. It often happens that in the course of the program is needed to generate reconfiguration manufacturing cell. In this case, it is necessary to start working again. When selecting the position of the robot is not possible sometimes modified the position or more positions. This is very dangerous

Content from this work may be used under the terms of the Creative Commons Attribution 3.0 licence. Any further distribution of this work must maintain attribution to the author(s) and the title of the work, journal citation and DOI.
Published under licence by IOP Publishing Ltd

because it can cause a collision. So it is with the robot to place the robot in the correct position will provide security jobs, providing access to all necessary positions [6, 7]. Now, the slot configuration is based on the experience of the designer. It often happens that in the course of the program is needed to generate reconfiguration manufacturing cell. In this case, it is needed to start working again. When selecting the position of the robot is not possible sometimes modified the position or more positions.

This is very dangerous because it can cause a collision.

For starters we need to remember that the robot does not set a point, but it is set in position (point has 3 coordinates, and the position has 6 coordinates) [4, 5]. Even if all the points of the robot work space available, it does not mean that all position in the workspace are available.

2. The search points robot assembly

Let we analyse where the robot can be installed if they are known only the operating position of the wrist. For example will show the difference between the position of the robot and the point of the robot workspace. Figure 1 shows the model of the robot from the robot wrist turned differently. Marked with the letter A workspace, marked with the letter B space in which the robot wrist can have direction as on the model.

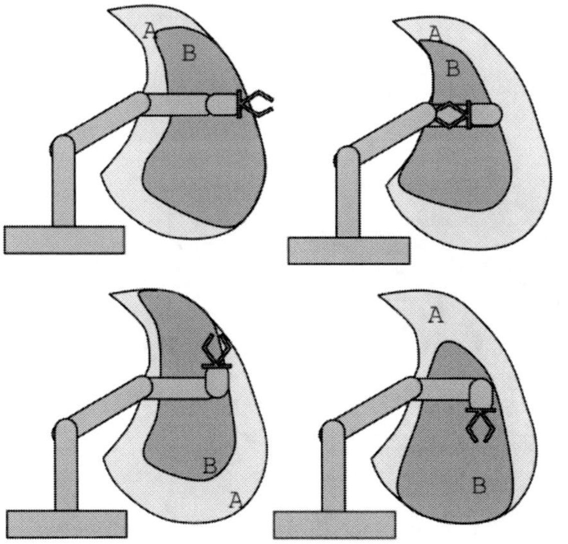

Figure 1. Comparison of working space, and space reachable in specific position of the robot wrist.

If we add equation (1) spaces reachable in each specific position of the robot wrist as a result we will have the entire workspace.

$$A=B_1\vee B_2\vee B_3\vee \ldots \vee B_n \quad (1)$$

where: A – robot workspace; B_1 … B_n – each space reachable in specific position of the robot wrist.

Workspace robot is presented everywhere because only depend on the construction of the robot and the rotation angles of the robot joints [8, 9].

The space in which the robot can be in any direction of the robot wrist is a conjunction equation (3) spaces reachable in each specific position of the robot wrist

$$C=B_1\wedge B_2 \wedge B_3 \wedge \ldots \wedge B_n \quad (2)$$

where: C – points reachable in any direction; B_1 … B_n – each space reachable in specific position of the robot wrist.

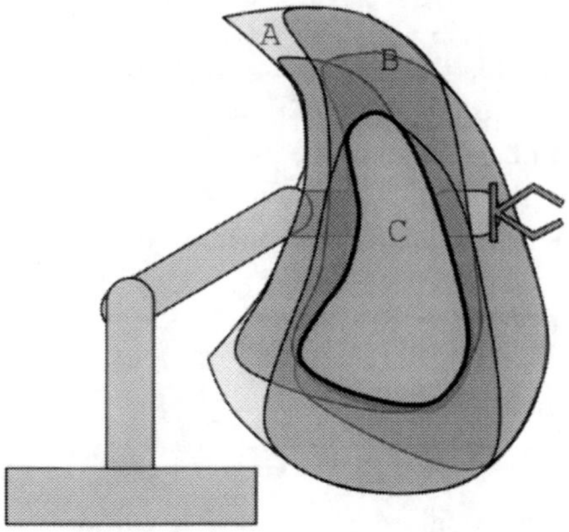

Figure 2. Points reachable in any direction.

Figure 2 shows what is the difference between workspace robot and space in which the wrist can be in any direction.

We choose any k position in Cartesian coordinates. In the traditional approach, we checked if our point is located in the workspace. But we will do otherwise, it will be fixed wrist and look for where it can stand a robot [10].

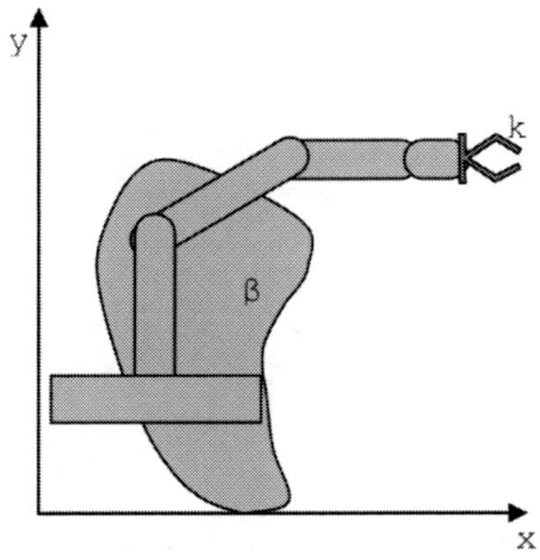

Figure 3. Possible place of the foundation of the wrist robot horizontally at the point k

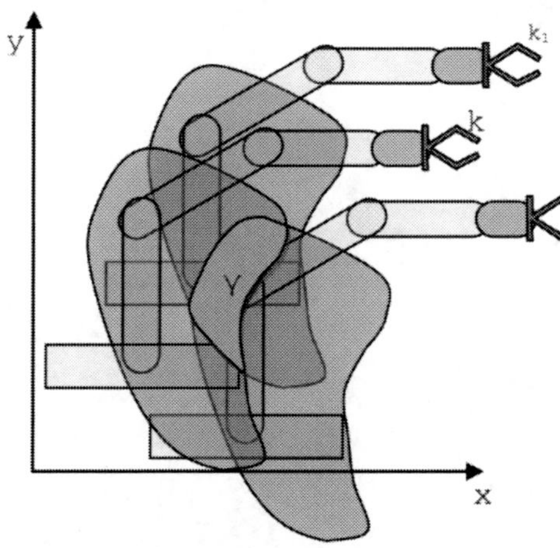

Figure 4. Possible place of the foundation of the wrist robot horizontally at the points k_1, k_2, k_3

In figure 3 the place of foundation robot the space labeled β.

The more complicated it is to find a place for the robot, which is expected to reach more position. Such an example is shown in figure 4. The space labeled γ, is the place for the robot. Easy to notice that the larger workspace robot, the more space to place the robot. Construction of a large workspace are more expensive and are usually smaller capacity. This method can be used to find the smallest robot that execute this task. When the distance between the points k_1 ... k_n increases the space in which the robot can be installed decreases.

$$\gamma=\beta_1 \wedge \beta_2 \wedge \beta_3 \wedge \ldots \wedge \beta_n \quad (3)$$

where: γ – place of installation robot to execution of all tasks; β_1 … β_n – Place of installation robot to execution of 1 … n tasks.

3. Place of installation of other components in the workspace

We need to know which direction the wrist is good for another component workspace. Figure 5 shows an example for placing another piece of the workspace. In figure 6 we can use when we know the desired direction of the wrist.

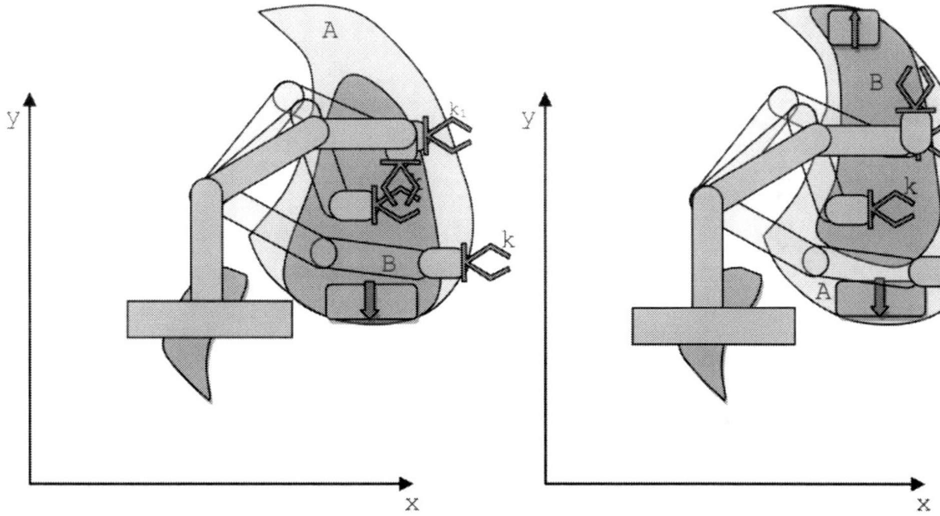

Figure 5. Place of installation of other components in the workspace.

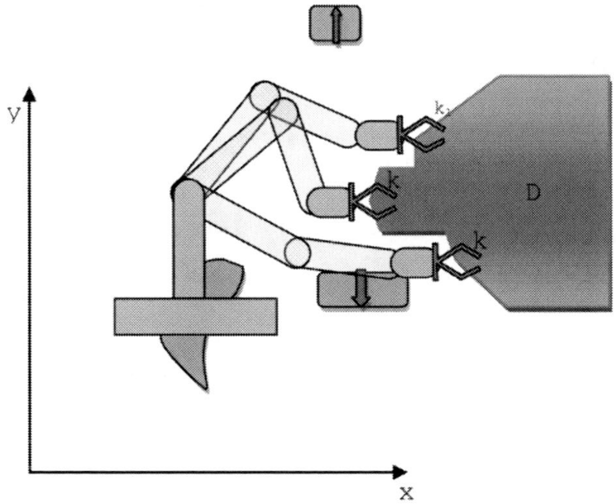

Figure 6. Final view of workspace.

Marked with the letter D is the place (part), in which the robot performs it's the task in positions k1, k2, k3. It happens that this part makes it difficult to reach the position [1].

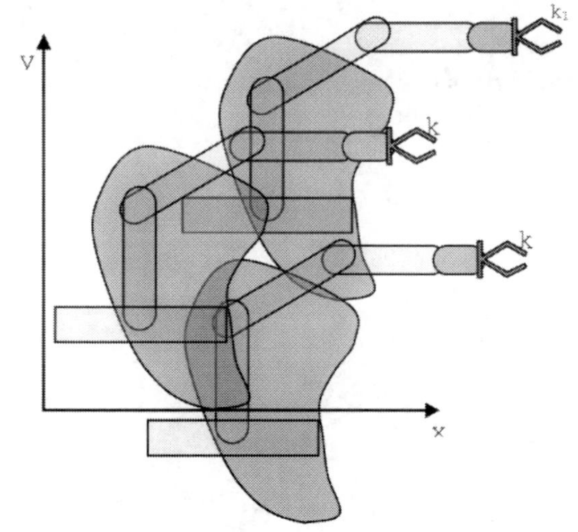

Figure 7. Example points k_1, k_2, k_3 whitout possible place of the foundation.

In a case shown on the figure 7 it is not possible to find a place from which all points are reachable. In this case, you can slightly modify the direction of the wrist and if it is unsuccessful, it is necessary to change the robot model or adding a second robot.

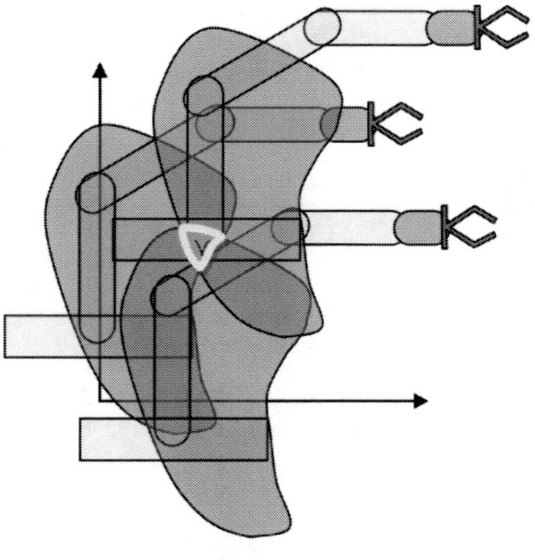

Figure 8. Example points k_1, k_2, k_3 with bigger robot.

Then it is needed to further reduce the space foundation work. Such an analysis is more complicated and will be described by the authors in later articles [2, 3].

4. Conclusions

The proposed method searches the foundation point of the robot is not yet complete requires an analysis of collision. It will be further developed for the analysis of collisions in two cases: robot carries the part to the place of assembly and lines that flow components are surrounded by a robot.

The presented method will make it easier to find the right foundation robot when its task is to reach a few positions. Use of this method will make it easier to find a place of foundation robot not by trial and error but more knowingly.

In more complex cell ought to be remember not only for proper flow of information especially when there are many components.

Acknowledgements
The work is realized within the project titled: "Modular automated production stand with instrumentation for non-invasive confirmation of product quality" funded by The National Centre for Research and Development, agreement No UOD-DEM-1-495/001.

References

[1] Ćwikła G, Sękala A and Woźniak M 2014 The expert system supporting design of the Manufacturing Information Acquisition System (MIAS) for production management *Advanced Materials Research* **1036** pp 852-857
[2] Hetmanczyk M P 2015 The Prediction Oriented Analysis of Mechatronic Machine Structures in Terms of the Signal Stream Flow *Solid State Phenomena* **220-221** pp 423-428
[3] Hetmanczyk M P and Michalski P 2014 The self-excitation phenomenon of quasi shielded inductive proximity switches *Advanced Materials Research* **837** pp 405-410
[4] Herbuś K, Kost G, Reclik D and Świder J 2014 Integration of a virtual 3D model of a robot manipulator with its tangible model (phantom) *Advanced Materials Research* **837** pp 582-587
[5] Gwiazda A, Herbuś K, Kost G and Ociepka P 2015 Designing mechatronics equipment based on the example of the Stewart platform *Solid State Phenomena* **220/221** pp 419-422
[6] Gwiazda A, Herbuś K, Kost G and Ociepka P 2015 Motion analysis of mechatronic equipment considering the example of the Stewart platform *Solid State Phenomena* **220/221** pp 479-484
[7] Ociepka P, Herbuś K and Gwiazda A 2014 Application of the CBR method for adding the design process of module manipulators *Advanced Materials Research* **1036** pp 1011-1016
[8] Herbuś K, Kost G, Reclik D and Świder J 2013 Integration of a virtual 3D model of a robot manipulator with its tangible model (phantom) *Advanced Materials Research* **837** pp 582-587
[9] Gwiazda A 2014 Construction development using virtual analysis on the example of a roof support *Applied Mechanics and Materials* **474**, pp 417-422
[10] Banaś W, Herbuś K, Kost G, Nierychlok A and Ociepka P Reclik D 2013 Simulation of the Stewart platform carried out using the Siemens NX and NI LabVIEW programs *Advanced Materials Research* **837** pp 537-542
[11] Paprocka W, Kempa M, Kalinowski K and Grabowik C 2014 A production scheduling model with maintenance *Advanced Materials Research* **1036** pp885-890
[12] Kalinowski W, Kempa M, Grabowik C and Paprocka I 2014 A survey on CAPP systems development methods *Advanced Materials Research* **837** pp 387-392
[13] Grabowik K, Kalinowski K, Paprocka and I Kempa W 2014 UML models of design and knowledge representation for technical production preparation needs *Advanced Materials Research* **837** pp 369-374
[14] Kempa W, Paprocka I, Kalinowski K and Grabowik C 2014 Estimation of reliability characteristics in a production scheduling model with failures and time-changing parameters described by Gamma and exponential distributions *Advanced Materials Research* **837** pp 116-121
[15] Sękala A, Banaś W and Gwiazda A 2014 Agent-based systems approach for robotic manufacturing cell integration *Advanced Materials Research* **1036** pp 721-725
[16] Sękala A and Dobrzańska-Danikiewicz A 2015 Possibilities of application of agent-based systems to support functioning of e-manufacturing environment *Mechatronics systems and materials* VI *Solid State Phenomena* **220/221** pp 781-784

Agent-based models in robotized manufacturing cells designing

A Sekala[1], A Gwiazda[1], K Foit[1], W Banas[1], P Hryniewicz[1] and G Kost[1]
[1]Silesian University of Technology, The Faculty Of Mechanical Engineering, Institute Of Engineering Processes Automation And Integrated Manufacturing Systems, ul. Konarskiego 18a 44-100 Gliwice, Poland

E-mail: agnieszka.sekala@polsl.pl

Abstract. The complexity of the components, presented in robotized manufacturing workcells, causes that already at the design phase is necessary to develop models presenting various aspects of their structure and functioning. These models are simplified representation of real systems and allow to, among others, systematize knowledge about the designed manufacturing workcell. They also facilitate defining and analyzing the interrelationships between its particular components. This paper proposes the agent-based approach applied for designing robotized manufacturing cells.

1. Introduction

Industrial robots are universal technological equipment. They are used in different work scenes depending on the utilized production system (figure 1). The most often they are utilized as equipment in manufacturing workcells. In this type of a production system robots realized some specialized tasks: servicing machine tools, parts reorientation and controlling the sequence of operations. Production lines are the other work scene, in which are utilized industrial robots. In comparison to workcells, production lines are external controlled. In this case industrial robots service particular manufacturing and assembly stands locate and position elements and are used as specific production buffers.

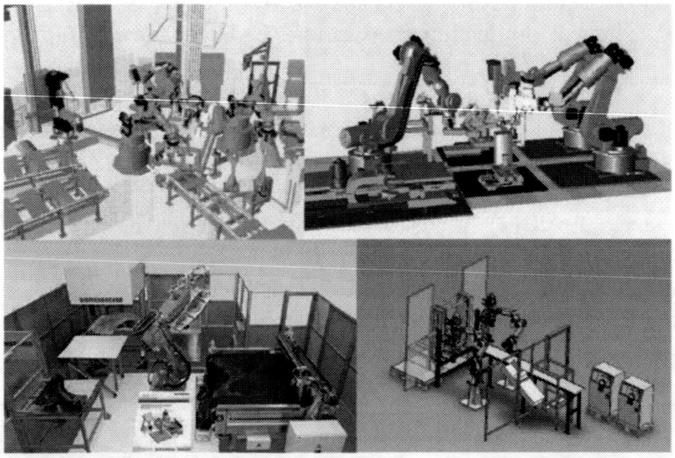

Figure 1. Different robot work scenes [1].

Content from this work may be used under the terms of the Creative Commons Attribution 3.0 licence. Any further distribution of this work must maintain attribution to the author(s) and the title of the work, journal citation and DOI.
Published under licence by IOP Publishing Ltd

Creating the robot work scenes is related with two main processes: the designing one and the selection one. During the designing process is determined the general structure of a workcell, including elements location, paths planning and area protection. During the selection process are chosen machine tools, auxiliary equipment and industrial robots. In this approach the workcell structure is the compromise between the designing and selection processes. One of the possibilities of changing this situation is to apply one of modern tools adding the heuristics elements of mentioned processes [2-4, 10-19].

2. Agent-based approach to the designing process

In the theory of artificial intelligence and automatics of the most popular tool is the concept of an agent. It is an autonomous unit which observes its environment, through sensors, and acts upon this environment using actuators [20, 21, 25]. The agent directs its activity towards achieving goals using embedded rules of rationality. Analyzing an agent, from the point of view of its construction, it could be stated that it is a system of associated elements.

An agent could be described as unit that characterizes with some properties [20-27]. Firstly it is an autonomous unit (figure 2). It means that an agent has a certain degree of control over their behavior and could operate without human or other program intervention. It could evaluate gathered facts and take proper action according to stored rules or acquired knowledge. Secondly, an agent is a communication unit. It could communicate both with others agents, and with other participants of its environment (including humans). Communication, together with autonomy and rationality allows the group of agents self-organizing. The next property of an agent is reactivity. It means that agent could respond to changes in its environment. Finally one should say that an agent is unit which is focusing on the goal. Generally speaking an agent not only reacts to changes in the environment, but it also takes the initiative to achieve determined goals. Taking into account the possibility of an agent to learn it could be described as an intelligent unit.

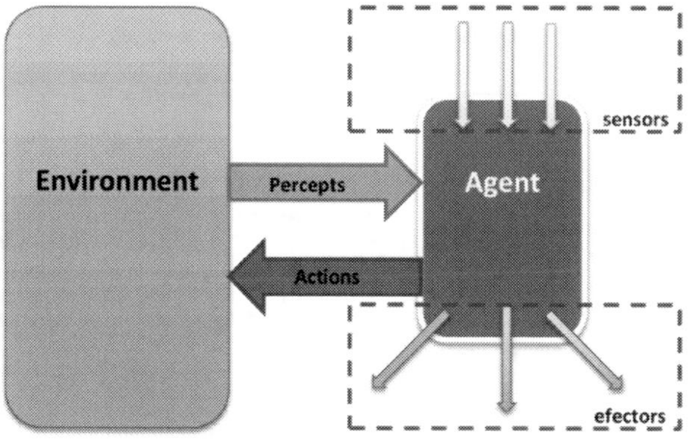

Figure 2. General model of an agent [27].

The designing process is the process of three sub-systems designing [5-9]: the structural one, the drive one and the control one. These sub-systems create a general structure of any complex technical mean. This structure corresponds to the structure of any mechatronics device [28], which includes: a mechanical sub-system (structural), an electric one (drive) and a computer one (control).

The designing process could be presented as a process of intelligent and heuristic processing of data, received from the environment in the form of technical requirements, to obtain an action taking the form of a technical solution.

The above diagram shows that the design process, from the special point of view, could be treated as a special action of an agent of a group of agents. In this form the design process is the agent-based one. Assuming the hierarchical structure of a system (including an agent) it is possible to state that a complex agent could be a system of less complex ones, what leads to the vision of a system of primary agents in the bottom of the hierarchy. Such structure is called a multi-agent system (MAS).

3. Workcell designing using MAS

A multi-agent system is a computerized system composed of multiple interacting intelligent agents within an environment [20, 27]. Multi-agent systems can be used to solve problems that are difficult or impossible for an individual agent. They are applied particularly for solving technical problems. The very important feature of the whole MAS is the self-organization. In the system of agents operation (agents and their environment) is programmable implemented the rule of self-organization of autonomous agents according to goal common for the whole system. This approach could help to elaborate the frameworks of a system for robots work scene designing, concerning the elements of that process as agents.

The MAS is frequently organized in a form of hierarchically controlled system. This hierarchical structure is called a holarchy what means a hierarchical system of agents. Groups of agents are called holons. In the same convention it could be designed a schematic structure of a holarchy that represents the system of a robotized workcell (figure 2). It also includes the supervising levels, responsible for self-organization, and the agent's level, representing devices in a workcell.

Taking into account presented considerations it is possible to elaborate the complex of a holarchy structure designated for designing robotized work scenes in a form of a workcell. This structure is presented in the figure 3.

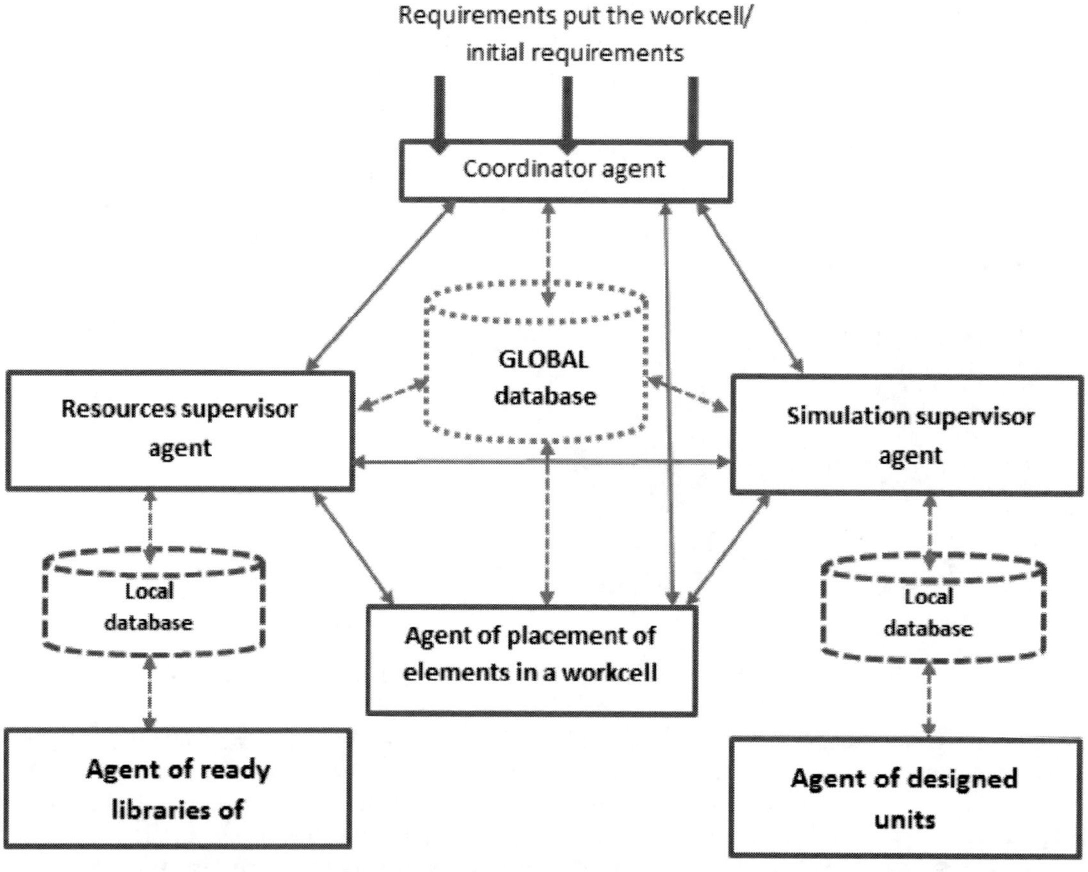

Figure 3. Holarchy for workcell designing as a self-organizing process.

The presented structure includes holons representing the mentioned previously processes of selection the workcell equipment (agents of ready libraries of elements (figure 4)) and of designing the workcell equipment (agents of designed elements (figure 5)).

IOP Conf. Series: Materials Science and Engineering **95** (2015) 012106 doi:10.1088/1757-899X/95/1/012106

Agent of ready libraries of elements	
Agent of robots	Agent of instrumentation–welding heads etc.
Agent of grippers	
Agent of storage equipment	Agent of transport systems
Agent of orientation devices	Agent of orientation devices
Agent of control systems selection	Agent of machine tools

Agent of designed units	
Agent of a material/parts flow etc.	Agent of program sequence
Agent of machines utilization	Agent of system effectiveness

Figure 4. Agents of ready libraries of elements. **Figure 5.** Agents of designed elements.

The supervising system consists of two levels of agents. The coordinator agent is aided with the global database. Agents on the lower level are aided with local databases.

4. Conclusions

In the presented work was discussed the concept of a multi-agent system designed for aiding the design process of robotized work scene (particularly a workcell). The system consists of two main groups of agents, having the possibility to learn. These groups represent the supervising levels and the operational level. Agents at the supervising level are aided with databases. The design is the result of self-organizing action of agents at the operational level.

Acknowledgements

The work is realized within the project titled: "Modular automated production stand with instrumentation for non-invasive confirmation of product quality" funded by The National Centre for Research and Development, agreement No UOD-DEM-1-495/001.

References

[1] Information on http://www.kaizenautomation.com/motosim.html, http://www.cardsplm solutions.nl/en/plm-software/tecnomatix/robcada-10/screenshots, http://www.compositesworld. com/articles/structural-preform-technologies-emerge-from-the-shadows, http://devlinksltd.com/ services/

[2] Foit K 2014 Mixed reality as a tool supporting programming of the robot *Advanced Materials Research* **1036** pp 737-742

[3] Foit K 2014 Introduction to solving task-level programming problems in logic programming language *Journal of Achievements in Materials and Manufacturing Engineering* **64/2** pp 78-84

[4] Banaś W and Sękala A 2014 Concepts of flexible production line, on the example of robotic cell *Advanced Materials Research* **1036** pp 749-754

[5] Gwiazda A, Sękala A, Monica Z and Banaś W 2014 Integrated approach to the designing process of complex technical systems *Advanced Materials Research* **1036** pp 1023-1027

[6] Gwiazda A, Sękala and Monica Z 2014 Integrated approach to the designing process of complex technical systems *Advanced Material Research* **1036** pp 1023-1027

[7] Sękala A, Gwiazda A and Banaś W 2014 Agent-based systems approach for robotic workcell integration *Advanced Materials Research* **1036** pp 721-725

[8] Gwiazda A 2014 System of designing complex technical means using fuzzy analysis *Applied Mechanics and Materials* **474** pp 147-152
[9] Gwiazda A 2014 Construction development using virtual analysis on the example of a roof support *Applied Mechanics and Materials* **474** pp 417-422
[10] Dymarek A, Dzitkowski T, Herbuś K, Kost G and Ociepka P 2014 Geometric analysis of motions exercised by the Stewart platform *Advanced Materials Research* **837** pp 351-356
[11] Ćwikła G, Sękala A and Woźniak M 2014 The expert system supporting design of the Manufacturing Information Acquisition System (MIAS) for production management *Advanced Materials Research* **1036** pp 852-857
[12] Grabowik Janik W 2013 The concrete casting matrixes inserts design preparation based on the master models *Advanced Materials Research* **702** pp 259-262
[13] Dobrzańska-Danikiewicz A 2006 The acceptation of the production orders for the realization in the manufacturing assembly systems *Journal of Materials Processing Technology* **175/1/3** pp 123-132
[14] Skołud B 2014 Market oriented approach to the production management on the operational level *Advanced Materials Research* **837** pp 663-668
[15] Grabowik A, Kalinowski K, Paprocka I and Kempa W 2014 UML models of design and knowledge representation for technical production preparation needs *Advanced Materials Research* **837** pp 369-374
[16] Dzitkowski T 2004 Computer - aided synthesis of discrete - continuous subsystems of machines with the assumed frequency spectrum represented by graphs *Journal of Material Processing Technology* **157-158** pp 144-149
[17] Buchacz A, Płaczek M and Wróbel A 2014 Modelling and analysis of systems with cylindrical piezoelectric transducers *Mechanika* **20/2** pp 183-189
[18] Dzitkowski T and Dymarek A 2014 Active reduction of identified machine drive system vibrations in the form of multi-stage gear units *Mechanika* **20/ 1** pp 87-91
[19] Dymarek A and Dzitkowski T 2013 Passive reduction of system vibrations to the desired amplitude value *J. Vibroeng.* **15/3** pp 1354-1364
[20] Botti V and Giret A 2008 A multi-agent methodology for holonic manufacturing systems (Berlin: Springer Verlag)
[21] Wooldridge M and Jennings N R 1995 Intelligent agents: theory and practice *Knowledge Engineering Review* **10/2** pp 115-152
[22] Ferreira P, Reyes V and Mestre J 2013 A web-based integration procedure for the development of reconfigurable robotic work-cells *Int. Journal of Advanced Robotic Systems* **10** pp 295-304
[23] Niazi M and Muaz H A 2011 Agent-based computing: from multi-agent systems to agent-based models. A visual survey *Scientometrics* **89/2** pp 479–499
[24] Babiceanu R F and Chen F F 2006 Development and applications of holonic manufacturing systems: a survey *Journal of Intelligent Manufacturing* **17** pp 111-131
[25] Rodriguez S, Gaud N, Hilaire V and Koukam A 2006 Modeling holonic systems with an organizational approach,A. Omicini, B. Dunin-Keplicz, J. Padget (Eds.) *Proceedings of EUMAS'06* Lisbon
[26] Monostori L, Váncza J and Kumara S 2006 Agent-based system for manufacturing *Annals of the CIRP* **55/2** pp 697-720
[27] Russell S and Norvig P 2005 Artificial intelligence: a modern approach (New York: Prentice-Hall)
[28] Silva W 2005 Mechatronics: An integrated approach (Boca Raton: CRC Press)

Modelling of cooperating robotized systems with the use of object-based approach

K Foit[1], A Gwiazda[1], W Banas[1], A Sekala[1] and P Hryniewicz[1]
[1]Silesian University of Technology, The Faculty Of Mechanical Engineering, Institute Of Engineering Processes Automation And Integrated Manufacturing Systems, ul. Konarskiego 18a 44-100 Gliwice, Poland

E-mail: krzysztof.foit@polsl.pl

Abstract. Today's robotized manufacturing systems are characterized by high efficiency. The emphasis is placed mainly on the simultaneous work of machines. It could manifest in many ways, where the most spectacular one is the cooperation of several robots, during work on the same detail. What's more, recently a dual-arm robots are used that could mimic the manipulative skills of human hands. As a result, it is often hard to deal with the situation, when it is necessary not only to maintain sufficient precision, but also the coordination and proper sequence of movements of individual robots' arms. The successful completion of this task depends on the individual robot control systems and their respective programmed, but also on the well-functioning communication between robot controllers. A major problem in case of cooperating robots is the possibility of collision between particular links of robots' kinematic chains. This is not a simple case, because the manufacturers of robotic systems do not disclose the details of the control algorithms, then it is hard to determine such situation. Another problem with cooperation of robots is how to inform the other units about start or completion of part of the task, so that other robots can take further actions. This paper focuses on communication between cooperating robotic units, assuming that every robot is represented by object-based model. This problem requires developing a form of communication protocol that the objects can use for collecting the information about its environment. The approach presented in the paper is not limited to the robots and could be used in a wider range, for example during modelling of the complete workcell or production line.

1. Introduction

Robots used in the production process getting better in implementing the activities carried out by worker. Newly designed manipulators are equipped with the pair of arms and be able to perform better in activities where the worker must use both hands. Some production lines are also equipped with multi-manipulator cells, which can perform the tasks that are beyond the capabilities of individual robot. Such cell realizes the task in a manner similar to the group of workers, so the robots should act in the way that ensures the correct accomplishment of work. The cooperation of manipulators may have a different character, but the most specific one is the collective manipulation with the same object [1-3] – the manipulators form a closed kinematic chain, analysis of which is more complicated. There are also other problems like load balancing between arms, force and position control etc. In some cases the kinematic chain may be closed for a while, but the manipulated object is motionless. It occurs when the one robot holds the object, while the other acts on it with some type of tools.

Content from this work may be used under the terms of the Creative Commons Attribution 3.0 licence. Any further distribution of this work must maintain attribution to the author(s) and the title of the work, journal citation and DOI.
Published under licence by IOP Publishing Ltd

Regardless of the case, the cooperation between robots requires the cooperation between its control systems.

The paper will discuss some issues connected with communication between the cooperating robots. In addition to the hardware solution, which means the use of appropriate connections (network) and communication protocols, it is also necessary to provide the right solutions at the level of program development. It is essential that there should exist some forms of messages that will be interpreted in the right way by all involved systems.

2. The Current State of Knowledge

2.1. Industrial and computer networks, topology of networks, hardware.

The one of the fast growing areas of industrial automation is the application of the computer networks. There exist many different solutions developed by many different manufacturers. Some of the network systems are specialized in a particular field, such as networks dedicated to the transmission of signals from the sensors, intended for control the drives or communication between robotic systems. Regardless of the application and manufacturer, network systems use of certain standards. The most important of these standards is the layered OSI model (ISO/IEC 7498-1), which structure is shown in figure 1. The networks are also characterized by topology, what is – in short – the organization of network nodes. The commonly used topologies are shown in figure 2.

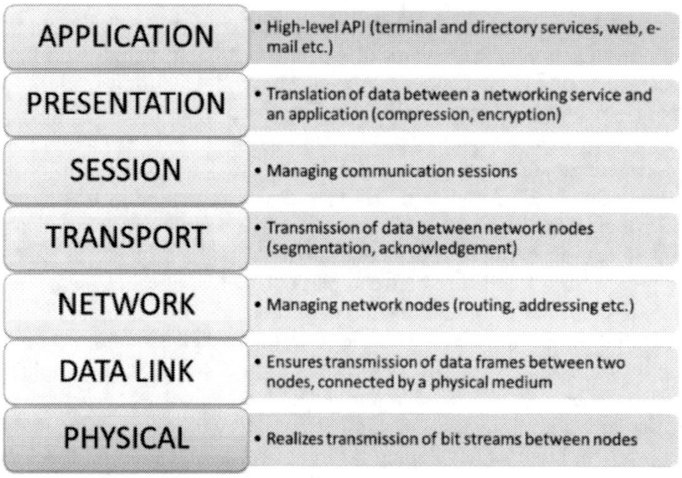

Figure 1. The layers of OSI model and its description.

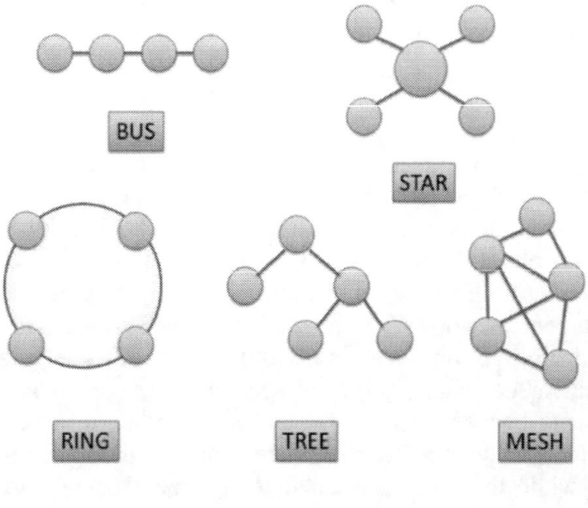

Figure 2. The commonly used topologies of computer networks.

Every protocol or standard is created in a way that respects the OSI model. It is important to take into consideration the fact that not every layer of OSI must be implemented.

In the case of industrial computer networks, the commonly used transmission medium is a copper wire or fiber optic. The wireless connections – although used – are not intended to connect network nodes of strategic importance, because this type of transmission is very susceptible to interferences. Wireless transmission of data is suited for less error-prone systems, like MES (Manufacturing Execution Systems) modules that collect the information about production process the PLC and save the data to the central database.

In the industrial network, a wide range of protocols are used, beginning from the simplest RS-232 protocol, through RS-485, Profibus, Ethernet, CAN etc. The type of used protocol is determined by communication requirements between network nodes (e.g. The AS-i protocol is best suited for sensors network, while the Ethernet is better for overall data communication). In the most of cases, the three basic structures can be distinguished: Master-slave, Multimaster and Ad-hoc.

The Master-slave structure is fairly widespread. It often uses the advanced properties of the PLC controllers for supervising and coordinating all the remained network devices. In the multimaster systems, every device has the same rights (for example in Ethernet or Profibus DP networks), while the ad-hoc networks are used for connecting two devices, like e.g. computer and robot controller. In such cases, the RS-232 or RS-485 standards are sufficient for this type of tasks.

The principles discussed above also apply to the networks where some of nodes are the controllers of robots. From the point of view of used physical links and protocols, the communication between the robots will occur in the same way as in the case of other automation devices.

2.2. Cooperating Robots: Communication and Programming Issues

Problems of cooperation between several robots are often referred to the mobile robotics [4, 5, 6]. Many authors use agent-based approach in order to take advantage of methods based on behavior of a community of living organisms. Jung and Zelinsky point out that "*Cooperation and communication are closely tied. [...] If we are to implement a concrete cooperative task that requires symbolic level communication, we must first identify the relationship between communication and cooperative behavior*" [5]. They use three levels of representation: iconic, indexical and symbolic to build the system of communication between two cooperating mobile robots that is based on living organisms communication abilities. As a starting point they use the results of researches on communication strategy in human society and animals' world. A slightly different approach is presented by Yan et al. [6] that advises to be careful when considering the Multi-Robot Systems (MRS) and Multi-Agent Systems (MAS). They refers to the model of Multi-Agent System as *"[...] traditional distributed computer system in which individual nodes are stationary*", while the Multi-Robot System "*[...] involves mobile robots that can move in the physical world and must interact with each other physically*". The authors also point out that there exist two types of strategies: cooperative or competitive. They emphasize the role of communication during realization of the cooperative strategy because of possible resource conflicts like sharing the communication medium, manipulate of the same object or sharing the same workspace.

The problems connected with communication medium sharing are discussed among others by Ye et al. [7] and Rybski et al. [8]. The first team examined the optimal use of communication bandwidth, while the other proposed the method of communication, developed for low-bandwidth networks. The mentioned method uses the scheduling that dynamically assigns the available resources to the robots in order to maximize the efficiency of data stream. The others publications that deals with the same problem are [9,10].

In the case of manipulation of the same object by more than one robot, the researchers concentrate mainly on the issues connected with kinematic chain [1-3]. The interesting discussion about self-organized group of robots is presented in [11], where robots mimic the behavior of the swarm of ants. In turn, Miyata et al. [12] proposed the system based on central motion and task planning, in order to achieve optimal effectiveness of the swarm of robots. The robots are divided into small teams that are delegated to the specific tasks.

Most of the issues concerning the cooperation between the robots are discussed in relation to mobile robots. The collaboration of typical, arm-equipped industrial robots is described in [13]. The method presented by Braun et al. refers to the group of robots that forms a workcell. The robots manipulate the same object at the time and the movement is supervised by common controller. This can provide the higher level of functionality, where all the robots, that form the workcell, could be treated as the single, "virtual" robot with several arms. In the other paper, Norman et al. [14] draw attention to an important aspect of manipulators' cooperation, which is the accuracy of movement. They proposed the so-called Indoor GPS system to control the position of each robot's arm.

As it could be seen, there is no standard approach to the problem of control of the robots that form a group and providing communication between them. At the highest level of generality it does not matter if we consider mobile robots or stationary ones, the used methods will be more or less the same, starting with the physical medium of transmission and eventually using the control algorithms for the specified group of the robots.

3. The Communication Between Object-Base Models of The Robots

The object-based model of the robot is the form that is not identified with particular type of a robot. It is possible to mention the brand and type in the object structure, but it is not necessary. In other words, the object-based model of robot could represent different brands and types of real robots that have similar properties (technical data, capabilities) and kinematic structure. In general, we could define the object as a set of features F (equation (1)):

$$O = \{F\} \Leftrightarrow F = \{R, A\} \tag{1}$$

where R is the set of routines (procedures, functions) and A the set of attributes. Objects will communicate with each other using the attributes and routines in order to set the appropriate attribute. In that manner we have three major ways of implementing the communication. The first one (figure 3a) is to use the blackboard or "scratchpad", where each object can write and read data. The blackboard is represented by the memory block, which is organized in a way that is recognized by all objects (robots). For example the blackboard could store the orientation of the subject of manipulation, coordinates of robots' effectors, tools' statuses etc. This is the implementation of public data structure that is the simplest method of storing the data, however very dangerous – in the case of writing improper data into the structure, entire system could be damaged. Therefore a slightly different solution should be used (figure 3b), wherein the table is a private property of one of the objects. This object supervises the correctness of the data written into the memory, by providing the read and write routines, that are available to the other objects. These routines could be either public or inherited, depending on the structure of objects. The third possibility bases on assumption that there is no public data structure available, but every object keeps the private copy of the data and shares it with the others (figure 3c). The communication model is therefore changed, because every object updates their own data table that then could have the read-only status for others. This solution is closer to the pattern of communication between agents in agent-based model [15], where the flow of information is regulated by sending requests to the other agents and receiving answers from them.

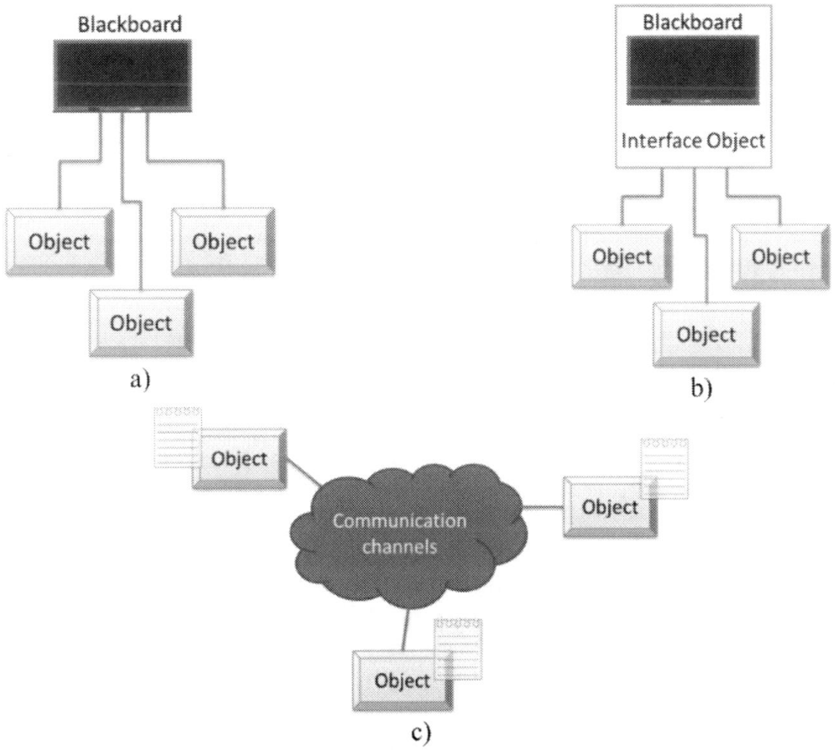

Figure 3. Communication models between objects, [16, 17].

Conclusions

The communication between objects in robotized systems is a complex problem. The main issue is the choice of appropriate hardware and communication protocols. The next step is to take into account the specific nature of the considered problem. Most of the research works describe the cases connected with mobile robotics and uses the agent-based description of the model. However many of conclusions could be also used in stationary robotics – like sharing of the communication medium, simultaneous manipulating of the same subject or planning the collision-free paths. Especially valuable is the idea of considering the management of multiple robots that form the workcell as the control of virtual, multi-arm robot.

The model of the robot in the form of the software object is most commonly used in all kinds of simulation programs. During performing the simulation in programs that allow advanced mathematical modeling (for example, Matlab, LabView, etc.), sometimes there is the need to deal with the representation of the object-based model in the form of source code. The methods of communication between objects that was described in this paper are only the basic solutions – they are the most universal, but not the most effective ones. On the other hand they could be implemented and elaborated in many programming environments that support procedural or object-oriented programming.

Acknowledgements

The work is realized within the project titled: "Modular automated production stand with instrumentation for non-invasive confirmation of product quality" funded by The National Centre for Research and Development, agreement No UOD-DEM-1-495/001.

IOP Conf. Series: Materials Science and Engineering **95** (2015) 012107 doi:10.1088/1757-899X/95/1/012107

References

[1] Hayati S 1986 Hybrid position/force control of multi-arm cooperating robots, Robotics and Automation *Proceedings of IEEE International Conference* **3**
[2] Al-Yahmadi A S, Abdo J and Hsia T C 2007 Modeling and control of two manipulators handling a flexible object *Journal of the Franklin Institute* **344.5** pp 349-361
[3] Bicchi A and Prattichizzo D 2000 Manipulability of cooperating robots with unactuated joints and closed-chain mechanisms *Robotics and Automation, IEEE Transactions* **16.4** pp 336-345
[4] Di Marco M, Garulli A, Giannitrapani A and Vicino A 2003 Simultaneous localization and map building for a team of cooperating robots: a set membership approach *Robotics and Automation, IEEE Transactions* **19(2)** pp 238-249
[5] Jung D and Zelinsky A 2000 Grounded symbolic communication between heterogeneous cooperating robots *Autonomous Robots* **8.3** pp 269-292
[6] Yan Z, Jouandeau N and Cherif A A 2013 A survey and analysis of multi-robot coordination *International Journal of Advanced Robotic Systems* **10**
[7] Ye W, Vaughan R T, Sukhatme G, Heidemann J, Estrin D and Mataric M J 2001 Evaluating control strategies for wireless-networked robots using an integrated robot and network simulation *Robotics and Automation, Proceedings ICRA. IEEE International Conference* 3 pp 2941-2947
[8] Rybski P E, Stoeter S A, Gini M, Hougen D F and Papanikolopoulos N P 2002 Performance of a distributed robotic system using shared communications channels *IEEE Transactions on Robotics and Automation* **18(5)** pp 713-727
[9] Nerurkar E D, Zhou K X and Roumeliotis S I 2011 A hybrid estimation framework for cooperative localization under communication constraints *Proceedings of IROS'11* pp 502-509
[10] Krontiris A and Bekris K E 2011 Using minimal communication to improve decentralized conflict resolution for non-holonomic vehicles *Proceedings of IROS'11* pp 3235-3240
[11] Kube C R and Bonabeau E 2000 Cooperative transport by ants and robot *Robotics and autonomous systems* **30.1** pp 85-101
[12] Miyata N, Ota J, Arai T and Asama H 2002 Cooperative transport by multiple mobile robots in unknown static environments associated with real-time task assignment *IEEE Transactions on Robotics and Automation* **18(5)** pp 769-780
[13] Braun B M, Starr G, Wood J E and Lumia R 2004 A framework for implementing cooperative motion on industrial controllers. Robotics and Automation *IEEE Transactions* **20(3)** pp 583-589
[14] Norman A R, Schönberg A, Gorlach I A and Schmitt R 2010 Cooperation of industrial robots with indoor-GPS *Proceedings of the international conference on competitive manufacturing–COMA* **10** pp 215-224
[15] Finin T, Fritzson R, McKay D and McEntire R 1994 KQML as an agent communication language *Proceedings of the third international conference on Information and knowledge management, ACM* pp 456-463
[16] Alves J Blackboard https://openclipart.org/detail/49363/Blackboard, Accessed: 28.02.2015
[17] Agone, Notes, CC Zero License, https://openclipart.org/detail/22037/Notes, Accessed: 28.02.2015

Object as a model of intelligent robot in the virtual workspace

K Foit[1], A Gwiazda[1], W Banas[1], A Sekala[1] and P Hryniewicz[1]
[1]Silesian University of Technology, The Faculty Of Mechanical Engineering, Institute Of Engineering Processes Automation And Integrated Manufacturing Systems, ul. Konarskiego 18a 44-100 Gliwice, Poland

E-mail: krzysztof.foit@polsl.pl

Abstract. The contemporary industry requires that every element of a production line will fit into the global schema, which is connected with the global structure of business. There is the need to find the practical and effective ways of the design and management of the production process. The term "effective" should be understood in a manner that there exists a method, which allows building a system of nodes and relations in order to describe the role of the particular machine in the production process. Among all the machines involved in the manufacturing process, industrial robots are the most complex ones. This complexity is reflected in the realization of elaborated tasks, involving handling, transporting or orienting the objects in a work space, and even performing simple machining processes, such as deburring, grinding, painting, applying adhesives and sealants etc. The robot also performs some activities connected with automatic tool changing and operating the equipment mounted on the wrist of the robot. Because of having the programmable control system, the robot also performs additional activities connected with sensors, vision systems, operating the storages of manipulated objects, tools or grippers, measuring stands, etc. For this reason the description of the robot as a part of production system should take into account the specific nature of this machine: the robot is a substitute of a worker, who performs his tasks in a particular environment. In this case, the model should be able to characterize the essence of "employment" in the sufficient way. One of the possible approaches to this problem is to treat the robot as an object, in the sense often used in computer science. This allows both: to describe certain operations performed on the object, as well as describing the operations performed by the object. This paper focuses mainly on the definition of the object as the model of the robot. This model is confronted with the other possible descriptions. The results can be further used during designing of the complete manufacturing system, which takes into account all the involved machines and has the form of an object-oriented model.

1. Introduction

Over the years, the role of industrial robots has changed in a meaningful way. Originally intended solely for manipulation, robots began to more actively participate in the production process, taking over the activities associated with work in hazardous environments (high temperature, fumes, etc.). This was mainly such works as painting of large surfaces (e.g., automobile sheets), welding and soldering, sandblasting etc. Nowadays the robots also carry out minor metal works like deburring, drilling, grinding or milling. Modern control systems allow using advanced vision systems, scanners, and tool changers. These entire features cause the robots increasingly replace a man at his workplace.

Content from this work may be used under the terms of the Creative Commons Attribution 3.0 licence. Any further distribution of this work must maintain attribution to the author(s) and the title of the work, journal citation and DOI.
Published under licence by IOP Publishing Ltd

For this reason, their model must be considered in a similar way as a "worker" on a production line – more universal than any other machine and performing with different tools.

One of the possible methods of description of the robot's model is to treat the robot as an object in the virtual space. The paper discusses the main assumptions of the mentioned method as well as compares it to the others, existing methods of modelling.

2. Basics of the method

At the beginning of the 21st century, the methods based on agent systems began to grow in popularity [1-3]. Although the concept of an agent was already known, it had not significant practical application. It was the dynamic development of information technology that has caused a significant growth of interest in the subject among researchers. Nowadays agent systems are common, especially in IT solutions [3]. They are also used to model phenomena in human or animals populations that may be further adapted to the needs of robotics – particularly for control of mobile robots group. In this context, the return to the concept of the object as a model of the robot seems a bit archaic, but it is a deliberate action, because agents are – more or less – the natural development of the object model.
One of the most important features that characterize an agent is its fit in a particular environment [4, 5]. Referring this property to the machines, we could say that the agent intended for cold environment cannot act in the hot one; the agent intended for terrestrial use cannot fly etc. The agent is also considered as an autonomous, so its interaction with the surrounding world depends on its internal "intelligence". As a contrast, the object is more universal and general description. It is often depended on the other objects, so it cannot be considered as autonomous. The communication between object and the environment is rather determined, but sometimes the object is a "dumb" formation, which does not interact with its surroundings. The more differences that distinguish objects from agents are shown in the table 1.

Table 1. Some differences between objects and agents.

Property	Agent	Object
Environment	specific	undefined
Level of abstraction	higher (strict description)	high (general description)
Inheritance	not precluded from some form of inheritance	commonly used
Autonomy	autonomous	related
Type/class notion	none	commonly used

The distinction between the agent and the object becomes clear when we ask about the role, which the object should to fulfill. For example, if someone has a transport company, he also has trucks. The trucks could be of different types, like dump truck, platform, container, refrigerator truck etc. He also could have some busses to transport people. Speaking broadly, we see that this man has some vehicles. If we want to describe these vehicles as an object, we could define the class *vehicle* and create the several objects that are distinguished by purpose (figure 1): for example *container_truck*, *bus* or *minibus*. Every object will be a child of *vehicle* class and will inherit its properties and methods – the examples of properties could be a color, engine type or vehicle mass, while the example methods could be drive, engine start etc. Treating the vehicle as an agent is slightly different. Considering the definition of the agent, we should treat every vehicle as an autonomous subject. In that way, it is easy to see what the agent could "say" about itself, or better – what the driver of the particular vehicle could say. We could collect the information such the position of vehicle, if it is duty or free, the amount of fuel, failures etc. The information could be stored in the management center, which – in turn – will collate them with the customers' demands (figure 2). The lines correspond to the communication abilities of each agent – it could send messages not only to the center, but also to the other agent.

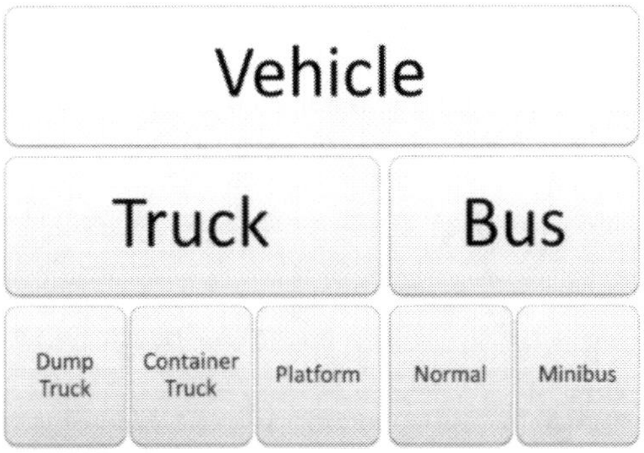

Figure 1. Simple illustration of object tree.

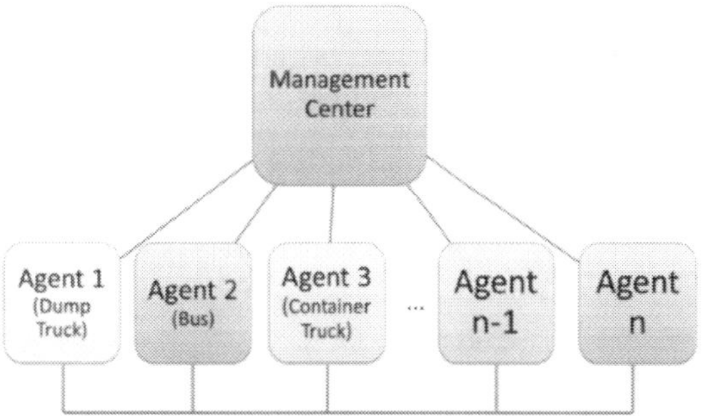

Figure 2. Simple organizational scheme of agent-based system.

As it can be seen, the basic distinction between the agent and the object depends primarily on approach [3, 6, 7]. It is therefore possible to treat an object as a base for creating the agent. This property will be used in further part of discussion.

3. The Object as a Model of a Robot

In the earlier discussion, the robot has been identified as the most complicated part of the production line. For this reason, it is difficult to find the right model that describes the robot in a universal way. Of course, the agent model could be taken into consideration: it contains all the information needed to specify the type of robot and control it. On the other hand this model applies only to the specific case and the concrete environment. In practice, the idea is to make the model of the robot as general as possible. The further upgrade of the model – making it more detailed – will be possible by adding more elements to an existing "skeleton". This approach corresponds to the application of model in the form of object [8-11], along with the implications in the form of inheriting of properties and methods.

Assuming that the object is basically the set of properties and methods (equation (1)),

$$O = \{P, M\}, \tag{1}$$

where the *P* stands for properties and *M* for methods, we could create the class named *Robot* in a similar way (equation (2)), adding the *name* attribute that distinguishes the particular robot from others :

$$\begin{aligned} &Robot = \{P, M\}, \\ &name \in P. \end{aligned} \tag{2}$$

The properties and methods can be used within the meaning of features that characterize a particular model of the robot, but it is difficult to use them at a high level of generality. It must be recognized that, in principle, it is impossible to choose all the features of an industrial robot at the stage of planning the production process. In this case, we can rely on the scope of activities that the robot will perform and use them as the basis to define the subsequent requirements. Speaking in a very general way, the kinematic chain of the robot could be the representation of the properties, while the control system could represent the methods.

4. Building an Object-based Model of a Robot

The first stage of designing of the production line, when the particular task is assigned to the particular machine (equation (3)):

$$f : T \rightarrow Mch \tag{3}$$

where T is the set of tasks and Mch the set of machines. Assuming that some tasks will be realized by the robot or by the group of robots, we should assign the task to the robot (equation (4)):

$$\exists t \in (T_1 \subset T) \wedge \exists r \in (R \subset Mch) : \varphi(t, r) \tag{4}$$

where T_1 is a subset of tasks that are realized by robot, R is a set of robots and $\varphi(t, r)$ means that the task t is realized by robot r.

At this moment, the tasks are assigned to the robot (or robots), but the sets of properties and method of the *Robot* class are empty. In order to accomplish the task we should fill the set of properties P and set of method M in the manner that (equation (5))

$$\forall[(r \in R) \wedge (t \in T_R)] \exists[(p \in P) \wedge (m \in M)] \Rightarrow \varphi(t, r) \tag{5}$$

where $T_R \in T$ is a subset of tasks realized by the particular robot. The equation 5 is based on assumption that the robot could realize the tasks from the T_R subset (see equation (4)).

5. Relationship Between Real Robot and its Object-based Model

As a result of conduction of the operations mentioned in the previous section, the complete model of the robot is obtained in the form of the object, but depending on the assumed level of details, the resulting models will differ.

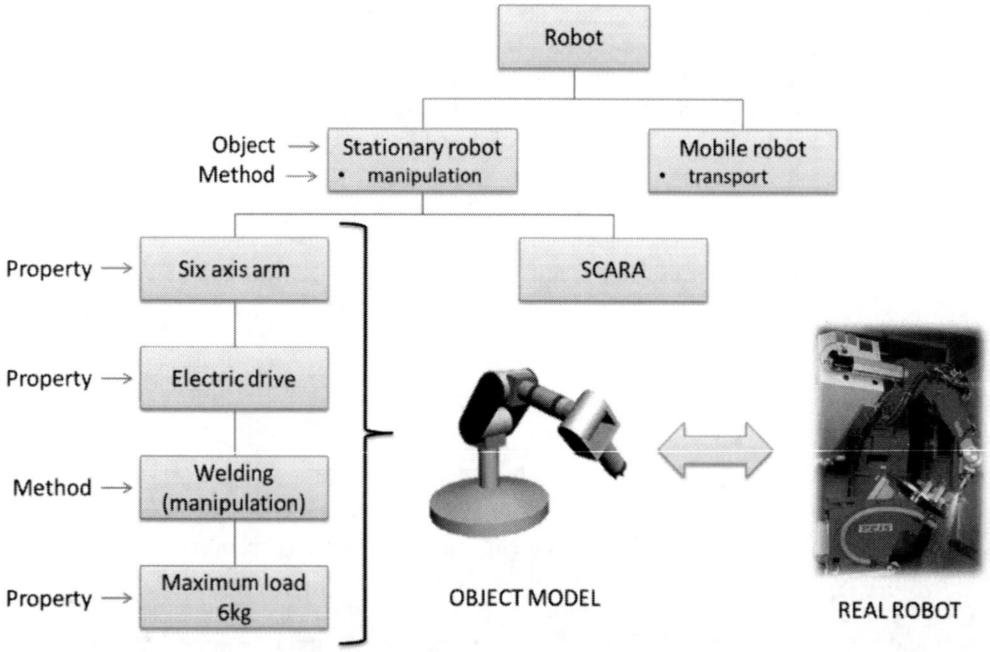

Figure 3. Simplified process of building the object-based model and its reference to real robot, [12-13].

Successive stages of assignment of another attributes leads to the complete characteristic of the robot. After completion of the process, the generalized model of the robot is obtained. In fact, due to its specific role and fit in a particular environment, this model can be seen as an agent. The final phase of the whole process is selection of a particular model of the robot from a specific manufacturer (figure 3).

6. Conclusions

Considering the production process we should pay attention to the things that are important on particular stage of designing the production line. First of all it is essential to define the types of machines that are involved in the whole process. Next the properties should be considered and eventually the way of communication (in the meaning of "language") between all of the devices. Because of its features, the robot should be seen as the very versatile machine that resembles the worker. For this reason, it is difficult to develop a model that will describe various application of the particular robot. In this paper the object-based approach was presented, as the one of possible ways of covering this problem.

The method shown in the paper, proposes the creation of abstract class named *Robot*, which could represent a real robot that acts in the production process. Further objects, which are created, are based on this class and equipped with the appropriate attributes and methods that allow fulfilling the specific tasks. The model, which is obtained in this manner, represents a hypothetical robot involved in the considered manufacturing process. Such a description is closer to the concept of an agent than an object, but the process leading to a final model is different – in the case of the object we are dealing with the evolution of the model and the creation of new forms based on previous results (inheritance).

The future use of the developed form of the robot's model could be connected with designing of manufacturing systems or robotized cells, using selected aspects of agent theory and object-oriented description.

Acknowledgements

The work is realized within the project titled: "Modular automated production stand with instrumentation for non-invasive confirmation of product quality" funded by The National Centre for Research and Development, agreement No UOD-DEM-1-495/001.

References

[1] Kubera Y, Mathieu P and Picault S 2010 Everything can be Agent! *Proceedings of the 9th International Conference on Autonomous Agents and Multiagent Systems* **1** pp 1547-1548

[2] Wood M F and DeLoach S A 2001 An overview of the multiagent systems engineering methodology *Agent-Oriented Software Engineering* **207-221** (Berlin: Springer Berlin Heidelberg)

[3] Jennings N R 2001 An agent-based approach for building complex software systems *Communications of the ACM* **44(4)** pp 35-41

[4] Franklin S and Graesser A 1997 Is it an Agent, or just a Program?: *A Taxonomy for Autonomous Agents, Intelligent agents III agent theories, architectures, and languages* (Berlin: Springer Berlin Heidelberg) pp 21-35

[5] Yan Z, Jouandeau N and Cherif A A 2013 A survey and analysis of multi-robot coordination *International Journal of Advanced Robotic Systems* **10**

[6] Sheremetov L B and Smirnov A V 1999 Component integration framework for manufacturing systems re-engineering: agent and object approach *Robotics and autonomous systems* **27(1)** pp 77-89

[7] Smimov A V and Sheremetov L B 1998 Agent & object-based manufacturing systems re-engineering: A case study *Changing the Ways We Work* pp 369-378

[8] Booch G 1986 Object-oriented development. Software Engineering *IEEE Transactions* **2** pp 211-221

[9] Rumbaugh J Blaha M Premerlani W Eddy F Lorensen W E 1991 Object-oriented modeling and design **199(1)** (Englewood Cliffs: Prentice-hall)
[10] Tummescheit H 2002 Design and implementation of object-oriented model libraries using modelica *Doctoral dissertation, Lund University*
[11] Otter M Elmqvist H Cellier F E 1996 Modeling of multibody systems with the object-oriented modeling language Dymola *Nonlinear Dynamics* **9(1-2)** pp 91-112
[12] RRU-Marketing, Laser Robot, CC-BY 3.0 Licence 2015 Retrieved from http://commons.wikimedia.org/wiki/File:Reis_Laser.jpg
[13] NeD80, 2015 3D Model of robotic arm, Public Domain, Retrieved from http://commons.wikimedia.org/wiki/File:Robot_arm_model_1.png

Virtual modelling of components of a production system as the tool of lean engineering

Z Monica[1]
[1]Silesian University of Technology, The Faculty Of Mechanical Engineering, Institute Of Engineering Processes Automation And Integrated Manufacturing Systems, ul. Konarskiego 18a 44-100 Gliwice, Poland

E-mail: zbigniew.monica@polsl.pl

Abstract. Between the most effective techniques of manufacturing management is considered the Lean Engineering. The term "lean engineering" was created by Japanese manufacturers. The high efficiency of this method resulted in a meaningful growth in concern in the philosophy of Lean among European companies, and consequently the use of its European markets. Lean philosophy is an approach to manufacturing to minimize the use of all resources, including time. These are resources that are used in the company for a variety of activities. This implies, first identify and then eliminate activities which does not generate added value in the field of design, manufacturing, supply chain management, and customer relations. The producers of these principles not only employ teams multi-professional employees at all levels of the organization, but also use a more automated machines to produce large quantities of products with a high degree of diversity. Lean Engineering is to use a number of principles and practical guidelines that allow you to reduce costs by eliminating absolute extravagance, and also simplification of all manufacturing processes and maintenance. Nowadays it could be applied the powerful engineering programs to realize the concept of Lean Engineering. They could be described using the term CAD/CAM/CAE. They consist of completely different packages for both the design of elements, as well process design. Their common feature is generally considered with their application area. They are used for computer programs assisting the design, development and manufacturing phases of a manufacturing process. The idea of the presented work is to use the Siemens NX software for aiding the process of Lean Engineering system creating. The investigated system is a robotized workcell. In the NX system are created the components of the designed workcell such as machine tools, as industrial robot, as conveyors and buffers. The system let to functionally link these components to simulate the work process and to introduce the rules of Lean Engineering. The purpose is also to determine the rules of Lean designing in such advanced design and simulation environments.

1. Introduction

One of the elements of a modern attitude towards the designing and optimization of complex technical systems is the use of advanced computer engineering environments [1,2]. They allow the betterment of technical solutions both in terms of geometric features and materials ones [3,4,5], which cause increasing of the quality level of technical solutions. The complementary approach is represented by solutions of specific problems that could be identified in complex technical systems [6,7]. This work presents a general approach based on the use of a 3D object of a manufacturing system, as a robotized

Content from this work may be used under the terms of the Creative Commons Attribution 3.0 licence. Any further distribution of this work must maintain attribution to the author(s) and the title of the work, journal citation and DOI.
Published under licence by IOP Publishing Ltd

work cell, to the issue of manufacturing process optimizing. The investigated system includes three devices: one lathe, one milling centre and one robot. This virtual environment was created using the Siemens NX 8.5 program

2. Creating the model of the lathe AFN TAE 45

The first element, of the created workcell, was the 3D model of the TAE AFN 45 lathe (figure 1). The model was developed using previously developed components (figure 2). It has a processor that allows executing instructions programmed in G-code.

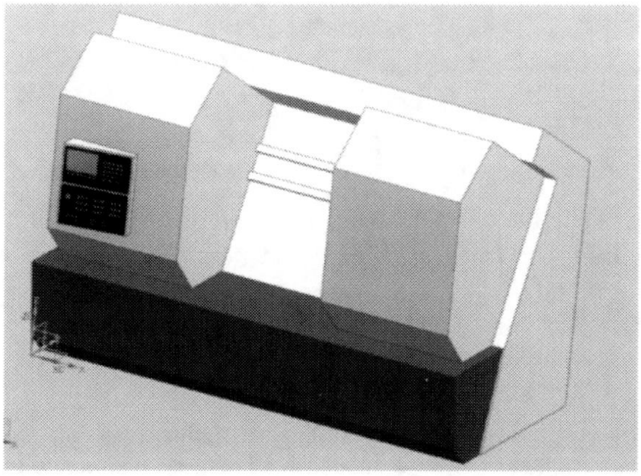

Figure 1. The 3D model of the lathe AFN TAE 45.

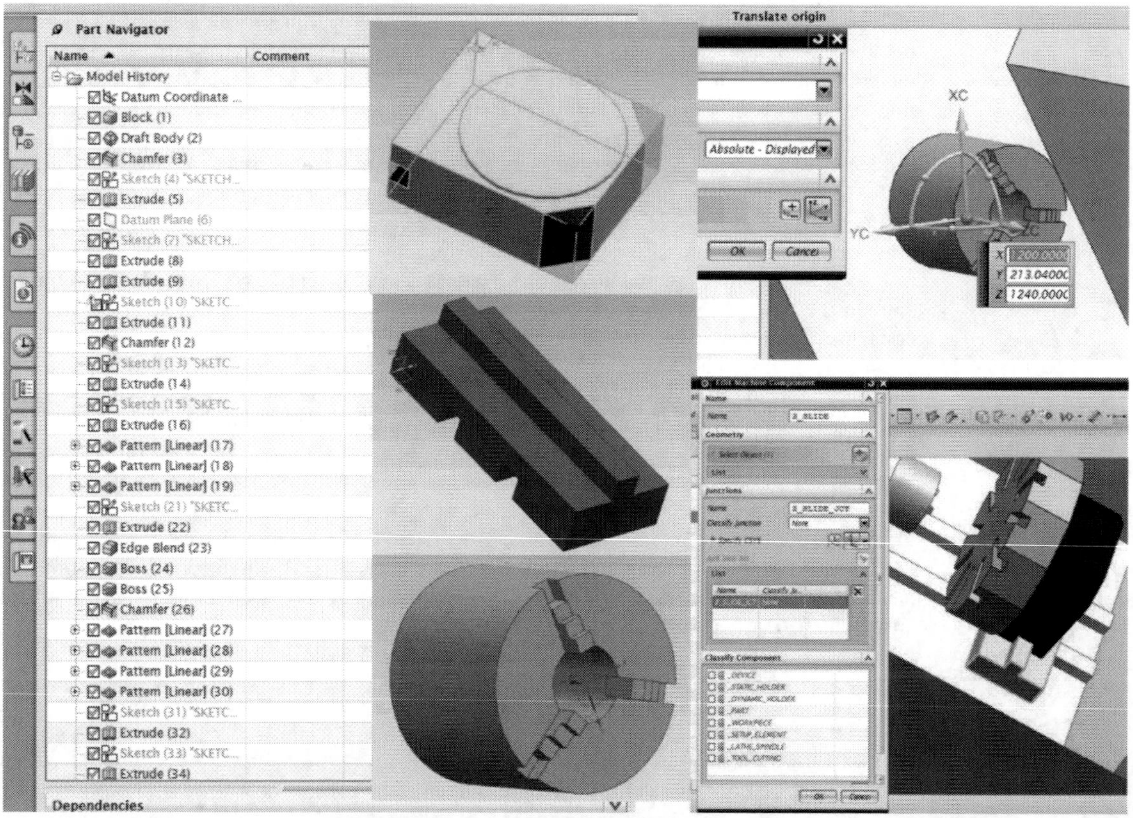

Figure 2. Stages of the lathe model creating.

The prepared model of the lathe could utilize 3D models of different actuators [8] or use information from real sensors to control the operation of the machine tool [9]. It also has a virtual input and output slots to link with other devices in the workcell.

3. Creating the model of the milling machine R1000 "Baca"

The second element, of the designed workcell, was created the R1000 "Baca" milling machine. Its 3D model is illustrated below (figure 3). Also the model of the milling machine was created as the assembly of the previously designed components (figure 4). It could be concluded additionally that the resulting base of component allows creating a broader base of models machine tools.

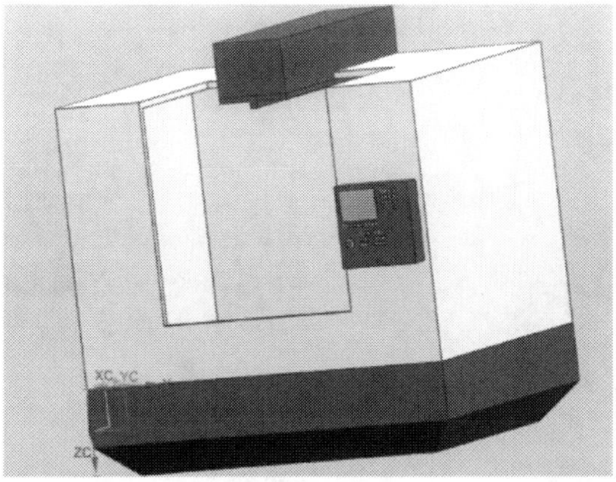

Figure 3. The 3D model of the milling machine R1000 "Baca".

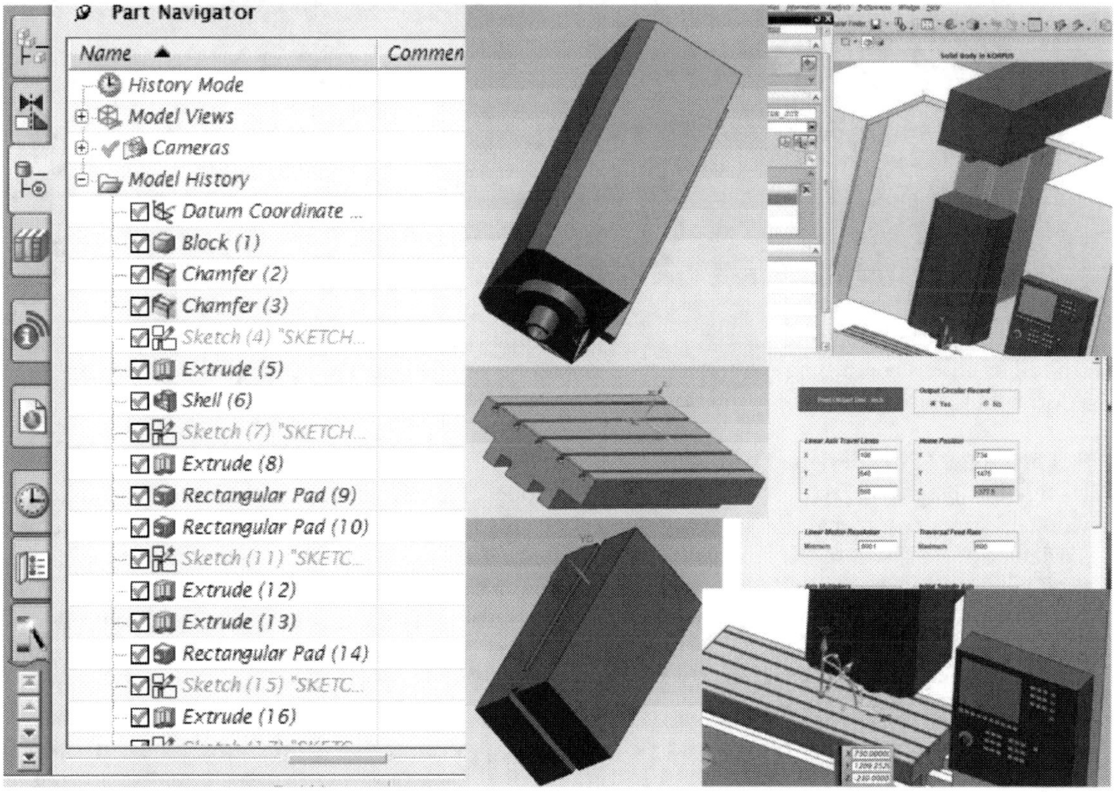

Figure 4. Stages of the milling machine model creating.

The 3D model of the milling machine was also completed with a program controller for the realization of programs prepared in G-code. It also has an input output slots for external communication and cooperation within the framework of the manufacturing system.

4. Creating the model of the robot M-16iiB

The last component of the manufacturing system was created the robot M-16iiB (Fanuc). Figure 5 shows the model and selected structural components of the robot. This model was selected as a pattern of the highest prevalent type of applied industrial robots.

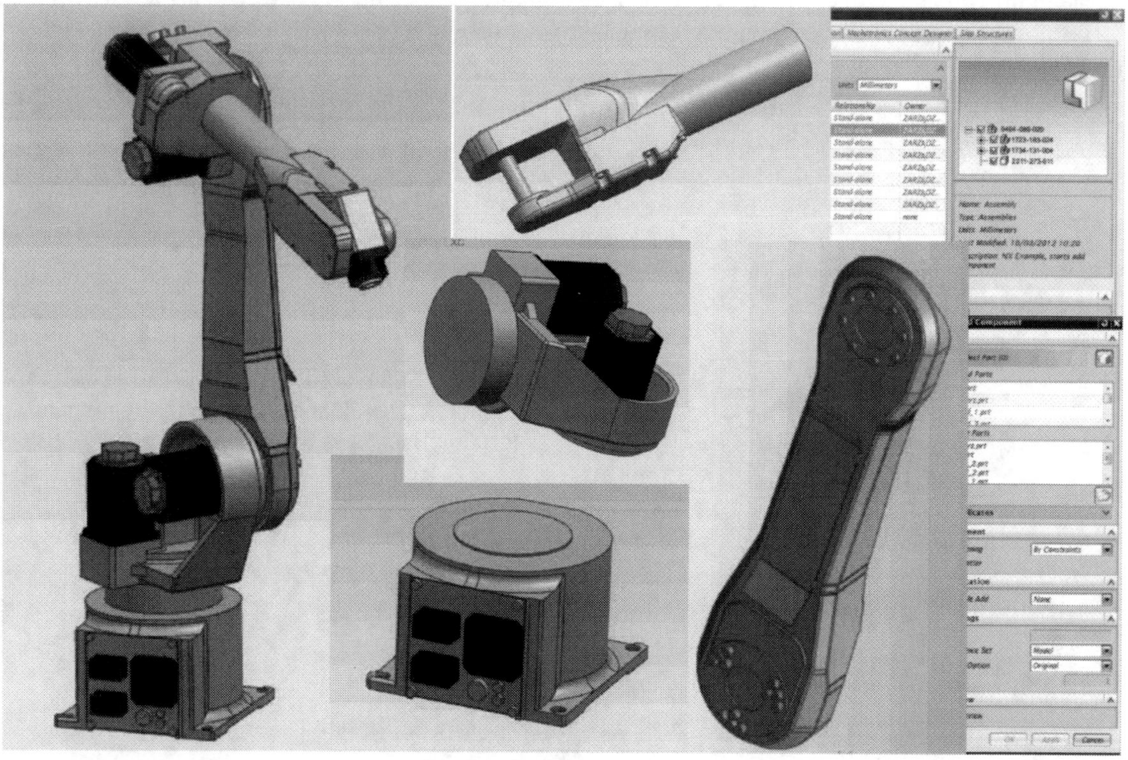

Figure 5. Creating the model of the robot M-16iiB.

The 3D model of the robot also is equipped input and output slots. Simultaneously, the robot controller allows controlling the sequence of operation of other devices. It could be purposeful to use information from real control systems [10,11].

5. The model of the designed workcell

Using the integrative approach [12] it was developed the 3D model of the entire manufacturing workcell (figure 6). In addition to the creation of the 3D model it was also identified relations between particular devices within the workcell control system. This model could be determined as a test one, used to investigate the overall dependences occurring in the manufacturing process.

The devices in the workcell cooperate with each other. Coordinating element is the controller of a robot. This order of operations includes the flow of a workpiece through the lathe and milling machine. Robot cooperates with particular machine tools (figure 7), as a device transporting and reorienting the workpiece. Robot controller also runs programs that control the particular machine tools and receives signals on the end of each procedure of realized programs. Thus, it could be observed, in a virtual environment, the work of the manufacturing workcell and analyzing the sequence of its operations in order to optimize the workcell work, including minimizing downtimes.

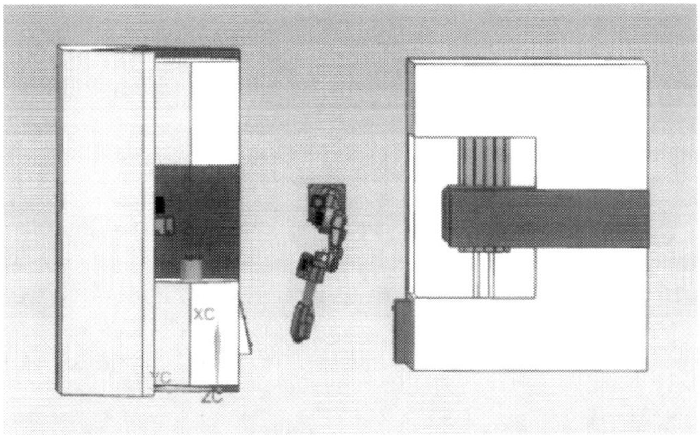

Figure 6. 3D model of the workcell

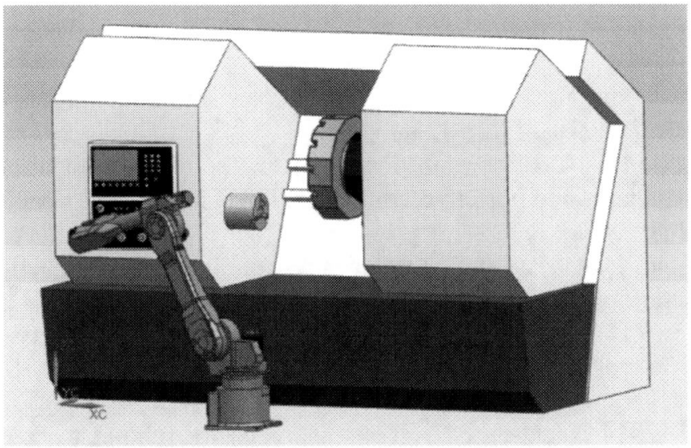

Figure 7. Cooperation of the robot with the lathe

Observations of the manufacturing process, and following analysis of the obtained schedules, let to locate bottlenecks in the process and to analyze them. In this way it is purposeful optimizing both the sequence of operations and the trajectory of robot moves. One would also connect the workcell with an external computer, on which can be implemented the program for automatic optimization, in real time, of the manufacturing process. This program cab be based on the deployment of multi-agent approach [13,14].

6. Conclusions

The solution is a part of studies on the controlling the manufacturing systems using modern technical solutions. In the most basic level, it uses the capabilities of modern systems for aiding engineering work and modeling of virtual environments [15,16]. They allow analyzing complex optimization problems without creating real technical systems.

The presented systems may be equipped with the procedures for controlling and optimization of operations on the base of modern information systems such as advisory programs [17]. This allows striving to develop methods of controlling the manufacturing systems, including the optimization procedures, which will operate in real time.

The other problem considered with creating 3D models of industry systems is to prepare the production plan that can optimize the resources utilization in the investigated system [18,19]. This task is the simpler that it could be realized also by the simulation approach [20].

References

[1] Dymarek A and Dzitkowski T 2012 Method of active synthesis of discrete fixed mechanical systems *J. of Vibroeng.* **14/2** pp 458-463

[2] Monica Z, Ociepka P and Herbuś K 2008 Symulacja procesu obróbki w oparciu o utworzony model obrabiarki w programie Unigraphics *Górnictwo Odkrywkowe* **4-5** pp 110-113
[3] Grajcar A, Płachcińska Topolska S and Kciuk M 2015 Effect of thermomechanical treatment on the corrosion behaviour of high-Mn austenitic steel with silicon and aluminum addition *J. Mat. in Tehn. / J. Mat. and Techn.* **6** pp 73-79
[4] Ćwikła G 2014 Real-time monitoring station for production systems *Adv. Mat. Res.* **837** pp 334-339
[5] Topolska S and Łabanowski J 2015 Duplex stainless steels toughness impact investigations *J. Mat. in Tehn. / J. Mat. and Techn.* **5** pp 114-119
[6] Dymarek A and Dzitkowski T 2013 Reduction vibration of mechanical systems *App. Mech. and Mat.* **307** pp 257-260
[7] Gwiazda A 2014 System of designing complex technical means using fuzzy analysis *App. Mech. and Mat.* **474** pp 147-152
[8] Płaczek M 2014 Computer aided analysis of systems with MFC actuators *Adv. Mat. Res.* **1036** pp 711-714
[9] Płaczek M 2013 Dynamic characteristics of a piezoelectric transducer with structural damping *Sol. St. Ph.* **198** pp 633-638
[10] Hetmańczyk M 2015 The prediction oriented analysis of mechatronic machine structures recorded by directed graphs *Sol. St. Ph.* **220-221** pp 429-434
[11] Michalski P, Hetmańczyk M and Świder J 2013 Simply-integrated method of judgments of expert knowledge collected in databases for objective computer-aided engineering systems *Hyb. Art. Int. Sys.* **8073** pp 262-268
[12] Sękala A, Gwiazda A, Monica Z and Banaś W 2014 Optimization of the Lean Production process using the virtual manufacturing cell *Adv. Mat. Res.* **1036** pp 858-863
[13] Sękala A and Dobrzańska-Danikiewicz A 2015 Possibilities of application of agent-based systems to support functioning of e-manufacturing environment *Sol. St. Ph.* **220/221** pp 781-784
[14] Sękala A, Gwiazda A and Banaś W 2014 Agent-based systems approach for robotic workcell integration *Adv. Mat. Res.* **1036** pp 721-725
[15] Monica Z 2011 Symulacja procesu obróbki na podstawie modelu obrabiarki utworzonego w programie NX *Wybrane Problemy Inżynierskie* **2** pp 267-272
[16] Gwiazda A 2014 System of designing complex technical means using fuzzy analysis *App. Mech. and Mat.* **474** pp 147-152
[17] Michalski P, Hetmańczyk M and Świder J 2013 Simply-integrated method of judgments of expert knowledge collected in databases for objective computer-aided engineering systems *Hyb. Art. Int. Sys.* **8073** pp 262-268
[18] Paprocka I, Kempa W M, Kalinowski K and Grabowik C 2014 A production scheduling model with maintenance *Adv. Mat. Res.* **1036** pp 885-890
[19] Paprocka I, Kempa W M, Kalinowski K and Grabowik C 2014 On Pareto Optimal Solution for Production and Maintenance Jobs Scheduling Problem in a Job Shop and Flow Shop with an Immune Algorithm *Adv. Mat. Res.* **1036** pp 875-880
[20] Krenczyk D and Olender M 2014 Production planning and control using advanced simulation systems Int. J. of Modern Manufacturing Technologies **VI/2** pp 38-43

Time-series pattern recognition with an immune algorithm

I Paprocka[1], **W M Kempa**[2], **C Grabowik**[1], **and K Kalinowski**[1]
[1]Silesian University of Technology, The Faculty of Mechanical Engineering, Institute of Engineering Processes Automation And Integrated Manufacturing Systems, ul. Konarskiego 18a 44-100 Gliwice, Poland
[2]Silesian University of Technology, Faculty of Applied Mathematics, Institute of Mathematics, 23 Kaszubska street, 44-100 Gliwice, Poland

E-mail: iwona.paprocka@polsl.pl

Abstract. In this paper, changes in sequences pattern describing damage-sensitive features of an object which undergoes a failure mode are recognized using an immune algorithm. A frequency response change is an effect for various failure modes occurrence. The objective of this paper is to present immune algorithm for pattern recognition which can discover dependencies between failure mode and effect - frequency response change. Changes in the effect are described with noise due to the fact that the object operates in external conditions. In the immune algorithm antibodies encode various changes in the effect after a given mode occurrence by a number of time. A pathogen encodes a noisy effect of the mode occurrence. Antibodies belonging to a given neighbourhood represent effects after a given type of failure mode occurrence. Antibodies from the neighbourhood undergo clonal selection and affinity maturation process. With the best matched antibody the type of failure mode is achieved.

1. Introduction

Genetic algorithms (GA) belong to the most popular evolutionary algorithms used for time-series pattern recognition problems. The example of application of GA for prediction of firms' stock prices is presented in [1], for making a buy and sell decision in [2], for prediction chaotic time series of sunspot data in [3]. In this paper an immune algorithm (IA) is proposed for time-series pattern recognition problem for failure mode detection. IA is an efficient multi-criteria solution search method [4]. Presented in this paper, time-series pattern recognition artificial immune system can be applied in various engineering problems for example for: the problem of design similarity evaluation [5] feature-based fault diagnosis [6], the classification of structural brain images [7] and fault diagnosis of hydraulic systems [8, 9].

In this paper, changes in sequences pattern describing damage-sensitive features of an object which undergoes a failure mode need to be recognized using an immune algorithm. Let us assume that a frequency response change is an effect we analyse for various failure modes occurrence. The objective of this paper is to present immune algorithm for pattern recognition which can discover dependencies between failure mode and effect - frequency response change. In the immune algorithm antibodies encode various changes in the effect after a given mode occurrence by a number of time. Changes in the effect can be described with noise due to the fact that the object operates in external conditions. In the immune algorithm a pathogen encodes a noisy effect of the mode occurrence. The degree of similarity between sequences of damage-sensitive features is computed in order to identify

Content from this work may be used under the terms of the Creative Commons Attribution 3.0 licence. Any further distribution of this work must maintain attribution to the author(s) and the title of the work, journal citation and DOI.
Published under licence by IOP Publishing Ltd

neighbourhoods. Antibodies (sequences) belonging to a given neighbourhood can represent effects after a given type of failure mode occurrence. Antibodies from the neighbourhood undergo clonal selection and affinity maturation process to achieve better affinity to the pathogen. With the best matched antibody the type of failure mode in the problem of frequency response change recognition is achieved.

The pattern recognition system which can discover dependencies between data entries in time-series is based on immunology thus the following analogies are adopted: a pathogen appearance fuels defending reaction of immune system by producing antibodies by *B-Lymphocytes* and receptors on surface of *T-Lymphocytes*. If the pathogen matches to antibodies or receptors, the Lymphocytes are stimulated to grow and clone. Offspring of Lymphocytes better adjusted to the pathogen, destroy it. Among offspring there are *memory T-Lymphocytes* that circulate in a blood for a long time and are called memory cells. If the same pathogen enters the immune system second time, the immune response is quicker and more effective. The immune system controls the diversity of generated antibodies. The small number of diverse antibodies means the defeat in the battle with a new pathogen. If *memory T-lymphocytes* re-match with the same pathogens, *memory T-lymphocytes* react more intensively involving more efficient proliferation of *effector T-lymphocytes* that kill the pathogens. *Suppressor T-lymphocytes* protect against excessive growth of *effector T-lymphocytes*.

2. Problem formulation

Let us consider an object working in a „normal" environment - without any disruption. The object without any failure is called a „healthy case" *H* and is described by a set of points $h(f^i, Pdf^i)$ achieved for time measurement points $i = 0..l$. Each point is described by frequency f^i and normalized function of probability density function Pdf^i, $i = 0..l$ (figure 1). The object is subjected to the influence of a failure mode occurrence by a number of times. After the mode occurrence frequency response change of the object is registered. After the mode occurrence the damaged object *A* is described by a set of points, each data is described by $a(f^i, Pdf^i)$ (frequency and normalized *Pdf*) in time point *i*, $i = 0..l$. The objective is to elaborate a pattern recognition system which can discover dependencies between data entries in time-series. The objective of the pattern recognition system is to detect a type of failure mode based on noisy frequency response change. The pathogen is described by noisy frequency response change $p(f^i, Pdf^i)$ (frequency and normalized *Pdf*) in time period *i*, $i = 0..l$.

3. Pattern recognition immune system

The phenomena mentioned in the introduction are used in the pattern recognition immune system: the pathogen represents noisy time-series data describing an effect of a type of failure mode occurrence, antibodies correspond to time-series data describing effects of possible types of failure modes occurrence and the best matched antibody to the pathogen corresponds to the solution of the time-series pattern recognition problem - the type of failure mode is achieved. A degree of affinity of an antibody to the pathogen represents a distance in a solution space.

The algorithms consists of following steps:

1. Antibody a_n coding;
2. Pathogen p coding;
3. Antibodies stimulation (affinity) degree D_n evaluation;
4. Antibodies that recognize the pathogen *p* create a Neighborhood of parents, $n_m \in N(p)$;

   By a number of generations:

5. The clonal proliferation process, each clone c_k^m of parent n_m undergoes hypermutation process and creates offspring o_k^m;
6. Offspring stimulation degree D_k^m evaluation;
7. Selection between parents n_m and offspring o_k^m;
8. Suppression process,
9. New randomly generated offspring is introduced instead of deleted ones to create $\tilde{o}_k^m$;
10. New offspring stimulation degree $\tilde{D}_k^m$ evaluation;
11. Selection between parents n_m and new offspring $\tilde{o}_k^m$;
12. The best matched antibody $\hat{n}_m$ is memorized in each generation;
13. Selection of recognized type of defect from the immune memory (IM).

Input parameters of the immune algorithm are: affinity threshold *At*, number of mutated genes *Nm*, suppression threshold *St*, number of iterations *Ino*.

Antibody a_n encodes an effect of a failure mode occurrence after *n* time. Genes of the antibody $a_n\left(f_n^i, Pdf_n^i\right)$ encodes frequency f response and normalized *Pdf* measured in point *l*, (*i* = *0..l*). Genes of each antibody $a_n\left(f_n^i, Pdf_n^i\right)$ are compared to genes of the "healthy case" $h\left(f^i, Pdf^i\right)$ (figure 1). Pathogen *p* encodes an effect of a failure mode occurrence but information is with noise. Genes of the pathogen $p\left(f^i, Pdf^i\right)$ encodes noisy frequency response and normalized *Pdf* measured in point *l*, (*i* = *0..l*) comparing to $h\left(f^i, Pdf^i\right)$.

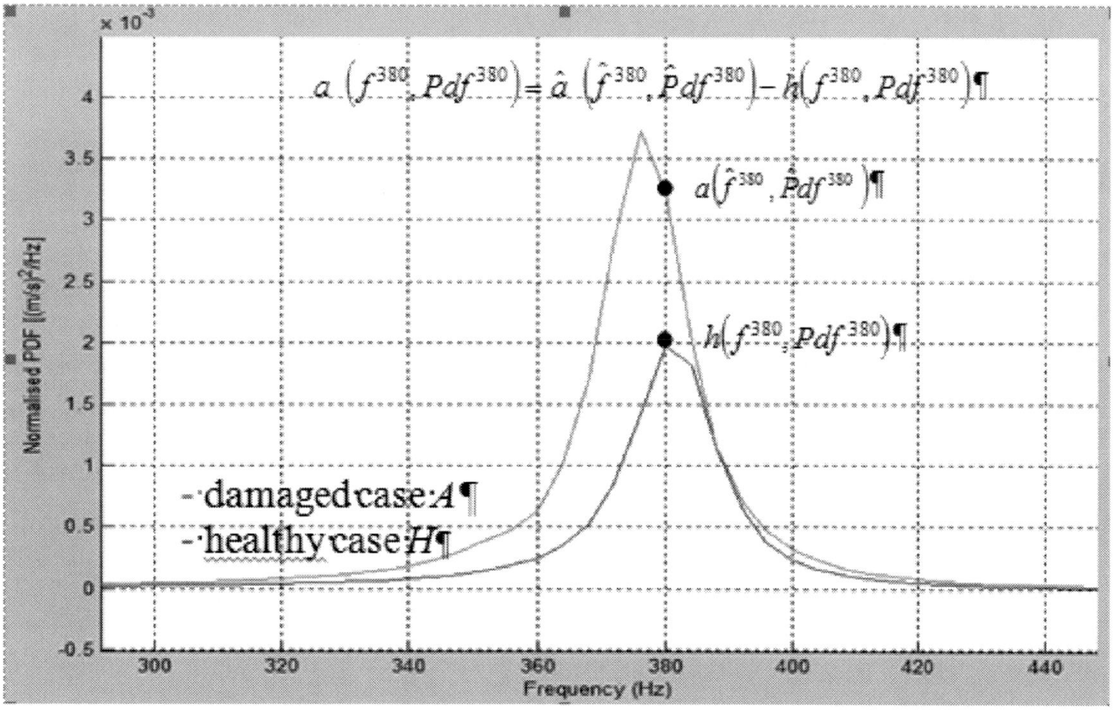

Figure 1. Coding of gene *i* = 380 of antibody chromosome *A* (*i* = 1..513).

The distance between antibodies and pathogen is computed using the following formula:

$$D_n(a_n, p) = \left(\sum_{i=1}^{l} \left(f_n^i - f^i\right)^2, \left(\sum_{i=1}^{l} \left(Pdf_n^i - Pdf^i\right)^2 \right) \cdot 10000 \right) = d_n(x_n, y_n) \tag{1}$$

In the stimulation phase, antibodies that recognize the pathogen p create a neighborhood of parents, $n_m(f^i, Pdf^i) \in N(p)$. Parameter x_n' states as the minimal distance in f and y_n' states as the minimal distance in Pdf obtained in any $d_n(x_n, y_n)$. Antibody a_n stimulates the pathogen p if the following formula is true:

$$x_n \leq (y_n' + At) \wedge y_n \geq (y_n' - At) \tag{2}$$

The higher value of At is, the more antibodies belongs to the Neighborhood $N(p)$ of the pathogen p. In the maturation process antibodies n_m that belongs to $N(p)$ are copied to create offspring c_k^m. A number of generated offspring m depends on the distance $D_m(n_m, p)$ of parent n_m to the pathogen p. The less distance $D_m(n_m, p)$ of parent n_m to pathogen p is, the higher number of offspring is produced. Offspring undergo hypermutation. A number of mutated genes of offspring c_k^m depends on the distance $D_m(n_m, p)$. Offspring is represented by a search point $d_m(x_m, y_m)$ in the solution space. The less distance $D_m(n_m, p)$ of parent n_m to pathogen p is, the less number of genes is mutated in clone c_k^m. In the hypermutation process new offspring are generated $o_k^m(f_i, Pdf_i)$.
In the offspring stimulation phase, the distance between new offspring and pathogen is calculated using $D_k^m(o_k^m, p)$ and obtained results are presented as $d_k^m(x_k^m, y_k^m)$. The similarity between offspring after hypermutation process and parents is compared. A single solution survives from the two: offspring after hypermutation process and parent, for which the following formula is true:

$$(x_m + y_m)^2 \geq (x_k^m + y_k^m)^2 \tag{3}$$

The suppression process allows the immune algorithm to maintain a high diversity of antibodies and to increase probability of recognizing the pathogen. In the suppression process the distance between offspring is calculated using $DS_k^m(o_k^m, o_k^m)$, the obtained measures state as $ds_k^m(x_k^m, y_k^m)$ in the solution space. Offspring o_k^m dies if the distance $ds_k^m(x_k^m, y_k^m)$ between any other offspring is less than St. Deleted offspring are replaced with new generated ones. Survived and new generated offspring state as $\tilde{o}_k^m$. The distance between $\tilde{o}_k^m$ and the pathogen p is calculated using $\tilde{D}_k^m(\tilde{o}_k^m, p)$, the results state as $\tilde{d}_k^m(\tilde{x}_k^m, \tilde{y}_k^m)$. A single solution survives from the two: $\tilde{o}_k^m$ and n_m if the following formula is true:

$$(x_m + y_m)^2 \geq (\tilde{x}_k^m + \tilde{y}_k^m)^2 \tag{4}$$

The antibody $\hat{n}_m$ with the shortest distance to the pathogen p, $\hat{D}_m(\hat{n}_m, p)$ is copied in each generation to create the immune memory. After a number of iteration the antibody with the shortest distance to the pathogen p from the immune memory is selected. With the best matched antibody $\hat{\hat{n}}_m$ the type of failure mode is obtained.

IOP Conf. Series: Materials Science and Engineering **95** (2015) 012110 doi:10.1088/1757-899X/95/1/012110

4. Computer example

An object is subjected to the influence of a failure mode occurrence 18 times. After each time of the mode occurrence the frequency response change of the object is registered $\hat{a}_n\left(\hat{f}_n^i, \hat{P}df_n^i\right)$ in time point $i = 1..513$ for $n = 1..18$. For example, the input data of the second antibody, after the second time of a mode occurrence are presented in Table 1. A randomly selected a_n undergoes noise in both f and Pdf characteristics to create pathogen $\hat{p}\left(\hat{f}^i, \hat{P}df^i\right)$ (figure 2). The objective is to detect a type of mode based on noisy frequency response change.

Table 1. The input data of the second antibody, after the second time of a damage occurrence.

i	1	2	3	4	...	510	511	512	513
f	0	4	8	12		2036	2040	2044	2048
Pdf	2.412e-06	4.629e-06	1.706e-06	6.729e-07		2.066e-07	2.058e-07	2.050e-07	1.023e-07

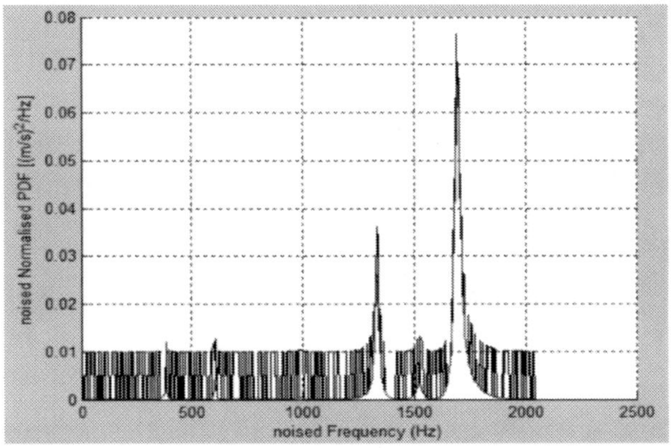

Figure 2. The pathogen $\hat{p}\left(\hat{f}^i, \hat{P}df^i\right)$ representation – the second antibody with the noise.

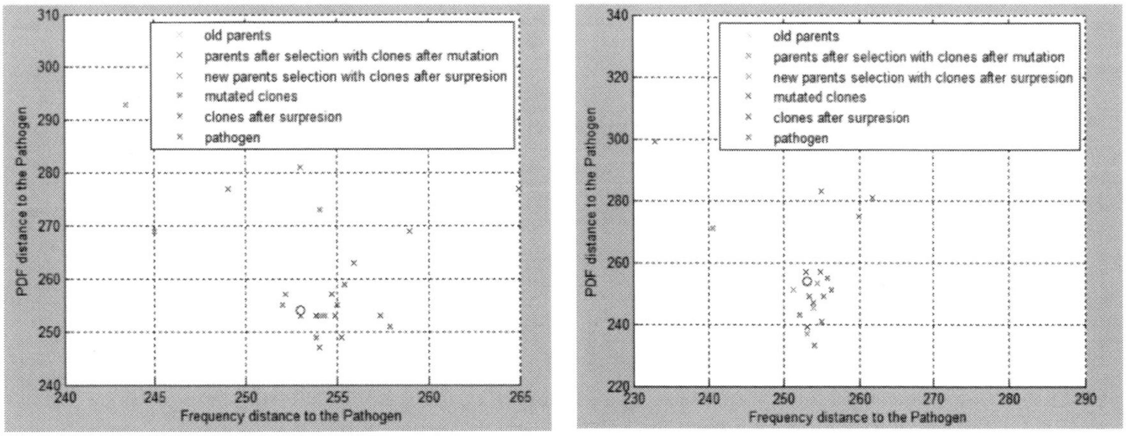

Figure 3. Solutions approaching to the pathogen in the maturation process.

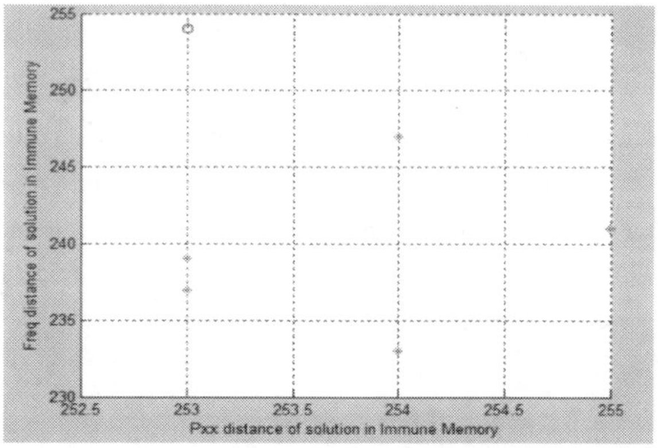

Figure 4. *f* and *Pdf* of the best solution $\hat{n}_m$ from the IM.

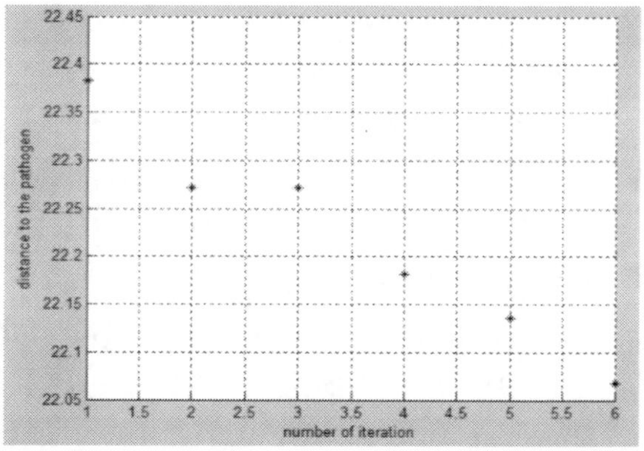

Figure 5. Distances of the best solutions $\hat{n}_m$ to the pathogen p, $\hat{D}_m(\hat{n}_m, p)$.

Input parameters of the immune algorithm are: $At = 1$, $Nm = 100$, $St = 300$, $Ino = 6$. Antibodies that recognize the pathogen p create a Neighborhood of old parents, $n_m \in N(p)$. Offspring o_k^m are produced in the clonal proliferation process. After selection and suppression processes new offspring $\tilde{o}_k^m$ are produced. In the figure 3 n_m, o_k^m and $\tilde{o}_k^m$ for the first and sixth generations are presented. Thanks to the suppression process a high diversity of generated antibodies is maintained. In the figure 4 frequency and normalized *Pdf* of the best matched antibodies $\hat{n}_m$ memorized in the IM are presented. The quality of achieved best antibodies $\hat{D}_m(\hat{n}_m, p)$ is improved in almost each generation as one can see in the figure 5. The second antibody has the shortest distance to the pathogen. The immune algorithm recognized the pathogen.

5. Conclusions

In the paper, the Immune Algorithm (IA) for time-series pattern recognition is presented. IA is used to identify similarity of sequences describing a frequency response change of the object which was undergone to the influence of failure mode (antibody) to a sequence describing also a frequency response but with a noise (pathogen). We assumed that the pathogen is a randomly selected antibody (solution) which sequence of a frequency response undergo noise, $p = \tilde{a}_2$. Solutions belonging to a given neighborhood represent a given type of mode. Four solutions belong to the neighborhood of the pathogen, $\{a_1, a_2, a_3, a_4,\} \in N(p)$. In the clonal selection and affinity maturation process solutions are approaching toward the pathogen. $\hat{n}_m$ which is the offspring of the parent a_2 obtained the shortest distance to the pathogen, $\hat{D}_2(\hat{n}_2, p)$=22.05. The best matched antibody encodes the type of mode in the time-series pattern recognition problem.

References

[1] Chih-Fong T and Yu-Chieh Hsiao 2010 Combining multiple feature selection methods for stock prediction: Union, intersection, and multi-intersection approaches *Decision Support Systems* **50** pp 258-269
[2] Wing W Y, Xue-Ling L and Trader L G 2014 Stock trading decision support based on feature selection by weighted localized generalization error model *Neurocomputing* **146** pp 104-112

[3] Ardalani-Farsa M and Zolfaghari S 2011 Residual analysis and combination of embedding theorem and artificial intelligence in chaotic time series forecasting *Applied Artificial Intelligence* **25** pp 45–73
[4] Paprocka I and Kalinowski K 2014 Pareto Optimality of Production Schedules in the Stage of Populations Selection of the MOIA Immune Algorithm *Applied Mechanics and Materials* **657** pp 869-873
[5] Rosenfeld A and Wechsler H 2000 Pattern Recognition: Historical Perspective and Future Directions *International Journal of Imaging Systems & Technology* **11** pp 101-116
[6] Tian H, Bo-Suk Yang and Zhong-Jun Yin 2007 Feature-based fault diagnosis system of induction motors using vibration signal *Journal of Quality in Maintenance Engineering* **13** pp 163-175
[7] Yudong Z, Shuihua W, Genlin Ji and Zhengchao D 2013 Genetic Pattern Search and Its Application to Brain Image Classification *Mathematical Problems in Engineering* pp 1-8
[8] Chen H.X, Chua Patrick S K and Lim G H 2006 Adaptive wavelet transform for vibration signal modelling and application in fault diagnosis of water hydraulic motor *Mechanical Systems & Signal Processing* **20** pp 2022-2045
[9] Duarte Jr Elias P and Nassu A 2010 Fault diagnosis of multiprocessor systems based on genetic and estimation of distribution algorithms: a performance evaluation *International Journal on Artificial Intelligence Tools* **19** pp 1-18

Semi-analytical computation of displacement in linear viscoelastic materials

S Spinu[1] **and D Gradinaru**[1]
[1]"Stefan cel Mare" University of Suceava, Department of Mechanics and Technologies, 13th University Street, 720229, Suceava, Romania

E-mail: sergiu.spinu@fim.usv.ro

Abstract. Prediction of mechanical contact performance based on elastic models is not accurate in case of viscoelastic materials; however, a closed–form description of the viscoelastic contact has yet to be found. This paper aims to advance a semi-analytical method for computation of displacement induced in viscoelastic materials by arbitrary surface tractions, as a prerequisite to a semi-analytical solution for the viscoelastic contact problem. The newly advanced model is expected to provide greater generality, allowing for arbitrary contact geometry and / or arbitrary loading history. While time-independent equations in the purely elastic model can be treated numerically by imposing a spatial discretization only, a viscoelastic constitutive law requires supplementary temporal discretization capable of simulating the memory effect specific to viscoelastic materials. By deriving new influence coefficients, computation of displacement induced in a viscoelastic material by a known but otherwise arbitrary history of surface tractions can be achieved via superposition authorized by the Boltzmann superposition theory applicable in the frame of linear viscoelasticity.

1. Introduction

The design of automotive belts and tires, seals or biomedical devices require extensive knowledge on the behavior of materials like elastomers or rubber, which fall into the category of viscoelastic materials. From a mechanical point of view, the tribological performances of the latter are particularly important in assessing the strength of the mechanical contact. When load is transmitted through a contact whose region is small compared to dimensions of the contacting bodies (i.e. a non-conforming contact), important gradients of stress and strain develop in the contact proximity. When the elastic limit of the material is surpassed, irreversible changes in material behavior occur, concluding with the destruction of the integrity of the involved machine element and thus leading to contact failure. The prediction of these stress and strain gradients based on the Hertz model is no longer accurate in case of viscoelastic materials, as the viscoelastic response to load depend explicitly on time and on the loading history.

The classic literature on the viscoelastic contact is based on the so-called correspondence principle between the elastic and the viscoelastic solution of a problem of stress analysis, stating that a viscoelastic problem has an associated elastic problem to which the former reduces after removal of time variable via transfer to Laplace transform domain. Indeed, the basic integral equations of the viscoelastic problem of stress analysis are identical, in form, when transferred to the Laplace domain, with the equations describing stresses in purely elastic materials. Consequently, if the boundary conditions are handled properly, a viscoelastic solution in the frequency domain is identical in form to

Content from this work may be used under the terms of the Creative Commons Attribution 3.0 licence. Any further distribution of this work must maintain attribution to the author(s) and the title of the work, journal citation and DOI.
Published under licence by IOP Publishing Ltd

its associated elastic solution. This remarkable property of the elastic and viscoelastic equations can thus be used to derive solutions of viscoelastic problems of stress analysis from their elastic counterparts, which are usually obtained more easily, e.g. from the work of Sneddon [1]. The applicability of this technique is however limited because the transient boundary conditions encountered in contact problems cannot be properly handled when transferring to Laplace domain, e.g. in generic contact problems there exist regions for which the boundary conditions are specified only for a short period of time, and thus the integral in the Laplace transform cannot be computed.

The first application of this principle to contact mechanics was reported by Lee and Radok [2], who derived the contact radius and the pressure distribution in a Hertz-type contact between linear viscoelastic materials. Displacement computation was achieved via integration of pressure over the contact area. Lee and Radok's result [2] is often referred to as a tentative solution, as it covers only the case when the contact radius increases monotonically with time. The generalization of this result was achieved later by Ting [3], who advanced a model in which, depending on the specifics of the loading history, up to five separate cases must be considered, leading to tedious algebraic manipulations of the emerging equations. Moreover, the resulting solutions, although explicit, involve numerical differentiation followed by numerical integration, as well as numerical resolution of transcendental equations. This mandatory numerical treatment often raises convergence issues, as opposed to the semi-analytical methods. The latter are based on manipulation of the so-called influence coefficients derived by closed-form integration of fundamental solutions in the theory of elasticity (i.e. the Green functions), and thus raise fewer convergence problems. Recent applications of this type of methods to viscoelastic contact analysis can be found in [4, 5].

This paper aims to advance a novel semi-analytical solution for the computation of displacement in viscoelastic materials, solution that is expected to assist the development of a semi-analytical method for the contact problem involving viscoelastic bodies.

2. Displacement computation

A surface distribution of normal or shear tractions, such as the ones resulting from a mechanical contact process, induce a displacement field whose knowledge is essential in solving the contact problem and in performing stress analysis in the contacting bodies. Displacement computation can only be accomplished after establishing the constitutive law of the material. While complete description of linear elastic and isotropic materials involves only two constants, e.g. the Young modulus and the Poisson's ratio ν, viscoelastic models employ various models and parameters. The assumptions described in [6] are adopted in this work to reduce the complexity of the arising equations to a computationally friendly form.

Firstly, the viscoelastic response is assumed to be linear (which is reasonable in the frame of small strain theory), thus allowing for the use of Boltzmann superposition theory. The latter states that that the stress or strain response to various strain or stress histories, acting simultaneously, is identical to the sum of responses when the same histories are applied separately. In mathematical form, the application of the Boltzmann hereditary integral operator is authorized in computation of viscoelastic response to various sequences of stress or strain.

Secondly, the viscoelastic material is assumed incompressible, which is reasonable for polymers, thus limiting the model complexity to a stress-strain law in shear, between the deviatoric stress s, the deviatoric strain e and the shear modulus G: $s = 2Ge$. In this framework, the strain response to a unit step change in stress is completely described by the creep compliance function $\Phi(t)$, and the stress response to a unit step change in strain by the relaxation modulus $\Psi(t)$. These two functions are interchangeable, but, unlike the purely elastic case, there is no reciprocity, i.e. $\Psi(t)\Phi(t) \neq 1$. By employing the Boltzmann hereditary integral together with these time-dependent functions, the viscoelastic response to any sequence of stress or strain in a window of observation $[0,t]$ can be expressed as:

$$s(t)=\int_0^t \Psi(t-t')\frac{\partial e(t')}{\partial t'}dt', \text{ or } e(t)=\int_0^t \Phi(t-t')\frac{\partial s(t')}{\partial t'}dt'. \tag{1}$$

Although the limiting boundary of a real solid is intrinsically rough, computational contact mechanics employs the half-space assumption, allowing for the use of fundamental solutions derived in the theory of linear elasticity for a semi-infinite body bounded by a plane surface. In a Cartesian coordinate system (x_1, x_2, x_3) having its origin on the half-space boundary and the x_3-axis along the direction of the normal to the half-space boundary, the normal displacement field u_3^e generated in a linear elastic and isotropic solid by a distribution of normal tractions $p(x_1, x_2)$ is computed using the Green function $G_e(x_1, x_2)$ for the elastic half-space derived by Boussinesq [7]:

$$u_3^e(x_1,x_2)=\int_{-\infty}^{\infty}\int_{-\infty}^{\infty} G_e(x_1-x_1', x_2-x_2')p(x_1',x_2')dx_1'dx_2' \tag{2}$$

where $G_e(x_1,x_2)=(1-\nu)/(2\pi G\sqrt{x_1^2+x_2^2})$ is the normal displacement induced at a point of coordinates (x_1, x_2) by a unity concentrated force acting in origin along direction of $\vec{x}_3$.

Lee and Radok [2] obtained the contact radius in the viscoelastic spherical contact problem by applying the hereditary integral operator of the type described in equation (1) to the Hertz (i.e. purely elastic) solution in which the elastic modulus $1/(2G)$ was replaced by the viscoelastic creep compliance. We are applying the same technique to equation (2), aiming to obtain the viscoelastic displacement generated by a known history of pressure $p(x_1', x_2', t')$ in a window of observation $[0, t]$:

$$u_3^{ve}(x_1,x_2,t)=\int_0^t \Phi(t-t')\frac{\partial}{\partial t'}\left[\int_{-\infty}^{\infty}\int_{-\infty}^{\infty}\frac{p(x_1',x_2',t')}{2\pi\sqrt{(x_1-x_1')^2+(x_2-x_2')^2}}dx_1'dx_2'\right]dt' \tag{3}$$

It should be remembered that the viscoelastic material is assumed incompressible, therefore $\nu = 0.5$. By interchanging differentiation and integration, an equivalent form can be obtained, as shown in Eq. 4. The semi-analytical treatment of these equation to attain a computationally friendly form is detailed in the following section.

$$u_3^{ve}(x_1,x_2,t)=\int_0^t\int_{-\infty}^{\infty}\int_{-\infty}^{\infty}\frac{\Phi(t-t')}{2\pi\sqrt{(x_1-x_1')^2+(x_2-x_2')^2}}\frac{\partial p(x_1',x_2',t')}{\partial t'}dx_1'dx_2'dt' \tag{4}$$

3. Semi-analytical method

The integral equations (2) (for the elastic case) and (3) (for the viscoelastic framework) cannot be solved analytically at this point. The principle of the semi-analytical method consist in considering all continuous distributions as piecewise constant on a uniformly spaced rectangular mesh established in the plane x_1x_2. Control points must be chosen for all the elementary cells of the grid (the centers of the cells are good candidates), and all continuous parameters are evaluated in these representing points, resulting in a digitized counterpart for each continuous distribution.

This discretization encourages a simplified notation taking as arguments the indexes of the cells rather than the continuous coordinates. For example, $p(i, j)$ denotes the pressure value computed in the center of the cell (i, j), and $p(i, j) = p(x_1^{(i)}, x_2^{(j)})$, where $x_1^{(i)}$ and $x_2^{(j)}$ are coordinates of the center of the cell (i, j). It follows that pressure is assumed uniform in the rectangular patch centered in $(x_1^{(i)}, x_2^{(j)})$, and therefore can be factored outside the integral operator in equation (2). The integral of the Green function $G_e(x_1, x_2)$ taken over the elementary patch of side lengths Δ_1 and Δ_2 along directions of $\vec{x}_1$ and $\vec{x}_2$, respectively, yield the influence coefficient for the elastic displacement IC_e:

$$IC_e(i-k, j-\ell) = \int_{x_2(\ell)-\Delta_2/2}^{x_2(\ell)+\Delta_2/2} \int_{x_1(k)-\Delta_1/2}^{x_1(k)+\Delta_1/2} G_e\left(x_1(i)-x_1', x_2(j)-x_2'\right) dx_1' \, dx_2', \quad (5)$$

which expresses the normal displacement induced in the observation cell (i, j) by a uniform pressure of magnitude $1/(\Delta_1\Delta_1)$ [Pa] acting in the cell (k, ℓ). The closed-form solution of the double integral in equation (5) was derived by Love [8]:

$$IC_e(i, j) = \frac{1-\nu}{2\pi G}\begin{pmatrix} f(x_1(i)+\Delta_1/2, x_2(j)+\Delta_2/2) + f(x_1(i)-\Delta_1/2, x_2(j)-\Delta_2/2)... \\ -f(x_1(i)+\Delta_1/2, x_2(j)-\Delta_2/2) - f(x_1(i)-\Delta_1/2, x_2(j)+\Delta_2/2) \end{pmatrix}, \text{ where}$$
$$f\left(x_1, x_2\right) = x_1 \ln\left(x_2 + \sqrt{x_1^2 + x_2^2}\right) + x_2 \ln\left(x_1 + \sqrt{x_1^2 + x_2^2}\right). \quad (6)$$

Within this framework, the semi-analytical counterpart of equation (2) results as:

$$u_3^e(i, j) = \sum_{k=1}^{N_1}\sum_{\ell=1}^{N_2} IC_e\left(i-k, j-\ell\right) p(k, \ell) \quad (7)$$

where N_1 and N_2 denotes the number of grids along directions of $\vec{x}_1$ and $\vec{x}_2$, respectively. The double convolution in equation (7) can be performed for any imposed pressure distribution. Optimum algorithmic efficiency is achieved using the Discrete Convolution Fast Fourier Transform (DCFFT) algorithm advanced by Liu, Wang and Liu [9]. The reduction of the order of computation comes from the convolution theorem, which states that the convolution operation reduces to an element-wise product in the Fourier transform domain. The semi-analytical displacement computation using equation (7) together with the DCFFT technique is now widely used in computational contact mechanics. In this paper, we intend to generalize this equation to the case of viscoelastic behavior.

As equations describing the purely elastic model are intrinsically time-independent, spatial discretization is adequate to circumvent the continuous integration in equation (2). An additional integration over the time span in which the body was loaded is present in equation (3), requiring an additional temporal mesh capable of simulating the memory effect specific to viscoelastic materials (i.e. the property that the current state depends upon all previous states achieved from the first loading). This temporal discretization should be chosen so that at $t = 0$ the body was undisturbed, and the time increment Δ_t should be small enough so that, during each step, the problem parameters can be assumed constant. A piecewise constant law is thus imposed along the temporal axis, adding a third parameter to the notation implemented in the purely elastic model. For example, $p(i, j, k)$ is the discrete counterpart of $p(x_1, x_2, t)$, denoting the pressure in the elementary cell (i, j) in the spatial mesh, achieved after k time increments, where $t = k\Delta_t$, with $k = 1\ldots N_t$. This assumption regarding the temporal variation of model parameters authorize the substitution of the partial derivative $\partial p(x_1', x_2', t')dt'/\partial t'$ in equation (4) with the finite difference $p(i, j, k) - p(i, j, k-1)$. The latter can be factored outside the spatial integral operator as in the purely elastic model. A viscoelastic influence coefficient can be defined similar to its elastic counterpart in equation (5), by including the viscoelastic material property, i.e. the creep compliance function:

$$IC_{ve}(i-\ell, j-m, k-n) = 2G\Phi\left((k-n)\Delta_t\right) IC_e(i-\ell, j-m) \tag{8}$$

This influence coefficient expresses the displacement observed after k time steps in the elementary patch (i, j) of the spatial mesh, due to a uniform pressure of $1/(\Delta_1\Delta_2)$ [Pa] that acted in the patch (ℓ, m) in the nth time step after the reference moment in which the body was undisturbed, with $n \le k$. The semi-analytical counterpart of Eq. 4 can thus be expressed as:

$$u_3^{ve}(i, j, k) = \sum_{n=1}^{N_t}\sum_{\ell=1}^{N_1}\sum_{m=1}^{N_2} IC_{ve}(i-\ell, j-m, k-n)\left(p(\ell, m, n) - p(\ell, m, n-1)\right), \tag{9}$$

where $i = 1\ldots N_1$, $j = 1\ldots N_2$, $k = 1\ldots N_t$. This equation clearly shows that the memory effect is considered explicitly in the displacement computation, as pressure distributions in all previous states, i.e. at all previous time increments, together with the current pressure, are needed to evaluate the current displacement. It is noteworthy that the contribution of the historical pressures can be separated from the contribution of the current pressure. This feature that will be used in the following section for validation purposes.

4. Program validation

Most results in the viscoelastic contact literature are given in pressure, not in displacement. Therefore, in order to validate this new displacement computation method, we need to link it to the corresponding pressure. In electrostatics, the one-to-one correspondence between pressure and displacement is guaranteed by the theorem of uniqueness of solutions, and this correspondence is at the base of the semi-analytical method [10] which solves the elastic contact problem by deriving both pressure and displacement from the contact interference equation. Provided displacement is computed according to the viscoelastic material model instead of the elastic one, this contact solver [10] should equally handle viscoelastic contact problems. Indeed, provided the memory effect of the viscoelastic material is properly considered in displacement computation, the same equilibrium and geometrical equations describe both elastic and viscoelastic contacts. However, while the contact solver can only be used to derive the current pressure, all historical pressures are needed in computation of displacement according to equation (9). To overcome this, we are using the Ting's partially analytical formulas [3] to obtain the historical pressures, and we subsequently use the viscoelastic displacement computed with the newly advanced semi-analytical method to obtain a new pressure. We are then comparing it with the one predicted by the Ting's formula for the same loading history. Based on the one-to-one correspondence between pressure and displacement, an agreement between the two pressure distributions would indicate that the performed semi-analytical displacement computation is indeed accurate.

A linear viscoelastic half-space described by a Maxwell rheological model having a τ relaxation time is indented by a rigid spherical punch in a step loading. Pressure distributions computed for different times t subsequent to the first contact, depicted in figure 1, match well the Ting's model, giving confidence in the capability of the newly advanced semi-analytical model to accurately predict viscoelastic displacement. Dimensionless pressure p, radial coordinate x_1 and normal displacement u_3 are defined as ratio to Hertz (i.e. purely elastic) central pressure p_H, contact radius a_H and rigid-body approach ω_{3H}, respectively, all corresponding to the maximum loading level.

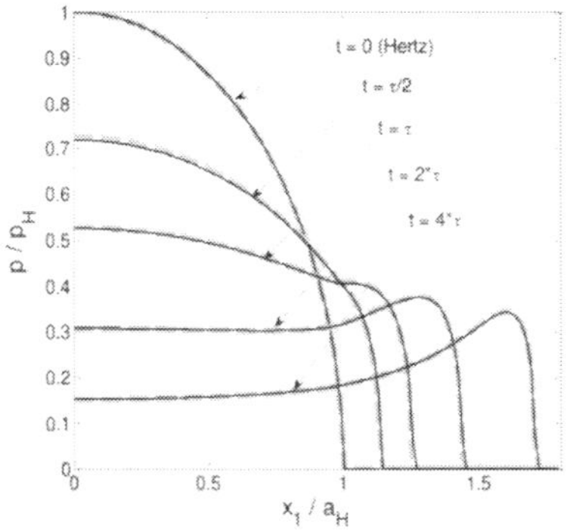

Figure 1. Contact pressure in a Maxwell viscoelastic contact: dashed grey lines—the novel semi-analytical method; continuous black lines—Ting's model.

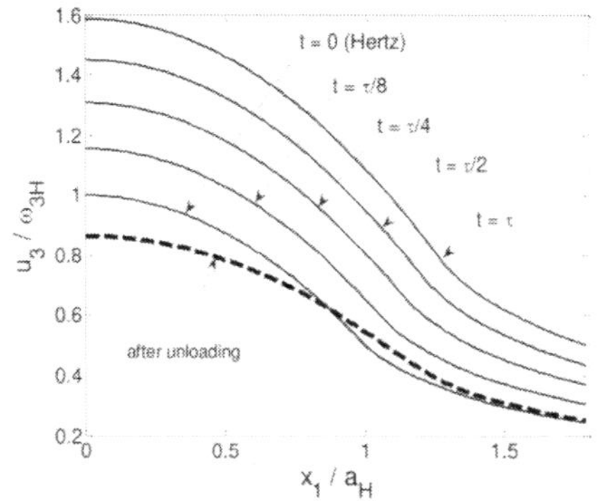

Figure 2. Displacement in a Maxwell viscoelastic contact loaded at time $t = 0$ and unloaded at $t = \tau$.

Equation (9) can also be used to predict the post-unloading displacement, by equating the post-unloading pressures to zero. This scenario can only be simulated using the semi-analytical formulation advanced herein. The displacement profiles prior and subsequent to indenter removal are depicted in figure 2. The elastic part due to the elastic spring in the Maxwell model is recovered instantaneously, while the dashpot displacement is predicted to persist indefinitely.

5. Conclusions

A semi-analytical method for the computation of displacement induced in a viscoelastic material by a known, but otherwise arbitrarily chosen history of surface tractions is advanced in this paper. The method is based on the correspondence principle between the elastic and the viscoelastic problems of stress analysis, authorizing the manipulation of the elastic solution in achievement of its viscoelastic counterpart. The linear viscoelastic model employs the Boltzmann hereditary integral to obtain the material response to various sequences of stress, in a directly additive manner. Both spatial and temporal discretizations are imposed in the semi-analytical formulation in order to circumvent analytical integration, which is eventually substituted by a discrete convolution product raising no convergence issues. The purely elastic influence coefficient for displacement is extended to a form that accounts for the time-dependent material property.

Simulations for a viscoelastic half-space described by a Maxwell rheological model, indented by a spherical rigid indenter in a step loading, confirm the validity of the newly advanced formula. Based on the latter, we are anticipating a semi-analytical method for the resolution of the viscoelastic contact problem, in which contact pressure and displacement are obtained for every new time increment in a step-by-step approach, by solving repeatedly the purely elastic contact problem with a modified initial contact geometry which accounts for the memory effect of the viscoelastic material.

References

[1] Sneddon I N 1965 The relation between load and penetration in the axisymmetric Boussinesq problem for a punch of arbitrary profile *Int. J. Engng Sci.* **3** pp 47-57

[2] Lee E H and Radok J R M 1960 The contact problem for viscoelastic bodies *ASME J. Appl. Mech.* **27** pp 438-444

[3] Ting T C T 1968 Contact problems in the linear theory of viscoelasticity *ASME J. Appl. Mech.* **35** pp 248-254

[4] Chen W W, Wang Q J, Huan Z and Luo X 2008 Semi-analytical viscoelastic contact modeling of polymer-based materials *ASME J. Tribol.* **133** pp 041404-1—041404-10

[5] Kozhevnikov I F, Duhamel D, Yin H P and Feng Z Q 2010 A new algorithm for solving the multi-indentation problem of rigid bodies of arbitrary shapes on a viscoelastic half-space *Int. J. Mech. Sci.* **52** pp 399-409
[6] Johnson K L 1985 Contact mechanics (Cambridge: University Press)
[7] Boussinesq J 1969 Application des potentiels á l'etude de l'equilibre et du mouvement des solides élastiques (Paris: Reed. A. Blanchard)
[8] Love A E H 1929 The stress produced in a semi-infinite solid by pressure on part of the boundary *Phil. Trans. Roy. Soc.* **228** pp 377-420
[9] Liu S B, Wang Q and Liu G 2000 A versatile method of discrete convolution and FFT (DC-FFT) for contact analyses *Wear* **243** pp 101-111
[10] Polonsky I A and Keer L M 1999 A numerical method for solving rough contact problems based on the multi-level multi-summation and conjugate gradient techniques *Wear* **231** pp 206-219

Digital prototyping technique applied for redesigning plastic products

A Pop[1] and A Andrei[2]
[1]University POLITEHNICA of Timisoara, Department of Mechatronics, P-ta Victoriei, No. 2, 300006 Timisoara, Romania
[2] University POLITEHNICA of Timisoara, Department of Material and Manufacturing Engineering, P-ta Victoriei, No. 2, 300006 Timisoara, Romania

E-mail: naskk_pop@yahoo.com

Abstract. After products are on the market for some time, they often need to be redesigned to meet new market requirements. New products are generally derived from similar but outdated products. Redesigning a product is an important part of the production and development process. The purpose of this paper is to show that using modern technology, like Digital Prototyping in industry is an effective way to produce new products. This paper tries to demonstrate and highlight the effectiveness of the concept of Digital Prototyping, both to reduce the design time of a new product, but also the costs required for implementing this step. The results of this paper show that using Digital Prototyping techniques in designing a new product from an existing one available on the market mould offers a significantly manufacturing time and cost reduction. The ability to simulate and test a new product with modern CAD-CAM programs in all aspects of production (designing of the 3D model, simulation of the structural resistance, analysis of the injection process and beautification) offers a helpful tool for engineers. The whole process can be realised by one skilled engineer very fast and effective.

1. Introduction

After products are on the market for some time, they often need to be redesigned. New products are generally derived from similar products. Redesigning is an important part of the product development process [1]. In industrial design is always a challenge to make the proper compromise between the most desirable parts shape, manufacturing cost and how resistant the part will be. As consequence, a better, more complex shape will increase the manufacturing cost. Using modern CAD-CAM programs for designing and simulating a new product can offer a manufacturing time reduction and a significantly lower production cost [2]. The basic role of CAD is to precisely define the geometry of a design, as it is critical to all the subsequent activities in the product cycle. Similarly, CAM is the technology concerned with the use of computer systems to plan, manage and control manufacturing operations through either direct or indirect computer interface with the plant's production resources so that a design can be materialized [3].

Redesigning products can be done using several methods:

- One of them involves using reverse engineering techniques. Reverse engineering is the opposite of forward engineering. It takes an existing product, and creates a CAD model, for

Content from this work may be used under the terms of the Creative Commons Attribution 3.0 licence. Any further distribution of this work must maintain attribution to the author(s) and the title of the work, journal citation and DOI.
Published under licence by IOP Publishing Ltd

modification or reproduction to the design aspect of the product [4]. Reverse engineering is usually undertaken in order to redesign the system for better maintainability or to produce a copy of a system without access to the design from which it was originally produced [5].

- Another way to redesign includes the integration of scanned images, drawings, in the modeling process. These images serve as a guide for the user when they are modeling the virtual object. The method is called Sketch based 3D modeling and we can say that is part of Top-Down methodology. One can say this because the designer is using the drawings to make a better image of the new product [6];
- The third method is to disassemble a product (or more) and it's re-designing orders to improve certain characteristics (new function) by adding new components [1].

2. Case study

In this paper we used modern technology based on digital prototyping and designed a new model of roulette from a model already existing on the market.

The chosen product is a tape measure device made of plastic components. Products on the market are relatively simple, they can be improved. For this we will use the existing model and redesign it to add some functional new elements that enhance the entire product.

To achieve this, we followed steps that are presented in this paper:

- Researching and choosing the reference product.

For this paper we are looking for a simple solution existing on the market today that requires and enables more features and functional components, all to contain more items in a single product making it more versatile, easy to use and inexpensive.

- Disassembling the product and studying the component.

At this stage we will use the existing product that will be disassembled and we study each component both in terms of functionality, but also in terms of the material and the degree of functionality.

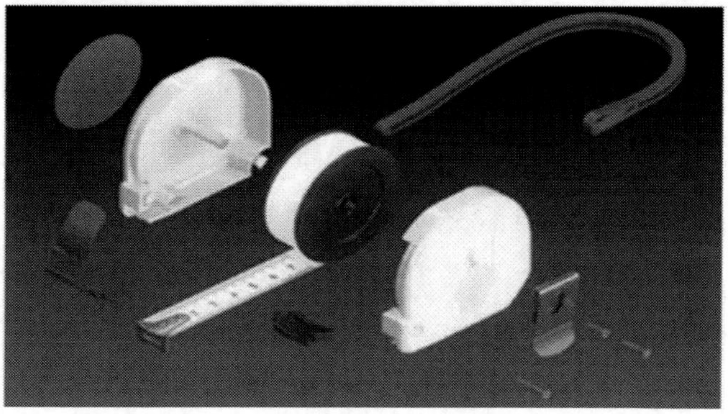

Figure 1. Isometric view of the exploded assembly.

We will create a CAD geometric model of each part of the studied component with the help of the CAD program SolidWorks. The figure below shows the assembly tape measure device (figure 1) in the exploded view which allows a better perspective on the components of which it is composed.

After we fully understand the functionality of the product, and we have studied each component we will draw, in close cooperation with the department of design and concept, a series of sketches (figure 2) with the purpose of changing, adapting and improving these components. These drawings are intended to cover the requirements of customers, their needs and to provide insight into new products. However, these drawings are not the final version of the products as they are just the reference to the client and design engineer.

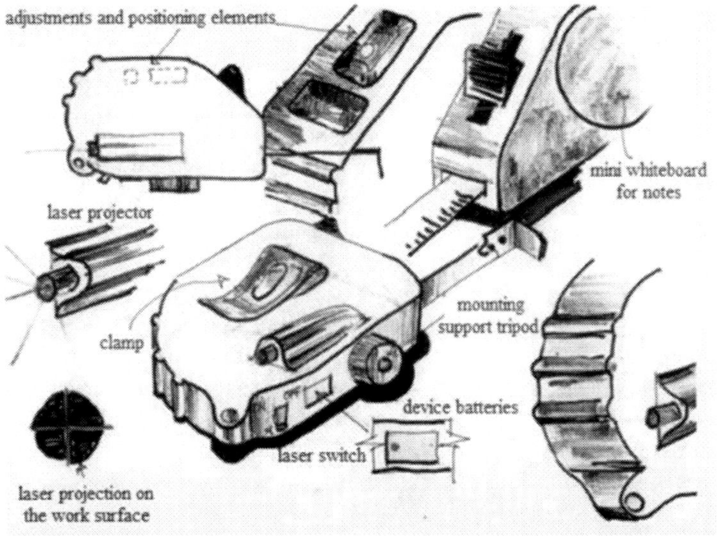

Figure 2. Concept sketches with different views and accessories.

3. Designing the product

In order not to substantially increase production costs for this type of product and also to not completely change the entire production line, the redesigning decision was reached only for a few components, namely housing tape measure device (figure 3). Along with these changes we will add other new components which will modify and enhance the functionality of the product.

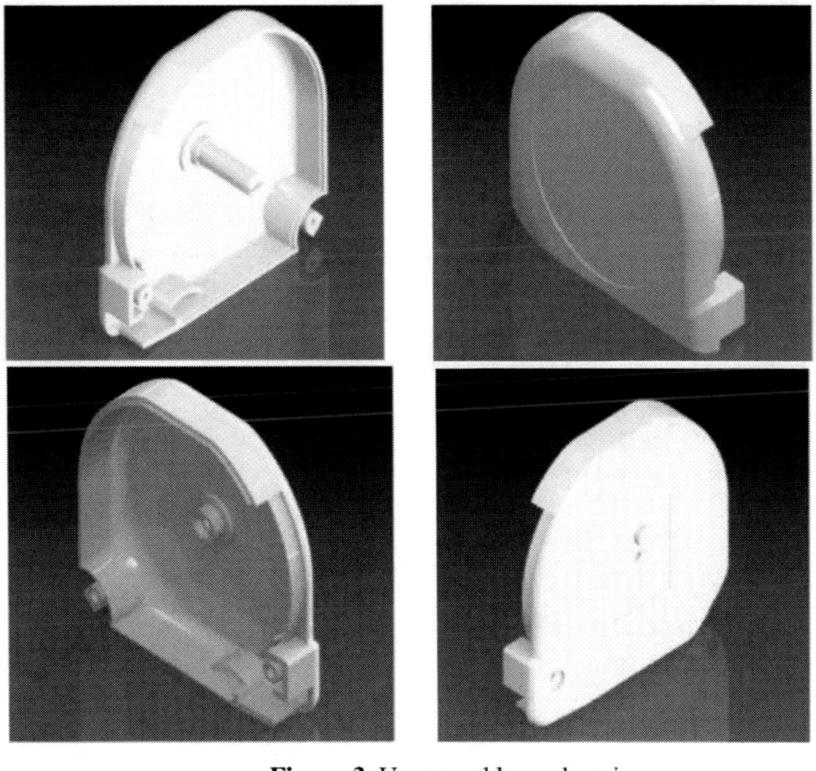

Figure 3. Upper and lower housing.

Components to be presented will be part of the new assembly. The first thing that is added to this new product is a laser sub-assembly. The assembly consists of the laser itself, a switch and batteries needed to power it. The purpose of this laser is to help draw a straight line on a wall, to help fixing of wall cabinets or shelves, etc. The following subset is a trip that serves to support and stabilize the tape measuring device when the laser is on. The tripod can tilt after any surfaces roulette can be placed in almost any location. Also, with tripod to achieve stabilization and flatness roulette are introduced bubble levels. To keep the batteries in place and to maintain permanent contact between them and close the circuit we will introduce a cap. Its role is not only to keep the batteries but also to allow quick and easy replacement thereof. Another element that is added to this new product is a magnetic support for the scale rule that comes attached to its end.

4. Structural analysis and simulation of the new product assembly

Simulation of a product offers a number of useful data for its design. The simulation results may determine a number of changes, such as the shape and material of the product. In order to simulate this assembly is required to follow certain steps to get the most accurate results compared with the reality. For the simulation we used the Simulation package provided by the CAD program SolidWorks.

Below is the assembly in the final version (figure 4) with all components placed in the operative position.

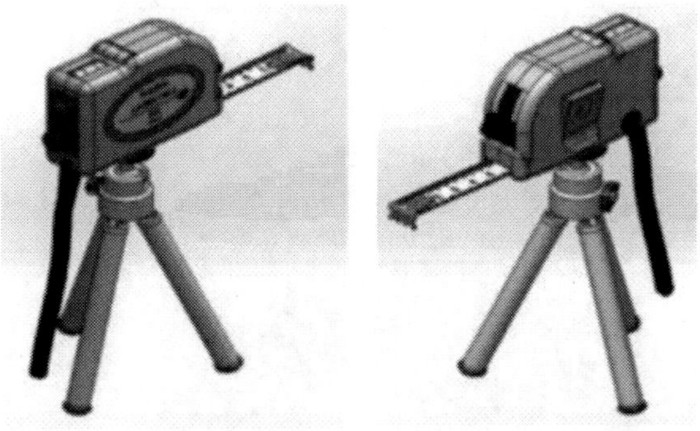

Figure 4. The final assembly of the redesigned product.

Simulation of a fall from 1.5 m

An analysis is made of the trop impact test for the plastic housing, as a first step towards the simulation of the impact event for design and development purposes. This simulation provides important data on the resistance assembly instant shock. We run the simulation on the plastic housing with the added mass of all the new components except the tripod.

To achieve this simulation we must follow these steps:

-The choice of simulation;

-The choice of material (ABS);

-Election of fall distance (h = 1.5m);

-Realization of the mesh (the more complex, the more the results will be accurate);

-Interpretation of data.

The maximum Von Misses stress resulted from the finite element simulations for the drop impact test of the housing tape was 18.952 [MPa] (figure 5). The maximum yield strength for ABS from the literature [7] is 42.5 [MPa] up to 44.2 [MPa].

Analyzing the data from figure 5 we conclude that upon impact, the housing tape measure does not suffer major damage leading to redesign it.

Analysis of the injection molding process

This analysis produces a set of parameters that are very important for the production of such components. The CAD software used was MoldFlow. The parameters are: the speed of injection, the

IOP Conf. Series: Materials Science and Engineering **95** (2015) 012112 doi:10.1088/1757-899X/95/1/012112

filling time of the mold, 1.09 sec. (figure 6), the cooling times, 13.39 sec. (figure 7). The material chosen for this product is ABS (acrylonitrile-butadiene-styrene) because of its higher injection load resistance and a significantly melt flow rate (MRF = 45 g/10min).

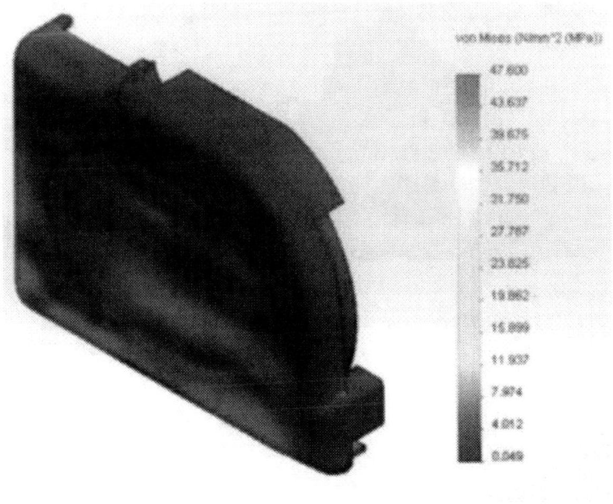

Figure 5. Interpretation of data.

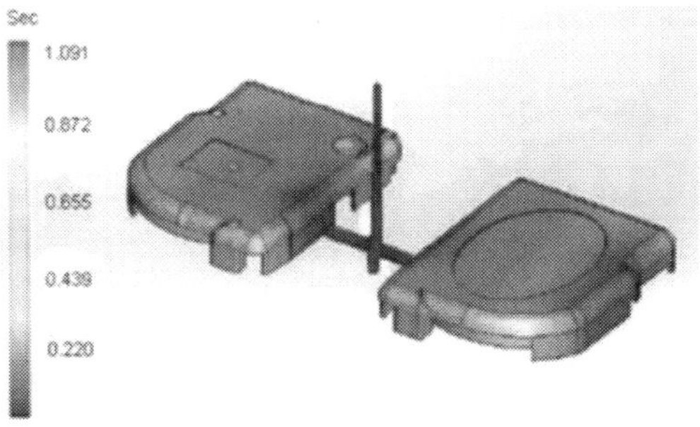

Figure 6. The time required to fill the mold (t=1.09 sec).

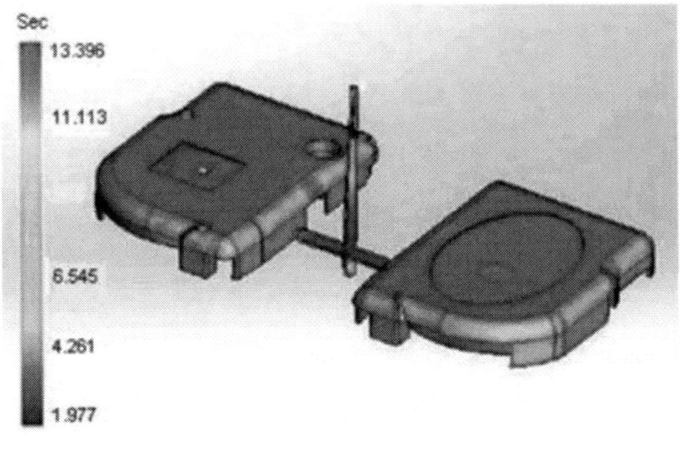

Figure 7. Product cooling time (13.39 sec).

4. Conclusions

In this paper we presented a modern and fast way of creating a new product from an existing one. There were used both known techniques for modeling parts and components and modern techniques for simulation and analysis of the product. Finally, we obtained a new product with new features that can be used in many fields.

The results of this paper show that using advance CAD-CAM techniques in designing a new product from an existing one available on the market mould offers a significantly manufacturing time and cost reduction. The ability to simulate and test a new product with modern CAD-CAM programs in all aspects of production (designing of the 3D model, simulation of the structural resistance, analysis of the injection process and beautification) offers a helpful tool for engineers. The whole process can be realised by one skilled engineer very fast and effective.

Acknowledgements
This paper is supported by the Sectoral Operational Programme Human Resources Development (SOP HRD), ID134378 financed from the European Social Fund and by the Romanian Govern.

References

[1] Shana J S, Gregory S and Shen Y T 2012 Redesign for product innovation *Design Studies* **33(2)** pp 160–184
[2] Zhang Y 2003 Research into the engineering application of reverse engineering technology *Journal of Materials Processing Technology* **139** pp 472–475
[3] Xu X W and He Q 2004 Striving for a totalintegration of CAD, CAPP, CAM and CNC *Robotics and Computer-Integrated Manufacturing* **20** pp 101–109
[4] Kolar V D 2008 Application of reverse engineering and rapid prototyping to casting *Doctoral thesis, Jawaharlal Nehru Technological University, India*
[5] Lee H and Soonhung H 2005 Reconstruction of 3D interacting solids of revolution from 2D orthographic views *Korea Advanced Institute of Science and Technology*
[6] Aiman Z 2013 Innovative Design for Active Disassembly and Sustainable Product Recovery *Electronic Theses and Dissertations University of Windsor*

IOP Conf. Series: Materials Science and Engineering 95 (2015) 012113 doi:10.1088/1757-899X/95/1/012113

Innovative methods for knowledge transfer

A M Goanta[1]
[1]"Dunarea de Jos" University of Galati, Engineering and Agronomy Faculty of Braila, Calea Calarasilor Street, 810017, Braila, Romania

E-mail: goanta_a_m@yahoo.com

Abstract. At this time, learning takes place, either with classic books on paper support or using books scanned or drawn and further converted into PDF or PPT files that are printed on type support CD / DVD. The latter modern means of learning, the study adds live on the Internet using search engines and not least e-learning method, which allows the study of bibliographic related materials in PDF or PPT, stacked and grouped on the basis of a curriculum imposed which can be accessed on a website via a user name and password. Innovative methods come to successfully use other file types than those mentioned above. The graphics in teaching technical subjects such as descriptive geometry can be achieved using animated PowerPoint files, allowing for visualization of steps to be taken, in the case of solving by drawing a descriptive geometry. Another innovative method relies on the use of HTML files, inspired by related sites help design software packages that can be used when teaching descriptive geometry that the technical design. Through this work, the author has proposed to present a new innovative method, which is inspired by the methods listed above, but involves using AVI files to teaching of computer-assisted type graphics or info graphics. In general this new author's method lends itself particularly well to the teaching of the use of software packages because the student actually see the place from where the delivered command is accessed and contextual options of right button of the mouse. These laboratory courses or mentoring can be freestanding cards support CD / DVD or can be posted on websites with restricted access based on user name and password. Practically paper presents the methodology of creating courses and tutorials in AVI format, and how to monitor the degree of accessing the website, on which there were posted mentioned teaching materials, using the tools offered by Google Analytics. The results consist of graphical work, about the degree of access to educational materials, made on the basis of various criteria, such as the: time of access, access location, age of those who accessed, chapters accessed, etc. Finally the author presents his findings on the benefits of this method.

1. Introduction

Current education is completely changed from that in the 80s-90s when a student was required to attend all course and laboratories to take notes further used for learning during examination sessions [1]. The entire bibliographic material was on paper support and the student could obtain it either from the library or by buying it from university libraries. There were, however, initiatives of non-conformist media such as sheets with formulas and demonstrations being made available to students, together with magnetic recording tapes that contained the texts. Together they formed an alternative to the classical method and the student used to listen to the explanation recorded on the magnetic tape and followed a logical mathematical demonstration on those sheets of paper. However, the most common method that has been perpetuated up to this day is the knowledge transfer on paper support. With the

Content from this work may be used under the terms of the Creative Commons Attribution 3.0 licence. Any further distribution of this work must maintain attribution to the author(s) and the title of the work, journal citation and DOI.
Published under licence by IOP Publishing Ltd

entrance into the 21st century the students' needs, generated by the century of speed and the new information technology, called for new modern methods of communication and learning on the educational market which subsequently experienced a strong expansion tendency. To this situation contributed the fact that the society we live in requires us to be active in as many directions as possible, even from different areas, to be able to reach that standard of life each of us wants. In this context the student, as an element of this modern society, is put under pressure from the society in which he lives and is forced to optimize his time dedicated to study so that he still has certain time resources to allow him to provide the desired living standard. In other words the student wishes to attend a university for a better future professional opening but he must keep his job as well, which often prevents him from being present at all the activities organized in the teaching process and leads him indirectly to poor results with adverse effects in the near future [2]. Given that there is extramural education it was introduced on larger scale the distance learning which is strongly supported [3] by the e-learning platforms method. These platforms have the incontestable advantage that are accessible 24/24 hours at any time of day or night, on a double condition, i.e. having an internet connection along with platform access data. Basically we succeeded in a very short time to move from the transfer of information from the paper support to electronic format. At this time almost all universities have educational platform for e-learning that contain information for all components of the teaching process, namely: lecture, seminars and laboratories [4]. In the last 15 years the author has been concerned about the ongoing upgrade of the teaching methods and transfer of information in the field of engineering education, focusing particularly on traditional formats of PDF, PPT with animation, or modern HTML and various video formats.

2. Innovative method's description and main stages

Since 2006 the author [5] has focused on the achievement of AVI courses as for the graphic engineering disciplines that use diversified software they are much more suitable for the transfer of information because it makes use of pictures and sound. Compared with this method, teaching such subjects using classic books require laborious expressions and a large number of explanatory images for each order separately, for every setting allowed by the software in question. Often, the author's method of expression in words, on paper, is not explicit enough so as to transmit to the student the entire procedure for implementing an order or instruction. Video format AVI / MPEG / FLV / SWF allow a more advanced explanation in graphic disciplines' case of engineering field because, while the teacher sets out the manner of implementing the orders, the student has the opportunity to see dynamically and in real time the effect of the command explained.

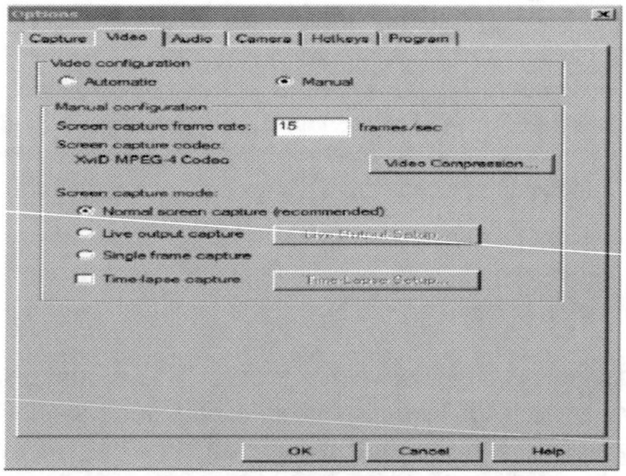

Figure 1. Setting format AVI and frame/sec.

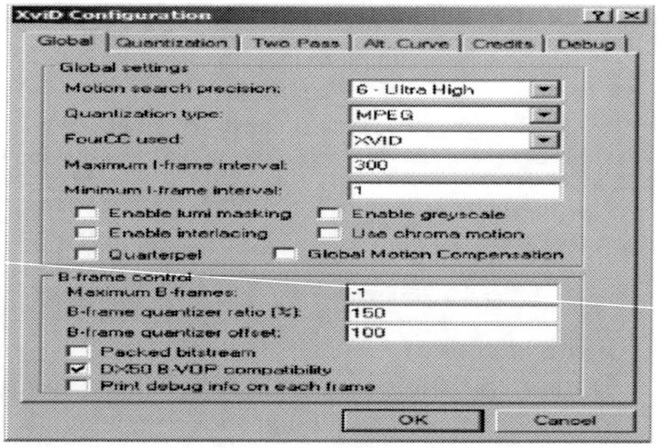

Figure 2. Setting the video codec.

The disciplines for which this method was applied: Infographic 2D/3D, Informatics systems of technological design, assisted graphic and parametric geometric modelling which are based on AutoCAD software, Solid Edge, Mechanical Desktop and NX Siemens. The method presented in this paper refers to the use of SWF files that are based on AVI and are finally posted on the internet in a certain logical sequence that follows a university curriculum.

For the management of the files SWF it is used [6] a Joomla platform. The generation method using Camtasia software [7], an AVI video file for WEB viewing, consists of the following steps:

A. Setting the recording parameters under Camtasia Recorder.

- **a.** Setting format AVI by imposing the number of frames/sec. and the video codec chosen (figure 1);
- **b.** Configuration of the video codec used to obtain maximum performance (figure 2);
- **c.** Setting the size of the recording area extracted from the entire surface of the screen in order to register and setting the audio parameters(figure 3);
- **d.** Setting the watermark type customization to ensure copyright

B. Generating the video file in format AVI with Camtasia Recorder.

C. Setting the parameters for transformation of AVI files SWF files with Camtasia Studio.

- **a.** Selection of AVI file generated to be further transformed into Flash file with SWF extension to be displayed on Internet.
- **b.** Preset Manager and select the project type: blog, CD, iPOD, WEB, Windows Media;
- **c.** Edit Production Preset (select the output file variant SWF/FLV/WMV/MOV/RM/CAMV/GIF animation);
- **d.** Settin Flash Template (Flash Options and Custom Size)
- **e.** Setting Video Options;
- **f.** Select a folder for the videos you are producing and Finish

D. Management and monitoring of information transfer through a JOOMLA platform.

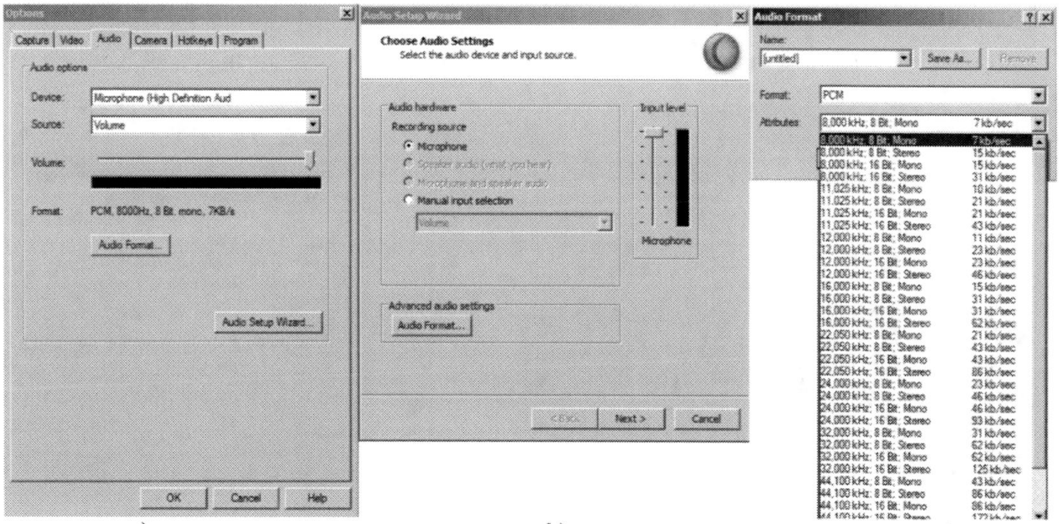

a) b) c)

Figure 3. Setting audio parameters.

Since the objective of the current work is to present an innovative method of knowledge transfer I shall present few windows associated to stage "C" which is made with Camtasia Studio. The file selection involves choosing the existing file in AVI format for which is desired the transformation by high-resolution rendering in SWF file. Specifically in the case presented it was selected the file

"Tutorial_gfile.avi" to transform by rendering. Finally we obtained an entire directory, viewed in figure 4, which contains a multitude of files among which it is found the "Tutorial_gfile.swf ".

It should be noted that all these files are required for posting on the JOOMLA platform. The software Camtasia Studio recommends, out of all extension variants, to use Flash SWF because this variant offers the highest quality relative to the minimum size of the file and on the other hand the resulting file can be opened under the optimized version with any browser. There is a small disadvantage though, namely the vast majority of the movements on the screen must be done at a relatively low speed.

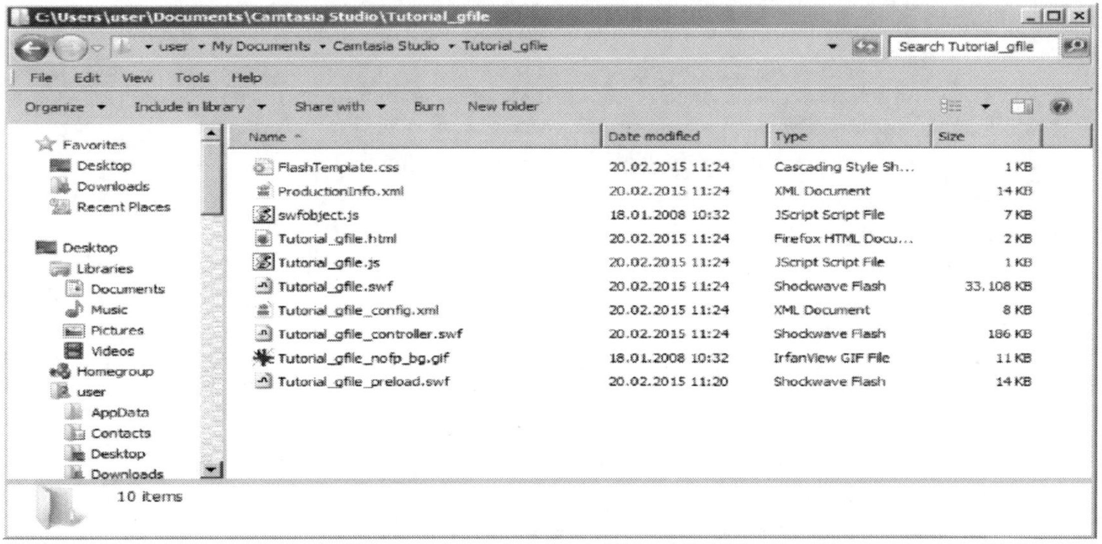

Figure 4. Directory with the resulting files.

At the stage visualized in figure 5, the author recommends the default setting on automatic variant and possibly preserve or increase the resolution after rendering to obtain a high quality image.

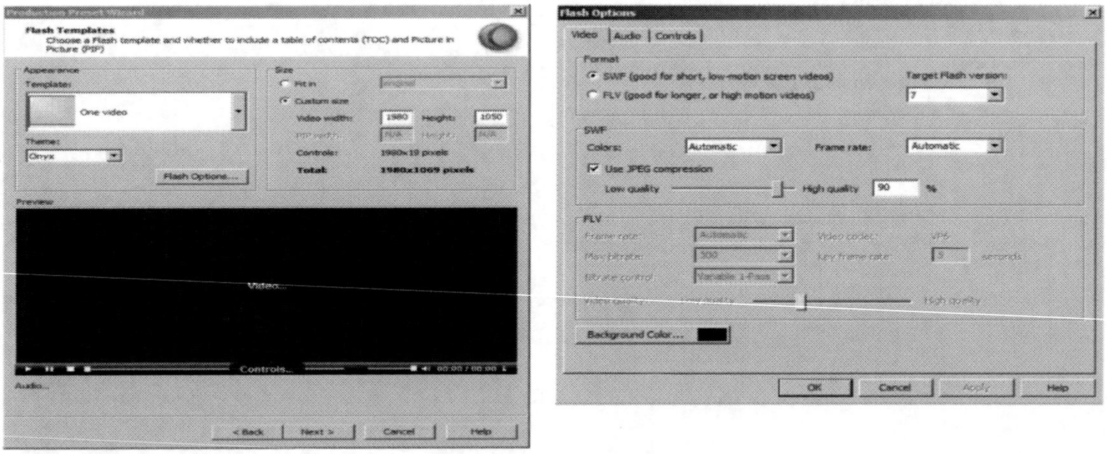

Figure 5. Setting Flash Template.

The main advantage of this method is given by the video file type that provides simultaneous sound and colour image at a resolution Full HD, while allowing real-time tracking of the dialogue mode with the PC in the explanation process of each instruction taught.

3. Monitoring results and knowledge transfer

By means of a JOOMLA platform, a site has been created at the address http://www.cursuriuniversitarebraila.ugal.ro where it has been posted since 2008 all the courses in video format graphics for technical engineering disciplines together with video recordings of the laboratory work. Also posted on the internet is the homework in all subjects with specific technical graphics. Monitoring the transfer of knowledge to students was performed on two distinct directions, namely: using JOOMLA capacity (figure 6) and indexing [8] website on Google Analytics (figure 7). Each of the methods of monitoring provides information complementary to that teacher who acts as administrator. Therefore Joomla platform provides information on whom of the students with access, entered the site to study and how long was connected, including in what time frame and what chapters he searched for.

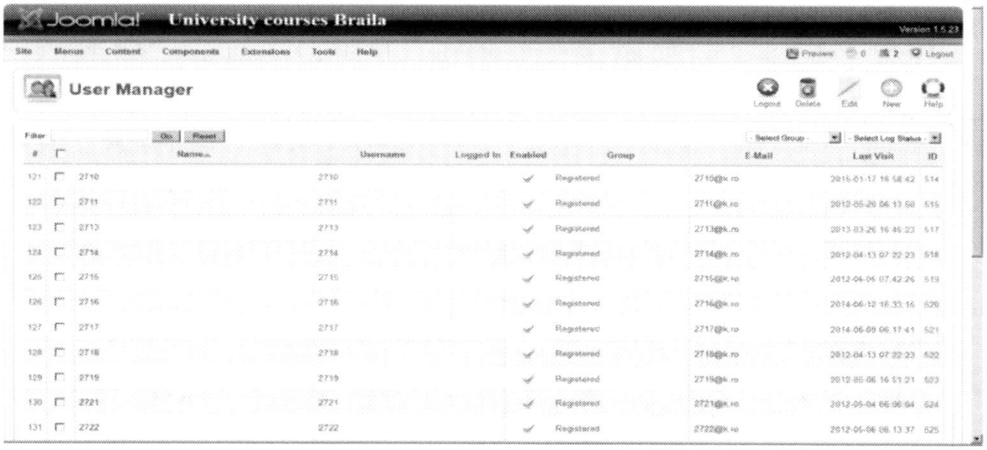

Figure 6. Monitoring by JOOMLA.

Figure 7. Monitoring by Google Analytics.

Monitoring with Google Analytics provides complementary information to JOOMLA variant without having the ability to determine nominal site users but having the opportunity to show the degree of access to the site within a certain period of time, how many work session have been in a period of time, how many pages were visited on average in each session, which was the average length of time, how many visitors were there in a certain period, which is the geographical positioning of visitors per continents, countries and cities, what operating systems had visitors on personal computers, what resolutions have visitors' monitors, what ISPs were using, what browser was used, the ratio of new users and those who returned to the site, etc. Graphical view of all this information can be done in hours, days, weeks or a month.

Figure 7 illustrates the degree of accessing the courses site from 19.12.2014 which is the beginning of the winter holiday until 02.20.2015 which is the last day of the first week of school after the holidays that followed the examination session. It can be seen easily that the two periods of vacations, winter holiday and after the winter session, the activity on the site is very low while in session, the degree of access increases significantly. It is also found that in the first week of school in February 2015, the degree of access to the site is slightly higher.

4. Conclusions

At the end of the paper it should be noted that the method presented has two types of monitoring, one made available by Joomla platform and one by indexing the site on Google Analytics. Due to the limited space of this paper, it could not be inserted all graphical representations resulting from the monitoring work, but I present in the conclusions:

-Monitoring by JOOMLA allows finding the last access made by a student or if he has never logged for study;

-Monitoring by indexing Google Analytics is more efficient in terms of who accesses the information, when accessed, where from access is made to the online courses

-Monitoring with Google Analytics is a perfect support consistent with the student psychology and behaviour in the sense that usually during vacation there is no study; during the semester there is a continuous increase of the study until the ninth week, then there is a slight decline due to both fatigue and lack of interest followed by sustained effort during the session.

-Monitoring with Google Analytics cannot distinguish between the visitors with user and password and those who stop at the moment of entering the access data

-The distribution of the degree of access by counties (Braila 76.44%, Bucuresti 5.46%, Galati 3.45%, Constanta 1.4% while the remaining districts have less than 1%) demonstrate that the site is intended almost exclusively to the students in Braila.

-The percentages related to the areas in Bucharest, Galati and Constanta is justified by the fact that students from Braila sent the access data to their friends in the university centers mentioned above.

References

[1] Sukumaran K and Srinath M V 2006 Mobilelearning: an innovative learning method of life long learning *paper presented at XXXVI ISTE Annual Convention & National Seminar* pp 14-16

[2] Hammond N 2003 Learning technology in higher education in the UK: Trends, drivers and strategies In M. Van der Wende and M. van der Ven (eds) *The use of ICT in Higher Education: A mirror of Europe. Lemma Publishers* Utrecht pp 109-122

[3] Goanta A M 2011 E-learning Method Applied to Technical Graphics Subjects *Journal of Industrial Design and Engineering Graphics*,pp13-16

[4] Matijevic M, Cvjetkovic V, Stefanovic M, Rankovic V and Stevanovic D 2007 Implementation of eLearning at the Faculty of Civil Engineering, Subotica, Serbia - Should the creators of long distance learning courses become aware of didactics guidelines for advanced training? *Proceedings of the eLearning Conference*, Belgrad, Republic of Serbia

[5] Goanta A M 2009 Communication Innovative Methods for Graphics Teaching On Technical Directions *Proceedings of the 5th International Vilnius Conference EURO-Mini Conference – KORDS* 2009. *Knowledge-Based Technologies and Methodologies for Strategic Decisions of Sustainable Development*", Lituania, Vilnius pp375-380

[6] Information on http://www.joomla.org/, Accessed: 17/01/2015

[7] Information on http://www.techsmith.com/camtasia.html, Accessed: 23/09/2014

[8] Information on http://www.google.com/analytics/, Accessed: 20/02/2015

IOP Conf. Series: Materials Science and Engineering **95** (2015) 012114 doi:10.1088/1757-899X/95/1/012114

Web platform for functional design

M R Dijmarescu[1] **and M C Dijmarescu**[2]
[1]University POLITEHNICA of Bucharest, Manufacturing Engineering Department, 313 Splaiul Independentei, 060042 Bucharest, Romania
[2] University POLITEHNICA of Bucharest, Materials Technology and Welding Department, 313 Splaiul Independentei, 060042 Bucharest, Romania

E-mail: manuela-d@live.com

Abstract. Today's global competitive trends, especially those related to industries, determine a much higher degree of pressure and demands for substantial innovation driven improvements, flexible and time sensitive solutions. Improving and optimizing the design activity by shortening its timeline and maintaining a high quality level for its output have become the main success factors. The evolution of design activity is strongly related to the evolution of education and research made in the design field. Thus, the development of web tools which can contain knowledge about mechanical products functionality and structure may be an important achievement for the education and industry. This paper presents a web platform which contains functional-constructive knowledge in the area of mechanical design field and was developed to support design activity. The proposed web tool can provide any user, even one without background in design theory, information about the functionality of products and the way it is related to the product structure.

1. Introduction

The demand for reducing the time-to-market for a new product and, at the same time, to provide it with a very good functional level, has become a main requirement in today's business environment. To support this requirement, the researchers have developed methods and tools which can help the designers to accomplish faster their design task. Most of these tools have as background design knowledge stored in databases. However, most of these databases contain only CAD data and no knowledge about products functionality [1].

It has been recognized that knowledge sharing in product design area influences both the development time and the product cost in a positive way [2]. Today, due to the evolution of the internet, easy access to design knowledge and design collaboration can be possible, and this makes the web-based design tools flexible and cost-effective.

If we look from the educational perspective, it is proven that today's students no longer spend their time writing by hand or collecting information from books in a library room. The functionality of these learning components has changed and computer and web-based applications made the knowledge more accessible and transportable. Web-based knowledge sharing is the start of new developments in education and especially in higher education area.

A web-tool which provides information about functionality and its applications is "production inspiration" [3]. Here, various solutions can be found for 37 primary functions linked to one of the states: solid, liquid, gas or field [3, 4]. This tool can be used as source of inspiration for products and

Content from this work may be used under the terms of the Creative Commons Attribution 3.0 licence. Any further distribution of this work must maintain attribution to the author(s) and the title of the work, journal citation and DOI.
Published under licence by IOP Publishing Ltd

manufacturing development, but the information regarding the solutions presented here is at minimum level and the designer work cannot be improved considerably by its use.

The most important achievement in the domain of sharing functional knowledge is the development of the web-based repository made by the University of Missouri–Rolla in collaboration with the University of Texas at Austin, Pennsylvania State University, Bucknell University, Virginia Polytechnic Institute and State University [5, 6]. However, using this tool implies a certain level of knowledge regarding Functional Basis functions and flows considered for products [7].

To improve the educational process in the design field, it is imperative to provide and share design knowledge based on products functionality. For this reason, it has been considered that the design of a web-platform with a simple and easy interface which contains functional-constructive knowledge which can be easy understood by any user with a technical background can provide a good basis for improving this process.

2. Objective

The objective of this research is to develop a web-platform to support students, teachers and engineers in the design activity by providing information about the functionality of products and the way it is related to their structure.

The main elements defining the research program have been: creating a design knowledge database with information about products functionality and its relationship to the constructive structure, documentation on the state of art regarding web tools available for supporting design activity, defining the platform structure, creating the platform interface and establishing its relationships to the knowledge database, testing the platform functionality.

3. Web platform design

The web platform was created having as a background a knowledge database developed for functional design [1]. One of its purposes is to be used as an interface between the potential users –teachers, engineers, and students – and the functional-constructive knowledge present in the database, for supporting users in design activities (table 1).

Using this web-tool, the information will be accessed easily because the user does not need to be connected at a device which has a fix location and can have the information at hand only if it has a device which allows him to have internet connection. The only problem which can appear is the use of CAD data, because this involves to have installed on his computer CAD software.

Table 1. Schematic extract of the functional-constructive knowledge [1].

Functional class		Specific functions		Sub-functions		Constructive units	
Code	Name	Code	Name	Code	Name	Code	Class
FC1	Motion	1SF1	Create			110CU1	Combustion engines
							…
		1SF2	Transfer	12F1	Rotary motion	121CU1	Rotary shafts
						121CU2	Bearings
							…
				12F2	Linear Motion	122CU1	Linear shafts
							…
					…		
FC2	Connect	2SF1	Assembly	21F1	Removable	211CU1	Fasteners
							…
				21F2	Permanent	212CU1	Welding
							…
					…		
				…			

For the design of the web-platform, the Hyper Text Mark-up Language, HTML, [8], and the Cascading Style Sheets, CSS, style [9] were used, as presented in table 2.

Table 2. Platform code and style examples.

Code	Style
<html xmlns="http://www.w3.org/1999/xhtml">	#container
<head>	{
<meta http-equiv="Content-Type" content="text/html;	margin: 0 auto;
charset=iso-8859-1" />	width: 1400px;
<title>Web - Platform for functional design</title>	background-color: #fff;
</head>	}
<link href="styles1.css" rel="stylesheet" type="text/css" />	

This work also incurs the need for a hardware environment. For the web-platform implementation, a data server will be necessary to contain the information from the database and an internet server on which the web application will be installed and from which the connection with the internet will be made. As an alternative, for reducing the costs with the hardware equipment, internet hosting can be bought from a specialised provider. The network relations of the web-platform, when hardware acquisition is considered, are briefly described in figure 1.

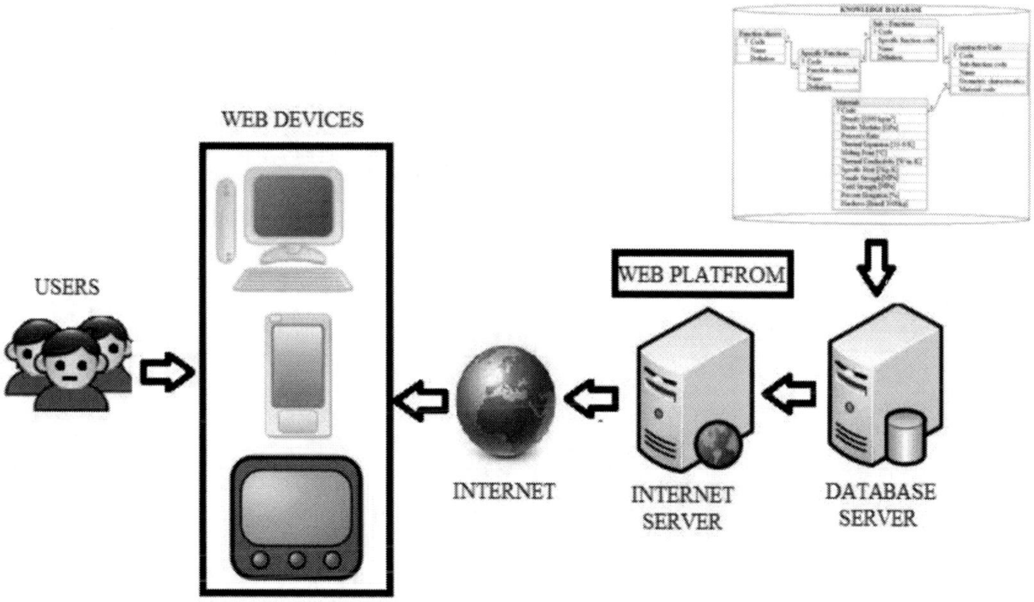

Figure 1. Relations between users and database content.

4. Web platform user interface

In general, the user interface design has as its main objective making the users interaction with the web-tool easy, efficient and enjoyable.

The web platform that has been created, Web Platform for Functional Design, denotes a simple and user friendly interface (figure 2). The main page contains a set of functional classes considered for the mechanical products – *motion, connect, control, support* and *convert* – represented by buttons which will guide the user to the specific functions, sub-functions, their definitions and the associated constructive-functional units assigned. For each constructive-functional unit, the user can retrieve information about its features and can download a CAD model which can be used as the component to be developed.

The platform includes also other sections which present general information related to its purpose, such as:

-*About* – general presentation of the web-tool and the reason for which it has been created;
-*News* – here news about research made in the design area will be listed;
-*Contact* – contact detail of the platform administrator;
and specific data and tools related with the design process, such as:
-*Design tools* – here tools for supporting design activity, as the tool for creating functional-constructive matrixes, FCM will be found;
-*CAD library* – here the CAD models for most of the considered design functions will be stored. These models will be accessible as .step format files or Autodesk Inventor format files. The CAD library will contain also a section on which students can exchange CAD data and also functional knowledge. The information inserted here will not influence or modify the functional-constructive knowledge presented in the main database, but the platform administrator can decide if the contents uploaded by users will be included there or not.
-*Materials library* – here information regarding material characteristics will be stored.

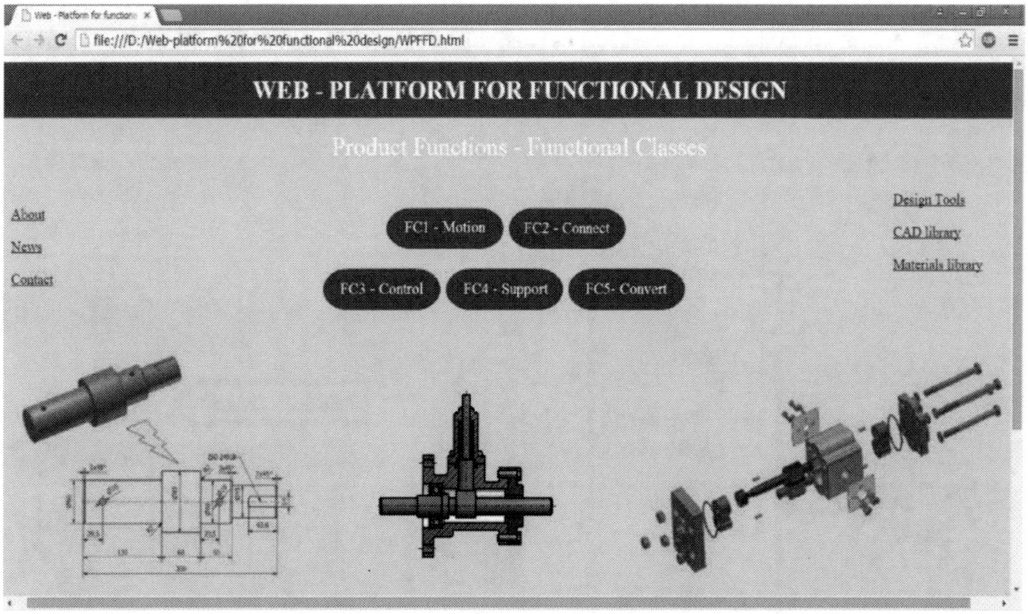

Figure 2. Web-platform main page.

Every functional class, specific function and sub-function name is linked to a web page which contains information related to that subject, and where, under their name or definition a button, *Constructive-functional units* is displayed, which guides the user to the different levels of constructive-functional solutions. By selecting one of the constructive-functional units displayed for a certain function, a page with the constructive characteristics of that specific unit is opened. If until this moment the user obtains only functional information, from this page the user can complete the design information with its constructive characteristics and download a CAD model with the desired dimensions by selecting them from the table present on this page. To obtain complete information related with the desired material from the list of possible options, the user should click on the material name and its characteristics will be displayed. A preview of what is described above is presented in figure 3. The path followed to access the presented platform pages is: *FC1 - Motion* (page content not shown) → *1SF2 - Transfer* → *12F1 - Rotary motion / Constructive-functional units* → *Rotary shafts*.

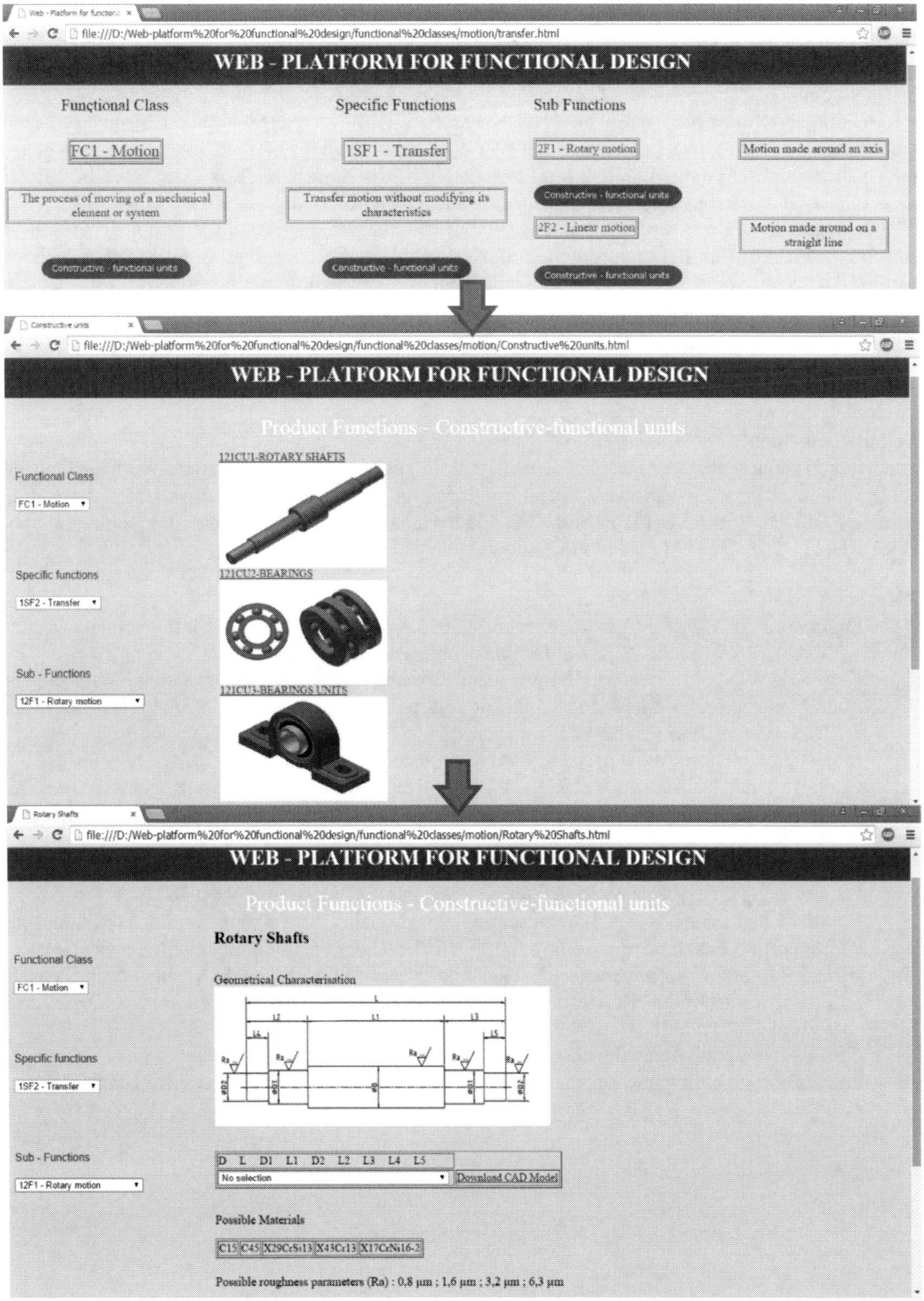

Figure 3. Preview of the web-platform interface.

5. Conclusions

This paper presents the design and the interface of a web-platform developed for supporting and sharing knowledge in the design process field.

The simple interface of this web-tool can provide, to any user which has a device which can allow internet connection, even without background on the design theory, the ability to consult and retrieve information about the functionality of products and the way this is related with their structure. Besides the functional-constructive knowledge, the platform also includes other sections presenting general information related to its purpose, such as news about research made in design area, specific data and tools related to the design process, a CAD library and a material library.

The implementation of this web-tool can shorten a lot the time necessary for design because for the elements retrieved from here the design is almost complete, the only missing information being the part/product drawings. But, to achieve the implementation of the proposed web-platform further work is needed in order to develop all the connections between the knowledge content, filling in complete information regarding geometrical and material characteristics for the considered constructive solutions, introduction of cost elements and tools, choosing the hardware support for making it accessible to the users and in order to introduce it in the design educational process.

Acknowledgement

This work is supported by the Sectorial Operational Programme Human Resources Development (SOP HRD), financed from the European Social Fund and the Romanian Government under the contract number POSDRU/159/1.5/S/137390 for Manuela-Roxana Dijmarescu and under the contract number POSDRU/159/1.5/S/134398 for Maria-Cristina Dijmarescu.

References

[1] Dijmarescu M D, Parvu C and Gheorghe M 2015 Knowledge Database for Functional Design of Mechanical Products *Applied Mechanics and Materials* **760** pp 15-20

[2] Chai K H, Wang Q ,Song M, Halman J and Brombacher A 2012 Understanding competencies in platform-based product development: antecedents and outcomes *J. of Product Innovation Management* **29-3** pp 452-472

[3] Information on http://www.productioninspiration.com, accessed: 10.08.2014.

[4] Chulvi V and Vidal R 2009 TRIZ on Design-oriented Knowledge-based Systems *The TRIZ Journal* available at http://www.triz-journal.com/triz-on-design-oriented-knowledge-based-systems

[5] Bohm M, Stone R and Szykman S 2005 Enhancing Virtual Product Representations for Advanced Design Repository Systems *J. Comput. Inf. Sci. Eng.* **5-4** pp 360-372

[6] Bohm M and Stone R 2004 Representing Functionality to Support Reuse: Conceptual and Supporting Functions *Proceedings of ASME DETC & CIE Conferences* Utah pp 411-419

[7] Hirtz J M, Stone R, McAdams D, Szykman S and Wood K 2002 A Functional Basis for Engineering Design: Reconciling and Evolving Previous Efforts *Research in Engineering Design* **13-2** pp 65-82

[8] Information on http://en.wikipedia.org/wiki/HTML, accessed: 10.02.2015

[9] Information on http://en.wikipedia.org/wiki/Cascading_Style_Sheets, accessed: 10.02.2015

IOP Conf. Series: Materials Science and Engineering **95** (2015) 012115 doi:10.1088/1757-899X/95/1/012115

Creating photo-realistic works in a 3D scene using layers styles to create an animation

A M Avramescu[1]
[1]University "Politehnica" of Bucharest, Department of Engineering Graphics and Industrial Design, 313 Splaiul Independentei, Sector 6, Bucharest, Romania

E-mail: avr_ana@yahoo.com

Abstract. Creating realist objects in a 3D scene is not an easy work. We have to be very careful to make the creation very detailed. If we don't know how to make these photo-realistic works, by using the techniques and a good reference photo we can create an amazing amount of detail and realism. For example, in this article there are some of these detailed methods from which we can learn the techniques necessary to make beautiful and realistic objects in a scene. More precisely, in this paper, we present how to create a 3D animated scene, mainly using the Pen Tool and Blending Options. Indeed, this work is based on teaching some simple ways of using the Layer Styles to create some great shadows, lights, textures and a realistic sense of 3 Dimension. The present work involves also showing how some interesting ways of using the illuminating and rendering options can create a realistic effect in a scene. Moreover, this article shows how to create photo realistic 3D models from a digital image. The present work proposes to present how to use Illustrator paths, texturing, basic lighting and rendering, how to apply textures and how to parent the building and objects components. We also propose to use this proposition to recreate smaller details or 3D objects from a 2D image. After a critic art stage, we are able now to present in this paper the architecture of a design method that proposes to create an animation. The aim is to create a conceptual and methodological tutorial to address this issue both scientifically and in practice. This objective also includes proposing, on strong scientific basis, a model that gives the possibility of a better understanding of the techniques necessary to create a realistic animation.

1. Introduction

The representation of lights, shadows and textures is realized with the help of software and hardware. A photo-realistic scene is a set of different objects which were defined geometrically in 3D space, together with viewing, shadow and lighting parameters. This paper briefly presents the different possibilities for representing a 3D scene have been explored in the literature and the way that all this given effects to a scene. This article has the aim to study how the shadows and the lights applied the renderings of a photo-realistic work and the big influence that they have in a 3D scene by using layers styles for the future animation.

3D Computer Graphics is a field where technology goes together with arts as a whole with lots of possibilities. In ours days, films and video industry have been the main target of 3D computer graphics. In this reason, there is hardly a movie that does not use special effects wholly or partially created on computer [1].

Content from this work may be used under the terms of the Creative Commons Attribution 3.0 licence. Any further distribution of this work must maintain attribution to the author(s) and the title of the work, journal citation and DOI.
Published under licence by IOP Publishing Ltd

IOP Conf. Series: Materials Science and Engineering **95** (2015) 012115 doi:10.1088/1757-899X/95/1/012115

When three-dimensional imagines are made interactive so that users feel involved in the scene, the experience is called virtual reality [2]. This makes it possible that scene created to look away from reality. 3D graphics is much more complex, working with textures, volumes, light sketches. Numerous sketch-based modelling techniques have been developed for meshes [3- 8], parametric surfaces [9], procedural surfaces [10], volumetric models [11] and implicit surfaces [12-17].

Figure 1. Light distribution and light reflection [18, 19].

Because of recent progress in sample techniques, the digital images, sounds and video are now part of our daily life. The world designing became increasingly integrated in people's activities each day. Recently, the development of 3D graphics opened the way to shaping complex objects or scenes in 3D [1, 5, 8].

The main advantage of the sketch-based modeling techniques over voxel-based models is their ability to provide a high-level surface paradigm based on a low-level volumetric representation [20]. Light can be used to focus a viewer's attention on a particular object within a scene [21, 22], figure 1.

2. Creating photo-realistic scene

Some volume sculpting applications use haptic feedback to give the user a sense of shaping a virtual material [2]. The representation of lights, shadows and textures is realized with the help of software and hardware, figure 2.

Figure 2. Source of the lights on the elements in the scene.

It is well known that we are able to see an object because light reflects from the object into our eyes. Just as in nature, illumination in a realistic scene happens through a complex interaction of lights, shadows and objects. The next step is an attempt to simulate the light effect that illuminates the scene, being perceived by the observer. The ways we implement this step represent a major influence for the quality of the image and for the quantity of time required.

Light can come from a great number of sources, the most being the sun, the source of natural lighting. The software Cinema 4D provides a number of lighting options that allow us to replicate natural and artificial lighting within scenes, see figure 3.

Figure 3. Source of the lights on the elements in the scene.

Rendering realistic illumination effects for presented animated scenes with many dynamic objects is a work that needs a lot of attention. It is well known how the important is the lighting for the overall perceived realism in a scene.

The representation of Layer Styles is great resources, and can really save heaps of time when designing something. The Layer Styles provide a great base of styles that are easily editable later on figure 5.

The aim of this work is based on teaching some simple ways of using the Layer Styles to create some great shadows, lights, textures and a realistic sense of 3 Dimension. The Layers Styles provide a unique spin on a simple circle, transforming it into an orb suitable for a realistic scene. The representation of Layer Styles is a great way to create fun and interesting photo-realistic effects and text effects without requiring a lot of effort, or even a lot of skill, figure 5.

The representation of Layer Styles helps by light and color theory to begin applying drop shadows, strokes, gradients, inner and outer glows and more to your images with layer styles, creating everything from subtle color effects to the wildest and craziest text effects anyone's ever seen. That, plus it also helps to have some free time on your hands, since layer styles can become a bit of an addiction. Layer styles are special effects that can be quickly and easily applied to individual layers in a realistic scene to drastically change the appearance of something in very little time.

The change or modification of camera parameters can be operated from accessing the 'Modify' command panel in the 'Parameters' field. At the same time, values for Lens, FOV (coverage angle) and Orthographic Projection can be operated here. Video clips and movies are very similar and, in spite of some minor differences, they merge into a single one [23].

They have the same underlying technology: high frequency rendering of similar frames in order to create the illusion of movement. Movies and animations are accompanied by sound so they can appear more real. The frame frequency most frequently used in the multimedia field is 25-30 FPS (frames per seconds). As the FPS value increases, the quality of the movie also increases but also demands more resources. This piece of work shows how dynamic rendering is applied to the scene. Playing the scene means the representation of images after modelling, structuring and animation of texts is done.

3. Future works

The present research tried to answer some existent questions through the lens of the objectives, but generated other matters that could be tackled through future research and analyzes. For future research, we propose the extension of comparative analysis for dynamic rendering types.

Simultaneously, as a new trend for future research, we suggest the creation of a more complex scene that includes knowledge of dynamic rendering influence on scene creation. This way, in order to optimize the projection solutions it becomes mandatory to perform a detailed study on the algorithms for light sources and shadows.

In the future works, we will focus our efforts on developing a new photo-realistic scene and to study how the lights and the shadows can change the way that audience will perceived a scene because, we think that the progress made in hardware and software fields, the renderings can be easily integrated in scenes of high complexity. A quality animated scene will be proposed. The quality of industrials scene is an extremely complex matter, difficult to describe objectively. This will be also a subject to discuss in our future works.

4. Conclusions

The graphic representation is the connection between imagination, reason and objective reality. This article presents how the shadows and the lights apply renderings of a scene. Today, for designing a photo-realistic scene, it is necessary to know, understand and anticipate the needs of the person that watch the scene. In the study in this paper it has found that due to the contribution of animated scene, a person compare information that he to his knowledge or experience. We have presented a series of articles relating to photo-realistic scene, trying to define how the lights and the shadows influence a 3D animation in a real evolution of the design field. One of the useful properties of Layer Styles are there relationship to the layer contents. Since the style is actually a separate entity that is just linked to the layer, it will continually update itself as a layers contents are edited or moved. Furthermore, the effects can easily be adjusted after applying them, making them non-destructive in nature.

This article aims to emphasize the approach of applying integrating methods for light sources and shadows in the proposed 3D scene. Rendering and dynamic 3D lighting play an important role for achieving a realistic scene, as close as possible to reality. Each type of rendering is specific for a certain characteristic depending on requirements for each user. In this regard, the comparative analysis led to a clear outcome. Considering the rendering for the chosen scene, we have concluded that Mental Ray rendering type is the one that gives more credibility to the scene and denotes an aspect of profound realism.

The results from the research comprise a scene presentation where rendering and dynamic 3D lighting play an essential role in the process of creating the animation. This research tackles the performance issues in the field of geometrical modelling, starting with the gaps from this sector, namely the insertion of profound realism in the process of creating animated 3D scenes. Thus, the research tries to approach this vast subject by interpreting the essential dynamic rendering features involved in the geometrical modelling field, highlighting the predictive valence of these features in relation to the performance of complex 3D scenes.

Acknowledgements

The author wishes to thank the Sectorial Operational Programme Human Resources Development 2007-2013 of the Romanian Ministry of Labour, Family and Social Protection through the Financial Agreement POSDRU/107/1.5/S/76813. The author also wishes to thank Mr. Cezar Pantelimon for all his support in this paper's work.

References

[1] Ghionea I 2004 Utilisation of knowledge-based parametric modeling in engineering design *Romanian Journal of Technical Sciences, Applied Mechanics* **49** pp 579-582

[2] Blanch R, Ferley E and Cani M P 2004 Gascuel J-D, Non-realistic haptic feedback for virtual sculpture *Technical Report RR-5090 INRIA, U.R. Rhone-Alpes*

[3] Zeleznik R and Herndon K 1996 Hughes J, Sketch: an interface for sketching 3D scenes *Proceedings of the SIGGRAPH conference* pp 63–70

[4] Igarashi T, Matsuoka S and Tanaka H 1999 Teddy: a sketching interface for 3-D free-form design *Proceedings of the SIGGRAPH conference* pp 409–16

[5] Igarashi T and Hughes J 2003 Smooth meshes for sketch-based free form modeling *Proceedings of the ACM symposium on interactive 3d graphics* pp 139–42
[6] Nealen A, Igarashi T, Sorkine O and Alexa M 2007 Fibermesh: designing free form surfaces with 3D curves *ACM Transactions on Graphics (SIGGRAPH)* **26(3)** pp 41
[7] Mori Y and Igarashi T 2007 Plushie: an interactive design system for plush toys *ACM Transactions on Graphics (SIGGRAPH)* **26(3)** pp 45
[8] Zimmermann J, Nealen A and Alexa M 2007 Silsketch: automated sketch-based editing of surface meshes *Proceedings of the 4th eurographics workshop on sketch-based interfaces and modeling* pp 23–30
[9] Cherlin J, Samavati F, Sousa M and Jorge J 2005 Sketch-based modeling with few strokes *Proceedings of the spring conference on computer graphics* pp 137–45
[10] Schmidt R and Singh K 2008 Sketch-based procedural surface modeling and compositing using surface trees *Computer Graphics Forum* **27(2)** pp 321–30
[11] Owada S, Nielsen F, Nakazawa K and Igarashi T 2003 A sketching interface for modeling the internal structures of 3D shapes *Proceedings of the 4th international symposium on smart graphics* pp 49–57
[12] Schmidt R, Wyvill B, Sousa M and Jorge J 2005 Shapeshop: sketch-based solid modeling with blobtrees *Proceedings of the 2nd eurographics workshop on sketch-based interfaces and modeling* pp 53–62
[13] Sugihara M, De Groot E, Wyvill B and Schmidt R 2008 A sketch-based method to control deformation in a skeletal implicit surface modeler *Proceedings of the 5th eurographics workshop on sketch-based interfaces and modeling* pp 65–72
[14] Karpenko O, Hughes J and Raskar R 2002 Free-form sketching with variational implicit surfaces *Computer Graphics Forum* 2002 **21(3)** pp 585–94
[15] Araujo B and Blobmaker J 2003 Free-form modelling with variational implicit surfaces *Proceedings of the 12 th Encontro Portugues de Computacao Graca* pp 17–26
[16] Tai C L, Zhang H and Fong JC-K 2004 Prototype modeling from sketched silhouettes based on convolution surfaces *Computer Graphics Forum* **23(1)** pp 71–83
[17] Alexe A, Gaildrat V and Barthe L 2004 Interactive modeling from sketches using spherical implicit functions *Proceedings of the international conference on computer graphics, virtual reality, visualization and interaction in Africa* pp 25–34
[18] Bogza A-M and Bogza P-C 2011 Illuminations and rendering scenes in 3D animation *Scientific Bulletin of the "Politehnica" Institute of Iasi, Romania,* **Tomul LVII (LXI), Fasc. 5** pp 141-148
[19] Avramescu A M 2011 Filming animated 3D scenes using dynamic lighting and rendering, *Dissertation, University "Politehnica" of Bucharest* pp 76
[20] Eyiyurekli M and Breen D 2010 Interactive free-form level-set surface-editing operators *Computers & Graphics* **34** pp 621–638
[21] Avramescu A M, Simion I and Enache I C 2013 Creating Realistic Images in a 3D Scene Using Illuminations and Renderings to Create an Animation *Scientific Bulletin of the "Politehnica" University of Timisoara, Romania*, **Tom 58(72), Fascicola 1** pp 95-100
[22] Roventa A and Avramescu A M 2013 Design issues of portable machines *Journal of Industrial Design and Engineering Graphics* **8(1)** pp 17-20
[23] Newman S D, Klatzky R L, Lederman S J and Just M A 2005 Imagining material versus geometric properties of objects: an fMRI study Cognitive *Brain Research* **23** pp 235 – 246

Special effects used in creating 3D animated scenes-part 1

A M Avramescu[1]
[1]University "Politehnica" of Bucharest, Department of Engineering Graphics and Industrial Design, 313 Splaiul Independentei, Sector 6, Bucharest, Romania

E-mail: avr_ana@yahoo.com

Abstract. In present, with the help of computer, we can create special effects that look so real that we almost don't perceive them as being different. These special effects are somehow hard to differentiate from the real elements like those on the screen. With the increasingly accesible 3D field that has more and more areas of application, the 3D technology goes easily from architecture to product designing. Real like 3D animations are used as means of learning, for multimedia presentations of big global corporations, for special effects and even for virtual actors in movies. Technology, as part of the movie art, is considered a prerequisite but the cinematography is the first art that had to wait for the correct intersection of technological development, innovation and human vision in order to attain full achievement. Increasingly more often, the majority of industries is using 3D sequences (three dimensional). 3D represented graphics, commercials and special effects from movies are all designed in 3D. The key for attaining real visual effects is to successfully combine various distinct elements: characters, objects, images and video scenes; like all these elements represent a whole that works in perfect harmony. This article aims to exhibit a game design from these days. Considering the advanced technology and futuristic vision of designers, nowadays we have different and multifarious game models. Special effects are decisively contributing in the creation of a realistic three-dimensional scene. These effects are essential for transmitting the emotional state of the scene. Creating the special effects is a work of finesse in order to achieve high quality scenes. Special effects can be used to get the attention of the onlooker on an object from a scene. Out of the conducted study, the best-selling game of the year 2010 was Call of Duty: Modern Warfare 2. This way, the article aims for the presented scene to be similar with many locations from this type of games, more precisely, a place from the Middle East, a very popular subject among game developers.

1. Introduction

3D Computer Graphics is somehow a new field that combines art and technology in a system that has multiple applications. Until recently, the main business target of 3D graphics was represented by movies and video industry. Nowadays, probably there is not a single movie (worth mentioning) made without special effects, partially or integrally generated on the computer [1, 2, 3].

Increasingly often, the majority of industries uses expressions containing 3D sequences (three-dimensional). Graphics can be represented in 3D, 3D sound is played by electronic components, commercials and special effects are designed in 3D [3].

The 3D field is increasingly accesible, gaining new areas of application. The 3D technology goes easily from architecture to product designing. Real like 3D animations are used as means of learning,

Content from this work may be used under the terms of the Creative Commons Attribution 3.0 licence. Any further distribution of this work must maintain attribution to the author(s) and the title of the work, journal citation and DOI.
Published under licence by IOP Publishing Ltd

for multimedia presentations of big global corporations, for special effects and even for virtual actors in movies [4].

The multifaceted possibilities of this field have started to be deeply explored. Most of the sketchers think that internet is the main element for the development of the aforementioned field. In the coming years, this will become a visual medium (like television) and hundreds of thousands of companies will need high quality graphics for Websites, animated presentations or virtual models of products [5, 6, 7]. The interest for 3D graphic programs grows as they are increasingly better designed. The evolution of special effects technology in 3D animations mirrors the technological development in the twentieth century, from industrial and mechanized to computerized and virtual. Even cartoons begin to leave behind those times when each image had to be drawn on paper, piece by piece. The place of countless pen drawings was undoubtedly replaced by computer and 3D professional graphics programs. This article aims to exhibit a game design from these days, using special effects and dynamic rendering in the process of scene creation.

2. Computer generated special effects (CGI)

In the year 1900, film producer George Melies was recording a scene on the street when his camera crashed while a cart was passing in front of the lens. While he tried to fix his camera, a hearse was passing on the street. When Melies processed the film frames, he saw, to his surprise, how the cart seemed to transform into a hearse. The freeze frame filming (aka cut-action) was a happy invention, one of the most effective tools for special effects, in the next century. The producers quickly learned how to manipulate the film in order to create illusions [8, 9, 10, 11].

In the year 1940, the process of combining many videos into one was made possible by perfecting the optical effects. Moreover, the creation of complex optical illusions was also possible. The producers could have more easily create matte videos where they were pausing one part of the lens while recording a scene. This way, another imagine could have been placed later on the film frame already left blank [11, 12].

During the past ten years, optical special effects have been replaced with digital processing. Even the bomb technicians who must participate in those scenes with fire and smoke, started to feel the competition of digital efects. Since 1990, it was obvious that movies with special effects should be divided in two parts: B.C. – before computers and A.D. – after digital. Computer generated images began to significantly interfere and alter the art of cinematography. Alongside with the developments in the computer industry, computer generated images will steadily evolve and shape the special effects of the future [13, 14].

3. Special effects in 3D animations

During the construction of special effects, there is a possibility to encounter effects and situations that are hard to make or animate using traditional key-frames. For example, falling leaves carried by the wind in a landscape or the glass that breaks when it reaches the ground, it may take hours to manually animate and the effect of the scene might not be as desired.

To simplify these effects, some techniques that tackle dynamics can be used to replicate the movement of an object influenced by different forces that are exerted. For example, a special effect can be represented by the motion of a leaf carried by the wind and tumbling to the ground. Using „Dynamics" techniques can be easy but frustrating at the same time. Anyway, for most of dynamics simulations, hundreds of options are available. This part from the research report will focus on important things underlying the creation of high complexity special effects.

The particle dynamics can be used in the creation of effects produced by countless elements including fire, fog, dust, explosions, mud or even a swarm of bees [4]. One particle is a point in the 3D space which can be altered by simulated forces, arranged as elements including spheres, clouds, images and undefined surfaces, even some objects like leaves or spaceships.

There are many prefab effects of the particles in order to simulate fire, emit smoke, produce lighting or, the creation of a river flow through a valley, image 1. In general, these scenes are constituted into a starting point for something more complex, so there is a need to adjust them in order to get the desired effect [4].

Figure 1. Smoke effect [4].

Fire produces a device that emits particles containing Gravity, Drag and Turbulence fields, attached to a particle of Cloud type. You can set Density, Radius, Direction and Intensity using the Options menu. The fire can be sending out from a surface, a curve or a directional point. This type of directional point will send out fire particles towards a specified direction of any point from that space.

4. Geometric modelling

If you follow the "checklist", your paper will comply to the requirements of the publisher and facilitate a problem-free publication process. Geometric modelling is about mathematical representation of curves, surfaces and solids, necessary to define the complex physical elements or the engineering objects. The computational geometry field is associated to geometric modelling and addresses the development, analysis and implementation of computational systems for specific algorithms.

The sphere of objects concerning the engineering field starts from simple machine parts to highly complex structures as ships, automobiles, airplanes, turbines and propeller blades, etc. In case of complex objects, the exterior shape of the surface is projected in order to reduce drag (for example, planes, ships, submarines) or to increase thrust. At the same time, these objects must satisfy other demands in order to accomplish some design requirements (for example, to handle a certain load or to be stable in case of disturbances, etc.). On the other side, there are cases where objects have significant esthetical demands (examples: automobiles, yachts, household products) [13]. Usually, one of engineers' tasks is to define complex models for engines, automobiles, planes, ships, submarines, underwater robots, offshore platforms, etc., models that are not entirely known beforehand. Consequently, the designing process is commonly iterative and involves:

-Creating the model based on design requirements;

-Analysis for performance assessment of the object;

-Changes for improving the model, followed by the analysis.

These steps may repeat like an iterative cycle till a satisfactory model comes out (even optimal in some cases) and meets all the design requirements, mitigating costs at the same time.

5. Creating the 3D scene

3D representations are obtained following a process of three-dimensional graphical creation using computers and special software programs. These creations can be represented as static images or video clips. The steps in obtaining a 3D project involve: modelling objects from the scene, scene illumination, texturing and rendering.

Further, a game was created using modelling, texturing and animation techniques, specific tools for video games. Considering that games must run in real time, limitations are quite harsh at least comparing to off-line rendering limits (as some rendering engines: Mental Ray, V-Ray, Brazil etc.). Although many times the optimization can be difficult, techniques used can be regarded as standard. That's why, an object will be formed using a more advanced technique called Sub-D modeling. Included in the final scene, this object is the tank.

As for the animation, from a technical standpoint a relatively simple infrastructure was kept. The turret of the tank was animated and the camera was moving in a three-dimensional way. In the post production stage, the explosion of the projectile destroying the track of the tank was added and synchronized to the scene.

From the artistically point of view, I have decided to create something that can be applied to a game from these days, Call of Duty: Modern Warfare 2. The scene presented in this article is similar with many of the locations from this type of games, like the Middle East place destroyed by this conflict.

Choosing this scene was determined by different technical and artistic challenges. From a technical standpoint, a destroyed building for example is much more difficult to create than a new one. This happens mostly because destroyed buildings have rugged, ruined details. Moreover, this type of destruction is more difficult to create in a 3D program, which is rigid; so any patchy or complex detail needs certain human intervention. The next step was the documentation. To efficiently create this scene, documentation was essential. The architectural style from these places, mood, level of building degradation, all contribute to a compelling scene. After research, the next step was the creation of a mood board, image 2. This term is used for a scene containing the most illustrative images found during the research process, to determine the style and mood. Traditionally, this image collection should be created on a big board, as it is called the moodboard [7, 8, 15].

Figure 2. Moodboards [16].

With the help of these mood boards, we have tried to identify the key elements of a scene created in a style already proposed at the beginning of this chapter. The evidence of destruction seemed extremely important for me. Knowing that I was not intending to present the scene in the middle of the action, those traces left from a possible conflict were essential for determining the atmosphere of the story. At the same time, the mood boards helped me to maintain the established visual path from the beginning and not to diverge from the original concept. Likewise, images gathered from the aforementioned game were very helpful, from the technical point of view. In some places, we can easily observe how problems have been solved (for example, concrete or grass).

Considering the shape of this object, starting from a simple box with the size of a sand bag was the easiest way. Going to Sub-Object Edge module, the margin ring was selected and the „Connect" command was pressed in order to unite the margins with a number specified by the user. The creation of two new margins was selected and with the help of Pinch parameter, the margins moved unequally one from another. This process has been repeated on the other margin ring, but this time, a single new row of margins was created.

Going to Vertez module and using Target Weld command, different vertices were united in order to obtain a shape closer to a sand bag. Yet, for adding something more to the shape without supplementing the number of poligons (optimization is very important in games), Cut command was used for creating new vertices, figure 3.

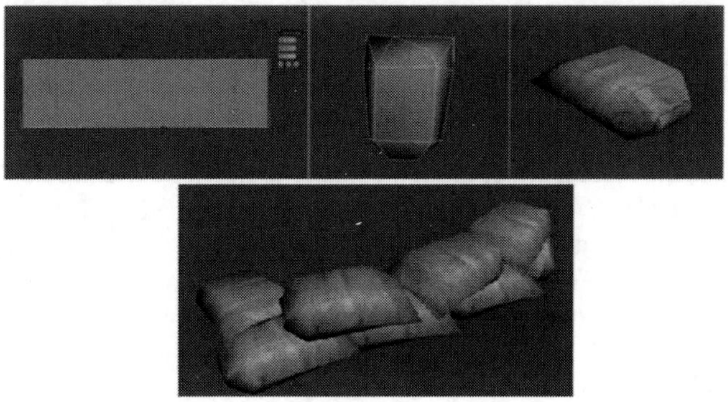

Figure 3. Sand bag modeling [16].

Thus, having the basic sand bag model, creating a wall was only about placing them (sand bags) one above the other in order to form a wall. Image 4 exhibits a scene of the barriers as they appear in the game (alongside with texture) and an image that shows their wireframe, to easily observe the modelling methods as described.

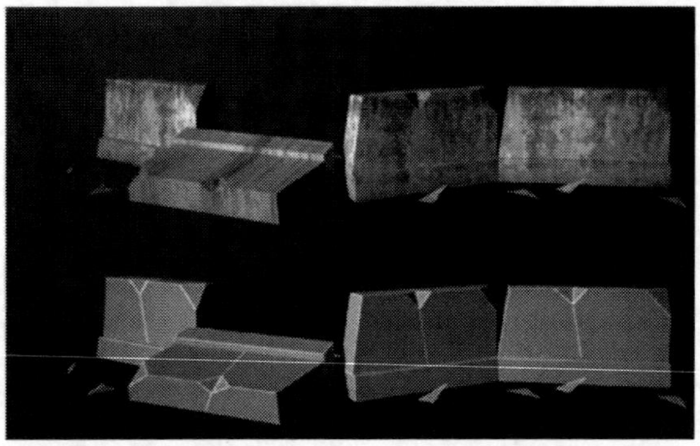

Figure 4. Barrier modeling [16].

Using „Move" and „Rotate" tools, these barriers have been arranged in a way to offer credibility for the scene. The barriers have been placed not only to block (their standard function) but also to defend, tying them to the central theme for this level.

6. Conclusions

There are many views stating that 3D animation will increasingly have more impact than it has today. 3D animated scenes offer a wide range of opportunities for direct camera recording. 3D graphics is much more complex using textures, volumes or light schemes. Everything is created using the computer and the intelligent mind of the graphics editor. 3D programs are spectacular considering the progress in both hardware and software. Illumination and dynamic rendering can be more easily integrated into complex scenes. The onlooker probably won't notice that some nature landscapes are generated on the computer by the graphics editor. The scope of this article is to concisely present the geometric modelling of 3D elements in the proposed 3D scene. The key for attaining real visual effects is to successfully combine various elements: characters, objects, images and video scenes; like all these elements represent a whole that works in perfect harmony.

Acknowledgements
The author wishes to thank engineer R. Atanasiu for his assistance in preparing the previous work and for the drawings. Also the author wishes to thank to the Operational Sectoral Program for Human Resources Development 2007-2013 of the Romanian Department of Labour, Family and Social Assistance through Financial Agreement **POSDRU/144/6.3/S/135227**.

References

[1] Fiell Ch and Fiell P 2001 Industrial Design A-Z *TASCHEN GmbH, KAhn*
[2] Jones A 1992 Art introduction (Bucharest: Lider Publishing House)
[3] Lee L 2008 Advanced MAYA. Texturing and lighting (USA: Autodesk Maya Press)
[4] Atanasiu R M and Avramescu A M 2011 Special effects in 3D animation *Scientific Bulletin-Politehnic Institut of Iași* **LVII (LXI), Fasc. 5** pp 153-158
[5] Lee L 2007 Maya Professional Tips and Techniques (USA: Sybex Publishing House)
[6] Oliverio G 2007 Maya 8 Character Modeling (USA: Wordware Applications Library Publisher)
[7] Alias Learning Tools, Learning Maya 7: The Modeling and Animation Handbook (USA: Alias Learning Tools Publishing House)
[8] Gray D 2004 Creating Striking Graphics with Maya and Photoshop (USA: Sybex Publisher)
[9] Demers O and Urszenyi C Digital Texturing & Painting (USA: New Riders Publishing)
[10] Vegter G 1999 The Apolar Bilinear form *Geometric Modeling, Mathematics of Computation*, **69(230)** pp 691-720
[11] Galyean T and Hughes J 1991 Sculpting: an interactive volumetric modeling technique *Proceedings of the SIGGRAPH conference* pp 267–274
[12] Chuang J H 1990 Surface Approximations in Geometric Modeling *PhD thesis, Purdue University, Computer Science*
[13] Bebevec P 2002 The Light Stage: Photorealistically Integrating Real Actors into Virtual Environments *Svenska Föreningen för Grafisk Databehandling* pp 17-22
[14] Kaushik V D, Pathak V K and Gupta P 2010 Geometric Modeling of 3D-Face Features and Its Applications *Journal Of Computers* **5(9)** pp 1305-1314
[15] Information on http://www.everything4maya.com/index.php?link=tutseries.
[16] Atanasiu R 2009 Specials effects in 3D animation *Dissertation* Bucharest pp 67

IOP Conf. Series: Materials Science and Engineering **95** (2015) 012117 doi:10.1088/1757-899X/95/1/012117

An automation of design and modelling tasks in NX Siemens environment with original software – generator module

M Zbiciak[1], **C Grabowik**[1] **and W Janik**[1]
[1]Silesian University of Technology, The Faculty of Mechanical Engineering, Institute Of Engineering Processes Automation And Integrated Manufacturing Systems, ul. Konarskiego 18a 44-100 Gliwice, Poland

E-mail: rafal.zbiciak@polsl.pl

Abstract. Nowadays the design constructional process is almost exclusively aided with CAD/CAE/CAM systems. It is evaluated that nearly 80% of design activities have a routine nature. These design routine tasks are highly susceptible to automation. Design automation is usually made with API tools which allow building original software responsible for adding different engineering activities. In this paper the original software worked out in order to automate engineering tasks at the stage of a product geometrical shape design is presented. The elaborated software works exclusively in NX Siemens CAD/CAM/CAE environment and was prepared in Microsoft Visual Studio with application of the .NET technology and NX SNAP library. The software functionality allows designing and modelling of spur and helicoidal involute gears. Moreover, it is possible to estimate relative manufacturing costs. With the Generator module it is possible to design and model both standard and non-standard gear wheels. The main advantage of the model generated in such a way is its better representation of an involute curve in comparison to those which are drawn in specialized standard CAD systems tools. It comes from fact that usually in CAD systems an involute curve is drawn by 3 points that respond to points located on the addendum circle, the reference diameter of a gear and the base circle respectively. In the Generator module the involute curve is drawn by 11 involute points which are located on and upper the base and the addendum circles therefore 3D gear wheels models are highly accurate. Application of the Generator module makes the modelling process very rapid so that the gear wheel modelling time is reduced to several seconds. During the conducted research the analysis of differences between standard 3 points and 11 points involutes was made. The results and conclusions drawn upon analysis are shown in details.

1. Introduction

Formerly a designed product came into being first of all and mainly in a designer imagination. In spite of a continuous technological progress this stage of the design-constructional process has not changed up today. Technological progress and continuous development of computer systems have led to creation of the full range of software which makes the process of design modelling in 3D space easy and efficient. In CAD software a 3D product model is represented by mathematical notation which unequivocally describes its design features. Base on this description a designer is able to prepare product views, cross-sections, make kinematic, dynamic calculation etc. Some of these activities have a routine character. In a daily practice of design and technological bureaus activity there could be

Content from this work may be used under the terms of the Creative Commons Attribution 3.0 licence. Any further distribution of this work must maintain attribution to the author(s) and the title of the work, journal citation and DOI.
Published under licence by IOP Publishing Ltd

observed problems with large time consumption of routine tasks performing, which in principle do not add an added value to a product [2,3,4].

On the other hand CAD/CAM/CAE systems users very seldom try to work out and next use their own software solutions which allow automating routine design tasks. These tasks, that taking into account their character, are especially prone to automation. In industry it is possible and highly recommended to apply software automation for this kind of design department activity which could be characterized by unchangeable, generally known algorithm for exactly defined input data and precisely defined result. For instance calculation that supports design and geometric modelling processes of gear wheels, shafts, sleeves.

Software adding and supporting of the design-constructional process, especially geometric modelling functions, in contemporary integrated CAD/CAM/CAE systems, is possible thanks to making available by software producers of OPEN/API libraries. In practice, it means the ability to create original software which preforms various tasks, in given system environment, specified by program author. These software applications could be made with high level programming languages such as C/C++, C#, Java, Object Pascal or Visual Basic in different programing environments for example Microsoft Visual Studio, Embaracadero RAD Studio. This functionality is offered inter alia by the following exemplary systems SIEMENS PLM NX, CATIA, ProEngineer, Inventor or Solid Works.

The original software developing process could be realized with the following methods:

-manual, in which a program structure is made by a programmer on their own without application of a system internal journaling mechanism. The journaling mechanism allows on automatic recording of actions performed by a system user during an element design geometry modelling. The manual method requires form a programmer to be very good at knowledge of a system capability to automation of considered routine task as well as excellent knowledge of a chosen programming language and programming environment;

-mixed, a part of a program structure is made with the journaling technique – a journal script is being recorded. The program script contains the subsequent geometric modelling steps; a reaming program code is made directly by a programmer.

The benefits of working out and using of original software, made by their own, in daily practice of design and technological departments could be as follows [1,4]:

-improvement and acceleration of the design-constructional process,

-enable of fast making of part families,

-reduce the risk of a design mistake making at the stage of a product design modelling stage by introducing of procedures of an automatic dimension selection for standard parts from relevant standards,

-constant supervision of a product being modelled according to, for instance the criterion of fatigue, dynamic properties etc.,

-provide possibility of a design optimization,

-automation of process planning,

-provide possibility of fast making of alternative machining routs,

-automatic generation of CNC programs,

-design knowledge storing.

Irrespective of the chosen CAD/CAM/CEA system environment the stage of original software development should be always preceded by an analysis of programming tools available in the system. This analysis and subsequent decision play a key role in the whole process of software development taking into consideration fact that particular programming tools have different abilities so programmer should find compromise between tools abilities and software development cost [3].

2. Problem formulation

Basis on a critical review and analysis conducted in the context of possibility of routine design and modelling tasks automation it was decided to work out original software supporting processes of gear wheel design and geometrical modelling. According to recommendations included in the introductory paragraph the software development stage was preceded by CAD/CAM/CAE system selection and a review of programming tools available in it. As a software development environment Siemens NX was selected. Siemens NX offers the following programing tools:

-NX Open it is a collection of APIs (application programming interfaces) that allows creating software applications for NX through an open architecture using high-level programming languages such as C/C++, Visual Basic, C#, and Java. It could be used for automation complex and routine-repetitive tasks;

-knowledge fusion which is an interpreter, object-oriented language that allows adding engineering knowledge to a part element by both design and geometric creating rules. The knowledge fusion language is declarative, rather than procedural one, which means that, in general, the rules are only evaluated when referenced or demanded. In KF it is possible to access external knowledge bases, such as databases or spreadsheets;

-SNAP (Simple NX Application Programming) is an easy-to-learn programming tool intended for mechanical designers and other typical NX users – not necessarily programmers. It can speed up work by automating simple routine tasks. SNAP is based on the Visual Basic (VB.Net) language and can be used with the NX Journal Editor or with IDEs, such as Visual Studio.

After a detailed analysis as custom-original software development tools SANP programming tool was chosen as a compromise between language abilities and development cost.

3. Initial program assumptions and the program structure

The program is designed as a tool that supports the processes of gear wheels parameters calculation, geometrical modelling and relative manufacturing cost evaluation. The range of designed gear wheels is limited to spur and helical gears. The original program consists of the two modules it is the generator module and cost module. The generator module is a main subject of this paper. In the figure 1 the general structure of the worked out program is presented.

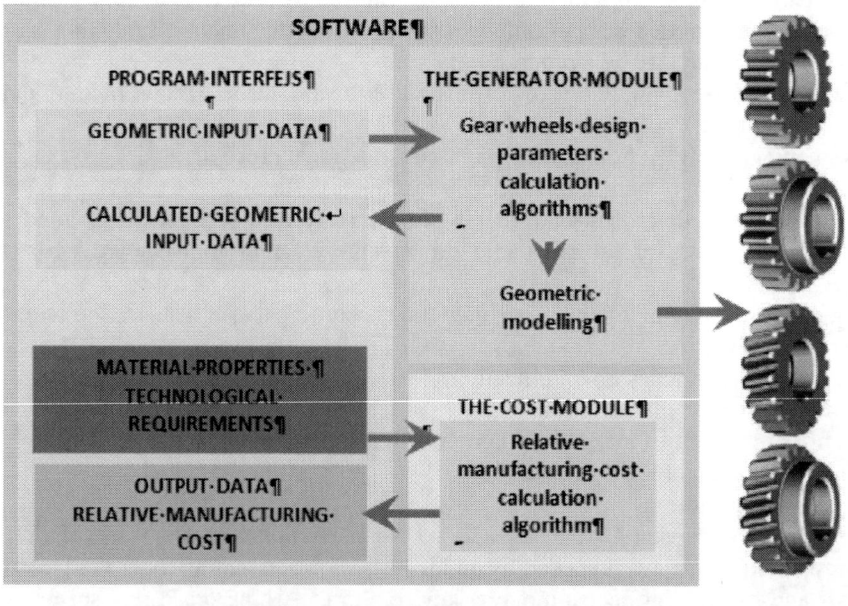

Figure 1. The original program structure.

A program user during the design session is obliged to introduce via program interface the following geometric input data: a teeth kind, wheel design shape (*A* or *B*), tooth module (m), a teeth number (z), a wheel rim width (B), a wheel internal diameter/mating shaft diameter (d_0), a tooth depth coefficient (y), a pressure angle (α), a helix angle (β), helix hand (*right* or *left*). Basis on the introduced input parameters the generator module calculates the rest of wheel parameters with gear wheels design parameters calculation algorithms. The calculated parameters are as follows (see the figure 2): a pitch diameter (D), a base diameter (D_b), an addendum/outside diameter (D_a), a root diameter (D_f), a tooth depth (H), addendum (H_a), dedendum (H_f). The slot parameters, rim geometrical shape and corresponding dimension set are selected according to appropriate standard.

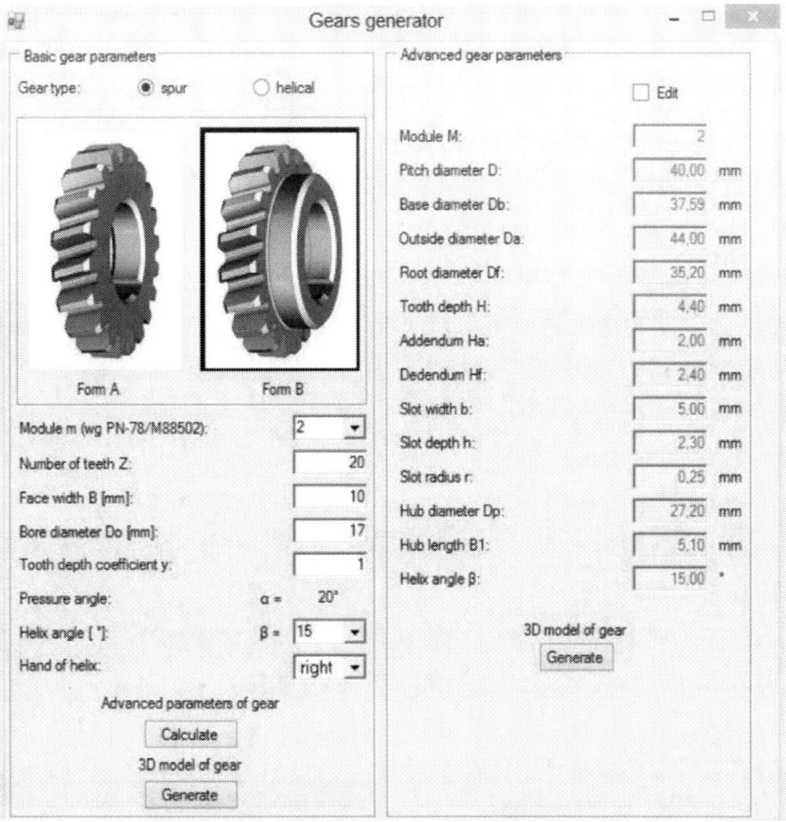

Figure 2. The generator model user interface.

Figure 3. The gear wheel example a) the real part model, b) model generated in the generator module: for the following design parameters $m = 2$, $z = 15$, $y = 1$.

4. Verification of the generator module correctness

The correctness of the generator module was carried out in the two directions. The first test was aimed at comparison of wheel design parameters calculated in the module with technical data given by industrial producers. Basis on this comparison it could be stated that these parameters are identical what proved the correctness of the generator calculation algorithms.

The aim of the second test was to compare geometry of the two involutes, it is the involute drawn by generator module with the involute drawn in a standard way. By the standard approach to involute drawing is understood this way in which the curve is drawn by the three points. These points are usually located at addendum, pitch and dedendum circles. This solution is commonly used in most CAD systems equipped with built-in gear wheel generator. In the presented research it was decided to support the involute curve being drawn on eleven points. This should assure better quality of the involute. Having in mind the test aim the default NX modelling preferences for vector graphics were changed. The distance tolerance value was changed from 0,0254 mm to 0,001 mm and angle tolerance form 0,5 mm to 0,01 mm. Involute curves were drawn in NX sketch environment for the following input parameters: modulus m = 1 mm, teeth number z = 20 and tooth depth coefficient y = 1. The curve drawn in a standard way it is by 3 points was denoted with the red colour whilst this drawn by 11point with the green one (see: the figure 4).

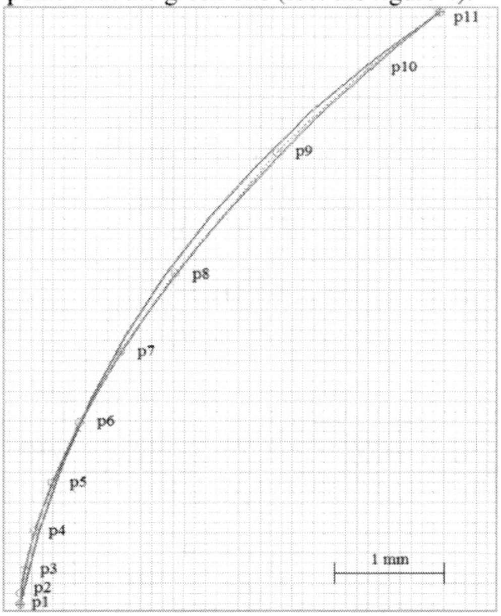

Figure 4. The involutes being compered.

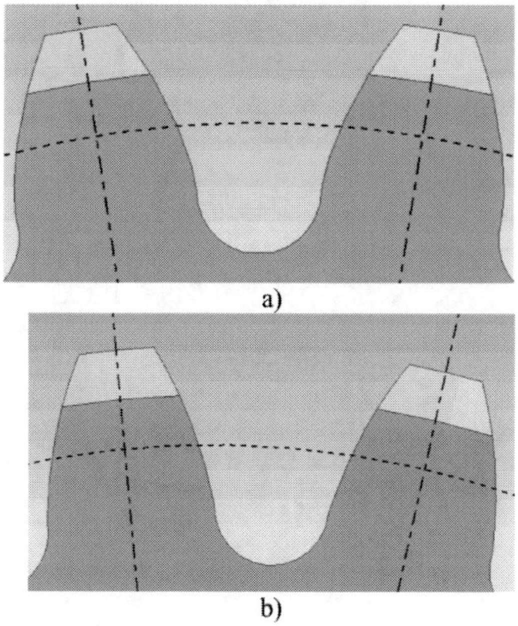

Figure 5. Tooth profile a) 11 pts. based, b) 3 pts. Based.

In the figure 4 the distinct difference between involutes profiles could be seen. In their lower part the involutes deviation is smaller than in the upper part in which the maximum deviations equals to 0,1 mm. It is expected that for bigger tooth modules this deviation would achieve greater values. In the figure the 4 the sides views of teeth for both involute profiles are show. The figure 5 analysis indicated that 11 pts. based tooth width measured at its base is wider than 3 pts. based.

In order to make a more precise analysis of involute profile shape influence on bending stress the FEM analysis in Siemens NX Design Simulation module was carried out. In the figure 6 results of the displacement analysis were shown. In case of the displacement analysis there is not difference between profile *a* and *b*.

The maximum displacement value equals to 0,001 mm for normal load 500 N applied to the tooth flank. The figure 7 presents results achieved for bending stress analysis. In case of the bending stress analysis difference between maximum (3 pts. based profile) and minimum value (11 pts. based profile) of bending stress equals to 2 MPa which seems to be not so high. In considered case lower value of bending stress for 11 pts. based involute profile could be explained by greater width of tooth measured at its base. Moreover on the basis of the stress distribution analysis it was found that in case of 3 pts.

based involute the area with increased stress level is higher than 11 pts. based curve. Presumably, it would have influence on fatigue strength especially for gear wheels of large dimensions.

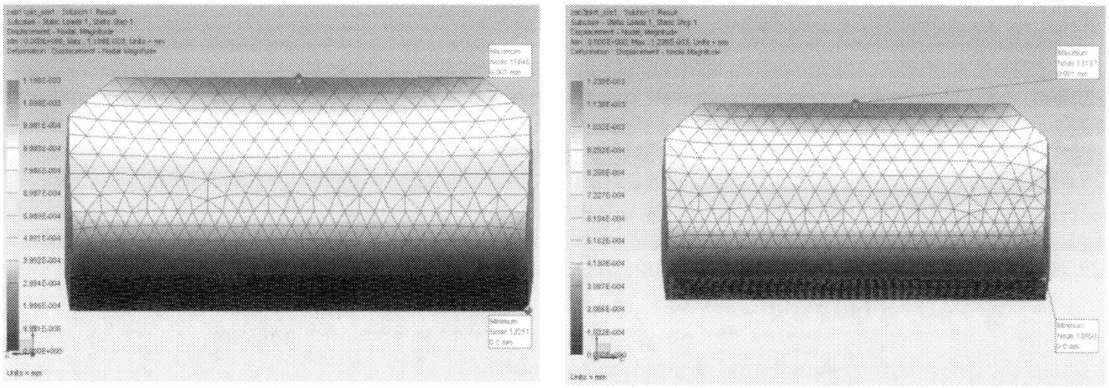

a) b)

Figure 6. Displacement analysis results a) 11 pts. based, b) 3 pts. Based.

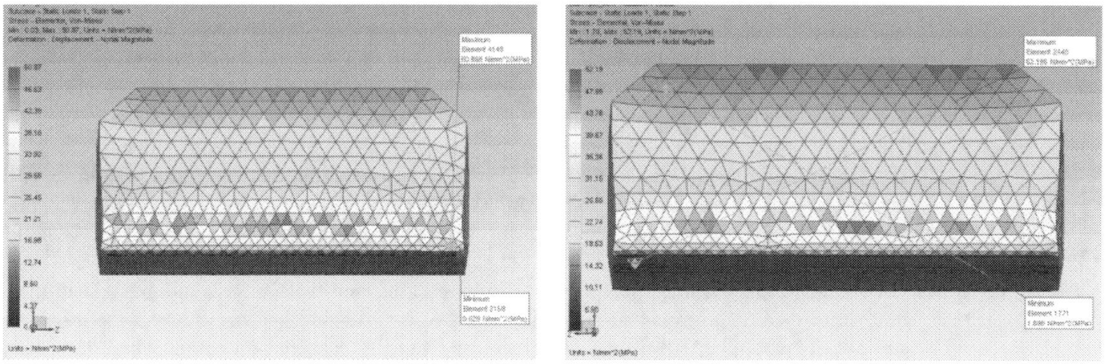

Figure 7. Bending stress analysis results a) 11 pts. based, b) 3 pts. Based.

5. Conclusions

Basis on the conducted research and work it could be found that original software facilitates the design-constructional process. In some cases this kind of software would give better results the CAD/CAM/CAE integrated systems built-in tools. In the considered work the gear wheel modelling time was reduce to particular minutes in opposite to not assisted gear wheel modelling process, realized in sketch module environment, in which modelling time is counted rather in hours. Moreover the software user has greater influence on final product model shape. This allows manufacturing product of better quality which very well meet a client requirements.

References

[1] Dzitkowski T and Dymarek A 2014 Active reduction of identified machine drive system vibrations in the form of multi-stage gear units *Mechanika* **20/2** pp 183-189

[2] Grabowik C and Janik W 2012 NX GRIP – CAD (Gliwice: The Silesian University Press)

[3] Ćwikła G 2014 Methods of manufacturing data acquisition for production management – a review *Advanced Materials Research* **837** pp 618-623

[4] Grabowik C, Kalinowski K, Kempa W and Paprocka I 2014 A survey on CAPP systems development methods *Advanced Materials Research* **837** 387-392

An automation of design and modelling tasks in NX Siemens environment with original software – cost module

R Zbiciak[1], **C Grabowik**[1] **and W Janik**[1]
[1]Silesian University of Technology, The Faculty of Mechanical Engineering, Institute of Engineering Processes Automation And Integrated Manufacturing Systems, ul. Konarskiego 18a 44-100 Gliwice, Poland

E-mail: rafal.zbiciak@polsl.pl

Abstract. The design-constructional process is a creation activity which strives to fulfil, as well as it possible at the certain moment of time, all demands and needs formulated by a user taking into account social, technical and technological advances. Engineer knowledge and skills and their inborn abilities have the greatest influence on the final product quality and cost. They have also deciding influence on product technical and economic value. Taking into account above it seems to be advisable to make software tools that support an engineer in the process of manufacturing cost estimation. The Cost module is built with analytical procedures which are used for relative manufacturing cost estimation. As in the case of the Generator module the Cost module was written in object programming language C# in Visual Studio environment. During the research the following eight factors, that have the greatest influence on overall manufacturing cost, were distinguished and defined: (i) a gear wheel teeth type it is straight or helicoidal, (ii) a gear wheel design shape A, B with or without wheel hub, (iii) a gear tooth module, (iv) teeth number, (v) gear rim width, (vi) gear wheel material, (vii) heat treatment or thermochemical treatment, (viii) accuracy class. Knowledge of parameters (i) to (v) is indispensable for proper modelling of 3D gear wheels models in CAD system environment. These parameters are also processed in the Cost module. The last three parameters it is (vi) to (viii) are exclusively used in the Cost module. The estimation of manufacturing relative cost is based on indexes calculated for each particular parameter. Estimated in this way the manufacturing relative cost gives an overview of design parameters influence on the final gear wheel manufacturing cost. This relative manufacturing cost takes values from 0.00 to 1,00 range. The bigger index value the higher relative manufacturing cost is. Verification whether the proposed algorithm of relative manufacturing costs estimation has been designed properly was made by comparison of the achieved from the algorithm results with those obtained from industry. This verification has indicated that in most cases both group of results are similar. Taking into account above it is possible to draw a conclusion that the Cost module might play significant role in design constructional process by adding an engineer at the selection stage of alternative gear wheels design. It should be remembered that real manufacturing cost can differ significantly according to available in a factory manufacturing techniques and stock of machine tools.

1. Introduction

It is assumed that almost 80% of product development process cost arises at the design stage [1, 2]. Therefore there is still a need for an appropriate methods and software tools which could help

Content from this work may be used under the terms of the Creative Commons Attribution 3.0 licence. Any further distribution of this work must maintain attribution to the author(s) and the title of the work, journal citation and DOI.
Published under licence by IOP Publishing Ltd

designers at proper design making [3, 4]. In the paper a method and software tool for manufacturing cost estimation is presented. In this method of gear wheels the relative manufacturing cost evaluation the following factors with the greatest influence on manufacturing cost were distinguished: (i) a gear wheel teeth type, (ii) a gear wheel design shape A, B with or without wheel hub, (iii) a gear tooth module, (iv) teeth number, (v) gear rim width, (vi) gear wheel material, (vii) heat treatment or thermochemical treatment, (viii) accuracy class. For each particular factor the corresponding relative manufacturing cost coefficient can be ascribed (see: the table 1):

k_{rz} – relative manufacturing cost coefficient on account of a gear wheel teeth type:

$$k_{rz} = \frac{wrz}{wrz_{\max}} \tag{1}$$

k_f – relative manufacturing cost coefficient on account of a gear wheel design shape:

$$k_f = \frac{wf}{wf_{\max}} \tag{2}$$

k_m – relative manufacturing cost coefficient on account of a gear tooth module:

$$k_m = \frac{wm}{wm_{\max}} \tag{3}$$

k_z – relative manufacturing cost coefficient on account of teeth number:

$$k_z = \frac{wz}{wz_{\max}} \tag{4}$$

k_w – relative manufacturing cost coefficient on account of gear rim width:

$$k_w = \frac{ww}{ww_{\max}} \tag{5}$$

k_t – relative manufacturing cost coefficient on account of gear wheel material:

$$k_t = \frac{wt}{wt_{\max}} \tag{6}$$

k_o – relative manufacturing cost coefficient on account of heat treatment or thermochemical treatment:

$$k_o = \frac{wo}{wo_{\max}} \tag{7}$$

k_k – relative manufacturing cost coefficient on account of accuracy class:

$$k_k = \frac{wk}{wk_{\max}} \tag{8}$$

Each relative manufacturing cost coefficient variant w_x can take value from the range of $wx = \{0,\ldots,n\}$ where n is an appropriate natural number (see: the table 1). There is a relation between a relative manufacturing cost value and manufacturing cost, greater relative manufacturing cost value means greater real manufacturing cost.

Relative manufacturing cost coefficients on account of a gear wheel teeth type and gear wheel design shape can take values from the range of {0,1}. The maximum values for coefficient variants *wx* were denoted in red (table 1).

Table 1. Gear wheels design parameters with corresponding values of coefficients variants *wx*.

Teeth number	*wz*	Gear rim width	*ww*	Gear wheel material	*wt*	Heat treatment or thermochemical treatment	*wo*	Accuracy class	*wk*
17-26	1	1-10	1	C35	1	Without	1	5	8
27-36	2	11-20	2	C45				6	7
37-46	3	21-30	3	C55		Single hardening	2	7	6
47-56	4	31-40	4	C10				8	5
57-66	5	41-50	5	C15		Double hardening	3	9	4
67-76	6	51-60	6	15Cr2	2			10	3
77-86	7	61-70	7	20Cr4		Carburizing	4	11	2
87-96	8	71-80	8	34CrMo4	3			12	1
97-106	9	81-90	9	37MnSi5	4			Modulus	*wm*
107-116	10	91-100	10	14HG				0,05	1
117-126	11	101-110	11	20MnCr5				0,14	10
127-136	12	111-120	12	18CrMo4	5	Wheel type	*wrz*	0,45	20
137-146	13	121-130	13	13CrMo4-5		Straight	1	1,375	30
147-156	14	131-140	14	18CrNi8		Helical	2	2,5	35
157-166	15	141-150	15	45HN	6			4,5	40
167-176	16			36CrNiMo4	7	Design shape	*wf*	14	50
177-186	17			34CrNiMo6	8	A type	1	25	55
187-200	18			30H2N2M		B type	2	90	66

The overall relative manufacturing cost for a given gear wheel can be calculated according to the following formula (value of each $k_{x_{\max}} = 1$):

$$k_c = \frac{k_{rz} + k_f + k_m + k_z + k_w + k_t + k_o + k_k}{k_{rz_{\max}} + k_{f_{\max}} + k_{m_{\max}} + k_{z_{\max}} + k_{w_{\max}} + k_{t_{\max}} + k_{o_{\max}} + k_{k_{\max}}} \tag{9}$$

2. Illustrative example

The presented example shows relative manufacturing calculations for the three gear wheels parts produced by an enterprise. The design parameters of the subjected parts in the table 2 are shown.

IOP Conf. Series: Materials Science and Engineering **95** (2015) 012118 doi:10.1088/1757-899X/95/1/012118

Table 2. Gear wheels design parameters.

	Gear wheel I	Gear wheel II	Gear wheel III
Teeth type	Straight	Straight	Straight
Design shape (A/B)	B	A	B
Tooth module	3	1	8
Teeth number	27	17	40
Gear rim width	30	15	80
Wheel material	13CrMo4-5	35	30H2N2M
Heat treatment	Carburizing	Single hardening	Without
Accuracy class	7	5	10

The detailed calculations for particular relative manufacturing cost coefficients in the table 3 are presented.

Table 3. Results of relative manufacturing cost coefficient calculations.

Relative manufacturing cost coefficient	Gear wheel I	Gear wheel II	Gear wheel III
$k_{rz} = \frac{wrz}{wrz_{max}}$	$k_{rz_1} = \frac{1}{2} = 0{,}5$	$k_{rz_2} = \frac{1}{2} = 0{,}5$	$k_{rz_3} = \frac{2}{2} = 1$
$k_f = \frac{wf}{wf_{max}}$	$k_{f_1} = \frac{2}{2} = 1$	$k_{f_2} = \frac{1}{2} = 0{,}5$	$k_{f3} = \frac{2}{2} = 1$
$k_m = \frac{wm}{wm_{max}}$	$k_{m_1} = \frac{27}{66} = 0{,}41$	$k_{m_2} = \frac{33}{66} = 0{,}5$	$k_{m_3} = \frac{45}{66} = 0{,}68$
$k_z = \frac{wz}{wz_{max}}$	$k_{z_1} = \frac{2}{18} = 0{,}11$	$k_{z_2} = \frac{1}{18} = 0{,}06$	$k_{z_3} = \frac{3}{18} = 0{,}17$
$k_w = \frac{ww}{ww_{max}}$	$k_{w_1} = \frac{3}{15} = 0{,}2$	$k_{w_2} = \frac{2}{15} = 0{,}13$	$k_{w_3} = \frac{8}{15} = 0{,}53$
$k_t = \frac{wt}{wt_{max}}$	$k_{t_1} = \frac{5}{8} = 0{,}63$	$k_{t_2} = \frac{1}{8} = 0{,}13$	$k_{t_3} = \frac{1}{1} = 1$
$k_o = \frac{wo}{wo_{max}}$	$k_{o_1} = \frac{4}{4} = 1$	$k_{o_2} = \frac{2}{4} = 0{,}5$	$k_{o_3} = \frac{1}{4} = 0{,}25$
$k_k = \frac{wk}{wk_{max}}$	$k_{k_1} = \frac{6}{8} = 0{,}75$	$k_{k_2} = \frac{8}{8} = 1$	$k_{k_3} = \frac{3}{8} = 0{,}38$

Finally the value of overall relative manufacturing cost for particular wheels is as follows:

- for the gear wheel I

$$k_{c_1} = \frac{0{,}5 + 1 + 0{,}41 + 0{,}11 + 0{,}2 + 0{,}63 + 1 + 0{,}75}{8} = \frac{4{,}6}{8} = 0{,}58 \tag{10}$$

- for the gear wheel II

$$k_{c_2} = \frac{0{,}5 + 0{,}5 + 0{,}5 + 0{,}06 + 0{,}13 + 0{,}13 + 0{,}5 + 1}{8} = \frac{3{,}19}{8} = 0{,}40 \tag{11}$$

– for the gear wheel III

$$k_{c_3} = \frac{1+1+0{,}68+0{,}17+0{,}53+1+0{,}25+0{,}38}{8} = \frac{5{,}01}{8} = 0{,}63 \quad (12)$$

It should be remembered that the proposed method of relative manufacturing cost calculation gives a general view on manufacturing cost. It arises form fact that the particular relative manufacturing cost coefficients was not defined with the high accuracy taking into account that they were defined in a subjective manner. Application of this method in an enterprise needs additional actions to be done. It could be necessary to change relative manufacturing cost values and in some case to introduce a system of coefficient weights for design parameters.

3. The cost module

The cost module is a part of original software designed to support an engineer during their work with gear wheels design modelling and fast relative manufacturing cost calculation (figure 1). Worked out software, developed in Microsoft Visual Studio 2012, works only in Siemens NX environment. The original software is composed of the two main modules it is the generator module and cost module. Generator module is used for gear wheels geometric parameters calculations and 3D modelling.

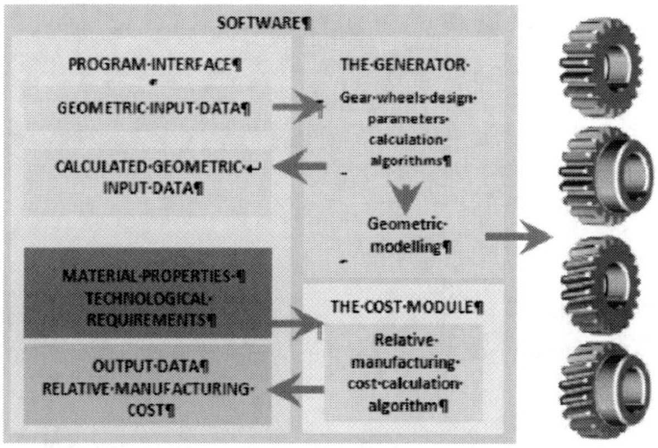

Figure 1. The original program structure.

The cost module is composed of computational procedures built on general idea of relative manufacturing cost calculation presented in this paper. The cost module gives a designer a general view of the impact of particular gear wheels design parameters on the final product economic efficiency so it supports the designer making their work easier. During conducted research the tests were carried out to verify the correctness of cost module. The research program included relative manufacturing cost calculations for 12 different gear wheel designs. The calculated relative costs were next compared with real manufacturing cost. Information about real manufacturing costs was achieved from the two independent producers. In the figure 2 the comparison between calculated relative manufacturing costs and real manufacturing costs is presented. The analysis of the figure 2 shows that the difference between real manufacturing costs in case of the same gear wheel design for both producers is small. The differences between real manufacturing costs for different gear wheels designs differs a lot but differences in calculated relative cost not. The small difference in relative manufacturing cost for particular gear wheels designs makes the evaluation of the influence of the designer decisions on design economic effectiveness difficult. In order to correct this phenomenon it was necessary to introduce weights system for relative manufacturing cost coefficients. The figure 3 shows the distinct improvement in relative manufacturing cost calculation. The overall relative manufacturing costs coefficients are almost at the same level as real manufacturing cost.

It is difficult to match relative and real manufacturing cost perfectly because during the research it was impossible to achieve information about producer' margins, they are probably different for each producer. Industrial application of the cost module needs the stage of weights system to be done. In such a case relative manufacturing cost results could by very useful for a designer helping them to choose correct design decisions.

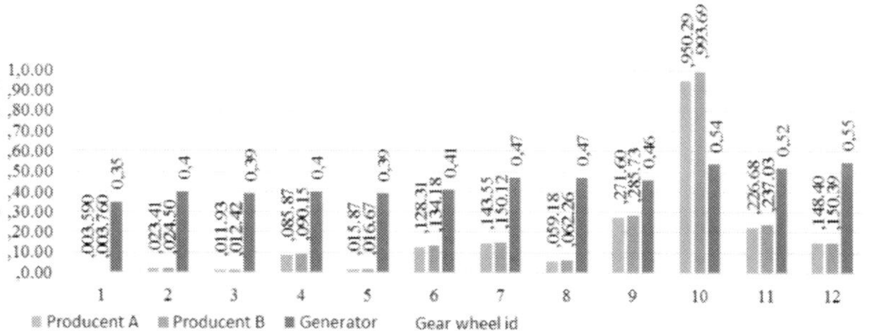

Figure 2. The comparison between real manufacturing cost and calculated relative manufacturing cost.

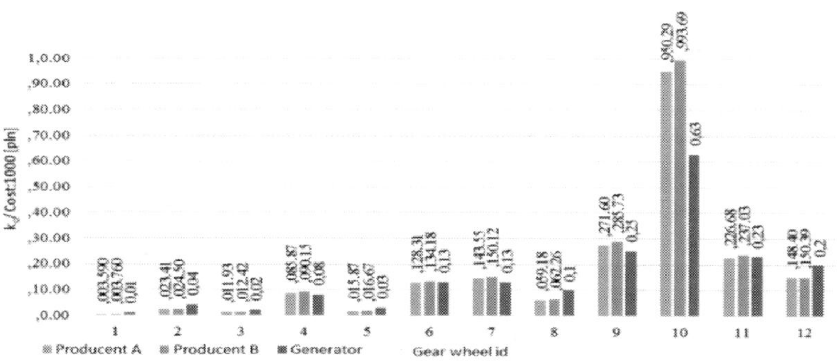

Figure 3. The comparison between real manufacturing cost and calculated relative manufacturing cost with weight system.

4. Conclusions

In contemporary industry cost estimation is considered as a strategic tool which can be applied to help decision making. It assumed that up to 80 percent of the product cost is determined by decisions taken by designers at the design-constructional stage. Therefore it is important to estimate manufacturing costs as early and as precisely as it is possible. Presented in the paper method and software tool is an answer to this problem.

References

[1] Chen X, Lai X and Gershenson J K 2010 Relative Design Cost Estimation at Design Stage Based on Design Features *Applied Mechanics and Materials* **26-28** pp 625-636

[2] In H P, Baik J, Kim S, Yang Y and Boehm B 2006 A quality-based cost estimation model for the product line life cycle *Communications of the ACM* **49/12** pp 85-88

[3] Cielniak M and Gendarz P 2013 Relative manufacturing costs of machines elements based on constructional similarity *Journal of Achievements in Materials and Manufacturing Engineering* **56/1** pp 20-28

[4] Saravi M, Newnes L, Mileham A R and Goh Y M 2008 Estimating Cost at the Conceptual Design Stage to Optimize Design in terms of Performance and Cost, Collaborative Product and Service Life Cycle Management for a Sustainable World *Advanced Concurrent Engineering* pp 123-130

IOP Conf. Series: Materials Science and Engineering **95** (2015) 012119 doi:10.1088/1757-899X/95/1/012119

Optimizing parameters of a technical system using quality function deployment method

M Baczkowicz[1] **and A Gwiazda**[1]
[1]Silesian University of Technology, The Faculty of Mechanical Engineering, Institute of Engineering Processes Automation And Integrated Manufacturing Systems, ul. Konarskiego 18a 44-100 Gliwice, Poland

E-mail: magdalena.baczkowicz@polsl.pl

Abstract. The article shows the practical use of Quality Function Deployment (QFD) on the example of a mechanized mining support. Firstly it gives a short description of this method and shows how the designing process, from the constructor point of view, looks like. The proposed method allows optimizing construction parameters and comparing them as well as adapting to customer requirements. QFD helps to determine the full set of crucial construction parameters and then their importance and difficulty of their execution. Secondly it shows chosen technical system and presents its construction with figures of the existing and future optimized model. The construction parameters were selected from the designer point of view. The method helps to specify a complete set of construction parameters, from the point of view, of the designed technical system and customer requirements. The QFD matrix can be adjusted depending on designing needs and not every part of it has to be considered. Designers can choose which parts are the most important. Due to this QFD can be a very flexible tool. The most important is to define relationships occurring between parameters and that part cannot be eliminated from the analysis.

1. Introduction

Nowadays the problem of optimizing parameters of technical systems is a common issue for many companies. The final product has to fulfil customer requirements which are becoming more and more complicated to translate into measurable technical requirements. The most important ones are called critical technical requirements and without them the final product cannot be completed. Critical requirements are the most difficult to implement. Quality Function Deployment (figure 1) can help with that problem by evaluating complete set of assumptions to determine the proper design parameters [1, 2].

QFD method was invented in the Mitsubishi Heavy Industry Ltd by two experts Shigeru Mizuno and Yoji Akao for better understanding and improvement the design processes. QDF was successfully used by Toyota which helped to reduce costs of introducing a new car into the market by 60% and shortening product development time by 1/3. This method allows increasing the customer satisfaction and improving the quality of every final product. A summary of the method is to create a QFD diagram often called the House of Quality. In essences basic general diagram shows (figure 1) customer requirements and answers how those needs are translated into design requirements. It also gives a view on how customers and design requirements correspond with each other, assigns weights and evaluate their relationships.

Content from this work may be used under the terms of the Creative Commons Attribution 3.0 licence. Any further distribution of this work must maintain attribution to the author(s) and the title of the work, journal citation and DOI.
Published under licence by IOP Publishing Ltd

The first step in the proposed method would be to select main groups of design parameters and define customer. The customer is understood not only as a person buying final product but also as a designer. In the article customers are represented by marketing department and designers by engineering department. Each department has different needs and the task of QFD is to combine both and choose their appropriate measured values. Most common requirements from customer point of view are: effective and practical use, appearance, easy operating, availability, physical proprieties (such as weight, temperature, type of material, etc.), reliability, environmental friendliness. The engineering department will have measurable requirements such as: dimensions, materials characterization, assembly features, power and energy consumption, ergonomic parameters, physical and chemical properties, gravity, cross-sectional shapes, cross-sectional dimensions and others depending on the type of the product.

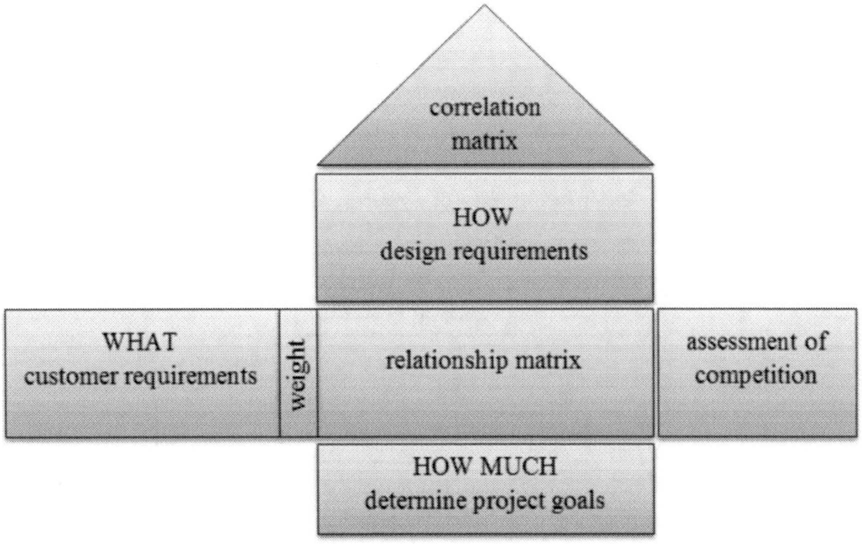

Figure 1. Quality Function Deployment diagram [1].

2. Optimization of a mining support

The mining mechanized support is the most important part of mining complexes (figure 2). It is utilized in hard coal mine for deep mining. The problems considered with this technical mean, results from the designed systems and adopted work parameters.

The role of a support is to protect the mining area against the rock falls what causes destruction of mining equipment, covering up a heading and, what is the most dangerous, jeopardizing human life. This is why this device should be proper designed and optimized. Moreover it should be prepared to assembly and disassembly under difficult mining conditions.

This analysis of parameters of a support could be realized according to the general model of the structure of a technical mean [3]. It consists of three main component systems: a "material" one, an "energetic" one and an "information" one (figure 3). The sequence of the component systems responds to their hierarchy. The material system includes such elements like: canopy, shields, lemniscate system, and foot piece. The energetic system includes main hydraulic props, hydraulic cylinder for canopy correction and hydraulic cylinder of a front shield. Finally the information system includes elements of a hydraulic system with control valves.

The presented structure, used as a decomposition tool, could be utilized to distinguished features that could be optimized using the QFD method.

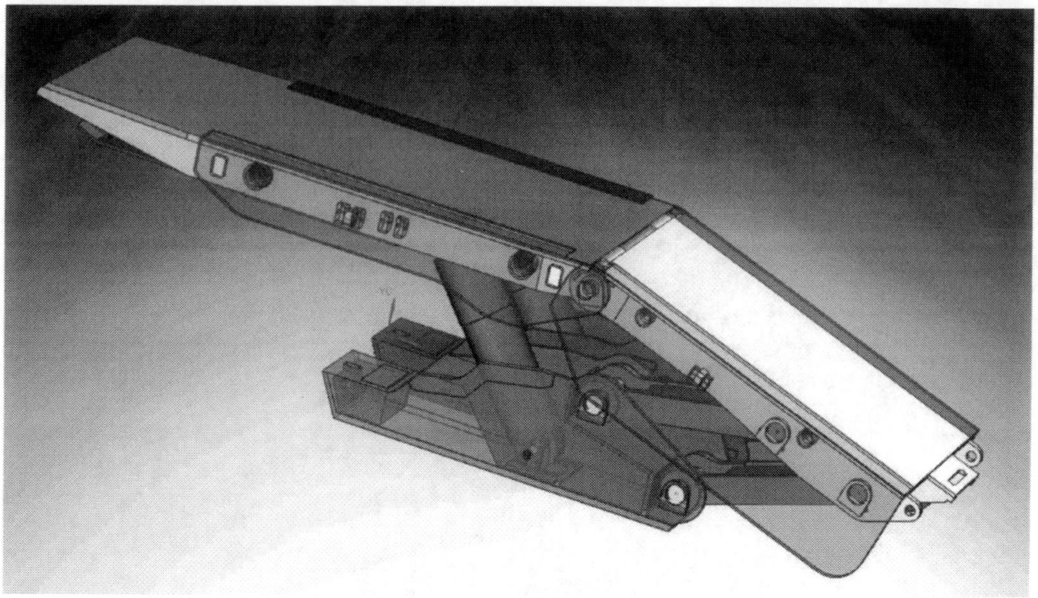

Figure 2. Virtual model of a mining mechanized support.

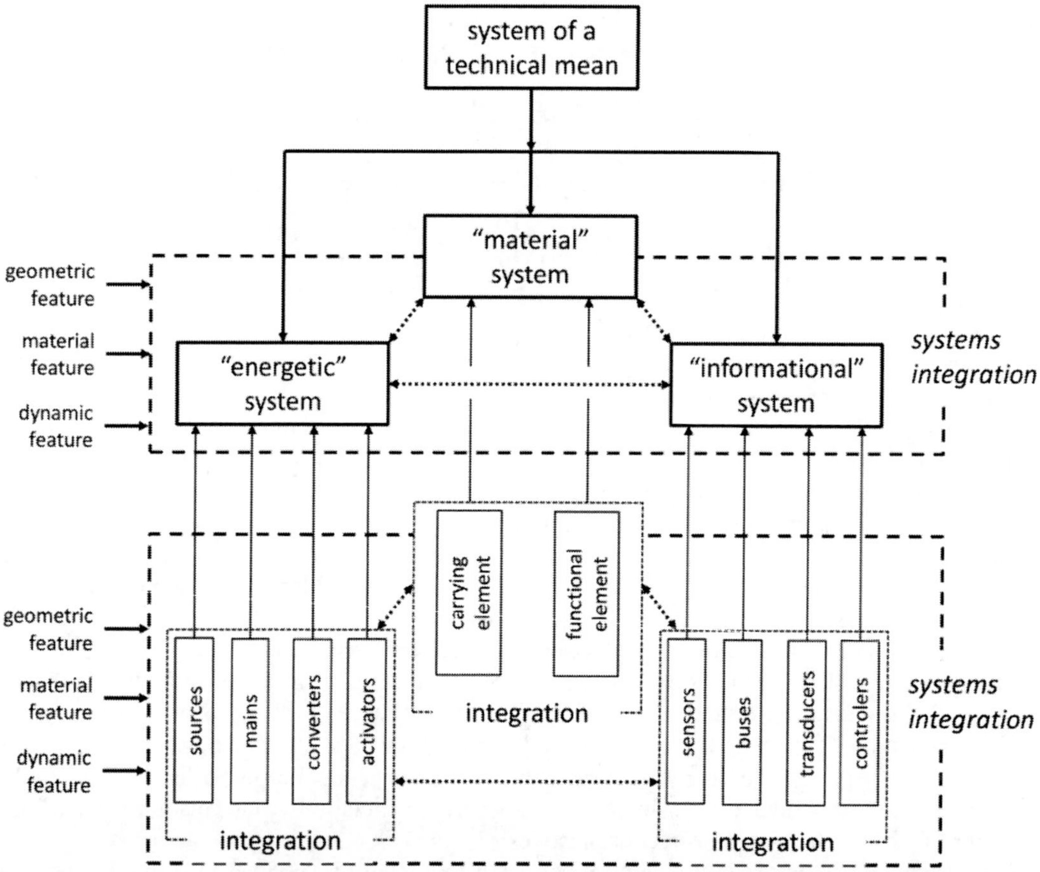

Figure 3. Model of the structure of a technical mean [3].

3. Quality Function Deployment methodology

The first step in determine construction parameters is to know how the test object is built. Traditional mechanized mining support has 9 main parts: front shield, props, base, lemniscate connectors, rear shield, correction prop, shield, side shields and canopy. Based on that, design requirements (construction requirements) and their weights are: compliance of direction of the load and support (0.06), compliance of direction of the load and canopy trajectory (0.06), the vulnerability of structures (0.14), versatility (0.12), continuous roof support (0.08), cover of workplace (0.07), standard components (0.11), simplicity (0.11), reuse of elements (0.15), workflow automation (0.09).

Than the requirements from the client point of view (marketing requirements) and also their weights are established such as:

- the use of the most accessible materials (0.13);
- costs (0.25);
- durability (0.21);
- prevention of destruction of the roof (0.42).

The second step is creating a QFD quality matrix (figure 2), were all the above requirements are put in, as well as theirs weights. All requirements are compared with each other and the assessment of existing relationships between them is made [4, 5]. Because there is not any relevant competition in the mining industry manufacturing mechanized mining support, analysis of the competition was omitted. Weights were analyzed and selected by experts (table 1).

Table 1. QFD matrix.

	rank features according to customer	*rank features according to designer*	*assessment of the validity of marketing requirements*	compliance of direction of the load and support	compliance of direction of the load and canopy trajectory	the vulnerability of structures	versatility	continuous roof support	cover of workplace	standard components	simplicity	reuse of elements	workflow automation
the use of the most accessible materials	1	2	0,13	0	0	1	2	0	0	3	2	2	0
costs	4	1	0,25	3	3	2	1	2	1	1	1	2	3
durability	2	3	0,21	1	2	2	1	1	0	1	2	2	1
prevention of destruction of the roof	3	4	0,42	2	3	1	0	3	1	0	0	0	1
			the validity of the technical parameters	0,06	0,06	0,14	0,12	0,08	0,07	0,11	0,11	0,15	0,09
			the importance of technical parameters	0,36	0,48	0,84	0,48	0,48	0,14	0,55	0,55	0,9	0,45
			critical features	c	c		c						c
			target technical parameters	↑	↑	↔	↑		↑		↑	↔	↔
			indicator of complexity	3	3	1	1	1	1	2	2	1	3

The gray area in the QFD matrix means future project goals and the correlation matrix is done using weights and establishment of the validity of marketing parameters [6, 7]. QFD diagram does not have to look like a typical house of quality. In some cases, as in this example it is easier to create a typical table and adjusted it, accordingly to a chosen technical system.

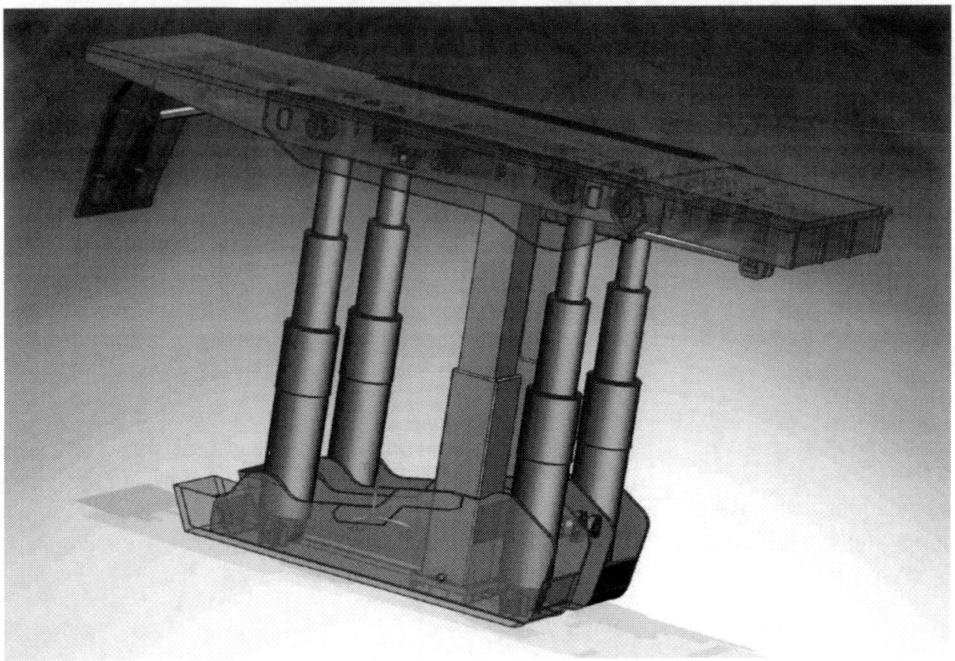

Figure 4. Model of exemplar support construction including optimized parameters.

4. Conclusions

The presented QFD approach helped to elaborate the set of constructional requirements that allowed designing the mechanized mining support that fulfil the most complex technical requirements indicated by the QFD analysis. The virtual model of this support is presented in figure 4. The model was used to conduct analysis considered with the assessment process of a new design [8]. The next analysis should be considered with material analysis [9,10] to select the most appropriate one taking into account the conditions in a hard coal wall.

The second direction of work is to consider with preparation the production scheme that allows decreasing manufacturing costs. To help with that it would be a good idea to use application of new intelligent tools [11,12] to realize this task as a multi-enterprise project.

The presented approach of the QFD technique to technical project realization shows that this multidisciplinary approach could bring synergetic effect in the designing process. Application of the QFD approach allows confronting customers' requirements (representatives of hard coal mines) and designers' problems. In comparison to other approaches, the QFD technique allows analyzing problem aspects not only in qualitative terms but also in quantitative ones. The last could be called as the main advantage of this engineering tool.

References

[1] Akao Y 1990 Quality function deployment: Integrating customer requirements into product design (Cambridge: Productivity Press)

[2] Akao Y and Mazur G H 2003 The leading edge in QFD: past, present and future *The Int. J. of Qua. & Rel. Man.* **20/1** pp 20-35

[3] Gwiazda A 2013 Designing of technical means using the integration methodological approach (Katowice: GIG Publishing) (in Polish)

[4] Chan L K and Wu M L 2002 Quality function deployment: A comprehensive review of its concepts and methods *Qua. Eng.* **15/1** pp 23-35
[5] Cohen L 1995 Quality function deployment: how to make QFD work for you *Addison-Wesley, Reading*
[6] Day R G 1993 Quality function deployment: linking a company with its customer (Milwaukee: ASQ Quality Press)
[7] Hunt R A and Xavier F B 2003 The leading edge in strategic QFD *The Int. J. of Qua. & Rel. Man.* **20/1** pp 56-73
[8] Gwiazda A 2014 Virtual analysis of the mining support under loads of the roof *Adv. Mat. Res.* **837** pp 393-398
[9] Topolska S and Łabanowski J 2015 Duplex stainless steels toughness impact investigations *J. Mat. in Tehn. / J. Mat. and Techn.* **5** pp 114-119
[10] Ćwiek J, Łabanowski J, Topolska S and Sozańska M 2012 Determination of failure causes of a steam turbine casting *Sol. St. Ph.* **183** pp 37-42
[11] Paprocka I, Kempa W, Kalinowski K and Grabowik C 2014 On Pareto optimal solution for production and maintenance jobs scheduling problem in a job shop and flow shop with an immune algorithm *Adv. Mat. Res.* **1036** pp 875-880
[12] Paprocka I and Kalinowski K 2014 Pareto optimality of production schedules in the stage of populations selection of the MOIA immune algorithm *App. Mech. and Mat.* **657** pp 869-873

A survey on methods of design features identification

C Grabowik[1], **K Kalinowski**[1], **I Paprocka**[1] **and W Kempa**[2]
[1]Silesian University of Technology, The Faculty of Mechanical Engineering, Institute of Engineering Processes Automation and Integrated Manufacturing Systems, ul. Konarskiego 18A 44-100 Gliwice, Poland
[2]Silesian University of Technology, The Faculty of Applied Mathematics, Institute of Mathematics, ul. Konarskiego 23A 44-100 Gliwice, Poland

E-mail: cezary.grabowik@polsl.pl

Abstract. It is widely accepted that design features are one of the most attractive integration method of most fields of engineering activities such as a design modelling, process planning or production scheduling. One of the most important tasks which are realized in the integration process of design and planning functions is a design translation meant as design data mapping into data which are important from process planning needs point of view, it is manufacturing data. A design geometrical shape translation process can be realized with application one of the following strategies: (i) designing with previously prepared design features library also known as DBF method it is design by feature, (ii) interactive design features recognition IFR, (iii) automatic design features recognition AFR. In case of the DBF method design geometrical shape is created with design features. There are two basic approaches for design modelling in DBF method it is classic in which a part design is modelled from beginning to end with application design features previously stored in a design features data base and hybrid where part is partially created with standard predefined CAD system tools and the rest with suitable design features. Automatic feature recognition consist in an autonomic searching of a product model represented with a specific design representation method in order to find those model features which might be potentially recognized as design features, manufacturing features, etc. This approach needs the searching algorithm to be prepared. The searching algorithm should allow carrying on the whole recognition process without a user supervision. Currently there are lots of AFR methods. These methods need the product model to be represented with B-Rep representation most often, CSG rarely, wireframe very rarely. In the IFR method potential features are being recognized by a user. This process is most often realized by a user who points out those surfaces which seem to belong to a currently identified feature. In the IFR method system designer defines a set of features and sets a collection of recognition process parameters. It allows to unambiguously identifying individual features in automatic or semi-automatic way directly in CAD system or in an external application to which the part model might be transferred. Additionally a user is able to define non-geometrical information such as: overall dimensions, surface roughness etc. In this paper a survey on methods of features identification and recognition is presented especially in context of AFR methods.

1. Introduction

Rule-based automatic design features recognition methods use a general principle which says that if in a product design it is possible to recognize a characteristic structure of a potential design feature this structure is compared, with system inference rules, with design features patterns stored in an AFR

Content from this work may be used under the terms of the Creative Commons Attribution 3.0 licence. Any further distribution of this work must maintain attribution to the author(s) and the title of the work, journal citation and DOI.
Published under licence by IOP Publishing Ltd

system knowledge base. During the process of an AFR system knowledge base development it is necessary to comply with rules uniqueness condition. In this connection it is not acceptable that two different design features could have the same corresponding pattern or more than one pattern structure definition for a particular design feature. Meeting of rule uniqueness condition allows making AFR systems which can recognize design features with a great precision. But in case of this kind of design feature recognition system is impossible to broaden a store of knowledge stored in a system knowledge base without user participation. As a result it is highly likely that in some cases it could be impossible to match a potential design feature structure with patterns stored in a knowledge base. In such case it is necessary to add appropriate rules and design features structure definitions manually.

Currently, *th*ere are many AFR systems worked out with application of design feature recognition logic, which first of all, differ in a manner of a product design representation. In such case the following approaches mostly are used: syntactic pattern recognition, predicate calculus also known as predicate logic, convex hull volumetric decomposition, cell-based volumetric decomposition, hint-based approach and hybrid.

2. Syntactic pattern recognition

In a method of syntactic pattern recognition an element model is made with application of sematic primitives. These primitives are stored with a certain descriptive language. First the set of language grammars, composed of the certain number of rules that allows defining particular syntactic patterns is built. The language grammar is responsible for delivering of detailed models which make distinguishing of characteristic chars sequence in a string possible. In a method of the syntactic pattern recognition a sentence (a string generated with a set of rule), a grammar (a set of rules) and parser are given. Parser is used in order to check whether a sentence was properly built according to language grammar. In the figure 1 AFR system architecture with syntactic pattern recognition is shown.

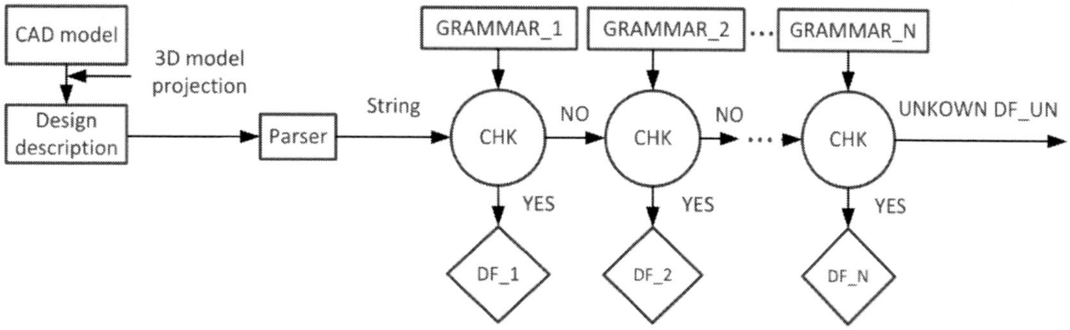

Figure 1. An AFR system with syntactic pattern recognition.

System form the figure 1 consists of: a design description module, parser and set of language grammars which represent particular design features patterns. At first these kinds of systems were able to recognize design features in 2D product design models, but their development has led to the state where design feature recognition in 3D product design model is also possible. In case of 3D models initial model preprocessing is necessary. Particular model faces have to be projected onto previously defined projection planes. The design product description module is responsible for a design description preparation. The product description is given in the shape of a string which is next examined by a system parser in order to check its correctness from language grammar point of view. If a product string is correct than the subsequent classification process is performed. During a classification process particular product design features are classified to appropriate pattern classes.

One of the latest syntactic pattern recognition methods is edge boundary classification method. In EBC method information about 3D models spatial addressability is used in order to distinguish outer and inner boundary object sides whilst identified edges loops are sought out in a product B-rep boundary representation. For each identified edges loop an EBC pattern is made. An EBC pattern is made as a result of a classification of points which belong to a test set. Initially the test set points are located in the vicinity of an edge that composes a given loop. A given point of the test set depending on its location according to a reference model is classified as an outer or inner element point. Basis on this information a special code is ascribed to the point. Thus formed for each loop code chain is a base for working out a given loop pattern. This loop pattern next might be used in a process of syntactic design feature recognition. The EBC method allows identifying the following class of objects: pockets, slots, steps, cylindrical and conical holes in rotational and prismatic parts. The most important drawback of syntactic pattern recognition systems is their limited range of application. They are able to identify design features in rotational part such as shafts, sleeves, shields and chosen designs of prismatic parts.

3. Predicate calculus – predicate logic

Predicate calculus design feature recognition systems are usually built with a set of production rules. A single production rule structure could be written in the following way: *If* C_1, C_2, C_3,…,C_n *Then* A, where A represents a particular design feature structure. If all conditions it is C_1, C_2, C_3,…,C_n of a production rule premise part are satisfied then a given structure identified in a part description is recognized as a structure compatible with A design feature structure. An example of the AFR system with predicate logic in [1] is presented. In the considered system a part model is represented with IGES format. The product 3D model is next transformed with original software in set of Prolog logic programming language facts. The process of automatic design features recognition begins with part faces extraction and base faces identification. From definition a base face should have at least one adjacent face and faces adjacency character should be a concave-type. In the next stage of the recognition process the product boundary end faces are identified. The main design features recognition criterion is type and the number of a product boundary end faces (excluding design features of the hole-type). In the figure 2 the general structure of the discussed AFR system (a) and the way of design features definition (b) are presented. This system allows recognizing of the following design features: slots, blind slots, holes, countersink holes, reliefs and chamfers.

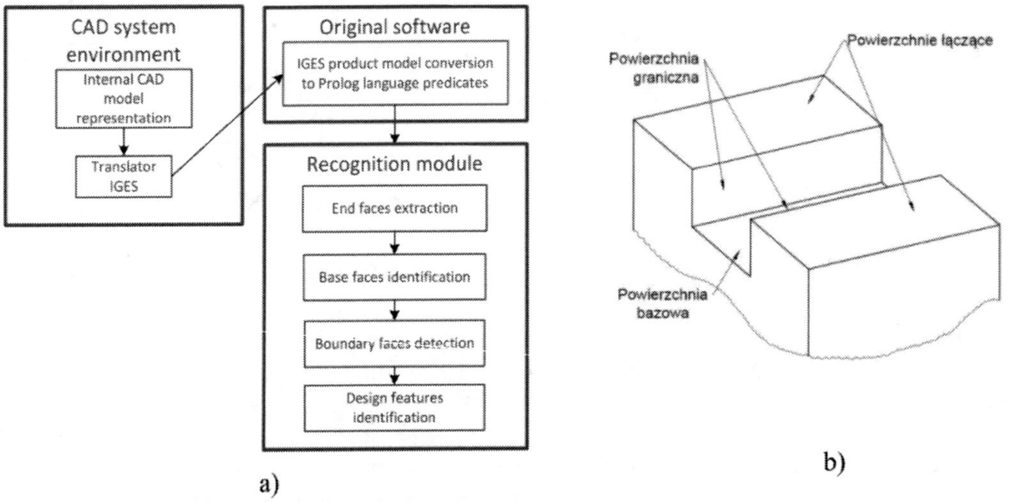

Figure 2. The AFR system structure (a), design feature definition (b).

The next example of the AFR system with predicate logic application in the [5] is presented. In this AFR system CSG product model representation is used. The identification process is carried out by a complete review of a text file with a program written in Visual Basic worked out with SolidWorks

API interface. The recognition process is being performed by comparing of features extracted from a product design with design features structures stored in Oracle database. Thanks to this AFR system it is possible to recognize design features in prismatic parts only excluding those features which intersect each other. The AFR system output information makes input information for a CAPP system planning activities.

4. Graph based design recognition methods

First studies on application of graph theory in design feature recognition were focused on working out of such design representation method which could allow representing both model topology and its geometry. Initially application of an attributed adjacency graph (AAG) in order to transformation of a B-rep product model made in any CAD system into a graph structure was presented. In the AAG graph for each particular graph edge 0 is ascribed if its vertexes satisfy concave adjacency relation and 1 for convex adjacency. In the figure 3 an example of a prismatic element for which design features are being recognized and its corresponding AAG graph structure are presented. A single design feature in the product design AAG graph is represented by the AGG subgraph structure. In this case the recognition process consist in searching, in the AGG graph of a product being processed, of structures which can be matched with those stored in an AFR system database. The subgraph structures are examined with logical rules. This approach known as the graphs isomorphism is characterized by large time-consuming and high computational complexity.

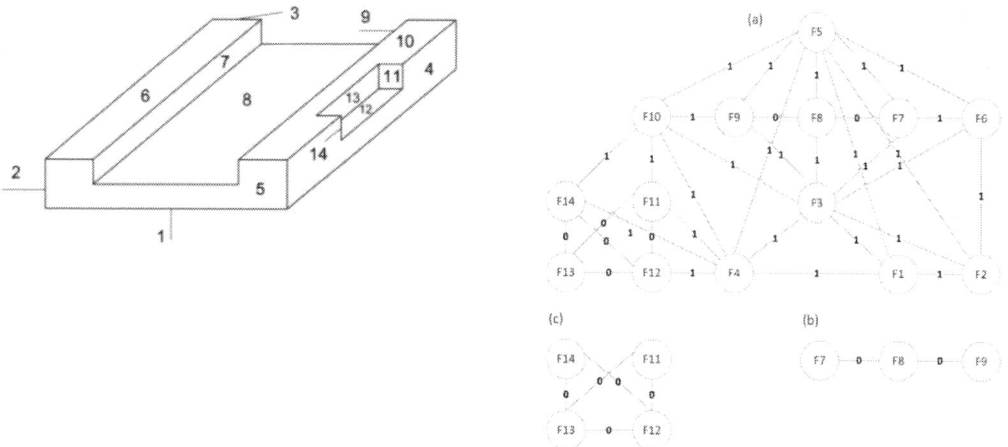

Figure 3. The prismatic part model (left), the product AGG graph (right) (a), the subgraph of the AAG graph representing design feature slot like (b), the subgraph of the AAG graph representing design feature blind slot like (c).

An alternative method to graphs isomorphism is a divided graphs isomorphism method in which the identification process is carried out by an AGG graph syntactic analysis. In such case syntactic analysis is carried out only for these graph vertices for which all product faces are convex.

An original idea of AAG graph application for automatic design features recognition purpose was characterized by identification ability of polyhedral design features only and inability of extracting boundary faces. AAG method limitations have given an impulse to AGG method modification. Based on original AGG graph concept a new Multi-Attributed Adjacency Graph method was worked out. The characteristic for MAAG method is that particular attributes values ascribed to graph edges give a model faces adjacency relation more precisely. For instance if a planar face makes with a curved face a convex angle then the attribute value equals 2. There is also a development of MAAG method for which a graph is presented by means of an adjacency matrix. In such case this AAG takes the name of multi-attributed adjacency matrix MAAM. Design features recognition process is then carried out with

MAAM matrix use, whilst particular MAAM matrix represent a singular design features structure. The most important drawback of graph based AFR systems was inability of design features recognition in case of their mutual interaction. This problem was solved by reconstructing of missing graph edges. These missing edges allow restoring information about mutual product faces connections lost as a result of appearing in a product model two or more design features which interact each other. Summarizing up, all AFR methods that use a graph theory can be characterized by a large number of preprocessing operations which results out of necessity of description of each part feature and geometric primitives. Moreover, most of graph based AFR systems allow recognize design features in prismatic parts only. What is more, there is no guarantee that this information, even if these systems are able to recognize design features efficiently, will be useful from a CAPP system process planning activities point of view.

5. Volumetric convex hull decomposition methods

Method of the volumetric convex hull decomposition consists in transformation of an input model into a set of intermediate volumes. Basis on these identified intermediate volumes design features are being search. This method uses a definition of ASV – alternating sum of volumes understood as the difference between a model volume and volumetric convex hull. On the basis of ASV definition in [2] a new partitioning procedure ASVP – alternating sum of volumes with partitioning was presented. In [4] on of the examples of ASVP use was presented. In this work basic design features and freeform design feature structures are considered. By freeform design features were assumed these features which contain freeform surfaces that could be machined with 2.5 or 3 axis milling operations. In the system [3] a B-rep product model made in any CAD system, exported with STEP or IGES neutral format, makes an input into a process of geometric information extraction. In that case of the ASVP method implementation particular model surfaces take different attribute values depending on whether they are a piece of a blank – SS, a piece of a finished product – MS or pieces of a product at particular stages of a manufacturing process – IS.

In such case a design feature is defined as a unique combination of the model surface with different attribute values it is SS, MS or IS. In the figure 4 the AFR system action schema is presented.

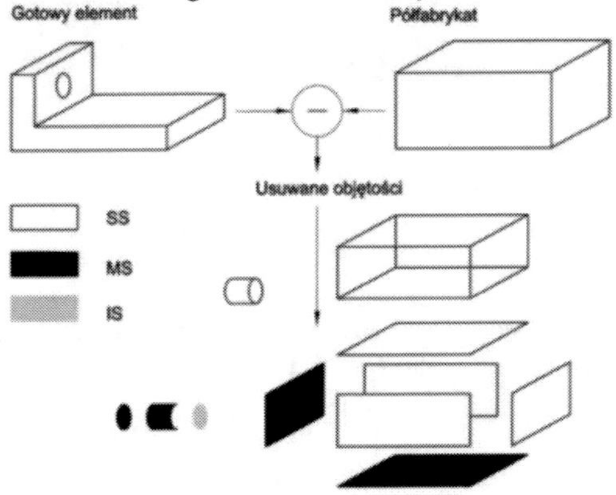

Figure 4. An illustration of design features recognition with ASVP method use [4].

6. Hint based methods

Unsatisfied need of the comprehensive solution of the problem of design features identification in case of features which interact each other has led to development of the new group of AFR methods it is hint-based methods. In a hint-based approach topological, geometric and heuristic information about product being analysed are used as hints on potential appearing of a certain class design features in the product design. There are some AFR systems which use hint approach but the most interesting are these presented in [6, 7]. In [6] system called OOFF from Object Oriented Feature Finder is presented. In this system an approach that consist in searching in a product design precisely defined patterns of geometric elements such as face, edge or vertex is used. As a matter of fat this approach is completely useless from a practical application angle. But this work introduced one important definition it is the presence rule definition. According to this definition manufacturing operation responsible for performing of a design feature with a machine tool should leave a mark on external surfaces of the machined product even if this design feature interacts with other product design features. The presence rule definition was used in later works giving foundations for working out of a new group of methods so called trace based methods. The introducing of the presence rule allowed working out patterns of minimal marks left by design features on a product external surfaces – boundary surfaces. So if it was possible to state a presence of a certain pattern on an external product surface it would suggest a potential possibility of existence of a certain class design feature in a product design. In [9] a modified hint-based method is presented. In this case as a hint is used additional information introduced by a user it is dimensional tolerances, surface roughness etc.

7. Conclusions

AFR systems were developed with the idea of a complete elimination of an expert presence in a process of design features identification. Despite the fact that AFR systems constantly evolve these systems still have lots of drawbacks and limitations. Among them we could distinguish high complexity of the recognition algorithms, long implementation time, limited application range, information ascribed to particular design features has a limited usefulness from process plan automation point of view, impossibility of design features dimension tolerance information acquiring, impossibility of a complete elimination of an expert presence in the identification process independently of applied method, almost complete lack of AFR methods industrial applications.

Taking into account above a problem of working out of a new efficient method of design features identification it is still valid. This method should be characterized by maximum simplicity and implementation ease. In spite of all the design feature method still seems to be one of the most attractive method that gives a methodological foundations for building of an integrated CAD/CAM/CAE system environment.

References

[1] Babić B and Miljkovic Z 1997 Feature recognition as the basis for integration of CAD and CAPP systems *Intelligent Manufacturing Processes and Systems* pp 596-601

[2] Kim Y S 1992 Recognition of form features using convex decomposition *Computer-Aided Design* **24** pp 461-476

[3] Markus A, Vancza J, and Horavath M 1997 Process planning by retrieval and adaptation *Computer in Industry* **33** pp 47-60

[4] Miao H K, Sridharan N and Shah J J 2002 CAD-CAM integration using machining features *International Journal of Computer Integrated Manufacturing* **15** pp 296-318

[5] Sadaiah M, Yadav D, Mohanram P and Radhakrishnan P 2002 A generative computer-aided process planning system for prismatic components *The I. J. of A. M. Techn.* **20** pp 709-719

[6] Vandenbrande J H and Requicha A A G 1993 Spatial reasoning for the automatic recognition of machinable features in solid models *IEEE Tran. on Patt. Ana. and Mach. Intell.*, **15** pp 265-285

[7] Waiyagan K and Bohez E L J 2009 Intelligent feature based process planning for five-axis mill-turn parts *Computers in Industry* **60** pp 296-316

[8] Xu H Li D 2009 Modelling of process parameter selection with mathematical logic for process planning *Robotics and Computer-Integrated Manufacturing* **25** pp 529-535

[9] Zheng Y, Taib J M and Tap M M 2011 Implementation of heuristic reasoning to recognize orthogonal and non-orthogonal inner loop features from boundary representation (B-rep) parts *Journal Mechanical* **33** pp 1-14

Recurrence plot for parameters analysing of internal combustion engine

O Alexa[1], C O Ilie[1], M Marinescu[1], R Vilau[1] and D Grosu[1]
[1]Military Technical Academy, Department of Military Automotive and Transportation, Blvd. George Cosbuc, No. 39-49, Sector 5, 050141 Bucharest, Romania

E-mail: alexa.octavian@gmail.com

Abstract. In many technical disciplines modern data analysis techniques has been successfully applied to understand the complexity of the system. The growing volume of theoretical knowledge about systems dynamic's offered researchers the opportunity to look for non-linear dynamics in data whose evolution linear models are unable to explain in a satisfactory manner. One approach in this respect is Recurrence Analysis - RA which is a graphical method designed to locate hidden recurring patterns, nonstationarity and structural changes. RA approach arose in natural sciences like physics and biology but quickly was adopted in economics and engineering. Meanwhile. The fast development of computer resources has provided powerful tools to perform this new and complex model. One free software which was used to perform our analysis is Visual Recurrence Analysis - VRA developed by Eugene Kononov. As is presented in this paper, the recurrence plot investigation for the analyzing of the internal combustion engine shows some of the RPA capabilities in this domain. We chose two specific engine parameters measured in two different tests to perform the RPA. These parameters are injection impulse width and engine angular speed and the tests are I11n and I51n. There were computed graphs for each of them. Graphs were analyzed and compared to obtain a conclusion. This work is an incipient research, being one of the first attempts of using recurrence plot for analyzing automotive dynamics. It opens a wide field of action for future research programs.

1. Introduction

Recurrence Plot Analysis (RPA) was introduced for the first time in 1987 in the paper published by Eckman, Kamhorst and Ruelle [1,2]. A recurrence plot (RP) is an advanced technique of nonlinear data analysis. It is a visualization (or a graph) of a square matrix, in which the matrix elements correspond to those times at which a state of a dynamical system recurs (columns and rows correspond then to a certain pair of times). The RP reveals all the times when the phase space trajectory of the dynamical system visits roughly the same area in the phase space [3].

The starting point is the idea that a measured time series is the result of the interaction between different relevant variables of a dynamic process. Each of that measured parameter is a one-dimension output of a multidimensional dynamic system. The method (RPA) infers characteristics of the original system and predicts its behaviour in the future given the history of its output. It analysed signal dynamics in reconstructed phase-space and reveals visual information about distance correlation between two points of attractor [3].

Content from this work may be used under the terms of the Creative Commons Attribution 3.0 licence. Any further distribution of this work must maintain attribution to the author(s) and the title of the work, journal citation and DOI.
Published under licence by IOP Publishing Ltd

For examples, an internal combustion engine is a highly complex system with a large number of parameters that influence its dynamics. Engine angular speed, injection impulse width or vehicle speed could be considered realizations of that dynamical system. Our research was motivated by the desire to identify more intimate dependencies between outputs and inputs of the internal combustion engine.
In this respect we've conducted a large number of tests using more than 10 cars with 1.6 l petrol engine with mileage ranging 13500 to 115000 km. The experimental research program was designed to obtain a variety of data for defining the vehicle dynamics. Most of the experimental data were acquired and stored in the tester. Those which could not be measured using the on-board system were computed using static characteristics of the engine that had been obtained on the test bench [4, 5].

2. Method

2.1. Recurrence Plot

In this method, a one-dimensional time series from a data file is expanded into a higher-dimensional space, in which the dynamic of the underlying generator takes place. This is done using a technique called "delayed coordinate embedding", which recreates a phase space portrait of the dynamical system under study from a single (scalar) time series [3].

Suppose that $x(i)$, $i=\overline{1,N}$ is one of the time series that describes a dynamical system. The recurrence plot is an array of dots in a N x N square (a matrix), where a dot is placed at (i, j) whenever $x(j)$ is sufficiently close to $x(i)$ [1].

The elements of the matrix are defined as follows:

$$R_{jk} = H(\varepsilon - d_{jk}) \tag{1}$$

$$d_{jk} = \left\| \vec{V}_j - \vec{V}_k \right\|_2 , \quad j,k = \overline{1,N} \tag{2}$$

Where: H is the Heaviside step function:

$$H(x) = \begin{cases} 0, & x < 0 \\ 1, & x \geq 0 \end{cases} \tag{3}$$

- ε is the threshold corridor, defined as a small percent from the standard deviation of initial time series (i.e. injection impulse width to be analysed, $i_j = i[j]$, $j = \overline{1,n}$);

- d_{jk} is the Euclidean distance between two vectors $\vec{V}_j - \vec{V}_k$;

$$R_{jk} = 1 \tag{4}$$

if and only if the state $\vec{V}_j$ it is inside a "sphere" of a radius ε, with $\vec{V}_k$ as a center [2].

The phase space vectors are reconstructed according to Taken's time-delay embedding theorem [2, 6], from univariate time series $i[1], i[2],...,i[1]$, (i.e., n – successive quantized samples of injection impulse width, measured from some appliance):

$$\vec{V}_j = \left(i_j, i_{j+\tau}, ..., i_{j+(m-1)\tau}\right) \tag{5}$$

where: j is the time index; m is the embedding dimension; τ is the time delay.

The number of vectors N, is computed as follow:

$$N = n - (m-1)\tau \tag{6}$$

where N is the horizontal/vertical dimension of recurrence matrix.

Embedding dimension m should be large enough to unfold the system trajectories from self-overlaps, but not too large, otherwise the noise will amplify. The rule of thumb is to set m to

$$m \leq 2N + 1, \tag{7}$$

where N is the number of operating variables, or degrees of freedom, in the dynamical system under study. One analytical method for estimating the embedding dimension is the False Nearest Neighbors method [7].

Choosing the optimal time delay is not an easy problem. If it is too long, the coordinates may become random in respect to each other, or it is too short, the coordinates used for each reconstructed vector will not be independent enough to carry any new information about the trajectory of the system in the state space. Usually, for estimating the time delay is used Mutual Information Function [7].

Each element of the square matrix, called recurrence matrix will receive a pixel, black or white, and will be displayed in a monochrome digital image, which is recurrence plot [2]. Also, the distances between all vectors are mapped to colors from the pre-defined color map and are displayed as colored pixels in their corresponding places. A recurrence plot is essentially a graphical representation of a correlation integral. The important distinction is that it, unlike the correlation integrals, preserve the temporal dependence in the time series, in addition to the spatial dependence [7].

The next step and most difficult and interesting in the same time, is to interpret the plot. Large-scale structures (typology) and small-scale patterns (texture) can be easily detected [1, 2, 8]. The topology can be characterized as homogeneous, periodic, drift, and disrupted. Texture can be classified in single/isolated dots, diagonal, horizontal/vertical lines, rectangular clusters [2]. If the underlying signal is truly random and has no structure, the distribution of colors over the RP will be uniform, and so there will not be any identifiable patterns. If, on the other hand, there is some determinism in the signal generator, it can be detected by some characteristic, distinct distribution of colors. The length of diagonal line segments of the same color on the RP can give you an idea about the signal predictability [7].

2.2. Recurrence plot for analysing internal combustion engine parameters

To be analysed, there were chosen injection impulse width and engine angular speed. Each of them is dependent of an unknown number of other parameters with a statistical variation. The first set of data was collected from a vehicle with a millage of 95000 km (I11n) and the second from one with a millage of 33800 km (I51n). The idea was to compare parameters of engines with different wear.

Table 1. Embedding dimension and Time delay.

		Embedding dimension *m*	Time delay *n*
Test I11n	Injection impulse width	5	6
	Engine angular speed	10	12
Test I51n	Injection impulse width	5	6
	Engine angular speed	7	5

First, we used the mutual information function and false nearest neighbors method to estimate the time delay and the embedding dimension for each parameter and test.

Figure 1 and figure 2 presents the results for choosing the embedding dimension and time delay of injection impulse width for test I11n, $m=5$ and $n=6$.

Table 1 shows the results for embedding dimension and time delay of injection impulse width and engine angular speed for tests I11n and I51n. As a first conclusion, injection impulse with seems to be more deterministic variable than the engine angular speed.

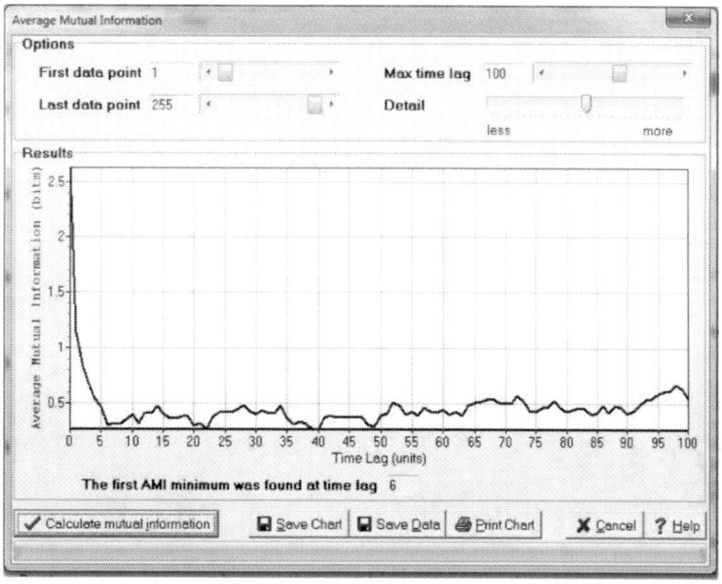

Figure 1. Average Mutual Information of injection impulse width, Test I11n.

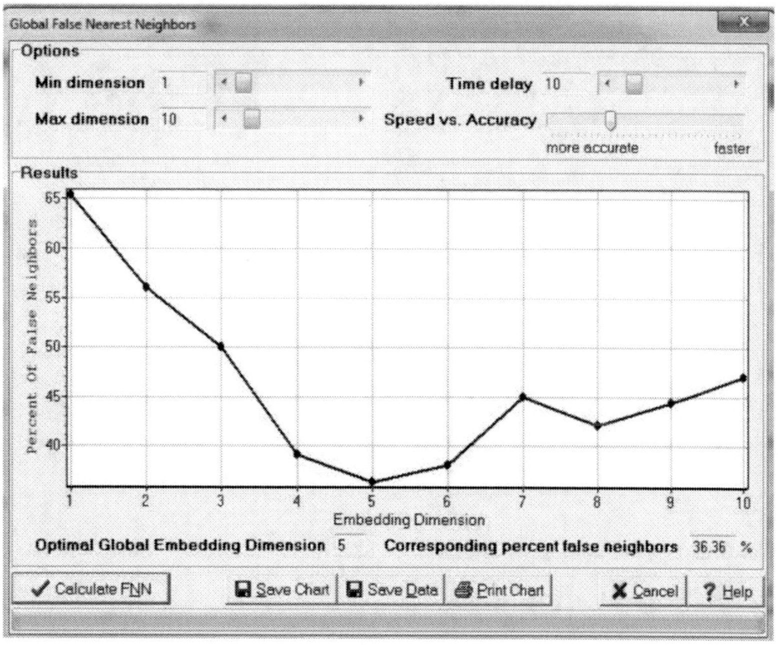

Figure 2. Global False Nearest Neighbors of injection impulse width, Test I11n.

Also, deterministic nature of injection impulse width is confirmed by the figure 3 and figure 4. There are some diagonal lines with evolution of states similar at different times that prove it. Similarities cannot be observed in recurrence plot of engine angular speed in figure 6 and figure 10. As it is known, electronic control unit sets the impulse injection width based on few deterministic

IOP Conf. Series: Materials Science and Engineering **95** (2015) 012121 doi:10.1088/1757-899X/95/1/012121

variables, while the angular engine speed is more influenced by stochastic sizes. All processes analysed are non-stationary, none of plots presents homogeneity.

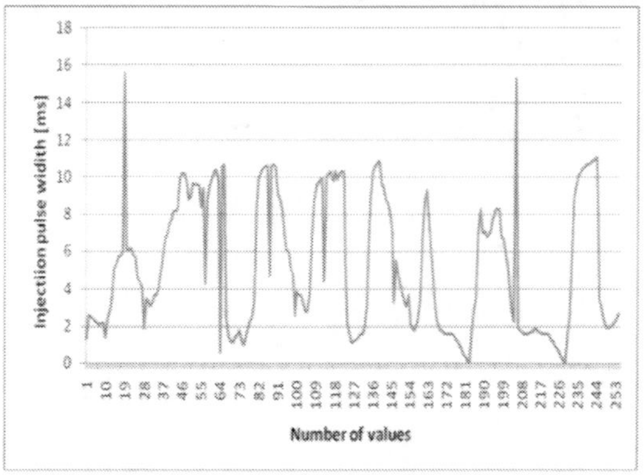

Figure 3. Injection impulse width, test I11n.

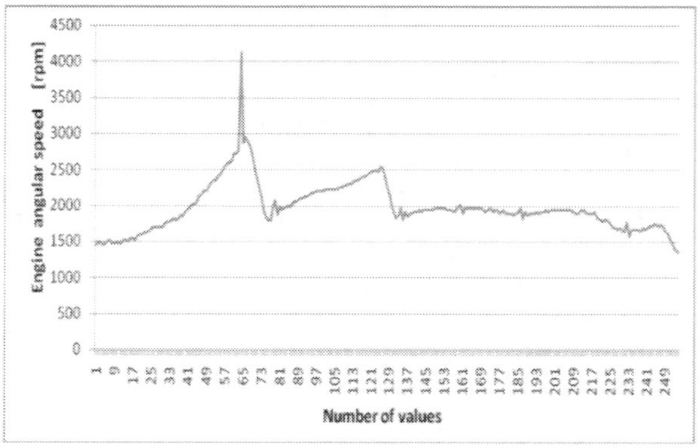

Figure 4. Engine angular speed, test I11n.

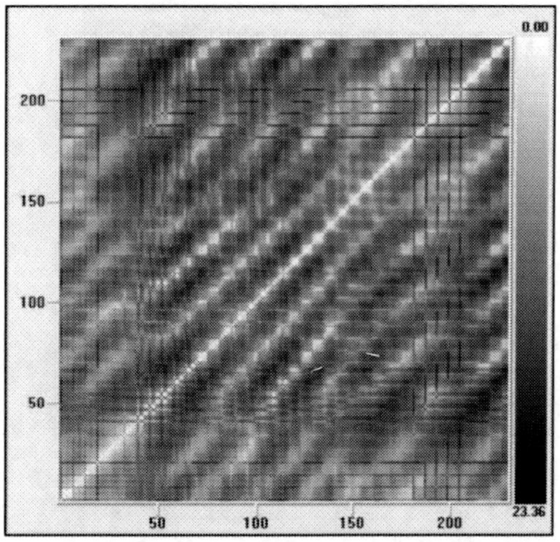

Figure 5. RP of injection impulse width, I11n.

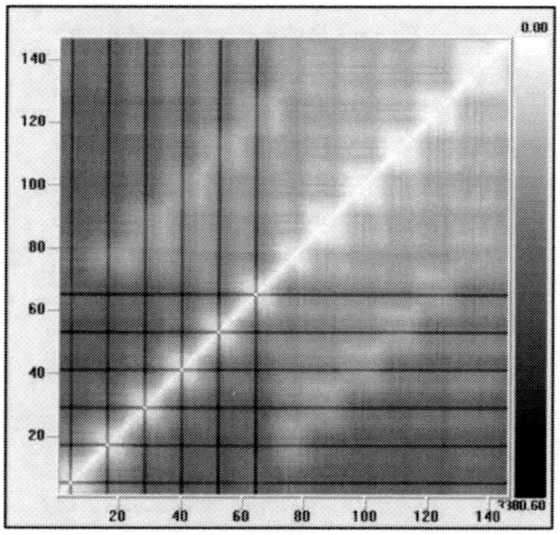

Figure 6. RP of engine angular speed, I11n.

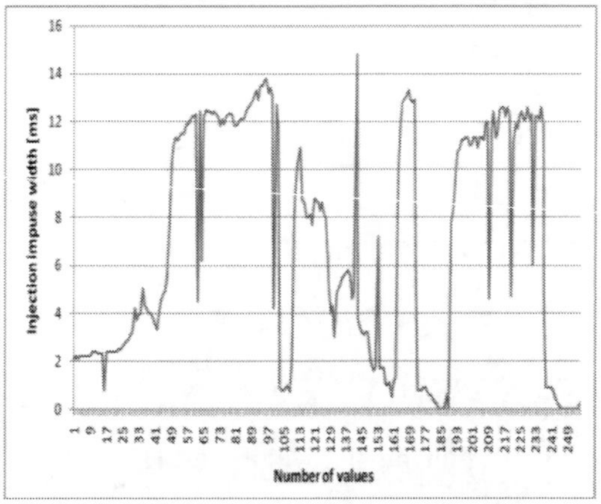

Figure 7. Injection impulse width, test I51n.

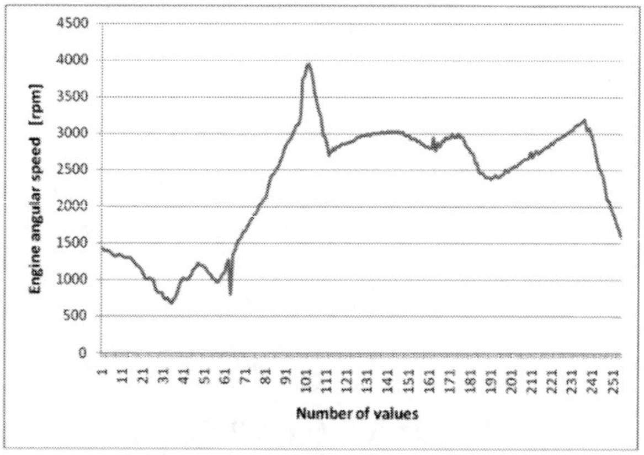

Figure 8. Engine angular speed, test I51n.

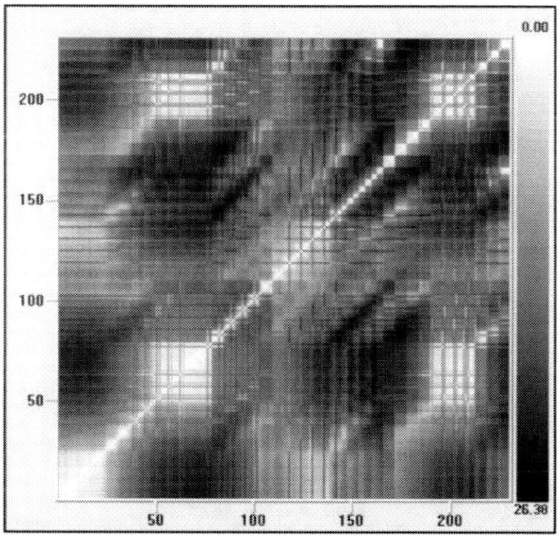

Figure 9. RP of injection impulse width, I51n.

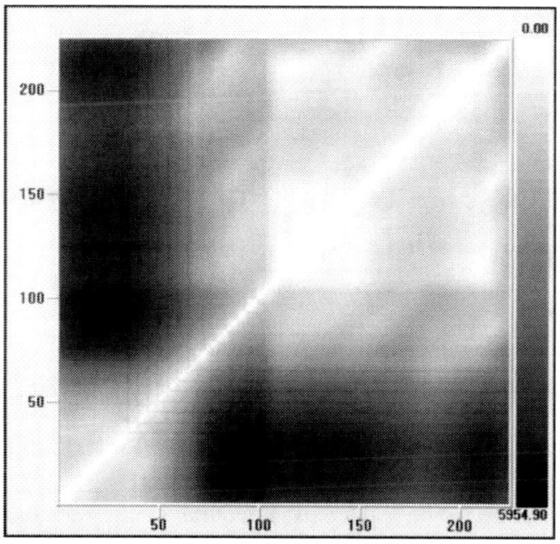

Figure 10. RP of engine angular speed, I51n.

3 Conclusions

This paper one of the first attempt of using recurrence plot for analyzing automotive dynamics. Yet, even if it is mostly a qualitative tool, recurrence plot is extremely powerful and useful in analysing dynamical systems. To accomplish a qualitative recurrence plot analysis, a quantitative approach has to consider. Future developments could lead to recognition of each engine based on graph obtained by this method. Also, it could be used to perform an extremely sensitive diagnose of vehicle systems. There is a vast field of applications for using this method in automotive.

References

[1] Ecmann J P, Kamphorst S O and Ruelle D 1987 Recurrence Plots of Dynamical Systems *Europhysics Letters* **5** pp 973-977
[2] Popescu F, Enache F, Vizitiu I C and Ciotârnaie P 2014 Recurrence Plot Analysis for characterization of appliance load signature *Proceedings of 10th International Conference on Communications - COMM* pp101-104
[3] Information on http://www.recurrence-plot.tk/glance.php
[4] Ilie C O 2014 Modeling of vehicle dynamics (Bucharest: Military Technical Academy)
[5] Ilie C O, Grosu D, Mocian O, Vilău R and Bartiş D 2014 Using statistically based modelling for vehicle dynamics *Advanced Materials Research* **1036** pp 564-567
[6] Takens F 1981 Detecting strange attractors in turbulence *Dynamical Systems and Turbulence* **898** (Berlin: Springer) pp 366-381
[7] Kononov E Visual Recurrence Analysis (VRA) software, version 4.9, available at http://nonlinear.110mb.com/vra/
[8] Marwan N and Kurths J 2005 Lines structures in recurrence plots *Physics Letters* **A336 (4-5)** pp 349-357

Aspects regarding at ^{13}C isotope separation column control using Petri nets system

M L Boca[1] **and M E Ciortea**[1]
[1]University "1 Decembrie 1918" of Alba Iulia, Faculty of Science and Engineering, Blvd. N. Iorga, No. 11-13, 515900, Alba Iulia, Romania

E-mail: loredana_boca1@yahoo.com

Abstract. This paper is intended to show that Petri nets can be also applicable in the chemical industry. It used linear programming, modeling underlying Petri nets, especially discrete event systems for isotopic separation, the purpose of considering and control events in real-time through graphical representations. In this paper it is simulate the control of ^{13}C Isotope Separation column using Petri nets. The major problem with ^{13}C comes from the difficulty of obtaining it and raising its natural fraction. Carbon isotopes can be obtained using many methods, one of them being the cryogenic distillation of carbon monoxide. Some few aspects regarding operating conditions and the construction of such cryogenic plants are known today, and even less information are available as far as the separation process modeling and control are concerned. In fact, the efficient control of the carbon monoxide distillation process represents a necessity for large-scale ^{13}C production. Referring to a classic distillation process, some models for carbon isotope separation have been proposed, some based on mass, component and energy balance equations, some on the nonlinear wave theory or the Cohen equations. For modeling the system it was used Petri nets because in this case it is deal with discrete event systems. In use of the non-timed and with auxiliary times Petri model, the transport stream was divided into sections and these sections will be analyzed successively.
Because of the complexity of the system and the large amount of calculations required it was not possible to analyze the system as a unitary whole. A first attempt to model the system as a unitary whole led to the blocking of the model during simulation, because of the large processing times.

1. Introduction

Petri-nets (PNs) are being grow up in aspects of their applications in the modeling, analysis and control of discrete manufacturing systems [1].

The aim of PNs is to model concurrent, synchronized interactions within a manufacturing system has contributed to their development as an important modeling tool.

PNs describe a manufacturing system graphically, and this contributes to a better understanding of the complex interactions within the system [1].

Since then, PNs have been modified, extensions proposed and detailed theoretical analysis performed. PNs are a "model for procedures, organization and devices where regulated flows play a role" [2].

Content from this work may be used under the terms of the Creative Commons Attribution 3.0 licence. Any further distribution of this work must maintain attribution to the author(s) and the title of the work, journal citation and DOI.
Published under licence by IOP Publishing Ltd

In modeling of the process using PN, independent events and causal dependencies are represented explicitly by introducing concurrent relations, which form the conceptual basis of net theory [3].

Systems can be represented at various levels of abstraction using the same descriptive language. PN representations allow for system-property verification in a systematic manner.

On the basis of supervisory model, the issues of safety verification, optimization of control parameters and scheduling of operating sequences are discussed.

At last, a distillation column is used as an example for modeling and analysis. Referring to a classic distillation process, some models for carbon isotope separation have been proposed, some based on mass, component and energy balance equations, some on the nonlinear wave theory or the Cohen equations.

2. Model of ^{13}C isotope separation column

The separation column used for experimentations one model can be configured in two ways: one - two columns of different diameters placed one within the other extension, and second way, one column with set diameter [4,5].

The main parts of the column are:

-The boiler of Primary column B1
-The boiler of Final column B2
-Vacuum jacket, M
-Aggregates vacuum, PVP (PD)
-Condenser- C
-First column C1
-Final column C2

In the column, using liquid nitrogen boiling at atmospheric pressure, the condenser (C) provides the reflux process to the top of the column with total condensation of carbon monoxide. The condenser from separation column is cylindrical and is provided with longitudinal fins.

It has the composition made of stainless steel, mono-block design. It measurements are: length is 800 mm and a diameter of 200mm.

Fluid flows gravitational from the condenser C1 on the column packing in counter current with monoxide carbon gas. In the first column, part of the liquid is vaporized and sent to the condenser and the other part enters into column C2. At the base of the column, the boiler B2 ensures the reflux process of carbon monoxide liquid [7, 8].

Column C1 has diameter 26mm and length 2500mm. Feed point divides the column in two enrichment sections: one of 1000 (1500) mm and second one dilution section of 1500 (1000) mm. It is filled with spiral triangular 2x2x0, 2mm stainless steel wire.

Column C2 has 16mm diameter and 4250mm length. It is filled with spiral triangular 1.8 x1, 8x0, 2mm stainless steel wire (figure 1).

Boilers, B1 and B2 are heated by electrical resistance, which can have a maximum power of 100W and 50W.

The main component of an isotope distillation process is separation column, because in the column, using the contact between the two phases - the liquid and vapor, takes place the isotopic enrichment [6,8].

Our ^{13}C experimental separation column is connected in its upper part, with liquid nitrogen-cooled condenser at atmospheric pressure (77 K) and at base on the boiler, which ensure the vaporization of the liquid. It operates at a temperature of about 79 K.

Column height is 7062 mm and is made also from stainless steel. In this model it was considered two columns: one with diameter of 16 mm and one with diameter of 26 mm, both the same height [8].

3. Model of a Petri Nets in industrial process

A Petri network type Location / Transition of the form [11, 12].

$$\sum = (P, T, F, W) \tag{1}$$

where:
- $F \subseteq (P \times T) \cup (T \times P)$ is a binary relation, called flux network relation,
- $W : F \rightarrow N$ is the weight function of the network $\sum(W(f))$ is called the weight of the element (f).

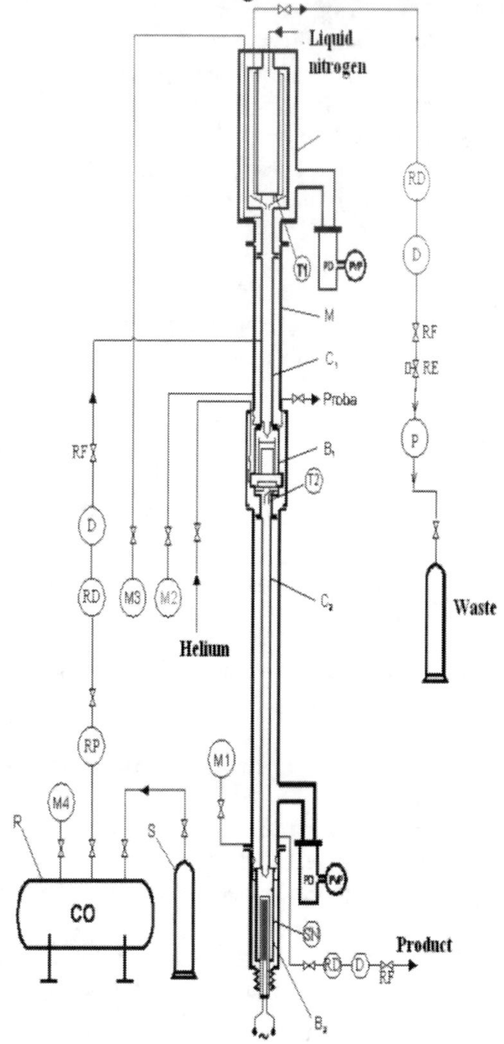

Figure 1. Model of separation column in distillation process [7].

Let (1) a network type location / transition, is called network marking $\sum$ any function $M : P \rightarrow N$ with the property $M(p) \leq K(p)$, for any $p \in P$, where $K : P \rightarrow N \cup \{\infty\}$ is the function of network capacity $\sum$.

$$\text{Let } \sum = (P, T, F, K, W) \tag{2}$$

The network P/T, where you get

$$N^P = \{M \mid M : P \rightarrow N \wedge (\forall p \in P)(M(p) \leq K(p))\} \tag{3}$$

If the network has only infinite capacity, then N^P coincides with the set of applications from P to N.

A marked P/T network is a pair $\gamma = (\Sigma, M_0)$, where Σ a support is network of the γ network, and M_0 is the initial marking of the γ network. [3, 11, 12].

For Σ network transitions, considered functions $t^-, t^+ : P \rightarrow N$ and $\Delta t : P \rightarrow Z$ defined by: $t^-(p) = W(p,t)$, $t^+(p) = W(t,p)$, $\Delta t(p) = t^+(p) - t^-(p)$, for any $p \in P$.

If w is known $(w \in T^*)$, is considered the function $\Delta w : P \rightarrow Z$ given through the relation:

$$|\Delta w(p) = \begin{cases} 0, \text{if } w = \lambda \\ \sum_{i=1}^{n} \Delta t_i(p), \text{if } w = t_1 t_2 \ldots t_n \ (n \geq 1), \text{for any } p \in P \end{cases} \tag{4}$$

In the dynamic evolution of a network there are known three types of semantics [11]:

1. sequential evolution of the network γ is given by the rule of transition, which consists of:

   -(RA – rule of applicability), transition t is possible by marking M in Σ and is noted $M[t\rangle_\Sigma$ if this relation takes place [11]:

$$W(p,t) \leq M(p), \forall p \in P \tag{5}$$

and

$$M(p) + W(t,p) \leq K(p), \forall p \in P \tag{6}$$

   -(RC – rule of calculation), M' marking is produced by the appearance of transition t at M marking and is noted $M[t\rangle M'$, if $M[t\rangle_\Sigma$ and

$$M'(p) = (M + \Delta t)(p), \forall p \in P \tag{7}$$

or

$$M' = M + O(t) - I(t). \tag{8}$$

2. concurrent evolution relative is the simplest way to capture the appearance of concurrent application of a transition. This method consists in generalizing the concept of "step" by moving from a transition to other transitions that can be applied concurrent to marking [10].

3. process type evolution, type subset doesn't point out the concurrently aspects like the concurrent application of a transition to itself. And so it is introduced the concept of process of a network using finite limited appearances networks.

Invariants are a formal method of analysis Petri networks which was introduced by Lautenbach, additional information [10, 11].

This method is used to demonstrate the dynamic properties of Petri networks and accessibility, limitation, reversibility and durability properties. An advantage of invariants is that they can be built during the system design phase, thus achieving efficient modelling. These methods are used to demonstrate certain structural and dynamic properties of the analysed system [12].

Using (1), network P/T. $|P| = m$, $|T| = n$. Matrix $m \times n$ defined by

$$A(p,t) = O(t,p) - I(p,t) \tag{9}$$

is the matrix of incidence of the network Σ.

$A(\cdot,t)$ defines the state change of all locations after the production transition t. Analogy is noted by $A(p,\cdot)$, defines induced changes in location p by the production of all transitions.

The markings can be regarded as m -dimensional column vectors. A marking produced by the production of transition t_j at the M marking can be calculated with:

$$M' = M + A \cdot f \quad (10)$$

where: f is a n dimensional vector that has 1 in line j and 0 everywhere else.

4. Model of a Petri Nets in ^{13}C isotope separation column. Structure of Petri model ^{13}C isotope separation column

It is used Petri nets in our ^{13}C Isotope Separation column for control the (figure 2):

-Condenser- C

-The boiler of final column B2

-Vacuum jacket, M [8, 9]

The set of concrete systems and the multitude of their conduction problems have created a variety of mathematical representations, called models (figure 3).

Their variety is determined mainly by the structure of "object" analysed, by the scope of the research and by information available (figure 4, 5).

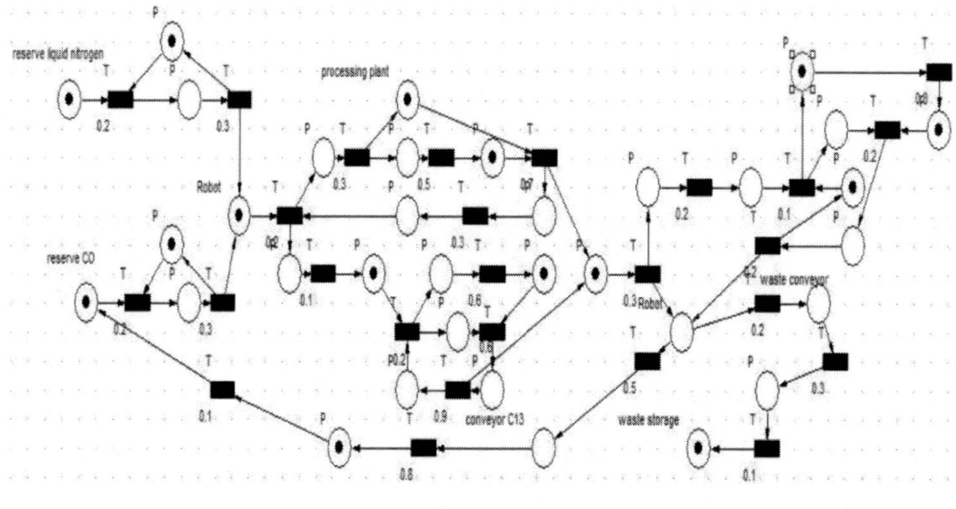

Figure 2. Petri net model applied on ^{13}C Isotope Separation column.

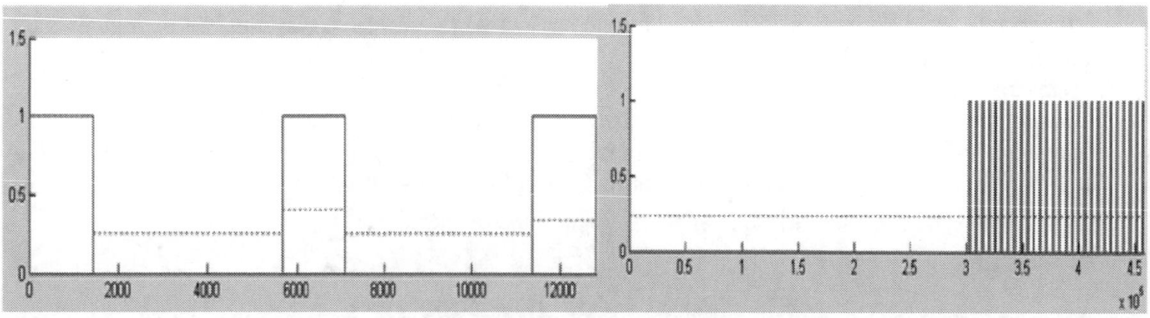

Figure 3. Evolution of the Queue Length indicators corresponding to the positions analysed (continues line – current value. dotted line - global value resulted from mediation within the simulation interval).

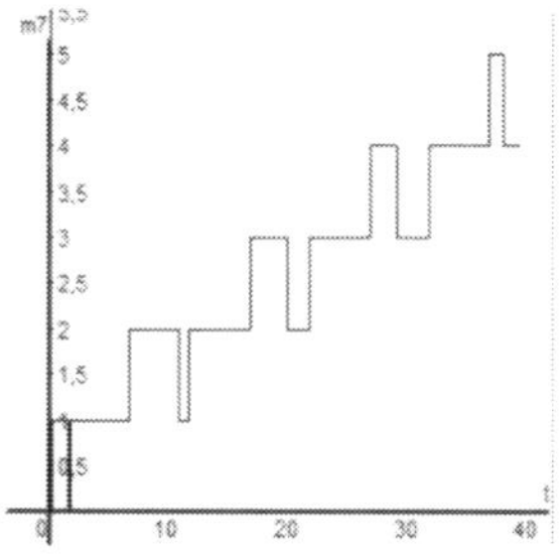

Figure 4. The graphical representation of the average time for the first transporter.

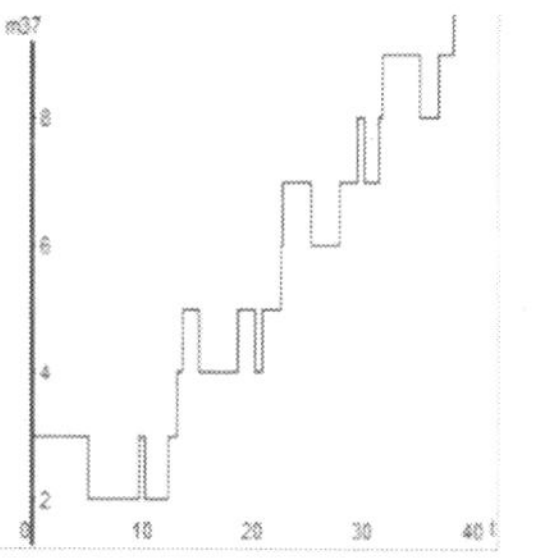

Figure 5. The graphical representation of the average value through the collection point.

5. Conclusions

This model system is using the Petri nets for discrete case event systems. In use of the non-timed and with auxiliary times Petri model, the transport stream was divided into sections and these sections was analyzed successively.

Due to the complexity of the system and the large amount of calculations required it was not possible to analyze the system as a unitary whole. A first attempt to model the system as a unitary whole led to the blocking of the model during simulation, because of the large processing times.

Like a future work it is intended to continue and make other model using Petri nets for distributed systems.

References

[1] Cecil J A, Srihari K and Emerson C R 1992 A Review of Petri-Net Applications in Manufacturing *The International Journal of Advanced Manufacturing Technology* pp 168-177

[2] Jucan T and Tiplea F L 1999 Reţele Petri. Teorie şi practică (Bucharest: Romanian Academy Publisher)

[3] Camerzan 2007 Proprietăţi structurale ale reţelelor Petri temporizate *Ph.D. Thesis* (Chisinau: Tiraspol State University) pp 12-15

[4] Axente D, Abrudean M and Bâldea A 1994 Separation of isotopes 15N, 18O, 10B, 13C through isotope exchange (Cluj Napoca: Casa Cărţii de Ştiinţă Publisher)

[5] Muntean M L, Unguresan A and Savu Abrudean M 2012 Modelling and composition estimation of an isotopic distillation column *Automation Quality and Testing Robotics* pp 109-113

[6] Pop C I and Dulf E H 2010 Control Strategies of the 13C Cryogenic Separation Column *CEAI* **12(2)** pp 36-43

[7] Boca M L, Secara M and Abrudean M 2013 Methods for control of some critical situations afferent of a ^{13}C isotope separation column *International Journal of Modern Manufacturing Technologies* **IV(1)** pp 21-26

[8] Boca M L and Secara M 2014 Analogy between "mission critical" detection in distributed systems and ^{13}C Isotope Separation Column *Proceedings of SPIE* **9258 92582N-1**

[9] Matjaz C and Wolfgang A H 2002 Architectural support for predictability in hard real-time systems *Control Engineering Practice* pp 51-59

[10] Mahulea C, Bârsan L and Păstrăvanu O 2001 Matlab tools for Petri-net-based approaches to flexible manufacturing systems Filip F G Dumitrache I Iliescu S (eds) *9th IFAC Symp. On Large Scale Systems LSS2001*, Bucharest

[11] Păstrăvanu O 1997 Sisteme cu evenimente discrete. Tehnici calitative bazate pe formalismul reţelelor Petri (Bucharest: MatrixRom Publisher)

[12] Păstrăvanu O, Matcovschi M and Mahulea C 2002 Aplicaţii ale reţelelor Petri în studierea Sistemelor cu evenimente discrete (Iasi: Gh. Asachi Publisher)

IOP Conf. Series: Materials Science and Engineering **95** (2015) 012123 doi:10.1088/1757-899X/95/1/012123

A method of computer aided design with self-generative models in NX Siemens environment

C Grabowik[1], K Kalinowski[2], W Kempa[2] and I Paprocka[2]
[1]Silesian University of Technology, The Faculty of Mechanical Engineering, Institute of Engineering Processes Automation and Integrated Manufacturing Systems, Konarskiego 18A Str., 44-100, Gliwice, Poland
[2]Silesian University of Technology, Faculty of Applied Mathematics, Institute of Mathematics, Konarskiego 23A Str., 44-100, Gliwice, Poland

E-mail: cezary.grabowik@polsl.pl

Abstract. Currently in CAD/CAE/CAM systems it is possible to create 3D design virtual models which are able to capture certain amount of knowledge. These models are especially useful in an automation of routine design tasks. These models are known as self-generative or auto generative and they can behave in an intelligent way. The main difference between the auto generative and fully parametric models consists in the auto generative models ability to self-organizing. In this case design model self-organizing means that aside from the possibility of making of automatic changes of model quantitative features these models possess knowledge how these changes should be made. Moreover they are able to change quality features according to specific knowledge. In spite of undoubted good points of self-generative models they are not so often used in design constructional process which is mainly caused by usually great complexity of these models. This complexity makes the process of self-generative time and labour consuming. It also needs a quite great investment outlays. The creation process of self-generative model consists of the three stages it is knowledge and information acquisition, model type selection and model implementation. In this paper methods of the computer aided design with self-generative models in NX Siemens CAD/CAE/CAM software are presented. There are the five methods of self-generative models preparation in NX with: parametric relations model, part families, GRIP language application, knowledge fusion and OPEN API mechanism. In the paper examples of each type of the self-generative model are presented. These methods make the constructional design process much faster. It is suggested to prepare this kind of self-generative models when there is a need of design variants creation. The conducted research on assessing the usefulness of elaborated models showed that they are highly recommended in case of routine tasks automation. But it is still difficult to distinguish which method of self-generative preparation is most preferred. It always depends on a problem complexity. The easiest way for such a model preparation is this with the parametric relations model whilst the hardest one is this with the OPEN API mechanism. From knowledge processing point of view the best choice is application of the knowledge fusion.

1. Introduction

Taking into account amount of work and time that have to be spent at a product design stage as well a designer intellectual effort, the following three types of a design process can be distinguished it is: (i) a creative design, (ii) innovative design and (iii) routine design. A creative design is a creative engineering activity that is aimed at searching of a completely new design of technical means. It might

Content from this work may be used under the terms of the Creative Commons Attribution 3.0 licence. Any further distribution of this work must maintain attribution to the author(s) and the title of the work, journal citation and DOI.
Published under licence by IOP Publishing Ltd

be characterized by a great complexity of problem decomposition into smaller design tasks. As a result of this kind of the design process a completely new, original and previously unheard-of solutions come into being [1, 2].

Innovative design is an engineering activity that consists in application of well-known technical means or design solution in order to find a new product design. During the innovative design some modifications of an existing product design are made. The result of these performed activities is so called an adopted product design. This adopted design is characterized by the unchanged system structure with some changes of partial solutions made on particular stages of the design process. In such case technical means functioning as a whole changes only slightly. In the innovative design only some pieces of detailed solution and their mutual location are being subjected to change. In conclusion it can be said that this design process course could be defined as process of new variants formation based on formerly designed technical solutions.

Routine design (multivariate design) is an engineer activity that consists in proper design information selection in order to identify a technical means structure which satisfied an identified need. In case of routine design, first a proper design documentation of formerly design product is selected; next the selected design documentation is being subjected to necessary changes. During this type of design activities similar tasks to those previously solved are being performed so pieces of a product design or even the whole product design are being subjected to changes. Product functioning and the principle of the technical solution stay unchanged. Characteristic for the routine design process is invariability of the product material, product manufacturing technology etc. for particular product components. The results of the routine design are different product design variants of formerly worked out designs. In the figure 1 different types of process design and product designs with respect to particular stages of the design process are shown.

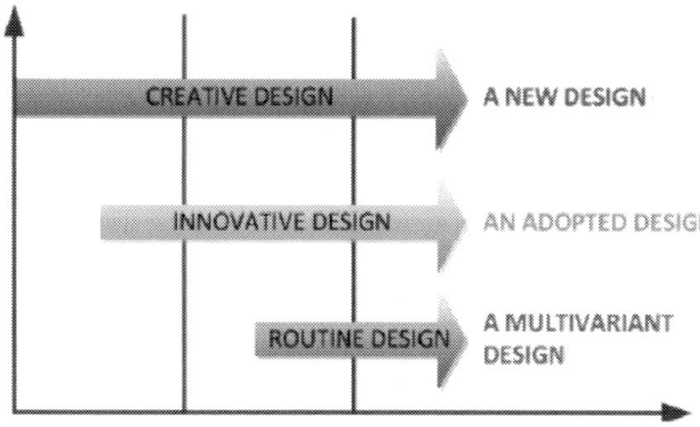

Figure 1. Types and stages of the design-constructional process.

The design process is an intellectual creative process whose results depend on a designer talent, imagination and creativity. Quality and effects of work of a designer depends inter alia on: (i) the designer intelligence and talent, (ii) acquired by the designer design knowledge and skills, (iii) the designer emotional involvement in the project (motivation, self-confidence), (iv) the amount of time spent on the project, (v) accessibility of devices that support designer work (easy access to information and computer equipment). Taking into account above, the following statement seems to be right "with unchanged other conditions, lack of talent could be compensated with an appropriate method". Usually a designer has very limited influence on some above outlined conditions and factors, for instance on factors that depend on people personality. Whereas factors such as technical and computer equipment, quality and efficiency of methods used in a design department could be shaped by the designer much easier. So the new more efficient methods of labour organization or the new computer systems that make the design process easier and faster could be applied. Observed, in the past few decade dynamic development of computer means supports this process. The contemporary processors

IOP Conf. Series: Materials Science and Engineering **95** (2015) 012123 doi:10.1088/1757-899X/95/1/012123

computing power increase, numerical and artificial intelligence methods development allow making more and more effective engineering tools. Current engineering tools development trends are mostly connected with application of artificial intelligence methods and tools. Thanks to this it is possible to capture and store skills and knowledge that arise during a new solution creation. Dynamic development of manufacturing systems requires application of new methods that support and make more efficient of a designer work. On the other hand recently could be observed increased interest of knowledge engineering domain. This domain includes issues of knowledge acquiring and storing as well its practical applications. One of the areas of knowledge engineering application is a computer aided design. Currently, CAD systems make storing of design-constructional knowledge and its integration with product geometrical models possible. Depending on the CAD system to be used in a design bureau, constructional-design knowledge can be store with parametric models, object models, design templates etc. [3, 4, 5].

2. Self-organizing models

In this work as a self-organizing model, a model that joins part model geometry with engineering knowledge and is capable of an intelligent behaviour in certain design condition is meant. In the other words self-organizing models make a full integration of design-constructional knowledge with solid product models possible. According to the introductory paragraph it could be noticed that especially prone to automation seems to be those engineering activities that have a routine character, so the potential application area of self-generative models is automation of the routine design. Thanks to application of this kind of models quick verification of many variants of a technical mean being designed is feasible. Taking into account that self-generative development process is labour and time-consuming they are improper for creative and innovative design [1, 2].

Self-generative model development process consists in increasing of a product solid model in design-constructional knowledge. Design-constructional knowledge can be introduced into a self-generative model by means of its different representation forms, such as:

- parameters that are usually defined by numerical or literal values;
- formulas that use formerly defined parameters – most often formulas take shape of math expressions;
- inference and production rules composed of If – Then structures. They are usually used for an automatic selection of design features standard dimensions that in a traditional design are performed by a designer;
- design tables that make controlling of parameters sets possible
- check-mate and checks, thanks to them it is feasible to test a product design quality on account of following of company design standards, adhering to best practices and finally meeting of modelling quality standards;
- knowledge templates used for knowledge management and sharing.

In the figure 2 a self-organizing model development process is shown. This development process consists of the five stages.

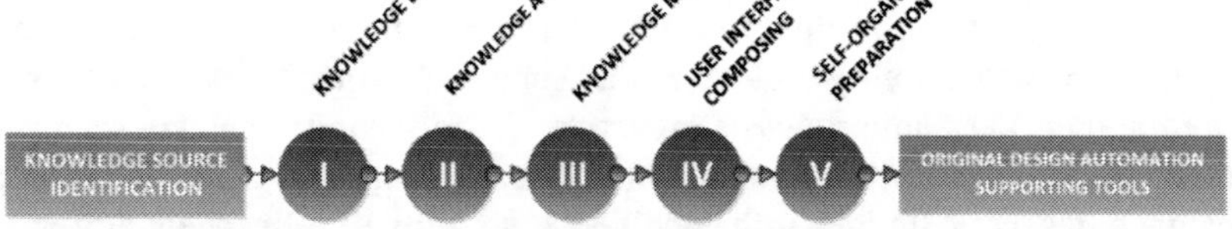

Figure 2. Self-organizing model development process.

The process of a self-organizing model development starts with identification of knowledge sources which are next subjected to decomposition process that is aimed at their complexity reducing.

At this development stage the self-organizing model system boundaries are defined. The second step includes knowledge acquisition which is usually preceded by working out of suitable knowledge acquisition tools. Next, the acquired constructional-design knowledge is modelled. It means that a computer interpretable model of knowledge is created. The resulting knowledge model might only be a computer interpretable when it is expressed in a knowledge representation language or data structure that enables the knowledge to be interpreted by software and to be stored in a database or some system of data exchange file (xml files for instance). In the fourth step a user interface structure is define. The last fifth stage includes activities connected with self-organizing model preparation; this also needs the CAD system environment to be chosen. In our research as a self-organizing models implementation environment a Siemens PLM NX 8.5 was chosen.

3. Methods of self-organizing models development in Siemens PLM NX 8.5 environment

Siemens NX 8.5 system makes the five method of self-generative models development available. They are as follows (see: the figure 3): (i) a fully parametric design model that constitute a combination of design features parameters and relations, (ii) a parametric model worked out with GRIP language, part families method, knowledge fusion, a parametric model worked out with application of chosen high-level programming language such as Visual Basic, C/C++. C# or Java.

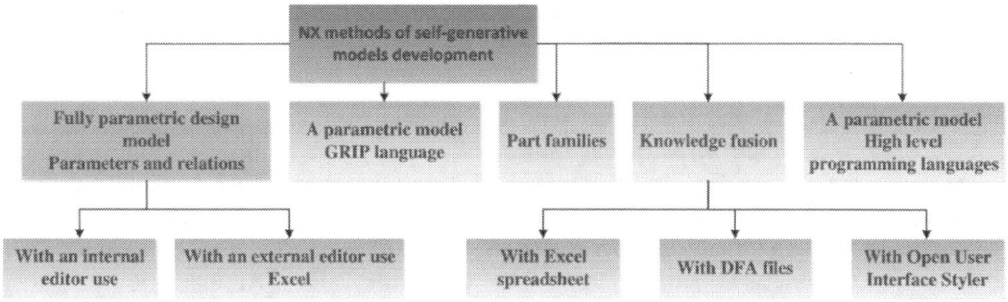

Figure 3. NX methods of self-organizing model development.

The choice of the most suitable method of a self-organizing model development depends primarily on the problem class. In case of complex engineering problems it is worth considering reaching for more advanced programing tools such as Visual Basic, Java, C/C++ or C#. For medium design problem class application of GRIP language is sufficient whilst simple problems could be supported by fully parametric models.

3.1. Fully parametric product model with parameters and relations

As mentioned above this method is used for relatively simple design problems. This method requires a good knowledge of the relationship that appears between particular technical means parameters and particular dimension values. It also requires knowledge of the mutual dimension relations. Thanks to this it is possible to change a product design shape by altering of these dimension values that were calculated in relational way to specific independent variable, for instance characteristic design features values or essential dimension etc. If the value of the independent value was changed than all connected dimension values are recalculated again. Next the product design shape is updated. Relation imposing is possible with an internal editor or Excel spreadsheet. In the figure 4 an example of this kind of method usage is shown.

3.2. A parametric model with GRIP language

Another method of self-organizing model development is a parametric model prepared with GRIP language. On account on GRIP language abilities it is possible to develop a medium hard parametric design models. Easy of learning is the main advantage of GRIP language, whilst the biggest drawback is incomplete coverage of all modelling features of Siemens NX system. So in some case is necessary to reach for more advanced programming tools for example SNAP or C#.

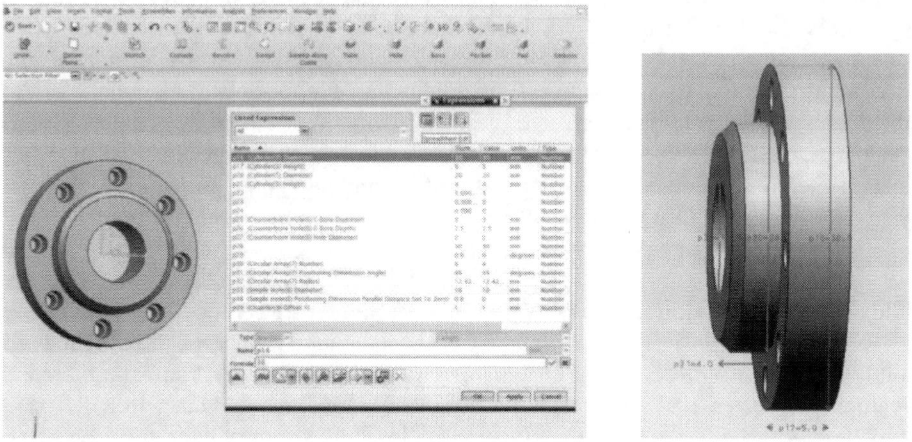

Figure 4. The method of self-organizing model development with fully parametric product model with parameters and relations.

3.3. Part families

Part family method allows making a self-organizing model with an idea of part family. From definition part family is a set of products with the same design, so particular part family members could differ each other only in particular dimension values. Part family development needs a part family template to be prepared – a solid model. Depending on design automation problem a part family template could represent either a stand-alone part or an assembly. In case of assembly all parameters that describe mutual assembly members location have to be precisely defined. Next Excel spreadsheet is used in order to store characteristic dimension values. An example of self-organizing model prepared with part family method in the figure 5 is shown.

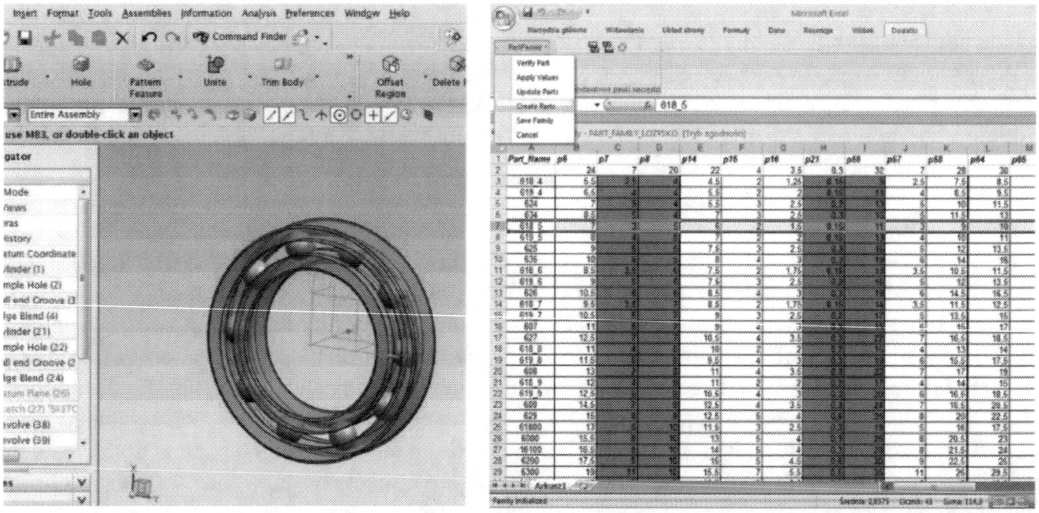

Figure 5. The method of self-organizing model development with fully parametric product model with parameters and relations.

3.4. Knowledge fusion
Knowledge fusion is an interpreter, object-oriented language that allows adding engineering knowledge to a part element by both design and geometric creating rules. The knowledge fusion language is declarative, rather than procedural one, which means that, in general, the rules are only evaluated when referenced or demanded. In KF it is possible to access external knowledge bases, such as databases or spreadsheets. Knowledge fusion allows making rules that are responsible for geometric modelling, performing of mathematical expressions, assemblies management, automatic generation of mechanical structures, performing of a dynamic and kinematic analyses, CNC controlled machine programming, different reports preparing. As a self-organizing model development tool knowledge fusion is slightly less powerful than a parametric model method with high-level programming languages. An example of self-organizing model prepared with knowledge fusion method in the figure 6 is shown.

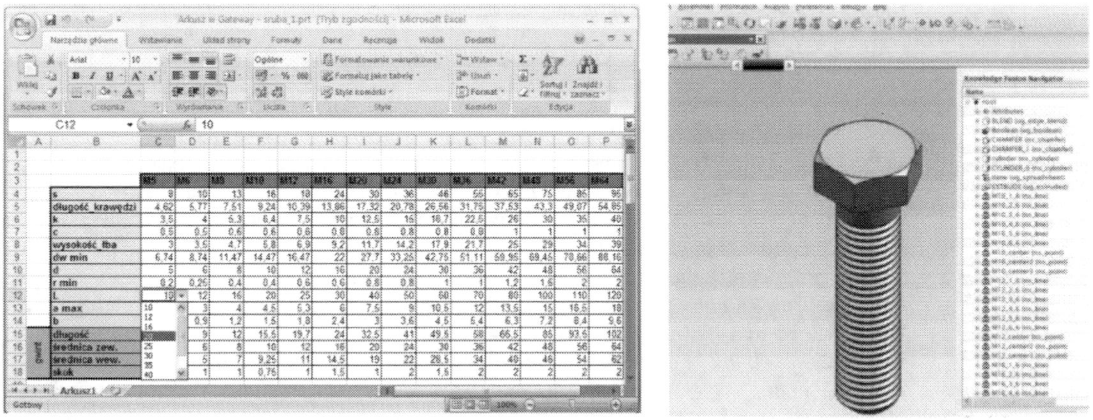

Figure 6. The method of self-organizing model development with knowledge fusion method.

A parametric model with high-level programing language use is the last method of self-generative method development. This is the most powerful tool of the self-generative model development. It is suggested to use this method for the most complicated design automation problems. Available in this method tools covers about 95% of Siemens NX functionality.

4. Conclusions
A self-organizing model is a modern tool that supports an engineer work. It allows integrating design-constructional knowledge in one coherent model. Thanks to application of this type of model it is possible to achieve a high degree of automation of routine tasks that are usually performed by designers. An extra advantage of self-generative model application is possibility of many design variant testing what allows improving the design-constructional process.

References
[1] Wronkowicz A and Wachla D 2014 *Model autogenerujący CAD zazębienia przekładni ślimakowej* (Zeszyty Naukowe Politechniki Śląskiej) seria Transport **82** pp 291-300

[2] Skarka W 2007 Modelowanie bazy wiedzy dla budowy modeli autogenerujacych w systemie Catia z zastosowaniem języka UML *Modelowanie Inżynierskie* **33** pp 145-152

[3] Hetmanczyk M and Michalski P 2014 The self-excitation phenomenon of quasi shielded inductive proximity switches *Advanced Materials Research* **837** pp 405-410

[4] Dzitkowski T and Dymarek A 2013 Active synthesis of discrete systems as a tool for stabilisation vibration *Applied Mechanics and Materials* **307** pp 295-298

[5] Sękala A, Banaś W and Gwiazda A 2014 Agent-based systems approach for robotic workcell integration *Advanced Materials Research* **1036** pp 721-725

IOP Conf. Series: Materials Science and Engineering **95** (2015) 012124 doi:10.1088/1757-899X/95/1/012124

Approximation method to compute domain related integrals in structural studies

E Oanta[1], C Panait[1], A Raicu[1], M Barhalescu[1] and T Axinte[1]
[1]Constanta Maritime University, Faculty of Naval Electro-Mechanics, 104 Mircea cel Batran Street, 900663, Constanta, Romania

E-mail: eoanta@yahoo.com

Abstract. Various engineering calculi use integral calculus in theoretical models, i.e. analytical and numerical models. For usual problems, integrals have mathematical exact solutions. If the domain of integration is complicated, there may be used several methods to calculate the integral. The first idea is to divide the domain in smaller sub-domains for which there are direct calculus relations, i.e. in strength of materials the bending moment may be computed in some discrete points using the graphical integration of the shear force diagram, which usually has a simple shape. Another example is in mathematics, where the surface of a subgraph may be approximated by a set of rectangles or trapezoids used to calculate the definite integral. The goal of the work is to introduce our studies about the calculus of the integrals in the transverse section domains, computer aided solutions and a generalizing method. The aim of our research is to create general computer based methods to execute the calculi in structural studies. Thus, we define a Boolean algebra which operates with 'simple' shape domains. This algebraic standpoint uses addition and subtraction, conditioned by the sign of every 'simple' shape (-1 for the shapes to be subtracted). By 'simple' shape or 'basic' shape we define either shapes for which there are direct calculus relations, or domains for which their frontiers are approximated by known functions and the according calculus is carried out using an algorithm. The 'basic' shapes are linked to the calculus of the most significant stresses in the section, refined aspect which needs special attention. Starting from this idea, in the libraries of 'basic' shapes, there were included rectangles, ellipses and domains whose frontiers are approximated by spline functions. The domain triangularization methods suggested that another 'basic' shape to be considered is the triangle. The subsequent phase was to deduce the exact relations for the calculus of the integrals associated to the transverse section problems. Thus we use a virtual rectangle which is framing the triangle, being generated supplementary right angled triangles. The sign of rectangle and the signs of the supplementary triangles are conditioned by the sign of the initial triangle. In this way, a generally located triangle for which we have direct calculus relations may be used to generate the discretization of any domain in transverse section associated integrals. A significant consequence of the paper is the opportunity to create modern computer aided engineering applications for structural studies, which use: intelligent applied mathematics background, modern informatics technologies and advanced computing techniques, such as calculus parallelization.

1. Introduction

Original models use analytic methods, as well as numerical solutions for the problems which have no exact mathematical solution, the computer being the instrument used to execute the calculi.

Content from this work may be used under the terms of the Creative Commons Attribution 3.0 licence. Any further distribution of this work must maintain attribution to the author(s) and the title of the work, journal citation and DOI.
Published under licence by IOP Publishing Ltd

There are several situations in which numerical solutions are 'embedded' in the specific technical solution, being applied the principles on which a certain numerical method is based, rather than a 'standard' method or a function stored into software libraries. Furthermore, computer based modelling integrates analytic and numeric studies with experiments, being created hybrid models, which may offer precise results to intricate phenomena.

The development of original computer based models is a permanent concern of the authors, along the time being created software instruments stored in software libraries, which were used in many research projects, [1, 2].

A paramount aspect at the initiation of a software project which is the background that supports the research models to be developed is the data management model to be conceived, useful to have a facile development of the components, [3]. Once this problem is resolved, the researchers may focus on the scientific problem to be solved.

Beside the data management, another vector in the sustainable development of the computer based instruments for scientific research projects is the creation of some specialized libraries of programs that solve a certain mathematical problem using several methods. In this way, the researcher is able to use the data management facilities, as well as the mathematical methods and the calculus techniques, to invent efficient instruments of numerical programming to be used in several fields of science.

2. Specific details regarding the calculus of the integrals

A mathematical problem which had to be repeatedly solved in different contexts of numerous projects using several specific methods and data processing techniques was the calculus of the integrals.

One of the projects which demanded the calculus of the integrals was regarded the calculus of the rotations and of the deflections of some distinct points along a structure, [4], using the strain energy methods. The structure was made of straight and curved intervals, every interval having its own geometrical properties and material constants. The general relation used to calculate the deflections, relation provided by the Mohr-Maxwell method is:

$$\begin{aligned}\delta = &\sum\int\frac{N\cdot n}{E\cdot A}ds+\sum\int k_Y\frac{T_Y\cdot t_Y}{G\cdot A_Y}ds+\sum\int k_Z\frac{T_Z\cdot t_Z}{G\cdot A_Z}ds+\\ &+\sum\int\frac{M_X\cdot m_X}{G\cdot I_P}ds+\sum\int\frac{M_Y\cdot m_Y}{E\cdot I_Y}ds+\sum\int\frac{M_Z\cdot m_Z}{E\cdot I_Z}ds\end{aligned} \tag{1}$$

In the integral, n, t_Y, t_Z, m_X, m_Y and m_Z are the functions of the internal forces and moments produced by the unit force or the unit moment, positioned in the section where the displacement or rotation must be computed.

In this case, we have a plane problem and relation (1) becomes:

$$\delta=\sum_{j=1}^{NJ}\left[\frac{1}{E\cdot A}\int(N\cdot n)\,ds\right]+\sum_{j=1}^{NJ}\left[k_Z\cdot\frac{1}{G\cdot A_Z}\int(T_Z\cdot t_Z)\,ds\right]+\sum_{j=1}^{NJ}\left[\frac{1}{E\cdot I_Y}\int(M_Y\cdot m_Y)\,ds\right] \tag{2}$$

where NJ is the total number of intervals of the system of straight and curved bars and j is the identifier of the current interval.

There was considered that the values of the deflections and of the rotations in every basic section of the structure, i.e. sections separating two intervals together with the sections in which there are located the external forces and moments, provide enough knowledge about the way how the structure is displaced.

Under these assumptions there was created a general model having in every basic section a vertical force, a horizontal force and a bending moment. Using these general loads there were deduced the general functions of the internal forces and moments. Substituting the real loads and considering the other ones to be zero, we obtain the functions of internal forces and moments, i.e. N , T_Z and M_Y . Next, we apply a single unit-force or unit-moment along the displacement or of the rotation to be computed, the other loads being zero. With this sets of loads we obtain the functions of the n , t_Z and m_Y internal forces and moment. The definition of the functions of the internal forces and moments was done algorithmically, therefore it resulted a flexible computation method, to be used in integrals shown in relation (2). These integrals had to be computed for both straight and curved intervals, so it was decided that a numerical approach is suitable. There was also conceived a method to minimize the number of calculi of the numerical computation of the integrals. A solver consisting of more than 6000 lines was developed for the automatic computations.

Another project in which a computer based method to compute integrals was employed, studied the behavior of a slender beam, [5]. According to the classic differential geometry, we have

$$\frac{1}{\rho_Z} = \frac{u_Z''}{\left[1 + \left(u_Z'\right)^2\right]^{\frac{3}{2}}} . \tag{3}$$

Because the small displacement assumption is used, u_Z' is small and $\left(u_Z'\right)^2$ is considered nil, so

$$\frac{1}{\rho_Z} = u_Z''(x). \tag{4}$$

But we have the relation

$$\frac{1}{\rho_Z} = \frac{M_Y(x)}{E \cdot I_Y}, \tag{5}$$

so, by respecting the general principle 'positive stresses are producing positive upper-level quantities (internal forces and moments, deflections aso.)', the differential equation of the neutral fiber is

$$\begin{cases} u_Z''(x) = -\dfrac{M_Y(x)}{E \cdot I_Y} \\ u_Y''(x) = \dfrac{M_Z(x)}{E \cdot I_Z} \end{cases} . \tag{6}$$

If we disregard the small displacements assumption, the general law of variation of the displacement is:

$$u_Z(x) = u_{Z\,ini} + \varphi_{Y\,ini} \cdot x + \int \left\{ \frac{-\int \dfrac{M_Y(x)}{E \cdot I_Y}}{\sqrt{1 - \left[-\int \dfrac{M_Y(x)}{E \cdot I_Y}\right]^2}} \right\} . \tag{7}$$

There was developed a program of more than 400 lines which computes the integrals. In this way we obtain the value of the displacement using the classic method, the displacement using the exact method, and the errors between these values. There was considered a cantilever having a rectangle transverse section. The vertical side of the rectangle was set to be a variable value, the beam being rigid for the first values and very flexible for the last ones. The results were analyzed to identify the extension of the small displacement hypothesis.

Another approach to calculate the integrals uses the Monte Carlo method, [6]. Thus, the random variables are extensively used in the computer based modeling of the economic events, as well as in electrical, mechanical and structural engineering. The level of randomness of some original pseudo-random numbers generators was evaluated using various criteria, i.e. the differences between the integration based on the Monte-Carlo method and the, so to say, classic computation methods.

3. Comparison between the methods employed to approximate the integrals

Automatic calculus of the geometrical properties was another project which demanded the calculus of the integrals.

There were conceived two methods to compute the integrals which define the geometrical properties.

First method, uses the general function

$$f(y,z) = y^u \cdot z^v, \tag{8}$$

where u, $v \in \mathrm{N}$ and $u + v \leq 2$. This function is included in the integral

$$I = \int_a^b \left(\int_{f_1(y)}^{f_2(y)} f(y,z)dz \right) dy, \tag{9}$$

the frontiers of the inner integral being modeled using spline functions.

This integral was resolved mathematically and the solution uses series. The same integral was solved using a numerical method.

This method precisely approximated the domain using spline functions and it was verified for various shapes of the transverse sections. Furthermore, the precision of the approximation of the curved frontiers of the domain was verified to select the optimum number of nodes. If the number of nodes is small, then the approximation may become unstable on some intervals. If we use too many nodes, the execution time of the according program will increase.

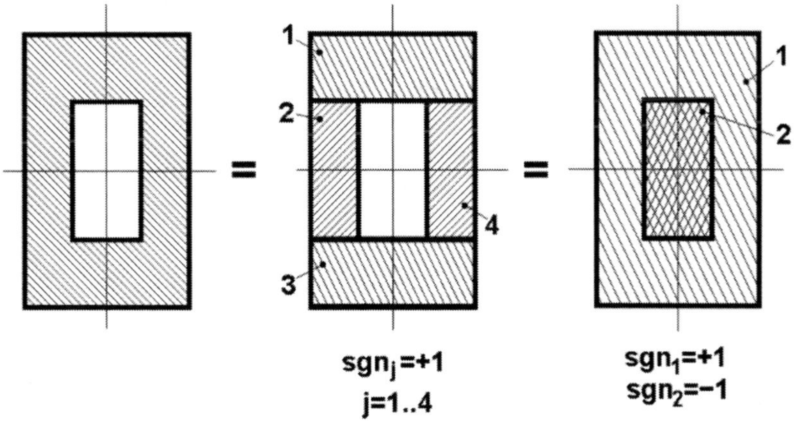

Figure 1. Division of a section in a set of hollow or solid 'simple' shapes.

The second method to compute the geometrical properties is to divide the surface of the transverse section in so-to-say 'basic' shapes, hollow or solid, [8]. We define the 'simple' shapes as shapes for which there are direct calculus relations of the geometrical properties. We assign $\mathrm{sgn}_j = -1$ if the j-th simple shape is hollow and $\mathrm{sgn}_j = +1$ if it is solid. In this way we create a Boolean algebra which operates using simple shapes. To conclude, the geometrical property of a section may be computed as the algebraic summation of the geometrical properties of the simple shapes. In reference [7] the libraries of simple shapes included only rectangles and ellipses.

4. Discussion

Based on the above mentioned methods, one can settle that the principle to divide the section in sub-domains may also include 'basic' shapes whose frontiers may be defined using spline functions.

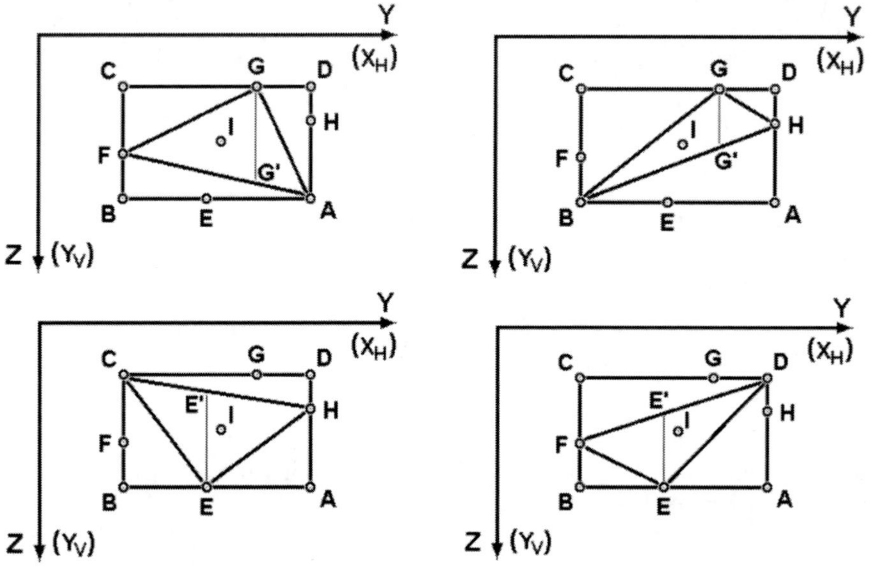

Figure 2. Calculus schemes for a generally located triangle.

Another simple shape which may be extensively used for the division of a transverse section is the triangle. Once the principal calculus relations are deduced, [8], there may be conceived a general algorithm which may be used for generally located triangle, figure 2. In this way, a triangle may be included in a rectangle. The surface between the rectangle and the initial triangle includes right angled triangles. For both, the rectangle and the right angled triangles, there are direct calculus relations. In this way it was conceived an algorithm which uses direct calculus relations for a general triangle. Furthermore, any transverse section may be divided in triangles. To conclude, the triangle is a general shape which may be included in the libraries of 'basic' shapes.

5. Conclusions

Calculus of the integrals in engineering has many particularities with respect to the specificity of the problem to be solved and the data processing technique to be used. Most of the solutions are based on approximations, the examples about the calculus of the geometrical properties being significant to identify methods to compute integrals, which may be based on: exact analytical methods, division of the domain in a set of simple shapes and direct calculus relations. If several discretizations of a domain are used, i.e. for a transverse section, there may be evaluated the precision of the solution for every division. A benefit of this approach is the general definition of the set of shapes of a domain, being possible to use advanced computing techniques, like calculus parallelization. Other strength is the aspect that the 'simple' shapes to be used for the calculus of the geometrical properties are associated to the set of nodes of the transverse section in which the significant stresses may be located. In this way, the choice about the 'simple' shapes in which the transverse section is divided may influence the time used to compute the results and their precision.

Acknowledgement
Concepts of computer aided instruments in engineering are originating in the research work of the MIEC2010 bilateral Ro-Md project, E. Oanta, C. Panait, L. Lepadatu, R. Tamas, M. Constantinescu, I. Odagescu, I. Tamas, G. Batrinca, C. Nistor, V. Marina, G. Iliadi, V. Sontea, V. Marina, V. Balan, V. (2010-2012), "Mathematical Models for Inter-Domain Research with Applications in Engineering and Economy", MIEC2010 - Bilateral Romania-Moldavia Scientific Research Project, under the guidance of the National Committee for Scientific Research, Romania, which is the continuing the studies of the ID1223 scientific research project: E. Oanta, C. Panait, B. Nicolescu, S. Dinu, A. Pescaru, A. Nita, G. Gavrila, (2007-2010), "Computer Aided Advanced Studies in Applied Elasticity from an Interdisciplinary Overview", under the guidance of the National University Research Council, Romania.

References
[1] Oanta E 1993 *Annual Scientific Session (Section: ICEs and Naval Equipment, Book of abstracts)* pp 33-34
[2] Oanta E and Comanita 1997 *International Congress of I.M.A.M. (IMAM '97)* (Istanbul: Turkey) 16.4.1-16.4.10
[3] Oanta E, Nicolescu B and Versatile A 1999 *ASME Design Engineering Technical Conference 1999 Proceedings of DETC99* (Las Vegas: Nevada)
[4] Oanta E and Nicolescu B 2001 *Proceedings of the XXV National Conference of Solid Mechanics–MECSOL* pp 103-110
[5] Oanta E and Nicolescu B 2003 *Constanta Maritime University Annals* **5** pp 53-58
[6] Oanta E *2007 Numerical Methods and Models in Economy (PhD Thesis)* p 246
[7] Oanta E 1998 *International Symposium on Marine Technologies and Management (TEHNONAV 1998)* **II** pp 159-166
[8] Oanta E, Constantinescu E, Lazaroiu G, Axinte T and Dascalescu A E *2013 Constanta Maritime University Annals* **20** pp 91-96

Extended precision data types for the development of the original computer aided engineering applications

A Pescaru[1], E Oanta[2], T Axinte[2] and A-D Dascalescu[3]
[1]Constanta Maritime University, Faculty of Navigation and Naval Transport, 104 Mircea cel Batran Street, 900663, Constanta, Romania
[2]Constanta Maritime University, Faculty of Naval Electro-Mechanics, 104 Mircea cel Batran Street, 900663, Constanta, Romania
[3] Politehnica' University of Bucharest, Faculty of Power Engineering, 313 Splaiul Independentei, Sector 6, 060042, Bucharest, Romania

E-mail: eoanta@yahoo.com

Abstract. Computer aided engineering is based on models of the phenomena which are expressed as algorithms. The implementations of the algorithms are usually software applications which are processing a large volume of numerical data, regardless the size of the input data. In this way, the finite element method applications used to have an input data generator which was creating the entire volume of geometrical data, starting from the initial geometrical information and the parameters stored in the input data file. Moreover, there were several data processing stages, such as: renumbering of the nodes meant to minimize the size of the band length of the system of equations to be solved, computation of the equivalent nodal forces, computation of the element stiffness matrix, assemblation of system of equations, solving the system of equations, computation of the secondary variables. The modern software application use pre-processing and post-processing programs to easily handle the information. Beside this example, CAE applications use various stages of complex computation, being very interesting the accuracy of the final results. Along time, the development of CAE applications was a constant concern of the authors and the accuracy of the results was a very important target. The paper presents the various computing techniques which were imagined and implemented in the resulting applications: finite element method programs, finite difference element method programs, applied general numerical methods applications, data generators, graphical applications, experimental data reduction programs. In this context, the use of the extended precision data types was one of the solutions, the limitations being imposed by the size of the memory which may be allocated. To avoid the memory-related problems the data was stored in files. To minimize the execution time, part of the file was accessed using the dynamic memory allocation facilities. One of the most important consequences of the paper is the design of a library which includes the optimized solutions previously tested, that may be used for the easily development of original CAE cross-platform applications. Last but not least, beside the generality of the data type solutions, there is targeted the development of a software library which may be used for the easily development of node-based CAE applications, each node having several known or unknown parameters, the system of equations being automatically generated and solved.

Content from this work may be used under the terms of the Creative Commons Attribution 3.0 licence. Any further distribution of this work must maintain attribution to the author(s) and the title of the work, journal citation and DOI.
Published under licence by IOP Publishing Ltd

1. Introduction

The definition of the most common data types as user defined distinct types is useful for the definition of the upper level sophisticated data types, as well as for the readily customization of a software library which may be included in a new application and that has specific requirements. Moreover, if all the programs use such definition it is easier to include these components in an upper level software, i.e. to integrate the components of a hybrid engineering model. The flexible definition of the basic data types is a long run concern of the authors and there are several aspects to be considered when a given type is conceived.

2. Criteria regarding the definition of flexible data types

When a major software project was started years ago, a basic concern was about the limited capacity of the memory where large arrays had to be stored. In order to solve this problem there was decided that the best choice was to store the data on the hard disk, as a random access file specific to a given programming language. Because various programming languages offered different data processing facilities, it was developed an interface based on constant structure text files. Moreover, various original software applications developed in different programming languages could be connected using this new interface, being possible to create multi-language applications. Improving this data processing method was a constant concern of the authors, [1, 2, 3].

```
TYPE
     KOD_TIP_LISTA   = 0..2;
     MATRIX_VALUE    = Double;
     INDEX_TYPE      = Integer;      { Byte/Word }
     P_REC_4         = ^INREG4;
     INREG4          = Record;
          Previous   : P_REC_4;
          X          : MATRIX_VALUE;
          CASE TIP_LISTA : KOD_TIP_LISTA of
               1 : (     Index          : INDEX_TYPE;
                         Total_Index    : INDEX_TYPE
                   )
          Next       : P_REC_4;
     End;
     VECTOR_LISTE    =    Array [1..MaxBlock] of P_REC_4;
```

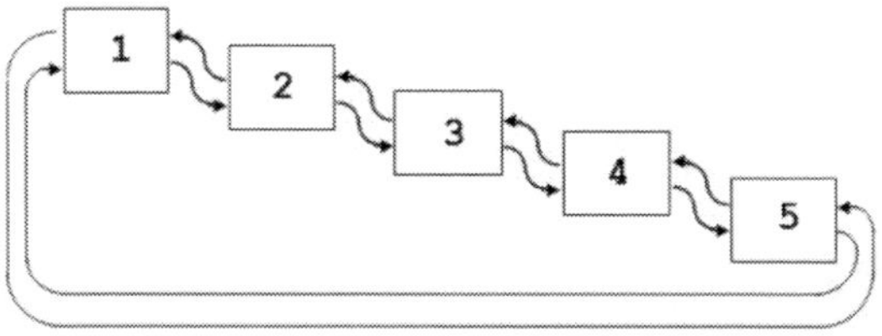

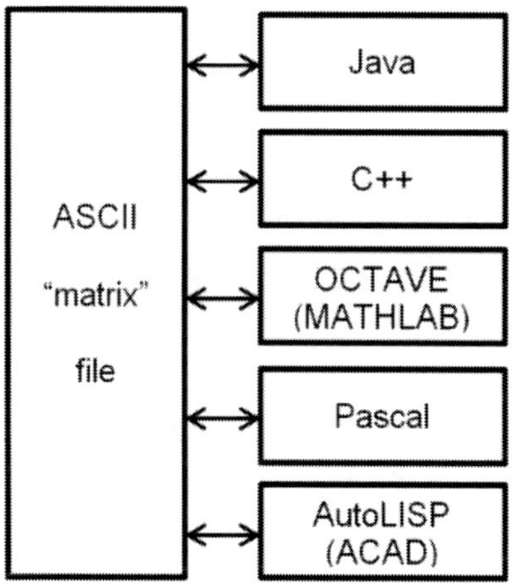

Figure 1. Data types, circular double connected list and interface between the programming languages.

An important objective was to decrease the access time, i.e. the data processing time. Several solutions were conceived and implemented: optimization of the memory use by defining a vector of double connected circular lists, creation of a RAM disk where the most frequently used files are stored, [4].

Another direction was to develop some libraries in order to offer a facile access to all the operations with the data types previously defined. Using this platform there were rapidly developed turnkey applications in various fields, [5], most of them solving numerical programming problems: analytic models in applied elasticity, physics, electro-technics; numerical methods; FEM and FDM applications; FEM input data generators; FEM output data processors; experimental data processors; visualization software instruments; hybrid models.

An important issue regarding the computer based models and in numerical programming is the accuracy of the results, [6]. There are several methods to improve the accuracy of the results, being very important to have a precise data type of the variables used to perform the calculi. For the example in figure 1, the 'real' type is designated as 'MATRIX_VALUE', while the 'integer' type used as an index is 'INDEX_TYPE'. If the real variables of the application to be developed are in a common range of values, there may be selected a simple precision type by modifying the 'MATRIX_VALUE' type. In this case it is possible to load a larger amount of data in the memory and the application to be faster. If accuracy is important, the 'MATRIX_VALUE' type may be defined to become an extended precision real type. All the upper level libraries and software applications are using this type, in this way being possible to modify the entire software application to be developed by performing minimum changes of the code.

However, there are special cases when a special approach is needed.

3. Discussions

There are certain cases when the numerical calculus results may become non-predictable.

Let us consider the calculus of the coefficients α and α', [7], (3), figure 1. The general forms of these coefficients are:

$$\alpha = -\chi \cdot \frac{\pi^2}{8} \cdot \frac{1}{\sum_{n=1,3,5,\dots}^{\infty} \left\{ \frac{(-1)^{\frac{n-1}{2}}}{n^2} \cdot \left[\frac{sh\left(\frac{\pi \cdot n \cdot z}{b}\right)}{ch\left(\frac{\pi \cdot n \cdot z}{2 \cdot b}\right)} \right] \cdot \cos\left(\frac{\pi \cdot n \cdot y}{b}\right) \right\}}, \quad (1)$$

$$\alpha' = -\chi \cdot \frac{\pi^2}{8} \cdot \frac{1}{\sum_{n=1,3,5,\dots}^{\infty} \left\{ \frac{(-1)^{\frac{n-1}{2}}}{n^2} \cdot \left[1 - \frac{ch\left(\frac{\pi \cdot n \cdot z}{b}\right)}{ch\left(\frac{\pi \cdot n \cdot z}{2 \cdot b}\right)} \right] \cdot \sin\left(\frac{\pi \cdot n \cdot y}{b}\right) \right\}}, \quad (2)$$

where

$$\chi = \frac{1}{3} \cdot \left\{ 1 - \frac{192}{\pi^5} \cdot \frac{b}{h} \cdot \sum_{n=1,3,5,\dots}^{\infty} \left[\frac{1}{n^5} \cdot th\left(\frac{n \cdot \pi \cdot b}{2 \cdot b}\right) \right] \right\} \quad (3)$$

and

$$\begin{cases} \tau_{ZX} = \dfrac{M_X}{\alpha \cdot h \cdot b^2} \\ \tau_{YX} = \dfrac{M_X}{\alpha' \cdot h \cdot b^2} \end{cases}. \quad (4)$$

The number of terms in the above sums must be carefully chosen because there may be generated large numbers which can lead to large round-off errors.

Same problem may be found in the calculus of the integral expressed as a summation, [8], (8):

$$I = v! \sum_{\substack{p_1,\dots,p_4=0 \\ p_1+\dots+p_4=v+1}}^{v+1} \frac{(b_3)^{p_1}\cdot(b_2)^{p_2}\cdot(b_1)^{p_3}\cdot(b_0)^{p_4}}{p_1!p_2!p_3!p_4!}\cdot\frac{b^{u+3p_1+2p_2+p_3+1}-a^{u+3p_1+2p_2+p_3+1}}{u+3p_1+2p_2+p_3+1} - \\ - v! \sum_{\substack{p_1,\dots,p_4=0 \\ p_1+\dots+p_4=v+1}}^{v+1} \frac{(a_3)^{p_1}\cdot(a_2)^{p_2}\cdot(a_1)^{p_3}\cdot(a_0)^{p_4}}{p_1!p_2!p_3!p_4!}\cdot\frac{b^{u+3p_1+2p_2+p_3+1}-a^{u+3p_1+2p_2+p_3+1}}{u+3p_1+2p_2+p_3+1} \tag{5}$$

In this way, there are extreme cases when, for some shapes of the calculus domain, the previous integral had different results if it was computed using a numerical method or if it was computed using (5). Moreover, in these extreme cases the same application used to compute the results of relation (5) offered different values on different computers or using different compilers.

It results that the calculus of the series requires a special attention in order to have an appropriate accuracy of the results. A solution is to design a user defined type having an extended precision. Moreover, this solution should be flexible in order to have a customized precision with respect to the problem to be solved.

From among many libraries that could solve extended precision and accuracy problems, two are to be noted for their versatility and support from the vast community that contribute to their development:

1. GNU MP - The Multiple Precision Arithmetic Library
2. GNU MPFR - The Multiple Precision Floating-Point Reliable Library

GNU MP, also designated as GMP, is a flexible library developed in C. It offers customizable precision for the basic data types: integer type and floating-point types. The main strength of this solution consists in the highest speed arithmetic for the applications which require more precision than the standard C types.

Many applications require a precision of some hundreds of bits. However, there are situations when thousands or millions of bits are required. This arbitrary precision solution offers high performance in both cases. In this way, the size of the operands leads to the selection of a given algorithm and meanwhile by minimizing the overhead.

Some of the most important solutions to maximize the speed are: defining basic arithmetic data types as full words, creating sophisticated algorithms, using high speed assembly code in the most often used loops and defining the speed as the most important target, consequently less attention being dedicated to the minimization of the complexity or to obtain a higher degree of refinement.

The library contains various classes of high-level types arithmetic functions:

- 'mpz' functions operate with high-level signed integers (≈150 arithmetic and logic functions).
- 'mpq' functions operate with high-level rational, being included almost 35 functions. The 'mpz' functions may be also utilized if they are separately used for the numerator, respectively for the denominator.
- 'mpf' functions operate with high-level floating-point (≈ 35 functions). These functions are useful if the 'double' type is not accurate enough.
- Classes of C++ functions which are interfaces to the previously mentioned functions.

Up-to-date information on GMP may be found in reference [9]. GMP is used in several computer algebra systems and also in 'Computational Geometry Algorithms Library', [10], where the use of the simple using ordinary floating point CPU math is not accurate enough and it even may lead to the so called 'explosion' of the geometry algorithms.

On the other hand, the main characteristics of MPFR, [11, 12], which makes the difference with respect to most customizable precision floating-point solutions, are:

- The results of the codes which use MPFR do not depend on the size of the word for a given machine (the mp_bits_per_limb is 64 on many processors), this means portability.

- High degree of customization of the precision, being possible to set the number of bits individually for each variable.
- Apart from the IEEE 754-1985 standard, the library includes four modes for the round-off operations for both simple calculi and for mathematical functions.

There may be noticed that the precision of the double-precision floating-point variables may be reached in MPFR using a 53 bits precision for the basic arithmetic operations and the square root.

The basic idea of the MPFR approach is that the operation is computed using the so called 'infinite accuracy' and, using the previously selected rounding mode, the result is rounded to the precision defined for the destination variable. The functions for floating-point operations are smoothly extending the IEEE 754 arithmetic. In this way the same result may be obtained on any computer, without any dependence on the size of the word on the current machine, the compiler type of the operating system.

Moreover, if two variables of lower precision are multiplied and the result is a variable having a large precision, the calculus being done using the previously mentioned 'infinite accuracy', it results a full precision number. This is an example regarding the fact that MPFR does not verify the accuracy during the computations. Interesting ideas regarding the loss of precision reasons and the use of the MPFR library to increase the accuracy are given in reference [13].

4. Conclusions

Using extended precision data representation was an important concern of the worldwide researchers. Using higher internal precision, there were created simpler, faster and more accurate algorithms for linear algebra problems [14, 15]. At present new ideas of research regarding the enhancement of the precision are followed, being created new algorithms for normalizing, dividing and square root calculus of the floating point expansions, [16].

In computer based engineering modelling there are certain cases, for instance the calculus of the series, when the necessary precision may be acquired only if extended precision types are used. Thus, in series (1), (2) and (5) there are subtracted two large numbers which might be nearly equal, this operation being known as catastrophic cancellation. Thus, the calculation may become dominated by errors. This loss of significance may be prevented using various mathematical particular methods or using scale factors. An IT general solution is to use arbitrary precision basic types which have a precision higher than the basic types offered by a commonly used programming language. So far, there were studied two solutions. Thus, GMP and MPFR offer important facilities regarding the use of the arbitrary precision types: flexible solutions, possibility to develop cross-platform applications, GNU license. A software project based on extended precision data types is under development. Special attention will be dedicated to the definition of some flexible data types and to the development of the according functions, all these being included in the basic libraries. After a thorough testing of the new basic types, the next stage will be the development of dedicated libraries which solve problems regarding: numerical methods and geometry data processing. The topmost level is dedicated to the development of turnkey applications for the computer based models.

Acknowledgement

Concepts Ideas regarding the computer based instruments in applied elasticity are the result of the models developed in the framework of the MIEC2010 bilateral Ro-Md research project, E. Oanta, C. Panait, L. Lepadatu, R. Tamas, M. Constantinescu, I. Odagescu, I. Tamas, G. Batrinca, C. Nistor, V. Marina, G. Iliadi, V. Sontea, V. Marina, V. Balan, V. (2010-2012), "Mathematical Models for Inter-Domain Approaches with Applications in Engineering and Economy", MIEC2010 - Bilateral Romania-Moldavia Scientific Research Project, under the supervision of the National Authority for Scientific Research (ANCS), Romania, that is the follow-up of the ID1223 scientific research project: E. Oanta, C. Panait, B. Nicolescu, S. Dinu, A. Pescaru, A. Nita, G. Gavrila, (2007-2010), "Computer Aided Advanced Studies in Applied Elasticity from an Interdisciplinary Perspective", under the supervision of the National University Research Council (CNCSIS), Romania.

References

[1] Kumbetlian G, Bocanete P, Oanta E, Fit M and Magardician N Matrices Described as Files. Pascal library. Interfaces with GWBasic, BasicCAD, C, AutoLISP languages. Demo Programs. Dedicated editor *The 2-nd National Conference on boundary and finite element* (University of Sibiu) **13/05/93-15/05/93** pp 37-43

[2] Oanta E 'Matrix' Variable Type Improvements and Applications, *The 3-rd National Conference on boundary and finite (*Merchant Marine Institute: Constanta) **25/05/95-27/05/95** pp 44-49

[3] Oanta E, Nicolescu B and Versatile A 1999 *ASME Design Engineering Technical Conference 1999 Proceedings of DETC99* (Las Vegas: Nevada)

[4] Oanta E 1993 Methods to optimize the automatic solving of the numerical calculus problems *Annual Scientific Session ICEs and Naval Equipment* pp 33-34

[5] Oanta E 1998 Advanced software for the data processing in the computational mechanics domain *International Symposium on Marine Technologies and Management* (Ovidius University of Constanta) **II** pp 151-158

[6] Oanta E 2012 Approximations in structural analytical studies *Constanta Maritime University Annals* **18** pp 129-134

[7] Oanta E, Panait C, Lazaroiu G and Dascalescu A E 2014 Computer aided instrument to be used as an automatic design component *Advanced Materials Research* **1036** pp 1017-1022

[8] Oanta E, Constantinescu E, Raicu A and Axinte T 2014 Analytic General Solution Employed to Calculate the Geometrical Characteristics in Structural Problems *Advanced Materials Research* **1036** pp 697-702

[9] https://gmplib.org/, Accessed: 05/01/2015

[10] http://www.cgal.org/, Accessed: 05/01/2015

[11] http://www.mpfr.org/, Accessed: 05/01/2015

[12] Fousse L, Hanrot G, Lefèvre V, Pélissier P and Zimmermann P 2007 A multiple-precision binary floating-point library with correct rounding *ACM Transactions on Mathematical Software* **33(2)**

[13] Ghazi K R, Lefèvre V, Théveny P and Zimmermann P 2010 Why and how to use arbitrary precision *Computing in Science & Engineering* **12(3)** p 5

[14] Li X, Demmel J, Bailey D, Henry G, Hida Y, Iskandar J, Kahan W, Kang S, Kapur A, Martin M, Thompson B, Tung T and Yoo D 2002 Design, Implementation and testing of Extended and Mixed Precision BLAS *ACM Transactions on Mathematical Software* **28**(2) pp 152-205

[15] Krougly Z L and Jeffrey D J Implementation and application of extended precision in Matlab *Mathematical Methods and applied Computing*, pp 103-108

[16] Joldes M, Marty O, Muller J and Popescu V 2015 Arithmetic algorithms for extended precision using floating-point expansions *LIP - ENS Lyon, LAAS-CNRS, ENS Cachan* (Research Report)

Innovation in engineering education through computer assisted learning and virtual university model

A Raicu[1] **and G Raicu**[2]
[1]Constanta Maritime University, Faculty of Naval Electro-Mechanics, 104 Mircea cel Batran Street, 900663, Constanta, Romania
[2]Constanta Maritime University, Faculty of Navigation and Naval Transport, 104 Mircea cel Batran Street, 900663, Constanta, Romania

E-mail: alexandra.raicu@cmu-edu.eu

Abstract. The paper presents the most important aspects of innovation in Engineering Education using Computer Assisted Learning. The authors propose to increase the quality of Engineering Education programs of study at European standards. The use of computer assisted learning methodologies in all studies is becoming an important resource in Higher Education. We intend to improve the concept of e-Learning using virtual terminals, online support and assisting special training through live seminars and interactive labs to develop a virtual university model. We intend to encourage computer assisted learning and innovation as sources of competitive advantage, to permit vision and learning analysis, identifies new sources of technology and ideas. Our work is based on our university datasets collected during last fifteen years using several e-Learning systems. In Constanta Maritime University (CMU), using e-Learning and Knowledge Management Services (KMS) is very important and we apply it effectively to achieve strategic objectives, such as collaboration, sharing and good practice. We have experience in this field since 2000 year using Moodle as KMS in our university. The term KMS can be associated to Open Source Software, Open Standards, Open Protocols and Open Knowledge licenses, initiatives and policies. In CMU Virtual Campus we have today over 12500 active users. Another experience of the authors is the implementation of MariTrainer Wiki educational platform based on Dokeos and DekiWiki under MARICOMP and MEP Leonardo da Vinci Project. We'll also present in this paper a case study under EU funded project POSDRU, where the authors implemented other educational platform in Technological High Schools from Romania used over 1000 teachers. Based on large datasets the study tries to improve the concept of e-Learning teaching using the revolutionary technologies. The new concept present in this paper is that the teaching and learning will be interactive and live. The new and modern techniques are the flexible learning courses, the production of learning demonstrators and testing. All the information from the virtual educational platform remain open space, communication between participants and continued after graduation, so we can talk about creating and maintaining a community of graduates, a partnership with them. Every European University must have a department which aims to provide computer assisted learning using knowledge creation through learning, capture and explication, sharing and collaborative communication, access, use and reuse and knowledge archiving.

Content from this work may be used under the terms of the Creative Commons Attribution 3.0 licence. Any further distribution of this work must maintain attribution to the author(s) and the title of the work, journal citation and DOI.
Published under licence by IOP Publishing Ltd

1. Introduction

Nowadays is particularly important to identify correctly all of modern teaching technologies used for development in university education. Learning concept has evolved over time leading to advanced forms currently based on knowledge engineering and artificial intelligence technologies.

Evolution of the communication technologies and especially the Internet technologies has initiated the concept of Virtual University development. A number of other advanced resources of collaborative information type (e.g. wiki concepts) may be useful as developing tools to platforms and software solutions to increase the quality of education using computerized assistive technologies. The new approach of education system is in a continuous learning process and changing globally and rapidly now, [1]. A Learning Management System (LMS) means a collection of dedicated software tools designed to manage user learning activity. Also, the term of Knowledge Management Services (KMS) can be associated to a plenty of meanings in Open Domain (e.g. Software, Standards, Protocols, Knowledge licenses, initiatives and policies).

Over the past fifteen years Constanta Maritime University initiated a series of experiments based on educational information technologies which have enabled new advanced statistics that can be important elements for infrastructure design in a virtual university concept.

Collaborations have been achieved in the framework of EU funded projects with partners from several maritime universities in countries such as Netherland, Germany, Norway, Ireland, Denmark, Finland, Estonia, Greece, Italy and Spain. We are jointly developed a series of specialized courses in maritime field designed from the start to use e-learning platform based on open source technologies such as Moodle, ATutor, Dokeos, ILIAS etc.

Use of the modern techniques as e-learning education, combined with classical teaching methods, represents the optimum solution for better and standing quality learning system inside of the lifelong learning concept.

2. Computer assisted learning in Constanta Maritime University

In Constanta Maritime University, applying computer assisted learning is important and we apply it effectively to achieve important and global objectives. We encourage learning and innovation in engineering education as sources of competitive concern, to allow vision and analysis, identifies new sources of technology and ideas. Using computer assisted learning in our university we explore creativity, communication between participants of the learning system and innovation, [2, 3].

You can see the practical Constanta Maritime University Campus approach, link: http://campus.e-shipping.ro. As Knowledge Management System in Constanta Maritime University we use Moodle as Open Source Software. We have today over 12500 active users in CMU Virtual Campus. In the meantime, we developed a new dedicated virtual education space, an upgrade one, link: http://campus.cmu-edu.eu/ using also Moodle. In the figure 1 is presented the main interface of the Constanta Maritime University Virtual Campus.

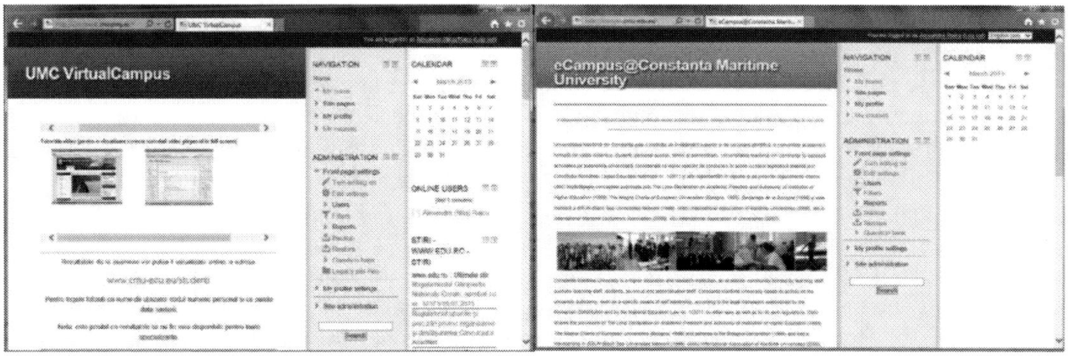

Figure 1. Constanta Maritime University Virtual Campus.

Using Moodle for our Virtual Campus we achieve strategic objectives as collaboration, encourage learning, sharing, creativity and innovation, through a free web application that teachers and instructors can use it for online learning. Moodle is a Course Management System (CMS), as well known as a Learning Management System (LMS) or a Virtual Learning Environment (VLE).

In Constanta Maritime University has introduced the concept of e-learning since 2000 years and from 2004 we made important investments in IT tool. An important approach that can make major contributions in increasing the overall quality of general education is based on the continuous process of improvement of teachers in high school using long life learning methods. In this regard were carried out a series of projects aimed to develop tools to increase the capacities for teaching of technical subjects using modern learning systems, computer aided type. Using classical capabilities of e-learning technologies in conjunction with a plenty of e-activities subsystems creates the premise of achieving a complete training using the most advanced learning techniques, [4].

The first e-learning EU project developed in CMU in 2007 achieved a Web based IMO Tanker Courses for distant simulation and tutorial systems on board. The developed Campus had the main goal to coordinate the development of a maritime flexible learning system.

Course management system ILIAS as knowledge management services was developed to support and enhance intensive knowledge processes, tasks or projects, creation, identification, organization, transfer, visualization, capturing, maintenance, distribution, evolution. You can see the practical approach using the link: http://training.e-shipping.ro, figure 2.

Another EU project that we developed in CMU was MariComp. The partner groups in this project consist of six MET institutions from different countries: Denmark, The Netherlands, Estonia, Ireland, Romania and Norway. The main objective of the project was to develop a European virtual learning space for maritime educational institutions and to create and evaluate a pilot e-learning course for maritime lecturers. In the figure 3, you can see the MariTrainer Wiki based on Dokeos and DekiWiki under MARICOMP Leonardo da Vinci Project, link: http://maricomp-web.eu.

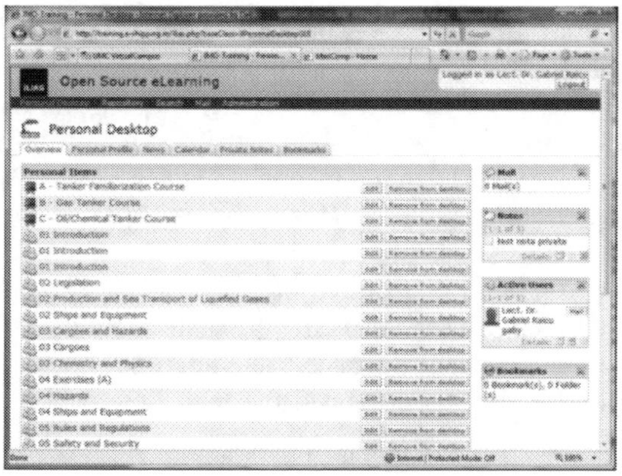

Figure 2. Web based IMO Tanker Courses.

Figure 3. MariTrainer Wiki.

The course developed under the project was a virtual learning space for maritime lecturers. The main objective of this platform was that European maritime lecturers with expert knowledge have better opportunities of developing their competencies and creating an international network of colleagues within the same maritime discipline. For the platform where lecturers inside and outside the partnership can seek and share knowledge was CMS Dokeos. The MariTrainer Wiki allowed us to gain access to a greater range of teaching and learning materials and methods, made us more prepared to enhance our skills, helped us to encourage a greater European dimension in the work of our organizations and met our continuing professional development needs.

Other EU project that we developed in Constanta Maritime University was Maritime Education Platform, you can see the link: http://mep.stc-r.nl. The main objective of this EU project was to improve the level of maritime education for lecturers in the EU, by producing and sharing the high quality educational material. The partners developed a virtual learning space called MEP, where they organised seminars, had a great cooperation between maritime educational institutions. That virtual dedicated space, MEP was an innovative virtual learning space filled with high quality e-learning courses, [5].

In 2011 we developed another large and important project under CMU using EU funds, with a target group of 1,000 teachers who are mainly focused on technological disciplines in training and development activities for the improvement of curriculum and professional development capabilities over several months. They address the complex issues being used in the LCMS software solutions, ERP/CRM and Knowledge Management Systems. The impact of information was high, being adapted to the needs of multidisciplinary training. Trends to use mobile terminals to change main aspects presentation of information in the educational activities generated virtualization premises of the educational process on the entire educational path, link: http://ciatm.forcon.cmu-edu.eu/moodle/login/index.php. You can see in the figure 4 screenshots of project portal.

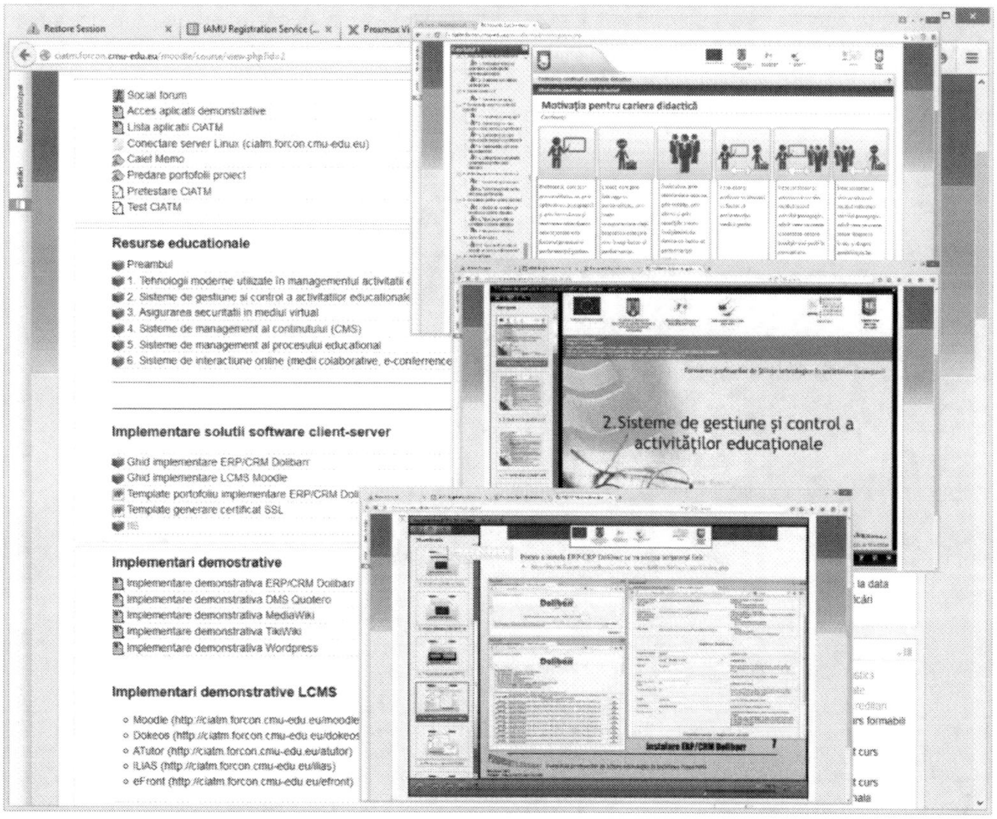

Figure 4. A new assisted e-learning approach: integrating LCMS, ERP/CRM & KMS.

3. Virtual University Model

During last decades there have been numerous attempts to develop some working models for adapting university education activities to new trends like e-learning and ICT technologies.

For a while it was possible to operate with mixed traditional learning models and e-learning methods. The emergence of new educational technologies and large scale adoption of mobile intelligent devices with high autonomy, permanently connected to the Internet, require reconsideration of the entire educational process, [6, 7].

Coexistence between traditional ways of teaching and learning and modern generated educational capabilities complement each other. Although any of the systems do not exclude other one, there was always one or the other pleadings regarding one of them. While narrowing is observed as the separation of the two ways of doing education, as technology becomes more mature - providing a high degree of naturalness in presenting information. Obviously, the difference between the two approaches is reduced by widespread acceptance of these mobile terminals in all sectors of social life.

The validity of the concept of true Virtual University has to rely on natural principles of evolution in the current ICT environment as education and classical techniques have made it in the natural environment before the appearance of information technologies.

The basic principle of Virtual University must derive from "naturalness" of interactions in virtual environment: advanced technologies for presentation and cognition, unrestricted access to knowledge, integrating social media, reputation and feedback. In almost every aspect the common principle preserve the concepts of classical education adding some specific elements of the virtual environment: responsiveness, widespread interaction, almost instantly feedback.

In a Virtual University all the materials, courses, information have to be open space, collaboration between participants and continued after graduation, to preserve a community of graduates, a partnership with them. Every European University must have a department to implement computer assisted learning using knowledge creation through learning, dissemination, preservation, capture and explication, sharing and collaborative communication, access, use and reuse and knowledge archiving, as you can see in the figure 5.

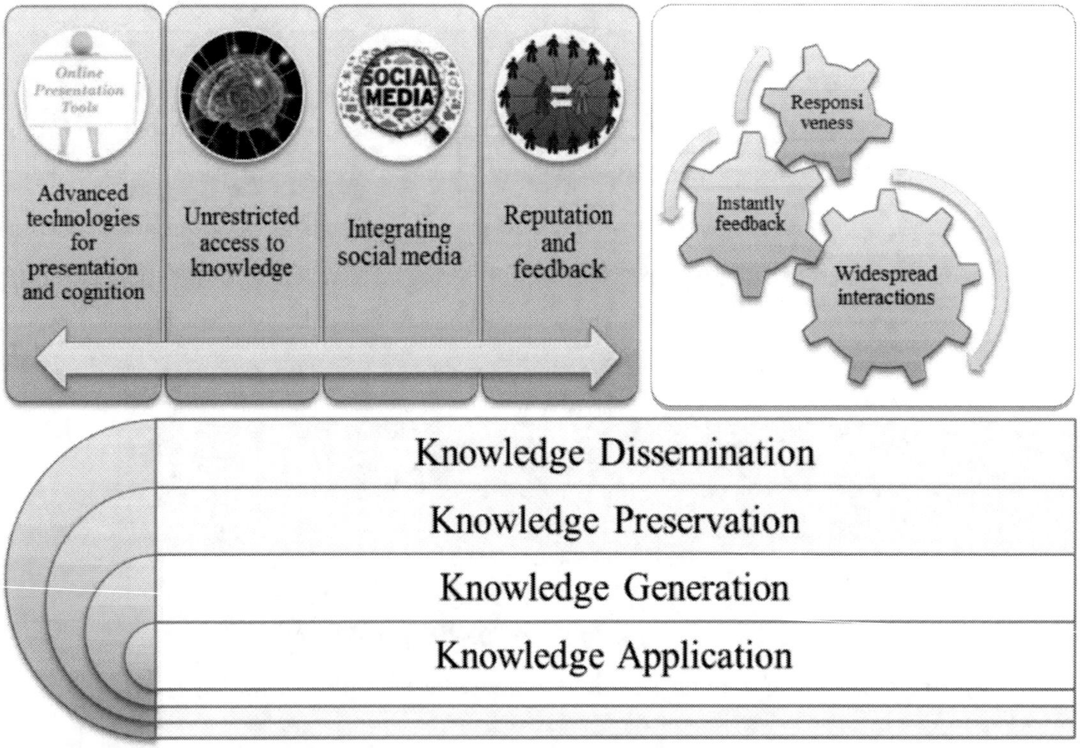

Figure 5. Virtual University abstraction layers.

4. Conclusions

This paper and discussions are based on our university datasets collected during last fifteen years using several e-learning systems. The quality of education will be improve day by day by various number of computer assisted learning software and tools.

Adapting of the educational process to the demands of virtual universities will raise discussions of specialists in education, which will help to define new advanced methods of education.

It is obvious that the future will impose major changes in the educational system beyond the current imaginary limits and the degree of integration of new technologies transforming the methods since become classics, in educational systems adapted to the needs of the age. The next evolutionary step in virtual education research is to offer a technical framework to negotiate between old fashion and new learning technologies in such a manner to obtain the best results of the both worlds.

Acknowledgement
Concepts Ideas regarding the computer assisted learning are the result of developed in the framework of the projects:

"Continuous training and development activities of the teachers from the technological disciplines in the knowledge society", POSDRU87/1.3/S/62647, partners ASE Bucharest, SIVECO S.A. and AltFactor S.R.L. Project coordination was by CMU, Raicu G., Panait C., Batranca G., Ciucur V. The period of the project was since 2010 until 2013.

„Maritime Education Platform – MEP", Raicu G. ,Barsan E., Niţă (Raicu) A., Hanzu R., Bejan R., Leonardo da Vinci project, Project coordination was by STC-Group Rotterdam, NL/08/LLP-LdV/TOI/12, – 2008-1-NL1-LEO05-00477, 2008 – 2010.

„Maricomp - Developing the Competencies of Maritime Lecturers", Raicu G., Niţă (Raicu) A., Acomi N., Arsenie P. Leonardo da Vinci project, Project coordination was by Svendborg International Maritime Academy – SIMAC, Contract No. DK/06/B/F/PP – 145 617, period 2006 – 2008.

References

[1] Raicu G and Nita (Raicu) A 2009 Evolutionary designs in e-Learning Virtual Engineering and Business Education, *5th Balkan Region Conference on Engineering and Business Education* (Lucian Blaga University Sibiu Romania) **I** and **II** pp 165-168

[2] Nita (Raicu) A and Oanta E 2014 Modern education facilities for CAD/CAM/CAE training of the future maritime engineers *Advanced Material Research* **837** pp 769-774

[3] Oanta E, Panait C, Lazaroiu G and Dascalescu A E 2014 Computer aided instrument to be used as an automatic design component *Advanced Materials Research* **1036** pp 1017-1022

[4] Nita (Raicu) A and Raicu G 2010 Increasing the CAD teaching quality through E-Learning *6th International Seminar on the Quality Management in Higher Education Location* (Tulcea: Romania) **2** pp 575-578

[5] Raicu A, Raicu G and Lamba D 2014 Improved the maritime online teaching by using the knowledge management concept *Constanta Maritime University Annals* **19** pp 65-68

[6] Raicu A and Raicu G 2010 Knowledge and quality management in universities *Ovidius University Annals* pp 291-294

[7] Raicu G and Nita (Raicu) A 2009 Impact of knowledge management in maritime education training *The International Scientific Session Challenges of the Knowledge Society* (Nicolae Titulescu Bucharest University) pp 1712-1732

IOP Conf. Series: Materials Science and Engineering **95** (2015) 012127 doi:10.1088/1757-899X/95/1/012127

The 5S lean method as a tool of industrial management performances

F C Filip[1] **and V Marascu-Klein**[1]

[1] "Transilvania" University of Brasov, Department of Engineering and Industrial Management, Mihai Viteazul Street, No. 5, 500174, Brasov, Romania

E-mail: florina-cristina.filip@unitbv.ro

Abstract. Implementing the 5S (seiri, seiton, seiso, seiketsu, and shitsuke) method is carried out through a significant study whose purpose to analyse and deployment the management performance in order to emphasize the problems and working mistakes, reducing waste (stationary and waiting times), flow transparency, storage areas by properly marking and labelling, establishing standards work (everyone knows exactly where are the necessary things), safety and ergonomic working places (the health of all employees). The study describes the impact of the 5S lean method implemented to storing, cleaning, developing and sustaining a production working place from an industrial company. In order to check and sustain the 5S process, it is needed to use an internal audit, called "5S audit". Implementing the 5S methodology requires organization and safety of the working process, properly marking and labelling of the working place, and audits to establish the work in progress and to maintain the improved activities.

1. Introduction

By permanent changing and improving of the current development conditions of the industrial companies, the implementation of framework management methods it is required in order to measure the performance of industrial processes [1]. One of the essential steps to set in motion and bring about a flourishing lean culture, as well as one of the most widely adopted method from the lean manufacturing toolbox, the 5S method is considered a basic lean concept, as it establishes the operational stability required for making and sustaining continuous improvements [2, 3, 4]. The 5S lean is a method to improve efficiencies and enhance the management performance [5]. The 5S is the methodology of creating and maintaining well organized, clean, high effective and high quality workplace. Its result is the effective organization of the workplace, reduction of work environment, elimination of losses connected with failures and breaks, improvement of the quality and safety of work [6, 7, 8, 9]. In Japanese, 5S is the short form of five words which represent the concept of good maintenance [10]. The 5S has the following significance [11, 12, 13, 14]:

• Sort (Seiri) – decide and separate or remove the unnecessary things by the necessary tools;

• Set in order (Seiton) – putting in an logical order each item most often used in a properly place and storage out those in a well-established location;

• Shine (Seison) – perform and maintaining a deep cleaning, eliminate sources of dirt and simplify cleaning process, quality can be obtained only in clean working environments;

• Standardize (Seiketsu) – establishing rules and storage areas by making simple visual rules along with training of maintain standards to reduce search time and avoid mistakes;

Content from this work may be used under the terms of the Creative Commons Attribution 3.0 licence. Any further distribution of this work must maintain attribution to the author(s) and the title of the work, journal citation and DOI.
Published under licence by IOP Publishing Ltd

• Sustain (Shitsuke) – all activities should be monitored, evaluated and continuously improved in order to respect and maintain the results achieved through the implementation of 5S method.

Implementing the 5S method should begin from trainings of productive workers in the range of the 5S's elements and advantages from their usage. Very important fact is that these method do not refer only to the productive positions, but also refer to the warehouse, office positions and others [6, 15, 16, 17, 18].

One of the main purposes of the 5S is to prepare the work environment to hold visual information. From that perspective, 5S is a method, while creating a visual workplace is the goal. So, 5S and visual management go hand in hand [2]. Visual management method assumes that by a simple observation, in maximum five minutes, it can be view the existing situation on which to establish a quick action plan to improve the entire production process [12, 19].

2. Tools of implementing the 5S lean method

The 5S audit is a standard tool used for periodical or permanent reviewing and verification of the 5S method. Development, implementation and continuous monitoring of the 5S implementation is the lean management responsibility and not the auditor. Following a 5S audit evaluation, it can examine if the situation has improved or has worsened against previous auditing [20].

Another tool used to implement the 5S lean method is a standard sheet created to be used in marking the production areas (table 1).

Table 1. Standard of marking the production areas.

Role	Delineates the walkways	Delineates the storage areas of the processing parts	Delineates the storage areas of finished parts	Delineates the storage areas of rework parts	Delineates the storage areas of rejected parts	Delineates the storage areas of trolleys, carriages, boxes, devices, molds	Delineates the storage areas of waste products	Delineates the dangerous areas	Delineates the fire extinguishers areas
Marking colors									

The 5S method implementation supposes establishing some tasks and responsibilities for the team involved in compliance the specific process activities. After establish the specific tasks and responsibilities of implementing the 5S, it was required to drawing the standardization form to do the process confirmation to all workplace (automated and manuals), as a tool in implementing the 5S.

3. Implementing the 5S method

The purpose of the study presented in this paper was to identify, analyze and implement the 5S method in an industrial company "A". To do this, in the company "A" held on a workshop where were attended the production area manager, the production preparation responsible, the foreman, the shop leader and workers. The 5S workshop was led in the lean management department of company "A".

Objectives of the 5S study imply:

• Preparing the 5S standard catalogue and implementing the standards;
• Clear tasks and responsibilities for the 5S team and the lean expert;
• Informing the team about the established tasks and responsibilities;
• Training the foreman and the shop leader regarding the implementation of the 5S standards;
• Achieving the 5S audit according to the established standard;
• Developing and continuous improvement of processes and workplaces;
• Drafting the standardization form of the process work;
• Performing the process confirmation of the workplaces according to the drawing standard form;
• Permanent update of the 5S standard catalogue.

3.1. Sorting

Sorting supposed to check all the area, even the most inaccessible, netherworld, behind, above the machines or tables etc., for separation the unnecessary things of the required and necessary tools. At the time of implementing sorting process, have addressed and asked the following questions: Why is needed of...?; Who needs the...?; How often is needed of...?; When was the last time I needed of...?

Based on answers of these questions, things were sorted on necessary and unnecessary tools. The unnecessary tools were removed and workplaces were released from the disturbing things.

3.2. Set in order

Set in order process required as the necessary things to be put in order at the working places. Once identified how often and why it is used each tools, was followed to have: all things needed and used constantly to be easily find and be easily accessible, all things that are not immediately needed are elsewhere stored, tools who are together used (by several operators), must stand together, and everything has its place.

Things used occasionally and seldom have been storage on the workplace but outside the direct using sphere. Their distance and location from the place of work depend on the frequency of using these materials or tools. Places of storage were marked in the manner to making possible their quick identification. For this, have been used colours from the standard sheet created to marking the production areas, and signs or tool boards. Once defined the places and methods of storage, these should be invariable. After implementing the second S's, all things and necessary tools were properly arranged for a quick usage and so, were shortened the time of preparing the workplace.

3.3. Shine

Quality can be achieved only in clean environments. Regular cleaning permits to identify and eliminate sources of disorder and to maintain the clean workplaces. During cleaning it is checked the cleanness of machine, workplace and floor, tightness of equipment, cleanness of lines, pipes, sources of light, current data, legibility and comprehensibility of delivered information etc. Indispensable is also taking care of and maintenance the personal tidiness of the operator [6].

Based on third S's implementation, all the work area and machines are maintained in cleanness and the work conditions are tidy and safe.

3.4. Standardize

Based on standard sheet of marking the production areas, has been achieved the standardization of (and not only): storage areas (make more transparent the stock levels), materials and working tools (decreases the spent time of their search), supporting materials (is reducing the possibility of their mixing and the searching time), tool trolleys (leads to an organized labour), personal things (maintain order and the employee's satisfaction), rejected and reprocessed parts (reduce errors that can occur by mixing parts), cleaning tools (improve access).

To sustain the standardization process was drawing the standard form by which is doing the process confirmation of all workplaces (figure 1). The standardization form helps keeping safety and quality rules, sequence of working steps, and process performances.

The 5S method and process confirmation connection involve some targets and principles:

- Go, see, understand the way, and how the working process is being executed;
- Observe the process for longer time, identify the waste, and the way to eliminate or reduce it;
- Identify deviations from standard and compare with the existing and trained standard;
- Inform the worker why you are there and discuss deviation before tells your observation;
- Give feedback what you observed (feedback is without blaming the worker);
- Agree with him on the next steps and on his role in it;
- What you promised, you should keep (implementation).

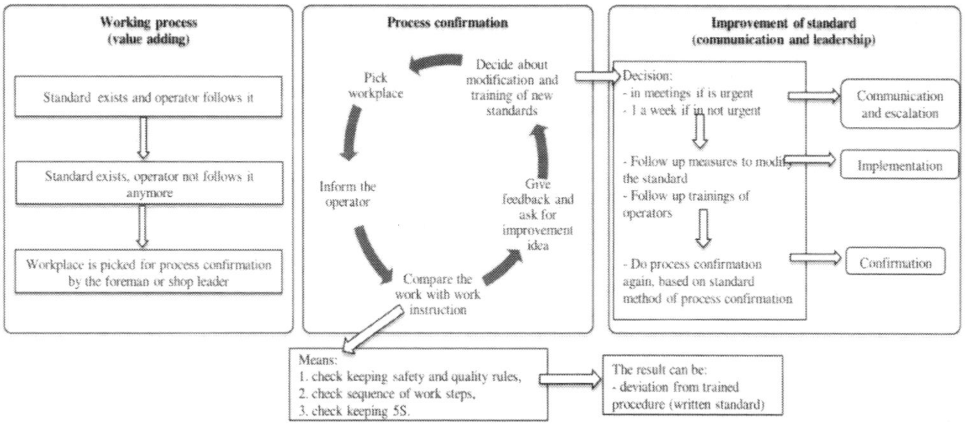

Figure 1. Connection between 5S and process confirmation.

3.5. Sustain

Sustaining suppose:

• Monitoring the compliance of rules;

• Periodical checking and updating the 5S standards;

• Immediately implement measures when are finding deviations from the 5S standard;

• Periodically control by specific 5S checklist and 5S audit, to tracking the current state of implementation;

• Leading forward the employees through consistent 5S audits, with standard questions to continuously improve results.

In the aim of check the 5S method and to sustain the process performance, was used the 5S audit. Figure 2 shows the evolution of the 5S audit, started in week 11 until week 26, in company "A".

The 5S audit percentage from week 11 until week 26

Stage	☹				😐				☺				☺			
Week	11	12	13	14	15	16	17	18	19	20	21	22	23	24	25	26
Target [%]	75	75	75	75	75	75	75	75	85	85	85	85	85	85	85	85
Value [%]	38	47	54	58	62	68	77	82	87	94	92	98	96	96	98	97

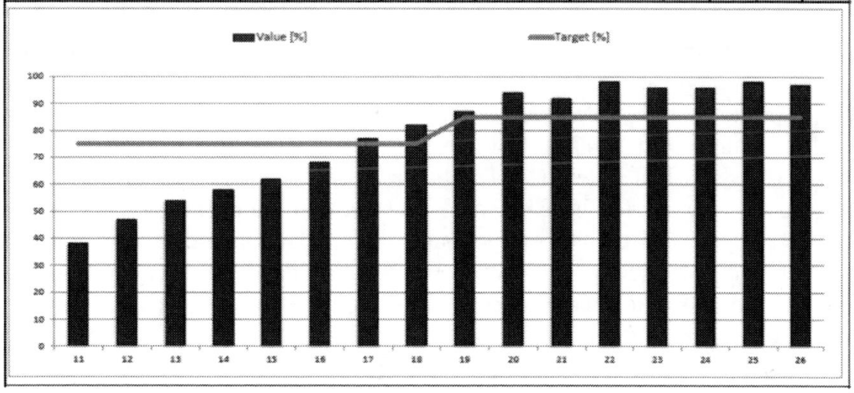

Figure 2. The 5S audit target.

As it can be seen in figure 2, by implementing and sustaining the specific actions and rules of 5S method is created a continuous improvement of the established targets. By continuous supporting of 5S method and checking the implemented process, it can measure the management performances of industrial processes.

4. Conclusions

The 5S study of this paper, describes the 5S (seiri, seiton, seiso, seiketsu, and shitsuke) method implemented to improve the workplaces, to increase the working environment quality, to eliminate or reduced mistakes, in order to sustain the industrial process performance. The advantages we can have by implementing the 5S method, are: development of a quality-conducive working environment, eliminate errors, mistakes and problems are visual, reduction of waste, reduction of waiting and searching time, transparency and clearness of workflow and workplaces, establish standards (everyone knows exactly where to find things), work safety and ergonomics of all employees.

After implementing the 5S method, have been achieved the following results (figure 3):

- Preparing workshops on 5S topics (agenda, participants, objectives, etc.);
- Moderating the 5S workshops;
- Drafting the 5S standard catalogue in 85% proportion and permanent update of its;
- Communicating the 5S standards catalogue;
- Supporting on training and implementation the new 5S standards;
- Supporting production workers in compliance standards;
- Supporting the development personal service on 5S topics;
- Drafting the standardization form of process work and the 5S audit form;
- Training for 5S audit and process confirmation;
- Achievement the process confirmation of workplaces and the 5S audit;
- Supporting the process confirmation achievement;
- Establishing the reporting mode of 5S activities;
- Performing and continuous improvement of the 5S method implementation.

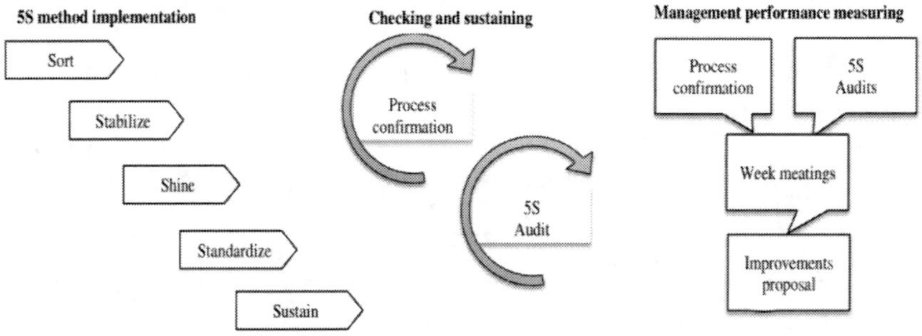

Figure 3. Phases of implementing the 5S method.

The 5S audit structure implies: responsible (production area manager and the foreman), participants (workers, production preparation responsible and the lean expert), fervency of doing (monthly by the foreman and quarterly by the production area manager), how long will take (30 minutes analyze and discussion of deviations and results), how to report (by an action plan and daily checking the implementation of established measures). The process confirmation main objective is to observe the work process and to identify the weaknesses. Responsible to doing the process confirmation is the shop leader whit support of the lean expert and the production preparation responsible. Process confirmation fervency is three workplaces per week, with a long between 20-30 minutes each workplace and he process confirmation results are reported each week in a meeting by establish the action plan, if are deviation from standards. The tools used in checking the process work has the advantage to allow the permanent checking of maintain and respect the standards, offer improvements who give a higher level of safety, order and efficiency to the workplaces, and give a permanent feedback and follow discipline in the process work. Implementing the 5S lean method gives the advantage of waste reducing, reducing the time access to the materials and needed tools, cleaning and orderly workplace, safety increase at the working place, productivity increase, improving discipline, stress reducing, and quick detection of problems.

Acknowledgement
This paper is supported by the Sectoral Operational Programme Human Resources Development (SOP HRD), financed from the European Social Fund and by the Romanian Government under the project number POSDRU/159/1.5/S/134378.

References
[1] Badea F and Burdus E 2009 Contributions on the lean management in the current evolution of a Company *J. Economia - Management serie* **12** pp 168-179
[2] 5S / Visual Workplace Handbook *Building the Foundation for Continuous Improvement, Product Automation Corporation* Information on http://www.gotopac.com
[3] Maharjan S 2011 Implementing the 5S methodology for the graphic communications management (University of Wisconsin-Stout)
[4] Cooper K, Keif M C and Macro K L 2007 *Lean Printing: Pathway to Success* (Sewickly: Pennsylvania PIAIGATF Press)
[5] Lean 2012 Increase efficiency and organization with Akro-Mils and 5S Lean *Akro-Mils/Myers Ind.*
[6] Michalska J and Szewieczek D 2007 The 5S methodology as a tool for improving the organization, *J. of Achievements in Mat. and Manufact. Engineering* **24** pp 211-214
[7] Urbaniak M 2004 Quality Management – Theory and Practice Difin (Warsaw)
[8] Ho S K 1996 TQM an Integrated Approaching – Implementing Total Quality through Japanese 5S and ISO 9000 (Kogan Page: London)
[9] Karkoszka T and Szewieczek D 2007 Risk of the processes in the aspect of quality, Natural Environment and Occupational Safety, *Journal of Achievements in Materials and Manufacturing Engineering* **20** pp 539-542
[10] Titu M A, Oprean C and Grecu 2010 D Applying the Kaizen method and the 5S technique in the activity of post-sale services in the knowledge-based organization *Proceedings of the International MultiConference of Engineers and Computer Scientists* **3** pp 1863-1867
[11] Black J and Miller D 2008 Toyota way to healthcare excellence: Increase efficiency and improve quality with Lean *Publisher Health Administration Press* (Chicago: U.S.A.)
[12] Burton T T and Boeder S M 2003 Lean Extended Enterprise: Moving Beyond the Four Walls to Value Stream Excellence *Publisher J. Ross Publishing* (Florida: U.S.A.)
[13] Liker J K 2004 The Toyota way: 14 Management Principles from the World's Greatest Manufacturer *Publisher McGraw-Hill Professional* (U.S.A.)
[14] Liker J K and Meier D 2005 Toyota way field book: A Practical Guide for Implementing Toyota's 4Ps *Publisher McGraw-Hill Professional* (U.S.A.)
[15] Michalska J 2006 The usage of the quality-cost analysis in a production process *Journal of Achievements in Materials and Manufacturing Engineering* **16** pp 190-198
[16] Dahlgaard J J, Kristensen K and Kanji G K 2000 Basis of quality management *PWN* (Warsaw)
[17] Karkoszka T and Roszak M 2005 Quality and environmental aspects in the technological process management *Proceedings of Conference Projecting & Managing of the Realization of the Production* (Zielona Góra) pp 63-68
[18] Michalska J 2005 Factors creating quality management in the company *Proceedings of the Conference INTELLECT 2005: Intellectual capital as a chance to improve the quality management in the global circumstances UMCS* (Lublin) pp 187-191
[19] Liff S and Posey P 2004 Seeing is Believing: How the new art of visual management can boost performance throughout your organization *Publisher AMACOM Books*, (New York: U.S.A.)
[20] Filip F C 2010 Theoretical researches regarding management methods and techniques for conducting the internal audit *Annals of DAAAM for 2010 & Proceedings of the 21st International DAAAM Symposium* **21** pp 1481-1482

IOP Conf. Series: Materials Science and Engineering **95** (2015) 012128 doi:10.1088/1757-899X/95/1/012128

A system model to integrate the "Green Manufacturing" concept in Romanian manufacturing organisation

D I Tilină[1], **M Zapciu**[1] **and C Mohora**[1]
[1]POLITEHNICA University of Bucharest, Faculty of Engineering and Management of Technological Systems, Machine and Production Systems Department, Splaiul Independentei nr. 313, Bucharest, Romania

E-mail: tilina.dana@gmail.com

Abstract. In Romania, the large majorities of the manufacturing companies consume natural resources and energy in an unsustainable manner. Over the years, the emissions of greenhouse gases have led not only to many environmental problems but also to important social and economic problems. A real solution to help the Romanian manufacturing companies to adapt to the new legislative requirements is the green manufacturing implementation. Considering the current situation, the purpose of this paper is to present a model that will integrate the green manufacturing concept at the organizational level based on the practices identified in the Romanian manufacturing companies at the operational level in the context of sustainable development.

1. Introduction

The EU environmental studies have shown that the manufacturing sector is the main consumer of energy and natural resources and release the largest amount of greenhouse gases. Because of this, all the manufacturing organisations are facing growing pressure to become "greener" or more environmentally friendly. Consequently, companies have had to review their production processes as a result of pressures from the customers, investors, community and governments [1].

The environmental experts and researchers believe that sustainable development and particularly sustainable manufacturing, is the solution for an industry to achieve balance with nature. Sustainable manufacturing combines the principles of Lean Manufacturing (waste reduction), Green Manufacturing (the 3R principle- reducing, recycling and reusing), TQM, and Corporate Social Responsibility [2].

Starting from the study that we made about the current situation of the Romanian's manufacturing organizations regarding the sustainable development/manufacturing; the purpose of this research is to develop a model to integrate the concept of "Green Manufacturing" in Romanian manufacturing industrial organizations. The aim of this model is to chart guidelines for the all the manufacturing enterprises who want to develop eco-innovative production systems.

2. Barriers for integrate Green manufacturing in the manufacturing organization

The term "Green manufacturing" is generically defined as "elimination of waste by re-defining the existing production process or system" [3].

Content from this work may be used under the terms of the Creative Commons Attribution 3.0 licence. Any further distribution of this work must maintain attribution to the author(s) and the title of the work, journal citation and DOI.
Published under licence by IOP Publishing Ltd

Even if we look at the data submitted by the EU on the benefits resulting from the application of this concept in the manufacturing organization however, the studies have shown that the implementation of Green Manufacturing is hampered by different barriers [4, 5].
A study of the relevant literature about the factors that prevent/ hinder the implementation of Green manufacturing is presented in table 1.

Table 1. List of Green Manufacturing barriers.

No.	Sources	Barriers
1.	M. Perron [6] Mittal et al. [7] Lee S.Y. [8] Walker et al. [9] Simpson et al. [10] Carley S. et al. [11]	1. *Lack of available resources* (financial and human resources, lack of time and of technical knowledge, High short-term costs, Uncertain benefits) 2. *Information barriers* (language barriers, lack of awareness of their business environmental impacts, lack of awareness of benefits etc.) 3. *Internal attitudes and perceptions* (inconsistent top management support, resistance to change within company culture, disbelief of the benefits of environmental initiatives, perceived cost, high perceived effort required etc.) 4. *Weak legislation, Low enforcement, Uncertain future legislation* 5. *Low public pressure, , Low customer demand* 6. *The firm's size* (a bigger size firm tends to be more willing to become "greener")
2.	Luthra et al. [12] Kumar S et al. [13]	15 barriers: *Systematic Economic Recycling, Lack of IT implementation, Market competition, The government support system, Lack of organization encouragement, Supplier reluctance to change towards GSCM, Poor quality of human resources; Unawareness of customers, Top management commitment; Government support and regulations, Cost implications; Systematic and Efficient Planning, Environment Management programs, Resistance to technology advancement adoption, Lack of implementing green practices*
3.	Kannan G. et al. [14] Mathiyazhagan K. et al. [5]	47 barriers dived in five categories: *Outsourcing* (6 barriers), *Technology* (9 barriers), *Knowledge* (11 barriers), *Financial* (9 barriers), *Involvement and support* (12 barriers)

As it can been seen the researchers has different or similar opinion regarding the type and the number of barriers from implementing the green concept. This difference is due to two important factors that may influence the outcome of the study: the country (USA, India, China, Germany etc.), the analyse level and the area of activity (automotive, manufacture of chemicals and chemical products, manufacture of basic pharmaceutical products, manufacture of plastic products, manufacture of electrical equipment etc.) where it was conducted.

3. A model to integrate the concept of "Green Manufacturing" in Romanian's manufacturing companies

The Sustainable manufacturing considers two separate, but related, topics: first, it seeks to support the development of sustainable products and second, it promotes the development of sustainable processes [15].

In a similar way Green Manufacturing (GM) can be looked at in two ways: first is the manufacturing of „green" products (for example those used in renewable energy systems and clean technology equipment) and second the "greening" of manufacturing (reducing pollution and waste by minimizing natural resource use, recycling and reusing what was considered waste, and reducing emissions) [16].

Starting from these, the research methodology contains three stages:

1. Collecting, analysing and interpretation of the data concerning the involvement of the Romanian manufacturing enterprises in the environmental issues.
2. The development of an integration model.
3. The development of measures to further improve production systems and the development of eco-innovative actions tracking their implementation within the organization.

3.1. First stage: Collecting, analysing and interpretation of the data concerning the involvement of the Romanian manufacturing enterprises in the environmental issues

The rapid industrialization, between 1960-1989, through infrastructure development in large manufacturing factories, has led to the over-exploitation of non-renewable and renewable natural resources. This contributed to the massively pollution of the environment. In all industrial companies the damage was resulting from absence of pollution-control installations but also from ineffective management programs. As a result, Romania has become more vulnerable to geomorphologic, hydrological and climate hazards [17].

Starting from the reality presented above, the Romanian Government enacted some specific regulations for the manufacturing companies, in order to ensure sustainable manufacturing in conformity with the EU objectives [17, 18]:

- The obligation to include the indicator "resource productivity" in the expenditure budgets and financial reports.
- The obligation to provide information's regarding the waste management, the environmental performance (processes and activities), and the application of green labelling (the number and relative share of products carrying environmental labels).
- The introduction of the "polluter-pays" principle according to which the polluter should bear the cost of measures to reduce pollution according to the extent of either the damage done to society or the exceeding of an acceptable level (standard) of pollution [19,20].

After 1989 many large manufacturing factories were closed and Romania's industry is today dominated by SME's. In 2015, from a total of 442.567 enterprises that operated in Romania, 441.206 are SME's and 1.361 are large enterprises [21]. The number and type of Romanian enterprises in the manufacturing industry are presented in figure 1.

Eurostat SBS database
Estimations

Unit: Number

section NACE Rev. 2	size class	2010	2011	2012	2013	2014	2015
Manufacturing	0 - 9	35.348	30.914	32.554	37.910	39.376	40.811
	10 - 49	9.845	10.184	9.789	10.175	10.201	10.203
	50 - 249	3.018	3.223	2.851	2.742	2.552	2.370
	250 +	722	731	650	638	585	535
	Total	48.933	45.052	45.845	51.465	52.713	53.918
	All SMEs	48.211	44.321	45.194	50.827	52.129	53.384

Figure 1. The number and the type of firms in Romanian manufacturing industry [21].

In the last "Flash Eurobarometer on SME's, Resource Efficiency and Green Markets" the survey shows that in Romania [22]:

- 42% of the SME's are complying with the environmental legislation but does not wish to go beyond these requirements and only 35% are contemplating doing more;

• only 25% are doing many action to be more resources efficient (the highest percentage is 64% in Spain) and for the future 34% are planning to implement many action;

• regarding the reasons why the enterprise is taking action to be more resources efficient the results are: cost savings (65%), demands from customers (13%), financial or fiscal incentives (12%), creation of a competitive advantage (10%);

• the biggest difficulty when they trying to set up resources efficiency plan is the complexity of administrative or legal procedures (29%) followed by cost of environmental action (17%), lack of specific environmental expertise (15%), difficulty to adapt legislation to the company (18%), difficulty in choosing the right actions for the company (9%).

In addition to all the above data, in the last survey published by ISO for 2013, Romania figures only with 1792 organization who have been certificated with ISO 14001 in the manufacturing industry, from a total of 8744 ISO 14001 certificate [23]. From a simple calculation result that only 3,5% of manufacturing companies had certified environmental management system.

3.2. Second stage: The development of an integration model

The proposed model has three stages like is shown in the figure 2.

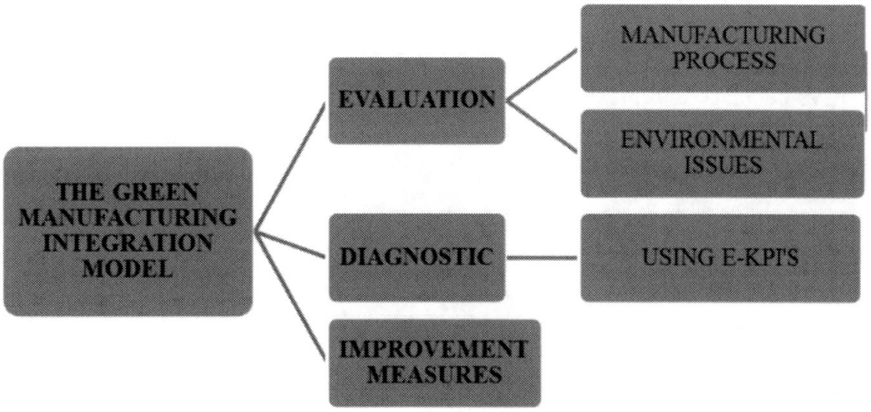

Figure 2. The integration model.

I. Evaluation. The assessment of the current situation of the manufacturing enterprise process: the description and analysis of the production process analyse the waste generation for each stage, identification of the green (environmental) tools used.

II. Diagnostic. The diagnostic stage is based on the use of Environmental Key Performance Indicators (E-KPI's).

Measuring direct and indirect impacts requires quantitative and qualitative indicators. Generally, environmental indicators are used for three major purposes [24]:

- to give information on environmental problems;
- to support policy development and priority setting;
- to monitor the effects.

Key performance indicators (KPI's) are a set of quantifiable measurements, which helps organizations in defining and measuring their progress towards their organizational goals.

F. John Reh, states that Environmental Key Performance Indicators provide organizations a tool for measuring their environmental performance: "E-KPI's are quantifiable metrics which reflects the environmental performance of an organizational performance in the context of achieving its wider goals and objectives" [25].

To identify the most appropriate E KPi's for the enterprise environment diagnosis we used the REM criteria:

- Relevant to the activity, operation and objectives;
- Easy to track over time;
- Measurable through quantitative metrics or qualitative surveys and interviews [26].

The Environmental Key Performance Indicators identified in the diagnosis research were established according to the following aspects: the environmental objectives of the organization, the period of time established for the diagnostic analysis (year, month), the periodicity evaluation indicators (daily, weekly, monthly) and finally the established assessment scale of the size indicators.

III. Improvement measures. Proposing improvement measures which take in account the continuous improvement and the impact on the production process, the degree adaptation of company departments, productivity, etc.

4. Conclusions

To produce in a sustainable way, the structural adjustment of the manufacturing process is the main way of increase resource productivity.

Regarding the environmental protection, on the whole, Romania scores was below average (compared to the EU Member States). Romanian SME's are less concerned with doing something to increase their resource efficiency and the measures taken in terms of environmental protection are due only because of the legislative constraints. Compared to the EU average of 26 %, only 22 % of companies have begun to exploit the opportunities offered by the demand for green products because they do not receive as much public support to do. In addition, the studies have shown that the implementation of GM in the manufacturing organization is not that easy because is hampered by different barriers.

Starting from these, the purpose of this research is to elaborate a model for helping the Romanian's manufacturing enterprises to develop eco-innovative production systems with the purpose of greening their products and processes while reducing the impact of their activities.

Acknowledgement
The work has been funded by the Sectorial Operational Program Human Resources Development 2007-2013 of the Ministry of European Funds through the Financial Agreement POSDRU/159/1.5/S/132397.

References

[1] E Bălan 2014 Proiectare conceptuală: teorii, metodologii şi aplicaţii *Modelare – simulare - proiectare în domeniul maşinilor-unelte şi sistemelor de maşini* ed Constantin G (Bucureşti: PRINTECH) chapter 5
[2] Jawahir I S 2008 Beyond the 3R's: 6R Concepts for Next Generation Manufacturing: Recent Trends and Case Studies *IIT MMAE Symposium on Sustainability and Product Development* (McCormick: Tribune Campus Centre Illinois Institute of Technology)
[3] Balan K 2008 Introduction to Green Manufacturing *The Shot Peener* **22(3)**
[4] Mittal V K and Sangwan K S 2014 Prioritizing Drivers for Green Manufacturing: Environmental, Social and Economic Perspectives *21st CIRP Conf. on Life Cycle Engineering* (*Procedia CIRP)* **15** pp 135-140
[5] Mathiyazhagan K, Govindan K., Noorul A and Geng Y 2013 An ISM approach for the barrier analysis in implementing green supply chain management *Journal of Cleaner Production* **47** pp 283-297
[6] Perron G M 2005 Barriers to Environmental Performance Improvements in Canadian SMEs (Dalhousie University: Canada)

[7] Mittal V K, Sangwan K S, Herrmann C and Egede P 2013 Comparison of drivers and barriers to Green Manufacturing: A Case of India and Germany *Re-engineering Manufacturing for Sustainability* (Springer: Singapore) pp 723-728
[8] Lee S Y 2008 Drivers for the participation of small and medium sized suppliers in green supply chain initiatives *Supply Chain Management International Journal* **13** pp 185-198
[9] Walker H, Di Sisto L and McBain D 2008 Drivers and barriers to environmental supply chain management practices: lessons from the public and private sectors *Journal of Purchasing and Supply Management* **14 (1)** pp 69-85
[10] Simpson M, Taylor N and Barker K 2004 Environmental responsibility in SMEs: does it deliver competitive advantage *Business Strategy and the Environment* pp 156-171
[11] Carley S, Jasinowski J, Glassley G, Strahan P, Attari S and Shackelford S 2014 Success Paths to Sustainable Manufacturing *School of Public and Environmental Affairs Indiana University*
[12] Luthra S, Kumar V, Kumar S and Haleem A 2011 Barriers to implement green supply chain management in automobile industry using interpretive structural modelling technique: An Indian perspective *Journal of Industrial Engineering and Management* **4(2)** pp 231-257
[13] Kumar S, Kumar S, Gahlot P and Haleem A 2013 Analysis of interdependence among the enablers of green concept implementation in Indian automobile supply chain *Journal of Engineering Research and Studies*
[14] Kannan G, Mathiyazhagan K, Devika K and Haq A N 2014 Barriers analysis for green supply chain management implementation in Indian industries using analytic hierarchy process *Int. J. Production Economics* **147** pp 555–568
[15] Sustainable Manufacturing: Preparing for a New Business Imperative 2014 American Society of Mechanical Engineers (ORF-2)
[16] Information on http://cleantechnica.com
[17] Government of Romania *United Nations Development Program* National Sustainable Development Strategy: Romania 2013-2020-2030
[18] Majstorovic V D 2014 Manufacturing Innovation and Horizon 2020 – Developing and Implement „New Manufacturing" *Proceedings in Manufacturing Systems* **9** pp 3–8
[19] Information on http://www.minind.ro/legislatie/politica_industriala_2005_2008.html
[20] United Nations 1997 Glossary of Environment Statistics *Studies in Methods* **F No. 67** (New York)
[21] Information on http://ec.europa.eu/enterprise/policies/sme/facts-figures-analysis/performance-review/index_en.htm#
[22] The European Commission 2013 Flash Eurobarometer on SMEs *Resource Efficiency and Green Markets*
[23] Information on http://www.iso.org/iso/iso-survey
[24] Smeets E and Weterings R 2010 Environmental indicators: Typology and overview, TNO Centre for Strategy *Technology and Policy* (Netherlands)
[25] Wajahat S and Rahul R 2013 Environmental KPI's for management and improvements in manufacturing: Increase employee sustainability commitment for Lean and Green production at ABB *Student thesis* (Mälardalen University: School of Innovation, Design and Engineering)
[26] Measuring Impact Framework methodology 2008 *World Business Council for Sustainable Development*

The evaluation of the environmental impact and the external factors of urban transport in Constanta

C Stanca[1], **V G Stîngă**[1], **S Georgescu**[1] **and O S Cupşa**[1]
[1]Constanta Maritime University, Faculty of Navigation, 104 Mircea cel Batran Street, 900663, Constanta, Romania

E-mail: ovidiu_cupsa@yahoo.com

Abstract. Transport activities are known to have a substantial negative environmental impact especially when referring to the urban transport. Studies have shown that external costs (as accidents, congestion, air emissions, climate change or noise) are an important subject of the European Union, that is why were carried out several research projects. This paper will highlight the current requirements and methodologies used by the European Union regarding the impact of the external costs of urban transport in most of the growth poles of Europe. Taking into consideration that Constanta is considered to be one of the seven major growth poles of Romania for the 2014-2020 period, this study aims at analyzing how the results of similar studies made in others centers of the European Union can be applied in Constanta, showing different methodologies and evaluations regarding the external costs and their impact. We will analyze how the conclusions obtained in previous projects are applicable to data collected by us throughout a field research on the technical description of the means of transport used it this city. As methodology, we will use one that was adopted by the European Union regarding the estimation of urban external costs, taking into consideration that each externality has a different method for estimating it. The results of this study may be useful in developing the sustainable urban mobility plan for Constanta, as a strategic plan design to reduce the impact of urban transport for a better quality of life at present and in the future. Through this paper we will get an insight into the urban transport in Constanta, but also data on external costs generated by the urban transport, given that road transport is considered to be the most polluting transport mode.

1. Introduction

Over the last years there has been a remarkable growth of interest in environmental impacts, due to the fact that they are significant and growing [1]. These impacts fall within three important categories: direct impacts (here we emphasize the immediate consequence of transport activities on the environment, impacts where the relationships between cause and effect are well understood), indirect impacts (the secondary (or tertiary) effects of transport activities on environmental systems, which involved relationships that are difficult to establish) and cumulative impacts (the multiplicative or synergetic consequences of transport activities, which involved effects that are often unpredicted) [2].

The evaluation of the environmental impact and the external costs of urban transport are an important element taken into consideration by the European Union within its strategic policy goals. The two most important documents of the European Commission, the Green Paper towards a new culture for urban mobility (COM(2007) 551) and the Action Plan on Urban Mobility (COM(2009) 490), support the need of a shift in transport, by reducing greenhouse gas emissions and pollution in

Content from this work may be used under the terms of the Creative Commons Attribution 3.0 licence. Any further distribution of this work must maintain attribution to the author(s) and the title of the work, journal citation and DOI.
Published under licence by IOP Publishing Ltd

order to achieve a smarter, safe, secure, more accessible and more environmentally friendly urban transport by using specialized vehicles with alternative fuels, but also using cleaner options like walking and cycling.

When referring to urban transport, it is clear that its environmental impact is higher due to the stop-start nature of many urban traffic conditions, which is exacerbated by growing congestion levels, despite recent improvement in vehicle fuel efficiency. We also need to highlight some points of the 2011 Transport White Paper concerning the environmental impact of urban transport that support the idea of a much cleaner urban transport: cities are the one that are most affected by this impact triggered by the high proportion of road accidents (almost 70 % of them occur within urban areas), congestion, noise exposure and poor air quality, cities being the ones that in the end will benefit from this switch while also taking into consideration the use of more specialized passenger vehicles that will use alternative propulsion systems and fuels [3].

We need to emphasize five activities related to transport that in the end affect the environment: the first one refers to infrastructure construction, the second one takes into consideration its maintenance and the last ones refer to vehicles: vehicle and parts manufacture, vehicle travel, vehicle maintenance and support and disposal of used vehicles and parts [4].

It is well known that transport contributes significantly to economic growth, but as we have mentioned earlier the transport also brings about negative impacts, most forms of transport giving rise to side effects that cannot be expressed in monetary terms [5]. When side effects of a certain activity impose a cost upon society, we may speak of such a cost as an external cost (see table 1), but it is important to point out that these costs aren't generally produced by the transport users, they are considered to be social costs imposed on others, but not paid for by the infrastructure user, as "internal costs" are.

Table 1. Total external costs for EU-27 (with the exemption of Malta and Cyprus, but including Norway and Switzerland) for passenger transport by cost category and transport mode.

Cost category	Road				Rail (passenger transport)	Aviation (passenger transport)
	Passenger cars	Buses & coaches	Motorcycles & mopeds	Total road passenger transport		
	€/(1,000 pkm*a)	€/(1,000 pkm*a)	€/(1,000 pkm*a)	€/(1,000 pkm*a)	€/(1,000 pkm*a)	€/(1,000 pkm*a)
Accidents	32.3	32.3	32.3	32.3	0.6	0.5
Air pollution	5.5	6.0	11.8	5.7	2.6	0.9
Noise	1.7	1.6	14.4	2.0	1.2	1.0
Biodiversity losses	0.2	0.4	0.1	0.2	0.0	0.1
Nature and landscape	0.6	0.3	0.5	0.6	0.2	0.6
Soil and water pollution	0.3	0.9	0.3	0.4	0.5	0.0
Urban effects	1.0	0.4	0.8	0.9	0.6	0.0

Source: CE Delft, Infras, Fraunhofer ISI, External Costs of Transport in Europe, Update Study for 2008, Delft, CE Delft, pp. 10 (2011)

Studies have shown that the current situation in urban areas could be regarded as a cycle of urban decline, due to the road transport which, despite of the fact that it brings lots of benefits to its users it also produces more traffic jams, leading to a decrease in the efficiency of public urban transport, but also in the quality of life in urban districts (due to accidents and to the high level of pollution). Nowadays there is a trend for the companies to move their headquarters to suburban areas, but this only leads to a higher level of pollution, because in those areas the provision of public transport is very low, so that workers are forced to travel by car [7]. The table below presents the relevant cost categories per urban transport mode showing that beside cycling, which is the most eco-friendly type of transport, train and tram are less polluting than urban road transport (car, trucks, bus, mopeds). Rail urban projects seem to be an optimal solution to road transport, as they reduce the environmental impacts, thanks to their benefits: regular, safe, with medium-high capacity.

The most important environmental impacts that need to be taken into account when referring to the urban transport are costs of congestion, noise, air pollution, climate change, accidents and the tear and wear of the use of the infrastructure.

Table 2. Relevant cost categories per urban transport mode.

Transport mode	Wear & tear	Climate	Air pollution	Noise	Accidents	Congestion
Car	X	X	X	X	X	X
Truck	X	X	X	X	X	X
Mopeds/ motorcycle	X	X	X	X	X	X
Cycling					X	
Bus	X	X	X	X	X	X
Train	X	X	X	X	X	
Tram	X	X	X	X	X	

Source: Bossche M Beekman R. Baanders B. Scholten B.2012 Study on Urban Aspects of the Internalisation of External Costs - MOVE/B4/310-1/2011, Final report, DG MOVE, Rotterdam, pp. 21

2. Methodologies

Different methodologies and evaluations of external costs and their impact have been developed, but only two of them have been analyzed within European Union's projects: Cost-Benefit Analysis (CBA) and Multicriteria Analysis (MCA) [7]. Even if the first one is more objective than the second one, due to the fact that it uses monetary values to measure all the effects, it cannot allow quantifying all the effects of external cost, because most environmental and social effects are difficult to account for.

The Multicriteria Analysis consists in an assessment that may be carried out under the objective of sustainable development, considering three important dimensions: economic, social and environmental sustainability. The analysis sets some objectives that need to be evaluated, each objective being measured within two scenarios (the reference scenario and the real scenario) by one or more specific criteria that will be measured through a combination of quantitative and qualitative indicators. It is important that every effect be aggregated in a final single value that will be individual social utility for each criterion. So, the final impact will be the weighted sum of all indicators multiplied by the weight assigned to their corresponding criterion as presented below:

$$\text{Socio - economic- environmental Utility} = \sum_{i=1}^{n} w_i \cdot x_i \qquad (1)$$

Where w_i are the weights and x_i the individual social utilities of each of the n indicators.

For example regarding the environmental improvement, the criteria will refer to external costs, which may establish the following indicators: air pollution, reduction of pollutant emissions (Tons/year of CO, SO2, NOx, etc.), percentage of persons that are less affected by noise, reduction of accident costs per year (Euro/car-km) [7].

As stated earlier, the most appropriate method of determining external costs is the Multicriteria Analysis. Through this method, adopted on an European level, there should be a clearer emphasis of external costs and also of the measures needed in the process of their internalization.

3. The external costs of urban transport in Constanta

The costs which affect the citizens directly are the noise level costs (due to the noise in the streets with a high daily average intensity), the sanitary costs- air pollution (due to the harming emissions of vehicles), the congestion costs (due to the increased number of cars, that cause losses of time), but also the costs of injuries or death due to accidents of the urban traffic [9], but two of them are considered to be the most important in Constanta: noise and air pollution costs.

a. Air pollution from urban transport

According to R. Gorham some factors can be identified that contribute to the air pollution from urban transport, influencing the amount of emissions, such as: the excessive use of vehicles, the age of the vehicle fleet and its technology used, poor vehicle maintenance and the unavailability or improper use of appropriate fuels [10].

According to the information provided by the Public Transportation Company of Constanta, 52,112,498 people were transported in 2012. When referring to the number of routes needed for urban road transport in Constanta, there are 26 routes, totaling 366 km, stopping at 549 bus stops. We also need point out that in February 2013 the total number of kilometers traveled daily by a single bus was of 174 km [13].

In 2013 Constanta's bus fleet was made up of 5 types of buses:

- MAZ 103-041: 112 buses;
- MAZ 103-065: 15 buses;
- MAZ 203-076: 15 buses;
- MAZ 107-468: 50 buses;
- AYATS: 10 buses.

Inner road network amounts in Constanta 410 km of streets, out of which 320 km of medium and light traffic streets and 90 km of heavy traffic streets [12]. The number of hours spent in traffic in Constanta totals 34,200 vehicles/ hour for cars, 1,300 vehicles/ hour for taxis and 6,100 passengers/ hour for public transport. The daily average travel speed of cars reaches 42 km/ h and for public transport an average speed of 32 km/ h is estimated.

According to assessments for the calculated year, motorized traffic in Constanta generates an amount of emissions per day as follows: 0.18 tons of NOx, 1.1 tons of CO, 0.09 tons/ particles of unburned hydrocarbons, 33.4 tons of SO2 and 102.5 tons of CO2.

b. The noise level from urban transport

Disorders of the human body functions due to noise are directly proportional to its intensity, frequency and duration [11].

- Acoustic intensity describes how the sound power of a source is spread through the environment and it is the power that passes through an imaginary surface element usually towards the direction of propagation of sound waves. Sound intensity level, measured in [dB] takes into account the physiological behavior of the ear. Propagation speed of the sound waves depends on the characteristics of the environment in which it propagates. Speed expression of the sound through air is:

$$c = \left(\frac{\gamma \cdot R \cdot T}{M} \right)^{1/2} \tag{2}$$

Where: γ is the adiabatic exponent of ideal gas; T – absolute temperature of the air; M – his molar mass. $R = 8.3 \cdot 10^3 [J \cdot Kmol^{-1} \cdot K^{-1}]$ - universal gas constant;

- Frequency which is the number of complete cycles of a sound in a unit of time, measured in Hertz (Hz). The ear perceives only audible sounds range between 16 and 16,000 Hz, but it does not perceive any ultrasound (acoustic oscillations with frequencies above 16,000 Hz) or infrasound frequency below 16 Hz.

Table 3. The level of acoustic intensity of different sources.

Noisy sound source	Intensity of different sources [dB]
Threshold of audibility	0
The ticking of a mechanic clock	30
Loud conversation	50-65
Usual road traffice	65-70
Railway traffic	56 -90
Intense traffic, symphonic orchestra	90
Thunder	90-100

Source: Gâtlan, A., Noise pollution in rail and urban rail transport, AGIR Journal (2012)

Regarding the number of people exposed to noise from motorized traffic in Constanta, according to information supplied by Constanta City Hall in the study "Development of strategic noise map for Constanta" [12] we can notice that:
- 99800 individuals, 32% of the population is exposed to noise levels between 55-59 dB;
- 56500 individuals, 18% of the population is exposed to noise levels between 60-64 dB;
- 52 600 individuals, 17% of the population is exposed to noise levels between 65-69 dB;
- 26200 persons, 9% of the population is exposed to noise levels between 70-74 dB;
- 700 people, less than 1% of the population is exposed to noise levels between 75 dB.

4. Conclusions

The analysis carried out in the research on the environmental effects of urban transport in Constanta emphasizes the fact that, due to the population density, and the fact that it is a harbour city, Constanta is among the top 5 cities in Romania affected by pollution caused by transport. As a further evidence of this we need to point out that Constanta was designated as one of the seven regional urban growth poles, for which the elaboration of a Sustainable Urban Mobility Plan is considered to be a priority.

As stated before the environmental impact of urban transport occurs mainly by air pollution from urban transport and its associated noise. So, we need to highlight that by 2011 public transport in Constanta included trolleybuses, trams and buses, but as shown in this study the buses are the only urban transport means that is still functional. Changing the urban public transport fleet led to a decrease in noise produced by trams, but at the same time is also led to traffic flows by eliminating trams. We should also emphasis the significant increase in the level of pollutants released into the atmosphere through road transport.

References

[1] Glasson J, Therivel R and Chadwick A 2013 Intro to environmental impact asses *Routledge* p 2
[2] Rodrigue J P 2013 The Environmental Impacts of Transportation (New York: Routledge) chapter 8
[3] Booz & Company 2012 Study on the financing needs in the area of sustainable urban mobility *Directorate-General for Mobility and Transport* (London) pp 22
[4] Corrales M, Grant M and Chan E 1996 Indicators of the environmental impacts of transportation- highway, rail, aviation, and maritime transport *United States Environmental Protection Agency, Office of Policy, Planning and Evaluation* pp 7
[5] Korzhenevych A, Dehnen N, Bröcker J, Holtkamp M, Meier H, Gibson G, Varma A and Cox V 2014 *Update of the Handbook on External Costs of Transport* pp 1
[6] CE Delft, Infras, Fraunhofer ISI 2011 External costs of transport in Europe *Update study for 2008* (Delft: CE Delft) pp 10

[7] Cascajo R 2005 Assessment of economic, social and environmental effects of rail urban projects *In Young Researchers Seminar* (The Hague: NL) pp 4-5
[8] Bossche M, Beekman R, Baanders B and Scholten B 2012 Study on urban aspects of the internalisation of external costs - Final report *DG MOVE* (Rotterdam) pp 21
[9] Barceloa I F and Campos-Cacheda J M 2012 Estimate of social and environmental costs for the urban distribution of goods. Practical case for the city of Barcelona *The Seventh International Conference on City Logistics (Procedia - Social and Behavioral Sciences)* **39** pp 818 – 830
[10] Gorham R 2002 Air pollution from ground transportation- An assessment of causes, strategies and tactics, and proposed actions for the international community *The Global Initiative on Transport Emissions* (A Partnership of the United Nations and the World Bank, United Nations) pp. 3
[11] Gâtlan A 2012 Noise pollution in rail and urban rail transport *AGIR Journal* pp 8-9
[12] http://www.primaria-constanta.ro, Accessed: 02/03/2015
[13] http://www.ratc.ro, Accessed: 02/03/2015

IOP Conf. Series: Materials Science and Engineering **95** (2015) 012130 doi:10.1088/1757-899X/95/1/012130

Two-Stage orders sequencing system for mixed-model assembly

M Zemczak[1], B Skołud[2] and D Krenczyk[2]
[1]University of Bielsko-Biala, Faculty of Mechanical Engineering and Informatics, Department of Production Engineering, Willowa 2, 43-309 Bielsko-Biala, Poland
[2]Silesian University of Technology, Faculty of Mechanical Engineering, Institute of Engineering Processes Automation and Integrated Manufacturing Systems, Konarskiego 18A, 44-100 Gliwice, Poland

E-mail: mzemczak@ath.bielsko.pl

Abstract. In the paper, the authors focus on the NP-hard problem of orders sequencing, formulated similarly to Car Sequencing Problem (CSP). The object of the research is the assembly line in an automotive industry company, on which few different models of products, each in a certain number of versions, are assembled on the shared resources, set in a line. Such production type is usually determined as a mixed-model production, and arose from the necessity of manufacturing customized products on the basis of very specific orders from single clients. The producers are nowadays obliged to provide each client the possibility to determine a huge amount of the features of the product they are willing to buy, as the competition in the automotive market is large. Due to the previously mentioned nature of the problem (NP-hard), in the given time period only satisfactory solutions are sought, as the optimal solution method has not yet been found. Most of the researchers that implemented inaccurate methods (e.g. evolutionary algorithms) to solving sequencing problems dropped the research after testing phase, as they were not able to obtain reproducible results, and met problems while determining the quality of the received solutions. Therefore a new approach to solving the problem, presented in this paper as a sequencing system is being developed. The sequencing system consists of a set of determined rules, implemented into computer environment. The system itself works in two stages. First of them is connected with the determination of a place in the storage buffer to which certain production orders should be sent. In the second stage of functioning, precise sets of sequences are determined and evaluated for certain parts of the storage buffer under certain criteria.

1. Introduction

For many years, studies have been conducted on the issue of the optimization of scheduling and resource allocation [1-3]. Optimization problems, both continuous and discrete groups are classified as NP-hard problems (problems for which finding a solution in polynomial time is not possible), due to their complexity, theoretical and computational. This problem is often encountered in the case of complex production systems, where on limited resources certain technical good must be produced within a given time. Scheduling is supposed to determine the sequence of production, which will utilize the production capacity remaining at the disposal of the company in the most of the advantageous way. Regardless of the specifics of the production system, its structure and organization,

Content from this work may be used under the terms of the Creative Commons Attribution 3.0 licence. Any further distribution of this work must maintain attribution to the author(s) and the title of the work, journal citation and DOI.
Published under licence by IOP Publishing Ltd

before the adoption of new orders review of the current state of the system is required, based on which the decision whether and when a new order can be accepted for implementation is taken [4].

Computational complexity and size of the practical problems clearly eliminates from consideration the exact algorithms (due to the fact that the solution must be given within the prescribed time window), leaving only the use of heuristic algorithms that solve the problems in a short, assumed time and with satisfactory accuracy. However, with assuming appropriate restrictions, a simpler set of rules controlling schedules can be selected to be used. The problem of scheduling jobs on the production lines has been formulated by [5] as MILPA (Mixed Integer Linear Programming), but at the outset the author observed that the complexity of the problem grows exponentially with the number of tasks to rank. For this reason, MILPA algorithms can be used only for small problems. In literature many attempts to solve scheduling problems, e.g. using genetic algorithms [6] can be found. Unfortunately, as any solution, they have their advantages and disadvantages. In most cases, obtaining the best effect requires a time-consuming "tuning" of the algorithm to a class of selected problems. Many authors carried out experiments with the use of algorithms based on heuristic methods [2, 7], wherein it should be noted that the studies were covering the small amount of data (for data with a greater extent, the solution couldn't be found within a reasonable time). Research on heuristic algorithms, providing solutions for problems where it is impossible or inefficient to use exact solutions, are rapidly developing field of science.

Suitable arrangement of the tasks leads to many benefits including better utilization of material resources (in this case, devices and tools), ensuring adequate plant production capacity, saturation of the production line, the appropriate management of employee time, as well as his more efficient operation. This translates directly into increased productivity and a reduction in unit costs, on the basis of economies of scale. Furthermore it should be noted that automotive companies are themselves interested in the search for solutions to the problem of scheduling, which can be seen, for example on the base of ROADEF Challenge 2005 competition announced by the French Association for Operations Research and Decision Support (fr. Société française de Recherche Opérationnelle et d'Aide à la Décision) and sponsored by the RENAULT group, whose aim was to develop a new universal method for scheduling.

2. Algorithms used for solving scheduling problems

Evolution Methods for solving scheduling problems can be generally divided into exact and approximate methods (figure 1).

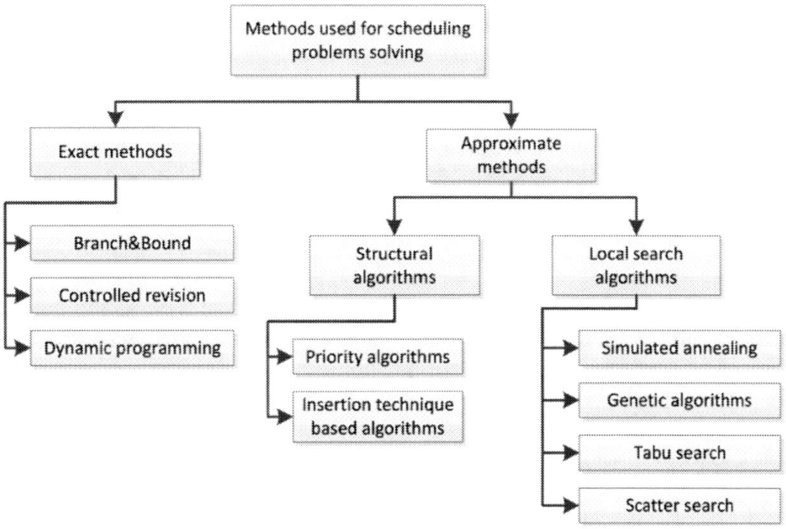

Figure 1. Division of methods used for scheduling problems solving.

The exact methods they can be divided into e.g. methods based on the principle of branch and bound, the method of controlled revision, methods based on dynamic programming scheme. Due to the aforementioned disadvantages and problems connected with their implementation to more complex tasks - the exact algorithms will not be further discussed. Among the approximate algorithms two major groups can be mentioned: structural algorithms that construct a solution to the problem and local search algorithms. Structural algorithms design is significantly dependent on the specifics of the problem. These algorithms can be divided into priority algorithms and algorithms that use the so-called insertions technique. The first rank the relative priorities of the tasks with the use of the rules of priority. The second group of algorithms, before ranking the tasks, creates a set of test solutions by inserting a task into a test position in current solution (constructed in the previous iteration of the algorithm). Among the selected test solutions, the best solution (in terms of objective function values) is chosen, which at the same becomes the base solution in the next iteration of the algorithm. Local search algorithms start operation from initial solution provided by the structural algorithm and work iteratively to improve the solution, using the techniques of reviewing the solution space. Among the most effective algorithms in this group simulated annealing, genetic algorithms, scatter search algorithms and tabu search algorithms can be specified. Presented methods are also implemented with the use of simulation software in order to visualize the solutions [8].

3. Scheduling problems description

Almost every scheduling problem can be described as a pair (X, F), where X is the set of feasible solutions to the problem (wherein $X \subseteq X^0$, where X^0 is the set of all solutions), F is the criterion of optimization (objective function). Solutions space is given explicitly only in a few cases, usually it is a set of independent variables, which can take the values of the specified ranges.

One of the basic concepts of local search algorithms is the movement. It can be described as an activity involving the transition between the two solutions and presented as a function of $v(x)\colon X \rightarrow X$, where $v(x)\colon x_v \in X$, $x_v \neq x$. Solution x_v, resulting from movement v is called the neighbouring solution of solution x. For each solution $x \in X$, a set of movements $V(x)$ which generates a neighbourhood $N(x)$ (1):

$$N(x) = \{x_v : v \in V(x)\} \tag{1}$$

of solution x can be defined. The definition of a set of movements and neighbourhood depends on the analysed problem and an algorithm used for its solutions.

Another important element is the local optimality of the solution. In the literature, in the context of local search algorithms the term "best solution" found by the algorithm often appears. This concept is the result of a mental shortcut pointing to the solution generating the smallest value of the objective function from all the solutions generated by the algorithm.

4. Structure of the system

For the purpose of the research, a structure of the system, corresponding to a real production system has been selected. The system consists of an input buffer, in which orders are stored before being admitted to assembly processes, and an assembly line, on which assembly operations are performed. Due to the fact, that assembly operations are different on each assembly station, their number for the purpose of the research has been limited to 50. The assembly stations on which the biggest differences in labor-consumption have been identified during the research in a company have the biggest impact on the production output. It should be noted, that even after solving major assembly line balancing problems, current assembly systems are still not flexible enough to adapt quickly to the exact version of the product that arrives, i.e the tact time is the same for all the stations in assembly line [3]. It could be observed, especially in car manufacturing companies, that in almost each and every single car the customer may order one of hundreds possible options. If the production plans were stable and not influenced by disturbances, the orders could be grouped and produced in batches. Each batch could have specific takt time, also dynamic assembly line balancing could be performed for each one. This condition, however is rarely met, due to the fact that each time a batch is changed the whole system would have to be emptied and adapted to the requirements of the next production batch. This would, however, decreased the systems efficiency, as usually single assembly line consists of more than a hundred assembly stations. Moreover, batch production usually limits the number of possible products versions. Plant managers usually seek to utilize production resources remaining at their disposal to the maximum extent. The assembly system, chosen for the purpose of presenting the orders sequencing system has been presented on figure 2.

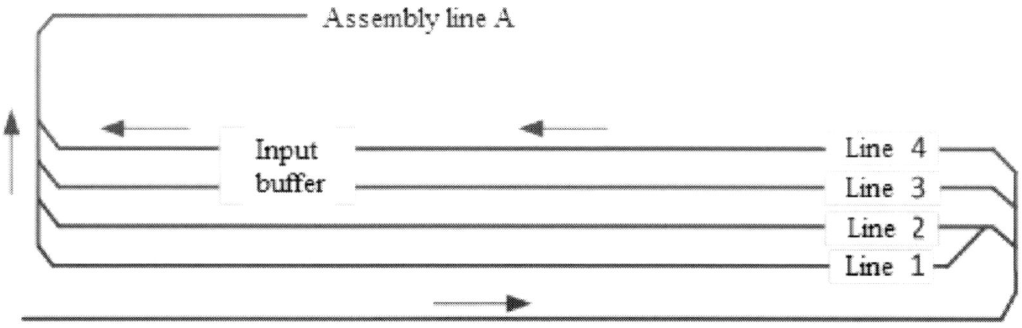

Figure 2. General overwiev of a selected assembly system.

As mentioned before, the system consists of an input buffer and a single assembly line. The input buffer is organized in four lines, on which the orders, transported via transporting system from the earlier stage of the manufacturing processes, are stored.

5. Orders sequencing software

For the purpose of sequencing and because of the fact that almost each car is different, data of each order is kept in a shared database. Data itself is created automatically after the order is taken from the client in the place of sale (usually a new record, containing information about the selected version, engine, options, etc.). The system itself performs sequencing in two stages (figure 3).

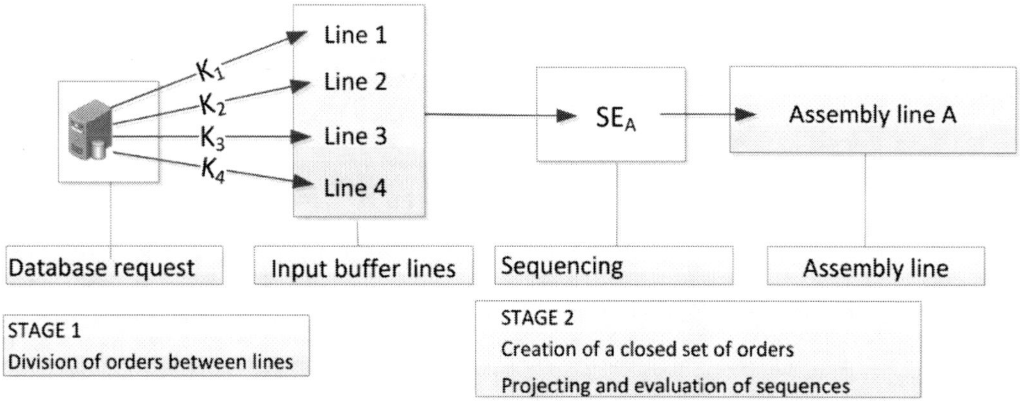

Figure 2. Main steps performed in orders sequencing system.

In the first stage the request for the data is sent, and according to the information obtained from the database orders are assigned to specific sets (K) and then sent to appropriate lines of the input buffer. The idea of the classification is quite simple, according to product labor-consumption matrix. Closed

IOP Conf. Series: Materials Science and Engineering **95** (2015) 012130 doi:10.1088/1757-899X/95/1/012130

set of orders, with the information about the lines of the buffer they are placed in has been presented in table 1.

Table 1. Closed set of orders stored in appropriate lines of the buffer.

Buffer line number	Orders
1	Z6, Z28, Z27, Z26, Z25, Z24, Z23, Z44, Z43, Z42, Z41, Z40
2	Z1, Z3, Z17, Z16, Z15, Z14, Z13, Z12, Z11, Z32, Z31
3	Z8, Z2, Z30, Z10, Z9, Z4, Z19, Z18, Z35, Z34, Z33
4	Z5, Z39, Z38, Z37, Z36, Z22, Z21, Z20, Z7, Z29, Z45

In the second stage, for orders in a closed set projecting and evaluation of sequences according to selected criterion is conducted. The effect of sequencing and evaluation in a form of best chosen sequence has been presented in table 2. The criteria used for the determination of the best sequence were the number of cycle times exceeded on the assembly line stations (ExC) and the average exceeding of the cycle time (AvEx).

Table 2. Best projected sequence according to selected criterion.

Best projected sequence	Criterion ExC	Criterion AvEx
Z5,Z39,Z6,Z8,Z2,Z38,Z37,Z36,Z22,Z30,Z1,Z21,Z3,Z17,Z20,Z7, Z29,Z10,Z16,Z45,Z15,Z14,Z28,Z27,Z13,Z26,Z12,Z11,Z25,Z24,Z 23,Z32, Z31,Z9, Z4,Z44,Z43,Z19,Z18,Z42,Z41,Z35,Z40,Z34,Z33	30	1,57

In the software there is also a possibility to generate schedules, on which assembly stations on which exceeding cycle times occurs may be observed. The example of such schedule has been presented on figure 4.

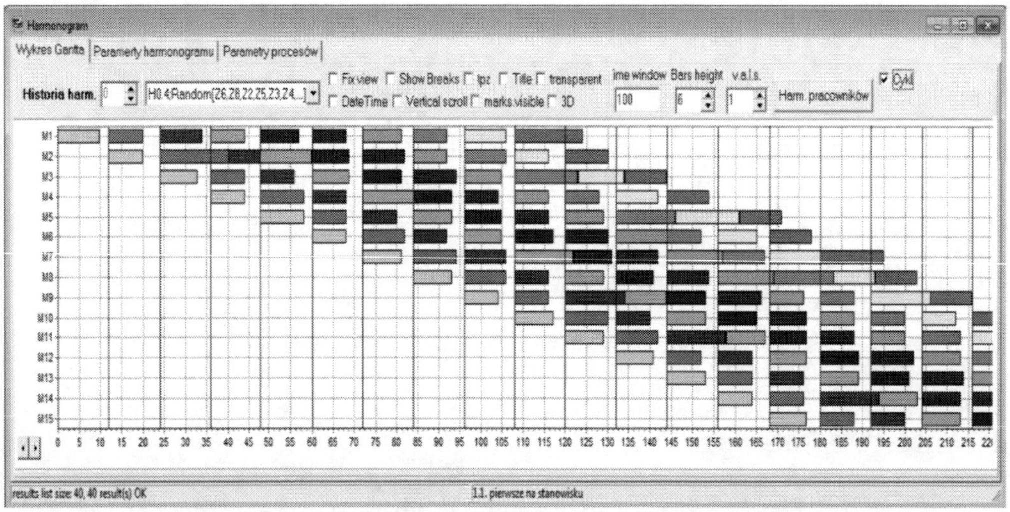

Figure 4. An example of schedule of projected sequence for a closed set of orders.

6. Conclusions

Application of advanced heuristic methods is not always practical, as there are many problems connected with the determination of quality of obtained solutions. However, the size and complexity of current scheduling problems eliminates the exact methods. One of the ways of solving these problems is connected with the possibility of joining both, exact and heuristic methods in problems solving. In presented paper, such solution has been presented. The first part of the system, basing on the simple rules determines the place in which order should be stored before admitting it to the assembly system. In the second step, basing on heuristic methods, the sequence is projected and evaluated according to selected criteria, the best is chosen and admitted to the assembly. This way, when problem is divided into smaller issues, the quality of solutions may be better. Many experiments are planned in order to prove the value of presented approach to scheduling problems solving.

Acknowledgement

This paper is a part of extensive work on tools for supporting production planning in mixed-model assembly systems.

References

[1] Janiak 1991 A Single machine scheduling problem with a common deadline and resource dependent release dates *European Journal of Operational Research* **53/3** pp 317-325
[2] McMullen P and Frazier G V 2000 A simulated annealing approach to mixed-model sequencing with multiple objectives on adjust in time line *IIE Transactions* **32/8** pp 679-686
[3] Zemczak M and Krenczyk D 2012 Formulation of a sequencing problem in a mixed-model production system *Journal of Machine Engineering* **12/3** pp 45-51
[4] Krenczyk D and Zemczak M 2014 Practical example of the integration of planning and simulation systems using the RapidSim software *Advanced Materials Research* **1036** pp 834-839
[5] Frey G 2000 Assembly line sequencing based on Petri-net T-invariants *Control Engineering Practice* **8/1** pp 63–69
[6] Srikanth K and Saxena B 2004 Improved genetic algorithm for the permutation flowshop scheduling problem *Computers & Operations Research* **31/4** pp 593-606
[7] Palshikar G K 2001 Simulated Annealing: A Heuristic Optimization Algorithm *Dr. Dobb's Journal* **26/9** pp 121-124
[8] Zemczak M 2012 Computer aided simulation of a mixed-model production system *Selected Engineering Problems* **3** pp 213-218

Theoretical and experimental researches on the operating costs of a wastewater treatment plant

M Panaitescu[1], F-V Panaitescu[1] and I-A Anton[1]
[1]Constanta Maritime University, Faculty of Naval Electro-Mechanics, 104 Mircea Street, PC 900663, Romania

E-mail: marianapan@yahoo.com

Abstract. Purpose of the work: The total cost of a sewage plants is often determined by the present value method. All of the annual operating costs for each process are converted to the value of today's correspondence and added to the costs of investment for each process, which leads to getting the current net value. The operating costs of the sewage plants are subdivided, in general, in the premises of the investment and operating costs. The latter can be stable (normal operation and maintenance, the establishment of power) or variables (chemical and power sludge treatment and disposal, of effluent charges). For the purpose of evaluating the preliminary costs so that an installation can choose between different alternatives in an incipient phase of a project, can be used cost functions. In this paper will be calculated the operational cost to make several scenarios in order to optimize its. Total operational cost (fixed and variable) is dependent global parameters of wastewater treatment plant. Research and methodology: The wastewater treatment plant costs are subdivided in investment and operating costs. We can use different cost functions to estimate fixed and variable operating costs. In this study we have used the statistical formulas for cost functions. The method which was applied to study the impact of the influent characteristics on the costs is economic analysis. Optimization of plant design consist in firstly, to assess the ability of the smallest design to treat the maximum loading rates to a given effluent quality and, secondly, to compare the cost of the two alternatives for average and maximum loading rates. Results: In this paper we obtained the statistical values for the investment cost functions, operational fixed costs and operational variable costs for wastewater treatment plant and its graphical representations. All costs were compared to the net values. Finally we observe that it is more economical to build a larger plant, especially if maximum loading rates are reached. The actual target of operational management is to directly implement the presented cost functions in a software tool, in which the design of a plant and the simulation of its behaviour are evaluated simultaneously.

1. Introduction

The total cost of a sewage plants is often determined by the present value method. All of the annual operating costs for each process are converted to the value of today's correspondence and added to the costs of investment for each process, which leads to getting the current net value. The operating costs of the sewage plants are subdivided, in general, in the premises of the investment and operating costs [1]. The costs can be stable or variables [2]. In this paper will be calculated the operational cost to make several scenarios in order to optimize its. Total operational cost (fixed and variable) is dependent global parameters of wastewater treatment plant [3].

Content from this work may be used under the terms of the Creative Commons Attribution 3.0 licence. Any further distribution of this work must maintain attribution to the author(s) and the title of the work, journal citation and DOI.
Published under licence by IOP Publishing Ltd

2. Research and Methodology

2.1. Fully parametric product model with parameters and relations

These costs are subdivided in investment and operating costs [3]. The investment costs (table 1) for a wastewater treatment plant can be quantified in comparison with the size of the process (area, volume flow) by using the laws of power or polynomial functions [4]. We choose Constanta North' waste water treatment plant (WWTP) for example. The functions for investment costs are depending on different parameters: flow rate (influent), capacity of oxygen (aeration system), volume of different equipments of each stage of WWTP (oxidation ditch, storage tank, thickener, aeration tank), area (settler), etc. We develop the cost functions for different scenarios (components of each stage of WWTP).

Example of investment cost (IC) functions may be found in the literature [5, 6]. Also, for mechanical stage of WWTP the investment costs functions (equation (1), equation (2), equation (3)) are:

$$IC= 2334\ Q^{0.637} \tag{1}$$

$$IC= 2123\ Q^{0.540} \tag{2}$$

$$IC= 3090\ Q^{0.349} \tag{3}$$

where, Q is the flow rate of influent, [m^3/h]. The values for these investment costs functions are presented in table 1:

Table 1. Investment costs functions for mechanical stage of Constanta North' WWTP.

Name of equipments	Cost function IC [Eu]	Parameter Q [m^3/h]	Range of parameter
Intake pumps	$2334\ Q^{0.637}$= 310561.697 $2123\ Q^{0.540}$= 134138.986 $3090\ Q^{0.349}$=45048.759	Q= 2160	Q= 250-4000

For second stage of WWTP the investment costs functions (equation (4), equation (5), equation (6), equation (7)) are:

$$IC= 2630\ A^{0.678} \tag{4}$$

$$IC=6338\ A^{0.325} \tag{5}$$

where: A= area of settler, [m^2], and

$$IC= 10304\ V^{0.477} \tag{6}$$

where: V= volume of aeration tank, [m^3],

$$IC= 8590\ OC^{0.433} \tag{7}$$

where: OC= oxygen capacity of aeration system, [kgO_2/h].

The values for these investment costs functions are presented in table 2.

Table 2. Investment costs functions for second stage of Constanta North' WWTP.

Name of equipments	Cost function IC [Eu]	Parameter, A[m^2],V[m^3], OC[kgO_2/h]	Range of parameter
Sedimentation system	$2630\ A^{0.678}$= 1212801.827 $6338\ A^{0.325}$= 119912.335	A= 8491, [m^2]	175-1250

Oxidation system	10304 $V^{0.477}$=1232104.617	V= 22680, [m^3]	1100-7300
Aeration system	8590 $OC^{0.433}$=14932.737	OC= 3.586, [[kgO_2/h]	30-630

For third stage of WWTP the investment costs functions (equation (8), equation (9), equation (10)) are:

$$IC=5038Q^{0.304} \tag{8}$$

where: Q= flow rate of sludge recirculation, [m3/h];

$$IC= 5559\ V^{0.473} \tag{9}$$

where: V= the volume of storage tank and thickener, [m^3];

$$IC= 3350\ Q^{0.363} \tag{10}$$

where: Q= flow rate of effluent, [m^3/h].

The values for these investment costs functions are presented in table 3.

Table 3. Investment costs functions for third stage of Constanta North' WWTP.

Name of equipments	Cost function IC [Eu]	Parameter, Q [m^3/h],V [m^3]	Range of parameter
Recirculation of mud	5038 $Q^{0.304}$= 17491.065	Q= 660, [m^3/h];	Q=35-2340;
Sedimentation tank and thickening agent	5559 $V^{0.473}$ =119850.805	V= 10.093, [m^3]	V=8-30
Effluent unit	13350 $Q^{0.363}$=37041.797	Q= 750, [m^3/h]	125-2130

Another investment costs functions are the costs for construction equation (11), infrastructures equation (12), electricity equation (13) and instrumentations equation (14). The formulas for these functions are [1, 6]:

$$IC= 6592\ Q^{0.498} \tag{11}$$
IC= 2 031 045 [Euro];

$$IC= 3873\ Q^{0.772} \tag{12}$$
IC= 14 989 033 [Euro];

$$IC= 16482\ Q^{0.383} \tag{13}$$
IC= 1 250 695 [Euro];

$$IC= 2438\ Q^{0.351} \tag{14}$$
IC= 158 364 [Euro];

where: Q= rate of flow, [m^3 /h].

The total investments costs are TIC = 18 429 137 [Eu].

2.2. Waste Water Treatment plant fixed and variable operating costs

We can use different cost functions to estimate fixed and variable operating costs (table 4, table 5).

IOP Conf. Series: Materials Science and Engineering **95** (2015) 012131 doi:10.1088/1757-899X/95/1/012131

Table 4. Fixed operating costs [2, 7].

Cost	Parameter, PE	Cost function	Symbols, units	Values/year [Euro]
Normal operation and maintenance	Population Equivalent, PE	$L = U_c * PE$ [1, 2]	L= labour, [man-hour/y]	231 111 111
			U_c=unit cost, [man-hour/y/PE]	925
Small equipments of second stage (N= 3)	Population Equivalent, PE	$C = N*U_c * PE$, [1, 2]	C = cost, [Euro/y] N= number of equipments'stage	693 333 333
Human costs	Population Equivalent, PE	$C = U_c *PE$, [1, 2]	C = cost, [Euro/y] U_c=unit cost, [man-hour/y/PE]	231 111 111

TOTAL fixed cost is ~1 155 556 480 [Euro].

Table 5. Variable operating costs [2, 8].

Cost	Parameter	Cost functions	Symbols and Units	Values/year
Pumping power	Flow rate Q [m^3/h]	$P = Q * \gamma * H / \eta$ [1, 2, 6, 8]	P= power, [kW]	4 610 758 [Euro]
			γ = specific weight of the liquid, [N/m^3]	1025 [N/m^3]
			H= dynamic head, [m]	2 [m]
			η = pump efficiency, [%]	85 [%]
Chemicals Consumption	C_n [kg]	$U_c * C_n$ [1],[2], [6]	U_c= unit cost, [Euro/kg chemicals]	1 664 000
Effluent taxes	COD (chemical oxygen demand), BOD5(biological oxygen demand), N(nitrogen), P (phosphorus), TSS (Total solids sediments)	$L=Uc (k_{org}.N_{org} + k_{nut}.N_{nut})$ [1, 2, 6]	U_c = unit cost, [Euro/kg] N_{org} = f (Q, BOD, TSS, COD) N_{nut} = f (Q, N, P)	73955555

TOTAL variable cost is ~ 80 230 314 [Euro].

2.3. *Waste Water Treatment plant total costs*

The total cost of a sewage plants is the sum of all cost, that means the value is TC= 5 577 642 876 [Euro/y]. All of the annual operating costs for each process are converted to the value of today's correspondence and added to the cost of investment for each process, which leads to getting net worth [2].

2.4. *The implementation of cost functions in a software tools*

For optimization the design of WWTP we can consider, firstly, to assess the ability of the smallest design to treat the maximum loading rates and, secondly, to compare the cost of the two alternatives for average and maximum loading rates [3].We can study the implementation of cost functions in a software tools. We can simulate with a modular, multi-purpose computer program a model of our plant, with graphical icons, select the process models, and assign parameters values. There are many sophisticated simulator, with on-screen sliders, swiches, and controls tools that help us to operating our plant in a virtual time (figure 1).

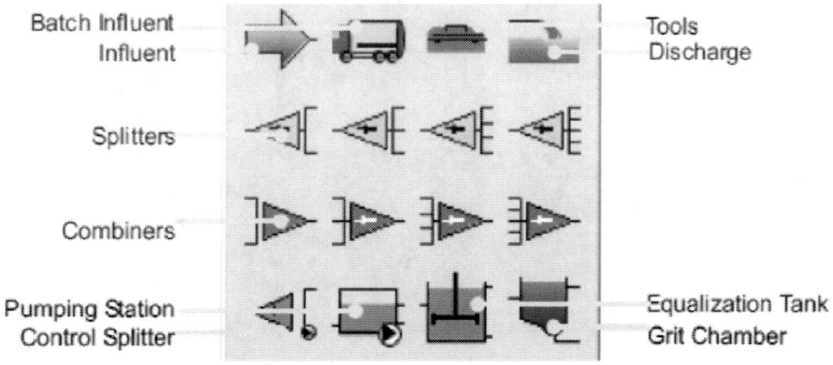

Figure 1. An example of software tools from a multi-purpose program [9].

With help of software tools and development of operation stages, we can perform automatic sensitivity analysis, which allows to clearly define the relationship between effluent quality and operational settings. Additionally, we can perform a multi-dimensional parameter search to find the best model inputs to achieve a desired outcome. On simulation, in technical literature [5] the results of this impact is analyzing on NPV (net present value) for different design of WWTPs, where the increase in the investment cost being compensated by a decrease of the operating cost. The sludge treatment and disposal represents the main operating cost, followed by the power consumption.

3. Conclusions

We can study the impact of the influent characteristics on the costs [1, 3, 5] by economic analysis. It is more economical to build a larger plant, especially if maximum loading rates are reached. The variable operating costs together fixed operating costs are very important values, too investment costs. If we want to optimize the design and operation of a plant, the definition of a standardized cost criterion is required to compare different treatment scenarios. The actual target of operational management is to directly implement the presented cost functions in a software tool, in which the design of a plant and the simulation of its behavior are evaluated simultaneously.

Acknowledgement

This This article is a result of the project "Increasing quality in marine higher education institutions by improving the teaching syllabus according to International Convention STCW (Standards of Training, Certification and Watchkeeping for Seafarers) with Manila amendments" POSDRU/156/1.2/G/141534. This project in co funded by European Social Fund through The Sectorial Operational Programme for Human Resources Development 2007-2013, coordinated by Constanta Maritime University.

The content of this scientific article does not reflect the official opinion of the European Union. Responsibility for the information and views expressed in the article lies entirely with the authors.

References

[1] Balmér P and Mattson B 1994 Wastewater treatment plant operation costs *Wat. Sci. Tech.* **30(4)** pp 7-15

[2] Panaitescu M and Panaitescu I I 2013 Evaluation of process parameters of the biological stage of wastewater treatment plant *Proceedings of 2nd International Conference on Energy and Environment Technologies and Equipment* (Brasov) pp 66-72

[3] Panaitescu I I, Robescu D N and Panaitescu M 2013 Operational management with application to the costs of a wastewater treatment plant (ARA'13) *Danube-Black Sea water regional forum* (Bucharest: Romania)

[4] Brett S W, Morse G K and Lester J N 1998 Operational expenditure in the water industry (I) : A methodology for estimating variable costs at an advanced water treatment works *European Water Management* **1(5)** pp 31-38

[5] Gillot S, De Clercq B, Defour D, Simoens F, Gernaey K and Vanrolleghem P A 2013 Optimization of wastewater treatment plant design and operation using simulation and cost analysis (Biomath Department, University of Gent: Belgium) *B-9000 Gent B-2610 Wilrijk*

[6] Vanrolleghem P A, Jeppsson U, Carstensen J, Carlsson B and Olsson G 1996 Integration of wastewater treatment plant design and operation - a systematic approach using cost functions *Wat. Sci. Tech.* **34(3-4)** pp 159-171

[7] Jacquet P 1999 Een globale kostenfunctie voor tuning en evaluatie van op respirometrie gebaseerde controle algoritmen voor actiefslibprocessen *Engineers Thesis* (Faculty of Agricultural and Applied Biological Sciences, University Gent, Belgium) pp. 122

[8] ASCE 1992 Standard measurement of oxygen transfer in clean water *American Society of Civil Engineers*

[9] www.hydromantis.com

Production scheduling with discrete and renewable additional resources

K Kalinowski[1], C Grabowik[1], I Paprocka[1] and W Kempa[2]
[1]Silesian University of Technology, The Faculty of Mechanical Engineering, Institute of Engineering Processes Automation and Integrated Manufacturing Systems, Konarskiego 18A Str., 44-100, Gliwice, Poland
[2]Silesian University of Technology, Faculty of Applied Mathematics, Institute of Mathematics, Konarskiego 23A Str., 44-100, Gliwice, Poland

E-mail: krzysztof.kalinowski@polsl.pl

Abstract. In this paper an approach to planning of additional resources when scheduling operations are discussed. The considered resources are assumed to be discrete and renewable. In most research in scheduling domain, the basic and often the only type of regarded resources is a workstation. It can be understood as a machine, a device or even as a separated space on the shop floor. In many cases, during the detailed scheduling of operations the need of using more than one resource, required for its implementation, can be indicated. Resource requirements for an operation may relate to different resources or resources of the same type. Additional resources are most often referred to these human resources, tools or equipment, for which the limited availability in the manufacturing system may have an influence on the execution dates of some operations. In the paper the concept of the division into basic and additional resources and their planning method was shown. A situation in which sets of basic and additional resources are not separable – the same additional resource may be a basic resource for another operation is also considered. Scheduling of operations, including greater amount of resources can cause many difficulties, depending on whether the resource is involved in the entire time of operation, only in the selected part(s) of operation (e.g. as auxiliary staff at setup time) or cyclic – e.g. when an operator supports more than one machine, or supervises the execution of several operations. For this reason the dates and work times of resources participation in the operation can be different. Presented issues are crucial when modelling of production scheduling environment and designing of structures for the purpose of scheduling software development.

1. Introduction

Formulating the problem of scheduling it is necessary to define the production resources first. The role of the decision maker at this stage is the proper selection of resources, which are directly involved in the execution of orders, and determination of the value of their parameters. The set of resources can be expanded iteratively, as a result of the feedback for further stages of the process of feasible schedule construction. This is due to the fact that production orders are specifying at the later stages of the scheduling, and that the decision-maker initially may not be sure which of the resources will be used for their realization. Modifications in the set of resources may also be associated with the production flow simulations concerning production capacity, the potential development of infrastructure and investment profitability.

Content from this work may be used under the terms of the Creative Commons Attribution 3.0 licence. Any further distribution of this work must maintain attribution to the author(s) and the title of the work, journal citation and DOI.
Published under licence by IOP Publishing Ltd

The general classification of production resources distinguishes three types: renewable, non-renewable and doubly constrained. Renewable resource is a resource that after the completion of one task can be load by the next one. Examples of this type of resources are machines, workers, tools etc. Limitations associated with the renewable resource concern the scope of its usage, e.g. capacity or availability at the given period of time. In such kind of system resources the number of resource units available at the every moment is usually limited. Non-renewable resources are not subject to regeneration process, and related constraints apply to their consumption. For a non-renewable resource its global quantity is mostly limited. Examples of non-renewable resources are the raw materials, funds or energy. In a production system doubly constrained resources may also be considered – for which both usage and consumption are limited. In this case limited availability at any given time, and total consumption are calculated. Double-limited resource can be e.g. energy (a medium) with limited power allocation and total consumption or funds, if the total capital of the entire project is defined and which is assigned to subsequent periods in a certain amount.

Scheduling problems, which take into account the limited availability of resources, are called Resource Constrained Scheduling Problems (RCSP) [1, 2, 3, 4]. In scheduling of manufacturing systems, in planning of activities the methods from the area of Resource Constrained Project Scheduling Problem (RCPSP) can also be used [1, 5].

From the scheduling point of view an important feature of resources is divisibility, according to which resources are distinguished discrete and continuous [6, 7]. Discrete resources are planned at fixed intervals and, in the planning period, they have a finite number of available and indivisible units of time to be used for tasks execution. Unlike the above, tasks can be assigned to the continuous resource at any time. Adoption of relatively small values of time unit and assumptions about considering the resources as a discrete, in fact, does not limit the availability of the resource and simplifies the planning of their work [8, 9, 10, 11].

In the rest of the paper only renewable and discreet type of resources are handled. The omission of non-renewable and double constrained resources has its justification in the practical problems of operation of manufacturing systems, among which the availability of positions, human resources, tools and accessories is essential for production planning.

2. Production resources

In most researches primary and often the only type of resource is a machine. A "machine" in this case can be understood as a literally machine but also as any kind of workstation, processor, device, workbench, a separate space, etc. In manufacturing systems, with a large number of resources, as a single resource can also be considered a production cell of a higher level, i.e. section, workcenter or production line, division, department or production plant in a specific location [12,13,14]. In this case, a schedule can be made in a hierarchical manner, by scheduling of higher level cells first, and then, more detailed, at lower level cells.

The production system is described by a set of resources. In a situation when some operations require more than one resource the set of resources $S=\{S_i, i=1,\ldots,m\}$ into two, not necessarily disjoint, sets $S = M \quad R$, including the main resources $M=\{M_a, a=1,\ldots,mm\}$ and additional resources $R = \{R_e, e=1,\ldots,mr\}$; where mm is the number of main resources, mr - the number of additional resources, and m - the number of all the resources, with relationship $m \leq mm + mr$.

In order to improve the process of scheduling of alternative resources - resource groups can be defined. $GRz = \{GRz_s, s = 0,\ldots,mg\}$, where GRz is a set of resources groups, $mg \geq 0$. A resource group indicates set of resources that can perform the selected operation or group of operations. The preference of choice (qualifications degree) of a i-th resource in the s-th group is expressed by the priority (weight) wzi,s. Resource membership of different groups represents the range of possibilities for its use, e.g. universality of production means, qualifications and skill level of the workers.

The production resource S_i (renewable, discrete) is described by:

$$S_i = (c_i, kz_i, Lgz_i, Ka_b) \tag{1}$$

where: c_i – the capacity of i-th resource, kz_i – the unit cost, Lgz_i – the list of resource groups, which include the i-th resource, Ka_b – b-th calendar, specifying the availability of the resource.

The resource capacity c_i refers to efficiency, and determines the maximum number of possible tasks, concurrently performed in the resource ($c_i \geq 1$). Resource with a capacity greater than 1 can process more tasks simultaneously. That resource can be treated as a set of resources with c = 1, assuming additional restrictions related to the synchronous loading/unloading, technological similarity etc. Unit cost of resource kz_i is a parameter used in calculation of the order cost in a given production route. List of groups of Lgz_i determines the ability to perform a given operation or group of operations, with the qualification degree expressed by priority. Membership of i-th resource is defined by resource list $Lgz_i \subset GRz$. The working time calendar (Ka_b) determines the availability of the resource for performing assigned operations. Resources in the production system work according to arbitrary generated calendars. The calendar is independent of the resource but resources can use the same calendar. The set Ka determines the system-defined calendars, $Ka = \{Ka_b, b=1,\ldots,mk\}$, where mk is the number of defined calendars. Working time calendar contains periods of availability $Ka_b = \{Od_{b,d}, d=1,\ldots,md_b\}$, where md_b is the number of availability periods in the b-th calendar. Each period is determined by the start date and finish date, $Od_{b,d} = (trd_{b,d}, tzd_{b,d})$.

3. Generation of calendar

Siemens Calendar time as the list of availability periods is generated from a set of calendar periods $Ok_b = \{Ok_{b,c}, c=1,\ldots,mc_b\}$, where mc_b represents the number of calendar periods defined in the b-this calendar. The calendar period is described by:

$$Ok_{b,c} = (tro_{b,c}, tzo_{b,c}, dost_{b,c}, cykl_{b,c}, typ_{b,c}), \tag{2}$$

where: $tro_{b,c}$ – starting date of the c-th calendar, $tzo_{b,c}$ – finishing date, $dost_{b,c}$ – availability of a resource, {available, unavailable},$cykl_{b,c}$ – periodicity, { $cykl_{b,c} = 0$ - non-cyclic, $cykl_{b,c} > 0$ - cyclic}, $typ_{b,c}$ – category of period {work, failure, maintenance, overtime, ...}.

On the basis of defined parameters of each period is assigned to one of four sets: cyclic periods of availability COD_b, cyclic periods of unavailability CON_b, non-cyclic periods of availability, NOD_b or non-cyclic periods of unavailability NON_b. The identification and aggregation of these periods enables creating the availability periods of a calendar (figure 1):

$$Ka_b = COD_b \setminus CON_b \cup NOD_b \setminus NON_b \tag{3}$$

4. Operations

The manufacturing process can be considered by the set of operations $P_k = \{O_{k,g}, g=1,\ldots,no_k\}$, where no_k is the number of operations in the k-th process. An operation $O_{k,g}$ is described by:

$$O_{k,g} = (Lwz_{k,g}, pre_{k,g}, re_{k,g}), \tag{4}$$

where: $Lwz_{k,g}$ – the list of resources requirements, $pre_{k,g}$ – the possibility of preemption, $re_{k,g}$ – the resumability (all applies to g-th operation in the k-th process).

The list of resources requirements ($Lwz_{k,g}$) determines the main and additional resources required for the operation executing. Alternative operations differ in this configuration of the required resources and are used for creating different routes of processes [15, 16, 17]. It was assumed that dates and work times of additional resources in the operation can be different (figure 2) so the list consists of set $Lwz_{k,g} = \{RR_{k,g,h}\}$, where $RR_{k,g,h}$ is a h-th requirement of $O_{k,g}$, and is described by:

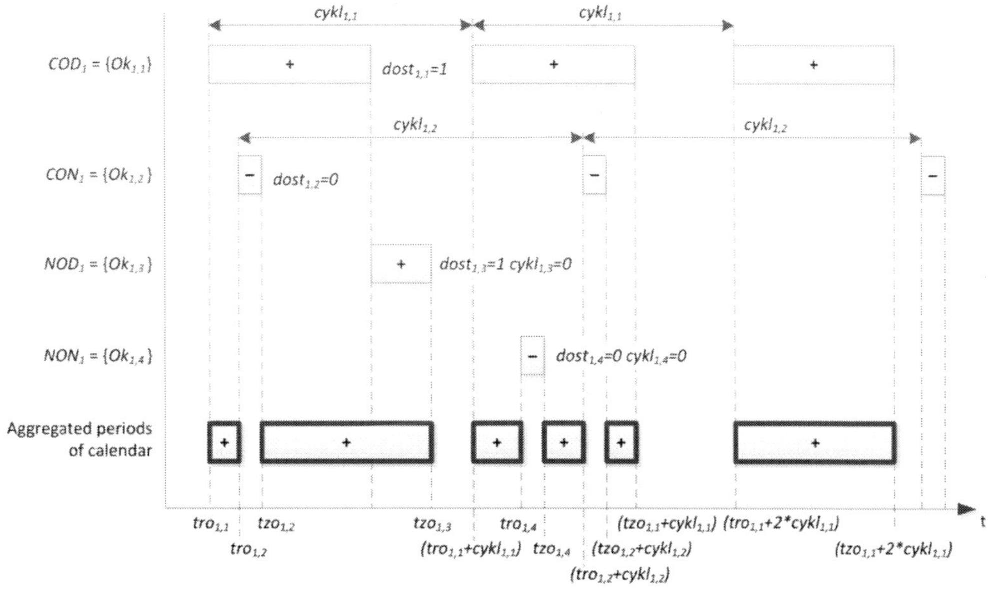

Figure 1. Construction of the resource work calendar.

$$RR_{k,g,h} = (So_{k,g,h}, Ls_{k,g,h}, cor_{k,g,h}, tj_{k,g,h}, tpz_{k,g,h}, ko_{k,g,h}), \tag{5}$$

where $So_{k,g}$ is a subset of the group of resources or specific resources required in g-th operation in the k-th process, $So_{k,g} \subset S$, $Ls_{k,g}$ – the list of quantitative requirements of particular resources in $So_{k,g}$, $cor_{k,g,h}$, - start date correction, $tj_{k,g,h}$ – the operation time, $tpz_{k,g,h}$ – the setup time, $ko_{k,g,h}$ – the cost.

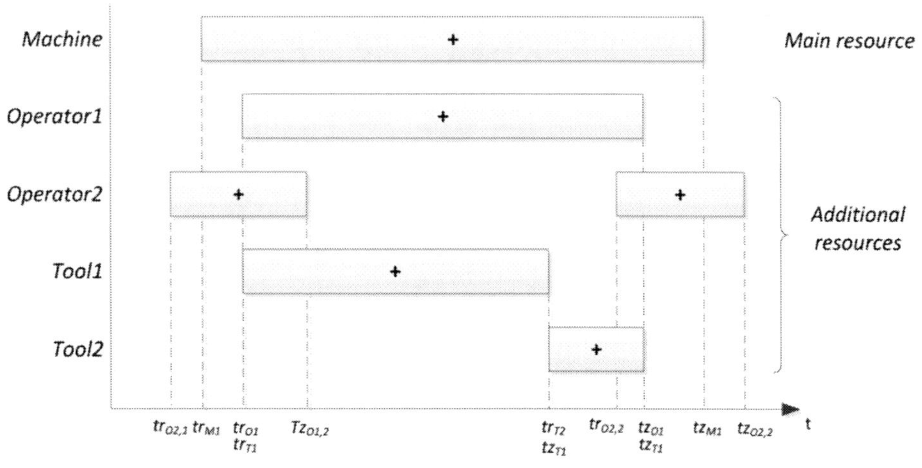

Figure 2. Resources requirements of an operation.

The setup time ($tpz_{k,g}$) can be fixed or variable, depending on the previous operation in the schedule. Taking into account the variable setup times allows precise scheduling of workstations, which setup times largely depends on the number of exchanged tools (if the capacity of tools magazine is not sufficient). This also allows minimizing setup times by sequencing of operations with similar tools requirements. The value of operations cost ($ko_{k,g}$) is used for calculation of the total cost of the given production order route. Preemption ($pre_{k,g}$) determines the admissibility of breaking execution of an operation, in order to carry out the other, e.g. with a higher priority. Resumability ($re_{k,g}$) determines if the operation may be resumed after the resource unavailability, resulting from the breakdown, calendar etc.

5. The method of operations scheduling with additional resources

The proposed method applies to operation scheduling with additional resources and is an extension of the method of scheduling presented in [17, 18]. Stages of the proposed method are presented in figure 3.

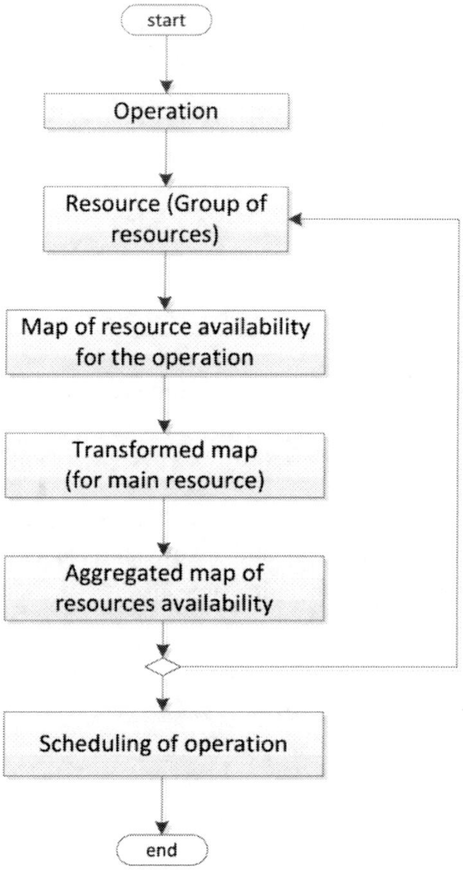

Figure 3. The prepartion to operation scheduling in case of additional resources.

At the first stage the identification of all the resources of operation to scheduling stored in $Lwz_{k,g}$ – the list of resources requirements is done. Then the procedure iteratively considers time constraints of each resource, starting from the main resource. Next, the map of resource(s) availability is created. If the item in the list indicates the resource group and more resources from the group – indicated is only the availability of required amount of resources. As was shown in figure 2 the working times of individual resources may vary from the time of main resource. In this case, particular availability maps should be transformed according to defined differences in dates and times of their work. Aggregation of all transformed maps of required resources availability enables scheduling of given operation.

6. Summary
Additional resources scheduling is one of the fundamental problems in most real manufacturing systems. Discussion is limited to systems with discrete production flow and considered resources are renewable (machines, tools, human resources). The paper presents the model of the production system and the general method for scheduling operations, as an extension of manufacturing system scheduling procedure, if more than one resource at the same time is required. The most important characteristics of resources, including time calendars are described. Such functionality greatly increases the usefulness of scheduling system for practical applications.

References
[1] Błażewicz J, Ecker K H, Pesch E, Schmidt G and Węglarz J 2007 *Scheduling computer and manufacturing processes* (Springer: Berlin)
[2] Van den Bergh J, Beliën J, De Bruecker P, Demeulemeester E and De Boeck L 2013 Personnel scheduling: A literature review *European Journal of Operational Research* **226** pp 367-385.
[3] Rattanatamrong P and Fortes J 2010 Real-time scheduling of mixture-of-experts systems with limited resources *Proceedings of the 13th ACM International Conference on Hybrid Systems: Computation and Control* (Stockholm) pp 71-80
[4] Kolisch R and Padman R 2001 An integrated survey of deterministic project scheduling *Omega* **29** pp 249-272
[5] Artigues C, Demassey S and Neron E 2010 Resource-Constrained Project Scheduling: Models, Algorithms *Extensions and Applications* (John Wiley & Sons: Portland)
[6] Edis E B, Oguz C and Ozkarahan I 2013 Parallel machine scheduling with additional resources: Notation, classification, models and solution methods *European Journal of Operational Research* **230** pp 449–463
[7] Slowinski R 1980 Two approaches to problems of resource allocation among project activities – a comparative study *Journal of the Operational Research Society* **31** pp711-723.
[8] Ćwikła G 2014 The methodology of development of the Manufacturing Information Acquisition System (MIAS) for production management *Applied Mechanics and Materials* **474** pp 27-32
[9] Zolkiewski S 2010 Dynamical Flexibility of Complex Damped Systems Vibrating Transversally in Transportation *Solid State Phenomena* **164** pp 339-342
[10] Dzitkowski T and Dymarek A 2013 Active synthesis of discrete systems as a tool for stabilisation vibration *Applied Mechanics and Materials* **307** pp 295-298
[11] Jurczyk-Bunkowska M 2013 Characteristics of Decision Problems In Innovation Process Planning *Proceedings of the 14th European Conference on Knowledge Management* (Kaunas) pp 795-804
[12] Sękala A, Banaś W and Gwiazda A 2014 Agent-based systems approach for robotic workcell integration *Advanced Materials Research* **1036** pp 721-725
[13] Hetmanczyk M and Michalski P 2014 The self-excitation phenomenon of quasi shielded inductive proximity switches *Advanced Materials Research* **837** pp 405-410
[14] Płaczek M 2014 Computer Aided Analysis of Systems with MFC Actuators *Advanced Materials Research* **1036** pp 711-714
[15] Kalinowski K, Grabowik C, Paprocka I and Kempa W 2014 The model of discrete production scheduling system in UML notation - classes diagrams *Advanced Materials Research* **837** pp 416-421
[16] Kalinowski K, Grabowik C, Paprocka I and Kempa W 2014 Interaction of the decision maker in the process of production scheduling *Advanced Materials Research* **1036** pp 830-833
[17] Kalinowski K, Grabowik C, Kempa W and Paprocka I 2014 The graph representation of multivariant and complex processes for production scheduling *Advanced Materials Research* **837** pp 422-427
[18] Kalinowski K, Krenczyk D and Grabowik C 2013 Predictive-reactive strategy for real time scheduling of manufacturing systems *Applied Mechanics and Materials* **307** pp 470-473

IOP Conf. Series: Materials Science and Engineering **95** (2015) 012133 doi:10.1088/1757-899X/95/1/012133

Multi-assortment rhythmic production planning and control

B Skolud[1], **D Krenczyk**[1] **and M Zemczak**[2]
[1]Silesian Silesian University of Technology, Faculty of Mechanical Engineering, Konarskiego 18A, 44-100 Gliwice, Poland
[2]University of Bielsko-Biala, Department of Production Engineering, Willowa 2, 43-309 Bielsko-Biala, Poland

E-mail: bozena.skolud@polsl.pl

Abstract. A method for production planning in a repetitive manufacturing system which allows for estimating the possibility of processing work orders in due time is presented. The difference between two approaches are presented; the first one one-piece flow elaborated in Toyota and the second one elaborated by authors that consists in defining sufficient conditions to filter all solutions and providing a set of admissible solutions for both the client and the producer. In the paper attention is focused on the buffer allocation. Illustrative examples are presented.

1. Introduction

Formulating Nowadays, two tendencies are observed in the industry, the first one: manufacturing of small quantity but in many varieties and the second one: manufacturing of small variety but in different quantity [1, 2, 3].

Both cases are characterized by small batches, what result with the need to shortening of the planning horizon.

It is not possible to elaborate a universal planning strategy suitable for every manufacturing condition. New approaches compete with each other in the context of the minimization of loses (lean manufacturing), which are related to the low level of resources utilization, minimization of production stock and storage costs. They reflect the producer's aim to increase profits and not to assure the satisfaction of manufacturing as such.

In this situation the most important thing is to be competitive. It means to assure lower price and precise time of production order completion then competitors. Because of that negotiation with clients about prices is impossible. To be better means to be cheaper. Costs of manufacturing decide about the profit [4, 5]. If the first estimating of production costs and terms of production order realization is not precise enough, then the profit is perceptible only in the phase of gaining of clients. In longer perspective it causes loose of the customer confidence as well as possibility of new orders gaining.

In the planning process the organizational variants, accomplishing limitations or subsetting the alternative solutions are determined, and then, the best variant is chosen basing on the additional criteria. Most cases of production planning and scheduling belong to a class of NP-hard problems. Many methods such as mathematical programming, computer simulation, application of artificial intelligence, are applied for solving these problems. A combinatorial explosion of possible variants makes it impossible to use an optimal solution in practice. Firstly, the feasible solution does not have

Content from this work may be used under the terms of the Creative Commons Attribution 3.0 licence. Any further distribution of this work must maintain attribution to the author(s) and the title of the work, journal citation and DOI.
Published under licence by IOP Publishing Ltd

IOP Conf. Series: Materials Science and Engineering **95** (2015) 012133 doi:10.1088/1757-899X/95/1/012133

to be an optimal one and the set of feasible solutions does not have to contain all the possible solutions.

The method presented in the paper guarantees meeting of the customer's demands. The constraint propagation technique is used. This approach proposes the creation of the sufficient conditions for filtering all solutions and it gives a set of admissible solutions for both the customer and the producer. The approach is presented with comparison to one piece flow philosophy proposed by Toyota.

2. Continuous flow

The methodology of continuous flow boils down to selection of products for simultaneous realization in the manufacturing system [6]. According to this idea product are classified into 4 groups. Products which are similar each other and characterized by high demand belong to the first group. In the contrary the 4th group contains products of low demand from the market and which are extremely differ from others (in design, size and technology). Products which are classified into the last (4th) group should not be allowed to perform in the considered system because of their big diversity and also because of time-consuming and expensive setup. Those elements should be commissioned in order to perform to another production cell or line or should be outsourced under cooperation to another company. In those systems an application of (Single Minute Exchange of Die) SMED techniques is recommended, in particular requalification of internal changeovers into external ones that are performed outside of the workstation, what, in practice, eliminated preset times.

With regard to products manufactured in the same system concurrently, one should be aware that their number does not have to be identical.

This approach is presented as an idea of the flow, but never is explained on the operational level. It is easy to imagine how it is carried out when one operational processes are analyzed. The problem appears when the process is multi operational and when processes have different production routes.

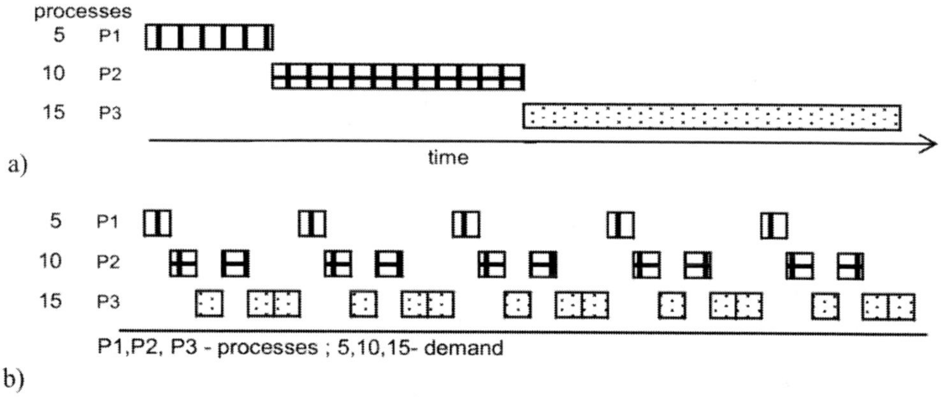

Figure 1. Production flow; a) batch production and b) one-piece flow [5].

In the figure 1 production flow is presented on the Gantt chart. It is visible how the flow is created starting from the longer period and batch production (figure 1a) and approaching to the single element flow (figure 1b).

2.1. Illustrative example

Below an example of the production flow on the operational level is presented. Given are processes described by matrices (1). First row of the matrix contains resources number from the production route, second one contains operations time respectively for these resources. Elements are manufactured one after one. Interoperations buffers are not provided in the system.

$$P1 = \begin{bmatrix} 1 & 2 & 3 \\ 4 & 3 & 3 \end{bmatrix}; P2 = \begin{bmatrix} 1 & 2 & 3 \\ 4 & 5 & 2 \end{bmatrix}; P3 = \begin{bmatrix} 1 & 2 & 3 \\ 2 & 3 & 3 \end{bmatrix} \quad (1)$$

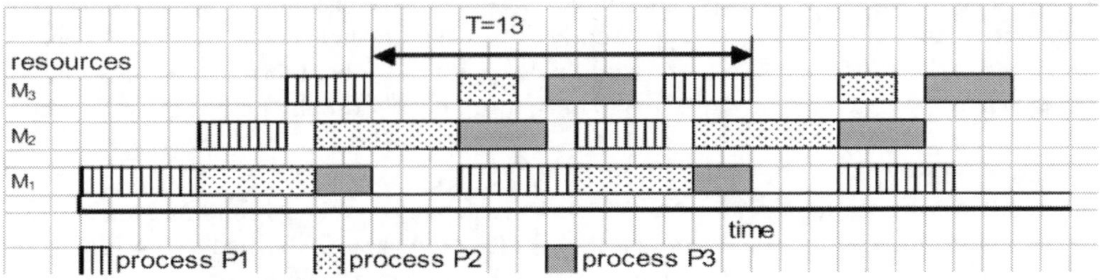

Figure 2. Production flow in the system without buffers.

The efficiency of resources utilisation in the repetitive period is calculated and is η=0, 74.

Let us assumed that between resources interoperational buffers are allocated. The production flow is presented in figure 3.

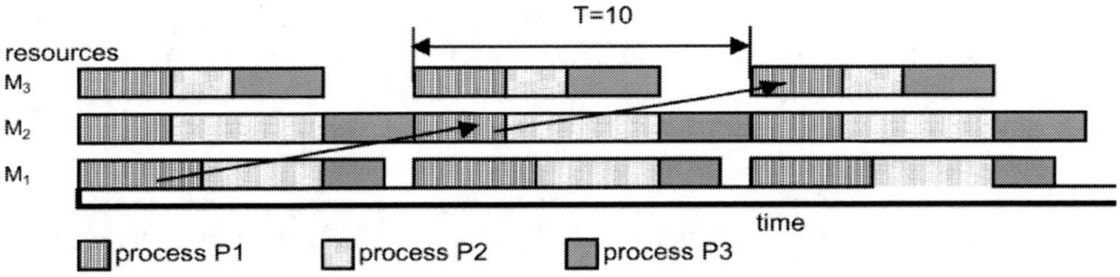

Figure 3. Production flow in the system with buffers allocated between resources.

The efficiency of the resources utilisation in the repetitive period is calculated and is =0,89 however this result can be obtained onlu if between resources buffers are allocated. This allows transfer the item to the buffer despite the fact that next machine in the route is busy. Interesting observations on buffers are presented in [7, 8].

3. Problem formulation

The production system is given and the set of production orders as well. Each production order is described by process with its route. Processes are concurrent, which means that before one finishes another process is already beginning. Each process can characterise different route with different number of operations and different flow direction.

The problem is: how to organise production flow to provide the best resources utilisation and order completion in due date (according to customer expectation) ?

The method bases on the synthesis of constraints. Theory of constraints (TOC) is applied in this approach. Constraints are used for creation the sequence of sufficient conditions that should be accomplish for a new production order acceptance for realisation in the given system. In the paper most of constraints are only mentioned. Authors of this paper concentrate mainly on constraints relating to the buffers capacity.

4. Buffers allocation in the context of the constraints propagation approach

The Buffers allocation and they capacities are one from list of constraints which should be satisfied in order to ensure deadlock free and starvation free functioning of the system and to ensure bottleneck utilisation according to theory of constraints (TOC) investigated and described by Goldratt.

Constraints in the production system are divided into two groups producers constraints and customers constraints.

One of the producer's constraints that are taken into account is the productivity of the system and limits of the possibility of the production order acceptance for realisation in the system, and permissible largest batch size, which may be processed by the system.

Other producer's constraints are buffers capacities that are available for realised processes. These constraints limit the batch sizes that may be processed in the given system. Sufficient conditions based on the producer's constraints are given in a form of theorems [9, 10, 11].

Accomplishing the producer demands assures that the system functions without deadlock and starvation. This fact does not guarantee the realisation of the production order in the time expected by the client. That is why other constrains like the time limit of the production order completion, should be taken into consideration.

If the production order cannot be realised in the expected time than the shortest possible realisation time is calculated.

The following question appears: what allocation of buffers capacity, for given batch production of order Z_{n+1}, providing the production orders (Z1, Z2, ..., Zn) realisation without disturbance?

The answer for this question is possible when the sufficient condition, determining buffers capacities, are known. Elements, executed in one cycle, are waiting for realisation in the next period. It causes that buffers capacity should be bigger than that one allocated for one batch production. This knowledge allows us to establish the condition for the synthesis of the system.

4.1. Illustrative example

Let's take the process P1 running by the M1 and M2 with times t1=1 and t2=4. The batch consists of p=3 elements.

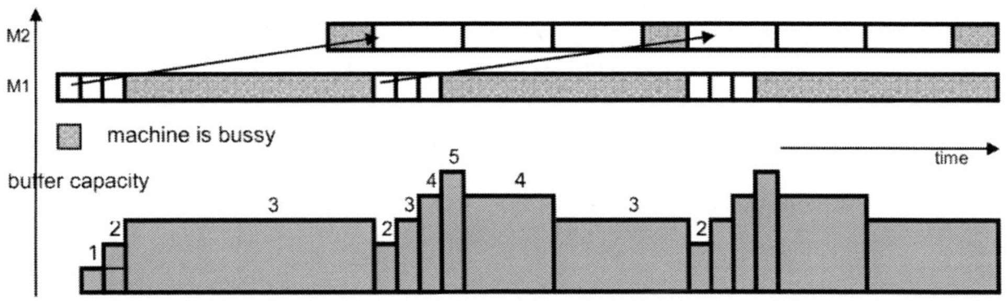

Figure 4. Buffers capacity allocation in the system.

Figure 5 present running processes and the capacity of the buffer indispensable for system functioning without disturbances. According to (2) required buffer capacity is 5 (figure 4).

4.2. Theorem

The system S is given. If the condition (3) concerning buffers capacity allocation holds, then realisation of the batch production, which consists of K elements, is possible without disturbance of production.

$$Cs(M_k, M_p) = 2*K-1 \tag{2}$$

where: Cs(Mk, Mp) - the capacity of the buffer allocated between k-th and p-th resource; K – the batch size.

Accomplishing of the condition from theorem assures completion of the batch of K elements.

If the realisation of two operations has place in the same period T, then sufficient buffer capacity allocated for this process between two neighbouring resources is equal to K?

In the system possible is a situation when one operation is executed in one period, and succeeding operation in the next one. In this situation buffer with capacity 2K-1 assures realisation of the production without disturbances.

5. Different flow direction

As was mentioned before, one-piece flow is always presented on the example of one- operational processes. The approach elaborated by authors of the paper allows for any production flow direction. The example presents multidirectional system of processes and also desired capacity of buffers which satisfied defined constraints.

5.1. Illustrative example

Let's assume a system of 4 resources (M_1, M_2, M_3, M_4). The following orders Z_1, Z_2, Z_3 are waiting for realization in the system. The routes of processes are presented in Fig. 5 and described by matrixes P_1, P_2, P_3. The first row of the matrix contains successive numbers of resources, the second row contains operations times on those resources, the third row contains pre-set time.

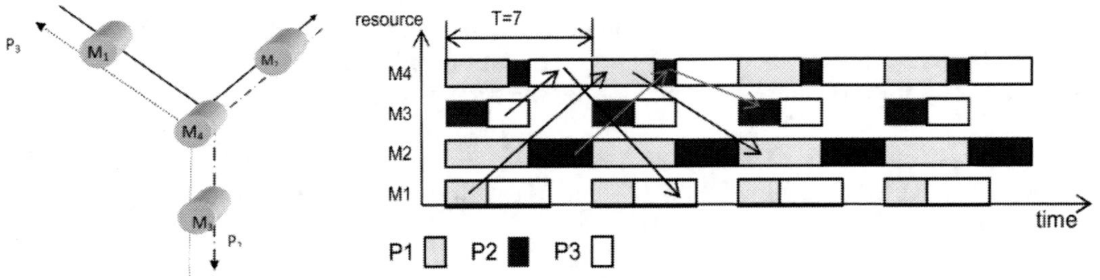

Figure 5. Flow of processes P1, P2, P3 in the system composed of 4 resources, a scheme of the flow and Gantt chart.

The coefficient of the resource utilisation is: $\eta = 0{,}821$. The required capacity of central storage is: Cs = 6 (or 1 place in each interoperational buffer), according to (2) .As was presented in the example the approach allows the analysis of the processes that have different routes.

6. Constraints sequencing

The constraints used for the synthesis of the system should be applied not randomly but in a precisely determined succession. To determine due time order completion (customer's demand), one should know the number of elements in one batch. This information is important only if it is possible to execute at least one element during one cycle. The qualitative functioning of the system means deadlock-free and starvation-free behaviour of the system. Considering these criteria it is possible to conduct the quality validation. This solution may not be the best one. It is very probable that the set of solutions does not contain an optimal one therefore the obtained solution is the best solution in a given sense of searching.

The conditions of qualitative system functioning determine the subset of admissible process realisations. Basing on this subset the satisfaction problem is formulated. This condition guarantees, the steady-state behaviour of the system and is a base of the analytical determination of the following parameters:

- the cycle time (repetitiveness);
- number of critical resources;
- capacity of storage system.

These parameters are the base of the determination of the solution which meets the elements that are definitely quantitative coefficient oriented (dedicated) to the producer or to the client.

7. Conclusion

In the paper the approach to the rhythmic production planning in the contrast of one-piece flow elaborated by Toyota is presented. Authors concentrate on constraints First of all the buffers allocation and its capacity was discuss as elements allowing critical resource creation according to TOC. Moreover it was presented an example which shows that this approach also relate to the system where processes are characterized by different flow directions.

The procedure of the synthesis can be applied in a computer system that aids decision-making. It can be used for assisting an engineer to give a quick answer to the question: Is it possible to realise the expected production order? And another question: Can the production order be executed in the time expected by the client? Presented approach was aplied in the The presented methodology constitutes for the "System of Production Order Validation" application. The system aids an engineer in decision making, allocating the dispatching rule, which co-ordinates the production flow.

References

[1] Banaszak Z, Skolud B and Zaremba M 2003 Computer-aided prototyping of production flows for a virtual enterprise *Journal of Intelligent Manufacturing* **14** Issue 1 pp 83-106

[2] Bocewicz G, Wójcik R and Banaszak Z 2014 Cyclic scheduling of multimodal concurrently flowing processes *Advances in Intelligent Systems and Computing* **240** pp 587-598

[3] Zemczak M and Krenczyk D 2014 A new procedure of production orders sequencing in mixed-model production systems *Advanced Materials Research* **1036** pp 864-868

[4] Hemmati S, Ebadian M and Nahvi A 2012 A new decision making structure for managing arriving orders in MTO environments *Expert Systems with Applications* **39 (3)** pp 2669-2676

[5] Skolud B 2014 Market oriented approach to the production management on the operational level *Advanced Materials Research* **837** pp 663-668

[6] Pool A, Wijngaard J and Van Der Zee D-J 2011 Lean planning in the semi-process industry, a case study *International Journal of Production Economics* **131(1)** pp 194-203

[7] Bahroun Z and Campagne J-P 2010 Medium Term Production Management for Cyclic Deliveries *Journal of Mathematical Modelling and Algorithms* **9** Issue 4 pp 311-342

[8] Bahroun Z, Campagne J-P and Moalla M 2006 Flow-shop cyclic scheduling for cyclic deliveries *12th IFAC Symposium on Information Control Problems in Manufacturing* (St Etienne: France) **3** pp 81-86

[9] Krenczyk D and Skołud D 2011 Production Preparation and Order Verification Systems Integration Using Method Based on Data Transformation and Data Mapping *Hybrid Artificial Intelligent Systems, Lecture Notes in Computer Science* **6679** pp 397–404

[10] Krenczyk D and Skołud B 2014 Transient states of cyclic production planning and control *Applied Mechanics and Materials* **657** pp 961-965

[11] Skołud B 2008 Approaches to the production flow scheduling and control. Intelligent manufacturing for industrial business process *Journal of Machine Engineering* **8/2** pp 5-13

IOP Conf. Series: Materials Science and Engineering **95** (2015) 012134 doi:10.1088/1757-899X/95/1/012134

The distributed agent-based approach in the e-manufacturing environment

A Sękala[1], G Kost[1], A Dobrzańska-Danikiewicz[1], W Banaś[1] and K Foit[1]
[1]Silesian University of Technology, Faculty of Mechanical Engineering,
Institute of Engineering Processes Automation and Integrated Manufacturing Systems, Konarskiego 18A, 44-100 Gliwice, Poland

E-mail: agnieszka.sekala@polsl.pl

Abstract. The deficiency of a coherent flow of information from a production department causes unplanned downtime and failures of machines and their equipment, which in turn results in production planning process based on incorrect and out-of-date information. All of these factors entail, as the consequence, the additional difficulties associated with the process of decision-making. They concern, among other, the coordination of components of a distributed system and providing the access to the required information, thereby generating unnecessary costs. The use of agent technology significantly speeds up the flow of information within the virtual enterprise. This paper includes the proposal of a multi-agent approach for the integration of processes within the virtual enterprise concept. The presented concept was elaborated to investigate the possible solutions of the ways of transmission of information in the production system taking into account the self-organization of constituent components. Thus it implicated the linking of the concept of multi-agent system with the system of managing the production information, based on the idea of e-manufacturing. The paper presents resulting scheme that should be the base for elaborating an informatics model of the target virtual system. The computer system itself is intended to be developed next.

1. Introduction

The complexity of the processes taking place within an enterprise causes that nowadays, in the conditions of global and most of all competitive economy, the efficient information flow between the enterprise components could have the decisive influence on its survival and development. The overriding goal of any enterprise should be the ability to adapt to the rapidly changing conditions of a competitive environment. Enterprises, which do not work comprehensively and put on the development of only certain areas of their activity, for example they focus only on costs decreasing, are on a hiding to nothing. Industrial practice shows that the much more effective strategy is to ensure the enterprise, as a whole, the possibility of rapid adaptation to function in changeable market conditions. This purpose could be achieved by integrating with each other manufacturing processes, technology, data about manufacturing resource and production systems, so that they cooperate together executing specified tasks. This integration should take place with the use of network solutions, treating as the priority to establish a closer relationship with the customer. This forces enterprises to better integrate systems in order to improve cooperation between the spheres of management, designing, production/manufacturing, engineering, supplying and distribution of a product [1, 3, 10-15].

Content from this work may be used under the terms of the Creative Commons Attribution 3.0 licence. Any further distribution of this work must maintain attribution to the author(s) and the title of the work, journal citation and DOI.
Published under licence by IOP Publishing Ltd

This method of system functioning is possible only if there is a suitable system for information exchange between existing levels of the enterprise organization and the resources existing at these levels. Nowadays, in the area of process control, the most frequently is performed the division of information systems in an enterprise on the so-called classes, in accordance with the role they perform [1, 4-7]. In classical organizational schemes it could be distinguished three basic levels that could be related with three management systems: strategic one, tactical ne and operational one. These systems, in the form of an organizational triangle are presented in figure 1.

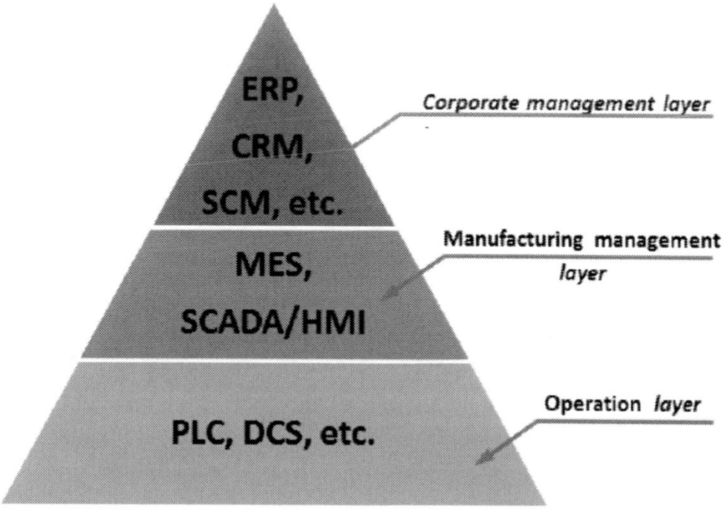

Figure 1. Classification of enterprise layers.

According to the presented classification it could be distinguished the following levels:

- ***Corporate management layer*** – layer concerning planning and enterprise resources management, and also relations with suppliers and customers (ERP-*enterprise resource planning*, CRM- *customer relationship management, SCM - supply chain management).*
- ***Manufacturing management layer*** - layer acting as an agent of information transferring between the operation layer and the corporate management layer (*MES - manufacturing execution system, SCADA - supervisory control and data acquisition, HMI - human–machine interface*).
- ***Operation layer*** – automatics systems responsible for controlling production units (*PLC - programmable logic controller*, DCS *-distributed control system).*

In the presented division it is not included the corporate layer, which is responsible for the laying out the direction of development of an enterprise. Modern business is a hybrid system in which coexist different solutions, as there are many different types of data sources concerning the state of a production system. The key element of the integration success of all or nearly all areas of an enterprise is the initial diagnosis and in-depth analysis of processes occurring in its interior, as well as their constituent elements and their proper modeling. Integration of the areas needs adequate distribution of tasks to dispersed subsystems cooperating with each other to collaborate together executing specified tasks [1].

2. E-manufacturing

The e-manufacturing philosophy is the concept of producing in accordance with changeable market needs. It was created as the response to current market needs, related with the dynamic development of the e-business strategy and the prevalent need for integration of activities within an enterprise, which are necessary for effective decisions making. The e-manufacturing consists of performing complete integration of production and business layers, basing on communication with the use of advanced information technology, using intelligent computer techniques (figure 2)[2, 8-9, 19]. Integration of the production layer and business one consists of creating a possibly rapid data transfer system from the production layer to the highest layers of enterprise management and transferring the decision from the management layer to the production one. The information exchange, in real time, enables dynamic making and implementation of strategic decisions. Therefore, the e-manufacturing environment could be defined as the system that integrates, in real time, production resources with suppliers and customers, using for this purpose the network-oriented computational tools and wireless technologies [16-21].

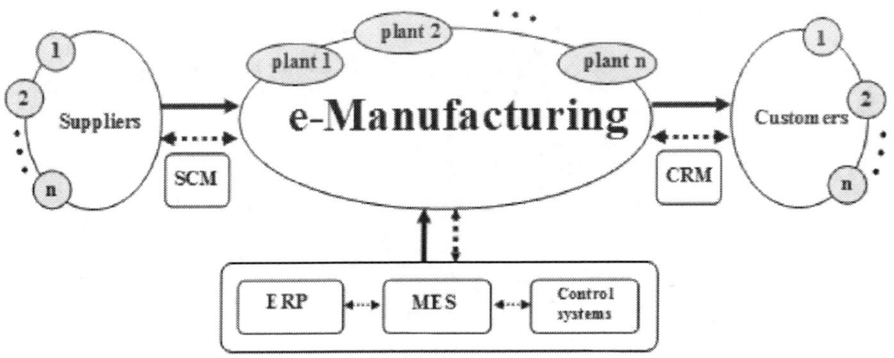

Figure 2. The e-manufacturing philosophy [19].

The most important element of the e-manufacturing concept is the common, multidimensional database that, in real time, gathers all necessary information, and which provides quick access to data and to various analyses. All operations carried out at the production level are referred to the management one.

The other reason for developing the e-manufacturing concept was the concept of a virtual enterprise or dynamic virtual manufacturing nets. In such production system the department organization exists only in the virtual reality. The real manufacturing departments are located in different factories, even in different countries. So the cooperation of such distributed production units is difficult and the profits of introducing and utilization of the informatics concept of e-manufacturing is the highest.

3. Application of multi-agent systems in an enterprise

The accessibility of a variety of technologically advanced equipment, computer applications while contributing to the technological progress, causes a number of inconveniences for an enterprise, both technical and business. Not always the communication between all levels/layers of an enterprise is well defined, with clearly designated borders of competency, but there is always a need for vertical communication with the process layer. This communication requires the solution of many problems, such as: lack of feedback in real time between the enterprise layers, costs associated with process implementation (employees hiring, tools and resources unification), existence of distortions in a data flow and the duration of each operation. Due to the universal and very general character the agent systems could be used in various scientific fields. Because of their ability to autonomously react to changes occurring in the environment, in which they are embedded, these systems are a convenient tool for analyzing, designing, and using in decentralized systems. In other words, multi-agent systems could be used wherever it operates a distributed resource of knowledge [22-27]. The complexity of the processes taking place in the e-manufacturing environment could be modeled as a multi-agent system, with generally defined objectives, which detailed specification may succeed during the process of system decomposition into smaller agent systems, which are, in some sense, independent. In figure 3 is shown the architecture of the proposed system, which is an extension of the approach proposed in [17].

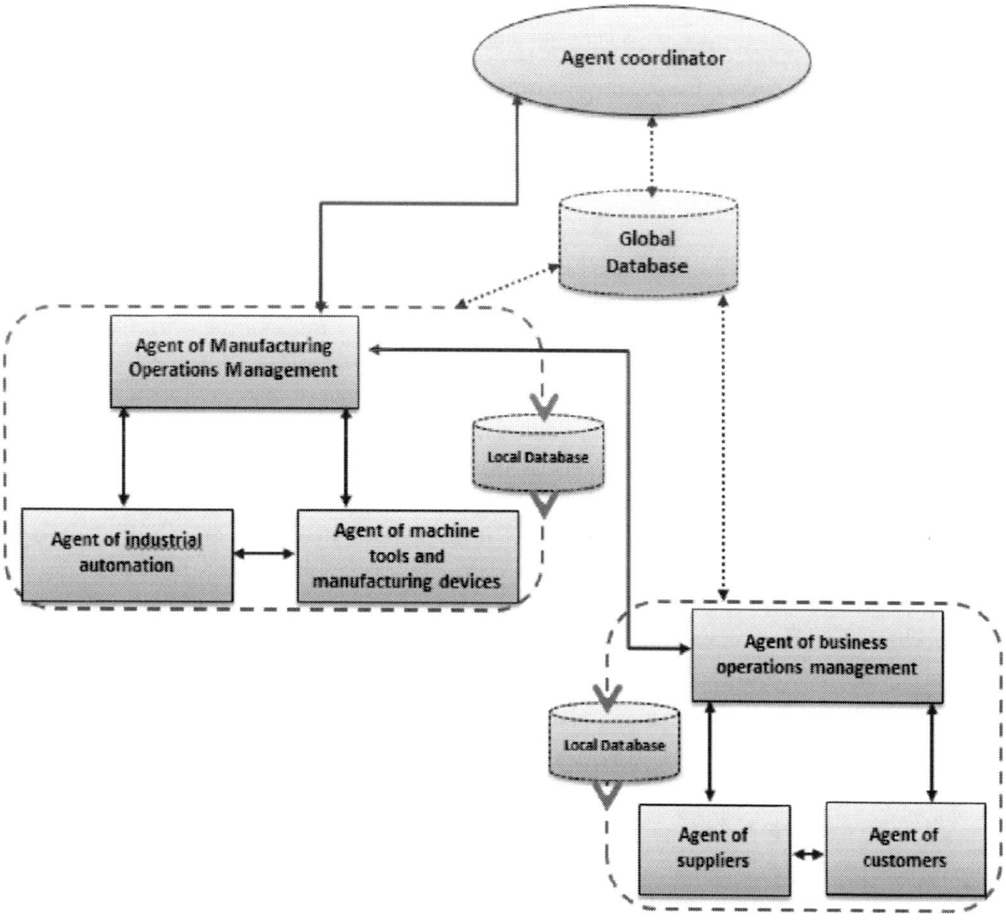

Figure 3. Architecture of the proposed system.

The proposed approach consists of a user interface, databases (local and global) and agents cooperating with each other. In the structure of the proposed system it could be observed the division on the so-called coordinating agents, which are responsible for the individual layers of an enterprise. The coordinators supervise the correct execution of the tasks by agent or group of agents assigned to them, plan tasks to perform and assign them to individual constituent elements (agents) within its own structure, communicate with the enterprise management system, as well as interact with each other. The global database collects information about the process state and the resources, whereas the agents have the ability to receive and analyze data, and to influence on the environment for example by reporting: the execution and/or the order of tasks, the current state of a process, the consumption of materials, the encountered problems, etc. Each of the agents contains detailed information on the tasks assigned to them. In addition, between all agents group there is cooperation, which is regarded as one of the most attractive features of the multi-agent approach and therefore, this approach opens up great opportunities related to the implementation of these systems in an e-manufacturing environment through direct data and information exchange between the components of the manufacturing process. The open structure of the system makes it possible to reconfigure the system, as well as add new control functions and modules.

4. Conclusions

The proposed approach is an extension of the concept presented in the work [17]. Multi-agent systems should be regarded as one of the tools supporting the activities of an enterprise, and not as an element necessary for its functioning. The main advantage of the MAS application is elimination of manual data processing, thereby reducing the cost of information acquisition. Through the use of multi-agent systems, it is possible to automate and optimize many processes that formerly required human intervention (e.g. equipment arrangement and equipment cooperation). These systems could aid or even replace a man first and foremost in many functions considered with decision making. They could decide faster and more accurately through taking into consideration more complex sets of data. Moreover, the open architecture provides data capturing, management of them and their comprehensive analysis in real time. The further stage of works will be determining the communication protocol between cooperating agents.

The elaborated concept could be characterized by some essential advantages. Firstly it is possible to use the existing, licensed software that is applied by our Institute. Secondly the integration process, in this approach, is based on the common data representation (in a global one and in local ones). The databases are the fundamental link allowing binding the system of separate program tools. It is however needed to elaborate proper interfaces for data format transferring. And finally the third advantage is related with the possibility of self-organization that could be conducted on the uniform data stored in the common databases.

References

[1] Ćwikła G 2014 Methods of manufacturing data acquisition for production management – a review *Advanced Materials Research* **837** pp 618-623
[2] Grabowik C, Kalinowski K, Paprocka I and Kempa W 2014 UML models of design and knowledge representation for technical production preparation needs *Advanced Materials Research* **837** pp 369-374
[3] Skołud B 2014 Market oriented approach to the production management on the operational level *Advanced Materials Research* **837** pp 663-668
[4] Grabowik C and Janik W 2013 The concrete casting matrixes inserts design preparation based on the master models *Advanced Materials Research* **702** pp 259-262
[5] Grabowik C, Ćwikła G and Janik W 2014 The new approach to design features identification *Applied Mechanics and Materials* **657** pp 750-754
[6] Hetmanczyk M P, Swider J and Wszołek G 2014 The analysis of the registration accuracy of distributed drives parameters *Advances in intelligent systems and computing. Recent advances in automation, robotics and measuring techniques* ed R Szewczyk C Zieliński M Kaliczyńska **267** (Springer: Berlin) pp 73-80
[7] Gwiazda A, Sękala A, Monica Z and Banaś W 2014 Integrated approach to the designing process of complex technical systems *Advanced Materials Research* **1036** pp 1023-1027
[8] Gwiazda A 2014 Construction development using virtual analysis on the example of a roof support *Applied Mechanics and Materials* **474** pp 417-422
[9] Gwiazda A 2008 Technical project management using quality methods *Journal of Achievements in Materials and Manufacturing Engineering* **31/2** pp 739-746
[10] Gwiazda A 2014 System of designing complex technical means using fuzzy analysis *Applied Mechanics and Materials* **474** pp 147-152

IOP Conf. Series: Materials Science and Engineering **95** (2015) 012134 doi:10.1088/1757-899X/95/1/012134

[11] Dzitkowski T and Dymarek A 2013 Active synthesis of machine drive systems *Applied Mechanics and Materials* **430** pp 178-183
[12] Dymarek A and Dzitkowski T 2013 Passive reduction of system vibrations to the desired amplitude value *J. Vibroeng.* **15/3** pp 1354-1364
[13] Foit K 2014 Introduction to solving task-level programming problems in logic programming language *Journal of Achievements in Materials and Manufacturing Engineering* **64/2** pp 78-84
[14] Buchacz A, Płaczek M and Wróbel A 2014 Modelling and analysis of systems with cylindrical piezoelectric transducers *Mechanika* **20/2** pp 183-189
[15] Białas K and Buchacz A 2014 The influence of changing the parameters of electrical components implementing the active reduction of vibration *Applied Mechanics and Materials* **657** pp 614-618
[16] Dobrzańska-Danikiewicz A 2006 The acceptation of the production orders for the realization in the manufacturing assembly systems *Journal of Materials Processing Technology* **175/1/3** pp 123-132
[17] Sękala A, Gwiazda A and Dobrzańska-Danikiewicz A 2015 Model of the e-Manufacturing Environment as the Multi-agent System *Applied Mechanics and Materials* **657** pp 854-858
[18] Sękala A and Dobrzańska-Danikiewicz A 2015 Possibilities of application of agent-based systems to support functioning of e-manufacturing environment *Solid State Phenomena* **220/221** pp 781-784
[19] Koc M, Ni J and Lee J 2002 Introduction of e-manufacturing *Proceeding of the International Conference on Frontiers on Design and Manufacturing* (Dalian: China) pp 43-47
[20] Lee J 2003 E-manufacturing-fundamental, tools, and transformation *Robotics and Computer - Integrated Manufacturing* **19/6** pp 501-507.
[21] Świder J, Sękala A and Michalski P 2007 e-F@ctory - The new style of manufacturing *Proceedings of the 9th Conference on Dynamical Systems: Theory and Applications* pp 805 -812
[22] Wooldridge M 2002 *An introduction to multi agent systems* (Wiley: Chichester)
[23] Wooldridge M and Jennings N R 1995 Intelligent agents, theory and practice *Knowledge Engineering Review* **10/2** pp 115-152
[24] Deen S M (ed) 2003 *Agent based manufacturing* (Springer Verlag: Berlin)
[25] Monostori L, Váncza J and Kumara S 2006 Agent-based system for manufacturing *Annals of the CIRP* **55/2** pp 697-720
[26] Rodriguez S, Gaud N, Hilaire V and Koukam A 2006 Modeling holonic systems with an organizational approach *Proceedings of EUMAS'06* A Omicini B Dunin-Keplicz J Padget Eds (Lisbon)
[27] Babiceanu R F and Chen F F 2006 Development and Applications of holonic manufacturing systems: A survey *Journal of Intelligent Manufacturing* **17** pp 111-131

IOP Conf. Series: Materials Science and Engineering **95** (2015) 012135 doi:10.1088/1757-899X/95/1/012135

Computer aided production planning – SWZ system of order verification

D Krenczyk [1] **and B Skolud** [1]
[1]Silesian University of Technology, Faculty of Mechanical Engineering, Konarskiego 18A, 44-100 Gliwice, Poland

E-mail: damian.krenczyk@polsl.pl

Abstract. SWZ (System of order verification) is a computer implementation of the methodology that support fast decision making on the acceptability of a production order, which allows to determine not the best possible solution, but admissible solution that is possible to find in an acceptable time (feasible solution) and acceptable due to the existing constraints. The methodology uses the propagation of constraints techniques and reduced to test a sequence of arbitrarily selected conditions. Fulfilment of all the conditions (the conjunction) provides the ability to perform production orders. In the paper examples of the application of SWZ system comprising the steps of planning and control is presented. The obtained results allowing the determination of acceptable production flow in the system - determination of the manufacturing system parameters those that ensure execution of orders in time under the resource constraints. SWZ also allows to generate the dispatching rules as a sequence of processing operations for each production resource, performed periodically during the production flow in the system. Furthermore the example of SWZ and simulation system integration is shown. SWZ has been enhanced with a module generating files containing the script code of the system model using the internal language of simulation and visualization system.

1. Introduction

Today's market is characterized by very high dynamics, recent trends can be observed to shorten the product life cycle, to reduce production lot sizes and shortening deadlines and time of the execution of order (figure 1). This is due to increased interest in a diverse range of products [1, 2]. A good example of the production areas are subject to such changes is electronic or automotive industry, in which the last years of was a period associated with the implementation of the concept of lean manufacturing.

The main principles of Lean Manufacturing are focusing on processes and flows of value streams, continuous improvement by eliminating the losses, which in turn leads to the continuous value flow. Implementation of Lean philosophy requires the widespread use of tools supporting the management process and the evaluation of changes in production, changes in the configuration of production systems, planning, scheduling, and control at the operational level.

The response to these needs is the concurrent production of multi-assortment, characterized by the production of a diversified range of products at the same planning period and high flexibility. Concurrent multi-assortment production systems make it possible to better adaptation of manufacturers to market expectations, allow the implementation of a short product lot sizes, and readily respond to changes in demand. These systems usually consist of a series of workstations, on which production processes are performed in an asynchronous manner.

Content from this work may be used under the terms of the Creative Commons Attribution 3.0 licence. Any further distribution of this work must maintain attribution to the author(s) and the title of the work, journal citation and DOI.
Published under licence by IOP Publishing Ltd

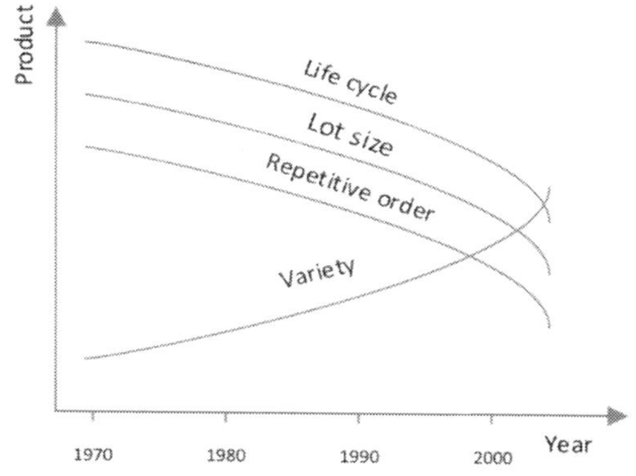

Figure 1. Market change [1].

Multi-assortment production systems can be characterized by cyclic production. In relation to the manufacturing process, it is characterized by the rhythmic character of the production flow and the simultaneous realization of various products [3, 4, 5, 6]. In this type of production, regular repetition of the manufacturing operations performed on the production resources, in which for every constant period T, the same sequence of operations is repeated for the resources and after the last operation the return to the first operation in the sequence. Because the characteristic feature of rhythmic production is the work of resources and equipment according to cyclic schedule of production flow, its important parameters (utilization of machines, means of transport, warehouses, production costs, punctuality, etc.) can be defined algebraically. This raises the possibility of forming the specified indicators of production such as the resources utilization, the level of inventory of work in progress on a high level. Therefore, it becomes necessary to support decision-making process in the production planning and control, which, depending on the level include, inter alia, the following tasks:

- division of a production lot size on the production batches;
- determine the timing and sequencing of production batches in conjunction with the required demand of material and the required efficiency of production resources;
- allocation of production system resources for each production batch;
- scheduling of production operations of each production batch;
- production flow control in the system;
- system monitoring and responding to the deviations from plans.

Realization of these tasks is related to the need to determine the allocation of system resources for specific tasks, resulting from the production plan. In the following sections of the paper authorial software to support production planning processes at the operational level is presented, including the above-mentioned functional areas. The functions and structure of the system of order verification SWZ and practical example of the capabilities is shown.

2. System of order verification SWZ

In previous works [5, 7, 8, 9] it has been presented decision-making methodology on the acceptability of a production order. Methodology is based on a sequence of selected conditions testing for generating sets of local dispatching meta-rules assigned to resources. The method of determining meta-rules is based on the constraint propagation methods described in [7, 10, 11]. The conjunction of all defined conditions (for the production system and production order identified constraints) guarantees the possibility of order execution (feasible solutions) (figure 2). The conditions, which are used in the methodology, include [5, 8]:

- system balance condition - the number of products entering into the system is equal to the number of processes leaving that system during one system cycle,

- buffer capacity condition - the capacity of the inter-resources buffer is equal or bigger than the realization number of the process during one system cycle.
- due time realization possibility condition – processes included in the production order will be executed within the due time required by the customer.

Lack of solution provides information about the necessary abandonment of specified conditions of the order, or having to meet the needs associated with an increase in available resource productivity, buffers capacity, etc.

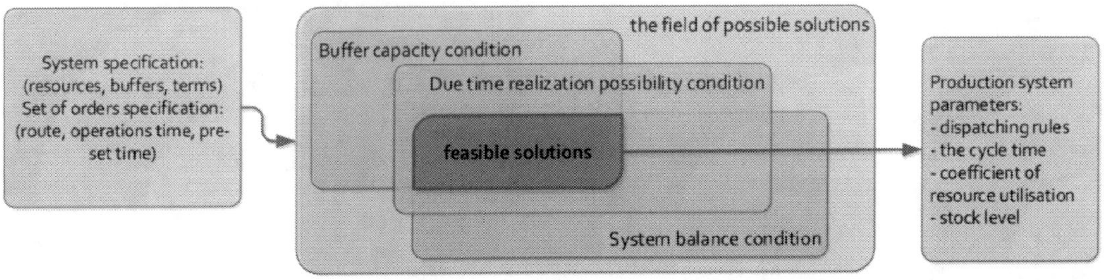

Figure 2. The conjunction of defined conditions.

This methodology has been implemented in orders verification computer system. The SWZ - computer production orders verification system, assisting in the decision-making processes of the acceptance or rejection of the production orders for production planning.

SWZ system enables determining the control procedures assigned to resources in the form of local dispatching rules for the steady-states of production flow and the transient dispatching rules for the starting-up and cease phases of production, used during the initiation and termination of orders in the production system. The generated rules are designed to ensure the acceptable production execution quality (without deadlocks and starvations) and timely execution of orders. SWZ also generates information about the necessity to reject the production order which do not have a chance to perform in due date with the identified resource constraints. Developed and implemented algorithms and methodology support operations of enterprises in different areas: area of works organization in the enterprise, during the development process of product and manufacturing processes design, in activities related to the maintenance and throughout the monitoring of works progress process and utilization of resources.

In the case of a positive decision on the acceptance of the order for execution, SWZ system determines the production flow control rules. However, if the solution does not guarantee the timeliness and currently realized production flow without disturbances, SWZ indicates the direct reasons for the rejection of newly introduced order.

2.1. Specification of the manufacturing system
The data specifying the system are: the number of orders planned for implementation, the number of resources in the manufacturing system, the type of warehouse: central or interop, available storage capacity. In addition, can add further information about the status of each of the stocks in each unit of cycle time.

2.2. Specification of the manufacturing orders
Orders specification for execution in the production system includes the following information: the lot size of the order, required (directive or planned) due dates, number of operations for each process, production routes, cycle times and setup times of operations.

Input data for SWZ are entered by filling the system and order data sheets (figure 3). The system state during the starting up of new jobs should be selected prior to entering the data into SWZ. Data on the setup and cycle time are entered into processes matrix (MPi) after the data on the number of operations for each production process.

2.3. Results of calculations

In case of order acceptance, a report is generated that contains:

- production batch size;
- job completion time for the accepted batch size;
- list of dispatching rules allocated to production resources, that coordinate the planned production flow in the system, including both a start-up rule and a cease rule for each phase of production,
- resources work schedule in the form of Gantt's charts,
- resources utilization level during the operation of the system.

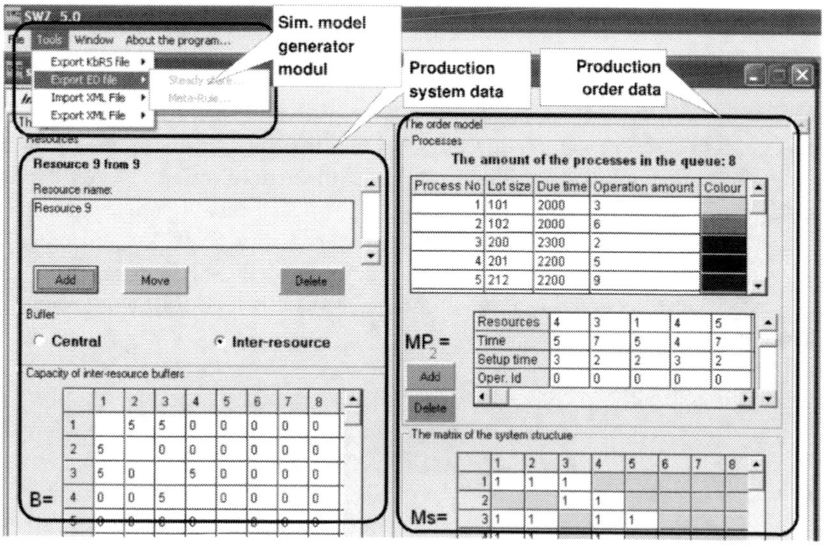

Figure 3. SWZ system interface.

Additionally SWZ system was equipped with a module of automatic generation of production system simulation models for simulation and visualization uses data transformation technique (figure 3). In the process of data exchange and transformation into the input file for the simulation software, Extensible Stylesheet Language Transformations - XSLT is used. The XSLT processor transforms data stored in SWZ automatically into file for the simulation system. The generated file containing the internal programming language code of simulation system, creating manufacturing system resources, i.e.: machines, inter-operational buffers, warehouses, input and output buffers, and information resources, i.e.: tables containing data about the setup and cycle times on production resources, tables containing the sequence of operations on resources, data on processes routes [12].

3. Illustrative example

Let's assume a production system, consisting of four resources M1 - M4. In the system three concurrent processes P1, P2 and P3 are performed (figure 4).

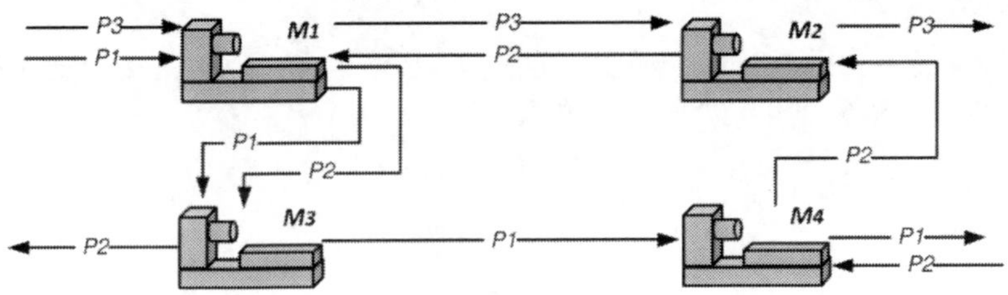

Figure 4. Production flow.

The production routes and times are recorded in the processes matrixes:

$$MP_1 = \begin{bmatrix} 1 & 3 & 4 \\ 4 & 2 & 3 \\ 0 & 0 & 0 \end{bmatrix},\ MP_2 = \begin{bmatrix} 4 & 2 & 1 & 3 \\ 5 & 6 & 5 & 8 \\ 0 & 0 & 0 & 0 \end{bmatrix},\ MP_3 = \begin{bmatrix} 1 & 2 \\ 4 & 3 \\ 0 & 0 \end{bmatrix} \quad (1)$$

The first row of the matrix corresponds to the resources over which the route of the process goes; in the second line cycle times on proper resources are given; the third line contains set-up times.

This data are complemented by the following values: the size of orders: P1 – 100 pcs, P2 – 100 pcs, P3 – 200 pcs, the due dates: tz1 = 2000, tz2 = 2000, tz3 = 2300, inter-operational buffers capacity: for all buffers 5 pcs was assumed. For the input data entered into SWZ system calculation has been executed. The system suggested change of the production batch of P3 process to 2. After accepting this change, all orders were accepted to realization in the system. The following report in SWZ system has been generated:

Local dispatching rules LRRKZ:
R1=(1,2,3,3)
R2=(2,3,3)
R3=(1,2)
R4=(1,2)

Trr (starting-up phase) = 50
Trw (cease phase)= 54

Repeatability of rules:
The rule allocated to the M1 resource - 1 time(s).
The rule allocated to the M2 resource - 1 time(s).
The rule allocated to the M3 resource - 1 time(s).
The rule allocated to the M4 resource - 1 time(s).
Buffers data:
The required capacity of the inter-resources buffer allocated between resources M2 and M1 is 2. Real capacity is 5
The required capacity of the inter-resources buffer allocated between resources M1 and M2 is 4. Real capacity is 5
The required capacity of the inter-resources buffer allocated between resources M4 and M2 is 2. Real capacity is 5
The required capacity of the inter-resources buffer allocated between resources M1 and M3 is 4. Real capacity is 5
The required capacity of the inter-resources buffer allocated between resources M3 and M4 is 2. Real capacity is 5

Rules realisation times at the resources :
Rule R1 = 17
Rule R2 = 12
Rule R3 = 10
Rule R4 = 8
System cycle : 17

Realisation times (steady state):
Process P1 = 1683
Process P2 = 1683
Process P3 = 1683

The coefficient of the system resources utilisation = 0,6911765

Rules containing the sequence of operation execution on resources in one cycle of the system have been generated. Realization times for the steady state and the maximum working time for starting-up and cease rules (*Trr, Trw*) were calculated. Resources utilization coefficient, which is in this case 69%, was determined.

In order to verify the correctness of the dispatching rules the simulation model of the production system for the Enterprise Dynamics simulation software has been automatically generated (figure 5).

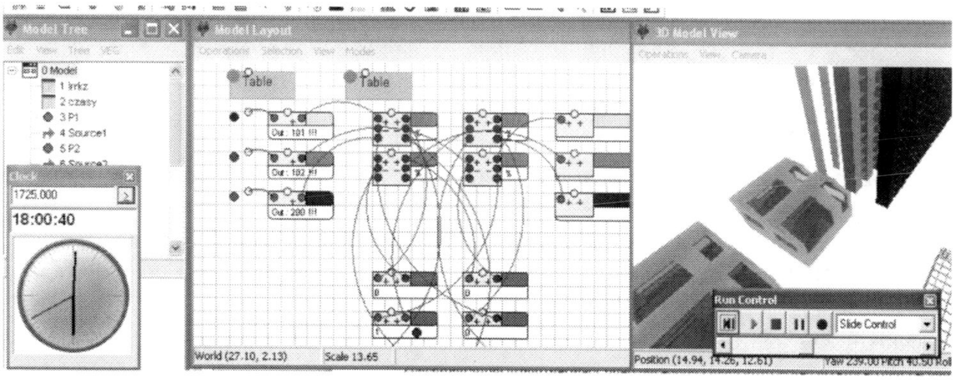

Figure 5. Simulation model.

The results obtained from the system operation simulation are consistent with the results obtained on the system SWZ. The total job completion time $Cmax$ = 1725 and is less than sum of realization times (for steady state) T=1683, max. time of starting-up phase Trr = 50 and max. time of cease phase Trw = 54.

4. Conclusions

A computer implementation of the support rapid decision-making methodology based on constraints propagation approach - SWZ system has been presented in the paper. The system aids an engineer in decision making, allocating the dispatching rule, which co-ordinates the production flow in the system (integrates the levels of planning and control). Two ways are possible: one for an empty system (usually when the set of orders is waiting for acceptance) and the other one for a system where other processes are performed (for a single order). The subject of further work in this field will be an extension of the methodology and SWZ system to transportation means, assembly processes and auxiliary operations.

References

[1] Wu C, Yu Shun F and Deyun X 2007 Computer integrated manufacturing *Handbook of Industrial Engineering* ed G Fellow (John Wiley & Sons: USA) pp 484–529

[2] Inman R A, Sale R S, Green K W and Whitten D 2011 Agile manufacturing: Relation to JIT, operational performance and firm performance *Journal of Operations Management* **29 4** pp 343–355

[3] Soltani S A and Karimi B 2015 Cyclic hybrid flow shop scheduling problem with limited buffers and machine eligibility constraints *The International Journal of Advanced Manufacturing Technology* **76(9-12)** pp 1739-1755

[4] Bocewicz G, Wójcik R and Banaszak Z 2014 Cyclic scheduling of multimodal concurrently flowing processes *Advances in Intelligent Systems and Computing* **240** pp 587-598

[5] Skolud B and Krenczyk D 2001 Flow synchronisation of the production systems – the distributed control approach *6th IFAC Workshop on Intelligent Manufacturing Systems* (Poznan: Poland) pp 127–132

[6] Kechadi M-T, Low K S and Goncalves G 2013 Recurrent neural network approach for cyclic job shop scheduling problem *Journal of Manufacturing Systems* **32(4)** pp 689-699

[7] Krenczyk D and Skolud B 2011 Production preparation and order verification systems integration using method based on data transformation and data mapping *Hybrid Artificial Intelligent Systems, Lecture Notes in Artificial Intelligence (subseries of Lecture Notes in Computer Science)* **6679** pp 397–404

[8] Krenczyk D and Dobrzanska-Danikiewicz A 2005 The deadlock protection method used in the production systems *Journal of Materials Processing Technology* **164** pp 1388–1394

[9] Krenczyk D, Kalinowski K and Grabowik C 2012 Integration Production Planning and Scheduling Systems for Determination of Transitional Phases in Repetitive Production *Hybrid Artificial Intelligent Systems, Lecture Notes in Artificial Intelligence (subseries of Lecture Notes in Computer Science)* **7209** pp 274–283

[10] Dobrzanska-Danikiewicz A and Krenczyk D 2015 The method of the production flow synchronisation using the meta-rule conception *Journal of Materials Processing Technology* **164** pp 1301-1308

[11] Lorterapong P and Ussavadilokrit M 2013 Construction scheduling using the constraint satisfaction problem method *J. Constr. Eng. Manage.-ASCE.* **139(4)** pp 414–422

[12] Krenczyk D 2014 Automatic generation method of simulation model for production planning and simulation systems integration *Advanced Materials Research* **1036** pp 825-829

IOP Conf. Series: Materials Science and Engineering **95** (2015) 012136 doi:10.1088/1757-899X/95/1/012136

Workload analyse of assembling process

L D Ghenghea[1]

[1]"Gheorghe Asachi" Technical University of Iasi-Romania, Department of Machine Manufacturing Technology, Blvd. Mangeron, No. 59A, 700050, Iasi, Romania

E-mail: dghenghea@yahoo.com

Abstract. The workload is the most important indicator for managers responsible of industrial technological processes no matter if these are automated, mechanized or simply manual in each case, machines or workers will be in the focus of workload measurements. The paper deals with workload analyses made to a most part manual assembling technology for roller bearings assembling process, executed in a big company, with integrated bearings manufacturing processes. In this analyses the delay sample technique have been used to identify and divide all bearing assemblers activities, to get information about time parts from 480 minutes day work time that workers allow to each activity. The developed study shows some ways to increase the process productivity without supplementary investments and also indicated the process automation could be the solution to gain maximum productivity.

1. Introduction

The assembling process could be considered the most important part of a technological process because it will put the manufacturer in position to go into the free market and take from it the monetary value of lot of work and resources introduced in the product. Few specialists could think that technological process relates with engineering, but nowadays is intensively used the term procedure for all profit or nonprofit activities, covered by quality management system. Speaking about technology like successive steps leading to fulfillment objectives, the humankind development could be seen in connection with technology starting with i) primary technologies till the first industrial revolution, ii) strong development of mature technologies that build modern society and finally with iii) emerging technologies, trying to solve the great problems of humanity. Technologies could use a great degree of automation but always the decisions and the control would be attributed to human that will have a great influence on efficiency of managed procedure for maximum efficiency in proper conditions for workers.

In this communication the work delay study method will be presented together with some results, obtain when it was applied for assembling radial bearings with cylindrical rollers made by teams of two or four qualified workers. This method of workload study has been introduced by L.H.C. Tippet [1] for textile industry in England and it is also known as: work sampling, random sampling, activity sampling, and ratio delay or observation ratio.

The mechanical shop where work sample study have been developed it is presented in figure 1 and here the technological process of bearings assembling it is carrying for a monthly figure around one million united states dollar and main type of bearings are radial and oscillate with small to big cylindrical or convex rollers, in steel or brass cages, in specific configuration like: NUP, NU, N type E for steel cage or type M for brass cage, dimensions related to radial cylindrical bearings or type C or M for oscillate bearings start with 50 mm inner diameter to 500 mm (example NU 210 E cylindrical

Content from this work may be used under the terms of the Creative Commons Attribution 3.0 licence. Any further distribution of this work must maintain attribution to the author(s) and the title of the work, journal citation and DOI.
Published under licence by IOP Publishing Ltd

IOP Conf. Series: Materials Science and Engineering **95** (2015) 012136 doi:10.1088/1757-899X/95/1/012136

rollers radial bearing with steel cage till NU 4100 M big cylindrical rollers radial bearing with brass cage) and for oscillate bearings from 22210 C with two steel cages to 22500 M with one brass cage). The production series had less than one thousand assembles for each shift and where no longer than two days of work, like a rule of the shop each assembling team is changed after a day of work to offer equal opportunity to all bearings assemblers in the shop.

Figure 1. Bearings assembling sector where the workload study took place.

The work-sampling method has been applied in the cylindrical rollers radial bearings sector for the bearings type presented in figure 2.

Figure 2. Bearings Radial bearings with cylindrical rollers type NU 210 E with steel cage (left side), N 210 M with brass cage (upper part of figure), NJ 214 M with brass cage (down part of the figure).

The structural difference between same number of cylindrical radial rollers bearings is presented in figure 3 and consists in type of materials for cage that is manufactured in steel or brass.

IOP Conf. Series: Materials Science and Engineering **95** (2015) 012136 doi:10.1088/1757-899X/95/1/012136

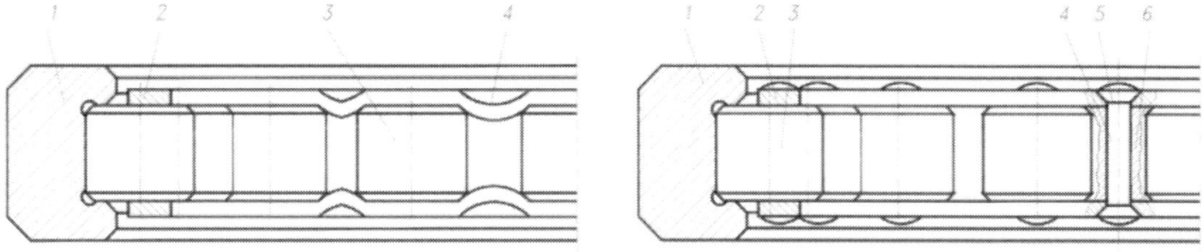

Figure 3. Cage type of cylindrical bearings (right side) the steel cage bearings, (left side) brass cage bearings, both figure: 1 outer ring, 2 cage, 3 cylindrical rollers, 4 rivet, 5 upper rivet head to be made in assembling process, 6 lid of the brass cage.

2. Phases of Workload Analyses

As method to analyse workload, work-sampling could be imagine that have been used by F. W. Taylor [2] the mechanical engineer considered founder of scientific management which used the observations in the mechanical shop he was leading, writing his communication at American Society of Mechanical Engineers titled The Shop Management and soon after that the fundamental book The Principles of Scientific Management. Taylor school of job description states these ideas: 1 eliminate idle times for workers and machines; 2 eliminate duplication of effort; 3 streamline the flow of work through the firm; 4 rearrange task sequences for more efficiency; 5 reduce jobs to short cycle, repetitive ones higher productivity and eventual replacement humans with machines; 6 systematic reduction of skill requirements for each job. Could be considered that Taylor's knowledge have been used by Ford in his automotive plant to manufacture the T Ford and to introduce assembling line and horizontal integration of parts production. In the 1930's Tippet based work-sampling techniques on the theory that percentage of observations for a particular activity, could reflect the average percentage of time spent accomplishing studied task.

The work-sampling study consists in a large number of observations made by an instructed technician, at random times, following a specific displacement trajectory and registering workers activities in predefined categories. Richardson [3] identifies utilizations of this measurement method as: 1 analysing an indirect or service activity that is largely unmeasured; 2 aiding in manpower requirements; 3 identifying problems within a work function; 4 establishing equipment usage problems; 5 aiding in supervision of employees; 6 identifying cycles within works and determining peak load variations. Smith [4] presented in 1978's main phases used in the majority of workload analyses based on work-sampling technique as follows: 1 determine objectives of the study; 2 define the scope of the study; 3 selection of the quantitative measures of output that reflect the condition being study; 4 establish the activity categories; 5 determine the number of observations needed; 6 schedule the randomized time for observations; 7 design the form for data collection; 8 decide who will conduct the study; 9 conduct an orientation meeting with workers to be observed; 10 conduct the study.

Assembling procedure is different and need teams with more members, just two for cylindrical roller radial bearings with steel cage (one to take rollers and cage to put them in a ring and other to press the assemble and stock products for future operations) and three for cylindrical rollers radial bearings with brass cage (one to take rollers and cage and put them in ring, other to introduce rivets to close brass cages and the last one to do hot forming of rivet head on a special press, for big cylindrical rollers bearings four members team are used because two people works to introduce rollers and cages into rings). The teams working for assembling cylindrical rollers bearings are specialized for each type and have lot of experience working together, in few cases when there lot of bearings in same type to made, the teams became larger generally working with two or three teams same bearing type. Assembling processes are very different for small and big cylindrical bearings and ask skills based on experience.

IOP Conf. Series: Materials Science and Engineering **95** (2015) 012136 doi:10.1088/1757-899X/95/1/012136

3. Workload study development and results

The first action in the workload study was to establish objectives, in a work meeting with assembling sector management when it have been decided that method could give important information's, have been establish type of assembling activities, the pathway to be followed by observers, random time establishment method and the template for data acquisition. The material flows in the sector have been established and those are presented in figure 4.

The bearings assemblers activities to be observed at random times where established as: 1 effective bearings assembling; 2 constitutive bearing parts transportation; 3 cleaning (demagnetize) constitutive parts; 4 dismantling bearings with deficiencies; 5 setting or testing installations or devices; 6 transport of assembled part of bearings to final quality control; 7 waiting constitutive parts of bearings; 8 receiving instructions for his chiefs; 9 refreshing breaks; 10 other activities (will be named by observer). These activities have been recorded by trained observers which work with templates presented in figure 5 at twenty randomized times in 480 minutes of an assembler workers day, registering specific types of assemblers established together with department management, observers followed same pathway for all observations and register activities.

After five days of observations made with observers in all three shifts working during twenty four hours in the Assembling and Mechanical Manufacturing shop for radial cylindrical rollers bearings in a meeting with the staff of that department the recorded activities have been once again put in classes presented in table 3, that reflected management interests about workload of bearings assembling workers. The results of work-sampling method applied to small radial cylindrical rollers bearings have been a surprise for department management that decided some future needed actions to achieve a greater effective assembling time, in the view of management have been measures like: new positions for assembling desks so to minimize movement of workers, bigger series of elements for assembling process, utilization of elements existing in rollers and rings from department warehouses, improvement of assembling elements transportation platforms.

Symbolization in table same in all rows: 1 current number of random observation during the day; JJ identification of bearing assembler or assembling team member; M activity number 1 from list meaning effective assembling bearings; T activity number 2 meaning transporting elements (rollers, cages or rings); V activity number 6 meaning verification of own work done till than; 30 symbolize all activities recorded in observation tour with short specification for those not closed by the list activities.

Table 1. Template for data acquisition used by observers.

No.	Names	Recorded activities from observations at fixed times														All obs/notes
		1	2	3	4	5	6	7	8	9	…	45	46	47	48	XXXXXX
1	JJ		M			T		M				T	V			30
2	KK		T			V		M				M	T			18
3	DG		W			O		T				M	Q			27
4	EF		M			T		R				C	T			10
5	AL		O			Q		C				T	M			30
…	…		…			…		…								
19	TD		M			V		T				M	T			29
20	VW		R			T		C				M	Q			29

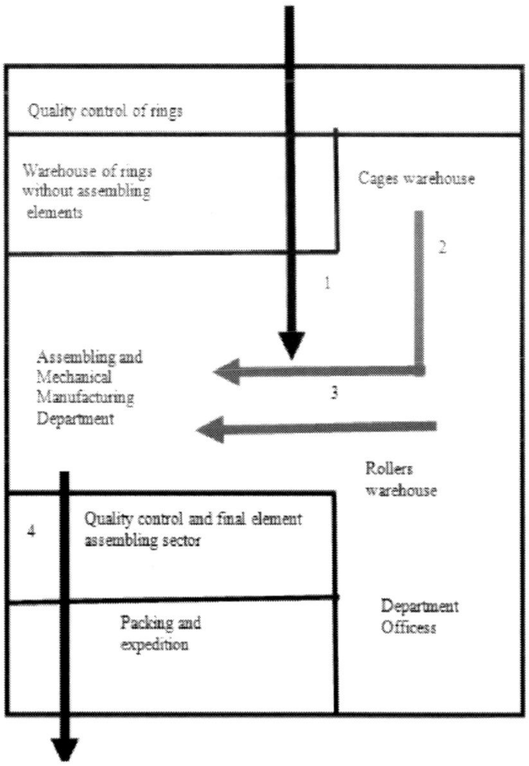

Figure 4. Production flow.

After five days of observations made with observers in all three shifts working during twenty four hours in the Assembling and Mechanical Manufacturing shop for radial cylindrical rollers bearings in a meeting with the staff of that department the recorded activities have been once again put in classes presented in Table 3, that reflected management interests about workload of bearings assembling workers. The results of work-sampling method applied to small radial cylindrical rollers bearings have been a surprise for department management that decided some future needed actions to achieve a greater effective assembling time, in the view of management have been measures like: new positions for assembling desks so to minimize movement of workers, bigger series of elements for assembling process, utilization of elements existing in rollers and rings from department warehouses, improvement of assembling elements transportation platforms.

Table 2. Final data after observations ending.

No.	Assembling activities	Observation from 100		Time from 480 min	
		Iron cage [%]	Brass cage [%]	Iron cage [%]	Brass cage [%]
1	Arrangement of ring and cages	10	9	28.8	25.9
2	Taking roller	8	9	23.4	25.9
3	Arrangements of roller	33	28	95.4	80.6
4	Cage deformation	29	15	83.5	43.2

5	Ensemble arrangement	10	9	28.8	25.9
6	Lid cage take out	-	3	-	8.6
7	Arrangement of lid cage	-	4	-	11.5
8	Arrangement of rivets	-	15	-	43.2

Table 3. Data interpretation for management objectives.

No.	Group of activities	Observations from 100%	Time from 480 min
1	Effective assembling bearings	46	220.80
2	Straight connected activities to bearings assembling operation	21	100.80
3	Not connected activities by bearings assembling operation	14	67.20
4	Activities which loose working time	16	76.80
5	Other activities	3	14.40
6	Total	100	480

4. Conclusions

This method use to study the workload have been used for the first time in that company and it underline an easy way to get important information about technological process and which way the work time is spend. It emphasizes some possibilities to increase productivity with low expenses. After this study between five to eight percent increase in productivity have been obtained with good collaborative suggestions from workers, also the problem of sorting rings operation have been solved totally, replacement of assembling shop was done and some improvements in warehouses for rings and rollers give a contribution to rise productivity cutting the waiting times.

This communication presents to readers and participants utilization of Tippet's work sampling method to estimate the workload of assembler bearings, also the data acquisition process and management appreciated results from Assembling and Mechanical Manufacturing sector from a big bearings manufacturing company.

References

[1] Information on http://en.wikipedia.org/wiki/L._H._C._Tippett, Accessed: 10.02.2015
[2] Information on http://en.wikipedia.org/wiki/The_Principles_of_Scientific_Management, Accessed: 25.01.2015
[3] Richardson W J 1976 *Cost Improvement Work Sampling and Shot Interval Scheduling* (Virginia: Reston Publishing)
[4] Smith G L 1978 *Work Measure, A Systems Approach* (Ohio: Grid Publishing)
[5] Disney S M 2009 Revisiting activity sampling: A fresh look at binomial proportion confidence intervals *Logistics Research Network Annual* (Cardiff University)
[6] Rawdon Y 2012 *Work Sample* (Software Engineering Institute Carnegie Mellon University: Pittsburgh)
[7] Vaccaro P A 2011 *Applied Management Science for Decision Making* (Pearson Prentice-Hall)

IOP Conf. Series: Materials Science and Engineering **95** (2015) 012137 doi:10.1088/1757-899X/95/1/012137

The relation between proactive environmental strategies and competitive advantage

A Butnariu[1] **and S Avasilcăi**[2]
[1] "Gheorghe Asachi" Technical University of Iasi-Romania, Department of Engineering and Management, Blvd. Mangeron, No. 59A, 700050, Iasi, Romania

E-mail: abutnariu@tex.tuiasi.ro

Abstract. There are two distinct orientations of the environmental management that companies may adopt: the model of compliance and the strategic model. The strategic model treats environmental expenses as investments that will lead to competitive advantage for the company. Nevertheless, there are few scientific works that prove the relation between corporate environmental investments and competitive advantage. Thereby, in order to bring clarifications about the profound implications of environmental investments, in the first stage of our research we have proposed the hypothesis that the environmental investments would probably lead to competitive advantage by creating capabilities that are mediators of this relation. In the second stage we have tested this hypothesis, using the research method of survey. A questionnaire was sent to managers in textile Romanian industry, and 109 answers were received. The data was analysed using the linear multiple regression method and the results confirm our hypothesis.

1. Introduction

From the moment the World Commission's Report on Environment and Development (known as the Bruntland Report) defined the term of sustainable development, in 1987, the managers of companies and management researchers searched reasons for "how" and "why" companies should incorporate environmental preoccupations in their strategic thinking. The relation between a proactive approach of environmental preoccupations and the firm's performance is a controversial issue in the literature. The reasons for that is because while some studies have proved a positive relation between the two [1], other studies do not identify a positive impact of environmental proactivity on financial performance, [2].

Our study is based on the resource based view of the firm and on the dynamic capabilities theory and its goal is to bring clarifications on the question "why" companies should consider adopting proactive environmental strategies in order to increase their competitive advantages and "what" is the mechanism that creates this advantage.

The resource based vision relates strategic advantages of an organisation to its resources, which are characterised by their strategic importance and by the organizations' ability to use them more efficiently than their competitors. The resources that offer added value to an organization are unique or valuable, difficult to imitate, not interchangeably. Among the most important resources, according this theory, are financial, physical, human and organizational resources [3].

The dynamic capabilities theory allows solving the deficiencies of the resource based vision of firm, with respect to considering the effects of time passing and evolution of capabilities. The dynamic

Content from this work may be used under the terms of the Creative Commons Attribution 3.0 licence. Any further distribution of this work must maintain attribution to the author(s) and the title of the work, journal citation and DOI.
Published under licence by IOP Publishing Ltd

capabilities are defined by Teece, [4] as the abilities to build, integrate and reconfigure the internal and external competencies in order to react to changing environments. The competitive benefits are perceived by the managers of proactive companies as the results of the strengths achieved by responsive environmental strategies. These strengths or capabilities are the coordinating mechanisms that facilitate the most efficient and competitive use of the firm's assets -tangible or intangible, [5].

The contribution of proactive environmental management to competitive advantage encompasses the cost and differentiation advantages, [6]. In this relation, the firm's capabilities should be considered a mediator variable. Wagner [2] identifies the importance for the firms to develop superior capabilities and resources, based on their relation with the natural environment as source of competitive advantage.

Table 1 presents a synthesis of the orientation of environmental management and sources of derived competitive advantage.

Table 1. The orientation of environmental management, sources and forms of competitive advantage.

The orientation of environmental management, sources and forms of competitive advantage		
Orientation of environmental management system	Sources of competitive advantage	Forms of advantage
Conformation model [7]	Environmental expenses are seen as costs, not as investments in competitive advantages	Potential advantage based on costs, but typically non-sustainable because this model suggests compliance with minimum environmental requirements
Strategic model	Environmental expenses are considered investments in firm's ability to create value for owners, clients and other stakeholders, through the development of superior competences or through adding basic resources, building a strategy based on competences [8]	Competitive advantage based on costs because of economies resulted from continuous improvements, "pollution prevention" programs, life cycle analysis, and potential lower expenses from litigations, lower insurance costs and lower energy costs And/or Competitive advantage based on a position of specialization/ differentiation which endorses the consumers sensitive to environmental problems or discovers/creates market performance as an entrance barrier. Enhanced image of the company could be a benefit of this positioning advantage [5].

According the resource based theory on firm, environmental strategies are put into practice by a group of simultaneous investments in several types of resources. Buysee and Verbeke [9] have empirically tested Hart's typology [7] and concluded that, by adopting a perspective of the resource based vision, an environmental strategy can be expressed as ungrouped investments in five distinct types of resources. These can be described as: 1.conventional environmental competences, which result from investments in productive products and processes. 2. training of employees in environmental issues; 3. formal management systems; 5. strategic planning that incorporates environmental issues. Buysee and Verbeke found that simultaneous investments in such resource domains are required to advance with respect to environmental proactivity. But they did not consider the profound consequences of these investments, like the development of the specific firm capabilities, which can lead to the achievement of competitive advantages.

Our study aims to complete the resource based theory of firm, by proposing the hypothesis that: "Investments in five types of resources (according Hart's theory) - employees' abilities, organizational competencies, formal management systems, strategic planning, will probably lead to the development of firm's capabilities".

2. Testing the hypothesis

In order to test our hypothesis we have used the survey method. A questionnaire was sent to 1000 managers of the textile firms in Romania. The companies were selected from the publicly available data bases. We have received 109 answers. The textile industry was chosen to test the hypothesis because this industry has evolved as a global, fragmented and complex system, that in order to reduce the social and environmental impacts along its products' life cycle uses an increasing number of corporate responsibility policies.

3. Measuring the environmental investments

The constructs that describe the investments in the five Hart type resource domains and their components are:

1. Investments in conventional environmental competences (products and productive processes) - resource domain 1:
- implementation of processes that reduce/eliminate the emissions of pollutants;
- elimination and treatment of hazardous waste;
- implementation of pollution control equipments;
- waste recycling;
- implementation of alarm/control systems for environmental accidents;
-systematic control of energy consumption;
- recycling of water used in the production process.

2. Investments in training and participation of employees in environmental aspects-resource domain 2:
- employees' training in environmental aspects.

3. Investments in organizational environmental competencies, which cross various functional areas of the company - resource domain 3:
- the company offers products and services that address consumers interested in environmental aspects;
- the products have been re-designed to reduce their impact on the environment.

4. Investment in management systems and formal procedures - resource domain 4:
- use of formal procedures to analyze environmental aspects for all the capital investments of the company;
- settling specific objectives for the environmental performance;
- Implementation of an environmental management system;
- auditing of the environmental activity;
- the existence of specific environmental standards to assess the suppliers.

5. Investments in strategic planning that incorporates environmental aspects - resource domain 5:
   - strategic environmental planning;
   - identification and assessment of new environmental aspects for their impact on the company on long term (five years or more).

The score assigned to every item has resulted from the answer to the next question: "How would you evaluate the effort of the company (investments, time allocated by the managers) regarding the next issues?" (on a scale of 1 to 5, where 1=there are no initiatives of this type, to 5 = very significant efforts)

The measure of every construct is the average of scores obtained by the items that compose every construct.

4. Measuring the organizational capabilities

In our study, we have used constructs based on the capabilities established by Sharma and Vredenburg [10], for which we have proposed component items [9]:

1. The capability to integrate the stakeholders

This capability implies the ability to establish collaboration relations based on trust with o wide variety of stakeholders, especially those with non-economic aims. These stakeholders can comprise local communities, environmental groups, regulation organizations, non-governmental organizations etc. The integration of stakeholders results not only from the product management, but also from the activities of habits conservation, resource management, waste reduction, energy conservation. The proactive companies engage the stakeholders in dialogues on the new locations, factory projects, and decisions on the waste recycling. The score of this construct was obtained by answering this question (on a Likert scale from 1 to 5): "What is the degree in which your firm is capable to collaborate with the stakeholders to find solutions for environmental problems?".

2. The capability of organizational learning is defined as the "development of knowledge, new senses and associations between past actions, the effectiveness of those actions and future actions". Learning in the organizations is described as the successful adaptation of the organization that evolves in an environment with rapid changes and as results of the behaviour based on common ideology and understanding of the changes that take place.

The capability to learn was calculated as the average of the scores obtained by these items (on a scale from 1 to 5):

-the degree in which the firm is capable to act before other firms in the sector;

-the degree in which the firm is capable to find opportunities in the environmental regulations shifts

3. The capability of continuous innovation

The learning processes, determined by environmental responsively strategies lead to changes of the organizational activities, procedures and objectives. Changes in technologies, processes, specifications, inputs and products can stimulate the achievement of internal capabilities and intangible assets based on knowledge. While environmental changes offer opportunities to the firm to be the first mover on a market, the probability that firm to benefit in a sustainable manner from this status will depend on the development of these capabilities. The scores of this construct are calculated as the average of scores obtained by the items (on a scale from 1 to 5):

-the degree in which the firm is capable to continuously innovate and enhance its operational activity while it reduces the environmental impact and

-the degree in which the firm has the ability to make continuous upgrading of its processes, products and systems.

5. The results of data analysis

In order to analyse the obtained data we have calculated the correlation coefficients between the two types of variables and then we have performed a linear multiple regression analysis with the SPSS software. The independent variables are the investments in the five resource domains (defined in the equation as X_1-X_5), whereas the dependent variables are the organizational capabilities (defined in the equation as Y_1-Y_3). We have performed three regression analyses, one for every capability as dependent variable. Table 2 presents the correlation coefficients between the analysed variables.

According to the values of R (Pearson correlation coefficient) in the table, we can conclude that there is a positive strong relation between the variables Investments in and the variables that describe the organizational capabilities. The R^2 is the coefficient of determination that indicates how well fits the data a linear model.

Table 2. Correlations between the environmental investments in resource domains and the capabilities that are mediators for achieving competitive advantages.

Correlations	Capability to integrate stakeholders		Capability to learn		Capability to innovate	
	R	R^2	R	R^2	R	R^2
Investments in resource domain 1	.509**	24.1%	.449**	20.2%	.547**	31.3%
Investments in resource domain 1	.500**	25.9%	.474**	22.4%	.499**	24.8%
Investments in resource domain 1	.447**	22.1%	.524**	27.5%	.519**	27.7%
Investments in resource domain 1	.494**	24.4%	.592**	34.7%	.606**	38.9%
Investments in resource domain 1	.472**	22.3%	.423**	17.9%	.569**	31.7%

***level of trust 99%*

The values of the Pearson coefficient for all the correlations are included in the interval 0.44-0.606, which shows a strong positive relation between the analysed variables.

We have used the multiple linear regressions with the SPSS software, a stepwise method, in order to eliminate the variables that have not an important influence on the dependent variable.

The model that describes the relation between the independent variables "Investments in resource domains" and the dependent variable "Integration of stakeholders" is statistically significant and is represented by the equation:

Integration of stakeholders (Y_1) = 1.788 + 0.398 * investments in resource domain 2 (X_2) + 0.287*investments in resource domain 3 (X_3)

The investments in resource domains 2 appear to have the strongest positive influence on the capability of stakeholder integration.

The model that describes the relation between independent variables "Investments in resource domains" and the dependent variable "Capability of organizational learning" is statistically significant and is represented by the equation:

Organizational learning (Y_2) = 1.405 + 0.437* investments in the resource domain 4 (X_4) + 0.213*investments in resource domain 3 (X_3)

It appears that the capability of learning is most influenced by the investments in resource domain 4 and investments in resource domain 3.

The model that describes the relation between independent variables "Investments in resource domains" and the dependent variable "Capability of continuous innovation" is statistically significant and is represented by the equation:

Capability of continuous innovation (Y_3) = 1.620 + 0.446* investments in resource domain 4 (X_4) + 0.224*investments in resource domain 5(X_5)

It appears that the investments in resource domain 4 and in resource domain 5 have the strongest influence on the organizational capability of continuous innovation.

Thus, the stated hypothesis is partially confirmed. Sustained investment in resource domains 2 and 3 have the strongest positive influence on the development of the capability to integrate stakeholders, while investments in resource domains 3 and 4 positively influence the capability of organizational learning. Also, investments in the resource domain 4 and investments in resource domain 5 favour the development of the innovation capability.

6. Conclusions

The sustained investments in all resource domains are highly correlated with each other. This is the reason for which the stepwise regression analysis eliminated from the models some types of investments. These correlations support the assertion that including the environmental aspects in strategic planning (resource domain 5) raises the employees' opportunities to practice and extend their knowledge on environmental protection (resource domain 2), and also to enhance the conventional environmental competencies of the company (resource domain 1) through all functional areas (resource domain 3). As a result, employees can reconfigure the formal management systems and procedures (resource domain 4).

We can conclude that not generally the environmental strategies influence the development of corporate strategies, but the investment in several resource domains. The capabilities are the result of sustained and simultaneous investments across these multiple resource domains, rather than of focusing on only one area.

The textile industry is characterized by a global competition, the shortening of products' life cycles and the decline of entry barriers. In this context, the resource based vision of the firm can offer an additional guidance to develop competitive strategies. The complex and comprehensive environmental strategies put into practice by investments in Hart type resource domains could be a new important competitive field to which the leader companies must pay attention, as they can positively influence the development of organizational capabilities that provide competitive advantages.

References

[1] Aragón-Correa J and Rubio-López E 2007 Proactive Corporate Environmental Strategies: Myths and Misunderstandings Long Range Planning **40 (3)** pp 357–381
[2] Wagner M 2005 How to reconcile environmental and economic performance to improve corporate sustainability: corporate environmental strategies in the European paper industry *Journal of Environmental Management* **76** pp 105–118
[3] Barney J 1991 Firm Resources and Sustained Competitive Advantage *Journal of Management* **17** pp 99-120
[4] Teece D, Pisano G and Shuen A 1997 Dynamic Capabilities and Strategic Management *Strategic Management Journal*, **18(7)** pp 509-533
[5] Hamel G and Prahalad C 1989 Strategic Intent *Harvard Business Review* **67(3)** pp 63–76
[6] González-Benito J and González-Benito O 2005 Environmental proactivity and business performance: an empirical analysis *Omega* **33(1)** pp 1-15
[7] Hart S 1995 *A natural-resource-based view of the firm Academy of Management Review* **20 (4)** pp 986-1014
[8] Nehrt C 1996 Timing and Intensity Effects of Environmental Investments *Strategic Management Journal* **17** pp 535–547
[9] Buysse K and Verbeke A 2003 Proactive environmental strategies: a stakeholder management perspective *Strategic Management Journal* **24** pp 453–470
[10] Sharma S and Vredenburg H 1998 Proactive corporate environmental strategy and the development of competitively valuable organizational capabilities *Strategic Management Journal* **19** pp 729–753

Poka Yoke system based on image analysis and object recognition

N Belu[1], L M Ionescu[2], A Misztal[3] and A Mazăre[2]
[1]University of Piteşti, Faculty of Mechanics and Technology, Department of Manufacturing and Industrial Management, 110040 Pitesti, Targu din Vale Street, no. 1, Argeş, Romania
[2]University of Pitesti, Faculty of Electronics, Communcations and Computers, Department of Electronics, Computers and Electrical Engineering, Str. Targu din Vale, nr.1, 110040, Pitesti, Romania
[3]Poznan University of Technology, Faculty of Engineering Management, 60-965 Poznan, ul. Strzelecka 11, Poland

E-mail: nadia.belu@upit.ro

Abstract. Poka Yoke is a method of quality management which is related to prevent faults from arising during production processes. It deals with "fail-safing" or "mistake-proofing". The Poka-yoke concept was generated and developed by Shigeo Shingo for the Toyota Production System. Poka Yoke is used in many fields, especially in monitoring production processes. In many cases, identifying faults in a production process involves a higher cost than necessary cost of disposal. Usually, poke yoke solutions are based on multiple sensors that identify some nonconformities. This means the presence of different equipment (mechanical, electronic) on production line. As a consequence, coupled with the fact that the method itself is an invasive, affecting the production process, would increase its price diagnostics. The bulky machines are the means by which a Poka Yoke system can be implemented become more sophisticated. In this paper we propose a solution for the Poka Yoke system based on image analysis and identification of faults. The solution consists of a module for image acquisition, mid-level processing and an object recognition module using associative memory (Hopfield network type). All are integrated into an embedded system with AD (Analog to Digital) converter and Zync 7000 (22 nm technology).

1. Introduction

Presently, in the competitive world any company has to manufacture high quality, fault free products at optimum cost in order to achieve customer satisfaction. According to today's opinions, the good quality can be achieved only by company, which implemented the Quality Management Systems, which used idea of continuous improvement of all processes and also which used the quality tools and quality methods, recycling technology inside of production process [1]. One of the important quality tools to add to any company's continuous improvement is Poka Yoke method. A Poka Yoke system is an in-process mistake proofing technique that ensures real-time 100 % inspections to prevent faults from occurring rather than fixing the problem afterward [2, 3].

Poka Yoke method was introduced by Shigeo Shingo in 1961, when this was one of engineers Toyota Motor Corporation. The name Poka Yoke Shigeo Shingo established in 1963, it is translated as "resistance to errors" (avoid (yoker) errors resulting from inattention (poka)) [1, 4]. The Poka Yoke

Content from this work may be used under the terms of the Creative Commons Attribution 3.0 licence. Any further distribution of this work must maintain attribution to the author(s) and the title of the work, journal citation and DOI.
Published under licence by IOP Publishing Ltd

method proposed initially by Shingo was to use simple, inexpensive, but effective techniques to avoid human error at work [1, 3]. Depending on the type of industry, the human factor is recognized as the cause of up to 80% of failures. This means that the human more or less directly contributes to a significant part of the failures [5]. In the competitive environment of our society, a safe working environment is a prerequisite to providing high quality products and services. Therefore the employers have begun to control the factors that may contribute to the growth of productivity in a specific workplace [6]. The optimal goal of the Poka Yoke systems is to benefit the industries with economical cost savings [3]. Tsou and Chen [7] developed a mathematic model that compared the economic returns of a system with and without mistake preventive activities, and further, the model was tested by using an automotive industry case study to confirm that mistake proofing devices were effective in reducing wastes. Being an effective way of catching errors, the Poka-Yoke method has even been adopted by construction, software, and other industries [8, 9], although the concept of mistake proofing was started from manufacturing. Our Poka Yoke solution system is based on image analysis and identification of faults. It consists of a module for image acquisition, mid-level processing and an object recognition module using associative memory (Hopfield network type). All are integrated into an embedded system.

2. Structure of the Pokayoke system

The identification of faults by image analysis is the least invasive solution. There are many works in this direction, however most of it has the image processing computer systems, hard or even impossible to use in some industrial environments. Our solution uses the image taken from a usual video camera, such as those used in security. The image is acquired, converted to digital format, then processed and converted into a binary matrix (monochrome image). Monochrome matrix reaches the entrance of associative memory which is able not only to identify the fault but classify the fault type. The system may then initiate the operation of means for correcting the fault or the subsequent elimination of the occurrence of a fault in the manufacturing process.

All the above mentioned operations are performed by an integrated system consisting of a set of high-speed ADC (Analog to Digital Converter) and an integrated circuit. The solution is so compact one, which can be easily integrated into the diagnostic system. It may be present in the vicinity of the video camera, resulting in a complete system for the identification of the faults and their diagnostic, system that can be placed anywhere on the production line. Due organization on the parallel computing blocks the system ensures fast processing and image diagnosis in real time.

In addition, by the ARM core (ARM core: processor integrated. Can run operating system based on Linux core and is used in our case as interface between acquisition system and operator) that is included in the integrated circuit is allowed to implement a web server and hosting a page that can be accessed in an intranet. Through this interface can be requested reports on the types of faults occurring. The following sections will present the structure of the system and then the results obtained by its use. The following section describes a block diagram of the system. Then, we discuss its three components placed on the flow image processing: acquisition and processing module of SVIDEO analog signal from the camera, mid-level digital image processing module and identification and classification of faults module.

2.1. Description of the system

The figure 1 illustrates a block diagram of the system. The central component is FPGA-ARM circuit (FPGA: is programmable logic device. In our case FPGA is used to implement entire image processing system: mid-level processing and pattern recognition) from Xilinx Company: Zync 7000.

This circuit is composed by ARM Cortex processor and an FPGA circuit. On the FPGA is implemented the digital image processing and classification of objects. These operations will be performed on hardware structures with parallel operation so that the speed of image acquisition from the camera will coincide with the speed of image processing and classification.

Image is captured from a video camera, converted in digital format by AD converter and then processed by ARM-FPGA integrated circuit (Zync 7000). As results we have indication of fault or good parts on Local monitor, commands to motor or servo for selection of fault part and logs files that can be read from remote devices using Ethernet. The parts can be identified by shape or by slot.

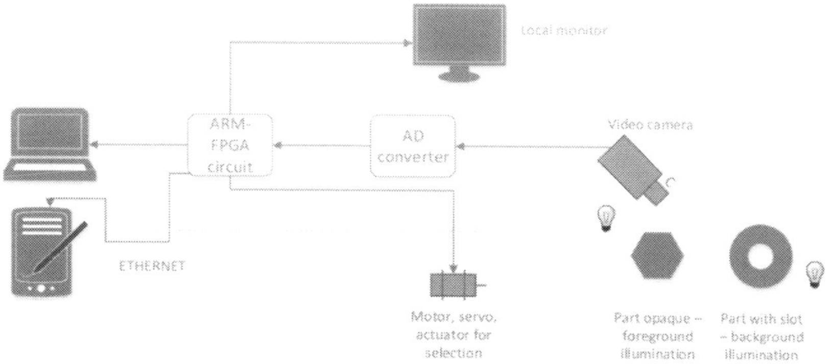

Figure 1. Block diagram of the system.

On ARM processor is installed an operating system based on Linux kernel (Xilinux) which allows installing and running a server will query the data from FPGA and will forward the request to a client via Ethernet. Digital image is provided by a circuit from Analog Device, specializing in the conversion of digital YCrCb SVIDEO format (16 bits per component).

2.2. Image acquisition and mid-level image processing

This Digital image, YCrCb (luminance, chrominance red, blue chrominance) reaches the entrance in FPGA as 16 x 3 bits size data. There have been two operations: format conversion from 48-bit YCrCb to monochrome image and storing in a bi-port configuration BRAM (Block RAM), figure 2.

In the end is obtained for each pixel acquired a value of 0 or logic 1. The information thus processed is stored in a internal memory in the bi-port configuration. Port A is organizing a bit (bit per pixel). Finally we have a full-frame stored. Arithmetic operations and comparisons take place with threshold frequency acquisition, from this point of view don't takes an additional processing time of acquisition and digital conversion, pixel by pixel, image.

2.3. Monochrome pattern recognition by image analysis

The Using BRAM memory in bi-port configuration has several advantages. Storage operations and reading operations are completely independent: storage can occur for a memory organization and with some frequency and reading for another organization and another frequency. The memory can operate at working speed of FPGA circuit. If storage is at a frequency of 50MHz (pixel acquisition from camera) at a bit reading can be done at 64bit 100MHz frequency. Thus, excluding the initial state when a frame has not yet been read completely, we have a placement rate to the network input independent of the refresh rate.

VDEC interface is used to receive image from AD converter – pixel by pixel. A pixel is processed, extracting luminance and chrominance and then converted in monochrome by levels comparator module. In the monochrome image areas of interest are highlighted.

The entire frame is stored on dual port memory (port A), other port is connected to the neural network input. The BRAM memory is organized in blocks, one block size is 16kb (more exactly18kb of which 16Kb data area and 2KB the parity used for certain configurations). An image with a resolution of 400x400 (160000 bytes) with one bit per pixel we need 10 memory blocks (circuit has a total of 16 blocks). Associative memory using 160,000 cells - one for each bit individually, grouped into 400 areas, each zone specialized in a specific area of image recognition. In total 20 different images are stored for each area. Image recognition is done entirely by assembling (using the principle of majority) recognized image in each area. Obviously, saving images was done previously in the network. For such types of networks based on the configuration Hopfield network - all interconnected - storage operation takes place in one step (as opposed to learn. The recognition (classification) of image object also occurs in a single step.

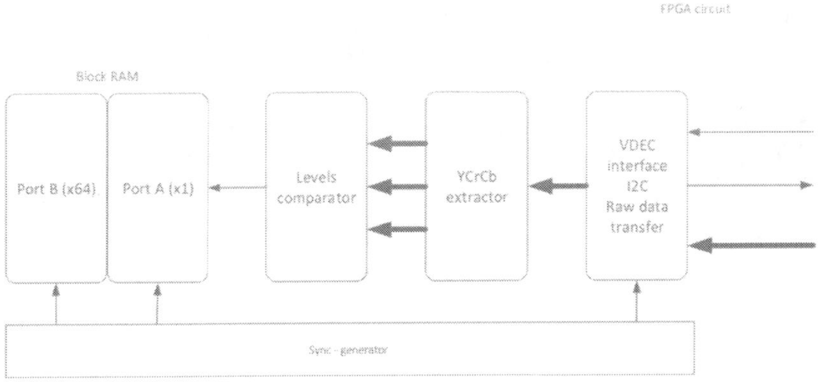

Figure 2. Image acquisition and mid-level processing modules.

3. Experimental results

Experiments were made using two types of components: T-shaped pieces and circle-shaped parts (with a slot (cavity) in the middle - Pretzel). T-shaped piece is painted with a specific colour (in this case red) that differentiates the black background. For these types of parts the lighting is behind the camera and falls perpendicularly on the part. The second type of piece is the ring-shaped with slit in the middle. The piece is placed with a light source behind.

The image analysis module converts the image in bit array. For the first type of parts we have a threshold for red chrominance in pixel (the threshold is chosen to 32767 half maximum chrominance). The module takes a pixel, which is in input format (Y – luminance, Cr, Cb – red and blue chrominance) and convert to 1 if Cr value is greater than threshold and to 0 otherwise. For the second type of parts we have 1 if Y is less than the threshold (half maximum luminance) else 0.

Whatever type of parts, image analysis for pattern recognition goes from a binary 2D array where the target object is highlighted (1 logic pixels belonging to the object, 0 the rest). Identification of the objects is the task artificial neural network. The network is learned with templates which contain shape for correct and damaged parts. For T-shaped piece we have 4 templates: one is for the correct part, other two defects in the left arm and right arm (missing or too small) and one with the bottom side to small.

Regarding the ring – shape parts we have 3 templates: one with the correct part, the other with interruption in the north and one in the south of interruption (the part has an "equator").

Our system identifies and classifies faults by analyzing the shape of the piece.

The system was tested using Sharp CCD camera with a resolution of 420 TV lines of which are captured images with 400x400 resolution. There are two types of target images tested: one with a part that has a slot and one with a part that has a certain shape (slit and form are T). Obviously can interpret and other parts. In the Fig.3 are illustrated images with Poka Yoke system.

Left images: Poka Yoke in action, bottom right acquisition board and camera, center and top local monitor with log file displayed – can see graphic interface of the Linux operating system (Ubuntu) which run on ARM processor, left and center a ring-shaped part with backlight source. Right picture:

AD converter extension board (left) and board with Zync circuit (right) - for experiment we use VDEC extension board and Zybo development board from Digilent. In both cases, there are 19 faults can be identified and classified alongside the correct part (the 20th picture). Acquisition speed and response time to recognize the shape shown in the table 1. Dual port BRAM (Block RAM) memory can be accessed independently on the two port, PORTA and PORTB. On port A one pixel is processed and stored in 342.5 ns and a pixel on PORTB can be read and interpreted in approximately 50 ns.

Figure 3. Images with the system.

Table 1. Results using our system.

Images from camera	AD sample & converting rate (max 50MHz)	Mid-level processing (including BRAM storing)	Neural network response (using parallel subnetworks)	Total time
20 frames / second	3.2MHz (312.5 ns per pixel)	~30ns / pixel	~50ns / subnetwork, 20ns areas integration 10ns memory read = 80 ns / pixel	Per pixel: Max (PORTA,PORTB) = 342.5 ns Entire image: **54.8 ms**

So I took maximum between these two times and results a response from entire image in 54.8 ms (approx. 1/20 second). This is maximum value – in practice response time / image varies between 40 – 50 ms. It can be seen that we have a system with real-time response, which can identify the types of failures that can occur only through image analysis. The performances of such a system are superior to other solutions on the market. Thus, we have a fully integrated system with low power consumption that can be placed anywhere in the production process and has a very low response time. A block diagram of the integrated system is presented in figure 4.

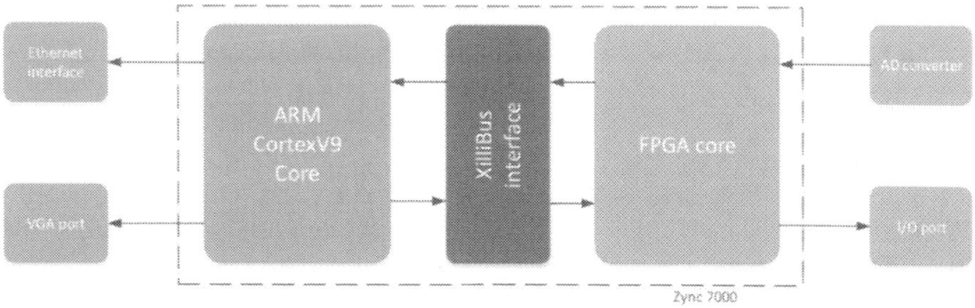

Figure 4. Block diagram of Zync integrated circuit and the interfaces used in project: FPGA core and ARM core.

FPGA core consist in digital acquisition system: AD interface, mid-level processing module and neural network. It uses AD converter port (synchronous serial interface) and can send to general purpose output ports commands to optional selection modules (motor, servo). ARM core runs operating system, can display status of Poka Yoke on local monitor or can send logs files to Ethernet. In addition, with ARM processor component, the system allows on-line query through a friendly web interface on a Web server.

4. Conclusions

This paper presented a fully integrated Poka Yoke system for determining faults by image analysis.

System performance is:

- Low cost, the system is based on a FPGA circuit, an AD acquisition mode and a video camera of a kind used in secure premises;
- Low power consumption by integrating an entire system circuit of acquisition: digital image processing, extraction and pattern recognition of the part;
- Response in real time by digital image processing with the same frequency acquisition and identification in a single step to the shape part.

The innovations introduced by the system are:

- Using an FPGA reconfigurable circuit that integrates all levels of processing and identification – the computer is replaced;
- Using an associative memory based on a Hopfield neural network type for rapid identification and classification of objects;
- The use of low resources to implement Poka Yoke.

As future directions for development of this Poka Yoke solutions can be the experimenting on the other faults types.

References

[1] Dudek-Burlikowska M and Szewieczek D 2009 The Poka-Yoke method as an improving quality tool of operations in the process *Journal of Achievements in Materials and Manufacturing Engineering* **36 (1)** pp 95-102

[2] Santos J, Wysk R and Torres J 2006 Improving production with lean thinking, *John Wiley& Sons* (Hoboken: NJ)

[3] Zhang J Z 2014 Development of an in-process Pokayoke system utilizing accelerometer and logistic regression modeling for monitoring injection molding flash *Int J Adv Manuf Technol* **71** pp 1793–1800

[4] Shinto S 1986 Zero Quality Control: Source Inspection and the Poka–Yoke System *Cambridge* (MA: Productivity Press)

[5] Butlewski M, Jasiulewicz-Kaczmarek M, Misztal A and Sławińska M 2014 Design methods of reducing human error in practice, Safety and Reliability: Methodology and Applications *Proceedings of the European Safety and Reliability Conference ESREL* 2014 ed T Nowakowski M Młyńczak A Jodejko-Pietruczuk S Werbińska-Wojciechowska (CRC Press:London) pp 1101-1106

[6] Ene A and Anghel D C 2013 A Neural Networks Application in Ergonomics *Proceedings of the Electronics, Computers and Artificial Intelligence* pp 107-110

[7] Tsou J C and Chen W J 2008 The impact of preventive activities on the economics of production systems: modeling and application *Appl Math Model* **32(6)** pp 1056–1065

[8] Patil P S and Burali Y N 2013 Review Paper On "Poka Yoke: The Revolutionary Idea In Total Productive Management" *International Journal Of Engineering And Science* **2(4)** pp 19-24

[9] Sadri R, Taheri P, Azarsa P and Ghavam H 2011 Improving productivity through mistake-proofing of construction process *Proc. Of CSIT* (IASCIT Press: Singapore)

IOP Conf. Series: Materials Science and Engineering **95** (2015) 012139 doi:10.1088/1757-899X/95/1/012139

An improved method for risk evaluation in failure modes and effects analysis of CNC lathe

N Rachieru[1], **N Belu**[1] **and D C Anghel**[1]
[1]University of Piteşti, Faculty of Mechanics and Technology, Department of Manufacturing and Industrial Management, 110040 Pitesti, Targu din Vale Street, no. 1, Argeş, Romania

E-mail: nrachieru@yahoo.com

Abstract. Failure mode and effects analysis (FMEA) is one of the most popular reliability analysis tools for identifying, assessing and eliminating potential failure modes in a wide range of industries. In general, failure modes in FMEA are evaluated and ranked through the risk priority number (RPN), which is obtained by the multiplication of crisp values of the risk factors, such as the occurrence (O), severity (S), and detection (D) of each failure mode. However, the crisp RPN method has been criticized to have several deficiencies. In this paper, linguistic variables, expressed in Gaussian, trapezoidal or triangular fuzzy numbers, are used to assess the ratings and weights for the risk factors S, O and D. A new risk assessment system based on the fuzzy set theory and fuzzy rule base theory is to be applied to assess and rank risks associated to failure modes that could appear in the functioning of Turn 55 Lathe CNC. Two case studies have been shown to demonstrate the methodology thus developed. It is illustrated a parallel between the results obtained by the traditional method and fuzzy logic for determining the RPNs. The results show that the proposed approach can reduce duplicated RPN numbers and get a more accurate, reasonable risk assessment. As a result, the stability of product and process can be assured.

1. Introduction

Quality planning is a process that leads to the effectiveness and efficiency of production processes. Correctness of quality planning lies in the fact that the individual stages and operations process should be designed so as to eliminate the possibility of non-compliance. However, in case of their occurrence, a high probability of detection should be ensured [1]. Failure Mode and Effects Analysis (FMEA) is one of the first structured, systematic and proactive techniques used for failure analysis. It is a widely used engineering technique for defining, identifying and eliminating known and/or potential failures, problems, errors and so on from system, design, process, and/or service before they reach the customer [2]. For analysing a specific product or system, a cross-functional expert team should be set up to conduct FMEA first. The first step in FMEA is to identify all possible failure modes of the product or system. Next, critical analysis is performed on the identified failure modes taking into consideration the risk factors: occurrence (O), severity (S), and detection (D). Conventionally, the ranking of failure modes for corrective actions is determined in terms of the risk priority number (RPN), which is the mathematical product of the S, O and D corresponding to the failure modes [3]. That is RPN = S x O x D, where O is the probability of the failure, S is the severity of the failure, and D is the probability of not detecting the failure. In order to obtain the RPN of a potential failure mode, the traditional FMEA uses an integer scale from 1 to 10 for evaluating the three risk factors. Generally, failure modes with

Content from this work may be used under the terms of the Creative Commons Attribution 3.0 licence. Any further distribution of this work must maintain attribution to the author(s) and the title of the work, journal citation and DOI.
Published under licence by IOP Publishing Ltd

higher RPN values are considered to be more important and are given higher priorities than those with lower RPN values [4]. However, it suffers from several shortcomings. It has been pointed out that the same RPN can be obtained from different combinations of different sets of S, O and D. Although the same RPN is obtained, the risk can be different and the relative importance of three risk factors is not taken into account. In other words, the risk factors are given to have the equal importance, which may not be the case in many practical applications of FMEA. The three risk factors are mostly difficult to be precisely determined. Much information in FMEA is often uncertain or vague and can be expressed by using linguistic terms such as *likely, important or very high* and so on [5,6,7,8]. In order to overcome the above shortcomings, a number of approaches have been suggested in the literature to enhance the FMEA methodology, such as grey theory [9], data envelopment analysis (DEA) [10], decision making trial and evaluation laboratory (DEMATEL) [11], Taguchi method, fuzzy logic and Genetic algorithms (GA) [12]. GA can be used to search for solutions difficult to obtain by other conventional methods, in different areas. They can be run on computer or can be accelerated on parallel hardware structures [13].

The aim of proposed paper is to present a new risk assessment system based on fuzzy theory to deal with the risk evaluation problems in FMEA. The paper presents a parallel between the typical and the fuzzy computation of RPNs of failure modes that could appear in the functioning of "Turn 55 Lathe CNC". Fuzzy set theory was initiated by Zadeh in the early 1960s. Fuzzy sets (FS) and fuzzy logic (FL) are powerful mathematical tools for modelling uncertain systems in industry, nature and humanity; and facilitators for common-sense reasoning in decision making in the absence of complete and precise information [8]. The three inputs (S) severity, (O) occurrence and (D) detecting are fuzzyfied and evaluated in a fuzzy inference engine built on a consistent base of IF-THEN rules. The fuzzy output is defuzzyfied to get the crisp value of the RPN that will be used for a more accurate ranking of the potential risks.

2. Application

2.1. Classical FMEA application

In the first part of the study a classical application of Design FMEA has been realized for Turn 55 Lathe CNC. The Concept Turn 55 is a desktop lathe driven by interchangeable CNC control Software running on a commercially available PC. This dual purpose turning center is the ideal solution for training students in further education when large industrial machines are not suitable. Its role in education is defined by its interchangeable control systems.

The evaluation of the failure modes is carried out by scoring the respective risk factors of occurrence, severity, and detection. For this purpose, usually 10-level scales are being used. The failure modes with higher RPNs are assumed to be more important and will be given higher priorities for correction. It is presented the failure with highest RPN values (54 and 72). Some of the data can be seen in table 1.

Table 1. Conventional FMEA for Lathe CNC.

Failure mode	Failure effect (s)	Cause (s)	S	O	D	RPN
F1. Wear of the mechanical components of tool machine	Poor quality of the surface of piece	C1. Overcoming life of the mechanic component	6	4	3	72
F2. Difficult processing (high energy consumption)	Deposition of material on the surface of the face of cutting tool	C2. Improper cutting regime	4	2	5	40
		C3.Electrical component can be damaged	4	2	3	24
F3. Incomplete	Nonconforming parts	C4. Writing error on NC	7	3	2	42

execution of the CNC program		program				
F4. CNC stopping	Cutting tool collision	C5. Errors on NC program	8	2	1	16
		C6. Incorrect installation of the blank on CNC	8	1	2	16
F5. Erosion of date cable	Data from computer are not transmitted to the tool machine	C7. Life cycle overflow	6	3	3	54
F6. Power LED no indication	Cannot determine whether the machine is on or off	C8. Faulty supply	2	2	1	4

2.2. Fuzzy FMEA Application

The fuzzy logic toolbox of Matlab software program has been used in calculating the values of RPN. A model was established for the FMEA technique having 3 inputs and 1 output variable, and given in figure 1. Four categories were associated to each fuzzy set: VL (Very Low), L, (Low), M (Moderate) and H (High). The output of the fuzzy system, FRPN, was scaled in the range 0...1000 in order to be compatible with the previous results. The occurrence, severity and not detection values of the failures were identified with the help of expert opinions and by using decision rules determined specifically. The rules were designed to take into account all possible situations. For Fuzzy FMEA application we present two case studies.

2.2.1. Case 1. In this case, for risk factors like severity, occurrence and priority number (RPN), the Gaussian membership function is selected because of its two important properties [14], viz., (i) it can lead to smooth, continuously differentiable hyper surfaces of a fuzzy model; (ii) it facilitates theoretical analysis of a fuzzy system because it is continuously and infinitely differentiable, i.e., it has derivatives of any grade (equation (1)). For Detection risk factor was used Cauchy membership function (generalized bell) (equation (2)).

$$\mu_A(x) = e^{-[x-c]^2/2\sigma^2} \quad (1) \qquad \mu_A(x) = 1/1 + \left|\frac{x-c}{a}\right|^{2b} \quad (2)$$

where c and σ parameterize the center and width of the Gaussian membership function, respectively; c represents the membership function center and a determines the membership function width, b a positive real parameter. The membership function for severity and detection are depicted in figure 2 - 3.

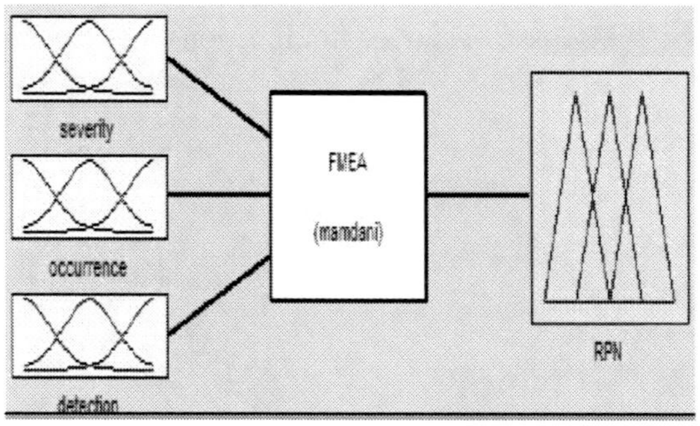

Figure 1. The Fuzzy FMEA model.

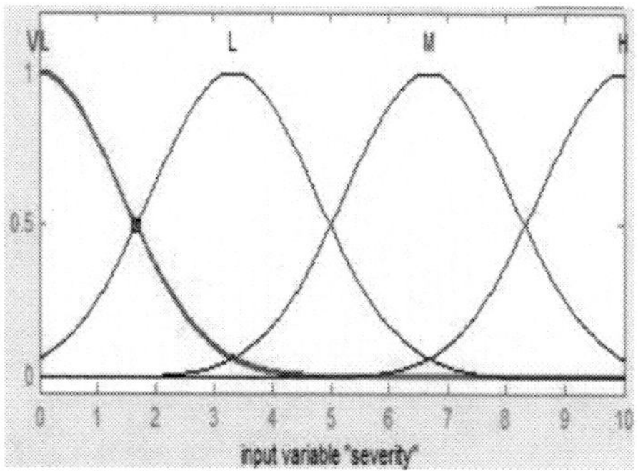

Figure 2. Membership functions for severity.

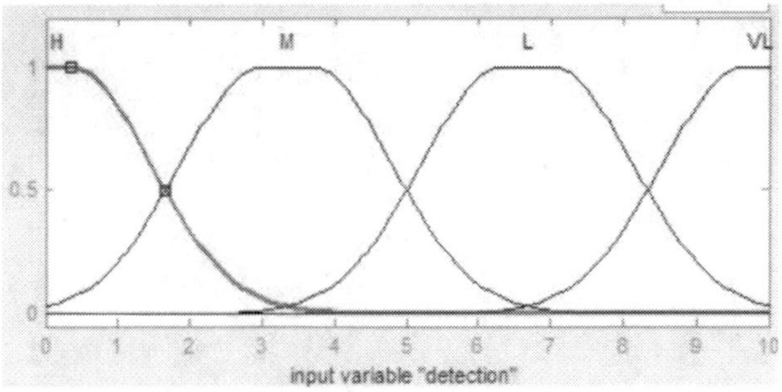

Figure 3. Membership functions for detection.

2.2.1. Case 2. In this case, for risk factors like severity, occurrence and priority number (RPN), the triangular membership function is selected (equation (3)) and for detection risk factor was used trapezoidal membership function (equation (4)). Triangular and trapezoidal fuzzy numbers are the most common used fuzzy numbers both in theory and practice. In fact, triangular fuzzy numbers are special cases of trapezoidal fuzzy numbers. When the two most promising values are the same number, the trapezoidal fuzzy number becomes a triangular fuzzy number. The membership functions are defined respectively as follows [15]:

$$\mu_A(x)=\begin{cases}(x-a)/(b-a), & \text{for } a\le x\le b\\ (c-x)/(c-b), & \text{for } b\le x\le c\\ 0 & \text{otherwise}\end{cases} \quad (3)$$

$$\mu_A(x)=\begin{cases}(x-a)/(b-a), & for\ a\le x\le b\\ 1 & for\ b\le x\le c\\ (d-x)/(d-c) & for\ c\le x\le d\\ 0 & otherwise\end{cases} \quad (4)$$

where $a \le b \le c \le d$ .The membership function for occurrence, detection and RPN are depicted in figure 4, figure 5 respectively figure 6.

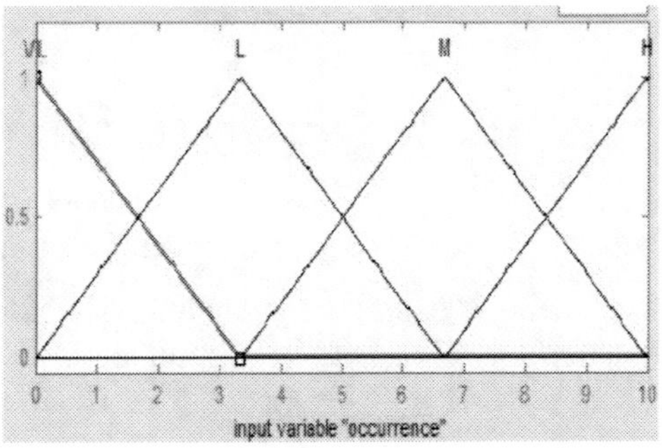

Figure 4. Membership functions for occurrence.

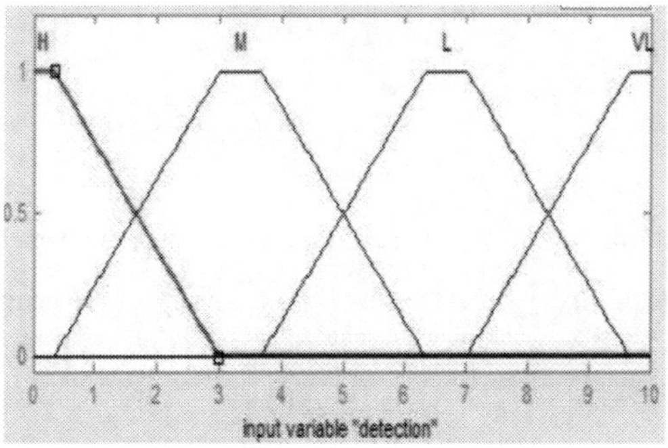

Figure 5. Membership functions for detection.

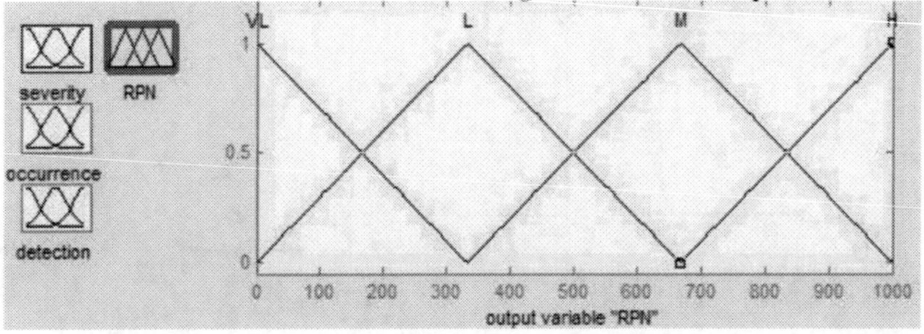

Figure 6. Membership functions for RPN.

The membership function derived from the expert is used to generate the fuzzy rule base. Table 2 presents the inference rules adopted for these applications.

Table 2. Inference rules.

Severity: VL					
FUZZY RPN		*Occurrence*			
		VL	*L*	*M*	*H*
Detection	*VL*	VL	VL	VL	L
	L	VL	VL	L	L
	M	VL	L	L	M
	H	L	L	M	M

Severity: L					
FUZZY RPN		*Occurrence*			
		VL	*L*	*M*	*H*
Detection	*VL*	VL	VL	L	L
	L	VL	L	L	M
	M	L	L	M	M
	H	L	M	M	H

Severity: M					
FUZZY RPN		*Occurrence*			
		VL	*L*	*M*	*H*
Detection	*VL*	VL	L	L	M
	L	L	L	M	M
	M	L	M	M	H
	H	M	M	H	H

Severity: H					
FUZZY RPN		*Occurrence*			
		VL	*L*	*M*	*H*
Detection	*VL*	L	L	M	M
	L	L	M	M	H
	M	M	M	H	H
	H	M	H	H	H

A total of 64 fuzzy rules $(4 \times 4 \times 4)$ are gathered from domain experts. These fuzzy rules are presented in an If–Then format and an example of two fuzzy rules is shown in figure 7.

Rule 1: If *Severity* is **Very Low**, *Occurrence* is **High** and *Detection* is **Moderate** then RPN is **Moderate**
Rule 2: If *Severity* is **Low**, *Occurrence* is **High** and *Detection* is **Moderate** then RPN is **High**
Rule 3: If *Severity* is **Moderate**, *Occurrence* is **High** and *Detection* is **Moderate** then RPN is **High**
Rule 4: If *Severity* is **High**, *Occurrence* is **Moderate** and *Detection* is **Moderate** then RPN is **High**

Figure 7. An example of four fuzzy rules.

Figure 8, presents the surface viewer a three-dimensional curve that represents the mapping from occurrence and severity.

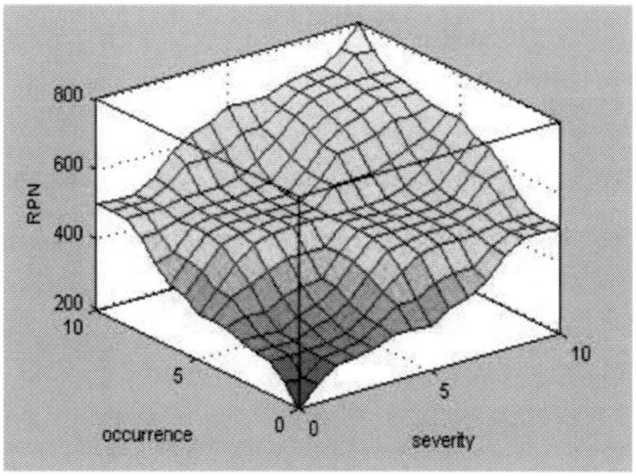

Figure 8. Surface viewer.

IOP Conf. Series: Materials Science and Engineering **95** (2015) 012139 doi:10.1088/1757-899X/95/1/012139

Minimum function was used in order to implement AND method and implication. Maximum function was used in order to implement OR method and aggregation. An important step in fuzzy modeling and fuzzy multi-criteria decision-making is the defuzzyfication task which transforms a fuzzy number into a crisp value.

Various techniques for this transformation are available, including the mean of maxima (MOM), center of area (COA), center of gravity (COG) or α - cut.

Different defuzzyfication methods extract different levels of information. Defuzzyfication is performed according to the membership function of the output variable. For our study the center of area was chosen for defuzzyfication. It finds the point where a vertical line would slice the aggregate set into two equal masses. The fuzzy logic system was simulated using MATLAB-SIMULINK.

As to the types of failure, the fuzzy RPN values provided in the model are given in table 3 in comparison with the RPN values of classical FMEA.

Table 3. Prioritization of failure modes.

Failure mode	Failure cause	RPN	Rank 1	Fuzzy RPN1	Rank 2	Fuzzy RPN2	Rank 3
F1	C1	72	**1**	598	**3**	586	**3**
F2	C2	40	**5**	393	**8**	399	**8**
	C3	24	**6**	412	**7**	423	**7**
F3	C4	42	**4**	615	**1**	619	**1**
F4	C5	16	**7**	601	**2**	591	**2**
	C6	16	**8**	478	**6**	492	**6**
F5	C7	54	**2**	527	**4**	513	**4**
F6	C8	4	**3**	527	**5**	513	**5**

3. Conclusions

Risk evaluation in FMEA, due to its intrinsic ambiguity and difficult formalization, is a particular complex task, usually accomplished by FMEA team members' experience and intuition. In this paper, a new risk evaluation methodology for FMEA based on fuzzy theory was proposed to deal with the risk factors and identify the most serious failure modes for corrective actions.

The results obtained by fuzzy inference in two case studies provide a hierarchy of potential risks that differs from the ranking established by classical FMEA. Also, two fuzzy proposed methodologies offer the same hierarchy of potential risks, although the RPNs value differs. The fuzzy inference does not allow identical values of RPNs to appear for different sets of risk factors. According to the analysis of the results produced by the traditional FMEA and the fuzzy FMEA methods, this research shows that a more accurate, reasonable ranking can be achieved by the application of FMEA based on fuzzy theory. More risk factors can be included if necessary. For further research, the results of our study can be compared with that of other results using Genetic Algorithm technique and Taguchi method.

References

[1] Misztal A and Bachorz S 2014 Quality planning of parts machine production based on housing of cylinder head milling machines *Applied Mechanics and Materials* **657** pp 986-990

[2] Stamatis H D 2003 Failure Mode and Effect Analysis *FMEA from Theory to Execution* **2 ed** (ASQC Press: New York)

[3] Chrysler Corporation, Ford Motor Company, General Motors Corporation 2008 Potential Failure Modes and Effects Analysis *Reference Manual* **4 edition**

[4] Arabian-Hoseynabadi H, Oraee H and Tavner P J 2010 Failure Modes and Effect Analysis (FMEA) for wind turbines *Electr. Power Energy Syst.* **32** pp 817–824

[5] Liu H C, Liu L and Liu N 2013 Risk evaluation approaches in failure mode and effects analysis: A literature review *Expert Systems with Application* **40** pp 828–838

[6] Seyed-Hosseini S M, Safaei N and Asgharpour M J 2006 Reprioritization of failures in asystem failure mode and effects analysis by decision making trial and evalua-tion laboratory technique *Reliab. Eng. Syst. Saf.* **91 (8)** pp 872–881
[7] Tay K M and Lim C P 2006 Fuzzy FMEA with a guided rules reduction system for prior-itization of failures *Int. J. Qual. Reliab. Manage.* **23 (8)** pp 1047–1066
[8] Liu H C, Liu L, Bian Q H, Lin Q L, Dong N and Xu P C 2011 Failure mode and effectsanalysis using fuzzy evidential reasoning approach and grey theory *ExpertSyst. Appl.* **38 (4)** pp 4403–4415
[9] Chang C L, Liu P H and Wei C C 2001 Failure mode and effects analysis using grey theory *Integrated Manufacturing Systems* **12(3)** p 211–216
[10] Chin K S, Wang Y M, Poon G K K and Yang J B 2009 Failure mode and effects analysis by data envelopment analysis *Decision Support Systems* **48(1)** pp 246–256
[11] Chang K H and Cheng C H 2011 Evaluating the risk of failure using the fuzzy OWA and DEMATEL method *Journal of Intelligent Manufacturing* **22(2)** pp 113–129
[12] Jaya M, Prakasha A, Senthilvelan T and Gnanadass R 2015 Optimization of process parameters through fuzzy logic and genetic algorithm – A case study in a process industry *Applied Soft Computing* **30** pp 94–103
[13] Mazare A, Ionescu L, Serban G and Barbu V 2011 Evolvable Hardware with Boolean Functions Network Implementation *Proceeding of International Conference on Applied Electronics*, **CFP1169A-PRT** pp 255-260
[14] Piegat A 2001 *Fuzzy Modeling and Control* (Springer-Verlag: New York, USA)
[15] Wang Y - M Chin K – S Poon G K K and Yang J B 2009 Risk evaluation in failure mode and effects analysis using fuzzy weighted geometric mean *Expert Syst. Appl.* **36** pp 1195–1207

The role of the production scheduling system in rescheduling

K Kalinowski[1], C Grabowik[1], W Kempa[2] and I Paprocka[1]
[1]Silesian University of Technology, The Faculty of Mechanical Engineering, Institute of Engineering Processes Automation and Integrated Manufacturing Systems, Konarskiego 18A Str., 44-100, Gliwice, Poland
[2]Silesian University of Technology, Faculty of Applied Mathematics, Institute of Mathematics, Konarskiego 23A Str., 44-100, Gliwice, Poland

E-mail: krzysztof.kalinowski@polsl.pl

Abstract. The paper presents the rescheduling problem in the context of cooperation between production scheduling system (PSS) and other units in an integrated manufacturing environment – decision makers and software systems. The main aim is to discuss the PSS functionality for maximizing automation of the rescheduling process, reducing the response time and improving the quality of generated solutions. PSSs operate in the meeting of tactical and operational level of planning and control, and play an important role in the production preparation and control. On the basis of information about orders, technology and production system state (e.g. resources availability) they prepare and/or update a detailed plan of production flow – a schedule. All necessary data for scheduling and rescheduling are usually collected in other systems both from organizational and technical production preparation, e.g. ERP, PLM, MES, CAPP or others, as well as they are entered directly by the decision-makers/operators. Data acquired in this way are often incomplete and inconsistent. Therefore the existing rescheduling software works according to interactive method – rather support but does not replace the human decision maker in tasks planning. When rescheduling, due to the limited amount of time to make a decision this interaction is particularly important. An additional problem arises in data acquisition, in the process of data exchanging between systems or in the identification of new data sources and their processing. Different approaches to rescheduling were characterized, including those solutions, where all these operations are carried out by an autonomous system and those in which scheduling is performed only upon request from the outside, for the newly created scheduling data representing the current state of the production system.

1. Introduction

The complexity of scheduling problems in real production systems requires a considerable amount of calculation in determining schedules. Therefore, solving of decision-making problems at all levels of planning and control is done by extensive use of dedicated software [1, 2]. Creating a schedule manually, by decision makers, even with a relatively small number of orders and operations, allows the construction of one or a small number of schedules, with unknown quality. Determined by this way a feasible solution is based largely on intuition and chance, and can hardly meet assumed goals. The use of appropriate scheduling software allows, among others, for searching larger area of solutions space in shorter time, and consequently for finding a better solution, closer to optimal. Despite numerous researches on production scheduling issues, describing detailed models and

Content from this work may be used under the terms of the Creative Commons Attribution 3.0 licence. Any further distribution of this work must maintain attribution to the author(s) and the title of the work, journal citation and DOI.
Published under licence by IOP Publishing Ltd

methods and techniques for obtaining solutions, very little space is dedicated to their practical implementation [3, 4].

From a computer system supporting scheduling of manufacturing systems a certain set of functions is expected. The scope of this functionality depends on many factors connected with particular implementation in a given a production environment, the nature of applied technologies, existing software and knowledge and experience of decision-makers. Scheduling system requirements may relate to: the scope and complexity of handled models of scheduling problem ($\alpha|\beta|\gamma$), generating and evaluating of solutions, interactivity and the requirements of the user and systems interfaces, possibilities for analyze of production capacity and throughput, operation in real-time conditions, connections with other software systems in the enterprise etc. [5]. In the paper the most important requirements of PSSs functionality in an integrated environment are discussed.

2. Integration and data exchange

Generating of correct solutions in a short time requires direct accessing to the appropriate data. These data can be imported from the external sources or entered directly by the user. Effective work organization in the enterprise requires a coherent information system, carried out mostly in the form of an integrated system based on a common, central or distributed database. Some companies, however, has not implemented this type of information systems and uses separated software with local databases or even only paper documentation. Manual entry or periodically copying data into/between software's databases of production systems, without adequate infrastructure of data exchange interfaces of resources, significantly reduces the usefulness of the scheduling system and contributes to delays in data processing and responding to the unexpected events. The lack of access to current data causes difficulties in making the right decisions, at every level of management.

The lack of an integrated environment requires acquiring data from independent systems. Then it is necessary to carry out the process of data mapping in order to accurately transforming of their internal structure of the system and testing the data integrity. Integration at the level of the exchange of data between systems can be realized in different ways [6]:

- by a common database, where one of the systems directly accesses the database of other system or new, common database is created,
- by the use of separate databases and the development of a common format for data exchanging.

In order to provide the access to current data the scheduling system should be integrated with other systems, both in the field of organizational and technical preparation and production control i.e.:

- *ERP (Enterprise Resource Planning), or less complex systems of MRP (Material Requirements Planning) or MRPII (Manufacturing Resources Planning)*, PLM (Product *Lifecycle Management)*, PDM *(Product Data Management)*, PMS *(Project Management Systems)*, BPM *(Business Processes Management)*,
- MES *(Manufacturing Execution System)*, SCADA *(Supervisory Control and Data Acquisition)* or systems of CAx class, e.g. CAPP *(Computer Aided Processes Planning)*.

Figure 1 shows the connections for the exchange of data in the ERP / MES / SCADA class of planning and control of production flow systems. Among the required input data for scheduling, acquired from the ERP class system, the following can be distinguished: information related to the description of parameters of the production system model containing resources availability, the actual packages of orders, criteria for schedules evaluation, as well as conditions and instructions for rescheduling [7]. The required data on the state of the production system are derived from the SCADA system [8, 9, 10, 11, 12].

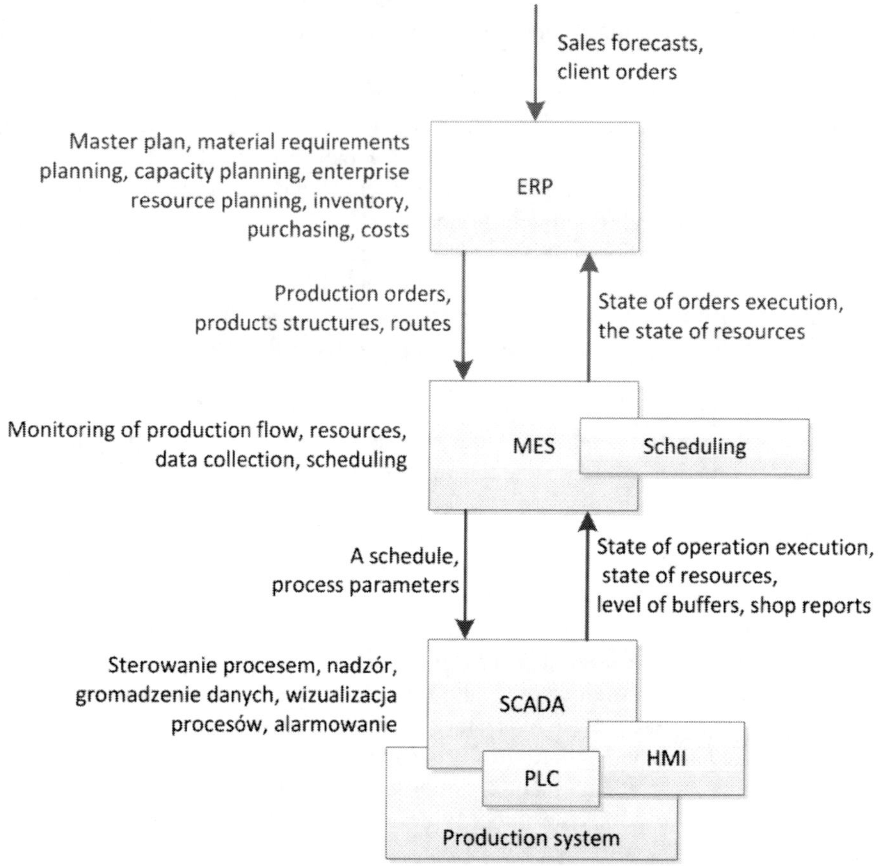

Figure 1. Data exchange in the integrated ERP/MES/SCADA environment.

The current development level of information systems used in enterprises, with built-in modules and the application of uniform standards for data exchange simplifies the process of integration between different systems, reducing both time and cost of implementation. The most commonly used data exchange standards are described by ISA (ANSI / ISA-95, ANSI / ISA-88), OPC, OAGIS, and MIMOSA standards [13, 14, 15, 16].

3. Custom and generic software

Siemens Computer-aided production scheduling can be implemented by dedicated software (custom software) or in the form of specialized modules/ components of standard system (generic software), the most common of ERP or MES class systems. A review of selected scheduling systems and to compare performance of their functions was presented in [17]. Beyond the classification of commercial dedicated and standard software an academic solutions are also described there. The generality level of standard systems allows their wide applications in various manufacturing companies and industries. The implementation of such kind of systems in the particular manufacturing plant often requires from the company carrying out a series of organizational changes, structures and implemented procedures. The universality of this approach, however, may have a weak side – created solutions may not include all the required constraints and evaluation criteria, resulting in the need for making significant corrections of generated solutions. An alternative to generic systems, which are popular rather in medium and large sized manufacturing systems, are dedicated, custom systems. This type of software is created from scratch or adapted from solutions used in another companies with a similar profile of production. The advantage of the generic software implementation is the ability to obtaining a significant range of required functions with little effort of required organizational changes and receiving the desired functionality of the software in less time. For creation or adaptation of dedicated solutions motivates: similar architecture of scheduling system, existing data exchange interfaces and the original user interface concept, large amount of non-standard functions and activities that are characteristic for a certain class of manufacturing systems, etc.

4. Expectations of the production scheduling software

Within the basic functionality of production scheduling system two functions can be distinguished: *scheduling*, that means the off-line schedule preparation, and *rescheduling* associated with on-line activities. The combination of both these functions is known as predictive-reactive scheduling. Therefore, these functions of the system should provide repairing of the adopted schedule, responding to any problems resulting from unexpected events in the production system. Among the most important decision-making problems in rescheduling the following can be mentioned: detection of disturbances and identification of their parameters, the decision about the schedule correction – immediate, postponed or not required, making proper changes in the scheduling model, determination of the rescheduling method, creation a set of operations to reschedule and a set of evaluation criteria, etc.

When creating software for rescheduling, it is important to ensuring data about the parameters of disturbances and the role of decision maker in modifying the input data for scheduling, determination of the rescheduling method, creation a set of operations to reschedule and a set of evaluation criteria. For this reason the 'a priori', 'a posteriori' or interactive method can be introduced [18]. The interactivity of the system allows the user taking an active part in the process of creating a solution. Depending on the applied method of interaction the participation of decision maker in the process of scheduling can significantly affecting the quality of the final solution by including additional, not included in the model conditions or preferences, e.g. informal evaluation criteria. Interactivity requires the presentation of the temporary results of calculations at the appropriate level of detail, in a clear and accepted (by a user) manner. Enabling interactivity is closely related with the development of a user interface. It should provide access to relevant information, depending on the user rights.

Rescheduling can be performed may be carried out at different intervals - at regular or irregular intervals. In the first case, the schedule is updated at predetermined time intervals. This method is often used in industrial enterprises, where data on the state of the production flow are not available and processed on-line, e.g. if are gathered in the traditional way, using paper forms. Prepared and revised schedule is valid for the entire cycle. The basic problem in the periodic rescheduling is the determination of an optimal rescheduling period, which should takes into account the frequency of disturbances occurring, their overall impact on the production flow and the system condition and also effectiveness parameters of resources. A main disadvantage of only periodic revised scheduling is outdating of the schedule in case of major disruption and shifting decisions on correction in time. The problem of continuing assigned tasks in presence of disturbances are solved locally, by lower-level manufacturing workers, dispatchers or operators. The undoubted advantage of this method is higher scheduling stability compared to rescheduling carried out after a single event.

Because of the strong dependence of certain scheduling algorithms on the production system configuration, the type of orders (structures of processes), given constraints and criteria a desirable feature of the system can be automatic model detection. The possibility of scheduling problem classification can greatly relieve the user of the system in the selection of appropriate algorithms and reduce the time to determine a solution. It is also important that in many cases, the given problem can be solved using different algorithms in parallel – the system should guarantee their execution and comparison.

A desirable feature of the system is the ability to carrying out scheduling simulations in various configurations of the production system, and analyses related to the resources load, i.e. the identification of "bottlenecks" and under loaded resources [3, 19].

In some cases, the scheduling system may be required to operate in real time [20, 21, 22]. The correctness of operation in real time depends on the result of the conducted calculations and also the time at which results are obtained. Therefore, in such systems two states are distinguished: internal and external. The calculations based on registered at a given moment, and entered into the scheduling system, the state of the production environment, are defined as the internal state. The current status of the manufacturing system, defined as external state, may change over time and make, obtained in the meantime solutions, out of date. This situation makes it necessary to establish criteria for limiting the time to determine a final solution. The sources of time type of constraints are different kinds of unforeseen events occurring in the production environment and affecting its state. The time requirements, depending on the application, may range from milliseconds, minutes / hours, or longer periods of time and may vary depending on the decision situations, even within the same environment. Another important issue is the limited and guaranteed time for creating a solution. According to the environmental requirements and potential consequences of missing the deadline for obtaining a solution by the system three basic types of real-time systems (RTS) can be characterized:

- *hard RTS* – systems with absolute time limits, in which exceeded the deadline invalidates the solution and is dangerous or catastrophic for the functionality of production environment,
- *firm RTS* – systems with absolute time limits, in which exceeded the deadline invalidates the solution without a strong impact on the operation of production environment,
- *soft RTS* – systems with "soft" time limits, in which the delay of due date reduces the usefulness of solutions but only after exceeded the deadline solution shall be cancelled.

The system requirements do not need to be associated with only one type of real-time system. The given system can handle a set of events, in which only some of them requires of operating in absolute limit conditions. Scheduling software in manufacturing systems in most cases can be treated as a soft real time system.

5. Conclusions

The role of production scheduling software in the integrated manufacturing environment is discussed in this paper. Possible ways of integration and the key software types, from the field of organizational and technical preparation and production control, supporting scheduling system are described. The basic functionality of the scheduling software for maximizing automation of the rescheduling process, reducing the response time and improving the quality of generated solutions were also identified. The correctness of the generated solutions by the scheduling system depends strongly on the availability of current data. In practice, the production systems in which data are collected manually or stored in separated systems are not capable to the proper response to disturbances. In relation to software development main advantages and disadvantages of using custom and generic software were mentioned. Particular attention was paid to expectations of the production scheduling and rescheduling software. Due to the dependence of unexpected events interactivity requirements relating to the system were not questioned here - conditioned by the need to participate the decision-maker in developing schedules. So the role of the system is still supporting but not replacing the human decision makers.

References

[1] Herrmann J W 2006 *Handbook of Production Scheduling* (Springer Science & Business Media: New York)

[2] Węglarz J 2013 *Project Scheduling: Recent Models, Algorithms, and Applications* (Springer-Verlag: New York)

[3] Beltz R and Mertens P 1996 Combining knowledge-based systems and simulation to solve a rescheduling problem *Decision Support Systems* **17** pp 141-157

[4] Jurczyk-Bunkowska M 2013 Innovation Process Planning: Aim, Scope and Constraints *Proceedings of the European Conference on Entrepreneurship and Innovation* **1** pp 347-354

[5] Liebowitz J and Potter W E1995 Scheduling Objectives, Requirements, Resources, Constraints, and Processes: Implications for a Generic Expert Scheduling System Architecture and Toolkit *Expert Systems with Applications* **9** pp 423-432
[6] Gwiazda A, Sękala A, Monica Z and Banaś W 2014 Integrated approach to the designing process of complex technical systems *Advanced Materials Research* **1036** pp 1023-1027
[7] Kalinowski K, Grabowik C, Kempa W and Paprocka I 2014 The procedure of reaction to unexpected events in scheduling of manufacturing systems with discrete production flow *Advanced Materials Research* **1036** pp 840-845
[8] Ćwikła G 2014 Real-time monitoring station for production systems *Advanced Materials Research* **837** pp 334-339
[9] Hetmanczyk M and Michalski P 2014 The aid of a mistake proofing with the use of mechatronic systems according to the Poka-Yoke methodology *Advanced Materials Research* **837** pp 399-404 Płaczek M 2014 Modelling of electric circuit's type influence on dynamic flexibility of system with passive vibration damping *Advanced Materials Research* **1036** pp 511-516
[10] Zolkiewski S 2010 Numerical Application for Dynamical Analysis of Rod and Beam Systems in Transportation *Solid State Phenomena* **164** pp 343-348
[11] Dzitkowski T and Dymarek A 2014 Active reduction of identified machine drive system vibrations in the form of multi-stage gear units *Mechanika 20* **2** pp 183-189
[12] Information on http://www.isa.org.
[13] Information on http://www.mimosa.org.
[14] Information on http://www.oagi.org.
[15] Information on http://www.opcfoundation.org.
[16] Pinedo M L 2012 *Scheduling. Theory, Algorithms and Systems* (Springer: New York)
[17] Kalinowski K, Grabowik C, Paprocka I and Kempa W 2014 Interaction of the decision maker in the process of production scheduling *Advanced Materials Research* **1036** pp 830-833
[18] Frantze´n M, Ng A and Moore P 2011 A simulation-based scheduling system for real-time optimization and decision making support *Robotics and Computer-Integrated Manufacturing* **27** pp 696-705
[19] Kalinowski K, Krenczyk D and Grabowik C 2013 Predictive-reactive strategy for real time scheduling of manufacturing systems *Applied Mechanics and Materials* **307** pp 470-473
[20] Buttazzo G C 2011 Predictable Scheduling Algorithms and Applications *Hard Real-Time Computing Systems* (Springer Science & Business Media)
[21] Cottet F, Delacroix J, Kaiser C and Mammeri Z 2002 *Scheduling in Real-Time Systems* (Wiley)

Using Teamcenter engineering software for a successive punching tool lifecycle management

F Blaga[1], **A-V Pele**[1], **I Stănăşel**[1], **T Buidoş**[1] **and V Hule**[1]
[1]University of Oradea, 1 Universităţii Street, 410087, Oradea, Romania

E-mail: fblaga@uoradea.ro

Abstract. The paper presents studies and researches results of the implementation of Teamcenter (TC) integrated management of a product lifecycle, in a virtual enterprise. The results are able to be implemented also in a real enterprise. The product was considered a successive punching and cutting tool, designed to materialize a metal sheet part. The paper defines the technical documentation flow (flow of information) in the process of constructive computer aided design of the tool. After the design phase is completed a list of parts is generated containing standard or manufactured components (BOM, Bill of Materials). The BOM may be exported to MS Excel (.xls) format and can be transferred to other departments of the company in order to supply the necessary materials and resources to achieve the final product. This paper describes the procedure to modify or change certain dimensions of sheet metal part obtained by punching. After 3D and 2D design, the digital prototype of punching tool moves to following lifecycle phase of the manufacturing process. For each operation of the technological process the corresponding phases are described in detail. Teamcenter enables to describe manufacturing company structure, underlying workstations that carry out various operations of manufacturing process. The paper revealed that the implementation of Teamcenter PDM in a company, improves efficiency of managing product information, eliminating time working with search, verification and correction of documentation, while ensuring the uniqueness and completeness of the product data.

1. Introduction

In last decade, there were a lot of references to Product Lifecycle Management (PLM) and their development. The general idea is the PLM systems implied "a single source strategy for gathering and managing product data in a single system" [1]. This objective is hard to be realized "due to the distributed nature of engineering work and systems" [1] and it seems to be impossible to develop a system which supports this way of working.

The paper [2] defines a framework for modelling the product system in the early development, which accompanies system design considering sustainability aspects in a prospective view. Also, the paper focuses on expanding modelling constructs by relevant behaviour elements capturing semantic links and information. In [3] it presents how PLM systems represent a very important foundation for achieving a more sustainable paradigm for life, a more sustainable development, engineering, manufacturing, use and disposal of products.

The paper [4] deal with the urgent problem to create integrated PLM solutions in today's European automotive industry. The different author's points of view using focus groups, blogs, and face-to-face meetings in a university community of practice are analyses and integrates in [5].

Content from this work may be used under the terms of the Creative Commons Attribution 3.0 licence. Any further distribution of this work must maintain attribution to the author(s) and the title of the work, journal citation and DOI.
Published under licence by IOP Publishing Ltd

IOP Conf. Series: Materials Science and Engineering **95** (2015) 012141 doi:10.1088/1757-899X/95/1/012141

2. Program and method

2.1. Assigning tasks

The term "technical data" of a product includes all data, both of the product and of the processes, used to design, manufacture, utilize, maintenance and dispose of it. The technical data are created and used throughout the product lifecycle. Some technical data, such as part geometry and NC programs, are created according to the technical conditions, some are created elsewhere, such as coming from the client, as a result of feedback obtained from the product testing.

Some of the data resulting from the material strength analysis of are used in the design study and others such as welding instructions in manufacturing, others such as installation instructions to customers or disassembly instructions at the end of the product lifecycle. Technical data are available in many places, on various storage supports and in large volumes.

Workflow is the sequence of technological operations, consisting of activities that creates or uses technical data. They intersect with the life cycle of the product. Flow can start the marketing department traditionally defined and can continue with design departments, manufacturing, services and so on. There is also a flow through other organizations that use technical data. Some of these activities take place outside the company.

Depending on the specific of company and the types of products designed, a flow of technical documentation can be initiated by different people, from design department, marketing department (where they settle and product requirements) to the high level management board of the company. If the product is a punching tool with successive punching and cutting operations the tasks circuit and flow may be defined in the Teamcenter Process Manager like in figure 1.

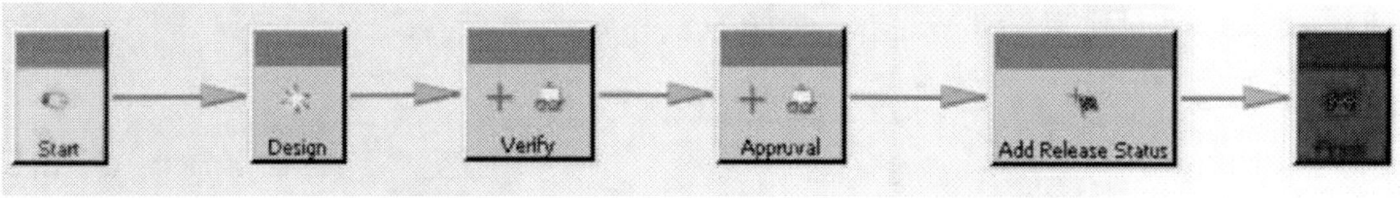

Figure 1. Documents workflow in design process.

After establishing the circuit of documents in design process on go to the allocation of tasks within the departments. Thus, on assign peoples to design the various components of the punch (figure 2). Also, on assign tasks to verify and check the documentation and for the approval of the whole process (figure 3).

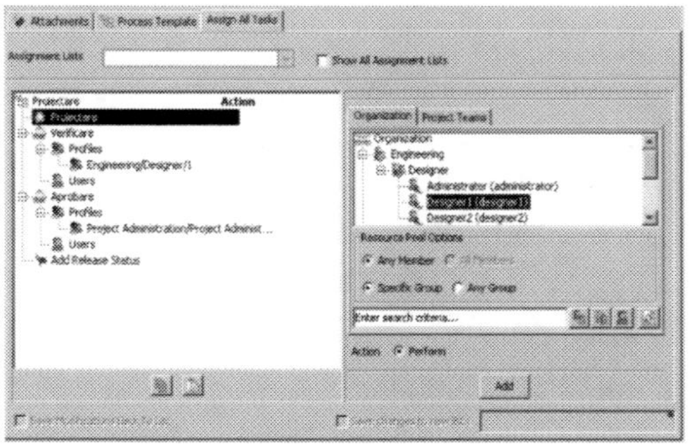

Figure 2. Assigning the design task.

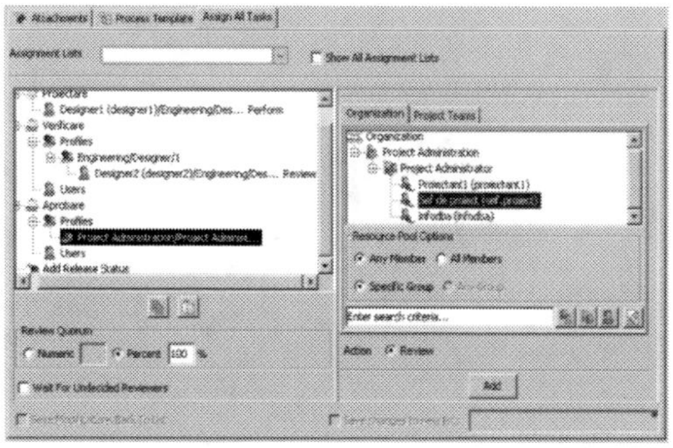

Figure 3. Assigning the approval task.

After assigning tasks on verify the start of the workflow. On can observe two tabs with two types of tasks: Task to Perform - where are listed the tasks and activities to be performed (figure 4) and to Task Track - where are listed the tasks to be pursued.

IOP Conf. Series: Materials Science and Engineering **95** (2015) 012141 doi:10.1088/1757-899X/95/1/012141

3. Design

Experiments Design activities will be performed by the team of specialists called Designers. In this team, the person named designer1 will find in the section Quick Links -> My Worklist the task of designing successive punching and cutting device (figure 4), as instructed by the Project Manager person. He has access to the available information, such as the drawings item, containing technical data about the product, other additional files and so on.

In the design activity are created items containing three-dimensional models of punching. In figure 5 is shown the window in which the item Upper package of punch is started to be created.

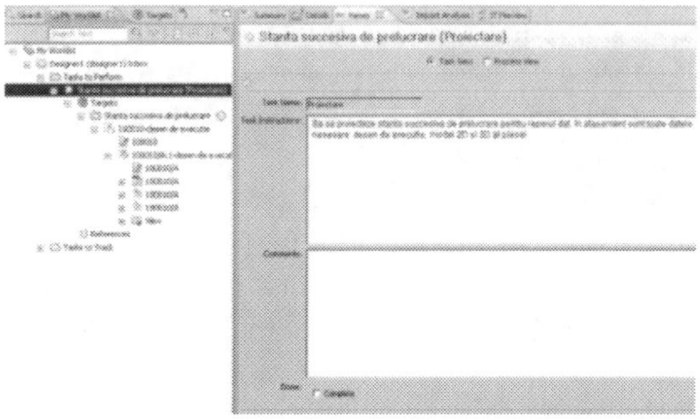

Figure 4. Task to do- Punch design.

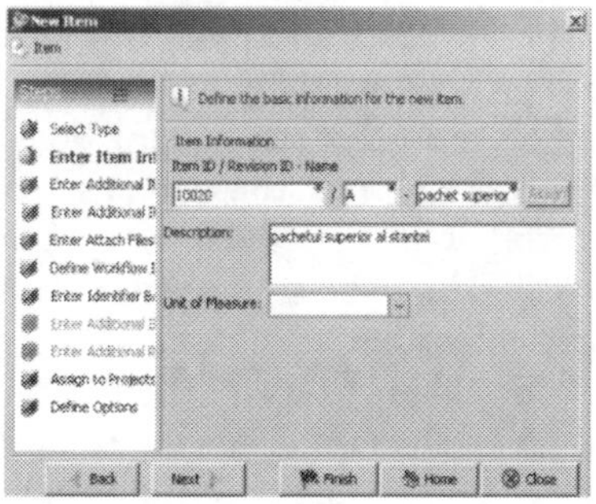

Figure 3. New item created for *Upper package.*

After defining items, in the Attached File will add files and import 3D model of the Upper package of the punching. When all parts are attached, the items form the tree structure of the product, a three-dimensional assembly with all parts included (figure 6).

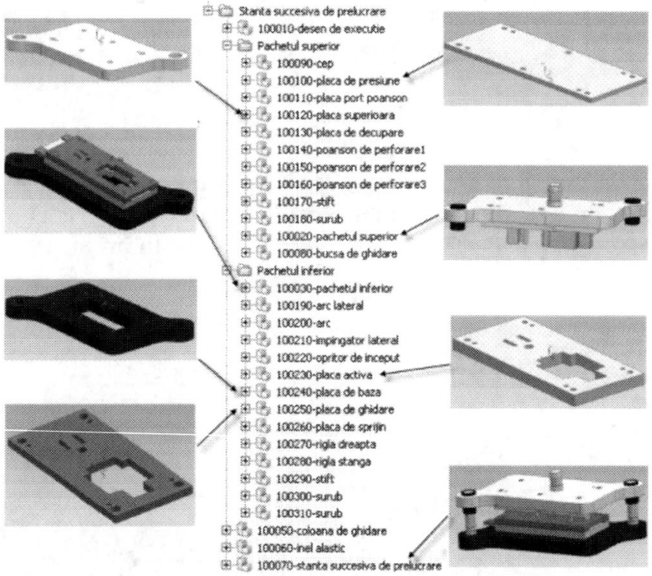

Figure 6. Product structure of punching.

3.1. Checking

Verification activity was attributed to user Designer2. It can access all the files associated with the project. The person appointed to verify will check the project according to the specification of the *Project Manager*. Depending on how the project meets the requirements imposed by, the verifier has three options: validates the project, reject the project, or takes no decision.

3.2. Approval

The state of the project seen in terms of the approving person Project Manager is shown in figure 7. The marked activities have been made. The *Project Manager* has three options: approve, reject, or decide anything. In the case of rejection (Reject), the project returns to the previous stage (Checking) for making changes.

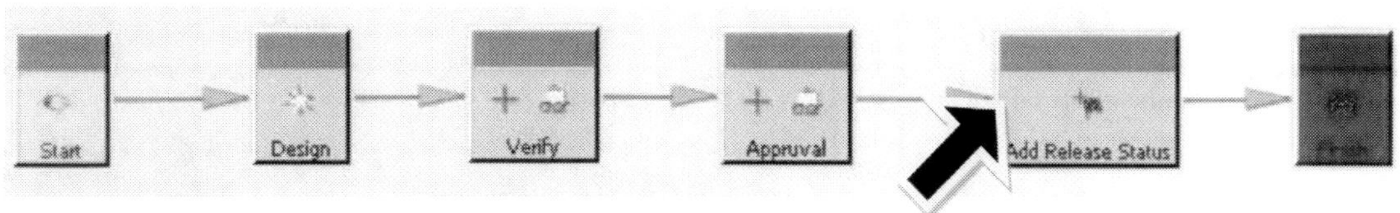

Figure 7. Workflow state, after Chavking task.

4. Creating product structure BOM (Bill of Materials)

BOM View Revision option defines constructive structure of the product design. After this operation, all standard and nonstandard components are inserted into the assembly of successive perforating and cut punching.

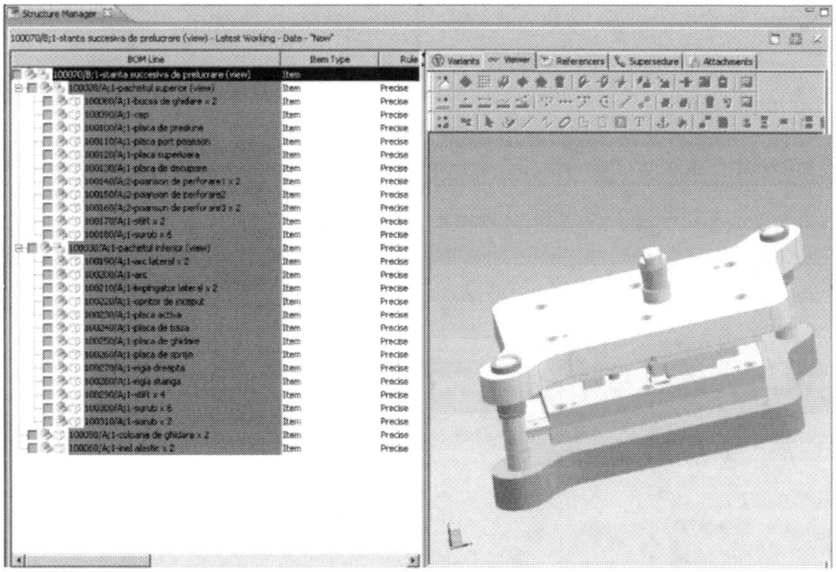

Figure 8. 3D visualization of product structure and BOM.

List of components in *.xls* format can be transferred to other departments of the company and used, for example, in order to supply the necessary materials to achieve the product.

5. Change Management in Teamcenter

If changes occur in the piece obtained by punching, these will cause changes in the components of the punch. Thus, if the initial piece (figure 9) has had a 20 mm hole, the modified piece will have a 25 mm hole (figure 10). This change will determine the 2D drawing item to have two revisions, A and B (figure 11).

This change must occur in *Workflow Process* module. Change must be described and approved). Also, the process of changing determines specific documentation flow.

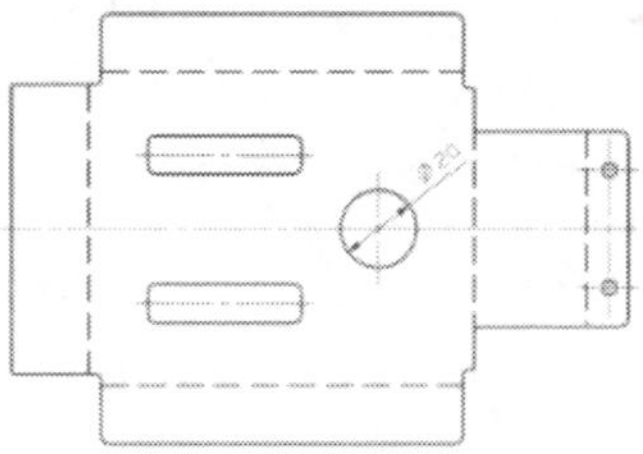

Figure 9. Initial dimension (Φ 20).

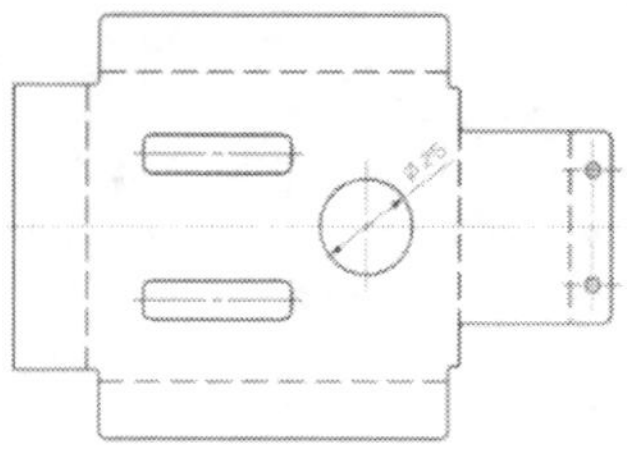

Figure 10. Modified dimension (Φ 25).

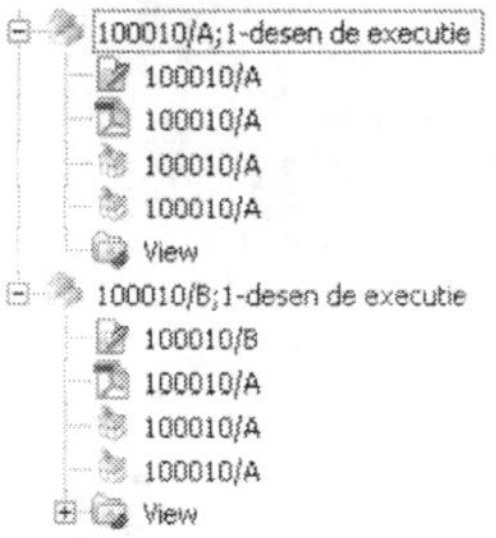

Figure 11. Item with two revisions.

This workflow involves several people: the person making the request for change (*Change Admin*), the person who make up technical report and recommendations regarding changes (*Author Recommendation*), the person that analyze the impact of change (Check Change Type), the person who analyzes change impact in terms of the customer (*Custom Preliminary Comments*) and the person who approves the change in the company (*Author Business Decision*).

After the analysis of required change, *Designer1* person will make all necessary and needed recommendations and will prepare documentation in this regard. This will create new revisions of modified items: 100070-successive processing punch item, item 10020-upper package, item 10030-lower package (with item 100250-guide plate, and item 100230-active plate).

6. Manufacturing Management in Teamcenter

Creating manufacturing processes is performed Manufacturing Process Planner application from the main menu. Defining a new manufacturing process of the punch is possible in the New Process window. In figure 12 is shown the process of "assembly of the successive perforating and cutting punch". Considering the assembly processes of components of lower and upper package, the configuration in figure 13 is resulted, corresponding to the manufacturing process of assembling the punch.

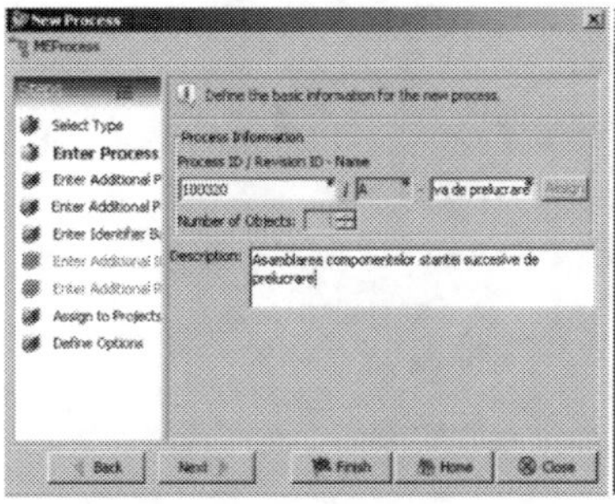

Figure 12. Defining the assembly process.

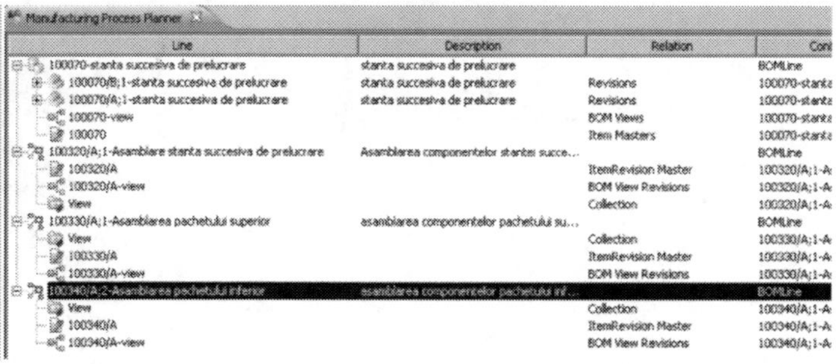

Figure 13. Manufacturing process of assembling the punch.

The next level is to establish operations necessary to manufacture of parts that go into the punch assembly. Thus, for the manufacture the part 100230/B-active plate is assigned unique code 100350/A- board manufacturing assets. Operations to manufacture active board part are: cutting, planar milling, contour milling, drilling (4xpinhole), drilling (4xscrew hole), drilling of punch hole (25 mm diameter hole punch), milling (2 x rectangular punch hole), milling (shape of the exterior).

In figure 14 is shown how to define the cutting operation. Setting the sequence of operations (technology) in the manufacturing process of the active board of successive perforating and cutting punch is shown in figure 15. Also, Teamcenter enables the design of manufacturing department's layout- (figure 16). After the design of manufacturing layout, the links among resources *product-process-manufacturing department* are created.

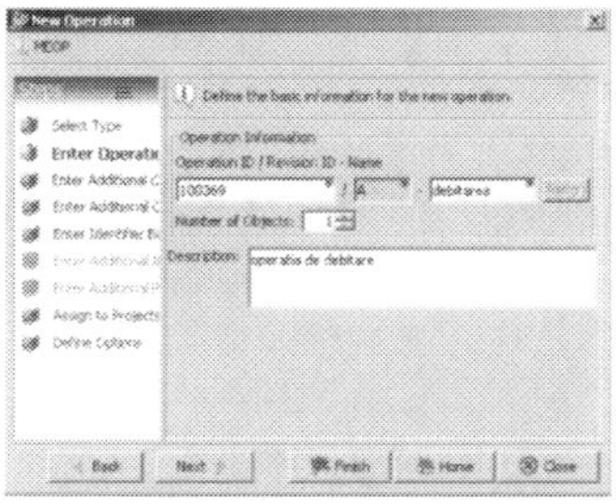

Figure 14. Defining cutting operation.

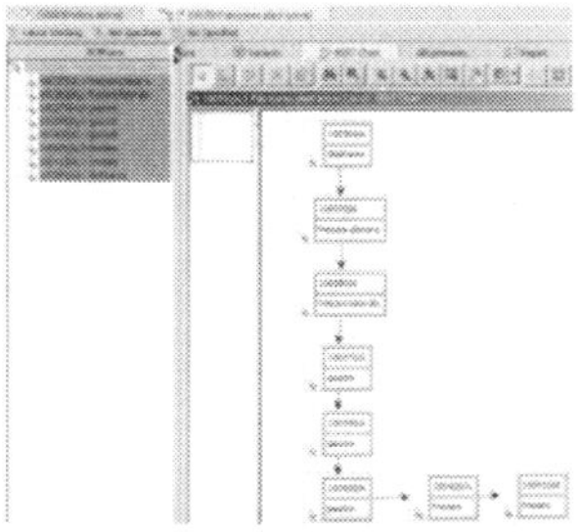

Figure 15. Technological workflow of active board manufacturing.

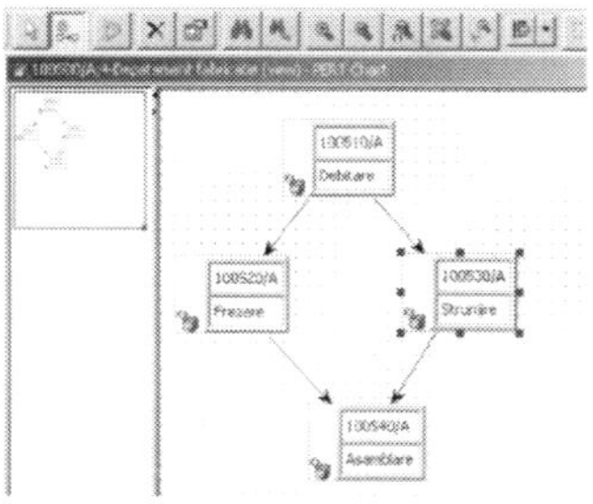

Figure 16. Layout of the manufacturing process

6.1. Generate reports and documentation production
If the information in digital format is not enough, the program provides users the option to convert digital information into the printed information on paper. After designing the processes, operations and activities, they should be described and detailed. The documentation that is generated is specific to each operation and activities for each manufactured or processed product.

7. Conclusions

The implementation of a PDM application, like Teamcenter, in a company, improves manufacturing capacity by managing product information, eliminating long time working with search, verification and correction of information, while ensuring the uniqueness and completeness of the data.

Teamcenter includes all processes and activities of the product life cycle and can create and generate reports at any stage thereof. The application significantly improves relations between different departments of the company through its interactive modules. Makes available a clear traceability of the product lifecycle management, helping to track changes that occur during the project development and also enables to reuse data.

Teamcenter manages projects and resources enabling the enterprise team selection. Provides the enterprise virtualization, enables the design and simulation of manufacturing errors and eliminates the waste of materials and resources.

References

[1] Stark J 2005 Product Lifecycle Management *21st century paradigm for product realization* (Springer: USA)

[2] Einger M, Dickopf Apostolov T, Schaefer P, Karl-Gerhard F and Keßer A 2014 Initial Approach for Sustainable Product Development Process based on Methods of Model Based systems Engineering *Product Lifecycle Management for a Global Market* pp 287- 300

[3] Duque Ciceri D, Garetti M and Terzi S 2009 Product Lifecycle Management Approach for Sustainability *Proceedings of the 19th CIRP Design Conference -Competitiv Design* pp 147-152

[4] Bergsjö D, Vielhaber M, Malvius D, Burr H and Malmqvist J 2007 Product Lifecycle Management for Cross-X Engineering Design *International Conference On Engineering Design* (Paris: France)

[5] Corallo A, Latino M E, Lazoi M, Lettera S, Marra M and Verardi S 2013 Defining Product Lifecycle Management: A Journey across, Features, Definitions, and Concepts *Industrial Engineering* (Hindawi Publishing Corporation) p 10

Some trends and proposals for the inclusion of sustainability in the design of manufacturing process

J Fradinho[1], D Nedelcu[2], A Gabriel-Santos[1], A Gonçalves-Coelho[1] and A Mourão[1]
[1]UNIDEMI & DEMI, Faculdade de Ciências e Tecnologia, Universidade Nova de Lisboa, Campus de Caparica, Caparica 2829-516, Portugal
[2]"Gheorghe Asachi" Technical University of Iaşi, Department of Machine Manufacturing Technology,Blvd. Mangeron, No. 59A, 700050, Iaşi, România

E-mail: jmf@fct.unl.pt

Abstract. Production processes are designed to meet requirements of three different natures, quality, cost and time. Environmental concerns have expanded the field of conceptual design through the introduction of sustainability requirements that are driven by the growing societal thoughtfulness about environmental issues. One could say that the major concern has been the definition of metrics or indices for sustainability. However, those metrics usually have some lack of consistency. More than ever, there is a need for an all-inclusive view at any level of decision-making, from the establishing of the design requirements to the implementation of the solutions. According to the Axiomatic Design Theory, sustainable designs are usually coupled designs that should be avoided. This raises a concern related to the very nature of sustainability: the cross effects between the actions that should be considered in the attempt to decouple the design solutions. In terms of production, one should clarify the characterization of the sustainability of production systems. The objectives of this paper are: i) to analyse some trends for approaching the sustainability of the production processes; ii) to define sustainability in terms of requirements for the design of the production processes; iii) to make some proposals based on the Axiomatic Design Theory, in order to establish the principles with which the guidelines for designing production processes must comply; iv) to discuss how to introduce this matter in teaching both manufacturing technology and design of production systems.

1. Introduction

Environmental sustainability has become a concern to the human society. The natural way in that a part of the world population already separates and discard the trash they produce at proper places is a good example of the collective consciousness. This kind of concerns also began to be reflected in the designing actions, mostly concerning to the type of raw materials and the energy consumption through the entire life cycle of the design objects, which includes decommissioning and disposal [1]. Thus, the concept of eco-design was born, which ultimately consists in seeking to design to minimize the environmental impacts throughout the products' life cycle. This characteristic turns out to become more important regarding the traditional balance performance vs. cost, as well as the available options for the products. Nowadays, the adoption of the development concept by corporations is already regarded as business value. Some authors developed design concepts on sustainability. According to them, the design of new products should also accomplish environmental aspects besides the economic

Content from this work may be used under the terms of the Creative Commons Attribution 3.0 licence. Any further distribution of this work must maintain attribution to the author(s) and the title of the work, journal citation and DOI.
Published under licence by IOP Publishing Ltd

and societal ones [2]. In this threefold concept, profit, planet and people are considered simultaneously (see figure 1). Thus, sustainable design is a kind of design that accomplishes the expected functionality that generates profit, that is socially acceptable and that uses reduced raw materials and energy, while generating minimal waste.

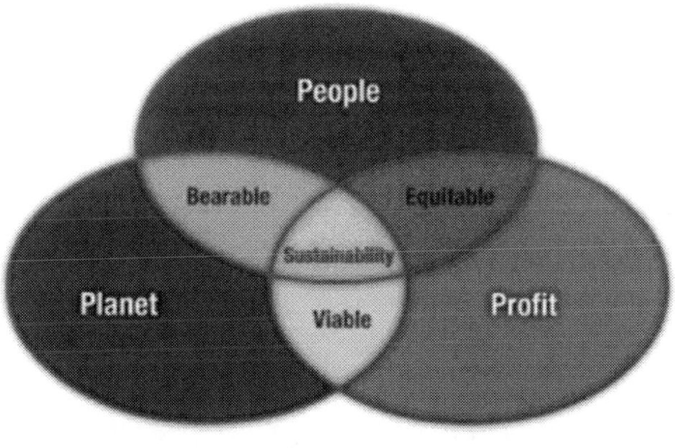

Figure 1. The triple bottom line of sustainability [2].

Sustainable design is applicable not only to product design but also to systems design, namely to industrial manufacturing processes. How do the companies know if they are using sustainable manufacturing processes? What should they do to move in the right direction? On which metrics should they support the sustainability assessments? This paper will try to contribute to the answers for those questions. First, some trends for sustainable manufacturing processes are presented and discussed (section 2), followed by the presentation of some metrics to assess sustainability (section 3), from which one concludes the coupled nature of sustainable design (section 4). Section 5 proposes an approach to deal with sustainability in the design of manufacturing processes. In the end, some suggestions are made for teaching sustainable manufacturing in engineering schools (section 6), followed by concluding remarks.

2. Some trends for approaching sustainability of manufacturing processes

Sustainability is a broad concept that involves materials, energy, waste and pollution. Considering all these aspects in manufacturing design is a hard task. So far, many authors have been working in the development of approaches for the inclusion of sustainability in the design of manufacturing processes. Those approaches usually have in common the use of suitable indicators in order to measure, compare or access sustainability.

Shuaib et. al. have built a global product sustainability index [3]. Their goal was to achieve a figure that describes the sustainability of manufactured products through their complete life cycle. They called it "Product Sustainability Index (ProdSI)". They used a top-down approach to identify which are the elements of the product that should be evaluated for sustainability. The approach seems more appropriate than the reverse one, where the metrics are established without knowing a priori how to aggregate them. They developed a five-level hierarchical structure:

- ProdSI – the overall aggregated product sustainability performance index;
- Subindex – the three aspects of TBL (economy, environment and society);
- Cluster – major elements or factors of product sustainability in the three TBL categories;
- Subcluster – decomposition of clusters to more-specific elements of product sustainability;
- Individual metrics – a quantifiable and measurable attribute or property related to a single parameter or indicator of product sustainability that is measured through the total product life cycle.

Table 1 partially depicts this structure. Each subcluster has one or more than one metrics that are not shown in the table. The procedure starts with the measurement of the individual metrics and the corresponding data collection, which is followed by data normalization (in a scale 0 - 10), the choice of weighting factors (based on experts' opinion that are gathered through surveys) and the data aggregation.

Table 1. Top down approach for individual metrics [5].

Subindex	Cluster	Subcluster
Economy	Industrial investment	Capital cost
		R&D cost
		Employee training
	Direct/indirect costs and overheads	Labor cost
		Materials
		Energy
		Logistics
		Product operational cost
		Legal costs
	Benefits and losses	Market value
		Quality losses
Environment	Material use and efficiency	Product material content
		Material utilization
		Regulations and certification
	Energy use and efficiency	Energy from renewable sources
		Energy from nonrenewable sources
		Energy regulations/certification
		Energy efficiency
	Other resources use and efficiency	Water use
		Recycled water use
		Other natural resources
		Natural resource regulations/certification
	Wastes and emissions	Gaseous emissions
		Solid waste
		Liquid waste
		Other waste and emissions
		Waste management/certification
	Product end of life (EOL)	EOL product/material recovery
		EOL product reuse
		EOL product remanufacturing
		EOL product recycling
		Product EOL regulations/certifications
Society	Product quality and durability	Product repair and maintenance
		Product reliability
		Return and recall
	Functional performance	Major product specifications
		Product customizability
		Product functional effectiveness
		Ease of operation
	Product EOL management	Ease of EOL activities
		Product EOL societal impact
	Product safety and health impact	Safety
		Health
	Product societal impact regulations and certifications	Product EOL regulation compliance
		Product EOL certification

In the end, they succeeded to create a representative index to compare product sustainability. The weakest point of the method is the use of weighting factors that induce subjectivity. Indeed, weighting is a very sensitive process that affects the result of the sustainability assessment.

In what to the sustainability characterization of manufacturing processes concerns, many works have been produced in recent years. Mani et. al identified the following challenges to develop the measurement for sustainable manufacturing [4]:

- uncertainties in manufacturing environment;
- dramatic changes in customer requirements;
- innovation in production technology;
- uncertainties in internal operating environment;
- inadequate approaches to overcome uncertainties;
- inadequate and unstructured information;
- inadequate decision models;
- undefined scope and boundaries within manufacturing unit processes;
- multiple unverified measurement methodologies.

The same authors presented a methodology for the evaluation and the improvement of sustainability in manufacturing industries. Based on unit manufacturing processes (UMP) (i.e., injection molding, casting and machining and surface treatment), modeling information is made by considering inputs, outputs, operation rules and resources, as well as the dataflow between operations. The goal is to make the UMP sustainable characterization, in order to create essential information to support decision-making related to sustainability. This characterization is made extracting key information of manufacturing processes from all the data available in handbooks, catalogues or metrics. Such information includes, but is not limited to, energy, emissions, pollutants, waste, scrap, alternative raw materials, cycle time and productivity. The main purpose of the characterization of the sustainability is to formulate strategies for evaluation and improvement of sustainability. The meth¬odology is depicted in figure 2.

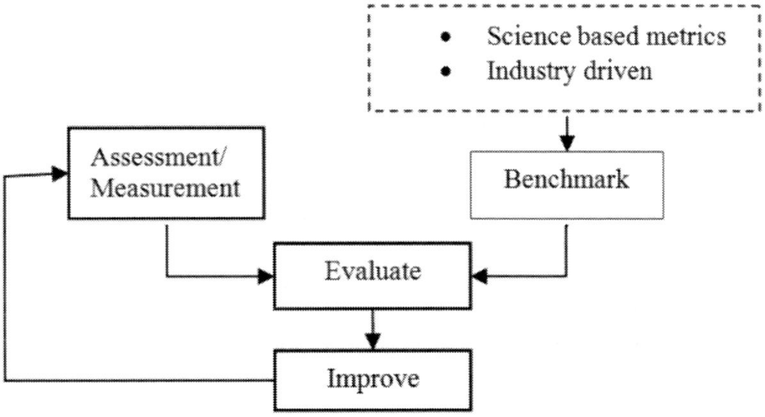

Figure 2. Sustainability evaluation and improvement methodology [4].

The two above cited works [3, 4] are representative of the state-of-art concerning to the sustainability approach to manufacturing processes. The current situation can be characterized by:

- General acceptance for the inclusion of sustainability concerns;
- Inexistence of a global and well-defined framework to characterize sustainability in manu-facturing;
- Use of representative indices as a basis to design decision-making.

3. Sustainability requirements and metrics for the design of production processes

Within According to the Triple Bottom Line concept, economic, environmental and societal requirements must be considered together in a sustainable design of any manufacturing process. Regardless the many works already carried out, sustainability stakeholders agree on the absence of a well-defined approach to characterize sustainability in manufacturing. The difficulty is related to the huge quantity of design requirements, as well as to the relations between them. Usually, the manner to deal with this drawback is to use indices to support decision-making.

Several indices have been proposed and used. Labor cost, energy consumption and return rate of faulty products are just a few examples.

In a sustainable design of any manufacturing process, the three types of sustainability requirements (economic, environmental and societal) should be considered together. Each type of requirement could generate many lower-level requirements, which turn sustainable designs into complex designs. Furthermore, many of the sustainable requirements depend on the same design variables. For instance, is easy to realize that economic and environmental requirements likely depend on the same design variables that govern the functional requirements. According to the Axiomatic Design terminology, most of these designs are of the coupled type [5].

This difficulty is often avoided by considering some design specifications as input constraints, and not as requirements, but this expeditious tactics usually hampers the designer's freedom.

4. How to deal with sustainability in design of manufacturing processes

Sustainable design of manufacturing processes is being studied trough several approaches and promising results have been achieved. However, an overall method for the sustainability assessment was not found so far.

Axiomatic Design theory (AD) provides a useful framework to deal with any type of design process because it allows distinguishing "good designs" from "poor designs". In addition, AD also allows identifying the best design among several "good designs". Therefore, AD seems to be an appropri¬ate tool to deal with sustainable designs.

Since the Brundtland Report [6], sustainability became a compulsory part of the development strategies of organizations. Whatever is the way that is chosen to achieve sustainability, its implementation unavoidably involves engineering decisions, in which one would have to cumulatively meet functionality, cost-effectiveness and safeguard the environment. Specifically, manufacturing technology and production systems are crucially important. Engineering professionals are actors of sustainable development, but as said Broman et. al., "an essential aspect of sustainable development is to include knowledge about it in the education of professionals" [7]. With an exception to what to the legal details concerns, an issue for discussion is if the development of an attitude toward sustainability in engineering education should be based on the inclusion of the theme in different disciplines, or if the matter should be regarded as an autonomous subject.

A possible approach to this theme is as follows: students should understand sustainability as an important goal to fulfill that should not be regarded as a constraint to the engineering solutions. Therefore, sustainability should be expressed by a set of functional requirements at each level of abstraction of the designing process, which in turn should be decomposed in a basis of suitable design parameters and process variables that cumulatively satisfy functionality, cost-effectiveness and environment. This implies that the evaluation of sustainability must be made in the conceptual stage of decision, instead at the detailed design. This approach implies that modern-day engineering education should encompass state-of-art technology, and should take into account that global sustainability entails concerns about "appropriate technology" [8]. The so-called "appropriate technology", as well as the corresponding design solutions, should therefore be taught. A basic concept that is comprised in the definition of sustainability is the concept of 'needs', which engineers must know to transform into functional requirements. As mentioned in the UNESCO Report, "the key factor is that all technology should be appropriate to economic, social and cultural contexts, as it once was when most technology was developed locally or when most situations were fairly similar in smaller scale pre-industrial days.

Problems arise when technology is transferred between different economic, social and cultural contexts, even within the same country, but especially between the developed and industrial to less developed and more traditional situations." [8]. Based on this reasoning, engineering education, specifically in the field of manufacturing engineering, should foster the attitude of students toward creating new sustainable solutions.
How to deal with sustainability in design of manufacturing processes

5. Conclusions

There is no global and well-defined framework to characterize sustainability in manufacturing and Axiomatic Design could yield to a suitable solution for design for sustainability, because AD allows identifying the existing couplings and trying to remove at least some of them. The current trends seem to be the inclusion of the theme sustainability in different disciplines, or as an autonomous subject. Whatever would be the solution for teaching, education in manufacturing engineering should foster the attitude of students toward creating sustainable solutions.

Acknowledgments

The authors gratefully thank the sponsorship of Fundação para a Ciência e Tecnologia through the Strategic Project PEst-OE/EME/UI0667/2011 – UNIDEMI.

References

[1] Giudice F, La Rosa G and Risitano A 2005 Materials selection in the Life-Cycle Design process: a method to integrate mechanical and environmental performances in optimal choice *Materials and Design* **26** pp 9-20
[2] Chiu M C and Chu C – H 2012 Review of Sustainable Product Design from Life Cycle Perspectives *International Journal of Precision Engineering and Manufacturing* **13(7)** pp 1259-1272
[3] Shuaib M, Seevers D, Xiangxue Z, Badurdeen F, Rouch K and Jawahir 2014 Product Sustainability Index (ProdSI) – A Metrics-based framework to Evaluate the Total Life Cycle Sustainability of Manufactured Products *Journal of Industrial Ecology* **18(4)** pp 491-507
[4] Mani M, Madan J, Lee J, Lyons K and Gupta S 2014 Sustainability characterization for manufacturing processes *International Journal of Production Research* **52(20)** pp 5895-5912
[5] Fradinho J, Gabriel-Santos A, Martinho A, Gonçalves-Coelho A and Mourão A 2014 Sustainability: how to deal with coupled designs *Proc. The 20th International Scientific Conference Knowledge-Based Organization* **3** pp 154-158
[6] Our Common Future 1987 Report of the World Commission on Environment and Development, Official Records of the General Assembly, Forty-second Session, Supplement **25** (United Nations: N.Y.)
[7] Broman G I, Byggeth S H and Robeárt K-H 2002 Integrating Environmental Aspects in Engineering Education *Int. J. Engng Ed.* **18(6)** pp 717-724
[8] *Engineering: Issues Challenges and Opportunities for Development* 2010 UNESCO Report (UNESCO: Paris)

IOP Conf. Series: Materials Science and Engineering 95 (2015) 012143 doi:10.1088/1757-899X/95/1/012143

Innovation management based on proactive engagement of customers: A case study on LEGO Group. Part I: Innovation Management at Lego Group

G Rusu[1] **and S Avasilcăi**[1]
[1]"Gheorghe Asachi" Technical University of Iasi, Department of Engineering and Management, Bd. D. Mangeron 29, Iasi, Romania

E-mail: grusu@tex.tuiasi.ro

Abstract. Customers' proactive engagement in the innovation process represents a business priority for companies which adopt the open innovation business model. In such a context, it is of outmost importance for companies to use the online environment and social media, in order to create an interactive and open dialogue with customers and other important external stakeholders, achieving to gather creative solutions and innovative ideas by involving them in the process of co-creating value. Thus, the current paper is based on a case study approach, which aims to highlight the open innovation business model of the LEGO Group, one of the most successful and active company in engaging customers in submitting ideas and creative solutions for developing new products and new technologies, through online platforms. The study then proceeds to analyze the innovation management at LEGO Group, emphasizing the most important elements regarding the management team, the success and failures, the evolution of the LEGO products focusing on the innovation efforts of the company, its mission, vision, and values, emphasizing the innovation terms which guide the actions and objectives of the LEGO Group. Also, the research based on the case study approach, outlines the most important policies and strategies of the company, the organizational structure consisting of flat structures which facilitate the orientation of the team management on the innovation process and the proactive involvement of consumers and other external stakeholders in product development, highlighting also the most important activities developed by the management team in exploring the new opportunities which may occur on the market, involving customers in sharing their ideas at festivals, participating to discussions of adult fans on web-based platforms and establishing partnerships with the external stakeholders in order to create value. Moreover, the paper is focused on identifying the company's concerns regarding the competitive strategies, new technologies developed for sustainable development and innovation, emphasizing also the concerns of the LEGO Group for a long-term orientation strategy which implies developing innovative products, made by ecological materials, reducing the negative impact on the environment. In addition, the company aims to foster innovation maintaining a clear vision, multiple contacts with the internal and external environment, and the flat structures which enable the communication between innovative management teams and top management team. Consequently, the study provides the insights regarding the innovation management of the LEGO Group, emphasizing the role of engaging customers and other external stakeholders in the co-creation of value.

Content from this work may be used under the terms of the Creative Commons Attribution 3.0 licence. Any further distribution of this work must maintain attribution to the author(s) and the title of the work, journal citation and DOI.
Published under licence by IOP Publishing Ltd

1. Introduction

In the context of changes that progressively appear in the modern business environment, companies should do the effort to adapt their R & D Department functioning to the new trends which require the involvement of the external stakeholders in the innovation process of the company products. Thus, recent literature in this area has recommended that companies should "think about their business from an open services perspective to discover new ways to generate profitable growth" [1]. Consistent with this trend of the new business models mentioned in the literature, the LEGO Group company managed to adapt to the dynamic evolution of the digital technologies, being a successful company by creating innovation possibilities for their products and by offering to their customers the opportunity to interact with company's site in order to choose by their own the playing and creation possibilities, making the LEGO products much more attractive.

In such context, it appears as of outmost importance to highlight the idea that engaging customers in the innovation process, by listening to their creative ideas, needs and expectations, creates the premises for a highly successful business, as LEGO Group is today. Thus, innovation management stays at the core of the LEGO Group business, placing the creation of innovative products for customers on a high level as this provides them the long-term success of their business.

2. Innovation Management in a New Era of Business

2.1. Co-creation in Innovation

LEGO Nowadays more and more successful companies choose to involve their customers in the innovation process, giving them the opportunity to express their needs and creative ideas for new product development, by co-creating with company's specialists innovative products or services adapted to their requests.

In this context, the perspective of co-creating which involves the customer in the creation of the new products emphasizes "the active role of the consumer as a key component of value co-creation in innovation activities" [2]. Based on this perspective, the customer is known as a prosumer, who can produce his own food, goods and services, the consumer becoming a co-producer by performing a part of the production process, as he is more informed, connected and active on the online environment in order to interact with companies for co-creating and co-designing new products and services [3].
Moreover, companies create online platforms, through which they communicate with customers, involving them in contests or challenges for those which are very creative and think they have innovative ideas to share regarding the creation of new products for the company. Thus, it is recommended that companies should co-create with customers, changing their role in the innovation process, by considering them active consumers, which are involved in the co-creation of new products and services [1]. [6].

2.2. The Open Innovation Model

The new model of business innovation requires that companies should be open to receive information for new product and service development from the internal but also from the external environment. Moreover, the open innovation model creates the basis for customers' proactive engagement in delivering information and creative ideas on the online environment, on special platforms companies create to maintain a continuous dialogue with their customers. Consequently, companies seek to come closer to their customers' needs, "leveraging platforms to create on-the-fly partnerships with large communities of programmers who use the common infrastructure and toolset to innovate and create value." [4]

Also, the open innovation paradigm outlines that, projects for new products or services can be launched from either internal or external technology sources and new technology or ideas can enter in the process from internal or external sources [5]. Thus, companies use the online environment to create interactions between innovators from the company and customers or other companies, who can bring new technology and ideas for creating innovative products or service for the market. This open dialogue is fostered by the new information technology and Internet which provide the premises for creativity and innovation to "take place over large amorphous networks of peers" [6], in order to "produce dynamic new goods and services that rival those of the world's largest and best-financed enterprises." [6]

In this context, the current paper, based on a case study approach, aims to investigate the case of innovation management on the successful toy company, LEGO Group, which adopted the open innovation business model and managed to cope with the failure periods encountered throughout its history. The case study is focused on analysing company's innovation management and its policies, concerns and interventions regarding open innovation.

3. Open Innovation Management: The Case of LEGO Group

3.1. LEGO Group: Management Team and History

LEGO LEGO Group is a privately held company, a property of the Kirk Kristiansen family, who founded it in Billund, Denmark, in 1932. The LEGO Group is owned by KIRKBI A/S (75%), the company for investments and strategic activities of Kirk Kristiansen family, and by LEGO Foundation (25%), a corporatist organization which has the same mission as the LEGO Group [7]. The object of activity of the company is to offer toys, experiences and learning materials for children from over 130 countries in the world, developing their creativity through play and learning [8].

Regarding the management team of the company, the most important actors at the global level are the members of the Management Board and the CEO of the company, Jorgen Vig Knudstorp. The Board of Directors of the company is represented by 7 members and the most important actors are Niels Jacobsen, the Chairman of the Board since 2008 and Kjeld Kirk Kristiansen, the Deputy Chairman of the Board since 1996 [9]. As for the Corporate Management, the most important actors from those twenty-one Vice Presidents, mainly involved in the innovation management of the firm, are Per Hjuler, Vice President for Innovation & Consumer Marketing, and Maarten Tibosch, the Vice President for Value Chain Innovation [10].

LEGO, the name of the company, is an abbreviation from the Danish words "leg godt", which means "play well", which is the ideal of the company. LEGO Group was founded in 1932 by Ole Kirk Kristiansen, and is proud of an ascending evolution, in the 80 years of activity becoming a modern global company and the third manufacturer of toys in the world, starting from a small carpenter's workshop. The LEGO brick remains the most important product of the company, although the business evolved, being named twice, "Toy of the Century", as this LEGO product offers unlimited building possibilities and develops creativity through playing [11].

The evolution of the company brings many important events in the company's history, as LEGO Group is a family business. The most successful periods in the company's history refer to the continuous expansion of the company, in 1987 being present with its products on 115 markets in the world. Also, LEGO company success registered in 1992 highlights a Guinness record in railway line construction and the introduction of LEGO products in China. In 1999, LEGO brick is named as one of the "Products of the Century", although the company is forced to restructure and the changes brought about the loss of 1000 jobs [12].

At the beginning of the third millennium LEGO encountered several changes of its mission and fundamental principles and a deficit of 1 billion DKK, failure which was overcome by obtaining profit due to success in Germany for LEGO My World product. However, year 2004 represented a turning point for LEGO as it registered a huge deficit, which determines the CEO, Kjeld Kirk Kristiansen, to create an action plan to turn around the company, appointing Jorgen Vig Knudstorp as the new CEO of the company. The company managed to overcome the failures, as LEGO makes the first steps towards co-creation in innovation, creating a dialogue with its customers re-launching DUPLO and founding LEGO Factory on www.LEGO.com site, which enables customers to create their own model to play and to buy LEGO bricks for building the model [13]. Strategic decisions regarding outsourcing of a part of production to external partners in 2006 placed the company in 2011 on the third place in the top of the world's toy manufacturers [14].

3.2. Product innovation efforts
LEGO products evolved, new toys being created according to technological changes and customers' needs. Also, LEGO created LEGO TECHNIC sets, PlayStation games, LEGO mini-figures, films, software like LEGO My World or Island 2, LEGO SCALA, LEGO Games, etc. Moreover, LEGO continued the product innovation efforts involving customers in the creation of the desired model online, and the most important products brought on the market after adopting this strategy of creation of new products taking into account customers' needs was the following: LEGO Designer sets, DINO 2010, LEGO Vikings, MINDSTORMS NXT, Parcheesi Board Game, electric building system Power Functions, Aqua Riders set, LEGO Indiana Jones, LEGO Games and Powerminers [13]. Also, according to the evolution of new technologies and customers' various needs who use more and more the Internet, LEGO launched in 2010 the online games LEGO Universe, LEGO Technic Unimog, Ninjago, etc. [14]. As one may observe, LEGO products were adapted to technological changes, customers' preferences, market tendencies and to new characters from films (e.g. The Lone Ranger, The Hobbit, etc.) and cartoons, taking into account the using of Internet and online platforms, but also the opportunity offered to customers from the world to create their own favorite game and to buy LEGO games more easy on the Internet. The LEGO online games enable children to create their own character, to play whatever they want, customizing the elements by choosing the color of the building bricks, deciding what they want to build and the place where they want to build [15].

Thus, LEGO adjustment to dynamic evolution of digital technologies fostered the company's success in the latest years, creating innovation possibilities for its products, offering to customers the opportunity to interact on company's site and to choose their possibilities of play and creation, this making the LEGO products more attractive and innovative.

3.3. Concept, values, policies and innovation strategies
LEGO's mission is to "inspire and develop the builders of tomorrow" [16], focusing on inspiring children to think creatively and to use their entire potential to build their own future [17]. The essential values of the company are: imagination, creativity, fun, learning, care and quality [16]. Of course, imagination and creativity are based on innovation, which is reflected in company's activities. The philosophy of the company consists of maximizing the positive impact on human beings and society and on offering to children creative high quality experiences of play [16].

Regarding the key policies of the company, LEGO emphasizes that their employees are the most important individuals who transfer the values of the company, and the organization is committed to developing competencies and creativity of their employees.

Another strategic direction of the company is the protection of the children right to develop, believing in the value of the play and learning by playing, for the society and children, sustaining the academic research in the play and learning and innovation field [16].

Also, the company strategy focuses on research and development activities which foster the achievement of a high level of innovation. Thus, the strategy of innovative product development through launching new products every year, such as LEGO® Ninjago and LEGO Friends, but also the strategy of placing production closer to the core markets, brought for LEGO Group revenue growths by 25%, almost triple the sales of 2007 [18].

Focusing mainly on environment protection, LEGO Group aims to explore creative and innovative ways to solve environment issues and to offer products according to the highest environment standards, reducing the energy, the impact produced by product transportation and the waste quantity, by recycling materials used in the production process and sourcing 100% from renewable energy, until 2020 [19].

On the other hand, LEGO strategy regarding the social responsibility is focused on creating a safe and fair workplace according to company's values, achieving with their internal and external stakeholders [20], and having a positive impact on the communities where they perform activities by developing opportunities for children to play and to grow in their local community [9].

The essential strategy of the company highlights its ambition to globalize and to innovate the LEGO play system, by supporting the commercial impulse, by covering the desired market segment, developing new products and launching new line products [16].

3.4. The organization of the company and key interventions of the team management
Starting from a vertical functional structure, applied to a family business, LEGO adjusted to the necessary changes to creating and developing new, innovative products, which encompasses teamwork and flat structures which could communicate efficiently and faster regarding the innovation product and launching new products line every year. In 2004, when Jorgen Van Knudstorp became the CEO of the company, the new approach of the manager focused on supporting innovation and teamwork, moving from the traditional organization to flatter structures, where teams from different fields collaborate and communicate directly about work activities [21]. Thus, the most important intervention of the team management of LEGO regarding the proactive engagement of consumers in developing innovative products was marked by the new manager, Jorgen Vig Knudstorp, which had a new vision which emphasizes the benefits of collaborating which „come from a dialogue with the community", focusing on openness, collaboration and innovative products launching [22].

Other important interventions of the team management in relation with proactive engagement of customers in developing new products are represented by BrickFest meetings, facilitated by the online platform LUGNET, between LEGO fans, who can share their construction techniques and their special creations, and LEGO managers who can found out information about the brand and the products features which users consider being the most appropriate, and they can look for new ideas for their products [23].

3.5. Exploring new market opportunities by involving customers in the innovation process
The exploration of new market opportunities was begun by LEGO since 2004 through creating online platforms of sharing ideas regarding future LEGO products. The first platform was LEGO Ambassador which represented the formalization of adult fans relations, which provided „a fast and direct way for LEGO managers to get in contact with adult fans, who may have new product or marketing ideas or be interested in providing feedback on products currently in development [22]. Focusing on creating web platforms for sharing creative ideas, LEGO managed to maintain a continuous dialogue with their customers, the LEGO Digital Designer platform facilitating new market opportunities exploration, through children engagement in uploading their models in a gallery, but also by accessing a virtual building application by means of which they can build creative models with LEGO bricks [24].

3.6. Company's innovation concerns
LEGO Group is focused on investments in research and development and innovation activities, as 60% of its sales come from these activities [9]. Also, one of the main concerns of the company is to create safe products for children, investing in product development, and to innovate in order to find sustainable materials to continue producing the LEGO brick [25]. Moreover, the company encourages the relations with the external environment, cooperating with educational institutions regarding research projects and new technologies [9]. Company's concerns for sustainable development, reflected also in products packaging, sustain innovation by rethinking product packaging in an innovative way, in smaller boxes, reducing the CO2 emissions by 10%, but also the raw materials for realizing packages and the energy for transportation [26]. The permanent dialogue with their suppliers aims to sustain the product innovation efforts, maximizing the positive impact on environment [27].

The innovation process is also sustained by providing stimulants for customers who bring new ideas materialized in new products launched on the market, innovators receiving 1% from net sales of the product [22], encouraging the launching of new ideas on online platforms.

4. Conclusions

To sum up, LEGO Group adjusted to changes occurred in the business environment, creating flatter organizational structures in order to sustain developing of innovative solutions for new products. Thus, the company adopted the open innovation business model and started a continuous dialogue with their stakeholders and customers on online platforms, in order to engage them in sharing creative ideas and developing innovative projects regarding new product development.

Moreover, the success of the company highlights the foreseen results of engaging customers in the innovation process of the organization, by creating the premises for innovators to send their innovative ideas and to collaborate with team management in order to create future products, according to customers' changing needs

Acknowledgement

This research was undertaken within the framework of the National Research Program PN II, financed by MEN – UEFISCDI, project PN-II-PT-PCCA-2013-4-1811.

References

[1] Chesbrough H W 2011 Open Services Innovation: Rethinking Your Business to Grow and Compete in a New Era *Jossey-Bass* (San Francisco)
[2] Russo-Spena T and Mele C 2012 "Five Co-s" in innovating: a practice-based view *Journal of Service Management* **23** pp 527-553
[3] Weber M E A 2012 Developing what customers really need: involving customers in innovations *International Journal of Innovation and Technology Management* **9** 1250018-1-1250018-15
[4] Tapscott D and Williams D A 2007 Opening up to collaboration *Business Week*
[5] Chesbrough H W, Vanhaverbeke W and West J 2006 Open Innovation: Researching a New Paradigm *Oxford University Press* (New York)
[6] Tapscott D and Williams A D 2007 Innovation in the age of mass collaboration *Business Week*
[7] Information on http://aboutus.lego.com/en-us/lego-group/ownership
[8] Information on http://aboutus.lego.com/en-us/lego-group
[9] Information on http://aboutus.lego.com/en-us/lego-group/annual-report
[10] Information on http://aboutus.lego.com/en-us/lego-group/management
[11] Information on http://aboutus.lego.com/en-us/lego-group/the_lego_history
[12] Information on http://aboutus.lego.com/en-us/lego-group/the_lego_history/1990
[13] Information on http://aboutus.lego.com/en-us/lego-group/the_lego_history/2000
[14] Information on http://aboutus.lego.com/en-us/lego-group/the_lego_history/2010
[15] Information on https://eu.chimaonline.com/en/create-customize
[16] Information on http://aboutus.lego.com/en-us/sustainability/our-approach
[17] Information http://aboutus.lego.com/en-us/lego-group/mission-and-vision
[18] Information on http://aboutus.lego.com/en-us/news-room/2013/february/annual-result-2012
[19] Information on http://aboutus.lego.com/en-us/sustainability/environment
[20] Information on http://aboutus.lego.com/en-us/sustainability/social-responsibility
[21] Information on http://pure.au.dk/portal-asb-student/files/36334745/BAMMC_Thesis_A_Case_Study_of_the_LEGO_Group.pdf
[22] Antorini Y M, Muniz Jr A M and Askildsen T 2012 Collaborating with customer communities: lessons from the LEGO Group *MIT Sloan Management Review* **53** pp 73-79
[23] Jo Hatch M and Schultz M 2010 Toward a theory of brand co-creation with implications for brand governance *Brand Management* **17** pp 590-604
[24] Information on http://ldd.lego.com/en-us/subpages/mindstorms
[25] Information on http://aboutus.lego.com/en-us/sustainability/product-safety
[26] Information on http://aboutus.lego.com/en-us/sustainability/sustainability-cases/smaller-boxes
[27] Information on http://aboutus.lego.com/en-us/sustainability/sustainability-cases/a-climate-savers-partner

Innovation management based on proactive engagement of customers: A case study on LEGO Group. Part II: Challenge of engaging the digital customer

S Avasilcăi[1] **and G Rusu**[1]
[1]"Gheorghe Asachi" Technical University of Iasi, Department of Engineering and Management, Bd. D. Mangeron 29, Iasi, Romania

E-mail: silvia.avasilcai@gmail.com

Abstract. To foster the development of innovative products and new technologies, nowadays companies use an open innovation system, encouraging stakeholders to contribute, using the companies' online platforms for open innovation or social media, bringing and sharing creative solutions and ideas in order to respond to challenging needs the company directly expresses. Accordingly, the current research continues the analysis of the LEGO Group innovation efforts, aiming to provide a case study approach based on describing the most important projects and online instruments company uses to interact with customers and other external stakeholders. Thus, by analysing the experience of the company in developing projects of involving stakeholders in the innovation processes, the article emphasizes the objective of these past projects developed by LEGO Group, outlining their objectives regarding the focus on the product or process innovation, the team management and stakeholders involved in the innovation actions and the results they obtained. Moreover, the case study highlights the features of the most important online instruments LEGO Group uses at the moment for engaging LEGO fans, children, parents, and other external stakeholders in developing new LEGO sets. Thus, LEGO online instruments provide the opportunity for customers to be creative and to respond to LEGO management team challenges. Accordingly, LEGO involve customers in bringing innovative ideas for LEGO sets through LEGO Ideas instrument, which aims to engage customers in submitting projects, voting and supporting ideas and also sharing them on social media. Also, the research emphasizes the role of supporting the open dialogue and interaction with customers and other external stakeholders through LEGO.com Create & Share Galleries instrument, using their creativity to upload innovative models in the public galleries. The continuous challenges LEGO launches for their fans create a long-term connection between company and its customers, supporting the value co-creation process, as the submitted ideas can materialize in new LEGO products which can be found on the market. As a consequence, customers' engagement in the co-creation process facilitated by the multiple online instruments provided by LEGO, resulted in positive outcomes for the company regarding new product development for the sets launched on the market to satisfy changing needs of their customers. The results provided by this case study approach can be useful for the business environment and academia as well in order to understand the role of engaging customers in the open innovation process, creating a competitive advantage on the market for companies.

Content from this work may be used under the terms of the Creative Commons Attribution 3.0 licence. Any further distribution of this work must maintain attribution to the author(s) and the title of the work, journal citation and DOI.
Published under licence by IOP Publishing Ltd

1. Introduction

The current case study approach continues the first part of the paper highlighting the LEGO Group company and the innovation management, where there were analysed the history of the company and management team, company's organizational products and services, the philosophy of the company, its mission and vision, the most important policies and strategies, the organizational structure, company's concerns for the innovation factors.

In the current paper we aim to analyse the co-creation in innovation process focusing on the experience of the LEGO Group and providing examples from company's history regarding the online platforms which facilitate the permanent dialogue with their customers and their creative ideas for new product development.

Thus, the second part of this case study regarding the innovation management on LEGO Group is focused on the following issues: the experience of the company in involving customers in the innovation processes by developing programs and projects, and the digitization innovation process, by using online instruments to interact with their customers. The case study presents the main instruments used by LEGO Group, identifying the target segment of customers and company's management of the relation with their customers.

2. Co-creation in innovation: Involving stakeholders in developing New LEGO products

2.1. The experience of LEGO in involving customers in the innovation processes

LEGO Group has a wide experience in creating new products as the number of projects developed in order to customize, develop and create new products is continuously increased.

Thus, the first project developed in 2005 was Ambassador, which provided a direct way for the company to connect with LEGO fans. The Ambassador program represented for the company discovering new ideas, new technologies, new business partnerships, and for adult fans, the collaboration with LEGO specialists enable them to influence the business decisions of LEGO and to encourage the company to develop new products [1]. The objective of the program was to recruit about 50 users committed to brand communities from the entire world in order to work for short periods of time within innovation facilities provided by LEGO [2]. The role of the ambassadors was to present requests and questions from the community to LEGO managers, and then managers asked ambassadors to look for ideas and feedback from users groups [3]. Also, managers participated into corporatist programs regarding product innovation and knowledge sharing activities with LEGO employees, this activity contributing to co-creating new products [2].

LEGO Kids Inner Circle is another LEGO program, created for children to comment about photographs with designers' concepts for the future toys and to criticize the sets prepared for launching on the market [3]. Accessing the program, the user can buy LEGO sets, mentioning at first his age, the most accessed plays in that day on LEGO site, how often he visits the LEGO site, the evaluation of the site, recommendation of the site to other users, the intention to come back on site and to buy LEGO products, demographic data regarding the user [4]. It may observed that this LEGO platform aims at this moment to obtain feedback regarding the site of the company, by marking answers on a questionnaire, and less to obtain ideas regarding the products which should be launched on the market by LEGO.

Another program focused on customization of the product, LEGO Factory (Design by Me) represented a real success for the company, by enabling users of the platform to create what they imagined on their personal computer, and then to order the customized product. Although this program, closed in January 2012, represented a great success for LEGO, Design by Me experience challenged the capacity of the company to accomplish the quality standards regarding LEGO service for the customized products [5]. However, the models can be found on co-creation platform called LEGO Digital Designer (LDD), which enable users to upload models in the public gallery, the success of the platform consisting in engaging millions of people in 6 years of building activities of customized models [5], when main actors from the company interacted with innovators from over the world.

Also, LEGO CUUSOO was another experiment for obtaining innovative ideas through the online environment. Launching this program in 2011 at the global level, LEGO Group managed to involve external stakeholders in a process focused on new product development, as the actors involved (designers for LEGO sets, marketing and business representatives) had the role of evaluating the new ideas for LEGO products and the projects obtaining over 10 000 supporting votes had the chance to become the new LEGO sets [6]. Though, at this moment LEGO CUUSOO team informs the users who access the platform that all projects, comments, profiles and the other data from LEGO CUUSOO were redirected to the new program of the LEGO Group, called LEGO Ideas [6].

3. Digitizing innovation by using the online environments

3.1. LEGO Ideas: the main instrument for interacting with LEGO customers

LEGO Among the numerous online instruments used by LEGO to interact with customers and to involve them in the co-creation in innovation process, LEGO Ideas is one of the most important at this moment. The homepage of LEGO Ideas is very suggestive, the visual identity being provided by LEGO logo, being present in the left side of the menu. The image representing the process followed by the company through interacting with customers expresses directly the question for users („Have an idea for a LEGO set?"), but also the stages of the process including sharing ideas, obtaining support, evaluating the idea by LEGO specialists and obtaining and launching a new LEGO product, as in the figure below [7].

Figure 1. The process of collecting creative ideas on LEGO Ideas [7].

Regarding the link with official website of the company, it may be observed that there is no link from the official website of the company to LEGO Ideas (https://ideas.lego.com/), which may represent a drawback if users are not familiar with LEGO CUUSOO official blog, website closed, but which provides the possibility to be redirected to the official website of LEGO Ideas program.

Though, it may be mentioned the externalization of this instrument, as the goal of the company is to receive as much ideas, being addressed to all users, excepting children below 13 years old. Moreover, LEGO encourages collaborations between members who build the model and those who promote the model, for the submitted projects [8].

LEGO Ideas presents to innovators on the homepage the possibility to create an account and to import data from LEGO CUUSOO. The button "Discover" enable LEGO fans to visualize the ongoing projects, to vote and to support ideas, mentioning the number of supporters, but also the number of days remained for voting during a year. Agreeing to support a project means that the user should respond to a questionnaire regarding the cost of the future product, the number of people buying it, the category of users for whom it is suitable, but also the difficulty of building the model [9].

The button "Submit" provides information to innovator regarding the project which will be sent, the way competition is conducted, the prizes and the rewards, the criteria for projects to be accepted, the quality standards of the project and rules for using the instrument and starting competition [10].

The interface of the instrument for submitting the project provides the stages for submitting the idea to LEGO specialists. The first step consists in giving the name of the project, followed by uploading an image of high quality reflecting the model [11]. In the section "Details", the innovator should offer a description of the project, mentioning its aim, having also the possibility to visualize the project uploaded on LEGO Ideas site. The last button ("Submit") offers the possibility for user to submit the project with the last changes, after marking that he agrees with terms and conditions and mentions if the idea of the project is original or it is based on the intellectual property of someone else.

Also, accessing "Community" the users of the platform can visualize the activities of the all users of the instrument and the votes of support. Pressing the button "How it works", the user receive information regarding all four stages which should be followed in order to submit an idea for developing a new product [12]. The next section, "My page", represents the profile of the user, who can access the realized activity, the submitted projects, the support offered to projects and to editing the profile. The last section of the menu, "Blog" comprise announcements regarding the results of the evaluation process, new ideas regarding the LEGO sets, the projects which are qualified for different evaluation rounds, etc. [13].

LEGO Ideas instrument provided by LEGO to interact with its customers addresses to a large segment of customers, LEGO adult fans, and children who are 13 years old or older, who can bring a new product idea for LEGO sets. Thus, the need defined by the company, which involves customers in developing new products using their creativity, is presented as a contest. This is indicated by the time period mentioned by the company (365 days) in which the innovator which submitted the project for the new LEGO set can gather votes, as support, passing into an evaluation process in those three periods from the year. Moreover, the creation process, innovation and competition are supported by providing rewards if the project submitted will become the new LEGO set launched on the market. Thus, innovators have the chance to receive 1% from the total net sales of the product, 5 complimentary copies of the LEGO Ideas set created by them, credit and bio in set materials as the LEGO Ideas set creators [14].

Regarding the intellectual property rights over the submitted ideas, the LEGO Group mentions the distinction made between "generic" ideas, which may be similar to other ideas existing in the world, but which may be different by the concept presentation, and "unique" ideas, original, such as a fictive vehicle, which benefits from the intellectual property right, as the company will not allow that the submitted idea to be used or modified by someone else [14].

LEGO Ideas platform encourages the social media presence, innovators being invited to share their ideas, projects, and to announce their friends that they participate to the contest and they need their support. The projects submitted for evaluation can be shared on Facebook, Twitter, Skype, blogs and other online communities to obtain support. Also, there other platforms which LEGO provide for sharing creations, ideas, contents, new models, such as LEGO.com Create & Share Galleries and ReBrick. The last one, ReBrick, is a social platform for bookmarking, where adult users can share, organize and discuss LEGO contents created by users of the LEGO digital platforms, teenagers and adults who use their creativity to build their own models, called MOCs (My Own Creation) [15].

3.2. LEGO.com Create & Share Galleries: the creativity trigger for LEGO customers

This online instrument aims the establishment of direct interaction with customers, by sharing ideas regarding new creations of LEGO fans. This platform contributes to a large extent to consolidation of the LEGO brand, as supports the interaction with customers of any age wanting to use their creativity to build new LEGO models. The visual identity of the brand is provided by the logo of LEGO brand illustrated to the page of interacting with customers and by an image which expresses a light bulb followed by green arrow directed toward the words "Create & Share". The LEGO.com Create & Share Galleries instrument has the advantage to be directly connected to the official website of the company, the connection being made by the main menu of the official page of the company. This direct connection from the official website menu highlights the importance that the company offers to activities of interaction with customers and to open innovation through the digital environment.

Thus, by accessing the official website of the LEGO Group (www.lego.com) and choosing the button Create & Share (http://www.lego.com/en-us/createandshare), the user opens the page which enables him to upload new models created in public galleries. The galleries provide different challenges for customers depending on occasion, as for example the celebration of Halloween opened the challenge "Spookify your LEGO sets" [16].

This LEGO instrument aims the open innovation and the externalization of the instrument, by involving all users, LEGO sets fans, who want to share their creative models with the others fans. By participating to LEGO challenges, the innovators use the LEGO bricks for building a model, take a photo of it and upload it on LEGO galleries, sharing their creations with the other users of the website. Also, if customers do not have enough LEGO bricks to build the models, the LEGO team provided the LEGO Digital Designer software which can be downloaded from the website and which enable them to create the models and to upload the creations from their personal computers on the LEGO website [17], supporting creativity and innovation of new LEGO models. After building the model using this application, the "Send" button enables the user to send and upload the new model on LEGO Digital Designer Gallery, to give a name to his creation, offering a description and selecting a category where to register the model [18].

By opening the LEGO Galleries, the LEGO.com Create & Share Galleries instrument offers information to LEGO customers regarding the current challenges, but also the challenges from the past, and the winners of the first 3 places of the LEGO contest. Each challenge which is launched has a certain theme, and in each gallery for each set, to which LEGO fans can upload creative models, there is a separate page which offers a menu with buttons for adding the creations ("Add creation"), for visualizing the personal creations ("My creations"), for adding the comments ("My comments"), or for searching the submitted models ("Search Creations"). The authors of the creative models have the opportunity to upload more than one image, video files, or files offered by the virtual application LEGO Digital Designer, with the model they build [19].

The segment of customers targeted by this instrument is very large, being addressed to both children and LEGO adult fans, who share their ideas for new LEGO products. The reward obtained for winning the contest is an intrinsic one, referring to the notoriety innovators gain by being known in the whole world for their creations awarded by the LEGO team. Regarding the intellectual property rights, the users who upload creations on LEGO website assign to LEGO Group the irreversible right to use, modify, to publish or to display the submitted material with different promotional or marketing aims [20].

3.3. Collecting, processing and analyzing customers' ideas

The process of collecting, processing and analysing customers' ideas involves the evaluation by the LEGO specialists of the submitted projects supported by more than 10 000 of supporters. If the project is selected for the production phase, the designers of the LEGO sets improve the original application in order to become a new LEGO product ready for launching on the market. The LEGO team decides which will be the main features of the product in the end, in 6 months being realized the final set for selling [21].

3.4. Assessing the results of involving customers in the innovation process

The involvement of customers in the co-creation process initiated by LEGO using the online platforms created positive results for the company, such as the profit for the year amounted to DKK 6.1 billion in 2013 and operating profit to DKK 8.3 billion [22]. Also, the company stated that new products generate 60% of the total sales for the company every year, LEGO mentioning that "an innovative and consumer-oriented development process is a fundamental parameter to the continued success" [22].

3.5. Future projects

The future projects of the company aims the globalization of company's presence on the market and continuous innovation of the LEGO play systems. Supporting the open innovation process, LEGO aims to launch new product lines, extending its presence to the global level and reaching to children from every country of the world [23]. Also, regarding the innovation efforts, the future projects of the company refer to creating the leverages for digitizing innovation on a wide range of subjects, such as integration of physical and digital play, marketing, e-commerce and business system [23].

4. Conclusions

LEGO Group developed a continuous dialogue with their customers, involving them in the open innovation process. The numerous open innovation platforms provided by the company in order to engage their customers in sharing their creative ideas about new LEGO sets enable customers to develop their ideas with LEGO specialists. The foreseen result of the co-creation process is the long-term success of the company as the customers' specific needs are satisfied by creating a permanent interaction with them in the online environment.

Acknowledgement

This research was undertaken within the framework of the National Research Program PN II, financed by MEN – UEFISCDI, project PN-II-PT-PCCA-2013-4-1811.

References

[1] Antorini Y M, Muniz Jr A M and Askildsen T 2012 Collaborating with customer communities: lessons from the LEGO Group *MIT Sloan Management Review* **53** pp 73-79
[2] Jo Hatch M and Schultz M 2010 Toward a theory of brand co-creation with implications for brand governance *Brand Management* **17** pp 590-604
[3] Robertson D C 2013 Brick by Brick. How LEGO Rewrote the Rules of Innovation and Conquered the Global Toy Industry *Crown Publishing Group* (New York)
[4] Information on http://www.legokidsinnercircle.com/legokids/eFocus/jsp/efocus
[5] Information on http://ldd.lego.com/en-us/subpages/designbyme/?domainredir=designbyme.lego.com
[6] Information on http://blog.lego.cuusoo.com/category/uncategorized/
[7] Information on https://ideas.lego.com/
[8] Information on https://ideas.lego.com/howitworks
[9] Information on https://ideas.lego.com/projects/85773
[10] Information on https://ideas.lego.com/projects/new
[11] Information on https://ideas.lego.com/projects/85915/edit/basics
[12] Information on https://ideas.lego.com/community
[13] Information on https://ideas.lego.com/blogs/blog
[14] Information on https://ideas.lego.com/guidelines
[15] Information on https://rebrick.lego.com/en-US/BottomMenu/About.aspx
[16] Information on http://www.us.lego.com/en-us/gallery?categoryid=B4359EC8-C31C-446C-AB19-CA755E5CDEB1
[17] Information on http://www.us.lego.com/en-us/gallery/upload?categoryId=64929898-C52F-49D8-B106-C0C36314093C
[18] Information on http://ldd.lego.com/en-us/application-pages/gallery-upload?id=dc37a520-6686-4059-8818-15f8ef20fc1c
[19] Information http://www.us.lego.com/en-us/disneyprincess/gallery/upload
[20] Information on http://www.us.lego.com/en-us/gallery/upload?categoryid=385a634c-63e8-42f1-a892-c092c4a03ebd
[21] Information on https://ideas.lego.com/guidelines#anchor-5
[22] Information on http://aboutus.lego.com/en-us/lego-group/annual-report
[23] Information on http://aboutus.lego.com/en-us/sustainability/our-approach

IOP Conf. Series: Materials Science and Engineering **95** (2015) 012145 doi:10.1088/1757-899X/95/1/012145

Innovation through developing consumers' community. Part I: Innovation in action

E Gălățanu (Avram)[1] and S Avasilcăi[1]
[1]"Gheorghe Asachi" Technical University of Iasi, Department of Engineering and Management, Bd. D. Mangeron 29, Iasi, Romania

E-mail: egalateanu@tex.tuiasi.ro

Abstract. Technological changes and need for innovation represents the main concerns for organizational growth and profitability. However the main priority is still about achieving high performance through product development and consumers' engagement activities. As implementation of open innovation applications increased and value co – creation became well known and major process, companies were engaged into value co – innovation activities. From this point of view the need for joint efforts with consumers in product development arose. Thus the primary condition for an organization to be consumer centric is to define clear the vision and mission which reflects the common efforts for co – creation and diffusion of innovation. As Research & Development processes evolved and interest for innovative concepts and products arose, companies started to implement the specific instruments for consumers' attraction and engagement into design and product development. The digitalized innovation became the main source for establishing the direct communication with the consumers. In order to achieve organization growth, profitability and recognition, the companies should be aware of the innovation importance and the need for internal change. From this point of view, there is necessary to assess the organizational structures, to implement new policies and to establish strategic targets. Basically it is justified the need for platform occurrence and development. Based on case study of BMW Group, recognised leader in automotive industry for innovative concepts, there will be analysed main features within organizational context which promotes the innovation implementation. There will be provided the review of the BMW Group experience of innovation activities, main consumers' engagement strategies, the values which promote the consumer – centric product development, new opportunities assessment, major policies and concerns. The foreseen result is to understand how companies are adapting to the technical and innovation changes as the main criteria for future product development and consumers' engagement motivation through the platform – based communication.

1. Introduction

Nowadays, the need for innovation became the main concern within economic and research context. The evolution from supply chain to value network stands out for new creative and innovative ways of doing business. From this point of view companies became aware that in order to achieve organizational growth and profitability, they should be open for collaboration with the final users. As

Content from this work may be used under the terms of the Creative Commons Attribution 3.0 licence. Any further distribution of this work must maintain attribution to the author(s) and the title of the work, journal citation and DOI.
Published under licence by IOP Publishing Ltd

result, social communities are seen as a part of product or service development process, valuable source of value for innovation processes [1]. Based on these aspects, the companies are adopting open innovation as mechanism for innovative instruments creation and implementation. Buur and Matthews identified the linkage between lead – users and innovation and defined the participatory design, as identification the "right" users and their involvement in product development processes [1]. According to their vision, the final users can be perceived as co – designers and product evaluators [1]. However, this requires considerable changes in organizational business models or structures through open innovation implementation and social communities' engagement.

2. Open innovation within organizational context

2.1. Open innovation and organizational change

LEGO Achieving profitability and growth through innovation implementation represents the key goals for organizational development. From this point of view open innovation is promoted as an important direction for "trial – and – learn" process development [2]. It has been defined as the mechanism for organizational accelerated growth [3], as necessary organizational change process [4], methods for knowledge sharing between the participants [5], techniques used by companies for opportunities' exploration [6], readiness for internal innovation [7], enterprise adaptation feature in terms of technological changes and globalization [8]. Gassman and Enkel presented open innovation as a process where internal organizational change represented the essential feature for value co - creation, as it is shown in figure 1 [9].

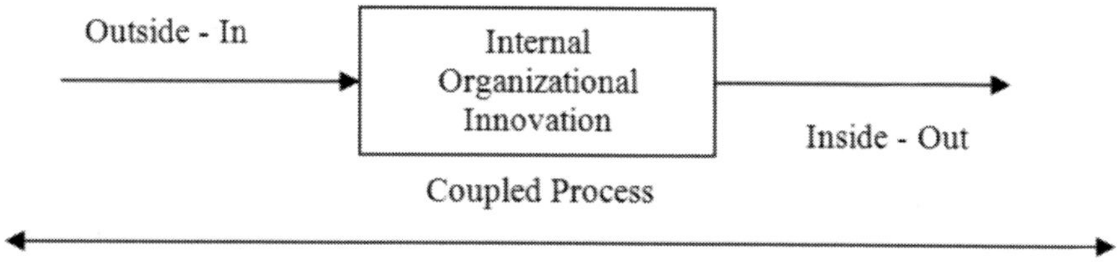

Figure 1. Open Innovation as a process [9].

However, the collaboration between the company and its end – users is highly explored by outside – in open innovation process [5, 9]. The inside – out component of open innovation emphasize innovation transfer outside the organizational boundaries, rather than its development inside the organization [7]. Coupled process of open innovation represents the joint effort for innovation creation, which is mostly defined as co – innovation [7]. Basically, the companies explore external knowledge, emphasize it into product development process and commercialize the innovation as joint – effort – based products or services [7].

2.2. Social communities as co – creators

Nowadays, collaborative networks became an important unit for innovation analysis, as they comprise not only the web of companies but also the social communities as co – creators. The final users' knowledge combined with organizational capabilities (resources and processes) represents a source of competitive advantage and an essential driver within coupled process of open innovation [7, 10]. Romero and Molina stated that in the era of "internet – of - things" it is important to explore virtual opportunities, so that the customers can access and create easier their own product, the companies' awareness about consumers' needs increases and co – innovation process is adopted in order to create significant value [10]. From this point of view, the consumers are engaged into innovation activities starting from idea generation until the final product development [10, 11]. However the motivation of social community as co – creators depends on the degree of consumers' voluntary engagement [12].

3. Case study: Innovation management at BMW Group

3.1. BMW Group: innovation history and development
Established in 1917, nowadays, BMW Group represents one of the largest vehicle manufacturer and global leader in automotive industry [13], general headquarter at Munich, Germany. As a public liability corporation, the company's main concern is to sustain the main strategy based on four essential pillars: growth, profitability, future development and the access to the technology and clients [14]. The image and trust were remarkably built due to the innovative solutions developed by highly specialized teams of designers and engineers. From this point of view BMW Group has known a successful development in terms of organizational growth and innovation implementation (table 1).

Table 1. Organizational growth and innovation implementation of BMW Group [15, 16].

Timeline	Company Development	Innovation Development
1917 - 1930	• BMW logo image • Establishment of Bayerische Motoren Werke • Launching the first motorcycle, BMW R32	• Airplane carburettor engine with aluminium primer and 185 [hp]
1930 - 1960	• Launching first airplane division and new plants at Allach, Spandau, Basdorf and Zuhldorf • Became the main manufacturer for German Air Force during the II World War • Loose power for own assets • First attempt for company's restructuring	• Launching the BMW VIIa airplane engine with 600 hp and 12 cylinders • First car designed and manufactured under BMW trademark • Usage of kidney type grid • New discoveries in aerodynamics and cylinders heads made off light alloy
1960 - 2000	• Gains strong position as successful automotive manufacturer • Acquisition of Hans Glas GmbH and creation of the biggest manufacturing facility in Dingolfing • Established the Foundation Herbert Quandt and offers sponsorships for knowledge transfer • The opening of BMW headquarter entitled "Four Cylinders and launching the first "Idea Plant" and Center of Research and Innovation in Munich	• BMW series 02 launching – original design, compact dimension, excellent quality • Boxer engine and crankshaft realized from light alloy • First digital electronic systems (Antilock Braking System) and board computer • Lightweight car construction, electric door system withdrawal, tubular frame and plastic drum
After 2000	• Concentrates on premium markets with brands: MINI, Rolls – Royce and BMW • Extended the Center for Research and Innovation by adding Design Division	• Internal combustion engine with hydrogen or traditional fuel

3.2. Innovation within BMW Group
Established The BMW Group highly promotes the innovation implementation as main driver for sustainable growth through clear mission and vision statements. From this point of view, the organization adopted a horizon 2020 mission: to become one of the leading automotive manufacturer and supplier for individual mobility [17]. "Connected Drive" became a synonym for intelligent connectivity, as BMW seeks to create a new form of interaction between the driver and the car through new technology implementation and intelligent solutions. The key aspect is related to sustainable resource consumption while offering a wide range of experience for drivers, which also is an

important strategic goal [18]. The company promotes the innovation through three main values: *design* – each product should contain group's signature, form and distinctive curves, a design for future [19], *engineering* – comprises a "what…if..." approach. A perspective to develop a wide range of industrial solutions for mobility, impressive technologies and driving experience [20], *innovation* – new technology implementation in product development process, as well as new technology creation in order to ensure the driver's experience [19].

However, these aspects are reflected mostly in product development process. BMW Group adopted a three – staged manufacturing process, which implies employees' engagement at each stage. Basically the Research & Development activities are based on the mix of highly specialized employees, innovative processes and future projects [21].

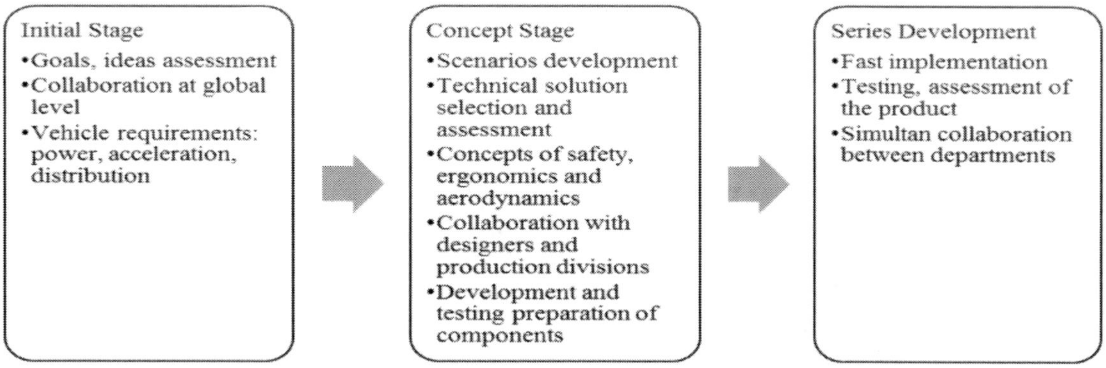

Figure 2. BMW Group product development process [22, 23, 24].

Each stage of product development is strictly controlled and based on collaboration between departments at national or global level, starting from the stage of concept or idea generation.

3.3. Innovation, market opportunities exploration and concerns

As BMW Group stated, in order to become leading automotive manufacturer, it is necessary to gain competitive advantage, to implement open innovation in product development process, but also to explore new market opportunities by using innovative ways. From this point of view, the company perceived the creation of specific instruments, such as BMW configurator. The potential customers can create their own vehicle according to their needs: model, colour, properties (acceleration, CO2 emissions, and consumption) [25]. Although virtual instruments gained a considerable awareness, BMW Group considers that the direct interaction with the specialists in domain and their plant is essential. From this point of view, the company organises "Junior Programme". It emphasize the learning of the most important concepts: sustainability, team work, environmental protection, vehicle design and technology, but also represents new way of exploring young potential. Workshop "Mobility Mission" represents a new way to explore young potential by given specific tasks and extracting ideas from the created sketches in terms of future development and prototyping future cars [26]. Another approach for innovation promotion is concretised through BMW Exhibition Houses: House of Design, Company House, Motorcycle House, Technology Exposition, Motorsport House, House of Brand and Series House. Each exposition has as objective to present and illustrate the successful implementation of innovation and to explore the consumers' enthusiasm for BMW trademark [27]. As main innovation concerns, The BMW Group concentrates on elaborating energy efficiency management, hydraulic power steering, ECO PRO Modules for efficient fuel consumption [28], incorporation of virtual features (Google Maps, BMW routes, ECO PRO analyser and other apps), renewable energy for individual mobility, reducing CO2 emissions, technology iDrive (intelligent mobility for electrical vehicles), usage of hydrogen as alternative fuel and ensuring risk assessment in order to achieve sustainable and innovation targets [29].

4. Conclusions

BMW Group represents a global leader in automotive industry. Their concerns for innovation implementation can be traced in the history of product development. However in order to achieve a greater balance and to gain leading position, the group had to adapt to the technological changes. The "Connected Drive" philosophy basically represents the mission and vision of the BMW Group, as they connected new virtual possibilities with their products and ensured the linkage between the driver, the car and external environment, aspects which were possible due to open innovation implementation.

Acknowledgement

This research was undertaken within the framework of the National Research Program PN II, financed by MEN – UEFISCDI, project PN-II-PT-PCCA-2013-4-1811.

References

[1] Buur J and Matthews B 2008 Participatory Innovation *International Journal of Innovation Management* **12** pp 255 – 273

[2] Gassmann O, Enkel E and Chesbrough H 2010 The Future of Open Innovation *R & D Management* (Blackwell Publishing) **40** pp 213 – 221

[3] Chesbrough H W, West J and Vanhaverbeke W 2006 Open Innovation: Researching a New Paradigm *Oxford University Press* (Oxford)

[4] Chiaroni D, Chiesa V and Frattini F 2011 The Open Innovation Journey: How Firms Dynamically Implement the Emerging Innovation Management Pradigm *Technovation* **31** pp 34 – 43

[5] Chesbrough W and Henry W 2011 Bringing Open Innovation to Services *MIT Sloan Management Review* **52** pp 85 – 90

[6] Ecevit Sati Z and Dursun F 2014 Using the E – Learning Methods for Managing Open Innovation Activities in Business Ecosystem *Journal of Social Sciences* **42** pp 349 – 358

[7] Enkel E, Gassmann O and Chesbrough H 2009 Open R & D and Open Innovation, Exploring the Phenomenon *R & D Management* **39** pp 311 – 316

[8] Zhang X, Ding L and Chen X 2014 Interaction of Open Innovation and Business Ecosystem *International Journal of U – and E – Service, Science and Technology* **7** pp 51 – 64

[9] Gassman O and Enkel E 2004 Towards a Theory of Open Innovation: Three Core Process Archetypes *R & D Management Conference* (Lisbon)

[10] Romero D and Molina A 2011 Value Co – Creation and Co – Innovation in the Networked Era *Journal of Production Planning and Control* **22** pp 447 – 472

[11] Reichwald R, Seifert S, Walcher D and Piller F 2004 IEEE Explore, Customers as Part of Value Webs: Towards a Framework for Webbed Customer Innovation Tools *Proceedings of 37th Annual Hawaii International Conference on System Sciences*

[12] Piller F Open Innovation with Customers: Crowdsourcing and Co – Creation at Threadless, *Social Science Research Network Electronic Journal*

[13] BMW Company Profile on http://www.bmwgroup.com/bmwgroup_prod/e/0_0_www_bmwgroup_com/unternehmen/unternehmensprofil/standorte/standorte/index.html

[14] BMW Company Profile. Strategy on http://www.bmwgroup.com/bmwgroup_prod/e/0_0_www_bmwgroup_com/unternehmen/unternehmensprofil/strategie/strategie.html

[15] BMW History Milestones on

http://www.bmwgroup.com/bmwgroup_prod/e/0_0_www_bmwgroup_com/unternehmen/historie/meilensteine/meilensteine.html
[16] BMW History Plants on: http://www.bmwgroup.com/bmwgroup_prod/e/0_0_www_bmwgroup_com/unternehmen/historie/werke/werke.shtml
[17] Mission Statement on http://www.bmwgroup.com/e/0_0_www_bmwgroup_com/investor_relations/fakten_zum_unternehmen/strategie.html
[18] BMW vision Connected Drive on http://www.bmw.com/com/en/insights/bmw_design_2012/visions/opener.html#row02-
[19] BMW Group values on http://www.bmwusa.com/InsideBMW.aspx
[20] BMW Group Engineering on http://www.bmwusa.com/Standard/Content/Innovations/Engineering/Default.aspx?from=/Standard/Content/Innovations/Engineering.aspx&return=/Standard/Content/Innovations/Engineering.aspx
[21] Research & Development pillars on http://www.bmwgroup.com/bmwgroup_prod/e/0_0_www_bmwgroup_com/forschung_entwicklung/ein_blicke_in_die_entwicklung/ueberblick.shtml
[22] Initial Stage of product development on http://www.bmwgroup.com/bmwgroup_prod/e/0_0_www_bmwgroup_com/forschung_entwicklung/ein_blicke_in_die_entwicklung/portal_zukunft/BMWGroupPortalZukunft143014.html
[23] Concept Stage of product development on http://www.bmwgroup.com/bmwgroup_prod/e/0_0_www_bmwgroup_com/forschung_entwicklung/ein_blicke_in_die_entwicklung/portal_zukunft/BMWGroupPortalZukunft143014.html
[24] Series Development Stage on http://www.bmwgroup.com/bmwgroup_prod/e/0_0_www_bmwgroup_com/forschung_entwicklung/ein_blicke_in_die_entwicklung/portal_zukunft/BMWGroupPortalZukunft143014.html
[25] BMW Configurator on http://www.bmw.co.uk/vc/ncc/xhtml/start/startWithModelSelection.faces?productType=1&brand=BM&country=GB&market=GB&locale=en_GB
[26] Workshop "Mobility Mission" on http://www.bmwwelt.com/en/junior_programm/junior_museum.html
[27] Seven Exhibition Houses on http://www.bmwwelt.com/en/exhibitions/museum/current/the_seven_houses.html
[28] Information on http://www.bmw.com/com/en/insights/technology/efficientdynamics/phase_2/technology/energy_management.html
[29] Information on http://www.bmw.com/com/en/insights/technology/cleanenergy/phase_2/cleanenergy.html

IOP Conf. Series: Materials Science and Engineering **95** (2015) 012146 doi:10.1088/1757-899X/95/1/012146

Creativity and innovation by empowering the customer: The case of Mulino Bianco. Part I: Organisational innovation context

A Bujor [1] **and S Avasilcăi**[1]
[1]"Gheorghe Asachi" Technical University of Iasi, Department of Engineering and Management, Bd. D. Mangeron 29, Iasi, Romania

E-mail: egalateanu@tex.tuiasi.ro

Abstract. The terms of creativity, co-creation, creative industries, innovation, and co-innovation are more and more used nowadays. While co-creation offers the possibility and encourages a more active involvement from the customers to create value rich experiences, innovation is responsible for the little improvements made for a better life, to grow a business, to improve products, services or company's productivity. Either customers, current and potential, or stakeholders' involvement into innovation activities, through their creativity, represent an important way of value creation, of actions' performance that increases the worth of goods, services, or business as a whole. More and more, different size businesses gather ideas for innovation from customers / stakeholders through their involvement into the early stages of the innovation process. Actually, it has been shown that their ideas sketch their needs and wishes, and have been described as "need information". Customers and stakeholders, in general, also offer ideas that have been called "solution information", which represents, not only need information, but also customer-based proposals that describe how ideas can be transformed into marketable products. The term of creative industries refers to those goods that can technically be reproduced, industrially produced, and commercially sold, this being one of the many definitions found in the literature. The classification of creative industries differs at the European Union level according to each country's development level and opening towards these industries' perception as an engine for economic growth. To better explore and identify the importance of innovation and creativity for new products' development or improvement a case study about Mulino Bianco was done. This methodology approach represents a part of a qualitative research approach within a broader research undertaken within the framework of the National Research Program PN II. Mulino Bianco was first launched in Italy, being one of the value brands of Barilla Group, which we can say, according to the Italian researches regarding their definition and classification, it belongs to creative industries: industry of food and taste. The foreseen result of this paper is highlighting the fact that the key strategy for the Group and for Mulino Bianco, by default, is mainly its customers' pro-active involvement in products' development or creation.

1. Introduction

Innovation, along with co-innovation, creativity and co-creation are processes that are found again in the top of the agenda of a corporate agenda. These processes have never been a fad, but have always been fashionable, out of fashion, being rediscovered as a growth factor in every generation of leadership. Too often, however, the relevant statements about innovation are followed of mediocre

Content from this work may be used under the terms of the Creative Commons Attribution 3.0 licence. Any further distribution of this work must maintain attribution to the author(s) and the title of the work, journal citation and DOI.
Published under licence by IOP Publishing Ltd

execution with anaemic results, and the groups of novation are quickly abolished in units of cost reduction [1].

The International Community has recently experienced a large variety of studies and measurements, depending on the particular definitions, regarding the concept of *cultural, creative industries and creative economy*. The use of the term *creative industries*, industry of which was identified as being part also the Barilla Group in Italy, varies from one country to another, being of relatively recent origin, developing in Australia, since 1994 with the launch of Creative Nation report. The term has seen a greater exposure in 1997, when the political decision makers of Department for Culture, Media and Sport (DCMS) in Great Britain established *Creative Industries Task Force* [2,3] It is noteworthy that the name of *creative industries* that have been developed since then has extended the scope of cultural industries beyond arts and marked a change of approach of potential commercial activities which until recently were considered pure or predominantly non-economic, [4].

Thus, in 1998, DCMS defined creative industries as "those activities which have their origin in individual creativity, artistry, talent and which have the potential to generate prosperity and jobs creation through the generation and exploitation of intellectual property", [5].

With all that in Italy there is no official definition of cultural industries, which is here meant by this is that represents cultural goods and services which can be "technically reproduced" or "industrially produced and commercially sold" according Edgar Morin. Thus in Italy we talk about [6]: books; press; broadcasting; cinema; music recordings; new mass-media. On the other hand it should be emphasized that the concept of creative industries in Italy was spread to other neighbouring industries, highly creative, among which an important place is occupied by the gastronomy, beside fashion, advertising etc.[6],becoming more popular lately.

2. Mulino Bianco – a valuable brand of Barilla Group

Mulino Bianco was launched in Italy, being one of the Barilla Group's brands of great value. Since its appearance, it quickly became the preferred brand of Italian cakes, currently being found in over 78% of Italian homes. It is already known the fact that Italy consumes more cakes than the United States, and that Italians serve cakes for breakfast, [7].

Barilla Group, which was founded in Parma, Italy, in 1877 initially as a store of bread and pasta, is ranked as one of the most important top groups in the food industry in Italy. Barilla Group leads the global business with pasta and pasta sauces in Europe, in the bakery business in Italy, and in business with crisp bread in Scandinavia. The Group has 41 production units, of which 13 are in Italy and 28 outside Italy, exporting to more than 100 countries. Every year, around 2.5 million tons of food produced by Barilla, Mulino Bianco, Voiello, Pavesi, Academia Barilla, Wasa, Harrys (France and Russia), Golden Toast (Germany), Misko (Greece) Filiz (Turkey), Yemin and Vesta (Mexico) brands are included by people from all over the world as part of their evening meals [8].

3. Barilla Group's process of products' creation

The company's attention given to quality and nutrition is complete and it is always united by the desire to surprise the ones who choose them and welcome them to their tables. The process, in which a new product that was born from an idea comes to life, is fascinating. This includes a series of interactions that can be unpredictable, between the development of prototypes on one hand and the permanent contact with the final consumers of the products in order to obtain the best quality of the final product.

Barilla Group launches approximately 50 new products, annually. This represents the ultimate stage of a long process that creates products in order to satisfy the necessities of those who buy them and bring them into their homes. The process behind the creation of a product is divided into several stages, also divided into four moments. At the end of each stage, the results are evaluated and it is decided whether or not the development should continue [9].

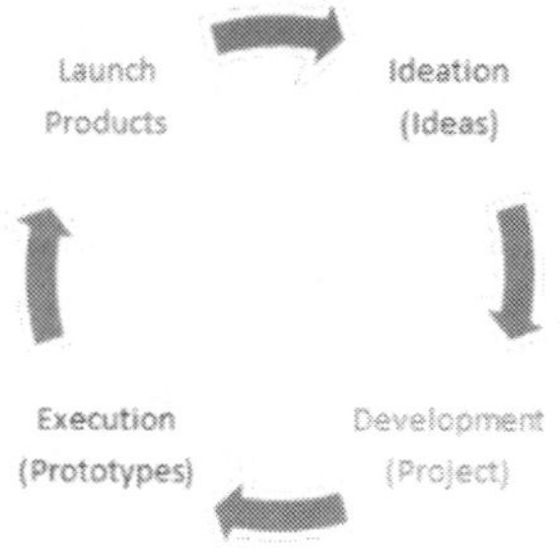

Figure 1. Barilla's moments of products 'creation [9].

The process is based on the interaction of different professionals, from the inside of the company, such as from the outside. These people are brought together by the close relationship of those who make the project of the products (design) and those for who the project is made (stakeholder). Each project/design team collaborates with a selected group of people that participates actively on creating new products. They are not simple tasters, but they are pro-active persons that are involved even in the first stages of the developing process [9].

4. Mulino Bianco's products

From the products production point of view, the evolution of the Mulino Bianco's products is divided into 4 big periods, between the years 1971-1975, 1976-1989, 1990-1999 and 2000-2011. In each of these periods, starting with the desire of being the number one choice of consumers in pastry production, the brand was permanently interested in developing innovative products and technological perfecting in order to always offer quality products and in conformity with their own environment politics. Thus, after years of experimenting, in October 1975 appeared in stores the first Mulino Bianco's biscuits packs, that were a great success [10].

The taste of the first Mulino Bianco biscuits is due to the experience of the talented English biscuit maker, George Maxwell, a real pastry maestro, who took care in secrecy of the experimental line, for several months, in an isle of the factory. He elaborated rigorously many natural recipes, without any additives or preservatives, inspired by the tradition from England [11].

In an era where industrial biscuits were button shaped, the idea of introducing symbolic iconography into designing a product meant an actual revolution and the innovation of using particular and irregular shapes with personalized drawings proved to be a winner, [11]. There were created hundreds of shapes, and then finally there were registered a number of 73 different shapes.

The years 1978 and 1979 have been very successful in which concerned the clients/the consumers, there was positive feed-back: Mulino Blanco was liked and the enthusiasm has doubled when the American firm Grace had withdrawn and the Italian firm returned to the family, into the hands of Pietro Barilla. Shortly, Mulino Bianco becomes the leader of the profile market, in the Italian pastry products domain and the incomes in the pastry domain have risen as the years passed by [12].

As the time passed, consumers appreciated the firm image, the quality of the products and the equilibrated price; the best results were highlighted above all, in the most innovative lines, opening new consuming areas, confirming the approach of Mulino Bianco to 40% incomes from the recorded total income in the pasta domain. [12] In the same year, 1983, the market quota in the pastry domain reached 17,2 % , the difference between it and the next competitor being of 8.4%.

The year 1986, when the incomes of Mulino Blanco have succeeded to equalize the incomes of Barilla pasta, the firm conquered the first position on the Italian market, reaching an elevation of 24%, it continued to create new products and lines of biscuits and sweets under the Mulino Bianco brand.

According to the data from the Mulino Bianco's site, the products fabricated under this brand are divided into 5 big categories, each of them being divided into assortments and types of products.

Figure 2. Mulino Bianco's products classification [13].

5. Research and Development at Mulino Bianco

For Mulino Bianco, as part of Barilla's Group, the products made are the main communication tools used to reach people from all over the Earth. Therefore, the main responsibility in Research and Development is to delight the ones for who are made Barilla's products. Only by establishing a direct and personal relationship with them, knowing their various life styles, also fully understanding their needs, expressed or not, can take to realize/offer new solutions of product that people can adopt in their routine [14]. The main tools to reach this goal are represented by the functional competencies, but by a knowledge of the alimentary and nutritional technologies too.

As part of Barilla's team, anyone is involved in the full cycle of life of making the product, of the first steps in production. Starting with generating the idea and personal creativity, throughout the ingredients, the nutritional science, technology, process definition and packaging, it works with experts of many different disciplines: from the process of baking, extruding the paste, drying, cooking, of the alimentary technologies of stabilization, of processing drinks, to chemistry-physics , microbiology, rheology, research of wheat and alimentary plants, science of ingredients, sensory analysis, evaluation and recommendation of the nutrients, to the techniques of market research, opinion surveys, statistics and projects, knowledge and people's management. All of these are necessary to take the product from the laboratory to the market, and this experience is "spiced" with a powerful incentive for innovation, a generous dose of passion and willing of working together to reach success, [14].

The involvement in the development of a new product can be realized even for a specific objective of consumption, or in thinking about new healthy food for our future, living daily in a world full of new ingredients, technologies, processes and alimentary experiences. An entire network of international experts and leading academic groups ensure the connection of the company with the most recent scientific discoveries and news related to nutrition. For Barilla, this means to be able to offer more and best opportunities than any other company that works with a great feeling of ethics and to people and the name Barilla which represents, [14]. The department of Research and Development is constantly using the knowledge and researches of the 250's researchers that hired and work in 6 of the offices from Italy, USA, Russia, France, Germany and Sweden.

Research and Development is a crucial stage in the process of transformation of an idea into a product, needing, most of the time, more time than the production phase. This can happen because it might take lot of time and effort to work with an idea and to find contributors. Therefore, the process of research and development is fundamental for what happens later, in the production phase. For example, the story that put the basis of the appearance of *Palicao, Fregoli* biscuits, it's a beautiful and interesting one, based on the experimentation and innovation. From a while, Mulino Bianco had in mind the idea to extend its presence for breakfast too, in the yet unexplored areas.

In the sector of the products that are consumed with milk, for years not improved, there were some interesting perspectives [15]. After many important tests on the products from the market, as the techniques from Mulino said, there was born the idea which took to the final product: "why don't we try a biscuit that soaks in milk? So there started the experiments to produce baby's biscuits. Every day there appeared new recipes and new tests were made: it was made the dough, it was baked and it was drinking. The process wasn't easy at all, and the questions and doubts have been many: Why wasn't it baking well? Why all the ingredients were soluble and yet they did not soften the biscuit? A thing was sure, though: the biscuits for babies wanted to be a child's food that included the bottle too. Analysing and eliminating the effects of those products, the first prototypes are adjusted, but they're only at the beginning. Only in this moment surprises started to appear.

Until that moment they tried to dissolute the product in milk at a low temperature. Then, one morning, inadvertently, it is thrown in hot milk. So what happens? Palicao starts to take density, by becoming chocolate in the cup. After one hour, the first surprise appears: in the cup there's a sort of pudding. Astonishment and curiosity are great. Enthusiasm grows and the question is: Why wouldn't they try to put it in the fridge mixer. And what if, after that, they put some sour cream too? The product, as a miracle, changes continuously under the experiments' eyes: from frappe to ice cream, to mousse, to cake. And so on, form the surprise till the *Sacher e alla bavarese*. The biscuit that melts into the milk was, actually, the ideal basis for all the receipts that imagination can offer: this is how Palicao gets birth, the cocoa miracle [15].

6. Mulino Bianco's Projects

Mulino Bianco's future projects are primarily related to the care for the environment. Since 2009 the brand has chosen to satisfy 100% of its energy plants using renewable energy sources, in order to always keep the care and attention devoted to nature. Because of this decision was possible to estimate the reduction of CO2 equivalent emissions by approximately 55.000 tons per year. Mulino Bianco's commitment to the environment has become concrete in several other top initiatives: -47% water used in bakeries (2004-2011 comparison), >90% recycled packages, and 100% eggs from free-range chicken [16].

One of the projects initiated in 2009 and which has achieved a sustainability management policy is "Mulino Bianco Tour". The project is an exciting itinerary and rich in content, addressing to families, schools and revealing, through fun and entertainment, *Mulino Bianco's Good World* and its many quality products, [16].

In 2014, to the full program of activities consolidated over the years, the laboratory "Recycling Adventure" was added, representing a workshop to raise awareness of the importance of respecting nature and simple gestures to protect it, the care for the "from farm to table" product [17].

Mulino Bianco future projects have as strategy the achieving of a good market segmentation and strategic positioning on the handicrafts market, healthy, wholesome and uncontaminated. Therefore the innovation strategy has always in view generating new products and packaging. Brand's website contains sections of traditional communication which describe history, products, promotions and events which are regularly organized, all these actually demonstrating the Mulino Bianco intend towards product sustainability. Such a blog is „Il Blog del Pane", which aims entertainment, big and small, about bread and discovers how such an ailment, rooted in Italian history and tradition, indispensable of Italian cuisine, can teach so many and is even considered a faithful „companion". [18]. There is no project announced related to bread, but still remains current on the platform the project „Ricordi di Pane", achieved collaborating with NABA (Nuova Accademia di Belle Arti, Milano).

7. Conclusions

The dream of Mulino Bianco brand is to provide a "Good World" (Mondo Buono) with deep roots, combining passion and experience, which is found in ingredients that are grown and selected with respect for nature and people, can be discovered at the table, in emotions of affection and the ones of being together and can be recognized in the pride of those who have sent their own know-how with the enthusiasm that only a good story can convey.

According Mulino Bianco the "Mondo Buono" products are born from a main principle: "To give people what we will give to our children", because to think of our children is as we think about the future, what we can do to build a world much better in all respects.

In order to achieve all these, the Mulino Bianco team give voice to all those who choose them, listening carefully to what they want, their suggestions and criticisms through an open. The road which the brand has chosen to have has no end, because never stops in building the future, and every achievement is for them a step closer to a better world.

Acknowledgement
This research was undertaken within the framework of the National Research Program PN II, financed by MEN – UEFISCDI, project PN-II-PT-PCCA-2013-4-1811.

References

[1] Kanter R B 2006 Innovation *The Classic Traps, Harvard Business Review*
[2] UNCTAD UNDP 2010 *Creative Economy Report: a Feasible Development Option*
[3] UNCTAD UNDP 2008 *Creative Economy Report*
[4] UNCTAD UNDP 2004 *Creative Industries and Development* (Eleventh session: São Paulo)
[5] Flew T 2012 The Creative Industries. Culture and Policy *SAGE Publications Ltd* (London)
[6] COMPENDIUM, Information on http://www.culturalpolicies.net/web/italy.php?aid=423
[7] Mulino Bianco, http://www.mulinobianco.com, accessed at 15/12/2014
[8] PR Newswire, http://www.prnewswire.com/news-releases/mulino-bianco-by-barilla-launches-in-the-united-states-bringing-italian-specialty-cookies-to-american-consumers-174923731.html accessed at 10/12/2014
[9] Barilla Group, http://www.barillagroup.com/corporate/en/home/cosafacciamo/come-nascono-i-prodotti-Barilla.html accessed at 10/12/2014
[10] Mulino Bianco, http://www.mulinobianco.it/storia-e-pubblicita/la-storia/1971-75/tutti-al-mulino accessed at 15/12/2014
[11] Mulino Bianco, http://www.mulinobianco.it/storia-e-pubblicita/i-prodotti/1971-75/nati-per-distinguersi accessed at 15/12/2014
[12] Mulino Bianco, http://www.mulinobianco.it/storia-e-pubblicita/la-storia/1976-89/mulino-bianco-cresce accessed at 15/12/2014
[13] Mulino Bianco, http://www.mulinobianco.it/home accessed at 10/12/2014
[14] Barilla Group, http://www.barillagroup.com/corporate/en/home/lavora-con-noi/organization-and-job-families/research-development-quality.html accessed at 30/11/2014
[15] Mulino Bianco, Information on http://www.mulinobianco.it/storia-e-pubblicita/i-prodotti/1990-99/essere-palicao-e-gli-altri accessed at 15/12/2014
[16] Barilla Group, www.barillagroup.com/saveas/corporate/en/home/media/press-kit/mulino-bianco/sidebar/0/items_files/file/Barilla_StoriaMulino_Eng.pdf11 accessed at 16/12/2014
[17] Mulino Bianco, http://www.mulinobianco.it/documents/10760/101307/Politica+di+Gestione+Sostenibile.pdf accessed at 16/12/2014.
[18] Mulino Bianco, http://www.mulinobianco.it/il-blog-del-pane accessed at 16/12/2014

Creativity and innovation by empowering the customer: The case of Mulino Bianco. Part II: The digital customer value added

A Bujor [1] **and S Avasilcăi** [1]
[1]"Gheorghe Asachi" Technical University of Iasi, Department of Engineering and Management, Bd. D. Mangeron 29, Iasi, Romania

E-mail: silvia.avasilcai@gmail.com

Abstract. The terms of creativity, co-creation, creative industries, innovation, and co-innovation are more and more used nowadays. While co-creation offers the possibility and encourages a more active involvement from the customers to create value rich experiences, innovation is responsible for the little improvements made for a better life, to grow a business, to improve products, services or company's productivity. Either customers, current and potential, or stakeholders' involvement into innovation activities, through their creativity, represent an important way of value creation, of actions' performance that increases the worth of goods, services, or business as a whole. More and more, different size businesses gather ideas for innovation from customers / stakeholders by involving them into the early stages of the innovation process. Actually, it has been shown that their ideas sketch their needs and wishes, and have been described as "need information". Customers and stakeholders, in general, also offer ideas that have been called "solution information", which represents, not only need information, but also customer-based proposals that describe how ideas can be transformed into marketable products. The term of creative industries refers to those goods that can technically be reproduced, industrially produced, and commercially sold, this being one of the many definitions found in the literature. Mulino Bianco was first launched in Italy, being one of the value brands of Barilla Group, which we can say, according to the Italian definition and classification, it belongs to creative industries: industry of food and taste. Even though Barilla Group's Research & Development department does its job very efficiently, developing and creating new products under different brands, lately a key strategy for the Group and for Mulino Bianco, by default, is customers' pro-active involvement in products' development or creation. One of the tools used for this is the Internet Toolkit and Web 2.0. by means of which customers are being invited to participate in competitions, being asked to design and share their concepts / views for new products, or to get involved in by voting those ideas they best like and would love to find them on markets. The aim of this paper is to explore and identify the involvement of stakeholders in Mullino Bianco's product development or improvement through creativity and innovation. As methodology approach, a case study about Mulino Bianco was done, and the foreseen result is highlighting the *Nel Mulino Che Vorrei* platform's features for consumer's engagement in the value creation and co-creation.

1. Introduction

Innovation gets in or out of fashion as a strategic factor for corporate growth, but with each wave of enthusiasm, managers make the same mistakes. Most often, they stumble in their research and

Content from this work may be used under the terms of the Creative Commons Attribution 3.0 licence. Any further distribution of this work must maintain attribution to the author(s) and the title of the work, journal citation and DOI.
Published under licence by IOP Publishing Ltd

development because they are engaged in a difficult balancing act: they need to protect existing revenue streams while coaxing along new ones. But "corporate entrepreneurship" needs not to be an oxymoron, and innovation can flourish if business executives heed lessons of the past.[1]
In order to innovate and to develop products taking into account the level of fat, sugar and salt, which have been given for every category, to reformulate already existent products, in order to reduce some nutrients, in 2013 Barilla invested 40 million Euros only in research and development activities that target both products and processes.
Mulino Bianco has always been concerned with its costumers' needs. Moreover, as time passes, the brand has been trying to decipher these needs and their desires. Today, through the new communication channels that facilitate relations, Mulino Bianco is convinced that the time for growth together with their customers is the best.

2. Mulino Bianco's Projects

Mulino Bianco was The initiated and implemented projects of Mulino Bianco generally target both participants' involvement, as they contribute with innovative ideas, and Mulino Bianco experts' involvement, who evaluate the ideas that participants have introduced and various interested categories, such as: the existing costumers, and the potential customers who contribute by voting the proposed ideas.

The projects / events that take place under Mulino Bianco's brand name, and which are dedicated to proactive involvement of clients / stakeholders in innovative processes are: *Nel Mulino Che Vorrei; I Talenti del Mulino, Buona Merenda; Una Sana Abitudine, Momenti di Pane Sapori Autentici; Il Tour del Mulino.*

Among the listed projects, Mulino Bianco has launched *Nel Mulino che vorrei (The Mulino that you wished for)* in order to collect ideas from those who want to get involved. All ideas are analyzed to confirm they are consistent with the mission, vision and values of Mulino Bianco, to finally be realized together [2] Without any prejudices, through this project the Mulino Bianco staff has entered the game, ready to listen to any proposal that is consistent with their daily work. They aim considering all ideas while they invite everyone to vote for the best of them, and in return, they promise transparency and involvement. All ideas receive a public feedback that include feasibility data and reasons why they have been chosen [2].

The *Nel Munino Che Vorrei* project functions according to several well-established rules, which provide organization and simplicity. There will be no prizes that involve money or objects and the participation is spontaneous and is based on relation parity [3]:

- those who participate as interested individuals have a wish, a need, an idea that they would like Mulino Bianco to bring into reality;
- Mulino Bianco wants to be closer to people's needs and to improve itself in a manner suggested by costumers.

Basically, the *Nel Mulino Che Vorrei* project came to life on the strength of a wish to collect and satisfy needs, wishes, suggestions and demands of improvement that clients and costumers expressed and thought that will help all of us [4].

The Mulino Bianco crew does not lack ideas, but the design process, as well as the building process needs a step-by-step approach and constant refinement activities in order to bring the best solutions. They are professionals, but their purpose is always the same: to satisfy the needs, demands, and wishes of their clients and customers in general. This is the reason why they consider that it is mainly important to know stakeholders' opinion before suggesting innovations, which might not meet clients' needs.

For this reason Mulino Bianco, through the *Nel Mulino Che Vorrei* project, is aiming to become a listener. The participation of all interested parts is very important: just by expressing opinions, by participating, by sending their own ideas, by commenting, and by voting other ideas which have already been published, we can help Mulino Bianco accomplish what we think is truly important and useful [4].

Those who are interested and wish to participate at this project are able to do it in three ways [5]:

- By sending the idea to the "Ideas for voting" areas. The more clear, descriptive and specific an idea is, the bigger the chances that it would be understood by more people in the community is, therefore it could be considered more valuable. The most voted ten ideas will be evaluated by Mulino Bianco, which can however consider other ideas if they prove valuable too after the evaluation.
- By voting other people's ideas, as each vote contributes to the process of determining the best ones. The most voted ten ideas will enter the evaluation process regardless of their nature, along with the ones chosen by Mulino Bianco.
- By observing upon other people's ideas, therefore by contributing with pieces of advice which may improve an idea or simply exaplins the vote.

All contributions are supervised by *Nel Mulino Che Vorrei* editors. They will publish posted ideas regardless of their connection to Mulino Bianco's activities in order to suppress any kind of discrimination. However, the contributions will be rejected if they do not meet the netiquette regulations. It is also very important that participants read the project's legal notes.

Lastly, all ideas that will be put into practice will be part of Mulino Bianco's world and will be published on the website. The author of the idea will be invited to the project release, and will write a note concerning his experience as a participant [6].

3. Mulino Bianco stakeholders' involvement in co-creation in innovation

The tool dedicated to interactions with the stakeholders is the Nel Mulino Che Vorrei platform through which interested parts can discover Mulino Bianco's projects, suggest their own ideas, vote other participants' ideas, visualize released projects, and enter the platform's blog or answer questions asked by Mulino Bianco in opinion polls. On the Nel Mulino Che Vorrei platform, the connection with the participants is carried out by individual accounts created by each participant.

Between the official Mulino Bianco website and the Nel Mulino Che Vorrei platform is a direct link. The platform's site is: http://www.nelmulinochevorrei.it, and can be directly accessed from the official Mulino Bianco website. Thanks to this situation, we can say that the visual identity of the project is, in fact, the visual identity of Mulino Bianco's platform. It is in fact about the Mulino Bianco brand:

Figure 1. Mulino Bianco's brand [7].

The process is based on the interaction of different professionals, from the inside of the company, such The preservation of the logo used by the pasta specialized brand, Barilla, as well as the blue font of Mulino Bianco, which is the specific color of Barilla packages is obvious. We can say that tradition is preserved using elements and colors which Barilla accustomed us to since it's very beginning. Yellow is the background color for the entire platform, therefore it is also used in Mulino Bianco's logo. This color combines well with the picture found in the logo, which suggestively represents a mill framed by wheat ears.

In this case we can bring into discussion the outsourcing possibility of this tool, as this platform, the first in Italy of this type, is used for idea exchanges between clients in order to underline their needs and wishes that can be brought to life by Mulino Bianco.

The *Nel Mulino Che Vorrei* project is part of an open innovation strategy that is linked to a process which starts and ends with the costumer (outside-in process) [8]. The *Il Mulino Che Vorrei* community was founded on 8th of March, 2009 as an example of a co-generation and crowd sourcing project, representing a workshop opened to everybody. People are able to contact the Mulino Bianco world, to suggest ideas and vote for the original ones. This type of community is a model of changing communication, previously implemented by Mulino Bianco, which does not aim to limit costumers to a passive, spectator level, but to transform them into active participants, capable of improving the brand with ideas, experiences, emotions and discussions [8] The platform can be directly accessed from the first page of Mulino Bianco's website and is divided in six areas:

- The area in which the project can be discovered – Scopri il progetto
- The area where people can suggest their ideas – Proponi la tua idea
- The area in which people can vote others' ideas – Vota la idee degli altri
- The area in which people can see realized projects – Gurda I progetti realizzati
- The area where customers can access the blog – Leggi il blog
- The area in which people can answer to Mulino Bianco's questions – Rispondi al Mulino.

By accessing each of these areas, which are located in the upper side of the website, right in the middle of it, new windows will open containing useful information about each of them. In fact, anyone can understand the topic right from the titles, which are quite suggestive.

As we have already seen, the community aims to collect ideas, to analyze and update them only if they are related to the brand's mission, vision and values. The company's management states that they will take into account any idea, even if it is not related to the current offer. Moreover, as transparency is the most important factor in this project, all ideas will be evaluated and a public answer will be posted in case of validity or not, along with the explained reasons why that choice was made.

In order to suggest an idea, people must firstly become a member of the Mulino Bianco community. When first launch, the platform registered 130.000 users, from which 75% of them were women, more precisely mothers, with an average age of 34. At the time of this study, the data recorded by the *Nel Mulino Che Vorrei* website are: 7567 ideas, 25.326 comments, and 395.060 votes. This data can be found on the right side of the platform, remaining visible when accessing the first three areas.

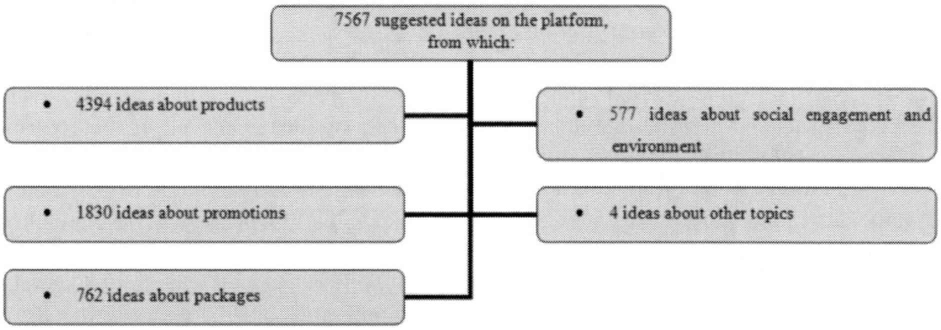

Figure 2. Suggested ideas on the platform [7].

The Mulino Bianco competitions/projects target not only current customers, but also potential customers of pastry products. The participation to the *Nel Mulino Che Vorrei* project is open to every adult person who is registered on the website and is an Italian citizen. We can therefore state that the targeted segment is quite restrictive.

4. The promotion of Mulino Bianco'projects

From Promoting the projects presented on *Nel Mulino Che Vorrei* platform is very important in order to collect as many ideas and votes as possible. In addition to the conventional methods, Barilla uses social media to develop creative initiatives. Social media can be defined as a group of Internet users who use the ideological and technological fundaments of Web 2.0 and allow building and sharing the generated content [9].

Social media is a new trend that changes the rules of communication with clients, allowing the company to get involved in direct contact with the final customer, at relatively low costs and high efficiency compared with traditional communication tools [10].

The Mulino Bianco initiative emphasizes, once again, how much social media, so used and loved by customers, enters more and more in the Italian corporation's environment. *Nel Mulino Che Vorrei* represents a radical approach of Social Media, as it affects both communication and the company [11]. By activating "Facebook Connect", we connect to our *Nel Mulino Che Vorrei* project account that uses the Facebook address. Until we connect, all actions that we do on the website through the Facebook application will be visible on our Facebook wall. Our votes and comments will be shared with all of our friends, who can also decide to suggest ideas or vote ours [12].

The project team's advice for all who meet the *Nel Mulino Che Vorrei* website is to take advantage and join, requiring being completed only few simple steps where we will be asked to authorize the publication of the votes and comments, and to accept the terms of use of Facebook and Nel Mulino Che Vorrei project [12].

5. The method of collecting, processing, and analyzing ideas

For The method of collecting, processing and analyzing the suggested ideas depends on the project. In order to participate at every project on the Mulino Bianco platform, registering on the website is mandatory. The user can register by writing down all required information, and after reading the Legal Notes and the Terms and Conditions of participation in competitions. The information must be real and correct in order to be processed according to the privacy policy published on the website. There is only one registration per e-mail allowed.

Once the registration process is complete, the user will receive an e-mail that will confirm the registration on the provided email address. Once registered, the user will be able to participate at any available project by sharing his/hers self-ideas, by voting others' ideas, by posting comments, and by participating in Mulino Bianco opinion polls that aim the improvement of products.

The Nel Mulino Che Vorrei editors supervise all contributions. They will publish all posted ideas, regardless of their connection with the brand's activity, in order to suppress any kind of discrimination. However, the contributions will be rejected if they do not meet the netiquette regulations. Also, it is very important that participants read the project's legal notes.

Mulino Bianco has set some internal rules that will guide the process of simultaneously evaluation the most 10 voted ideas and a variable number of options of new ideas from all of the contributors. All ideas entered in the process of evaluation will be vote blocked. Once evaluated, positively or negatively, the ideas will be eliminated from the evaluation list, and subsequently will be published. At the same time, the most voted idea will enter the evaluation process. The evaluation will be performed on two levels [13]:

On the first level of evaluation, the idea involves the Mulino Bianco Directorate, which gives an opinion on the business idea: such an idea could be eliminated if it is very similar to an already analyzed project in the immediate past, and is not feasible. This phase lasts approximately six weeks.

On the second level of evaluation, the idea will be evaluated from a cost/benefit point of view, and will enter into the business processes to adapt to periods of development of other projects. If they are positively evaluated, will enter the analysis cycle, and in depth building capacity. This phase requires different times, sometimes very long, depending on the type of idea that is processed.

The actual implementation will be announced on the website, and the required realization times will be identified, procedure that is from time to time evaluated. It is very important to know that timetable implementation for a food industry might take a long time, given the complexity of the tests needed before communication / marketing. For example, starting to produce a new type of biscuit might take 24 months of work. Periodically, Mulino Bianco team will provide information concerning that work stage.[13] Mulino Bianco is committed to providing a complete explanation for each idea that has entered the evaluation process, but that will not be materialized.

6. Conclusions

The In order to innovate and to develop products taking into account the level of fat, sugar and salt, which have been given for every category, to reformulate already existent products, in order to reduce some nutrients, in 2013 Barilla invested 40 million Euros only in research and development activities that target both products and processes.

Nel Mulino Che Vorrei project came to life on the strength of a wish to collect and satisfy needs, wishes, suggestions and demands of improvement that clients and costumers expressed and thought that will help all of us.

Through *Nel Mulino Che Vorrei* project, Mulino Bianco is aiming to become a listener. The participation of all interested parts is very important: just by expressing opinions, by participating, by sending their own ideas, by commenting, and by voting other ideas which have already been published, we can help Mulino Bianco accomplish what we think is truly important and useful.

Acknowledgement
This research was undertaken within the framework of the National Research Program PN II, financed by MEN – UEFISCDI, project PN-II-PT-PCCA-2013-4-1811.

References

[1] R B 2006 Innovation *The Classic Traps, Harvard Business Review*
[2] Nel Mulino che Vorrei, http://www.nelmulinochevorrei.it/scopri.html, Accessed: 20/12/2014.
[3] Nel Mulino che Vorrei, http://www.nelmulinochevorrei.it/index.php?ctl=index&cmd=come, Accessed: 20/12/2014.
[4] Nel Mulino che Vorrei, http://www.nelmulinochevorrei.it/index.php?ctl=index&cmd=perche, Accessed: 20/12/2014.
[5] Nel Mulino che Vorrei, http://www.nelmulinochevorrei.it/index.php?ctl=index&cmd=cosa, Accessed: 20/12/2014.
[6] Nel Mulino che Vorrei, http://www.nelmulinochevorrei.it/index.php?ctl=index&cmd=poi, Accessed: 21/12/2014.
[7] Nel Mulino che Vorrei, http://www.nelmulinochevorrei.it/index.html, Accessed: 19/12/2014.
[8] Di Guardo M and Castriota M 2013 The challenge and opportunities of crowdsourcing web communities: an Italian case study *International Journal of Electronic Commerce Studies* pp 79-92
[9] Kaplan A and Haenlein M 2010 Users of the worlds, Unite! The challenges and opportunities of social media *Business Horizons* **53(1)** pp 59-68
[10] Martini A, Massa S and Testa S 2014 Customer co-creation projects and the social media: the case of Barilla of Italy *Business Horizons, forthcoming*
[11] Maggi S Blog, Information on http://www.steblog.net/2009/03/nel-mulino-che-vorrei-good-sign-for.html
[12] Nel Mulino che Vorrei, http://www.nelmulinochevorrei.it/connetti-con-facebook.html, Accessed: 21/12/2014.
[13] Nel Mulino che Vorrei, http://www.nelmulinochevorrei.it/index.php?ctl=index&cmd=fara, Accessed: 21/12/2014.

IOP Conf. Series: Materials Science and Engineering **95** (2015) 012148 doi:10.1088/1757-899X/95/1/012148

Innovation through developing consumers communities. Part II: Digitalizing the innovation processes

S Avasilcai[1] **and E Galateanu (Avram)**[1]
[1]"Gheorghe Asachi" Technical University of Iasi, Department of Engineering and Management, Bld. D. Mangeron no. 29, 700050, Iasi, Romania

E-mail: egalateanu@tex.tuiasi.ro

Abstract. The current research recognises the concept of innovation as the main driver for organisational growth and profitability. The companies seek to develop new ways to engage consumers and customers into co – creation value through the product design, development and distribution processes. However the main concern is manifested for new and creative ways of customization products based on consumers' requirements and needs. Thus the need for innovative virtual instruments arose as the demand from social communities for personalised products or services increased. Basically companies should develop own innovative platforms, where consumers can participate, with ideas, concepts or other relevant contributions, and interact with designers or engineers for product development. This paper aims to present the most important features of platform development within BMW Group as a concept and as innovative instrument. From this point of view it is important to enhance past experiences of the company in the field of co – creation projects. There will be highlighted the dual consumers' character as co – creator and co – evaluator based on their involvement in the proposed and developed projects and platform structure. The significant impact on platform functioning it has the diversity of company's concerns for Research & Development and innovation activities. From this point of view there will be assessed the platform structure, the main proposed themes and the evaluation process. The main outcome is to highlight the significance of platform development as innovative tool for consumers' communities' enhancement. Based on the analysis of "BMW Co-Creation Lab", there will be revealed the main consumers concerns in terms of safety, comfort and appearance of the products. Thus it is important to understand the evaluation process of gathered ideas and intellectual property policy. The importance of platform development and implementation will be highlighted by company's results in terms of Research & Development investments and future projects which will be proposed, assessed and implemented by BMW Group in order to show the responsibility for their products and consumers.

1. Introduction

The current research emphasize the need of open innovation as the main driver for organizational growth and profitability. However it is also stated that in order to implement the open innovation outside – in component it is necessary to create a virtual environment for interactions with social communities [1]. This kind of organizational openness resulted as consequence of the switch from traditional supply chain to value network, and is concretized by value creation. Therefore, the evolving value networks seek to engage into collaborative processes in order to gain competitive advantage [1]. From this point of view, co – creation became the basis for consumers' engagement into

Content from this work may be used under the terms of the Creative Commons Attribution 3.0 licence. Any further distribution of this work must maintain attribution to the author(s) and the title of the work, journal citation and DOI.
Published under licence by IOP Publishing Ltd

organizational activities and is achieved through exploitation of those interactions [1]. Based on online communication technologies and the need for collaboration, the manufacturers or service providers improved their offerings by including social communities as value creation engine [1]. Although innovation management has major importance as it justifies the need of consumers' engagement, the main concern is still about the locus of innovation, the integration into the product development process [2], especially at early stage – ideas generation and exploitation [2]. Despite the fact that innovation is a key concept for product development process, its sources are highly assessed in order to create new product. From this point of view, the final users are seen as sources of innovation [3], as they contribute with skills, knowledge, creativity and motivation [2] in the product development. The main interest is concentrated on the mechanism for ideas collection and their exploitation from the producer point of view. Hence the development of virtual instruments and social community's involvement became an unavoidable condition in terms of innovation implementation and interaction with final users. This user – driven approach requires empowerment and connectedness of customers or consumers with product producers or service providers via virtual platforms, as instrument for ideas collection and generation [4].

2. Social communities' customization and digitalized interactions

As consumers' buying power increased, the need for product personalization arose [5]. The integration at the early stages of product development, design and development, of the social communities implies the companies' adaptation to market trends [5]. Hence there is important to define the consumers' actions in order to understand the level of product' customization [6], which is comprised from:

Consumers' data evaluation – collecting data from different indirect channels, such as sales, reports, internet logging data,

Consumers' response – observe and ask the consumers directly: surveys, interviews, focus group,

Co – create with consumers – consumers' involvement into product creation by introducing specific developed tools [6].

From this point of view, companies adopt a platform thinking approach based on the main interests: differentiation, relations and efficiency of the products, which represents the engine of value co - creation and communities' customization [7]. However, Piller considered that a social community' customization is successful only when in covers all necessary information and provide an efficient communication between those who are involved [7]. The degree of covered information and efficient communication between manufacturers or producers and their final users is covered by platforms' implementation, which represents an important instrument for customization distribution, implicitly reflects the degree of interaction digitalization [7].

3. Case Study: BMW platform and implications

3.1. Co – Creation with stakeholders

According to the BMW Group vision and philosophy the product innovation should involve first of all customers' engagement. However, the company's previous experience is relevant for platform development research. As the group's vision is to provide innovative and connected solutions, the main concern is related to product development at early stages, consolidated afterwards through collaboration between the departments [8]. From this point of view the group implemented Virtual Innovation Agency, which represents an interface for ideas or concepts collection related to different problems solving solutions or other relevant design aspects in order to ensure The BMW Group's leadership in automotive industry worldwide [4, 8]. Previous experience in co – creation innovation and stakeholders' engagement is concertized through several successful projects delivered on "BMW Co – Creation Lab", as it is shown in table 1.

The "BMW Co – Creation Lab" is designed for car owners or other customers with special attention for automotive industry, who jointly with the group's experts participate in co – creation activities through assessments, commentaries and suggestions [15]. The main subjects addressed, but not limited to engineering, electronics, performance and motorsport, design and life style, marketing and sales [15]

Table 1. Previous developed projects on "BMW Co- Creation Lab".

Projects	Type	Objective	Key aspects
Interior Design Contest [9]	Ideas contest	Functionality, design and experience	Vehicle interior design Creative ideas for vehicle personalization
Mobility Services [10]	Ideas contest and discussion forum	Urban mobility services and solutions	General mobility Electric vehicles Network and communications standardization
Customer innovation Lab [11]	Co-creation lab	Electric and electronics, life style and accessories, sales and marketing	Online services Future driver's assistance systems
BMW on internet [12]	Workshop	Accessories and life style, internal and external design	Virtual configuration of BMW products
CleanEnergy Community motoring [13]	Discussion forum	Sustainable energy and BMW Hydrogen 7 in virtual communities	BMW Hydrogen 7 technology Renewable and sustainable energy use
BMW Idea Lab [14]	Event	Product innovation process	Use of multimedia virtual instrument Performance and engineering

3.2. "BMW Co – Creation Lab": digitalized innovation

As it was previously stated, the digitalized innovation requires the creation of a virtual interface in order to ensure efficient communication and interaction between the social communities and producers. From this point of view, the BMW Group implemented "BMW Co – Creation Lab" as a virtual meeting environment, where the participants get to propose, assess and discuss major subjects related to product innovation [15].

The instrument's affiliation to the BMW Group can be traced through brand's identification, the BMW logo, comprised from Bavarian colors – white and blue. The importance of brand visualization is reflected by the fact that it represents an essential design element [16].

The platform is structured as virtual community, where potential co – creators can participate at a various developed projects by BMW Group's experts. The access to the platform is granted based on individual registration virtual form. The potential co – creators should fill special registration form with [17]:

Account information – e- mail address, user name and password

Personal information – age, occupation, educational qualification, industry, family status, country

Survey – motivation for involvement into "BMW Co – Creation Lab" activities

Vehicle Information – car type and specification [17].

The access to the platform is granted based on account validation, which represents a direct link to the participant's e-mail address and comprises also information about the latest developed projects. The participants have the opportunity to explore the virtual world of co – creation and to meet all the contributors at "Co - Creators" section [18]. This section comprises also statistical data concerning the type of vehicles that other co – creators possess [18]. The co – creators listing is represented by a 3D sphere, comprised from other contributors accounts [18].

The access to the specific project is realized by pressing the "Participate at this project!" button. However it will redirect the participant to the Terms of Use and Data Protection page, where at the bottom it can be found the direct link to the project website "Proceed to the project website" [19].

The visibility of the platform is ensured by the possibility of sharing the contest information through social networks, such as Facebook or Twitter, as well as mailing option, through Gmail, Hotmail and others, and images networks, Pinterest, Instagram and other.

3.3. Data collection and analysis
The ideas submitted to the "Co – Creation Lab" are assessed and reviewed by the BMW Group's experts in order to discover the current trends, ideas and needs according to the topics of the proposed and developed project [20]. After the first review, the most appealing and applicable ideas will be presented to the experts non – member to the BMW Group in an anonymous way, such as discussion forums (figure 1). The Intellectual Property policy ensures the personal data protection for each participant of "BMW Co- Creation Lab" [20].

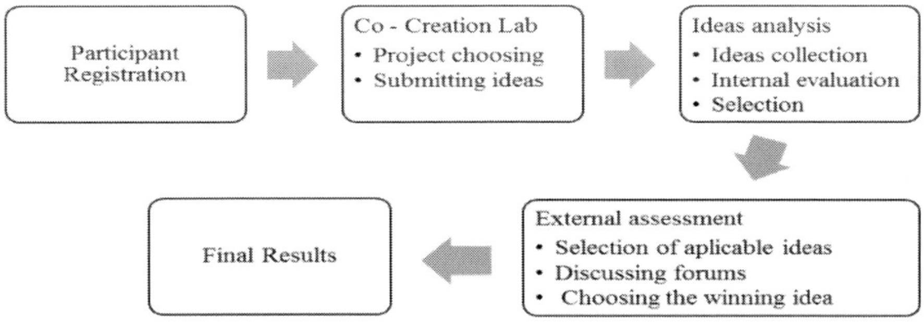

Figure 1. Data collection and analysis process in "BMW Co – Creation Lab" [20].

3.4. Company' results and future projects
The main objective of platform implementation is to explore social communities' ideas and needs in order to create sustainable and innovative products. According to BMW Group's vision the platform use facilitates the interactions between the company and the potential consumers, the implementation of ideas in product development process at concept and design levels. From this point of view, the success of co – creation activities is concretized by major financial results derived from the use of new technologies and innovation [21]. The Research and Development activities are reflected by new models launching and increased number of innovative products which ensured the group's future through increasing investment into research and development.

In future, The BMW Group seeks to respect their commitment for innovative and sustainable manufacturing. There is still present the company's concern for lightweight construction, innovative design and connected drive vision, which are pillars of their 2020 strategy [22]. Sustainable mobility represents an important energy management issue in terms of efficient dynamics [23]. The use of renewable energies and alternative fuels, BMW Hydrogen 7 project, as well as led technologies represents the major interest in innovative product development process [24].

4. Conclusions
Open innovation as well as co – creation with social communities expands the companies' horizons in terms of product development and innovation implementation. However, each economic entity organizes its own activities according to their main strategies.

Nowadays, the reflection of innovation implementation is supported by platform development activities. Based on the case study in BMW Group it can be seen that innovative products require innovative ideas, gathered in a systematic and creative ways, and assessment of consumers' needs based on their own developed solutions.

The created virtual instruments became well known for the interactive projects development, which are implemented in products 'design, engineering or services. From this point of view, The BMW Group perceive to identify not only innovative ideas, but also to create the new meanings for life style and driving. "Connected Drive" reflects the combination of innovative engineering solutions, creative design and drivers' experience.

Acknowledgement
This research was undertaken within the framework of the National Research Program PN II, financed by MEN – UEFISCDI, project PN-II-PT-PCCA-2013-4-1811

References

[1] Romero D, Molina A, Taylor and Francis 2011 Collaborative Networked Organisations and Customer Communities: Value Co – Creation and Co – Innovation in the Networked Era *J. of Production Planning and Control* **22** pp 447 – 472
[2] Haavisto P 2012 Social media discussion forums and product innovation – The way forward? *First Monday* **17** pp 1 – 13
[3] Gambardella A, Raasch C and von Hippel E 2014The user innovation paradigm: impacts on markets and welfare, *Social Sci. Res. Network Electronic J.* available at http://ssrn.com/abstract=2079763
[4] Russo – Spena T and Mele C 2012 Five Co-s in innovating: a practice – based view *J. of Service Manag.* **23** pp 527 – 553
[5] Berger C and Piller F 2003 Customer as co – designers *IEE Manuf. Eng.* pp 42 – 45
[6] Piller F, Ihl C and Vossen A A typology of customer co – creation in the innovation process *Social Sci. Res. Network Electronic J.* available at http://ssrn.com/abstract=1732127
[7] Piller F 2002 Customer interaction and digitizability – a structural approach to mass customization *Moving towards mass customization* edRautenstrauch et.al (Heidelberg/Berlin/New York: Springer)
[8] Information on on http://www.bmwgroup.com/e/0_0_www_bmwgroup_com/forschung_entwicklung/ein_blicke_in_die_entwicklung/via/via.html
[9] Information on https://www.bmwgroup-cocreationlab.com/cocreation/project/interior-ideacontest
[10] Information on https://www.bmwgroup-cocreationlab.com/cocreation/project/tomorrows-urban-mobility-services-ideacontest
[11] Information on https://www.bmwgroup-cocreationlab.com/cocreation/project/customer-innovation-lab
[12] Information on https://www.bmwgroup-cocreationlab.com/cocreation/project/bmw-on-the-internet
[13] Information on https://www.bmwgroup-cocreationlab.com/cocreation/project/clean-energy-comunity-monitoring
[14] Information on https://www.bmwgroup-cocreationlab.com/cocreation/project/m-power-tour-event-idea-lab
[15] Information on https://www.bmwgroup-cocreationlab.com/about
[16] Information on http://www.bmwusa.com/Standard/Content/Innovations/Design/SignatureElements.aspx
[17] Information on https://www.bmwgroup-cocreationlab.com/users/register2
[18] Information on https://www.bmwgroup-cocreationlab.com/cocreators
[19] Information on https://www.bmwgroup-cocreationlab.com/users/participate/trunk-contest
[20] Information on https://www.bmwgroup-cocreationlab.com/terms
[21] BMW Group, Quarterly Report (2014), 1 – 59 available at http://www.bmwgroup.com/e/0_0_www_bmwgroup_com/investor_relations/finanzberichte/zwischenberichte/2014/0914/_pdf/Q3_2014_BMW_Group_engl_Online.pdf
[22] Information onhttp://www.bmwgroup.com/bmwgroup_prod/com/en/responsibility/whats_next/lightweight_construction/index.html
[23] Information onhttp://www.bmwgroup.com/bmwgroup_prod/com/en/responsibility/whats_next/sustainable_mobility/index.html
[24] Information onhttp://www.bmwgroup.com/bmwgroup_prod/com/en/responsibility/whats_next/renewable_energies/index.html

A case study on open innovation on Procter & Gamble. Part I: Innovation strategy over years

I G Agafitei[1] **and S Avasilcai**[1]
[1]Gheorghe Asachi Technical University of Iasi-Romania, Department of Engineering and Management, 28 Mangeron St., 700050, Iasi, Romania

E-mail: iuliana.agafitei@gmail.com

Abstract. The purpose of the present paper is to analyse how product innovation process has been organized by a multinational company, such as Procter & Gamble. Since open innovation is a new approach on the local market, we have chosen to conduct a research in form of a case study showing the steps that the company has taken in order to involve other organizations to contribute to the development of new products. The paper will first describe what innovation management theory is suggesting. Following this, the mission and vision of Procter & Gamble will be analysed emphasizing its innovation efforts in time. The approach that the company has nowadays, related to development of new products, has its roots in several complex initiatives in the past, initiatives that have in their central attention the customer. Finally, it provides the research limitations and implications for future analysis on the case of Procter & Gamble open innovation approach.

1. Introduction

An increasing interest has been shown in the last few years for all that is connected to innovation: new products and services, management of innovation, innovation models and innovation tools. What modern economic theories have proven for sure is that innovation has an effect on a company's capability to differentiate itself. Paul Romer has further developed the theories of Schumpeter, the so called neo-Schumpeterian economic growth theory, suggesting the main idea that firms will try to increase their profits by directing all their efforts into developing new products and services [1]. At the opposite side of neo-classical economics that looks at industry and economic performance without taking into consideration the differences between firms, the Schumpeterian view strongly supports that the manner in which a company manages its resources over time and develops its capabilities will influence its innovation performance [2].

2. Introduction

A mutual consideration on innovation concept places this process in the area of new products and services, which are produced by using new information and that take place in two phases: invention and exploitation [1, 2]. Looking at innovation as a management process allows us to see the distinction between an innovation and a product, which is the output of innovation [2]. Another key aspect regarding innovation as a process is the distinction between innovation and invention. Others have concluded that innovation has to do more with the commercial and practical application of an invention. The latter is the conception of an idea and intellectual effort of putting together scientific information comparing to innovation being the subsequent result of invention into the economy [2].

Content from this work may be used under the terms of the Creative Commons Attribution 3.0 licence. Any further distribution of this work must maintain attribution to the author(s) and the title of the work, journal citation and DOI.
Published under licence by IOP Publishing Ltd

Since the benefits of welcoming innovation into a firm have been proven by economic theories, the next step would be to see how those interested can influence successfully the innovation process. One way of listing the specifics is to do empirical studies as this to see how other companies organize themselves. In table 1 below we can see the most important key features that can easily transpose into the vision and mission of a company in order to sustain an innovation process [2].

Table 1. Summary of characteristics that facilitate the innovation process.

Organizational requirement	Characterized by:
Growth orientation	A commitment to long term growth rather than short term profit
Organisational heritage and innovation experience	Widespread recognition of the value of innovation
Vigilance and external links	The ability of the organisation to be aware of its threats and opportunities
Commitment to technology and R&D intensity	The willingness to invest in the long term development of technology
Acceptance of risks	The willingness to include risky opportunities in a balanced portfolio
Cross-functional cooperation and coordination within organisational structure	Mutual respect among individuals and a willingness to work together across functions
Receptivity	The ability to be aware of, to identify and to take effective advantage of externally developed technology
Space for creativity	An ability to manage the innovation dilemma and provide room for creativity
Strategy towards innovation	Strategic planning and selection of technologies and markets
Coordination of a diverse range of skills	Developing a marketable product requires combining a wide range of specialised knowledge

Still, in order to differentiate itself a company has to design an innovation model that has the role of defining that specific structure, resources, target customers, value proposition and other details that eventually will help the company focus on what they want to accomplish. As Vargo el al. appreciated [3], value is created when the customers reach the context and integrate the products or services offered. In other words, based on a service-dominant logic, the focus is not on the offering per se but on the customers' value-creation process [3]. Year 2000 has brought a new perspective over the interaction of firms and consumers showing the customers are taking active roles expressing what they expect from companies experiences. Later on, in 2004, the official debut of the term co-creation has been made by Prahalad and Ramaswamy, the authors explaining that co-creation of value is an initiative of the customers who are "dissatisfied with available choices" [4].

All companies, similar in size with Procter & Gamble and even those who are smaller face the pressure of reducing costs of production and constantly improving their operations. Repetitive tasks and a well-organized environment allow the firm to deliver their product under the right specifications but at the same time, in the same environment there is a need for creativity in order to make room for new ideas, approaches and products [2]. To sum up, it is the firm's challenge to balance this kind of environment if they want to allow innovation in their house. Precisely due to this heavy mission that a firm may have, balancing its resources and activities, open innovation has taken a lot of attention in the last years. Adding the input of customers and integrating outside company knowhow on top of all other resources that the company has, has proven to be a success in the case of many companies. Open innovation, implies that "firms can and should use external ideas as well as internal ideas, and internal and external paths to market", in their pursue to advance their technology. One view of the actual author who coined the term is that open innovation can be thought of as the antithesis of the traditional vertical integration model where internal research and development activities lead to internally developed products that are to be distributed by the firm [5].

3. Case study: Procter & Gamble innovation in time

In order to draw conclusion on how did Procter & Gamble organized its activities over time, we have conducted an analysis of its evolution pointing to the most important moments. Founded in 1837, the company has its headquarter in Cincinnati, United States and gathers an impressive portfolio of products with 23 brands with annual sales of $1 billion to more than $10 billion, and 14 with sales of $500 million to $1 billion - many of those with billion-dollar potential [6].

3.1. Procter & Gamble: company evolution and product development

After 50 years since it was established, in 1887, P&G comes up with a new way of involving its employees in the company activity. It was the first time the company initiated a special program that offered the opportunity for employees to take shares in the company and define their role in what its success was. In 1919 William Cooper Procter, the son of William Procter continues with his efforts to institutionalize the relationship between the company and its employees. In 1924 it was for the first time in the United States when a company has taken the decision to grant all resources for a research study with a strong focus on its consumers. The goal of this activity was to offer a better understanding on what consumers needs are, anticipate it and respond by creating products that would improve the quality of their day by day life [7]. Following the same strategic direction, in 1941, P&G develops a customer relationship department to handle all inquiries in a formal way. In the years to follow, by 1980 there were several toll free numbers assigned and email correspondence was growing, all for the same purpose to be close to the consumers and, allow them to contact the company by all means. All the strategic movements the company has done over time have created a tremendous advantage of being one of the companies that best knows its consumers. Along the way Procter & Gamble has grown through many acquisitions and product developments which made its portfolio to be one of the most complex at that time and even today. In table 2 we can see just a few of the most important events in the history of the company indicating the desire to improve its production and research capabilities, be the first to use the technological breakthroughs and connect with its consumers.

Presently the company has several research facilities in different parts of the world, from Western Europe – 8 research centres, to Asia, where the Singapore Innovation is considered the third high-tech innovation centre in the Asian region [9].

With a declared purpose of improving the life of its consumers, the company states that the results will be that satisfied consumers will turn the company into a market leader [10]. When it comes to its values the company mentions integrity, leadership, ownership, passion for winning and trust. One of the key attitudes that strongly suggest the desire to innovate is the fact that all employees are encouraged to act as owners and also constantly challenge the status quo. Since employees are truly considered to be an asset, The Clay Street project has been initiated with the purpose for creative thinking to deliver great innovative product and change the culture. Teams inside P&G were exposed to diverse ideas and methods that caused them to think differently about their challenge and themselves [11]. Among the eight principles that lay at the heart of P&G foundation there is also innovation. This principle articulates the fact that the company places great value on new consumer innovations and this is why challenges convention and wishes to reinvent the way it does business [12]. Leadership is an important coordinate placed by the company as one of its values. The importance of leadership is crucial to the innovation process as it allows all ideas to come to the surface and not to be blocked at a certain hierarchical level or by internal procedures. So the company is committed to eliminate organizational barriers in order to deliver its strategy [12].

Table 2. Important historical events over time [8].

Year	Event
1890	P&G sets up an analytical lab at Ivorydale to study and improve the soap-making process. It is one of the earliest product research labs in America.
1896	P&G's first colour print advertisement – an ad for Ivory – appears in Cosmopolitan magazine picturing this "Ivory Lady."
1915	The Company builds its first manufacturing facility outside the United States, in Canada. Employing 75 people, the plant produces Ivory soap and Crisco.
1931	Procter & Gamble's brand management system is born.
1939	Just five months after the introduction of television in the U.S., P&G airs its first TV commercial (for Ivory Soap) during the first televised major league baseball game.
1943	The Company creates its first division – the Drug Products Division – to sell its growing line of toilet goods.
1952	A new research facility, Miami Valley Laboratories, opens in Cincinnati. MVL is the Company's first facility dedicated solely to upstream research.
1965	The first paper plant built by P&G opens in Mehoopany, Pennsylvania.
1988	The company starts to produce its goods also in China, 1st operation in the world largest consumer world.
1995	P&G opens its Health Care Research Centre in Cincinnati to serve as the worldwide hub for the Company's health care business. The Health Care Research Centre is designed to promote innovation and efficiency in bringing new health care products to market faster.
1998	P&G announces Organization 2005, a new global organizational design to drive innovative ideas to world markets faster.

3.3. Open innovation approach: working with external partners & employees

By analysing its capacity to launch new product on the market, Procter & Gamble has been considered for quite some years now one of the most innovative consumer products company. But this performance it's not the result of chance but more of a systematic and disciplined approach [13]. When using The Initiative Diamond approach based on a stage gate model, the company stated to have aligned the entire organization on how to use Stage Gate success criteria, portfolio and resources to deliver better innovations. SIMPL (Success Initiative Management & Product Launch) process designed in 1990 was a rigorous one and it explained the steps to take the idea through launching it and post-launch. There are four drivers of performance that worked in the case of P&G during the usage of stage-gate model: a product innovation and technology strategy for the business; an effective and efficient idea-to-launch system; resource commitment, focusing on the right projects and portfolio management; a positive climate and environment for innovation [13]. Beginning of the year 2000 the company has changed its focus from research and development since the growth generated by it was not efficient anymore. The budgets allocated to innovation were under pressure due to the high costs of the new technologies, at the same time only 35% of the new products were achieving their financial objective. The new focus has become developing new products by enhancing external relations and the company's new goal was to reach 50% of the innovations from collaborating with external parties [14]. In the following year P&G had a designated team in charge to share the principles of disruptive innovation. Quite often A.G. Lafley, the CEO of P&G, has been mentioned as one of the leaders who master the skills of disruptive innovators. For example, by asking a lot of questions, he is considered to have changed the way the company worked. Another example supporting this is the fact the all employees are encouraged to devote 25% of their time on finding ways to improve what they do [15]. One of the new directions set by the company's CEO was to increase the organizational support in order to develop new business. For this, new teams were appointed, with main purpose to grow the business, teams with resources kept separate from the core business.

At the same time that Connect & Develop platform has been launched to support the strategy mentioned above, TAG (Technology Acquisition Group) was one of the initiatives that have been defined with the purpose to search, outside the company, for technology [16]. In table 3 we can see examples of, internal & external connections, used by the company in order to leverage its technical capabilities into new as well as existing products [17]

Table 3. External and internal activities supporting open innovation [17].

Procter & Gamble's connections for innovation	
External connections	Internal connections
Joint Developments	Innovation Net
National Laboratories	Smart Learning Reports
Technology Acquisition Group	Communities of Practice
Complementary Technology Innovations	Global Technology Council
Critical Supplier Partnerships	Corporate Innovation Fund
Endorsements	
Patent/IP Licensing & Donation	
University Research Fellow Program	

The group TAG has become a primary portal for soliciting external technologies and allowed inventors to submit their ideas for evaluation, supporting this way the new approach that P&G had for licensing technology. The strategy was to improve the innovation process and turn it into a much simpler approach: license all technology at the latest three years after market introduction or five years after patent approval [17].

4. Conclusion

In a very official manner the company stated that has gone from being „protective" with its technology resources to being „proactive" with the clear intention to open the door to see their innovation in the marketplace, not only in their products. The purpose was to generate revenue for their stakeholders and any type of collaboration, from selling, donating, trading with suppliers was taken into account [17]. A good example to understand the change in the company's approach, that will create the right environment for great product innovation, is to look at the new strategy based on licensing. As Sakkab mentioned [17], previously the company had been under the negative impact of the Cold War when no type of collaboration was happening. Starting the year 2000 P&G representative declares „will sell, donate, swap or collaborate with suppliers for capital avoidance, or trade for lower pricing" as long as it can be measured and also produce value for all stakeholders. The present paper looks at key decisions taken by Procter & Gamble related to the innovation approach with the purpose to show the importance of the mentality and internal efforts of all organization members. It look like the open innovation direction came from top – down creating tools such as TAG or InnovationNet, the latter one allowing P&G researchers to connect with thousands of innovators across the globe and exchange information on latest novelty. Part two of the present study will look into the features and impact of Connect & Develop platform on new product development inside P&G. The goal of this case study was to emphasize all the initiatives that supported innovation and even earlier – the general interest for consumer delight which subsequently became conversion for profit. The limitations of the present study are related to the qualitative approach taken that doesn't allow statistical inference. A series of other such analysis would be necessary to see whether companies with successful products fall under the same approach.

Acknowledgement

This research was undertaken within the framework of the National Research Program PN II, financed by MEN – UEFISCDI, project PN-II-PT-PCCA-2013-4-1811.

References

[1] Afuah A 1998 *Innovation Management: Strategies, Implementation and Profits* (New York: Oxford University Press)
[2] Trott P 2012 *Innovation management and new product development* fifth ed. (Pearson Education)
[3] Gustafsson A, Kristensson P and Witell L 2012 Customer co-creation in service innovation: a matter of communication?, *J. Service Manag.* **(23)3**
[4] Prahalad C K, Ramaswamy V 2004 Co-Creation Experiences: The next practice in value creation *J. of Interactive Marketing* **18** (Wiley InterScience)
[5] Chesbrough H 2006 *Open innovation: Researching a new paradigm* (Oxford University press)
[6] Procter & Gamble Leadership Brands, http://www.pgbalkans.com/en_US/brands/index.shtml, accessed 22 January 2015
[7] Information on P&G History, http://www.pg.com/ro_RO/company/heritage.shtml, accessed 22 January 2015
[8] Information on P&G History, http://www.pg.com/translations/history_pdf/english_history.pdf, accessed 22 January 2015
[9] Information on P&G latest innovations, http://www.pg.com/en_US/downloads/innovation/factsheet_Singapore.pdf, accessed 22 January 2015
[10] Why Innovation Matters and How to Make it Work, https://www.pg.com/en_ US/downloads/company/purpose_people/touching_lives_improving_life.pdf, accessed 28 March 2015
[11] Information on P&G latest innovations, http://www.pg.com/en_US/downloads/innovation/factsheet_ClayStreet_FINAL.pdf, accessed 22 January 2015
[12] Information on P&G Purpose, Values and Principles, http://www.pg.com/en_US/company/purpose_people/pvp.shtml, accessed 13 February 2015
[13] Information on P&G New Product Development strategy – "Innovation Diamond" http://www.stage-gate.com/downloads/wp/wp_21.pdf, accessed 13 February 2015
[14] Information on P&G new model for innovation, https://hbr.org/2006/03/connect-and-develop-inside-procter-gambles-new-model-for-innovation, accessed 13 February 2015
[15] Dyer J, Gregersen H and Christensen C M 2011 *The innovator's DNA* (Harvard Business Review Press)
[16] Dodgson M, Gann D and Salter A 2006 The role of technology in the shift towards open innovation: the case of Procter & Gamble *R&D Management* **36** (Blackwell Publishing Ltd)
[17] Sakkab N Y 2009 Connect & Develop Complements research & Develop at P&G *Industrial Research Institute*

A case study on open innovation on Procter & Gamble. Part II: Co-creation and digital involvement

I G Agafitei[1] and S Avasilcai[1]
[1]Gheorghe Asachi Technical University of Iasi-Romania, Department of Engineering and Management, 28 Mangeron St., 700050, Iasi, Romania

E-mail: iuliana.agafitei@gmail.com

Abstract. The first part of the present study case has pointed to the efforts directed by Procter & Gamble over time to keep close to its customers and the manner in which management has supported innovation within the company. Following the principles of disruptive innovation, the CEO of the company has set the goal for the company as to produce at least 50% of the innovations from external ideas. So, the aim of this paper is to continue to analyse the tools and the approach that P&G has taken in order to launch new products by partnering with other companies. Innovating by collaborating with external entities holding a highly specialized level of technology and engineering knowhow has been appreciated to contribute to the value creation process. In order to understand how open innovation strategy has worked for Procter & Gamble and which was the designated tool used to enrich the company with valuable input in order to develop products, we have taken a look at P&G's open innovation platform Connect + Develop.

1. Concept introduction

Although not considered to be a new approach, open innovation has captured more attention in the last years especially from companies such as P&G, PepsiCo, Lego, Barilla or Heineken. The manner in which these companies have chosen to work together with their customers, suppliers, distributors or other partners is different and supported by different tools. The principles behind open innovation are dating since the 60s, for example: the idea that there are people outside the organization who have the expertise to contribute is supported by the "gatekeeper" role a person could have within the company [1]. The term of co-creation has been coined by C. K. Prahalad and Venkat Ramaswamy with the occasion of an article in Harvard Business Review in 2000 supporting the idea that co-creation was high related to the desire of the customer to define the product he or she will eventually buy and not wait for companies to decide for them the characteristics of the goods [2]. In order to support the idea that co-creating has high chances to revealing information that it's not obvious or that comes up by taking a classical approach – like market research, information that needs to be gathered from all stakeholders, V. Ramaswamy and F. Gouillart have come up with the following separation between a classical strategy and strategy by co-creating [3] – table 1.

By analyzing the platform and communication approach of Procter& Gamble we will see that the stakeholders of the company are not only the customers and that the company is looking to joint its resources with those of other businesses, help them grow but also come up with new products under P&G brand.

Content from this work may be used under the terms of the Creative Commons Attribution 3.0 licence. Any further distribution of this work must maintain attribution to the author(s) and the title of the work, journal citation and DOI.
Published under licence by IOP Publishing Ltd

Table 1. Traditional strategy vs. Co-creation strategy [3].

Traditional strategy	Co-creation strategy
Value: Creates value by delivering defined customer experience to targeted customer set.	Value: Creates value by constantly enhancing experiences for all stakeholders.
Goal: Establishes strategic goals at the outset and doesn't significantly change them.	Goal: Uses the initial strategic goal as a starting point and lets the full strategy emerge over time.
Traditional strategy	Co-creation strategy
Key Focus: Focuses on the interests of the firm: that is, how the firm can maximize its share of the created value relative to the shares of its industry competitors and the other members of its value chain.	Key Focus: Focuses on the interests of all stakeholders and how the ecosystem can maximize the size of the pie; maximizing the share of value captured by the firm is secondary.
Advantage: Achieves advantage through realizing economies of scale before competitors do and making big, bold moves (such as acquisitions and investments in proprietary assets).	Advantage: Achieves advantage through the increased engagement of stakeholders and by continually building new interactions and experiences, which lead to higher productivity, higher creativity, and lower costs and risks.

2. Procter & Gamble: strategy background

With a powerful research and development team of scientists and researchers located all over the world and a strong brand, Procter & Gamble has encountered a slowdown in sales through the late 1990s. The business model created along the years by the company has been considered a closed one, meaning that there were no exchanges with outside partners regarding development of new products and even improvements. Technology owned was considered „corporate secret" and was used less than 10 percent. For example, licensing was done seldom and the policy that was followed was quite rigid. Precisely due to the large team of professionals that P&G hold and the increasing costs of new technology, determined the company strategists to shift to a different model where proactivity was practiced [4].

3. Procter & Gamble platform: overview

Connect and Develop platform, launched in 2001, is considered one of the official tools that help the company collect and manage ideas to supports its open innovation approach. In order to make everything clear for the visitor, the first section of this platform deals with the benefits and motivations of the company when using this approach. The main reasons behind it are: creating an opportunity for people outside the organization to improve P&G products; avoid stale and repetitive ideas from the employees. The benefit is that, this way the company manages to avoid investing in research and development much more than necessary. The name of the platform is intended to suggest exactly the approach that will rule over the company in the years to come: research and development is becoming connect and develop in order to connect the technologies much better so that the development process will become shorter and efficiency will increase [4]. The platform can be accessed at the following URL http://www.pgconnectdevelop.com/ and also using a new window link on several pages on the official website of the company www.pgbalkans.com. The present version of the website was launched by the company in 2013 allowing for several new types of activities: user submissions are directly feed into the business category leaders which allowed responsible inside P&G to speed up the evaluation process. The designated team can also search by topic, share with other colleagues and track progress of all submitted information.

4. Data collection and analysis by dedicated Connect & Develop team

By gathering a large set of ideas from companies and individuals, P&G can easily get insight from a diversity of professionals. The only request that is mentioned on Connect & Develop, just as introducing the concept, is intellectual property; P&G requires all ideas to follow information on the granted patent or the published patent application [5].

Similar to other companies, P&G knows its areas where improvements are required and communicates the list of present needs. Grouped into categories under the section called Our needs, these are connected to the product categories manufactured by the company: personal and family care innovations; beauty and grooming innovations, household care innovations; packing innovations; industrial and manufacturing innovations; digital, business and retail innovations; other innovations [6]. One example of specific request that the company has is within the category of laundry disinfection in professional business products. The interest is to find technologies cost effectively in disinfecting or sanitizing laundry. The information sheet that can only be accessed after creating an account, explaining technical aspects for the solution is divided into few sections as it follows [7]:

Need description. P&G explains what is looking for.

Background. Specific details regarding products in the sanitizing/disinfecting laundry are given in order to underline the standards that the solution has to comply with.

We are looking for. Details regarding the criteria that technology has to comply with.

We are not looking for. The company states very clear the type of solutions that it's not interested in.

The reasoning behind using an online platform such as Connect & Develop to manage new ideas and proposals is the volume of open innovation submissions. So that all these submissions to be reviewed, the online platform help the company to manage the process efficiently, collect all the information they need and also respond. The submissions are reviewed by a specific team member of Connect and Develop who will pass the innovation to the relevant business in the company [8].

Using the same online space – Connect & Develop, P&G states its interest in collaborating with companies - small or large, entrepreneurs, academia and research groups. Despite the size of the company, P&G wants to reassure those who are interested in submitting ideas that, internally they work as a „small company". The cores in charge of Connect & Develop have the capacity of analysing the information submitted, review and evaluate and finally share it with an Expert Panel. The role of the Experts Panel is to be connected to all P&G businesses and share their knowledge on the company's strategy, needs and plans [9].

When talking about its success stories on Connect & Develop, P&G classifies its open innovations initiatives into main categories: academia partnerships, joint ventures, trademark licensing, in & out licensing. Improvements to already existing and popular products are listed in Other case studies section mentioning the collaboration between P&G and other companies. Below are some examples of collaboration that the company has developed using Connect + Develop – table 2.

Table 2. Open collaboration managed on Connect & Develop [10].

Company objective	Type of collaboration
Sponsoring high-potential start-ups who can demonstrate innovative technology and/or business model ideas to dramatically accelerate quality and speed of critical business opportunities for P&G brands across the globe.	Partnership & support: start up sponsoring
Identify strategic research themes across a wide range of academic disciplines including engineering, maths and physics, design, computing, chemistry, psychology, business and the arts. Link the work done at Leeds to the value of the company's commercial products.	Academia partnerships
Enter a new growth area in the health-care business	Joint Venture
Expand brand equity into new areas	Trademark Licensing
Introduce a new product category in portfolio	In & out licensing

IOP Conf. Series: Materials Science and Engineering **95** (2015) 012150 doi:10.1088/1757-899X/95/1/012150

5. Procter & Gamble successful products resulting from open innovation

Tide Pods, Olay, Always Infinity or Crest are just some of the brands that have benefit from open innovation partnerships. Crest Whitestrips Advanced Seal teeth whitening product is the result of the collaboration with Corium International, Inc. a privately-owned, clinical stage company engaged in the research, development and manufacture of advanced transdermal drug delivery technologies and products. Through the combination of its proprietary delivery technologies and its development and manufacturing expertise, Corium is developing a range of novel active and passive transdermal products as well as consumer products. Corium's innovative "form fit, no slip, clean removal" Corplex technology has been of great interest for P&G's technicians that decided to integrate it into the product. A commercially proven versatile polymer technology that can be delivered in liquid, semi-liquid and solid formats for many drug delivery, medical and consumer product applications, Corplex consists of composite blends of polymers with unique adhesive, physical/chemical, and water-absorbing properties and offers a range of hydrophilic to hydrophobic properties, enabling excellent adhesion to both dry and wet surfaces and a variety of wear times. Corium has been awarded due to this collaboration that produced outstanding business results. Crest Whitestrips Advanced Seal teeth-whitening product was launched in early 2009 and has received great reviews from consumers who were offered a new solution to whitening their teeth [11]. The partnership is a good example of how the company has succeeded to create value for all involved stakeholders in the process, proving the benefits of an open innovation approach. This partnership helped P&G to avoid investing time and research resources at the same time as Corium, the result was that both companies have contributed and built on each other's capabilities.

To emphasize the flexibility and determination that P&G has to develop great products, while managing more than one partner in their efforts to develop a new product category, below are the different companies that P&G worked with that led to a successful product range under the same brand name – Febreze (table 3).

Table 3. External partners who helped build Febreze products [10].

Febreze Trademark Licensing Program		
Product	Licensing Partner	Outcome for P&G
Febreze air filters	Imagine One Resources	Trademark for use with residential air filters
Febreze vacuum filters and carpet powders	Bissell	A new line of vacuum bags and filters; cleaning carpet powders, including specialty products such as pet odour remover
Febreze spot and stain removers	Home Care Industries, Inc	A new line of carpet, rug and upholstery spot and stain removers
Honeywell Febreze fan	Kaz	Two speed fan freshener

6. Connect and Develop results

With more than 10 years of success from 2000 global partnerships and more than 4000 ideas a year C +D, focus is to deliver discontinuous innovations. The open innovation programme reached its initial goal in 2005 – to have more than 50% of product innovations as a partnership with external entities. In 2010 the company has set two new goals: to triple the contribution through the open innovation platform by delivering 30 billion dollars in annual sales and become a Partner of Choice when it comes to innovation collaboration [12].Judging by the impact or part of the product that will include the innovation elements, the company is performing the following types of innovation – table 4.

Table 4. Types of innovation resulting from C+D* collaborations [7] [13].

P&G's posted need on the platform / What are they looking for?	Type of innovation according to the OECD** classification
Need: Refillable Dispensing Package for Wet-Wipes What is P&G looking for: innovative packaging design to improve ease of use and overall consumer experience of wet-wipe product; an IP protected existing solution; novel packaging design solution.	Marketing/product innovation
Need: Digestive Wellness/products, technologies and packaging that improve benefit areas of treat, manage and/or enhance in Digestive Wellness What is P&G looking for: technologies capable of providing new or augmenting established health claims, increasing consumer satisfaction, enhancing consumer compliance, or enabling greater convenience. Opportunities may consist of new product forms, packaging innovations, delivery technologies, or new actives/ingredients or actives'/ingredients' combinations.	Product/Process innovation
Need: pseudoephedrine technologies What is P&G looking for: formulations that will reduce/inhibit the in vitro extraction/use of PSE, yet maintain clinical bioavailability, effectiveness and safety.	Product/Process innovation

* *Connect + Develop*
** *Organisation for Economic Co-operation and Development*

7. Conclusion

At a closer look, the success coming from the various products the company has launched is not strictly related to just the platform Connect + Develop, it is the mixture of all tools and types of collaborations, mentioned in the first part of this case study, that creates opportunities for specialists to develop new solutions. In the last financial report issued by the company innovation is still mentioned as "company lifeblood" [14].

The official purpose of the platform is clearly aligned with the mission of the company „delighting the consumer with innovations that can make every day better" [15]. In comparison with the manner in which other consumer product companies have organized their open innovation process, P&G keeps a high control over all partners' input. There are rules stated in order to communicate precisely the areas where new ideas are welcome. The fuzzy front end of a product development process is considerate to be the „messy getting started" period of new products and although it doesn't require capital investment it can take 50% of development time [16]. Most of the featured needs presented on Connect & Develop are related to products already existing, the purpose of the input would be to help the company change the performance of the product.

A future research idea would be to analyse in depth, taking into consideration other innovation theories besides OECD classification, the type of innovation that results from using an online solution such as the platform Connect + Develop. In the examples mentioned above it is not clear whether they should qualify for process innovation also since the technologies are not owned by P&G – the company just integrates it. The limitations of the present study are connected to the fact that much of the information regarding this platform has been desk researched and most of the scientific research available is mainly related to the open innovation strategy inside the company. The goal of the present case study was to show the results of the change from a closed strategy to a new, open innovation strategy leading to numerous partnerships and success products.

Acknowledgement
This research was undertaken within the framework of the National Research Program PN II, financed by MEN – UEFISCDI, project PN-II-PT-PCCA-2013-4-1811.

IOP Conf. Series: Materials Science and Engineering **95** (2015) 012150 doi:10.1088/1757-899X/95/1/012150

References

[1] Trott P and Hartman D Why "open innovation" is old wine in new bottles *Int. J. Innovation Management* **13(4)** pp 715-736

[2] Prahalad C K and Ramaswamy V 2004 Co-Creation Experiences: The next practice in value creation *J. of Interactive Marketing* **18** (Wiley InterScience)

[3] Information on product development by co-creation: https://hbr.org/2010/10/building-the-co-creative-enterprise, accessed 10 March 2015

[4] Sakkab N Y 2009 *Connect &Develop Complements research & Develop at P&G* (Industrial Research Institute)

[5] What is open innovation, http://www.pgconnectdevelop.com/home/open_innovation.html, accessed 10 February 2015

[6] Information on P&G's needs, http://www.pgconnectdevelop.com/home/needs.html, accessed 10 February 2015

[7] Information on household care innovations, https://pgconnectdevelop.inovasuite.com/pg/, accessed 10 February 2015

[8] About submitting to P&G, http://www.pgconnectdevelop.com/home/frequently_asked_questions/about-submitting-to-pg.html, accessed 10 February 2015

[9] About partnering with P&G, http: //www.pgconnectdevelop.com/home/frequently_asked_questions/about-partnering-with-pg.html, accessed 10 February 2015

[10] Open innovation stories, http://www.pgconnectdevelop.com/home/stories.html, accessed 20 March 2015

[11] Corium International, Inc. Honored as P&G's External Business Partner of the Year http://www.coriumgroup.com/PreseRelease_HonoredPandGsExternalBusinessPartner.html, accessed 13 March 2015

[12] Connect + Develop Fact Sheet, http://www.pg.com/en_US/downloads/innovation/C_D_factsheet.pdf, accessed 10 March 2015

[13] OECD/Eurostat 2005 *Oslo Manual: Guidelines for Collecting and Interpreting Innovation Data, 3rd Edition* The Measurement of Scientific and Technological Activities (Paris: OECD Publishing)

[14] Annual Report 2014, http://www.pginvestor.com/interactive/lookandfeel/ 4004124/PG_Annual_Report_2014.pdf, accessed 20 March 2015

[15] Why innovation matters and how to make it work – Speech Transcript, https://www.pg.com/en_US/downloads/company/purpose_people/touching_lives_improving_life.pdf, accessed 15 March 2015

[16] Nambisan S "Designing virtual customer environments for new product development: Toward a theory" *The academy of Management Review* **27(3)** pp 392-413

Coastline planning and management through digital mapping systems

M Hysenaj[1]
[1]University of Shkodër, Albania

E-mail: mhysenaj@unishk.edu.al

Abstract. Albania is a country with a coastline of 316 km. The potentiality offered turns into a determinant factor for the Albanian economy. However specific issues need a solution. One of them remains the shoreline pollution. It affects mostly foreign visitors, also local population which recently tends to avoid attending these areas, instead they frequent foreign places. The importance of GIS technology in the water sector is undisputed. This paper will present a full set of digital maps representing a complete picture of the Albanian shoreline situation. The entire coastline is divided into the major frequented areas with a spatial extension based mainly on district level.

1. Introduction

Many coastal ecosystems still remain adversely affected by current land-use practices and the ever-growing demand for coastal lands from urbanization. And unfortunately, the recent economic recession is expected to provide only temporary respite in the trend of coastal land take [1]. In both developed and developing countries, the coastal zone is likely to undergo the most profound change in the near future [2]. As already known the two primary factors for coastline change remains natural process and human intervention. Natural factors remain great forces that are hard to manage, while human processes belong to the classified category which can be totally controlled and well managed.

The study involves measures on 70 check point over the Albanian coastline [3] with the aim of identifying problematic zones with high percentage of pollution values. The results reveal 36 problematic areas with high risk of bacteria, translated in other words half of the coastline is unsuitable to be frequented by tourists or native population.

The main reason that induces the verified situation is the uncontrolled urbanization phenomenon near these areas followed by unplanned investments, eager to benefit at the expense of coastal space leading to the degradation of natural resources which will close development options.

It is generally expected that Remote Sensing and GIS techniques and instruments will be of increasing importance in coastline and coastal zone management [4]. Through GIS technologies we intend to estimate the coastline situation creating a full picture overview of the current condition. This knowledge will enable coastal planners, inhabitants and decision makers to better understand and evaluate the current situation with the goal of improving the list of necessary steps to be undertaken producing a result to the benefit of the Albanian coastline.

2. Materials and methods

Coliform bacteria have been used to evaluate the general quality of water. Two other groups of bacteria that are present in feces are: fecal streptococci (FS) and Clostridium (FC) [5]. Our

Content from this work may be used under the terms of the Creative Commons Attribution 3.0 licence. Any further distribution of this work must maintain attribution to the author(s) and the title of the work, journal citation and DOI.
Published under licence by IOP Publishing Ltd

investigation consists by taking in consideration the values obtained by the Public Health Institute which has set a limit rate of 100. The paper reveals the current situation of the whole Albanian coastline. A set of digital maps have been structured cutting the coastline into specific areas according to the most visited beaches and their geographical position. Small beaches lying on close spatial areas have been grouped for study reason. As source map we have used the digital map of Albania under the GCS_WGS_1984 projection will a scale of 1:400.000.

The research goal is to define the most polluted areas and their risk factor. The results will show us the problematic of the studied area expressed through digital maps. From a visual impact the population awareness could increase and the responsible authorities' management process could improve.

As we notice from the digital maps (figure 1, figure 2) the northern coast which covers two primary beaches Velipoja and Shëngjin generally is considered an acceptable area to be frequented by the population. From the 12 measured check points only Shëngjin beach contain a critical area that exceeds the allowed limit (table 1, 2).

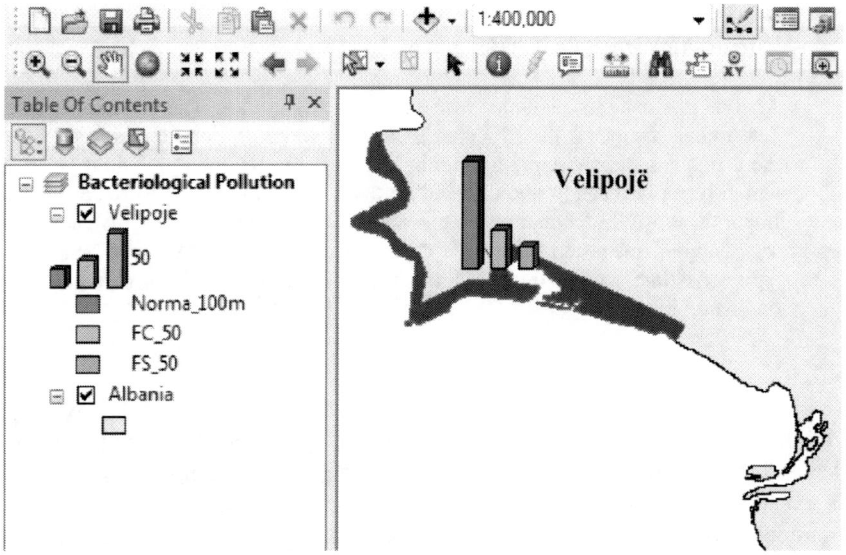

Figure 1. Albania Coastal Pollution, Study zone: Velipojë, Software: Package ArcGIS, ArcMap 10.
Source: Ministry of Environment, Forestry and Water Administration.

Table 1.

Ch. P.	1	2	3	4	5	6	7	Average	Norma/100 ml
FC-50	40	35	33	29	48	34	33	36	100
FS-50	23	24	20	14	27	22	20	21	100

If we continue geographically down the territory basically in the central zone of the Albanian coastline we face a problematic situation. The values generated induce us in categorizing Durrës and many parts of Kavaja beach among the dangerous areas to be strictly avoided in any period of the year. As we can obviously see from the map and the relevant table (figure 3), most of the values not only exceed the defined normative but boost by multiple times the limit of 100. All the 21 check points that belong to Durrës beach are considered inappropriate to be visited (table 3). As for Kavaja shoreline the situation seems to be more optimistic in half of the coast but still the other half needs urgent intervention (table 4). According to these values we can rank these two beaches as the most polluted areas in the whole Albanian coastline.

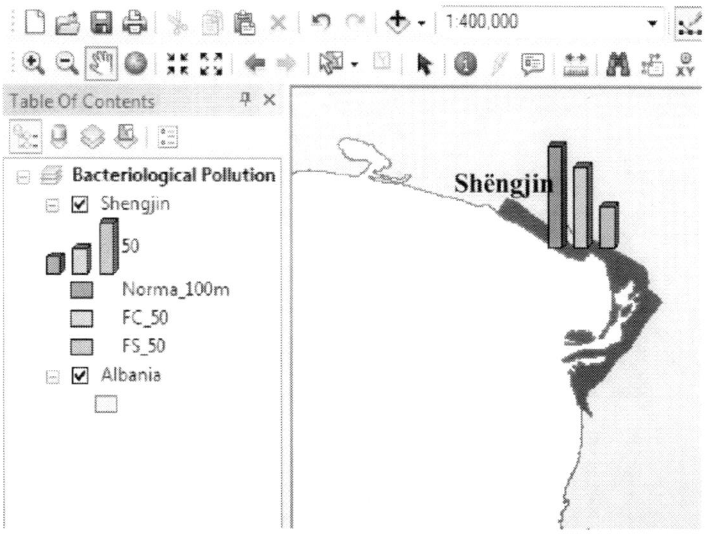

Figure2. Albania Coastal Pollution, Study zone: Shëngjin, Software: Package ArcGIS, ArcMap 10
Source: Ministry of Environment, Forestry and Water Administration.

Table 2.

Ch. P.	8	9	10	11	12	Average	Norma/100 ml
FC-50	85	128	56	83	46	80	100
FS-50	45	87	20	36	19	41	100

Moving into the south the situation seems to change gradually. Vlora beach faces areas that are adaptable to be frequented with parameters below the allowed normative (figure 4). From 11 check points, 7 of them seem to have good quality water (table 5). Still if we notice carefully there is especially one zone that exceeds the limit by 15 times and three more by 3 or 4 times.

Not by chance Dhërmi, Himarë and Borsh are the most frequented beaches during summer and in general. According to the map and the corresponding values (figure 5) the average of the measuring parameters (table 6), FC = 9.8 and FS = 7.5, ranks them among the most required places by native and especially foreign tourists. They remain the referring point for a nice and pleasant vacation period.

The southern beach of Saranda closes the Albanian coastline. Based on the map (figure 6) and the relevant table of measured check points (table 7) we conclude that the situation doesn't seem to improve joining the pessimistic overview created by the overall panorama of the whole country shoreline.

Regardless of the influencing factors, whether natural or human, tougher policy must be followed. Local governments must collect and maintain updated sets of digital maps concerning coastline issues. In addition, a GIS approach to coastal management provides a process to bring together groups that cut across professional disciplines to work together toward a common goal and more effectively communicate information about coastal management issues to decision-makers [6].

The mapping of the coastline situation is a mandatory input of vital importance for further decision-making progress [7]. We insist in leading further digital mapping projects to investigate environmental issues for the benefit of population welfare. The application of GIS and remote sensing techniques to monitor coastline change should be carried out jointly with routine observatory work in order to improve our understanding of the environmental problematic [8].

IOP Conf. Series: Materials Science and Engineering **95** (2015) 012151 doi:10.1088/1757-899X/95/1/012151

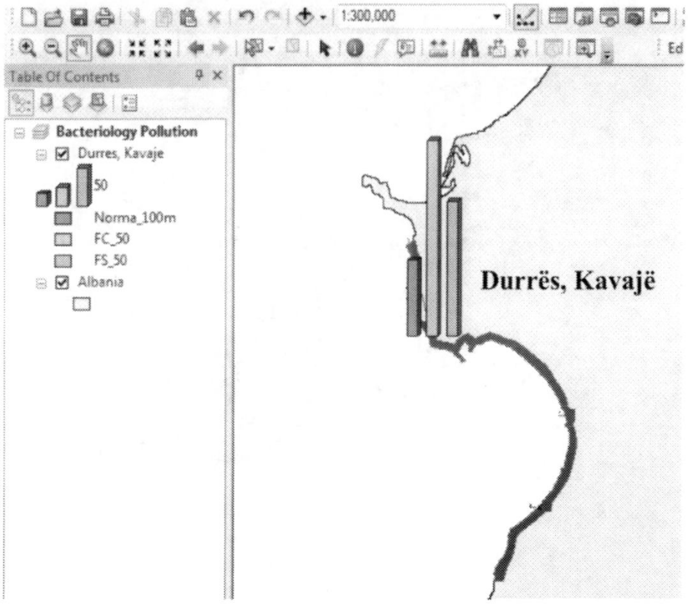

Figure 3. Albania Coastal Pollution, Study zone: Durrës, Kavajë
Source: Ministry of Environment, Forestry and Water Administratio.

Table 3.

FC-50, FS-50 values

Ch. P.	FC-50	FS-50
1	1073	669
2	906	687
3	679	435
4	640	591
5	455	424
6	281	206
7	247	194
8	199	161
9	186	142
10	171	135
11	162	124
12	162	107
13	154	118
14	151	104
15	142	118
16	140	93
17	130	95
18	121	92
19	118	88
20	115	72
21	114	81
Average	271	226
Norma 100 ml	100	100

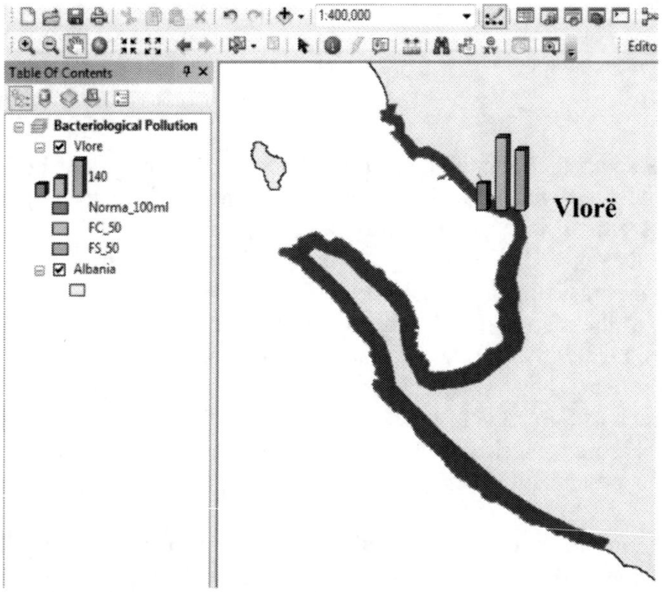

Figure4. Albania Coastal Pollution, Study zone: Vlorë; Software: Package ArcGIS,
ArcMap 10; Source: Ministry of Environment, Forestry and Water Administration.

Table 4.

Ch. P.	22	23	24	25	26	27	28	29	30	31
FC-50	54	91	95	112	112	159	318	325	404	460
FS-50	20	38	39	56	50	56	256	177	296	326

Table 5.

Ch. P.	1	2	3	4	5	6	7	8	9	10	11	Average	Norma/100 ml
FC-50	355	437	1519	466	41	33	28	23	24	38	23	271	100
FS-50	317	318	1304	429	23	23	16	12	14	19	14	226	100

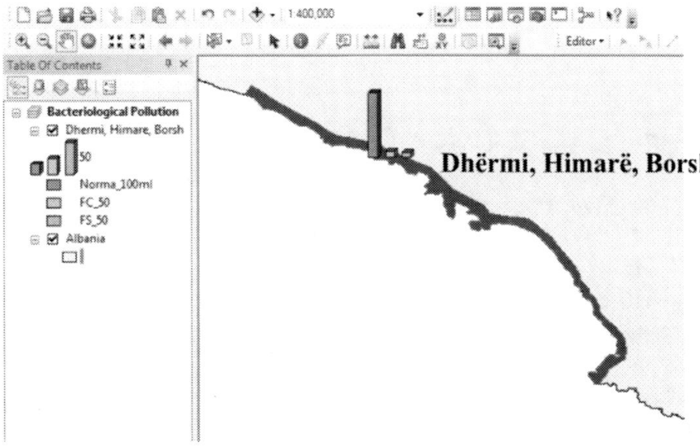

Figure 5. Albania Coastal Pollution, Study zone: Dhërmi, Himarë, Borsh; Software: Package ArcGIS ArcMap 10; Source: Ministry of Environment, Forestry and Water Administration.

Table 6.

Ch. P.	1	2	3	4	5	6	7	8	9	10	Average	Norma/100 ml
FC-50	9.1	11	12.3	13	8	5	9	8	11	12	9.8	100
FS-50	6.5	7.1	9.5	18	6	3	7	5	7	7	7.5	100

Table 7.

Ch. P.	1	2	3	4	5	6	Average	Norma/100 ml
FC-50	29	30	124	146	69	670	178	100
FS-50	16	19	103	121	58	580	150	100

3. Conclusions

The coastline issue in Albania remains a problematic topic due to its particular importance in the economical and management sector. Due to remote sensing we can easily gain access to spatial data which integrated into the proper software can produce important outcome for the benefit of coastline planners and decision-makers.

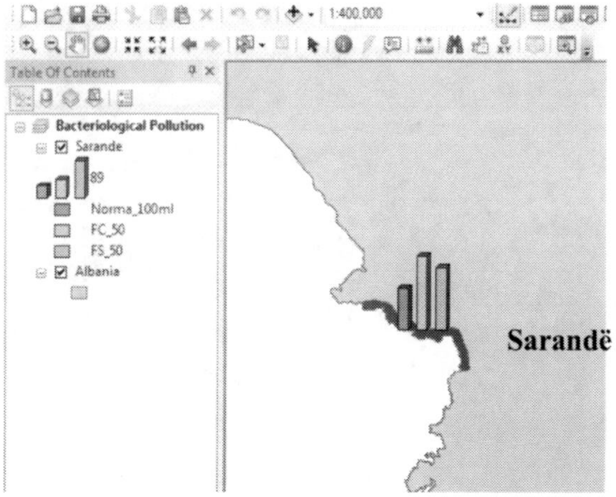

Figure 6. Albania Coastal Pollution, Study zone: Sarandë; Software: Package ArcGIS,
ArcMap 10; Source: Ministry of Environment, Forestry and Water Administration.

The current shoreline situation seems to be pessimistic leading to the importance of urgent intervention from responsible authorities. We induce in exploiting maximally digital mapping as the right solution for issues that are spatially connected.

Analyzing data from a visual perspective turns to be more productive than just rough data. Digital maps show us that half of the coastline suffers from high level of pollution which makes them inappropriate for any kind of exploitation. The other half still need correction in many areas which are considered somewhat appropriate. If some of them are still under pollution limits, it does not come because of their careful management, but simply because of non massive frequentation [9].

The paper outline the importance of data management where digital maps tend to induce both categories, coastline exploiters including native inhabitants and tourist, also responsible entities including local authorities and government institutions and organizations.

References

[1] Czerniak P, Pickaver A, Ferreira M, Steijn R and Devilee E 2011 *Integrated Coastal Zone management information and technology in ICZM practices around Europe*

[2] Niya A K 2013 Shoreline Change Mapping Using Remote Sensing and GIS *Int. J. Remote Sensing Applications* **3(3)**

[3] *Source data for digital maps* (Ministry of Environment Forestry and Water Administration)

[4] van Heuvel T and Hillen R *Coastline management with GIS in the Netherlands* (National Institute for Coastal and Marine Management/RIKZ)

[5] Oram B *J. Bacteria in the Environment and Drinking Water* PG

[6] Hart D A 2000 *Acquisition and Integration of Digital Parcel Mapping to Support Coastal Management Along the Lake Michigan Coast of Wisconsin*

[7] Hysenaj M 2011 *Geographical Information Systems*

[8] Yanli T *The Application of GIS and RS for Coastline Change Detection and Risk Assesment to Enhanced Sea Level Rise* ITC

[9] Quality of beaches, Fecal Coliforms and Intestinal Enterococcus Indicators 2011 (Open Data Albania)

Digital development of products with NX9 for academical areas

A M Goanta[1]

[1]"Dunarea de Jos" University of Galati, Engineering and Agronomy Faculty of Braila, Research Centre for Mechanics of the Machines and Technological Equipments, Calea Calarasilor Street, 810017, Braila, Romania

E-mail: goanta_a_m@yahoo.com

Abstract. International competitiveness forced the manufacturing enterprises to look for new ways to accelerate the development of digital products through innovation, global alliances and strategic partnerships. In an environment of global research and development of distributed geographically, all members of the joint teams made up of companies and universities need to access updated and accurate information about products created by any of the type employed, student, teacher. Current design processes involve more complex products consisting of elements of design created by multiple teams, disciplines and suppliers using independent CAD systems. Even when using a 3D CAD mature technology, many companies fail to significantly reduce losses in the process, improve product quality or product type to ensure successful innovations to market arouse interest. These challenges require a radical rethinking of the business model, which belongs to the field of design, which must be based on digital development of products based on integrated files. Through this work, the author has proposed to provide both synthesis and transformations brought news of the integrated NX [1, 2, 3] from Siemens PLM Software 9, following a news results detailed documentary study, and personal results obtained by applying the same version, the digital and integrated development of a product type device test beams. Based on educational license received for NX 9 was made a detailed study of the innovations made by this release, and the application of some of them went to graphical modelling and getting all the documentation of a test device bearing beams. Also, were synthesized in terms of methodology, the steps to take to obtain graphical documentation. The results consist of: 3D models of all parts and assembly 3D model of the three-dimensional constraints of all component parts and not least respectively all drawings and assembly drawing. The most important consequence of the paper is the obtaining of integrated files that can be subjected to further analysis type CAE / CAM / PDM software components by the same company. Additional advantages related files by the synthesis of integrated CAD / CAE / CAM / PDM.

1. Digital Integration and PLM Aspects – "Product Lifecycle Management"

Digital integration is the stage a community tends to in order to enter the global network [4]. When talking about digital integration and development we refer to many aspects of the daily life starting with electronic commerce until getting digital products needed in any field including health and education. Electronic commerce and applications in information and communication technology (ICT) as components of the digital development of a society as a whole have become the engines of the economic development, significantly influencing the world we live in. It should be noted that from this

Content from this work may be used under the terms of the Creative Commons Attribution 3.0 licence. Any further distribution of this work must maintain attribution to the author(s) and the title of the work, journal citation and DOI.
Published under licence by IOP Publishing Ltd

impact on the daily life generated by technological breakthroughs the developed countries are the first to benefit. The increasing discrepancy between developed and poor countries, especially visible economically and socially, is worrying also from the perspective of information and communications technology. The rapid expanding the Internet and "digitization" of society among developed countries creates in the less developed countries real problems of integration according to the technological changes. For the majority of less developed countries or there has been or there is an optimum moment to join global networks of information able to propel them towards welfare However, it is not important only to be ready on time; without sustained and consistent effort, developing countries will find it much more difficult to bridge the already existing gap between them and the rest of the rich world and the use of new technologies will remain a phenomenon characteristic only to economically prosperous countries. Digital integration creates new market opportunities for companies and individuals, removes barriers that traditionally suppresses the flow of information and allows movement of goods to and from developing countries, promoting efficiency as the ultimate goal. Students can learn more about the world and about themselves using the global network. Business men can learn about new opportunities available on the market and management techniques likely to improve the company performance. Central and local government offices can provide better public services. Individuals can communicate with their friends and family and can inform virtually about any area of interest. Digital integration can offer developing countries new ways to improve their social, economic and political life. Ideas and opportunities for a better life are becoming easier to access as information and communication technologies are becoming cheaper and cheaper. All multinational companies wishing to expand business area and optimize profit have increasingly sought to focus more on less developed countries too.

Responding to the general industry trends dictated by environmental compliance, management complexity caused by globalization, shortening the time for launching a product on the market, the increasing levels of personalization and innovation of the designed products, Siemens Industry Software [5] provides an integrated portfolio of software solutions for product development, fundamentally oriented towards ensuring a high level of efficiency in production, increasing profitability and meeting the expectations of the product delivery. The main challenges industry faces today are related to:

- managing the growing complexity of design and production;
- improving both processes and outcomes in relation to changes in terms of complexity and diversity of product performance, quality and innovativeness;
- reducing costs and optimizing resources without compromising quality;
- reducing product delivery times.

Among the most important results obtained by companies using SIEMENS design products, we mention:

- Issues identified and resolved more quickly;
- Greater production flexibility;
- Less waste, more products;
- Investment decisions validated;
- 80% fewer problems after production of machinery or equipment designed;
- design modification reduced from 90% to 60%;
- Simplified production processes;
- Reduced product development time;
- Improved collaboration between departments and external suppliers.

One of the key phases of product development is to produce prototypes [6]. This phase follows the design and construction phase after the technological processing of the production process, as final stage in product development and as testing for mass production preparation. For this reason, the information appearing during production of the prototype is one of the key factors in the decision making with large series of products or giving it up. The pace of developing new products today and the global market demands require companies to adapt quickly to new conditions and constant adoption of new technologies. In recent years, the rapid prototyping technology, supported by computer-aided design (CAD) has become an inevitable engineering tool. Reduction of product development because of innovation and constant fight on the market and the increasing complexity of products and the need of real models for a qualitative assessment such as shape, size and control functions are all factors that influenced the rapid development of this technology and its implementation in many areas of engineering. Studies on industry made by analysts suggest [7] that the vast majority of products which actually represents 70% of company revenue today will be withdrawn within a period of 5 years. Most concepts do not reach the market, and among those available on the market between 50 and 70% fail. Thus, new product development has become essential for the existence and success of any company. To face market pressures every company should create a platform for continuous innovation. This platform is an integrated business approach based on information and consists of people, processes / practices and technology known in industry as Product Lifecycle Management (PLM) [8]. For example, Teamcenter produced by SIEMENS manages complete information about this product from marketing, production concept, design (CAD), analysis (CAE), processing (CAM), digital manufacturing (DM), sale, maintenance, service, to withdrawal and recycling. Figure 1 illustrates the influence of PLM concept influence on the incubated stages of business, development, maturity, decline and end of life of the product. It is clearly seen that the time to launch the product on the market is much smaller when applying PLM concept, which leads to lower launch costs in manufacturing. Also, the periods of development, maturity and decline, profit is maximized and the lifetime of the product is growing significantly.

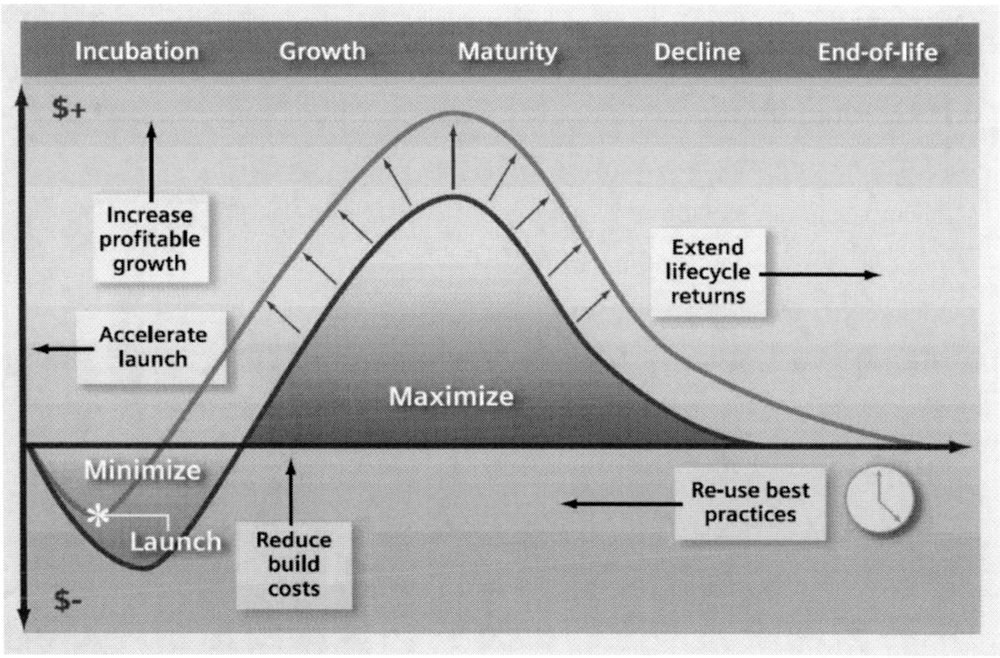

Figure 1. Long-term business benefit of PLM concept.

2. Present day evolution and performance of NX SIEMENS product

The software package NX currently under SIEMENS property, appeared in 1969 and was originally known as Unigraphics, then much later, in 2002 acquired the name of NX. Since 2007, the NX has introduced the Synchronous technology that comes to revolutionize the 3D modelling work because allows designer to act directly using the mouse over certain parts of the generated body and easily modify a shape, which has not been seen up to that moment. In the NX 9 version 9, the synchronous technology was extended to the 2D changes in that it allows operations of „shooting" the edges, under

IOP Conf. Series: Materials Science and Engineering **95** (2015) 012152 doi:10.1088/1757-899X/95/1/012152

the name of „Synchronous Sketch". The supremacy of all these tools is because they can be used in the same file in a single interface. The key benefits of this integrated software are: immediately propagate changes into all components of the project, there are no transfer files because all the information is found in a single file, it is possible to purchase and subsequently use complementary modules, with the development of the company without affecting interpretation unit of the designed project. We should not forget that the Parasolid modelling is based on the Parasolid nucleus which is very powerful and flexible, being used even by competitors. In the process of reverse engineering, NX takes the points cloud resulted from a 3D scan, it evaluates it, possibly fixes it and turns it on a surface or solid, which subsequently transform with editing commands in order to obtain new complex geometries. NX 9 allows the opening [9] of both IGES Step transfer files and other files created by other CAD systems. Basically Siemens NX became a fully integrated solution for the conception and design of complex systems with specialized components such as those below, for the following industry sectors:

- Automotive industry - reverse engineering, concurrent engineering, multidisciplinary design (mechanical and electrical), CAE analysis (fluid flow, mechanical strength and impact analysis) complex assemblies, realistic visualization, etc.
- Aero and marine industry - parts and complex assemblies, CAE analysis (fluid flow) and multi-axis machining.
- Consumer goods industry - areas with unique design, complex molds, visualization and rendering.
- Machinery and equipment manufacture industry- complex assemblies and multidisciplinary engineering

This integrated design solution with history and experience of over 40 years offered the highest level tools for modelling and analyzing multidisciplinary specialized industrial fields as mentioned above.

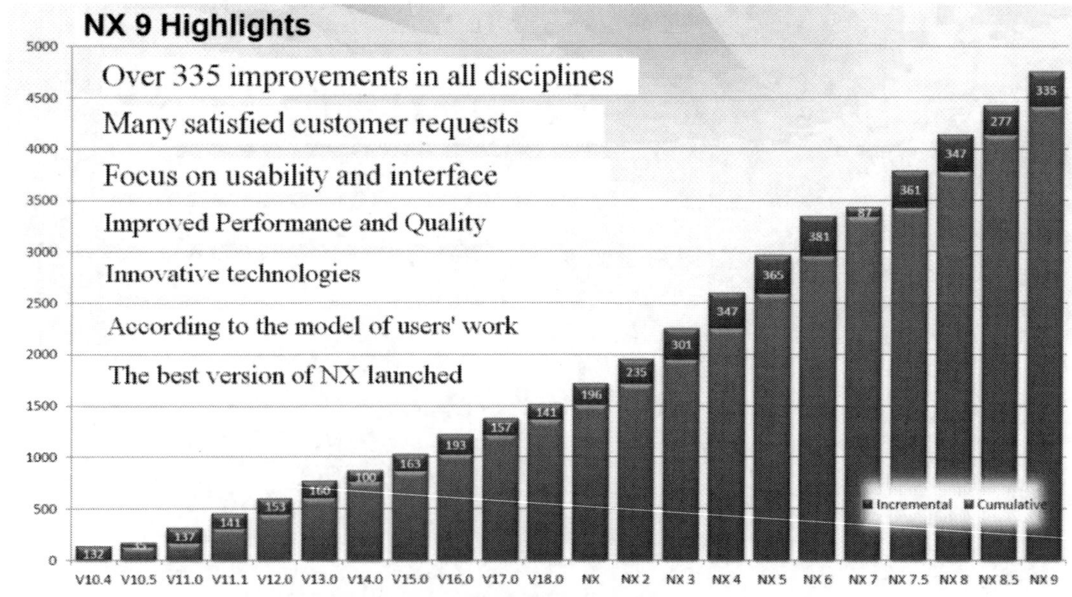

Figure 2. NX evolution.

NX 9 version has an improved module of "Distributed Memory Parallel" (DMP) and a new method of response at processor core level, significantly reducing computation times of the operation type input / output within the photorealistic visualization and analysis with finite elements.

3. 3D and 2D results

To obtain overall three-dimensional results, each item was achieved three-dimensionally starting from parameterized sketches which were applied commands for generating 3D features. Steps to be taken for modelling the items are [10]:

- Choosing the drawing plan by command "Sketch" and setting the position and orientation of the coordinate system
- Drawing a sketch by controls like „Profile", Line", „Arc", „Circle", „Rectangle".
- Editing the sketch by controls like „Fillet", „Chamfer", „Offset Curve", „Pattern Curve", „Mirror Curve", „Project Curve", „Quick Trim", „Quick Extend", „Make Corner".
- Setting 2D constrains 2D which keeps the form of the sketch as a geometric family.
- Insert all required sizes to complete the definition of the sketch.
- Based on drawings it will be generated 3D features to which Boolean operations of "Unite", "Subtract", "Intersect" type are applied
- Attaching materials of well defines physical – mechanical properties.
- Opening an assembly file type and insertion, based on 3D constraints, of each item in order to generate a file that contains the 3D model of the entire assembly.

As a case study it was considered the generation of a test device for load-bearing beams for which all parts were modelled geometrically and all the execution drawings were obtained, the overall drawing. Figure 3 and figure 4 illustrate the hook subassembly and whole device less the anchoring cable and fastening hook.

Regarding the results of 2D it should be noted that the following steps must be run, namely:

- Switching to mode "Drafting".
- Defining the paper size and representation scale.
- Establishment of European or American type of projection
- Imposing the baseline projection and getting other screenings.
- Technical sizing of the drawing obtained and adding tolerances of form and position, and size tolerances respectively.

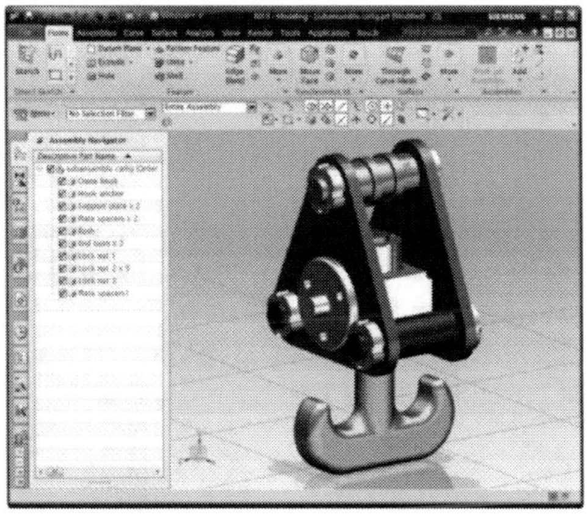

Figure 3. Hook subassembly.

Figure 4. Device assembly.

Figure 5 and figure 6 shows the dialog box where it is set the paper size and type and the projection systems, i.e. the 2D result obtained.

4. Conclusions

Mechanical design capabilities of NX 9 are unmatched in terms of computing power, versatility, flexibility and productivity digital product development environment usually provide. The latest version of NX 9 includes new features and technological innovations that provide increased flexibility and product development up to 5 times more productivity, removing old paradigms related to flexibility and productivity when working with 2D data and large assemblies.

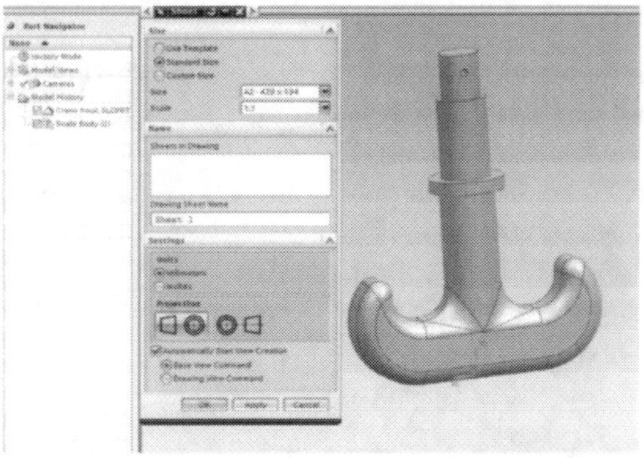

Figure 5. Setting box for the 2D drawing.

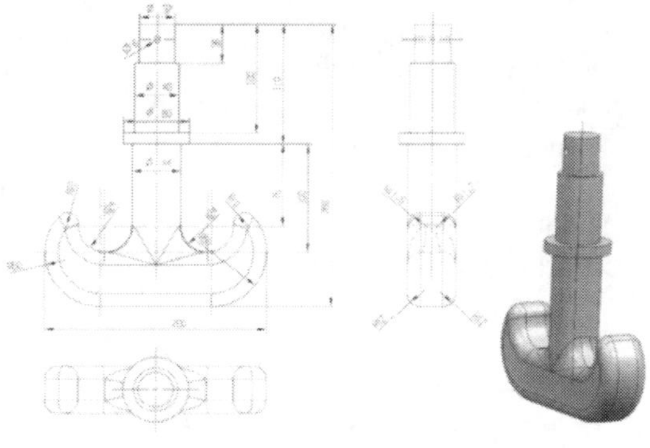

Figure 6. Execution drawing for the hook.

NX 9 offers the widest range of CAD / CAM / CAE integrated applications (figure 2) and fully associative in industry. NX is found in the full range of development processes related to product design, manufacturing and simulation that enable companies to encourage the use of best practices, by capturing and reusing the knowledge on product and process.

The case study demonstrates the device beams tried all these additional capabilities of NX 9 software. Ribbon interface of version NX9 helps users find, understand and use directly and efficiently controls through a small number of clicks without having to resort systematically to the Help page. The ribbon previews very well most common commands and in addition, unlike other similar interface, allows complete customization. Also to be mentioned as a novelty is the functionality "Multi-Touch" which allows the use of tablets and touch type Display sites and gestures interpretation of mouse or stylus type. The benefits of integrated files are as follows:

- using a single database and a common software infrastructure to provide information in real time, allowing managers to take better decisions;
- eliminating transfer files that can lead to errors;
- increase the quality and speed of data access is provided by single database covering all functional areas of the company;
- integrated data provides better efficiency in terms of decision support and assistance to users.

References

[1] Leu C M, Thomas A and Kolan K 2014 *NX 9.0 for Engineering Design* (Missouri, USA: University of Science&Technology)

[2] Shih R 2014 *Parametric Modelling with NX 9* (Mission, Kansas, USA: SDC Publications)

[3] Ticko S 2014 *NX 9.0 for Designers* (Hammond, Indiana USA: Pardue University Calumet)

[4] Information on: http://cyber.law.harvard.edu/readinessguide/Readiness-translation %20 Romanian.doc. Accessed: 12/10/2014

[5] Information on: http://economie.hotnews.ro/stiri-companii-16862301-eveniment-dezvoltarea-digitala-produselor-din-industria-auto.htm Accessed: 14/09/2014

[6] Information on: http://www.link-academy.com/Academy-Digital-Prototyping_992_2_103_237 Accessed: 23/09/2014

[7] Information on: http://www.marketwatch.ro/articol/7168/Cum_dezvoltam_noi_produse_de_s ucces/ Accessed: 15/11/2014

[8] Manole G, Oprea E and Iosip M 2010 *Conceptia si proiectarea produselor* (Bucharest: Publisher Qual Media)

[9] Goanta A M and Nedelcut F 2014 NX Siemens - Integrated Solution of PLM Mechanical Design *Ann. Eng. and Agronomy Faculty of Braila* **2** pp 15-18

[10] Goanta A M 2014 Siemens PLM Solution Applied to the Design of Agriculture Facilities and Equipment *Ann. Eng. and Agronomy Faculty of Braila* **1** pp 43-46

IOP Conf. Series: Materials Science and Engineering **95** (2015) 012153 doi:10.1088/1757-899X/95/1/012153

Application of the MIAS methodology in design of the data acquisition system for wastewater treatment plant

G Ćwikła[1], D Krenczyk[1], A Kampa[1] and G Golda[1]
[1]Silesian University of Technology, Faculty of Mechanical Engineering, Institute of Engineering Processes Automation and Integrated Manufacturing Systems, Konarskiego 18A, Gliwice, Poland

E-mail: grzegorz.cwikla@polsl.pl

Abstract. This paper presents application of MIAS (Manufacturing Information Acquisition System) methodology to develop customized data acquisition system supporting management of the Central Wastewater Treatment Plant (CWWTP) in Gliwice, Poland, being example of production systems leading continuous flow, automated production processes. Access to current data on the state of production system is a key to efficient management of a company, allowing fast reaction or even anticipation of future problems with equipment and reduction of waste. Overview of both analysis and synthesis of organisational solutions, data sources, data pre-processing and communication interfaces, realised according to proposed MIAS methodology, had been presented. The stage of analysis covered i.e.: organisational structure of the company, IT systems used in the company, specifics of technological processes, machines and equipment, structure of control systems, assignments of crew members, materials used in the technological processes. This paper also presents results of the stage of synthesis of technical and organisational solutions of MIAS for CWWTP, including proposed solutions covering MIAS architecture and connections with other IT systems, data sources in production system that are currently available and newly created, data pre-processing procedures, and necessary communication interfaces.

1. Introduction

Contemporary market challenges mostly connected with worldwide competition between manufacturers calls for new methods of improving the operation and management of companies. Everyone agrees that efficient management of a company is currently possible thanks to the software supporting almost every area of a company operation, including resources planning [1], human resources management, accounting, relations with customers, production preparation [2] (development of product's design [3] and technological processes planning [4]), advanced methods of production scheduling [5] (including maintenance operations [6]), down to the systems supporting automated control of technological processes. Competition between software producers results in development of complex software platforms, integrating many aspects of a shopfloor operation and management, there are new classes of software arising, fulfilling the needs of different branches of industry. Almost every IT system (ERP, MES, etc.) supporting management of different areas should be provided with data on current state of production system [7].

In that context, acquisition of data on the state of production system is crucial in the management of company, because access to real-time information describing the state of production system (including completion of production orders, state of crew, machinery and equipment, OEE, flow of material and semi-finished products, etc.) is the basis for decision making in every company. Detailed

Content from this work may be used under the terms of the Creative Commons Attribution 3.0 licence. Any further distribution of this work must maintain attribution to the author(s) and the title of the work, journal citation and DOI.
Published under licence by IOP Publishing Ltd

issues of data acquisition and data sources, as well as fundamentals of the MIAS (Manufacturing Information Acquisition System) methodology itself was described in prior publications [8], thus only main rules will be reminded here. Availability of data in any given company strongly depends on specific of its production processes (discrete, batch, continuous), branch of industry, level of automation and share of manual operations, etc. In modern company acquisition of data on the state of the production system should be carried-on automatically, mostly without involvement of workers. The MIAS methodology is an example of multi-stage, multi-level methodology, where process of development of data acquisition system, understood as a set of cooperating hardware and software solutions, is divided into 3 main stages: analysis of various aspects of the company in which MIAS is applied, synthesis of technical and organisational solutions, and evaluation of proposed solutions. There are 4 proposed levels of analysis: level of company as a system, level of data sources, level of data pre-processing, and level of communication [8].

This paper presents example of application of the MIAS methodology in the company leading continuous, automated production processes – Central Wastewater Treatment Plant (CWWTP) in Gliwice, Poland. Overview of both analysis and synthesis of organisational solutions, data sources, data pre-processing and communication interfaces had been presented.

The stage of analysis covered i.e.: organisational structure of the company, IT systems used in the company, specifics of technological processes, machines and equipment, structure of control systems, assignments of crew members, materials used in the technological processes. During the stage of synthesis of technical and organisational solutions of MIAS for CWWTP, the solutions covering MIAS architecture and connections with other IT systems, data sources in production system that are currently available and newly created, data pre-processing procedures, and necessary communication interfaces had been proposed. The main reason to create information acquisition system for CWWTP was to provide better information on usage of equipment and assignments of crew members in order to achieve better management of production system, as well as to create complete database containing data on operation of wastewater treatment plant, necessary for fulfilling law requirements concerning compulsory information on the chemical composition of intake wastewater and outtake water cleaned during mechanical and biological treatment processes.

2. Dependence of data acquisition possibilities on a level of technological processes automation

Characteristics of a company and type of production processes strongly influences availability of data on the state of production system and realized tasks. Types of production classification can be based on products variety or volume and branch of industry. Production systems/processes can be broadly classified as continuous production systems (process/flow production, mass production) or intermittent production systems (batch production systems, job-shop production systems, project production systems).

Despite other characteristics of a production system, level of automation of technological processes is one of the most important factors in context of data acquisition [9]. This is caused by the fact that automatic process control requires the installation of hardware (PLCs, sensors and measuring devices) allowing control over production processes. Selected data obtained from automatic control systems can be extracted and reused for tasks that are not directly connected with the control of the processes. It can be stated that presence of industrial control systems can guarantee easy access to information on the state of manufacturing system (automated data acquisition). Data obtained from automated production system often do not carry all the information needed because usually not all activities in the factory are automated (i.e. maintenance).

In the production systems characterized by a predominance of tasks performed manually the basis of information retrieval is the method called manual data acquisition (mainly direct, verbal communication between workers and managers, sometimes formalised - forms or reports).

Requirements of IT systems supporting management (ERP, MES) calls for more reliable and faster solutions. This is a reason to define semi-automatic (assisted-manual) acquisition method, in which technologies like automatic identification systems (barcodes, RFID, machine vision systems), mobile or stationary data collectors, GPS, and other methods (3D scanning, photogrammetry, sound and voice analysis, triangulation) can be used. Semi-automatic data acquisition method can provide us with data, that cannot be collected from automatic data sources. All devices and software responsible for data acquisition should constitute integrated data acquisition system, combining data sources in the shop floor, middle-layer sub-systems responsible for the pre-processing and reduction of the amount of data, as well as communication interfaces and industrial database responsible for data archiving, aimed at providing information for IT systems supporting management.

3. Data acquisition in continuous flow processes production system

The application of MIAS methodology to create data acquisition system for a company, in which continuous flow production processes are main part of operation, is presented on the example of Central Wastewater Treatment Plant (CWWTP) in Gliwice, Poland.

CWWTP is modern facility, completed in 2005, main technological processes of wastewater treatment are automated, it receives wastewater from about 200,000 population city and cleans it in continuous, relatively slowly pacing biotechnological flow process, that cannot be interrupted. Characteristics (operation and structure of CWWTP) has been analysed according to MIAS methodology. CWWTP is a part of the Water and Sewerage Company (WSC) in Gliwice, using the PerfectExpert/ERP as an IT system integrating management of the whole company.

However, there is an information gap between WSC and CWWTP, because there is no direct connection between WSC's ERP and CWWTP IT systems. Most of the data from CWWTP are exported in standard MS Office (doc, xls) documents and then entered into ERP terminals. Communication between WWTP and the rest of company (ERP) should be necessarily improved. There are no MES or other middle-layer IT systems used. Wastewater treatment plant is specific example – this is not a typical profit-oriented company, the main objective of CWWTP is to run technological processes in a way that meet the quality requirements of cleaned wastewater at minimal operating costs. These quality standards requires collecting and archiving of data on chemical composition (content of the particular substances) and quality of input sewage, sludge and treated wastewater.

3.1. Technological processes of wastewater treatment, sludge processing and co-generation

The CWWTP plant has been designed for the removal of organic matter and nitrogen by simultaneous denitrification and nitrification, and removal of phosphorus using biological dephosphatation and chemical precipitation with iron salts [10]. Technological scheme of CWWTP consist of two main branches (figure 1): wastewater treatment line (bar screens station, aerated vertical grit chambers, circular primary settling tanks, activated sludge tanks, pre-denitrification tank, anaerobic tank and aerobic-anoxic tank, biological reactors, etc.), and sludge processing line (gravity and mechanical thickeners, anaerobic digestion chambers, filter presses, etc.) [11]. Additional devices are responsible for processing gases (e.g. methane) produced in biotechnological processes. Gases are used to generate thermal and electrical energy.

3.2. Control system of CWWTP

Most of the technological devices are tanks of various shapes and sizes, open or closed, often equipped with a agitators and aerators. Devices are often large, and distributed over a vast area, that creates problem of long distance communication. The flow of wastewater and sludge between objects is possible through pipelines and forced by gravity or pumps. Valves and adjustable shunts allow control over the flow. Control system consists of PLCs, sensors and measurement systems (liquid levels, temperature, chemical composition, etc.) and frequency inverters, motors, electrovalves, etc.

The core of the control system consists of 8 Siemens S7-300 PLCs, located in separate areas nearby the group of controlled devices (figure 1). Each of PLCs has 2 fieldbus interfaces: close-range copper wired Profibus, responsible for control over devices in its areas, and long-range fiber-optic link, configured in closed-loop redundant topology, connecting all area's PLCs. One of the PLCs is used for controlling of HMI synoptic table in the control room, which is currently used as a secondary HMI device. Primary HMI system is based on the computer with HMI/SCADA PRO-2000 system, allowing supervisory control over technological wastewater treatment processes and data archiving.

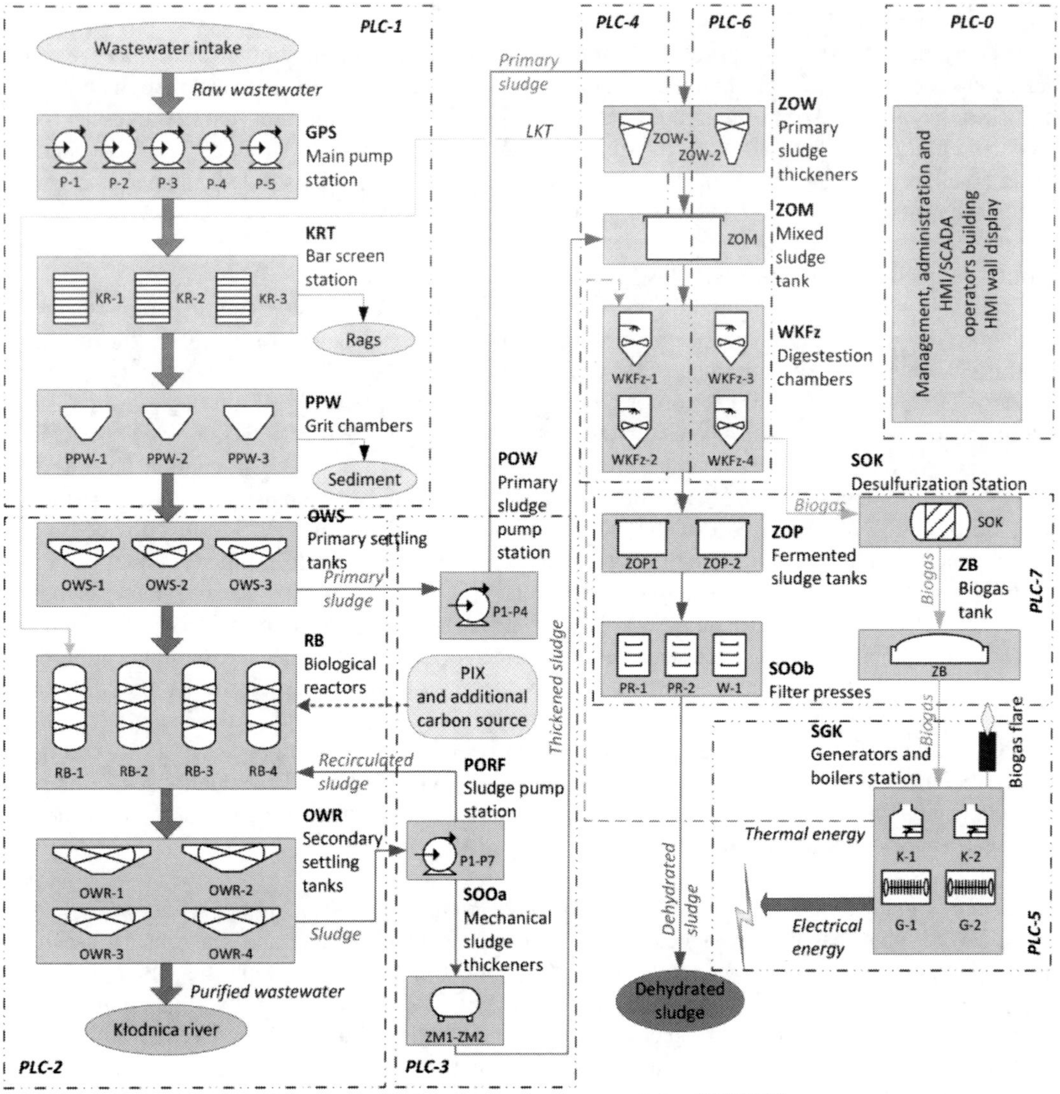

Figure 1. Technological scheme of the CWWTP.

Except signals from controlled devices (state of machinery, frequency inverters, softstarts, valves), control system processes data from measuring devices (e.g. wastewater/sludge flowmeters, wastewater temperature and pH sensors, chemical substances sensors – ammonia nitrogen, nitrate nitrogen, phosphorus – measured every few seconds or minutes, depending on the substance), allowing applying automatic control algorithms to obtain proper progress of wastewater mechanical, chemical and biological treatment processes, as well as providing data required because of reporting wastewater treatment results to the relevant institutions.

3.3. Objects and activities not included in existing data acquisition system

Control system of CWWTP has been designed only for control over wastewater treatment processes, not covering arising needs of data exchange with ERP system and only containing data on the state of main technological processes. Activity of workers (mainly technological processes operators and maintenance crew) is not reported and cannot be managed from ERP system, as well as handling and usage of materials, side products and means of local transport. Activity of crew (about 30 workers) is currently managed and reported manually. Workers are divided into 3 process operator teams (plus 3

HMI/SCADA operators), responsible for uninterrupted operation of CWWTP on 3 shifts, and one team of maintenance workers (electricians, mechanics). CWWTP is supervised by plant manager, chief process engineer and maintenance foreman.

Materials handling in CWWTP are relatively not complicated because material needs are low – only small amount of reagents, allowing better control over biochemical processes is necessary. However, in a modern production plant, usage of this substances, as well as handling of wastewater treatment residues, should be, without delays, reported to ERP system.

3.4. Proposed MIAS for CWWTP

As a result of MIAS analysis stage, it can be stated that newly designed data acquisition system for WWTP should cover information needs in three main areas:

- improved, direct ERP reporting of automated control system state, augmented with data from additional sensors, archived and processed in advanced event precognition software, allowing better control in case of weather changes, and rationalised maintenance,
- reporting of workers activities of to the ERP system – it should be based on newly installed local HMI panels in selected plant areas, and automatic identification systems (mobile data collectors with barcode scanners), supported by organisational solutions,
- reporting of materials and residues (waste) flow to the ERP system, based on automatic identification systems (mobile data collectors with barcode scanners) and SCADA-connected vehicle scales, supported by organisational solutions.

The core of the new data acquisition system (figure 2) is based on industrial database (Proficy Historian), equipped with collectors for available data sources. Architecture of industrial database divided into main database engine and data collector modules, allows purchase of necessary modules only. OPC collector is the most important in case of CWWTP, allowing access to data from SCADA and new data sources. SCADA system will remain the main source of data on state of technological processes, including data from newly created sources. Proficy Historian contains calculation collector, allowing development of scripts responsible for real-time data pre-processing, reduction and interpretation. Data is presented as pre-defined reports and in operator dashboards, and exported to ERP system. Additionally, advanced fault and problems anticipation software [12] (e.g. Proficy Troubleshooter), detecting coincidence of signals [13], leading to problems in CWWTP operation in the past (problems can appear in specific weather conditions, melting of snow causing wastewater temperature drop, or composition of wastewater) will be used. Full acquisition of data on activities of workers and materials handling will be possible thanks to additional HMI panels in specific plant areas (allowing reporting of faults and tasks completion), and portable data collectors. Labels will be attached to material containers or tanks (liquid materials), also workers will get personal barcode identifiers and typical tasks barcode charts.

4. Conclusions

Design of integrated data acquisition system for continuous flow automated production system, based on MIAS methodology, is easier than in case of non-automated production system, thanks to availability of control system and SCADA, providing access to existing data sources and allowing further expansion. Only activities of workers, flow of materials and transport system have to be included in expanded data acquisition system. Concept of MIAS for CWWTP, allowing reporting of this additional data has been developed. Newly designed data acquisition system should provide information on state of technological processes, machines, OEE, KPIs, crew activities, material flow, etc. MIAS is currently in phase of development and initial application, based on Proficy Historian, allowing easy data collection (from SCADA and new data sources through OPC server), pre-processing, presentation and interfacing with ERP. Data on activity of workers and material are collected using HMI panels and mobile data collectors, that should provide more detailed information on the state of CWWTP. Full evaluation of results of operation of MIAS will be possible after complete introduction, but first impressions shows improvement in the data exchange between CWWTP and the ERP system – there is less need for time-consuming manual data entry and availability of detailed data allows better planning and management of CWWTP.

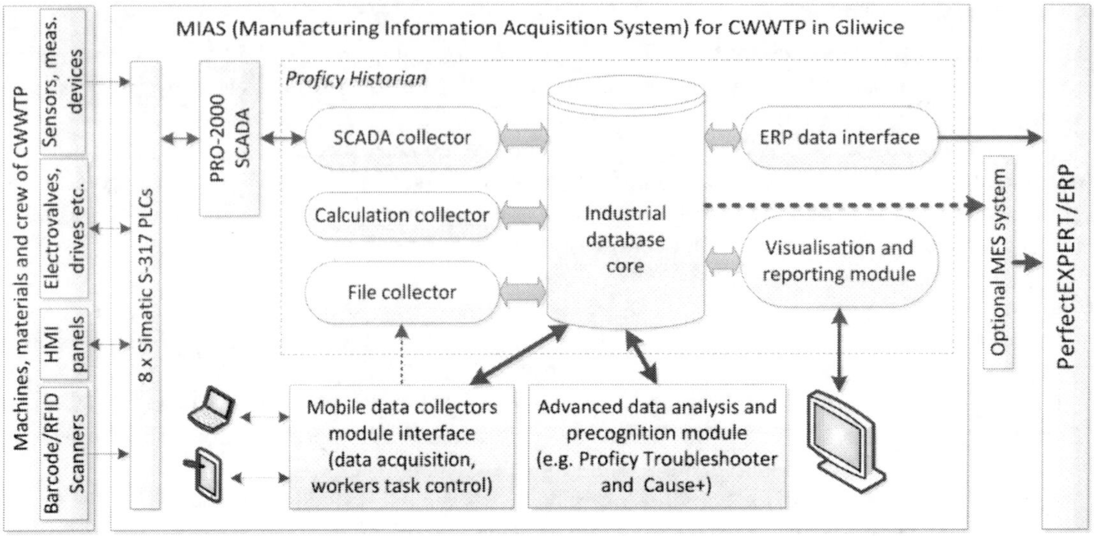

Figure 2. Proposed structure of data acquisition system for CWWTP.

References

[1] Krenczyk D and Skołud B 2011 Production Preparation and Order Verification Systems Integration Using Method Based on Data Transformation and Data *Mapping Lecture Notes in Artificial Intelligence, Hybrid Artificial Intelligent Systems*. **6679** pp 397-404

[2] Chwastyk P and Kołosowski M 2012 Integration CAD/CAPP/CAM Systems in Design Process of Innovative Products. Ann. DAAAM for 2012 & Proc. 23rd Int. DAAAM Symp. **(23)1** (Vienna, Austria)

[3] Gwiazda A, Sękala A and Monica Z. 2014 Integrated approach to the designing process of complex technical systems *Advanced Material Res.* **1036** pp 1023-1027

[4] Grabowik C, Kalinowski K, Kempa W and Paprocka I 2014 A survey on CAPP systems development methods *Advanced Materials Res.* **837** pp 387-392

[5] Paprocka I, Kempa W M, Grabowik C and Kalinowski K 2014 Predictive and reactive scheduling for a critical machine of a production system *Advanced Materials Res.* **1036** pp 909-914

[6] Paprocka I, Kempa W M, Kalinowski K and Grabowik C 2014 A production scheduling model with maintenance *Advanced Materials Res.* **1036** pp 885-890

[7] Sękala A and Dobrzańska-Danikiewicz A 2015 Possibilities of application of agent-based systems to support functioning of e-manufacturing environment *Mechatronics systems and materials* VI. edA V Valiulis *Solid State Phenomena* **220/221** pp 781-784

[8] Ćwikła G 2014 The methodology of development of the Manufacturing Information Acquisition System (MIAS) for production management *Applied Mechanics and Materials* **474** pp 27-32

[9] Ćwikła G 2014 Methods of manufacturing data acquisition for production management – a review Advanced Materials Res. **837** pp 618-623

[10] Ćwikła J and Konieczny K 2011 Treatment of Sludge Water With Reverse Osmosis *Environment Protection Eng.* **(37)4** pp 21-34

[11] Sochacki A, Kubiatowicz J, Surmacz-Górska J and Ćwikła J 2013 Plant-wide modelling and simulation using steady-state data: a case study of the Gliwice WWTP *Water Practice and Technology* **(8)1** pp 142-150

[12] Hetmanczyk M, Swider J and Wszołek G 2014 The systematized data structures oriented towards diagnosis and prediction *Advances in Intelligent Systems and Computing. Recent advances in automation, robotics and measuring techniques* **267** (Springer) pp 63-72

[13] Hetmańczyk M 2013 The multilevel prognosis system based on matrices and digraphs methods *Solid State Phenomena* **199** (Stafa-Zurich: Trans Tech Publications) pp 79-84

IOP Conf. Series: Materials Science and Engineering **95** (2015) 012154 doi:10.1088/1757-899X/95/1/012154

Fuzzy methods in decision making process - A particular approach in manufacturing systems

A M Coroiu[1]
[1]Babes-Bolyai University, Faculty of Mathematics and Computer-Science, Mihail Kogalniceanu Street, No. 1, Cluj-Napoca 400084, Cluj, Romania

E-mail: adrianac@cs.ubbcluj.ro

Abstract. We are living in a competitive environment, so we can see and understand that the most of manufacturing firms do the best in order to accomplish meeting demand, increasing quality, decreasing costs, and delivery rate. In present a stake point of interest is represented by the development of fuzzy technology. A particular approach for this is represented through the development of methodologies to enhance the ability to managed complicated optimization and decision making aspects involving non-probabilistic uncertainty with the reason to understand, development, and practice the fuzzy technologies to be used in fields such as economic, engineering, management, and societal problems. Fuzzy analysis represents a method for solving problems which are related to uncertainty and vagueness; it is used in multiple areas, such as engineering and has applications in decision making problems, planning and production. As a definition for decision making process we can use the next one: result of mental processes based upon cognitive process with a main role in the selection of a course of action among several alternatives. Every process of decision making can be represented as a result of a final choice and the output can be represented as an action or as an opinion of choice. Different types of uncertainty can be discovered in a wide variety of optimization and decision making problems related to planning and operation of power systems and subsystems. The mixture of the uncertainty factor in the construction of different models serves for increasing their adequacy and, as a result, the reliability and factual efficiency of decisions based on their analysis. Another definition of decision making process which came to illustrate and sustain the necessity of using fuzzy method: the decision making is an approach of choosing a strategy among many different projects in order to achieve some purposes and is formulated as three different models: high risk decision, usual risk decision and low risk decision - some specific formulas of fuzzy logic. The fuzzy set concepts has some certain parameterization features which are certain extensions of crisp and fuzzy relations respectively and have a rich potential for application to the decision making problems. The proposed approach from this paper presents advantages of fuzzy approach, in comparison with other paradigm and presents a particular way in which fuzzy logic can emerge in decision making process and planning process with implication, as a simulation, in manufacturing - involved in measuring performance of advanced manufacturing systems. Finally, an example is presented to illustrate our simulation.

1. Introduction and related work

The concept of fuzzy sets developed by Zadeh [1] handle with impreciseness and incertitude, uncertainty, tangled in all decision making aspects. The fuzzy set theory has been derived from the probability theory and is more suitable to modeling any real life phenomena.

Content from this work may be used under the terms of the Creative Commons Attribution 3.0 licence. Any further distribution of this work must maintain attribution to the author(s) and the title of the work, journal citation and DOI.
Published under licence by IOP Publishing Ltd

Starting with '80, a large number of researchers applied fuzzy set concepts to solve many problems in Engineering and Management.

One of the most representative concept of fuzzy theory is represented by the adaptive model based on free estimation.

Some techniques related that the Intelligent Systems adaptively estimate continuous functions from data without specifying mathematically how the results are depend on initial dates are presented in [2]. In this example, this refers to the ability of fuzzy systems to map an initial dates domain X e.g., denote to an results dates range Y e.g., judgments/decisions without denoting the function f: X → Y. Anyway, mathematically it has been demonstrated that fuzzy systems are universal approximations of continuous functions of an approximately general class [3]. Due to the universal approximations, modeling techniques based on fuzzy models have a native freedom from a priori assumption of the type of relationships that may exist through different variables. Usually an universal approach places no theoretical limits on the modeling capabilities of fuzzy systems, in practice, although to optimally construct a model for a given data set to achieve the full modeling strength of the approach is still an open question [10]. Convert and Crager [4] show how fuzzy sets can be used to capture linguistic values such as high and low in variables related to human performance such as job or task experience and performance. According to this, the variable performance can be captured by specifying a finite universal set that consists in levels of performance using three fuzzy sets such as high, moderate, and limited efficiency.

Noteworthy, the concept of fuzzy set is equivalent with early psychometric ideas; for instance Thurstone [5] has the idea that an individual's opinion could be characterized by more than a single point estimate response as suggested by Hesketh and Hesketh [6].

Newell and Simon [7] have been demonstrated that much of human problem solving could be expressed as if-then types of production rules.

Moreover, expert and other Intelligent systems have been implemented to model, capture, and support human decision processes. These usually rule-based systems allow different points, including the fact that human experts are often needed to articulate propositional rules, that the symbolic processing normally used prevents direct application of mathematics and that traditional rule-based systems require a large number of rules that are often friable and thus they are not rough to the required novel set of data inputs.

The advantages of modeling decision making process are numerous, nevertheless a short analysis of traditional modeling papers, for instance linear regression shows a number of factors suggesting that inquiry on alternative methods is warranted.

These factors include the use of idealistic, orthogonal judgment response, arising from the difficulty in analyzing inter-correlated responses with multiple regression [8]; reliance on linear models even though the cited pervasiveness of linearity may be more reflective of a lack of research on alternative models [9]; and limited selection of methods for eliciting participants verbal descriptions of their judgment policies.

2. Aspects of decision making process

One of the application of the theory of decision is to determine some optimal strategies where a decision is faced with different decision alternatives and an uncertain pattern of next events.

For example, a manufacturer of a new product which will be launched (or will be used in different build) can think to manufacture a wide number of pieces of the product if the consumer acceptance and consequently demand for the product intent to increase. Withal, the producer would like to produce a small number of pieces if the market requirement for that product are going to be low.

Unfortunately, the launching of a certain product which will use the product discussed require to the producer to create a production quantity decision before the demand is actually know. The new acceptance of actual consumer for a certain product will not be determined until the items have been placed in the stores and the consumers have had the opportunity to purchase them.

The selection of the best production volume decision from among several production volume alternatives when the decision Filter is faced with the uncertainty of future demand is a problem for the theory of decision. The theory of decision commences with the assumption that regardless of the type of decision involved, all the Decision making process problems have certain common characteristics which are briefly enumerated below:

The filter of decision. This refers to individual or a group of individuals responsible for making the choice of an appropriate course of action amongst the available course of action. Courses of Action: The courses of action or strategies are the acts that are available to the decision Filter. The Decision Analysis involves a selection among two or more courses of action and the problem is to choose the best of these variants, as a final purpose to achieve an objective.

Many different people have take a decision about the similar situation and then trade potential cognitive interference aimed at improving decision making process outcomes. Some the commonly argued cognitive biases are selective search for evidence; premature stops of search for evidence; inertia; selective perception; wishful thinking or optimism bias; choice supportive bias; Recency; repetition bias; anchoring and adjustment; source credibility bias; incremental Decision making process; attribution symmetry; role fulfillment; underestimating uncertainty and the illusion of control.

Nowadays, in our fields as business or industry, we need rational decision making process. If we have to define a decision, we can use a simple one, an if-then action or the selection from two or more courses of action.

Decision making process can be regarded as a consequence of mental processes that are fundamental cognitive in nature leading to the selection of a course of action among several alternatives. Every decision making process causes a final selection [8]. The result can be an action or an view of choice. Decision making process is vital for all categories of problems which may be either long-range or short-range in nature; or the problem may be at relatively high or low level managerial responsibility. The inherent analysis in the theory of decision is a discipline providing various tools for modeling different decision situations in view of explaining them or prescribing actions increasing the coherence between the possibilities offered by the situation, and goals and values systems of agents involved. Decision making process for human performance represented a topic of active research from several point of views. From a psychological point of view, it is necessary to examine individual decisions in the context of preferences and needs that an individual had and values he looking for. From a cognitive point of view, the decision making process process must be regarded as a continuous process integrated in the interaction with the environment. From a normative point of view, the analysis of individual decisions is thoughtful with the logic of decision making process and rationality and the invariant choice it leads to [10]. At another level, it might be regarded as a problem solving activity which is finished when a satisfactory solution is set out. Therefore, decision making process is a reasoning or emotional process which can be rational or irrational, for the main based on explicit hypothesis or tacit hypothesis.

3. A case study: The distribution of tasks application

The term fuzzy suggests an image of a boundary zone, rather than an abrupt frontier. The set relations are being considered as relations composed of crisp sets, to distinguish them from fuzzy set relations. As with set relations, we are only guided by intuition in deciding which objects are members and which are not; a formal basis for how to determine the membership grade of a fuzzy set relation is absent. The membership grade is a precise, but arbitrary measure as it rests on personal opinion, not reason. The range of values of membership grade is $0 \leq \mu \leq 1$, the higher the value, the higher the membership grade.

The members of fuzzy relations are taken from a universe of discourse which comprises of all objects that can be taken into consideration and generally depends on the context.

There are two ways to represent a membership function viz., continuous or discrete. A continuous fuzzy relation A is defined by means of a continuous membership function, $\mu A(x)$.

A trapezoidal membership function is a linear parts, and a continuous function, directed by four parameters such as, a, b, c, d [11]:

$$\mu_{trapezoid}(x;a,b,c,d)=\begin{cases}0, & x\le a\\ \dfrac{x-a}{b-a}, & a\le x\le b\\ 1, & b\le x\le c;\quad x\in\Re\\ \dfrac{d-x}{d-c}, & c\le x\le d\\ 0, & d\le x\end{cases} \quad (1)$$

To achieve a discrete triangular membership function from the trapezoid membership function let us assume that the universe is a vector U of 7 elements. Each membership value corresponds to one element of the universe, with the universe in the bottom row, and the value for a membership for the top row. As a rule of thumb, the continuous form is more computing increased, but a storage demanding smaller than the discrete form.

In real application, the problem is often intricate and elaborate due the fact that resources may be limited or the set of tasks may not be known exactly, and the set of tasks may change in time. Generally resource allocation algorithms could not handle in a proper manner the constrained situations, especially the uncertainty in the environment and their dynamics.

For an example we consider a repartition of tasks between the members of a team to accomplish a mission in an activity. We now consider another decision-making problem of allocating a particular job to the best possible person who fulfills the requirements of the job.

Let

U = {p1, p2, p3, p4, p5, p6}

be the crisp set of five persons for the job.

Let

E = {enterprising, average, confident, confused, yes risks, no risks}

be the set of parameters.

Let

(F1, A1) = {F1 (enterprising) = {p1/ 0.5, p2/ 0.7, p3/ 0.3, p4/ 0.1, p5/ 0.8, p6/ 0.9}, F1 (average) = {p1/ 0.3, p2/ 0.1, p3/ 0.5, p4/ 0.8, p5/ 0.05, p6/ 0.7}

be the set describing the enterprising qualities of the person.

Again

(F2, A2) = {F2 (confident) = {p1/ 0.6, p2/ 0.8, p3/ 0.5, p4/ 0.2, p5/ 0.9, p6/ 0.8}, F3 (confused) = {p1/ 0.3, p2/ 0.1, p3/ 0.7, p4/ 0.9, p5/ 0.5, p6/ 0.6} be the set describing the confidence level of the person.

Similarly,

(F3, A3) = {F3 (yes risks) = {p1/ 0.7, p2/ 0.8, p3/ 0.5, p4/ 0.2, p5/ 0.6, p6/ 0.5}, F1 (no risks) = {p1/ 0.3, p2/ 0.07, p3/ 0.65, p4/ 0.95, p5/ 0.1, p6/ 0.6}

be the set describing the willingness level of the person.

Let us assume that the particular job requires an enterprising, confident person who is yes risks.

Our problem (Table 1) is to find the candidate who best suits the requirements of the job. To solve this problem we use the definition of a fuzzy relation (R, C) of the fuzzy sets (F1, A1), (F2, A2), (F3, A3) of all candidates who are enterprising, confident, yes risks.

By definition, (R, C) is given by (R, C) = {p1/ 0.21, p2/ 0.45, p3/ 0.08, p4/ 0.05, p5/ 0.43, p6/ 0.36}.

From the relation it is evident that the most suitable candidate for the job is p2 who possesses the greatest membership value in the relation (R, C).

Now, we present the probability and possibility distributions for the above problem corresponding to one specific parameter. Considering the risk taking parameter, eg. yes risks from the parameter set E we have the following probability distribution prob for the persons PTi; i = 1,………,6 from the set.

Table 1. Probability risk taking parameter (PT).

Tasks Person	Prob (PT)
1,00	0,25
2,00	0,55
3,00	0,10
4,00	0,10
5,00	0,00
6,00	0,00

Probability values of person PTi with respect to risk taking parameter. Again, a fuzzy set expressing the risk taking attitude of the persons pi; i = 1,.........,6 from the set U may be the expressed using the following possibility distribution π,

Table 2. Probability values of person Pti.

Tasks Person	Π (PTi)
1,00	1
2,00	1
3,00	1
4,00	1
5,00	0,8
6,00	0,7

Possibility values of person Pti with respect to risk to take attitude.

It is to be noted that each possibility is at least as high as the corresponding probability. Further, the sum of prob (Pti – table 2) is always equal to 1 but π (pi) may be equal to, greater or less than 1. As it is obvious from the above discussion that it is much easier to represent a large amount of information viz., different parameters using the fuzzy relations which is the prime requirement in most decision making situations, because the final decision to the problem is dependent on various associated parameters. Besides this, using probability and possibility distributions only a partial representation of the information is possible, which leads to final decision results which are inaccurate and incomplete. Finally, the solution to the problem is obtained with minimal computational effort using fuzzy relations.

4. Conclusions

As a conclusion, the paper use some notions important for planning and decision: we are using the fuzzy concepts and process of decision making.

Firstly, the utility of fuzzy logic is due to the inference structure that enables the human reasoning capabilities to be applied to artificial knowledge-based systems. Fuzzy logic theory plays an important role in inference mechanism, learning, adaptation, fault tolerance, parallelism, and generalization, is generally used in cognitive uncertainty or computational neural networks.

Secondly, the process of making decisions is the fundamental activity of human beings. In other words, decision making represent an important field in all aspects of life. In any decision process we consider the information about the outcome and choose among two or more alternatives for subsequent action. If good decisions are made, then we may get a good expected output. Decision making is defined to include any choice or selection alternatives. A decision is said to be made under certainty, where the outcome for each action can be determined precisely. A decision is made under risk. When the only available knowledge concerning the outcomes consists of their conditional probability distributions.

The uncertainty existing is the prime domain for fuzzy decision making.

Nowadays, our world is furthermore based on the use of electronics and computers to control the behavior of real-world resources, and in this case, we can see the utility of fuzzy aspects in real-life, usually in process of decision.

Related to the future work, we want to investigate the distribution of tasks application considering and another value which is important when we are talking about uncernatinty, the time. Another ideea of future work is to use intelligent agents.

References

[1] Zadeh L A 1965 Fuzzy Sets *Information and Control* **8** pp 338-353
[2] Kosko B 1992 *Neural Networks and Fuzzy Systems* (New Jersey: Engelwood Cliffs, Prentice Hall)
[3] Kiir G and Yuan B 1995 *Fuzzy Sets and Fuzzy Logic: Theory and Applications* (Prentice Hall)
[4] Craiger P and Coovert M D 1994 Modeling dynamic social and psychological processes with fuzzy cognitive maps *Proc. of 3rd IEEE Conf. on Fuzzy Systems*
[5] Thurstone L L 1973 *The Nature of Intelligence* (London: Routledge)
[6] Hesketh T and Hesketh B 1994 Computerized Fuzzy Ratings: The concept of a fuzzy class. Behavior Research Methods *Instruments and Computers* **26** pp 272-277
[7] Newell A and Herbert S 1972 *Human Problem Solving* (N.J.: Prentice-Hall Englewood Cliffs)
[8] Hammond K R 1980 Introduction to Brunswikian theory and methods: New Directions for Methodology of Social and Behavioral Science pp 1-11
[9] Ganzach Y and Schul Y 1995 The influence of quantity of information and valence framing on decision *Acta Psychologica* **89** pp 23-36
[10] Kahneman D and Tversky A 2000 *Choice, Values, Frames* (The Cambridge University Press)
[11] Jang S, Sun C T and Mizutani E 1997 *Neuro-Fuzzy and Soft Computing* (Prentice Hall)

IOP Conf. Series: Materials Science and Engineering **95** (2015) 012155 doi:10.1088/1757-899X/95/1/012155

Estimation of overall equipment effectiveness using simulation programme

I Paprocka[1], W Kempa[2], K Kalinowski[1] and C Grabowik[1]
[1]Silesian University of Technology, The Faculty of Mechanical Engineering, Institute of Engineering Processes Automation and Integrated Manufacturing Systems, Konarskiego 18A Str., 44-100, Gliwice, Poland
[2]Silesian University of Technology, Faculty of Applied Mathematics, Institute of Mathematics, Konarskiego 23A Str., 44-100, Gliwice, Poland

E-mail: iwona.paprocka@polsl.pl

Abstract. A queuing system of the M/M/1/N type with cyclic failure-free and repair times and cyclic defective production is used as a model of a single machine production system. Production tasks arrive according to an exponential process and are served with normally disturbed processing time. Successive working (failure-free) times, repair times and defect-free production have exponential distributions. Defective products can be recirculated and back to the production line according to a Bernoulli distributions. The objective is to evaluate effectiveness of the production system using Overall Equipment Effectiveness before and after introducing improvement initiatives.

1. Introduction

According to the words of Dave Logozzo: "If you're sensitive to the flow, you will easily notice wastes" companies aim to achieve a production without downtime in their cells or production lines [1]. Manufacturers aim to eliminate wastes, i.e. anything that does not add a value to a product [1]. Manufacturers try to establish close cooperation with a client and to produce products of a high quality and reasonable prices. The following methodologies are applied for reducing a cycle-time: six-sigma for value-added operations and lean for non-value-added operations. Six losses are distinguished: downtime losses (breakdown losses, set-up and adjustment losses), speed losses (idling and minor stoppage losses, speed losses) and quality losses (quality defects, rework and reduced yield during start-up) according to [2]. Overall Equipment Effectiveness (OEE) is proposed to measure improvement initiatives.

Production in manufacturing cells is realized in a tact time in order to overcome overproduction. The production capacity is adjusted by decreasing or increasing the tact time up to the maximal utilisation of given resources. Companies add overtime for existing workers or increase a number of production cells to meet the capacity constraints. In the paper, the OEE is proposed to evaluate problems with availability (breakdown or equipment failure, set-up and adjustment), performance efficiency (idling and minor stoppage, reduced speed) and product quality (defects and reworks, reduced yield). The OEE is used to maximize performance of existing capacity. A. Ron and J. Rooda [3] compare the OEE to equipment effectiveness E in order to indicate the influence of downtime and rework. P. Muchiri and L. Pintelon [2] investigate how the OEE indicator has evolved with time depending on individual needs of industries, indicating: overall factory effectiveness OFE, overall

Content from this work may be used under the terms of the Creative Commons Attribution 3.0 licence. Any further distribution of this work must maintain attribution to the author(s) and the title of the work, journal citation and DOI.
Published under licence by IOP Publishing Ltd

plant effectiveness OPE, overall throughput effectiveness OTE, production equipment effectiveness PEE, overall asset effectiveness OAE and total equipment effectiveness performance TEEP. They underlined that the OEE identifies losses reducing a measure of effectiveness of equipment over a period of time. For other performance measurements see [4, 5, 6].

In the paper we deal with the M/M/1/N-type queuing system, in which the arrival stream of jobs is described by exponential process with intensity λ, and jobs are being served individually by a single machine, with the service speed normally distributed with μ and α and setup time β. The total system capacity equals N i.e. we have N-1 places in the magazine and one place in service. If the arriving job finds all places occupied, it is lost. Arriving production tasks form a single waiting line and are served in the order of their arrivals. At time t = 0 the first working (failure-free) period begins that ends with a machine failure after an exponentially distributed time with mean γ. Successively, the repair time with mean η begins immediately at the end of failure-free time. Time to defective product occurrence is described by Bernoulli distribution with mean ϖ. Defective number of products can be recirculated and back to the production line according to a Bernoulli distribution with mean ρ. The production line is modeled in the Enterprise Dynamics System (ED). Wastes due to unplanned events in the machine's work on the OEE are summed up running experiments in the ED.

2. Problem formulation

The assumption of the TPM method is to maintain machines in the state of productivity. The responsibility of each maintenance employee is to identify, monitor and remove causes of waste: breakdowns, little downtimes, work below a nominal performance, retooling of a machine and inadequate quality. Breaks for technical service are included in the production schedule. Each unplanned disturbance causes an increase in waste. The objective is to assess waste due to unplanned events over the OEE indicator, such as: (1) unplanned downtimes because of equipment failure occurring after failure-free time with mean γ and with repair time with mean η, (2) set-up and adjustment losses β during changing over from requirement of one item to another, (3) reduced speed losses with mean μ and deviation α due to differences between equipment designed speed and actual operating speed, (4) quality defects with mean ϖ due to malfunctioning production equipment, (5) reworks with mean ρ due to malfunctioning production equipment, (6) reduced yield during start-up that is losses that occur from machine start-up to stabilization.

The OEE indicator compares the real usage of the machine to the ideal usage, according to plan. The available time is the time duration of a shift. The schedule includes predicted downtimes: machine failures, changing of workers' shift, retooling, etc. A time of efficient production is the time of a product production. Components of the OEE: Availability (A), Performance (P) and Quality (Q) represent an impact of a waste in the production systems efficiency [2]. It is good when the OEE indicator reaches around 80% to make full use of machines and equipment. Below we study in details the influence of key system parameters λ , γ , η , β , μ , α , ϖ and ρ on the OEE and his components A, P and Q in different scenarios of system operation.

3. Simulation model creation

The model is created using a graphical user interface by selecting the component objects of simulation model from the available predefined atoms. Following atoms are used to model the manufacturing line: production process, source, buffer, machine for setup time, critical machine, machine for close down time, control stand (starting from the left hand side in the figure 1). Good quality products are sent to the upper sink. If a bad quality product can be reworked is sent back to the production line, if the product cannot be reworked is sent to waste (bottom sink). In the system P1 manufacturing process is executed. Atoms used in the model are subjected to the parameterization process. Possible alternative input parameters λ, γ, η, β, μ, α, ϖ and ρ for the process are presented in Sections describing the experiments and results. The input parameters are modeled in eight tables, the example of table for interarrival times is in figure 1. Two data recorder atoms are used to collect information about λ and μ, and atoms: Gantt Initialize and Gantt Chart are used to generate Gantt charts after a simulation.

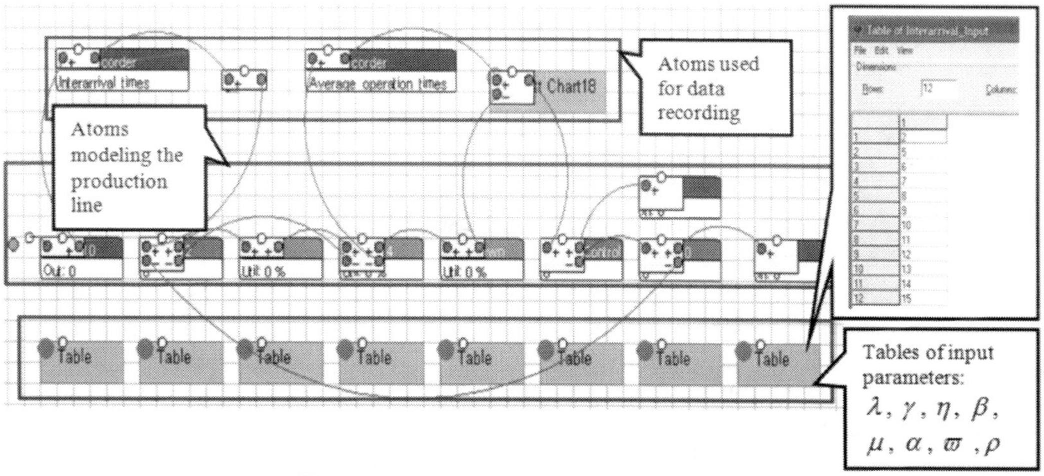

Figure 1. Atoms of the simulation model.

4. Simulation Experiment

Simulation experiments for the variable parameters of the production system (λ, γ, η, β, μ, α, ϖ, ρ) are made with the application of the Experimental Wizard Module. The experimental module allows to perform various experiments in an automatic way for many simulations without the interaction of an analyst. The option: Separate runs was selected as the method of Simulation. Number of independent simulations (of observations) is 10. Duration time of observation is 4800 mins in each simulation. In each simulation input parameter values are read from tables. In order to read automatically a value of λ in each simulation following codes are written in 4DScript: On start of simulation run: *if(Interarrival_Input(1,1)>11,set Interarrival_Input (1,1,2))* and On end of simulation run: set *Interarrival_Input (1,1, Interarrival_Input (1,1)+1)*. After each run of simulation a value of is read from a vector indicated in the cell (1,1) of *Interarrival_Input Table*. Each time of simulation value of a counter from the cell (1,1) is increased by 1. The counter receives the value of 2 if the value of counter is more than 11. In the Source atom, parameters: inter-arrival time and time till first product arrives are described by: *mins(Interarrival_Input(Interarrival_Input (1,1),1))*.

The discrete-event simulation run for analyzing the behavior of the model. Results of production system work are generated in reports. In tabel 1 the example of the report generated for the input data described in the Section " Effect of speed of processing on the OEE" is presented.

5. Effect of speed of processing on the OEE

Let us observe the OEE after T=4800[min], taking interarrival time, mean values for operating speed, setup time, failure-free and repair times, defects-free times and rework occurrence times equal to 9, 8, 5, 340 and 16[min], 95 and 50 [number] respectively. A production task is stored in a finite-capacity buffer if it arrives and the machine is busy or setup activities are done. The level of buffer saturation n =6. The machine needs a setup time before providing the service of the first task after the idle time.

Let us analyze now the impact of the processing rate on the OEE. Observe the mean A, P and Q for the system described above and for 10 values of mean processing rate μ of jobs, namely 2, 5, 8, 10, 12, 14, 16, 18, 20, 22, 24 [min] and standard deviation equals to 2[seconds]. The results of the simulation are presented in the table 1 and geometrically in the figure 2. For low processing speed (2-9) the OEE is below 80% which is unacceptable. For processing speed (2-9) differences between results obtained for the OEE are significant. As the service speed increases, the differences are getting smaller. The explanation of this phenomenon is in that, at a very high processing rate, the production capacity of the production system is not used since the intensity of arrivals equals 9. Improvements in equipment effectiveness over a period of time is economically reasonable only if the interarrival time of a new task decreases as well.

Table 1. Raw data and mean OEE computing as a function of processing rate after simulation time T=288000[seconds].

Raw Data										
Server	Average staytime (Ideal Cycle Time) [in seconds]									
	121,807	240,09	361,62	481,589	602,088	722,241	842,855	963,312	1083,585	1204,003
Server14	Departures									
	556	544	554	545	476	397	340	298	265	238
Server	Utilization (%)									
Idle	0.764	0.543	0.304	0.086	0.002	0.002	0.002	0.002	0.002	0.002
Busy	0,231	0,453	0,692	0,909	0,993	0,993	0,993	0,993	0,993	0,993
Down	0,003	0,003	0,003	0,003	0,003	0,003	0,003	0,003	0,003	0,003
Sink	Good Units									
	514	512	521	510	454	372	324	288	255	234
Source	Units Started									
	533	533	533	533	481	400	342	303	273	247
Availability (A) = Availability Time / (Simulation Time /60)										
Availability Time = (Simulation Time - (Busy x Simulation Time))/60										
	4784	4784	4784	4784	4784	4784	4784	4784	4784	4784
(A)	0,996	0,99	0,9967	0,9967	0,9967	0,9967	0,996	0,996	0,996	0,996
Performance = (Good Units* Average staytime) / Available Time										
	0,218	0,428	0,656	0,855	0,952	0,936	0,951	0,966	0,962	0,981
Quality = Good Units / Units Started x 100										
	0,964	0,960	0,977	0,956	0,943	0,93	0,947	0,95	0,934	0,947
OEE = Availability (A) x Performance x Quality										
	0,209	0,41	0,639	0,816	0,895	0,867	0,898	0,915	0,896	0,926

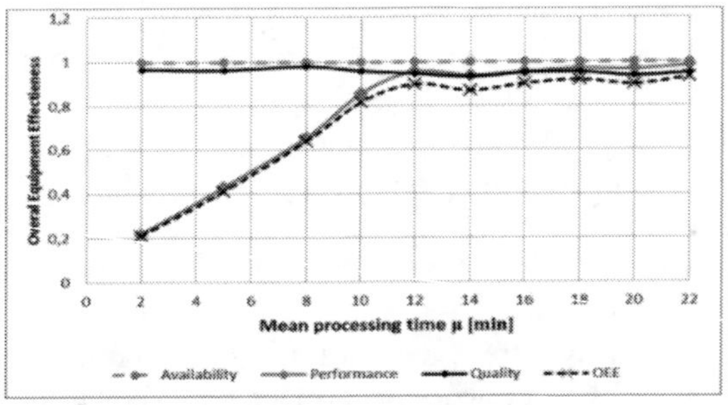

Figure 2. Mean OEE as a function of processing rate after T=288000[seconds].

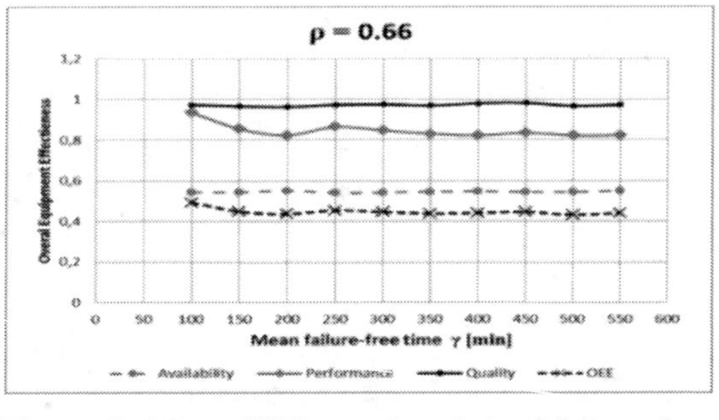

Figure 3. Mean OEE as a function of failure-free time and traffic load equal to 0.66.

6. Impact of the failure-free period duration on the OEE

Let us study now the possible impact of the failure-free period duration γ on the OEE before the fixed time T. Let us take into consideration the system in which mean service rate equals $\mu = 9$ [min], the repair time is long with mean $\eta = 16$ [min] and for 10 mean values of γ : 100, 150… 600 [min].

Accepting means of processing times: $\lambda = 6$, 9 and 12 [min], we get the values of the traffic load $\rho = \lambda / \mu$ equal to 0.66, 1.0 (critical loading) and 1.33 (overloading), respectively. For the initial condition n=6 we obtain the results given for $\rho = 0.66$ in the figure 3 and for $\rho = 1$ in the figure 4. The OEE increases for the increasing ρ to the critical loading level. A slight decrease of this characteristic is connected with the increasing γ. Comparing the OEE for critical loading and the case of the overloaded system, the OEE for $\rho = 1.33$ is slightly worst (figure 5), in practice, the only time period during which the machine does not process the jobs is its repair time.

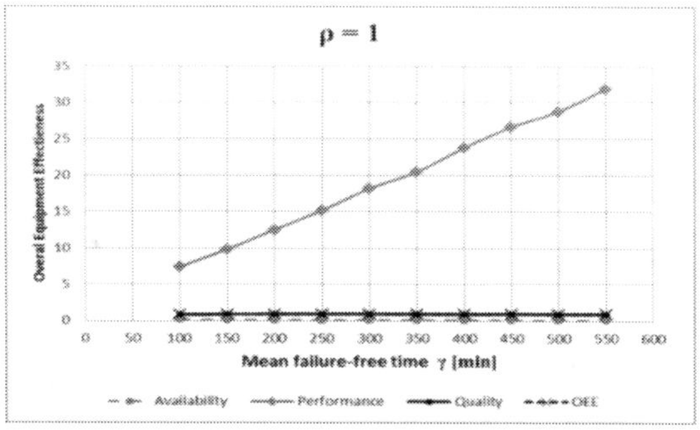

Figure 4. Mean OEE as a function of failure-free time and traffic load equal to 1.

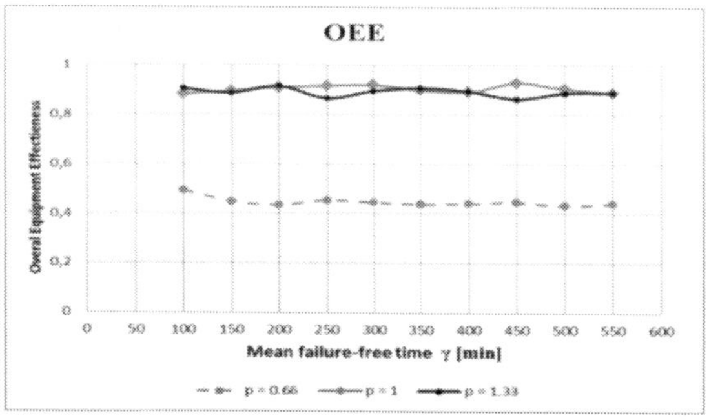

Figure 5. Mean OEE as a function of failure-free time and traffic load equal to 0.66, 1 and 1.33.

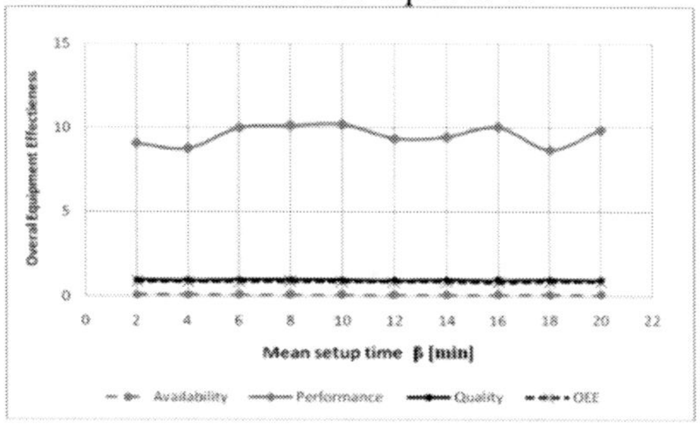

Figure 6. Mean OEE as a function of setup time after T=4800 [min].

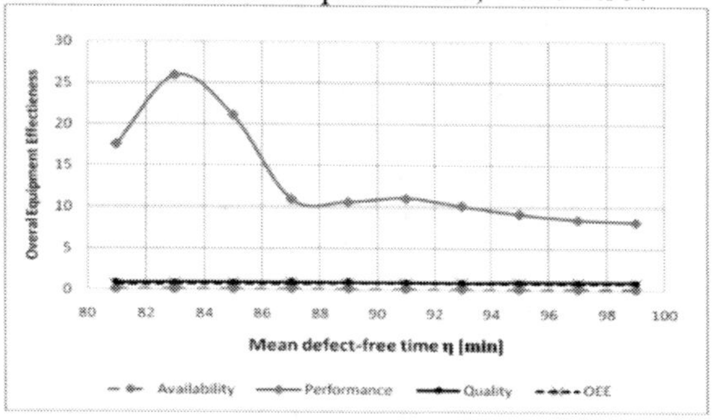

Figure 7 Mean OEE as a function of defect-free time after T=4800 [min].

7. Impact of the setup time on the OEE

Let us investigate what is the impact of setup time duration α on the OEE. Observe the system for proportion: repair time/failure free-time: 8/500 [min] and for 10 mean values of α, namely 2, 4... 20[min]. For $\lambda = 9$ and $\mu = 8$ [min], we obtain results visualized in figure 6. As one can observe, the mean α increases, however, the OEE decreases very little. The explanation of this phenomenon is in that, the intensity of arrivals equals 9 is close to the ideal cycle time thus no setup times activities are done for this data in the most scenarios.

8. Impact of the defects-free time on the OEE

Next we investigate what is the impact of defects-free time duration on the OEE. Take α =22, and observe the system for 10 mean values of η, namely 81, 83,..., 99[min]. We obtain results visualized in the figure 7 and 8. As one can observe, the OEE increases as η increases. For $81<\eta<87$ the performance value of the system changes very rapid.

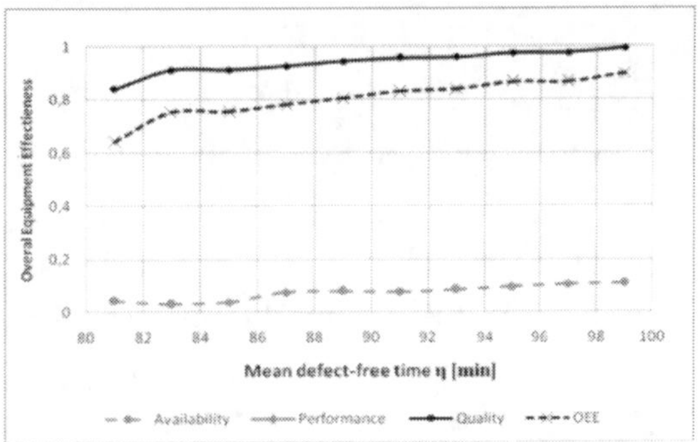

Figure 8. Mean OEE as a function of defect-free time with the range 0 to 1 for Y axis.

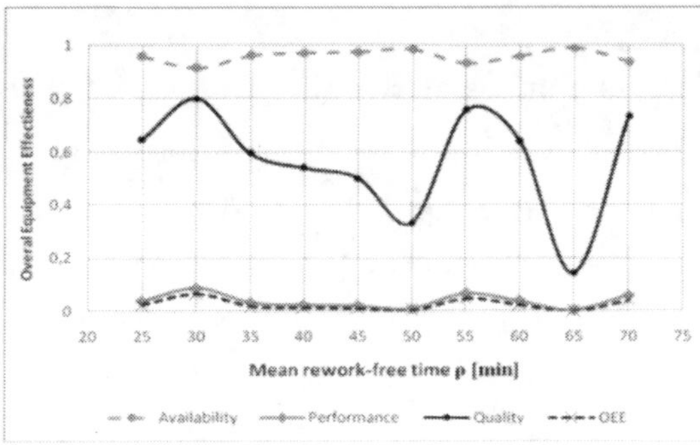

Figure 9. Mean OEE as a function of rework-free time after T=4800 [min].

9. Impact of the rework-free time on the OEE

Finally, investigate what is the impact of rework-free time duration on the OEE. Observe the system for $\eta = 89$ and for 10 different mean values of ρ, namely 25, 30... 70[min]. We obtain results visualized in figure 9. As one can observe, the impact of rework-free time on the OEE is the highest. We can see the burstiness of the OEE as ρ increases.

10. Conclusions

In the paper the effectiveness of production system of the M/M/1/N type with cyclic failure-free, repair times and cyclic defective production is evaluated. In the production system defective products can be recirculated and back to the production line. The Overall Equipment Effectiveness is used to evaluate improvement initiatives. Differences between results obtained for the OEE are significant for processing speed from 2 to 9. Increasing the processing speed over 9 is economically unreasonable since the intensity of arrivals equals 9. Increasing the traffic load to the critical loading level increases the OEE as well. The setup time has no effect on the OEE. The OEE increases with the defects-free time increasing. Increasing rework-free time a high burstiness of the OEE has been noticed.

References

[1] Rother M and Shook J 2003 Learning to see *Value stream mapping to create value and eliminate muda* (Brookline Massachusetts USA The lean enterprise institute)

[2] Muchiri P and Pintelon L 2008 Performance measurement using overall equipment effectiveness (OEE): literature review and practical application discussion *Int. J. of Production Research* **46(13)** pp 3517-3535.

[3] Ron Aand Rooda J 2006 OEE and equipment effectiveness: an evaluation *Int. J. of Production Research* **44(23)** pp 4987-5003

[4] Kempa W M, Paprocka I, Kalinowski K and C. Grabowik 2014 On transient queue-size distribution in a single-machine production system with breakdowns *Advanced Materials Research* **1036** pp 505-510

[5] Paprocka I, Kempa W M, Grabowik C and Kalinowski K 2014 Predictive and reactive scheduling for a critical machine of a production system *Advanced Materials Research* **1036** pp 909-914

[6] Gwiazda A 2008 Technical project management using quality methods *Journal of Achievements in Materials and Manufacturing Engineering* **31(2)** pp 739-746

Institute of Physics
Dirac House, Temple Back
Bristol BS1 6BE UK

ISSN: 1757-8981
ISBN 978-1-5108-1466-0

CURRAN ASSOCIATES INC.
proceedings.com

Thermal Performance of the Exterior Envelopes of Whole Buildings XII International Conference

Clearwater, Florida, USA
1 - 5 December 2013

OAK RIDGE
National Laboratory